SECOND EDITION

Organic Chemistry

Ralph J. Fessenden
Joan S. Fessenden

University of Montana

WG Willard Grant Press
Boston, Massachusetts

PWS PUBLISHERS

Prindle, Weber & Schmidt · ✗ · Willard Grant Press · **wg** · Duxbury Press · ♠

Statler Office Building · 20 Providence Street · Boston, Massachusetts 02116

Library of Congress Cataloging in Publication Data

Fessenden, Ralph J.
    Organic chemistry.

    Includes index.
    1. Chemistry, Organic.   I. Fessenden, Joan S.
II. Title.
QD251.2.F49   1982   547   81-13496

ISBN 0-87150-752-8   AACR2
ISBN 0-534-98009-0   (International edition)

Printed in the United States of America.
82 83 84 85 86 — 10 9 8 7 6 5 4 3 2 1

Production editor: David M. Chelton
Text designer: Megan Brook
Cover designer: Martucci Studio
Text compositor: Composition House Ltd.
Art studio: J & R Services
Cover printer: Lehigh Press Lithographers
Text printer and binder: Rand McNally & Company

Preface to the
Second Edition

When we wrote the first edition of *Organic Chemistry,* our goal was to produce a manageable text for the one-year introductory organic chemistry course—a text designed to guide students in their studies and not simply provide a compilation of organic chemical knowledge. Throughout our writing we tried always to keep the student in mind. For the most part, we think we were successful in achieving our goal. The response from users of the first edition, professors and students alike, has been gratifying.

In this second edition, we have retained the same overall organization as in the first. We have also kept the amount of chemistry, as well as the length of the book, about the same. The revisions, most of them based on the experiences of users of the first edition, have been made primarily to improve clarity, correctness, and logical flow of material.

Changes that occur throughout this edition include the addition of *preparation tables,* which list the reactions used to prepare the various classes of compounds; short sections on how each class of compounds can be used in *synthesis;* and *reaction summary tables,* which are incorporated into the synthesis sections. *Chapter summaries* have been retained to emphasize the important points covered. These tables and sections are included for the student's review and convenience. Other recurring changes include greater emphasis on electron shifts in mechanisms and more extensive use of dimensional (wedge) formulas.

In addition to these changes, the *problem sets* at the ends of the chapters have been improved. We have replaced some of the more repetitive drill problems with a greater number of problems that require some thought. Some of these problems are moderately demanding, while others, toward the ends of the problem sets, should be quite challenging.

The text is organized into three parts: concepts of structure and bonding; organic reactions and mechanisms; and topics of more specialized interest.

Introductory material.   Chapters 1 and 2 are primarily reviews of atomic and molecular structure, along with electronegativity, hydrogen bonding, acid–base reactions (expanded somewhat in this edition), and molecular orbitals (presented for the most part in a pictorial way). Bonding in some simple nitrogen and oxygen compounds is included as a way of introducing the concept of functional groups, but this topic is emphasized less than in the first edition. A brief introduction to resonance theory, with a few more examples than in the first edition, is also presented here.

The student's introduction to structural isomerism and nomenclature comes in Chapter 3. Besides describing the nomenclature of alkanes, we briefly introduce the naming of a few other classes of compounds that will be encountered early in the book. Chapter 4 on stereochemistry contains discussions of structure: geometric isomerism, conformation, and chirality. Fischer projections have been moved from the carbohydrates chapter to Chapter 4 for the convenience of instructors who use these projections early in the course. Resolution, formerly in Chapter 15, has also been included here, partly to demonstrate the important difference between enantiomers and diastereomers and partly as an extension of acid–base reactions.

Organic reactions.   Mechanisms are introduced in Chapter 5 with the substitution and elimination reactions of alkyl halides. We have several reasons for taking this approach. First, the typical S_N2 reaction path is a concerted reaction with a single transition state and is thus ideal for introducing transition-state diagrams and reaction kinetics. Second, the S_N1 path follows logically from the S_N2 mechanism and allows us to introduce steric hindrance and carbocations early in the course. Finally, ionic reactions allow us to apply the stereochemical principles just covered in Chapter 4.

Chapter 5 has been extensively revised and tightened. Because the E1 mechanism is more important in alcohol chemistry than in alkyl halide chemistry, the principal discussion of the E1 path has been moved to Chapter 7. The topic of solving synthesis problems has been moved to Chapter 6 on free-radical halogenation and organometallic compounds, by which time students will have learned enough chemistry for retrosynthetic analysis to be presented. Otherwise, Chapter 6 has been revised only slightly.

Chapter 7 (alcohols and ethers) has been tightened considerably by rearranging the order of topics. In this edition, all alcohol chemistry is presented first, followed by ether and epoxide chemistry. The discussions of inorganic esters and phenols have been shortened; however, phenols are now discussed at greater length in the chapter on benzene and substituted benzenes.

Infrared and nmr spectroscopy, presented in Chapter 8, provide a break in the organizational pattern of organic reactions and are discussed later in this preface.

In Chapter 9 on alkenes and alkynes, the topic of hydroboration has been expanded because of the increasing interest in this area. Although the Diels–Alder reaction is still presented in Chapter 9, and has been expanded upon in this edition, the general topic of pericyclic reactions has been moved to a new Chapter 17. Chapter 10 (benzene and substituted benzenes) has been expanded to include aryldiazonium salts, which previously appeared in Chapter 15, and phenols (formerly in Chapter 7); both are discussed in somewhat greater detail than in the previous edition.

The chemistry of carbonyl compounds in Chapters 11–14 has undergone only minor revision. Chapter 15 (amines) has been shortened by transferring resolution and aryldiazonium salts to earlier chapters; however, a brief discussion of phase-transfer catalysis has been added.

Topics of specialized interest.   Chapters 16–21 cover polycyclic and heterocyclic compounds; pericyclic reactions; carbohydrates; proteins; lipids; and uv spectroscopy, color, and mass spectrometry. Except for minor revisions and updating, the principal changes in these chapters include the addition of a simple alkaloid synthesis to Chapter 16 and the coverage of pericyclic reactions using the frontier molecular orbital approach in the new Chapter 17.

Other topics of special interest, such as carbenes, polymers, and the metabolism of ethanol, appear within chapters where they logically follow from the chemistry being discussed. Wherever possible, these subjects are placed in separate sections so that they can be dealt with as the instructor deems best.

Although we have not changed our basic philosophy concerning the presentation of nomenclature, spectroscopy, synthesis, bio-organic material, and problems in this second edition, our approach and the minor changes made are worth stating.

Nomenclature.   In keeping with current trends in nomenclature, we stress IUPAC names in this text, more so than in the first edition; however, we also include some trivial names (such as acetone and *t*-butyl chloride) that are part of every organic chemist's vocabulary. As in the first edition, our presentation begins with a brief survey of systematic nomenclature in Chapter 3. The names presented there are those that the student will encounter again in chapters immediately following. The nomenclature for each class of compounds is then discussed in more detail in later chapters. An appendix is included for those who wish additional material or a quick source of reference.

Spectroscopy.   Spectroscopy is discussed as early as we think feasible—Chapter 8. By this time, the student has a working knowledge of structure, a few functional groups, and a few reactions. However, those who wish to do so may cover the spectroscopy chapter right after Chapter 4, as soon as the students are familiar with organic structures.

We have included infrared and nuclear magnetic resonance spectroscopy in Chapter 8 because of their importance in structure determination. Sufficient background in the principles behind infrared and nmr spectroscopy is presented so that students can appreciate why spectra and structures are related, but the emphasis is on structure. Wherever appropriate after Chapter 8, we have included sections on the infrared and nmr characteristics of the compound classes being discussed. Structure-determination problems involving infrared and nmr spectra are included at the ends of many of these later chapters.

Revisions in Chapter 8 include the use of infrared spectra with cm^{-1} as the principal scale, because this is the type of spectrum students will likely encounter in the laboratory, and the addition of a short section on carbon-13 nmr spectroscopy.

Ultraviolet and mass spectra are covered in Chapter 21. These topics are designed to stand alone; therefore, either or both can be presented along with infrared and nmr spectra if the instructor wishes.

Synthesis: We have placed our formal discussions of synthesis in separate sections—many of them new—at the ends of appropriate chapters so that they may be emphasized or de-emphasized in the lecture presentation. The purposes of these sections are to provide additional review of material covered previously and to give students an opportunity to apply their knowledge.

Bio-organic material.   Many students in the introductory organic course are majoring in medical or biological fields. Therefore, numerous sections and problems that are biological in nature have been included. We have selected material that is appropriate to the chemistry under discussion and that requires application of organic logic. Our intention is to show the close relationship between organic chemistry and the biological sciences.

Problems.   We are firm believers in problem solving as an important part of learning organic chemistry, and we have included more than 1150 unsolved problems in the text. Within each chapter, a number of worked-out sample problems are included to illustrate the approach to problem solving and to provide further information. Often these sample problems are followed directly by study problems with answers at the end of the book. Some of these study problems are designed to relate previous material to the present discussion. Others are designed to test students on their mastery of new material.

The problems at the end of each chapter are of two types: drill problems and thought problems. Although their order of presentation tends to follow the chapter organization, they are graded in difficulty. The last several problems in each chapter should challenge even the best students. As mentioned, we have changed the mixture of problems to include a somewhat greater number of intermediate and more-challenging problems, generally of the synthesis type, but not exclusively so. The *Study Guide with Solutions* that accompanies this text contains the answers to the chapter-end problems and also provides further explanation where appropriate.

Finally, we have prepared several supplements in addition to the *Study Guide* just mentioned. Many figures in the text, especially the spectra, have been included in a set of *Overhead Transparencies* available from the publisher. We have written an *Instructor's Guide* containing what we hope is useful information for instructors using this text. It also contains a table correlating every end-of-chapter problem in the book with the text section on which it is based.

Acknowledgments

We have appreciated the many suggestions and corrections sent to us by users of the first edition, especially Lee Clapp (Brown Univ.), Frank Guziec (New Mexico State Univ.), Edward Hoganson (Edinboro St. Coll.), David Todd (Worcester Polytechnic Inst.), Roy Upham (St. Anselm Coll.), and many others too numerous to mention.

We are very grateful to our colleagues who have reviewed the manuscript for this second edition and have contributed many excellent suggestions: Robert R. Beishline (Weber State Univ.); Robert Damrauer (Univ. of Colorado, Denver); Slayton A. Evans, Jr. (Univ. of North Carolina, Chapel Hill); A. Denise George (Nebraska Wesleyan Univ.); John Jacobus (Tulane Univ.); Allen Schoffstall (Univ. of Colorado, Colorado Springs); Malcolm P. Stevens (Univ. of Hartford); and Leroy G. Wade, Jr. (Colorado St. Univ., Fort Collins). Above all, we are indebted to Ronald Kluger (Univ. of Toronto) for his early encouragement and suggestions, as well as for his review and re-review of the entire manuscript for this second edition.

We also thank Sadtler Research Laboratories, Inc., for providing the two actual, or ''real,'' spectra used in Chapter 8 and our typist Laurie Palmer for her careful and prompt work.

The staff at Willard Grant Press has been exceptionally supportive. The notes and comments from the sales representatives have been extremely valuable in formulating our plans for the second edition. For the second time, our special thanks go to our enthusiastic editor Bruce Thrasher and to David Chelton for his careful production work, art and design coordination, copyediting, and many suggestions for improvement.

Ralph J. Fessenden
Joan S. Fessenden

University of Montana
Missoula, Montana

Contents

CHAPTER 3

Structural Isomerism, Nomenclature, and Alkanes   81

CHAPTER 4

Stereochemistry   110

CHAPTER 5

Alkyl Halides; Substitution and Elimination Reactions   161

CHAPTER 6

Free-Radical Reactions; Organometallic Compounds   215

CHAPTER 7

Alcohols, Ethers, and Related Compounds   255

CHAPTER 8

Spectroscopy I: Infrared and Nuclear Magnetic Resonance   312

CHAPTER 9

Alkenes and Alkynes   375

CHAPTER 10

Aromaticity, Benzene, and Substituted Benzenes   450

CHAPTER 11

Aldehydes and Ketones   509

CHAPTER 12

Carboxylic Acids   569

CHAPTER 13

Derivatives of Carboxylic Acids   605

CHAPTER 14

Enolates and Carbanions: Building Blocks for Organic Synthesis   663

CHAPTER 15

Amines   706

CHAPTER 16

Polycyclic and Heterocyclic Aromatic Compounds   **742**

CHAPTER 17

Pericyclic Reactions   **781**

CHAPTER 18

Carbohydrates   **806**

APPENDIX

CHAPTER 1

Atoms and Molecules— A Review

Around 1850, organic chemistry was defined as **the chemistry of compounds that come from living things**—hence the term *organic*. This definition was well-outgrown by about 1900. By that time, chemists were synthesizing new organic compounds in the laboratory, and many of these new compounds had no link with any living thing. Today, organic chemistry is defined as **the chemistry of the compounds of carbon**. This definition too is not entirely correct, because a few carbon compounds, such as carbon dioxide, sodium carbonate, and potassium cyanide, are considered to be inorganic. We accept this definition, however, because all organic compounds do contain carbon.

Carbon is but one element among many in the periodic table. What is so unique about carbon that its compounds justify a major subdivision in the study of chemistry? The answer is that carbon atoms can be covalently bonded to other carbon atoms and to atoms of other elements in a wide variety of ways, leading to an almost infinite number of different compounds. These compounds range in complexity from the simple compound methane (CH_4), the major component of natural gas and marsh gas, to the quite complex nucleic acids, the carriers of the genetic code in living systems.

A knowledge of organic chemistry is indispensable to many scientists. For example, because living systems are composed primarily of water and organic compounds, almost any area of study concerned with plants, animals, or micro-organisms depends on the principles of organic chemistry. These areas of study include medicine and the medical sciences, biochemistry, microbiology, agriculture, and many others. However, these are not the only fields that depend on organic chemistry. Plastics and synthetic fibers are also organic compounds. Petroleum and natural gas consist mostly of compounds of carbon and hydrogen that have been formed by the decomposition of plants. Coal is a mixture of elemental carbon combined with compounds of carbon and hydrogen.

Where do we start? The cornerstone of organic chemistry is the covalent bond. Before we discuss the structure, nomenclature, and reactions of organic compounds in detail, we will first review some aspects of atomic structure and bonding (Chapter 1) and then molecular orbitals (Chapter 2) as these topics apply to organic compounds.

Electron Structure of the Atom

The most important elements to organic chemists are carbon, hydrogen, oxygen, and nitrogen. These four elements are in the first two periods of the periodic table and their electrons are all found in the two electron shells closest to the nucleus. Consequently, our discussion of the electron structures of atoms will center mainly on elements with electrons only in these two electron shells.

Each electron shell is associated with a certain amount of energy. Electrons close to the nucleus are more attracted by the protons in the nucleus than are electrons farther away. Therefore, the closer an electron is to the nucleus, the lower is its energy. The electron shell closest to the nucleus is the one of lowest energy, and an electron in this shell is said to be at the **first energy level**. Electrons in the second shell, at the **second energy level**, are of higher energy than those in the first shell. Electrons in the third shell, at the **third energy level**, are of higher energy yet.

A.   Atomic orbitals

We cannot accurately determine the position of an electron relative to the nucleus of an atom. Instead, we must rely upon quantum theory to describe the most likely location of an electron. Each electron shell of an atom is subdivided into **atomic orbitals**, an atomic orbital being a region in space where the probability of finding an electron of a specific energy content is high (90–95%). **Electron density** is another term used to describe the probability of finding an electron in a particular spot; a higher electron density means a greater probability, while a lower electron density means a lesser probability.

The first electron shell contains only the spherical 1s orbital. The probability of finding a 1s electron is highest in this sphere. The second shell, which is slightly farther from the nucleus than the first shell, contains one 2s orbital and three 2p orbitals. The 2s orbital, like the 1s orbital, is spherical.

Figure 1.1 shows a graph of electron density in the 1s and 2s orbitals as a function of distance from the nucleus. It may be seen from the graph that the 1s and 2s orbitals do not have sharply defined surfaces, but rather the electron density increases and decreases over a range of distances from the nucleus. The result is that the 1s and 2s orbitals overlap each other.

The electron density–distance curve for the 2s orbital reveals two areas of high electron density separated by a zero point. This zero point is called a **node**, and represents a region in space where the probability of finding an electron (the 2s electron in this case) is very small. All orbitals except the 1s orbital have nodes. Pictorial representations of the 1s and 2s orbitals are shown in Figure 1.2.

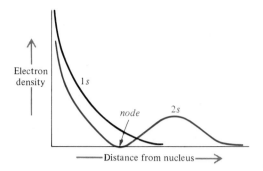

FIGURE 1.1. Graphic relationship between the 1s and 2s atomic orbitals.

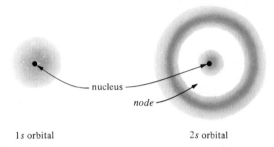

1s orbital   2s orbital

FIGURE 1.2. Pictorial representations of the 1s and 2s atomic orbitals.

The second energy level also contains three 2p atomic orbitals. The 2p orbitals are at a slightly greater distance from the nucleus than the 2s orbital and are of slightly higher energy. The p orbitals are shaped rather like dumbbells; each p orbital has two lobes separated by a node (a nodal plane in this case) at the nucleus (see Figure 1.3).

A sphere (an s orbital) is nondirectional; that is, it appears the same when viewed from any direction. This is not the case with a p orbital, which can assume different orientations about the nucleus. The three 2p orbitals are at *right angles* to each other—this orientation allows maximum distance between the electrons in the three p orbitals and thus minimizes repulsions between electrons in different p orbitals. The mutually perpendicular p orbitals are sometimes designated p_x, p_y, and p_z. The subscript letters refer to the x, y, and z axes that may be drawn through pictures of these p orbitals, as in Figure 1.3.

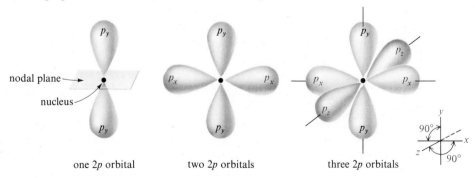

one 2p orbital   two 2p orbitals   three 2p orbitals

FIGURE 1.3. The shapes and orientations of the 2p orbitals.

TABLE 1.1. Atomic orbitals in the first three energy levels

Energy level	Atomic orbitals
1	$1s$
2	$2s\ 2p_x\ 2p_y\ 2p_z$
3	$3s\ 3p_x\ 3p_y\ 3p_z$ plus five $3d$

Since the three $2p$ orbitals are equivalent in shape and in distance from the nucleus, they have equal energies. Orbitals that have the same energy, such as the three $2p$ orbitals, are said to be **degenerate**.

The third electron shell contains one $3s$ orbital, three $3p$ orbitals, and also five $3d$ orbitals. The numbers of atomic orbitals at each of the first three energy levels are summarized in Table 1.1.

B.   Filling the orbitals

Electrons have spin, which can be either clockwise or counterclockwise. The spin of a charged particle gives rise to a small magnetic field, or **magnetic moment**, and two electrons with opposite spin have *opposite magnetic moments*. The repulsion between the negative charges of two electrons with opposite spin is minimized by the opposite magnetic moments, allowing two such electrons to become *paired* within an orbital. For this reason, any orbital can hold a maximum of two electrons, but those electrons must be of opposite spin. Because of the number of orbitals at each energy level (one at the first energy level, four at the second, and nine at the third), the first three energy levels can hold up to two, eight, and 18 electrons, respectively.

The **aufbau principle** (German, "building up") states that as we progress from hydrogen (atomic number 1) to atoms of successively higher atomic number, orbitals become filled with electrons in such a way that the *lowest-energy orbitals are filled first*. A hydrogen atom has its single electron in a $1s$ orbital. The next element, helium (atomic number 2), has its second electron also in the $1s$ orbital. The two electrons in this orbital are paired.

TABLE 1.2. Electron configurations of the elements in periods 1 and 2

Element	Atomic number	Electron configuration
H	1	$1s^1$
He	2	$1s^2$
Li	3	$1s^2\ 2s^1$
Be	4	$1s^2\ 2s^2$
B	5	$1s^2\ 2s^2\ 2p^1$
C	6	$1s^2\ 2s^2\ 2p^2$
N	7	$1s^2\ 2s^2\ 2p^3$
O	8	$1s^2\ 2s^2\ 2p^4$
F	9	$1s^2\ 2s^2\ 2p^5$
Ne	10	$1s^2\ 2s^2\ 2p^6$

A description of the electron structure for an element is called its **electron configuration**. The electron configuration for H is $1s^1$, which means one electron (superscript 1) in the $1s$ orbital. For He, the electron configuration is $1s^2$, meaning *two* electrons (superscript 2) in the $1s$ orbital. Lithium (atomic number 3) has two electrons in the $1s$ orbital and one electron in the $2s$ orbital; its electron configuration is $1s^2\ 2s^1$.

The electron configurations for the first- and second-period elements are shown in Table 1.2. In carbon and the succeeding elements, each $2p$ orbital receives one electron before any $2p$ orbital receives a second electron. This is an example of **Hund's rule**: In filling atomic orbitals, pairing of two electrons in degenerate orbitals does not occur until each degenerate orbital contains one electron. Therefore, an atom of carbon has an electron configuration of $1s^2\ 2s^2\ 2p_x{}^1\ 2p_y{}^1$.

SECTION 1.2.

Atomic Radius

The **radius of an atom** is the distance from the center of the nucleus to the outermost electrons. The atomic radius is determined by measuring the **bond length** (the distance between nuclei) in a covalent compound such as Cl—Cl or H—H and dividing by two. Therefore, atomic radii are often called **covalent radii**. Values for atomic radii are usually given in Angstroms (Å), where 1 Å = 10^{-8} cm.

For H$_2$:

bond length = 0.74 Å

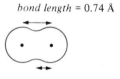

atomic radius = 0.37 Å

Atomic radii vary depending on the extent of attraction between the nucleus and its electrons. The greater the attraction, the smaller is the atomic radius. What factors affect this attraction? The most important factors are *the number of protons in the nucleus* and *the number of shells containing electrons*.

A nucleus with a greater number of protons has a greater attraction for its electrons, including the outermost electrons. Consider the elements of the second row of the periodic table (lithium to fluorine). An atom of any of these elements has electrons in only the first two electron shells. As we progress stepwise from lithium to fluorine, a proton is added to the nucleus. At each step, the nucleus has a greater attraction for the electrons, and the atomic radius decreases (refer to Figure 1.4).

	Li	Be	B	C	N	O	F
atomic number:	3	4	5	6	7	8	9

decreasing atomic radius

H 0.37						
Li 1.225	Be 0.889	B 0.80	C 0.771	N 0.74	O 0.74	F 0.72
Na 1.572	Mg 1.364	Al 1.248	Si 1.173	P 1.10	S 1.04	Cl 0.994
						Br 1.142
						I 1.334

FIGURE 1.4. Atomic radii of some of the elements (in Angstroms, Å, where $1Å = 10^{-8}$ cm).

As we proceed from top to bottom within a group in the periodic table, the number of electron shells increases and, therefore, so does the atomic radius.

**increasing
atomic radius**

H   (1 shell)
Li   (2 shells)
Na   (3 shells)
K   (4 shells)

In organic chemistry, atoms are bonded together in close proximity to one another by covalent bonds. We will find the concept of atomic radii useful in estimating the attractions and repulsions between atoms and in discussing co-valent bond strengths.

SECTION 1.3.

Electronegativity

Electronegativity is a measure of the ability of an atom to attract its outer, or valence, electrons. Since it is the outer electrons of an atom that are used for bonding, electronegativity is useful for predicting and explaining chemical re-activity. Like the atomic radius, electronegativity is affected by the number of protons in the nucleus and by the number of shells containing electrons. A greater number of protons means a greater positive nuclear charge, and thus an increased attraction for the bonding electrons. Therefore, electronegativity increases as we go from left to right in a given period of the periodic table.

Li Be B C N O F
increasing electronegativity

Attractions between oppositely charged particles increase with decreasing distance between the particles. Thus, electronegativity increases as we proceed from bottom to top in a given group of the periodic table because the valence electrons are closer to the nucleus.

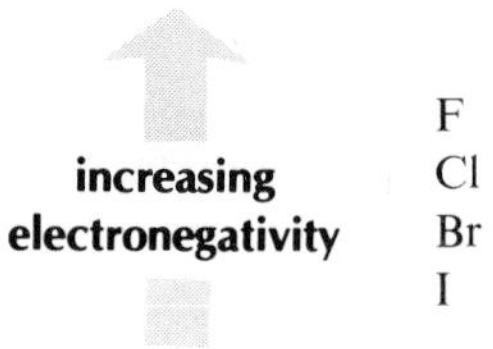

↑	F
increasing	Cl
electronegativity	Br
	I

The **Pauling scale** (Figure 1.5) is a numerical scale of electronegativities. This scale is derived from bond-energy calculations for different elements joined by covalent bonds. In the Pauling scale, fluorine, the most electronegative element, has an electronegativity value of 4. Lithium, which has a low electronegativity, has a value of 1. An element with a very low electronegativity (such as lithium) is sometimes called an **electropositive** element. Carbon has an intermediate electronegativity value of 2.5.

H 2.1						
Li 1.0	Be 1.5	B 2.0	C 2.5	N 3.0	O 3.5	F 4.0
Na 0.9	Mg 1.2	Al 1.5	Si 1.8	P 2.1	S 2.5	Cl 3.0
						Br 2.8
						I 2.5

FIGURE 1.5. Electronegativities of some elements (Pauling scale).

SECTION 1.4.

Introduction to the Chemical Bond

Because of their different electron structures, atoms can become bonded together in molecules in different ways. In 1916, G. N. Lewis and W. Kössel advanced the following theories:

1. An **ionic bond** results from the transfer of electrons from one atom to another.

2. A **covalent bond** results from the sharing of a pair of electrons by two atoms.

3. Atoms transfer or share electrons so as to gain a **noble-gas electron configuration**. This configuration is usually eight electrons in the outer shell, corresponding to the electron configuration of neon and argon. This theory is called the **octet rule**.

An ionic bond is formed by electron transfer. One atom donates one or more of its outermost, or bonding, electrons to another atom or atoms. The atom that loses electrons becomes a positive ion, or **cation**. The atom that gains the electrons becomes a negative ion, or **anion**. The ionic bond results from the electrostatic attraction between these oppositely charged ions. We may illustrate electron transfer by using dots to represent the bonding electrons.

$$Na\cdot + \cdot \ddot{C}l: \longrightarrow Na :\ddot{C}l: \quad or \quad Na^+ \; Cl^-$$

A covalent bond is produced by the sharing of a pair of bonding electrons between two atoms. Shared electrons result from the merging of the atomic orbitals into shared orbitals called **molecular orbitals**, a topic that we will discuss in Chapter 2. For now, we will use dots to represent bonding electrons. With the dot formulas, called **Lewis formulas**, we can easily count electrons and see that the atoms attain noble-gas configurations: two electrons (helium configuration) for hydrogen and eight electrons for most other atoms.

$$H\cdot + H\cdot \longrightarrow H\!:\!H$$
$$:\ddot{C}l\cdot + \cdot\ddot{C}l: \longrightarrow :\ddot{C}l\!:\!\ddot{C}l:$$

covalent bond

$$\cdot\dot{C}\cdot + 4\,H\cdot \longrightarrow \begin{array}{c} H \\ H\!:\!\ddot{C}\!:\!H \\ H \end{array} \longleftarrow \textit{four covalent bonds}$$

The sharing of one pair of electrons between two atoms is called a **single bond**. Two atoms can share two pairs or even three pairs of electrons; these multiple bonds are called **double bonds** and **triple bonds**, respectively.

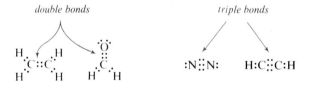

double bonds   *triple bonds*

:N::N: H:C::C:H

SAMPLE PROBLEMS

Circle the eight bonding electrons associated with the carbon atom in each of the following structures:

$$\begin{array}{c} \ddot{O} \\ \| \\ H\!:\!C\!:\!H \end{array} \qquad :\!\ddot{O}\!::\!C\!::\!\ddot{O}: \qquad H\!:\!C\!:\!:\!N:$$

Solution:

$$\begin{array}{c} \ddot{O} \\ H\!:\!\widehat{C}\!:\!H \end{array} \qquad :\!\ddot{O}\,\widehat{::\!C\!::}\,\ddot{O}: \qquad H\,\widehat{:\!C\!::}\,N:$$

For the structures in the preceding problem, circle the two electrons associated with each hydrogen atom and the eight electrons associated with each oxygen or nitrogen atom.

Solution:

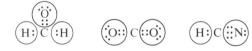

When do atoms form ionic bonds and when do they form covalent bonds? Ionic bonds are formed when the electronegativity difference between two atoms is large (greater than about 1.7). For example, a sodium atom (electronegativity 0.9), with little attraction for its bonding electron, readily loses this electron to a chlorine atom (electronegativity 3.0). On the other hand, the electronegativity difference between two carbon atoms is zero; between carbon and hydrogen, only 0.4; and between carbon and chlorine, 0.5. Because carbon has an electronegativity of 2.5, intermediate between the extremes of high and low electronegativity, it almost never forms ionic bonds with other elements. Instead, *carbon forms covalent bonds with other carbon atoms and with atoms of other elements.*

A.   Valence

The **valence** of an atom is the number of electrons that the atom loses, gains, or shares. In a covalent molecule, the valence of each atom is the number of covalent bonds that the atom forms. Carbon has four bonding electrons and forms four covalent bonds to attain an octet. Therefore, we say that carbon has a valence of four. Table 1.3 lists typical valences of elements commonly found in organic compounds.

$$
\cdot \overset{\cdot}{\underset{\cdot}{C}} \cdot \quad + \quad 4\,H\cdot \quad \longrightarrow \quad H : \overset{\cdot\cdot}{\underset{H}{\overset{H}{C}}} : H
$$

valence   valence
of 4   of 1

$$
\cdot \overset{\cdot}{\underset{\cdot}{C}} \cdot \quad + \quad 4\,:\!\overset{\cdot\cdot}{\underset{\cdot\cdot}{Cl}}\!\cdot \quad \longrightarrow \quad :\!\overset{\cdot\cdot}{\underset{\cdot\cdot}{Cl}}\!:\overset{:\overset{\cdot\cdot}{\underset{\cdot\cdot}{Cl}}:}{\underset{:\overset{\cdot\cdot}{\underset{\cdot\cdot}{Cl}}:}{C}}:\!\overset{\cdot\cdot}{\underset{\cdot\cdot}{Cl}}\!:
$$

valence   valence
of 4   of 1

TABLE 1.3. Most common valences of some elements typically encountered in organic compounds

Element	Valence
H	1
C	4
O	2
N	3
Cl	1
I	1
Br	1

For simple structures, we can often deduce the Lewis formula for a compound of known composition from valence rules alone.

SAMPLE PROBLEMS

Write the Lewis formulas for H_2O and C_2H_6.

Solution:

1. Determine the number of valence electrons of each atom: $H = 1$; $O = 6$; $C = 4$.

2. Draw the skeleton of the molecule following the rules of valence: H can form one covalent bond; O can form two; and C can form four.

$$\begin{array}{ccc} & & \text{H H} \\ \text{H O H} & & \text{H C C H} \\ & & \text{H H} \end{array}$$

3. Distribute the valence electrons in such a way that each H has two electrons and each other atom has an octet.

$$\begin{array}{cc} & \text{H H} \\ \text{H:}\ddot{\text{O}}\text{:H} & \text{H:}\ddot{\text{C}}\text{:}\ddot{\text{C}}\text{:H} \\ & \text{H H} \end{array}$$

Write the Lewis formula for C_3H_8 and two Lewis formulas for C_2H_6O.

Solution:

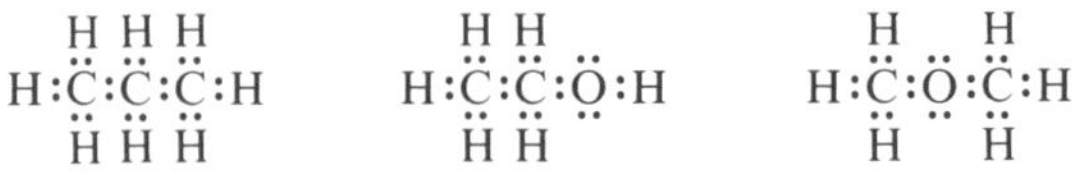

$$\begin{array}{ccc} \text{H H H} & \text{H H} & \text{H \quad H} \\ \text{H:}\ddot{\text{C}}\text{:}\ddot{\text{C}}\text{:}\ddot{\text{C}}\text{:H} & \text{H:}\ddot{\text{C}}\text{:}\ddot{\text{C}}\text{:}\ddot{\text{O}}\text{:H} & \text{H:}\ddot{\text{C}}\text{:}\ddot{\text{O}}\text{:}\ddot{\text{C}}\text{:H} \\ \text{H H H} & \text{H H} & \text{H \quad H} \end{array}$$

STUDY PROBLEM

1.1.   Each of the following structures contains a double or triple bond. Give the Lewis formulas.

(a) Cl_2CO (b) C_2Cl_4 (c) C_2H_3Br (d) C_2HF

B.   Formal charge

In some structures, the rules of valence do not seem to be followed. One such case occurs when one atom in a molecule provides both electrons for a covalent bond.

$$\text{H:}\ddot{\text{O}}\text{:N::}\ddot{\text{O}}\qquad \textit{both } e^- \textit{ from } N$$

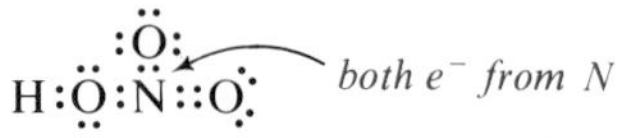

nitric acid, HNO_3

If one atom provides both electrons for a covalent bond, the bond is called a **coordinate covalent bond**. In such a case, some atoms within the covalent structure carry a positive or negative charge called a **formal charge**. In the coordinate covalent bond, the donating atom has a formal charge of $+1$, and the accepting atom has a formal charge of -1. Since both charges are found within the same structure, the opposite charges cancel, and the molecule as a whole may have no net ionic charge. (If the charges do not cancel, the structure represents an ion with a net ionic charge.)

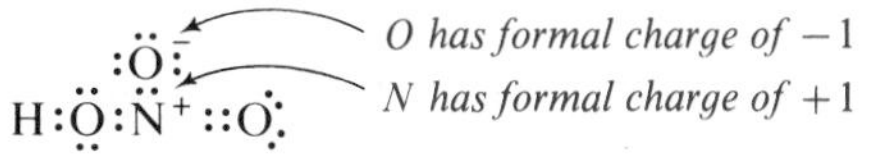

$$H : \ddot{O} : \overset{..}{N^+} :: \ddot{O} :$$

O has formal charge of −1
N has formal charge of +1

The formal charge for an atom may be calculated by the following equation:

formal charge = (number of valence e^- in a neutral atom)
$-\frac{1}{2}$ (number of shared e^-) − (number of unshared valence e^-)

Example Using the formula, we can calculate the formal charges on N and each O in HNO_3. (H does not carry a formal charge in covalent molecules.)

$$H : \ddot{O} \overset{\ddot{O}}{N} \overset{..}{O} :$$

no. of valence e^- for N = 5
$\frac{1}{2}$(no. of shared e^- for N) = 4
no. of unshared e^- for N = 0
formal charge for N = 5 − 4 − 0 = +1

For each O, the same technique is used.

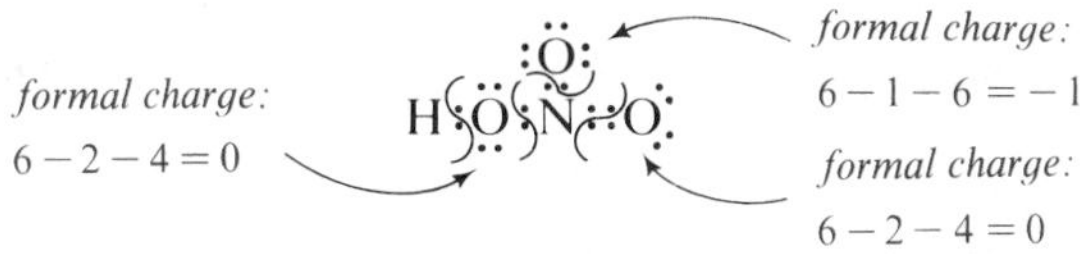

formal charge:
$6 - 2 - 4 = 0$

$$H \overset{\ddot{O}}{O} \overset{}{N} \overset{..}{O} :$$

formal charge:
$6 - 1 - 6 = -1$
formal charge:
$6 - 2 - 4 = 0$

Some other examples of calculation of formal charge follow:

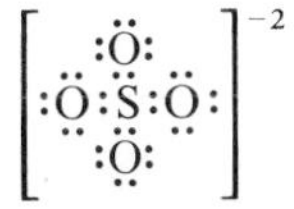

$$\left[: \ddot{O} : \overset{\ddot{O}}{S} : \ddot{O} : \right]^{-2}$$

for S: $6 - 4 - 0 = +2$
for each O: $6 - 1 - 6 = -1$

(Because the formal charges do not cancel,
the formula represents an ion: SO_4^{-2}.)

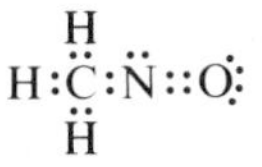

$$\begin{array}{c} H \\ H : \overset{..}{C} : \overset{..}{N} :: \ddot{O} : \\ H \end{array}$$

for C: $4 - 4 - 0 = 0$
for N: $5 - 3 - 2 = 0$
for O: $6 - 2 - 4 = 0$

STUDY PROBLEMS

1.2.   Calculate the formal charges on C and N in the following formulas, and determine whether each represents a neutral molecule or an ion.

(a) 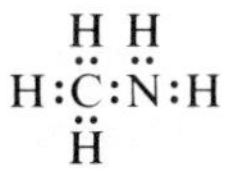 $\begin{array}{c} H \ H \\ H : \overset{..}{C} : \overset{..}{N} : H \\ H \end{array}$ (b) 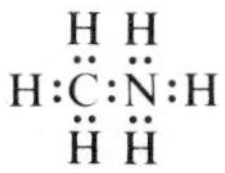 $\begin{array}{c} H \ H \\ H : \overset{..}{C} : \overset{..}{N} : H \\ H \ H \end{array}$

1.3.   Calculate the formal charge on each atom in the following structures:

(a)  $H : \ddot{O} : \overset{..}{N} :: \ddot{O} :$ (b)  $\begin{array}{c} H \\ H : \overset{..}{C} :: N :: \overset{..}{N} : \end{array}$ (c) $\begin{array}{c} H \ \ \ddot{O} : \\ H : \overset{..}{C} : \ddot{O} : \overset{\ddot{O}}{S} : \ddot{O} : H \\ H \ \ \ddot{O} : \end{array}$

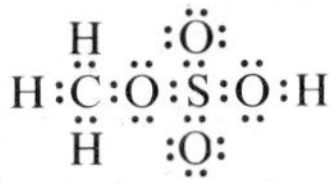

Chemical Formulas in Organic Chemistry

Lewis formulas are useful for keeping track of bonding electrons, but organic chemists rarely use true Lewis formulas. Let us consider the types of chemical formulas that are more frequently encountered.

An **empirical formula** tells us the types of atoms and their numerical ratio in a molecule. For example, a molecule of ethane contains carbon and hydrogen atoms in a ratio of 1 to 3; the empirical formula is CH_3. A **molecular formula** tells us the actual number of atoms in a molecule, not simply the ratio. The molecular formula for ethane is C_2H_6. A **structural formula** shows the *structure* of a molecule—that is, the order of attachment of the atoms. In order to explain or predict chemical reactivity, we need to know the structure of a molecule; therefore, structural formulas are the most useful of the different types of formulas.

$$CH_3 \qquad\qquad C_2H_6 \qquad\qquad \begin{array}{c} H \quad H \\ | \quad\; | \\ H-C-C-H \\ | \quad\; | \\ H \quad H \end{array}$$

empirical formula   *molecular formula*   *structural formula*
for ethane   *for ethane*   *for ethane*

A.   Structural formulas

Lewis formulas are one type of structural formula. However, chemists usually represent a covalent structure by using a dash for each shared pair of electrons, and rarely show unshared pairs of valence electrons. Formulas with dashes for bonds are called **valence-bond formulas**. In this text, we will also refer to them as **complete structural formulas**.

$$H:H \quad \text{becomes} \quad H-H$$

$$:\ddot{C}l:\ddot{C}l: \quad \text{becomes} \quad Cl-Cl$$

$$\begin{array}{c} H \\ \cdot\cdot \\ H:C:H \\ \ddot{H} \end{array} \quad \text{becomes} \quad \begin{array}{c} H \\ | \\ H-C-H \\ | \\ H \end{array}$$

$$\begin{array}{c} H \quad\;\; H \\ \cdot \quad\;\; \cdot \\ :C::C: \\ H\cdot \quad \cdot H \end{array} \quad \text{becomes} \quad \begin{array}{c} H \qquad\quad H \\ \diagdown \qquad \diagup \\ C=C \\ \diagup \qquad \diagdown \\ H \qquad\quad H \end{array}$$

Although unshared pairs of electrons are not usually shown in valence-bond formulas, we will sometimes show these electrons when we want to emphasize their role in a chemical reaction.

All represent the same molecule:

— *unshared pair of* e^-

$$\begin{array}{c} \cdot\cdot \\ H:N:H \\ \ddot{H} \end{array} \quad \text{or} \quad \begin{array}{c} H-N-H \\ | \\ H \end{array} \quad \text{or} \quad \begin{array}{c} \cdot\cdot \\ H-\ddot{N}-H \\ | \\ H \end{array}$$

SAMPLE PROBLEM

Write the valence-bond formula for each of the following Lewis formulas:

$$
\begin{array}{ccc}
\text{H} \quad \text{H} & \text{H} \text{H} & \overset{\ddot{\text{O}}}{:} \\
\text{H}:\ddot{\text{C}}:\ddot{\text{O}}:\ddot{\text{C}}:\text{H} & \text{H}:\ddot{\text{C}}:\ddot{\text{C}}:\ddot{\text{O}}:\text{H} & \text{H}:\ddot{\text{C}}:\text{H} \\
\text{H} \quad \text{H} & \text{H} \text{H} &
\end{array}
$$

Solution:

$$
\begin{array}{ccc}
\text{H} \quad\quad \text{H} & \text{H} \quad \text{H} & \text{O} \\
| \quad\quad | & | \quad | & \| \\
\text{H}-\text{C}-\text{O}-\text{C}-\text{H} & \text{H}-\text{C}-\text{C}-\text{O}-\text{H} & \text{H}-\text{C}-\text{H} \\
| \quad\quad | & | \quad | & \\
\text{H} \quad\quad \text{H} & \text{H} \quad \text{H} &
\end{array}
$$

B.   Condensed structural formulas

Complete structural formulas are frequently condensed to shorter, more convenient formulas. In **condensed structural formulas**, bonds are not always shown, and atoms of the same type bonded to one other atom are grouped together. The structure of a molecule is still evident from a condensed structural formula as long as the rules of valence are taken into consideration.

$$
\text{CH}_3\text{CH}_3 \quad \text{is the condensed structural formula for} \quad
\begin{array}{c}
\text{H} \quad \text{H} \\
| \quad | \\
\text{H}-\text{C}-\text{C}-\text{H} \\
| \quad | \\
\text{H} \quad \text{H}
\end{array}
$$

$$
\text{CH}_3\text{CH}_2\text{OH} \quad \text{is the condensed structural formula for} \quad
\begin{array}{c}
\text{H} \quad \text{H} \\
| \quad | \\
\text{H}-\text{C}-\text{C}-\text{O}-\text{H} \\
| \quad | \\
\text{H} \quad \text{H}
\end{array}
$$

SAMPLE PROBLEM

Write (a) the complete structural formula (showing all bonds as dashes), and (b) the condensed structural formula, for each of the following Lewis formulas:

$$
\begin{array}{cc}
\text{H} \text{H} & \text{H} \quad \text{H} \quad \text{H} \\
\text{H}:\ddot{\text{C}}:\ddot{\text{C}}:\ddot{\text{C}}\text{l}: & \text{H}:\ddot{\text{C}} : \ddot{\text{C}} : \ddot{\text{C}}:\text{H} \\
\text{H} \text{H} & \text{H} \quad :\ddot{\text{C}}\text{l}: \quad \text{H}
\end{array}
$$

Solution:

$$
\begin{array}{cc}
& \text{H} \quad \text{H} \quad\quad \text{H} \quad \text{H} \quad \text{H} \\
& | \quad | \quad\quad\quad | \quad | \quad | \\
\text{(a)} \quad \text{H}-\text{C}-\text{C}-\text{Cl} \quad\quad \text{H}-\text{C}-\text{C}-\text{C}-\text{H} \\
& | \quad | \quad\quad\quad | \quad | \quad | \\
& \text{H} \quad \text{H} \quad\quad \text{H} \quad \text{Cl} \quad \text{H}
\end{array}
$$

$$
\text{(b)} \quad \text{CH}_3\text{CH}_2\text{Cl} \quad\quad\quad \text{CH}_3\text{CHClCH}_3
$$

Structural formulas may be condensed even further if a molecule has two or more identical groups of atoms. In these cases, are used to parentheses enclose a repetitive group of atoms. The subscript following the second parenthesis indicates the number of times the entire group is found at that position in the molecule.

$$(CH_3)_2CHOH \quad \text{is the same as} \quad CH_3-\overset{\overset{\displaystyle CH_3}{|}}{\underset{\underset{\displaystyle H}{|}}{C}}-OH$$

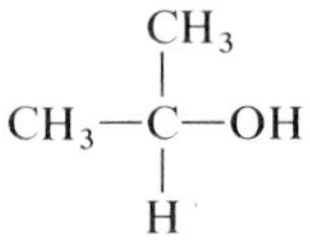

$$(CH_3)_3CCl \quad \text{is the same as} \quad CH_3-\overset{\overset{\displaystyle CH_3}{|}}{\underset{\underset{\displaystyle CH_3}{|}}{C}}-Cl$$

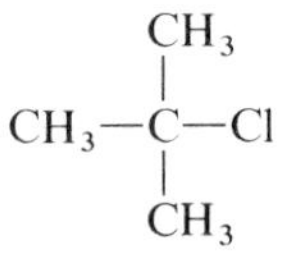

$$CH_3(CH_2)_3CH_3 \quad \text{is the same as} \quad CH_3CH_2CH_2CH_2CH_3$$

For the sake of clarity, double or triple bonds are usually shown in a condensed structural formula.

$$CH_3CH{=}CH_2 \quad \text{is the same as}$$

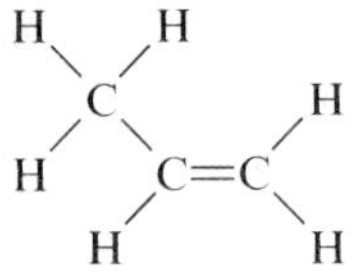

$$CH_3C{\equiv}CH \quad \text{is the same as} \quad H-\overset{\overset{\displaystyle H}{|}}{\underset{\underset{\displaystyle H}{|}}{C}}-C{\equiv}C-H$$

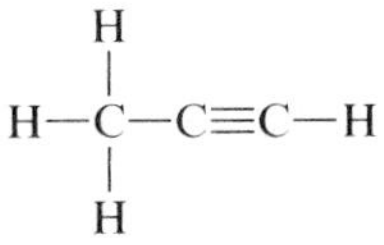

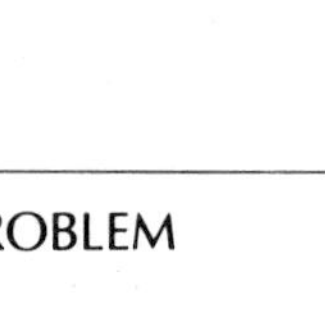

$$\overset{\displaystyle O}{\overset{\displaystyle \|}{CH_3CCH_2CH_3}} \quad \text{is the same as} \quad H-\overset{\overset{\displaystyle H}{|}}{\underset{\underset{\displaystyle H}{|}}{C}}-\overset{\overset{\displaystyle O}{\|}}{C}-\overset{\overset{\displaystyle H}{|}}{\underset{\underset{\displaystyle H}{|}}{C}}-\overset{\overset{\displaystyle H}{|}}{\underset{\underset{\displaystyle H}{|}}{C}}-H$$

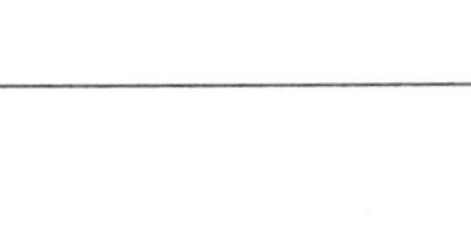

STUDY PROBLEM

1.4.   For each of the following formulas, write a more condensed formula:

(a) 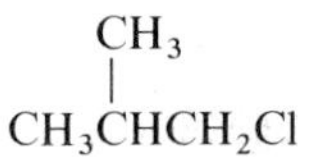 $CH_3\overset{\overset{\displaystyle CH_3}{|}}{C}HCH_2Cl$ (b) $CH_3\overset{\overset{\displaystyle Cl}{|}}{C}HCl$ (c) $CH_3CH_2CH_2CH_2\overset{\overset{\displaystyle Cl}{|}}{C}HCH_2Cl$

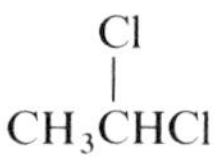

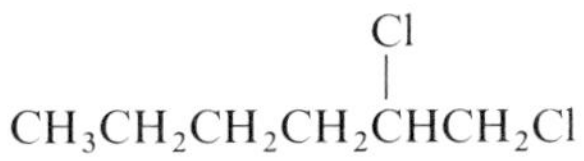

(d) 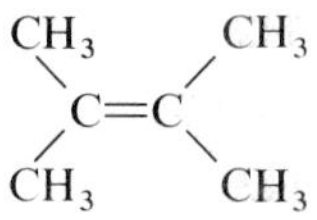 $\underset{\underset{\displaystyle CH_3}{}}{\overset{\overset{\displaystyle CH_3}{}}{C}}{=}\underset{\underset{\displaystyle CH_3}{}}{\overset{\overset{\displaystyle CH_3}{}}{C}}$ (e) $N{\equiv}C-CH_2-C{\equiv}N$

C.   Cyclic compounds and polygon formulas

A compound such as $CH_3CH_2CH_2CH_3$ is said to have its carbon atoms connected in a chain. Carbon atoms can be joined together in rings as well as in chains; a compound containing one or more rings is called a **cyclic compound**.

Cyclic structures are usually represented by **polygon formulas**, which are another type of condensed structural formula. For example, a triangle is used to represent a three-membered ring, while a hexagon is used for a six-membered ring.

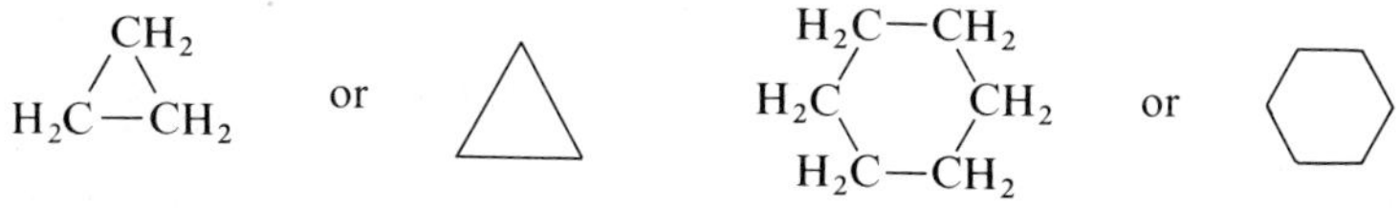

$$CH_2 \diagdown \diagup \quad H_2C-CH_2 \qquad \text{or} \qquad \triangle \qquad\qquad H_2C\diagup^{H_2C-CH_2}\diagdown CH_2 \\ H_2C-CH_2 \qquad \text{or} \qquad \hexagon$$

In polygon formulas, a corner represents a carbon atom along with its hydrogens; the sides of the polygon represent the bonds joining the carbons. If an atom or group other than hydrogen is attached to a carbon of the ring, the number of hydrogens at that position is reduced accordingly.

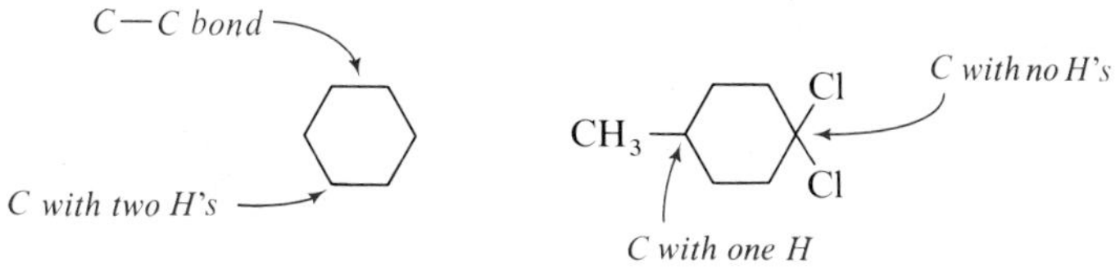

C—C bond

C with two H's

C with no H's

C with one H

CH_3 Cl Cl

Rings can contain atoms other than carbon; these atoms and any hydrogens attached to them must be indicated in the polygon formula. Double bonds must also be indicated.

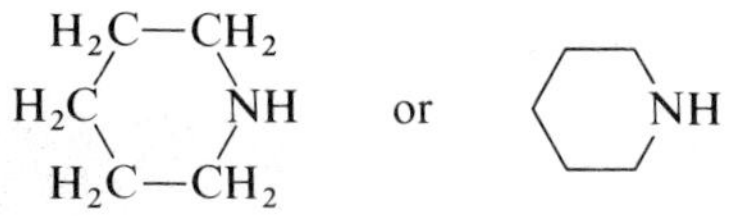

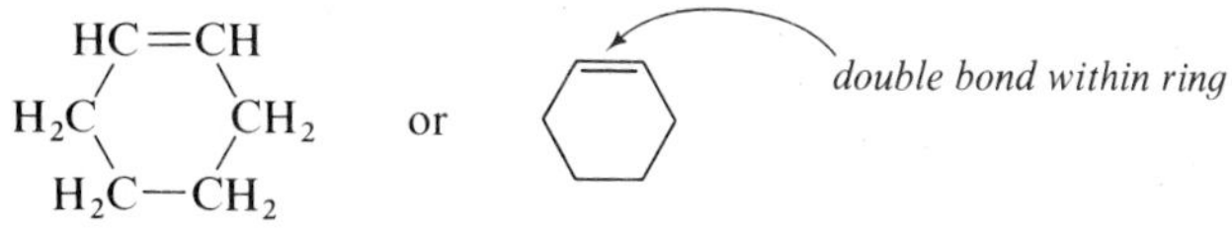

$$H_2C\diagup^{H_2C-CH_2}\diagdown NH \\ H_2C-CH_2 \qquad \text{or} \qquad \hexagon NH$$

$$H_2C\diagup^{HC=CH}\diagdown CH_2 \\ H_2C-CH_2 \qquad \text{or} \qquad \hexagon \quad \text{—}\quad \textit{double bond within ring}$$

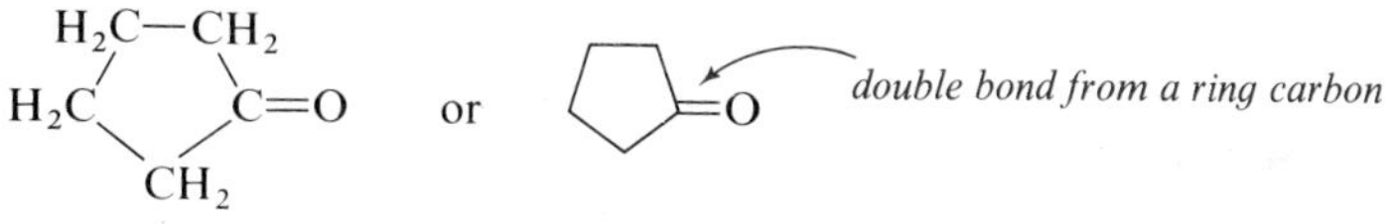

$$H_2C\diagup^{H_2C-CH_2}\diagdown C{=}O \\ CH_2 \qquad \text{or} \qquad \pentagon{=}O \quad \textit{double bond from a ring carbon}$$

STUDY PROBLEMS

1.5. Draw complete structural formulas for the following structures, showing each C, each H, and each bond.

(a) 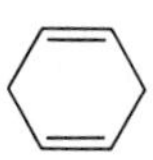 $\hexagon$ (b)  $\hexagon NCH_3$ (c) 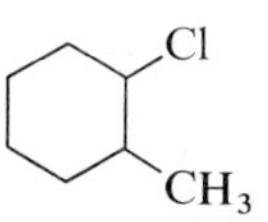 $\hexagon$ with Cl and CH_3

1.6. Draw polygon formulas for the following structures:

(a) 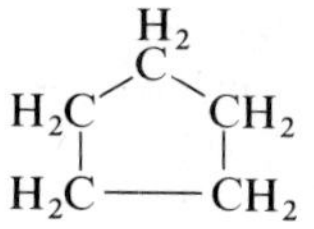
$$H_2C\diagup^{\overset{H_2}{C}}\diagdown CH_2 \\ H_2C{-\!-}CH_2$$

(b) 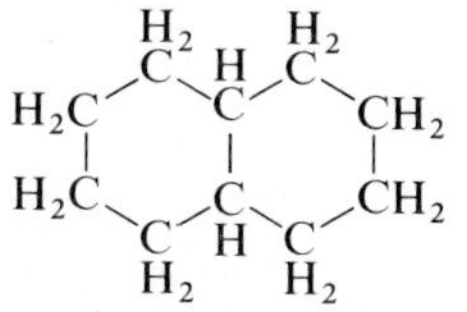
$$H_2C\diagup^{\overset{H_2}{C}-\overset{H}{C}-\overset{H_2}{C}}\diagdown CH_2 \\ H_2C\diagdown_{\underset{H_2}{C}-\underset{H}{C}-\underset{H_2}{C}}\diagup CH_2$$

(c)
$$H_2C{-\!-\!-}CH_2 \\ H_2C\diagdown_{O}\diagup CH_2$$

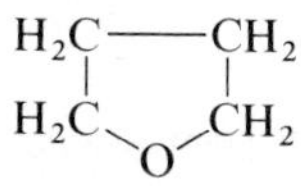

(d) 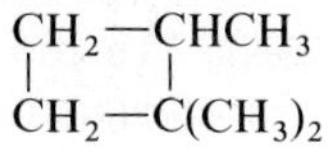
$$CH_2{-}CHCH_3 \\ CH_2{-}C(CH_3)_2$$

(e) 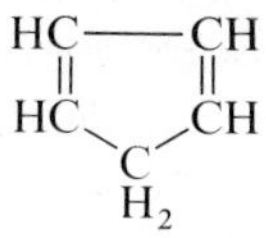
$$HC{-\!-\!-}CH \\ \| \qquad \| \\ HC\diagdown_{\underset{H_2}{C}}\diagup CH$$

SECTION 1.6.

Bond Lengths and Bond Angles

We have discussed how chemists represent covalent compounds. Now let us consider some of the properties of covalent bonds. The distance that separates the nuclei of two covalently bonded atoms is called the **bond length**. Covalent bond lengths, which can be measured experimentally, range from 0.74 Å to 2 Å.

If there are more than two atoms in a molecule, the bonds form an angle, called the **bond angle**. Bond angles vary from about 60° up to 180°.

Most organic structures contain more than three atoms, and are three-dimensional rather than two-dimensional. The preceding structural formula for ammonia (NH_3) illustrates one technique for representing a three-dimensional structure. A line bond (—) represents a bond in the plane of the paper. The solid wedge (——) represents a bond coming out of the paper toward the viewer; the H at the wide end of the solid wedge is in front of the paper. The broken wedge (⸳⸳⸳⸳⸳⸳) represents a bond pointing back into the paper; the H at the small end of the broken wedge is behind the paper.

SECTION 1.7.

Bond Dissociation Energy

When atoms bond together to form molecules, energy is liberated (usually as heat or light). Thus, for a molecule to be dissociated into its atoms, energy must be supplied.

There are two ways a bond may dissociate. One way is by **heterolytic cleavage** (Greek *hetero*, "different"), in which both bonding electrons are retained by one of the atoms. The result of heterolytic cleavage is a pair of ions.

We use a curved arrow (⌒) in these equations to show the direction in which the pair of bonding electrons moves during bond breakage. In the heterolytic cleavage of HCl or H_2O, the bonding electrons are transferred to the more electronegative Cl or O.

The other process by which a bond may dissociate is **homolytic cleavage** (Greek *homo*, "same"). In this case, each atom involved in the covalent bond receives one electron from the original shared pair. Electrically neutral atoms or groups of atoms result.

Homolytic cleavage: $H{-}H \longrightarrow H\cdot + H\cdot$

$H{-}Cl \longrightarrow H\cdot + Cl\cdot$

$H_3C{-}H \longrightarrow H_3C\cdot + H\cdot$

Note that the curved arrows in these equations have only half an arrowhead. This type of arrow (⌒), called a fishhook, is used to show the direction of shift of *one* electron, whereas the curved arrow with a complete head (⌒) is used to show the direction of shift of a *pair* of electrons.

Another new symbol is the single dot, as in $Cl\cdot$. This dot stands for a lone, unshared, unpaired electron. Other outer electrons are ignored in this symbolism. The symbol $Cl\cdot$ really means $:\ddot{C}l\cdot$. An atom such as $H\cdot$ or a group of atoms such as $H_3C\cdot$ that contains an unpaired electron is called a **free radical**. Free radicals are usually electrically neutral; therefore, there are no electrostatic attractions between free radicals, as there are between ions. Also, most free radicals are of high energy; consequently, they are unstable and very reactive.

Homolytic cleavage is more useful than heterolytic cleavage in determining the energies required for bond dissociations because calculations are not complicated by ionic attractions between the products. From measurements of the components of dissociating gases at high temperatures, the **change in enthalpy** ΔH (change in heat content, or energy) has been calculated for a large number of bond dissociations. For the reaction $CH_4 \rightarrow CH_3\cdot + H\cdot$, ΔH equals 104 kcal/mole. In other words, to cleave one hydrogen atom from each carbon atom in one mole of CH_4 requires 104 kcal. This value (104 kcal/mole) is the **bond dissociation energy** for the $H_3C{-}H$ bond.

The bond dissociation energies for several types of bonds are listed in Table 1.4. To break a more stable bond requires a higher energy input. For example, cleavage of HF to $H\cdot$ and $F\cdot$ (135 kcal/mole) is difficult compared with cleavage of the $O{-}O$ bond in hydrogen peroxide, HOOH (35 kcal/mole).

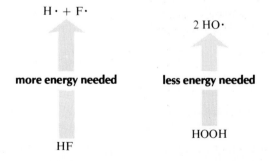

Note in Table 1.4 that atoms joined by multiple bonds require more energy for dissociation than the same atoms joined by single bonds ($CH{\equiv}CH$, 230

TABLE 1.4. Selected bond dissociation energies (in kcal/mole)

Miscellaneous bonds		C—H bonds		C—X bonds[a]		C—C bonds	
H—H	104	CH_3—H	104	CH_3—Cl	83.5	CH_3—CH_3	88
N≡N	226	CH_3CH_2—H	98	CH_3CH_2—Cl	81.5	CH_2=CH_2	163
F—F	37	$(CH_3)_2CH$—H	94.5	$(CH_3)_2CH$—Cl	81	CH≡CH	230
Cl—Cl	58	$(CH_3)_3C$—H	91	$(CH_3)_3C$—Cl	78.5		
Br—Br	46	CH_2=CH—H	108	CH_2=CH—Cl	84		
I—I	36						
H—F	135						
H—Cl	103			CH_3—Br	70		
H—Br	87			CH_3CH_2—Br	68		
H—I	71			$(CH_3)_2CH$—Br	68		
HO—OH	35			$(CH_3)_3C$—Br	67		

[a] X refers to a halogen.

kcal/mole, versus CH_3—CH_3, 88 kcal/mole). Also note that other parts of a molecule may affect the bond dissociation energy:

$$H_3C\overset{\frown\frown}{—}H + 104 \, \text{kcal/mole} \xrightarrow{\text{more difficult}} H_3C\cdot + H\cdot$$

$$(CH_3)_3C\overset{\frown\frown}{—}H + 91 \, \text{kcal/mole} \xrightarrow{\text{easier}} (CH_3)_3C\cdot + H\cdot$$

Bond dissociation energies allow a chemist to calculate relative stabilities of compounds and to predict (to an extent) the courses of chemical reactions. For example, one reaction we will discuss later in this text is the chlorination of methane, CH_4:

$$CH_4 + Cl_2 \longrightarrow CH_3Cl + HCl$$

Will this reaction be **exothermic** (energy-releasing) or **endothermic** (energy-absorbing)? We may break down the reaction into its component parts and calculate from the individual bond dissociation energies whether energy will be liberated or required. The greater the amount of energy liberated, the more favorable is the reaction. (Note that in these equations, $+\Delta H$ indicates energy put into a reaction, while $-\Delta H$ indicates energy that is liberated.)

		ΔH
Cl—Cl + 58 kcal/mole $\longrightarrow$ Cl$\cdot$ + Cl$\cdot$		+ 58 kcal/mole
H_3C—H + 104 kcal/mole $\longrightarrow$ $H_3C\cdot$ + H$\cdot$		+ 104 kcal/mole
$H_3C\cdot$ + Cl$\cdot$ $\longrightarrow$ H_3C—Cl + 83.5 kcal/mole		− 83.5 kcal/mole
H$\cdot$ + Cl$\cdot$ $\longrightarrow$ H—Cl + 103 kcal/mole		− 103 kcal/mole

net reaction:

$$Cl_2 + CH_4 \longrightarrow CH_3Cl + HCl + 24.5 \, \text{kcal/mole} \qquad \text{net } \Delta H = -24.5 \, \text{kcal/mole}$$

We calculate that this reaction should be exothermic. When we run this reaction in the laboratory, we find that it is indeed exothermic.

STUDY PROBLEM

1.7. Using bond dissociation energies from Table 1.4, predict which of the following reactions liberates more energy:

(a) $(CH_3)_3CH + Cl_2 \longrightarrow (CH_3)_3CCl + HCl$

(b) $CH_4 + Cl_2 \longrightarrow CH_3Cl + HCl$

SECTION 1.8.

Polar Covalent Bonds

Atoms with equal or nearly equal electronegativities form covalent bonds in which both atoms exert equal or nearly equal attractions for the bonding electrons. This type of covalent bond is called a **nonpolar bond**. In organic molecules, carbon–carbon bonds and carbon–hydrogen bonds are the most common types of non-polar bonds.

Some compounds containing relatively nonpolar covalent bonds:

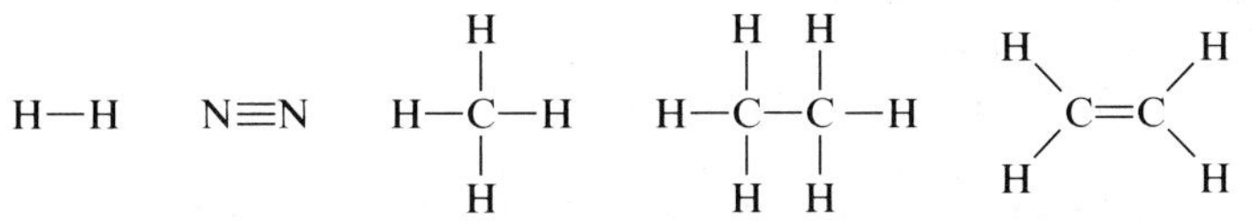

In covalent compounds like H_2O, HCl, CH_3OH, or $H_2C=O$, one atom has a substantially greater electronegativity than the others. The more electro-negative atom has a greater attraction for the bonding electrons—not enough of an attraction for the atom to break off as an ion, but enough so that this atom takes the larger share of electron density. The result is a **polar covalent bond**, a bond with an uneven distribution of electron density. The degree of polarity of a bond depends partly on the difference in electronegativities of the two atoms bonded together and partly on other factors, such as the size of the atoms. We may think of chemical bonds as a continuum from nonpolar covalent bonds to ionic bonds. Within this continuum, we speak of the increasing **ionic character** of the bonds.

$$H-H \qquad CH_3-O-CH_3 \qquad H-O--H \qquad H-Cl \qquad Na^+ \ Cl^-$$

increasing ionic character of bonds

The distribution of electrons in a polar molecule may be symbolized by **partial charges**: $\delta+$ (partial positive) and $\delta-$ (partial negative). Another way of representing the different electron densities within a molecule is by a crossed arrow ($\longmapsto$) that points from the partially positive end of a molecule to the partially negative end.

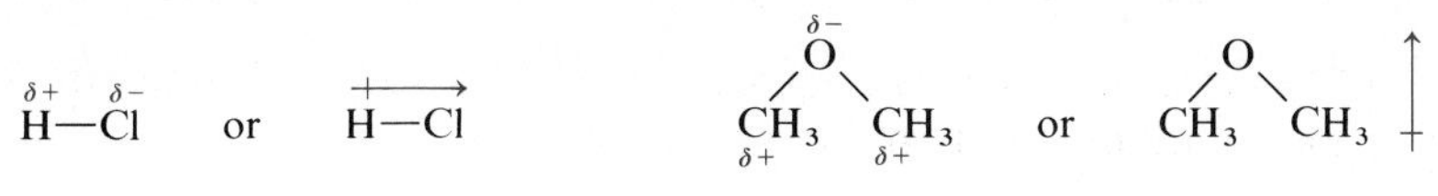

STUDY PROBLEM

1.8. Using partial charges, indicate the polarity of the following compounds:

$$\overset{\displaystyle O}{\overset{\displaystyle \|}{}}$$

(a) CH_3Br (b) CH_3COH

A.   Bond moments

If a polar bond, such as an O—H bond, is subjected to an electric field, the bond feels a certain amount of "turning force." This force is simply the push of the electric field to align the bond in the field. A more polar bond feels more force than a less polar bond. The **bond moment**, a measure of the polarity of a bond, may be calculated from the value of the force felt by that bond.

The bond moment is defined as $e \times d$, where e is the charge (in electrostatic units) and d is the distance between the charges (in Å), and is reported in units called *Debyes* (D). Bond moments range from 0.4 D for the nonpolar C—H bond to 3.5 D for the highly polar C≡N bond (see Table 1.5). The bond moment for a particular bond is relatively constant from compound to compound.

TABLE 1.5. Bond moments for selected covalent bonds

Bond[a]	Bond moment, D	Bond[a]	Bond moment, D
H—C	0.4	C—Cl	1.46
H—N	1.31	C—Br	1.38
H—O	1.51	C—I	1.19
C—N	0.22	C=O	2.3
C—O	0.74	C≡N	3.5
C—F	1.41		

[a] In each case, the more positive atom is on the left.

STUDY PROBLEMS

1.9. Which of the indicated bonds in each pair of compounds is more polar?

(a) $CH_3—NH_2$ or $CH_3—OH$ (b) $CH_3—OH$ or $CH_3O—H$
(c) $CH_3—Cl$ or $CH_3—OH$

1.10. Use a crossed arrow to show the approximate direction of the bond moment (if any) of the double or triple bond in each of the following structures:

$$\overset{\displaystyle O}{\overset{\displaystyle \|}{}}$$

(a) $CH_3C≡N$ (b) CH_3CCH_3 (c) $CH≡CH$ (d) 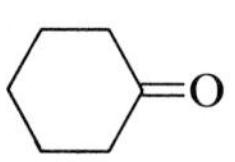 ⬡=O

B.   Dipole moments

The **dipole moment** μ is the *vector sum of the bond moments* in a molecule. Because vector addition takes into account the direction as well as the magnitude of the bond moments, the dipole moment is a measure of the polarity of the molecule as a whole.

The dipole moments for a few organic compounds are listed in Table 1.6. Note that the dipole moment of CCl_4 is zero, even though each C—Cl bond has a moment of 1.46 D. The reason for this apparent anomaly is that the CCl_4 molecule is symmetrical around its central atom; thus, the bond moments cancel and result in a vector sum of zero. Carbon dioxide is another molecule with bond moments but no dipole moment. Again, this is a case of a symmetrical molecule in which the bond moments cancel. On the other hand, the bond moments of a water molecule do not cancel, and water has a net dipole moment. From this observation, we can deduce that the water molecule is not symmetrical around the oxygen. Dipole moments can thus be used to help determine molecular geometry.

TABLE 1.6. Dipole moments of selected compounds

Compound	Dipole moment, D	Compound	Dipole moment, D
H_2O	1.84	CH_3OCH_3	1.3
NH_3	1.46	$\underset{\displaystyle CH_3CH}{\overset{\displaystyle O}{\|\|}}$	2.7
CH_3Cl	1.86		
CCl_4	0		
CO_2	0	$\underset{\displaystyle CH_3CCH_3}{\overset{\displaystyle O}{\|\|}}$	2.8

SECTION 1.9.

Attractions Between Molecules

A.   Dipole–dipole interactions

Except in a highly dispersed gas, molecules attract and repel each other. These attractions and repulsions arise primarily from molecular dipole–dipole interactions. For example, in the liquid state, molecules of CH_3I can either attract or repel each other, depending on the orientation of the molecules. Two CH_3I molecules are attracted to each other because of the attraction between the partially negative iodine of one molecule and the partially positive carbon of the other molecule.

$$\begin{array}{c} \text{H} \\ \diagdown\,\underset{\text{H}^{\diagup}\,\text{H}}{\overset{\delta+\quad\delta-}{\text{C}-\text{I}}}----\,\underset{\text{H}^{\cdots}\,\text{H}}{\overset{\delta+\quad\delta-}{\text{C}-\text{I}}}----\,\underset{\text{H}^{\cdots}\,\text{H}}{\overset{\delta+\quad\delta-}{\text{C}-\text{I}}} \end{array}$$

attractions

When the iodine ends of two CH_3I molecules approach closely, the two molecules repel each other.

$$\begin{array}{c} \text{H} \qquad\qquad \text{H} \\ \diagdown\,\underset{\text{H}^{\diagup}\,\text{H}}{\overset{\delta+\quad\delta-}{\text{C}-\text{I}}}\,\bigg)\bigg(\,\underset{\text{H}}{\overset{\delta-\quad\delta+}{\text{I}-\text{C}}}\diagdown\text{H} \end{array}$$

repulsion

Nonpolar molecules are attracted to each other by weak dipole–dipole interactions called **London forces**. London forces arise from dipoles *induced* in one molecule by another. In this case, electrons of one molecule are weakly attracted to a nucleus of a second molecule; then the electrons of the second molecule are repelled by the electrons of the first. The result is an uneven distribution of electrons and an induced dipole. Figure 1.6 depicts how an induced dipole can arise when two molecules approach each other.

The various dipole–dipole interactions (attractive and repulsive) are collectively called **van der Waals forces**. The distance between molecules has an important effect on the strength of van der Waals forces. The distance at which attraction is greatest is called the **van der Waals radius**. If two atoms approach each other more closely than this distance, repulsions develop between the two nuclei and between the two sets of electrons. When the distance between two molecules becomes larger than the van der Waals radius, the attractive forces between the molecules decrease.

Continuous-chain molecules, such as $CH_3CH_2CH_2CH_2CH_3$, can align themselves in zigzag chains, enabling the atoms of different molecules to assume positions that match the van der Waals radii. Maximal van der Waals attractions can develop between such long-chain molecules. Branched molecules cannot approach one another closely enough for all the atoms to assume optimal van der Waals distances. Because more energy is necessary to overcome van der Waals attractions and to free molecules from the liquid state, continuous-chain com-

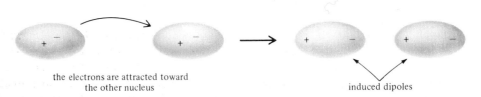

the electrons are attracted toward
the other nucleus

induced dipoles

FIGURE 1.6. Nonpolar molecules can induce dipoles in each other.

pounds have higher boiling points than branched compounds of the same molecular weight and otherwise similar structures.

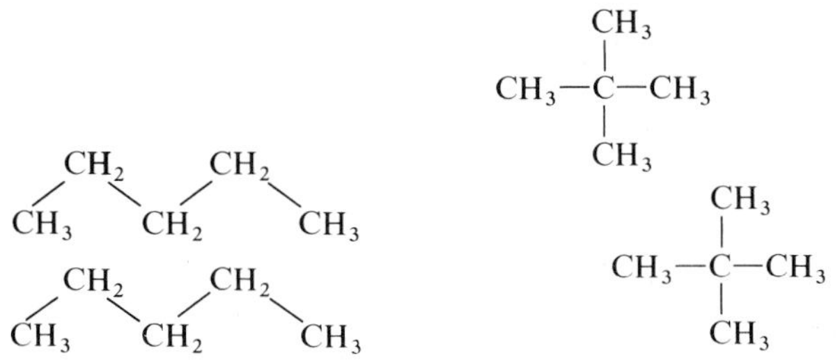

$$CH_3-\overset{\overset{\displaystyle CH_3}{|}}{\underset{\underset{\displaystyle CH_3}{|}}{C}}-CH_3$$

bp 36°

$$CH_3-\overset{\overset{\displaystyle CH_3}{|}}{\underset{\underset{\displaystyle CH_3}{|}}{C}}-CH_3$$

bp 9.5°

B. Hydrogen bonding

An especially strong type of dipole–dipole interaction occurs between molecules containing a hydrogen atom bonded to nitrogen, oxygen, or fluorine. Each of these latter elements is electronegative and has unshared valence electrons. Some typical compounds that contain an NH, OH, or FH bond are:

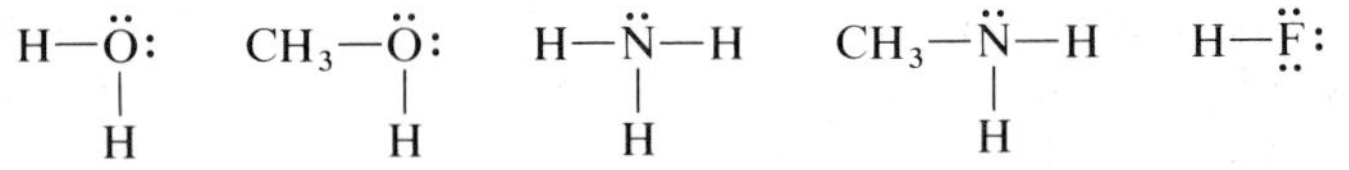

$$H-\overset{..}{\underset{|}{O}}: \quad CH_3-\overset{..}{\underset{|}{O}}: \quad H-\overset{..}{\underset{|}{N}}-H \quad CH_3-\overset{..}{\underset{|}{N}}-H \quad H-\overset{..}{\underset{..}{F}}:$$
$$\quad H \qquad\quad H \qquad\quad H \qquad\qquad H$$

In the liquid state, the molecules of any one of these compounds have strong attractions for one another. A partially positive hydrogen atom of one molecule is attracted to the unshared pair of electrons of the electronegative atom of another molecule. This attraction is called a **hydrogen bond**.

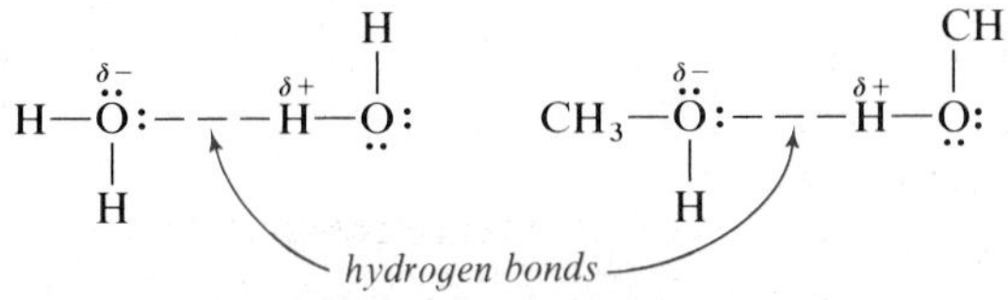

Compounds or groups containing only carbon and hydrogen cannot undergo hydrogen bonding. As an example, consider methane, CH_4. Methane cannot undergo hydrogen bonding for two reasons:

1. Because the CH bond is relatively nonpolar, a CH_4 molecule does not have a partially positive H.
2. The carbon atom in CH_4 has no unshared electrons to attract a hydrogen atom.

The dissociation energy of a hydrogen bond is only 5–10 kcal/mole, much lower than the bond dissociation energy of a typical covalent bond (80–100 kcal/mole) but substantially stronger than most dipole–dipole attractions. The reason for this difference is the size of the atoms involved. A hydrogen atom is small compared to other atoms and can occupy a position very close to the unshared electrons of an electronegative atom. A strong electrostatic attraction results.

Atoms larger than hydrogen cannot occupy positions so near to each other; consequently, dipole–dipole attractions between other atoms are weaker.

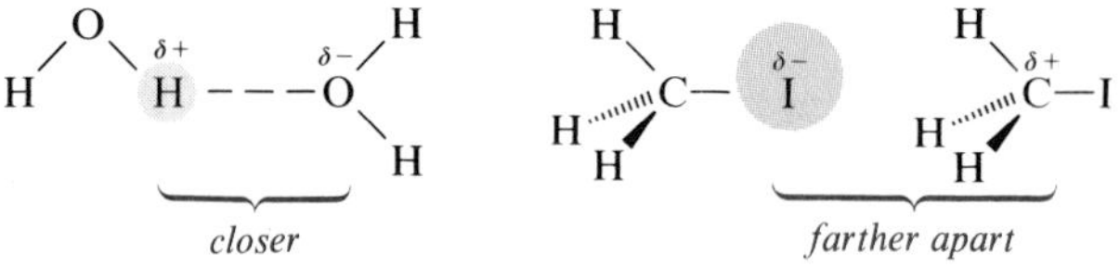

closer   farther apart

SAMPLE PROBLEM

Show the hydrogen bonding between two molecules of $CH_3CH_2NH_2$.

Solution:

1. Look for one or more partially positive hydrogens.

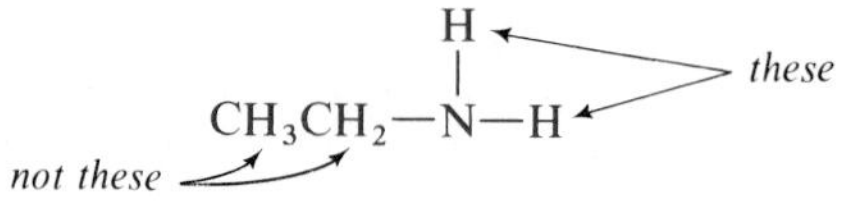

2. Look for an electronegative atom (N, O, or F) with unshared electrons.

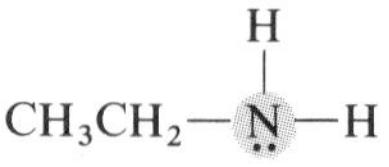

3. Draw two molecules with a hydrogen bond between one partially positive H and the N of the other molecule.

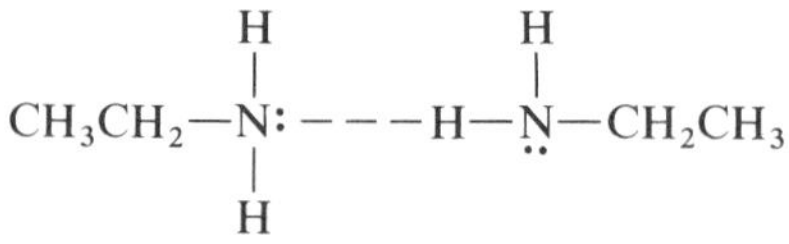

Hydrogen bonds are not all the same strength. An $O---HO$ hydrogen bond is stronger than an $N---HN$ hydrogen bond. Why is this true? Oxygen is more electronegative than nitrogen; therefore, the O—H group is more polar and has a more positive H. This more positive H is more strongly attracted by a negative center.

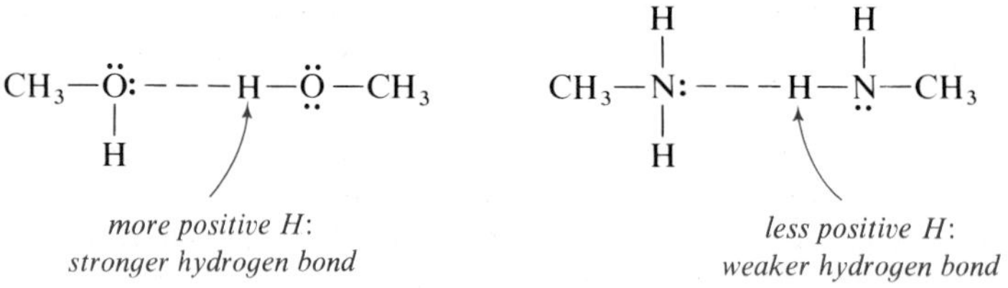

more positive H: less positive H:
stronger hydrogen bond   weaker hydrogen bond

Hydrogen bonds may form between two different compounds, such as between CH_3OH and H_2O or between CH_3NH_2 and H_2O. In these cases, there is often more than one possibility for hydrogen bonding. The following structures show two types of hydrogen bonds between CH_3NH_2 and H_2O. (In a mixture of these two compounds, hydrogen bonds can also form between two molecules of H_2O and between two molecules of CH_3NH_2.)

$$
\begin{array}{cc}
\underset{\substack{\\ \text{H}}}{\overset{\text{H}}{\underset{|}{\overset{|}{\text{CH}_3-\text{N:}}}}}\text{---H}-\overset{\text{..}}{\underset{\text{..}}{\text{O}}}\text{:} & \overset{\text{H}}{\underset{|}{\text{CH}_3-\text{N}}}-\text{H---}\overset{\text{H}}{\underset{|}{\overset{..}{\underset{..}{\text{O}}}-\text{H}}}
\end{array}
$$

| *more positive H:* | *less positive H:* |
| *stronger hydrogen bond* | *weaker hydrogen bond* |

Table 1.7 shows the amount of energy needed to break some different types of hydrogen bonds. Note that the OH-- --N hydrogen bond is the strongest of the group. Because nitrogen is less electronegative than oxygen, its electrons are more loosely held and more easily attracted by another atom. The combination of nitrogen's loose electrons and the more positive hydrogen of an OH group leads to a quite strong hydrogen bond.

TABLE 1.7. Approximate dissociation energies of some hydrogen bonds

Type of hydrogen bond	Approximate dissociation energy (*kcal/mole*)
$-\text{O}-\text{H}---:\overset{\|}{\underset{\|}{\text{N}}}-$	7
$-\text{O}-\text{H}---:\overset{\|}{\underset{\|}{\text{O}}}-$	5
$-\overset{\|}{\underset{\|}{\text{N}}}-\text{H}---:\overset{\|}{\underset{\|}{\text{N}}}-$	3
$-\overset{\|}{\underset{\|}{\text{N}}}-\text{H}---:\overset{\|}{\underset{\|}{\text{O}}}-$	2

C.   Effects of hydrogen bonding

Hydrogen bonds are rather like glue between molecules. Although a single hydrogen bond by itself is weak, all the molecules taken together may form a great many hydrogen bonds.

For all substances, boiling points increase with molecular weight because of increased van der Waals attractions. However, a hydrogen-bonded compound has a *higher boiling point* than would be predicted from molecular weight considerations alone. For a hydrogen-bonded liquid to be volatilized, additional energy must be supplied for breaking all the intermolecular hydrogen bonds.

Ethanol (CH_3CH_2OH) and dimethyl ether (CH_3OCH_3) have the same molecular weight. However, ethanol has a much higher boiling point than does dimethyl ether—ethanol is a liquid at room temperature while dimethyl ether is a gas. The difference in boiling points between these two compounds can be directly attributed to the fact that ethanol molecules are joined by hydrogen bonds, while dimethyl ether molecules cannot form hydrogen bonds among themselves. Note that hydrogen bonding affects the boiling point to a much greater extent than branching does.

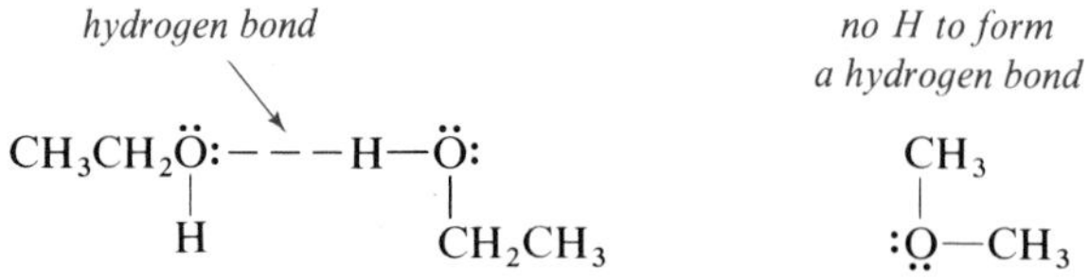

hydrogen bond   *no H to form*
                                                            a hydrogen bond

$$CH_3CH_2\overset{\cdot\cdot}{\underset{|}{O}}: - - - H - \overset{\cdot\cdot}{O}:$$

$$\underset{H}{\overset{|}{}} \qquad \underset{CH_2CH_3}{\overset{|}{}}$$

ethanol
bp 78.5°

$$CH_3$$
$$|$$
$$:\underset{\cdot\cdot}{O} - CH_3$$

dimethyl ether,
bp −23.6°

Solubility of covalent compounds in water is another property affected by hydrogen bonding. A compound that can form hydrogen bonds with water tends to be far more soluble in water than a compound that cannot. Sugars, such as glucose, contain many —OH groups and are quite soluble in water. Cyclohexane, however, cannot form hydrogen bonds and cannot break the existing hydrogen bonds in water; therefore, cyclohexane is water-insoluble.

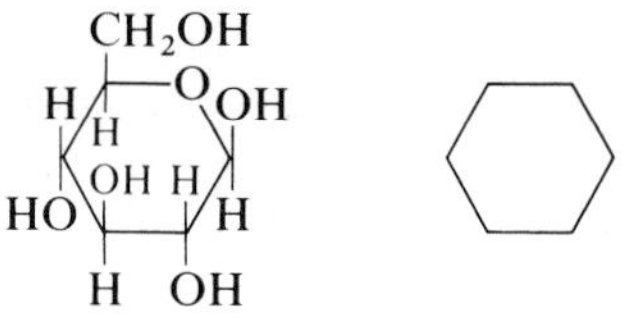

glucose

cyclohexane

soluble in water

insoluble in water

STUDY PROBLEM

1.11.   Show all the types of hydrogen bonds (if any) that would be found in:

(a)   liquid $CH_3CH_2CH_2NH_2$ (b)   a solution of CH_3OH in H_2O
(c)   liquid $CH_3CH_2OCH_2CH_3$ (d)   a solution of CH_3OCH_3 in H_2O

SECTION 1.10.

Acids and Bases

According to the **Brønsted–Lowry concept** of acids and bases, an **acid** is a substance that can donate a positively charged hydrogen ion, or proton (H^+). Two examples of Brønsted–Lowry acids are HCl and HNO_3. A **base** is defined as a substance that can accept H^+; examples are OH^- and NH_3. (A typical source of hydroxide ions is Na^+ OH^-.) Although we speak of "proton donors" and "proton acceptors," curved arrows are used in organic chemistry to indicate the actions of *electrons*, not protons. Therefore, in the following equation, a curved arrow is drawn from the unshared electrons of the base to the proton that it is accepting.

$$H\overset{\cdot\cdot}{\underset{\cdot\cdot}{O}}:^- \;\; + \;\; H - \overset{\cdot\cdot}{\underset{\cdot\cdot}{Cl}}: \;\;\longrightarrow\;\; H\overset{\cdot\cdot}{\underset{\cdot\cdot}{O}} - H \;\; + \;\; :\overset{\cdot\cdot}{\underset{\cdot\cdot}{Cl}}:^-$$

a base   *an acid*
(H^+ *acceptor*)   (H^+ *donor*)

Recall from your general chemistry course that a **strong acid** is an acid that undergoes essentially complete ionization in water. Representative strong acids are HCl, HNO_3, and H_2SO_4. The ionization of these strong acids is a typical acid–base reaction. The acid (HCl, for example) donates a proton to the base (H_2O). The equilibrium lies far toward the right (complete ionization of HCl) because H_2O is a stronger base than Cl^- and HCl is a stronger acid than H_3O^+.

$$H-\overset{..}{\underset{..}{O}}-H \; + \; H-\overset{..}{\underset{..}{Cl}}: \; \longrightarrow \; H-\overset{\overset{\textstyle H}{|}}{\underset{..}{O}}{}^{\!+}\!-H \; + \; :\overset{..}{\underset{..}{Cl}}:^-$$

stronger base   *stronger acid*
than Cl⁻   *than H₃O⁺*

A **weak acid**, by contrast, is only partially ionized in water. Carbonic acid is a typical inorganic weak acid. The equilibrium lies to the left because H_3O^+ is the stronger acid and HCO_3^- is the stronger base.

$$H-\overset{..}{\underset{..}{O}}-H + H-\overset{..}{\underset{..}{O}}\overset{O}{\overset{||}{C}}OH \; \rightleftharpoons \; H-\overset{\overset{\textstyle H}{|}}{\underset{..}{O}}{}^{\!+}\!-H \; + \; {}^-:\overset{..}{\underset{..}{O}}\overset{O}{\overset{||}{C}}OH$$

carbonic acid   bicarbonate ion

stronger acid   *stronger base*
than H₂CO₃   *than H₂O*

Also, recall that *bases* are classified as strong (such as OH^-) or weak (NH_3), depending on their affinity for protons.

$$:NH_3 \; + H-\overset{..}{\underset{..}{O}}-H \; \rightleftharpoons \; \overset{+}{N}H_4 \quad + \quad {}^-:\overset{..}{\underset{..}{O}}H$$

ammonia   ammonium ion   hydroxide ion

a weak base   *stronger acid*   *stronger base*
                        than H₂O   *than NH₃*

Let us now consider some organic compounds that act as acids and bases. **Amines** are a class of organic compounds structurally similar to ammonia; an amine contains a nitrogen atom that is covalently bonded to one or more carbon atoms and that has an unshared pair of electrons.

Some common amines:

$$CH_3-\overset{..}{\underset{\underset{\textstyle H}{|}}{N}}-H \qquad CH_3-\overset{..}{\underset{\underset{\textstyle CH_3}{|}}{N}}-H \qquad CH_3-\overset{..}{\underset{\underset{\textstyle CH_3}{|}}{N}}-CH_3$$

Amines, like ammonia, are weak bases and undergo reversible reactions with water or other weak acids. (The use of a strong acid drives the reaction to completion.)

$$CH_3\overset{..}{N}H_2 + H-\overset{..}{\underset{..}{O}}H \; \rightleftharpoons \; CH_3\overset{+}{N}H_3 + :\overset{..}{\underset{..}{O}}H^-$$

$$CH_3\overset{..}{N}H_2 + H-\overset{..}{\underset{..}{Cl}}: \; \longrightarrow \; CH_3\overset{+}{N}H_3 + :\overset{..}{\underset{..}{Cl}}:^-$$

An organic compound containing a **carboxyl group** ($-CO_2H$) is a weak acid. Compounds that contain carboxyl groups are called **carboxylic acids**.

Acetic acid, CH_3CO_2H, is an example. One of the reasons for the acidity of carboxylic acids is the polarity of the O—H bond. (Another reason for this acidity will be discussed in Section 2.9.)

The carboxyl group:

usually written $-COH$, $-COOH$, or $-CO_2H$

In the presence of a base, H^+ is removed from the carboxyl group and the carboxylate anion is formed. Because carboxylic acids are only weakly acidic, these reactions do not proceed to completion unless a stronger base than water is used, as the reaction arrows in the following equations indicate.

acetic acid   acetate ion

a carboxylic acid   *a carboxylate ion*

STUDY PROBLEMS

1.12. Which of the following compounds or ions act as acids and which act as bases in H_2O?

(a) $^-NH_2$ (b) $CH_3CH_2CH_2COH$ (with O double bonded above C)   (c) $^-OCH_2CH_3$

(d) (cyclohexane ring with N^+ bearing two H)   (e) (cyclohexane ring with NH)

1.13. Rewrite and complete the following equations for acid–base reactions. Include in your answer (1) electron dots, and (2) reaction arrows that show the direction of the equilibrium. (If the reaction proceeds essentially to completion, use a single arrow.)

(a) $CH_3CH_2CO_2H + H_2O$ (b) (benzene ring)$-CO_2H + OH^-$

(c) $(CH_3)_2NH + H_2O$ (d) (cyclohexane ring)$NH + CH_3CO_2H$

A.   Acidity Constants

A chemical reaction has an **equilibrium constant** K that reflects how far the reaction proceeds toward completion. For the ionization of an acid in water, this constant is called an **acidity constant** K_a. An equilibrium constant is determined by the following general equation, with concentration values given in molarity, M:

$$K = \frac{\text{concentrations of products in } M}{\text{concentrations of reactants in } M}$$

For acetic acid: $CH_3CO_2H \; \rightleftharpoons \; CH_3CO_2^- + H^+$

$$K_a = \frac{[CH_3CO_2^-][H^+]^*}{[CH_3CO_2H]}$$

where $[H^+]$ = molar concentration of H^+

$[CH_3CO_2^-]$ = molar concentration of $CH_3CO_2^-$

$[CH_3CO_2H]$ = molar concentration of CH_3CO_2H

The more ionized an acid is, the larger is the value for K_a because the values in the numerator are larger. *A stronger acid has a larger K_a value.* Any acid with a $K_a > 10$ is considered a strong acid. (For HCl, $K_a \cong 10^7$.) By contrast, typical carboxylic acids, such as acetic acid, have K_a values much smaller than 1. (For CH_3CO_2H, $K_a = 1.75 \times 10^{-5}$.)

$$K_a = \frac{[H^+][\text{anion}]}{[\text{un-ionized acid}]} \quad \begin{array}{l} \longleftarrow \text{ as numerator increases,} \\ \quad K_a \text{ increases} \end{array}$$

Just as pH is the negative logarithm of hydrogen ion concentration, pK_a is the negative logarithm of K_a. We will use pK_a values in this text for comparison of acid strengths. (The K_a values and pK_a values for some carboxylic acids are listed in Table 1.8.)

$$pH = -\log[H^+]$$

$$pK_a = -\log K_a$$

Examples: If $K_a = 10^{-3}$,   then $pK_a = 3$

If $K_a = 10^2$,   then $pK_a = -2$

---

* More correctly, the *activity*, or *effective concentration*, should be used, rather than molarity. Since activities of ions approach their molarities in dilute solution, molarity may be used for the sake of simplicity. In addition, the equilibrium expression should contain the hydrogen acceptor, water:

$$CH_3CO_2H + H_2O \; \rightleftharpoons \; CH_3CO_2^- + H_3O^+$$

$$K_a' = \frac{[CH_3CO_2^-][H_3O^+]}{[CH_3CO_2H][H_2O]}$$

For all practical purposes, the molar concentration of water remains fixed at 55.5. This constant factor is generally grouped with the equilibrium constant K_a, and the $[H_3O^+]$ term is simplified to $[H^+]$. Even though "naked" protons do not exist as such in solution, we will frequently use the symbol H^+ instead of H_3O^+ in this book to represent an aqueous acid.

$$K_a = K_a'[H_2O] = \frac{[CH_3CO_2^-][H_3O^+]}{[CH_3CO_2H]} = \frac{[CH_3CO_2^-][H^+]}{[CH_3CO_2H]}$$

TABLE 1.8. Acidity constants and pK_a values for some acids

Formula	K_a	pK_a
strong:		
HCl	$\sim 10^7$	~ -7
H_2SO_4	$\sim 10^5$	~ -5
moderately strong:		
H_3PO_4	7.52×10^{-3}	2.12
weak:		
HCO_2H	17.5×10^{-5}	3.75
CH_3CO_2H	1.75×10^{-5}	4.75
$CH_3CH_2CO_2H$	1.34×10^{-5}	4.87
very weak:		
HCN	4.93×10^{-10}	9.31
H_2O	2.00×10^{-16}	15.7

As K_a gets larger (stronger acid), pK_a gets smaller. *The smaller the value for* pK_a, *the stronger the acid.*

$$
\begin{array}{lcccc}
K_a: & 10^{-10} & 10^{-5} & 10^{-1} & 10^2 \\
pK_a: & 10 & 5 & 1 & -2
\end{array}
$$

increasing acid strength

SAMPLE PROBLEM

Calculate the pK_a of an acid with K_a equal to 136×10^{-5}.

Solution:
$$
\begin{aligned}
pK_a &= -\log K_a \\
&= -\log(136 \times 10^{-5}) \\
&= -\log(1.36 \times 10^{-3}) \\
&= -(\log 1.36 - 3) \\
&= 3 - \log 1.36 \\
&= 3 - 0.133 \quad \longleftarrow \textit{from log table or calculator} \\
&= 2.87
\end{aligned}
$$

B.   Basicity constants

The reversible reaction of a weak base with water, like the reaction of a weak acid with water, results in a small but constant concentration of ions at equilibrium. The **basicity constant** K_b is the equilibrium constant for this reaction. As in the case of K_a, the value for $[H_2O]$ is included in K_b in the equilibrium expression.

$$
NH_3 + H_2O \;\rightleftharpoons\; NH_4^+ + OH^-
$$

$$
K_b = \frac{[NH_4^+][OH^-]}{[NH_3]}
$$

$$
pK_b = -\log K_b
$$

TABLE 1.9. Basicity constants and pK_b values for ammonia and some amines

Formula	K_b	pK_b
NH_3	1.79×10^{-5}	4.75
CH_3NH_2	45×10^{-5}	3.34
$(CH_3)_2NH$	54×10^{-5}	3.27
$(CH_3)_3N$	6.5×10^{-5}	4.19

With an increase in base strength, the value for K_b increases and the pK_b value decreases. *The smaller the value for* pK_b, *the stronger the base.*

$$
\begin{array}{llll}
K_b: & 10^{-10} & 10^{-7} & 10^{-5} \\
pK_b: & 10 & 7 & 5
\end{array}
$$

increasing base strength ⟶

STUDY PROBLEMS

1.14. List the following compounds in order of increasing basicity (weakest first). See Table 1.9 for pK_b values.

(a) NH_3 (b) CH_3NH_2 (c) $(CH_3)_2NH$

1.15. List the following anions in order of increasing basicity:

(a) CH_3O^-, $pK_b = -1.5$ (b) $CH_3CO_2^-$, $pK_b = 9.25$
(c) Cl^-, $pK_b = 21$

C. Conjugate acids and bases

The concept of conjugate acids and bases is useful for comparisons of acidities and basicities. The **conjugate base** of an acid is the ion or molecule that results after the loss of H^+ from the acid. For example, the chloride ion is the conjugate base of HCl. The **conjugate acid** of a base is the protonated form of the base. Thus, the conjugate acid of NH_3 is NH_4^+.

If an acid is strong, its conjugate base is a weak base:

$$
\underset{\substack{\textit{strong acid} \\ \textit{(loses } H^+ \textit{ readily)}}}{HCl} \quad + \quad H_2O \quad \longrightarrow \quad H_3O^+ \quad + \quad \underset{\substack{\textit{very weak base} \\ \textit{(has little attraction for } H^+\textit{)}}}{Cl^-}
$$

On the other hand, if an acid is weak or very weak, its conjugate base is moderately strong or strong, depending on the affinity of the conjugate base for H^+.

$$
\underset{\textit{weak acid}}{CH_3\overset{\overset{O}{\|}}{C}OH + H_2O} \quad \rightleftharpoons \quad \underset{\textit{moderate base}}{CH_3\overset{\overset{O}{\|}}{C}O^-} \quad + \quad H_3O^+
$$

$$
\underset{\textit{very weak acid}}{2\,H_2O} \quad \rightleftharpoons \quad H_3O^+ \quad + \quad \underset{\textit{strong base}}{OH^-}
$$

Thus, as the acid strengths of a series of compounds increase, the base strengths of their conjugate bases decrease.

$$H_2O \quad HCN \quad CH_3CO_2H \quad H_3PO_4 \quad HCl$$

increasing acidity; decreasing basicity of conjugate base

D.   Lewis acids and bases

Although many acid–base reactions involve the transfer of a proton from an acid to a base, some acid–base reactions do not involve proton transfer. For this reason, the more general Lewis concept of acids and bases was developed. A **Lewis acid** is a substance that can **accept a pair of electrons**. Any species with an electron-deficient atom can act as a Lewis acid; for example, H^+ is a Lewis acid. Most Lewis acids other than H^+ that we will encounter in this text are anhydrous metal salts (for example, $ZnCl_2$, $FeCl_3$, and $AlBr_3$).

$$FeBr_3 \quad + \quad :\ddot{Br}-\ddot{Br}: \quad \rightleftharpoons \quad FeBr_4^- + \ddot{Br}:^+$$

a Lewis acid
(electron acceptor)

A **Lewis base** is a substance that can **donate a pair of electrons**. Examples of Lewis bases are NH_3 and ^-OH, each of which has an unshared pair of electrons. (Most Lewis bases are bases by the Brønsted–Lowry theory as well.)

$$\ddot{N}H_3 \quad + \quad H-\ddot{\underset{..}{C}l}: \quad \longrightarrow \quad NH_4^+ + :\ddot{\underset{..}{C}l}:^-$$

a Lewis base
(electron donor)

SAMPLE PROBLEM

Methylamine (CH_3NH_2) undergoes a Lewis acid–base reaction with boron trifluoride (BF_3) to yield $CH_3NH_2-BF_3$.   (a) Write the equation for this reaction, showing the complete structural formula for the product and the formal charges on N and B.   (b) Identify each reactant as a Lewis acid or Lewis base.

Solution:

$$CH_3\ddot{N}H_2 \quad + \quad BF_3 \quad \longrightarrow \quad \underset{\overset{|}{H}\ \overset{|}{H}\ \overset{|}{F}}{\overset{\overset{|}{H}\ \overset{|}{H}\ \overset{|}{F}}{H-C-N^+-B^--F}}$$

Lewis base   *Lewis acid*

STUDY PROBLEM

1.16.   Identify the reactants in the following equations as Lewis acids or Lewis bases:

(a) $$CH_3\overset{\cdot\ddot{O}\cdot}{\overset{\|}{C}}CH_3 + H^+ \rightleftharpoons CH_3\overset{\overset{+}{:\ddot{O}H}}{\overset{\|}{C}}CH_3$$

(b) $(CH_3)_3C^+ + :\ddot{\underset{..}{Cl}}:^- \longrightarrow (CH_3)_3C\ddot{\underset{..}{Cl}}:$

(c) $CH_3\overset{\overset{O}{\|}}{C}OCH_3 + {}^-:\ddot{\underset{..}{O}}CH_3 \rightleftharpoons {}^-:CH_2\overset{\overset{O}{\|}}{C}OCH_3 + H\ddot{\underset{..}{O}}CH_3$

Summary

The probable location (relative to the nucleus) of an electron with a particular energy is called an **atomic orbital**. The first electron shell (closest to the nucleus, lowest energy) contains only the spherical $1s$ orbital. The second shell (higher energy) contains a spherical $2s$ orbital and three mutually perpendicular, two-lobed $2p$ orbitals. Any orbital can hold a maximum of two paired (opposite spin) electrons.

The **atomic radius** equals half the distance between nuclei bonded by a non-polar covalent bond, such as in H—H. The atomic radius increases as we go down any group in the periodic table and decreases as we go from left to right across a period. **Electronegativity** is a measure of the pull of the nucleus on the outer electrons of the atom. It decreases as we go down any group and increases as we go from left to right in the periodic table.

A chemical bond results from electron transfer (**ionic bond**) or electron sharing (**covalent bond**). The number of bonds an atom can form (the **valence**) is determined by the number of bonding electrons. Carbon has four bonding electrons and forms four covalent bonds.

An **empirical formula** tells us the *relative number* of different atoms in a molecule, and the **molecular formula** tells us the *actual number* of different atoms in a covalent molecule.

$$C_2H_5 \qquad C_4H_{10}$$

empirical   molecular
formula   formula

In **structural formulas**, which depict the structures of molecules, pairs of electrons may be represented by dots or by lines. Unshared valence electrons are not always shown in structural formulas.

Lewis formula   complete structural formula   condensed structural formula

An atom may share two, four, or six electrons with another atom—that is, two atoms may be joined by a **single bond**, a **double bond**, or a **triple bond**.

$$H-C\equiv C-\overset{\overset{H}{|}}{C}=CH_2$$

A **formal charge** arises from a **coordinate covalent bond**, a bond in which both electrons are supplied by one atom.

$$CH_3{-}\overset{+}{N}\underset{\ddot{\underset{\cdot\cdot}{O}}{:}^-}{\overset{\displaystyle\ddot{O}{:}}{\diagup\diagdown}}\qquad both\ e^-\ from\ N$$

The **bond length** is the distance between nuclei of covalently bonded atoms. The **bond angle** is the angle between two covalent bonds in a molecule. The **bond dissociation energy** (ΔH in the following equation) is the amount of energy needed to effect **homolytic cleavage** of a covalent bond.

$$H_3C{-}H \longrightarrow H_3C\cdot + H\cdot \qquad \Delta H = +104\ \text{kcal/mole}$$

A **polar covalent bond** is a covalent bond between atoms with substantially different electronegativities. The **bond moment** is a measure of the polarity of the bond. The **dipole moment** is a measure of the polarity of the entire molecule.

Dipole–dipole attractions between molecules (**van der Waals attractions**) are generally well under 5 kcal/mole except for **hydrogen bonds** (attractions between a partially positive H and an unshared pair of electrons of N, O, or F), which require 5–10 kcal/mole for their dissociation. Hydrogen bonding leads to an increase in boiling point and water solubility of a compound.

A **Brønsted–Lowry acid** is a substance that can donate H^+; a **Brønsted–Lowry base** is a substance that can accept H^+. The strength of an acid or a base is reported as K_a (or pK_a) or as K_b (or pK_b), respectively. A stronger acid has a larger value for K_a (and a smaller pK_a); a stronger base has a larger K_b (and a smaller pK_b). (See Tables 1.8 and 1.9.)

$$K_a = \frac{[H^+][\overset{anion}{A^-}]}{[HA]} \quad \text{and} \quad pK_a = -\log K_a$$

$$K_b = \frac{[\overset{+}{B}H][OH^-]}{[\underset{base}{B{:}}]} \quad \text{and} \quad pK_b = -\log K_b$$

strong acids ($pK_a < -1$):   HCl, HNO_3, H_2SO_4

weak acids: ($pK_a > 3$): CH_3CO_2H, HCN, H_2O

strong bases: ^-OH, $^-OCH_3$

weak bases: NH_3, CH_3NH_2

The **conjugate base** of a strong acid is a weak base, while the conjugate base of a very weak acid is a strong base. In the following equation, as HA decreases in acid strength, A^- increases in base strength.

$$H{-}A \quad \rightleftharpoons \quad H^+ \quad + \quad {:}A^-$$

conjugate acid of A^-   *conjugate base of HA*

A **Lewis acid** is a substance that can *accept* a pair of electrons, while a **Lewis base** is a substance that can *donate* a pair of electrons.

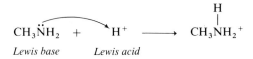

STUDY PROBLEMS

1.17. Without referring to the text, write the electron configuration (for example, $1s^2\ 2s^1$) for a free, nonbonded atom of:

(a) carbon (atomic no. 6)   (b) silicon (14)
(c) phosphorus (15)   (d) sulfur (16)

1.18. What element corresponds with each of the following electron configurations?

(a) $1s^2\ 2s^2\ 2p^6\ 3s^1$ (b) $1s^2\ 2s^2\ 2p^6\ 3s^2\ 3p^5$ (c) $1s^2\ 2s^2\ 2p^6$

1.19. From the electron configurations, which pairs of elements would you expect to exhibit similar chemical behavior? Why?

(a) $1s^2\ 2s^1$ and $1s^2\ 2s^2$
(b) $1s^2\ 2s^1$ and $1s^2\ 2s^2\ 2p^6\ 3s^2\ 3p^6\ 4s^1$
(c) $1s^2\ 2s^1$ and $1s^2\ 2s^2\ 2p^6\ 3s^1$

1.20. If an atom used two p atomic orbitals to form single covalent bonds with two hydrogen atoms, what would be the expected bond angle?

1.21. Which element in each of the following lists has the largest atomic radius? (Do not refer to the text, but make your prediction on the basis of relative locations in the periodic table inside the back cover of this book.)

(a) Si, C, O   (b) B, C, F   (c) H, C, O

1.22. Which element in each of the following lists is the most electronegative? (Refer to Figure 1.5.)

(a) C, H, O   (b) C, H, N   (c) C, H, Mg   (d) C, Cl, O

1.23. Which of the following compounds would you expect to have (1) *only ionic bonds*; (2) *only covalent bonds*; or (3) *both ionic and covalent bonds*?

(a) CH_3CO_2Na (b) CH_3I (c) LiOH   (d) CH_3ONa
(e) CH_3OH (f) $Mg(OH)Br$ (g) H_2S (h) $CHCl_3$

1.24. Without referring to the text, list the valences for the elements in period 2.

1.25. What is the expected *maximum* number of covalent bonds that an atom of any second-period element could form? Why?

1.26. Give the *Lewis formula* for each of the following structures:

(a) 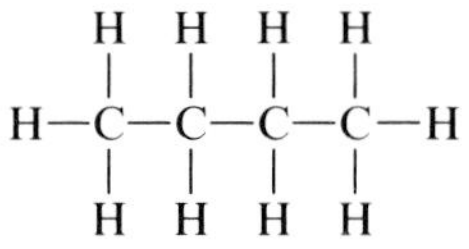
$$\begin{array}{ccccccc} & H & H & H & H \\ & | & | & | & | \\ H- & C- & C- & C- & C & -H \\ & | & | & | & | \\ & H & H & H & H \end{array}$$

(b)
$$\begin{array}{ccc} H & H & H \\ | & | & | \\ H-C\!\!\!-\!\!\!-\!\!\!-C\!\!\!-\!\!\!-\!\!\!-C-H \\ | & | & | \\ H & H-C-H & H \\ & | & \\ & H & \end{array}$$

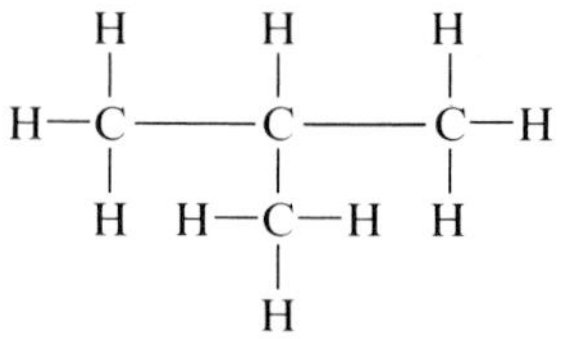

(c) $CH_3CHClCH(CH_3)_2$

(d) H_2O_2

(e) 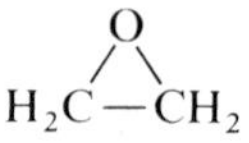 $H_2C\overset{O}{\overset{/\backslash}{-}}CH_2$

1.27. Calculate the formal charges of all atoms except H in each of the following structures:

(a) 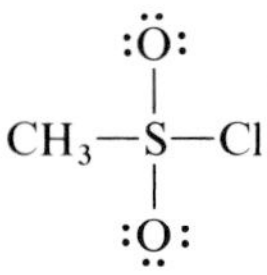
$$CH_3-\overset{:\ddot{O}:}{\underset{:\underset{\cdot\cdot}{O}:}{\overset{|}{S}}}-Cl$$

(b) $CH_3-C\equiv N:$

(c) 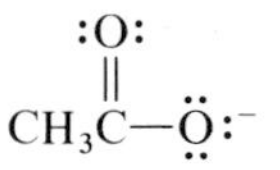
$$CH_3\overset{:O:}{\overset{\|}{C}}-\ddot{\underset{\cdot\cdot}{O}}:^-$$

(d) 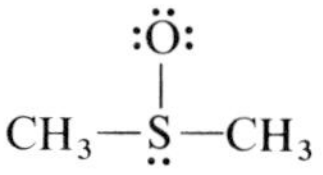
$$CH_3-\overset{:\ddot{O}:}{\underset{\cdot\cdot}{\overset{|}{S}}}-CH_3$$

1.28. Give the *complete structural formula* (showing each atom and using lines for bonds) for each of the following condensed formulas:

(a) $(CH_3)_2CHCHBrCH_3$

(b)  $(CH_3)_2CHC\overset{O}{\overset{\|}{}}NHCH(CH_3)CH\!=\!CH_2$

(c) $(CH_3)_2C\!=\!C(CH_3)C\equiv CCH_3$

1.29. Write the *condensed structural formula* for each of the following structures:

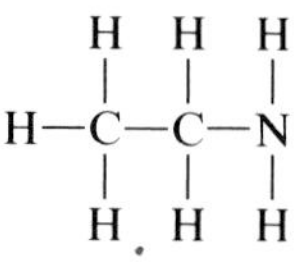

(a) 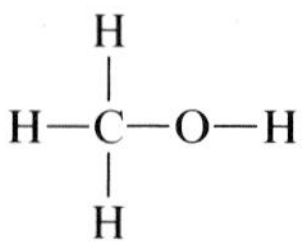
$$\begin{array}{c} H \\ | \\ H-C-O-H \\ | \\ H \end{array}$$

(b)
$$\begin{array}{ccc} H & H & H \\ | & | & | \\ H-C-C-N \\ | & | & | \\ H & H & H \end{array}$$

(c)
$$\begin{array}{ccc} & H & H & H \\ H:\!\ddot{C}\!:\!\ddot{N}\!:\!\ddot{C}\!:\!H \\ & H & & H \end{array}$$

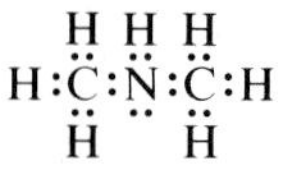

1.30. Write the *molecular formula* for each of the following condensed structural formulas:

(a) $CH_3CH_2CH_2CH_2OH$

(b) 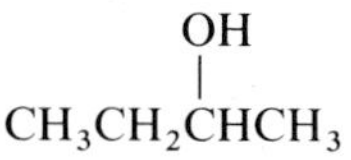 $CH_3CH_2\overset{OH}{\overset{|}{C}}HCH_3$

(c) $(CH_3)_3COH$

(d) $(CH_3)_2CHCH_2OH$

1.31. Give the *complete structural formula* for each of the following compounds:

(a) $H_2C\!=\!CHCH\!=\!CHCN$

(b) CH_3COCH_3

(c) CH_3CO_2H

(d) $OHCCH_2CHO$

(*Hint:* Each structure contains at least one double or triple bond that is not shown. Use rules of valence to find these.)

1.32. Show any *unshared pairs of valence electrons* (if any) in the following formulas:

(a) CH_3NH_2 (b) $(CH_3)_3N$ (c) $(CH_3)_3NH^+$ (d) CH_3OH

(e) $(CH_3)_3COH$ (f) $CH_2{=}CH_2$ (g) $H_2C{=}O$

1.33. Draw a polygon formula for each of the following cyclic structures:

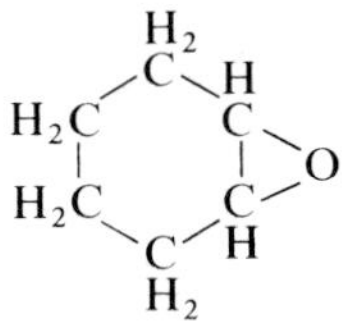

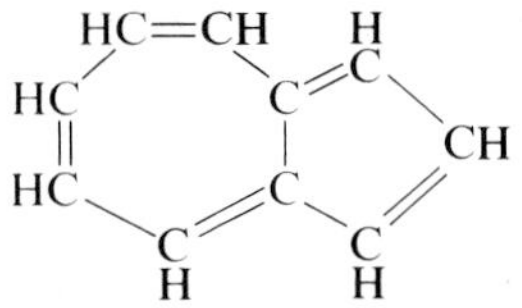

1.34. Convert each of the following polygon formulas to a complete structural formula, showing each atom and each bond. Show also any unshared pairs of valence electrons.

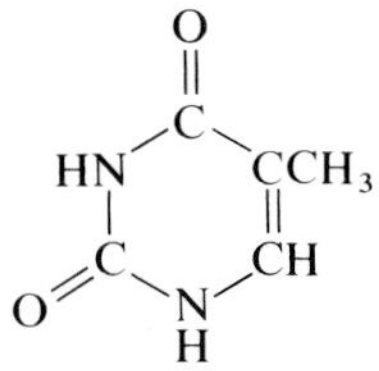

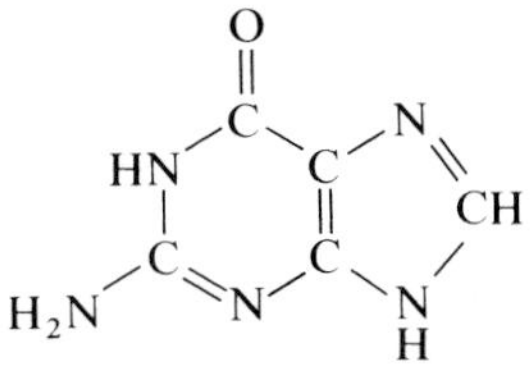

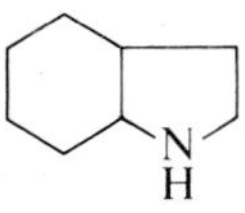

1.35. Draw the polygon and the structural formulas for a carbon ring system containing:

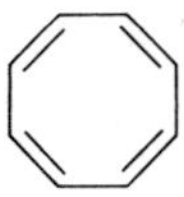

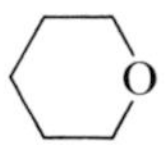

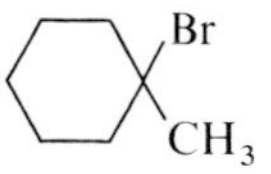

(a) six ring carbons and a double bond;

(b) five ring carbons, one of which is part of a carbonyl group ($C{=}O$).

1.36. The bond dissociation energy of the carbon–halogen bond of CH_3F is 108 kcal/mole; for CH_3Cl, it is 83.5 kcal/mole; for CH_3Br, it is 70 kcal/mole; and for CH_3I, it is 56 kcal/mole. Calculate the net ΔH for each of the following reactions:

(a) $CH_4 + F_2 \longrightarrow CH_3F + HF$

(b) $CH_4 + Cl_2 \longrightarrow CH_3Cl + HCl$

(c) $CH_4 + Br_2 \longrightarrow CH_3Br + HBr$

(d) $CH_4 + I_2 \longrightarrow CH_3I + HI$

1.37. Write chemical equations for (1) the *homolytic cleavage*, and (2) the *heterolytic cleavage*, of each of the following compounds at the indicated bond. (Apply your knowledge of electronegativities in the heterolytic cleavages.)

(a) $CH_3CH_2{-}Cl$ (b) $H{-}OH$ (c) $H{-}NH_2$

(d) $CH_3{-}OH$ (e) $CH_3O{-}H$

1.38. Which is the positive end and which is the negative end of the dipole in each of the following bonds?

(a) $C{-}Mg$ (b) $C{-}Br$ (c) $C{-}O$

(d) $C{-}Cl$ (e) $C{-}H$ (f) $C{-}B$

1.39. Circle the most electronegative element in each of the following structures, and show the direction of polarization of its bond(s):

(a) CH_3OH (b) $CH_3\overset{\overset{\displaystyle O}{\|}}{C}CH_3$

(c) FCH_2CO_2H (d) $(CH_3)_2NCH_2CH_2OH$

1.40. Arrange each of the following series of compounds in order of increasing polarity (least polar first):

(a) $CH_3CH_2CH_2NH_2$, $CH_3CH_2CH_3$, $CH_3CH_2CH_2OH$
(b) $CH_3CH_2CH_2Br$, $CH_3CH_2CH_2I$, $CH_3CH_2CH_2Cl$

1.41. Draw structures to show the hydrogen bonding (if any) you would expect in the following compounds in their pure liquid states:

(a) $(CH_3)_2NH$ (b) $CH_3CH_2OCH_3$ (c) CH_3CH_2F

(d) $(CH_3)_3N$ (e) $(CH_3)_2C{=}O$ (f) $CH_3OCH_2CH_2OH$

1.42. Which of the following compounds would form hydrogen bonds with itself? With water?

(a) $CH_3CH_2CH_2OH$ (b) $CH_3\overset{\overset{\displaystyle O}{\|}}{C}OH$ (c) $CH_3\overset{\overset{\displaystyle CH_3}{|}}{C}HCH_3$

(d) $(CH_3)_2CHOCH_3$ (e) 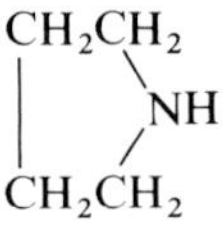 $\begin{matrix} CH_2CH_2 \\ | \quad\quad\; \diagdown \\ \quad\quad\quad NH \\ | \quad\quad\; \diagup \\ CH_2CH_2 \end{matrix}$ (f) 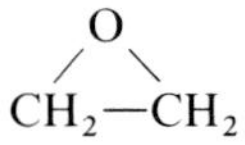 $CH_2\overset{\displaystyle O}{\underset{\diagup\diagdown}{-}}CH_2$

1.43. Show all types of hydrogen bonds in an aqueous solution of $(CH_3)_2NH$. Which is the strongest hydrogen bond?

1.44. Complete the following equations for acid–base reactions:

(a) $CH_3O^- + H_2O \;\rightleftharpoons\;$ (b) $CH_3NH_2 + HCl \;\rightleftharpoons\;$

(c) $HO_2CCO_2H + excess\ OH^- \;\rightleftharpoons\;$ (d)  $\bigcirc\!\!\!\!NH + H^+ \;\rightleftharpoons\;$

(e) $CH_3CO_2^- + H^+ \;\rightleftharpoons\;$ (f) $CH_3NH_2 + CH_3CO_2H \;\rightleftharpoons\;$

(g) $CH_3CO_2H + CH_3O^- \;\rightleftharpoons\;$ (h) $CH_3\overset{+}{N}H_3 + CH_3O^- \;\rightleftharpoons\;$

(i) $HO\overset{\overset{\displaystyle O}{\|}}{C}OH + 2CH_3O^- \;\rightleftharpoons\;$

1.45. Calculate the pK_a of each of the following compounds, and arrange in order of increasing acidity (weakest acid first):

Structure	K_a
(a) CH_3CO_2H	1.75×10^{-5}
(b) 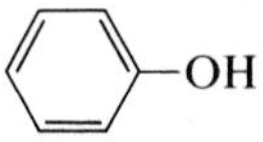 $\bigcirc\!\!-OH$	1.0×10^{-10}

(c) 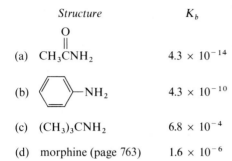 5.2 × 10⁻⁵

(d) CH_3CH_2OH ~10^{-16}

(e) CH_3CH_3 ~10^{-43}

1.46. Calculate the pK_b of the following bases, and arrange in order of increasing base strength:

Structure	K_b
(a) $CH_3\overset{O}{\overset{\|}{C}}NH_2$	4.3 × 10⁻¹⁴
(b) ⬡—NH_2	4.3 × 10⁻¹⁰
(c) $(CH_3)_3CNH_2$	6.8 × 10⁻⁴
(d) morphine (page 763)	1.6 × 10⁻⁶

1.47. Write the formulas for (a) the conjugate acid of $(CH_3)_2NH$, and (b) the conjugate base of $CH_3CH_2CH_2CO_2H$.

1.48. From the pK_a values for the conjugate acids in Table 1.8 (page 30), list the following anions in order of increasing base strength:

(a) $CH_3CO_2^-$ (b) HCO_2^- (c) Cl^- (d) $H_2PO_4^-$

1.49. If you mix equal volumes of equimolar solutions of NaBr and LiCl, you will obtain the same solution as a similar mixture prepared from LiBr and NaCl. Why? Would this be true for solutions of CH_3Cl + NaBr and CH_3Br + NaCl? Why?

1.50. Lead(IV) chloride, $PbCl_4$, is a liquid at room temperature (mp −15°), while lead(II) chloride, $PbCl_2$, is a high-melting solid (mp 501°). What do these properties suggest about the bonding in these two compounds?

1.51. Assign the proper ionic charge to each of the following ions:

(a) $:C{\equiv}N:$ (b) $:C{\equiv}CH$ (c) $CH_2\ddot{O}H$ (d) $CH_3\ddot{O}:$

(e) $CH_2{=}CHCH_2$ (f) $(CH_3)_3C$ (g) $:\ddot{N}H_2$ (h) $CH_3\overset{\cdot\ddot{O}\cdot}{\overset{\|}{C}}\ddot{O}:$

1.52. None of the following compounds contains double bonds. Can you devise a condensed structural formula for each? More than one answer may be possible. (Remember the valences.)

(a) C_3H_6 (b) C_2H_4O (c) C_4H_8O

1.53. Assuming that each of the following compounds contains a double bond, write a condensed structural formula for each. (There may be more than one correct answer.)

(a) C_3H_6 (b) C_2H_4O (c) C_4H_8O

1.54. Using Table 1.4 (page 18), calculate the heats of reaction for:

(a) $CH_3CH_3 + Cl_2 \longrightarrow CH_3CH_2Cl + HCl$
(b) $CH_3CH_3 + Br_2 \longrightarrow CH_3CH_2Br + HBr$

Which reaction liberates more energy?

1.55. Write an equation to show the heterolytic cleavage (at the most likely position) of each of the following molecules. (Remember to take electronegativities into consideration.)

(a) $(CH_3)_2CHBr$ (b) CH_3CH_2Li

(c) $(CH_3)_2CHOCH(CH_3)_2$ (d) 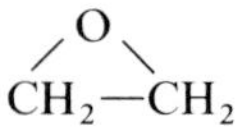 $CH_2\overset{O}{\diagup \diagdown}CH_2$

1.56. BF_3 has a dipole moment of zero. Suggest a shape for the BF_3 molecule.

1.57. Arrange the following compounds according to increasing solubility in water, least soluble first:

(a) CH_3CO_2H (b) CH_3CH_3 (c) $CH_3CH_2OCH_2CH_3$

1.58. Diethyl ether, $CH_3CH_2OCH_2CH_3$, and 1-butanol, $CH_3CH_2CH_2CH_2OH$, are equally soluble in water, but the boiling point of 1-butanol is 83° higher than that of diethyl ether. What explanation can you give for these observations?

1.59. In the following reactions, which of the reactants is the Lewis acid and which is the Lewis base?

(a) $(CH_3)_3CCl + AlCl_3 \rightleftharpoons (CH_3)_3C^+ + AlCl_4^-$
(b) $(CH_3)_3C^+ + CH_2=CH_2 \rightleftharpoons (CH_3)_3CCH_2\overset{+}{C}H_2$
(c) $(CH_3)_3C^+ + H_2O \rightleftharpoons (CH_3)_3C\overset{+}{O}H$ $\underset{H}{|}$

(d) $CH_2=CH_2 + Br_2 \rightleftharpoons CH_2\overset{\overset{+}{Br}}{\diagup \diagdown}CH_2 + Br^-$

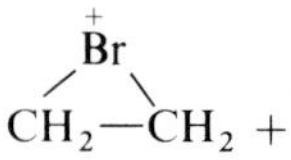

1.60. Indicate the most likely position of attack by a Lewis acid on each of the following structures:

(a)  $HN\diagup\diagdown NH$ (b) 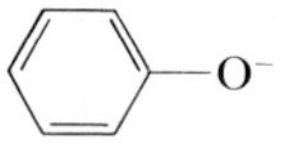 ⬡–O^- (c) CH_3OH (d) CH_3CH_2Cl

1.61. Like water, an alcohol can act as a weak acid or a weak base. Write equations for the reaction of methanol (CH_3OH) with (a) concentrated sulfuric acid, and (b) sodamide ($NaNH_2$), an extremely strong base.

1.62. In Chapter 6, we will discuss the following reaction:

$$CH_3I + Mg \longrightarrow CH_3MgI$$

(a) The product of this reaction (methylmagnesium iodide) is a much stronger base than even the hydroxide ion. Can you suggest a reason for this?
(b) If water is added to CH_3MgI, a vigorous reaction ensues. Predict the products.

CHAPTER 2

Orbitals and Their Role
in Covalent Bonding

In Chapter 1, we took a brief look at atomic orbitals and at covalent bonding. In this chapter, we will discuss how covalent bonds are produced by the formation of molecular orbitals. Various approaches to molecular orbitals have been developed. The **molecular orbital (MO) theory** provides mathematical descriptions of orbitals, their energies, and their interactions. The **valence-shell electron-pair repulsion (VSEPR) theory** is based on the premise that valence electrons or electron pairs of an atom repel each other. These repulsions can be used to explain observed bond angles and molecular geometry. In the **valence-bond theory**, valence-bond formulas are used to describe covalent bonds and their interactions.

Within their limitations, all of these theories are successful and are often in agreement. However, no single one of them is practical to use in all discussions of organic compounds and their reactions. For this reason, we will present some of the highlights of each theory without necessarily attempting to differentiate among them. You will find that a knowledge of molecular orbitals and molecular shapes is invaluable in a study of organic reactions—whether they occur in a flask in the laboratory or in the cells of an animal.

SECTION 2.1.

Properties of Waves

Until 1923, chemists assumed that electrons were nothing more than negatively charged particles whirling about the atomic nuclei. In 1923, Louis de Broglie, a French graduate student, proposed the revolutionary idea that electrons have properties of waves as well as properties of particles. De Broglie's proposal met with skepticism at first, but his idea was the seed that grew into today's quantum-mechanical concept of electron motion and the molecular orbital theory.

Quantum mechanics is a mathematical subject. For our understanding of covalent bonds, we need only the results of quantum-mechanical studies, rather than the mathematical equations themselves. With this in mind, let us survey some of the basic concepts of wave motion as they pertain to the current theories of covalent bonds.

We will begin with some simple **standing waves** (Figure 2.1), the type of wave that results when you pluck a string, like a guitar string, that is fixed at both ends. This type of wave exhibits motion in only one dimension. By contrast, the standing waves caused by beating the head of a drum are two-dimensional, and the wave system of an electron is three-dimensional. The height of a standing wave is its **amplitude**, which may be up (positive value) or down (negative value) in relation to the resting position of the string. (Note that the + or − sign of amplitude is a mathematical sign, not an electrical charge.) A position on the wave at which the amplitude is zero is called a **node**, and corresponds to a position on the guitar string that does not move as the string vibrates.

Two standing waves can be either **in phase** or **out of phase** in reference to each other. Intermediate states in which waves are only partially in phase are also possible. We can illustrate these terms by two wave systems on two identical vibrating strings. If the positive and negative amplitudes of the two waves correspond to each other, the two waves are *in phase*. If the mathematical signs of the amplitudes are opposite, the waves are *out of phase* (see Figure 2.2).

If two in-phase waves on the same string overlap, they **reinforce** each other. The reinforcement is expressed by addition of the mathematical functions of the same sign describing the waves. Conversely, a pair of overlapping waves that are

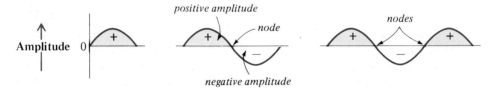

FIGURE 2.1. Some standing waves of a vibrating string with fixed ends (positive amplitudes shaded).

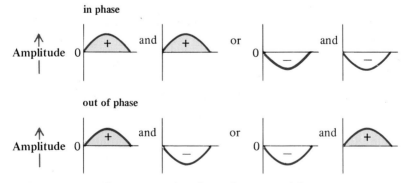

FIGURE 2.2. Two standing waves may be either in phase or out of phase.

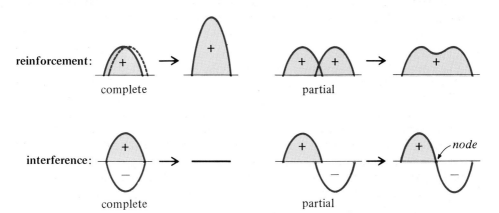

FIGURE 2.3. Reinforcement and interference of waves.

out of phase **interfere** with each other. The process of interference is represented by addition of two mathematical functions of opposite sign. Complete interference results in the cancelling of one wave by another. Partial overlap of two out-of-phase waves gives rise to a node. Figure 2.3 illustrates reinforcement and interference.

Although a three-dimensional electron wave system is more complicated than a one-dimensional string system, the principles are similar. Each atomic orbital of an atom behaves as a wave function and may have a positive or negative amplitude. If the orbital has both positive and negative amplitudes, it has a node. Figure 2.4 depicts the 1s, 2s, and 2p orbitals, including their signs of amplitude and their nodes.

One atomic orbital can overlap an atomic orbital of another atom. (Mathematically, the wave functions describing each overlapping orbital are added together. These calculations are referred to as the **linear combination of atomic orbitals**, or LCAO, theory.) When the overlapping orbitals are in phase, the result is reinforcement and a **bonding molecular orbital**. On the other hand, interaction between atomic orbitals that are out of phase results in interference, creating a node between the two nuclei. Interference leads to an **antibonding molecular orbital**. We will expand upon these definitions of bonding and antibonding orbitals in Section 2.2B.

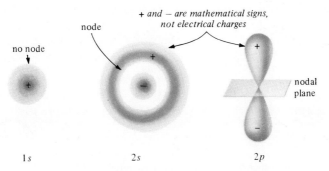

FIGURE 2.4. The 1s, 2s, and 2p orbitals with their signs of amplitude.

SECTION 2.2.

Bonding in Hydrogen

Hydrogen (H_2) is the simplest molecule. We will look at the covalent bond of H_2 in some detail because many features of this bond are similar to those of more-complex covalent bonds.

Let us consider two isolated hydrogen atoms, each with one electron in a $1s$ atomic orbital. As these two atoms begin bond formation, the electron of each atom becomes attracted by the nucleus of the other atom, as well as by its own nucleus. When the nuclei are at a certain distance from each other (the bond length, 0.74 Å for H_2), the atomic orbitals merge, or overlap, to reinforce each other and form a bonding molecular orbital. This molecular orbital encompasses both hydrogen nuclei and contains two paired electrons (one from each H). Both electrons are now equally attracted to both nuclei. Because a large portion of the negatively charged electron density of this new orbital is located between the two positively charged nuclei, repulsions between the nuclei are minimized. This molecular orbital results in the covalent bond between the two hydrogen atoms in H_2 (see Figure 2.5).

A.   The sigma bond

The molecular orbital that bonds two hydrogen atoms together is *cylindrically symmetrical*—that is, symmetrical about a line, or axis, joining the two nuclei. Think of the axis as an axle, and rotate the orbital around this axis. If the appearance of the orbital is not changed by the rotation, the orbital is symmetrical around that axis (see Figure 2.6).

Any molecular orbital that is symmetrical about the axis connecting the nuclei is called a **sigma (σ) molecular orbital**; the bond is a **sigma bond**. The bond in H_2 is only one of many sigma bonds we will encounter. (We will also encounter molecular orbitals that are not sigma orbitals—that is, orbitals that are not symmetrical about their nuclear axes.)

B.   The bonding orbital and the antibonding orbital

When a pair of waves overlap, they either reinforce each other or interfere with each other. Addition of two in-phase $1s$ atomic orbitals of two H atoms results in

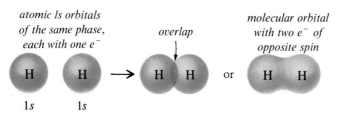

FIGURE 2.5. The formation of the bonding molecular orbital in H_2 from in-phase $1s$ orbitals.

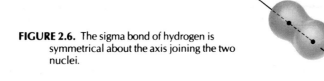

FIGURE 2.6. The sigma bond of hydrogen is symmetrical about the axis joining the two nuclei.

reinforcement and produces the σ bonding molecular orbital with a high electron density between the bonded nuclei.

If two waves are out of phase, they interfere with each other. Interference of two out-of-phase atomic orbitals of two hydrogen atoms gives a molecular orbital with a *node between the nuclei*. In this molecular orbital, the probability of finding an electron between the nuclei is *very low*. Therefore, this particular molecular orbital gives rise to a system where the two nuclei are not shielded by the pair of electrons, and the nuclei repel each other. Because of the nuclear repulsion, this system is of *higher energy* than the system of two independent H atoms. This higher-energy orbital is the **antibonding orbital**, in this case, a "sigma star," or σ^*, orbital (the * meaning "antibonding"). Figure 2.7 compares the shapes of the σ and σ^* orbitals for H_2.

The energy of the H_2 molecule with two electrons in the σ bonding orbital is *lower* by 104 kcal/mole than the combined energy of two separate hydrogen atoms. The energy of the hydrogen molecule with electrons in the σ^* antibonding orbital, on the other hand, is *higher* than that of two separate hydrogen atoms. These

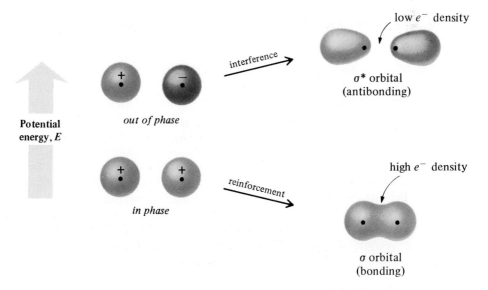

FIGURE 2.7. Reinforcement and interference of two 1s orbitals. (The + and − refer to phases of the wave functions, not electrical charges.)

relative energies may be represented by the following diagram:

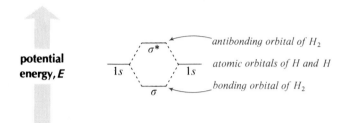

A molecular orbital, like an atomic orbital, can hold no electrons, one electron, or two paired electrons. The two electrons in a hydrogen molecule go into the lowest-energy orbital available, the σ bonding orbital. In the following diagram, we use a pair of arrows (one pointing up and one pointing down) to represent a pair of electrons of opposite spin.

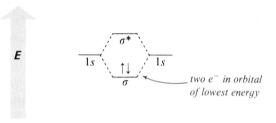

We said in Chapter 1 that electrons in different atomic orbitals differ in energy because of the various distances of these electrons from the nucleus. The higher energy of a molecule with electrons in an antibonding orbital (compared to the energy of the molecule with electrons in a bonding orbital) does not arise from the electrons being farther from the nuclei. Instead, the higher energy arises from the presence of the node between the nuclei.

For the hydrogen molecule, the σ and σ^* orbitals are in the same general region of space. Although two particles of matter cannot occupy the same space at the same time, two orbitals may. Remember, orbitals are not matter, but are simply regions of space where the probability of finding an electron with a particular energy is high.

All bonding molecular orbitals have antibonding orbitals associated with them. In each case, a molecule with electrons in a bonding molecular orbital has a *lower energy* than the energy of the nonbonded atoms, and a molecule with electrons in an antibonding orbital has a *higher energy* than that of the nonbonded atoms. Because the antibonding orbitals are of high energy, the electrons are not generally found there. Almost all the chemistry in this text will deal with molecules in the **ground state**, the state in which the electrons are in the lowest-energy orbitals. However, we will encounter a few situations where energy absorbed by a molecule is used to promote an electron from a low-energy orbital to a higher-energy orbital. A molecule is said to be in an **excited state** when one or more electrons are not in the orbital of lowest energy.

SECTION 2.3.

Some General Features of Bonding and Antibonding Orbitals

Let us summarize some general rules that apply to all molecular orbitals, not only the molecular orbitals of H_2:

1. Any orbital (molecular or atomic) can hold a maximum of two electrons, which must be of opposite spin.

2. The number of molecular orbitals equals the number of atomic orbitals that went into their formation. (For H_2, two $1s$ atomic orbitals yield two molecular orbitals: σ and σ^*.)

3. In the filling of molecular orbitals with electrons, the lowest-energy orbitals are filled first. If two orbitals are degenerate (of equal energies), each gets one electron before either is filled.

SECTION 2.4.

Hybrid Orbitals of Carbon

When a hydrogen atom becomes part of a molecule, it uses its $1s$ atomic orbital for bonding. The situation with the carbon atom is somewhat different. Carbon has two electrons in the $1s$ orbital; consequently, the $1s$ orbital is a filled orbital that is not used for bonding. The four electrons at the *second energy level* of carbon are the bonding electrons.

There are four atomic orbitals at the second energy level: one $2s$ and three $2p$ orbitals. However, carbon does not use these four orbitals in their pure states for bonding. Instead, carbon blends, or **hybridizes**, its four second-level atomic orbitals in one of three different ways for bonding:

1. *sp³* **hybridization,** used when carbon forms four single bonds.

2. *sp²* **hybridization,** used when carbon forms a double bond.

3. *sp* **hybridization,** used when carbon forms a triple bond or *cumulated* double bonds (two double bonds to a single carbon atom).

Why does a carbon atom form compounds with hybrid orbitals rather than with unhybridized atomic orbitals? The answer is that hybridization gives stronger bonds because of greater overlap, and therefore results in more-stable, lower-energy molecules. As we discuss each type of hybridization, note that the *shape* of each hybrid orbital is favorable for maximum overlap with an orbital of another atom. Also note that the *geometries* of the three types of hybrid orbitals allow attached groups to be as far from each other as possible, thus minimizing their repulsions for each other.

A. sp^3 Hybridization

In methane (CH_4), the carbon atom has four equivalent bonds to hydrogen. Each C—H bond has a bond length of 1.09 Å and a bond dissociation energy of 104 kcal/mole. The bond angle between each C—H bond is 109.5°. From this experimental evidence alone, it is evident that carbon does not form bonds by means of one *s* atomic orbital and three *p* atomic orbitals. If that were the case, the four C—H bonds would not all be equivalent.

According to present-day theory, these four equivalent bonds arise from complete hybridization of the four atomic orbitals (one 2s orbital and three 2p orbitals) to yield four equivalent sp^3 orbitals. For this to be accomplished, one of the 2s electrons must be promoted to the empty 2p orbital. This promotion requires energy (about 96 kcal/mole), but this energy is more than regained by the concurrent formation of chemical bonds. The four sp^3 orbitals have equal energies —slightly higher than that of the 2s orbital, but slightly lower than that of the 2p orbitals. Each of the sp^3 orbitals contains one electron for bonding.

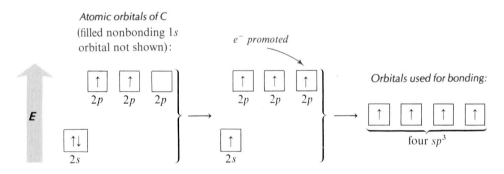

The preceding diagram is called an **orbital diagram**. Each box in the diagram represents an orbital. The relative energies of the various orbitals are signified by the vertical positions of the boxes within the diagram. Electrons are represented by arrows, and the direction of electron spin is indicated by the direction of the arrow.

The sp^3 orbital, which results from a blend of the 2s and 2p orbitals, is shaped rather like a bowling pin: it has a large lobe and a small lobe (of opposite amplitude) with a node at the nucleus. Figure 2.8 shows one isolated sp^3 orbital. The small end of the hybrid orbital is not used for bonding because overlap of the large end with another orbital gives more complete overlap and results in a stronger bond.

Four sp^3-hybrid orbitals surround the carbon nucleus. Because of repulsions between electrons in different orbitals, these sp^3 orbitals lie as far apart from each

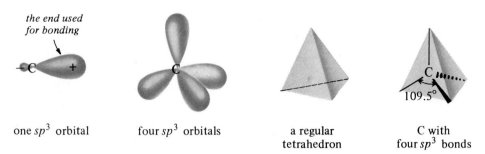

the end used for bonding

one sp^3 orbital   four sp^3 orbitals   a regular tetrahedron   C with four sp^3 bonds

109.5°

FIGURE 2.8. The four sp^3-hybrid orbitals of carbon point toward the corners of a regular tetrahedron.

other as possible while still extending away from the same carbon nucleus—that is, the four orbitals point toward the corners of a regular tetrahedron (Figure 2.8). This geometry gives idealized bond angles of 109.5°. An sp^3 carbon atom is often referred to as a **tetrahedral carbon atom** because of the geometry of its bonds.

When an sp^3 carbon atom forms bonds, it does so by overlapping each of its four sp^3 orbitals (each with one electron) with orbitals from four other atoms (each orbital in turn containing one electron). In methane (Figures 2.9 and 2.10), each sp^3 orbital of carbon overlaps with a $1s$ orbital of hydrogen. Each of the resultant sp^3–s molecular orbitals is symmetrical around the axis passing through the nuclei of the carbon and the hydrogen. The covalent bonds between C and H in methane, like the H—H covalent bond, are sigma bonds.

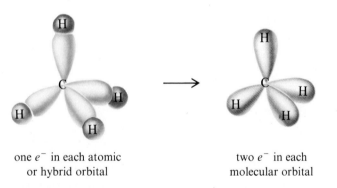

one e^- in each atomic or hybrid orbital

two e^- in each molecular orbital

FIGURE 2.9. Formation of C—H sigma bonds in methane, CH_4. (The small lobes of the sp^3 orbitals are not shown.)

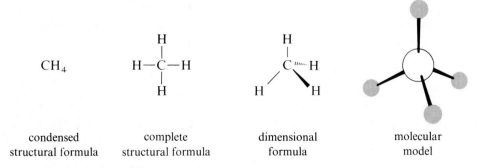

CH_4

$$H—\overset{\displaystyle H}{\underset{\displaystyle H}{C}}—H$$

condensed structural formula   complete structural formula   dimensional formula   molecular model

FIGURE 2.10. Some different ways of representing methane.

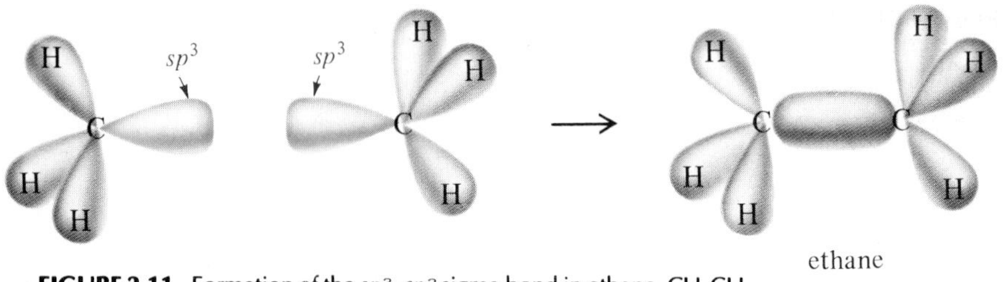

FIGURE 2.11. Formation of the sp^3–sp^3 sigma bond in ethane, CH_3CH_3.

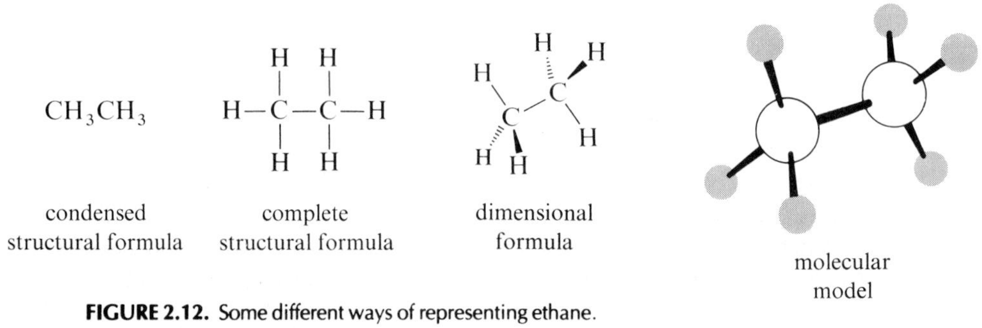

CH_3CH_3

condensed
structural formula

complete
structural formula

dimensional
formula

molecular
model

FIGURE 2.12. Some different ways of representing ethane.

Ethane (CH_3CH_3) contains two sp^3 carbon atoms. These two carbon atoms form a C—C sigma bond by the overlap of one sp^3 orbital from each carbon (sp^3–sp^3 sigma bond). Each carbon atom has three remaining sp^3 orbitals, and each of these overlaps with a $1s$ orbital of a hydrogen atom to form a C—H sigma bond. Each carbon atom in ethane is tetrahedral (see Figures 2.11 and 2.12).

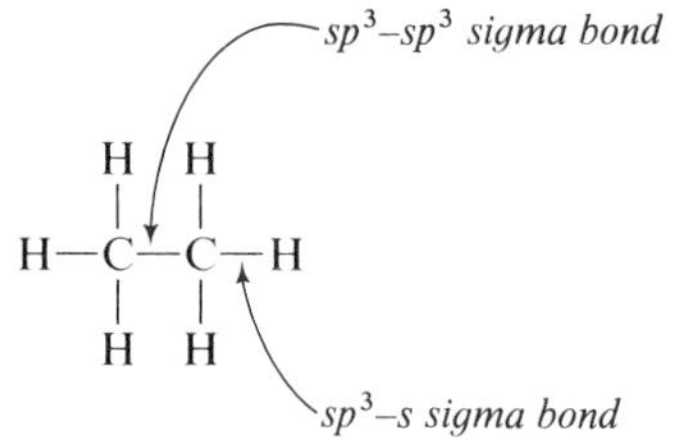

In any molecule, any carbon atom bonded to four other atoms is in the sp^3-hybrid state, and the four bonds from that carbon are sigma bonds. When carbon is bonded to four other atoms, the sp^3 hybridization allows maximal overlap and places the four attached atoms at the maximum distances from each other. If possible, the sp^3 bond angles are 109.5°; however, other factors, such as dipole–dipole repulsions or the geometry of a cyclic compound, can cause deviations from this ideal bond angle.

Examples of structures with sp^3 carbons (each C has four sigma bonds):

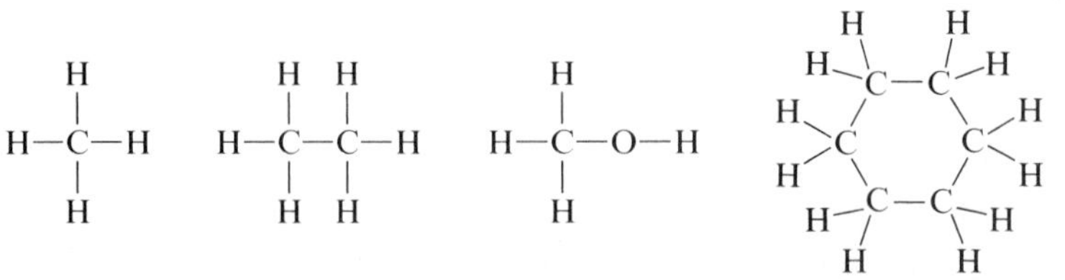

SAMPLE PROBLEM

Give the complete structural formula (showing all atoms and bonds) for propane ($CH_3CH_2CH_3$). Which types of orbitals overlap to form each bond?

Solution:

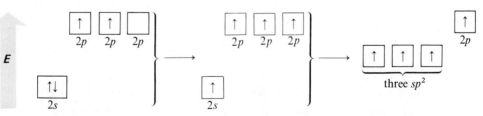

STUDY PROBLEM

2.1.   Write the complete structural formula for each of the following compounds. Which types of orbitals overlap to form each bond?

(a) $(CH_3)_3CH$ (b)   ⬠

B.   *sp²* Hybridization

When carbon is bonded to another atom by a double bond, the carbon atom is in the sp^2-hybrid state.

Examples of compounds with sp^2 carbons:

$$\underset{\text{ethylene}}{\overset{H}{\underset{H}{>}}C=C\overset{H}{\underset{H}{<}}} \qquad \underset{\text{formaldehyde}}{\overset{H}{\underset{H}{>}}C=O}$$

To form sp^2 bonding orbitals, carbon hybridizes its $2s$ orbital with only two of its $2p$ orbitals. One p orbital remains unhybridized on the carbon atom. Because three atomic orbitals are used to form the sp^2 orbitals, three sp^2-hybrid orbitals result. Each sp^2 orbital has a shape similar to that of an sp^3 orbital and contains one electron that can be used for bonding.

The three sp^2 orbitals around a carbon nucleus lie as far apart from one another as possible—that is, the sp^2 orbitals lie in a plane with angles of 120° (ideally) between them. An sp^2-hybridized carbon atom is said to be a **trigonal** (triangular) carbon. Figure 2.13 shows a carbon atom with three sp^2 orbitals and the one unhybridized p orbital, which is perpendicular to the sp^2 plane.

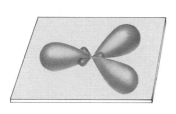

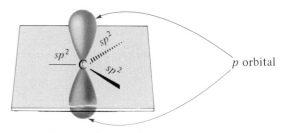

p orbital

a trigonal carbon; three sp^2
orbitals in a plane with $120°$
angles between them

p orbital at a right angle
to the plane

FIGURE 2.13. Carbon in the sp^2-hybrid state.

In ethylene ($CH_2{=}CH_2$), two sp^2 carbons are joined by a sigma bond
formed by the overlap of one sp^2 orbital from each carbon atom. (This sigma bond
is one of the bonds of the double bond.) Each carbon atom still has two sp^2
orbitals left for bonding with hydrogen. (Each carbon atom also has a *p* orbital,
which is not shown in the following structure.)

Planar sigma-bond structure of ethylene (p orbitals not shown):

$$\begin{array}{ccc} H & {\scriptstyle 121°} & H \\ \diagdown & & \diagup \\ & C{-}C & {\scriptstyle 118°} \\ \diagup & & \diagdown \\ H & & H \end{array}$$

What of the remaining *p* orbital on each carbon? Each *p* orbital has two lobes,
one above the plane of the sigma bonds and the other (of opposite amplitude)
below the plane. Each *p* orbital contains one electron. If these *p* electrons become
paired in a bonding molecular orbital, then the energy of the system is lowered.
Since the *p* orbitals lie side by side in the ethylene molecule, the *ends* of the orbitals
cannot overlap, as they do in sigma-bond formation. Rather, the two *p* orbitals
overlap their *sides* (see Figure 2.14). The result of this side-to-side overlap is the
pi (π) bond—a bonding molecular orbital joining the two carbons and located above
and below the plane of the sigma bonds. The pi bond is the second bond of the
double bond.

Any carbon atom that is bonded to three other atoms is in the sp^2-hybrid
state. In stable compounds, the *p* orbital on the sp^2 carbon must overlap with a

The π bond is a two-lobed molecular
orbital containing one pair of e^- and
having a node at the site of the sigma
bond.

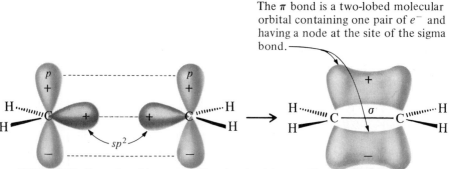

FIGURE 2.14. Formation of the sp^2–sp^2 sigma bond and the *p*–*p* pi bond in ethylene,
$CH_2{=}CH_2$. (The + and – refer to phases of the wave functions, not electrical charges.)

p orbital of an adjacent atom, which can be another carbon atom or an atom of some other element.

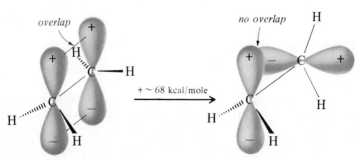

sp²ᵃ carbons

C. Some features of the pi bond

Each *p* orbital contributing to a pi bond has two lobes and has a node at the nucleus. It is not surprising that the pi orbital also is two-lobed and has a node. Unlike a sigma orbital, a pi orbital is not cylindrically symmetrical. However, just like any molecular orbital, a pi orbital can hold a maximum of two paired electrons.

A $2p$ orbital of carbon is of slightly higher energy than an sp^2 orbital. For this reason, a pi bond, which is formed from two $2p$ orbitals, has slightly higher energy and is slightly less stable than an sp^2–sp^2 sigma bond. The bond dissociation energy of the sigma bond of ethylene's carbon–carbon double bond is estimated to be 95 kcal/mole, while that of the pi bond is estimated to be only 68 kcal/mole.

The more-exposed pi electrons are more vulnerable to external effects than are electrons in sigma bonds. The pi bond is polarized more easily—we might say that the pi electrons are more mobile. The pi electrons are more easily promoted to a higher-energy (antibonding) orbital. Also, they are more readily attacked by an outside atom or molecule. What does this vulnerability mean in terms of the chemistry of pi-bonded compounds? In a molecule, the pi bond is a site of chemical reactivity.

Another property of the pi bond is that its geometry holds a portion of its molecule in a rigid shape. For the carbon atoms to rotate around their bonds, the pi bond must first be broken (see Figure 2.15). In chemical reactions, molecules may have sufficient energy (about 68 kcal/mole) for this bond to break. In a flask at room temperature, however, molecules do not have enough energy for this bond breakage to occur. (Approximately 20 kcal/mole is the maximum energy available to molecules at room temperature.) The significance of the rigidity of pi bonds will be discussed in Chapter 4.

FIGURE 2.15. The portion of a molecule surrounding a pi bond is held in a planar structure unless enough energy is supplied to break the pi bond.

In a structural formula, a double bond is indicated by two identical lines. Keep in mind that the double bond is not simply two identical bonds, but that the double line represents one strong sigma bond and one weak pi bond.

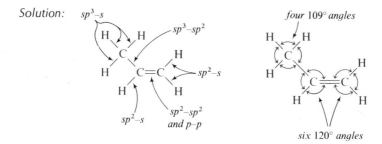

SAMPLE PROBLEM

What type of overlap (sp^3–s, for example) is present in each bond of $CH_3CH=CH_2$? What is each bond angle (approximately)?

Solution:

STUDY PROBLEMS

2.2. Give the complete structural formula for each of the following compounds. Indicate which types of orbitals are used to form each bond.

(a) $CH_2=C(CH_3)_2$ (b) $CH_2=CHCH=CH_2$ (c)

2.3. For compound (b) in Problem 2.2, draw the structure showing the pi bonds with their correct geometry in reference to the sigma bonds. (Use lines to represent the sigma bonds.)

D.   The bonding and antibonding orbitals of ethylene

The carbon–carbon sigma bond in ethylene results from overlap of two sp^2 orbitals. Because *two* sp^2 orbitals form this bond, *two* molecular orbitals result. The other molecular orbital, arising from interference between the two sp^2 orbitals, is the antibonding σ^* orbital. This antibonding orbital is similar to the σ^* orbital of hydrogen: it has a node between the two carbon nuclei and is of high energy. The two electrons of the C—C sigma bond in ethylene are usually found in the lower-energy σ orbital.

In the pi bond joining the two carbons of ethylene, each carbon atom contributes one p orbital for a total of two p orbitals; therefore, two pi molecular orbitals result. One of these is the π bonding orbital that arises from the overlap of two in-phase p orbitals. The other orbital is the π^* antibonding orbital, which arises from interference between two p orbitals of opposite phase. These two orbitals are frequently designated π_1 for the bonding orbital and π_2^* for the antibonding orbital. Some molecules contain several π orbitals, which we number in order of increasing energy; the subscript numbers are useful for differentiating the π orbitals. (We do not usually use subscript numbers for σ orbitals because σ^* orbitals are usually of minor importance to the organic chemist.)

Figure 2.16 shows the orbital representations of the π_1 and π_2^* orbitals of ethylene. Note that besides the node at the σ-bond site, the π_2^* orbital has an additional node *between the two carbon nuclei*. A minimum of the pi electron density is located between the nuclei in this orbital; thus, the π_2^* orbital is of higher energy than the π_1 orbital. In the ground state of ethylene, the pi electrons are found in the lower-energy π_1 orbital.

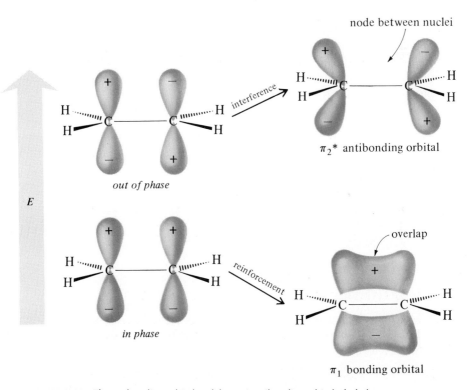

FIGURE 2.16. The π_1 bonding orbital and the π_2^* antibonding orbital of ethylene.

The following diagram comparing energies of the σ, σ^*, π_1, and π_2^* orbitals shows that the σ^* orbital is of higher energy than the π_2^* orbital. The amount of energy required to promote an electron from the σ orbital to the σ^* orbital is therefore greater than the energy required to promote a π electron to the π^* orbital.

Ground state of C=C in ethylene:

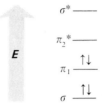

Why is more energy required to promote a σ electron than a π electron? Sigma bonding electrons are found close to the nucleus and, for the most part, directly between the nuclei. Promotion of one of these sigma electrons results in severe nuclear repulsion. However, the π bonding electrons are farther from the nucleus. Promotion of one of these pi electrons does not result in such severe nuclear repulsions. (Also, when a pi electron is promoted, the nuclei are still shielded from each other by σ bonding electrons.)

Because of the large amount of energy required to promote a sigma electron, electron transitions of the $\sigma \rightarrow \sigma^*$ type are rare and relatively unimportant to the organic chemist. However, $\pi \rightarrow \pi^*$ electron transitions, which require less energy, are important. For example, $\pi \rightarrow \pi^*$ transitions are responsible for vision, a topic that will be mentioned in Chapter 21, and for the energy capture needed for photosynthesis.

Two possible excited states of C=C in ethylene:

$\sigma \rightarrow \sigma^*$, higher ΔE

$\pi_1 \rightarrow \pi_2^*$, lower ΔE

E. *sp* Hybridization

When a carbon atom is joined to only two other atoms, as in acetylene (CH≡CH), its hybridization state is *sp*. One 2*s* orbital blends with only one 2*p* orbital to form two *sp*-hybrid orbitals. In this case, two unhybridized 2*p* orbitals remain, each with one electron.

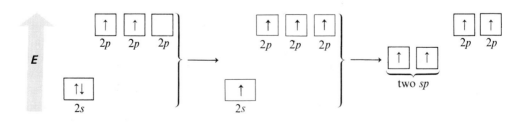

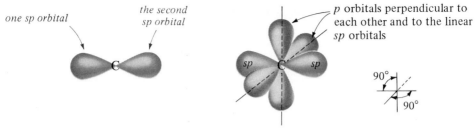

one *sp orbital*

the second sp orbital

p orbitals perpendicular to each other and to the linear *sp* orbitals

sp   C   *sp*

90°

90°

the two *sp* orbitals are linear

the two *p* orbitals

FIGURE 2.17. Carbon in the *sp*-hybrid state.

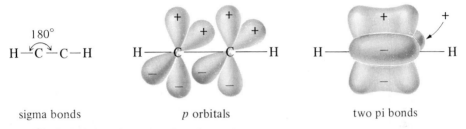

180°

H—C—C—H

H—C——C—H

H——C——H

sigma bonds

p orbitals

two pi bonds

FIGURE 2.18. Bonding in acetylene, CH≡CH.

The two *sp* orbitals lie as far apart as possible, in a straight line with an angle of 180° between them. The *p* orbitals are perpendicular to each other and to the line of the *sp* orbitals (see Figure 2.17).

In CH≡CH, the two carbon atoms are joined by an *sp–sp* sigma bond. Each carbon is also bonded to a hydrogen atom by an *sp–s* sigma bond. The two *p* orbitals of one carbon then overlap with the two *p* orbitals of the other carbon to form *two* pi bonds. One pi bond is above and below the line of the sigma bonds, as shown in Figure 2.18; the other pi bond is located in front and back.

As you might guess, the chemical reactions of a compound containing a triple bond are not too different from those of a compound containing a double bond. Instead of one pi bond, there are two.

one σ bond, two pi bonds

H—C≡C—H   CH₃—C≡C—H   H—C≡N

STUDY PROBLEM

2.4.   What type of overlap is present in each carbon–carbon bond of CH_3—C≡CH? What are the approximate bond angles around each carbon?

F.   Effects of hybridization on bond lengths

A 2*s* orbital is of slightly lower energy than a 2*p* orbital. On the average, 2*s* electrons are found closer to the nucleus than 2*p* electrons. For this reason, a hybrid orbital with a greater proportion of *s* character is of lower energy and is closer to the nucleus than a hybrid orbital with less *s* character.

An *sp*-hybrid orbital is one-half *s* and one-half *p*; we may say that the *sp* orbital has 50% *s* character and 50% *p* character. At the other extreme is the *sp*³ orbital, which has only one-fourth, or 25%, *s* character.

	hybridization of carbon	percent s character
$CH{\equiv}CH$	*sp*	50
$CH_2{=}CH_2$	*sp*²	$33\frac{1}{3}$
CH_3CH_3	*sp*³	25

Because the *sp* orbital contains more *s* character, it is closer to its nucleus; it forms shorter and stronger bonds than the *sp*³ orbital. The *sp*² orbital is intermediate between *sp* and *sp*³ in its *s* character and in the length and strength of the bonds it forms.

Table 2.1 shows the differences in bond lengths among the three C—C and C—H bond types. Note that the *sp–s* CH bond in CH≡CH is the shortest, while the *sp*³*–s* CH bond is the longest. We see an even wider variation in the C—C bonds because these bond lengths are affected by the *number* of bonds joining the carbon atoms as well as by the hybridization of the carbon atoms.

STUDY PROBLEM

2.5. For each of the following structures, list the numbered bonds in order of increasing bond length (shortest bond first):

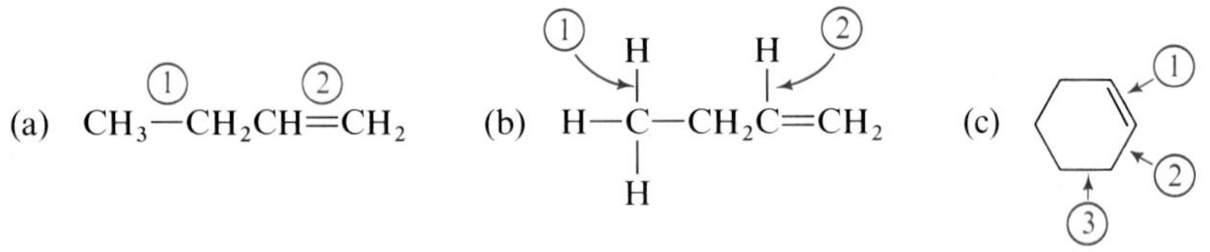

(a) CH₃—CH₂CH=CH₂   (b) H—C—CH₂C=CH₂   (c)

TABLE 2.1. Effect of hybridization on bond length

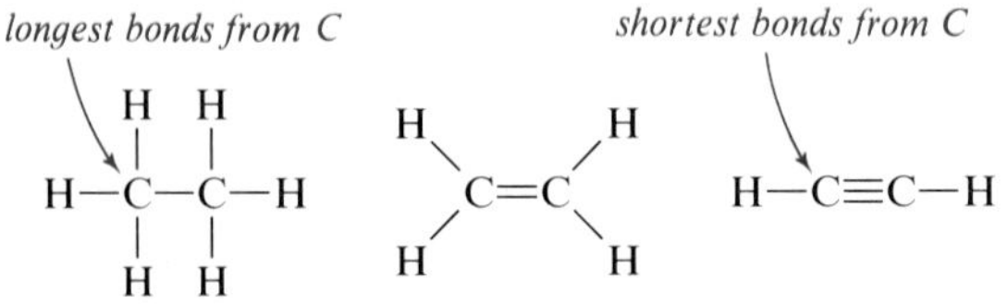

	longest bonds from C		shortest bonds from C
percent s character:	25	$33\frac{1}{3}$	50
C—C bond length:	1.54 Å	1.34 Å	1.20 Å
C—H bond length:	1.09 Å	1.08 Å	1.06 Å

G. Summary of the hybrid orbitals of carbon

1. When a carbon atom is bonded to *four other atoms*, the bonds from the carbon atom are formed from four equivalent sp^3 orbitals. The sp^3 carbon is **tetrahedral**.

tetrahedral   *Examples:* CH_4, $CHCl_3$, (hexagon) sp^3

2. When carbon bonds to *three other atoms*, the bonds from the carbon atom are formed from three equivalent sp^2 orbitals with one *p* orbital remaining. The sp^2 orbitals form three sigma bonds; the *p* orbital forms a pi bond. The sp^2 carbon is **trigonal**.

σ and π

trigonal   *Examples:* $CH_3CH=CH_2$, $H_2C=O$, (ring) sp^2

3. When carbon bonds to *two other atoms*, the bonds from the carbon atom are formed from two equivalent *sp* bonds, with two *p* orbitals remaining. The two *p* orbitals overlap with two *p* orbitals of another atom to form two pi bonds. The *sp* orbitals form two equivalent and **linear** sigma bonds.

one σ, two π

$$H-C\equiv C-H$$

linear   *Examples:* $CH_3C\equiv CH$, $HC\equiv N$

sp carbons

SECTION 2.5.

Functional Groups

Although sp^3–sp^3 carbon–carbon bonds and sp^3–*s* carbon–hydrogen bonds are common to almost all organic compounds, surprisingly, these bonds do not usually play a major role in organic reactions. For the most part, it is the presence of either pi bonds or other atoms in an organic structure that confers reactivity. A site of chemical reactivity in a molecule is called a **functional group**. A pi bond or an electronegative (or electropositive) atom in an organic molecule can lead to chemical reaction; either one of these is considered a functional group or part of a functional group.

Some functional groups (circled):

CH_3 $CH=C H_2$ $CH_3CH_2 NH_2$ $CH_3CH_2 OH$

STUDY PROBLEM

2.6.   Circle the functional groups in the following structures:

(a) 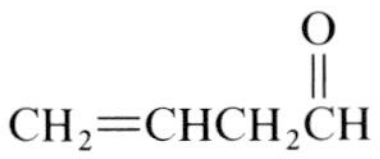 $CH_2{=}CHCH_2\overset{\overset{\displaystyle O}{\|}}{C}H$ (b) 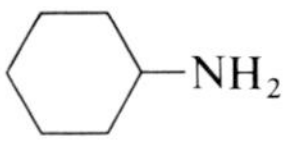   ⬡—NH_2 (c)

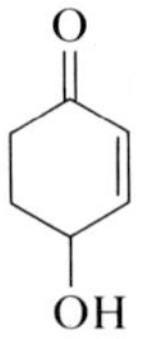

Compounds with the same functional group tend to undergo the same chemical reactions. For example, each of the following series of compounds contains a **hydroxyl group** (—OH). These compounds all belong to the class of compounds called **alcohols**, and all undergo similar reactions.

Some alcohols:

$$CH_3CH_2OH \qquad (CH_3)_3COH \qquad ⬡{-}OH$$

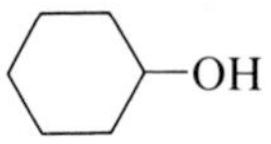

Because of the similarities in reactivity among compounds with the same functional group, it is frequently convenient to use a general formula for a series of these compounds. We usually use R to represent an **alkyl group**, a group that contains only sp^3 carbon atoms plus hydrogens. By this technique, we may represent an alcohol as ROH. Table 2.2 shows some functional groups and some classes of compounds with generalized formulas.

R—   means an alkyl group, such as $CH_3{-}$, $CH_3CH_2{-}$, or   ⬡—

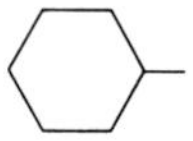

ROH   means an alcohol, such as CH_3OH, CH_3CH_2OH, or   ⬡—OH

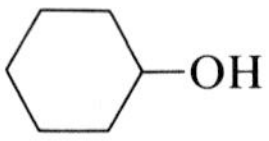

TABLE 2.2.   Some functional groups and compound classes

Functional group		Class of compound	
Structure	*Name*	*General formula*	*Class name*
C=C	double bond	$R_2C{=}CR_2$	alkene
C≡C	triple bond	$RC{\equiv}CR'^a$	alkyne
—NH_2	amino group	RNH_2	amine
—OH	hydroxyl group	ROH	alcohol
—OR	alkoxyl group	$R'OR^a$	ether

^{a} R' refers to an alkyl group that may be the same as or different from R.

SAMPLE PROBLEM

The following compounds are all **amines**. Write a general formula for an amine.

$$CH_3NH_2 \qquad CH_3CH_2NH_2 \qquad (CH_3)_2CHCH_2NH_2$$

Solution: RNH_2

STUDY PROBLEM

2.7. The following compounds are all **carboxylic acids**. Write the general formula for a carboxylic acid.

$$CH_3CO_2H \qquad CH_3CH_2CH_2CO_2H \qquad \text{⬡}{-}CO_2H$$

SECTION 2.6.

Hybrid Orbitals of Nitrogen and Oxygen

A.   Amines

Many important functional groups in organic compounds contain nitrogen or oxygen. Electronically, nitrogen is similar to carbon, and the atomic orbitals of nitrogen hybridize in a manner very similar to those of carbon:

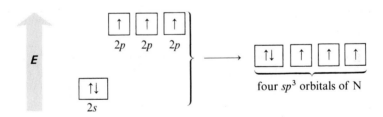

As these orbital diagrams show, nitrogen can hybridize its four second-level atomic orbitals to four equivalent sp^3 bonding orbitals. However, note one important difference between nitrogen and carbon. While carbon has four electrons to distribute in four sp^3 orbitals, nitrogen has *five* electrons to distribute in four sp^3 orbitals. One of the sp^3 orbitals of nitrogen is filled with a pair of electrons, and nitrogen can form compounds with only three covalent bonds to other atoms.

A molecule of ammonia contains an sp^3 nitrogen atom bonded to three hydrogen atoms. An **amine** molecule has a similar structure: an sp^3 nitrogen atom bonded to one or more carbon atoms. In either ammonia or an amine, the nitrogen has one orbital filled with a pair of unshared valence electrons. Figure 2.19 shows the geometry and the filled orbitals of ammonia and two amines: the similarities in structure are evident in this figure.

The unshared pair of electrons in the filled orbital on the nitrogen of ammonia and amines allows these compounds to act as bases. When an amine is treated with

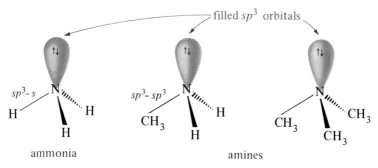

FIGURE 2.19. Bonding in ammonia and in two amines.

an acid, these unshared electrons are used to form a sigma bond with the acid. The product is an **amine salt**.

$$CH_3 - \overset{..}{N} - CH_3 + HCl \longrightarrow CH_3 - \overset{\overset{\displaystyle H}{|}}{\underset{\underset{\displaystyle CH_3}{|}}{N^+}} - CH_3 \ Cl^-$$

an amine   *an amine salt*

By analogy with carbon, we would expect the H—N—H bond angle in NH_3 to be 109.5°. Experiments have shown that this is not so; the bond angles in NH_3 are 107.3°. One explanation for this is that the bond angles are compressed by the large size of the filled orbital with its unshared electrons. (Since the electrons in this filled orbital are attracted to only one nucleus rather than to two nuclei, they are less tightly held; therefore, the filled orbital is larger than an N—H sigma orbital.) When atoms other than hydrogen are bonded to an sp^3 nitrogen, the bond angles are observed to be closer to the tetrahedral angle of 109.5° because of repulsions between these larger groups.

107.3°
ammonia

108°
trimethylamine

Like carbon, nitrogen is also found in organic compounds in the sp^2- and sp-hybrid states. Again, the important difference between nitrogen and carbon is that one of the orbitals of nitrogen is filled with a pair of unshared electrons.

sp^2
$$CH_3CH{=}\overset{..}{N}{\overset{\displaystyle \cdot}{}}$$
CH_3

sp
$$CH_3CH_2C{\equiv}N\text{:}$$

STUDY PROBLEM

2.8. Give the complete structural formula for each of the following compounds and tell which types of orbitals overlap to form each bond:

$$\overset{\overset{\displaystyle NH}{\|}}{}$$

(a) $(CH_3)_2NCH_2NH_2$ (b) $H_2NCNHCH_2CN$ (c) [hexagon]NH

B.   Water, alcohols, and ethers

Like carbon and nitrogen, oxygen forms bonds with sp^3-hybrid orbitals. Because oxygen has six bonding electrons, it forms two covalent bonds and has two filled orbitals.

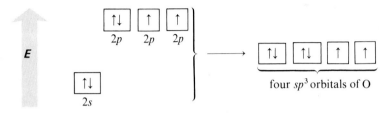

Water is an example of a compound containing an sp^3 oxygen. The bond angle in water is measured to be 104.5°, not the idealized 109.5°. It is believed that the size of the filled orbitals with unshared electrons compresses the H—O—H bond angle, just as the filled orbital in ammonia compresses the H—N—H bond angles.

There are a number of classes of organic compounds that contain sp^3 oxygen atoms. For the present, let us consider just two, alcohols and ethers: ROH and ROR′. The bonding to the oxygen in alcohols and ethers is directly analogous to the bonding in water. In each case, the oxygen is sp^3 hybridized and has two pairs of unshared valence electrons, as is shown in Figure 2.20.

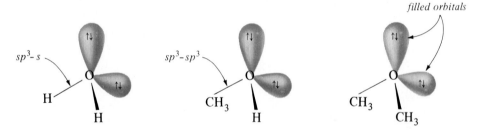

FIGURE 2.20. Bonding in water, in the alcohol CH_3OH, and in the ether CH_3OCH_3.

C.   Carbonyl compounds

The **carbonyl group** (C=O) contains an sp^2 carbon atom connected to an oxygen atom by a double bond. It is tempting to think that a carbonyl oxygen is in the sp^2-hybrid state just as the carbonyl carbon is; however, chemists are not truly sure of the hybridization of a carbonyl oxygen because there is no bond angle to measure.

The geometry of a carbonyl group is determined by the sp^2 carbon. The carbonyl group is *planar* around the trigonal sp^2 carbon. The carbon–oxygen bond contains a pair of *exposed pi electrons*. The oxygen also has *two pairs of unshared electrons*. Figure 2.21 shows the geometry of a carbonyl group.

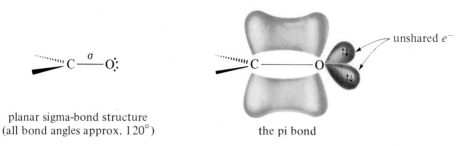

planar sigma-bond structure
(all bond angles approx. $120°$)

the pi bond

FIGURE 2.21. Bonding in the carbonyl group.

The carbonyl group is more polar than the C—O group in an alcohol or ether. The probable reason for this enhanced polarity is that the mobile pi electrons are more easily drawn toward the electronegative oxygen than are the C—O sigma electrons.

The carbonyl group itself is not considered to be a functional group; rather, it is part of a variety of functional groups. The functional group and class of compound are determined by the other atoms bonded to the carbonyl carbon. If one of the atoms bonded to the carbonyl carbon is a hydrogen, then the compound is an **aldehyde**. If two carbons are bonded to the carbonyl carbon, then the compound is a **ketone**. Of course, in Chapter 1 you already encountered **carboxylic acids**, in which an OH group is bonded to the carbonyl group. We will mention other classes of carbonyl compounds later in the book.

$$\underset{\text{an aldehyde}}{\overset{\displaystyle O \atop \displaystyle \|}{\text{RCH}} \ \ \text{or} \ \ \text{RCHO}} \qquad \underset{\text{a ketone}}{\overset{\displaystyle O \atop \displaystyle \|}{\text{RCR}}} \qquad \underset{\text{a carboxylic acid}}{\overset{\displaystyle O \atop \displaystyle \|}{\text{RCOH}} \ \ \text{or} \ \ \text{RCO}_2\text{H}}$$

SAMPLE PROBLEM

Classify each of the following compounds as an aldehyde, a ketone, or a carboxylic acid:

(a) ⬡—CO$_2$H   (b) $CH_3CH_2\overset{\displaystyle O \atop \displaystyle \|}{\text{CH}}$

(c) ⬡—$\overset{\displaystyle O \atop \displaystyle \|}{\text{CCH}_3}$ (d) HCHO

Solution: (a) carboxylic acid; (b) aldehyde; (c) ketone; (d) aldehyde

STUDY PROBLEM

2.9.   Write condensed structural formulas for four-carbon compounds that exemplify (a) an aldehyde; (b) a ketone; (c) a carboxylic acid; and (d) a carboxylic acid that is also a ketone. (There may be more than one correct answer for each part.)

SECTION 2.7.

Conjugated Double Bonds

An organic molecule may contain more than one functional group. In most polyfunctional compounds, each functional group is independent of another; however, this is not always the case. Let us consider some compounds with more than one carbon–carbon double bond.

There are two principal ways double bonds can be positioned in an organic molecule. Two double bonds originating at adjacent atoms are called **conjugated double bonds**.

Conjugated double bonds:

$$CH_2=CH-CH=CH_2$$

adjacent carbon atoms

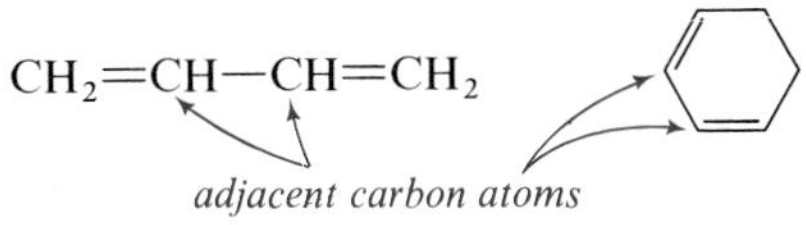

Double bonds joining atoms that are not adjacent are called **isolated**, or **nonconjugated**, double bonds.

Isolated double bonds:

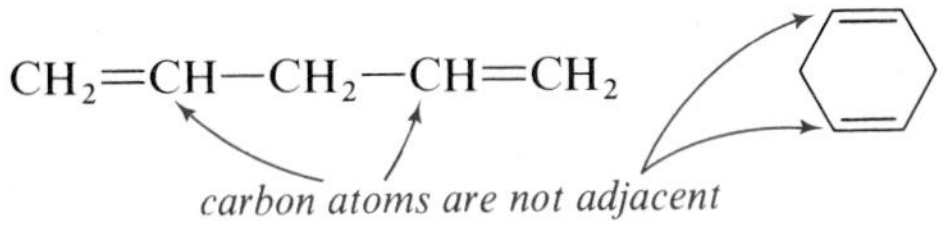

$$CH_2=CH-CH_2-CH=CH_2$$

carbon atoms are not adjacent

STUDY PROBLEMS

2.10. Vitamin A$_1$ has the structure shown below. How many conjugated double bonds does it contain? How many isolated double bonds?

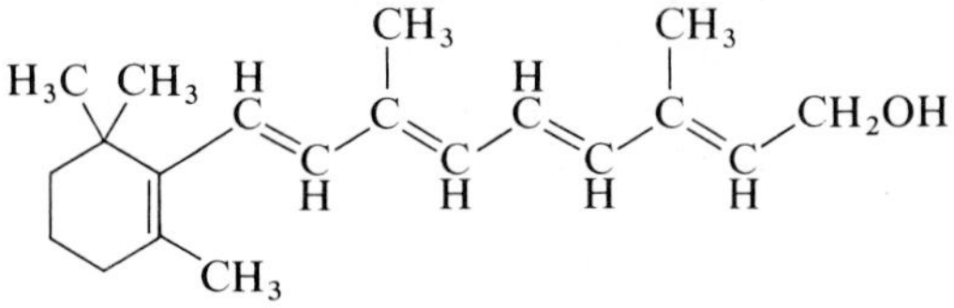

2.11. Draw the structure of an eight-carbon compound that has (a) three conjugated double bonds; (b) two conjugated double bonds and one isolated double bond; and (c) three isolated double bonds. (There may be more than one correct answer.)

Isolated double bonds behave independently; each double bond undergoes reaction as if the other were not present. Conjugated double bonds, on the other hand, are not independent of each other; there is electronic interaction between them. Let us choose the simplest of the conjugated systems, $CH_2=CH-CH=CH_2$, called 1,3-butadiene, to discuss this phenomenon. Figure 2.22 illustrates the *p*-orbital overlap in 1,3-butadiene.

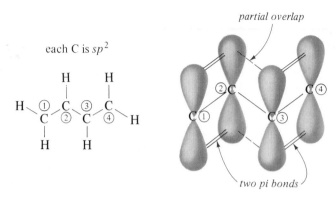

each C is sp^2

top view of planar σ bonds
(p orbitals not shown)

side view showing p orbitals
(hydrogens not shown)

FIGURE 2.22. The p-orbital picture of 1,3-butadiene, $CH_2=CH-CH=CH_2$.

We have numbered the carbon atoms of 1,3-butadiene in Figure 2.22 for reference. There are two pairs of p orbitals that form two pi bonds: one pi bond between carbons 1 and 2, and one pi bond between carbons 3 and 4. However, the p orbitals of carbons 2 and 3 are also adjacent, and *partial overlap* of these p orbitals occurs. Although most of the pi-electron density is located between carbons 1–2 and 3–4, some pi-electron density is also found between carbons 2–3. (Figure 2.22 shows a composite of the bonding orbitals in 1,3-butadiene. For a more-detailed discussion of the π molecular orbitals of this compound, see Section 17.1. Also, refer to Figure 17.1, page 783, for a representation of the four π molecular orbitals.)

We use a number of terms to describe this pi-bond interaction in conjugated systems. We may say that there is **partial p-orbital overlap** between the central carbons. We may also say that the bond between carbons 2 and 3 in 1,3-butadiene has **partial double-bond character**. Yet another way of describing the system is to say that the pi electrons are **delocalized**, which means that the pi-electron density is distributed over a somewhat larger region within the molecule. By contrast, **localized electrons** are restricted to two nuclei; nonconjugated double bonds contain localized pi electrons.

STUDY PROBLEMS

2.12. Draw the p orbitals of each of the following compounds, using lines to show pi-bond overlap and dotted lines to show partial overlap, as we have done in Figure 2.22.

(a) $CH_2=CHCH_2CH=CHCH=CH_2$ (b)

2.13. Write the formula for an open-chain four-carbon ketone in which the carbonyl group is in conjugation with a carbon–carbon double bond.

SECTION 2.8.

Benzene

Benzene (C_6H_6) is a cyclic compound with six carbon atoms joined in a ring. Each carbon atom is *sp²* hybridized, and the ring is planar. Each carbon atom has one hydrogen atom attached to it, and each carbon atom also has an unhybridized *p* orbital perpendicular to the plane of the sigma bonds of the ring. Each of these six *p* orbitals can contribute one electron for pi bonding (see Figure 2.23).

 With six *p* electrons, benzene should contain three pi bonds. We could draw three pi bonds in the ring one way (formula **A**), or we could draw them another way (formula **B**). However, we might also wonder—could this type of *p*-orbital system lead to *complete* delocalization of all six *p* electrons (formula **C**) instead of just partial delocalization?

<pre>
      H   H
       \   /
        C — C
       /   \
  H — C   C — H
       \   /
        C — C
       /   \
      H   H
</pre>

⬡ or ⬡ or maybe ⬡
  A B   C

sigma-bond structure
(Each C is *sp²* and
has one *p* electron.)

placement of pi bonds
(The circle represents complete delocalization.)

 It is known that all carbon–carbon bond lengths in benzene are the same, 1.40 Å. All six bonds are longer than C—C double bonds, but shorter than C—C single bonds. If the benzene ring contained three localized double bonds separated by three single bonds, the bonds would be of different lengths. The fact that all the carbon–carbon bonds in the benzene ring have the same length suggests that the benzene ring does not contain alternate single and double bonds.

 From the bond lengths plus a body of other evidence that will be presented in later chapters, chemists have concluded that benzene is a symmetrical molecule and that each of the six ring bonds is like each of the other ring bonds. Instead of alternate double and single bonds, the six pi electrons are *completely delocalized* in a cloud of electronic charge shaped rather like a pair of donuts. This cloud of pi electrons is called the **aromatic pi cloud** of benzene. Figure 2.23 depicts the lowest-energy bonding orbital of benzene, the one commonly used to represent the

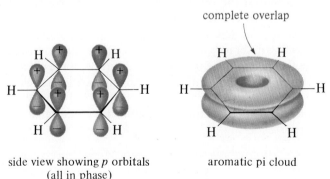

side view showing *p* orbitals
(all in phase)

aromatic pi cloud

FIGURE 2.23. The *p*-orbital picture of benzene, C_6H_6.

aromatic pi cloud. (Figure 10.6, page 461, shows the six π orbitals of benzene.) For most purposes in this text, we will use a hexagon with a circle inside it to represent benzene; the circle represents the aromatic pi cloud.

Benzene is just one member of a class of **aromatic compounds**, compounds that contain aromatic pi clouds. Historically, the term "aromatic" derives from the fact that many of these compounds have distinctive odors. We will discuss aromatic compounds in detail in Chapters 10 and 16.

SECTION 2.9.

Resonance

Methane (CH_4) and ethylene ($CH_2{=}CH_2$) are examples of organic compounds with structures that may reasonably be described using single valence-bond formulas (that is, using lines for pairs of bonding electrons). In each case, a line joining two atomic symbols represents a covalent bond between two atoms.

Benzene is an example of an organic compound that cannot be accurately represented by a single valence-bond formula. The delocalization of the pi electrons results in a system in which the pi electrons encompass more than two atoms. Classical valence-bond notation does not cover this circumstance. (The circle in the hexagon representing the aromatic pi cloud in benzene is a fairly recent addition to organic symbolism.) In order to describe the pi-electron distribution in benzene using classical valence-bond formulas, *two* formulas must be used.

resonance structures for benzene
(Kekulé formulas)

These two valence-bond formulas for benzene are called **Kekulé formulas** in honor of Friedrich August Kekulé, who first proposed them in 1872. Kekulé's original proposal was brilliant, but unfortunately incorrect. His idea was that the two structures for benzene shifted back and forth so fast that neither could be isolated independently of the other. The two Kekulé structures were said to be in **resonance** with each other; for this reason, the Kekulé structures are also called the **resonance symbols**, or **resonance structures**, for benzene. We now know that benzene does not shift between two different structures; the real structure of benzene is a composite of the two resonance structures. We say that benzene is a **resonance hybrid** of the two resonance structures.

Whenever we can describe a molecular structure by two or more valence-bond formulas that differ *only in the positions of the electrons* (usually pi electrons), none of these formulas will be in complete accord with the chemical and physical properties of the compound. If different resonance structures can be written for a compound, we can assume a delocalization of electron density. These statements are true for all aromatic structures, as well as for some other structures we will mention shortly.

The real structure of naphthalene is a composite of the resonance structures:

The real structure of pyridine is a composite of the resonance structures:

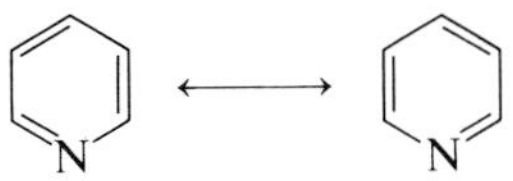

    The important thing to keep in mind is that resonance symbols are not *true* structures; the true structure is a composite of all the resonance symbols. Many chemists have made the analogy that a rhinoceros (a real animal) might be described as a resonance hybrid of a unicorn (imaginary) and a dragon (imaginary). A rhinoceros does not shift back and forth from unicorn to dragon, but is simply an animal with characteristics of both a unicorn and a dragon.

    To show that two or more formulas represent resonance structures (imaginary) and not real structures in equilibrium, we use a double-headed arrow ($\leftrightarrow$). By contrast, we indicate an equilibrium by *two* arrows ($\rightleftarrows$).

Resonance: rhinoceros: unicorn $\longleftrightarrow$ dragon
              (*real*)   (*imaginary*)

                benzene:

                (*real*)   (*imaginary*)

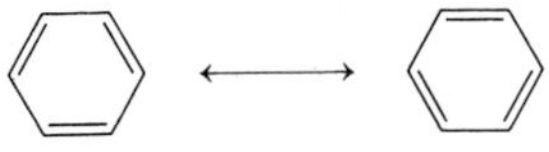

Equilibrium: $2\,H_2O$ $\rightleftarrows$ $H_3O^+ + OH^-$
                   (*real*)   (*real*)   (*real*)

    Aromatic compounds are not the only compounds for which single valence-bond formulas are inadequate. The **nitro group** ($-NO_2$) is a group of atoms that we can best describe using resonance structures. A single valence-bond structure for the nitro group shows two types of N—O bond. However, it is known that the two N—O bonds are the same length. Two valence-bond structures are needed; the true structure of the NO_2 group is somewhere in between the two. Resonance structures for nitromethane, CH_3NO_2, are shown below. To show that the bonds from the nitrogen to the oxygens in the NO_2 group are the same, some chemists represent the nitro group with dotted lines for the partial double bonds.

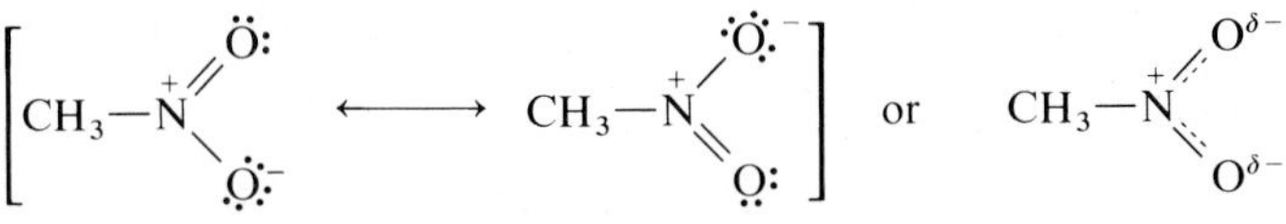

    The **carbonate ion** ($CO_3{}^{2-}$) is an ion that cannot be represented by a single valence-bond structure. Each C—O bond in the carbonate ion is the same length. We must use *three* resonance structures to describe the real structure.

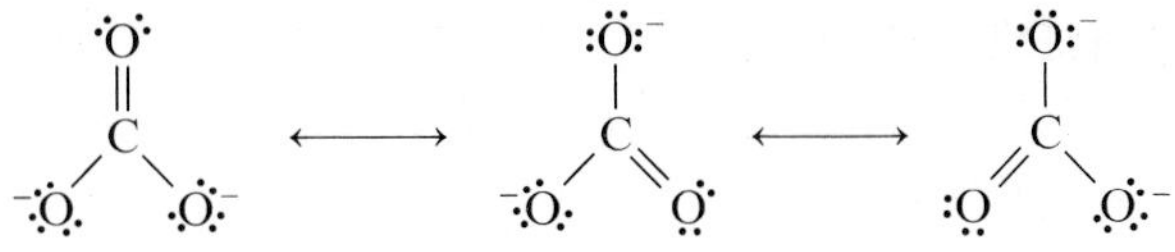

A. Electron shifts

In a series of resonance symbols, we often show shifts of pi-bond electrons by small curved arrows. The value of electron-shift arrows is that they allow us to progress systematically from one resonance symbol to another. These electron shifts are purely artificial because the pi electrons do not truly shift, but are delocalized. Note that electron-shift arrows may be drawn only to an *adjacent atom* or an *adjacent bond position.*

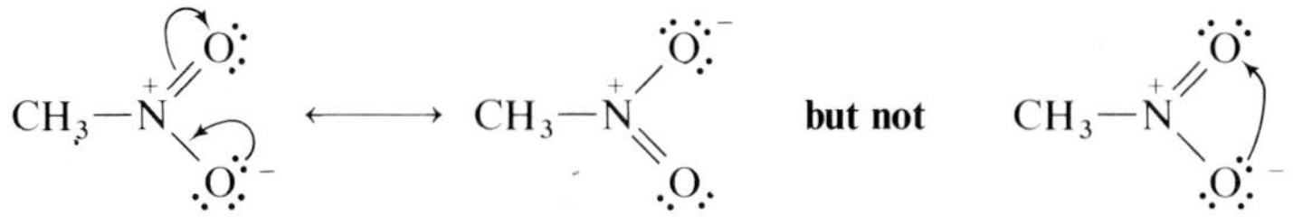

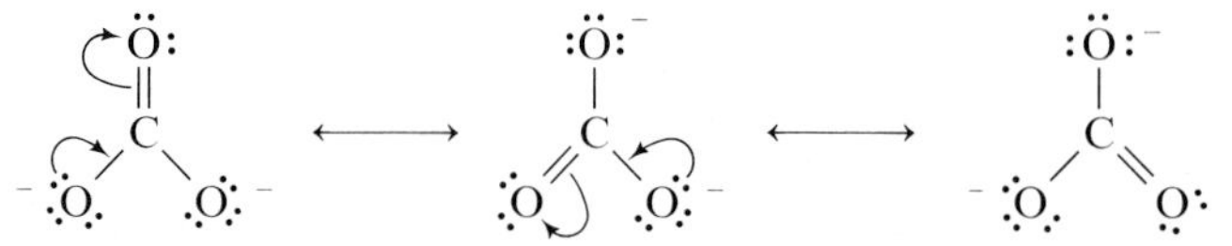

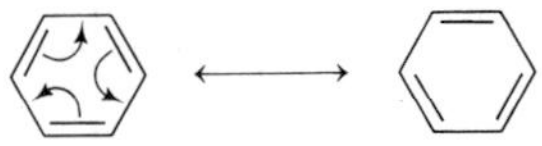

When using electron-shift arrows, we must pay special attention to valence rules. An atom of the period-2 elements can accommodate a maximum of eight valence electrons. For purposes of keeping track of these electrons, it is often advisable to show the unshared electrons in resonance structures.

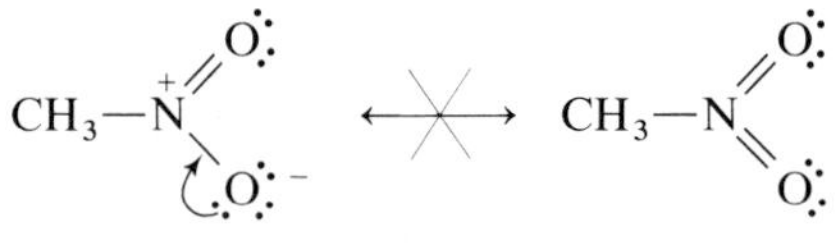

not possible: N cannot accommodate ten e⁻

SAMPLE PROBLEM

Write resonance structures with electron-shift arrows for pyrimidine and for the acetate ion, $CH_3CO_2{}^-$.

pyrimidine

Solution:

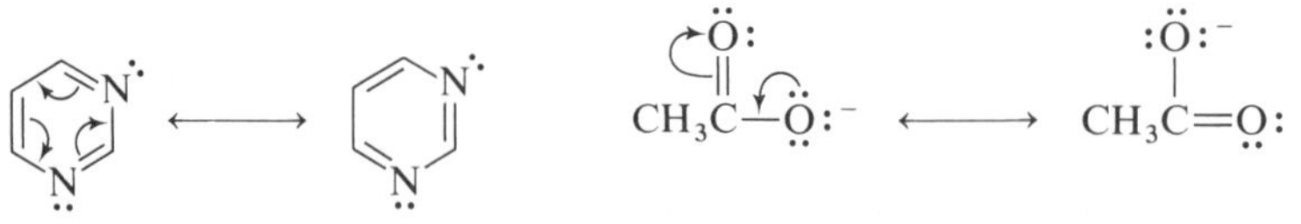

STUDY PROBLEM

2.14. The nitrate ion, $NO_3{}^-$, contains three equivalent N—O bonds. Write resonance structures for the nitrate ion.

B.   Major and minor contributors

In each example we have shown so far, the bonding has been the same in each resonance symbol. When the bonding is the same, the resonance structures are of equal energy and are equivalent to each other. *Equivalent resonance structures contribute equally to the real structure.*

Resonance structures for a compound or ion may not all contribute equally to the real structure. Consider the following example:

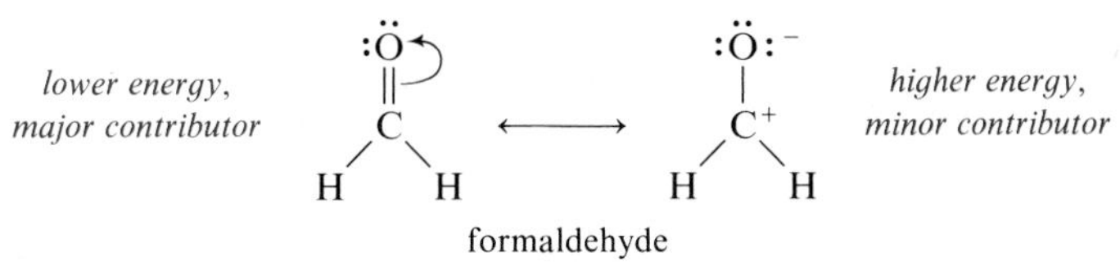

lower energy,   *higher energy,*
major contributor   *minor contributor*

formaldehyde

This pair of resonance structures for formaldehyde shows one structure (on the left) in which all atoms have an octet of valence electrons and no charge separation. This resonance structure is of lower energy than the right-hand structure, in which the carbon atom has only six electrons surrounding it and the carbon–oxygen bond has a charge separation.

The lower-energy structure is more like the real structure than is the higher-energy structure. We say that the left-hand structure is a *more important resonance structure*, or a *major contributor* to the real structure, while the right-hand structure is a *less important resonance structure*, or a *minor contributor*. (This is similar to saying that a rhinoceros is like a dragon with just a little unicorn thrown in.)

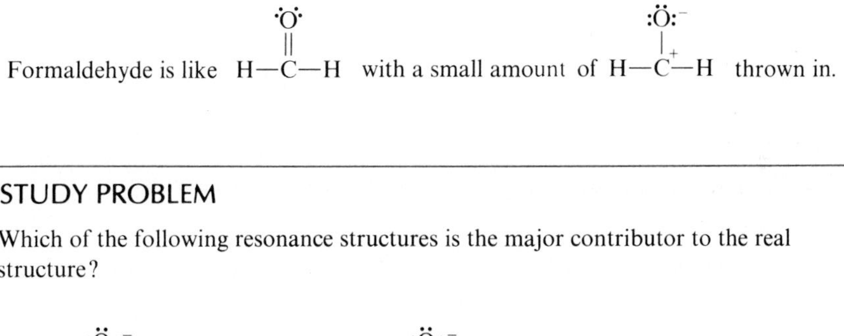

Formaldehyde is like H—C—H with a small amount of H—C—H thrown in.

STUDY PROBLEM

2.15.   Which of the following resonance structures is the major contributor to the real structure?

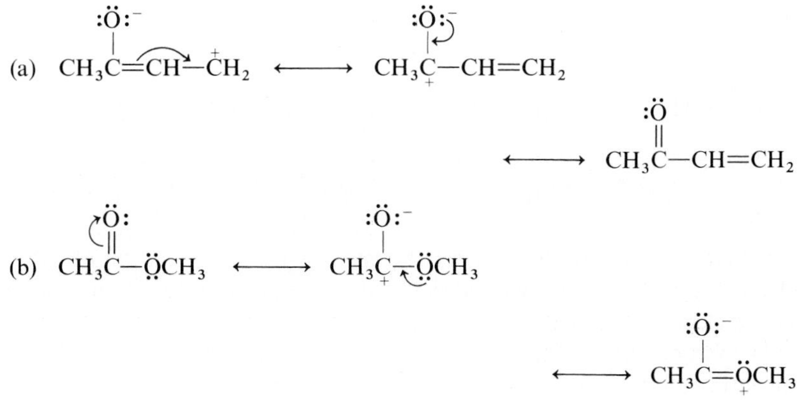

C.   Rules for writing resonance structures

We will encounter resonance structures frequently in the coming chapters. Let us summarize the rules for writing resonance structures:

1.   *Only electrons (not atoms) may be shifted*, and they may be shifted only to adjacent atoms or bond positions.

2.   Resonance structures in which an atom carries more than its quota of electrons (eight for the period-2 elements) are not contributors to the real structure.

3.   The more important resonance structures show each atom with a complete octet and as little charge separation as possible.

| more important because | less important because |
| each atom has an octet | *N* has only six electrons |

D.   Resonance stabilization

If a structure is a resonance hybrid of two or more resonance structures, *the energy of the real structure is lower than that of any single resonance structure.* The real structure is said to be **resonance-stabilized**. In most cases, the energy difference is slight, but for aromatic systems, like benzene or naphthalene, the energy difference is substantial. (The reasons for aromatic stabilization will be discussed in Chapter 10.)

Resonance stabilization is most important if two or more resonance structures for a compound are *equivalent or nearly equivalent in energy.* A high-energy, minor-contributing resonance structure adds little stabilization.

The reason for the energy differences between hypothetical resonance structures and the real structure of a compound is not entirely understood. Certainly, part of the reason is that a delocalized electron is attracted to more than one nucleus. It is generally true that a system with delocalization of electrons or of electronic charge is of lower energy and of greater stability than a system with localized electrons or electronic charge.

Although carbon atoms do not usually form ionic bonds, resonance-stabilization allows some of these ionic compounds to exist (even if for just a fleeting moment). For example,

$$CH_3CH_2\overset{+}{C}H_2 \ Cl^-$$
cannot be formed under
usual laboratory conditions

$$CH_2{=}CH\overset{+}{C}H_2 \ Cl^-$$
highly reactive and
short-lived, but can exist
under ordinary conditions

Resonance structures for $CH_2{=}CH\overset{+}{C}H_2$:

$$CH_2{=}\overset{\frown}{CH}{-}\overset{+}{C}H_2 \longleftrightarrow \overset{+}{C}H_2{-}CH{=}CH_2$$
equal contributors

The principal reason that carboxylic acids are acidic is that the carboxylate ion is stabilized by delocalization of the negative charge after removal of the proton.

$$CH_3\overset{\overset{\cdot\cdot}{\ddot O}}{\overset{\|}{C}}\overset{\cdot\cdot}{\ddot O}{\frown}H + H_2O \rightleftharpoons \left[CH_3\overset{:\overset{\cdot\cdot}{O}}{\overset{\|}{C}}\overset{\cdot\cdot}{\ddot O}:^- \longleftrightarrow CH_3\overset{:\ddot O:^-}{\overset{|}{C}}{=}\overset{\cdot\cdot}{\ddot O}: \right] + H_3O^+$$

In the carboxylate ion, the − charge
is shared by both O atoms.

SAMPLE PROBLEM

Which of the following compounds are resonance-stabilized? (*Note*: Because the bonding electrons of sulfur are in the third energy level, it can accommodate more than eight electrons.)

(a) $CH_3CH_2CH_3$

(b)
$$\overset{H}{\underset{H}{\diagdown}}C{=}C\overset{CH_3}{\underset{H}{\diagup}}$$

(c)
$$CH_3O{-}\overset{\overset{O}{\|}}{\underset{\|}{S}}{-}O^-\overset{}{\underset{O}{}}$$

Solution: (a) is not resonance-stabilized because it contains no pi electrons.

(b) is not resonance-stabilized because there is no position in the molecule where the pi electrons can be delocalized.

(c) is resonance-stabilized:

$$CH_3O\overset{:\ddot O:^-}{\underset{\cdot\overset{\cdot\cdot}{O}\cdot}{S}}{=}\ddot O: \longleftrightarrow CH_3O\overset{\overset{\cdot\cdot}{O}}{\underset{\cdot\overset{\cdot\cdot}{O}\cdot}{S}}{-}\ddot O:^- \longleftrightarrow CH_3O\overset{\overset{\cdot\cdot}{O}}{\underset{:O:^-}{S}}{=}\ddot O:$$

STUDY PROBLEM

2.16. (1) Write resonance structures for the following ions. (2) Indicate which, if any, of the resonance structures is the major contributor or whether the structures are of similar energy.

(a) $CH_3CH{=}CH\overset{+}{C}HCH_3$

(b) $\bigcirc{-}\overset{+}{C}H_2$

(c) $CH_3\overset{\overset{\cdot\cdot}{\ddot O}}{\overset{\|}{C}}\overset{-}{\ddot C}H_2$

(d) $:\bar CH_2C{\equiv}N:$

(e) $\bigcirc{-}\ddot O:^-$

(f) $\bigcirc{-}\overset{\overset{\cdot\cdot}{\ddot O}}{\overset{\|}{C}}\overset{\cdot\cdot}{\ddot O}:^-$ (Be careful!)

Summary

A covalent bond is the result of two atomic orbitals overlapping to form a lower-energy **bonding molecular orbital** in which two electrons are paired.

The four atomic orbitals of C undergo **hybridization** in bond formation:

sp^3 **hybridization:**   four equivalent single bonds (*tetrahedral*).

sp^2 **hybridization:**   three equivalent single bonds (*trigonal*) plus one unhybridized p orbital.

sp **hybridization:**   two equivalent single bonds (*linear*) plus two unhybridized p orbitals.

Overlap of a hybrid orbital with an orbital of another atom (end-to-end overlap) results in a **sigma bond**. Overlap of a p orbital with a parallel p orbital of another atom (side-to-side overlap) results in a **pi bond**.

$$\sigma \text{ and } \pi \qquad \sigma \text{ and two } \pi$$

$$
\begin{array}{c}
\text{H} \quad \text{H}\\
|\quad\;\; |\\
\text{H}-\text{C}-\text{C}-\text{H}\\
|\quad\;\; |\\
\text{H} \quad \text{H}
\end{array}
\qquad
\begin{array}{c}
\text{H} \qquad\quad \text{H}\\
\diagdown \quad\;\; \diagup\\
\text{C}=\text{C}\\
\diagup \quad\;\; \diagdown\\
\text{H} \qquad\quad \text{H}
\end{array}
\qquad
\text{H}-\text{C}\equiv\text{C}-\text{H}
$$

$$sp^3, all\;\sigma \qquad\qquad sp^2 \qquad\qquad sp$$

Each bonding molecular orbital has an **antibonding molecular orbital** associated with it. Antibonding orbitals are of higher energy than either bonding orbitals or the atomic orbitals that went into their formation. In the ground state, molecules have their electrons in the lowest-energy orbitals, usually the bonding orbitals.

A **functional group** is a *site of chemical reactivity* in a molecule and arises from a *pi bond* or from *differences in electronegativity* between bonded atoms. A double or triple bond is a functional group. The following general formulas show examples of other functional groups:

$$R\ddot{N}H_2 \qquad\qquad R\ddot{O}H \qquad\qquad R\ddot{O}R'$$

an amine \qquad\qquad *an alcohol* \qquad\qquad *an ether*

$$
\begin{array}{c}
\overset{\displaystyle\cdot\cdot}{\underset{\displaystyle\cdot\cdot}{\text{O}}}\\
\|\\
\text{RCH}
\end{array}
\qquad\qquad
\begin{array}{c}
\overset{\displaystyle\cdot\cdot}{\underset{\displaystyle\cdot\cdot}{\text{O}}}\\
\|\\
\text{RCR}'
\end{array}
\qquad\qquad
\begin{array}{c}
\overset{\displaystyle\cdot\cdot}{\underset{\displaystyle\cdot\cdot}{\text{O}}}\\
\|\;\;..\\
\text{RC}\ddot{\text{O}}\text{H}
\end{array}
$$

an aldehyde \qquad\qquad *a ketone* \qquad\qquad *a carboxylic acid*

Pi bonds extending from adjacent carbon atoms are called **conjugated double bonds**. In the case of benzene and other aromatic compounds, complete delocalization of pi electrons results in the **aromatic pi cloud**.

$$CH_2=CH-CH=CH_2$$

some pi-electron density here \qquad\qquad *pi electrons completely delocalized*

Resonance symbols may be used to show delocalization of *p* electrons.

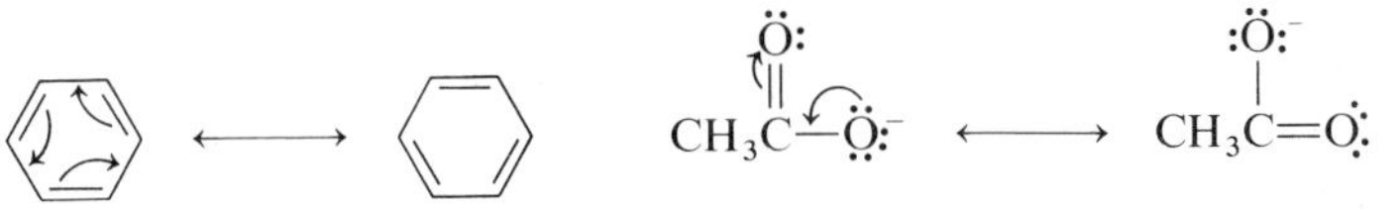

If all resonance symbols are not equivalent, the lowest-energy resonance symbol is the **major contributor**.

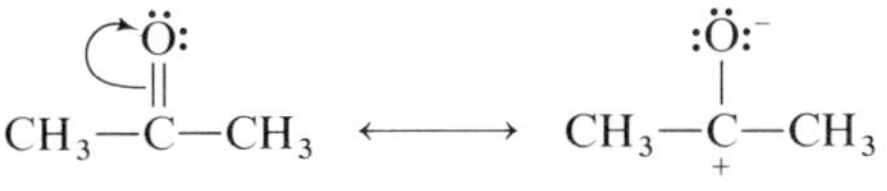

major contributor

Delocalization of pi electrons in a molecule, or of electronic charge in an ion, results in a slight increase in the stability of the system. However, for aromatic compounds, the increase in stability is substantial.

STUDY PROBLEMS

2.17. Do the following structures represent the same or different compounds? Explain.

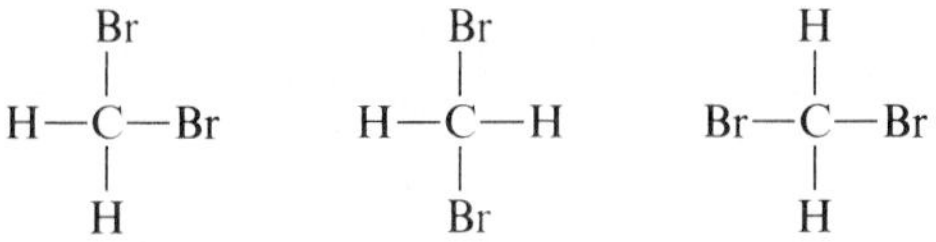

2.18. Which of the following molecular formulas could represent a real compound?

(a) C_2H_7 (b) C_3H_6 (c) C_3H_8 (d) C_4H_9

2.19. Draw a *p*-orbital picture for each of the following compounds. (Use lines for sigma bonds.)

(a) $(CH_3)_2C=C(CH_3)_2$ (b) $CH_3C\equiv CCH_3$

2.20. Indicate the hybridization of each carbon in the following structures:

(a) 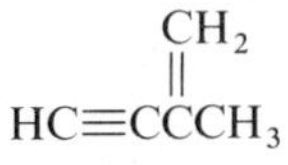 $HC\equiv C\overset{\overset{\displaystyle CH_2}{\|}}{C}CH_3$ (b) $CH_3OCH_2CH_3$ (c)

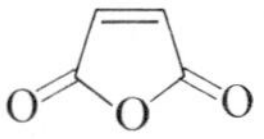

2.21. In each of the following structures, does the pair of indicated carbon atoms lie in the same plane or not?

(a) 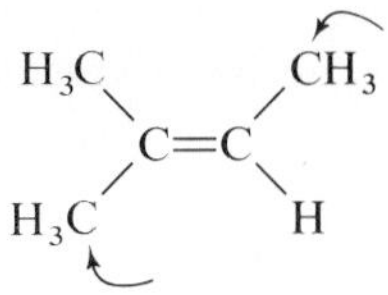   (b) 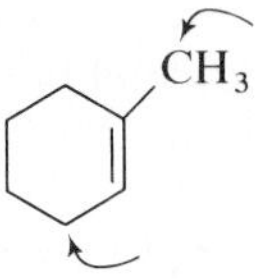   (c)

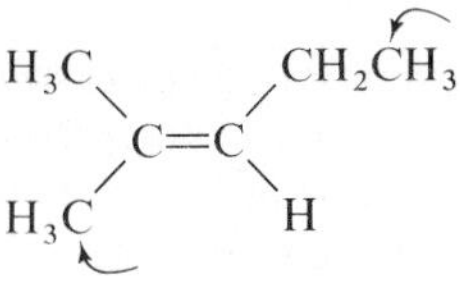

2.22. What would be the expected *sigma-bond angles* in the following structures?

2.23. Draw an orbital picture for each of the following formulas. Indicate the types of orbitals from each atom, and show any orbitals containing unshared valence electrons. (Use lines for sigma bonds.)

(a) H_3O^+ (b) HCN   (c) OH^- (d) ⬡—CH=CH$_2$

(e) NH_2NH_2 (f) $(CH_3)_4N^+$ (g) $^-OCH_3$

2.24. The molecular-orbital diagram that follows is for the carbon–carbon double bond in cyclohexene.

cyclohexene

$$\sigma^* \; \underline{\quad}$$
$$\pi_2^* \; \underline{\quad}$$
$$\pi_1 \; \underline{\quad}$$
$$\sigma \; \underline{\quad}$$

E

(a) Indicate by arrows which orbitals contain the double-bond electrons when cyclohexene is in the ground state.

(b) Absorption of ultraviolet light by a double bond results in the promotion of an electron to a higher-energy orbital. Which electron transition requires the least energy when cyclohexene goes to an excited state?

2.25. In each of the following structures or pairs of structures, which of the indicated bonds (1 or 2) is the shorter one?

(a) Cl—①CH$_3$,   ⬡—②Cl   (b) $\overset{H}{\underset{H}{>}}$C=CH—C≡C—①②H

(c) ⬡—①CH$_3$,   ⬡—②CH$_3$ (d) CH$_3$—①OCH$_3$,   CH$_3$$\overset{O②}{\overset{||}{C}}CH_3$

2.26. *Piperazine* is the drug of choice for treatment of patients with roundworms or pinworms. *Pyridine* is a pungent-smelling aromatic compound used as a solvent and reagent in organic laboratories.

(a) HN⬡NH   (b) ⬡N

piperazine   pyridine

(1) Draw the complete structural formulas of these compounds, showing each atom and using a line for each covalent bond.

(2) Indicate unshared electron pairs by dots.

(3) What is the hybridization of each ring atom?

2.27. Show by electron dots any unshared valence electrons in the following structures:

(a) CH_3Cl (b) 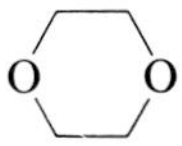   (c) CH_3NHCH_3 (d) $CH_3\overset{\overset{\displaystyle O}{\|}}{C}OH$

2.28. Which compound of each pair is more polar? Explain.

(a) CH_3CH_2OH, CH_3OCH_3 (b) CH_3OCH_3, $(CH_3)_2C{=}O$
(c) CH_3NH_2, $(CH_3)_3N$ (d) CH_3CH_3, CH_3CH_2Cl

2.29. Circle and name the functional groups in the following structures:

(a) $CH_3\overset{\overset{\displaystyle O}{\|}}{C}H$ (b) $CH_3\overset{\overset{\displaystyle O}{\|}}{C}CH_2CH_3$ (c) $CH_3CH{=}CH\overset{\overset{\displaystyle O}{\|}}{C}OH$

(d) 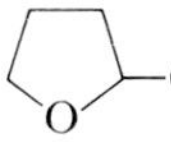 —CH_2OH (e) 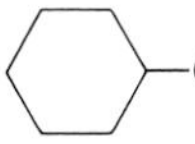 —CHO (f) $CH_3\overset{\overset{\displaystyle O}{\|}}{C}CH_2OH$

(g)

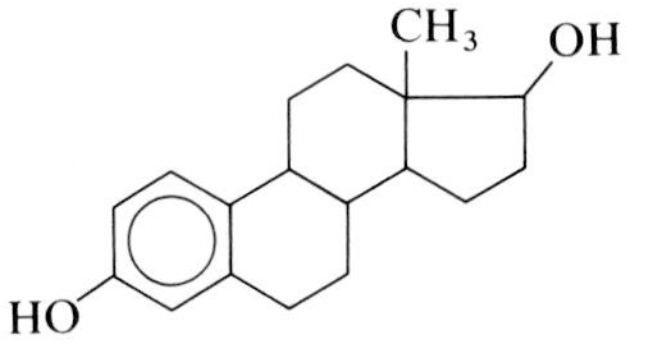

estradiol

(*a female hormone*)

(h)

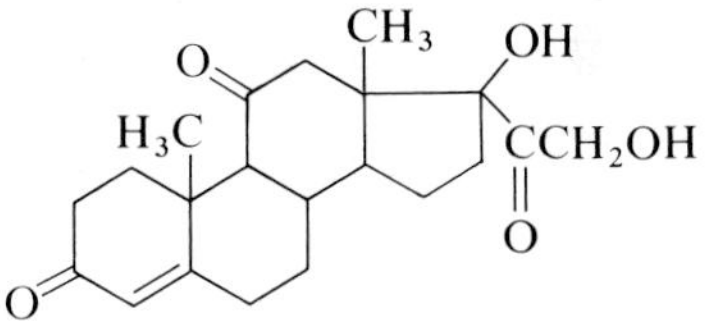

cortisone

(*an adrenal steroid used to treat arthritis*)

2.30. Write a general formula, using R and R′, for each of the following compounds. (*Example*: $CH_3OH = ROH$.)

(a) 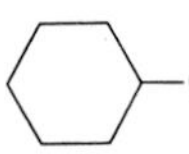 —OH (b) $CH_3CH_2CH_2\overset{\overset{\displaystyle O}{\|}}{C}OCH_3$ (c) $CH_3CH_2NHCH_3$

2.31. Give a three-carbon open-chain structure to illustrate each of the following compound types: (a) alkene; (b) alkyne; (c) ether; (d) alcohol; (e) amine; (f) ketone; (g) aldehyde; (h) carboxylic acid. (There may be more than one correct answer in each case.)

2.32. Redraw each structure to emphasize the carbonyl group:

(a) $CH_3CH_2CH_2CO_2H$ (b) 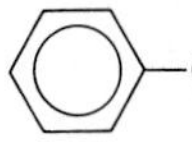 —CHO (c) 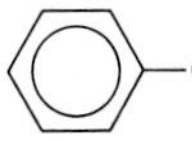 —CO_2CH_3

2.33. Write a condensed structural formula for each of the following compounds. (There may be more than one answer.)

(a) a ketone, C_4H_8O (b) a cyclic alcohol, $C_6H_{12}O$
(c) an alkene, C_5H_{10} (d) an aldehyde, C_3H_6O

2.34. Indicate whether each of the following compounds contains *delocalized* or *localized* pi electrons:

(a) 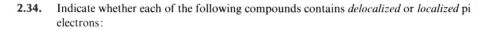   (b)   (c)

(d) 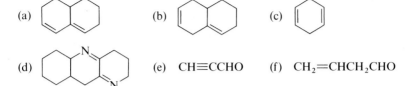   (e) $CH\equiv CCHO$ (f) $CH_2=CHCH_2CHO$

2.35. Give the structure of an open-chain five-carbon compound with   (a) two double bonds in conjugation;   (b) two isolated double bonds. (More than one answer may be possible.)

2.36. In the following structure for cyclohexanone,   (a) show the placement of a C—C double bond in conjugation with the carbonyl group, and (b) indicate the hybridization of each carbon atom in the structure with the C—C double bond.

cyclohexanone

2.37. Tell whether each of the following pairs of structures are resonance symbols or compounds (or ions) in equilibrium:

(a) CH_3CO_2H and $CH_3CO_2^- + H^+$

(b) $CH_3CH_2N=O$ and $CH_3CH=N-OH$

(c) $CH_3\overset{O}{\overset{||}{C}}CH_2^-$ and $CH_3\overset{O^-}{\overset{|}{C}}=CH_2$

(d) $CH_3\overset{+}{C}HCH=CH_2$ and $CH_3CH=CH\overset{+}{C}H_2$

2.38. Which of the following pairs of structures are resonance symbols of each other?

(a) $(CH_3)_2CH\overset{O}{\overset{||}{C}}O^-$, $(CH_3)_2CH\overset{O^-}{\overset{|}{C}}=O$ (b) $CH_3\overset{+OH}{\overset{||}{C}}CH_3$, $CH_3\overset{OH}{\overset{|}{C}}CH_3$

(c) $CH_3\overset{O}{\overset{||}{C}}CH_3$, $CH_3\overset{OH}{\overset{|}{C}}=CH_2$ (d)

(e) $CH_2=C=O$, $CH\equiv C-OH$

(f) $CH_2=CHCH=CH\overset{+}{C}H_2$, $\overset{+}{C}H_2CH=CHCH=CH_2$

2.39. Give the important resonance contributors for each of the following structures:

(a) $H\overset{O}{\overset{||}{C}}O^-$ (b) $HO\overset{O}{\overset{||}{C}}O^-$ (c)

(d)   (e)

2.40. Which of the following resonance structures would be the major contributor to the real structure?

(a) 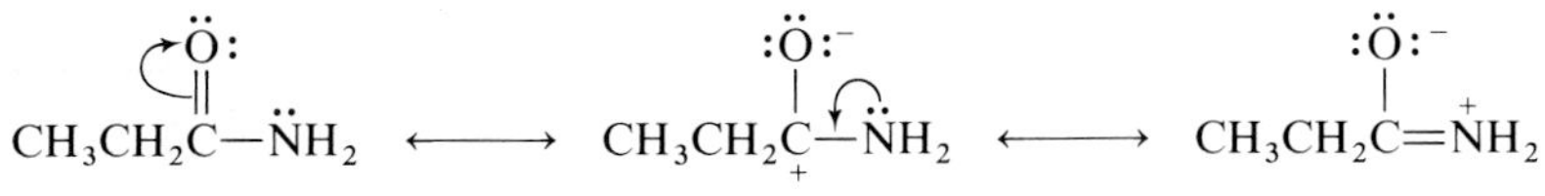 $CH_3CH_2\overset{\ddot{O}:}{\underset{\parallel}{C}}-\ddot{N}H_2$ ⟷ $CH_3CH_2\overset{:\ddot{O}:^-}{\underset{\mid}{C}}-\ddot{N}H_2$ ⟷ $CH_3CH_2\overset{:\ddot{O}:^-}{\underset{\mid}{C}}=\overset{+}{N}H_2$

(b) 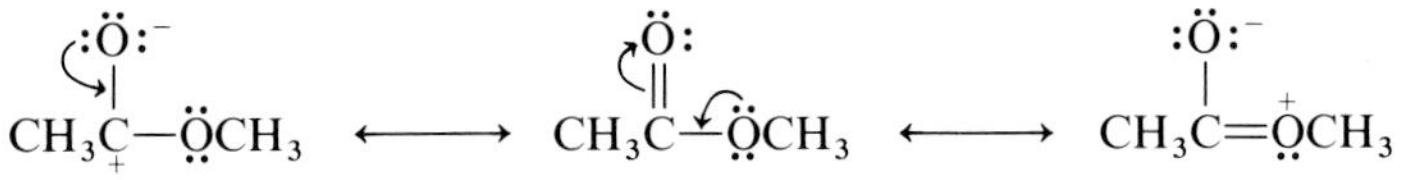 $CH_3\overset{:\ddot{O}:^-}{\underset{\mid}{\overset{+}{C}}}-\ddot{O}CH_3$ ⟷ $CH_3\overset{\ddot{O}:}{\underset{\parallel}{C}}-\ddot{O}CH_3$ ⟷ $CH_3\overset{:\ddot{O}:^-}{\underset{\mid}{C}}=\overset{+}{O}CH_3$

(c) 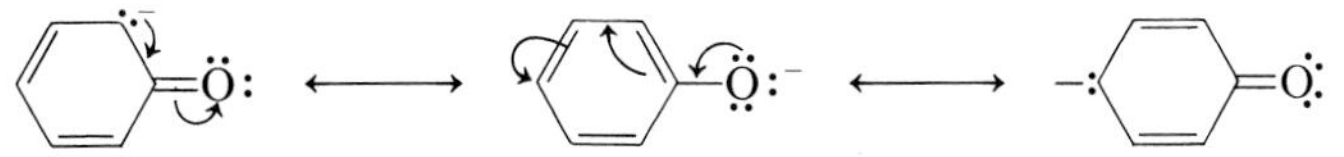 ⟷ ⟷

2.41. Write the important Kekulé structures for each of the following aromatic compounds. (*Hint*: First, determine the number of pi electrons. Second, determine the number of double bonds formed by this number of pi electrons. Third, follow the rules of valence and distribute the double bonds in as many ways as possible.)

(a) 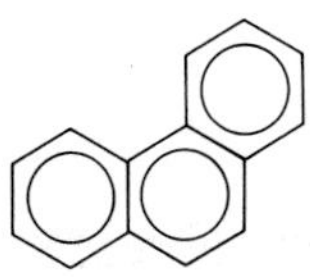   (b) 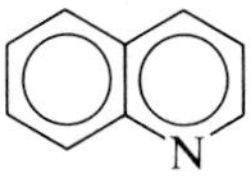   (c)

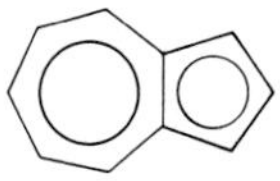

2.42. Which ion in each of the following pairs would be more stabilized? Explain your answers with structures.

(a) 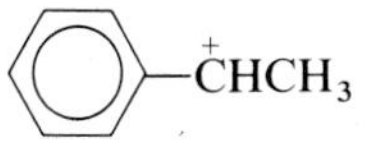 and

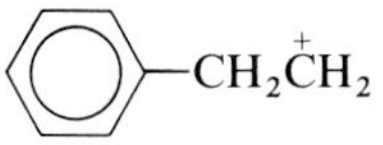

(b) $CH_3CH_2\overset{+}{C}HCH_3$ and $CH_2{=}CH\overset{+}{C}HCH_3$

(c) $:\bar{C}H_2CH_3$ and $:\bar{C}H_2NO_2$

2.43. *Allene* has the following structure: $CH_2{=}C{=}CH_2$. (a) What is the hybridization of each carbon? (b) What would be the expected angle between the two pi bonds? (c) Are the two pi bonds conjugated? (d) Can electronic charge be delocalized throughout both pi bonds?

2.44. The $O-C-O$ bond angle in CO_2 is $180°$. Write the Lewis structure and draw the *p*-orbital picture. What is the hybridization of the carbon?

2.45. *Guanidines* and *amidines* are classes of compounds that react as bases. Draw the resonance structures for the conjugate acids—that is, the cations.

(a) $CH_3\ddot{N}H-\overset{:NH}{\underset{}{\overset{\parallel}{C}}}-\ddot{N}H_2 + H^+$ ⇌ $CH_3\ddot{N}H-\overset{+NH_2}{\underset{}{\overset{\parallel}{C}}}-\ddot{N}H_2$
a guanidine

(b) $CH_3-\overset{:NH}{\underset{}{\overset{\parallel}{C}}}-\ddot{N}H_2 + H^+$ ⇌ $CH_3-\overset{+NH_2}{\underset{}{\overset{\parallel}{C}}}-\ddot{N}H_2$
an amidine

2.46. What are the electron distributions in the bonding and antibonding orbitals of $H_2{}^+$ and $H_2{}^-$?

2.47.   *Triplet methylene* is a very unstable species with the formula $:CH_2$. The H—C—H bond angle is 180°. Draw the orbital picture (using lines for sigma bonds). Indicate the hybridization state of the carbon atom.

2.48.   The hydrogen atoms circled in the following formulas are unusually acidic compared to other hydrogens bonded to carbon. Suggest a reason for this acidity. (*Hint*: Consider the resonance structures of the anions that result from loss of the circled hydrogens.)

(a) $CH_3\overset{\displaystyle O}{\overset{\|}{C}}CH\overset{\displaystyle O}{\overset{\|}{C}}CH_3$ (b)   NC—CH—CN

                              (H)   (H)

2.49.   One of the few known ionic compounds of carbon in which the organic ion contains only C and H is triphenylmethyl perchlorate. Why is it that this compound can be isolated, while a phenylmethyl cation (called the *benzyl cation*) is too unstable to yield isolable salts? (Use structures in your answer.)

        triphenylmethyl perchlorate   benzyl cation

CHAPTER 3

Structural Isomerism, Nomenclature, and Alkanes

In Chapter 2, we discussed the bonding of several compounds that contain only carbon and hydrogen. A compound containing only these two elements is called a **hydrocarbon**. Methane (CH_4), ethylene ($CH_2{=}CH_2$), and benzene (C_6H_6) are all examples of hydrocarbons.

Hydrocarbons with only sp^3 carbon atoms (that is, only single bonds) are called **alkanes** (or **cycloalkanes** if the carbon atoms are joined in rings). Some typical alkanes are methane, ethane (CH_3CH_3), propane ($CH_3CH_2CH_3$), and butane ($CH_3CH_2CH_2CH_3$). All these alkanes are gases that are found in petroleum deposits and are used as fuels. Gasoline is primarily a mixture of alkanes. At the end of this chapter, we will discuss some of these aspects of hydrocarbon chemistry.

Alkanes and cycloalkanes are said to be **saturated hydrocarbons**, meaning "saturated with hydrogen." These compounds do not undergo reaction with hydrogen. Compounds containing pi bonds are said to be **unsaturated**; under the proper reaction conditions, they undergo reaction with hydrogen to yield saturated products.

Saturated hydrocarbons:

$$CH_4 \ + H_2 \ \longrightarrow \ \text{no reaction}$$

methane

an alkane

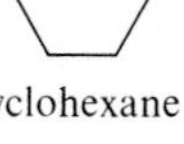

$$\hexagon \ + H_2 \ \longrightarrow \ \text{no reaction}$$

cyclohexane

a cycloalkane

Unsaturated hydrocarbons:

$$CH_2{=}CH_2 + H_2 \xrightarrow{\text{Ni catalyst}} CH_3CH_3$$

ethylene   ethane

$$CH_3C{\equiv}CH + 2\,H_2 \xrightarrow{\text{Ni catalyst}} CH_3CH_2CH_3$$

propyne   propane

$$\text{benzene} + 3\,H_2 \xrightarrow[\text{heat, pressure}]{\text{Ni catalyst}} \text{cyclohexane}$$

benzene   cyclohexane

In this chapter, we will discuss primarily alkanes and cycloalkanes—the saturated hydrocarbons. Because saturated hydrocarbons lack a true functional group, their chemistry is not typical of most organic compounds; however, these compounds provide the carbon skeletons of organic compounds that do contain functional groups. Therefore, the discussion of saturated hydrocarbons provides an excellent opportunity for introducing the variations in the structures of organic compounds and the naming of organic compounds.

SECTION 3.1.

Structural Isomers

Variations in the structures of organic compounds can arise from different numbers of atoms or types of atoms in molecules. However, variations in structure can also arise from *the order in which the atoms are attached to each other in a molecule*. For example, we can write two different structural formulas for the molecular formula C_2H_6O. These two structural formulas represent two different compounds: dimethyl ether (bp $-23.6°$), a gas that has been used as a refrigerant and as an aerosol propellant, and ethanol (bp $78.5°$), a liquid that is used as a solvent and in alcoholic beverages.

$$CH_3OCH_3 \qquad\qquad CH_3CH_2OH$$

dimethyl ether   ethanol

a gas at room temperature   *a liquid at room temperature*

Two or more different compounds that have the same molecular formula are called **isomers** of each other. If the compounds with the same molecular formula have their atoms attached in different orders, they have different structures and are said to be **structural isomers** of each other. (We will encounter other types of isomerism later.) Dimethyl ether and ethanol are examples of a pair of structural isomers.

Alkanes that contain three or fewer carbons have no isomers. In each case, there is only one possible way in which the atoms can be arranged.

No isomers: CH_4 CH_3CH_3 $CH_3CH_2CH_3$

methane   ethane   propane

The four-carbon alkane (C_4H_{10}) has two possibilities for arrangement of the carbon atoms. As the number of carbon atoms increases, so does the number of isomers. The molecular formula C_5H_{12} represents three structural isomers; C_6H_{14} represents five; and $C_{10}H_{22}$ represents 75!

Structural isomers for C_4H_{10}:

$$CH_3CH_2CH_2CH_3$$

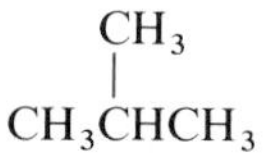

butane, bp $-0.5°$

a continuous-chain alkane

$$CH_3 \atop | \atop CH_3CHCH_3$$

methylpropane, bp $-12°$

a branched-chain alkane

SAMPLE PROBLEM

Write formulas for the three structural isomers of C_5H_{12}.

Solution:

(a) $CH_3CH_2CH_2CH_2CH_3$ (b) 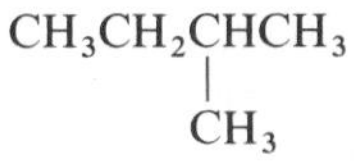 $CH_3CH_2CHCH_3 \atop \qquad\qquad | \atop \qquad\qquad CH_3$ (c) 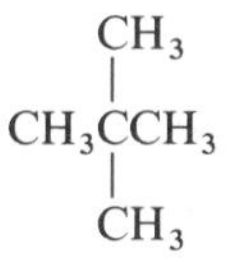 $CH_3 \atop | \atop CH_3CCH_3 \atop | \atop CH_3$

You might have listed $CH_3CHCH_2CH_3 \atop | \atop CH_3$. This structure is the same as (b).

Different positions of attachment of a functional group in a molecule also lead to structural isomerism. The alcohols 1-propanol and 2-propanol are structural isomers with slightly different properties. The alkenes 1-butene and 2-butene are also structural isomers with different properties.

$$CH_3CH_2CH_2OH$$

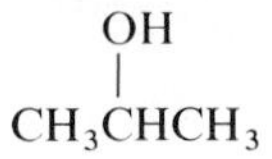

1-propanol
bp 97°

$$OH \atop | \atop CH_3CHCH_3$$

2-propanol
bp 82°

$$CH_2{=}CHCH_2CH_3$$

1-butene
bp $-6.3°$

$$CH_3CH{=}CHCH_3$$

2-butene
bp 3.7°

STUDY PROBLEMS

3.1. Draw a structural isomer for each of the following compounds. (There is more than one answer for each.)

(a) 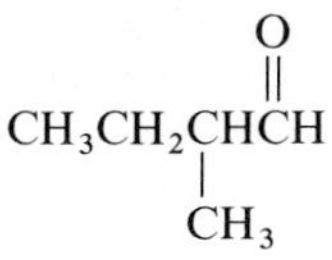 $CH_3CH_2CHCH \atop \quad\quad O \atop \quad\quad \| \atop \qquad | \atop \qquad CH_3$ (b) 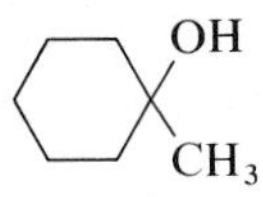   (c) $CH_2{=}CHCH_2OH$

3.2. For each of the following compounds, write the structural formula of an isomer that has a *different functional group*. (There may be more than one correct answer.)

(a) CH_3COH
$\overset{O}{\overset{\|}{}}$

(b)   ⬡—OCH_3

(c) $CH_3CH_2CCH_2CH_3$
$\overset{O}{\overset{\|}{}}$

Example: An isomer of $CH_3CH_2CH_2OH$ that has a different functional group is $CH_3OCH_2CH_3$, but not $(CH_3)_2CHOH$, which is an isomer with the *same* functional group.

A.   Isomers or not?

Molecules can move around in space and twist and turn in "snake-like motion," as Kekulé once described it. (Kekulé, you will recall, was the chemist who proposed a structure for benzene.) We may write the same structure on paper in a number of ways. The *order of attachment* of the atoms is the factor that determines if two structural formulas represent isomers or the same compound. For example, all the following formulas show the same order of attachment of atoms; they all represent the same compound and do not represent isomers.

All represent the same compound:

$$CH_3CH_2\overset{\overset{\displaystyle OH}{|}}{C}HCH_3 \qquad CH_3\overset{\overset{\displaystyle CH_3CH_2}{|}}{C}HOH \qquad \overset{\displaystyle CH_3CH_2}{\underset{\displaystyle CH_3}{>}}CHOH \qquad CH_3\overset{\overset{\displaystyle OH}{|}}{C}HCH_2CH_3$$

SAMPLE PROBLEM

Which of the following pairs of formulas represent isomers and which represent the same compound?

(a) $CH_3\overset{\overset{\displaystyle Cl}{|}}{C}HCHCl_2$ and $Cl_2CH\overset{\overset{\displaystyle Cl}{|}}{C}HCH_3$

(b)   ⬠   and   ⬡

(c) Cl—⬡—Cl and   ⬡ with Cl above and Cl below

(d) $(CH_3)_3C\overset{\overset{\displaystyle OH}{|}}{C}HCH_3$ and $(CH_3)_3CCH_2CH_2OH$

Solution: (c) and (d) represent pairs of structural isomers. In (a) and (b) the structures are oriented differently, but the order of attachment of the atoms is the same; therefore, the formulas in (a) and (b) represent the same compound.

B.   A ring or unsaturation?

Given a molecular formula for a hydrocarbon, we can often deduce a reasonable amount of information about its structure. For example, all acyclic (non-cyclic) alkanes have the general formula C_nH_{2n+2}, where $n =$ the number of carbon atoms in the molecule. Propane ($CH_3CH_2CH_3$, or C_3H_8) has three carbon atoms ($n = 3$). The number of hydrogen atoms in propane is $2n + 2$, or eight. Try the formula on the following alkanes:

$$CH_4 \qquad CH_3\overset{\underset{|}{CH_3}}{C}HCH_3 \qquad CH_3CH_2CH_2\overset{\underset{|}{CH_3}}{C}HCH_3$$

The presence of a ring or a double bond reduces the number of hydrogens in the formula by two for each double bond or ring; that is, a compound with the general formula C_nH_{2n} contains either one double bond or one ring.

$$CH_3CH_2CH_2CH_3 \qquad CH_3CH_2CH{=}CH_2 \qquad \square$$
$$C_4H_{10} \qquad\qquad C_4H_8 \qquad\qquad C_4H_8$$
$$C_nH_{2n+2} \qquad\qquad C_nH_{2n} \qquad\qquad C_nH_{2n}$$

A compound with the general formula C_nH_{2n-2} might have one triple bond, two rings, two double bonds, or one ring plus one double bond.

Three of many possible structural isomers for C_8H_{14}:

$$CH_3(CH_2)_5C{\equiv}CH$$

one triple bond   *two rings*   *one double bond and one ring*

STUDY PROBLEM

3.3. What can you deduce about the possible structures represented by each of the following molecular formulas?

(a) $C_{16}H_{34}$ (b) $C_{10}H_{20}$ (c) C_5H_8

SECTION 3.2.

How Organic Nomenclature Developed

In the middle of the 19th century, the structures of many organic compounds were unknown. At that time, compounds were given names that were illustrative of their origins or their properties. Some compounds were named after friends or relatives of the chemists who first discovered them. For example, the name *barbituric acid* (and hence the common drug classification *barbiturates*) comes from the woman's name Barbara. At one time, the carboxylic acid HCO_2H was obtained from the distillation of red ants. This acid was given the name *formic acid* from the Latin *formica*, which means "ants." These names are both called **trivial names** or **common names**. In many respects, these trivial names are like nicknames; both these compounds have more formal (but seldom-used) names.

Faced with the specter of an unlimited number of organic compounds, each with its own quaint name, organic chemists in the late 19th century decided to systematize organic nomenclature to correlate the names of compounds with their structures. The system of nomenclature that has been developed is called the **Geneva system** or the **IUPAC system**. The term *Geneva* comes from the fact that the first nomenclature conference was held in Geneva, Switzerland. The initials IUPAC stand for the *International Union of Pure and Applied Chemistry*, the organization responsible for the continued development of organic nomenclature.

In the next section, we will present a brief survey of the IUPAC nomenclature system along with some frequently used trivial names. In future chapters we will expand our discussion of nomenclature as it becomes necessary. A more complete outline of organic nomenclature is presented in the appendix.

SECTION 3.3.

A Survey of Organic Nomenclature

A.   Continuous-chain alkanes

The IUPAC system of nomenclature is based upon the idea that the structure of an organic compound can be used to derive its name and, in turn, that a unique structure can be drawn for each name. The foundations of the IUPAC system are the names of the continuous-chain alkanes. The structures and names of the first ten continuous-chain alkanes are shown in Table 3.1.

The compounds in Table 3.1 are arranged so that each compound differs from its neighbors by only a methylene (CH_2) group. Such a grouping of compounds is called a **homologous series**, and the compounds in such a list are called **homologs**.

From Table 3.1, you can see that all the alkane names end in **-ane**, which is the IUPAC ending denoting a saturated hydrocarbon. The first parts of the names of the first four alkanes (methane through butane) are derived from the traditional trivial names. The higher alkane names are derived from Greek or Latin numbers; for example, pentane is from *penta*, "five."

TABLE 3.1. The first ten continuous-chain alkanes

Number of carbons	Structure	Name
1	CH_4	methane
2	CH_3CH_3	ethane
3	$CH_3CH_2CH_3$	propane
4	$CH_3(CH_2)_2CH_3$	butane
5	$CH_3(CH_2)_3CH_3$	pentane
6	$CH_3(CH_2)_4CH_3$	hexane
7	$CH_3(CH_2)_5CH_3$	heptane
8	$CH_3(CH_2)_6CH_3$	octane
9	$CH_3(CH_2)_7CH_3$	nonane
10	$CH_3(CH_2)_8CH_3$	decane

Let us briefly consider the derivations of the names for the first four alkanes. *Methane* (CH_4) is named after methyl alcohol (CH_3OH). *Methyl*, in turn, is a combination of the Greek words *methy* (wine) and *hyle* (wood). Methyl alcohol can be prepared by heating wood in the absence of air. Even today, this alcohol is sometimes referred to as wood alcohol.

The name *ethane* (CH_3CH_3) is derived from the Greek word *aithein*, which means "to kindle or blaze." Ethane is quite flammable. The name for *propane* ($CH_3CH_2CH_3$) is taken from the trivial name for the three-carbon carboxylic acid, propionic acid ($CH_3CH_2CO_2H$). *Propionic* is a combination of the Greek *proto* (first) and *pion* (fat). Propionic acid is the first (or lowest-molecular-weight) carboxylic acid to exhibit properties of fatty acids, which are acids that can be obtained from fats. *Butane* ($CH_3CH_2CH_2CH_3$) is named after butyric acid—the odorous component of rancid butter (Latin *butyrum*, "butter").

B. Cycloalkanes

Cycloalkanes are named according to the number of carbon atoms in the ring, with the prefix **cyclo-** added.

cyclopentane   cyclohexane   cyclooctane

C. Side chains

When alkyl groups or functional groups are attached to an alkane chain, the continuous chain is called the **root**, or **parent**. The groups are then designated in the name of the compound by prefixes and suffixes on the name of the parent.

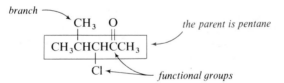

A **side chain**, or **branch**, is an alkyl group branching from a parent chain. A continuous-chain alkyl group is named after its own alkane parent with the **-ane** ending changed to **-yl**. (CH_4 is methane; therefore, the $CH_3—$ group is the *methyl group*. CH_3CH_3 is ethane; therefore, the $CH_3CH_2—$ group is the *ethyl group*.) The names for the first five continuous-chain alkyl groups are listed in Table 3.2.

TABLE 3.2. The first five continuous-chain alkyl groups

Structure	Name
$CH_3—$	methyl
$CH_3CH_2—$	ethyl
$CH_3CH_2CH_2—$	propyl
$CH_3(CH_2)_2CH_2—$	butyl
$CH_3(CH_2)_3CH_2—$	pentyl

How is the name of a side chain incorporated into the name of a compound? To illustrate the technique, we will use hexane as the parent. If there is a methyl group on the second carbon of the hexane chain, the compound is named 2-methylhexane; *2-* for the position of attachment on the parent, *methyl* for the branch at this position, *hexane* for the parent. Methylhexane is one word.

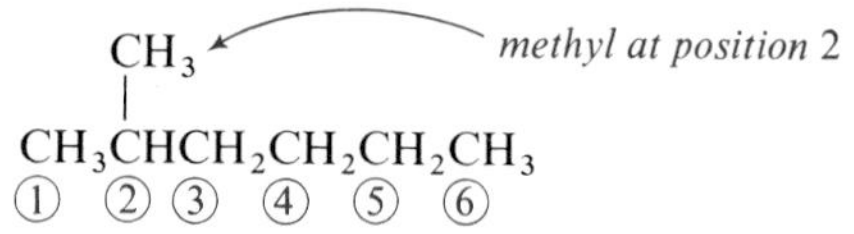

$$CH_3 \quad \text{methyl at position 2}$$
$$CH_3CHCH_2CH_2CH_2CH_3$$
$$① \; ② \; ③ \; ④ \; ⑤ \; ⑥$$

2-methylhexane

SAMPLE PROBLEM

Why is the following compound *not* named 1-methylhexane? What is its correct name?

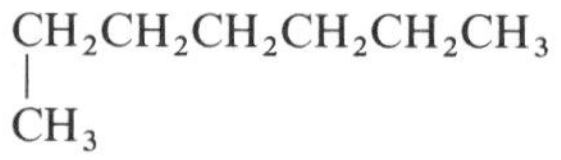

$$CH_2CH_2CH_2CH_2CH_2CH_3$$
$$|$$
$$CH_3$$

Solution: The structure contains a continuous chain of seven carbons; it is called *heptane*.

The general procedure to be followed in naming branched alkanes follows:

1.   Number the carbon atoms in the longest continuous chain, starting at the end closer to the position of the branch so that the prefix number will be as low as possible. (The longest continuous chain in an alkane is the parent, even if the chain is not shown in a straight line.)

2.   Identify the branch and its position.

3.   Attach the number and the name of the branch to the name of the parent.

SAMPLE PROBLEMS

Name the following compound:

$$CH_3CH_2CHCH_2CH_2CH_2CH_3$$
$$|$$
$$CH_2CH_2CH_3$$

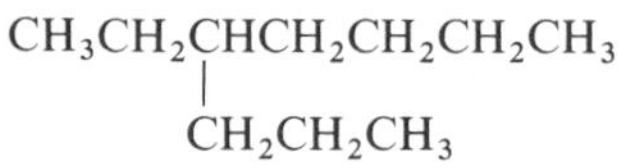

Solution: The longest continuous chain is eight carbons. The parent is *octane*.

$$
\begin{array}{c}
④ \; ⑤ \; ⑥ \; ⑦ \; ⑧ \\
\textit{numbering:} \quad C-C-C-C-C-C-C \\
| \\
C-C-C \\
③ \; ② \; ①
\end{array}
$$

Start at the end closer to the branch

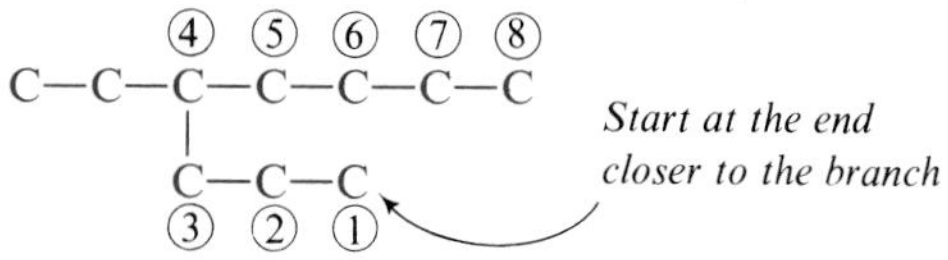

alkyl group: ethyl at carbon 4

name: 4-ethyloctane

Write the structure for 3-methylpentane.

Solution: the parent chain: C—C—C—C—C   (pentane)

> *numbering:*   C—C—C—C—C
> ① ② ③ ④ ⑤

$$CH_3$$
$$|$$
the alkyl group: C—C—C—C—C   (3-methyl)

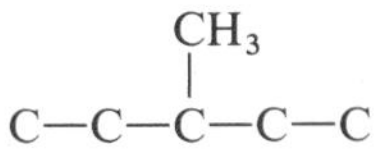

> *add the hydrogens (each carbon must have four bonds):*

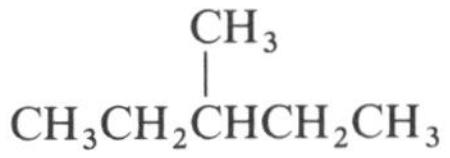

$$CH_3$$
$$|$$
$$CH_3CH_2CHCH_2CH_3$$

Name the following structure. In this name, a prefix number is not needed. Why not?

$$(CH_3)_2CHCH_2CH_3$$

Solution: Methylbutane. A prefix number is not needed because there is only one methylbutane:

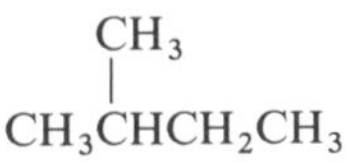

$$CH_3 \qquad\qquad\qquad\qquad CH_3$$
$$| \qquad\qquad\qquad\qquad\qquad |$$
$$CH_3CHCH_2CH_3 \quad \text{is the same as} \quad CH_3CH_2CHCH_3$$

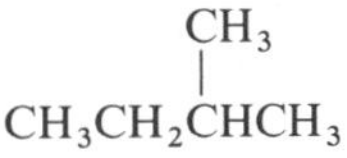

The only other place to attach the methyl group would be on an end carbon; however, this compound is pentane, not methylbutane:

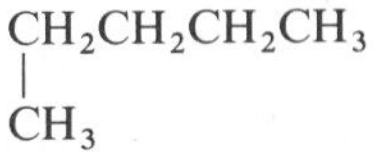

$$CH_2CH_2CH_2CH_3$$
$$|$$
$$CH_3$$

D. Branched side chains

An alkyl group may be branched, rather than a continuous chain. The following examples show branched side chains on a cyclohexane ring and on a heptane chain, respectively.

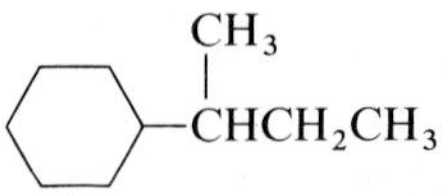

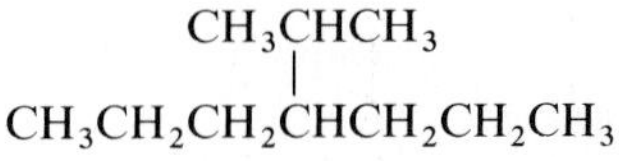

$$CH_3 \qquad\qquad\qquad CH_3CHCH_3$$
$$| \qquad\qquad\qquad\qquad\qquad |$$
$$\text{—CHCH}_2CH_3 \qquad CH_3CH_2CH_2CHCH_2CH_2CH_3$$

Common branched groups have specific names. For example, the two propyl groups are called the **propyl group** and the **isopropyl group**.

$$\qquad\qquad\qquad\qquad\qquad CH_3$$
$$\qquad\qquad\qquad\qquad\qquad |$$
$$CH_3CH_2CH_2\text{—} \qquad CH_3CH\text{—}$$
$$\text{propyl (or }n\text{-propyl)} \qquad \text{isopropyl}$$

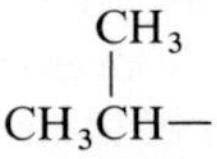

To emphasize that a side chain is *not* branched, a prefix *n*- (for *normal-*) is often used. (The *n*- is redundant since the absence of a prefix also indicates a continuous chain.) The prefix *iso-* (from *isomeric*) is used to indicate a <u>methyl branch at the end of the alkyl side chain</u>.

A four-carbon side chain has four structural possibilities. The **butyl**, or **n-butyl**, group is the continuous-chain group. The **isobutyl group** has a methyl branch at the end of the chain. These two butyl groups are named similarly to the propyl groups.

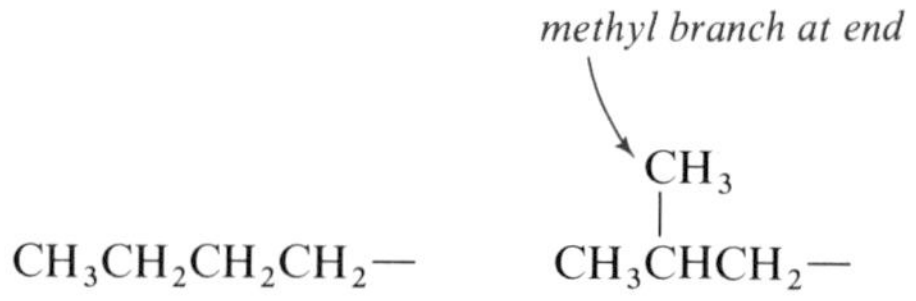

methyl branch at end

$$CH_3CH_2CH_2CH_2—$$ $$CH_3CHCH_2—$$
$$CH_3$$

butyl (or *n*-butyl)   isobutyl

The *secondary*-**butyl group** (abbreviated *sec*-butyl) has two carbons bonded to the head (attaching) carbon. The *tertiary*-**butyl group** (abbreviated *tert*-butyl or *t*-butyl) has three carbons attached to the head carbon.

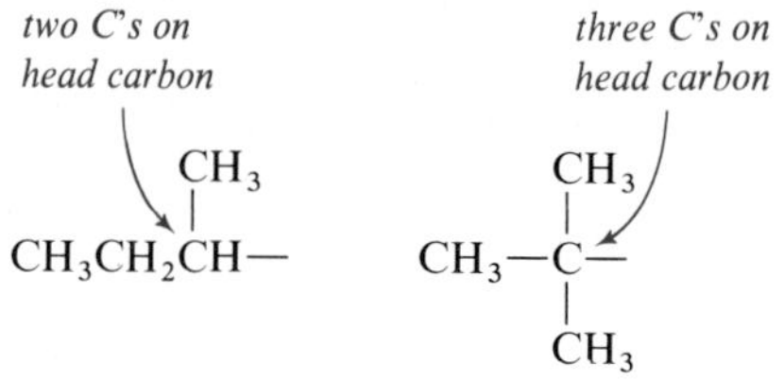

two C's on head carbon   *three C's on head carbon*

$$CH_3CH_2CH—$$ $$CH_3—C—$$
$$CH_3$$ $$CH_3$$
          $$CH_3$$

sec-butyl   *t*-butyl

SAMPLE PROBLEM

Name the following compounds:

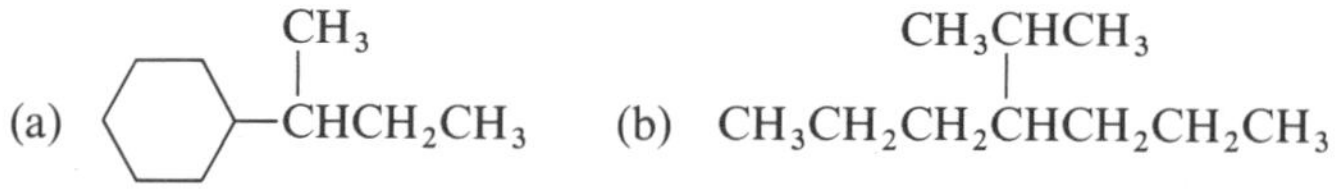

(a) ⬡—CHCH$_2$CH$_3$ (b) CH$_3$CH$_2$CH$_2$CHCH$_2$CH$_2$CH$_3$
       CH$_3$ CH$_3$CHCH$_3$

Solution: (a) *sec*-butylcyclohexane;   (b) 4-isopropylheptane

STUDY PROBLEM

3.4.   Give structures for   (a) *n*-propylcyclohexane;   (b) isobutylcyclohexane;
(c) 4-*t*-butyloctane.

E.   Multiple branches

If two or more branches are attached to a parent chain, more prefixes are added to the parent name. The prefixes are placed alphabetically, each with its number indicating the position of attachment to the parent.

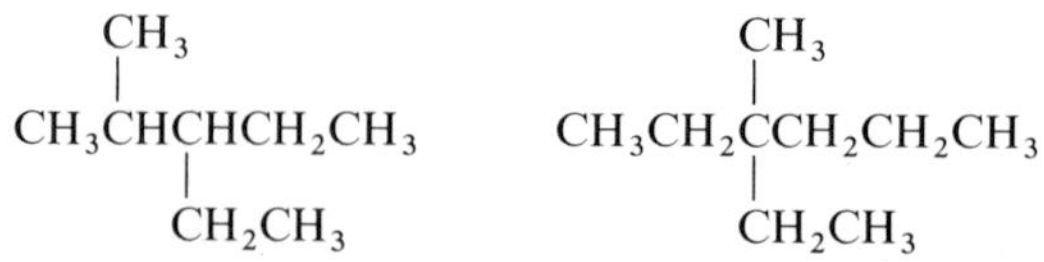

$$CH_3CHCHCH_2CH_3$$ $$CH_3CH_2CCH_2CH_2CH_3$$
$$CH_3$$ $$CH_3$$
$$CH_2CH_3$$ $$CH_2CH_3$$

3-ethyl-2-methylpentane   3-ethyl-3-methylhexane

TABLE 3.3. Prefixes for naming multiple substituents

Number	Prefix
2	di-
3	tri-
4	tetra-
5	penta-
6	hexa-

If two or more substituents on a parent are the same (such as two methyl groups or three ethyl groups), these groups are consolidated in the name. For example, *dimethyl* means "two methyl groups" and *triethyl* means "three ethyl groups." The prefixes (*di-*, *tri-*, etc.) that denote number are listed in Table 3.3. In a name, the *di-* or *tri-* prefix is preceded by position numbers. If *di* is used, two numbers are required; if *tri* is used, three numbers must be present. Note the use of commas and hyphens in the following examples:

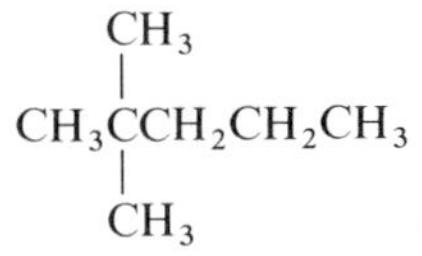

$$CH_3$$
$$|$$
$$CH_3CCH_2CH_2CH_3$$
$$|$$
$$CH_3$$

2,2-dimethylpentane

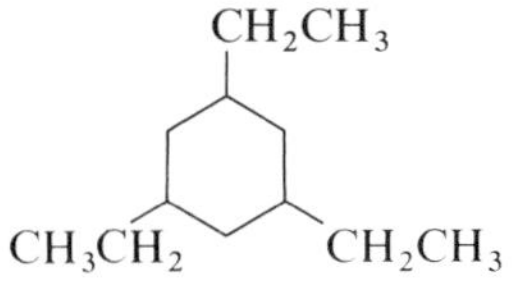

1,3,5-triethylcyclohexane

SAMPLE PROBLEM

Name the following compounds:

(a)

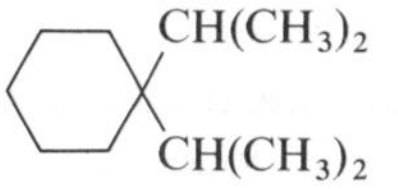

(b) 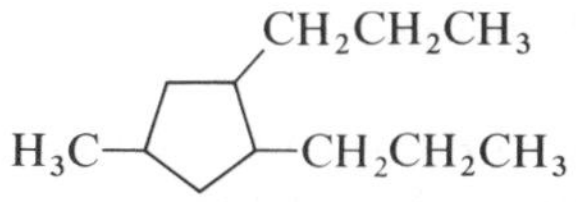 H_3C—

(c) 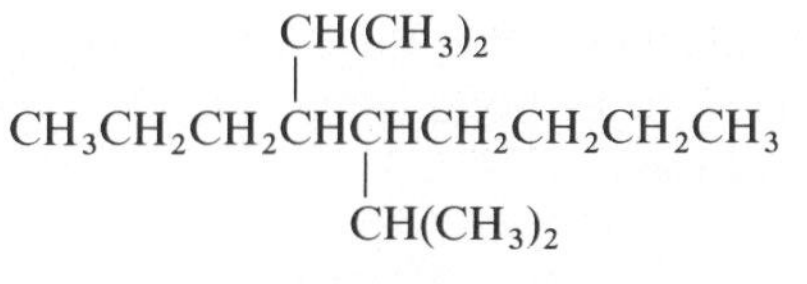 $CH_3CH_2CH_2CHCHCH_2CH_2CH_2CH_3$

Solution: (a) 1,1-diisopropylcyclohexane; (b) 4-methyl-1,2-dipropylcyclopentane;   (c) 4,5-diisopropylnonane. (We number a ring so that the prefix numbers are as low as possible. We alphabetize *dipropyl* as *propyl*.)

F.   Other prefix substituents

Like alkyl branches, a few functional groups are named as prefixes to the parent name. Some of these groups and their prefix names are listed in Table 3.4.

TABLE 3.4. Some substituents named as prefixes

Substituent	Prefix name
$-NO_2$	nitro-
$-F$	fluoro-
$-Cl$	chloro-
$-Br$	bromo-
$-I$	iodo-

The rules governing the use of these prefixes are identical to those for the alkyl groups except that the parent is the longest continuous chain *containing the functional group*. The position of the functional group is specified by a number (as low as possible), and identical groups are preceded by *di-* or *tri-*.

$$\overset{\displaystyle NO_2}{\underset{\displaystyle |}{}}\ \overset{\displaystyle Br}{\underset{\displaystyle |}{}}$$
$$CH_3CH_2CHCH_2CH_2$$

$$CH_3-\overset{\displaystyle Cl}{\underset{\displaystyle Cl}{C}}-\overset{\displaystyle Cl}{\underset{\displaystyle Cl}{C}}-CH_3$$

1-bromo-3-nitropentane   2,2,3,3-tetrachlorobutane

STUDY PROBLEM

3.5.   Give structures for:

(a)   1,1,2-trichloroethane   (b)   1,2-dichloro-4-nitrocyclohexane

G.   Alkenes and alkynes

In IUPAC nomenclature, carbon–carbon unsaturation is always designated by a change in the *ending* of the parent name. As we have indicated, if the parent hydrocarbon contains no double or triple bonds, the suffix **-ane** is used. If a double bond is present, the **-ane** ending is changed to **-ene**; the general name for a hydrocarbon with a double bond is **alkene**. A triple bond is indicated by **-yne**; a hydrocarbon containing this group is an **alkyne**.

$$CH_3CH_3 \qquad\qquad CH_2{=}CH_2 \qquad\qquad HC{\equiv}CH$$

ethane   ethene   ethyne

an alkane   (*trivial name*: ethylene)   (*trivial name*: acetylene)

                     an alkene   *an alkyne*

When the parent contains four or more carbons, a prefix number must be used to indicate the position of the double or triple bond. The chain is numbered so as to *give the double or triple bond as low a number as possible*, even if a prefix group must then receive a higher number. Only a single number is used for each double and triple bond; it is understood that the double or triple bond *begins at this numbered position and goes to the carbon with the next higher number.* Thus, a prefix number 2 means the double or triple bond is between carbons 2 and 3, not between carbons 2 and 1.

$C{=}C$ *starts at carbon 1* $C{=}C$ *starts at carbon 2*

$CH_2{=}CHCH_2CH_3$ $CH_3CH{=}CHCH_3$

1-butene   2-butene

a branch precedes the name

CH_3

$CH_3CHCH_2CH_2C{\equiv}CH$

5-methyl-1-hexyne

If a structure contains more than one double or triple bond, the name becomes slightly more complex. The following example of a **diene**, a compound with two double bonds, shows the placement of the numbers and the *di*. (Note that an *a* is inserted before *di* to ease the pronunciation.)

$$CH_2{=}CHCH{=}CH_2$$

1,3-butadiene

— *two double bonds*

four C's

SAMPLE PROBLEM

Name the following compounds:

(a) $(CH_3)_2CHCH_2CH_2CH{=}CH_2$ (b) $CH_2{=}CHCH{=}CHCH{=}CH_2$
(c) $CH{\equiv}CC{\equiv}CH$

Solution: (a) 5-methyl-1-hexene   (b) 1,3,5-hexatriene   (c) 1,3-butadiyne

STUDY PROBLEMS

3.6. Name the following compounds:

(a) $CH_3CH{=}CH_2$ (b)  ⬡   (c) 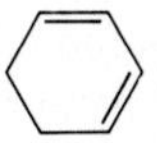 ⬡   (d) $CH_3CH{=}CHCl$

3.7. Give structures for:

(a) cyclopentene   (b) 1,3-pentadiyne   (c) 1-methyl-1-cyclobutene

H.   Alcohols

In the IUPAC system, the name of an alcohol (ROH) is the name of the parent hydrocarbon with the final -e changed to **-ol**. Prefix numbers are used when necessary; the hydroxyl group (—OH) receives the lowest prefix number possible.

$$CH_3CH_2CH_3 \qquad CH_3CH_2CH_2OH \qquad CH_3\overset{\overset{\displaystyle OH}{|}}{C}HCH_3$$

propane   1-propanol   2-propanol

$$CH_3\overset{\overset{\displaystyle CH_3}{|}}{C}HCH_2CH_3 \qquad HOCH_2\overset{\overset{\displaystyle CH_3}{|}}{C}HCH_2CH_3 \qquad CH_3\overset{\overset{\displaystyle CH_3}{|}}{C}HCH_2CH_2OH$$

methylbutane   2-methyl-1-butanol   3-methyl-1-butanol

cyclopentane   cyclopentanol—OH

I.   Amines

Simple amines (RNH_2, R_2NH, or R_3N) are usually named by the name of the alkyl group followed by the suffix **-amine**. A substituent on the nitrogen is sometimes preceded with the prefix *N*- (not *n*-, the abbreviation for *normal-*).

$$CH_3NH_2 \qquad\qquad (CH_3)_2NH \qquad\qquad (CH_3)_2NCH_2CH_3$$

methylamine   dimethylamine   ethyldimethylamine
                                            or *N,N*-dimethylethylamine

$$CH_3\overset{\overset{\displaystyle NH_2}{|}}{C}HCH_2CH_2CH_3$$

2-pentylamine

STUDY PROBLEM

3.8.   (a)   Name the following compounds:

$$CH_3CH_2CH_2\overset{\overset{\displaystyle OH}{|}}{C}HCH_3 \qquad CH_3-\!\!\!\bigcirc\!\!\!-NH_2$$

(b)   Give the structures of 3-methyl-1-cyclohexanol and *n*-butylisopropylamine.

J.   Aldehydes and ketones

Because an aldehyde (RCHO) contains a carbonyl group bonded to a hydrogen atom, the aldehyde group must be at the beginning of a carbon chain. The aldehyde carbon is considered carbon 1; therefore, no number is used in the name to indicate the position. The ending of an aldehyde name is **-al**.

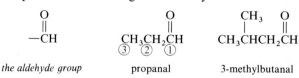

the aldehyde group   propanal   3-methylbutanal

A keto group, by definition, cannot be at the beginning of a carbon chain. Therefore, except for propanone and a few other simple ketones, a prefix number is necessary. The chain should be numbered to give the carbonyl group the lowest possible number. The ending for a ketone name is **-one**.

$$\underset{\text{the keto group}}{\overset{\displaystyle \overset{O}{\underset{\|}{}}}{-C-}} \qquad \underset{\substack{\text{propanone} \\ (\textit{trivial name}: \text{acetone})}}{\overset{\displaystyle \overset{O}{\underset{\|}{}}}{CH_3CCH_3}} \qquad \underset{\text{3-methyl-2-hexanone}}{\overset{\displaystyle \overset{O}{\underset{\|}{}}}{CH_3CCHCH_2CH_2CH_3}}$$

In the third structure, a CH_3 branch is shown below the carbon bearing the H.

K.   Carboxylic acids

A carboxyl group, like an aldehyde group, must be at the beginning of a carbon chain and again contains the first carbon atom (carbon-1). Again, a number is not needed in the name. The ending for a carboxylic acid name is **-oic acid**.

$$\underset{\text{the carboxyl group}}{\overset{\displaystyle \overset{O}{\underset{\|}{}}}{-COH}} \qquad \underset{\substack{\text{ethanoic acid} \\ (\textit{trivial name}: \text{acetic acid})}}{\overset{\displaystyle \overset{O}{\underset{\|}{}}}{\underset{②~①}{CH_3COH}}} \qquad \underset{\text{5,5-dimethylhexanoic acid}}{\overset{\displaystyle \overset{O}{\underset{\|}{}}}{CH_3CCH_2CH_2CH_2COH}}$$

with a CH_3 group above and below the second carbon.

STUDY PROBLEM

3.9.   Write structures and names for an aldehyde, a ketone, and a carboxylic acid, each containing five carbons in a continuous chain.

L.   Esters

An ester is similar to a carboxylic acid, but the acidic hydrogen has been replaced by an alkyl group.

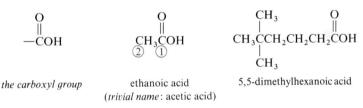

a carboxylic acid   an ester — R instead of H

The name of an ester consists of two words: (1) the name of the ester alkyl group, and (2) the name of the carboxylic acid with the **-ic acid** ending changed to **-ate**. (The ester alkyl group is always the group attached to the oxygen, while the carboxylic acid portion is always that portion containing the carbonyl group.)

$$\underset{\substack{\text{methyl ethanoate} \\ (\textit{trivial name}: \text{methyl acetate})}}{\overset{\displaystyle \overset{O}{\underset{\|}{}}}{\underset{}{CH_3CO-CH_3}}}$$

from ethanoic acid ⟶   ⟵ *methyl group*

SAMPLE PROBLEM

Name the following ester: $\overset{\displaystyle O}{\overset{\displaystyle \|}{CH_3CH_2CH_2COCH_2CH_3}}$

Solution: *Step 1:*   the ester alkyl group = ethyl
             Step 2:   the acid = butanoic acid
             Step 3:   the name = ethyl butanoate

STUDY PROBLEM

3.10.   Write the structures for   (a) cyclohexyl hexanoate and   (b) *t*-butyl butanoate.

M. Benzene compounds

The benzene ring is considered to be a parent in the same way that a continuous-chain alkane is. Alkyl groups, halogens, and the nitro group are named as prefixes to benzene.

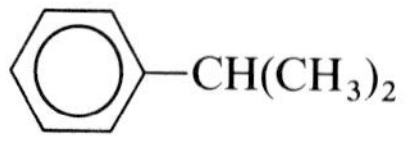

isopropylbenzene 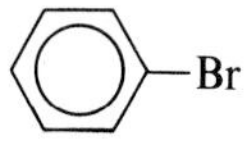   bromobenzene 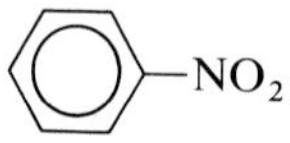   nitrobenzene

When a benzene ring is attached to an alkane chain with a functional group or to an alkane chain of seven or more carbon atoms, benzene is considered a substituent instead of a parent. The name for a benzene substituent is **phenyl**.

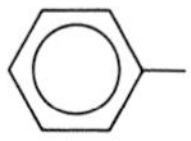

the phenyl group   2-phenyl-1-ethanol

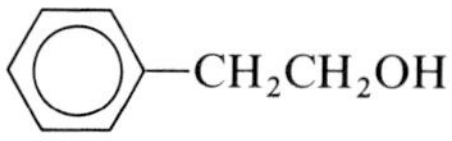

(C_6H_5-)

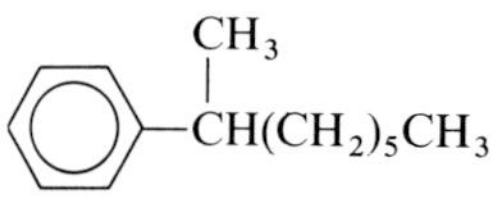

2-phenyloctane

N. Conflicts in numbering

A structure that has more than one type of substituent can sometimes be numbered in more than one way. Should the name for $ClCH_2CH{=}CH_2$ be 1-chloro-2-propene or 3-chloro-1-propene? To cover such situations, a *system of priorities* for prefix numbers has been developed (Table 3.5); the higher-priority substituent receives the lower number. (A more complete list of nomenclature priorities is found in Table A8 in the appendix.)

TABLE 3.5. Nomenclature priorities of selected functional groups

	Partial structure	*Name*
	$-CO_2H$	-oic acid
	$-\overset{\overset{\displaystyle O}{\|\|}}{C}H$	-al
increasing priority	$-\overset{\overset{\displaystyle O}{\|\|}}{C}-$	-one
	$-OH$	-ol
	$-NR_2$	-amine
	$\diagdown C{=}C \diagup$	-ene
	$R-$, C_6H_5-, $Cl-$, $Br-$, $-NO_2$, etc.	prefix substituents

From Table 3.5, we can see that a double bond is higher in priority than Cl. When numbering a carbon chain, we give the double bond the lowest possible number. The name of $ClCH_2CH{=}CH_2$ is therefore 3-chloro-1-propene, and not 1-chloro-2-propene. Similarly, the following compounds are numbered to give the higher-priority groups the lower prefix numbers.

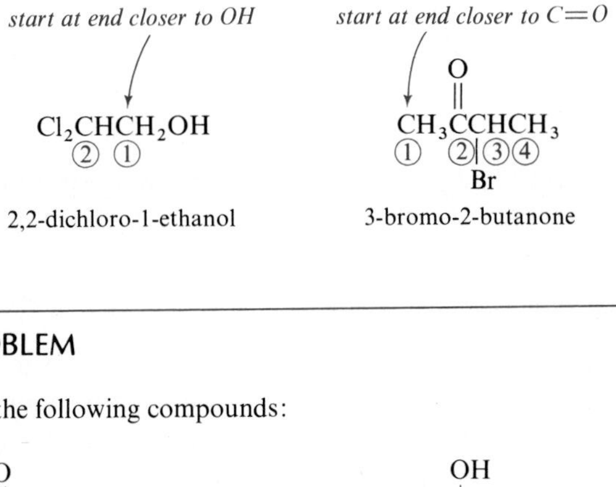

start at end closer to OH

Cl_2CHCH_2OH
② ①

2,2-dichloro-1-ethanol

start at end closer to C=O

$CH_3\overset{\overset{\displaystyle O}{\|\|}}{C}CHCH_3$
① ②③④
Br

3-bromo-2-butanone

STUDY PROBLEM

3.11. Name each of the following compounds:

(a) $CH_3CH_2\overset{\overset{\displaystyle O}{\|\|}}{C}CH_2CH_2CH_2NO_2$

(b) (benzene ring)$-\overset{\overset{\displaystyle OH}{\|}}{C}HCH_2CH_2Br$

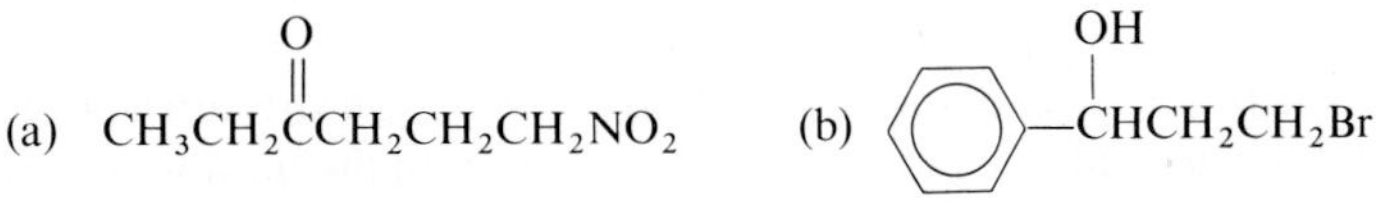

SECTION 3.4

Alkanes

Most organic compounds have a portion of their structures composed of carbon atoms and hydrogen atoms. A fat is one example of an organic compound with ester groups and with long hydrocarbon chains, which may be alkyl or alkenyl (containing a double bond).

$$
\begin{array}{l}
\overset{\displaystyle O}{\underset{\displaystyle \parallel}{}} \\
CH_2OC(CH_2)_{14}CH_3 \\
\overset{\displaystyle O}{\underset{\displaystyle \parallel}{}} \\
CHOC(CH_2)_{14}CH_3 \\
\overset{\displaystyle O}{\underset{\displaystyle \parallel}{}} \\
CH_2OC(CH_2)_7CH{=}CH(CH_2)_7CH_3
\end{array}
$$

long hydrocarbon chains

a typical animal fat

Early chemists did not know the molecular structure of a fat, but they did know that many compounds containing long hydrocarbon chains behave similarly to fats. (For example, most of these compounds are water-insoluble and are less dense than water.) For this reason, compounds with hydrocarbon chains are re- ferred to as **aliphatic compounds** (Greek *aleiphatos*, "fat"). (The term *aliphatic compound* is usually contrasted to *aromatic compound*, such as benzene or a sub- stituted benzene.)

Some aliphatic compounds:

$$
\begin{array}{l}
\overset{\displaystyle CH_3}{\underset{\displaystyle |}{}} \\
CH_3CHCH_2CH_3 \quad\quad CH_3CH_2CH_2OH \quad\quad CH_3CH_2NHCH_2CH_3
\end{array}
$$

Some of the physical and chemical properties of an aliphatic compound arise from the alkyl part of its molecules. Therefore, much of what we have to say about alkanes and cycloalkanes is true for other organic compounds as well. Of course, the properties of a compound are also greatly determined by any functional groups that may be present. For example, a hydroxyl group in a molecule leads to hydrogen bonding and a large change in physical properties. Ethane (CH_3CH_3) is a gas at room temperature, while ethanol (CH_3CH_2OH) is a liquid.

A.   Physical properties of alkanes

Alkanes are nonpolar compounds. As a result, the attractive forces between molecules are weak. The continuous-chain alkanes through butane are gases at room temperature, while the C_5 to C_{17} alkanes are liquids (see Table 3.6). The continuous-chain alkanes with 18 or more carbon atoms are solids.

The boiling point of a compound depends, in part, on the amount of energy needed by the molecules of that compound to escape from the liquid into the vapor phase. The boiling points of the compounds of a homologous series, such as the alkanes in Table 3.6, increase by about 30° for each additional methylene (CH_2) group. This increase in boiling point is due principally to an increase in the van der Waals attractions between longer and longer molecules. Other homologous series show similar effects.

$$CH_3I \quad\quad\quad CH_3CH_2I \quad\quad\quad CH_3CH_2CH_2I$$

iodomethane   iodoethane   1-iodopropane
bp 43°   bp 72°   bp 102°

TABLE 3.6. Boiling points of some alkanes

Structure	Bp, °C	Structure	Bp, °C
CH_4	-162	$CH_3(CH_2)_4CH_3$	69
CH_3CH_3	-88.5	$CH_3(CH_2)_5CH_3$	98
$CH_3CH_2CH_3$	-42	$CH_3(CH_2)_6CH_3$	126
$CH_3(CH_2)_2CH_3$	0	$CH_3(CH_2)_7CH_3$	151
$CH_3(CH_2)_3CH_3$	36	$CH_3(CH_2)_8CH_3$	174

As we mentioned in Section 1.9A, branching in the hydrocarbon portion of a molecule lowers the boiling point from the expected value because of interference with van der Waals attractions between the molecules in the liquid state.

Because they are nonpolar, alkanes are soluble in nonpolar or slightly polar solvents such as other alkanes, diethyl ether ($CH_3CH_2OCH_2CH_3$), or benzene. The solubility arises from van der Waals attractions between the solvent and the solute. Alkanes are insoluble in water.

All the alkanes have densities less than that of water, a fact easy to remember because we know that gasoline and motor oil (which are principally alkanes) float on water.

B.   Chemical properties of alkanes

Alkanes and cycloalkanes are chemically unreactive compared to organic compounds with functional groups. For example, many organic compounds undergo reaction with strong acids, bases, oxidizing agents, or reducing agents. Alkanes and cycloalkanes generally do not undergo reaction with these reagents. Because of their lack of reactivity, alkanes are sometimes referred to as **paraffins** (Latin *parum affinis*, "slight affinity").

There are two principal reactions of alkanes that we will discuss in this text. One is the *reaction with halogens*, such as chlorine gas. We will present this reaction in detail in Chapter 6. The other important reaction of alkanes is *combustion*. The remainder of this chapter will be concerned primarily with the combustion of alkanes and their use as a source of energy.

Halogenation:

$$CH_4 + Cl_2 \xrightarrow{\text{ultraviolet light}} CH_3Cl + HCl$$

methane   chloromethane

$$CH_3CH_2CH_3 + Br_2 \xrightarrow{\text{ultraviolet light}} \overset{\displaystyle Br}{\underset{}{CH_3CHCH_3}} + HBr$$

propane   2-bromopropane

Combustion:

$$CH_4 + 2\,O_2 \xrightarrow{\text{spark}} CO_2 + 2\,H_2O$$

carbon dioxide

$$CH_3CH_2CH_3 + 5\,O_2 \xrightarrow{\text{spark}} 3\,CO_2 + 4\,H_2O$$

C.   Combustion

Combustion is the process of burning—that is, the rapid reaction of a compound with oxygen. Combustion is accompanied by the release of light and heat, two forms of energy that humans have sought since they first built a fire and found that it kept them warm. Although we will present the subject of combustion under the heading of alkanes, keep in mind that almost all organic compounds can burn.

Combustion of organic mixtures, such as wood, is not always a simple conversion to CO_2 and H_2O. Instead, combustion is the result of a large number of complex reactions. One type of reaction that occurs is **pyrolysis**, the thermal fragmentation of large molecules into smaller molecules in the absence of oxygen. Pyrolysis of large molecules in wood, for example, yields smaller gaseous molecules that then react with oxygen above the surface of the wood. This reaction with oxygen gives rise to the flames. On the surface of the wood, a slow, but very hot, oxidation of the carbonaceous residue takes place. Most of the heat from a wood or coal fire results from this slow oxidation, rather than from the actual flames.

Complete combustion is the conversion of a compound to CO_2 and H_2O. If the oxygen supply is insufficient for complete combustion, **incomplete combustion** occurs. Incomplete combustion leads to carbon monoxide, or sometimes carbon as carbon black or soot.

Incomplete combustion:

$$2 CH_3CH_2CH_3 + 7 O_2 \longrightarrow \underset{\substack{\text{carbon} \\ \text{monoxide}}}{6 CO} + 8 H_2O$$

$$CH_3CH_2CH_3 + 2 O_2 \longrightarrow \underset{\text{carbon}}{3 C} + 4 H_2O$$

D.   Heat of combustion

The energy released when a compound is oxidized completely to CO_2 and H_2O is called the **heat of combustion** ΔH. Under controlled laboratory conditions, ΔH may be measured. (As we described in Section 1.7, the value for ΔH is negative when energy is liberated because the molecules have lost energy.)

$$\underset{\text{methane}}{CH_4} + 2 O_2 \longrightarrow CO_2 + 2 H_2O + 213 \text{ kcal/mole}$$

$$\underset{\text{butane}}{2 CH_3CH_2CH_2CH_3} + 13 O_2 \longrightarrow 8 CO_2 + 10 H_2O + 688 \text{ kcal/mole}$$

Values for heats of combustion (Table 3.7) depend primarily upon the number of carbon and hydrogen atoms in a molecule. In a homologous series, the energy

TABLE 3.7. Heats of combustion for some hydrocarbons

Name	*$-\Delta H$, kcal/mole*	*Name*	*$-\Delta H$, kcal/mole*
methane	213	cyclopropane	500
ethane	373	cyclobutane	656
propane	531	cyclopentane	794
butane	688	cyclohexane	944

liberated increases by about 157 kcal/mole for each additional methylene group. The heat of combustion of a compound may also reflect unusual bonding characteristics. For example, cyclohexane has a ΔH of -944 kcal/mole, or -157 kcal per methylene group, the same value observed for open-chain alkanes. However, cyclopropane has a ΔH of -167 kcal per methylene group, a higher value than that for an open-chain alkane. (We will discuss the reason for this in Section 4.4A.)

$$\text{cyclohexane} + 9\,O_2 \longrightarrow 6\,CO_2 + 6\,H_2O + 157\ \text{kcal/mole per } CH_2 \text{ group}$$

cyclohexane

$$2\ \text{cyclopropane} + 9\,O_2 \longrightarrow 6\,CO_2 + 6\,H_2O + 167\ \text{kcal/mole per } CH_2 \text{ group}$$

cyclopropane

SECTION 3.5.

The Hydrocarbon Resources

A. Natural gas and petroleum

Natural gas, which is 60–90% methane depending on its source, has been formed by the anaerobic decay (decay in the absence of air) of plants. The other components of natural gas are ethane and propane, along with nitrogen and carbon dioxide. Natural gas found in the Texas panhandle and in Oklahoma also is a source of helium. Deposits of natural gas are usually found with petroleum deposits.

Petroleum has been formed by the decay of plants and animals, probably of marine origin. Crude petroleum, called *crude oil*, is a complex mixture of aliphatic and aromatic compounds, including sulfur and nitrogen compounds (1–6%). In fact, over 500 compounds have been detected in a single sample of petroleum. The actual composition varies from deposit to deposit.

Because of its complexity, crude oil itself is not very useful. Separating the crude oil into useful components is called **refining**. The first step in refining is a fractional distillation, called **straight-run distillation**. The fractions that are collected are listed in Table 3.8.

TABLE 3.8. Fractions of straight-run distillation

Boiling range, °C	Number of carbons	Name	Use
under 30	1–4	gas fraction	heating fuel
30–180	5–10	gasoline	automobile fuel
180–230	11–12	kerosene	jet fuel
230–305	13–17	gas oil	diesel fuel, heating fuel
305–405	18–25	heavy gas oil	heating fuel

Residue: (1) Volatile oils: lubricating oils, paraffin wax, and petroleum jelly.
(2) Nonvolatile material: asphalt and petroleum coke.

The gasoline fraction of straight-run distillation is too scanty for the needs of our automobile-oriented society, and straight-run gasoline is usually of poor quality. To increase both quantity and quality of the gasoline fraction, the higher-boiling fractions are subjected to *cracking* and *reforming*.

Catalytic cracking is the process of heating the high-boiling material under pressure in the presence of a catalyst (finely divided, acid-washed aluminum silicate clay). Under these conditions, large molecules are cracked, or broken, into smaller fragments.

Steam cracking is a technique for converting alkanes to alkenes, and **catalytic reforming** converts aliphatic compounds to aromatic compounds. The alkenes and aromatics formed in these cracking and reforming procedures are used as starting materials for making plastics and other synthetic organic compounds. The following represents just a sample of the many reactions that can occur in cracking and reforming.

$$CH_3CH_2CH_2CH_2CH_2CH_2CH_3$$
heptane

$$CH_3CH_2CH_2CH_2CH_3 + CH_2=CH_2$$
pentane   ethene
                    (ethylene)

$$\langle\bigcirc\rangle-CH_3 + 4\,H_2$$
methylbenzene
(toluene)

The high-compression engines in automobiles are relatively efficient for their weight, but in these engines continuous-chain hydrocarbons burn unevenly and cause knocking, the ticking noise heard when a car accelerates uphill. Knocking decreases the power output of the engine and decreases the life of the engine through wear and tear. Quality automobile fuels contain branched alkanes and aromatic compounds, which burn more evenly than continuous-chain compounds. Fortunately, cracking procedures provide both branched alkanes and aromatics.

At one time, isooctane (a trivial name) was the alkane with the best anti-knock characteristics for automobile engines, and heptane was the poorest. These two compounds were used to develop an octane rating of petroleum fuels.

$$\begin{matrix} & CH_3 & CH_3 \\ & | & | \\ CH_3CCH_2CHCH_3 & & CH_3(CH_2)_5CH_3 \\ & | \\ & CH_3 \end{matrix}$$

2,2,4-trimethylpentane   heptane
("isooctane")

To rate the quality of a gasoline, the fuel is compared with a mixture of iso-octane and heptane and given an **octane number**. An octane number of 100 means that the gasoline is equivalent in burning characteristics to pure isooctane. Gasoline with an octane number of 0 is equivalent to pure heptane. An octane number of 75 is given to gasoline that is equivalent to a mixture of 75% isooctane and 25% heptane.

Additives are also added to gasoline to decrease engine knock and increase octane ratings. The best-known additive is *Ethyl fluid*, which contains approximately 65% tetraethyllead; 25% 1,2-dibromoethane; and 10% 1,2-dichloroethane. The halogenated hydrocarbons are essential for conversion of the lead to the volatile lead bromide, which is removed from the cylinder in the exhaust.

$(CH_3CH_2)_4Pb$ $BrCH_2CH_2Br$ $ClCH_2CH_2Cl$

tetraethyllead   1,2-dibromoethane   1,2-dichloroethane

(ethylene dibromide)   (ethylene dichloride)

A gasoline engine puts forth a variety of pollutants: unburned hydrocarbons, carbon monoxide, and nitrogen oxides. The presence of tetraethyllead in the gasoline adds lead compounds to the list of pollutants. Catalytic converters have been installed in many automobiles to convert nonoxidized and partially oxidized compounds to more-highly oxidized and acceptable forms of exhaust. For example, a catalytic converter oxidizes unburned hydrocarbons and carbon monoxide to carbon dioxide and water. The platinum catalyst used in these converters is "poisoned" (made nonfunctional) by lead compounds; therefore, leaded gasoline should not be used in cars equipped with catalytic converters.

The octane number of gasoline can be increased by the addition of other compounds besides Ethyl fluid. Benzene (octane number 106) is routinely added to gasoline for this purpose. Ethanol; *t*-butyl alcohol, $(CH_3)_3COH$; and *t*-butyl methyl ether, $(CH_3)_3COCH_3$, may also be used to increase the octane rating. (A mixture of approximately 90% gasoline–10% ethanol is popularly called "gasohol.")

B. Coal

Coal is formed by the bacterial decomposition of plants under varying degrees of pressure. Coal is classified by its carbon content: *anthracite*, or hard coal, contains the highest carbon content, followed by *bituminous* (*soft*) *coal*, *lignite*, and, finally, *peat*. Because some coals also contain 2–6% sulfur, the burning of coal can lead to severe air pollution and "acid rain."

When coal is subjected to heat and distillation in the absence of air, a process called **destructive distillation**, three crude products result: *coal gas* (primarily CH_4 and H_2), *coal tar* (the condensable distillate), and *coke* (the residue). Both coal gas and coke are useful fuels. (Today, coke is used primarily in the manufacture of steel.) Coal tar is rich in aromatic compounds, which are formed in the destructive distillation.

Some of the aromatic compounds found in coal tar:

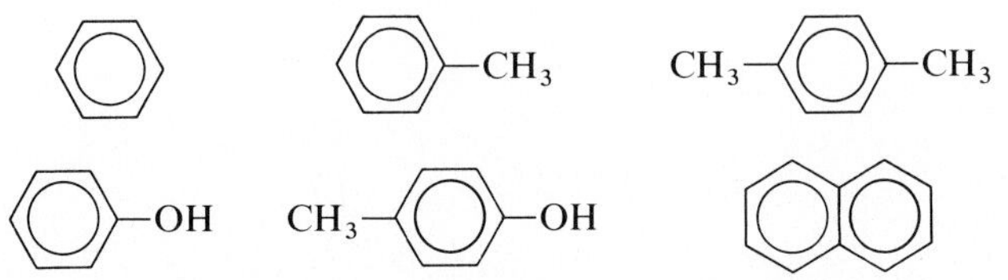

Until petroleum became plentiful and cheap in the 1940's, coal was a primary source of synthetic organic compounds. Currently, however, over 90% of the organic chemicals produced in the United States are synthesized from petroleum. The reasons for the switch to petrochemicals are that petroleum-refining processes are less costly and also that these processes give rise to less pollution than most coal-refining processes. Unfortunately, the world's petroleum resources are rapidly being depleted and becoming more expensive, while the coal reserves of the world are still very extensive. Economical ways to convert coal to useful fuels and chemicals (with a minimum of air pollution) are under investigation at the present time.

The conversion of coal to gaseous or liquid fuels, called synthetic fuels or "syn fuels," is referred to as **coal gasification** or **coal liquefaction**, respectively. Many gasification plants use the German-developed **Lurgi process**, or modifications of this technique, in which a bed of coal is treated with steam at high temperatures to yield *synthesis gas* ($CO + H_2$). Synthesis gas itself is only a moderately efficient fuel, and carbon monoxide is highly toxic. Thus, synthesis gas is treated with additional hydrogen to yield methane.

Coal gasification:

$$C + H_2O \xrightarrow{\text{heat}} \underbrace{CO + H_2}_{\text{"synthesis gas"}} \xrightarrow[\text{Ni catalyst}]{2H_2} CH_4 + H_2O$$

coal   steam   methane

The liquefaction of coal is its conversion to liquid alkanes. The classical process for accomplishing this conversion is the **Fischer–Tropsch synthesis**, which was developed in Germany during World War II. The Republic of South Africa synthesizes most of its gasoline and organic chemicals by this process.

Coal liquefaction (Fischer–Tropsch synthesis):

$$C + H_2O \xrightarrow{\text{heat}} CO + H_2 \xrightarrow[\text{heat, pressure}]{\overset{H_2}{\text{Fe catalyst}}} \text{alkanes} + H_2O$$

coal   steam

Because coal reserves, too, are finite, other sources of hydrocarbons are being investigated and developed. These include oil shale in the United States; tar sands in the Athabasca basin in Alberta, Canada; and biological sources, such as agricultural wastes and plants of the *Euphorbia* genus (succulents with a high percent of hydrocarbons in their sap).

Summary

Compounds with the same molecular formula, but different structures (order of attachment of the atoms), are **structural isomers** of each other.

The **IUPAC system of nomenclature** is based on the names of the *continuous-chain alkanes* as parents. If a hydrocarbon chain forms a ring, the prefix *cyclo-* is added to the alkane name. Branches and functional groups are indicated in a name by prefixes or suffixes.

The longest continuous chain containing the functional group (if any) is the parent. The chain is numbered from the end nearer to the branches or functional groups. (The functional group of highest priority, as listed in Table 3.5, receives the lowest number.) Positions of substitution on the chain are then specified by these numbers.

$$\text{Cl}$$
$$\overset{\text{Cl}}{\underset{\underset{CH_3}{|}}{\overset{(5)(4)|(3)}{CH_3CCH}}}\overset{(2)(1)}{=CHCH_2Cl}$$

1,4-dichloro-4-methyl-2-pentene:

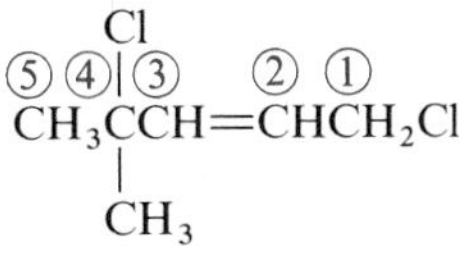

The principal reactions of alkanes are **halogenation** and **combustion**. The **heat of combustion** of an alkane (or other compound) is the result of a decrease in bond energies between the original compound and the products $CO_2 + H_2O$.

Petroleum is the world's principal source of gasoline and organic chemicals today. Alkanes, alkenes, and aromatic compounds are obtained by **refining**: straight-run distillation, cracking, and reforming. In the future, **coal-gasification** or **coal-liquefaction** processes may be the principal sources of methane and other alkanes.

STUDY PROBLEMS

3.12. Which of the following compounds are unsaturated?

(a) CH_3CN

(b)

(c)

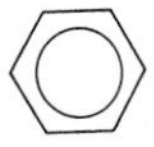

(d) $CH_3CH_2CH{=}CH_2$ (e) $CH_3CH_2CH_2OH$ (f) CH_3CH_2CHO

3.13. Write a formula for one structural isomer of each of the following compounds. (If no structural isomer exists, indicate this fact.)

(a)    (b) $BrCH_2CH_2Br$ (c) CH_3OH

(d)  $\overset{O}{\overset{\|}{HCH}}$ (e) $ClCH_2Br$ (f)  $\overset{O}{\overset{\|}{HCOH}}$

3.14. Which of the following pairs of structures represent structural isomers?

$$\text{(a)} \quad CH_3CH_2CH_2CH_2CH_2OH, \quad CH_3\overset{\overset{\displaystyle CH_3}{|}}{\underset{\underset{\displaystyle CH_3}{|}}{C}}CH_2CH_2OH$$

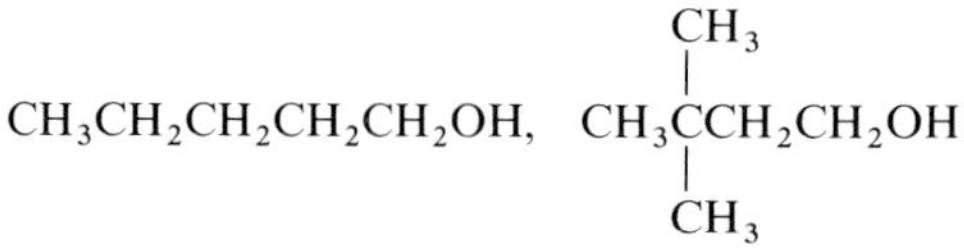

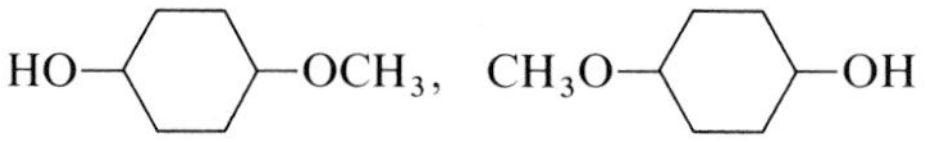

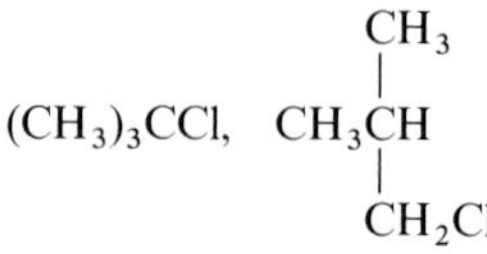

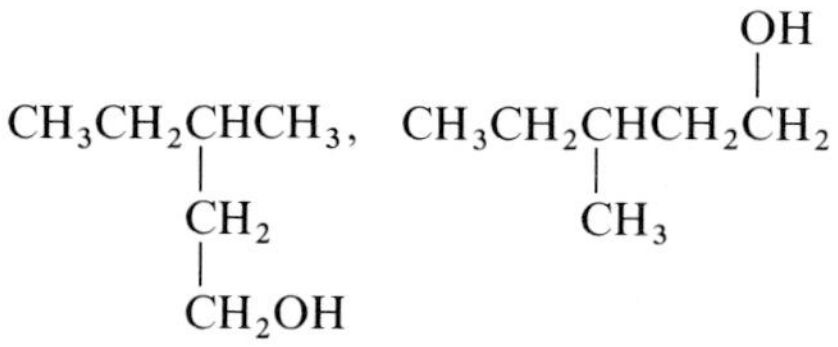

(b) HO—⬡—OCH_3, CH_3O—⬡—OH

$$\text{(c)} \quad (CH_3)_3CCl, \quad CH_3\overset{\overset{\displaystyle CH_3}{|}}{\underset{\underset{\displaystyle CH_2Cl}{|}}{CH}}$$

$$\text{(d)} \quad CH_3CH_2\underset{\underset{\underset{\displaystyle CH_2OH}{|}}{\underset{\displaystyle CH_2}{|}}}{CH}CH_3, \quad CH_3CH_2\underset{\underset{\displaystyle CH_3}{|}}{\overset{\overset{\displaystyle OH}{|}}{CH}}CHCH_2CH_2$$

$$\text{(e)} \quad CH_3\overset{\overset{\displaystyle OH}{|}}{CH}CH_2CH_3, \quad CH_3CH_2\overset{\overset{\displaystyle OH}{|}}{CH}CH_3$$

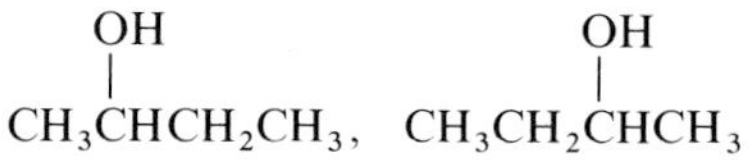

(f) 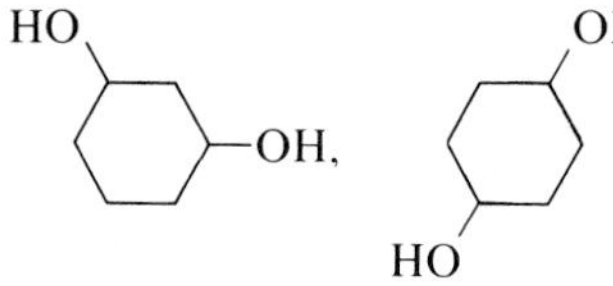 $\overset{\displaystyle HO}{⬡}$—$OH$, $\underset{\displaystyle HO}{\overset{\displaystyle OH}{⬡}}$

3.15. Write structural formulas for the indicated compounds:

    (a)   five structural isomers for C_6H_{14}
    (b)   all the isomeric alcohols for $C_4H_{10}O$
    (c)   all the isomeric amines for $C_4H_{11}N$
    (d)   all the structural isomers for C_3H_6BrCl
    (e)   all the structural isomers for C_4H_6

3.16. In the following list of structures, which are structural isomers of each other and which are the same compound?

    (a) $CH_3CCl_2CH_2CH_3$ (b) $CH_3CH(CH_2Cl)CH_2CH_3$

    (c) $CH_3C(CH_3)_2CH_2CH_2OH$ (d) $CH_3CHClCH_2CH_3$

    (e) $CH_2ClCH(CH_3)CH_2CH_3$ (f) $CH_3CHClCHClCH_3$

3.17. What is the molecular formula and the general formula (e.g., C_nH_{2n}) for each of the following compounds?

$$\text{(a)} \quad CH_3\overset{\overset{\displaystyle CH_3}{|}}{\underset{\underset{\displaystyle CH=CH_2}{|}}{C}}CH_3 \qquad \text{(b)} \quad ⬡⬜$$

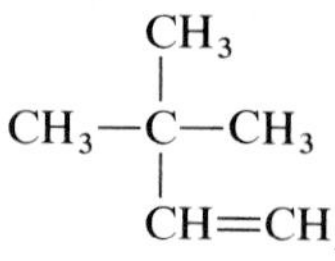

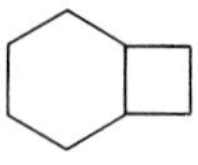

(c) $CH_2=CHCH=CHCH_2CH(CH_3)_2$ (d)

3.18. Write the formula for one structural isomer of each of the following compounds:

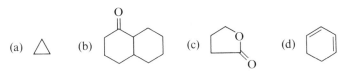

(a)   (b)   (c)   (d)

3.19. For each of the following compounds, write the structural formula of an isomer that has a *different functional group* (see Problem 3.2, page 84).

(a)   (b) $HOCH_2\overset{\overset{\displaystyle O}{\|}}{C}H$ (c) $H\overset{\overset{\displaystyle O}{\|}}{C}CH_2CH_3$

3.20. Arrange the following structures into a homologous series:

(a) $CH_3CH_2CH_2CH_2CH_2OH$ (b) CH_3OH (c) $CH_3CH_2CH_2OH$

(d) CH_3CH_2OH (e) $CH_3CH_2CH_2CH_2OH$

3.21. Write structural formulas for a homologous series of alkyl bromides (monobromoalkanes) from C_5 to C_{10}.

3.22. Give the condensed structural (or polygon) formula of each of the following compounds:
(a) 2,2-dimethyloctane;   (b) 3,4-diethylheptane;   (c) 4-ethyl-2,4-dimethylnonane;
(d) 1,3-diisopropylcyclohexane;   (e) *sec*-butylcyclopentane;   (f) *t*-butylbenzene;
(g) isobutylcycloheptane;   (h) 1-methyl-3-pentylcyclohexane;   (i) 4-isopropylheptane.

3.23. Write the IUPAC name for each of the following compounds:

(a) $(CH_3CH_2)_2CHCH_2CH_3$ (b) $(CH_3)_2CHCH(CH_3)_2$

(c) CH_3CH_2—   —CH_3 (d) $(CH_3)_3C$—

(e) $CH_3CH_2CH_2C(CH_3)_2CH_2C(CH_3)_3$ (f) $[(CH_3)_3C]_2CHCH_3$

3.24. Each of the following names is *incorrect*. Give a more suitable name in each case.
(a) 3-propylpentane;   (b) 6-methyloctane;   (c) *t*-butylmethane;
(d) 3-methyl-3-ethyl-4-methyldecane.

3.25. Write structural formulas for the following compounds, each of which contains a total of six carbons in its structure. (There may be more than one correct answer.)

(a)   a ketone with a methyl branch
(b)   a continuous-chain ketone
(c)   a cyclic ketone with a methyl branch
(d)   a cyclic ketone with no branches

3.26. Draw structures for five-membered carbon rings with the following branches: (a) propyl; (b) isopropyl; (c) butyl; (d) isobutyl; (e) *sec*-butyl; (f) *t*-butyl; (g) cyclobutyl; (h) pentyl.

3.27. Circle and name the functional groups in the following structures:

(a) $CH_2{=}CHCHC{\equiv}CCH(CH_3)_2$ (with OH on the third carbon)

(b) $BrCH_2\overset{O}{\underset{||}{C}}{-}\overset{O}{\underset{||}{C}}OH$

(c)

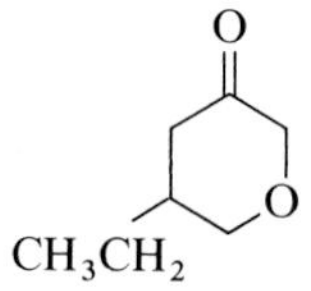

(d)

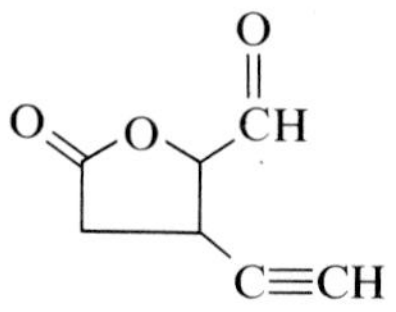

3.28. Give structures for all the monochloro isomers of the following alkanes: (a) *n*-pentane; (b) cyclopentane; (c) 2,2-dimethylbutane; (d) 2,2-dimethylpropane.

3.29. Write an IUPAC name for each of the following structures:

(a)

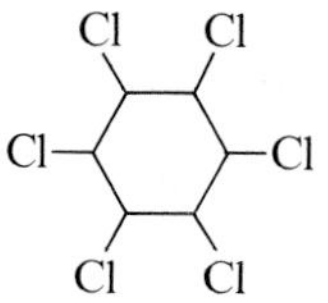

(b) $CH_2ClCHCl_2$

(c) $(CH_3)_2CBrCCl(CH_3)_2$

(d) CH_3NO_2

3.30. Write the formula for each of the following names: (a) 1-bromo-1,2-diphenylpropane; (b) hexachloroethane; (c) 2-iodo-1-octanol; (d) 1,1-dichloro-3-methylcyclohexane.

3.31. Write IUPAC names for the following compounds:

(a) $CH_3CH_2CH{=}CHCH_3$

(b) $(CH_3)_2C{=}CHCH_3$

(c)

(d) $CH_2{=}CHC(CH_3){=}CHCH{=}CH_2$

(e) $CH_3C{\equiv}CCH_3$

(f) $Cl_2C{=}CHCl$

(g)

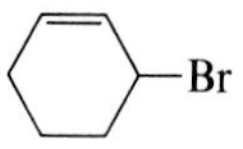

(h)

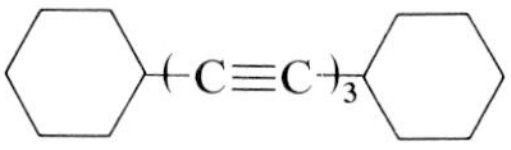

3.32. Give the IUPAC name and structure for (a) a continuous-chain seven-carbon aldehyde; (b) a two-carbon carboxylic acid that contains two Cl atoms; (c) a continuous-chain, nine-carbon ketone that is symmetrical (the carbonyl group is in the center).

3.33. Write the IUPAC name for each of the following structures:

(a) $CH_3CH{=}CHCHCCl_3$ (with CH_3 branch)

(b) $CH_2{=}CHCH_2NO_2$

(c) 〈◯〉—CH$_2$CH$_2$NH$_2$ (d) CH$_3$CHCOCH$_2$CH$_3$

with O double-bonded above the carbonyl carbon and Cl below the CHC carbon:

(d) $$CH_3\overset{\displaystyle Cl}{\underset{}{C}}H\overset{\displaystyle O}{\overset{\|}{C}}OCH_2CH_3$$

3.34. Write the formula for each of the following compounds: (a) isopropyl propanoate; (b) 4,4-diethyl-3-methyloctanal; (c) 1-cyclohexyl-1-ethanol.

3.35. (a) From the data in Table 3.7 (page 100), predict the heat of combustion of octane.
(b) From these data, could you also extrapolate the heat of combustion of 2-methylheptane? Explain.

3.36. Which one of each of the following pairs of structural isomers would you expect to have the higher boiling point: (a) hexane or 2,2-dimethylbutane? (b) 2-butene or methylpropene? (c) 1-pentanol or 2,2-dimethyl-1-propanol? Why?

3.37. Methanol is often added to automobile gas tanks in the winter to prevent the gas line from freezing. The freezing point of gasoline is somewhere around $-50°C$, far below the temperature of any anticipated cold spell. Suggest: (a) why a gas line might freeze, and (b) how methanol could help prevent this.

3.38. A chemist determines that a compound with the molecular formula C_4H_8O is a ketone. What are the possible structures for this compound?

3.39. An alcohol with no carbon–carbon double bonds has the formula C_4H_8O. What are the possible structures?

3.40. What are the possible structures for a carboxylic acid with the molecular formula $C_4H_8O_2$?

3.41. A four-carbon compound contains an aldehyde group and a carboxyl group. What are two possible structures for this compound?

3.42. A compound with the formula $C_4H_6O_2$ contains at least one carbonyl group. The compound does not contain a hydroxyl group; an alkoxyl group (—OR in an ether or ester); or a carbon–carbon double bond. The compound is not acidic. What are the possible structures for this compound?

3.43. (a) Complete, but do not balance, the following equation, showing all possible monochlorinated products:

$$CH_3CH_2CH_3 + Cl_2 \xrightarrow{\text{ultraviolet light}}$$

(b) For the equation in (a), show all possible *dichlorinated* products.
(c) If the substitution of Cl for H were completely random for the reaction in (a), what would be the ratio of isomeric products?

3.44. On page 102, we show the cracking of heptane to yield pentane and ethylene. What other cracking products could be obtained? (Use equations in your answer.)

3.45. A mixture of *n*-hexane, *n*-heptane, and *n*-octane is subjected to **dehydrogenation** (loss of H_2) by treatment with heat and an appropriate catalyst. What aromatic compounds are likely to result? (Used balanced equations in your answer.)

CHAPTER 4

Stereochemistry

Stereochemistry is the study of molecules in three dimensions—that is, how atoms in a molecule are arranged in space relative to one another. The three aspects of stereochemistry that will be covered in this chapter are:

1. **Geometric isomers:** how rigidity in a molecule can lead to isomerism;
2. **Conformations of molecules:** the shapes of molecules and how they can change;
3. **Chirality of molecules:** how the right- or left-handed arrangement of atoms around a carbon atom can lead to isomerism.

It is often difficult to visualize a three-dimensional molecule from a two-dimensional illustration. Therefore, in our discussions of stereochemistry here and in subsequent chapters, we strongly urge you to use a set of molecular models.

SECTION 4.1.

Geometric Isomerism in Alkenes

In Chapter 3, we defined structural isomers as compounds with the same molecular formula but with different orders of attachment of their atoms. Structural isomerism is only one type of isomerism. A second type of isomerism is **geometric isomerism**, which results from rigidity in molecules and occurs in only two classes of compounds: alkenes and cyclic compounds.

Molecules are not quiet, static particles. They move, spin, rotate, and flex. Atoms and groups attached only by sigma bonds can rotate so that the overall shape of a molecule is in a state of continuous change. However, groups attached by a double bond cannot rotate around the double bond without the pi bond being broken. The amount of energy needed to break a carbon–carbon pi bond (about 68 kcal/mole) is not available to molecules at room temperature. Because of the rigidity of a pi bond, groups attached to pi-bonded carbons are fixed in space relative to one another.

We usually write the structure for an alkene as if the sp^2 carbon atoms and the atoms attached to them are all in the plane of the paper. In this representation, we can visualize one lobe of the pi bond as being in front of the paper and the other lobe of the pi bond as being underneath the paper, behind the front lobe (see Figure 4.1).

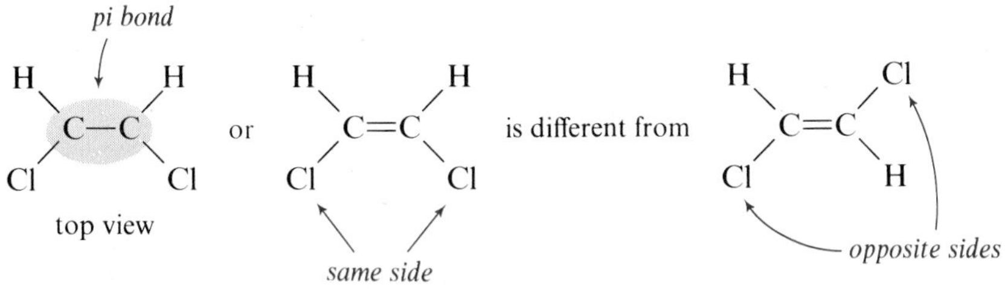

FIGURE 4.1. The groups attached to sp^2 carbons are fixed in relation to one another.

In Figure 4.1, we show a structure with two Cl atoms (one on each sp^2 carbon) on one side of the pi bond and two H atoms on the other side. Because the double bond is rigid, this molecule is not readily interconvertible with the compound in which the Cl atoms are on opposite sides of the pi bond.

Two groups on the *same side of the pi bond* are said to be ***cis*** (Latin, "on the side"). Groups on the *opposite sides* are said to be ***trans*** (Latin, "across"). Note how the *cis* or *trans* designation is incorporated into the name.

cis-1,2-dichloroethene
bp 60°

trans-1,2-dichloroethene
bp 48°

The *cis*- and *trans*-1,2-dichloroethenes have different physical properties (such as boiling points); they are different compounds. However, these two compounds are not structural isomers because the order of attachment of the atoms and the location of the double bond are the same in each compound. This pair of isomers falls into the general category of **stereoisomers**: different compounds that have the same structure, differing only in the *arrangement of the atoms in space*. This pair of isomers falls into a more specific category of **geometric isomers** (also called ***cis, trans*-isomers**): stereoisomers that differ by groups being on the same side or on opposite sides of a site of rigidity in a molecule.

The requirement for geometric isomerism in alkenes is that each carbon atom involved in the pi bond have two different groups attached to it, such as H and Cl, or CH_3 and Cl. If one of the carbons of the double bond has two identical groups, such as two H atoms or two CH_3 groups, then geometric isomerism is not possible. (We urge you to make molecular models and verify for yourself this requirement of geometric isomerism.)

Geometric isomers:

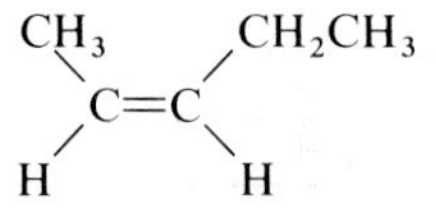

and

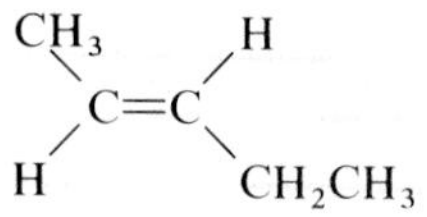

cis-2-pentene *trans*-2-pentene

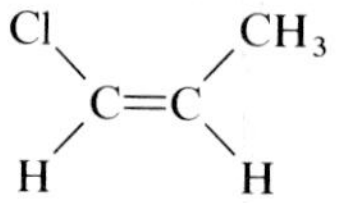

and

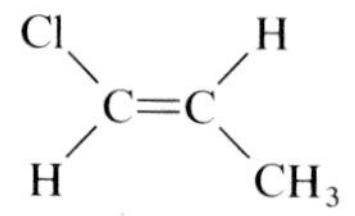

cis-1-chloro-1-propene *trans*-1-chloro-1-propene

Not geometric isomers:

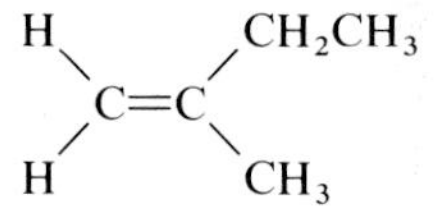

is the same as

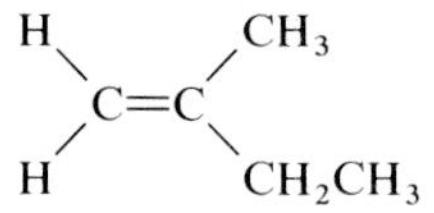

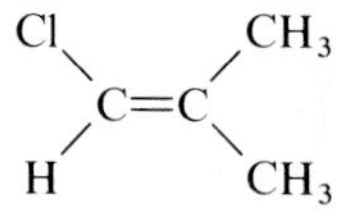

is the same as

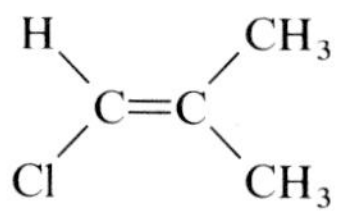

SAMPLE PROBLEMS

Label each of the following pairs of structures as *structural isomers* of each other, as *geometric isomers* of each other, or as the *same compound*:

(a)

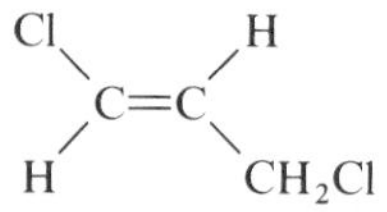

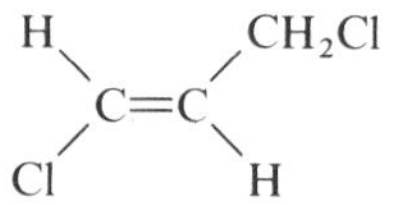

(b)

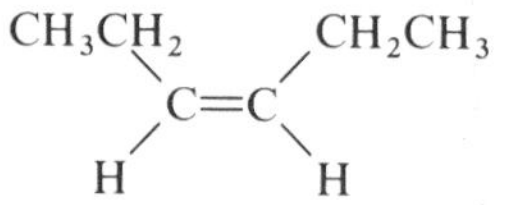

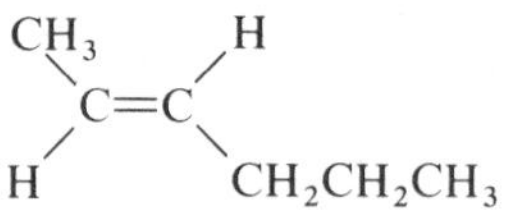

(c)

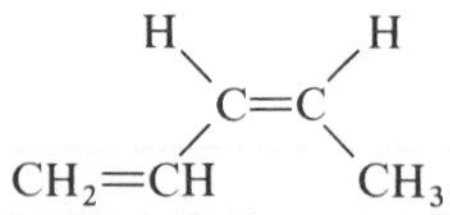

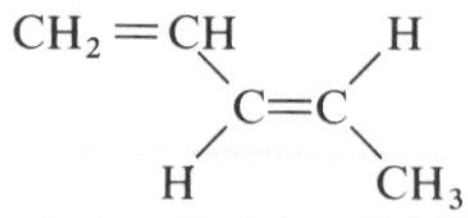

Solution:

(a) *same compound* (H's *trans* to each other in each);
(b) *structural isomers* (position of double bond is different; each structural isomer has a geometric isomer);
(c) *geometric isomers* (the first is *cis*; the second is *trans*).

Why does the dry-cleaning solvent trichloroethene ($Cl_2C=CHCl$) *not* have geometric isomers?

Solution: Exchanging any two groups in the structure does not give a different isomer. For geometric isomers to exist, there must be two different groups attached to each carbon of the double bond.

$$\begin{array}{c} Cl \\ \diagdown \\ \diagup \\ Cl \end{array} C=C \begin{array}{c} H \\ \diagup \\ \diagdown \\ Cl \end{array} \quad \text{is the same as} \quad \begin{array}{c} Cl \\ \diagdown \\ \diagup \\ Cl \end{array} C=C \begin{array}{c} Cl \\ \diagup \\ \diagdown \\ H \end{array}$$

Is geometric isomerism around a triple bond possible?

Solution: No. The groups attached to an *sp* carbon lie in a line. There is no "same side" or "opposite side."

A.   (*E*) and (*Z*) system of nomenclature

When there are three or four different groups attached to the carbon atoms of a double bond, a pair of geometric isomers exists, but it is sometimes difficult to assign *cis* or *trans* designations to the isomers.

$$\begin{array}{c} Br \\ \diagdown \\ \diagup \\ I \end{array} C=C \begin{array}{c} F \\ \diagup \\ \diagdown \\ Cl \end{array}$$

cis or *trans*?

   In our example, we can say that Br and Cl are *trans* to each other, or that I and Cl are *cis* to each other. However, we cannot name the structure in its entirety as being either the *cis* or the *trans* isomer. Because of the ambiguity in cases of this type, a more general system of isomer assignment has been devised, called the **(*E*) and (*Z*) system.** In practice, geometric isomers frequently are named by the *cis* and *trans* system if possible; the (*E*) and (*Z*) system is generally used only for those compounds that cannot be designated *cis* or *trans*.
   The (*E*) and (*Z*) system is based on an assignment of priorities (not to be confused with nomenclature priorities) to the atoms or groups attached to each carbon of the double bond. If the higher-priority atoms or groups are on *opposite sides* of the pi bond, the isomer is (*E*). If the higher-priority groups are on the *same side*, the isomer is (*Z*). (The letter *E* is from the German *entgegen*, "across"; the letter *Z* is from the German *zusammen*, "together.")
   If the two atoms on each double-bond carbon are different, priority is based on the atomic numbers of the single atoms directly attached to the double-bond carbons. *The atom with the higher atomic number receives a higher priority*. In our

example, I has a larger atomic number than does Br; I is of higher priority. On the other carbon of the double bond, Cl is of higher priority than F.

	F	Cl	Br	I
atomic number:	9	17	35	53

increasing priority

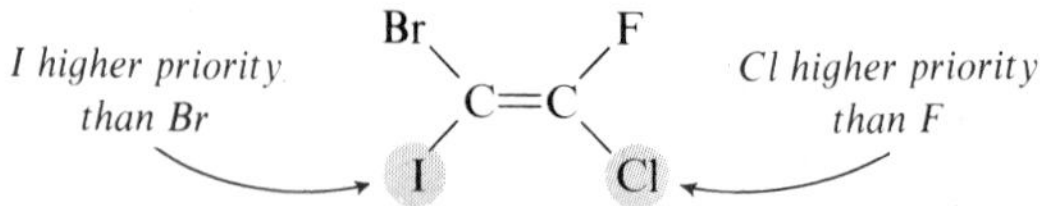

(Z)-1-bromo-2-chloro-
2-fluoro-1-iodoethene

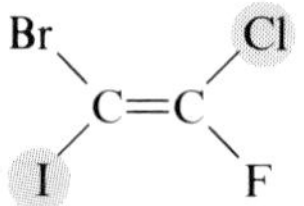

(E)-1-bromo-2-chloro-
2-fluoro-1-iodoethene

SAMPLE PROBLEMS

Is the following structure *(E)* or *(Z)*?

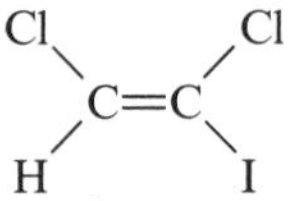

Solution: On one carbon of the double bond, the Cl has a higher priority than H. On the other carbon, I is of higher priority than Cl. The higher-priority atoms are on opposite sides; therefore, the isomer is *(E)*. Its name is *(E)*-1,2-dichloro-1-iodoethene.

Name the following compound by the *(E)* and *(Z)* system:

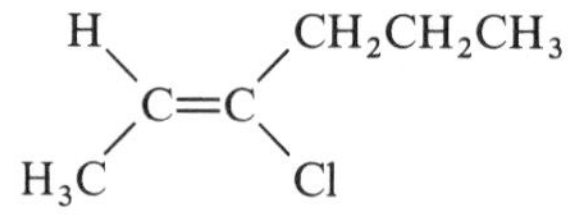

Solution: The left-hand carbon of the double bond has an H and a C attached to it; the C has the higher priority. The right-hand carbon has a Cl and a C attached to it; the Cl has a higher priority. (Look at the *single atoms* directly attached to the double-bond carbon: C, and not the entire —CH₂CH₂CH₃ group.) The higher-priority atoms are on the same side. The compound is named *(Z)*-3-chloro-2-hexene.

STUDY PROBLEM

4.1.   Name the following compounds by the *(E)* and *(Z)* system:

(a)

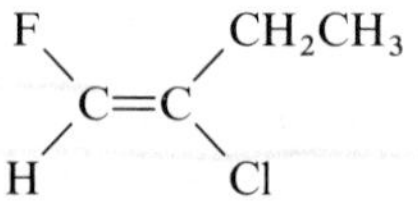

(b)

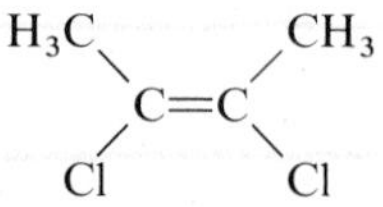

B.   Sequence rules

Determination of priorities by atomic number alone cannot handle all cases. For example, how would we name the following compound by the (*E*) and (*Z*) system?

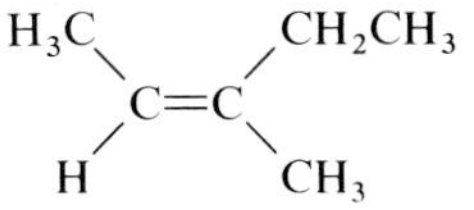

$$H_3C \diagdown \atop{C=C} \diagup {CH_2CH_3} $$

$$ H \diagup \qquad \diagdown CH_3 $$

To handle such a case, and others like it, a set of *sequence rules to determine order of priority* has been developed. These priority rules form the basis of the **Cahn–Ingold–Prelog nomenclature system,** named in honor of the chemists who developed the system.

✳   *Sequence rules for order of priority:*

1.   If the atoms in question are different, the sequence order is by atomic number, with the atom of highest atomic number receiving the highest priority.

F   Cl   Br   I

increasing priority

2.   If two isotopes of the same element are present, the *isotope of higher mass* receives the higher priority.

^1_1H, or H ^2_1H, or D
hydrogen   deuterium

increasing priority

3.   If two atoms are identical, the atomic numbers of the *next atoms* are used for priority assignment. If these atoms also have identical atoms attached to them, priority is determined at the first point of difference along the chain. The atom that has attached to it an atom of higher priority has the higher priority. (Do not use the sums of the atomic numbers, but look for the single atom of highest priority.)

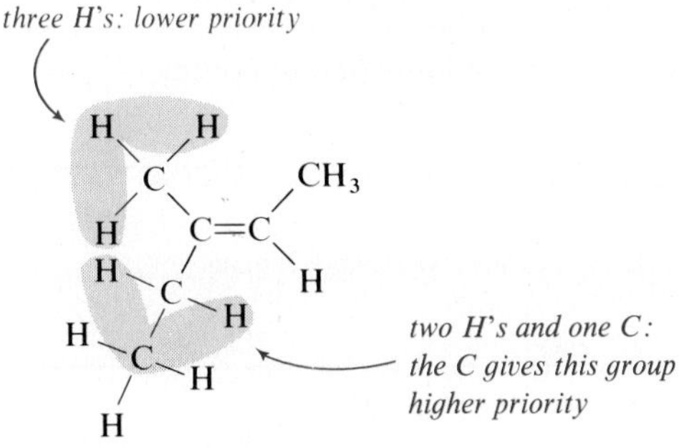

three H's: lower priority

*two H's and one C:
the C gives this group
higher priority*

*Cl gives this group
higher priority*

(*E*)-3-methyl-2-pentene

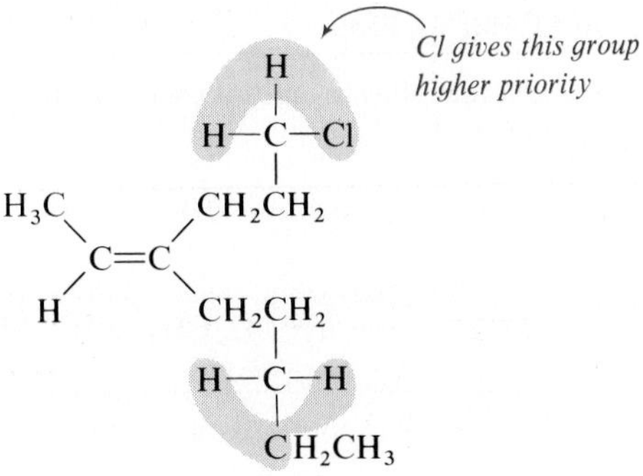

(*Z*)-6-chloro-3-pentyl-2-hexene

4.   Atoms attached by double or triple bonds are given single-bond *equivalencies* so that they may be treated like single-bonded groups in determining priority. Each doubly bonded atom is duplicated (or triplicated for triple bonds), a process better seen in examples.

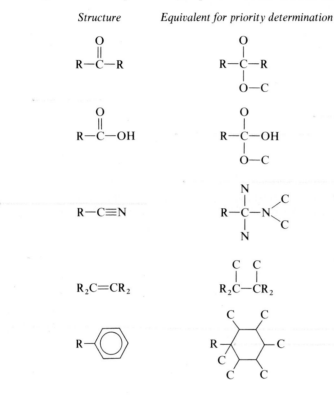

By this rule, we obtain the following priority sequence:

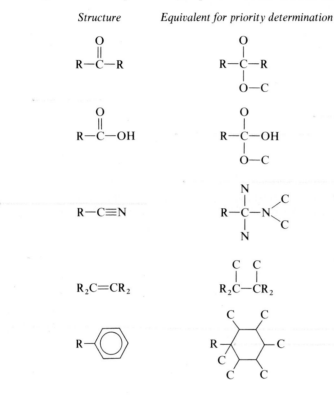

SAMPLE PROBLEMS

List the following atoms or groups in order of increasing priority (lowest priority first):

$$-NH_2 \quad -H \quad -CH_3 \quad -Cl$$

Solution: Increasing atomic number of the attached atom (N, H, C, Cl) gives increasing priority: H, CH_3, NH_2, Cl.

List the order of priority of the following groups (lowest to highest):

$$-CO_2H \quad -CO_2CH_3 \quad -CH_2OH \quad -OH \quad -H$$

Solution: Following the sequence rules:

$$-H \qquad -CH_2OH \qquad -CO_2H \qquad -CO_2CH_3 \qquad -OH$$

Which group is of higher priority, the *n*-butyl group or the isobutyl group?

Solution: The first carbon is the same in each (two H's and one C attached).

$$-CH_2CH_2CH_2CH_3 \qquad\qquad -CH_2CH(CH_3)_2$$

Therefore, we proceed to the second carbon and we find that the isobutyl group is of higher priority.

$$-CH_2\overset{\displaystyle H}{\underset{\displaystyle H}{C}}CH_2CH_3 \qquad\qquad -CH_2\overset{\displaystyle H}{\underset{\displaystyle CH_3}{C}}CH_3$$

The *n*-butyl group has one C and two H's at the first point of difference.	The isobutyl group has two C's and one H at the first point of difference.

STUDY PROBLEM

4.2. Tell whether each of the following compounds is (*E*) or (*Z*):

(a)
$$\overset{D}{}\diagdown\!\!\underset{H}{}\diagup C=C\overset{H}{}\diagup\!\!\underset{D}{}\diagdown$$

(b)
$$\overset{H_3C}{}\diagdown\!\!\underset{H}{}\diagup C=C\overset{CH_2CH_3}{}\diagup\!\!\underset{CH_3}{}\diagdown$$

(c)
$$\overset{Cl}{}\diagdown\!\!\underset{H}{}\diagup C=C\overset{CH_2CH_3}{}\diagup\!\!\underset{CH(CH_3)_2}{}\diagdown$$

(d)
$$\overset{CH_3\overset{O}{\overset{\|}{C}}}{}\diagdown\!\!\underset{ClCH_2}{}\diagup C=C\overset{\overset{O}{\overset{\|}{C}}CH_2Cl}{}\diagup\!\!\underset{Cl}{}\diagdown$$

SECTION 4.2.

Geometric Isomerism in Cyclic Compounds

We have seen how restricted rotation around a double bond can lead to geometric isomerism. Let us now consider restricted rotation in cyclic compounds.

Atoms joined in a ring are not free to rotate around the sigma bonds of the ring. Rotation around the ring sigma bonds would require that attached atoms or groups pass through the center of the ring, and van der Waals repulsions prevent this from happening unless the ring contains ten or more carbon atoms. The most common rings in organic compounds are five- and six-membered rings; therefore, we will concentrate our discussion on rings of six and fewer carbon atoms.

groups cannot
rotate completely
around ring bonds CH_2CH_3 *a group can*
rotate completely
around this bond

H

For the moment, we will assume that the carbon atoms of a cyclic structure such as cyclohexane form a plane. (While this is not strictly correct, as we will show later in this chapter, it is often convenient to assume that they do lie in a plane.) For the present discussion, we will view the plane of the ring as being almost horizontal. The edge of the ring projected toward us is shaded more heavily.

away from viewer

toward viewer

Each carbon atom in the cyclohexane ring is joined to its neighboring ring carbon atoms and also to two other atoms or groups. The bonds to these two other groups are represented by vertical lines. A group attached to the top of a vertical line is said to be *above the plane of the ring*, and the group attached to the bottom of a vertical line is said to be *below the plane of the ring*.

CH_3 ——— *above the plane*

OH ←——— *below the plane*

In this symbolism, hydrogen atoms attached to the ring and their bonds are not always shown.

CH_3 *is the same as*   H   H
OH   H CH_3
                                              H   H
                                              H   H
                                              H   OH

Another way of showing how groups are attached to the ring is by using a broken wedge to indicate a group below the plane of the ring, and a solid line bond or a wedge to represent a group above the plane.

above the plane

CH_3 H_3C

OH   HO

below the plane

Descriptions of substituents as being "above the plane" or "below the plane" are correct for only a particular representation of a structure. A molecule can be flipped over in space and the descriptions reversed.

The important point is that, in all of the preceding formulas, the methyl group and the hydroxyl group are on *opposite sides* of the plane of the ring. When the two groups are on opposite sides of the ring, they are *trans*; when they are on the same side, they are *cis*. These designations are directly analogous to *cis* and *trans* in alkenes. The *cis*- and *trans*-compounds are geometric isomers of each other, just as *cis*- and *trans*-alkenes are.

trans-2-methyl-1-cyclohexanol   *cis*-2-methyl-1-cyclohexanol

SAMPLE PROBLEM

Tell whether each of the following compounds is *cis*, *trans*, or neither:

Solution: (a) *trans*; (b) neither (because the benzene ring and its substituents are in the same plane, the substituents cannot be up or down); (c) *cis*.

STUDY PROBLEM

4.3. Draw formulas for the geometric isomers of 2-isopropyl-5-methyl-1-cyclohexanol (commonly called *menthol*), which is used in cigarettes and throat lozenges.

SECTION 4.3.

Conformations of Open-Chain Compounds

In open-chain compounds, groups attached by sigma bonds can rotate around these bonds. Therefore, the atoms in an open-chain molecule can assume an infinite number of positions in space relative to one another. Ethane is a small molecule, but even ethane can assume different arrangements in space, called **conformations**.

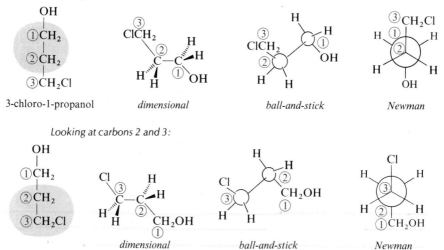

FIGURE 4.2. A dimensional formula, a ball-and-stick formula, and a Newman projection of ethane.

To represent conformations, we will use three types of formulas: **dimensional formulas, ball-and-stick formulas**, and **Newman projections**. (We suggest that you also use models to compare different conformations.) A ball-and-stick formula and a dimensional formula are three-dimensional representations of the molecular model of a compound (see Figure 4.2). A Newman projection is an end-on view of only *two carbon atoms* in the molecule. The bond joining these two carbons is hidden. The three bonds attached to the front carbon appear to go to the center of the projection, and the three bonds of the rear carbon are only partially shown.

bonds from the front carbon   *bonds from the rear carbon*   *Newman projection for CH_3CH_3*

Newman projections may be drawn for molecules with more than two carbon atoms. Because only two carbon atoms at a time can be shown in the projection, more than one Newman projection may be drawn for a molecule. For example, we can show two Newman projections for 3-chloro-1-propanol.

Looking at carbons 1 and 2:

3-chloro-1-propanol   *dimensional*   *ball-and-stick*   *Newman*

Looking at carbons 2 and 3:

dimensional   *ball-and-stick*   *Newman*

Because of rotation around its sigma bonds, a molecule can assume any number of conformations. However, certain conformations are more stable than others. These preferred conformations are called **conformers**. Conformers are not isomers because they are interconvertible—they are only different spatial orientations of the same molecule.

In our formulas of ethane and 3-chloro-1-propanol, we have shown **staggered conformers,** in which the hydrogen atoms or the attached groups are as far apart from one another as possible. Because the C—C bond can undergo rotation, the hydrogen atoms might also be **eclipsed,** or as close as possible, one behind the other in the Newman projection. We will show them not quite eclipsed so they can be seen.

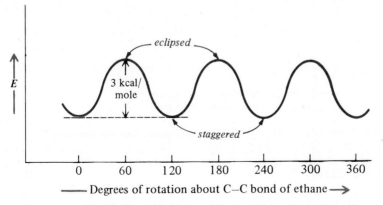

The rotation around sigma bonds is often called **free rotation**, but it is not entirely free. The eclipsed conformation of ethane is about 3 kcal/mole less stable (of higher energy) than the staggered conformer because of minor repulsions between the bonding electrons to the hydrogen atoms. To undergo rotation from a staggered to an eclipsed conformation, a mole of ethane molecules would require 3 kcal. Since this amount of energy is readily available to molecules at room temperature, the rotation can occur easily; this is why the different conformations are not isomers. However, even though the conformations of ethane are interconvertible at room temperature, at any given time we would expect a greater percentage of ethane molecules to be in the staggered conformation because of its lower energy. A diagram showing the ebb and flow in potential energy with rotation around the C—C bond in ethane is presented in Figure 4.3.

Butane ($CH_3CH_2CH_2CH_3$), like ethane, can exist in eclipsed and staggered conformations. In butane, there are two methyl groups, which are relatively large, attached to the center two carbons. Viewing butane from the center two carbons, the presence of these methyl groups gives rise to two types of staggered conformations that differ in the positions of the methyl groups in relation to each other. The staggered conformation in which the methyl groups are at the maximum distance

FIGURE 4.3. Energy changes involved in rotation around the carbon–carbon sigma bond of ethane.

apart is called the ***anti*** conformer (Greek *anti*, "against"). The staggered conformations in which the methyl groups are closer are called *gauche* conformers (French *gauche*, "left" or "crooked"). Newman projections for one-half a complete rotation follow.

Partial rotation around the carbon 2–carbon 3 bond of butane (rear carbon rotating):

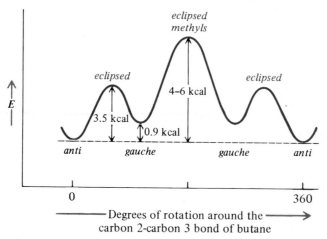

The larger the groups attached to two carbon atoms, the greater is the energy difference between the molecule's conformations. It takes more energy to push together two bulky groups than two small groups. While it takes only 3 kcal/mole for ethane to rotate from staggered to eclipsed, it takes 4–6 kcal/mole for butane to rotate from *anti* to the conformation in which the methyls are eclipsed. The energy relationships of the complete rotation around the carbon 2–carbon 3 bond of butane are shown in Figure 4.4.

FIGURE 4.4. Energy relations (in kcal/mole) of the different conformations of butane.

STUDY PROBLEM

4.4. Draw Newman projections for the *anti* and *gauche* conformations of
(a) 1-bromo-2-chloroethane, and   (b) 3-hydroxypropanoic acid.

SECTION 4.4.

Shapes of Cyclic Compounds

A.   Ring strain

In 1885, Adolf von Baeyer, a German chemist, theorized that cyclic compounds form planar rings. Baeyer further theorized that all cyclic compounds except for cyclopentane would be "strained" because their bond angles are not close to the tetrahedral angle of 109.5°. He proposed that, because of the abnormally small bond angles of the ring, cyclopropane and cyclobutane would be more reactive than an open-chain alkane. According to Baeyer, cyclopentane would be the most stable ring system (because its bond angles are closest to tetrahedral), and then reactivity would increase again starting with cyclohexane.

Bond angles according to Baeyer:

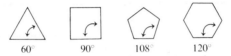

| 60° | 90° | 108° | 120° |

Baeyer's theory was not entirely correct. Cyclohexane and larger-sized rings are not more reactive than cyclopentane. We now know that cyclohexane is not a flat ring with bond angles of 120°, but rather a puckered ring with bond angles close to 109°, the normal sp^3 bond angles.

bond angles $\sim 109°$

However, there is indeed what we call **ring strain** in the smaller ring systems. Cyclopropane is the most reactive of the cycloalkanes. Its heat of combustion is higher per CH_2 group than that of other alkanes (Table 4.1; also see Section 3.4D).

TABLE 4.1. Strain energies from heat of combustion data

	$-\Delta H$ per $CH_2{}^a$	Strain energy per $CH_2{}^b$	Strain energy, total^c
cyclopropane	167 kcal/mole	10 kcal/mole	30 kcal/mole
cyclobutane	164	7	28
cyclopentane	159	2	10
cyclohexane	157	0	0

^{a} $-\Delta H$/mole divided by number of CH_2 groups.

^{b} The difference between (1) the value of $-\Delta H/CH_2$ for that compound, and (2) the value for cyclohexane, the assumption being that cyclohexane is not strained.

^{c} Strain energy/CH_2 × number of CH_2 groups.

When treated with hydrogen gas, cyclopentane does not react, but cyclopropane undergoes ring opening.

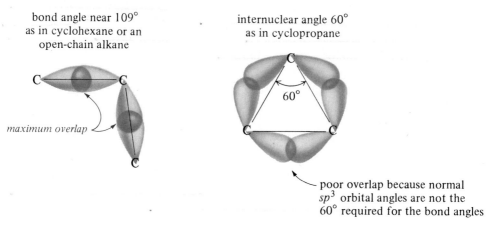

cyclopentane

cyclopropane   propane

Today we would say that the sp^3 orbitals of the carbon atoms in cyclopropane cannot undergo complete overlap with each other because the angles between the carbon atoms of cyclopropane are geometrically required to be 60° (see Figure 4.5). The ring sigma bonds of cyclopropane are of higher energy than sp^3 sigma bonds that have the normal tetrahedral angle. The cyclopropane bonds are more easily broken than most other C—C sigma bonds and, in comparable reactions, more energy is released.

bond angle near 109°
as in cyclohexane or an
open-chain alkane

internuclear angle 60°
as in cyclopropane

60°

maximum overlap

poor overlap because normal
sp^3 orbital angles are not the
60° required for the bond angles

FIGURE 4.5. Maximum overlap cannot be achieved between the ring carbon atoms in cyclopropane.

Cyclobutane is less reactive than cyclopropane but more reactive than cyclopentane. Following along reasonably with Baeyer's theory, the cyclopentane ring is stable and is far less reactive than the three- and four-membered rings.

With cyclohexane and larger rings, Baeyer's predictions fail. Cyclohexane and rings larger than cyclohexane are found in puckered conformations rather than as flat rings, and are not particularly reactive. Larger rings are not commonly found in naturally occurring compounds, as are five- and six-membered rings. Baeyer felt the rarity was due to ring strain. We now realize that the rarity of larger rings is not due primarily to unusually high bond energies. Instead, the scarcity of these compounds arises from the decreasing probability that the ends of longer molecules will find each other to undergo reaction and form a ring. (The problem is one of *entropy*, or randomness, and not one of *enthalpy*, or energy change.)

SAMPLE PROBLEM

Would you expect the benzene ring to be flat or puckered?

Solution: The benzene ring is flat because the carbon atoms are sp^2 (not sp^3) hybridized. The normal positioning of sp^2 bonds is planar and with angles of 120° between them, corresponding to a regular planar hexagon.

STUDY PROBLEM

4.5.   Considering ring sizes, which of the following compounds would you expect to suffer from substantial amounts of ring strain?

(a) 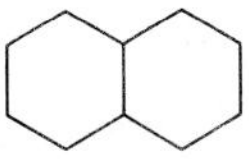
decalin

(b) 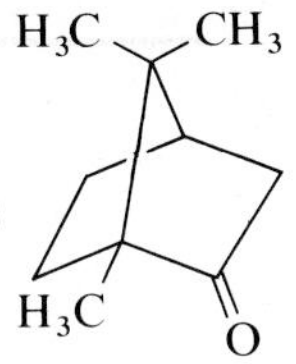 H_3C CH_3 H_3C O
camphor

(c) 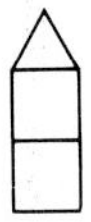
apartmenthousane

(d) 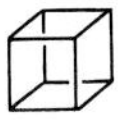
cubane

B.   Ring puckering and hydrogen–hydrogen repulsions

If the cyclohexane ring were flat, all the hydrogen atoms on the ring carbons would be eclipsed. In the puckered conformer that we have shown, however, all the hydrogens are staggered. The energy of this puckered conformer of cyclohexane is lower than the energy of flat cyclohexane, both because of more-favorable sp^3 bond angles and also because of fewer hydrogen–hydrogen repulsions.

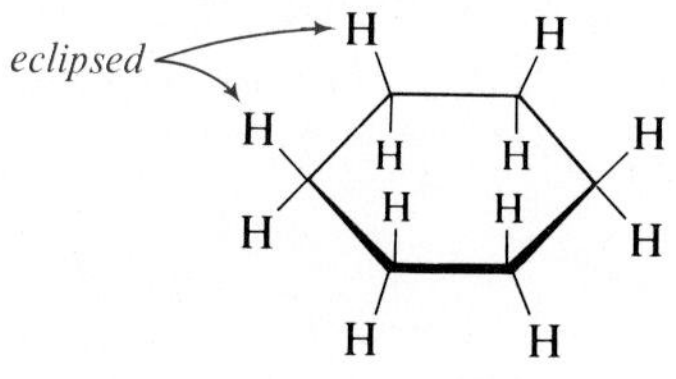

flat

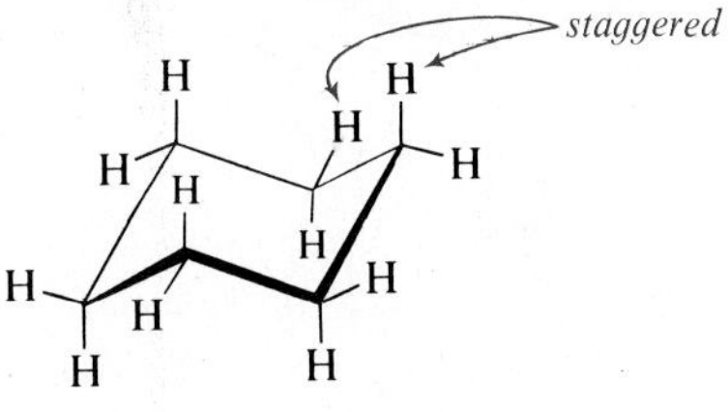

puckered

What of the other cyclic compounds? Cyclopentane would have near-optimal bond angles (108°) if it were flat, but cyclopentane also is slightly puckered so that the hydrogen atoms attached to the ring carbons are staggered. Cyclobutane (flat bond angles of 90°) also is puckered, even though the puckering causes more-strained bond angles. Cyclopropane must be planar; geometrically, three points (or three carbon atoms) define a plane. The hydrogen atoms in cyclopropane necessarily are eclipsed.

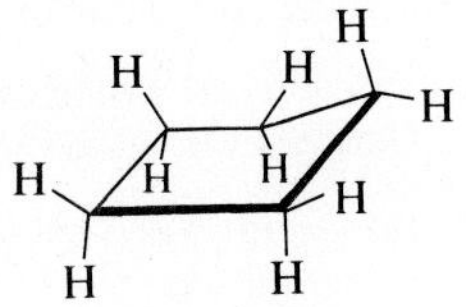

envelope form
of cyclopentane

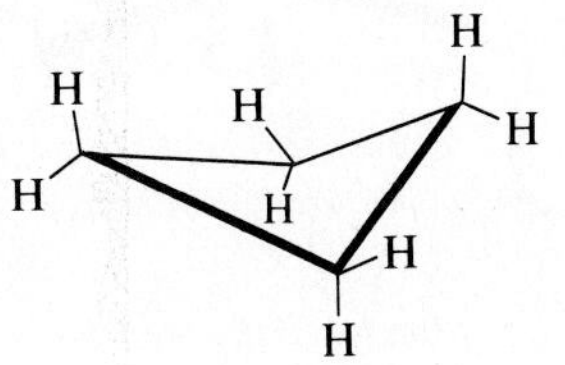

butterfly form
of cyclobutane

The Conformers of Cyclohexane

The cyclohexane ring, either alone or in fused-ring systems (ring systems that jointly share carbon atoms), is the most important of all the ring systems. In this section, we will study the conformations of cyclohexane and substituted cyclohexanes. In Chapter 20, we will discuss the conformations of fused-ring systems.

There are many shapes that a cyclohexane ring can assume, and any single cyclohexane molecule is in a continuous state of flexing into different shapes. (Molecular models are invaluable for showing the relationships among the various conformations.) So far, we have shown the **chair form** of cyclohexane. Some other shapes the cyclohexane molecule can assume are as follows:

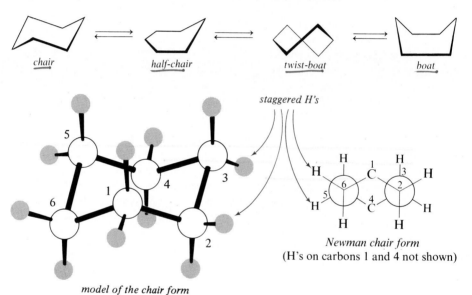

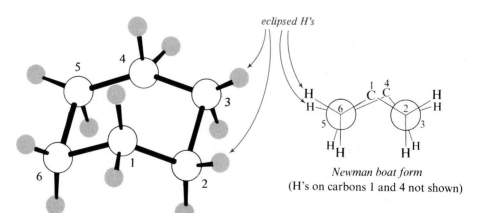

FIGURE 4.6. Molecular models and Newman projections of the chair and boat forms of cyclohexane.

None of these other conformations has the <u>favorable, staggered-hydrogen</u> <u>structure of the chair form</u>. The eclipsing of hydrogens, as in the boat form, adds to the energy of the molecule. Figure 4.6 shows models and Newman projections of the chair form and boat form; the staggered and eclipsed hydrogens are apparent in these representations.

The energy requirements for the interconversion of the different conformations of cyclohexane are shown in Figure 4.7. We can see that the <u>chair form has</u> <u>the lowest energy</u>, while the <u>half-chair (which has an almost-planar structure) has</u> <u>the highest energy.</u> At any given time, we would expect most cyclohexane molecules to be in the chair form. Indeed, it has been calculated that about 99.9% of cyclohexane molecules are in the chair form at any one time.

Equatorial and axial substituents.   The carbon atoms of the chair form of cyclohexane roughly form a plane. For purposes of discussion, an axis may be drawn perpendicular to this plane. These operations are shown in Figure 4.8.

Each ring carbon of cyclohexane is bonded to two hydrogen atoms. The bond to one of these hydrogens is in the <u>rough plane of the ring</u>; this hydrogen atom is called an **equatorial hydrogen**. The bond to the other hydrogen <u>atom is parallel to</u> the axis; this is an **axial hydrogen**. Each of the six carbon atoms of cyclohexane

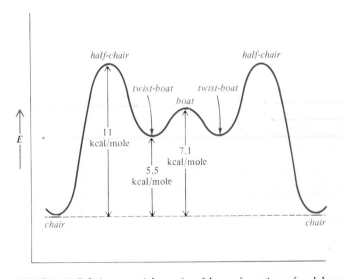

FIGURE 4.7. Relative potential energies of the conformations of cyclohexane.

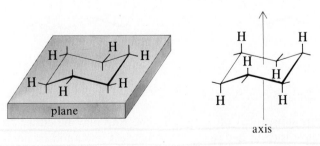

equatorial H's (in plane)   axial H's (parallel to axis)

FIGURE 4.8. The equatorial and axial hydrogens of cyclohexane.

has one equatorial and one axial hydrogen atom. (Again, refer to Figure 4.8.) In the flipping and reflipping between the conformers, <u>axial becomes equatorial,</u> while equatorial becomes axial.

A <u>methyl group is bulkier than a hydrogen atom</u>. When the methyl group in <u>methylcyclohexane is in the axial position</u>, the <u>methyl group and axial hydrogens</u> <u>on the same side of the ring repel each other</u>. Interactions between axial groups are called **axial–axial interactions**. When the methyl group is in the equatorial position, the repulsions are minimized. Thus, the energy of the conformer with an equatorial methyl is lower. At room temperature, about 95% of methylcyclohexane molecules are in the conformation in which the methyl group is equatorial.

The <u>bulkier the group,</u> the greater is the <u>energy difference between axial</u> <u>and equatorial conformers</u>. In other words, a cyclohexane ring with a bulky substituent is <u>more likely to have that group in the equatorial position</u>. When the size of the substituent group reaches *t*-butyl, the difference in energies between the conformers becomes quite large. *t*-Butylcyclohexane is often said to be "frozen" in the conformation in which the *t*-butyl group is equatorial. The ring is not truly frozen, but the energy difference (5.6 kcal/mole) between the equatorial and the axial positions of the *t*-butyl group means that only 1 in 10,000 molecules has the *t*-butyl group in an axial position at any given time.

favored

B.   Disubstituted cyclohexanes

Two groups substituted on a cyclohexane ring may be either *cis* or *trans*. The *cis*- and *trans*-disubstituted rings are geometric isomers and are not interconvertible at room temperature; however, either isomer may assume a variety of conformations. For example, consider some chair forms of *cis*-1,2-dimethylcyclohexane.

Some different representations of cis-1,2-dimethylcyclohexane,

both "down"   *both "down"*   *both "up"*

Because this is the *cis*-isomer, the methyl groups must both be on the same side of the ring, regardless of the conformation. In each chair conformation that we can draw, one methyl is axial and the other is equatorial. For any *cis*-1,2-disubstituted cyclohexane, one substituent must be axial and the other substituent must be equatorial. (Refer back to Figure 4.8 or use molecular models and verify this statement for yourself.)

When *cis*-1,2-dimethylcyclohexane changes from one chair conformer to the other, the two methyl groups reverse their equatorial–axial status. The energies of the two conformers are equal because their structures and bonding are identical. Therefore, this compound exists primarily as a 50 : 50 mixture of these two chair-form conformers.

Conformers of cis-1,2-dimethylcyclohexane:

axial, equatorial (or a,e)   *equatorial, axial (or e, a)*

In *trans*-1,2-dimethylcyclohexane, the methyl groups are on *opposite* sides of the ring. In the chair form of the *trans*-isomer, one group must be attached to an "uppermost" bond, while the other is attached to a "lowermost" bond.

Some different representations of trans-1,2-dimethylcyclohexane,

No matter how the two adjacent *trans* groups are shown, they are *both axial* (*a, a*) or they are *both equatorial (e, e)*. There is no way for two groups to be *trans* and 1,2 on the chair form of cyclohexane without assuming either the *a,a* or the *e,e* conformation.

Conformers of trans-1,2-dimethylcyclohexane:

a,a   e,e
favored

A single methyl group on a cyclohexane ring assumes the equatorial position preferentially. *Two* methyl groups on a cyclohexane ring also assume the equatorial positions preferentially, if possible. In *trans*-1,2-dimethylcyclohexane, the *e,e* conformer is the preferred conformer and is of lower energy than the *a,a* conformation. The *trans e,e* conformer is also of lower energy (by 1.87 kcal/mole) than either conformer of the *cis*-compound, which must be *a,e* or *e,a*.

In the case of a 1,2-disubstituted cyclohexane, the *trans*-isomer is more stable than the *cis*-isomer because both substituents can be equatorial. However, when the two substituents are 1,3 to each other on a cyclohexane ring, the *cis*-isomer is more stable than the *trans*-isomer. The reason is that both substituents in the *cis*-1,3-isomer can be equatorial. In the *trans*-1,3-isomer, one group must be axial.

cis-1,3-Dimethylcyclohexane:

a,a   e,e
favored

trans-1,3-Dimethylcyclohexane:

e,a   a,e

SAMPLE PROBLEM

What are the possible equatorial–axial relationships for the *cis*- and *trans*-1,4-dimethylcyclohexanes? In each case, which conformer is of lower energy?

Solution: cis-1,4: *a,e; e,a*
              trans-1,4: *a,a; e,e*

The *cis*-conformers are of equal energy. The *e,e trans*-conformer is of lower energy than the *a,a trans*-conformer (and of lower energy than either *cis*-conformer).

STUDY PROBLEM

4.6. Label each of the following disubstituted rings as *cis* or *trans* and as *a,a*; *e,e*; or *a,e*. (If the structure has no *cis* or *trans* isomers, indicate this.)

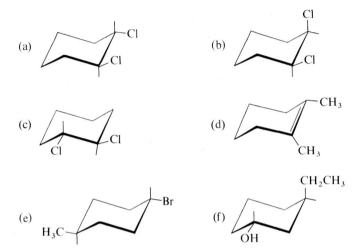

SECTION 4.6

Chirality

A.   Chirality of objects and molecules

Consider your left hand. Your hand <u>cannot be superimposed on its mirror image</u>. If you hold your left hand up to a mirror, the image looks like a right hand. If you do not have a mirror available, hold your hands together with the palms facing each other—you can see that they are mirror images. Try to superimpose your hands (both palms down)—you cannot do it (see Figure 4.9). This right- and

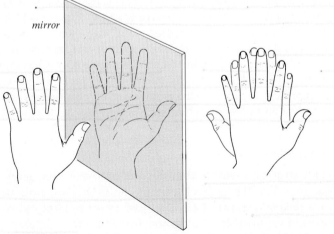

FIGURE 4.9. A *chiral* object *cannot* be superimposed on its mirror image.

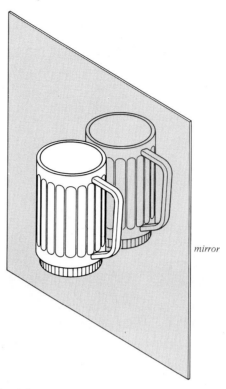

FIGURE 4.10. An *achiral* object *is* superimposable on its mirror image.

left-handedness is also encountered in shoes and gloves. (Try wearing a left-handed glove on a right hand!)

Any object that *cannot be superimposed on its mirror image* is said to be **chiral** (Greek *cheir*, "the hand"). Hands, gloves, and shoes are all chiral. Conversely, a plain cup or a cube is **achiral** (not chiral); these can be superimposed on their mirror images. Figure 4.10 shows a cup being superimposed on its mirror image. The same principles of right- and left-handedness also apply to molecules. A molecule that can be superimposed upon its mirror image is *achiral*. A molecule that *cannot* be superimposed on its mirror image is *chiral*. Figures 4.11 and 4.12 show an achiral molecule and a chiral molecule along with their mirror images.

An achiral molecule and its superimposable mirror-image molecule are the same compound; they are not isomers. But a chiral molecule is *not* superimposable on its mirror image; this molecule and its mirror-image molecule are different compounds, and represent a pair of stereoisomers called **enantiomers**. A pair of enantiomers is simply a pair of isomers that are *nonsuperimposable mirror images*.

B.   Chiral carbon atoms

The most common structural feature (but not the only one) that gives rise to chirality in molecules is that the molecule contains an sp^3 carbon atom with *four different groups* attached to it (Figure 4.12). Such a molecule is chiral and exists as a pair of enantiomers. For this reason, a carbon atom with four different groups attached is usually called a **chiral carbon atom** (although, technically, it is the mole-

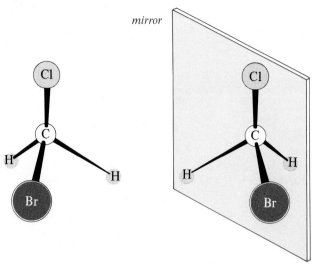

mirror image is the same compound

FIGURE 4.11. A molecule with a single carbon atom that has two identical substituents (H in this case) is achiral and can be superimposed upon its mirror image. (Try it with models.)

cule and not the carbon atom that is chiral). By learning to recognize chiral carbons within a formula, we can greatly simplify the problem of identifying structures that can exist as enantiomers.

In order to identify a chiral carbon, we must determine that all four groups attached to the sp^3 carbon are different. In many cases, the problem is trivial; for example, if the carbon is attached to two or more H atoms ($-CH_2-$ or $-CH_3$), then the carbon cannot be chiral. However, in a few cases the problem can be more

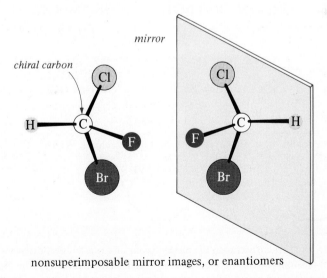

nonsuperimposable mirror images, or enantiomers

FIGURE 4.12. A molecule that has four different groups attached to one single carbon atom is chiral and not superimposable on its mirror image. (Try it with models.)

challenging. In these cases we must inspect each *entire group* attached to the carbon in question, not just the atoms bonded directly to that carbon.

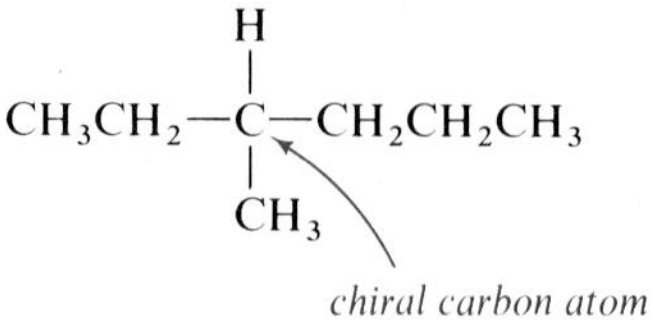

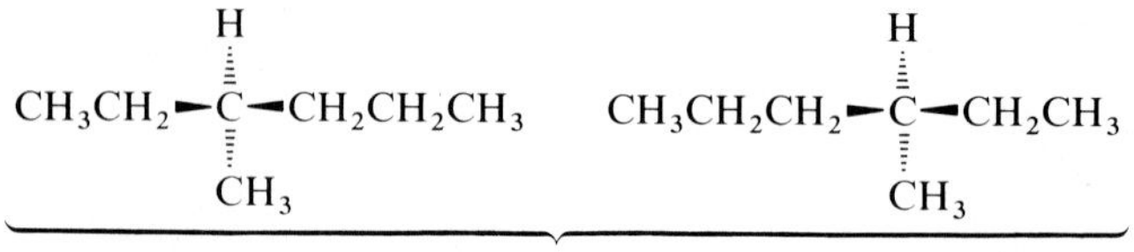

$$CH_3CH_2-\overset{\overset{\displaystyle H}{|}}{\underset{\underset{\displaystyle CH_3}{|}}{C}}-CH_2CH_2CH_3$$

chiral carbon atom

$$CH_3CH_2-\overset{\overset{\displaystyle H}{\vdots}}{\underset{\underset{\displaystyle CH_3}{|}}{C}}-CH_2CH_2CH_3 \qquad CH_3CH_2CH_2-\overset{\overset{\displaystyle H}{\vdots}}{\underset{\underset{\displaystyle CH_3}{|}}{C}}-CH_2CH_3$$

enantiomers

dimensional formulas

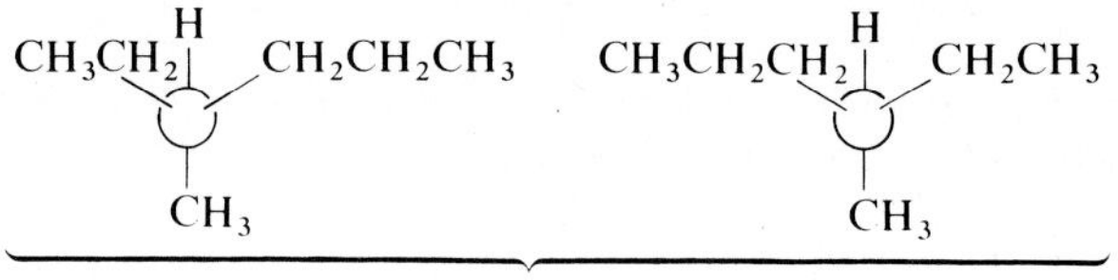

$$CH_3CH_2 \overset{\displaystyle H}{|} \quad CH_2CH_2CH_3 \qquad CH_3CH_2CH_2 \overset{\displaystyle H}{|} \quad CH_2CH_3$$

$$CH_3 \qquad\qquad CH_3$$

enantiomers

ball-and-stick formulas

SAMPLE PROBLEMS

Chiral carbon atoms are often starred for emphasis. Star the chiral carbons in the following structures:

$$CH_3\overset{\overset{\displaystyle Cl}{|}}{\underset{\underset{\displaystyle Br}{|}}{C}}CH_2CH=CHCH_3 \qquad CH_3\overset{\overset{\displaystyle O}{\|}}{\underset{\underset{\displaystyle CH_2CH_3}{|}}{C}H}COCH_3$$

Solution: $CH_3\overset{*}{\overset{\overset{\displaystyle Cl}{|}}{\underset{\underset{\displaystyle Br}{|}}{C}}}CH_2CH=CHCH_3 \qquad CH_3\overset{*}{\underset{\underset{\displaystyle CH_2CH_3}{|}}{C}H}\overset{\overset{\displaystyle O}{\|}}{C}OCH_3$

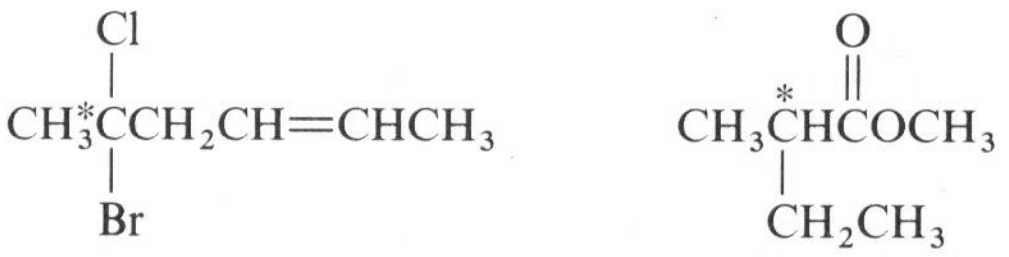

One of the following molecules is chiral, and one is achiral. Which is the chiral molecule?

(a) 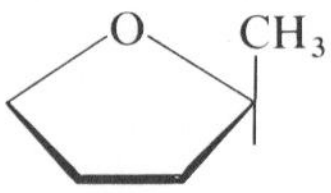   (b)

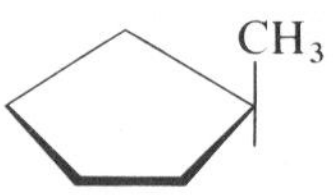

Solution: The ring carbon with the methyl group in (a) is a chiral carbon because it has four different groups attached: CH_3, H, O, and CH_2; therefore, the structure in (a) is chiral. The corresponding carbon in (b) has two identical groups; it is achiral.

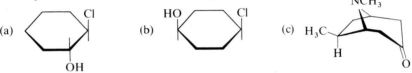

four different groups: the two attachments with arrows
this molecule is chiral   are the same; this molecule is achiral

STUDY PROBLEMS

4.7.   Star the chiral carbons (if any) in the following formulas:

(a) CH_3CHBr
with CH_3 above

(b) CH_3CH_2CHBr
with CH_3 above

(c) benzene ring—CH_2CH—benzene ring
with F above

4.8.   Star any chiral carbon atoms in the following formulas:

(a) ring with Cl and OH

(b) ring with HO and Cl

(c) H_3C— ring with NCH_3, H, and O

Drawing structures of a pair of enantiomers for a molecule with one chiral carbon is relatively easy. The *interchange of any two groups* around the chiral carbon results in the enantiomer. The following examples show two ways to convert the formula for one compound (**A**) to the formula for its enantiomer (**B**). (Use models to verify that the two formulas shown for **B** actually do represent the same molecule.)

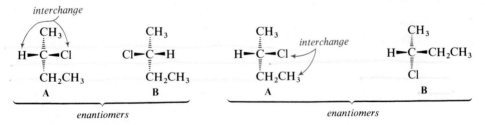

enantiomers   *enantiomers*

STUDY PROBLEM

4.9.   Draw the formulas (either dimensional or ball-and-stick) for the two enantiomers of each of the following compounds:

(a) benzene ring—$CHCH_3$
with Br above

(b) $CH_3CH_2CHCH_3$
with OH above

C. Fischer projections

In the late 1800's, the German chemist Emil Fischer introduced projection formulas for showing the spatial arrangement of groups around chiral carbon atoms. These projection formulas are called **Fischer projections**. Because Fischer devised these formulas for representing sugar molecules, we will illustrate the type of Fischer projection commonly used today with the simplest of sugars: 2,3-dihydroxypropanal (usually called glyceraldehyde) and 2,3,4-trihydroxybutanal (erythrose). Glyceraldehyde has one chiral carbon atom (carbon 2), while erythrose has *two* chiral carbons (carbons 2 and 3). As you can see from the following representations for these two compounds, a Fischer projection is simply a shorthand way to represent a ball-and-stick or dimensional formula.

For glyceraldehyde:

$$
\begin{array}{ccc}
\text{①CH} & \text{CH} & \text{CH} \\
\text{H——OH} & \text{H——C——OH} & \text{H——OH} \\
\text{②} & & \\
\text{③CH}_2\text{OH} & \text{CH}_2\text{OH} & \text{CH}_2\text{OH}
\end{array}
$$

or   becomes

Fischer projection

For erythrose:

$$
\begin{array}{ccc}
\text{①CH} & \text{CH} & \text{CH} \\
\text{H——OH} & \text{H——C——OH} & \text{H——OH} \\
\text{②} & & \\
\text{H——OH} & \text{H——C——OH} & \text{H——OH} \\
\text{③} & & \\
\text{④CH}_2\text{OH} & \text{CH}_2\text{OH} & \text{CH}_2\text{OH}
\end{array}
$$

or   becomes

Fischer projection

In drawing a Fischer projection, we assume that the molecule is *completely stretched out in the plane of the paper with all its substituents eclipsed,* regardless of any preferred conformation. The preceding formulas for erythrose show the conformation used for the Fischer projection. By convention, the carbonyl group (or group of highest nomenclature priority) is placed at or near the top. Thus, *the top carbon is carbon 1.* Each intersection of the horizontal and vertical lines represents a chiral carbon atom. *Each horizontal line represents a bond coming toward the viewer,* while *the vertical line represents bonds going back, away from the viewer.*

away from viewer — CHO — *toward viewer*

$$
\begin{array}{c}
\text{CHO} \\
\text{H——OH} \\
\text{CH}_2\text{OH}
\end{array}
$$

A pair of enantiomers is easily recognized when Fischer projections are used.

$$
\begin{array}{cc}
\overset{1}{\text{CHO}} & \text{CHO} \\
\text{H}\overset{2}{——}\text{OH} & \text{HO——H} \\
\text{H}\overset{3}{——}\text{OH} & \text{HO——H} \\
\text{HO}\overset{4}{——}\text{H} & \text{H——OH} \\
\overset{5}{\text{CH}_2\text{OH}} & \text{CH}_2\text{OH}
\end{array}
$$

Carbons 2, 3, and 4 are chiral.   *The enantiomer shows all groups on chiral carbons transposed from left to right.*

A Fischer projection may be rotated 180° in the plane of the paper, but it may not be flipped over or rotated by any other angle. Either of these last two operations would take the formula out of the Fischer projection and lead to an incorrect structure.

Correct:

$$\text{H} \underset{\text{CH}_2\text{OH}}{\overset{\text{CHO}}{\underset{|}{\overset{|}{\text{—}}}}\text{OH}} \quad \xrightarrow{\text{rotate } 180°} \quad \text{HO} \underset{\text{CHO}}{\overset{\text{CH}_2\text{OH}}{\underset{|}{\overset{|}{\text{—}}}}\text{H}}$$

same as left-hand structure

Incorrect:

$$\text{H} \underset{\text{CH}_2\text{OH}}{\overset{\text{CHO}}{\underset{|}{\overset{|}{\text{—}}}}\text{OH}} \quad \xrightarrow{\text{rotate } 90°} \quad \text{HOCH}_2 \underset{\text{OH}}{\overset{\text{H}}{\underset{|}{\overset{|}{\text{—}}}}\text{CHO}}$$

NOT THE SAME
(To verify the difference, draw
this projection as a dimensional
formula with the horizontal bonds
coming toward you.)

Fischer projections are a convenient shorthand way of representing chiral molecules. Because of the limitations placed upon them, such as the preceding one of rotation, Fischer projections must be used carefully. We suggest that you convert Fischer projections to ball-and-stick or dimensional formulas (or use models) when performing any spatial manipulations. In future chapters, we will stress dimensional formulas rather than Fischer projections until we discuss carbohydrates (Chapter 18).

STUDY PROBLEM

4.10. (a) Convert the following Fischer projection to a dimensional formula:

$$\text{HO} \underset{\text{CH}_2\text{OH}}{\overset{\text{CO}_2\text{H}}{\underset{|}{\overset{|}{\text{—}}}}\text{H}}$$

(b) Convert the following dimensional formula for alanine (an amino acid found in proteins) to a Fischer projection:

$$\text{CH}_3\overset{\overset{\textstyle \text{NH}_2}{|}}{\underset{\underset{\textstyle \text{H}}{|}}{\text{C}}}\text{CO}_2\text{H}$$

2-aminopropanoic acid
(alanine)

Rotation of Plane-Polarized Light

With the exception of chirality, the structures of a pair of enantiomers are the same. Therefore, almost all of their physical and chemical properties are the same. For example, each pure enantiomer of a pair has the same melting point and the same boiling point. Only two sets of properties are different for a single pair of enantiomers:

1.   interactions with other chiral substances, and

2.   the direction of rotation of the plane of polarization of plane-polarized light.

Ordinary light travels in waves, and the waves are at right angles to the direction of travel. **Plane-polarized light** is light in which all wave vibrations have been filtered out except for those in one plane. The plane polarization is effected by passing ordinary light through a pair of calcite crystals ($CaCO_3$) or a polarizing lens. (The same principle is used in Polaroid sunglasses.) Figure 4.13 shows a simplified diagram of the plane polarization of light.

If plane-polarized light is passed through a solution containing a single enantiomer, the plane of polarization of the light is *rotated* either to the right or to the left (see Figure 4.14). The rotation of plane-polarized light is referred to as **optical rotation**. A compound that rotates the plane of polarization of plane-polarized light is said to be **optically active**. For this reason, enantiomers are sometimes referred to as **optical isomers**.

A **polarimeter** is an instrument designed for polarizing light and then showing the angle of rotation of the plane of polarization of the light by an optically active compound. The amount of rotation depends on (1) the structure of the molecules; (2) the temperature; (3) the wavelength; and (4) the number of molecules in the path of the light.

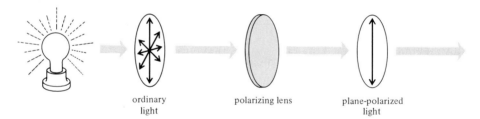

ordinary   polarizing lens   plane-polarized
light   light

FIGURE 4.13.   The plane polarization of light.

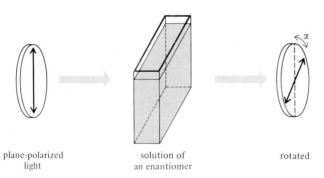

plane-polarized   solution of   rotated
light   an enantiomer

FIGURE 4.14.   The plane of polarization of plane-polarized light is rotated by a single enantiomer.

Specific rotation is the amount of rotation by 1.00 gram of sample in 1.00 mL of solution in a tube with a path length of 1.00 decimeter (dm), at a specified temperature and wavelength. The commonly used wavelength is 589.3 nm (the D line of sodium), where 1.0 nm = 10^{-9} m. The specific rotation for a compound (at 20°, for example) may be calculated from the observed rotation by the following formula:

$$[\alpha]_D^{20} = \frac{\alpha}{lc} \quad \text{✳}$$

where $[\alpha]_D^{20}$ = specific rotation of sodium D line at 20°
α = observed rotation at 20°
l = cell length in dm (1.0 dm = 10 cm)
c = concentration of sample solution in g/mL

The fact that some compounds rotate the plane of polarization of plane-polarized light was discovered in 1815 by the French physicist Jean-Baptiste Biot. But it was Louis Pasteur who in 1848 made the momentous discovery that there are two types of sodium ammonium tartrate crystals and that these two types are mirror images of each other. (We will discuss the structure of tartaric acid in Section 4.9C.)

Pasteur painstakingly separated the "left-handed" crystals from the "right-handed" crystals with a pair of tweezers. Imagine his amazement when he found that (1) a solution of the original mixture of crystals did not rotate plane-polarized light; (2) a solution of left-handed crystals did rotate plane-polarized light; and (3) a solution of right-handed crystals also rotated plane-polarized light to exactly the same extent, but in the opposite direction. Pasteur's experiment and later experiments of innumerable other scientists lead us to the following conclusions:

1.   A pair of pure enantiomers each rotates the plane of polarization of plane-polarized light the same number of degrees, but in the opposite directions (one to the left; one to the right).

2.   A mixture of equal parts of a pair of enantiomers does not rotate plane-polarized light.

A.   Some terms used in discussing optical rotation

The enantiomer of any enantiomeric pair that rotates plane-polarized light to the *right* is said to be **dextrorotatory** (Latin *dexter*, "right"). Its mirror image, which rotates plane-polarized light to the *left*, is said to be **levorotatory** (Latin *laevus*, "left"). The direction of rotation is specified in the name by (+) for dextrorotatory and (−) for levorotatory. (In older literature, *d* for dextrorotatory and *l* for levorotatory are sometimes encountered.)

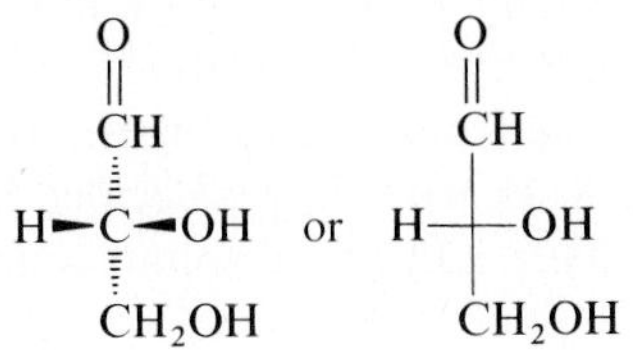

(+)-glyceraldehyde
$[\alpha]_D^{20} = +8.7°$

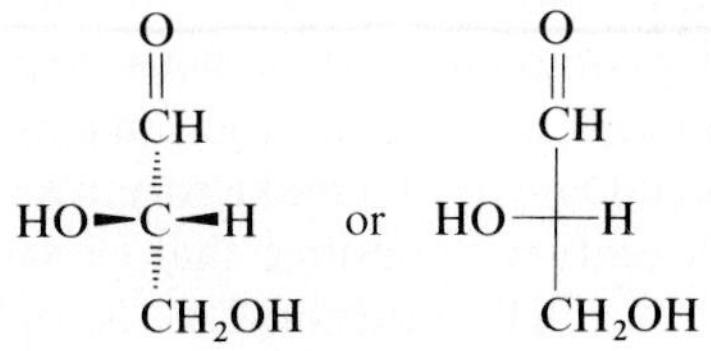

(−)-glyceraldehyde
$[\alpha]_D^{20} = -8.7°$

A mixture of equal parts of any pair of enantiomers is called a **racemic mixture**, or **racemic modification**. A racemic mixture may be indicated in the name by the prefix ($\pm$). (In older literature, *dl* is used to signify a racemic mixture.) Thus, racemic glyceraldehyde is called ($\pm$)-glyceraldehyde. (The term *racemic* comes from the Latin *racemis*, "a bunch of grapes." The reason for this unusual derivation is that *racemic* was first used to describe racemic tartaric acid, which was isolated as a by-product of wine-making.)

$$50\% \ (+)\text{-glyceraldehyde} + 50\% \ (-)\text{-glyceraldehyde} = (\pm)\text{-glyceraldehyde}$$
$$[\alpha]_D^{20} = +8.7° \qquad\qquad [\alpha]_D^{20} = -8.7° \qquad\qquad [\alpha]_D^{20} = 0$$
$$\textit{a racemic mixture}$$

A racemic mixture does not rotate plane-polarized light because the rotation by each enantiomer is cancelled by the equal and opposite rotation by the other. A solution of either a racemic mixture or of an achiral compound is said to be **optically inactive**, but the causes of the optical inactivity are different.

SECTION 4.8.

Relative and Absolute Configurations

The order of arrangement of four groups around a chiral carbon atom is called the **absolute configuration** around that atom. (Do not confuse configurations with *conformations*, which are shapes arising from rotation around bonds.) A pair of enantiomers have opposite configurations. For example, ($+$)- and ($-$)-glyceraldehyde have opposite configurations. But which formula represents the dextrorotatory enantiomer and which represents the levorotatory one? Until 1951, chemists did not know! Prior to that time, it was known that ($+$)-glyceraldehyde and ($-$)-glyceric acid (2,3-dihydroxypropanoic acid) have the same configuration around carbon 2, even though they rotate plane-polarized light in opposite directions. But it was not known whether the OH ar carbon 2 was to the right or to the left in the formulas as shown:

O ||	O ||	*same configuration,*
COH	CH	*but is the −OH on*
H─C─OH	H─C─OH	*the right (as shown)*
CH₂OH	CH₂OH	*or on the left?*
(−)-glyceric acid	(+)-glyceraldehyde	

Even though chemists did not know the absolute configurations of ($-$)-glyceric acid and other optically active compounds, they could speak of the **relative configurations** of these compounds. Usually, the configurations of optically active compounds were related to ($+$)-glyceraldehyde. The relative configurations were said to be either the same as, or opposite from, the configuration of ($+$)-glyceraldehyde. To make formulas easier to work with, it was decided in the late 19th century to assume that ($+$)-glyceraldehyde had the absolute configuration with the OH on carbon 2 to the right.

In 1951, x-ray diffraction studies by J. M. Bijvoet at the University of Utrecht in Holland showed that the original assumption was correct. ($+$)-

Glyceraldehyde does indeed have the absolute configuration that chemists had been using for 60 years. If the early chemists had guessed wrong, the chemical literature would have been in a state of confusion—all pre-1951 articles would show configurations the *reverse* of those in more modern articles. The original assumption was indeed a lucky guess!

The direction of rotation of plane-polarized light by a particular enantiomer is a physical property. The absolute configuration of a particular enantiomer is a characteristic of its molecular structure. There is no simple relationship between the absolute configuration of a particular enantiomer and its direction of rotation of plane-polarized light. As we have said, the enantiomer of glyceric acid with the same absolute configuration as (+)-glyceraldehyde is levorotatory, not dextrorotatory.

A.   Assignment of configuration: the (*R*) and (*S*) system

We have shown how the direction of rotation of plane-polarized light can be indicated by (+) or (−). However, a system is also needed to indicate the absolute configuration—that is, the actual arrangement of groups around a chiral carbon. This system is the **(*R*) and (*S*) system**, or the **Cahn–Ingold–Prelog system**. The letter (*R*) comes from the Latin *rectus*, "right," while (*S*) comes from the Latin *sinister*, "left." Any chiral carbon atom has either an (*R*) configuration or an (*S*) configuration; therefore, one enantiomer is (*R*) and the other is (*S*). A racemic mixture may be designated (*R*)(*S*), meaning a mixture of the two.

In the (*R*) and (*S*) system, groups are assigned a priority ranking using the same set of rules as are used in the (*E*) and (*Z*) system; however, the priority ranking is used in a slightly different manner. To assign an (*R*) or (*S*) configuration to a chiral carbon:

1.   Rank the four groups (or atoms) attached to the chiral carbon in order of priority by the Cahn–Ingold–Prelog sequence rules (pages 115–116).

2.   Project the molecule with the group of *lowest priority* to the rear.

3.   Select the group of *highest priority* and *draw a curved arrow to the group of next highest priority.*

4.   If this arrow is *clockwise*, the configuration is (*R*). If the arrow is *counter-clockwise*, the configuration is (*S*).

Let us illustrate the use of this procedure by assigning (*R*) and (*S*) to the enantiomers of 1-bromo-1-chloroethane.

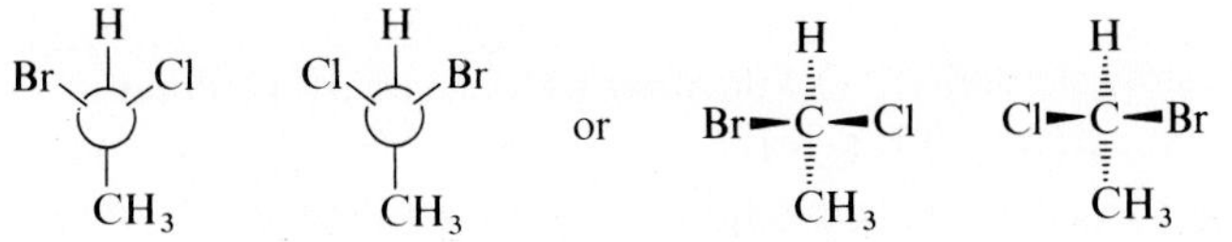

enantiomers of 1-bromo-1-chloroethane

1.   Rank the four groups. In the case at hand, the order of priority of the four atoms by atomic number is Br (highest), Cl, C, H (lowest).

2.   Draw projections with the lowest-priority atom (H) in the rear. (This atom is hidden behind the carbon atom in the following projections.)

3.   Draw an arrow from highest priority (Br) to second-highest priority (Cl).
4.   Assign (*R*) and (*S*).

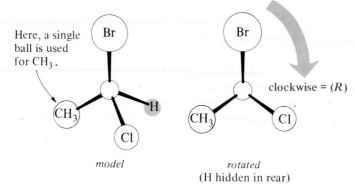

clockwise = (*R*)   counterclockwise = (*S*)
(*R*)-1-bromo-1-chloroethane   (*S*)-1-bromo-1-chloroethane

Placing a structure in the correct position for an (*R*) or (*S*) assignment is easily done with a molecular model. Construct the model, grasp the group of lowest priority with one hand, and turn the model so that the remaining three groups face you. Determine if the structure is (*R*) or (*S*) in the usual way. (See Figure 4.15.)

Here, a single
ball is used
for CH₃ .

Br   Br

CH₃   clockwise = (*R*)
H
Cl   CH₃   Cl

model   *rotated*
(H hidden in rear)

FIGURE 4.15. Using a molecular model to determine (*R*) or (*S*) configuration.

SAMPLE PROBLEMS

Name the following compound, including an (*R*) or (*S*) designation.

HO   CH₃   HO—C—CH₃
H   H
CH₂CH₃   or   CH₂CH₃

Solution:

1.   order of priority: OH (highest), CH₂CH₃, CH₃, H (lowest)
2.   projection with H in rear:

OH   OH

CH₃CH₂   CH₃   or   CH₃CH₂   CH₃
C

3.   counterclockwise = (*S*)
4.   name: (*S*)-2-butanol

Draw the structure of (*R*)-(−)-2-butanol.

Solution:

1. Write the formula of the compound without regard to configuration of the chiral carbon.

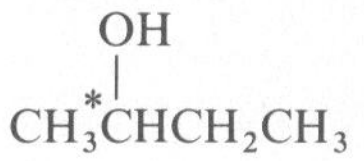

$$CH_3\overset{*}{C}HCH_2CH_3$$

with OH above the starred carbon.

2. Assign priorities around the chiral carbon: OH (highest), CH_2CH_3, CH_3, H (lowest).

3. Draw the projection with the lowest-priority group (H) in the rear. Place the remaining groups in such a way that from OH to CH_2CH_3 is clockwise for the (*R*) configuration.

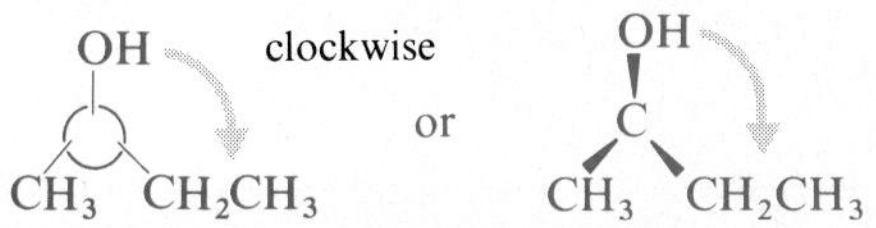

OH   clockwise   OH

CH₃   CH₂CH₃   or   CH₃   CH₂CH₃

4. Redraw the structure as a ball-and-stick or dimensional formula. (Models make this task easier.)

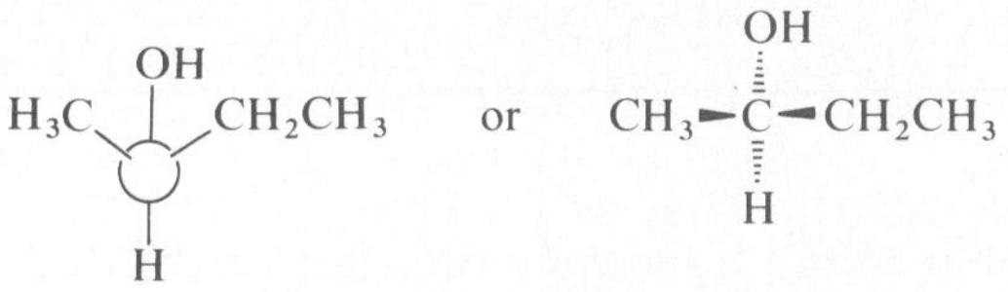

OH
H₃C   CH₂CH₃   or   CH₃—C—CH₂CH₃

H   H

Note that in the solution of this problem, the sign of rotation in the name is not used. The (−) is the indication of direction of rotation of plane-polarized light, a physical property of (*R*)-2-butanol, and is not used in determining its configuration.

STUDY PROBLEMS

4.11. Assign each of the following molecules an (*R*) or (*S*) configuration:

(a) 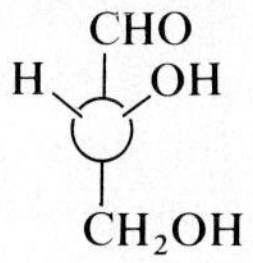
CHO
H   OH
CH₂OH

(b) 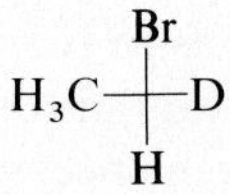   H₃C—D
Br

H

(c) 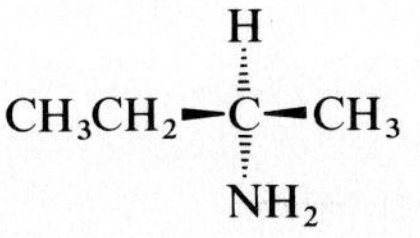   CH₃CH₂—C—CH₃
H

NH₂

(d)   H₂N—H
CO₂H

CH₂OH

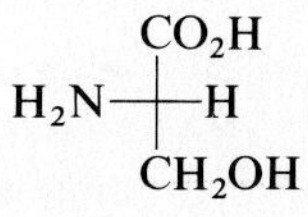

4.12. Draw formulas for   (a) (*R*)-3-bromoheptane and   (b) (*S*)-2-pentanol that show the absolute configurations around the chiral carbons.

SECTION 4.9.

More Than One Chiral Carbon Atom

Our discussion so far has concentrated primarily on compounds that contain only one chiral carbon, but compounds can have more than one chiral atom. Consider a compound with two different chiral carbons. Each of these two chiral carbons can be either (R) or (S); consequently, there are four different ways in which these configurations can be arranged within a molecule. A molecule with two different chiral carbons can therefore have four different stereoisomers.

		total molecular
chiral carbon 1	*chiral carbon 2*	*configuration*
(R)	(R)	($1R,2R$)
(R)	(S)	($1R,2S$)
(S)	(R)	($1S,2R$)
(S)	(S)	($1S,2S$)

SAMPLE PROBLEM

How many stereoisomers could exist for a compound that has three different chiral carbons?

Solution: eight: ($1R,2R,3R$); ($1R,2R,3S$); ($1R,2S,3R$); ($1R,2S,3S$); ($1S,2R,3R$); ($1S,2R,3S$); ($1S,2S,3R$); ($1S,2S,3S$).

The *maximum number of optical isomers* for a compound is 2^n, where n is the number of chiral atoms. If there are two chiral carbons, then there can be up to four stereoisomers ($2^2 = 4$); when there are three such carbon atoms, there can be up to eight stereoisomers ($2^3 = 8$).

STUDY PROBLEM

4.13. What is the maximum number of stereoisomers for each of the following compounds?   (a) 1,2-dibromo-1-phenylpropane;
(b) 1,2-dibromo-2-methyl-1-phenylpropane;   (c) 2,3,4,5-tetrahydroxypentanal

A.   (R) and (S) system for a compound with two chiral carbon atoms

To assign (R) or (S) configurations to two chiral carbon atoms in one molecule, we consider each chiral carbon atom in turn. We will use the sugar erythrose to illustrate the technique. (For convenience, we do not usually use broken wedges in a formula that shows more than one chiral carbon.)

For carbon 2:

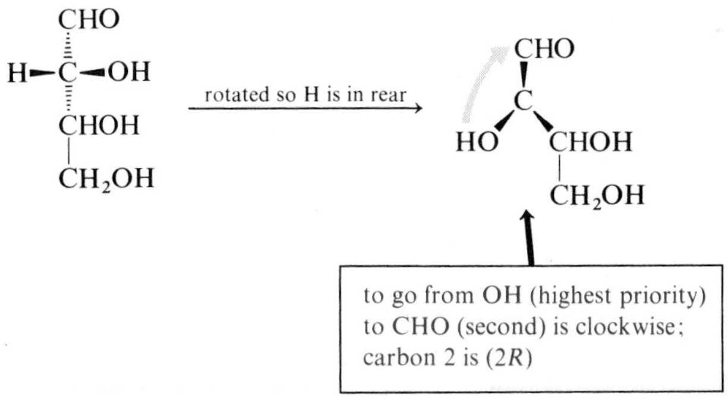

CHO
H—C—OH
CHOH
CH₂OH

$\xrightarrow{\text{rotated so H is in rear}}$

CHO
C
HO   CHOH
CH₂OH

to go from OH (highest priority)
to CHO (second) is clockwise;
carbon 2 is (2R)

For carbon 3:

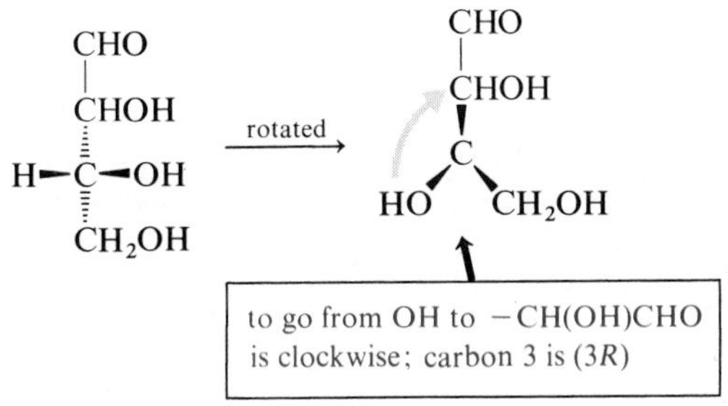

CHO
CHOH
H—C—OH
CH₂OH

$\xrightarrow{\text{rotated}}$

CHO
CHOH
C
HO   CH₂OH

to go from OH to −CH(OH)CHO
is clockwise; carbon 3 is (3R)

The IUPAC name for this stereoisomer is therefore (2*R*,3*R*)-2,3,4-tri-hydroxybutanal. Note that the numbers and letters in the single set of parentheses refer to the configurations around two different chiral carbons in one molecule. Contrast this notation with (*R*)(*S*), which means a racemic mixture.

B.   Diastereomers

When a molecule has more than one chiral carbon, not all of the optical isomers are enantiomers. By definition, enantiomers (mirror images) come in pairs.

The four stereoisomers of 2,3,4-trihydroxybutanal:

CHO	CHO	CHO	CHO
H—C—OH	HO—C—H	H—C—OH	HO—C—H
H—C—OH	HO—C—H	HO—C—H	H—C—OH
CH₂OH	CH₂OH	CH₂OH	CH₂OH
(2R,3R)	(2S,3S)	(2R,3S)	(2S,3R)

enantiomers   *enantiomers*

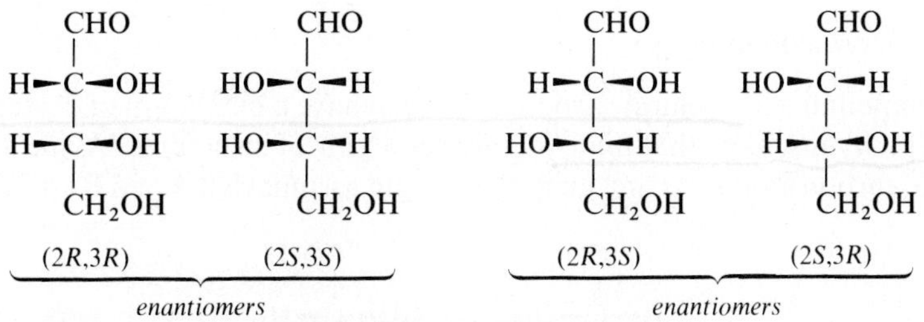

Note that the (2*R*,3*R*)- and (2*S*,3*S*)-stereoisomers are enantiomers. The (2*R*,3*S*)- and (2*S*,3*R*)-stereoisomers are also enantiomers. However, the (2*S*,3*S*)- and (2*R*,3*S*)-stereoisomers are *not* enantiomers. (What other nonenantiomeric pairings can be arranged?) Any pair of stereoisomers that are not enantiomers are called **diastereomers**, or **diastereoisomers**. Thus the (2*S*,3*S*)- and (2*R*,3*S*)-stereo-isomers are diastereomers. (*Geometric isomers* are also diastereomers by this definition.)

$$
\begin{array}{cc}
\text{CHO} & \text{CHO} \\
| & | \\
\text{HO}-\text{C}-\text{H} & \text{H}-\text{C}-\text{OH} \\
| & | \\
\text{HO}-\text{C}-\text{H} & \text{HO}-\text{C}-\text{H} \\
| & | \\
\text{CH}_2\text{OH} & \text{CH}_2\text{OH}
\end{array}
$$

a pair of diastereomers:
stereoisomers that are not enantiomers

A pair of enantiomers have identical physical and chemical properties except for interactions with other chiral molecules and for the direction of rotation of plane-polarized light. Diastereomers, however, are chemically and physically different. They have different melting points, different solubilities, and often undergo chemical reactions in a different fashion.

SAMPLE PROBLEM

Identify each pair of molecules as *structural isomers, enantiomers,* or *diastereomers.*

(a)
$$
\begin{array}{cc}
\text{CO}_2\text{H} & \text{CO}_2\text{H} \\
| & | \\
\text{H}_2\text{N}-\text{C}-\text{H} & \text{H}-\text{C}-\text{NH}_2 \\
| & | \\
\text{CH}_2\text{OH} & \text{CH}_2\text{OH}
\end{array}
$$

(b)
$$
\begin{array}{cc}
\text{CH}_3 \quad \text{CH}_3 & \text{CH}_3 \quad \text{H} \\
\diagdown \quad \diagup & \diagdown \quad \diagup \\
\text{C}=\text{C} & \text{C}=\text{C} \\
\diagup \quad \diagdown & \diagup \quad \diagdown \\
\text{H} \quad \text{H} & \text{H} \quad \text{CH}_3
\end{array}
$$

(c)
$$
\begin{array}{cc}
\text{CHO} & \text{CHO} \\
\text{H} \diagdown \diagup \text{OH} & \text{H} \diagdown \diagup \text{OH} \\
\text{H} \diagup \diagdown \text{OH} & \text{HO} \diagup \diagdown \text{H} \\
| & | \\
\text{CH}_2\text{OH} & \text{CH}_2\text{OH}
\end{array}
$$

(d)
$$
\begin{array}{cc}
\text{CHO} & \text{CHO} \\
| & | \\
\text{H}-\text{C}-\text{OH} & \text{HO}-\text{C}-\text{H} \\
| & | \\
\text{H}-\text{C}-\text{OH} & \text{HO}-\text{C}-\text{H} \\
| & | \\
\text{H}-\text{C}-\text{OH} & \text{HO}-\text{C}-\text{H} \\
| & | \\
\text{CH}_2\text{OH} & \text{CH}_2\text{OH}
\end{array}
$$

Solution: (a) and (d) are enantiomers. (b) and (c) are diastereomers.

C.   *Meso* compounds

A compound with *n* chiral carbon atoms can have a maximum of 2^n stereoisomers, but it may not have that many. Consider a pair of structures (**A** and **B**) with two chiral carbon atoms. At first glance, we might assume that **A** and **B** are enantiomers.

$$
\begin{array}{cc}
\text{CO}_2\text{H} & \text{CO}_2\text{H} \\
| & | \\
\text{H}-\text{C}-\text{OH} & \text{HO}-\text{C}-\text{H} \\
| & | \\
\text{H}-\text{C}-\text{OH} & \text{HO}-\text{C}-\text{H} \\
| & | \\
\text{CO}_2\text{H} & \text{CO}_2\text{H}
\end{array}
$$

A   **B**

Let us take compound **B** and rotate it 180° in the plane of the paper. We can see that compound **B** is identical to compound **A**! Indeed, **A** and **B** are mirror images, but the mirror images are superimposable; therefore, **A** and **B** are the same compound.

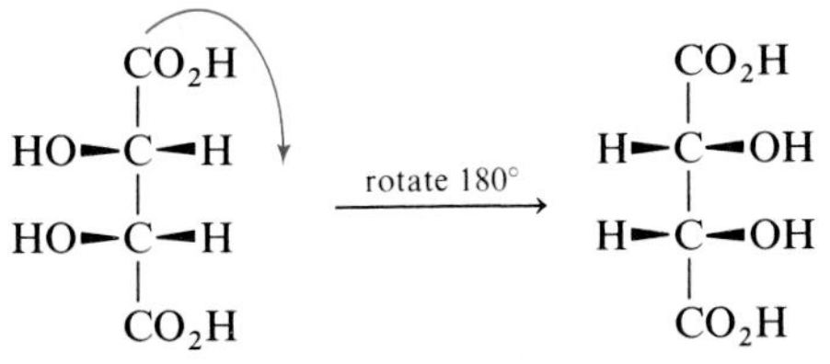

B **B** *superimposable on* **A**

How is it possible that a molecule with two chiral carbons is superimposable on its mirror image? The answer is that, in at least one conformation, this molecule has an **internal plane of symmetry**. The "top" half of the molecule is the mirror image of the "bottom" half. We might say that the two halves of the molecule cancel each other so far as chirality is concerned. Therefore, the molecule as a whole is achiral and does not cause a rotation of plane-polarized light. Figure 4.16 illustrates an internal plane of symmetry.

A stereoisomer that contains chiral carbons but can be superimposed on its mirror image is called a ***meso* form**. The compound we have been discussing is *meso*-tartaric acid.

A compound with two chiral carbons can have up to four stereoisomers. For tartaric acid, we have looked at two possibilities, and these two added up to only one isomer, *meso*-tartaric acid. What about the other two stereoisomers of tartaric acid? Do they have an internal plane of symmetry? No, they do not. The top halves are not mirror images of the bottom halves. Rotation of either structure by 180° does not result in the other structure or in the *meso*-isomer. The following two stereoisomers of tartaric acid are enantiomers. They are both optically active

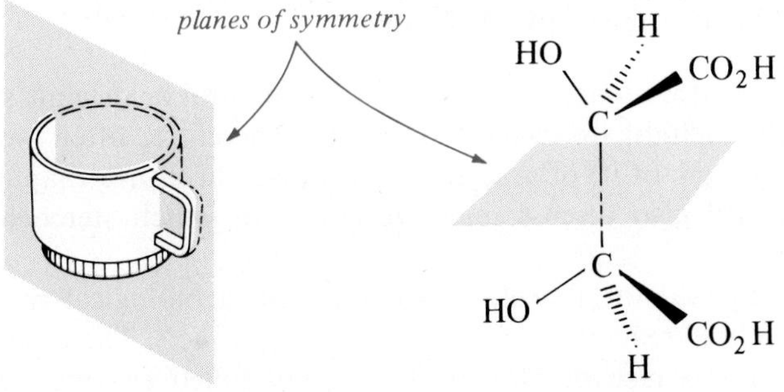

meso-tartaric acid
$[\alpha] = 0$

FIGURE 4.16. An internal plane of symmetry is a hypothetical plane that bisects an object into two identical, mirror-reflective halves. An object with an internal plane of symmetry is achiral (can be superimposed on its mirror image).

and rotate the plane of polarization of plane-polarized light equally, but in opposite directions.

$$
\begin{array}{cc}
\text{CO}_2\text{H} & \text{CO}_2\text{H} \\
| & | \\
\text{H—C—OH} & \text{HO—C—H} \\
\text{-----------} & \text{-----------} \quad NOT\ a\ plane \\
\text{HO—C—H} & \text{H—C—OH} \quad of\ symmetry \\
| & | \\
\text{CO}_2\text{H} & \text{CO}_2\text{H}
\end{array}
$$

(2R,3R)-(+)-tartaric acid   (2S,3S)-(−)-tartaric acid

What are our conclusions about tartaric acid? Because of the internal plane of symmetry in *meso*-tartaric acid, there is a total of only *three* stereoisomers for tartaric acid, rather than the four stereoisomers predicted by the 2^n rule. These three stereoisomers are a pair of enantiomers and the diastereomeric *meso* form. The *meso* forms of some other compounds follow:

$$
\begin{array}{ccc}
& \text{CHO} & \text{CH}_2\text{OH} \\
\text{Cl} & | & | \\
| & \text{H—C—OH} & \text{H—C—OH} \\
\text{H}_3\text{C—C—H} & | & | \\
\text{-----------} & \text{---H—C—OH---} & \text{H—C—OH} \\
\text{H}_3\text{C—C—H} & | & \text{-------- plane of symmetry} \\
| & \text{H—C—OH} & \text{H—C—OH} \\
\text{Cl} & | & | \\
& \text{CHO} & \text{H—C—OH} \\
& & | \\
& & \text{CH}_2\text{OH}
\end{array}
$$

STUDY PROBLEM

4.14. Draw all possible stereoisomers of 2,3-dichlorobutane. Indicate any enantiomeric pairs.

SECTION 4.10

Resolution of a Racemic Mixture

In most laboratory reactions, a chemist uses achiral or racemic starting materials and obtains achiral or racemic products. Therefore, often we will ignore the chirality (or lack of it) of reactants or products in future chapters. On the other hand, we will also discuss many reactions in which stereochemistry is quite important.

As opposed to laboratory reactions, most biological reactions start with chiral or achiral reactants and lead to chiral products. These biological reactions are possible because of biological catalysts called **enzymes**, which are chiral themselves. Recall that a pair of enantiomers have the same chemical properties *except* for their interactions with other chiral substances (Section 4.7). Because enzymes are chiral, they can act very selectively in their catalytic action. For example, when an organism ingests a racemic mixture of alanine, only the (S)-alanine becomes incorporated into protein structures. The (R)-alanine is not used in proteins; rather, it is oxidized to a keto acid with the aid of other enzymes and enters other metabolic schemes.

$$\begin{array}{cc}
CO_2H & CO_2H \\
\mid & \mid \\
H_2N-C-H & H-C-NH_2 \\
\mid & \mid \\
CH_3 & CH_3 \\
\textit{(S)-alanine} & \textit{(R)-alanine}
\end{array}$$

In the laboratory, the physical separation of a racemic mixture into its pure enantiomers is called the **resolution**, or **resolving**, of the racemic mixture. Pasteur's separation of racemic sodium ammonium tartrate was a resolution of that mixture. It is a very rare occurrence for enantiomers to crystallize separately; therefore, the method that Pasteur used cannot be considered a general technique. Because a pair of enantiomers exhibit the same chemical and physical properties, they cannot be separated by ordinary chemical or physical means. Instead, chemists must rely on chiral reagents or chiral catalysts (which, one way or another, almost always have their origins in living organisms).

The most general technique for resolving a pair of enantiomers is to subject them to a reaction with a chiral reagent to obtain a pair of **diastereomeric products**. Diastereomers, you will remember, are different compounds with different physical properties. Thus, a pair of diastereomers may be separated by ordinary physical means, such as crystallization.

Let us illustrate a general procedure for the laboratory resolution of $(R)(S)$-RCO_2H, a racemic mixture of a carboxylic acid, where (R)-RCO_2H and (S)-RCO_2H represent the two enantiomers. A carboxylic acid will react with an amine to yield a salt (Section 1.10):

$$\underset{\textit{a carboxylic acid}}{RC\ddot{O}-H} + \underset{\textit{an amine}}{R'\ddot{N}H_2} \longrightarrow \underset{\textit{a salt}}{RC\ddot{O}:^- \; R'\overset{+}{N}H_3}$$

Reaction of the $(R)(S)$ carboxylic acid with an amine that is a pure enantiomer results in a pair of diastereomeric salts: the amine salt of the (R) acid and the amine salt of the (S) acid.

$$\begin{array}{c}
(R)\text{-}RCO_2H \\
\text{and} \\
(S)\text{-}RCO_2H
\end{array} + (S)\text{-}R'NH_2 \longrightarrow \left\{ \begin{array}{c}
(R)\text{-}RCO_2^- \; (S)\text{-}R'NH_3^+ \\
\text{and} \\
(S)\text{-}RCO_2^- \; (S)\text{-}R'NH_3^+
\end{array} \right\}$$

racemic mixture   *an enantiomeric amine*   *The (R,S)-salt and the (S,S)-salt are not enantiomers: they are diastereomers and may be separated.*

In this reaction, the only possible products are the (R,S)-salt and the (S,S)-salt, which are not enantiomers of each other. The enantiomers of these two salts are the (S,R)-salt and the (R,R)-salt, respectively. Neither of these latter enantiomers can be formed in this reaction because only the (S)-amine was used as a reactant.

After separation, each diastereomeric salt is treated with a strong base to regenerate the amine. The amine and the carboxylate ion may be separated by extraction with a solvent such as diethyl ether (in which the amine is soluble, but the

carboxylate salt is not). Acidification of the aqueous phase yields the free enantiomeric carboxylic acid.

$$(R)\text{-RCO}_2^-\ (S)\text{-RNH}_3^+ \xrightarrow{\ \text{OH}^-\ } (R)\text{-RCO}_2^- + (S)\text{-RNH}_2$$

one of the pure
diastereomeric salts

$$\xrightarrow{\ \text{HCl}\ } (R)\text{-RCO}_2\text{H}$$

as a pure enantiomer

This resolution of a racemic acid depends on salt formation with a single
enantiomer of a chiral amine. Commonly used amines are amphetamine, which is
commercially available as pure enantiomers, and the naturally occurring strychnine (page 733).

$$\text{C}_6\text{H}_5\text{—CH}_2\overset{\overset{\displaystyle \text{CH}_3}{|}}{\text{CH}}\text{NH}_2$$

amphetamine

STUDY PROBLEM

4.15. (a) Write Fischer projections for the enantiomers of amphetamine, and assign
the (R) and (S) configurations.
(b) Write an equation to show the products of the reaction of (R)-amphetamine
with (R)(S)-lactic acid.

$$\text{CH}_3\overset{\overset{\displaystyle \text{OH}}{|}}{\text{CH}}\text{CO}_2\text{H}$$

2-hydroxypropanoic acid
(lactic acid)

Summary

Stereoisomerism is isomerism resulting from different spatial arrangements of
atoms in molecules. **Geometric isomerism**, one form of stereoisomerism, results
from groups being *cis* (same side) or *trans* (opposite sides) around a pi bond
or on a ring. Geometric isomers of alkenes may also be differentiated by the letter
(E), opposite sides, or (Z), same side.

Rotation of groups around sigma bonds results in different **conformations**,
such as the **eclipsed**, *gauche*, **staggered**, and *anti* conformations. Lower-energy
conformers predominate. Conformers are interconvertible at room temperature
and therefore are not isolable isomers. A cyclic compound assumes puckered
conformations to relieve strain of unfavorable bond angles and, more important,
to minimize repulsions of substituents. For the cyclohexane ring, the chair-form
conformer with substituents **equatorial** instead of **axial** is favored.

A **chiral molecule** is a molecule that is nonsuperimposable on its mirror image.
The pair of nonsuperimposable mirror images are called **enantiomers** and represent
another type of stereoisomerism. Each member of a pair of enantiomers rotates

the plane of polarization of plane-polarized light an equal amount, but in opposite directions. An equimolar mixture of enantiomers, called a **racemic mixture**, is optically inactive.

Chirality usually arises from the presence of a carbon with four different atoms or groups attached to it. The arrangement of these groups around the chiral carbon is called the **absolute configuration** and may be described as (R) or (S). **Fischer projections** are often used to depict chiral molecules.

A molecule with more than one chiral carbon has more stereoisomers than a single enantiomeric pair. Stereoisomers that are not enantiomers are **diastereomers**. If a molecule has more than one chiral carbon and can be superimposed on its mirror image, it is optically inactive and is called a *meso* **form**.

The different types of isomerism may be summarized:

A. **Structural isomers** differ in order of attachment of atoms:

$$(CH_3)_2CHCH_3 \quad \text{and} \quad CH_3CH_2CH_2CH_3$$

B. **Stereoisomers** differ in arrangement of atoms in space.

    1. **Enantiomers:** nonsuperimposable mirror images

(2R,3R)   (2S,3S)

    2. **Diastereomers:** nonenantiomeric stereoisomers

Containing chiral carbons:

(2R,3R)   *meso*

Achiral:

cis, or (Z)   *trans*, or (E)

also called geometric isomers

A pair of enantiomers have the same physical and chemical properties except for the direction of rotation of the plane of polarization of plane-polarized light and their interactions with other chiral substances. Enantiomers may be resolved by (1) treatment with a chiral reagent to yield a pair of diastereomers; (2) separation of the diastereomers, which do *not* have the same properties; and (3) regeneration of the separated enantiomers.

STUDY PROBLEMS

4.16. Give structural formulas for each of the following compounds (both geometric isomers, if any) and label each as *cis*, *trans*, or *no geometric isomer*: (a) 1-hexene; (b) 2-hexene; (c) 2-methyl-1-butene; (d) 1-chloro-2-butene; (e) 1,3-diethylcyclohexane.

4.17. Write structural formulas for the alkenes of molecular formula C_5H_{10} that exhibit geometric isomerism. Indicate the *cis-* and *trans-*structures.

4.18. Which of the following compounds exhibit geometric isomerism?

(a) 1,2-diphenylethene
(b) 1-butene-3-yne, $CH_2\!\!=\!\!CHC\!\!\equiv\!\!CH$
(c) 2-pentene-4-yne, $CH_3CH\!\!=\!\!CHC\!\!\equiv\!\!CH$
(d) 2,3-dimethyl-2-pentene
(e) ethyl 2-butenoate, $CH_3CH\!\!=\!\!CHCO_2CH_2CH_3$

4.19. Assign (*E*) or (*Z*) to each of the following alkenes. (*Note:* $C_6H_5\!\!-$ = phenyl.)

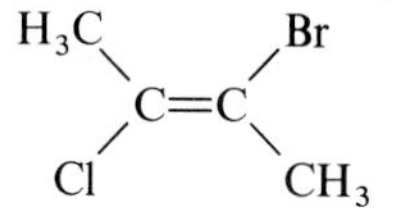

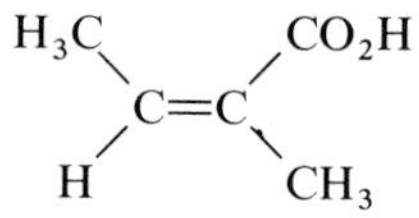

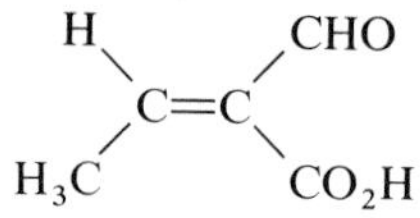

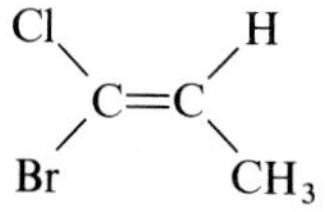

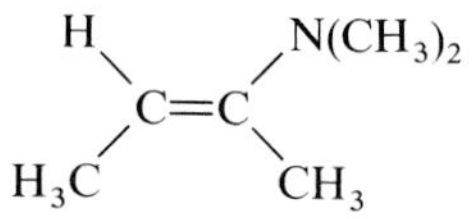

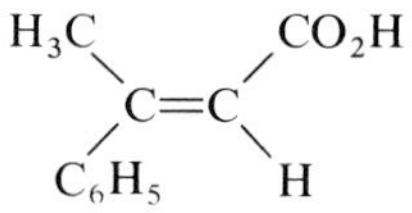

4.20. Draw the structure of each of the following compounds, showing its stereochemistry:

(a) (*E*)-2-chloro-2-butene
(b) (2Z,4Z)-nonadiene
(c) (*Z*)-2-pentene
(d) (*E*)-2-bromo-1-nitro-2-butene

4.21. Name and give formulas for the geometric isomers of 2,4-hexadiene, using the *E* and *Z* system.

4.22. Draw the formulas and label the geometric isomers (if any) for the following compounds:

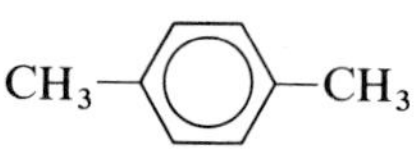

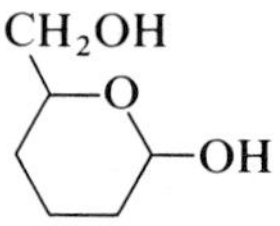

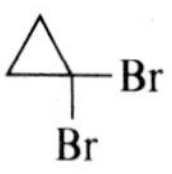

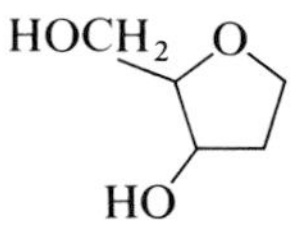

4.23. Draw the Newman projections for the *anti* and two eclipsed conformations of 1,2-diiodoethane. Of the two eclipsed conformations, which is of higher energy?

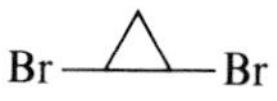

4.24. Draw the Newman projection for an *anti* conformer (if any) for each of the following compounds. Use the circled carbons as the center of the Newman projection.

(a) 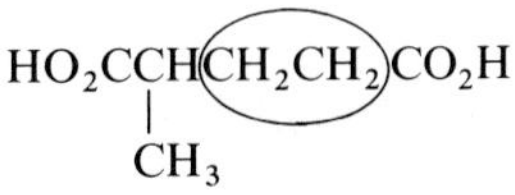 $HO_2CCH(CH_2CH_2)CO_2H$
    $|$
    CH_3

(b) 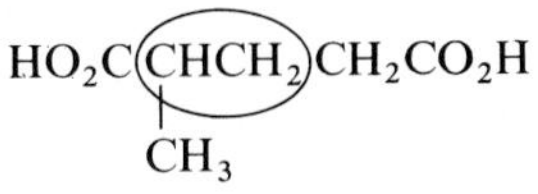 $HO_2C(CHCH_2)CH_2CO_2H$
    $|$
    CH_3

(c) 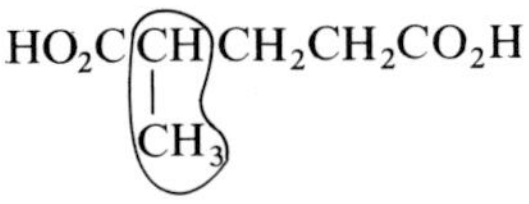 $HO_2C(CH)CH_2CH_2CO_2H$
    $|$
    CH_3

4.25. Which of the following compounds contains one or more strained rings?

(a) 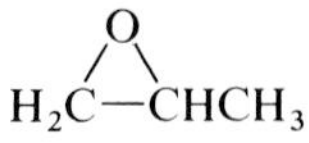 $H_2C—CHCH_3$ with O bridge

propylene oxide

(b) 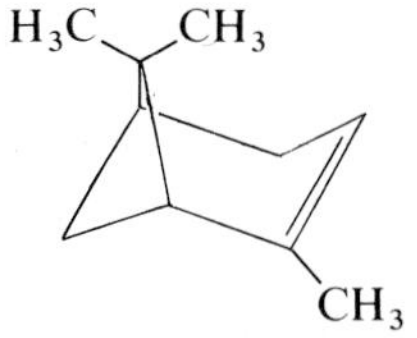 α-pinene structure with H_3C, CH_3 and CH_3

α-pinene

in turpentine

(c) 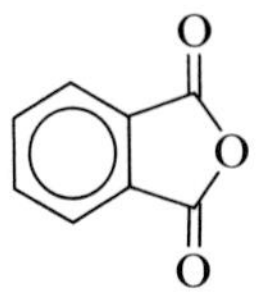 phthalic anhydride structure

phthalic anhydride

(d) 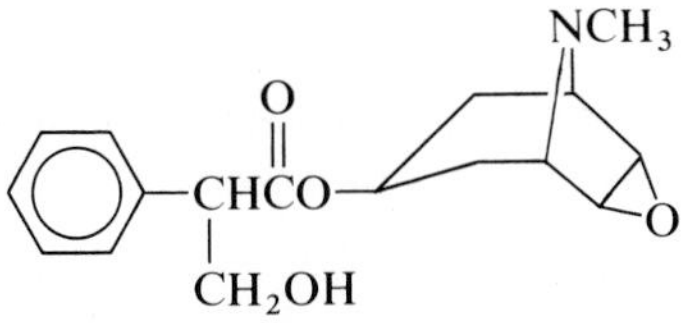 scopolamine structure with NCH_3, $—CHCO—$, CH_2OH

scopolamine

a preoperative anesthetic

4.26. Label each of the following positions as axial or equatorial:

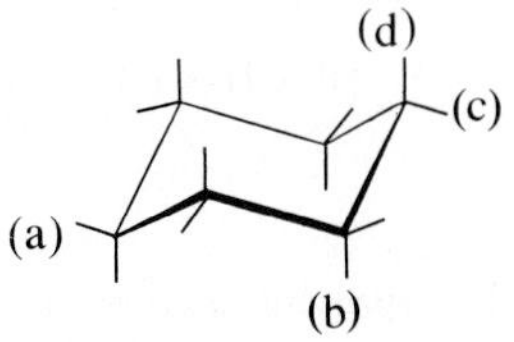

cyclohexane chair with positions (a), (b), (c), (d)

4.27. Which of the following conformations is most stable? (More than one compound is shown.)

(a) 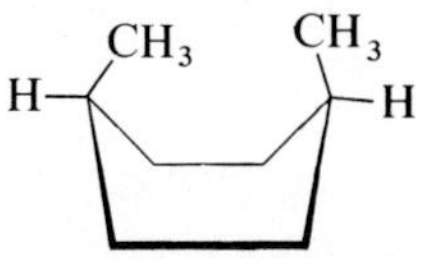 chair with CH_3, CH_3, H—, —H

(b) 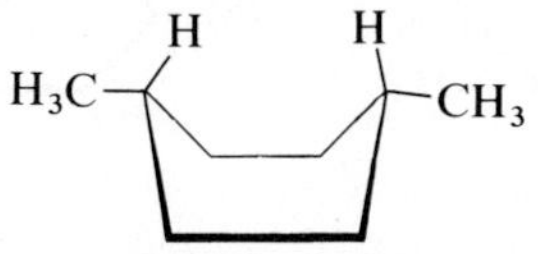 chair with H, H, H_3C—, —CH_3

(c) 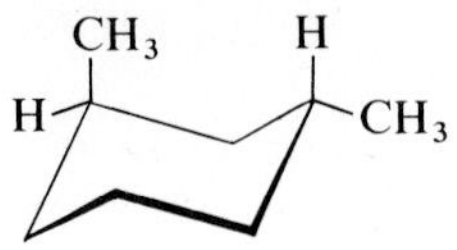 chair with CH_3, H, H—, —CH_3

(d) 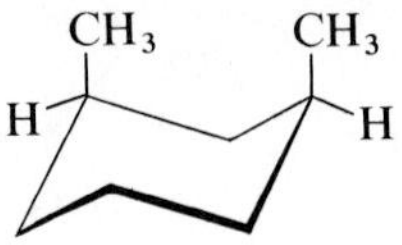 chair with CH_3, CH_3, H—, —H

4.28.   Which of the following conformations is more stable?

(a)   (b)

4.29.   (a)   Why is a *cis*-1,3-disubstituted cyclohexane more stable than the corresponding *trans*-structure?
(b)   Is the *cis*-1,2-isomer more stable than the *trans*-1,2-disubstituted cyclohexane?

4.30.   Draw the following compound in the chair form with all of the ring hydrogen atoms in axial positions:

$$HOCH_2$$
$$O$$
$$HO \quad\quad OH$$
$$OH \quad OH$$

4.31.   Draw the structure of the preferred conformation of   (a) 1-methyl-1-propylcyclohexane; (b) *cis*-1-methyl-2-propylcyclohexane;   (c) *trans*-1-methyl-3-propylcyclohexane.

4.32.   Which of the following compounds contain chiral molecules?
(a) 2-methyl-2-phenylbutane;   (b) 1,2-dibromo-1,2-diphenylpropane.

4.33.   Star the chiral carbons (if any):

(a) $(CH_3)_2CHCHBrCH_3$

(b) $CH_3CH_2CH_2\overset{\overset{\displaystyle CH_3}{|}}{C}HOH$

(c) $H_2NCH_2CO_2H$

(d) $CH_3CHBrCHBrCH_2CH_3$

(e) $H_2N\overset{\underset{\displaystyle CH_3}{|}}{C}HCO_2H$

(f) $HOCH_2\overset{\underset{\displaystyle OH}{|}}{C}HCH_2OH$

4.34.   Which of the following formulas represents a chiral molecule?

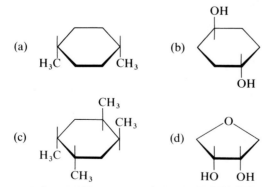

4.35. Show the configuration of the enantiomer of each of the following compounds:

(a)
$$
\begin{array}{c}
CHO \\
| \\
H\!-\!C\!-\!OH \\
| \\
H\!-\!C\!-\!OH \\
| \\
H\!-\!C\!-\!OH \\
| \\
CH_2OH
\end{array}
$$

(b)
$$
\begin{array}{c}
CO_2H \\
| \\
H_2N\!-\!C\!-\!H \\
| \\
CH_2\!-\!\!\bigcirc\!\!-\!OH
\end{array}
$$

4.36. Give the structures and configurations for all the stereoisomers of five-carbon alcohols (with one OH) that have stereoisomers, and indicate any enantiomeric pairs.

4.37. A carboxylic acid of the formula $C_3H_5O_2Br$ is optically active. What is its structure?

4.38. Give the structure of the lowest-molecular-weight alcohol (containing only C, H, and O) that is chiral.

4.39. Convert the following dimensional formulas to Fischer projections. (Molecular models may be helpful.)

(a)
$$
\begin{array}{c}
CO_2H \\
| \\
H_2N\!-\!C\!-\!H \\
| \\
CH_2CH(CH_3)_2
\end{array}
$$

(b)
$$
\begin{array}{c}
CO_2H \\
| \\
C \\
H \quad OH \\
| \\
CH_3
\end{array}
$$

(c)
$$
\begin{array}{c}
CH_3 \\
| \\
C\cdots Br \\
CH_3CH_2 \quad CH \\
\| \\
O
\end{array}
$$

(d)
$$
\begin{array}{c}
H_3C \qquad H \\
\diagdown \qquad \diagup \\
C\!-\!C \quad OH \\
H\diagup \quad \diagdown \\
NH_2 \qquad CH \\
\| \\
O
\end{array}
$$

(e)
$$
\begin{array}{c}
CO_2H \quad Br \\
\diagdown \quad | \quad H \\
C\!-\!C \\
H\diagup \quad \diagdown \\
Br \qquad CO_2H
\end{array}
$$

4.40. Convert the following Fischer projections to dimensional formulas (not necessarily the lowest-energy conformers):

(a)
$$
\begin{array}{c}
CO_2H \\
H_2N\!-\!\!|\!-\!H \\
CH_2OH
\end{array}
$$

(b)
$$
\begin{array}{c}
CHO \\
H\!-\!|\!-\!OH \\
HO\!-\!|\!-\!H \\
CH_2OH
\end{array}
$$

(c)
$$
\begin{array}{c}
CH_2OH \\
| \\
C\!=\!O \\
H\!-\!|\!-\!OH \\
CH_2OH
\end{array}
$$

4.41. For each Fischer projection in Problem 4.40, draw the Fischer projection of the enantiomer.

4.42. Match the compound on the left with its stereoisomer, if any, on the right.

(a)
$$
\begin{array}{c}
HOCH_2 \\
\diagup\!-\!O \\
\bigcirc
\end{array}
$$

(1)
$$
\begin{array}{c}
CO_2H \\
| \\
H_2N\cdots C\cdots H \\
| \\
CH_3
\end{array}
$$

(b) 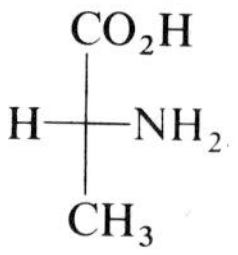
$$
\begin{array}{c}
CO_2H \\
H\!\!-\!\!\!\!\!\!-\!\!NH_2 \\
CH_3
\end{array}
$$

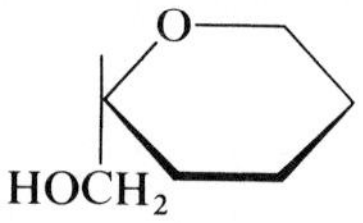

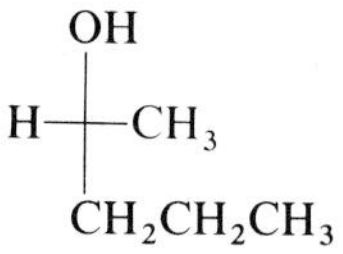

(2)
HOCH₂

(c)
$$
\begin{array}{c}
OH \\
H\!\!-\!\!\!\!\!\!-\!\!CH_3 \\
CH_2CH_2CH_3
\end{array}
$$

(3) $CH_3\!\!-\!\!\overset{H}{\underset{OH}{C}}\!\!-\!\!CH_2CH_3$

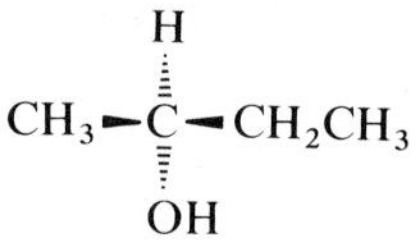

(d) 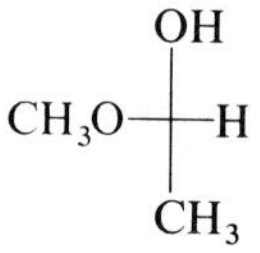
$$
\begin{array}{c}
OH \\
CH_3O\!\!-\!\!\!\!\!\!-\!\!H \\
CH_3
\end{array}
$$

(4) 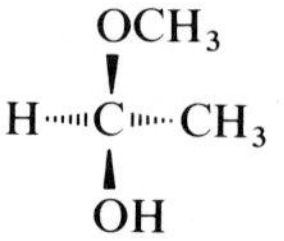 $H\cdots\overset{OCH_3}{\underset{OH}{C}}\cdots CH_3$

4.43. Which of the following pairs of formulas represent enantiomeric pairs?

(a) 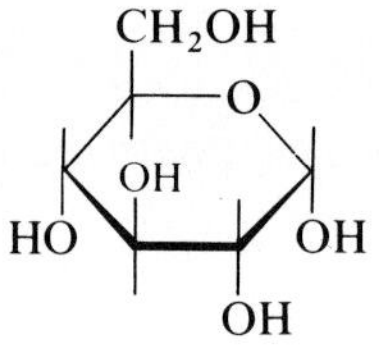 and

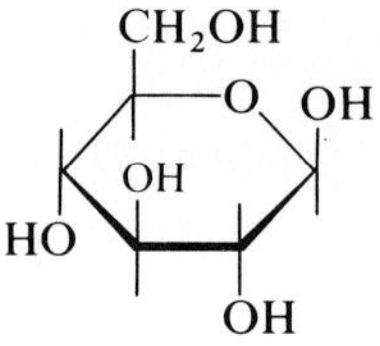

(b) 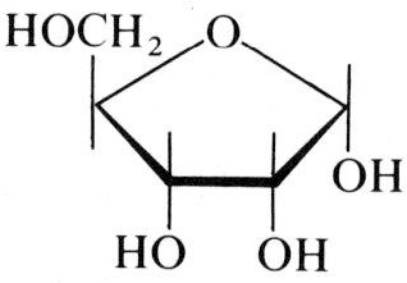 and

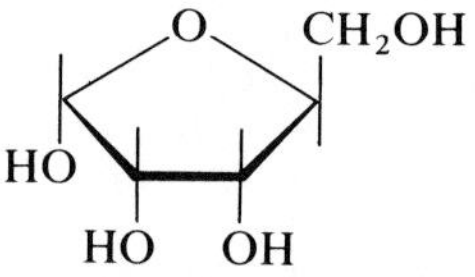

(c) 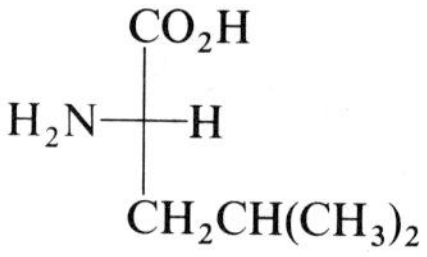
$$
\begin{array}{c}
CO_2H \\
H_2N\!\!-\!\!\!\!\!\!-\!\!H \\
CH_2CH(CH_3)_2
\end{array}
$$
and
$$
\begin{array}{c}
CO_2H \\
H_2N\!\!-\!\!\!\!\!\!-\!\!H \\
(CH_3)_2CHCH_2
\end{array}
$$

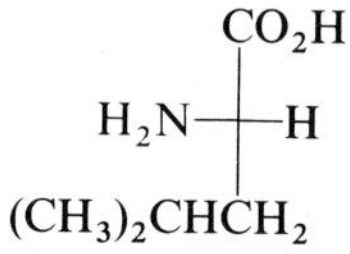

(d) 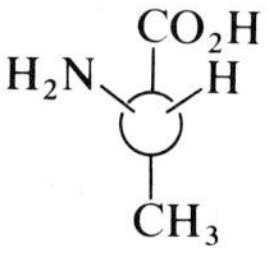 and

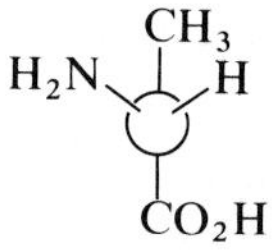

4.44. A pure sample of (S)-2-butanol has an $[\alpha]_D^{25}$ of $+13.5°$. What is the specific rotation of (R)-2-butanol?

4.45. Which of the following formulas represent *meso* compounds?

(a)

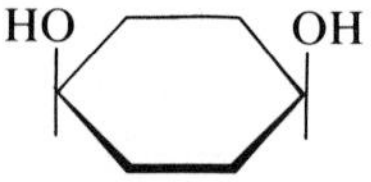

(b) 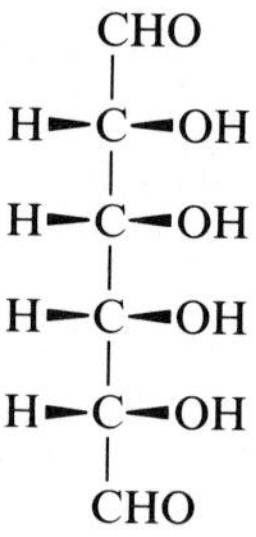
$$
\begin{array}{c}
CHO \\
H\!\!-\!\!C\!\!-\!\!OH \\
H\!\!-\!\!C\!\!-\!\!OH \\
H\!\!-\!\!C\!\!-\!\!OH \\
H\!\!-\!\!C\!\!-\!\!OH \\
CHO
\end{array}
$$

(c)

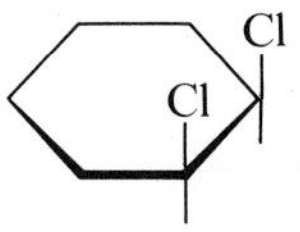

4.46. How many chiral carbons do each of the following compounds contain?

(a)

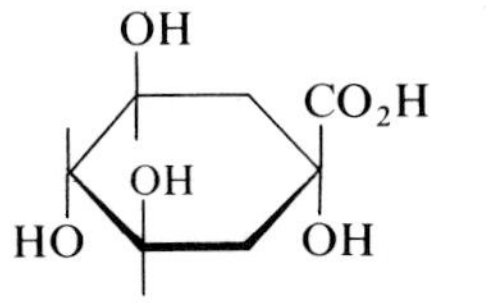

OH
CO$_2$H
OH
HO
OH

(b)

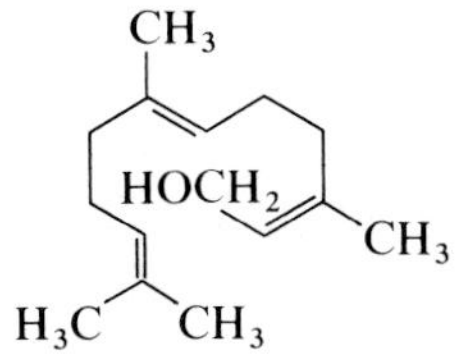

CH$_3$
HOCH$_2$
CH$_3$
H$_3$C   CH$_3$

(c)

CH$_3$
O
=C(CH$_3$)$_2$
CH$_3$

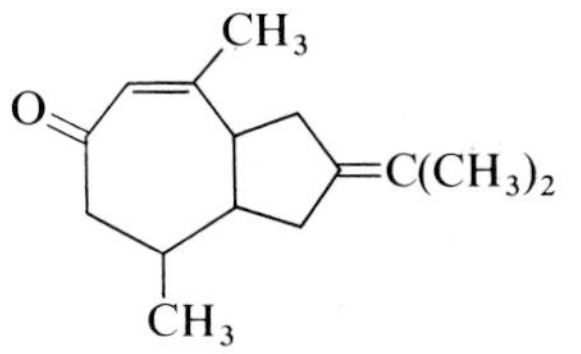

(d)

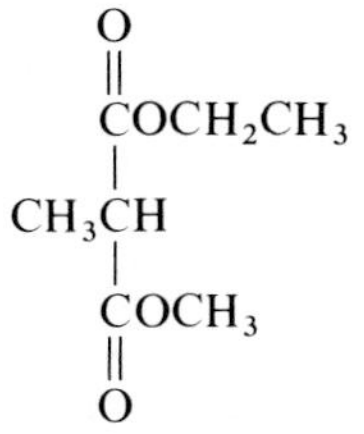

$$O$$
$$\overset{\|}{C}OCH_2CH_3$$
CH$_3$CH
$\overset{|}{C}OCH_3$
$\overset{\|}{O}$

(e) 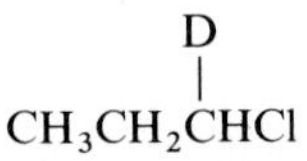 CH$_3$CH$_2$CHCl
      $\overset{D}{|}$

(f)

H$_3$C   CH$_3$
HO$_2$C   CO$_2$H

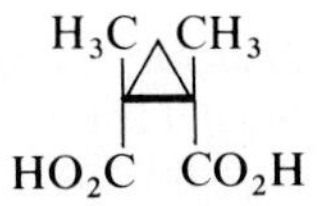

4.47. Draw dimensional or ball-and-stick formulas for the following compounds. (Use models.)

(a) (*R*)-2-bromopropanoic acid, and

(b) ethyl (2*R*,3*S*)-3-amino-2-iodo-3-phenylpropanoate, C$_6$H$_5$CHCHICOCH$_2$CH$_3$
    $\overset{NH_2}{|}$ $\overset{O}{\|}$

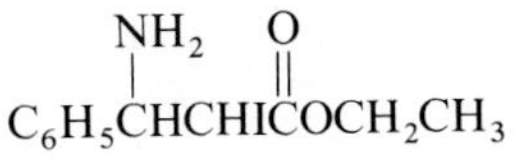

4.48. What is the *maximum number* of stereoisomers possible for
(a) 3-hydroxy-2-methylbutanoic acid, and   (b) 2,4-dimethyl-1-pentanol?

4.49. (a)   Which of the following Fischer projections represent enantiomers?
       (b)   Which are diastereomers?
       (c)   Which is a *meso* form?

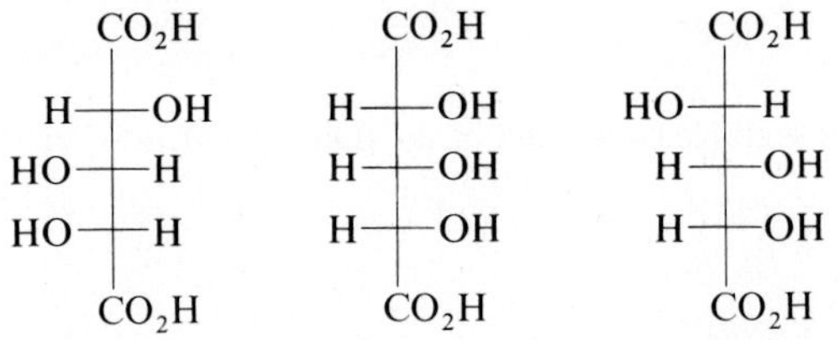

CO$_2$H
H——OH
HO——H
HO——H
CO$_2$H
I

CO$_2$H
H——OH
H——OH
H——OH
CO$_2$H
II

CO$_2$H
HO——H
H——OH
H——OH
CO$_2$H
III

4.50. Write Fischer projections for all possible configurations of 2,3,4-pentanetriol. Indicate enantiomeric pairs. Are there any *meso* forms?

4.51. Draw dimensional or ball-and-stick formulas for   (a) any enantiomeric pairs,   (b) a *meso* form, and   (c) any diastereomeric pairs for HO$_2$CCHBrCH$_2$CHBrCO$_2$H.

4.52. Draw a dimensional or ball-and-stick formula for (2*S*,3*S*)-2,3-dichlorobutanedioic acid, HO$_2$CCHClCHClCO$_2$H.

4.53. Assign an (R) or (S) designation to each of the following structures:

(a)

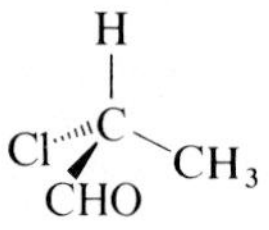

(b)

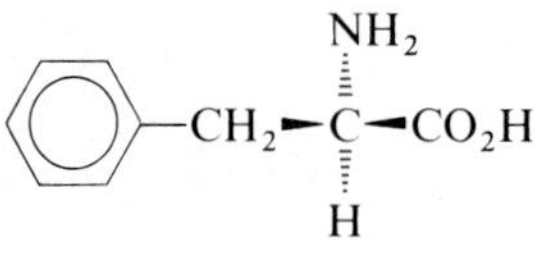

(c)

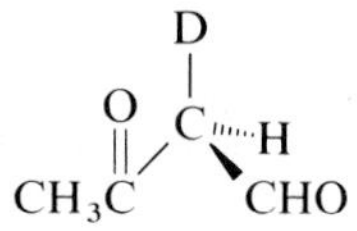

(d)

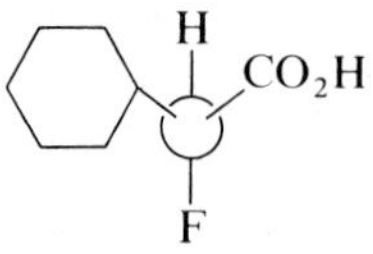

(e)

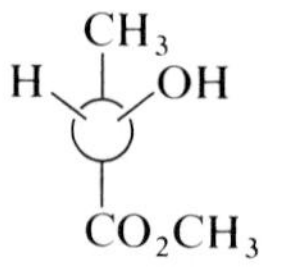

(f)

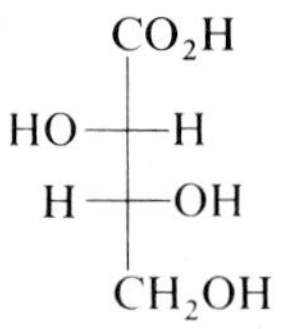

$$\begin{array}{c} CO_2H \\ HO \!-\!\!\!-\!\!\!-\! H \\ H \!-\!\!\!-\!\!\!-\! OH \\ CH_2OH \end{array}$$

4.54. (a) Draw formulas for all the geometric isomers of 1,2,3,4-cyclobutanetetracarboxylic acid. (b) Which of these geometric isomers contain no chiral carbons?   (c) Which are *meso* forms?   (d) Which can exist as enantiomers?

4.55. In the middle of the 19th century, it was proposed that a carbon atom with four groups attached is tetrahedral and not flat or pyramidal. Place yourself in the position of a chemist of that era and answer the following questions.   (1) In each of the following systems, how many stereoisomers would exist for a compound in which carbon is attached to four different groups?   (2) In each system, how many pairs of enantiomers are possible?

(a)   Carbon forms four planar bonds with bond angles of 90°.
(b)   Carbon forms four tetrahedral bonds with bond angles of 109.5°.
(c)   Carbon forms four regular (square) pyramidal bonds (carbon at the apex).

4.56. Draw the most stable chair-form conformations for the following structures. (Note that *cis* and *trans* relationships may be deduced by the broken wedges or lack of them.)

(a)

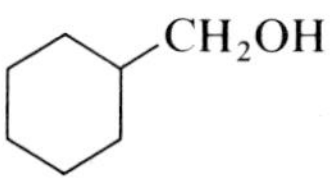

(b)

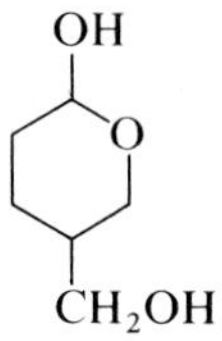

(c)

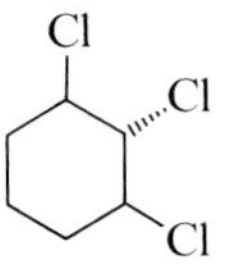

4.57. (a)   Draw the structure for the most stable conformer of the following menthyl chloride:

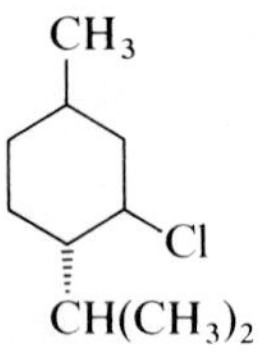

(b)   Draw the structure of the chair-form conformer in which Cl is *anti* to an H on one of the adjacent carbon atoms. Would you expect a substantial percentage of the menthyl chloride molecules to be in this latter conformation?

4.58. *Inositols* are 1,2,3,4,5,6-hexahydroxycyclohexanes that are found in all cells. *scyllo*-Inositol is the most stable of all the inositols. Draw its structure in the chair form.

4.59. Draw Newman projections of carbons 2 and 3 in the following structures, showing all possible staggered conformations:

(a)

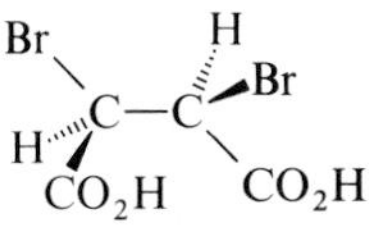

(b)

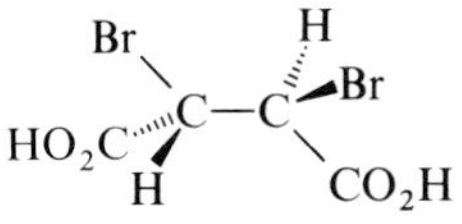

4.60. Draw Newman projections for the most stable and the least stable conformations of (1S,2R)-1,2-dibromo-1,2-diphenylethane.

4.61. Draw Newman projections for all stereoisomers of 1-bromo-1,2-diphenylpropane in which the H on carbon 2 and the Br are *anti*.

4.62. Assign (R) and (S) configurations to the stereoisomers of 1-bromo-1,2-diphenylpropane.

4.63. (S)-2-Iodobutane has an $[\alpha]_D^{24}$ of $+15.9°$.   (a) What is the observed rotation at 24° of an equimolar mixture of (R)- and (S)-2-iodobutane?   (b) What is the observed rotation (at 24°, 1-dm sample tube) of a solution (1.0 g/mL) of a mixture that is 25% (R)- and 75% (S)-2-iodobutane?

4.64. What is the specific rotation of each of the following solutions at 20°, Na D-line?

(a)   1.00 g of sample is diluted to 5.00 mL. A 3.00-mL aliquot is placed in a tube that is 1.0 cm long. The observed rotation is $+0.45°$.

(b)   A 0.20-g sample is diluted to 2.0 mL and placed in a 10-cm tube. The observed rotation is $-3.2°$.

4.65. Each of the following compounds is dissolved in an optically inactive solvent. Which would cause a rotation of the plane of polarization of plane-polarized light?

$$\underset{\quad}{NH_2}\ \underset{\quad}{NH_2}$$
$$\quad\quad |\quad\ \ |$$

(a)   (2S,3R)-butanediamine, $CH_3CH-CHCH_3$
(b)   (2S,3S)-butanediamine
(c)   an equimolar mixture of (a) and (b)
(d)   an equimolar mixture of (b) and (2R,3R)-butanediamine
(e)   the principal constituent of oil of balsam (page 912).

4.66. Calculate the specific rotation of   (a) an equimolar mixture of $(-)$-tartaric acid and *meso*-tartaric acid, and   (b) an equimolar mixture of $(-)$- and $(+)$-tartaric acid. (The $[\alpha]_D^{20}$ for $(+)$-tartaric acid is $+12.0°$.)

4.67. $(+)$-Tartaric acid (page 148) is produced commercially as a by-product of wine-making. Using flow equations, show how you would resolve $(\pm)$-amphetamine (1-phenyl-2-propylamine).

4.68. The compound $CH_3CH=C=CHCH_3$ belongs to a class of compounds called *allenes*. Although this compound has no chiral carbon, it does exist as a pair of enantiomers. Explain. (*Hint*: Consider the bonding and the resultant geometry. Use models.)

4.69.   Which of the following compounds has an enantiomer? [Use models for (b) and (c).]

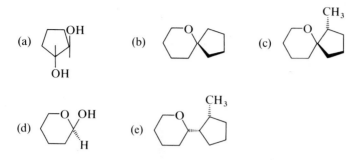

4.70.   (a)   If we consider the cyclohexane ring to be planar, we would predict that
cis-1,2-dimethylcyclohexane would be achiral, even though it contains two chiral
carbons, because it has an internal plane of symmetry. However, if we consider the
more realistic *chair form* of this compound, we see that it does *not* contain such a plane
of symmetry. Yet, *cis*-1,2-dimethylcyclohexane cannot be resolved into a pair of
enantiomers. Explain. (*Hint*: Consider the conformational equilibrium. Use models.)

  (b)   Could *trans*-1,2-dimethylcyclohexane be resolved? Explain.

CHAPTER 5

Alkyl Halides;
Substitution and
Elimination Reactions

Organohalogen compounds are used extensively in modern society. Some are used as solvents, some as insecticides, and some as intermediates in the synthesis of other organic compounds. Most organohalogen compounds are synthetic. Naturally occurring organohalogen compounds are rather rare. Thyroxine, a component of the thyroid hormone thyroglobulin, is a naturally occurring iodine compound. Another naturally occurring organohalogen compound is Tyrian purple, obtained in tiny amounts from a rare species of snail found in Crete. Tyrian purple was used as a dye by the Phoenician royalty and later by the Romans. (We still hear the phrases "royal purple" or "born to the purple.") The structure of this compound is shown on page 917.

As a class of compounds, organohalogen compounds are toxic and should be used with caution. For example, the solvents carbon tetrachloride (CCl_4) and chloroform ($CHCl_3$) both cause liver damage when inhaled in excess. Insecticides that contain halogens (such as DDT) have been widely used in agriculture; however, their use has declined in recent years due to their detrimental effects upon the environment.

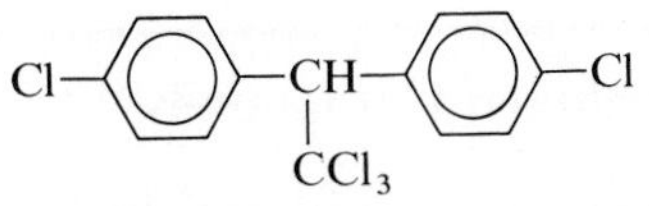

$$Cl\!-\!\!\langle\bigcirc\rangle\!-\!\underset{\underset{CCl_3}{|}}{CH}\!-\!\langle\bigcirc\rangle\!-\!Cl$$

DDT

an insecticide

Compounds containing only carbon, hydrogen, and a halogen atom fall into one of three categories: **alkyl halides**, **aryl halides** (in which a halogen is bonded to a carbon of an aromatic ring), and **vinylic halides** (in which a halogen is bonded to a double-bonded carbon). A few examples follow:

Alkyl halides (RX): CH_3I CH_3CH_2Cl

iodomethane   chloroethane

Aryl halides (ArX):

bromobenzene   a polychlorinated biphenyl (PCB)

*a toxic compound that
has been used as a
cooling fluid in transformers*

Vinylic halides:

$CH_2{=}CHCl$

$CH_3CH{=}CCH_3$ with Br substituent

chloroethene
(vinyl chloride)   2-bromo-2-butene

*the starting material for
polyvinyl chloride (PVC),
a plastic used for piping,
house siding, phonograph
records, and trash bags*

We have already defined R as the general symbol for an alkyl group. In a similar manner, Ar is the general symbol for an aromatic, or **aryl**, group. A halogen atom (F, Cl, Br, or I) may be represented by X. Using these general symbols, an alkyl halide is RX, and an aryl halide such as bromobenzene (C_6H_5Br) is ArX.

A halogen atom in an organic compound is a functional group, and the C—X bond is a site of chemical reactivity. In this chapter, we will discuss reactions of only the alkyl halides. *Aryl halides and vinylic halides do not undergo the reactions that we will present in this chapter*, partly because a bond from an sp^2 carbon is stronger than a bond from an sp^3 carbon (Section 2.4F). Because this is the first chapter devoted to compounds containing functional groups, we will also use this chapter as an introduction to organic chemical reactions.

SECTION 5.1.

Bonding in Organohalogen Compounds

The carbon–halogen sigma bond is formed by the overlap of an orbital of the halogen atom and a hybrid orbital of the carbon atom. We cannot be sure of the hybridization (or lack of it) of the halogen atom in an organic halide because a halogen forms only one covalent bond and therefore has no bond angle around it. However, carbon uses the same type of hybrid orbital to bond to a halogen atom as it does to bond to a hydrogen atom or to another carbon atom.

A halogen atom is electronegative with respect to carbon; therefore, alkyl halides are polar, as illustrated by the following dipole moments. (Dipole moments were discussed in Section 1.8B.)

Dipole moments of the methyl halides:

$$CH_3F \qquad CH_3Cl \qquad CH_3Br \qquad CH_3I$$
$$\mu = \quad 1.81 \text{ D} \qquad 1.86 \text{ D} \qquad 1.78 \text{ D} \qquad 1.64 \text{ D}$$

A carbon atom bonded to a halogen atom has a *partial positive charge*. The positive charge renders this particular carbon atom in an organic molecule susceptible to attack by an anion. As we will see, attack at this positive carbon is part of the general pattern of reactions of alkyl halides.

this carbon has no partial positive charge and is not attacked by anions

$$CH_3CH_2CH_2\overset{\delta+}{CH_2}\!\!-\!\!\overset{\delta-}{Br}$$

this carbon has a partial positive charge and can be attacked by an anion

SECTION 5.2.

Physical Properties of Halogenated Alkanes

The names, boiling points, and densities of several halogenated alkanes are listed in Table 5.1. Except for fluorine, halogen atoms are heavy compared to carbon or hydrogen atoms. The increase in molecular weight as halogen atoms are substituted into hydrocarbon molecules causes an increase in the boiling points of a series of compounds. Compare, for example, the boiling points of CH_3Cl, CH_2Cl_2, $CHCl_3$, and CCl_4.

Again, because of the mass of a halogen atom, the densities of liquid alkyl halides are often greater than those of other comparable organic compounds. Whereas most organic compounds are lighter than water, many common halogenated solvents, such as chloroform or dichloromethane, are heavier than water

TABLE 5.1. Physical properties of some halogenated alkanes

IUPAC name	Trivial name	Formula	Bp, °C	Density, g/cc at 20°
chloromethane	methyl chloride	CH_3Cl	−24	gas
dichloromethane	methylene chloride	CH_2Cl_2	40	1.34
trichloromethane	chloroform	$CHCl_3$	61	1.49
tetrachloromethane	carbon tetrachloride	CCl_4	77	1.60
bromomethane	methyl bromide	CH_3Br	5	gas
iodomethane	methyl iodide	CH_3I	43	2.28

(densities greater than 1.0 g/cc). These compounds sink to the bottom of a container of water rather than floating on top as do most organic compounds. Halogenated hydrocarbons do not form hydrogen bonds and are insoluble in water.

Nomenclature and Classification of Alkyl Halides

In the IUPAC system, an alkyl halide is named with a **halo-** prefix. Many common alkyl halides also have trivial **functional-group names**. In these names, the name of the alkyl group is given, followed by the name of the halide.

$$\begin{array}{c} Cl \\ | \\ CH_3CHCH_2CH_3 \end{array} \qquad \langle\ \rangle\!-\!Br$$

IUPAC:	2-chlorobutane	bromocyclohexane
trivial:	*sec*-butyl chloride	cyclohexyl bromide

In chemical reactions, the structure of the alkyl portion of an alkyl halide is important. Therefore, we need to differentiate the four types of alkyl halides: **methyl**, **primary**, **secondary**, and **tertiary**.

A **methyl halide** is a structure in which one hydrogen of methane has been replaced by a halogen.

The methyl halides:

CH_3F	CH_3Cl	CH_3Br	CH_3I
fluoromethane	chloromethane	bromomethane	iodomethane

The *head carbon* of an alkyl halide is the carbon atom bonded to the halogen. A **primary (1°) alkyl halide** (RCH_2X) has *one alkyl group* bonded to the head carbon. In the following examples, the head carbons and their hydrogens are circled.

Primary alkyl halides (one alkyl group attached to head):

$$CH_3\!-\!CH_2Br \qquad (CH_3)_3C\!-\!CH_2Cl \qquad \langle\ \rangle\!-\!CH_2I$$

bromoethane	1-chloro-2,2-dimethylpropane	(iodomethyl)cyclohexane
(ethyl bromide)	(neopentyl chloride)	

A **secondary (2°) alkyl halide** (R_2CHX) has *two* alkyl groups attached to the head carbon, and a **tertiary (3°) alkyl halide** (R_3CX) has *three* alkyl groups attached to the head carbon. (Note that a halogen attached to a cycloalkane ring must be either secondary or tertiary.)

Secondary alkyl halides (two alkyl groups attached to head):

$$\begin{array}{c} CH_3 \\ | \\ CH_3CH_2\ CH\!-\!Br \end{array} \qquad \begin{array}{c} H_2C\!-\!CH_2 \\ H_2C \qquad\ \ CH\!-\!Cl \\ C \\ H_2 \end{array} \text{ or } \langle\ \rangle\!-\!Cl$$

2-bromobutane	chlorocyclopentane
(*sec*-butyl bromide)	(cyclopentyl chloride)

Tertiary alkyl halides (three alkyl groups attached to head):

$$CH_3-\overset{\overset{\displaystyle CH_3}{|}}{\underset{\underset{\displaystyle CH_3}{|}}{C}}-Cl$$

2-chloro-2-methylpropane
(*t*-butyl chloride)

1-bromo-1-methylcyclopentane

SAMPLE PROBLEM

Classify each of the following alkyl halides as 1°, 2°, or 3°:

(a) $CH_3\overset{\overset{\displaystyle CH_3}{|}}{C}HCH_2Cl$ (b)   (c) $CH_3\overset{|}{C}HCl$
$CH_2CH_2CH_3$

Solution: (a) 1°;   (b) 3°;   (c) 2°

STUDY PROBLEMS

5.1.   Write two names for   (a) $(CH_3)_3CI$;   (b) $(CH_3)_2CHCl$;   (c) $(CH_3)_2CHCH_2I$.

5.2.   Write structural formulas for   (a) 1,1-dibromobutane;   (b) 3-chloro-1-butene;
and   (c) 2-fluoro-1-ethanol.

SECTION 5.4.

A Preview of Substitution and Elimination Reactions

A.   Substitution reactions

The head carbon atom of an alkyl halide has a partial positive charge. This carbon is particularly susceptible to attack by an anion or by any other species that carries an unshared pair of electrons in the outer shell. The result is a **substitution reaction**—a reaction in which one atom, ion, or group is substituted for another.

the σ-bond electrons
leave with the halogen

$$H\ddot{O}:^- \; + \; CH_3\overset{\delta+}{C}H_2\overset{\delta-}{\ddot{B}}r: \; \longrightarrow \; CH_3CH_2-\ddot{O}H \; + \; :\ddot{B}r:^-$$

hydroxide ion   bromoethane   ethanol

$$CH_3\ddot{O}:^- \; + \; CH_3CH_2CH_2\overset{}{\ddot{C}}l: \; \longrightarrow \; CH_3CH_2CH_2-\ddot{O}CH_3 \; + \; :\ddot{C}l:^-$$

methoxide ion   1-chloropropane   methyl *n*-propyl ether

In substitution reactions of the alkyl halides, the halide is called the **leaving group**, a term meaning any group that can be displaced from a carbon atom.

Halide ions are *good* leaving groups because they are very weak bases. Strong bases, such as ⁻OH, are very poor leaving groups. In this chapter, we will discuss only the halides as leaving groups; we will introduce other leaving groups as we encounter them in subsequent chapters.

In substitution reactions of alkyl halides, the iodide ion is the halide most easily displaced, followed by the bromide ion and then the chloride ion. Because F⁻ is a stronger base than the other halide ions, it is not as good a leaving group. From a practical standpoint, only Cl, Br, and I are good enough leaving groups to be useful in substitution reactions. For this reason, when we refer to RX, we usually mean alkyl chlorides, bromides, and iodides.

$$RF \quad RCl \quad RBr \quad RI$$
increasing reactivity

The species that attacks an alkyl halide in a substitution reaction is called a **nucleophile** (literally, "nucleus lover"), often abbreviated Nu⁻. In the preceding equations, OH⁻ and CH_3O^- are the nucleophiles. Generally, a nucleophile is any species that is attracted to a positive center; thus, a nucleophile is a Lewis base. Most nucleophiles are anions; however, some neutral polar molecules, such as H_2O, CH_3OH, and CH_3NH_2, can also act as nucleophiles. These neutral molecules all contain unshared electrons that can be used to form sigma bonds. Substitutions by nucleophiles are called **nucleophilic substitutions**, or **nucleophilic displacements**.

The opposite of a nucleophile is an **electrophile** ("electron lover"), often abbreviated E⁺. An electrophile is any species that is attracted toward a negative center—that is, an electrophile is a Lewis acid, such as H⁺ or $ZnCl_2$. Electrophilic reactions are common in organic chemistry; you will encounter many of these reactions in later chapters.

B.   Elimination reactions

When an alkyl halide is treated with a *strong base*, an **elimination reaction** can occur. An elimination reaction is one in which a molecule loses atoms or ions from its structure. The organic product of an elimination reaction of an alkyl halide is an alkene. In this type of elimination reaction, the elements H and X are lost from the alkyl halide; therefore, these reactions are also called **dehydrohalogenation reactions**. (The prefix **de-** means "minus" or "loss of.")

$$CH_3CH-CH_2 + {}^-\!\ddot{O}H \longrightarrow CH_3CH{=}CH_2 + H_2\ddot{O} + Br^-$$

2-bromopropane   propene
(isopropyl bromide)   (propylene)

$$CH_3-\underset{\underset{CH_3}{|}}{\overset{\overset{CH_3}{|}}{C}}-Cl + {}^-OH \longrightarrow CH_3-\underset{\underset{CH_3}{|}}{\overset{\overset{CH_2}{\|}}{C}} + H_2O + Cl^-$$

2-chloro-2-methylpropane   methylpropene
(*t*-butyl chloride)   (isobutylene)

C.   Competing reactions

A hydroxide ion or alkoxide ion (RO^-) can react as a *nucleophile* in a substitution reaction or as a *base* in an elimination reaction. Which type of reaction actually occurs depends on a number of factors, such as the structure of the alkyl halide ($1°$, $2°$, or $3°$), the strength of the base, the nature of the solvent, and the temperature.

   Methyl and primary alkyl halides tend to yield substitution products, not elimination products. Under equivalent conditions, *tertiary alkyl halides* yield principally elimination products, and not substitution products. *Secondary alkyl halides* are intermediate in their behavior; the relative proportion of the substitution product to the elimination product depends to a large extent upon the experimental conditions.

$1°$: $CH_3CH_2Br + CH_3CH_2O^- \xrightarrow[25°]{CH_3CH_2OH} CH_3CH_2OCH_2CH_3$
 almost 100%

$2°$: $(CH_3)_2CHBr + CH_3CH_2O^- \xrightarrow[25°]{CH_3CH_2OH} (CH_3)_2CHOCH_2CH_3 + CH_2{=}CHCH_3$
 20%   80%

$3°$: $(CH_3)_3CBr + CH_3CH_2O^- \xrightarrow[25°]{CH_3CH_2OH} (CH_3)_3COCH_2CH_3 + CH_2{=}C(CH_3)_2$
 5%   95%

   Because more than one reaction can occur between an alkyl halide and a nucleophile or base, substitution reactions and elimination reactions are said to be **competing reactions**. Competing reactions are common in organic chemistry. Because mixtures of products are the rule rather than the exception when competing reactions occur, we will not balance most of the organic equations in this book.

   In Sections 5.5–5.10, we will discuss two different types of substitution reactions (called S_N1 and S_N2 reactions) and two types of elimination reactions (E1 and E2). We will discuss each type of reaction individually, and then we will summarize the factors that can help us predict which of these reactions will predominate in a given case.

D.   Nucleophilicity versus basicity

Before proceeding with the details of substitution and elimination reactions, let us briefly consider the similarities and the differences between bases and nucleophiles. Under the proper circumstances, all bases can act as nucleophiles. Conversely, all nucleophiles can act as bases. In either case, the reagent reacts by donating a pair of electrons to form a new sigma bond.

   Basicity is a measure of a reagent's ability to accept a proton in an acid–base reaction. Therefore, the relative base strengths of a series of reagents are determined by comparing the relative positions of their equilibria in an acid–base reaction, such as the degree of ionization in water.

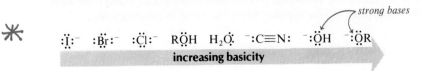

strong bases

$:\ddot{I}:^-$ $:\ddot{Br}:^-$ $:\ddot{Cl}:^-$ $R\ddot{O}H$ $H_2\ddot{O}:$ $^-:C{\equiv}N:$ $:\ddot{O}H^-$ $^-:\ddot{O}R$

increasing basicity

In contrast to basicity, **nucleophilicity** is a measure of a reagent's ability to cause a substitution reaction. The relative nucleophilicities of a series of reagents are determined by their relative rates of reaction in a substitution reaction, such as a substitution reaction with bromoethane.

$$H_2\ddot{O}\quad R\ddot{O}H\quad :\ddot{C}l:^-\quad :\ddot{B}r:^-\quad ^-:\ddot{O}H\quad ^-:\ddot{O}R\quad :\ddot{I}:^-\quad ^-:CN:$$

increasing nucleophilicity

A list of relative nucleophilicities does not exactly parallel a list of base strengths; however, a stronger base is usually a better nucleophile than a weaker base. For example, OH^- (a strong base) is a better nucleophile than Cl^- or H_2O (weak bases). We will discuss the factors affecting nucleophilicity in greater detail in Section 5.10.

SAMPLE PROBLEM

Give the structures of the substitution products (if any) of the following reactions:

(a) $CH_3CH_2CH_2CH_2I + {}^-CN \longrightarrow$

(b)  ⬡—$Cl + {}^-OH \longrightarrow$

(c) Cl—⬡—$CH_2Cl + {}^-OH \longrightarrow$

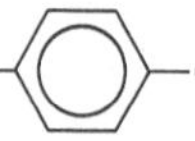

Solution: (a) $CH_3CH_2CH_2CH_2CN$; (b) no reaction for an aryl halide;

(c) Cl—⬡—CH_2OH

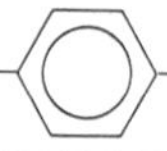

STUDY PROBLEMS

5.3. Write equations (not necessarily balanced) for the reactions of the following compounds with CH_3O^-. Show both substitution and elimination products (if any).

(a) $(CH_3CH_2)_2CHCl$ (b) 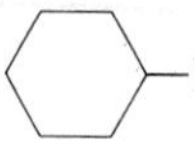 ⬡—Br (c) 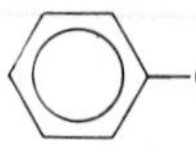 ⬡—CH_2I

5.4. Show by equations how you would prepare the following compounds from alkyl halides and appropriate nucleophiles. Show two paths to compound (a).

(a) $CH_3CH_2CH_2OCH_2CH_2CH(CH_3)_2$ (b) 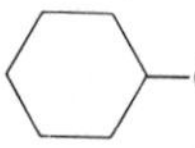 ⬡—CN

SECTION 5.5.

The S_N2 Reaction

The reaction of bromoethane with hydroxide ion to yield ethanol and bromide ion (page 165) is a typical **S_N2 reaction**. (S_N2 means "substitution, nucleophilic, bimolecular." The term *bimolecular* will be defined in Section 5.5B.) Virtually any methyl or primary alkyl halide undergoes an S_N2 reaction with any relatively strong nucleophile: ⁻OH, ⁻OR, ⁻CN, and others that we have not yet mentioned. Methyl or primary alkyl halides also undergo reaction with weak nucleophiles, such as H_2O, but these reactions are too slow to be of practical value. Secondary alkyl halides can also undergo S_N2 reactions; however, tertiary alkyl halides do not.

A. Reaction mechanism

The detailed description of how a reaction occurs is called a **reaction mechanism**. A reaction mechanism must take into account all known facts. For some reactions, the number of facts known is considerable, and the particular reaction mechanisms are accepted by most chemists. The mechanisms of some other reactions are still quite speculative. The S_N2 reaction is one that has been studied extensively; there is a large amount of experimental data supporting the mechanism that we will present.

For molecules to undergo a chemical reaction, they must first collide. Most collisions between molecules do not result in a reaction; rather, the molecules simply rebound. To undergo reaction, the colliding molecules must contain enough *potential energy* for bond breakage to occur. Also, the *orientation* of the molecules relative to each other is often an important factor in determining whether a reaction will occur. This is particularly true in an S_N2 reaction. In this section, we will first discuss the stereochemistry of the S_N2 reaction, then we will discuss the energy requirements.

B. Stereochemistry of an S_N2 reaction

In the S_N2 reaction between bromoethane and hydroxide ion, the oxygen of the hydroxide ion collides with the rear of the head carbon and displaces the bromide ion.

Overall S_N2 reaction:

rear attack

When a nucleophile collides with the backside of a tetrahedral carbon atom bonded to a halogen, two things occur simultaneously: (1) a new bond begins to form, and (2) the C—X bond begins to break. The process is said to be a one-step, or **concerted**, process. If the potential energy of the two colliding species is high enough, a point is reached where it is energetically more favorable for the new bond to form and the old C—X bond to break. As the reactants are converted to products, they must pass through an in-between state that has a high potential energy relative to the reactants and the products. This state is called the **transition state**, or the **activated complex**. Because the transition state involves two particles (Nu⁻ and RX), the S_N2 reaction is said to be **bimolecular**. (The "2" in S_N2 indicates bimolecular.)

partial bonds

$$HO^- + \quad \overset{H}{\underset{H}{\overset{CH_3}{\underset{|}{C}}}}{-}Br \longrightarrow \left[HO - - - \overset{H}{\underset{H}{\overset{CH_3}{\underset{|}{C}}}} - - - Br \right]^- \longrightarrow HO - \overset{H_3C}{\underset{H}{\overset{H}{C}}} + Br^-$$

transition state:
high potential energy,
equally able to go to reactants
or products

A transition state in any reaction is the fleeting high-energy arrangement of the reactants as they go to products. We cannot isolate a transition state and put it in a flask. The transition state is simply a description of "molecules in a state of transition." We will often use square brackets in an equation to show any temporary, nonisolable structure in a reaction. Here, we use brackets to enclose the structure of a transition state. Later, we will sometimes use brackets to indicate unstable products that undergo further reaction.

For the S_N2 reaction, the transition state involves a temporary rehybridization of the head carbon from sp^3 to sp^2 and finally back to sp^3 again. In the transition state, the carbon atom has three planar sp^2 bonds, plus two half-bonds using the p orbital.

p orbital

$$- - - \overset{H \quad CH_3}{\underset{|}{\overset{|}{C}}} - - -$$

sp² carbon

As the nucleophile attacks from the rear of the molecule, relative to the halogen atom, the other three groups attached to the carbon flatten out in the transition state, then flip to the other side of the carbon atom, much as an umbrella blowing inside out. (Models would be useful to help you visualize this.) This flipping is called **inversion**, or **Walden inversion** after the chemist who discovered it.

The existence of inversion as part of the mechanism of an S_N2 reaction has been beautifully demonstrated by reactions of pure enantiomers of chiral secondary alkyl halides. For example, the S_N2 reaction of (R)-2-bromooctane with ⁻OH yields almost exclusively (S)-2-octanol.

$$HO^- + \quad \overset{H}{\underset{CH_3}{\overset{CH_2(CH_2)_4CH_3}{\underset{|}{C}}}}{-}Br \quad \xrightarrow{S_N2} \quad HO - \overset{CH_2(CH_2)_4CH_3}{\underset{CH_3}{\overset{H}{C}}} \quad + Br^-$$

(R)-2-bromooctane

(S)-2-octanol

96% inversion

Most reactions involving chiral molecules are carried out with racemic mixtures—that is, equal mixtures of (*R*) and (*S*) reactants. In these cases, the products also are racemic mixtures. Even though inversion occurs, we cannot observe the effects because half the molecules go one way and half the molecules go the other way.

STUDY PROBLEM

5.5. Write an equation (showing the stereochemistry by using dimensional formulas) for the S_N2 reaction of (*S*)-2-bromobutane with CN^-.

C.   Energy in an S_N2 reaction

We have mentioned that colliding molecules need energy to undergo reaction. We will now look at these energy requirements in more detail.

Molecules moving around in a solution contain a certain amount of potential energy in their bonds and a certain amount of kinetic energy from their movement. Not all molecules in solution have exactly the same amount of potential or kinetic energy; however, we may speak of the *average energy of the molecules*. The total energy of the reaction mixture may be increased, usually by heating the solution. When heated, the molecules gain kinetic energy, collide more frequently and more energetically, and exchange some kinetic energy for potential energy.

Before a reaction can begin to occur, some of the colliding molecules and ions in the flask must contain enough energy to reach the transition state upon collision. Reaching the potential-energy level of the transition state is rather like driving an old car to a mountain pass. Does the car have enough energy to make the top? Or will it stall and slide back down the mountain? Once you reach the top, which way do you go—back the way you came or on down the other side? Once you are descending the far side, the choice is easy—you can relax and let the car roll to the bottom.

Figure 5.1 shows an energy diagram for the progress of an S_N2 reaction. The potential energy required to reach the transition state forms an energy barrier;

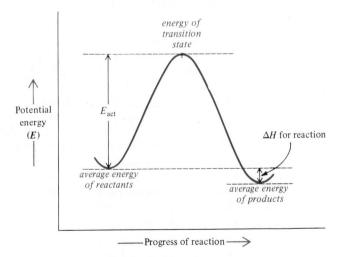

FIGURE 5.1. Energy diagram of an S_N2 reaction.

it is the point of maximum energy on the graph. For a colliding alkyl halide and nucleophile to reach the transition state, they need a certain minimum amount of energy called the **energy of activation E_{act}**. At the transition state, the molecules find it just as easy to go back to reactants or on to products. But, once over the top, the path of least resistance is that of going to products. The difference between the average potential energy of the reactants and that of the products is the change in enthalpy ΔH for the reaction (see Section 1.7).

D.   Rate of an S_N2 reaction

Each molecule that undergoes reaction to yield product must pass through the transition state, both structurally and energetically. Since the energies of all the molecules are not the same, a certain amount of time is required for all the molecules present to react. This time requirement gives rise to the **rate of a reaction**. The rate of a chemical reaction is a measure of how fast the reaction proceeds; that is, how fast reactants are consumed and products are formed. **Reaction kinetics** is the term used to describe the study and measurement of reaction rates.

The rate of a reaction depends on many variables, some of which may be held constant for a given experiment (temperature and solvent, for example). In this chapter, we will be concerned primarily with two variables: (1) the concentrations of the reactants, and (2) the structures of the reactants.

Increasing the concentration of reactants undergoing an S_N2 reaction increases the rate at which products are formed because it increases the frequency of molecular collisions. Typically, the rate of an S_N2 reaction is proportional to the concentrations of both reactants. If all other variables are held constant and the concentration of either the alkyl halide or the nucleophile is doubled, the rate of product formation is doubled. If either concentration is tripled, the rate is tripled.

$$Nu^- + RX \longrightarrow RNu + X^-$$

$$S_N2 \text{ rate} = k\,[RX][Nu^-]$$

In this equation, $[RX]$ and $[Nu^-]$ represent the concentrations in moles/liter of the alkyl halide and the nucleophile, respectively. The term k is the proportionality constant, called the **rate constant**, between these concentrations and the measured rate of product formation. The value for k is constant for the same reaction under identical experimental conditions (solvent, temperature, etc.)

SAMPLE PROBLEM

What would be the effect on the rate of the S_N2 reaction of CH_3I with CH_3O^- if the concentrations of *both* reactants were doubled and all other variables were held constant?

Solution: If the concentrations of both CH_3I and CH_3O^- were doubled, the rate would quadruple—the reaction would proceed four times as fast.

Because the rate of an S_N2 reaction depends on the concentrations of two particles (RX and Nu^-), the rate is said to be **second order**. The S_N2 reaction is said to follow **second-order kinetics**. (Although the S_N2 reaction is also bimolecular, not all bimolecular reactions are second-order and not all second-order reactions are bimolecular. For example, see Problem 5.64, page 213.)

E.   Effect of E_{act} on rate and on products

The effect of the energies of activation on the relative rates of reaction may be stated simply: *Under the same conditions, the reaction with the lower E_{act} has a faster rate.* The reason for this relationship is that, if less energy is required for reaction, a greater number of molecules have enough energy to react.

Let us consider a case in which one starting material can undergo two different irreversible reactions leading to two different products. (A reaction in which the E_{act} of the reverse reaction is substantially greater than the E_{act} of the forward reaction is exothermic and essentially irreversible.) When the starting material can undergo two such reactions, *the product of the faster reaction (the one with the lower E_{act}) predominates.* Figure 5.2 shows energy curves of two such reactions of the same starting material.

The E_{act} is the energy of the transition state relative to that of the reactants. Therefore, relative rates of reaction are related to the energies of the transition states. In competing reactions of the same starting material, *the reaction with the lower-energy transition state is the faster reaction.* From Figure 5.2, it is evident that the reaction with the lower-energy transition state has the lower E_{act}.

A species of low potential energy is more stable than one of high energy. Therefore, we can also say that *the reaction with a more stabilized transition-state structure is the faster reaction.* This concept is useful in analyzing competing reactions to determine which reaction predominates.

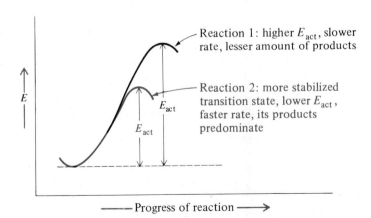

FIGURE 5.2. In competing reactions of a single starting material, the reaction with the lower E_{act} is the faster one. If the reactions are essentially irreversible, the products of the faster reaction predominate.

F.   Effect of structure on rate

Reaction kinetics provide a valuable tool for exploring the effects of structure upon reactivity. Consider the following two reactions:

(1) $OH^- +$ CH_3Br ⟶ $CH_3OH + Br^-$

                        bromomethane   methanol

                        a methyl halide

(2) $OH^- +$ CH_3CH_2Br ⟶ $CH_3CH_2OH + Br^-$

                        bromoethane   ethanol

                        a 1° alkyl halide

Both are S_N2 reactions and both yield alcohols. The two reactions differ only in the alkyl portion of the alkyl halides. We may ask the following question: "Does this difference in the alkyl group have an effect upon the rate of the S_N2 reaction?" To answer such a question, the rates of the two reactions are measured under the same reaction conditions (the same solvent, the same concentrations, and the same temperature). Then, either the two rate constants (k_1 and k_2) are determined or, more commonly, the *relative rates* are determined.

$$CH_3Br \xrightarrow{\;OH^-\;} CH_3OH \qquad rate_1 = k_1[CH_3Br][OH^-]$$

$$CH_3CH_2Br \xrightarrow{\;OH^-\;} CH_3CH_2OH \qquad rate_2 = k_2[CH_3CH_2Br][OH^-]$$

$$\textit{relative rates of reaction of } CH_3Br \textit{ compared to } CH_3CH_2Br = \frac{rate_1}{rate_2}$$

Under the experimental conditions used in one study, bromomethane undergoes reaction 30 times faster than bromoethane. (If it takes one hour for the bromoethane reaction to reach 50% completion, the bromomethane reaction would take about 1/30 as long, or only two minutes, to reach 50% completion!) We conclude that there is indeed a big difference in how the methyl and ethyl groups affect the rate of the reaction.

In a similar fashion, the relative rates of a variety of S_N2 reactions of alkyl halides have been determined. Table 5.2 shows some average relative rates (compared to ethyl halides) of S_N2 reactions of alkyl halides.

TABLE 5.2. Average relative rates of some alkyl halides in a typical S_N2 reaction

Alkyl halide	Relative rate
CH_3X	30
CH_3CH_2X	1
$CH_3CH_2CH_2X$	0.4
$CH_3CH_2CH_2CH_2X$	0.4
$(CH_3)_2CHX$	0.03
$(CH_3)_3CX$	~0

G.   Steric hindrance in S$_N$2 reactions

In S$_N$2 reactions of the alkyl halides listed in Table 5.2, methyl halides show the fastest rate, followed by primary alkyl halides, then secondary alkyl halides. Tertiary alkyl halides do not undergo S$_N$2 reactions.

※   3°RX   2°RX   1°RX   CH$_3$X
increasing rate of S$_N$2 reaction

As the number of alkyl groups attached to the head carbon increases (CH$_3$X → 1° → 2° → 3°), the transition state becomes increasingly crowded with atoms. Consider the following examples of reactions of alkyl bromides with the methoxide ion (CH$_3$O$^-$) as the nucleophile (CH$_3$O$^-$ + RBr → CH$_3$OR + Br$^-$):

CH$_3$O$^-$ ⟶ (bromomethane)

bromomethane

very fast

CH$_3$O$^-$ ⟶ (bromoethane)

bromoethane

moderate rate

CH$_3$O$^-$ ⟶ (2-bromopropane)

2-bromopropane

very slow

CH$_3$O$^-$ ⟶ (t-butyl bromide)

t-butyl bromide

no S$_N$2 reaction

Spatial crowding in structures is called **steric hindrance**. When large groups are crowded in a small space, repulsions between groups become severe and therefore the energy of the system is high. In an S$_N$2 reaction, the energy of a crowded transition state is higher than that of a transition state with less steric hindrance. For this reason, the rates of reaction become progressively slower in the series methyl, primary, secondary, and tertiary (see Figure 5.3). The energy of the S$_N$2 transition state of a tertiary alkyl halide is so high relative to other possible reaction paths that the S$_N$2 reaction does not proceed.

SAMPLE PROBLEM

The rate of S$_N$2 reaction of neopentyl bromide, (CH$_3$)$_3$CCH$_2$Br, with sodium ethoxide, Na$^+$ $^-$OCH$_2$CH$_3$, proceeds about 0.00001 times as fast as the reaction of bromoethane. Explain.

Solution: Although neopentyl bromide is a primary alkyl halide, the alkyl group attached to the head carbon atom is very bulky. The steric hindrance in the transition state is considerable. Therefore, the E_{act} is high and the rate is slow.

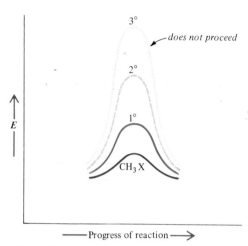

FIGURE 5.3. Energy diagram for S_N2 reactions of different types of alkyl halides.

STUDY PROBLEM

5.6.   Which member of each of the following pairs undergoes the faster S_N2 reaction?

(a) ⬡—CH_2Cl and CH_3—⬡—Cl

(b) $(CH_3)_2CHCHCH_2CH_3$ and $(CH_3)_2CHCH_2CHCH_3$
            | |
            Cl   Cl

(c) ⬡—CH_2Cl and CH_3—⬡—Cl

SECTION 5.6.

The S_N1 Reaction

Because of steric hindrance, *t*-butyl bromide and other tertiary alkyl halides do not undergo S_N2 reactions. Yet, if *t*-butyl bromide is treated with a nucleophile that is a *very weak base* (such as H_2O or CH_3CH_2OH), substitution products are formed, along with elimination products. Because H_2O or CH_3CH_2OH is also used as the solvent, this type of substitution reaction is sometimes called a **solvolysis reaction** (from *solvent* and *-lysis,* "breaking").

$$(CH_3)_3CBr \xrightarrow[\quad 25°\quad]{CH_3CH_2OH} (CH_3)_3COCH_2CH_3 + CH_2{=}C(CH_3)_2$$

t-butyl ethyl ether   methylpropene
          80%   20%

$$\xrightarrow[\quad 25°\quad]{H_2O} (CH_3)_3COH \;+\; CH_2{=}C(CH_3)_2$$

t-butyl alcohol   methylpropene
          70%   30%

If tertiary alkyl halides cannot undergo S_N2 reactions, how are the substitution products formed? The answer is that tertiary alkyl halides undergo substitution by a different mechanism, called the S_N1 **reaction** (substitution, nucleophilic, unimolecular.) The experimental results obtained in S_N1 reactions are considerably different from those obtained in S_N2 reactions. Typically, if a pure enantiomer of an alkyl halide containing a chiral C—X carbon undergoes an S_N1 reaction, *racemic* substitution products are obtained (not the inverted products observed in an S_N2 reaction). It has also been determined that the concentration of nucleophile generally has little effect upon the overall rate of an S_N1 reaction. (In contrast, the rate of an S_N2 reaction is directly proportional to the concentration of the nucleophile.) To explain these experimental results, we will discuss the mechanism of the S_N1 reaction using *t*-butyl bromide and water. For the moment, we will ignore the elimination product.

$$(CH_3)_3CBr \ + \ H_2O \ \longrightarrow \ (CH_3)_3COH \ + \ H^+ + Br^-$$

  t-butyl bromide   *t*-butyl alcohol

A.   The S_N1 mechanism

The S_N1 reaction of a tertiary alkyl halide is a *stepwise reaction*. Step 1 is the cleavage of the alkyl halide into a pair of ions: the halide ion and a **carbocation**, an ion in which a carbon atom carries a positive charge. Because S_N1 reactions involve ionization, these reactions are aided by polar solvents, such as H_2O, that can solvate, and thus stabilize, ions.

Step 1:

$$(CH_3)_3C{-}\overset{..}{\underset{..}{Br}}: \ \longrightarrow \ \left[(CH_3)_3C\overset{\delta+}{-}{-}{-}\overset{\delta-}{\underset{..}{\overset{..}{Br}}}: \right] \ \longrightarrow \ [(CH_3)_3C^+] \ + \ :\overset{..}{\underset{..}{Br}}:^-$$

         transition state 1   *unstable carbocation*
                                                  intermediate

Step 2 is the combining of the carbocation with the nucleophile (H_2O) to yield the initial product, a protonated alcohol.

Step 2:

$$[(CH_3)_3\overset{\frown}{C^+}] + H_2\overset{..}{O}: \ \longrightarrow \ \left[(CH_3)_3\overset{\delta+}{C}{-}{-}{-}\overset{H}{\underset{\delta+}{\overset{|}{O}H}} \right] \ \longrightarrow \ (CH_3)_3C{-}\overset{H}{\overset{|}{O}H}$$

                   transition state 2   *protonated*
                                                       t-butyl alcohol

The final step in the sequence is the loss of H^+ by the protonated alcohol in a rapid, reversible acid–base reaction with the solvent.

Step 3:

$$(CH_3)_3C\overset{\overset{\displaystyle H}{\overset{|}{\underset{+}{}}}}{O}H + H_2\overset{..}{O}: \ \rightleftharpoons \ (CH_3)_3C\overset{..}{O}H \ + \ H_3\overset{+}{O}:$$

            excess   *t*-butyl alcohol

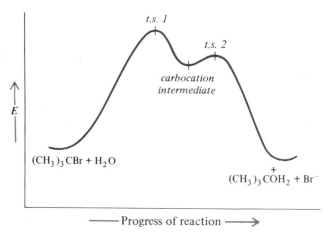

FIGURE 5.4. Energy diagram for a typical S_N1 reaction.

The overall reaction of *t*-butyl bromide with water is thus actually composed of two separate reactions: the S_N1 reaction (ionization followed by combination with the nucleophile) and an acid–base reaction. The steps may be summarized:

$$(CH_3)_3CBr \xrightarrow[\text{slow}]{-Br^-} [(CH_3)_3C^+] \xrightarrow[\text{fast}]{H_2O} (CH_3)_3C\overset{+}{O}H_2 \xrightarrow[\text{fast}]{-H^+} (CH_3)_3COH$$

$\frown$ *means loss of Br$^-$*

S_N1 *reaction*   *acid–base reaction*

Let us consider the energy diagram for an S_N1 reaction (Figure 5.4). Typically, Step 1 (the ionization) has a high E_{act}; it is the *slow step* in the overall process. Enough energy must be supplied to the tertiary alkyl halide to break the C—X sigma bond and yield the carbocation and the halide ion.

The carbocation is an **intermediate** in this reaction, a structure that is formed during the reaction and then undergoes further reaction to products. An intermediate is not a transition state. An intermediate has a finite lifetime; a transition state does not. At the transition state, molecules are undergoing bond-breaking and bond-making. The potential energy of a transition state is a high point on a potential-energy curve. By contrast, an intermediate is a temporary, reactive product. No bond-breaking or bond-making is occurring in an intermediate. An intermediate is of lower energy than transition states that surround it, but is of higher energy than the final products. The energy diagram in Figure 5.4 shows a dip for the carbocation formation; the dip is not a large one because the carbocation is a high-energy, reactive species.

Step 2 in the sequence of an S_N1 reaction is the reaction of the carbocation with a nucleophile. The two combine in a reaction having a low E_{act}—a fast reaction.

SAMPLE PROBLEM

Write equations for the steps in the S_N1 reaction of 2-chloro-2-methylbutane with methanol. (Include the deprotonation step in your answer.)

Solution:

Step 1 (ionization):

$$(CH_3)_2\overset{\overset{\displaystyle :\ddot{C}l:}{|}}{C}CH_2CH_3 \longrightarrow [(CH_3)_2\overset{+}{C}CH_2CH_3] + :\ddot{C}l:^-$$

a carbocation

Step 2 (combination with Nu⁻):

$$[(CH_3)_2\overset{+}{C}CH_2CH_3] + CH_3\ddot{O}H \longrightarrow (CH_3)_2\overset{\overset{\displaystyle H\overset{+}{O}CH_3}{|}}{C}CH_2CH_3$$

a protonated ether

Step 3 (loss of H⁺ to solvent):

$$CH_3\ddot{O}H + (CH_3)_2\overset{\overset{\displaystyle H-\overset{+}{O}CH_3}{|}}{C}CH_2CH_3 \longrightarrow (CH_3)_2\overset{\overset{\displaystyle :\ddot{O}CH_3}{|}}{C}CH_2CH_3 + CH_3\overset{+}{\ddot{O}}H_2$$

an ether

STUDY PROBLEM

5.7.   Complete the following equations for solvolysis reactions:

(a) $(CH_3CH_2)_3CI + H_2O \longrightarrow$

(b) [cyclohexane ring with CH_3 and Cl substituents] $+ CH_3OH \longrightarrow$

B.   Stereochemistry of an S_N1 reaction

A carbocation (or **carbonium ion**, as it is also called) is a carbon atom with only three groups attached, instead of the usual four. Because there are only three groups, the bonds to these groups lie in a plane, and the angles between the bonds from the positive carbon are approximately 120°. To attain this geometry, the positive carbon is sp^2-hybridized and has an empty p orbital.

empty p orbital

sigma-bond angles around C⁺ are approximately 120°

Let us consider the S_N1 reaction of a chiral alkyl halide. When (S)-3-bromo-3-methylhexane is treated with water, it undergoes racemization to yield the (R)(S) alcohol. The S_N1 mechanism can be used to explain this racemization. The first step in this S_N1 reaction is ionization of the alkyl halide to a carbocation and a halide ion:

Step 1:

$$CH_3\text{-}\underset{\underset{CH_3CH_2CH_2}{|}}{\overset{\overset{Br}{|}}{C}}\text{-}CH_2CH_3 \xrightarrow{-Br^-} \left[\underset{CH_3CH_2CH_2}{CH_3\text{-}\overset{+}{C}\text{-}CH_2CH_3} \right]$$

(S)-3-bromo-3-methylhexane   *planar carbocation*

In the second step, H_2O attacks the carbocation to form two protonated alcohols. (We show the alcohols *after* deprotonation in the following equations.) If the H_2O molecule attacks the empty *p* orbital from the top side, as we have shown the structure, the carbocation yields the (S)-enantiomer of the alcohol. However, if the H_2O attacks from the bottom side, the (R)-enantiomer is the product. Since there is equal probability of attack from either side, equal amounts of the (R)- and (S)-alcohols are produced. Thus, inversion is not observed in a typical S_N1 reaction, as it is in a typical S_N2 reaction.

Step 2 and 3:

$$\underset{CH_3CH_2CH_2}{\overset{\overset{H_2O}{\searrow}}{\underset{}{CH_3\text{-}\overset{+}{C}\text{-}CH_2CH_3}}} \xrightarrow{-H^+} \underset{CH_3CH_2CH_2 \quad CH_2CH_3}{CH_3\text{-}\overset{\overset{OH}{|}}{C}}$$

attack from top   (S)-3-methyl-3-hexanol

$$\underset{\underset{H_2O}{\nearrow}}{CH_3CH_2CH_2}\overset{\overset{CH_3}{|}}{\overset{+}{C}}\text{-}CH_2CH_3 \xrightarrow{-H^+} \underset{\underset{OH}{|}}{\overset{CH_3CH_2CH_2 \quad CH_3}{C}}\text{-}CH_2CH_3$$

attack from bottom   (R)-3-methyl-3-hexanol

C. Rate of an S_N1 reaction

We mentioned earlier that the rate of a typical S_N1 reaction does not depend on the concentration of the nucleophile, but depends only on the concentration of alkyl halide.

$$S_N1 \text{ rate} = k\,[RX]$$

The reason for this behavior is that the reaction between R^+ and Nu^- is very fast, but the concentration of R^+ is very low. The fast combination of R^+ and Nu^- occurs only when a carbocation is formed. Therefore, the rate of the overall reaction is determined entirely by how fast RX can ionize and form R^+ carbocations. This ionization step (Step 1 of the overall reaction) is called the **rate-determining step**, or the **rate-limiting step**. In any stepwise reaction, the slowest step in the entire sequence is the rate-determining step.

An S_N1 reaction is characteristically **first order** in rate because the rate is proportional to the concentration of only one reactant (RX). It is a **unimolecular reaction** because only one particle (RX) is involved in the *transition state of the rate-determining step.* (The "1" in S_N1 refers to unimolecular.)

Rate-determining step:

$$(CH_3)_3C—Br \longrightarrow [(CH_3)_3\overset{\delta+}{C}---\overset{\delta-}{Br}] \longrightarrow [(CH_3)_3C^+] + Br^-$$

transition state l
(from one particle)

D. Relative reactivities in S_N1 reactions

Table 5.3 lists the relative rates of reaction of some alkyl bromides under typical S_N1 conditions (solvolysis in water). Note that a secondary alkyl halide undergoes substitution 11.6 times faster than a primary alkyl halide under these conditions, while a tertiary alkyl halide undergoes reaction a million times faster than a primary halide!

Methyl:

$$CH_3Br \quad + H_2O \xrightarrow[S_N1]{\text{negligible rate}} CH_3OH \quad + Br^- + H^+$$

1°:

$$CH_3CH_2Br \quad + H_2O \xrightarrow[S_N1]{\text{negligible rate}} CH_3CH_2OH \quad + Br^- + H^+$$

2°:

$$(CH_3)_2CHBr + H_2O \xrightarrow[S_N1]{\text{very slow}} (CH_3)_2CHOH + Br^- + H^+$$

3°:

$$(CH_3)_3CBr \quad + H_2O \xrightarrow[S_N1]{\text{fast}} (CH_3)_3COH \quad + Br^- + H^+$$

The rates at which different alkyl halides undergo S_N1 reaction depend on the relative energies of activation leading to the different carbocations. In this reaction, the energy of the transition state leading to the carbocation is largely determined by the stability of the carbocation, which is already half-formed in the transition state. We say that the transition state has **carbocation character**. Therefore, the reaction leading to a more stabilized, lower-energy carbocation has the faster rate. A tertiary alkyl halide yields a carbocation that is more stabilized than the carbocation from a methyl halide or a primary alkyl halide and, consequently, this reaction has the fastest rate.

TABLE 5.3. Relative rates of some alkyl bromides under typical S_N1 conditions

CH_3Br	1.00[a]
CH_3CH_2Br	1.00[a]
$(CH_3)_2CHBr$	11.6
$(CH_3)_3CBr$	1.2×10^6

[a] The observed reaction of the methyl or primary bromide probably occurs by a different path (S_N2, not S_N1).

E.   Stability of carbocations

A carbocation is unstable and quickly undergoes further reaction. However, we may still speak of *relative stabilities* of carbocations. The different types of carbocations that concern us here are the **methyl cation** (the carbocation resulting from ionization of a methyl halide), **primary carbocations** (from 1° alkyl halides), **secondary carbocations** (from 2° alkyl halides), and **tertiary carbocations** (from 3° alkyl halides). Some examples follow:

$$^+CH_3 \qquad CH_3\overset{+}{C}H_2 \qquad (CH_3)_2\overset{+}{C}H \qquad (CH_3)_3C^+$$
$$\text{methyl} \qquad\quad 1° \qquad\qquad 2° \qquad\qquad 3°$$

What increases the stability of a positively charged carbon atom? The answer is: anything that can *disperse the positive charge*. In alkyl cations, the principal phenomenon that disperses the positive charge is the **inductive effect**, a term used to describe the polarization of a bond by a nearby electronegative or electropositive atom. In a carbocation, the positively charged carbon is an electropositive center. The electron density of the sigma bonds is shifted toward the positive carbon. We will use arrows in place of line bonds to show the direction of this attraction.

$$H_3C \searrow \overset{+}{C} \swarrow CH_3$$
$$\uparrow$$
$$CH_3$$

sigma-bond e⁻ are drawn toward the positive charge

This shift of electron density creates a partial positive charge on the adjacent atoms. These partial positive charges, in turn, polarize the next sigma bonds. In this way, the positive charge of the carbocation is somewhat dispersed, and the carbocation is stabilized to some extent.

All the atoms help disperse the positive charge.

Alkyl groups contain more atoms and electrons than does a hydrogen atom. When there are more alkyl groups attached to a positively charged carbon atom, there are more atoms that can help share the positive charge and help stabilize the carbocation. Because of the lack of stabilization, methyl and primary halides do not normally form carbocations.

$$^+CH_3 \qquad CH_3\overset{+}{C}H_2 \qquad (CH_3)_2\overset{+}{C}H \qquad (CH_3)_3C^+$$
$$\text{methyl} \qquad\quad 1° \qquad\qquad 2° \qquad\qquad 3°$$

**increasing carbocation stability;
increasing S_N1 rate of RX**

STUDY PROBLEM

5.8. List the following carbocations in order of increasing stability (least stable first):

(a) ⬡$^+$ (b) ⬡—$\overset{+}{C}H_2$ (c) ⬡—$\overset{+}{C}(CH_3)_2$

Another factor that may increase the stability of tertiary carbocations is **steric assistance.** Repulsions between groups in an alkyl halide add energy to the neutral molecule. Therefore, the ground-state energy of a tertiary alkyl halide is higher than that of a comparable primary or secondary alkyl halide. The groups attached to the head carbon are farther apart in the planar carbocation than in the alkyl halide, and repulsions are minimized. The result of the higher ground-state energy of RX due to steric hindrance is that *less additional energy* is needed for RX to form a carbocation.

repulsions

fewer repulsions

$H_3C-\overset{+}{C}\overset{CH_3}{\underset{CH_3}{\cdots}}$

a t-butyl halide   *the t-butyl cation*

Another theory proposed to explain the relative stabilities of carbocations is **hyperconjugation,** the partial overlap of an sp^3-s orbital (a C—H bond) with the empty p orbital of the positively charged carbon.

sp^3-s *empty p orbital*

An ethyl cation has only three C—H bonds that can overlap the empty p orbital, but the *t*-butyl cation has *nine* C—H bonds that can help disperse the charge in this fashion. Therefore, a tertiary carbocation is stabilized by a greater dispersal of the positive charge.

F.   Rearrangements of carbocations

The following secondary alkyl chloride can undergo S_N1 reaction with bromide ion as the nucleophile. However, in addition to the expected product, a *second* substitution product is observed.

$(CH_3)_3\overset{Cl}{\underset{|}{C}}CHCH_3$ $+ Br^-$

2-chloro-3,3-dimethylbutane

$\xrightarrow{-Cl^-}$ $(CH_3)_3\overset{Br}{\underset{|}{C}}CHCH_3$

2-bromo-3,3-dimethylbutane

expected product

$\xrightarrow{-Cl^-}$ $(CH_3)_2\overset{Br}{\underset{|}{C}}CH(CH_3)_2$

2-bromo-2,3-dimethylbutane

unexpected product

Let us examine the intermediate carbocation in this reaction more carefully. The expected carbocation is a *secondary* carbocation.

$$CH_3-\underset{\underset{CH_3}{|}}{\overset{\overset{CH_3}{|}}{C}}-\underset{}{\overset{\overset{Cl}{|}}{C}}HCH_3 \quad \xrightarrow{-Cl^-} \quad \left[CH_3-\underset{\underset{CH_3}{|}}{\overset{\overset{CH_3}{|}}{C}}-\overset{+}{C}HCH_3 \right]$$

expected carbocation; 2°

A secondary carbocation is of much higher energy than a tertiary carbocation. The energy of this particular carbocation can be lowered by the shift of a methyl group with its bonding electrons from the adjacent carbon atom. The result is the rearrangement of the secondary carbocation to a more stable tertiary carbocation.

1,2-Shift of a methyl group:

$$CH_3-\underset{\underset{CH_3}{|}}{\overset{\overset{CH_3}{|}}{C}}-\overset{+}{C}HCH_3 \quad \longrightarrow \quad CH_3-\underset{\underset{CH_3}{|}}{\overset{+}{C}}-\overset{\overset{CH_3}{|}}{C}HCH_3$$

a 2° carbocation   *a more stable 3° carbocation*

The shift of an atom or of a group from an adjacent carbon is called a **1,2-shift**. (The numbers 1,2 used in this context have nothing to do with nomenclature numbers, but refer to the positive carbon and the adjacent atom.) The 1,2-shift of a methyl group is called a **methyl shift**, or a **methide shift**. (The *-ide* suffix is sometimes used because ⁻:CH_3 is an anion; however, the 1,2-shift is a concerted reaction step and no methide anion is actually formed.)

The presence of both secondary and tertiary carbocations in solution leads to the two observed products, the so-called "normal" product and the **rearrange-** **ment product**, a product in which the skeleton or the position of the functional group is different from that of the starting material.

$$[(CH_3)_3C\overset{+}{C}HCH_3] \quad \xrightarrow{Br^-} \quad (CH_3)_3C\underset{}{\overset{\overset{Br}{|}}{C}}HCH_3$$

2° carbocation   2-bromo-3,3-dimethylbutane

$$[(CH_3)_2\overset{+}{C}CH(CH_3)_2] \quad \xrightarrow{Br^-} \quad (CH_3)_2\underset{}{\overset{\overset{Br}{|}}{C}}CH(CH_3)_2$$

3° carbocation   2-bromo-2,3-dimethylbutane

If an alkyl group, an aryl group, or a hydrogen atom (each with its bonding electrons) on an adjacent carbon atom can shift and thereby create a more stable carbocation, rearrangement can occur. Rearrangement can also occur when a pair of carbocations are of equivalent stabilities. The extent of rearrangement that will be observed in a reaction is often hard to predict and depends on a number of factors, including the relative stabilities of the carbocations in question and the reaction conditions (solvent, etc.). The following rearrangements exemplify 1,2-shifts and the formation of more-stable carbocations.

A methide shift:

$$CH_3-\overset{\overset{\displaystyle CH_3}{|}}{\underset{\underset{\displaystyle CH_3}{|}}{C}}-\overset{+}{C}HCH_2CH_3 \longrightarrow CH_3-\overset{\overset{\displaystyle CH_3}{|}}{\underset{\underset{\displaystyle CH_3}{|}}{\overset{+}{C}}}-CHCH_2CH_3$$

      a 2° carbocation   *a more stable 3° carbocation*

A hydride (H⁻) shift:

$$CH_3-\overset{\overset{\displaystyle H}{|}}{\underset{\underset{\displaystyle CH_3}{|}}{C}}-\overset{+}{C}HCH_3 \longrightarrow CH_3-\overset{+}{\underset{\underset{\displaystyle CH_3}{|}}{C}}-CH_2CH_3$$

      a 2° carbocation   *a more stable 3° carbocation*

STUDY PROBLEMS

5.9. Although many carbocations can form more-stable carbocations by 1,2-shifts, not all carbocations have structures that can yield more-stable carbocations by rearrangement. Indicate which of the following cations is likely to undergo rearrangement. Show by an arrow the shift of an alkyl group or a hydrogen, and give the structure of the rearranged carbocation.

(a) $CH_3CH_2\overset{+}{C}HCH_3$ (b) $(CH_3)_2CH\overset{+}{C}HCH_2CH_3$

5.10. What S_N1 products would be formed by the following reactants?

$$(CH_3)_2CH\overset{\overset{\displaystyle Cl}{|}}{C}HCH_2CH_3 + Br^- \longrightarrow$$

SECTION 5.7.

Substitution Reactions of Allylic Halides and Benzylic Halides

Two special types of halides behave differently in S_N1 and S_N2 reactions from the alkyl halides we have been discussing. These are the **allylic halides** and the **benzylic halides.**

$$CH_2=CH-CH_2- \qquad CH_2=CHCH_2Cl$$

    the allyl group   3-chloro-1-propene
                       (allyl chloride)

$$\text{⟨O⟩}-CH_2- \qquad \text{⟨O⟩}-CH_2Br$$

    the benzyl group   benzyl bromide

An atom or group that is attached to the carbon atom *adjacent to one of the sp^2 carbon atoms* is said to be in the **allylic position** or the **benzylic position**, respectively. The halogen atoms in our previous examples and also in the following examples are in allylic or benzylic positions.

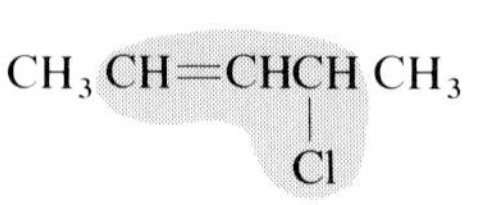

$$CH_3 \, CH=CHCH \, CH_3$$
$$| $$
$$Cl$$

4-chloro-2-pentene
an allylic chloride

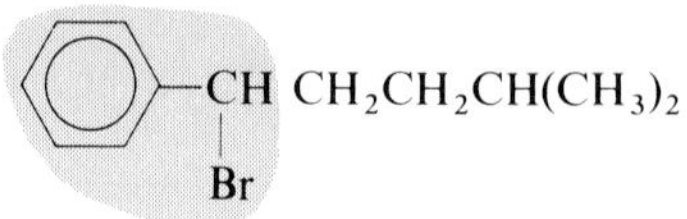

$$-CH \, CH_2CH_2CH(CH_3)_2$$
$$|$$
$$Br$$

1-bromo-4-methyl-1-phenylpentane
a benzylic bromide

SAMPLE PROBLEM

Circle the allylic or benzylic halide grouping in each of the following structures. In addition, classify each as being 1°, 2°, or 3°.

(a) 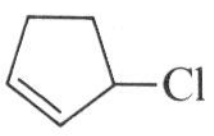 [cyclopentene]—Cl   (b) 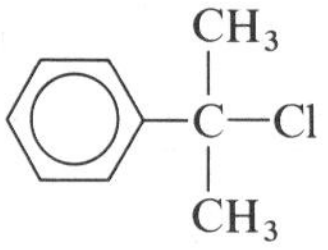 [benzene ring]—C(CH$_3$)(CH$_3$)—Cl   (c) [cyclohexane with =CH$_2$ and Cl]

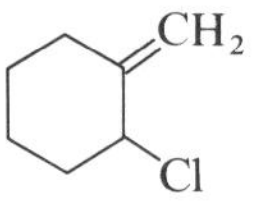

Solution:

(a) 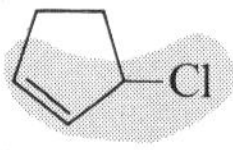 [cyclopentene]—Cl   (b) 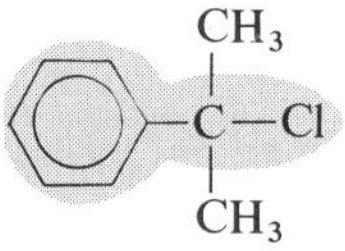 [benzene ring]—C(CH$_3$)(CH$_3$)—Cl   (c) [cyclohexane with =CH$_2$ and Cl]

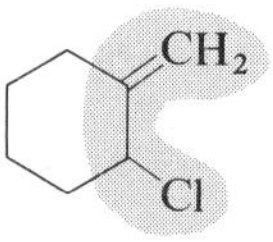

    2°   3°   2°

STUDY PROBLEMS

5.11.   Circle the benzylic or allylic halide grouping in each of the following structures:

(a) 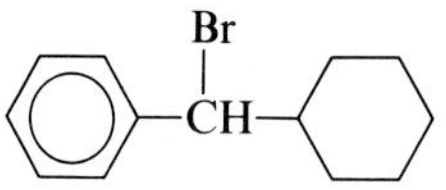 [phenyl]—CH(Br)—[cyclohexane]   (b) 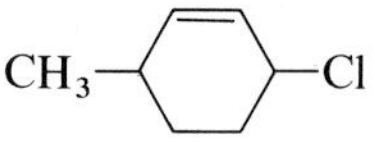 CH$_3$—[cyclohexene]—Cl

5.12.   Classify each of the following compounds as a vinylic halide, an allylic halide, a benzylic halide, or an aryl halide:

(a) 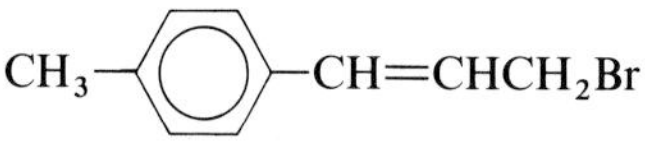 CH$_3$—[benzene]—CH=CHCH$_2$Br   (b) 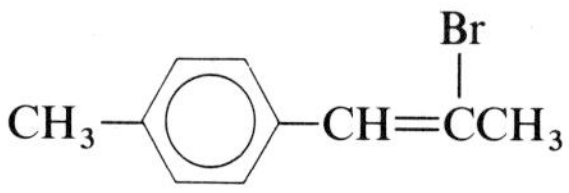 CH$_3$—[benzene]—CH=CCH$_3$ (with Br)

(c) 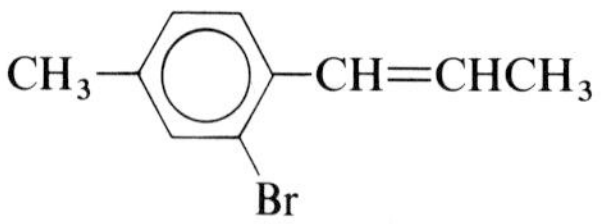 CH$_3$—[benzene (with Br)]—CH=CHCH$_3$ (d) 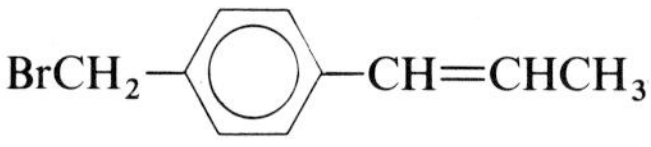 BrCH$_2$—[benzene]—CH=CHCH$_3$

A. S_N1 reactions

Most primary alkyl halides undergo substitution by the S_N2 path exclusively and do not undergo S_N1 reactions. However, a primary allylic halide or benzylic halide is very reactive in both S_N1 and S_N2 reactions. Table 5.4 lists the relative reactivities of some halides under typical S_N1 conditions. You can see from the table that an allyl halide is more than 30 times more reactive than an ethyl halide, and a benzyl halide is almost 400 times as reactive. If *two* phenyl groups are present, the halide is 100,000 times as reactive!

increasing rate of S_N1 reaction

$$CH_2{=}CHCH_2Cl + H_2O \longrightarrow CH_2{=}CHCH_2OH + Cl^- + H^+$$

allyl chloride   2-propen-1-ol
                                      (allyl alcohol)

$$\langle\!\bigcirc\!\rangle{-}CH_2Cl \ + \ H_2O \longrightarrow \langle\!\bigcirc\!\rangle{-}CH_2OH + Cl^- + H^+$$

benzyl chloride   benzyl alcohol

a lachrymator:
a compound that causes
tears to flow

The reason for the enhanced reactivity of these two types of halides in an S_N1 reaction lies in the *resonance-stabilization of the carbocation and of the transition state leading to the carbocation*. Carbocations are stabilized by dispersal of the positive charge. *Inductive stabilization* involves dispersal of the positive charge through *sigma bonds*. We have used the inductive effect to explain the relative stabilities of 1°, 2°, and 3° carbocations. *Resonance-stabilization* involves the dispersal of the positive charge by *pi bonds*.

Let us consider the S_N1 reaction of allyl chloride with H_2O:

$$CH_2{=}CHCH_2Cl \xrightarrow[S_N1]{-Cl^-} [CH_2{=}CH\overset{+}{C}H_2] \xrightarrow[-H^+]{H_2O} CH_2{=}CHCH_2OH$$

allyl cation

Recall from Section 2.9 that structures differing only in the position of pi electrons are resonance structures. If resonance structures can be drawn for a molecule or ion, the resonance hybrid (the real structure) has less energy than if delocalization of electrons or electrical charges could not occur. The two resonance structures for the allyl cation are identical in structure and bonding; therefore, they have the same energy content and contribute equally to the structure of the actual allyl cation. Because the allyl cation is resonance-stabilized, the energy of

---

TABLE 5.4. Relative rates of some organic halides under typical S_N1 conditions

Halide	*Relative rate*
CH_3CH_2X	1.0^a
$CH_2{=}CHCH_2X$	33
$C_6H_5CH_2X$	380
$(C_6H_5)_2CHX$	$\sim 10^5$

^{a} The observed reaction probably proceeds by an S_N2 path.

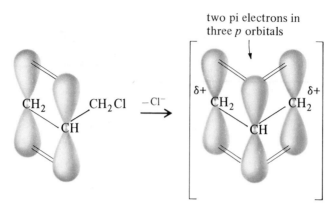

FIGURE 5.5. Formation of the allyl cation from allyl chloride.

the transition state leading to its formation is relatively low. Consequently, the rate of its S_N1 reaction is fairly fast. A *p*-orbital picture (a composite of the π bonding orbitals) of the allyl cation is shown in Figure 5.5.

$$[CH_2{=}CH{-}\overset{+}{C}H_2 \quad \longleftrightarrow \quad \overset{+}{C}H_2{-}CH{=}CH_2]$$

resonance structures for the allyl cation
(equal contributors)

Both terminal (end) carbons in the allyl cation have an equal amount of positive charge. Which atom is attacked by the nucleophile? The answer is—either of them! Let us use another allylic system to illustrate this fact. The S_N1 reaction of 1-chloro-2-butene with H_2O leads to two products. These two products arise from attack of H_2O at either of the two partially positive carbon atoms.

$$CH_3CH{=}CHCH_2Cl \quad \xrightarrow[S_N1]{-Cl^-} \quad [CH_3CH{=}CH{-}\overset{+}{C}H_2 \quad \longleftrightarrow \quad CH_3\overset{+}{C}H{-}CH{=}CH_2]$$

1-chloro-2-butene

$$\Big\downarrow \begin{array}{l} H_2O \\ -H^+ \end{array}$$

$$\begin{array}{cc} CH_3CH{=}CHCH_2 & CH_3CHCH{=}CH_2 \\ \quad\quad\quad | & \quad\quad | \\ \quad\quad\quad OH & \quad\quad OH \\ \text{2-buten-1-ol} & \text{3-buten-2-ol} \end{array}$$

Benzyl halides also show enhancement of the S_N1 rate because of resonance-stabilization of the transition state leading to the carbocation. In this case, the pi electrons in the aromatic pi cloud of the benzene ring help disperse the charge.

benzyl chloride   benzyl cation   benzyl alcohol

We usually symbolize the aromatic pi cloud of benzene by a circle in the ring. However, in discussions of delocalization of pi electrons, Kekulé formulas are more convenient. With Kekulé formulas, we can count the pi electrons in the ring and readily see which atoms are electron-deficient. Note the similarity between the resonance strctures for the benzyl cation and those for the allyl cation. The benzyl cation has four resonance structures similar to allylic resonance structures.

similar to allyl cation

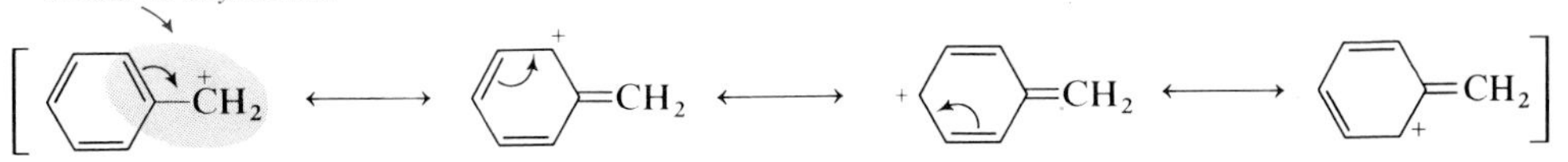

major contributor

The first resonance structure shown is the major contributor because this structure has aromatic stabilization. Therefore, the most positive carbon in the intermediate is the benzyl carbon. This is the carbon attacked by the nucleophile.

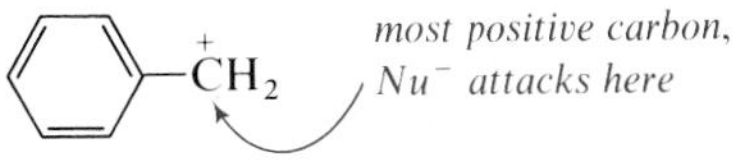

most positive carbon,
Nu⁻ attacks here

STUDY PROBLEM

5.13. Predict the products of the solvolysis in water of:

(a) 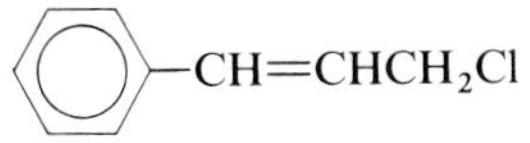 $\bigcirc$—CH=CHCH$_2$Cl   (b)  CH$_3$—$\bigcirc$—Br

---

B. S$_N$2 reactions

Allylic halides and benzylic halides also undergo S$_N$2 reactions at faster rates than primary alkyl halides or even methyl halides. Table 5.5 lists the average relative rates of some halides in a typical S$_N$2 reaction.

The reason for the greater S$_N$2 reactivity of allylic and benzylic halides is that the allylic pi bond or the aromatic pi cloud reduces the energy of the transition state of an S$_N$2 reaction. In the transition state, the carbon undergoing reaction changes from the sp^3-hybrid state to the sp^2-hybrid state and has a p orbital. This p orbital forms partial bonds with both the incoming nucleophile and the leaving group. The entire grouping of atoms carries a negative charge. Adjacent p orbitals, as in an allylic or benzylic group, undergo partial overlap with the transitional p orbital. In this way, adjacent p orbitals help delocalize the negative charge and thus lower the energy of the transition state. Figure 5.6 shows the p orbitals of the allylic case; the benzylic case is similar.

TABLE 5.5. Average relative S$_N$2 rates for some organic halides

Halide	Relative rate
CH$_3$X	30
CH$_3$CH$_2$X	1
(CH$_3$)$_2$CHX	0.03
CH$_2$=CHCH$_2$X	40
C$_6$H$_5$CH$_2$X	120

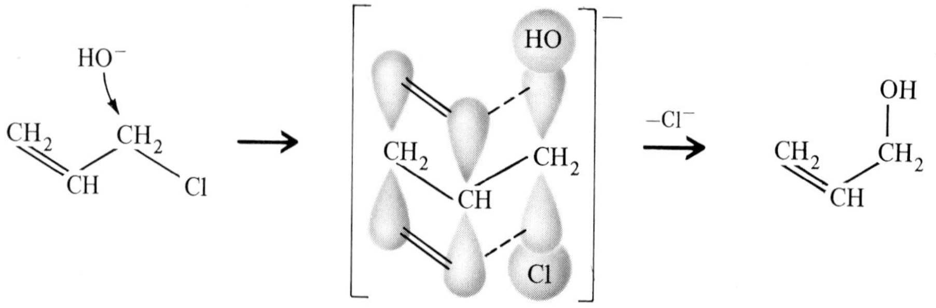

S_N2 *transition state*
showing p orbitals

FIGURE 5.6. Stabilization of the transition state in an S_N2 reaction of allyl chloride.

For increased stabilization to occur in either S_N1 or S_N2 reactions of compounds with pi systems, the pi system must be *adjacent* to the reacting carbon. If it is farther away, it cannot overlap and cannot help stabilize the transition state.

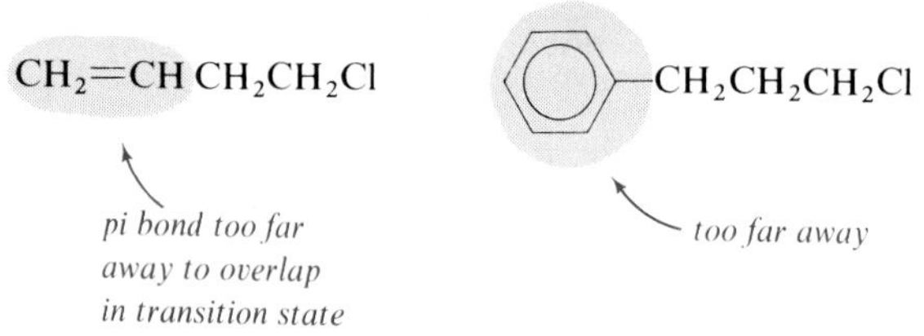

$$CH_2=CH\ CH_2CH_2Cl$$

pi bond too far
away to overlap
in transition state

$-CH_2CH_2CH_2Cl$

too far away

SAMPLE PROBLEM

Which of the following compounds exhibit enhanced reactivities in S_N1 and S_N2 reactions because of resonance-stabilization or partial *p*-orbital overlap?

(a) $CH_3CH=CHCHBrCH=CH_2$ (b) $CH_3CH=CHCH_2CHBrCH_2CH_3$
(c) $C_6H_5CH=CHCH_2I$ (d) $CH_2=CHCH_2CHBrCH=CH_2$

Solution: (a), (c), (d)

SECTION 5.8.

The E1 Reaction

A carbocation is a high-energy, unstable intermediate that quickly undergoes further reaction. One way a carbocation can reach a stable product is by combining with a nucleophile. This, of course, is the S_N1 reaction. However, there is an alternative: the carbocation can *lose a proton to a base in an elimination reaction*, an **E1 reaction** in this case, and become an alkene.

Substitution (S_N1):

$$(CH_3)_3CBr \xrightarrow{-Br^-} [(CH_3)_3C^+] \xrightarrow[-H^+]{H_2O} (CH_3)_3COH$$

t-butyl bromide   *t*-butyl cation   *t*-butyl alcohol

Elimination (E1):

$$(CH_3)_3CBr \xrightarrow{-Br^-} \left[(CH_3)_2\overset{+}{C}-CH_2 \atop H\right] \xrightarrow{H\ddot{O}H} (CH_3)_2C{=}CH_2 + H_3\overset{+}{O}{:}$$

t-butyl cation   methylpropene

The first step in an E1 reaction is identical to the first step in an S_N1 reaction: the ionization of the alkyl halide. This is the slow step and thus the rate-determining step of the overall reaction. Like an S_N1 reaction, a typical E1 reaction shows first-order kinetics, with the rate of the reaction dependent on the concentration of only the alkyl halide. Since only one reactant is involved in the transition state of the rate-determining step, the E1 reaction, like the S_N1 reaction, is unimolecular.

Step 1 (slow):

$$(CH_3)_3C\ddot{B}r{:} \longrightarrow \left[{H_3C \atop H_3C}\overset{+}{C}{-}CH_3\right] + {:}\ddot{B}r{:}^-$$

carbocation
intermediate

In the second step of an elimination reaction, the base removes a proton
✳ from a carbon that is *adjacent to the positive carbon*. The electrons of that carbon–hydrogen sigma bond shift toward the positive charge, the adjacent carbon rehybridizes from the sp^3 state to the sp^2 state, and an alkene is formed.

Step 2 (fast):

$$\left[{H_3C \atop H_3C}\overset{+}{C}{-}CH_2 \atop H\right] + H_2\ddot{O}{:} \longrightarrow \left[{H_3C \atop H_3C}\overset{\delta+}{C}{=}CH_2 \atop H{-}\overset{\delta+}{\ddot{O}}H_2\right]$$

transition state

$$\longrightarrow {H_3C \atop H_3C}C{=}CH_2 + H_3\overset{+}{O}{:}$$

Because an E1 reaction, like an S_N1 reaction, proceeds through a carbocation intermediate, it is not surprising that tertiary alkyl halides undergo this reaction far more rapidly than other alkyl halides. E1 reactions of alkyl halides occur under the same conditions as S_N1 reactions (polar solvent, very weak base, etc.); therefore, S_N1 and E1 reactions are competing reactions. Under the mild conditions required for these carbocation reactions of alkyl halides, the S_N1 product generally predominates over the E1 product. For this reason, E1 reactions of alkyl halides are relatively unimportant. You will learn in Chapter 7 that E1 reactions of alcohols, however, are quite important. We will discuss these reactions in greater detail at that time.

STUDY PROBLEM

5.14. Some tertiary alkyl halides yield mixtures of alkenes as well as a substitution product when they are subjected to S_N1 conditions. Predict all likely products from the reactions of 2-bromo-2-methylbutane with ethanol under S_N1 conditions.

SECTION 5.9.

The E2 Reaction

The most useful elimination reaction of alkyl halides is the **E2 reaction** (bimolecular elimination). E2 reactions of alkyl halides are favored by the use of strong bases, such as $^-$OH or $^-$OR, and high temperatures. Typically, an E2 reaction is carried out by heating the alkyl halide with K^+ $^-$OH or Na^+ $^-OCH_2CH_3$ in ethanol.

$$
\underset{\substack{\text{2-bromopropane} \\ \text{(isopropyl bromide)}}}{CH_3\overset{\overset{\displaystyle Br}{|}}{C}HCH_3} + CH_3CH_2O^- \xrightarrow[\substack{\text{heat} \\ \text{E2}}]{CH_3CH_2OH} \underset{\text{propene}}{CH_3CH=CH_2} + CH_3CH_2OH + Br^-
$$

The E2 reaction does not proceed by way of a carbocation intermediate, but is a **concerted reaction**—that is, it occurs in one step, just as an S_N2 reaction does.

$$
RO^- + CH_3\overset{\overset{\displaystyle Br}{|}}{C}HCH_3 \longrightarrow R\overset{..}{O}:^- \overset{(1)}{\frown} H\overset{(2)}{\frown} CH_2 - \overset{:\overset{..}{B}r:}{\underset{(3)}{|}}{C}HCH_3 \longrightarrow
$$

$$
R\overset{..}{O}H + CH_2=CHCH_3 + :\overset{..}{\underset{..}{B}}r:^-
$$

$*\begin{cases} (1) & \text{The base is forming a bond with the hydrogen.} \\ (2) & \text{The C—H electrons are forming the pi bond.} \\ (3) & \text{The bromine is departing with the pair of electrons} \\ & \text{from the carbon–bromine sigma bond.} \end{cases}$

The preceding equation shows the mechanism with arrows representing "electron-pushing." The structure of the transition state in this one-step reaction follows:

$$
\begin{array}{c}
R\overset{..}{O}:^{\delta-} \\
| \\
H \\
| \\
CH_2 \text{---} CHCH_3 \\
| \\
:\overset{..}{B}r:^{\delta-}
\end{array}
$$

E2 transition state

In E2 reactions, as in E1 reactions, tertiary alkyl halides undergo reaction fastest, and primary alkyl halides react slowest. (When treated with a base, primary alkyl halides usually undergo substitution so readily that little alkene is formed.)

$$1°RX \quad 2°RX \quad 3°RX$$

increasing rate of E2

A. Kinetic isotope effect

One piece of experimental evidence that supports our understanding of the E2 mechanism is the difference in the rates of elimination of deuteriated and non-deuteriated alkyl halides. A difference in the rates of reaction between compounds containing different isotopes is called a **kinetic isotope effect.**

Deuterium ($_1^2H$, or D) is an isotope of hydrogen with a nucleus containing one proton and one neutron. The C—D bond is stronger than the C—H bond by 1.2 kcal/mole. We have postulated that the breaking of the CH bond is an integral part of the rate-determining step (the *only* step) of an E2 reaction. What happens when the H that is eliminated is replaced by D? The stronger CD bond requires more energy to be broken. For this reason, the E_{act} is increased (Figure 5.7), and the rate for the elimination reaction is slower.

When the following 2-bromopropanes are subjected to an E2 reaction with $CH_3CH_2O^-$ as the base, it has been observed that the deuteriated compound undergoes reaction at *one-seventh* the rate of ordinary 2-bromopropane, a fact that supports the E2 reaction mechanism we have described.

$$CH_3CH_2O^- + CH_3\overset{\overset{\displaystyle Br}{|}}{C}HCH_3 \xrightarrow{\text{fast}} CH_3CH_2OH + CH_2{=}CHCH_3 + Br^-$$

$$CH_3CH_2O^- + CD_3\overset{\overset{\displaystyle Br}{|}}{C}HCD_3 \xrightarrow{\text{slow}} CH_3CH_2OD + CD_2{=}CHCD_3 + Br^-$$

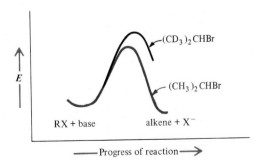

FIGURE 5.7. Energy diagrams for the E2 reactions of 2-bromopropane and a deuteriated 2-bromopropane.

SAMPLE PROBLEM

Why is a kinetic isotope effect not observed with $(CH_3)_3CBr$ and $(CD_3)_3CBr$ in E1 reactions?

Solution: The cleavage of a C—H bond is not involved in the rate-determining step of an E1 reaction.

B.   Mixtures of alkenes

Often, E1 and E2 reactions are referred to as **beta (β) eliminations**. This term reflects which hydrogen atom is lost in the reaction. Different types of carbon and hydrogen atoms in a molecule may be labeled as α, β, and so forth, according to the Greek alphabet. The carbon atom *attached to the principal functional group* in a molecule is called the **alpha (α) carbon**, and the adjacent carbon is the **beta (β) carbon**. The hydrogens attached to the α carbon are called α hydrogens, while those attached to the β carbon are β hydrogens. In a β elimination, a β hydrogen is lost when the alkene is formed. (Of course, an alkyl halide with no β hydrogen cannot undergo a β elimination.)

β carbons and hydrogens circled:

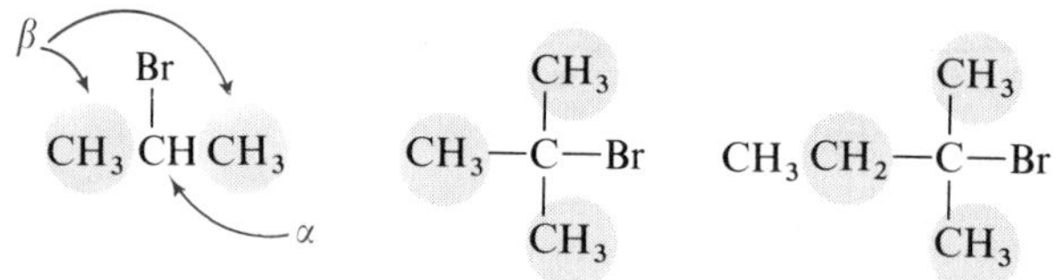

If 2-bromopropane or *t*-butyl bromide undergoes elimination, there is only one possible alkene product. However, if the alkyl groups around the α carbon are different and there are more than one type of β hydrogens, then more than one alkene can result. The E2 reaction of 2-bromobutane yields two alkenes because two types of hydrogen atoms can be eliminated: a hydrogen from a CH_3 group or a hydrogen from a CH_2 group.

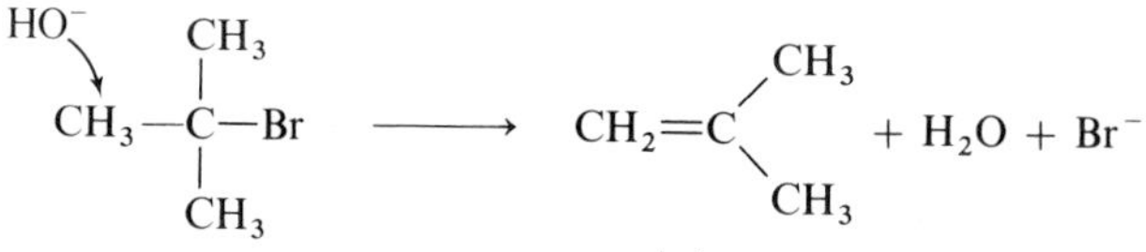

t-butyl bromide   methylpropene

only one type of β H   *only possible alkene*

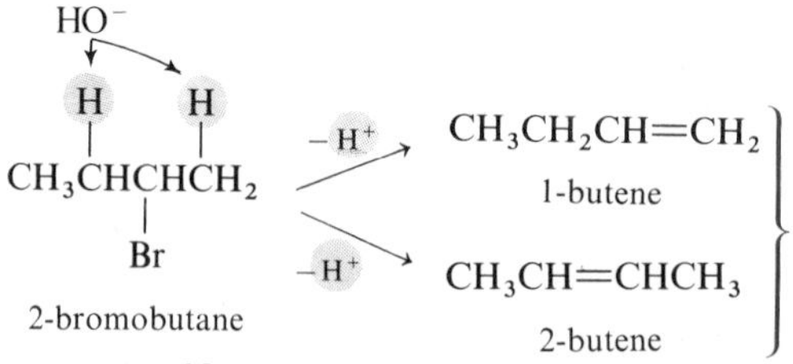

2-bromobutane

two types of β H

SAMPLE PROBLEMS

Circle the β carbons and hydrogens in the following structures:

(a) $CH_3CH_2CHCH_2CH_2CH_3$ (b)

$\quad\quad\quad\quad\quad\quad |$
$\quad\quad\quad\quad\quad\quad Br$

$\quad\quad\quad\quad\quad\quad\quad\quad\quad\quad\quad\quad\quad CH_3$
$\quad\quad\quad\quad\quad\quad\quad\quad\quad\quad\quad\quad\quad Br$

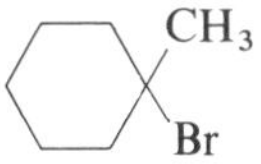

Solution: (a) CH_3 (CH$_2$) CH (CH$_2$) CH_2CH_3 (b)

$\quad\quad\quad\quad\quad\quad\quad\quad\quad |$
$\quad\quad\quad\quad\quad\quad\quad\quad\quad Br$

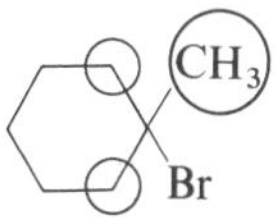

In the preceding problem, tell how many different *types* of β hydrogens are in each structure.

Solution: (a) two types;   (b) two types (the ring CH_2 groups are equivalent to each other; the CH_3 is different).

STUDY PROBLEM

5.15.   Write the structures of the alkenes that could result from the E2 reaction of each of the preceding alkyl bromides.

C.   Which alkene is formed?

In 1875, the Russian chemist Alexander Saytseff formulated the following rule, now called the **Saytseff rule**: *In elimination reactions, the alkene with the greatest number of alkyl groups on the doubly-bonded carbon atoms predominates in the product mixture.* We will refer to this alkene as the *more highly substituted alkene.* The Saytseff rule predicts that 2-butene would predominate over 1-butene as a product in the E2 reaction of 2-bromobutane. This indeed is what occurs. In the following reaction, the mixture of alkene products consists of 80% 2-butene and only 20% 1-butene.

$$\underset{\text{2-bromobutane}}{CH_3CH_2\overset{\overset{\displaystyle Br}{|}}{C}HCH_3} \xrightarrow[\text{CH}_3\text{CH}_2\text{OH}]{Na^+ \ ^-OCH_2CH_3} \underset{\underset{80\%}{\text{2-butene}}}{CH_3 CH\!=\!CH CH_3} + \underset{\underset{20\%}{\text{1-butene}}}{CH_3CH_2 CH\!=\!CH_2}$$

two R's on C=C, more highly substituted   *one R on C=C*

It has been determined that *more highly substituted alkenes are more stable than less substituted alkenes* (this will be discussed further in Chapter 9). Therefore, an E2 elimination leads to the *more stable alkene.*

$$CH_2\!=\!CH_2 \quad CH_3CH\!=\!CH_2 \quad CH_3CH\!=\!CHCH_3 \quad (CH_3)_2C\!=\!C(CH_3)_2$$

increasing stability

To see why the more stable alkene (2-butene) is formed in preference to the less stable alkene (1-butene), let us consider the transition states leading to these two butenes. In either transition state, the <u>base is removing a proton</u> and a double bond is being formed. We say that this <u>transition state has some **double-bond character**,</u> which we represent as a dotted line in the formula.

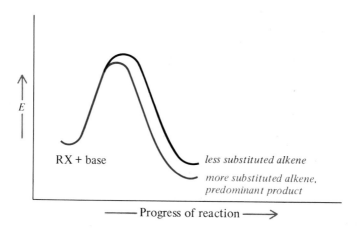

Because both transition states leading to the alkenes have some double-bond character, the transition state leading to the more stable alkene is itself more stabilized and of lower energy. The reaction with the lower-energy transition state proceeds at a faster rate; therefore, the more stable alkene is the predominant product (see Figure 5.8).

FIGURE 5.8. Energy diagram for a typical E2 reaction, showing why the more substituted alkene predominates.

SAMPLE PROBLEMS

Which is the more stable alkene,   (a) $(CH_3)_3CCH=CHCH_3$, or (b) $CH_3CH=C(CH_3)_2$?

Solution: (b) with three R's is more stable than (a) with only two R's.

Predict the major alkene product of the E2 dehydrohalogenation of $CH_3CH_2C(CH_3)_2Cl$.

Solution: The two possible alkenes are $CH_3CH=C(CH_3)_2$ and

$$CH_3CH_2C=CH_2 .$$
$$| $$
$$CH_3$$

The first predominates (3 R's versus 2 R's).

STUDY PROBLEM

5.16.   Predict the major alkene product of the E2 dehydrohalogenation of 1-chloro-1-methylcyclohexane.

Part of our question about which alkene would be formed in dehydro-halogenation has been answered: the most highly substituted alkene predominates in the product mixture. The most highly substituted alkene can often exist as *cis* and *trans* diastereomers (geometric isomers). Is there a selectivity as to which diastereomer is formed? Experimentally, it has been determined that *trans* alkenes are generally more stable than their *cis* isomers, presumably because of less steric hindrance in the *trans* isomers. Therefore, it is not surprising that *trans* alkenes predominate as products of E2 reactions. Again, the reason is a more stabilized transition state. The following equation shows the results of an E2 reaction of 2-bromopentane.

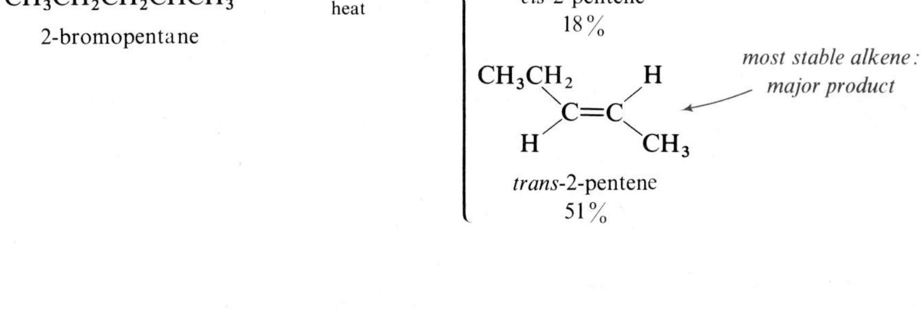

D.   Stereochemistry of an E2 reaction

In the transition state of an E2 elimination, the attacking base and the leaving group are generally as far apart as possible, or *anti*, For this reason, the E2 elimination is often referred to as ***anti*-elimination**.

anti-Elimination:

dimensional	*ball-and-stick*	*Newman*

The interesting feature about *anti*-elimination is that the *anti*-positioning of the H and Br that are lost determines the stereochemistry of the product alkene. To see how this happens, let us look at the E2 reactions of some stereoisomeric halides. The compound 1-bromo-1,2-diphenylpropane has two chiral carbon atoms (carbons 1 and 2) and four stereoisomers.

The four stereoisomers of $\overset{③}{CH_3}\overset{②}{CH}{-}\overset{①}{CH}Br$:
$$C_6H_5 \quad C_6H_5$$

(1*R*,2*R*)	(1*S*,2*S*)	(1*R*,2*S*)	(1*S*,2*R*)

enantiomers	enantiomers

Because there is only one β hydrogen in the starting halide, any one of these stereoisomers yields $C_6H_5(CH_3)C{=}CHC_6H_5$. However, geometric isomerism is possible in this product.

only one β hydrogen

$$CH_3CH{-}CHBr + {^-}OR \xrightarrow{\text{E2}} CH_3C{=}CH + ROH + Br^-$$
$$C_6H_5 \ C_6H_5 C_6H_5 \ C_6H_5$$

When either (1*R*,2*R*)-1-bromo-1,2-diphenylpropane or its (1*S*,2*S*)-enantiomer undergoes E2 reaction, the (*Z*)-alkene is formed exclusively; no (*E*)-alkene is formed.

(1*R*,2*R*)	(*Z*)-1,2-diphenyl-1-propene

The reason for all (*Z*) and no (*E*) product is that there is only one conformation of either of these enantiomers in which the Br and the beta hydrogen are *anti*. In either the (1*R*,2*R*)- or the (1*S*,2*S*)-enantiomer, the *anti* alignment of H and Br puts the phenyl groups on the same side of the molecule, and the (*Z*)-alkene results. If the elimination could occur regardless of the conformation of the enantiomers, then some (*E*)-alkene would also be observed.

anti H and Br means cis-phenyls

$$\xrightarrow[\text{E2}]{-\text{HBr}}$$

(1*R*,2*R*)   (*Z*)-alkene

STUDY PROBLEM

5.17. Write equations for the *anti*-elimination of the (1*S*,2*S*)-enantiomer, as we have done for the (1*R*,2*R*)-enantiomer.

Just the opposite situation prevails with the (1*R*,2*S*)- or (1*S*,2*R*)-enantiomers. Either of these isomers yields the (*E*)-alkene and not the (*Z*)-alkene. The reason, once again, is that there is only one conformation for each of these enantiomers in which the Br and the single beta hydrogen are in an *anti* relationship. In these conformations, the phenyl groups are on opposite sides of the molecule.

anti H and Br means trans-phenyls

(1*R*,2*S*)   (1*S*,2*R*)

A reaction in which different stereoisomers of the reactant yield stereo-isomerically different products is said to be a **stereospecific reaction**. The E2 reaction is an example of a stereospecific reaction.

Halocycloalkanes, such as chlorocyclohexane, can also undergo E2 reactions. In these cases, the conformations of the ring play an important role in the course of the reaction. In order to be *anti* on a cyclohexane ring, the leaving group (such as chlorine) and a β hydrogen must be 1,2-*trans* and *diaxial*. No other conformation places the H and Cl *anti* to each other. (Try it with models.) Even though this conformation is not the favored one, a certain percent of halocycloalkane molecules are in this conformation at any given time and can thus undergo elimination.

Cl is equatorial and not anti to any β hydrogens   *Cl is axial and anti to two β hydrogens*   *OH⁻ could attack either H shown*

$$\begin{array}{c} -\text{H}_2\text{O} \\ -\text{Cl}^- \end{array}$$

STUDY PROBLEMS

5.18.   Draw dimensional formulas for the conformers of
(1*R*,2*S*)-1-bromo-1,2-diphenylpropane and (1*S*,2*R*)-1-bromo-1,2-diphenylpropane
that undergo E2 reaction.

5.19.   1,2-Dibromo-1,2-diphenylethane contains two chiral carbon atoms and has a
pair of enantiomers plus a *meso* diastereomer. Any of the stereoisomers of this
compound can undergo an E2 reaction to yield 1-bromo-1,2-diphenylethene.
The *meso* form yields one geometric isomer of the alkene, while the racemic
mixture of enantiomers yields the other geometric isomer. Predict the
stereochemistry of the products of these two reactions.

5.20.   Write an equation using conformational formulas that illustrates the E2 reaction
of *trans*-1-chloro-2-methylcyclohexane in aqueous sodium hydroxide. (Be sure to
show the required conformation of the ring for *anti*-elimination.)

E.   Hofmann products

Most dehydrohalogenations follow the Saytseff rule and the more substituted
alkene predominates. However, under some circumstances, the major product of
an E2 dehydrohalogenation is the *less substituted, less stable alkene*. When the
less substituted alkene is the predominant product, we say that the reaction yields
the **Hofmann product**.

   When is the less substituted alkene likely to be the predominant product? A
common phenomenon leading to the less substituted alkene is **steric hindrance** in
that transition state leading to the most substituted alkene. Steric hindrance can
raise the energy of this transition state so substantially that the reaction follows a
different course and yields the less substituted alkene. The steric hindrance may be
caused by any one of three factors. First of all, it may be caused by the **size of the
attacking base.** In the elimination reaction of 2-bromobutane with the small
ethoxide ion, the more substituted alkene predominates. With the bulky *t*-butoxide
ion, the 1- and 2-butenes are formed in equal amounts.

$$CH_3CH_2CHCH_3 \quad (2\text{-bromobutane})$$

with $CH_3CH_2O^-$ (small):

$$CH_3CH=CHCH_3 + CH_3CH_2CH=CH_2$$

2-butene (80%)   1-butene (20%)

with $(CH_3)_3CO^-$ (bulky):

$$CH_3CH=CHCH_3 + CH_3CH_2CH=CH_2$$

2-butene (50%)   1-butene (50%)

Reaction with the t-butoxide ion:

$(CH_3)_3CO^-$

*attack at carbon 3:
more steric hindrance*

$$CH_3CH\overset{\displaystyle H}{\underset{\displaystyle Br}{C}}HCH_3 \longrightarrow CH_3CH=CHCH_3$$

2-butene

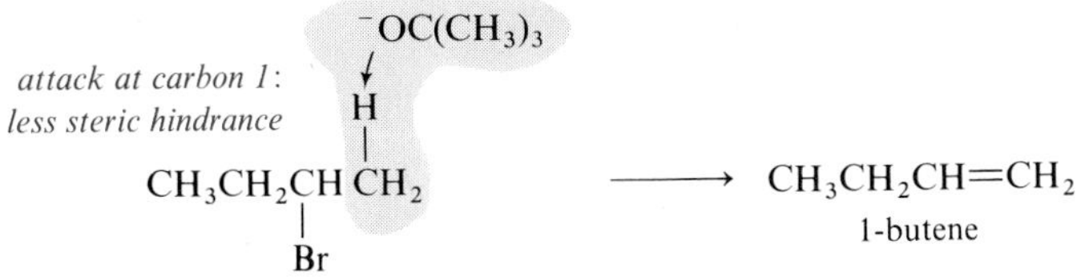

attack at carbon 1:
less steric hindrance

$$^-OC(CH_3)_3$$

$$CH_3CH_2CH\,CH_2 \longrightarrow CH_3CH_2CH{=}CH_2$$

$$\underset{Br}{|}$$

1-butene

Second, steric hindrance might be caused by the **bulkiness of groups surrounding the leaving group** in the alkyl halide. The hindered 2-bromo-2,4,4-trimethylpentane yields the less substituted alkene in an E2 reaction, even with a small base like the ethoxide ion.

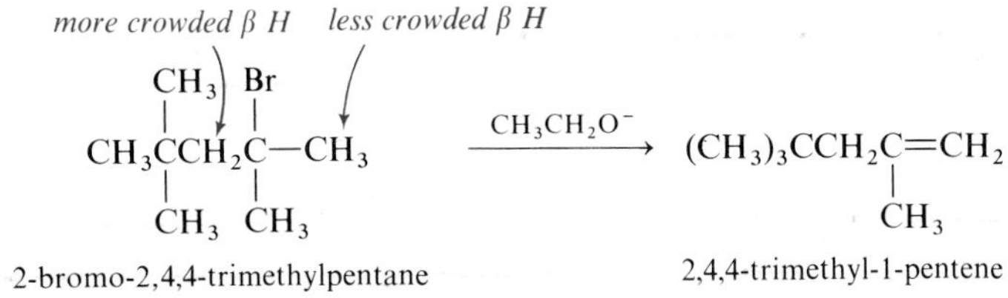

more crowded β H   *less crowded β H*

$$CH_3CCH_2C{-}CH_3 \xrightarrow{\ CH_3CH_2O^-\ } (CH_3)_3CCH_2C{=}CH_2$$

2-bromo-2,4,4-trimethylpentane   2,4,4-trimethyl-1-pentene

Third, if the **leaving group itself is large and bulky**, the Hofmann product may predominate. This type of reaction will be discussed in Section 15.10.

STUDY PROBLEMS

5.21. Write formulas for both the Hofmann and Saytseff products of the E2 reactions of   (a) 3-bromo-2-methylpentane and   (b) 1-chloro-1-methylcyclohexane.

5.22. Predict the major alkene products of the following E2 reactions:

(a) $CH_3CH_2CH_2CHBrCH_3$ + $^-OC(CH_3)_3$ $\longrightarrow$
(b) $CH_3CH_2CHBrCH_3$ + OH^- $\longrightarrow$

SECTION 5.10.

Factors Governing Substitution and Elimination Reactions

At the start of this chapter, we mentioned that S_N1, S_N2, E1, and E2 are *competing reactions*. A single alkyl halide could be undergoing substitution, elimination, and rearrangements all in the same reaction flask. If this happens, a mixture of a large number of products can result. However, a chemist can control the products of the reaction to a certain extent by a proper choice of the reagents and reaction conditions.

What, then, are the factors that affect the course of substitution and elimination reactions of alkyl halides? These factors are:

1. the structure of the alkyl halide;
2. the nature of the nucleophile or base;
3. the nature of the solvent;
4. the concentration of the nucleophile or base;
5. the temperature.

A.   The alkyl halide

We have mentioned that the type of alkyl halide affects the mechanism of the reaction. Now that we have looked at the four principal mechanisms by which an alkyl halide can undergo reaction with a nucleophile or base, we can summarize how the different alkyl halides act.

Methyl and primary alkyl halides tend to undergo S_N2 reactions. They do not form carbocations and thus cannot undergo S_N1 or E1 reactions. Primary alkyl halides undergo E2 reactions slowly, if at all.

Secondary alkyl halides can undergo reaction by any path, but S_N2 and E2 are more common than E1 or S_N1. The reactions of secondary alkyl halides are more subject to control by conditions in the reaction flask (concentration of nucleophile, solvent, etc.) than are reactions of other alkyl halides.

Tertiary alkyl halides undergo primarily E2 reactions with a strong base (such as ¯OH or ¯OR), but undergo the S_N1 reaction and some E1 reaction with a very weak base (such as H_2O or ROH).
Table 5.6 shows general equations to summarize the reactions of the different types of alkyl halides.

B.   The nucleophile or base

The difference between nucleophilicity and basicity was discussed in Section 5.4D. As we mentioned in that section, a strong base is generally also a good nucleophile. Two other factors can affect the relative nucleophilicities of reactants, sometimes dramatically. One of these factors is the *solvent* used for the reaction. Solvent effects will be discussed in Section 5.10C.
The *polarizability* of an ion or molecule is another factor that affects its nucleophilicity. The outer electrons of larger atoms are farther from the nucleus and less tightly held than those of smaller atoms. The outer electrons of larger atoms are therefore more easily distorted by attraction to a positive center and can attack a partially positive carbon atom more readily. For example, the iodide ion is usually a better nucleophile than the chloride ion.

TABLE 5.6. The principal reactions of the different types of alkyl halides[a]

	Halide		*Products*
Methyl and primary:	$RCH_2X + Nu^-$	$\xrightarrow{\;S_N2\;}$	RCH_2Nu
Secondary:	$R_2CHX + Nu^-$	$\xrightarrow{\;S_N2 + E2\;}$	R_2CHNu + alkenes
Tertiary:	$R_3CX\;\; + H_2O$	$\xrightarrow{\;S_N1 \text{ and } E1\;}$	R_3COH + alkenes
	$R_3CX\;\; + R'OH$	$\xrightarrow{\;S_N1 \text{ and } E1\;}$	R_3COR' + alkenes
	$R_3CX\;\; + \text{base}$	$\xrightarrow{\;E2\;}$	alkenes

[a] For an E1 or E2 reaction to occur, the alkyl halide must contain at least one β hydrogen.

The degree of nucleophilicity versus basicity can affect the course of a reaction. The reaction of a *primary alkyl halide* with a strong nucleophile (see page 168) follows an S_N2 path, even if the nucleophile is also a strong base. However, for a *tertiary alkyl halide*, any moderately strong base favors E2 reaction. Only the weakest bases (H_2O, ROH) result in substitution (by an S_N1 path).

For *secondary alkyl halides*, strong nucleophiles (such as CN^-) favor S_N2 reactions, while weak nucleophiles (such as H_2O) favor carbocation reactions, primarily S_N1 with some E1. Strong bases (such as ^-OH or ^-OR) favor E2 reactions.

> strong nucleophile: S_N2
> weak nucleophile: S_N1
> strong base: E2

STUDY PROBLEM

5.23. Predict which is generally the better nucleophile: CH_3S^- or CH_3O^-. Explain the reason for your answer.

C.   The solvent

The solvent exerts its influence on substitution and elimination reactions by its ability or inability to solvate ions: carbocations, nucleophiles or bases, and leaving groups. The ability of a solvent to solvate ions is determined by its polarity, which is usually reported as a **dielectric constant**. Whereas a dipole moment is a measure of the polarity of a single molecule, the dielectric constant is a measure of the polarity of a liquid (many molecules with interactions between them). A highly polar solvent has a high dielectric constant. Table 5.7 lists some common organic solvents, their dielectric constants, and the relative rates of a typical S_N1 reaction in that solvent.

While dielectric constants can provide a guide for solvent selection, there are no firm rules about how to predict the best solvent for a given reaction. (The solubilities of the reactants must be considered too!) In general, a *very polar solvent* (such as water) encourages S_N1 reactions by helping stabilize the carbocation through solvation. Conversely, a less polar solvent (such as acetone) favors S_N2 and E2 reactions because it does not aid ionization.

TABLE 5.7. Relative rates of typical S_N1 reactions in various solvents

Solvent	*Formula*	*Dielectric constant*	*Approximate relative rate*
formic acid	HCO_2H	58	15,000
water	H_2O	78.5	4000
80% aqueous ethanol	$CH_3CH_2OH–H_2O$	67	185
ethanol	CH_3CH_2OH	24	95
acetone	$CH_3\overset{\overset{\displaystyle O}{\|\|}}{C}CH_3$	21	0.5

In addition to solvation of a carbocation, the solvation of the nucleophile is very important. The choice of solvent can actually change the ranking of nucleophilicity within a group of nucleophiles. A solvent that can solvate (and thus stabilize) an anion reduces its nucleophilicity. By contrast, a solvent that cannot solvate an anion enhances its nucleophilicity. The chloride ion is a far better nucleophile in dimethylformamide (DMF), where it is not solvated, than in ethanol, where it is solvated.

D.   Concentration of the nucleophile or base

By controlling the concentration of nucleophile or base, a chemist has direct control over the rates of S_N2 and E2 reactions. Increasing the concentration of nucleophile generally has no effect on the rates of S_N1 or E1 reactions, but increases S_N2 or E2 reaction rates proportionally. Therefore, a high concentration of nucleophile or base aids S_N2 or E2 reaction, respectively; a low concentration favors S_N1 or E1.

> **high concentration of Nu⁻ or base:** S_N2 or E2
>
> **low concentration of Nu⁻ :** S_N1 or E1

E.   Temperature

An increase in temperature increases the rates of all substitution and elimination reactions. However, an increase in temperature usually leads to a greater increase in elimination products. (The reason for this is that elimination reactions usually have higher E_{act}'s than do substitution reactions, and higher temperatures enable a greater number of molecules to reach the elimination transition state.)

SECTION 5.11.

Synthesizing Other Compounds from Alkyl Halides

From a practical standpoint, only S_N2 and E2 reactions are useful for synthesizing other compounds from alkyl halides. S_N1 and E1 reactions usually yield mixtures of products.

A large number of functional groups can be obtained by S_N2 reactions. So far, we have presented only a few nucleophiles, but many others can be used. For example, an *ester* is the product if a *carboxylate salt* is used as a nucleophile. (A carboxylate ion is a weak base and a fairly weak nucleophile; therefore, the reaction proceeds best with the most reactive halides such as benzylic or allylic halides.)

$$CH_3C\overset{O}{\overset{\|}{C}}\overset{..}{\underset{..}{O}}{:}^- \quad + \quad C_6H_5CH_2\overset{..}{\underset{..}{Br}}: \quad \longrightarrow \quad C_6H_5CH_2O\overset{O}{\overset{\|}{C}}CH_3 + :\overset{..}{\underset{..}{Br}}{:}^-$$

acetate ion   benzyl bromide   benzyl acetate

a carboxylate ion   *an ester*

An *amine salt* can be prepared if *ammonia* or an *amine* (weak bases, moderate nucleophiles) is used as a nucleophile. (This reaction will be discussed in Section 15.5A.)

$$(CH_3)_3N: \quad + \quad CH_3CH_2CH_2\overset{..}{\underset{..}{Br}}: \quad \longrightarrow \quad CH_3CH_2CH_2\overset{+}{N}(CH_3)_3 \ :\overset{..}{\underset{..}{Br}}{:}^-$$

trimethylamine   trimethyl-*n*-propylammonium bromide

an amine   *a salt*

An *alkene* can be prepared by heating a secondary or tertiary alkyl halide with a *strong base* such as potassium hydroxide or the alkali metal salt of an alcohol in an alcohol solvent. Generally, the more highly substituted, *trans* alkene is the product. The less substituted alkene can sometimes be prepared if a bulky base, such as K^+ $^-OC(CH_3)_3$, is used.

Table 5.8 summarizes the types of products that can be obtained by S_N2 and E2 reactions of alkyl halides.

TABLE 5.8. Some types of compounds that can be synthesized from alkyl halides

Reactants[a]		Principal product		Typical reagents
1° RX	+ $^-OR'$ $\longrightarrow$	ROR'	an ether	Na^+ $^-OCH_2CH_3$, Na^+ $^-OC_6H_5$
1° RX	+ ^-OH $\longrightarrow$	ROH	an alcohol	Na^+ ^-OH, K^+ ^-OH
1° or 2° RX	+ ^-CN $\longrightarrow$	RCN	a nitrile	Na^+ ^-CN
1° or 2° RX	+ $^-SR'$ $\longrightarrow$	RSR'	a sulfide, or thioether	Na^+ $^-SCH_2CH_3$
1° or 2° RX	+ $^-O\overset{O}{\overset{\|}{C}}R'$ $\longrightarrow$	RO$\overset{O}{\overset{\|}{C}}$R'	an ester[b]	Na^+ $^-O_2CCH_3$
1° or 2° RX	+ I^- $\longrightarrow$	RI	an alkyl iodide	Na^+ ^-I
1° or 2° RX	+ NR'_3 $\longrightarrow$	$R\overset{+}{N}R'_3$ X^-	an ammonium salt	$(CH_3)_3N$
2° or 3° R_2CHCXR_2	+ $^-OR'$ $\longrightarrow$	$R_2C{=}CR_2$	an alkene	K^+ ^-OH, Na^+ $^-OCH_2CH_3$, K^+ $^-OC(CH_3)_3$

[a] Where 1° RX is specified, methyl halides, allylic halides, and benzylic halides may also be used.

[b] A reactive halide must be used.

STUDY PROBLEM

5.24. Write equations to show how you would synthesize the following compounds from organic halides and other appropriate reagents. If there are two routes to the compound, choose the better one. If two routes are equivalent, show both.

(a) $CH_3CH_2CH_2O$—⬡   (b) ⬡—CH_2SCH_3

(c) $CH_3CH_2OCH_2CH_2CH_3$ (d) ⬡—$\overset{\overset{\displaystyle O}{\|}}{C}OCH_2CH{=}CH_2$

(e) ⬡—OCH_2CH_3 (f) $(CH_3)_2CHCH{=}CHCH_3$

Summary

An alkyl halide contains a good **leaving group** (X^-) and is readily attacked by **nucleophiles** (Nu^-). Reaction occurs by one or more of four possible paths: S_N1, S_N2, E1, E2.

An S_N1 or E1 reaction proceeds through a **carbocation intermediate**:

$$RX \xrightarrow{\;-X^-\;} [R^+] \begin{cases} \xrightarrow[S_N1]{Nu^-} & RNu \\[2mm] \xrightarrow[E1]{-H^+} & \text{alkene} \end{cases}$$

A carbocation intermediate usually leads to a mixture of products: *a substitution product, an alkene*, and also *rearrangement products*. Rearrangement products occur if the carbocation can form a more stable carbocation by a 1,2-shift of H, Ar, or R. If RX is optically active, racemization can occur in an S_N1 reaction.

The rate of a typical S_N1 or E1 reaction depends on the concentration of only RX; thus, these reactions are said to be **first order**. The **rate-determining step** (slow step) in an S_N1 or E1 reaction is the formation of R^+. The stability of R^+ determines the **energy of the transition state** (E_{act}) in this step because the transition state has carbocation character. The *order of stability* of carbocations is $3° > 2° \gg 1° \gg CH_3^+$. For this reason, the likelihood of RX to undergo S_N1 or E1 reaction is $3° > 2° \gg 1° \gg CH_3X$. Allylic and benzylic halides undergo S_N1 reactions readily because of resonance-stabilization of the intermediate carbocation.

An S_N2 reaction is a **concerted reaction** that leads to *inversion*. Inversion can be observed if RX is optically active. An E2 reaction is also a concerted reaction that results by *anti*-elimination of H^+ and X^-.

$$\text{Nu}^- + \text{R--X} \xrightarrow{\text{S}_\text{N}2} \text{NuR} + \text{X}^-$$

$$\text{R}_2\text{C--CR}_2 \xrightarrow{\text{E2}} \text{ROH} + \text{R}_2\text{C}{=}\text{CR}_2 + \text{X}^-$$

Both S_N2 and E2 reactions follow **second-order kinetics**: the rate is dependent on the concentrations of both RX and Nu$^-$ because both are involved in the transition state. Because of steric hindrance, the order of reactivity of RX in S_N2 reactions is $CH_3X > 1° > 2° \gg 3°$.

Because the transition state has double-bond character, the order of reactivity of RX in E2 reactions is $3° > 2° \gg 1°$, the same order as in the E1 reaction. The *most substituted alkene* usually predominates in E2 reactions (**Saytseff rule**). The *trans* alkene usually predominates over the *cis* alkene. If steric hindrance inhibits the formation of the most substituted alkene, then the *least substituted alkene* predominates (**Hofmann product**).

STUDY PROBLEMS

5.25. Name each of the following compounds by the IUPAC system:

(a) $CCl_3CH{=}CH_2$ (b) $CH_3CHCH_2CH_2Br$ (with Br on second carbon)   (c) [cyclohexane with Br and CH$_3$]

(d) [cyclopentane with OH and Cl]   (e) [structure with I, H, CO$_2$H, CH$_3$]

5.26. Give the structure for each of the following compounds: (a) isobutyl iodide;
(b) 1-iodo-2-methylpropane;   (c) *cis*-1,3-dichlorocyclohexane;
(d) 2-bromo-3-methyl-1-butanol;   (e) (2R,3R)-2-bromo-3-chlorobutane.

5.27. Classify the following organohalogen compounds as methyl, 1°, 2°, or 3°, and, if applicable, allylic, benzylic, or vinylic.

(a) $(CH_3)_3CCH_2Cl$ (b) [benzene]$-CH_2Br$ (c) [benzene]$-CHCl-$[benzene]

(d) $(CH_3CH_2)_3CCl$ (e) [cyclohexane with CH$_3$ and Cl]   (f) $CH_3CH_2CH{=}CHCl$

5.28. Which compound in each of the following pairs would undergo more rapid S_N2 reaction?

(a) $(CH_3)_3CI$ or $(CH_3CH_2)_2CHI$ (b) $(CH_3)_2CHI$ or $(CH_3)_2CHCl$

(c) ⬡—Cl or ⬡—CH$_2$Cl   (d) ⬡—Cl or ⬡(CH$_3$)(Cl)

5.29. Complete the following equations for S_N2 reactions.

(a) ⬡—Cl + NaSCH$_3$ ⟶

(b) O⬡—Br + CH$_3$CH$_2$CH$_2$ONa ⟶

(c) $ICH_2CH_2CH_2I + 2\,NaOH$ ⟶

(d) ⬡N: + CH$_3$I ⟶

(e) $(CH_3CH_2\overset{\overset{\displaystyle O}{\|}}{C})_2\ddot{C}H^- + CH_3I$ ⟶

5.30. Predict the likely substitution and elimination products when the following compounds are treated with NaCN:

(a) naphthalenyl—CHCH$_3$ with Br   (b) $CH_3CH_2SCH_2CH_2Cl$ (c) O⬡—Cl

5.31. Suggest reagents for the preparation of each of the following compounds by an S_N2-type reaction of an organohalogen compound with a nucleophile:

(a) $(CH_3)_2CHOC_6H_5$ (b) cyclopropane oxide   (c) $CH_3\overset{\overset{\displaystyle O}{\|}}{C}O$—⬡

5.32. Write equations showing the structures of the transition state and the product for the S_N2 reaction of $^-OCH_3$ with each of the following alkyl halides:

(a) (R)-2-bromobutane;   (b) *trans*-1-chloromethyl-4-methylcyclohexane;
(c) (R)-2-bromo-3-methylbutane;   (d) (S)-2-bromo-3-methylbutane.

5.33. Predict the product of an S_N2 reaction of each of the following nucleophiles with (S)-2-iodohexane:

(a) $^-SCH_2CH_3$ (b) $CH_3C{\equiv}C{:}^-$ (c) (R)-$CH_3CH_2\overset{\overset{\displaystyle O^-}{|}}{C}HCH_2CH_2CH_3$

5.34. Predict the S_N2 product:

(a) H$_3$C—⬡—Cl + OH$^-$ ⟶   (b) Cl—⬡—Cl + 2OH$^-$ ⟶

5.35. Which of the following two syntheses for *t*-butyl ethyl ether would be preferred and why?

(a) $(CH_3)_3CO^- + CH_3CH_2Br$ (b) $(CH_3)_3CBr + CH_3CH_2O^-$

5.36. If Reaction 1 and Reaction 2 are two different irreversible reactions of a single starting material, and if Reaction 1 has a higher-energy transition state than Reaction 2,

(a) which reaction has the higher E_{act}?
(b) which reaction has the faster rate?
(c) which products will predominate in the product mixture?

5.37. What is the effect on the rate of the S_N2 reaction of CH_3I and OH^- when:

(a) The concentration of CH_3I is tripled and that of OH^- is doubled?
(b) The concentration of OH^- is halved?
(c) The temperature is increased?
(d) The ratio of solvent to reactants is doubled?

5.38. Which of the following carbocations is the most stable? Which is the least stable?

(a) 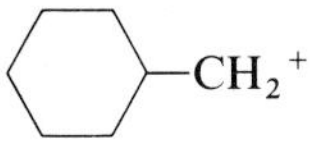 $-CH_2^+$ (b) $(CH_3CH_2)_2CH^+$ (c) $(CH_3CH_2)_3C^+$

5.39. (1) Show the products of the solvolysis of the following compounds in *aqueous ethanol*.
(2) Which halide would react the fastest?

(a) $(CH_3)_2CHBr$ (b) $(CH_3)_3CBr$ (c) $(CH_3)_3CI$

5.40. Predict the product of the aqueous solvolysis of the following halides:

(a) *cis*-1-iodo-3-methylcyclohexane   (b) (*R*)-2-iodooctane

(c) (2*R*,4*S*)-2-iodo-4-methylhexane   (d)

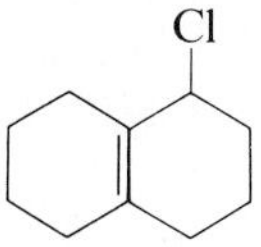

5.41. Each of the following carbocations is capable of undergoing rearrangement to a more stable carbocation. Suggest a structure for the rearranged carbocation.

(a) $(CH_3)_3C\overset{+}{C}HCH_2CH_3$ (b) $CH_2{=}CHCH_2\overset{+}{C}HCH_3$

(c) $(CH_3)_2CH\overset{+}{C}HCH_2CH(CH_3)_2$ (d) 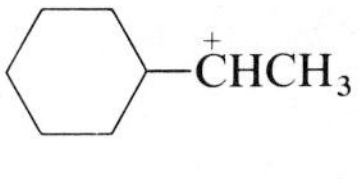 $-\overset{+}{C}HCH_3$

5.42. The following reactions proceed with rearrangement. Show the initial carbocation, the rearranged carbocation, and the major rearrangement product in each case.

(a) $(CH_3)_3CCHICH_3 + H_2O \longrightarrow$
(b) $(CH_3)_2CHCHICH_2CH_3 + CH_3CH_2OH \longrightarrow$

5.43. Reaction (a), which follows, is known to proceed readily. What is the organic product of reaction (b)?

(a) $HO-\overset{\displaystyle CH_3O}{\underset{\displaystyle CH_3O}{\bigcirc}}-CHO + CO_3^{2-} \rightleftharpoons {}^-O-\overset{\displaystyle CH_3O}{\underset{\displaystyle CH_3O}{\bigcirc}}-CHO + HCO_3^-$

(b) $HO-\overset{\displaystyle CH_3O}{\underset{\displaystyle CH_3O}{\bigcirc}}-CHO + \bigcirc-CH_2Cl \xrightarrow{CO_3^{2-}}$

5.44. Suggest a synthesis for each of the following compounds from an organic halide:

(a)   [structure: naphthalene-type ring system with CH_2OH and $HOCH_2$ substituents]

(b) $CH_2{=}CHCH_2SCH_3$

5.45. Give the important resonance structures of each of the following cations:

(a) $CH_2{=}CHCH{=}C\overset{+}{H}CH_2$

(b) $\bigcirc-\overset{+}{C}H-\bigcirc$

(c) $\bigcirc-\overset{+}{C}H{=}CH_2$

(d) $\bigcirc-\overset{+}{C}HCH_3$

5.46. Complete the following equations, showing structures of all likely substitution products:

(a) [cyclohexane ring]$=CH_2 + CN^- \longrightarrow$ (with Br substituent)

(b) [naphthalene ring with CH_2Cl] $+ SH^- \longrightarrow$

(c) $\bigcirc-O^- + \bigcirc-CH_2Cl \longrightarrow$

(d) $CH_3CH{=}CHCH{=}CHCH_2Cl + {}^-OCH_3 \longrightarrow$

(e) [phthalimide-type structure] $NCHCO_2^- + O_2N-\bigcirc-CH_2Br \longrightarrow$ (with CH_2OH)

5.47. (a) Write the equations for the steps of the E1 reaction of 2-iodohexane with H_2O, showing only the major alkene product.
   (b) Which step determines the rate of reaction?
   (c) What other alkene products could be formed?
   (d) Which step determines the ratio of alkene products?

5.48. Which is the more stable alkene: (a) 1-butene or 2-butene?
(b) 2,3-dimethyl-1-butene or 2,3-dimethyl-2-butene?   (c) 2-methyl-2-pentene or
4-methyl-2-pentene?   (d) 1-methyl-1-cyclohexene or 3-methyl-1-cyclohexene?

5.49. What would be the predominant product of E1 reaction of each of the following
compounds?

5.50. What would be the predominant product of E2 reaction (NaOH as the base) of each of the
compounds in Problem 5.49?

5.51. Each of the following alkyl halides can undergo rearrangement in an E1 reaction. For each,
show the initial carbocation, the rearranged carbocation, and the anticipated major
rearrangement product.

(a) $(CH_3)_2CHCHClCH(CH_3)_2$ (b) $(CH_3)_3CCHBrCH_2CH_2CH_3$

5.52. Which compound in each of the following pairs undergoes E2 reaction more rapidly?

(a) $(CH_3)_2CHCHBrCH_2CH_3$ or $(CH_3)_2CBrCH_2CH_2CH_3$

(b) $(CH_3)_2CHCHICH_3$ or $(CH_3)_2CHCH_2CH_2I$

(c) CH_3CHICH_3 or $CH_3CHBrCH_3$

5.53. Which hydrogen(s) must be replaced by deuterium if a maximum kinetic isotope effect is to
be observed in an E2 reaction of each of the following compounds?

(a) $(CH_3)_2CClCH_3$ (b)   (c)

5.54. For each of the following reactions, indicate whether the product is mainly the Saytseff
product or the Hofmann product:

(a) $CH_3CHClCH_2CH_2CH_3$ $\longrightarrow$ $CH_2{=}CHCH_2CH_2CH_3$

(b) $CH_3CHClCH_2CH_2CH_3$ $\longrightarrow$ $CH_3CH{=}CHCH_2CH_3$

(c) $\langle\!\!\!\bigcirc\!\!\!\rangle{-}CH_2CHICH(CH_3)_2$ $\longrightarrow$ $\langle\!\!\!\bigcirc\!\!\!\rangle{-}CH_2CH{=}C(CH_3)_2$

5.55. Predict the principal E2 product of the reaction of each of the following compounds with
Na^+ $^-OCH_3$:   (a) (*S*)-2-bromopentane;   (b) 2,6-dichloroheptane;
(d) (1*S*,2*S*)-1-bromo-1,2-diphenylbutane;   (d) (*S*)-1-chloro-1-cyclohexylethane.

5.56. Complete the following equations, showing only the major organic product, and predict which reaction mechanism (S_N1, S_N2, E1, E2) is the most likely:

(a) $(CH_3)_2CHBr + KI \xrightarrow{\text{acetone}}$

(b) $(CH_3)_2CHBr + KOH \xrightarrow[\text{heat}]{CH_3CH_2OH}$

(c) $(CH_3)_2CHBr + CH_3CH_2OH \xrightarrow{\text{heat}}$

(d) $(CH_3)_3CI + CH_3OH \longrightarrow$

(e) 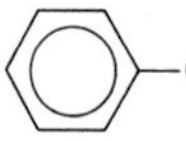 $-O^- + CH_3I \longrightarrow$

(f) $(CH_3)_2CClCH_2CH_3 + Na^+ \ ^-OCH_3 \xrightarrow{CH_3OH}$

(g) $CH_3CHBrCH_2CH_2CH_3 + KOH \xrightarrow[\text{heat}]{CH_3CH_2OH}$

(h) $(CH_3)_2CBrCH_2CH_3 + K^+ \ ^-OC(CH_3)_3 \longrightarrow$

5.57. Which reaction would be more likely to give a rearrangement product:
(a) 2-bromo-3-methylbutane with 10% aqueous KOH, or   (b) 2-bromo-3-methylbutane with 0.1% aqueous KOH?

5.58. Write equations to show how you could prepare the following compounds from organic halides:

(a) $CH_2=CHCH_2OH$

(b) $(CH_3)_2CHCH_2SH$

(c) *trans*-$C_6H_5CH=CHC_6H_5$

(d)

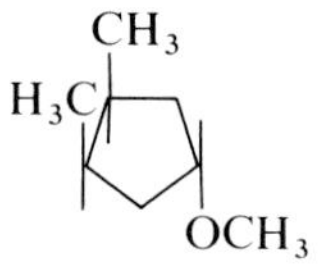

(e)

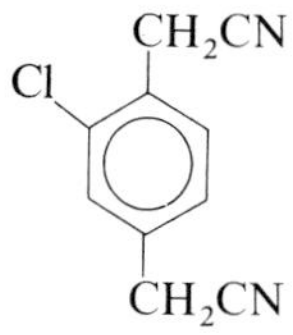

(f) $C_6H_5CH_2OCC_6H_5$ with carbonyl O above the C

(g) $(CH_3)_3\overset{+}{N}CH_2CH_2CH_2\overset{+}{N}(CH_3)_3 \ 2\,Br^-$

(h) $(2S,3R)$-$C_6H_5\overset{|}{C}HCHCH_3$ with CN substituents

5.59. Explain the following observations:

(a)   The following halide is inert toward nucleophiles under either S_N1 or S_N2 conditions.

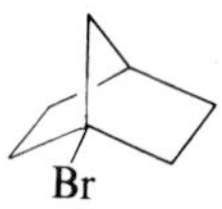

(b) *Quinuclidine* undergoes reaction with iodomethane about 50 times faster than does triethylamine, even though these two amines have almost equal basicities.

quinuclidine

(c) When heated in a solvent, (S)-4-bromo-*trans*-2-pentene undergoes racemization.
(d) Chloromethyl methyl ether ($ClCH_2OCH_3$) undergoes rapid S_N1 reaction.
(e) When (R)-2-iodooctane is treated with radioactive iodide ion, the rate of racemization is found to be twice the rate of incorporation of radioactive iodine.

5.60. A compound with the formula C_4H_9Cl, upon treatment with a strong base, yields three isomeric alkenes. What is the structure of this alkyl halide?

5.61. How would you distinguish (chemically) between each of the following pairs of compounds?

(a) Cl—⟨◯⟩—CH_3 and ⟨◯⟩—CH_2Cl

(b) ⟨◯⟩—Br and ⟨◯⟩—Br

(c) (2S,3R)-2-chloro-3-methylpentane and its (2S,3S)-diastereomer

5.62. (2R,3S)-2-Bromo-3-deuteriobutane undergoes an E2 reaction when treated with $NaOCH_2CH_3$ in ethanol. What would be the principal product(s)?

5.63. (2S,3S)-3-Bromo-2-methoxybutane undergoes an S_N2 reaction with CH_3O^- to yield an optically inactive product. Explain. (Use an equation in your answer.)

5.64. Consider the following hypothetical two-step reaction:

$$A + B \xrightleftharpoons{fast} [AB] \xrightarrow{slow} products$$

(a) Would this reaction show overall *first-order* or *second-order* kinetics? Explain.
(b) Is this reaction *unimolecular* or *bimolecular*? Explain.

5.65. Suggest a synthesis for 1,4-dioxane from 1,2-dibromoethane and no other organic reagents.

(*Hint*: $ROH \xrightarrow{Na} RO^- \ Na^+ + \frac{1}{2}H_2$.)

1,4-dioxane

5.66.   Suggest a mechanism for each of the following reactions:

(a) $(CH_3)_2CHCl + AgNO_3 + H_2O \longrightarrow (CH_3)_2CHOH + AgCl$

(b) $+ AgNO_3 + H_2O \longrightarrow$ $+ AgCl$

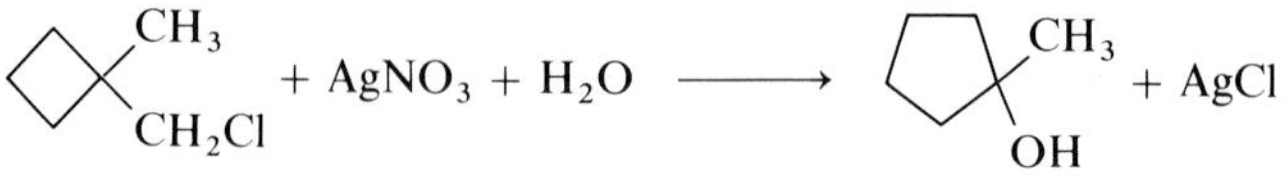

(c) $-CH_2Cl + H_2O \longrightarrow$

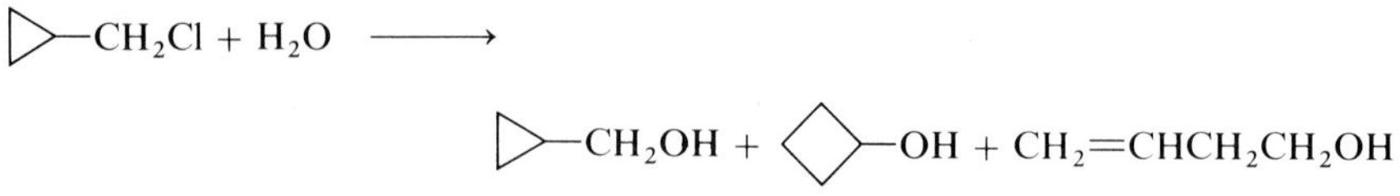

$-CH_2OH +$ $-OH + CH_2{=}CHCH_2CH_2OH$

(d) $CH_2{=}CHCHCH_3 \xrightarrow[H_2O]{LiBr} CH_3CH{=}CHCH_2Br$
with Br above the CHCH₃ carbon

(e) $CH_3CH_2CH_2Cl + NO_2^- \longrightarrow CH_3CH_2CH_2NO_2 + CH_3CH_2CH_2ONO + Cl^-$

(f) $-CH_2Cl + CN^- \xrightarrow{H_2O} CH_3-$$-CN + Cl^-$

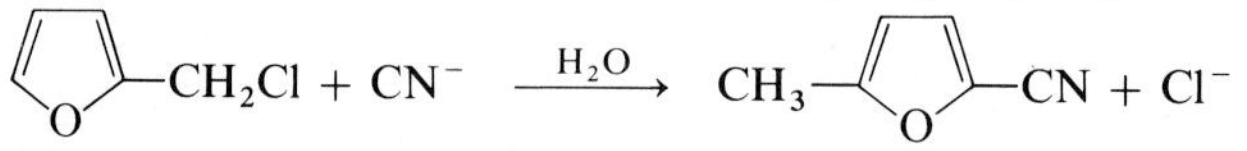

5.67.   Formulas for two of the stereoisomers of menthyl chloride follow. One of these menthyl chlorides undergoes rapid E2 reaction when treated with a base to yield two alkenes: 75% A and 25% B. The other menthyl chloride undergoes a slow E2 reaction and yields only one alkene.

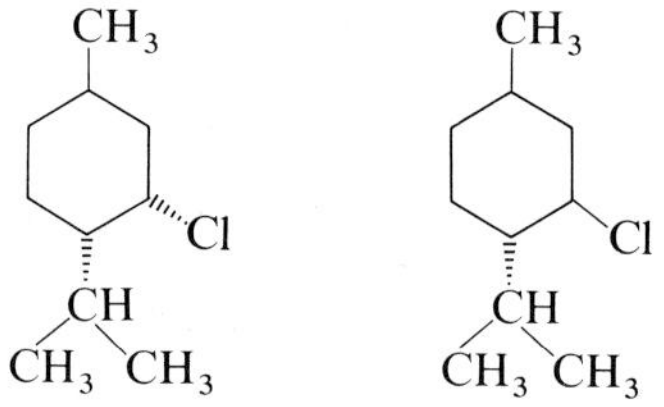

(a)   Which menthyl chloride undergoes the more rapid reaction? Why?
(b)   What two alkenes are formed in this reaction, and which one predominates? Explain.
(c)   Which menthyl chloride undergoes the slow reaction, and what is the structure of the single alkene product? Explain why the reaction is slow and yields only one alkene.

CHAPTER 6

Free-Radical Reactions;
Organometallic Compounds

Many organohalogen compounds are prepared industrially by the reaction of hydrocarbons and halogens, two inexpensive starting materials. Direct halogenation reactions often proceed explosively and, as a general rule, give mixtures of products. For these reasons, direct halogenation is used only occasionally in the laboratory.

$$CH_4 + Cl_2 \xrightarrow{\text{ultraviolet light}} CH_3Cl + CH_2Cl_2 + HCl + \text{other products}$$
methane

$$CH_3CH_3 + Cl_2 \xrightarrow{\text{ultraviolet light}} CH_3CH_2Cl + ClCH_2CH_2Cl + HCl + \text{other products}$$
ethane

$-CH_3 + Cl_2 \xrightarrow{\text{ultraviolet light}}$ $-CH_2Cl + HCl$

toluene

Direct halogenation reactions proceed by a **free-radical mechanism**, which is different from the mechanisms discussed in Chapter 5 for nucleophilic substitutions and eliminations. Free-radical reactions have biological and practical importance. For example, organisms utilize atmospheric oxygen by a sequence of reactions that begins with a free-radical oxidation–reduction. Butter and other fats become rancid partly by free-radical reactions with oxygen.

Organometallic compounds are the second topic we will introduce in this chapter. Reactions of these compounds appear periodically throughout the rest of this book. It is appropriate to mention them here because most organometallic compounds are prepared from alkyl halides.

$$4\,CH_3CH_2Cl \;+\;\; 4\,Na/Pb \;\longrightarrow\; (CH_3CH_2)_4Pb \qquad +\; 3\,Pb\; +\; 4\,NaCl$$

sodium–lead
alloy

tetraethyllead

an organometallic compound

phenylmagnesium bromide

an organometallic compound

SECTION 6.1

A Typical Free-Radical Reaction: Chlorination of Methane

The term **free radical** refers to any atom or group of atoms that has an *odd number of electrons*. Because the number of electrons is odd, the electrons in a free radical cannot all be paired. Although a free radical usually has no positive or negative charge, such a species is highly reactive because of its unpaired electron. A free radical is usually found as a high-energy, nonisolable reaction intermediate.

We usually symbolize a free radical with a single dot representing the unpaired electron.

Lewis formulas for typical free radicals:

$:\!\ddot{C}\!l\cdot \qquad :\!\ddot{B}\!r\cdot \qquad H\!:\!\overset{\displaystyle H}{\underset{\displaystyle H}{\ddot{C}}}\cdot$

Usual formulas for free radicals:

$Cl\cdot \qquad Br\cdot \qquad H_3C\cdot \;$ or $\; CH_3\cdot \;$ or $\; \cdot CH_3$

STUDY PROBLEM

6.1. Write Lewis formulas for the following free radicals:

(a) $HO\cdot$ (b) $CH_3O\cdot$ (c) $CH_3CH_2\cdot$

The chlorination of methane in the presence of ultraviolet light (symbolized *hv*; see page 314) is a classical example of a free-radical reaction. The result of the reaction of Cl_2 with CH_4 is the *substitution* of one or more chlorine atoms for hydrogen atoms on the carbon.

$$CH_4 \;\; + Cl_2 \;\xrightarrow{\;hv\;}\; CH_3Cl \quad + \quad CH_2Cl_2$$

methane   chloromethane   dichloromethane
                       (methyl chloride)   (methylene chloride)

$$+ \quad CHCl_3 \quad + \quad CCl_4 \quad + HCl$$

trichloromethane   tetrachloromethane
(chloroform)   (carbon tetrachloride)

Although methane is the simplest alkane, four organic products can be formed in its chlorination. Small amounts of higher alkanes, such as ethane, and their chlorinated products may also be formed. We will discuss first the reactions leading to CH_3Cl, then we will expand the discussion to the formation of other products.

The mechanism of a free-radical reaction is best thought of as a series of stepwise reactions, each step falling into one of the following categories: (1) **initiation** of the free-radical reaction; (2) **propagation** of the free-radical reaction; and (3) **termination** of the free-radical reaction.

A. Initiation

As the term implies, the initiation step is the initial formation of free radicals. In the chlorination of methane, the initiation step is the homolytic cleavage of Cl_2 into two chlorine free radicals. The energy for this reaction step is provided by ultraviolet light or by heating the mixture to a very high temperature.

Step 1 (initiation):

$$Cl{-}Cl + 58 \text{ kcal/mole} \xrightarrow{\textit{hv or heat}} 2\,Cl\cdot$$

free radicals

B. Propagation

After its formation, the chlorine free radical starts a series of reactions in which new free radicals are formed. Collectively, these reactions are called the **propagation steps** of the free-radical reaction. In effect, the initial formation of a few free radicals results in the propagation of new free radicals in a self-perpetuating reaction called a **chain reaction**.

As the first propagation step, the reactive chlorine free radical abstracts a hydrogen atom from methane to yield a methyl free radical and HCl.

$$Cl\cdot + H{:}CH_3 + 1 \text{ kcal/mole} \longrightarrow H{:}Cl + \cdot CH_3$$

The methyl free radical is also reactive. In the second propagation step, the methyl free radical abstracts a chlorine atom from Cl_2.

$$\cdot CH_3 + Cl{:}Cl \longrightarrow CH_3Cl + Cl\cdot + 25.5 \text{ kcal/mole}$$

chloromethane

This step yields one of the products of the overall reaction, chloromethane. This step also regenerates a new chlorine free radical that can abstract a hydrogen atom from another methane molecule and start the propagation sequence over again.

The overall sequence so far is:

Initiation: $Cl_2 \xrightarrow{\textit{hv or heat}} 2\,Cl\cdot$

Propagation:

$$CH_4 + Cl\cdot \longrightarrow \cdot CH_3 + HCl$$

$$\cdot CH_3 + Cl_2 \longrightarrow CH_3Cl + Cl\cdot$$

can undergo reaction with CH_4

Because a single Cl· causes reaction and a Cl· is also formed, the process could, in theory, continue indefinitely. However, as you might imagine, the reaction does not continue indefinitely. The *number of cycles* (that is, the number of passes through the propagation steps) is called the **chain length**. The chain length of a free-radical reaction depends partly upon the energies of the radicals involved in the propagation. (We will discuss this subject shortly.) For a free-radical chlorination of a hydrocarbon, the chain length is about 1000.

C.   Termination

The propagation cycle is broken by **termination reactions**. Any reaction that results in the destruction of free radicals or in the formation of stable, nonreactive free radicals can terminate the free-radical propagation cycle. The chlorination of methane is terminated principally by free radicals combining with other free radicals; this is a process of destruction of free radicals. In Section 6.7, we will mention termination by formation of stable, nonreactive free radicals.

Termination steps:

$$Cl· + ·CH_3 \longrightarrow CH_3Cl$$

$$·CH_3 + ·CH_3 \longrightarrow CH_3CH_3$$

The second termination step shown is an example of a **coupling reaction**: the joining together of two alkyl groups.

STUDY PROBLEM

6.2.   Write equations for the initiation, propagation, and termination reactions leading to the formation of chlorocyclohexane from cyclohexane and chlorine.

D.   Why free-radical reactions yield mixtures of products

Free-radical reactions are often characterized by a multitude of products. For example, the chlorination of methane can yield four organic products. The reason for the formation of these mixtures is that the high-energy chlorine free radical is not particularly selective about which hydrogen it abstracts during the propagation step.

While chlorine is undergoing reaction with methane, chloromethane is being formed. In time, the chlorine free radicals are more likely to collide with chloromethane molecules than with methane molecules, and a new propagation cycle is started. In this new cycle, chloromethyl free radicals ($·CH_2Cl$) are formed. These undergo reaction with chlorine molecules to yield dichloromethane (CH_2Cl_2). As in the previous cycle leading to CH_3Cl, another chlorine free radical is regenerated in the process.

Propagation steps leading to dichloromethane:

$$Cl· + CH_3Cl \longrightarrow HCl + ·CH_2Cl$$

$$·CH_2Cl + Cl_2 \longrightarrow CH_2Cl_2 + Cl·$$

dichloromethane

SAMPLE PROBLEM

Write the propagation steps leading to the formation of trichloromethane (chloroform) from dichloromethane.

Solution:

$$\cdot Cl + CH_2Cl_2 \longrightarrow HCl + \cdot CHCl_2$$

$$\cdot CHCl_2 + Cl_2 \longrightarrow Cl\cdot + CHCl_3$$

STUDY PROBLEM

6.3. Write the propagation steps leading to the formation of tetrachloromethane (carbon tetrachloride) from trichloromethane.

The free-radical chlorination of methane can yield four organic products (or more, if coupling products are considered). Higher alkanes can produce even larger numbers of products because there are more hydrogens available that can enter into propagation reactions.

STUDY PROBLEM

6.4. How many chloroalkanes could be produced in the chlorination of ethane?

SAMPLE PROBLEM

A chemist wishes to make chloroethane from chlorine and ethane. If he wants to avoid higher chlorination products, would he use: (a) an equimolar mixture of CH_3CH_3 and Cl_2; (b) an excess of Cl_2; or (c) an excess of CH_3CH_3?

Solution: (c) By using an excess of CH_3CH_3, the chemist increases the probability of collisions between $Cl\cdot$ and CH_3CH_3 and decreases the probability of collisions between $Cl\cdot$ and CH_3CH_2Cl.

SECTION 6.1.

Relative Reactivities of the Halogens

The halogens vary dramatically in their reactivity toward alkanes in free-radical reactions. Fluorine undergoes explosive reactions with hydrocarbons. Chlorine is next in terms of reactivity, followed by bromine. Iodine is nonreactive toward alkanes.

$$I_2 \quad Br_2 \quad Cl_2 \quad F_2$$

increasing reactivity as free-radical agents →

The relative reactivity of the halogens toward alkanes is *not* due to the ease with which X_2 molecules are cleaved into free radicals. From the bond dissociation energies for the halogens, we can see that the relative ease of homolytic cleavage is almost the reverse of their reactivity in halogenation reactions.

	F_2	Cl_2	Br_2	I_2
bond dissociation energy (kcal/mole):	37	58	46	36

The order of reactivity is primarily a result of the ΔH of the propagation steps in free-radical halogenation. The propagation steps of fluorination are highly exothermic. More than enough energy is produced to rupture additional F—F bonds and cause an extremely rapid, explosive reaction.

$$F\cdot + CH_4 \longrightarrow HF + CH_3\cdot + 31\ kcal/mole \qquad \Delta H = -31\ kcal/mole$$
$$CH_3\cdot + F_2 \longrightarrow CH_3F + F\cdot + 71\ kcal/mole \qquad \Delta H = -71\ kcal/mole$$

$$CH_4 + F_2 \longrightarrow CH_3F + HF + 102\ kcal/mole \qquad net\ \Delta H = -102\ kcal/mole$$

Just the reverse situation is encountered with iodine: the propagation steps are *endothermic*—that is, the products are of higher energy than the reactants. Most important, the energy required by $I\cdot$ to abstract hydrogen from a C—H bond is *substantially* endothermic. The result is that the iodine free radical does not enter into a chain reaction; $I\cdot$ is an example of a *stable free radical*, a free radical that does not abstract hydrogens.

$$I\cdot + CH_4 + 33\ kcal/mole \longrightarrow HI + CH_3\cdot \qquad \Delta H = +33\ kcal/mole$$
$$CH_3\cdot + I_2 \longrightarrow CH_3I + I\cdot + 20\ kcal/mole \qquad \Delta H = -20\ kcal/mole$$

$$CH_4 + I_2 + 13\ kcal/mole \longrightarrow CH_3I + HI \qquad net\ \Delta H = +13\ kcal/mole$$

Chlorine and bromine are intermediate between fluorine and iodine in ΔH of the propagation steps and therefore are also intermediate in reactivity. Figure 6.1 shows energy diagrams for the reactions of Cl_2 and Br_2 with methane.

$$Cl\cdot + CH_4 + 1\ kcal/mole \longrightarrow HCl + CH_3\cdot \qquad \Delta H = +1\ kcal/mole$$
$$CH_3\cdot + Cl_2 \longrightarrow CH_3Cl + Cl\cdot + 25.5\ kcal/mole \qquad \Delta H = -25.5\ kcal/mole$$

$$CH_4 + Cl_2 \longrightarrow CH_3Cl + HCl + 24.5\ kcal/mole \qquad net\ \Delta H = -24.5\ kcal/mole$$

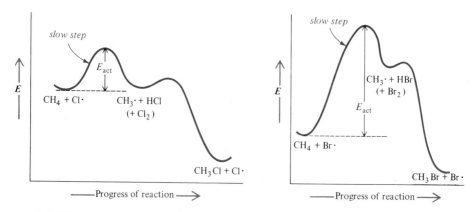

FIGURE 6.1. Energy diagrams for the free-radical chlorination and bromination of methane.

$$Br\cdot + CH_4 + 17\ kcal/mole \longrightarrow HBr + CH_3\cdot \qquad \Delta H = +17\ kcal/mole$$
$$CH_3\cdot + Br_2 \longrightarrow CH_3Br + Br\cdot + 24\ kcal/mole \qquad \Delta H = -24\ kcal/mole$$

$$CH_4 + Br_2 \longrightarrow CH_3Br + HBr + 7\ kcal/mole \qquad net\ \Delta H = -7\ kcal/mole$$

In summary, we find that only chlorine and bromine are useful as free-radical halogenating agents. Fluorine is too reactive toward alkanes, and iodine is not reactive enough.

SECTION 6.3.

Stereochemistry of Free-Radical Halogenation

An alkyl free radical is a species in which a carbon atom has three groups attached to it and a single, unpaired electron. We will look at the structure of the methyl free radical; other alkyl free radicals have similar bonding around the free-radical carbon.

Because there are only three attachments to the free-radical carbon, this carbon is in the sp^2-hybrid state. The three sp^2 orbitals are planar and the unpaired electron is in the p orbital. The structure is very similar to that of a carbocation except that the p orbital of a carbocation is empty.

methyl free radical, $CH_3\cdot$

When a pure enantiomer of a chiral alkyl halide undergoes S_N1 reaction at the chiral carbon, racemization is observed. As we discussed in Chapter 5, racemization arises from the nucleophile being able to attack either lobe of the empty p orbital of the carbocation. If a hydrogen is abstracted from the chiral carbon of a pure enantiomer in a free-radical reaction, racemization also occurs.

(S)-1-chloro-2-methylbutane

$(R)(S)$-1,2-dichloro-2-methylbutane

(*racemic*)

The preceding reaction can lead to a number of products; there are five carbon atoms in the molecule that can lose a hydrogen and gain a chlorine. We would also expect to find as products trichlorinated alkanes, tetrachlorinated alkanes, and so forth. But we are interested only in the one product that has been chlorinated at the chiral carbon. When we isolate this specific product, we find that it is a racemic mixture of the (R) and (S) enantiomers. Just as in an S_N1 reaction, this evidence leads us to believe that the free radical is flat (sp^2 hybrid) and that a chlorine atom can add to either lobe of the *p* orbital.

planar free radical

(*one e⁻ in p orbital*)

STUDY PROBLEM

6.5.   Another dichloroalkane formed in the chlorination of (*S*)-1-chloro-2-methylbutane is $CH_3CHClCH(CH_3)CH_2Cl$. Is this dichloroalkane racemic or not?

SECTION 6.4.

Hydrogen Abstraction: The Rate-Determining Step

Unlike the kinetics of substitution and elimination reactions, the kinetics of a free-radical reaction are quite complex. Simple rate expressions, such as first-order or second-order, are not encountered in free-radical chemistry. The reason for this complexity is that the steps in a free-radical reaction are enmeshed in a cyclical process of varying chain lengths. However, evidence does point to the **hydrogen-abstraction step** as the step governing the overall rate at which products are formed. For example, methane (CH_4) undergoes free-radical chlorination 12 times faster than perdeuteriomethane (CD_4), indicating that the CH bond is broken in the rate-determining step of the reaction.

H (or D) abstraction is the rate-determining step:

$$CH_4 + Cl\cdot \xrightarrow{\text{faster}} CH_3\cdot + HCl$$

$$CD_4 + Cl\cdot \xrightarrow{\text{slower}} CD_3\cdot + DCl$$

A.   Which hydrogen is abstracted?

The hydrogen atoms in organic compounds may be classified as **methyl** (CH_4), **primary** (attached to a 1° carbon), **secondary** (attached to a 2° carbon), **tertiary** (attached to a 3° carbon), **allylic** (on a carbon adjacent to a double bond), or **benzylic** (on a carbon adjacent to an aromatic ring).

2° hydrogens
↓
$CH_3CH_2CH_3$ $CH_2=CHCH_3$ 〈benzene〉—CH_3 ✗
↑   ↑   ↑
1° hydrogens   allylic hydrogens   benzylic hydrogens

SAMPLE PROBLEM

Classify each circled H as 1°, 2°, 3°, allylic, or benzylic:

(a) $(CH_3)_3CC$ Ⓗ$_3$ (b) $(CH_3)_3C$ Ⓗ   (c)

(d)   (e)

Solution: (a) 1°;   (b) 3°;   (c) 2°;   (d) allylic and 3°;   (e) benzylic and 2°

These different types of hydrogen atoms are not abstracted at identical rates by free radicals. Instead, there is a degree of selectivity in hydrogen abstraction. The reaction of propane with a small amount of chlorine under free-radical conditions yields two monochlorinated products, 1-chloropropane and 2-chloropropane, with the 2-chloropropane predominating.

$$CH_3CH_2CH_3 + Cl_2 \xrightarrow{h\nu} CH_3\overset{\underset{|}{Cl}}{C}HCH_3 \; + \; CH_3CH_2CH_2Cl$$

propane

2-chloropropane   1-chloropropane
(isopropyl chloride)   (*n*-propyl chloride)
55%   45%

There are *six* primary hydrogens and *two* secondary hydrogens altogether in propane; the ratio of primary to secondary hydrogens in propane is 6/2, or 3/1. If all the hydrogens underwent abstraction at equal rates, we would observe three times more 1-chloropropane than 2-chloropropane in the product mixture. This is *not* what is observed when the reaction is carried out; instead, slightly *more* 2-chloropropane is formed. We conclude that statistical H abstraction does not occur, and that secondary hydrogens are abstracted faster than primary hydrogens.

Another example of how the relative rates of hydrogen abstraction affect the product ratio follows. (We will discuss the greater selectivity of Br_2 in Section 6.5.)

$\overset{\underset{|}{CH_3}}{C}H_3CHCH_3$
methylpropane
(isobutane)

$\xrightarrow{Cl_2}$

$(CH_3)_2CHCH_2Cl$ + $(CH_3)_3CCl$ + other products
1-chloro-2-methylpropane
(isobutyl chloride)   *t*-butyl chloride
50%   30%   20%

$\xrightarrow{Br_2}$

$(CH_3)_3CBr$
t-butyl bromide
almost 100%

STUDY PROBLEM

6.6.   In the chlorination of methylpropane, what would be the expected product ratio of isobutyl chloride to t-butyl chloride if all the hydrogens were abstracted at equal rates?

Through these and similar experiments, the order of reactivity of hydrogens toward free-radical halogenation has been determined. The relative rates for halogenation reactions of a few compounds are given in Table 6.1.

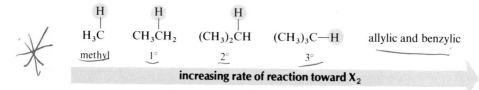

$$\begin{array}{ccccc} \overset{\text{H}}{\underset{|}{}} & \overset{\text{H}}{\underset{|}{}} & \overset{\text{H}}{\underset{|}{}} & & \\ \text{H}_3\text{C} & \text{CH}_3\text{CH}_2 & (\text{CH}_3)_2\text{CH} & (\text{CH}_3)_3\text{C}-\text{H} & \text{allylic and benzylic} \\ \text{methyl} & 1° & 2° & 3° & \end{array}$$

increasing rate of reaction toward X$_2$

TABLE 6.1. Average relative rates of hydrogen abstraction

Hydrocarbon	Reagent[a]	
	Br_2	Cl_2
CH_3-H	0.0007	0.004
CH_3CH_2-H	1	1
$(CH_3)_2CH-H$	220	4.3
$(CH_3)_3C-H$	19,400	6.0
$C_6H_5CH_2-H$	64,000	1.3
$(C_6H_5)_2CH-H$	6.2×10^5	2.6
$(C_6H_5)_3C-H$	1.14×10^6	9.5

[a] The two columns contain data from two separate studies of relative rates. The chlorination of ethane proceeds much more rapidly than bromination under the same conditions.

B.   Relative stabilities of alkyl free radicals

To understand why some hydrogens are abstracted more easily than others, we must look at the transition states of the hydrogen-abstraction steps. The following equations show the hydrogen-abstraction steps in the chlorination of methane and methylpropane. (The symbol $\delta \cdot$ is used to show that both the chlorine atoms and the carbon atoms have partial free-radical character in the transition states.)

$$\text{Cl}\cdot + \text{H}-\overset{\text{H}}{\underset{\text{H}}{\overset{|}{\text{C}}}}{\overset{\text{H}}{}} \longrightarrow \left[\overset{\delta\cdot}{\text{Cl}}\text{----}\text{H}\text{----}\overset{\text{H H}}{\underset{\text{H}}{\text{C}}}{}^{\delta\cdot} \right] \longrightarrow \text{Cl}-\text{H} + \overset{\text{H H}}{\underset{\text{H}}{\text{C}}}\cdot$$

methane	transition state	planar
		methyl free radical

methylpropane   transition state   planar
                                                                     t-butyl free radical

In the transition state, the CH bond is breaking. Part of the reason for the reactivity sequence $3° > 2° > 1° > CH_4$ may be attributed to the increasing CH bond strength as we go from 3° to methyl. It is easier to break a 3° CH bond than a 2° CH bond.

$$CH_3—H \qquad CH_3CH_2—H \qquad (CH_3)_2CH—H \qquad (CH_3)_3C—H$$

**bond dissociation
energy (kcal/mole):**   104   98   94.5   91

decreasing bond strength

However, the relative CH bond strengths are probably <u>not the only reason</u> for the reactivity differences. The transition-state structure for hydrogen abstraction has some free-radical character; therefore, the energy of the transition state is largely determined by the stability of the alkyl free radical being formed. The order of free-radical stability, just like that of carbocation stability, increases as we proceed from methyl to tertiary. It is thought that the free-radical intermediates are stabilized by interaction with neighboring sigma bonds, possibly by hyperconjugation (see Section 5.6E).

$$·CH_3 \qquad ·CH_2CH_3 \qquad (CH_3)_2\overset{·}{C}H \qquad (CH_3)_3C· \qquad \text{allylic and benzylic}$$

methyl   1°   2°   3°

increasing stability

As in the case of carbocation reactions, we find enhanced free-radical reactivity at allylic and benzylic positions because of resonance-stabilization of the intermediate.

allylic

$$CH_2=CHCH_3 \xrightarrow[500°]{Cl_2} CH_2=CHCH_2Cl$$

propene   3-chloro-1-propene
                           (allyl chloride)
                           90%

benzylic

ethylbenzene   (1-chloroethyl)benzene   (2-chloroethyl)benzene
                              56%   44%

STUDY PROBLEMS

6.7.   Write resonance structures for the following free radicals:

(a) $CH_2{=}CH\dot{C}H_2$ (b) 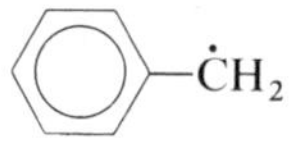 $-\dot{C}H_2$

6.8.   List the following free radicals in order of increasing stability (least stable first):

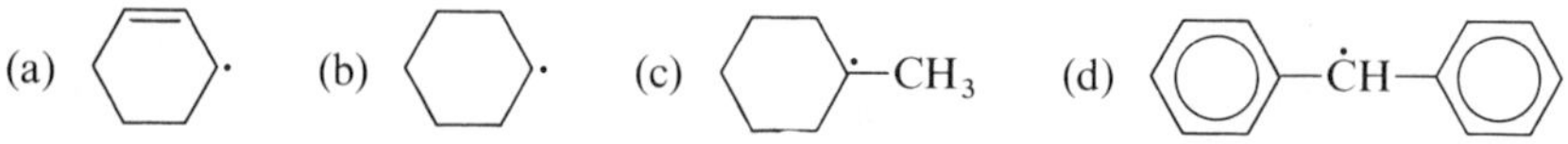

(a)   (b)   (c) $-CH_3$ (d) $-\dot{C}H-$

C.   Rearrangements of free radicals

Alkyl free radicals are, in many respects, similar to carbocations. Both are sp^2 hybrids; both undergo racemization if reaction occurs at a resolved chiral carbon; and both show the same order of stability in terms of structure. Carbocations tend to undergo rearrangement to more-stable carbocations. Does the same hold true for free radicals? No, this is one of the differences between free radicals and carbocations. While free-radical rearrangements are not unknown, they are not common.

Rearrangement:

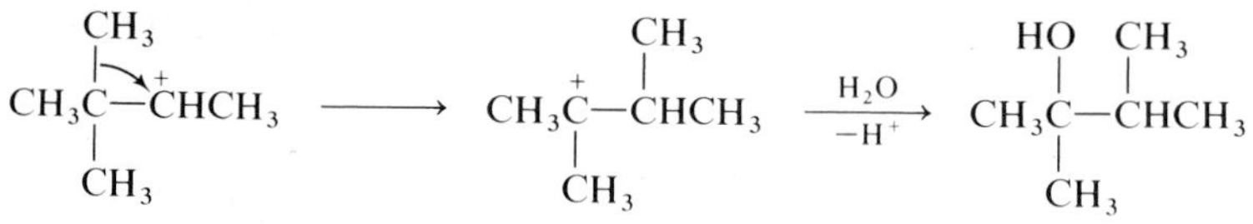

No rearrangement:

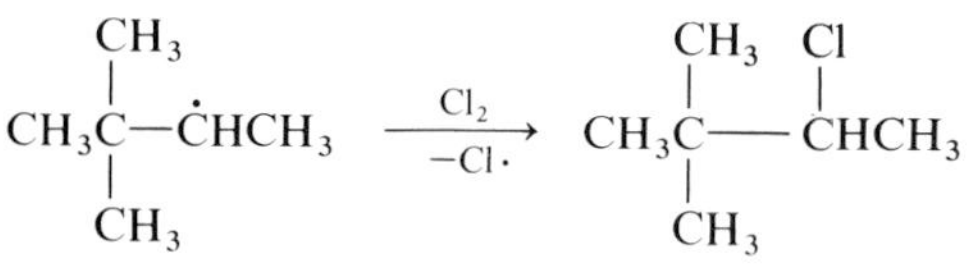

SECTION 6.5.

Selective Free-Radical Halogenations

A.   Bromine versus chlorine

Although free-radical halogenations often lead to mixtures of products, good yields of single products may be obtained in some cases. Compare the product ratios of the chlorination and the bromination of propane:

$$\text{CH}_3\text{CH}_2\text{CH}_3 \xrightarrow[hv]{\text{Cl}_2} \underset{\substack{\text{2-chloropropane} \\ 55\%}}{\text{CH}_3\overset{\text{Cl}}{\underset{|}{\text{CH}}}\text{CH}_3} + \underset{\substack{\text{1-chloropropane} \\ 45\%}}{\text{CH}_3\text{CH}_2\text{CH}_2\text{Cl}}$$

$$\xrightarrow[hv]{\text{Br}_2} \underset{\substack{\text{2-bromopropane} \\ 98\%}}{\text{CH}_3\overset{\text{Br}}{\underset{|}{\text{CH}}}\text{CH}_3} + \underset{\substack{\text{1-bromopropane} \\ 2\%}}{\text{CH}_3\text{CH}_2\text{CH}_2\text{Br}}$$

We can see that bromine, which yields 98% 2-bromopropane, is more selective about abstracting a secondary hydrogen than is chlorine. The selectivity of bromine arises from the fact that bromine is less reactive than chlorine in free-radical halogenations. To see why this is so, we will consider a pair of hypothetical energy diagrams (Figure 6.2).

Reaction 1 in Figure 6.2 is an *exothermic reaction with a low E_{act}*. Note that the structure of the transition state in Reaction 1 is *very close to that of the reactants*. Like Reaction 1, the hydrogen-abstraction step in the chlorination of propane is exothermic and has a low E_{act}. (You could calculate the E_{act} from Table 1.4, page 18.) Therefore, the transition state in this reaction step resembles the reactants more than it does the products:

$$\underset{\text{propane}}{\overset{\text{CH}_3}{\underset{\text{CH}_3}{\text{H}\cdots\text{C}-\text{H}}}} \xrightarrow{\text{Cl}\cdot} \underset{\substack{\text{transition state} \\ \text{resembles reactants}}}{\left[\overset{\text{CH}_3}{\underset{\text{CH}_3}{\text{H}\cdots\text{C}^{\delta\cdot}\text{---H}\text{---------}\overset{\delta\cdot}{\text{Cl}}}}\right]} \longrightarrow \underset{\substack{\text{isopropyl} \\ \text{free radical}}}{\overset{\text{H}\quad\text{CH}_3}{\underset{\text{CH}_3}{\text{C}\cdot}}} + \text{HCl}$$

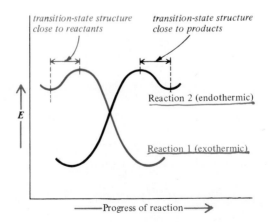

FIGURE 6.2. Energy diagrams showing the relationship of transition-state structure to the exothermic or endothermic nature of a reaction.

Now let us look at Reaction 2 in Figure 6.2. This is an *endothermic reaction with a high* E_{act}. The structure of the transition state in Reaction 2 is *close to that of the products*. The hydrogen-abstraction step in free-radical bromination is more endothermic and has a higher E_{act} than in chlorination. The structure of the transition state in bromination has a greater resemblance to the product alkyl free radical.

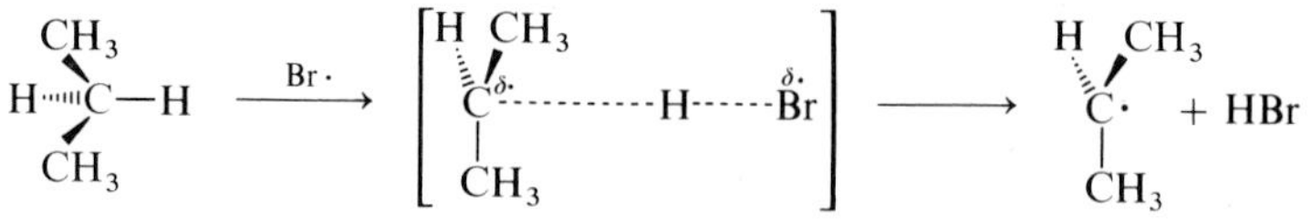

*transition state
resembles products*

Because this transition state in bromination resembles the alkyl free radical, it is highly influenced by the stability of the alkyl free radical. The reaction proceeds through the lower-energy transition state to yield the lower-energy, more stable free radical: $CH_3\dot{C}HCH_3$ is highly favored over $CH_3CH_2\dot{C}H_2$. By contrast, the transition state in chlorination is less influenced by the stability of the alkyl free radical: $CH_3\dot{C}HCH_3$ is only slightly favored over $CH_3CH_2\dot{C}H_2$. Therefore, chlorine is more likely to yield product mixtures.

B.   Benzylic and allylic halogenations

Toluene may be selectively halogenated at the benzylic position with either chlorine or bromine. If more than one alkyl position on a benzene side chain is open to attack, as in ethylbenzene, the more selective bromine is the reagent of choice for halogenation at the benzylic position.

toluene

a benzyl halide
100%

where X = Cl or Br

(1-chloroethyl)benzene
56%

(2-chloroethyl)benzene
44%

ethylbenzene

(1-bromoethyl)benzene
100%

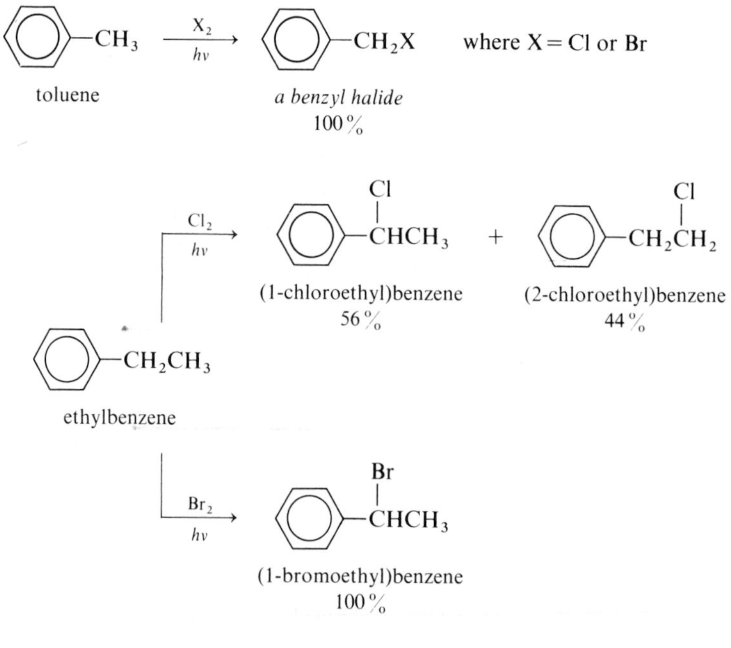

A convenient laboratory reagent for selective halogenation is **N-bromosuc-cinimide (NBS)**, which introduces bromine at the allylic and benzylic positions, but not at other positions. An NBS reaction is catalyzed by light or by some source of free radicals.

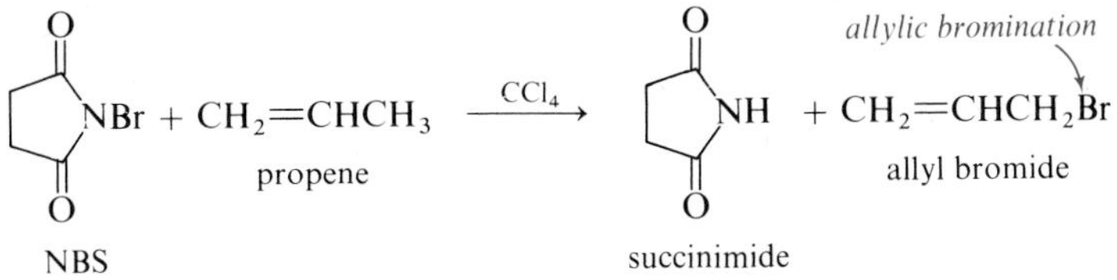

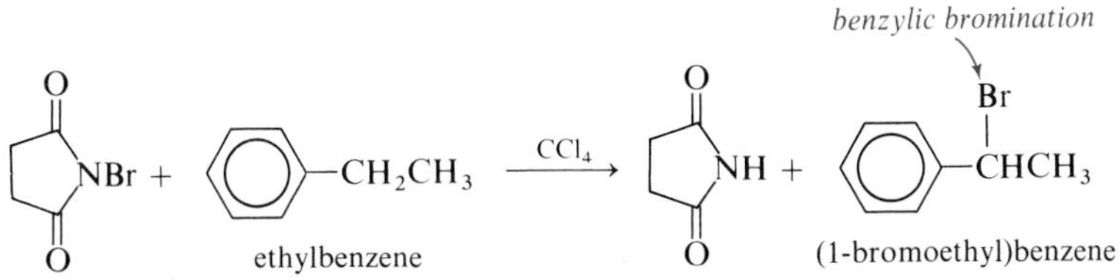

allylic bromination

NBS + propene →(CCl$_4$) succinimide + allyl bromide CH$_2$=CHCH$_2$Br

benzylic bromination

NBS + ethylbenzene →(CCl$_4$) succinimide + (1-bromoethyl)benzene

The selective action of NBS depends partly on its ability to provide a low, but constant, concentration of Br$_2$, which is the halogenating agent. The Br$_2$ is generated by the reaction of HBr (a product of the halogenation) and NBS. Therefore, as Br$_2$ is consumed, more is formed.

Br$_2$ consumed:

$$CH_2=CHCH_3 + Br_2 \longrightarrow CH_2=CHCH_2Br + HBr$$

Br$_2$ generated:

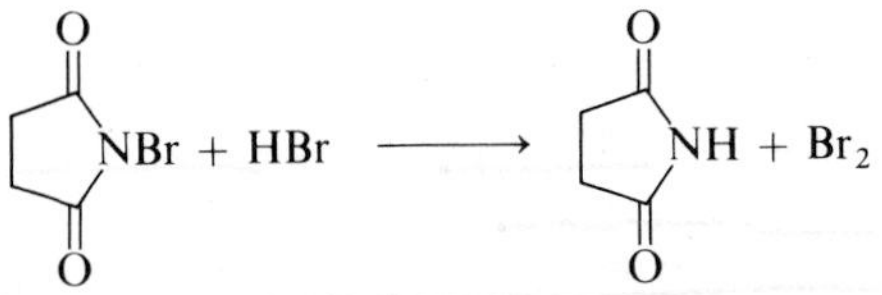

NBr + HBr ⟶ NH + Br$_2$

If the two carbon atoms in an allylic system are not equivalent, mixtures of products result; however, the products may not necessarily be formed in equal amounts. If groups are added to the allylic system, the energies of the resonance structures may not be equivalent; one resonance structure may be a greater contributor.

$$CH_3CH_2CH=CH_2 \xrightarrow{NBS} [CH_3\overset{\cdot}{C}HCH=CH_2 \longleftrightarrow CH_3CH=CH\overset{\cdot}{C}H_2]$$

1-butene   *resonance structures*

$$\underset{\text{3-bromo-1-butene}}{CH_3\overset{\displaystyle Br}{\overset{|}{C}}HCH=CH_2} \qquad \underset{\text{1-bromo-2-butene}}{CH_3CH=CHCH_2Br}$$

SAMPLE PROBLEM

. In the NBS bromination of 1-butene, which of the two products would you
expect to predominate?

Solution: The intermediate free radical has two resonance structures. Both are
allylic radicals, but one is 1° and one is 2°. A 2° free radical is more stable
than a 1° free radical; therefore, the 2° structure is the major contributing
resonance structure. The major product is 3-bromo-1-butene.

STUDY PROBLEM

6.9.   Predict the major halogenation products of NBS halogenation of the following
compounds: (a) *n*-butylbenzene;   (b) cyclohexene;   (c) 1-phenyl-1-propene.

SECTION 6.6.

Other Free-Radical Reactions

As we have mentioned, free-radical reactions are not limited to the halogenation of hydrocarbons, but are encountered in many areas of organic chemistry.
For example, free radicals add readily to double bonds (another complication in
allylic halogenation). However, we will limit our present discussion to sigma-
bond free-radical reactions and consider only a few of the many processes known
to proceed by a free-radical mechanism.

A.   Pyrolysis

In Section 3.4C, we defined **pyrolysis** as the *thermal decomposition of organic
compounds in the absence of oxygen.* When organic molecules are heated to high
temperatures, carbon–carbon sigma bonds rupture and the molecules are broken
into free-radical fragments. (The temperature required depends on the bond
dissociation energies.) This fragmentation step, called **thermally induced homolysis**
(homolytic cleavage caused by heat), is the initiation step for a series of free-
radical reactions. The following equations illustrate some possible pyrolysis
reactions of pentane. (There are other possible positions of cleavage and subsequent reaction.)

$$CH_3CH_2CH_2\!-\!CH_2CH_3 \quad \xrightarrow{\text{heat}} \quad CH_3CH_2CH_2\cdot + CH_3CH_2\cdot$$

Once formed, the free radicals can enter into typical propagation reactions
yielding new free radicals. For example, a free radical can abstract a hydrogen
atom from another pentane molecule.

$$CH_3CH_2\cdot + CH_3CH_2\overset{\textstyle(H)}{C}HCH_2CH_3 \quad \longrightarrow \quad CH_3CH_3 + CH_3CH_2\dot{C}HCH_2CH_3$$

The free radicals can also undergo **β cleavage**, a reaction that, in our example, yields an alkene and a methyl free radical.

$$CH_3CH_2\dot{C}H\!-\!CH_2\!-\!CH_3 \longrightarrow CH_3CH_2CH\!\!=\!\!CH_2 + CH_3\cdot$$

β to free radical

The termination steps in pyrolysis reactions may be either the joining together (coupling) of two free radicals or the **disproportionation** of two radicals. Disproportionation is a general term used to describe a reaction in which two equivalent species react with each other so that one is oxidized and one is reduced. Disproportionation between two alkyl free radicals involves the transfer of a hydrogen atom from one free radical to the other. Two stable products, an alkane (the reduced product) and an alkene (the oxidized product) are formed.

$$CH_3\dot{C}H_2\cdot + \overset{\textstyle\frown}{C}H_2\!-\!\dot{C}H_2 \longrightarrow CH_3CH_3 + CH_2\!\!=\!\!CH_2$$

two ethyl radicals   ethane   ethylene

reduced   oxidized

Controlled pyrolysis has been used industrially for the cracking of high-molecular-weight compounds to lower-molecular-weight compounds, which are often more useful. Until about 1925, pyrolysis of wood provided the major source for methanol (wood alcohol). The thermal cracking of high-boiling petroleum fractions into lower-boiling gasoline fractions was once the only method available for obtaining more gasoline from petroleum. Both of these techniques have been replaced by more sophisticated chemical reactions. The cracking of petroleum is now accomplished with the help of catalysts (Section 3.5), while methanol is largely produced by the catalytic hydrogenation of carbon monoxide ($CO + 2H_2 \rightarrow CH_3OH$).

Pyrolysis is studied today primarily for gaining knowledge about the mechanism of combustion—for example, about the processes involved in the burning of wood or coal. The gaseous molecules that undergo combustion arise from free-radical pyrolysis of molecules on the surface of wood or coal. Fire retardants function not by directly reducing the flames, but by inhibiting the surface and vapor-phase free-radical reactions.

B.   Free radicals in biological systems

Free-radical reactions are an integral part of the chemistry of living systems. Let us consider one example. Animals use food partly for energy. Carbohydrates, for example, are converted to glucose, which then can be converted to carbon dioxide, water, and energy.

$$C_6H_{12}O_6 \quad + 6O_2 \xrightarrow{\text{many steps}} 6CO_2 + 6H_2O + 686\,\text{kcal/mole}$$

glucose

from sugars and starches

The oxidation of glucose is not a direct oxidation like combustion. In an animal cell, a lengthy series of oxidation–reduction reactions is required for the conversion of glucose to CO_2 and H_2O. In the final steps of the oxidation, the electrons needed for the reduction of O_2 to H_2O are supplied by the iron(II) ion.

$$\tfrac{1}{2}O_2 + 2H^+ + 2Fe^{2+} \xrightarrow{\text{cell}} H_2O + 2Fe^{3+}$$

A key structure in this process is a **hydroquinone** (a dihydroxybenzene), a type of compound that is easily oxidized to a **quinone** (a dicarbonyl compound). Conversely, a quinone is easily reduced to a hydroquinone. The following equation depicts the reversible oxidation–reduction reaction between the simplest members of these classes, hydroquinone and quinone.

an oxidizing agent

[O] / [H]

hydroquinone   quinone

a reducing agent

The iron(II) ion in cells is regenerated from the iron(III) ion by reaction with a hydroquinone. The product quinone, which is found in all cells, is called **coenzyme Q**, or **ubiquinone** (for "ubiquitous quinone"). The ring of ubiquinone is substituted with two methoxyl groups (CH_3O—), a methyl group, and a long alkenyl hydrocarbon chain (R) that varies with the source (yeast, mammals, etc.).

$+ 2Fe^{3+} \longrightarrow$ $+ 2H^+ + 2Fe^{2+}$

dihydroubiquinone   ubiquinone
                                    (coenzyme Q)

The conversion of one Fe^{3+} ion to Fe^{2+} is a one-electron change. The action of dihydroubiquinone depends upon the ability of a hydroquinone to lose a single electron at a time, the intermediate being a relatively stable **semiquinone** free radical.

$\xrightarrow[-e^-]{-H^+}$ $\xrightarrow[-e^-]{-H^+}$

semiquinone

a free radical

STUDY PROBLEM

6.10. The semiquinone free radical is resonance-stabilized. Draw the resonance structures.

Let us now consider an example of undesirable free-radical reactions in living systems. Certain types of radiation (α, β, and γ radiation, and x-rays) are called **ionizing radiations** and are known to damage living cells by causing molecules to cleave into ions and free radicals. These cleavages can cause cellular damage by one of two routes: (1) direct destruction of cellular components, or (2) formation of radicals or ions that undergo abnormal reaction with other cellular components.

The nucleic acids are compounds that we will discuss in Chapter 16. These compounds carry the genetic code and, in this capacity, are responsible for cellular multiplication, reproduction of an organism, and the biosynthesis of proteins. When exposed to radiation, the nucleic acids are subject to **depolymerization**— that is, fragmentation of large molecules into smaller molecules. Mitotic (reproducing) cells are more vulnerable to radiation-caused damage than other cells. This fact is used to advantage in radiation treatment of cancer; cancer cells, which reproduce at an abnormally high rate, are more susceptible to radiation damage than are normal cells.

C. Oxygen as a free-radical reagent

Molecular oxygen is different from the compounds we have been studying so far because a stable molecule of O_2 in the ground state has two unpaired electrons; oxygen is said to be a **diradical.** The structure of O_2 cannot be adequately explained by valence-bond formulas because one pair of $2s$ electrons is in an antibonding orbital. An orbital diagram is shown in Figure 6.3. For our purposes, we will represent molecular oxygen as $\cdot O{-}O\cdot$ or simply O_2.

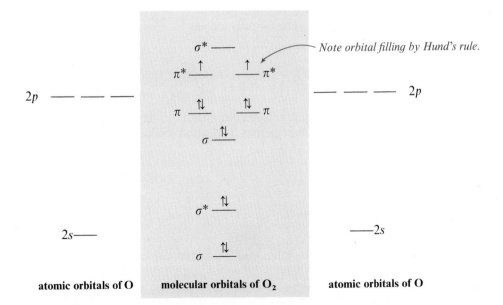

FIGURE 6.3. Orbital diagram for O_2. The lower-energy σ and σ^* orbitals arise from the two $2s$ atomic orbitals of O. Two $2p$ atomic orbitals overlap to form a higher-energy σ orbital. Each oxygen atom in O_2 also has a pair of unshared electrons in a π orbital (from a $2p$ orbital) and one unshared electron in a π^* orbital.

Oxygen is a stable diradical and therefore is a selective free-radical agent. A compound that contains double bonds, allylic or benzylic hydrogens, or tertiary hydrogens is susceptible to **air oxidation**, also called **auto-oxidation** or **autoxidation**. Compounds with only primary and secondary hydrogens are not as susceptible. (From our discussion of free-radical halogenation reactions, the relative reactivities of these hydrogens should not be surprising.)

Fats and vegetable oils often contain double bonds. Auto-oxidation of a fat yields a mixture of products that includes low-molecular-weight (and foul-smelling) carboxylic acids. For example, rancid butter contains the odorous butanoic acid.

Linseed oil and other vegetable oils, which contain many double bonds, are used as drying oils in paint and varnish. These compounds are purposely allowed to undergo air oxidation because the molecules join together, or *polymerize*, into a tough film on the painted surface.

Auto-oxidation initially leads to **hydroperoxides**, compounds containing the —OOH group, which are readily converted to mixtures of alcohols, ketones, and other products. Because mixtures are the usual result, auto-oxidation is rarely used as an organic synthetic technique.

Auto-oxidation:

$$R_3CH + \cdot O{-}O\cdot \longrightarrow \underset{a\ hydroperoxide}{R_3COOH} \longrightarrow \text{mixture of products}$$

In the laboratory, a chemist is most likely to encounter auto-oxidation products as undesirable impurities in ethers and aldehydes. In ethers, the carbon adjacent to the oxygen is the position of attack. Peroxides of ethers explode when heated. For example, diethyl ether is a common laboratory solvent that is purified by distillation. Unless the peroxides have been removed prior to distillation (by a reducing agent, for example), they will become concentrated in the distilling flask as the ether is boiled away. The result could easily be an explosion.

$$\underset{\text{diethyl ether}}{CH_3CH_2OCH_2CH_3} + O_2 \longrightarrow \underset{a\ hydroperoxide}{CH_3CH_2O\overset{\overset{\displaystyle OOH}{|}}{C}HCH_3}$$

The product of aldehyde auto-oxidation is a carboxylic acid, which is formed by reaction of the intermediate peroxy acid with the aldehyde.

$$\underset{\substack{\text{ethanal}\\ \text{(acetaldehyde)}}}{CH_3\overset{\overset{\displaystyle O}{||}}{C}H} + O_2 \longrightarrow \underset{\text{peroxyacetic acid}}{CH_3\overset{\overset{\displaystyle O}{||}}{C}OOH}$$

$$CH_3\overset{\overset{\displaystyle O}{||}}{C}OOH + CH_3\overset{\overset{\displaystyle O}{||}}{C}H \longrightarrow \underset{\text{acetic acid}}{2\ CH_3\overset{\overset{\displaystyle O}{||}}{C}OH}$$

STUDY PROBLEM

6.11. When exposed to air, diisopropyl ether forms peroxides at a more rapid rate than diethyl ether does. Suggest a reason for this behavior.

SECTION 6.7.

Free-Radical Initiators and Inhibitors

A **free-radical initiator** is anything that can initiate a free-radical reaction. The action of ultraviolet light to bring about free-radical halogenation is the action of an initiator. There are several types of compounds that may be added to a reaction mixture to initiate free-radical reactions. These compounds are sometimes erroneously called **free-radical catalysts.** They are not truly catalysts because they are often consumed in the reaction.

Any compound that can easily decompose into free radicals can act as an initiator. **Peroxides.** (ROOR) are one example. They form free radicals easily because the RO—OR bond dissociation energy is only about 35 kcal/mole, lower than that for most other bonds. Benzoyl peroxide and peroxybenzoic acid are two peroxides that are commonly used in conjunction with NBS brominations.

benzoyl peroxide

peroxybenzoic acid

As the name implies, a **free-radical inhibitor** inhibits a free-radical reaction. An inhibitor is sometimes referred to as a **free-radical " trap."** The usual action of a free-radical inhibitor is to undergo reaction with reactive free radicals to form relatively stable and nonreactive free radicals.

An inhibitor used to control auto-oxidation is called an **antioxidant,** or, in the food industry, a **preservative.** *Phenols,* compounds with an —OH group attached to an aromatic ring carbon, are effective antioxidants. Recall that hydroquinone is a natural storehouse for unpaired electrons (Section 6.6B). The inhibitory powers of phenols are directly analogous. The free-radical products of these compounds are resonance-stabilized and thus nonreactive compared to most other free radicals.

phenol

the simplest of
the phenols

" *trapped* "

Resonance structures of the phenol free radical:

The food preservative BHT is a synthetic phenol. (BHA, another preservative, is closely related to BHT; instead of a methyl group on the ring, BHA has an —OCH$_3$ group.) A naturally occurring preservative found in vegetable oils, especially wheat germ oil, is α-tocopherol, or vitamin E.

phenol groups

3,5-bis(*t*-butyl)-4-hydroxytoluene
"BHT"

α-tocopherol
(vitamin E)

STUDY PROBLEMS

6.12. *Azobisisobutyronitrile* (AIBN) is often used as a free-radical initiator because it yields free radicals (along with N$_2$) readily upon heating. (a) Write an equation with fishhook arrows (⌒) for this thermal decomposition. (b) Suggest a reason why AIBN yields free radicals readily.

$$(CH_3)_2\overset{\overset{\textstyle CN}{|}}{C}-N{=}N-\overset{\overset{\textstyle CN}{|}}{C}(CH_3)_2$$

AIBN

6.13. Arylamines, like phenols, can act as antioxidants. For example, *N*-phenyl-2-naphthylamine is added to rubber articles to prevent free-radical degradation of the rubber. Write an equation that shows how this amine can function as an antioxidant.

N-phenyl-2-naphthylamine

SECTION 6.8.

Organometallic Compounds

An **organometallic compound** is defined as a compound in which *carbon is bonded directly to a metallic atom* (such as mercury, zinc, lead, magnesium, or lithium) or to certain *metalloids* (such as silicon, arsenic, or selenium).

CH$_3$CH$_2$CH$_2$CH$_2$Li   (CH$_3$)$_4$Si   CH$_3$ONa

n-butyllithium   tetramethylsilane (TMS)   sodium methoxide

organometallic   *organometallic*   *not considered organometallic*
(no carbon–metal bond)

Organometallic compounds are named in one of two ways:

1. They are named as **alkylmetals** (one word):

$$CH_3CH_2CH_2Li \qquad (CH_3CH_2)_4Pb$$

   n-propyllithium   tetraethyllead

If the metal is bonded to an inorganic anion as well as to a carbon atom, the compound is named as a derivative of the inorganic salt.

$$CH_3MgBr$$

methylmagnesium bromide

⟨○⟩—HgCl

phenylmercuric chloride

2. Compounds of silicon and some other metalloids are named as **derivatives of the hydrides.**

$$SiH_4 \qquad (CH_3)_2SiH_2 \qquad (C_6H_5)_2Si(CH_3)_2$$

silane   dimethylsilane   dimethyldiphenylsilane

SECTION 6.9.

Organomagnesium Halides: Grignard Reagents

One of the most useful classes of reagents in organic synthesis is that of the organo-magnesium halides (RMgX). These compounds are called **Grignard reagents** after the French chemist Victor Grignard, who received the Nobel Prize in 1912 for work in this area of organometallic chemistry. A Grignard reagent is the product of a free-radical reaction between magnesium metal and an organohalogen compound in an ether solvent.

$$R{:}X + Mg \longrightarrow [R{\cdot} + {\cdot}MgX] \longrightarrow R{-}MgX$$

                                                  a Grignard reagent

The reaction is general and does not depend to any great extent upon the nature of the R group. Primary, secondary, and tertiary alkyl halides, as well as allylic and benzylic halides, all form Grignard reagents.

$$CH_3I \quad + Mg \longrightarrow \quad CH_3MgI$$

iodomethane   methylmagnesium iodide

$$(CH_3)_3CBr \quad + Mg \longrightarrow \quad (CH_3)_3CMgBr$$

t-butyl bromide*   *t*-butylmagnesium bromide

⟨○⟩—CH_2Cl + Mg ⟶ ⟨○⟩—CH_2MgCl

benzyl chloride   benzylmagnesium chloride

Aryl and vinylic halides (X on the doubly bonded carbon) are generally quite inert toward nucleophilic substitution and elimination. These compounds are not as reactive as alkyl halides toward magnesium, but their Grignard reagents may still be prepared.

$$\text{C}_6\text{H}_5\!-\!\text{Br} + \text{Mg} \longrightarrow \text{C}_6\text{H}_5\!-\!\text{MgBr}$$

bromobenzene   phenylmagnesium bromide

an aryl Grignard reagent

$$\text{CH}_2\!\!=\!\!\text{CHI} + \text{Mg} \longrightarrow \text{CH}_2\!\!=\!\!\text{CHMgI}$$

iodoethene   ethenylmagnesium iodide
(vinyl iodide)   (vinylmagnesium iodide)

a vinylic Grignard reagent

Organomagnesium halides are unstable unless they are solvated. The usual solvent for a Grignard reagent is diethyl ether ($CH_3CH_2OCH_2CH_3$), which is nonreactive toward Grignard reagents, but can donate unshared electrons to the empty *d* orbitals of Mg. The ethyl groups provide a hydrocarbon environment that acts as the solvent for the alkyl portion of a Grignard reagent.

$$\begin{array}{ccc} \text{CH}_3\text{CH}_2 & & \text{CH}_2\text{CH}_3 \\ & \ddot{\text{O}} & \\ & | & \\ \text{CH}_3\!-\!&\text{Mg}\!-\!\text{I} \\ & | & \\ & \ddot{\text{O}} & \\ \text{CH}_3\text{CH}_2 & & \text{CH}_2\text{CH}_3 \end{array}$$

STUDY PROBLEM

6.14. Which of the following compounds would be a suitable solvent for a Grignard reagent?

(a) $CH_3CH_2CH_2CH_2CH_3$ (b) $CH_3OCH_2CH_2CH_3$

(c)

(d) O ⬡ O

(e) $CH_3OCH_2CH_2Cl$

A.   Reactivity of Grignard reagents

What is unique about a Grignard reagent? In most organic compounds, carbon carries either no partial charge or a partial positive charge. In a Grignard reagent, carbon is bonded to an electropositive element and consequently carries a *partial negative charge.*

$$\overset{\delta+}{\text{CH}_3\text{CH}_2}\!-\!\overset{\delta-}{\text{Br}} + \text{Mg} \xrightarrow{\text{diethyl ether}} \overset{\delta-}{\text{CH}_3\text{CH}_2}\!-\!\overset{\delta+}{\text{Mg}}\!-\!\text{Br} \quad \textit{carbon is } \delta-$$

It is generally true that a carbon atom bonded to a metallic atom is the more electronegative of the two atoms and carries a partial negative charge. An ion with a negatively charged carbon atom is called a **carbanion**. A carbon bonded to a metallic atom therefore has **carbanion character**.

a carbanion   has carbanion character

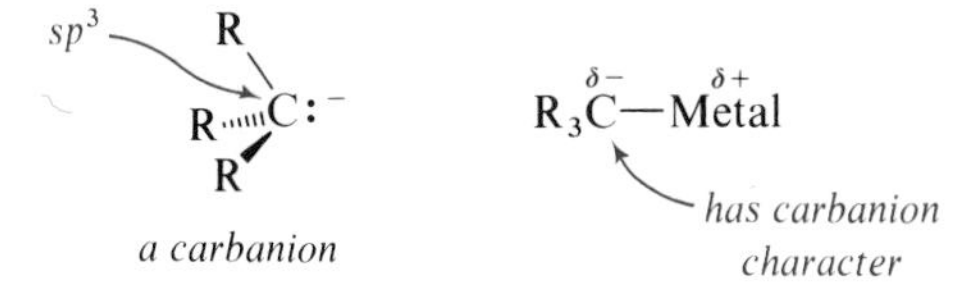

Carbanions are one of the strongest classes of bases encountered in the laboratory. Because a Grignard reagent has a partially negative carbon, (1) it is an *extremely strong base*, and (2) the alkyl or aryl portion of the Grignard reagent can act as a *nucleophile*. We will discuss the action of a Grignard reagent as a nucleophile here and examine its basicity in Section 6.11.

The most important reactions of Grignard reagents are those with carbonyl compounds. In a carbonyl group (C=O), the electrons in the carbon–oxygen bonds (sigma and pi) are drawn toward the electronegative oxygen. The carbon of the carbonyl group, which has a partial positive charge, is attacked by the nucleophilic carbon of the Grignard reagent. The following equations show how Grignard reagents undergo reactions with ketones. (Note that these reactions are not free-radical reactions. When a Grignard reagent attacks a carbonyl group, the electrons move in pairs, not singly.)

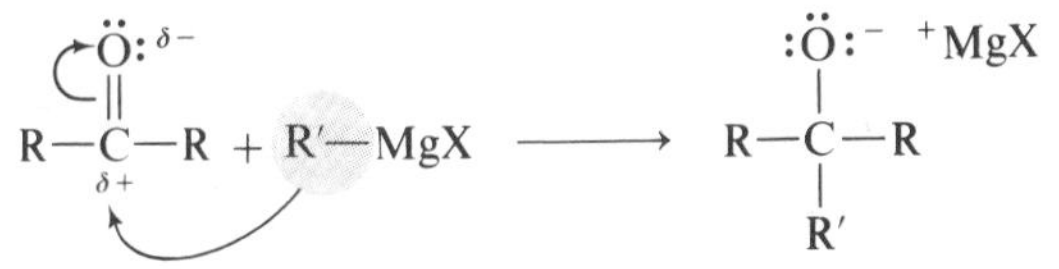

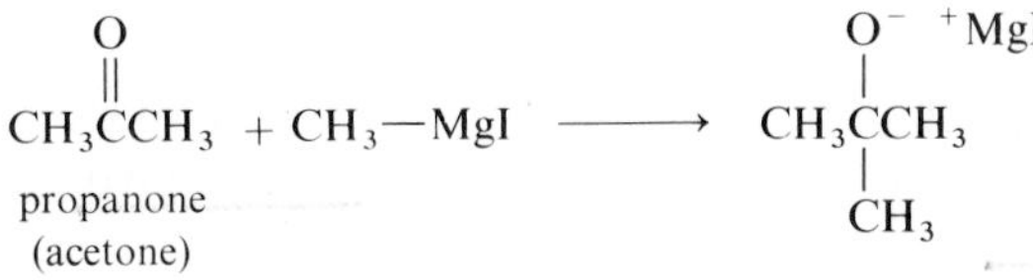

Reactions of RMgX with ketones:

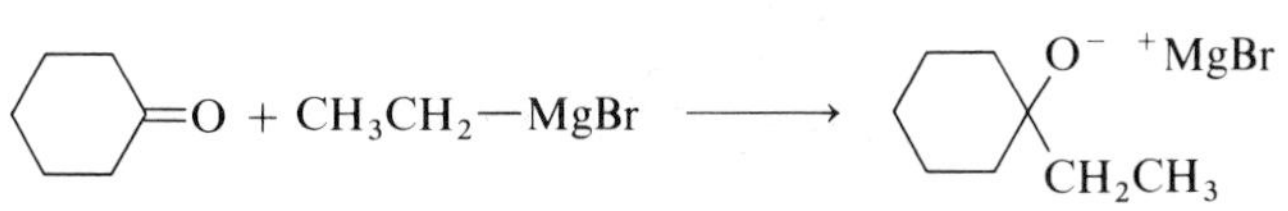

The product of the reaction of RMgX with a ketone is the magnesium salt of an alcohol. When treated with water or aqueous acid, this magnesium salt yields the alcohol and a mixed inorganic magnesium salt. The hydrolyzed product of the reaction of a ketone with a Grignard reagent is a *tertiary alcohol*.

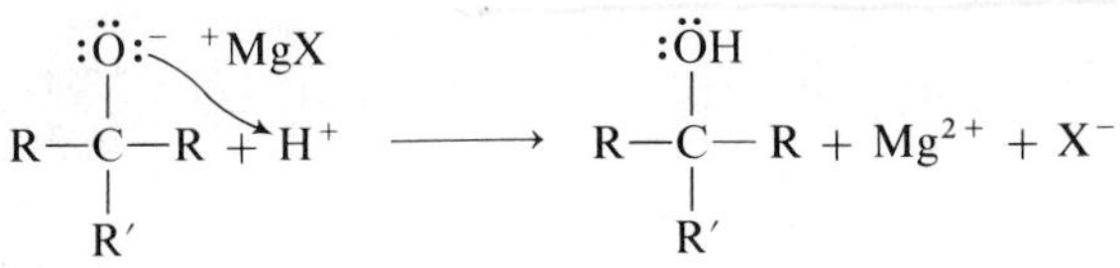

a 3° alcohol

The two steps of a Grignard reaction are usually combined into one equation:

$$
\underset{\text{acetone}}{CH_3\overset{\displaystyle O}{\overset{\|}{C}}CH_3}
\xrightarrow[\text{(2) H}_2\text{O, H}^+]{\text{(1) CH}_3\text{MgI}}
\underset{\substack{\displaystyle | \\ CH_3 \\[2pt] \textit{t-butyl alcohol}}}{CH_3\overset{\displaystyle OH}{\overset{|}{C}}CH_3}
$$

Not only ketones, but almost all compounds containing carbonyl groups (aldehydes, esters, carbon dioxide, etc.) undergo reaction with Grignard reagents. For this reason, Grignard reactions are invaluable to the synthetic organic chemist for building up complicated carbon skeletons from simpler skeletons. Some examples of reactions of Grignard reagents with aldehydes to yield *secondary alcohols* follow:

Reactions of RMgX with aldehydes:

$$
\textit{General:} \quad
\underset{\substack{\displaystyle \\ \textit{an aldehyde}}}{R\overset{\displaystyle O}{\overset{\|}{C}}H}
\xrightarrow[\text{(2) H}_2\text{O, H}^+]{\text{(1) R}'\text{MgX}}
\underset{\substack{\displaystyle \\ \textit{a 2° alcohol}}}{R\overset{\displaystyle OH}{\overset{|}{C}}HR'}
$$

$$
\underset{\substack{\text{ethanal} \\ \text{(acetaldehyde)}}}{CH_3\overset{\displaystyle O}{\overset{\|}{C}}H}
\xrightarrow[\text{(2) H}_2\text{O, H}^+]{\text{(1) CH}_3\text{MgI}}
\underset{\substack{\text{2-propanol} \\ \text{(isopropyl alcohol)}}}{CH_3\overset{\displaystyle OH}{\overset{|}{C}}HCH_3}
$$

$$
\underset{\substack{\text{propanal} \\ \text{(propionaldehyde)}}}{CH_3CH_2\overset{\displaystyle O}{\overset{\|}{C}}H}
\xrightarrow[\text{(2) H}_2\text{O, H}^+]{\text{(1) C}_6\text{H}_5\text{MgBr}}
\underset{\text{1-phenyl-1-propanol}}{CH_3CH_2\overset{\displaystyle OH}{\overset{|}{C}}H-\text{C}_6\text{H}_5}
$$

An exception to the general rule that a Grignard reaction with an aldehyde yields a secondary alcohol is the reaction of formaldehyde. This aldehyde yields a *primary alcohol* in the Grignard reaction.

$$
\textit{General:} \quad
\underset{\substack{\text{methanal} \\ \text{(formaldehyde)}}}{H\overset{\displaystyle O}{\overset{\|}{C}}H}
\xrightarrow[\text{(2) H}_2\text{O, H}^+]{\text{(1) RMgX}}
\underset{\textit{a 1° alcohol}}{RCH_2OH}
$$

The reaction of a Grignard reagent with carbon dioxide (often as dry ice) does not yield an alcohol but a *magnesium carboxylate salt.* The magnesium salt is insoluble in the ether solvent used in a Grignard reaction; therefore, only one of the two pi bonds of CO_2 reacts. Treatment of the insoluble magnesium salt with aqueous acid liberates the *carboxylic acid.*

$$
\textit{Step 1:} \quad \ddot{O}{=}C{=}\ddot{O} + R{-}MgX \longrightarrow \ddot{O}{=}C{-}\ddot{O}{:}^- \;\; ^+MgX
$$
$$
\underset{\substack{\displaystyle | \\ R \\[2pt] \textit{a carboxylate} \\ \textit{(insoluble in ethers)}}}{}
$$

Step 2:

$$\underset{\substack{\text{|} \\ \text{RCO}^- \ ^+\text{MgX} + \text{H}^+}}{\text{O}} \longrightarrow \underset{\substack{\text{|} \\ \text{RCOH} \\ \textit{a carboxylic acid}}}{\text{O}} + \text{Mg}^{2+} + \text{X}^-$$

Table 6.2 summarizes the Grignard reactions that we have mentioned here.

TABLE 6.2. Some products from Grignard syntheses[a]

Carbonyl compound	$\xrightarrow[\text{(2) H}_2\text{O, H}^+]{\text{(1) RMgX}}$	Product	
$\underset{\text{HCH}}{\overset{\text{O}}{\parallel}}$		RCH_2OH	**a 1° alcohol**
$\underset{\text{R'CH}}{\overset{\text{O}}{\parallel}}$		$\underset{\text{RCHR'}}{\overset{\text{OH}}{\vert}}$	**a 2° alcohol**
$\underset{\text{R'CR''}}{\overset{\text{O}}{\parallel}}$		$\underset{\underset{\text{R}}{\vert}}{\overset{\text{OH}}{\underset{\text{R'CR''}}{\vert}}}$	**a 3° alcohol**
CO_2		RCO_2H	**a carboxylic acid**

[a] Other Grignard reactions will be discussed in Sections 7.16B, 11.13, 13.3C, and 13.5C.

SAMPLE PROBLEM

A chemist (a) treats iodobenzene with magnesium metal and diethyl ether; (b) adds acetone; and finally, (c) adds a dilute solution of HCl. Write an equation to represent each reaction.

Solution: (a) $\text{C}_6\text{H}_5\text{I} + \text{Mg} \xrightarrow{\text{diethyl ether}} \text{C}_6\text{H}_5\text{MgI}$

(b) $\underset{\text{CH}_3\text{CCH}_3}{\overset{\text{O}}{\parallel}} + \text{C}_6\text{H}_5\text{MgI} \longrightarrow \underset{\underset{\text{C}_6\text{H}_5}{\vert}}{\overset{\text{OMgI}}{\underset{\text{CH}_3\text{CCH}_3}{\vert}}}$

(c) $\underset{\text{(CH}_3)_2\text{CC}_6\text{H}_5}{\overset{\text{OMgI}}{\vert}} + \text{H}^+ \longrightarrow \underset{\text{(CH}_3)_2\text{CC}_6\text{H}_5}{\overset{\text{OH}}{\vert}} + \text{Mg}^{2+} + \text{I}^-$

STUDY PROBLEM

6.15. Suggest a Grignard synthesis for 1-cyclohexyl-1-ethanol starting with bromocyclohexane and an aldehyde.

SECTION 6.10.

Other Organometallics

Grignard reagents are but one type of a large number of useful organometallic compounds. **Lithium reagents,** another type of organometallic compound, are prepared by the reaction of lithium metal with an alkyl halide in a hydrocarbon or ether solvent.

General: $RX + 2 Li \longrightarrow R{-}Li + LiX$

an alkyllithium

$CH_3CH_2CH_2Br + 2 Li \longrightarrow CH_3CH_2CH_2Li + LiBr$

1-bromopropane
(*n*-propyl bromide)   *n*-propyllithium

A lithium reagent is similar to a Grignard reagent in many ways and undergoes similar reactions. However, the C—Li bond has more ionic character than the C—Mg bond because lithium is more electropositive than magnesium. Lithium reagents are more reactive as nucleophiles than Grignard reagents because the carbon involved in a C—Li bond is more negative.

General:

$$R{-}\underset{a\ ketone}{\overset{\overset{\displaystyle O}{\|}}{C}}{-}R \xrightarrow[\text{(2) } H_2O,\ H^+]{\text{(1) } R'Li} R{-}\underset{\underset{\displaystyle R'}{|}}{\overset{\overset{\displaystyle OH}{|}}{C}}{-}R$$

a 3° alcohol

$$CH_3CH_2CH_2\overset{\overset{\displaystyle O}{\|}}{C}CH_2CH_3 + CH_3CH_2\overset{\delta-}{C}H_2\overset{\delta+}{-}Li \longrightarrow$$

3-hexanone

$$CH_3CH_2CH_2\underset{\underset{\displaystyle CH_2CH_2CH_3}{|}}{\overset{\overset{\displaystyle O^-\ Li^+}{|}}{C}}CH_2CH_3 \xrightarrow{H_2O,\ H^+} CH_3CH_2CH_2\underset{\underset{\displaystyle CH_2CH_2CH_3}{|}}{\overset{\overset{\displaystyle OH}{|}}{C}}CH_2CH_3$$

4-ethyl-4-heptanol

Lithium dialkylcopper reagents, also called **cuprates,** are synthesized from an alkyllithium and a copper(I) halide, such as CuI.

$2 CH_3Li + CuI \longrightarrow (CH_3)_2CuLi + LiI$

a cuprate

These reagents are especially useful in synthesizing *unsymmetrical alkanes of the type* R—R′, where R comes from the cuprate and R′ from an alkyl halide. Best yields are obtained when R′X is a primary alkyl halide, but the R group in R_2CuLi can be almost any alkyl or aryl group.

General: $R_2CuLi + R'X \xrightarrow{0-25°} R{-}R'$

a cuprate   *a 1° alkyl halide*   *an unsymmetrical alkane*

$$(CH_3)_2CHBr \xrightarrow{\text{Li}} (CH_3)_2CHLi \xrightarrow{\text{CuI}} [(CH_3)_2CH]_2CuLi$$

2-bromopropane   *a cuprate*

$$\xrightarrow{CH_3CH_2CH_2Br} (CH_3)_2CH-CH_2CH_2CH_3$$

2-methylpentane

bromocyclohexane   *a cuprate*

methylcyclohexane

STUDY PROBLEM

6.16. Suggest syntheses for

(a) [cyclohexyl]—$CH_2CH_2CH_3$ (b) $(CH_3)_3CCH_2CH_2CH_3$

from compounds containing six or fewer carbon atoms. (There may be more than one correct answer.)

SECTION 6.11.

Reaction of Organometallics with Acidic Hydrogens

We have shown how some organometallic compounds can act as nucleophiles. Many organometallic compounds, such as Grignard reagents and lithium reagents, are also extremely strong bases. A hydrogen that can be abstracted from a compound by a Grignard reagent is said to be an *acidic hydrogen* in relation to the Grignard reagent. Grignard reagents and lithium reagents undergo rapid reaction with compounds that have acidic hydrogens, yielding hydrocarbons and metal salts.

$$\overset{\delta-}{R}-MgX + H-\overset{..}{\underset{..}{O}}H \longrightarrow RH + XMg\overset{..}{\underset{..}{O}}H$$

$$\overset{\delta-}{R}-Li + H-\overset{..}{\underset{..}{O}}H \longrightarrow RH + Li\overset{..}{\underset{..}{O}}H$$

Grignard reagents and lithium reagents are stronger bases than ^-OH, ^-OR, $^-NH_2$, or $RC\equiv C^-$. Therefore, the type of hydrogen that is acidic toward RMgX or RLi is any H bonded to *oxygen, nitrogen,* or an *sp-hybridized carbon*. (The reasons why $RC\equiv CH$ can lose a proton will be discussed in Chapter 9.)

Some structures that contain hydrogens acidic toward RMgX or RLi:

HO H	H₂N H	ArO H
RO H	R₂N H	RCO₂ H
RC≡CH		

SAMPLE PROBLEM

What are the products of the reaction of methylmagnesium iodide with each of the seven structures shown that contain acidic hydrogens?

Solution:

$$CH_3MgI + H_2O \longrightarrow CH_4 + HOMgI$$

$$CH_3MgI + ROH \longrightarrow CH_4 + ROMgI$$

$$CH_3MgI + RC \equiv CH \longrightarrow CH_4 + RC \equiv CMgI$$

$$CH_3MgI + NH_3 \longrightarrow CH_4 + NH_2MgI$$

$$CH_3MgI + R_2NH \longrightarrow CH_4 + R_2NMgI$$

$$CH_3MgI + ArOH \longrightarrow CH_4 + ArOMgI$$

$$CH_3MgI + RCO_2H \longrightarrow CH_4 + RCO_2MgI$$

The principal ramifications of the reactivities of RMgX and RLi toward acidic hydrogens are that: (1) water, alcohols, and other compounds with acidic hydrogens must be excluded from a Grignard or lithium reaction mixture (unless an alkane is the desired product), and (2) the presence of certain functional groups in an organohalogen compound precludes the formation of stable Grignard or lithium reagents.

SAMPLE PROBLEM

A student maintains that a Grignard reagent undergoes reaction with water to yield an alcohol. What is wrong with this statement?

Solution: The carbon of the Grignard reagent is partially *negative* and undergoes reaction with a *positive group* (such as H$^+$), not with a negative group (such as $^-$OH).

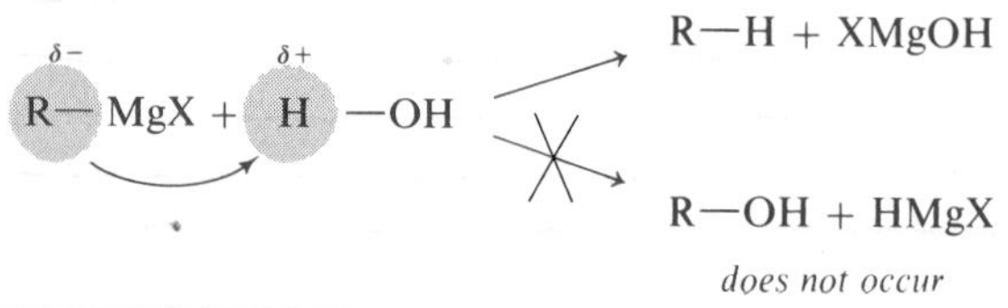

$$R-H + XMgOH$$

$$R-MgX + H-OH$$

$$R-OH + HMgX$$

does not occur

STUDY PROBLEMS

6.17. Which of the following compounds could *not* be used to prepare a Grignard reagent? Explain.

(a) $\underset{\underset{\displaystyle NH_2}{|}}{CH_3CHCH_2CH_2Br}$ (b) 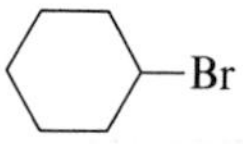 —Br   (c)

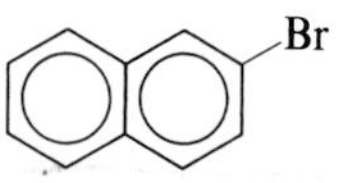

(d) $BrCH_2\overset{\displaystyle O}{\overset{\displaystyle \|}{C}}OH$ (e) $CH_3C \equiv CCH_2CH_2I$

6.18. Suggest reagents for the following conversion:

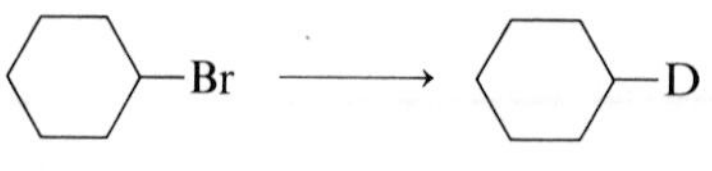

—Br $\longrightarrow$ —D

SECTION 6.12.

Synthesis Problems

Organic chemists often synthesize compounds in the laboratory. The syntheses may be simple and straightforward (for example, the preparation of a particular simple alcohol for a rate study), or they may be very involved (for example, the laboratory synthesis of a complex biological molecule). Even if you do not become a laboratory chemist, designing synthetic schemes on paper is a valuable way to learn to think in the language of organic chemistry.

In this text, you will encounter many __synthesis problems__: problems in which you are asked to show by equations how you would prepare a particular compound. In some of these problems (but not all), a starting material will be specified. Often, there will be more than one correct solution to a synthesis problem. However, only one correct answer to each synthesis problem is usually given in this text or the accompanying study guide. If your answer is significantly different from the one given, you should verify its correctness with your instructor.

The synthesis problems posed in this text are not intended to be valid in a laboratory sense. The solution to a true laboratory synthesis problem would include a complete search of the chemical literature to see if a particular compound or sequence of reactions has already been studied by other chemists. Then, various possible pathways to the desired compound are drawn up. Each pathway is evaluated from a practical laboratory standpoint (likelihood of success, cost in terms of reagents and time, availability of starting materials, hazards, etc.). Of the various proposed pathways, one is selected. Finally, the synthetic sequence is tested in the laboratory. In solving synthesis problems in this text, you need consider only reactions or reaction sequences that are reasonable, based on the information previously presented in the text.

A. Solving synthesis problems

The following suggestions may help you answer synthesis problems correctly.

1. A typical study problem gives the reactants and asks for the product or products:

$$A + B \longrightarrow \quad ?$$

In synthesis problems, the reverse question is asked: Given the product, what are the reactants?

$$? + ? \longrightarrow \quad C$$

Thus, as you study organic reactions, you must learn them both ways. You must be able to answer such questions as "What reactions yield alcohols?" as well as "The reaction of a Grignard reagent with formaldehyde yields what?"

2. Use only reactions that give the product in reasonable yield. If one reaction gives 100% of the desired product, use that reaction. However, if no such reaction is available, use a reaction that gives 50 or 60% yield. Do not use reactions that give very low yields (under 25%).

3.   It is <u>acceptable to use flow equations</u>, with reagents and reaction conditions above or below the arrows. This technique, while not necessary with a one-step synthesis, saves considerable time in writing the solutions to multistep syntheses.

$$A \xrightarrow{\text{xy}} B \xrightarrow[100°]{\text{yz}} C$$

4.   <u>Unless you are asked</u> to do so, yo<u>u need not balance equations</u> nor indicate <u>minor products</u>.

5.   If the text shows a reaction of a simple compound, you may usually extrapolate the reaction to more-complex, but similar, structures.

   Example:

$$CH_3\overset{\displaystyle O}{\overset{\displaystyle \|}{C}}CH_3 \xrightarrow[\text{(2) } H_2O, H^+]{\text{(1) } CH_3MgI} (CH_3)_3COH$$

   This reaction may be extrapolated to other ketones and Grignard reagents.

$$CH_3CH_2\overset{\displaystyle O}{\overset{\displaystyle \|}{C}}CH_2CH_3 \xrightarrow[\text{(2) } H_2O, H^+]{\text{(1)} \langle\text{—}\rangle\text{—MgI}} CH_3CH_2\overset{\displaystyle OH}{\underset{\displaystyle \bigcirc}{C}}CH_2CH_3$$

6.   In a synthesis problem, do not be intimidated by the complexity of a structure, but <u>focus your attention on the functional groups.</u> Inspect the structure for its important features and worry only about the small portion that undergoes reaction. When you have a larger vocabulary of organic reactions, you will want to inspect a complex structure for other functional groups that might also undergo reaction under the same conditions.

B.   Multistep synthetic problems

Most syntheses in the laboratory (and many synthesis problems in this book) require more than a single step from commercially available starting materials to desired products. When confronted with a synthesis that will require two or more steps and a reaction sequence is not immediately evident, do *not* choose a likely starting material and try to convert it to the product. Instead, <u>*start with the*</u> <u>*product and work backwards, one step at a time, to the starting material.*</u> This procedure, called **retrosynthetic analysis**, is best explained with an example.

   Example.   Show by flow equations how you would prepare 3-deuterio-1-propene from a nondeuteriated hydrocarbon, standard inorganic reagents, and appropriate solvents.

1.   <u>Write the structure of the product.</u>

$$CH_2\text{=}CHCH_2D$$

2.   Consider, not a possible starting hydrocarbon, but a reaction that leads directly to this given product. A Grignard reagent, treated with D_2O, is a way to introduce deuterium into a structure. Write the equation for this reaction,

$$CH_2=CHCH_2MgBr \xrightarrow{D_2O} CH_2=CHCH_2D$$

3.   Next, what reagents are needed to prepare the allyl Grignard reagent? (Again, you are working "backwards.")

$$CH_2=CHCH_2Br \xrightarrow[\text{diethyl ether}]{Mg} CH_2=CHCH_2MgBr$$

4.   Finally, what reaction could be used to prepare allyl bromide?

$$CH_2=CHCH_3 \xrightarrow{NBS} CH_2=CHCH_2Br$$

5.   Because we have worked our way backwards to a nondeuteriated hydrocarbon, we have solved the problem. The answer is now written forward, rather than backwards.

$$CH_2=CHCH_3 \xrightarrow{NBS} CH_2=CHCH_2Br \xrightarrow[\text{diethyl ether}]{Mg}$$

$$CH_2=CHCH_2MgBr \xrightarrow{D_2O} CH_2=CHCH_2D$$

Example.   Show how you could convert diphenylmethane to diphenylmethanol.

1.   Write the structures.

2.   Consider the starting material and the product. They have the same carbon skeleton and differ only by an OH group at the benzylic position. Your first question should be: "Is there a one-step reaction that converts a benzylic H to a benzylic OH?' The answer is "no"; therefore, more than one step will be necessary and the problem should be approached by retrosynthetic analysis.

3   Consider the product and ask, "What reactions yield alcohols?" At this point, you have been presented with only two: substitution reactions of alkyl halides and Grignard reactions of carbonyl compounds. A Grignard reaction can be ruled out in this case because the starting material and the product have the same carbon skeleton. Thus, a substitution reaction must be used.

4.   Your next question should be: "What halide and what nucleophile would I need to obtain the desired product?" The answer is:

$$C_6H_5\overset{\displaystyle X}{\underset{|}{C}}HC_6H_5 \xrightarrow{OH^-} C_6H_5\overset{\displaystyle OH}{\underset{|}{C}}HC_6H_5 \qquad \text{where } X = Cl, Br, \text{ or } I$$

5. The synthesis has now been simplified to the conversion of $C_6H_5CH_2C_6H_5$
    to $C_6H_5\underset{\underset{X}{|}}{C}HC_6H_5$.

    The logic process is now repeated. "Is there a one-step reaction by
    which a benzylic H can be converted to Br, Cl, or I?" For Br and Cl, the
    answer is "yes." There are a number of reactions that will work:
    free-radical chlorination or bromination, or an NBS bromination.
    Choosing bromination, for example, we can write the equation:

    $$C_6H_5CH_2C_6H_5 \xrightarrow[hv]{Br_2} C_6H_5\underset{\underset{Br}{|}}{C}HC_6H_5$$

6. Combine the equations into a final synthetic sequence.

    $$C_6H_5CH_2C_6H_5 \xrightarrow[hv]{Br_2} C_6H_5\underset{\underset{Br}{|}}{C}HC_6H_5 \xrightarrow{OH^-} C_6H_5\underset{\underset{OH}{|}}{C}HC_6H_5$$

STUDY PROBLEMS

6.19. Show how you would make the following conversions. Use any reagents or
other starting materials required.

(a) diphenylmethane to 1,1-diphenyl-2-propanol
(b) toluene (methylbenzene) to $C_6H_5CH_2OCH_3$

(c) 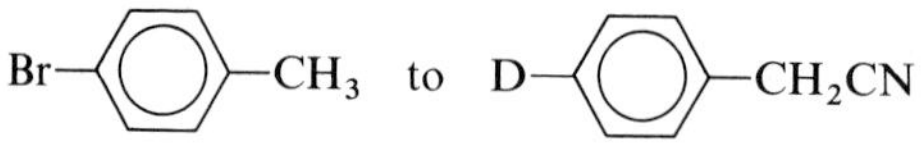 Br—⟨◯⟩—CH_3 to D—⟨◯⟩—CH_2CN

6.20. Suggest syntheses for the following compounds from organic compounds
containing six or fewer carbon atoms and any other required reagents:

(a) 3,5-dimethyl-3-hexanol   (b) cyclohexylmethanol

Summary

A **free radical** is an atom or group of atoms with an unpaired electron. Free-radical reactions are **chain reactions** that involve **initiation** (formation of free radicals); **propagation** (reactions in which new free radicals are formed); and **termination** (coupling, disproportionation, or the formation of stable free radicals). Racemization occurs at a reacting, resolved, chiral carbon during a free-radical reaction.

The order of reactivity of H's toward free-radical abstraction is $CH_4 <$ $1° < 2° < 3° <$ allylic or benzylic. The order of reactivity is the result of the relative stabilities of the free-radical intermediates.

Cl_2 is more reactive and less selective than Br_2 in free-radical halogenations. *N*-**Bromosuccinimide** (NBS) is a selective brominating agent for *allylic* and *benzylic positions*.

$$CH_3CH_2CH_3 \xrightarrow[hv]{Cl_2} CH_3CHClCH_3 + CH_3CH_2CH_2Cl + HCl$$

$$\xrightarrow[hv]{Br_2} CH_3CHBrCH_3 + HBr$$

$$CH_2{=}CHCH_3 \xrightarrow{NBS} CH_2{=}CHCH_2Br$$

$$\text{(phenyl)}{-}CH_2R \xrightarrow{NBS} \text{(phenyl)}{-}CHBrR$$

Other free-radical reactions besides halogenation include **pyrolysis**, a thermal free-radical decomposition of organic compounds; the **biological reduction** of O_2; and **auto-oxidation**, a free-radical oxidation by O_2 that results in decomposition of fats, oils, rubber, ethers, and aldehydes.

Free-radical initiators are substances that cause the formation of free radicals. Ultraviolet light and peroxides (which contain the easily broken $—O—O—$ bond) are examples. **Free-radical inhibitors**, such as I_2 or phenols, are substances that form nonreactive free radicals.

Organometallic compounds are compounds that contain a carbon–metal bond. A **Grignard reagent** (RMgX) is reactive because of the nucleophilic carbon bonded to Mg. (For a summary of some Grignard reactions, see Table 6.2, page 241.)

$$\underset{\delta+}{\overset{\delta-}{-C}}{=}O + \overset{\delta-}{R}{-}MgX \longrightarrow \underset{R}{-}\overset{OMgX}{\underset{|}{C}}{-} \xrightarrow{H_2O, H^+} \underset{R}{-}\overset{OH}{\underset{|}{C}}{-}$$

An **alkyllithium reagent** (RLi) undergoes reactions similar to those of Grignard reagents. **Cuprates** (R_2CuLi) are used to prepare hydrocarbons.

Grignard reagents and lithium reagents both undergo reaction with compounds that contain acidic hydrogens to yield alkanes.

$$RMgX + H{-}OH \longrightarrow RH + HOMgX$$

$$RLi + H{-}OH \longrightarrow RH + LiOH$$

STUDY PROBLEMS

6.21. Write Lewis formulas for (a) $CH_3CH_2O\cdot$ and (b) $CH_2{=}CH{-}\dot{C}H_2$.

6.22. Label the following reactions as *initiation*, *propagation*, or *termination* steps:

(a) $(CH_3)_3C\cdot + CH_2{=}CH_2 \longrightarrow (CH_3)_3C{-}CH_2CH_2\cdot$

(b)
$$C_6H_5\overset{\displaystyle O}{\overset{\displaystyle \|}{C}}OOH \longrightarrow C_6H_5\overset{\displaystyle O}{\overset{\displaystyle \|}{C}}O\cdot + \cdot OH$$

(c) $2\,CH_3CH_2CH_2\cdot \longrightarrow CH_3CH_2CH_3 + CH_3CH{=}CH_2$

(d) $2\,CH_3CH_2\cdot \longrightarrow CH_3CH_2CH_2CH_3$

(e) $Br\cdot + CH_2{=}CH_2 \longrightarrow \cdot CH_2CH_2Br$

6.23. Write equations for the steps in the free-radical dichlorination of cyclopentane to yield 1,2-dichlorocyclopentane.

6.24. If all H's were abstracted at equivalent rates, what would be the ratio of monochlorination products of an equimolar mixture of $CH_3CH_2CH_3$ and cyclohexane?

6.25. List the products that would be obtained from the free-radical monochlorination of each of the following compounds. (Do not forget to indicate stereoisomers.)

(a) (*R*)-1,2-dichloropropane   (b) (*R*)-2-chlorobutane

6.26. Only one monochlorination product is obtained from an alkane with the molecular formula C_5H_{12}. What is the structure of the alkane?

6.27. Rank the following free radicals in order of increasing stability (least stable first):

(a) $CH_3CH_2\dot{C}(CH_3)_2$ (b) $CH_3\dot{C}HCH_3$ (c) $C_6H_5\dot{C}HCH_3$

(d) $CH_3CH{=}CHCH_2\dot{C}H_2$ (e) $C_6H_5\dot{C}HCH{=}CH_2$

6.28. Rank the following hydrocarbons in order of increasing ease of free-radical bromination:

(a) $CH_3CH_2CH(CH_3)_2$ (b) $CH_3CH_2CH_3$ (c) $C_6H_5CH_2CH_3$

6.29. Draw all the important resonance structures for the following free radicals:

(a) 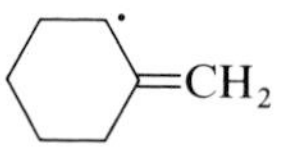   (b) 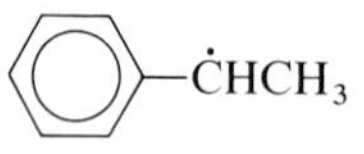   (c)

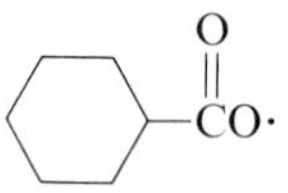

6.30. In each of the following structures, circle the position (or positions) that you would expect to be attacked by a low-energy free radical. Explain your choices.

(a) 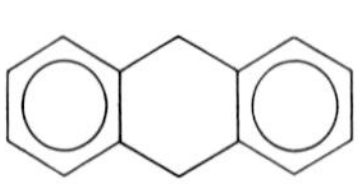   (b)

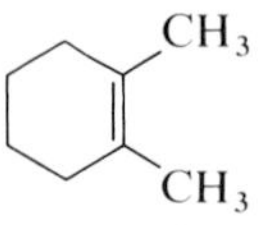

(c) CH₃—[cyclohexene ring]—CH(CH₃)₂   (d)

$$
\begin{array}{c}
CH_3 \\
| \\
CH \\
CH_2 \quad CH \\
| \qquad\quad || \\
CH_2 \quad CH_2 \\
CH \\
|| \\
C \\
H_3C \qquad CH_3
\end{array}
$$

6.31. Complete the following equations, showing only the major organic products:

(a) [cyclohexyl]—CH₃ + Br₂ $\xrightarrow{hv}$

(b) [phenyl]—CH₂CH₂CH₃ + [succinimide-type ring with two C=O and N—Br] $\xrightarrow[CCl_4]{hv}$

(c) [indene bicyclic structure] + Br₂ $\xrightarrow{hv}$

(d) [cyclohexane with OH and CH₂C₆H₅] + [N-bromo cyclic imide with two C=O, N—Br] $\xrightarrow[CCl_4]{hv}$

(e) [quinoline fused ring with N and CH₃] + Br₂ $\xrightarrow{hv}$

6.32. Suggest a mechanism that explains the following observation. (The * represents an isotopic label: ^{14}C.)

[cyclohexene with * label] $\xrightarrow{NBS}$ [Br-substituted cyclohexene with *] + [Br-substituted cyclohexene with *] + [Br-substituted cyclohexene with *]

6.33. What would be the disproportionation products of (a) CH₃CH₂CH₂CH₂· and (b) CH₃ĊHCH₃?

6.34. What would be the coupling product for each of the examples in Problem 6.33?

6.35. Predict the β-cleavage and disproportionation products of the following free radicals:

(a) $(CH_3)_2\dot{C}CH_2CH_2CH_3$ (b) [cyclohexyl radical] ·

6.36. Which of the following compounds would form a hydroperoxide (ROOH) readily upon exposure to air?

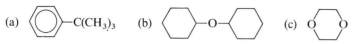

(a)   (b)   (c)

6.37. Which of the following compounds are organometallic compounds?

(a) CH_3HgOH (b) $(C_6H_5O)_2Mg$ (c) $(CH_3)_3CNa$

6.38. Name the following organometallic compounds:

(a) C_6H_5MgI (b) $CH_3CH_2CH_2MgCl$ (c) $CH_3(CH_2)_5CH_2Li$

6.39. Complete the following equations:

(a) $+ Mg \xrightarrow{\text{diethyl ether}}$

(b) $-I + Mg \xrightarrow{\text{diethyl ether}}$

(c) $-Br + Mg \xrightarrow{\text{diethyl ether}}$

6.40. Predict the organic products:

(a) $=O \xrightarrow[(2) H_2O, H^+]{(1) CH_3MgI}$

(b) $CH_3CH_2\overset{O}{\overset{\|}{C}}H \xrightarrow[(2) H_2O, H^+]{(1)\quad \text{—MgBr}}$

(c) $C_6H_5\overset{O}{\overset{\|}{C}}H \xrightarrow[(2) H_2O, H^+]{(1) CH_3CH_2MgBr}$

(d) $=O \xrightarrow[(2) H_2O, H^+]{(1)\quad \text{—MgCl}}$

6.41. How could you prepare each of the following compounds, starting with 2-bromopropane, magnesium, and other appropriate reagents? (a) 2,3-dimethyl-2-butanol; (b) 3-methyl-2-butanol.

6.42. (1) Which of the following compounds contain acidic hydrogens that would be removed by CH_3MgI? (2) Write equations showing the products (if any) of these reactions, assuming that an excess of CH_3MgI is used.

(a) $C_6H_5C\equiv CCH_3$ (b) $HOCH_2CH_2OH$

(c) $(CH_3CH_2)_2NH$ (d) $HO_2C-\overset{CO_2H}{\overset{|}{C}H}-CO_2H$

6.43. Complete the following equations, showing all principal products:

(a) —Br + Li $\longrightarrow$

(b) Li——CH$_2$CH$_2$CO$_2^-$ + H$_3$O$^+$ $\longrightarrow$

(c) —Li + NH $\longrightarrow$

6.44. Starting with any organic halide of four carbons or less, suggest a method for the preparation of each of the following compounds:

(a) CH$_3$CH$_2$CH$_2$CH$_2$CH$_3$ (b) CH$_3$(CH$_2$)$_6$CH$_3$

(c) (CH$_3$)$_2$CHCH$_2$CH$_2$CH$_2$CH$_3$ (d) CH$_2$=CHCH$_2$D

(e) CH$_3$CHDCH$_3$ (f) (CH$_3$)$_2$CHCH$_2$CH=CH$_2$

6.45. The free-radical bromination of C$_6$H$_5$CH$_2$D with Br$_2$ shows a kinetic isotope effect. Suggest a reason for the fact that, if HBr is removed from the reaction mixture as soon as it is formed, the isotope effect is even greater.

6.46. Explain why Compound (a) decomposes to free radicals about 75 times faster than (b).

(a)

(b)

6.47. A chemist treated CH$_3$CH=CHCH$_2$Cl with magnesium in anhydrous ether and then added acetone (propanone). After hydrolysis, instead of a single alkenyl alcohol as a product, the chemist obtained *two* alcohols. What are the structures of the two alcohols?

6.48. The aqueous electrolysis of salts of carboxylic acids (called the **Kolbe electrolysis**) yields carbon dioxide and hydrocarbons by a free-radical path. What would be the products of the electrolysis of sodium acetate (CH$_3$CO$_2$Na)?

6.49. Reaction of ethylsodium with (R)-2-chlorooctane yields (R)-3-methylnonane.

(a) Has the chiral carbon been inverted?

(b) Draw a transition-state structure that would explain the observed stereochemistry.

6.50. Upon free-radical bromination, *n*-pentane yields almost exclusively two monobromo compounds, A and B. Upon treatment with NaOCH$_3$ under E2 conditions, A and B yield predominantly the same product, C. What are the structures of A, B, and C?

6.51. Suggest one (or more, if possible) synthetic route to each of the following alcohols, starting with an organic halide and other needed organic reagents:

(a) $(CH_3)_3CCHCH_3$ with OH

(b) [decalin ring with CH_2OH substituent]

(c) [ring structure with HO and CH_3]

(d) CH_3CH_2C— [cyclohexane ring] with OH and CH_3

(e) $(CH_3)_2CHCH$— [cyclopentane ring] with OH

(f) $(C_6H_5)_3COH$

6.52. Show how you would synthesize the following compounds from the suggested starting materials. Use any reagents or other starting materials necessary unless otherwise specified.

(a) [anthracene-type ring] ⟶ [ring with CN groups top and bottom]

(b) [fused ring system] ⟶ [fused ring system]

(c) CH_3CH=$CHCH_3$ ⟶ $CH_3\overset{O}{\overset{||}{C}}OCH_2CH$=$CHCH_2O\overset{O}{\overset{||}{C}}CH_3$

(d) $C_6H_5CH_2CH_2CH_3$ ⟶ C_6H_5CH=$CHCH_2\overset{OH}{\underset{|}{C}}HCH_3$
(Also show the structure of the principal by-product, if any.)

(e) $C_6H_5CH_3$ (as the only organic reactant) ⟶ $C_6H_5CH_2CH_2C_6H_5$

(f) [bicyclic ring] ⟶ [bicyclic ring with $CH_2CH_2CH_3$]

(*Hint*: Consider the Kekulé formulas of the product.)

CHAPTER 7

Alcohols, Ethers, and Related Compounds

Alcohols (ROH) and ethers (ROR) are so much a part of our everyday lives that even laymen are familiar with the terms. Diethyl ether (ether) is used as an anesthetic. Ethanol (ethyl alcohol, grain alcohol, or just "alcohol") is used in beverages. 2-Propanol (isopropyl alcohol, or rubbing alcohol) is used as a bacteriocidal agent. Methanol (methyl alcohol, or wood alcohol) is used as an automobile gas-line antifreeze. In the laboratory and in industry, all these compounds are used as solvents and reagents.

In this chapter we will discuss alcohols, ethers, and epoxides (which are a special type of ether). We will also briefly mention phenols and some sulfur analogs of alcohols and ethers.

CH_3CH_2OH ⬡—OH $CH_3CH_2OCH_2CH_3$ $CH_3\overset{\displaystyle O}{\overset{\diagup\diagdown}{CH-CH_2}}$

ethanol	phenol	diethyl ether	propylene oxide
an alcohol	*a phenol*	*an ether*	*an epoxide*

SECTION 7.1.

Bonding in Alcohols and Ethers

The bonding in alcohols and ethers was mentioned in Chapter 2. Both types of compounds have bonding similar to that in water. In all three cases, the oxygen is in the sp^3-hybrid state. Two of the sp^3 orbitals of the oxygen atom are bonded to other atoms, and the remaining two orbitals are filled with two electrons each (see Figure 2.20, page 63).

$$\overset{..}{\underset{..}{O}} \quad \overset{..}{\underset{..}{O}} \quad \overset{..}{\underset{..}{O}}$$
$$\text{H} \qquad \text{H} \qquad \text{R} \qquad \text{H} \qquad \text{R} \qquad \text{R}$$

water   *an alcohol*   *an ether*

Alcohols and ethers are composed of polar molecules. In either type of compound, the oxygen carries a partial negative charge. However, an alcohol molecule is more polar than an ether molecule. The reason for this is that hydrogen is more electropositive than carbon, and therefore an O—H bond is more polar than an O—R bond. The dipole moments of the following compounds show decreasing polarity in the series H_2O, ROH, and ROR.

$$\overset{\delta -}{O} \qquad \overset{\delta -}{O} \qquad \overset{\delta -}{O}$$
$$\overset{\delta +}{H} \qquad \overset{\delta +}{H} \qquad \overset{(\delta +)}{H_3C} \qquad \overset{\delta +}{H} \qquad \overset{(\delta +)}{H_3C} \qquad \overset{(\delta +)}{CH_3}$$

μ: 1.8 D   1.7 D   1.3 D

Ethers can be either open-chain or cyclic. When the ring size (including the oxygen) is five or greater, the chemistry of the ether may be extrapolated from that of open-chain counterparts. (There are some differences in rates of reaction because the oxygen in a cyclic ether is less sterically hindered—its alkyl substituents are tied back in a ring.) Epoxides contain three-membered ether rings. Epoxides are more reactive than other ethers because of ring strain.

Some cyclic ethers:

$$\overset{O}{\underset{CH_2-CH_2}{\diagup \diagdown}}$$

ethylene oxide   tetrahydrofuran   1,4-dioxane
an epoxide   THF

SECTION 7.2.

Physical Properties of Alcohols and Ethers

A.   Boiling points

Because alcohols can form hydrogen bonds with other alcohol molecules, they have higher boiling points than alkyl halides or ethers of comparable molecular weights. Table 7.1 compares the boiling points of some alcohols and organic halides with the same carbon skeletons.

B.   Solubility in water

Alcohols of low molecular weight are miscible with water, while the corresponding alkyl halides are water-insoluble. This water solubility is directly attributable to hydrogen bonding between alcohols and water.

TABLE 7.1. Comparison of the boiling points of some alcohols and chloroalkanes

Alcohol	Bp, °C	Chloroalkane	Bp, °C
CH_3OH	64.5	CH_3Cl	−24
CH_3CH_2OH	78.3	CH_3CH_2Cl	13
$CH_3CH_2CH_2OH$	97.2	$CH_3CH_2CH_2Cl$	46
$HOCH_2CH_2OH$	197	$ClCH_2CH_2Cl$	83.5
$\underset{\displaystyle HOCH_2CHCH_2OH}{\overset{\displaystyle OH}{}}$	290	$ClCH_2CHClCH_2Cl$	157

The hydrocarbon portion of an alcohol is **hydrophobic**—that is, it repels water molecules. As the length of the hydrocarbon portion of an alcohol molecule increases, the water solubility of the alcohol decreases. When the hydrocarbon chain is long enough, it overcomes the **hydrophilic** (water-loving) properties of the hydroxyl group. The three-carbon alcohols, 1- and 2-propanol, are miscible in water, while only 8.3 grams of 1-butanol dissolves in 100 grams of water. (These solubilities are summarized in Table 7.2.)

Branching increases water solubility. Although 1-butanol is only slightly soluble, *t*-butyl alcohol, $(CH_3)_3COH$, is miscible with water. The reason for this is that the *t*-butyl group is more compact and less hydrophobic than the *n*-butyl group. An increase in the number of OH groups also increases hydrophilicity and solubility. Sucrose (table sugar, page 838) has twelve carbons, but it also has eight hydroxyl groups and is readily soluble in water.

Ethers cannot form hydrogen bonds with themselves because they have no hydrogen attached to the oxygen. However, ethers can form hydrogen bonds with water, alcohols, or phenols. Because of hydrogen bonding with H_2O, the solubilities of the four-carbon compounds diethyl ether and 1-butanol (Tables 7.2 and 7.3) are about the same.

C. Solvent properties

Water is an excellent solvent for ionic compounds. The OH bond is polar and provides the dipole necessary to solvate both cations and anions. Alcohols also can dissolve ionic compounds, but to a lesser extent. (Ethers cannot dissolve

TABLE 7.2. Physical properties of some alcohols

IUPAC name	Trivial name	Formula	Bp, °C	Density, g/cc at 20°C	Solubility in H_2O
methanol	methyl alcohol	CH_3OH	64.5	0.79	∞
ethanol	ethyl alcohol	CH_3CH_2OH	78.3	0.79	∞
1-propanol	propyl alcohol	$CH_3CH_2CH_2OH$	97.2	0.80	∞
2-propanol	isopropyl alcohol	$(CH_3)_2CHOH$	82.3	0.79	∞
1-butanol	butyl alcohol	$CH_3(CH_2)_3OH$	117	0.81	8.3 g/100 cc

TABLE 7.3. Physical properties of some ethers and epoxides

Name	Formula	Bp, °C	Density, g/cc at 20°C	Solubility in H_2O
dimethyl ether	CH_3OCH_3	−24	gas	∞
diethyl ether	$CH_3CH_2OCH_2CH_3$	34.6	0.71	8 g/100 cc
tetrahydrofuran	(ring structure with O)	66	0.89	∞
oxirane (ethylene oxide)	CH_2CH_2 (with O bridge)	13.5	0.88 (at 10°)	∞
methyloxirane (propylene oxide)	CH_3CHCH_2 (with O bridge)	34.3	0.86	∞

TABLE 7.4. Solubility of sodium chloride in water and in some alcohols

Solvent	Dielectric constant	Solubility of NaCl, g/100 cc at 25°C
H_2O	78	36.2
CH_3OH	32	1.4
CH_3CH_2OH	24	0.06
$CH_3CH_2CH_2OH$	20	0.01

ionic compounds.) Table 7.4 lists the solubility of sodium chloride in water and in a few alcohols. Note that the solubility of NaCl decreases as the hydrocarbon chain of the alcohol increases in length.

SECTION 7.3.

Nomenclature of Alcohols and Ethers

A. IUPAC names of alcohols

The IUPAC names of alcohols are taken from the names of the parent alkanes, but with the ending **-ol**. A prefix number, chosen to be as low as possible, is used if necessary.

		OH
CH_3OH	$CH_3CH_2CH_2OH$	CH_3CHCH_3
IUPAC: methanol	1-propanol	2-propanol

More than one hydroxyl group is designated by *di-*, *tri-*, etc., just before the -ol ending.

$$OH$$
$$|$$
$$CH_3CHCH_2CH_2OH$$

1,3-butanediol
a diol

STUDY PROBLEMS

7.1.   Name the following compounds:

$$OH$$
$$|$$
(a) $(CH_3)_2CHCHCH(CH_3)_2$ (b)

7.2.   Write the structures for   (a) 3-ethyl-3-methyl-2-pentanol, and (b) 2,2-dimethyl-1,4-hexanediol.

A hydroxyl group is often found in a molecule that contains other functional groups. In the IUPAC system, the numbering and the suffix in the name of a multi-functional compound are determined by nomenclature priority (Section 3.3N).

$$-R, -X, \text{etc.} \qquad \overset{\diagdown}{\underset{\diagup}{C}}=\overset{\diagup}{\underset{\diagdown}{C}} \qquad -OH \qquad \overset{O}{\overset{||}{-C-}} \qquad \overset{O}{\overset{||}{-CH}} \qquad -CO_2H$$

increasing nomenclature priority   ✶

Carboxylic acids, aldehydes, and ketones have higher nomenclature priority than the hydroxyl group; one of these groups receives the lowest nomenclature number and is also given the suffix position in the name. The lower-priority OH group is then named by the prefix **hydroxy-**, as may be seen in the following examples:

$$\overset{HO\ \ O}{\underset{|\ \ \ ||}{CH_3CHCOH}} \qquad \overset{O}{\overset{||}{HOCH_2CH_2CH}} \qquad \overset{O}{\overset{||}{HOCH_2CH_2CCH_3}}$$

2-hydroxypropanoic acid   3-hydroxypropanal   4-hydroxy-2-butanone
(lactic acid)

In a compound that contains an OH group and also has a double bond or a group usually named as a prefix, the hydroxyl group has the higher nomenclature priority. In these cases, the OH receives the lowest prefix number and is given the -ol ending. Note in the following examples how a double-bond suffix is inserted into the name of an unsaturated alcohol.

3,3-dichloro-1-cyclohexanol $CH_2=CHCH_2CH_2OH$ 4-methyl-2-cyclohexen-1-ol

3-buten-1-ol

STUDY PROBLEM

7.3.   Name the following compounds by the IUPAC system:

(a) ⬡—OH   (b) HO—⬡=O   (c) $BrCH_2CH_2OH$

B.   Trivial names of alcohols

Just as CH_3I may be called methyl iodide, CH_3OH may be called methyl alcohol. This type of name is a popular way of naming alcohols with common alkyl groups.

$$(CH_3)_3COH \qquad (CH_3)_2CHOH$$

t-butyl alcohol   isopropyl alcohol

A diol (especially a 1,2-diol) is often referred to as a **glycol**. The trivial name for a 1,2-diol is that of the corresponding **alkene** followed by the word **glycol**. Epoxides and 1,2-dihalides are often named similarly. The naming of a saturated compound as a derivative of an alkene is unfortunate; however, the practice arose quite innocently in the early years of organic chemistry because all these compounds can be prepared from alkenes.

	$CH_2=CH_2$	$\underset{\displaystyle CH_2-CH_2}{\overset{\displaystyle OH \quad OH}{}}$	$\underset{\displaystyle CH_2-CH_2}{\overset{\displaystyle Br \quad Br}{}}$	$\underset{\displaystyle CH_2-CH_2}{\overset{\displaystyle O}{}}$
IUPAC:	ethene	1,2-ethanediol	1,2-dibromoethane	oxirane
trivial:	ethylene	ethylene glycol	ethylene dibromide	ethylene oxide

C.   Classification of alcohols

Alcohols, like alkyl halides, may be classified as **methyl**, **primary**, **secondary**, or **tertiary**, as well as **allylic** or **benzylic**.

CH_3OH	CH_3CH_2OH	$(CH_3)_2CHOH$	$(CH_3)_3COH$
methyl	1°	2°	3°

$$CH_3CH=CHCH_2OH \qquad \text{⬡—}\overset{\displaystyle OH}{\underset{}{C}}HCH_3$$

an allylic alcohol   a benzylic alcohol
(and 1°)   (and 2°)

D. Ethers

Simple open-chain ethers are named almost exclusively by their trivial names, as **alkyl ethers**.

$$CH_3CH_2OCH_2CH_3 \qquad (CH_3)_2CHOCH(CH_3)_2 \qquad CH_3OCH_2CH_3$$

diethyl ether   diisopropyl ether   methyl ethyl ether
(or ethyl ether, or
simply "ether")

The names of more-complex ethers follow systematic nomenclature rules. An **alkoxy-** prefix is used when there is more than one alkoxyl (RO—) group or when there is a functional group of higher priority. (Note that a hydroxyl group has priority over an alkoxyl group.)

1,2-dimethoxycyclohexane   1-isopropoxy-2-   5-ethoxy-2-pentanol
                             methoxycyclohexane

In the IUPAC system, epoxides are called **oxiranes**. In the numbering of these rings, the oxygen is always considered position 1.

2-ethyloxirane

SAMPLE PROBLEM

Name the following compounds:

(a) $(CH_3)_3COCH_3$ (b)   (c)   (d)

Solution: (a) *t*-butyl methyl ether;   (b) *trans*-4-methylcyclohexanol;
(c) *cis*-cyclohexene glycol, or *cis*-1,2-cyclohexanediol;
(d) 2-hydroxycyclohexanone.

STUDY PROBLEM

7.4.   Write structures for   (a) ethyl phenyl ether; (b) 2-butanol; (c) *sec*-butyl alcohol;   (d) 2,2-dimethyloxirane.

SECTION 7.4.

Preparation of Alcohols

We have already discussed two reactions that yield alcohols as products: (1) $RX + OH^-$ (Chapter 5), and (2) Grignard reactions (Section 6.9A). In this section, we will present a survey of these reactions plus a few others that are commonly used to synthesize alcohols.

A.   Nucleophilic substitution reactions

The reaction of an alkyl halide with hydroxide ion is a nucleophilic substitution reaction. When primary alkyl halides are heated with aqueous sodium hydroxide, reaction occurs by an S_N2 path. Primary alcohols may be prepared in good yields by this technique. Because secondary and tertiary alkyl halides are likely to give elimination products, they are not generally as useful for preparing alcohols.

$$ CH_3CH_2CH_2Br + OH^- \xrightarrow{\text{heat}} CH_3CH_2CH_2OH + Br^- $$

1-bromopropane   1-propanol

a 1° alkyl halide   a 1° alcohol

B.   Grignard reactions

Grignard reactions provide an excellent route to alcohols with complex carbon skeletons. A Grignard reaction:

1.   with formaldehyde yields a *primary alcohol*;
2.   with any other aldehyde yields a *secondary alcohol*; and
3.   with a ketone yields a *tertiary alcohol*.

These reactions are summarized in Table 6.2 (page 241) and in Table 7.5 (page 264).

　　　Some other Grignard reactions also lead to alcohols. The reaction of a Grignard reagent with *ethylene oxide* yields a *primary alcohol*. This reaction will be discussed later in this chapter (Section 7.16). The reaction of a Grignard reagent with an *ester* leads to a *tertiary alcohol*. (If a formate ester is used, a secondary alcohol is the product.) The reactions of Grignard reagents with esters will be discussed in Section 13.5C.

Some additional Grignard reactions that yield alcohols (see also Tables 6.2 and 7.5):

　　1° Alcohols from ethylene oxide:

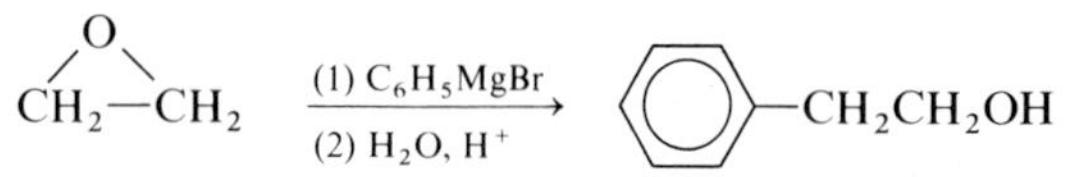

$$ \underset{\text{ethylene oxide}}{\overset{O}{\underset{}{CH_2-CH_2}}} \xrightarrow[\text{(2) } H_2O,\ H^+]{\text{(1) } C_6H_5MgBr} \underset{\text{2-phenyl-1-ethanol}}{C_6H_5-CH_2CH_2OH} $$

2° Alcohols from formate esters:

$$\underset{\text{methyl formate}}{\underset{||}{\overset{O}{HCOCH_3}}} \xrightarrow[\text{(2) H}_2\text{O, H}^+]{\text{(1) 2 CH}_3\text{CH}_2\text{MgBr}} \underset{\text{3-pentanol}}{\overset{\overset{OH}{|}}{\underset{\underset{CH_2CH_3}{|}}{HC-CH_2CH_3}}}$$

from RMgX

3° Alcohols from other esters: *from RMgX*

$$\underset{\text{ethyl acetate}}{\underset{||}{\overset{O}{CH_3COCH_2CH_3}}} \xrightarrow[\text{(2) H}_2\text{O, H}^+]{\text{(1) 2 CH}_3\text{CH}_2\text{MgBr}} \underset{\text{3-methyl-3-pentanol}}{\overset{\overset{OH}{|}}{\underset{\underset{CH_2CH_3}{|}}{CH_3C-CH_2CH_3}}}$$

STUDY PROBLEM

7.5. Write equations to show how the following conversions could be made:

(a) ⟨◯⟩—Br ⟶ ⟨◯⟩—CH₂OH

(b) $CH_3CH_2CH_2CH_2Cl \longrightarrow CH_3CH_2CH_2CH_2OH$

(c) $CH_3CH_2Br \longrightarrow$ ⟨◯⟩—$\underset{\underset{OH}{|}}{CHCH_2CH_3}$

C. Reduction of carbonyl compounds

Alcohols may be prepared from carbonyl compounds by **reduction reactions** in which hydrogen atoms are added to the carbonyl group. For example, *reduction of a ketone* by catalytic hydrogenation or with a metal hydride yields a *secondary alcohol*. Yields are often 90–100%. These reactions will be discussed in more detail in Section 11.14.

$$\underset{\text{acetone}}{\underset{||}{\overset{O}{CH_3CCH_3}}} \xrightarrow[\text{(2) H}_2\text{O, H}^+]{\text{(1) NaBH}_4} \underset{\text{2-propanol}}{\overset{\overset{OH}{|}}{CH_3CHCH_3}}$$

$$\underset{\text{cyclohexanone}}{\overset{\bigcirc=O}{}} \xrightarrow[\text{heat, pressure}]{\text{H}_2, \text{Ni catalyst}} \underset{\text{cyclohexanol}}{\overset{\bigcirc-OH}{}}$$

D. Hydration of alkenes

When an alkene is treated with water plus a strong acid, which acts as a catalyst, the elements of water (H^+ and OH^-) add to the double bond in a **hydration**

reaction. The product is an alcohol. Many alcohols, such as laboratory ethanol, are made commercially by the hydration of alkenes. Limitations and variations of hydration reactions, as well as the mechanism, will be discussed in Chapter 9.

$$CH_2{=}CH_2 + H_2O \xrightarrow{\ H^+\ } CH_3CH_2OH$$

ethylene   ethanol

$$\text{cyclohexene} + H_2O \xrightarrow{\ H^+\ } \text{cyclohexanol}\text{—OH}$$

cyclohexene   cyclohexanol

Table 7.5 summarizes the various ways to prepare alcohols.

STUDY PROBLEM

7.6. Write equations that show how each of the following alcohols can be prepared from (1) an alkene, and (2) a ketone:

(a) 2-butanol   (b) 2,4-dimethyl-1-cyclopentanol

TABLE 7.5. Summary of laboratory syntheses of alcohols

Reaction		Section reference
Primary alcohols:		
$RCH_2X + {}^-OH \xrightarrow{\ S_N2\ } RCH_2OH$		5.5
HCH (O) $\xrightarrow[\text{(2) } H_2O,\ H^+]{\text{(1) RMgX}} RCH_2OH$		6.9
CH_2CH_2 (O) $\xrightarrow[\text{(2) } H_2O,\ H^+]{\text{(1) RMgX}} RCH_2CH_2OH$		7.16
$R_2C{=}CH_2 \xrightarrow[\text{(2) } H_2O_2,\ OH^-]{\text{(1) } BH_3} R_2CHCH_2OH$		9.10
RCH (O) $\xrightarrow{\ [H]\ } RCH_2OH$		11.14
Secondary alcohols:		
RCH (O) $\xrightarrow[\text{(2) } H_2O,\ H^+]{\text{(1) R'MgX}} RCHR'$ (OH)		6.9
$HCOR$ (O) $\xrightarrow[\text{(2) } H_2O,\ H^+]{\text{(1) 2 R'MgX}} R'CHR'$ (OH)		13.5C
RCR' (O) $\xrightarrow{\ [H]\ } RCHR'$ (OH)		11.14
$RCH{=}CHR^a \xrightarrow{\ H_2O,\ H^+\ } RCHCH_2R$ (OH)		9.8

TABLE 7.5. (*continued*)

Reaction	Section reference

Tertiary alcohols:

$$\underset{RCR'}{\overset{O}{\parallel}} \quad \xrightarrow[\text{(2) H}_2\text{O, H}^+]{\text{(1) R''MgX}} \quad \underset{\underset{R''}{\overset{OH}{\mid}}}{RCR'} \qquad 6.9$$

$$\underset{RCCl}{\overset{O}{\parallel}} \quad \text{or} \quad \underset{RCOR'}{\overset{O}{\parallel}} \quad \xrightarrow[\text{(2) H}_2\text{O, H}^+]{\text{(1) R''MgX}} \quad \underset{}{\overset{OH}{\mid}}{RCR''_2} \qquad 13.3C, \ 13.5C$$

1,2-Diols:

$$RCH{=}CHR \quad \xrightarrow[25°]{\text{MnO}_4^-, \text{ OH}^-} \quad \underset{}{\overset{OH \ \ \ OH}{\mid \ \ \ \ \mid}}{RCH{-}CHR} \qquad 9.14A$$

$$RCH{=}CHR \quad \xrightarrow[\text{(2) H}_2\text{O, H}^+]{\text{(1) C}_6\text{H}_5\text{CO}_3\text{H}} \quad \underset{\overset{}{\underset{OH}{\mid}}}{\overset{\overset{OH}{\mid}}{RCHCHR}} \qquad 9.14B$$

[a] Other preparations of 2° alcohols from alkenes include oxymercuration–demercuration (Section 9.9) and hydroboration–oxidation (Section 9.10).

E. Ethanol by fermentation

The ethanol used in beverages is obtained by the enzyme-catalyzed **fermentation of carbohydrates (sugars and starches).** One type of enzyme converts carbohydrates to glucose, then to ethanol; another type leads to vinegar (acetic acid), with ethanol as an intermediate.

$$\underset{\substack{\text{glucose}\\ \textit{a sugar}}}{C_6H_{12}O_6} \quad \xrightarrow{\text{enzymes}} \quad \underset{\text{ethanol}}{CH_3CH_2OH}$$

The source of the carbohydrates used for fermentation depends on availability and on the purposes of the alcohol. In the U.S., carbohydrates are obtained primarily from corn and from the molasses residue of sugar refining. However, potatoes, rice, rye, or fruit (grapes, blackberries, etc.) may also be used.

Fermentation of any of these fruits, vegetables, or grains ceases when alcoholic content reaches 14 to 16 percent. If a higher concentration of alcohol is desired, the mixture is distilled. The distillate is an azeotrope of 95 % ethanol–5 % water. (An azeotrope is a mixture that boils at a constant boiling point as if it were a pure compound.) This distillate may then be used to fortify the fermentation mixture, or it may be diluted with water to the desired strength.

Because alcoholic beverages are taxed in almost all countries of the world, most ethanol sold for laboratory or industrial purposes (and not taxed as a liquor) is **denatured**—that is, small amounts of toxic impurities are added so that the ethanol cannot be diverted from the laboratory or factory into illegal beverages.

SECTION 7.5.

Reactivity of Alcohols

An alcohol can lose its hydroxyl proton to a sufficiently powerful base in an acid–base reaction. The product is an alkoxide. This reaction will be discussed in Section 7.10.

O—H bond broken:

$$CH_3O\!\!-\!\!H + Na^+ \; \ddot{N}H_2^- \longrightarrow CH_3O^- \; Na^+ + NH_3$$

methanol   sodamide   sodium methoxide

a strong base

Alcohols can also undergo substitution reactions and elimination reactions in which the C—O bond is broken.

C—O bond broken:

$$CH_3CH_2\!\!\!\!+\!\!OH + HBr \longrightarrow CH_3CH_2\!\!-\!\!Br + H_2O$$

ethanol   bromoethane

$$\underset{heat}{\overset{H_2SO_4}{\longrightarrow}} CH_2\!\!=\!\!CH_2 + H_2O$$

$$\overset{H}{\underset{|}{CH_2}}\!\!-\!\!CH_2\!\!-\!\!OH$$

ethylene

These substitution and elimination reactions are similar to the substitution and elimination reactions of alkyl halides. Unlike alkyl halides, however, alcohols do not undergo substitution or elimination reactions in neutral or alkaline solution; therefore, treatment of an alcohol with a nucleophile such as CN^- in neutral or alkaline solution does not result in substitution products or alkenes. Why not? The answer is that, in general, a leaving group must be a fairly weak base. In Chapter 5, you learned that Cl^-, Br^-, and I^- are good leaving groups and are readily displaced from alkyl halides. These ions are very weak bases. But ^-OH, which would be the leaving group of an alcohol in neutral or alkaline solution, is a *strong base* and thus is a very poor leaving group.

good leaving group

$$CH_3CH_2\!\!-\!\!Br + {}^-OH \longrightarrow CH_3CH_2OH + Br^-$$

poor leaving group

$$CH_3CH_2\!\!-\!\!OH + Br^- \longrightarrow \text{no reaction}$$

Substitution and elimination reactions of alcohols do proceed in acidic solution. We will consider what happens initially to alcohols in acidic solution and then consider the further reactions of these compounds.

SECTION 7.6.

Substitution Reactions of Alcohols

In acidic solution, alcohols are protonated. This reaction is an acid–base equilibrium with the alcohol acting as a base. It is the same type of reaction that occurs between water and a proton.

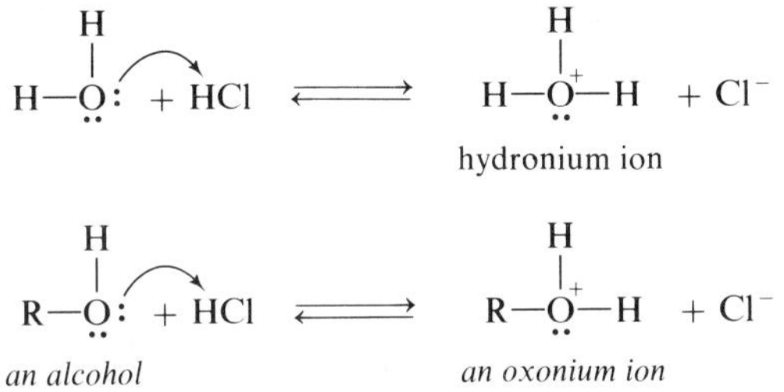

$$H-\underset{\cdot\cdot}{O}\overset{H}{:} + HCl \;\rightleftharpoons\; H-\underset{\cdot\cdot}{\overset{H}{\overset{|}{O}}}-H + Cl^-$$

hydronium ion

$$R-\underset{\cdot\cdot}{O}\overset{H}{:} + HCl \;\rightleftharpoons\; R-\underset{\cdot\cdot}{\overset{H}{\overset{|}{O}}}-H + Cl^-$$

an alcohol   *an oxonium ion*

In each case, an empty $1s$ orbital of H^+ overlaps with one of the filled valence orbitals of the oxygen, and an O—H sigma bond is formed (see Figure 7.1). The product of the reaction with water is the protonated water molecule, or the **hydronium ion**. A protonated alcohol molecule is called an **oxonium ion**.

Protonation reactions are rapid and reversible. Because any hydrogen attached to the oxygen may be lost in the reverse reaction, the OH protons of alcohols undergo rapid exchange with other acidic protons. For example, if methanol is treated with DCl, the solution soon contains CH_3OD and HCl as well as CH_3OH and DCl.

$$CH_3OH \;\xrightarrow{D^+}\; CH_3\overset{+}{\underset{D}{\overset{|}{O}}}H \;\xrightarrow{-H^+}\; CH_3OD$$

Although —OH is a poor leaving group, $-OH_2^+$ is a good leaving group because it is lost as water, a very weak base.

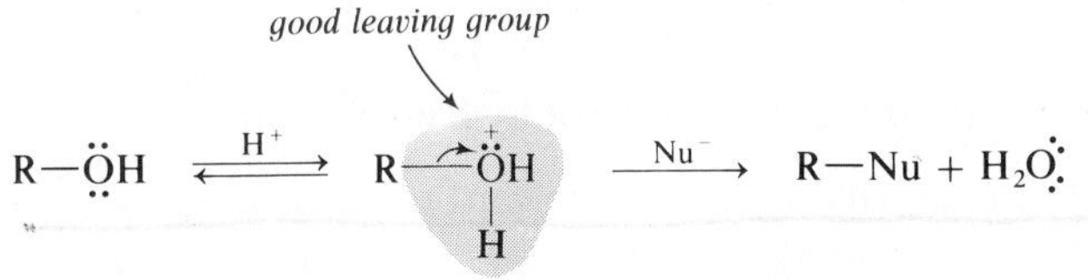

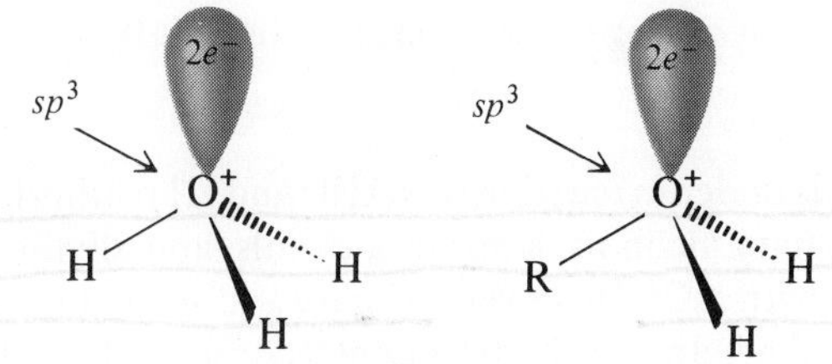

good leaving group

$$R-\underset{\cdot\cdot}{\overset{\cdot\cdot}{O}}H \;\xrightleftharpoons{H^+}\; R-\overset{+}{\underset{H}{\overset{|}{O}}}H \;\xrightarrow{Nu^-}\; R-Nu + H_2\underset{\cdot\cdot}{\overset{\cdot\cdot}{O}}:$$

FIGURE 7.1. Bonding in the hydronium ion (H_3O^+) and in an oxonium ion (ROH_2^+).

The most useful reagents in substitution reactions of alcohols are the hydrogen halides. The product of the reaction of an alcohol with HX is an alkyl halide.

$$CH_3CH_2CH_2OH + HBr \longrightarrow CH_3CH_2CH_2Br + H_2O$$

1-propanol   1-bromopropane

$$(CH_3)_2CHOH + HI \longrightarrow (CH_3)_2CHI + H_2O$$

2-propanol   2-iodopropane

$$(CH_3)_3COH + HCl \longrightarrow (CH_3)_3CCl + H_2O$$

t-butyl alcohol   *t*-butyl chloride

A.   Reactivity of the hydrogen halides

In alcohol substitution reactions, the reactivity of the hydrogen halides is as follows:

	HF	HCl	HBr	HI
pK_a:	3.2	-7	-9	-9.5

✳ increasing acid strength; increasing reactivity toward ROH ⟶

Although HI, HBr, and HCl are all considered strong acids (almost completely ionized in water), HI is the strongest acid of the group. HF is a weak acid. (Some reasons for this order of acid strength will be discussed in Chapter 12.) The order of reactivity of these acids toward alcohols simply parallels the relative acid strengths.

⬆ increasing rate of reaction

$$ROH + HI \longrightarrow RI + H_2O$$

$$ROH + HBr \longrightarrow RBr + H_2O$$

$$ROH + HCl \longrightarrow RCl + H_2O$$

B.   Reactivity of alcohols toward hydrogen halides

The order of reactivity of alcohols toward the hydrogen halides follows:

methyl   1°   2°   3°   benzylic   allylic

✳ increasing reactivity of ROH toward HX ⟶

All alcohols undergo reaction with HBr and HI readily to yield alkyl bromides and iodides. Tertiary alcohols, benzylic alcohols, and allylic alcohols also undergo reaction readily with HCl. However, primary and secondary alcohols are less reactive and require the help of anhydrous $ZnCl_2$ or a similar catalyst before they can undergo reaction with the less reactive HCl in a reasonable period of time.

increasing
reactivity
toward HX

$3°$: $(CH_3)_3COH$ $+ HCl$ $\xrightarrow{25°}$ $(CH_3)_3CCl$ $+ H_2O$

$2°$: $(CH_3)_2CHOH$ $+ HCl$ $\xrightarrow{ZnCl_2}$ $(CH_3)_2CHCl$ $+ H_2O$

$1°$: CH_3CH_2OH $+ HCl$ $\xrightarrow[\text{heat}]{ZnCl_2}$ CH_3CH_2Cl $+ H_2O$

The function of the zinc chloride is similar to that of H^+. Anhydrous zinc chloride is a powerful Lewis acid with empty orbitals that can accept electrons from the oxygen. The formation of a complex of $ZnCl_2$ with the alcohol oxygen weakens the C—O bond and thus enhances the leaving ability of the oxygen group.

$$CH_3CH_2\overset{..}{\underset{H}{O}}: + ZnCl_2 \longrightarrow CH_3CH_2\overset{+}{\underset{H}{\overset{..}{O}}}-\bar{Z}nCl_2$$

the complex

$$:\overset{..}{\underset{}{C}}l:^- + \overset{}{\underset{CH_3\ \ H}{C}H_2}-\overset{+}{\overset{..}{O}}-\bar{Z}nCl_2 \xrightarrow{S_N2} :\overset{..}{\underset{}{C}}l-\overset{}{\underset{CH_3}{C}H_2} + H\overset{..}{O}ZnCl + Cl^-$$

displacement by Cl^-   *the product*

C. S_N1 or S_N2?

It has been observed that secondary alcohols and tertiary alcohols sometimes undergo rearrangement when treated with HX. Most primary alcohols do not. The conclusion is that secondary and tertiary alcohols undergo reaction with hydrogen halides by the S_N1 path (through a carbocation), while primary alcohols undergo reaction by the S_N2 path (backside displacement).

✳ *Methyl and primary alcohols, S_N2:* ✳.

$$CH_3CH_2\overset{..}{O}H \overset{H^+}{\rightleftharpoons} CH_3CH_2-\overset{+}{\overset{..}{O}}H_2 \xrightarrow{X^-}$$

protonated

$$\left[:\overset{\delta-}{\overset{..}{X}}---\overset{}{\underset{CH_3}{C}H_2}---\overset{\delta+}{\overset{..}{O}}H_2\right] \longrightarrow :\overset{..}{X}-\overset{}{\underset{CH_3}{C}H_2} + H_2\overset{..}{O}:$$

S_N2 *transition state*

Other alcohols, S_N1:

$$(CH_3)_2CH\overset{..}{O}H \overset{H^+}{\rightleftharpoons} (CH_3)_2CH-\overset{+}{\overset{..}{O}}H_2 \overset{-H_2O}{\rightleftharpoons} [(CH_3)_2\overset{+}{C}H] \xrightarrow{X^-} (CH_3)_2CHX$$

protonated   *carbocation*
          intermediate

SAMPLE PROBLEM

The only alkyl halide formed in the reaction of 3-methyl-2-butanol with HBr is a product of rearrangement. Give the structure of the product and show how it is formed.

Solution:

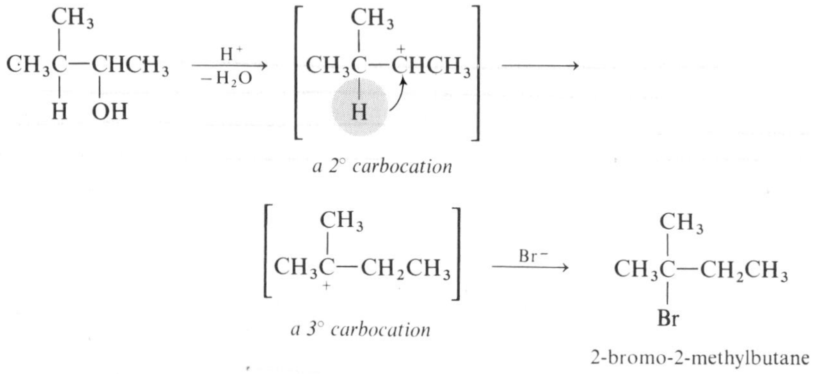

$$
\underset{\substack{|\quad\;|\\ \text{H}\quad\text{OH}}}{\text{CH}_3\overset{\text{CH}_3}{\overset{|}{\text{C}}}-\text{CHCH}_3}
\xrightarrow[-\text{H}_2\text{O}]{\text{H}^+}
\left[\underset{\substack{|\\ \text{H}}}{\text{CH}_3\overset{\text{CH}_3}{\overset{|}{\text{C}}}-\overset{+}{\text{C}}\text{HCH}_3}\right]
\longrightarrow
$$

a 2° carbocation

$$
\left[\text{CH}_3\overset{\text{CH}_3}{\underset{+}{\overset{|}{\text{C}}}}-\text{CH}_2\text{CH}_3\right]
\xrightarrow{\text{Br}^-}
\underset{\text{Br}}{\text{CH}_3\overset{\text{CH}_3}{\overset{|}{\text{C}}}-\text{CH}_2\text{CH}_3}
$$

a 3° carbocation

2-bromo-2-methylbutane

SECTION 7.7.

Other Reagents Used to Convert Alcohols to Alkyl Halides

Other halogenating reagents, such as PX_3, $SOCl_2$, and PX_5, can be used to convert alcohols to alkyl halides without rearrangement. We will discuss only two of these reagents, phosphorus trichloride and thionyl chloride.

Both these reagents undergo reaction with alcohols to form *inorganic esters,* a topic that will be discussed in Section 7.12. What is germane to the present discussion is that the resulting inorganic ester groups, like $-OH_2{}^+$, are *good leaving groups*. The following reactions are excellent methods for the preparation of alkyl chlorides from primary and secondary alcohols. (Tertiary alcohols generally undergo elimination reactions under these conditions.)

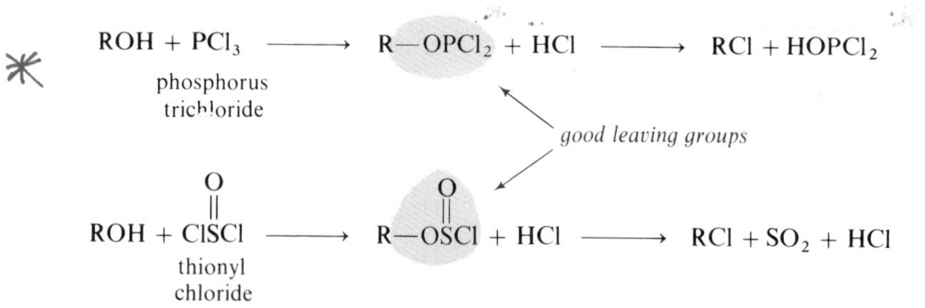

$$
\text{ROH} + PCl_3 \longrightarrow R-OPCl_2 + \text{HCl} \longrightarrow RCl + \text{HOPCl}_2
$$

phosphorus
trichloride

good leaving groups

$$
\text{ROH} + \underset{\text{thionyl chloride}}{\overset{\text{O}}{\overset{\|}{\text{ClSCl}}}} \longrightarrow R-\overset{\text{O}}{\overset{\|}{\text{OSCl}}} + \text{HCl} \longrightarrow RCl + SO_2 + \text{HCl}
$$

Phosphorus trichloride first undergoes reaction with an alcohol to yield a phosphite ester and HCl. This initial reaction step does not involve cleavage of the C—O bond. No racemization of a pure enantiomeric alcohol is observed, as would happen if the reaction went through a carbocation. Of course, if the starting alcohol is either achiral or racemic, so is the product.

Step 1:

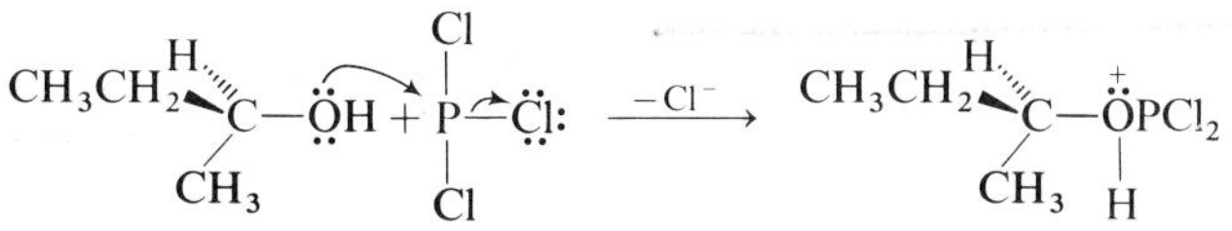

$$CH_3CH_2\overset{H_{\cdots}}{\underset{CH_3}{\diagdown}}C-\overset{\cdots}{O}H + \overset{Cl}{\underset{Cl}{\overset{|}{P}}}{\diagdown}\overset{\cdots}{Cl}: \quad \xrightarrow{-Cl^-} \quad CH_3CH_2\overset{H_{\cdots}}{\underset{CH_3}{\diagdown}}\underset{H}{\overset{+}{C}}-\overset{+}{O}PCl_2$$

(S)-2-butanol   *a protonated*
                                          (S)-chlorophosphite ester

The second step in the reaction is the S_N2 attack by Cl^-. As in any S_N2
reaction, inversion of configuration is observed if the starting material is a single
enantiomer.

Step 2:

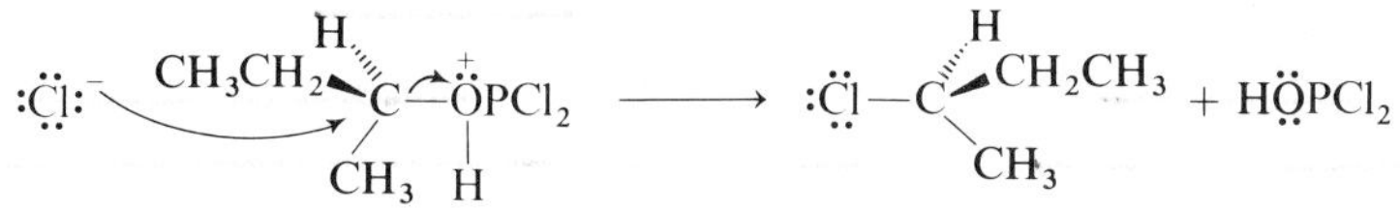

$$:\overset{\cdots}{\underset{\cdots}{Cl}}:^- \quad CH_3CH_2\overset{H_{\cdots}}{\underset{CH_3\ H}{\diagdown}}\overset{+}{C}\overset{}{\diagup}\overset{\cdots}{O}PCl_2 \quad \longrightarrow \quad :\overset{\cdots}{\underset{\cdots}{Cl}}-C\overset{H}{\underset{CH_3}{\diagup}}\overset{CH_2CH_3}{} + H\overset{\cdots}{O}PCl_2$$

                                          (R)-2-chlorobutane
                                              (inverted)

Each of the three P—Cl bonds can undergo reaction; the end product of the
phosphorus trihalide is phosphorous acid (H_3PO_3).

$$3\,ROH + PCl_3 \quad \longrightarrow \quad 3\,RCl + H_3PO_3$$

STUDY PROBLEM

7.7.   Would you expect deuterium and chlorine to be *cis* or *trans* in the product of the
following reaction?

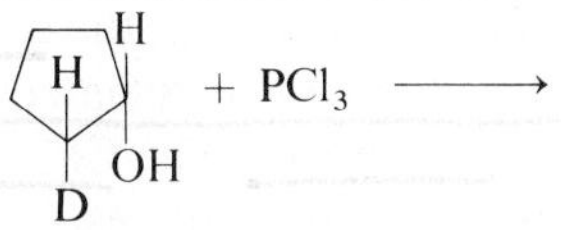

$$\underset{\underset{D}{OH}}{\overset{H}{\underset{H}{\diagup}}} + PCl_3 \quad \longrightarrow$$

The reactions of alcohols with **thionyl chloride** are interesting. If an *amine
solvent* is used for the reaction, a chiral, resolved alcohol yields the alkyl chloride
with the inverted configuration. By contrast, if an *ether solvent* is used, the alkyl
chloride that is obtained has the same configuration as that of the starting alcohol.
In this latter case, we say that the reaction proceeds with **retention of configuration**.

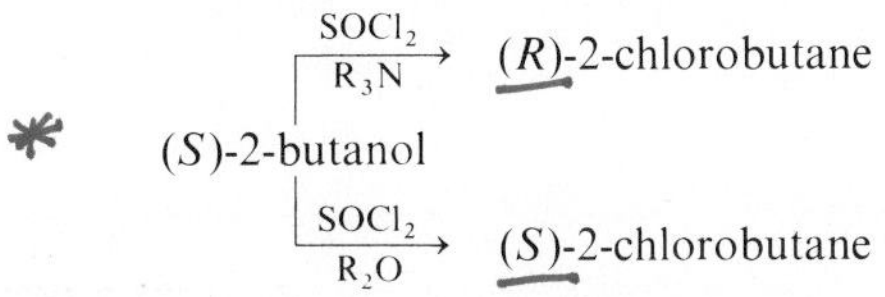

$$\ast \qquad (S)\text{-2-butanol} \overset{\overset{SOCl_2}{R_3N} \longrightarrow (R)\text{-2-chlorobutane}}{\underset{\underset{R_2O}{SOCl_2} \longrightarrow (S)\text{-2-chlorobutane}}{\vert\qquad\qquad\qquad}}$$

In either an amine solvent or an ether solvent, the first step in the reaction
sequence is analogous to that of the reaction with phosphorus trichloride—the
formation of an inorganic ester. Again, the C—O bond is not broken in this first

step. If the starting alcohol is a pure enantiomer, the chlorosulfite ester has the same configuration as the alcohol.

Step 1:

(S)-2-butanol   an (S)-chlorosulfite

An amine solvent reacts with the HCl formed in this reaction to yield an amine salt:

$$R_3N: + HCl \longrightarrow R_3NH^+ \ Cl^-$$

The chloride ion from this acid–base reaction attacks the chlorosulfite ester in a typical S_N2 reaction, which results in an alkyl chloride with an inverted configuration.

Step 2 in an amine solvent, an S_N2 reaction:

(R)-2-chlorobutane

An ether solvent cannot solvate and stabilize ions. The HCl formed in Step 1 (the formation of the chlorosulfite ester) escapes into the atmosphere. Thus, in an ether solution, the chloride ion for the alkyl chloride must come from the chlorosulfite ester grouping. Evidence shows that the chlorosulfite dissociates into ions. These ions cannot become completely dissociated in an ether solution, but remain very close to each other as an **ion pair**. The Cl^- then attacks the positive carbon from the *same side* as the C—O bond; the result is a retained configuration. This type of reaction is called an **S_Ni reaction**, where "*i*" stands for *internal return*. S_Ni reactions are quite rare in organic chemistry.

Step 2 in an ether solvent, an S_Ni reaction:

ion pair   (S)-2-chlorobutane

STUDY PROBLEM

7.8.   List three reagents that could be used to prepare 2-chlorooctane from (R)-2-octanol. Tell whether each would lead to *racemization*, *inversion*, or *retention of configuration*.

SECTION 7.8.

Elimination Reactions of Alcohols

Alcohols, like alkyl halides, undergo elimination reactions to yield alkenes. Because water is lost in the elimination, this reaction is called a **dehydration reaction**.

$$(CH_3)_3COH \xrightarrow[60°]{conc.\ H_2SO_4} (CH_3)_2C{=}CH_2 + H_2O$$

t-butyl alcohol
a 3° alcohol

methylpropene
(isobutylene)

$$(CH_3)_2CHOH \xrightarrow[100°]{conc.\ H_2SO_4} CH_3CH{=}CH_2 + H_2O$$

2-propanol
a 2° alcohol

propene
(propylene)

$$CH_3CH_2OH \xrightarrow[180°]{conc.\ H_2SO_4} CH_2{=}CH_2 + H_2O$$

ethanol
a 1° alcohol

ethene
(ethylene)

Although sulfuric acid is often the acid of choice for a dehydration catalyst, any strong acid can cause dehydration of an alcohol. Note the comparative ease with which a tertiary alcohol undergoes elimination—simply warming it with concentrated H_2SO_4 leads to the alkene. Elimination is a prevalent side reaction in substitution reactions of tertiary alcohols with HX.

1° ROH   2° ROH   3° ROH

increasing ease of dehydration

For secondary and tertiary alcohols, dehydration follows an **E1 path**. The hydroxyl group is protonated, a carbocation is formed with loss of a water molecule, and then a proton is eliminated to yield the alkene. (Primary alcohols probably undergo dehydration by an E2 path.) A secondary or tertiary alcohol yields little, if any, substitution product in hot sulfuric acid; however, primary alcohols may yield ethers or sulfates. These competing reactions are discussed in Section 7.12B.

Let us consider the dehydration of 2-pentanol, a secondary alcohol that undergoes a typical E1 reaction.

Step 1 (protonation and loss of water):

$$CH_3CH_2CH_2\overset{\displaystyle :\overset{\cdot\cdot}{O}H}{\underset{|}{C}}HCH_3 \underset{\xleftarrow{\hspace{1cm}}}{\overset{H^+}{\underset{fast}{\xrightarrow{\hspace{1cm}}}}} CH_3CH_2CH_2\overset{\displaystyle \overset{+}{\overset{\cdot\cdot}{O}}H_2}{\underset{|}{C}}HCH_3 \underset{\xleftarrow{\hspace{1cm}}}{\overset{-H_2O}{\underset{slow}{\xrightarrow{\hspace{1cm}}}}} [CH_3CH_2CH_2\overset{+}{C}HCH_3]$$

protonated

a carbocation

Step 2 (loss of H⁺):

$$CH_3CH_2\overset{\displaystyle H}{\underset{|}{C}}H{-}\overset{+}{C}HCH_3 \underset{\xleftarrow{\hspace{1cm}}}{\overset{fast}{\xrightarrow{\hspace{1cm}}}} \left[CH_3CH_2\overset{H^{\delta+}}{\underset{|}{C}}H{\cdots}\overset{\delta+}{C}HCH_3 \right] \underset{\xleftarrow{\hspace{1cm}}}{\overset{-H^+}{\underset{fast}{\xrightarrow{\hspace{1cm}}}}} CH_3CH_2CH{=}CHCH_3$$

transition state

2-pentene

In this reaction, the formation of the carbocation is the slow step and thus the rate-determining step, just as carbocation formation is the rate-determining step in an S_N1 or E1 reaction of an alkyl halide. In the second step, which is the faster step, the carbocation loses H^+ (to H_2O, HSO_4^-, or another molecule of alcohol). In this second step, the double bond is partially formed in the transition state. For this reason, when more than one alkene can be formed, a typical E1 reaction yields predominantly the *more substituted, more stable alkene* (Saytseff rule). Recall from Chapter 5 that E2 reactions also usually yield the more stable alkene.

$$\underset{\text{2-pentanol}}{CH_3CH_2CH_2\overset{\overset{\displaystyle OH}{\displaystyle |}}{C}HCH_3} \quad \xrightarrow[\text{E1}]{\overset{\displaystyle H_2SO_4}{\text{heat}}} \quad \underset{\underset{80\%}{\text{2-pentene}}}{CH_3CH_2CH{=}CHCH_3} + \underset{\underset{5\%}{\text{1-pentene}}}{CH_3CH_2CH_2CH{=}CH_2}$$

STUDY PROBLEM

7.9.   Would you expect to find *cis-* or *trans-*2-pentene as the predominant product in the dehydration of 2-pentanol? Explain.

In a typical E1 reaction of an alcohol, Step 1 determines the overall rate of the reaction, but *Step 2 determines the product ratio*. Figure 7.2 shows an energy diagram that exemplifies this statement.

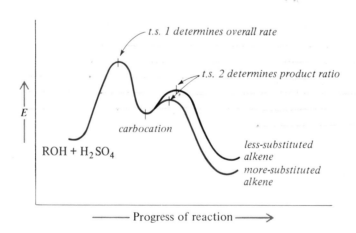

FIGURE 7.2. Energy diagram for a typical E1 reaction of a secondary or tertiary alcohol.

In any elimination reaction in which the double bond can be in *conjugation with a benzene ring*, the conjugated product is formed in preference to the non-conjugated product. The conjugated alkene is of lower energy, as is the transition state leading to its formation. In fact, it is often difficult to isolate alcohols in which the hydroxyl group is one or two carbons away from a benzene ring; dehydration is usually spontaneous under acidic conditions.

OH
|
⬡—CH$_2$CHCH$_2$CH$_3$ $\xrightarrow{-H_2O}$

1-phenyl-2-butanol

in conjugation with benzene ring

⬡—CH=CHCH$_2$CH$_3$ and not ⬡—CH$_2$CH=CHCH$_3$

1-phenyl-1-butene   1-phenyl-2-butene

Recall from Section 5.6F that carbocations can undergo rearrangements. Are rearrangements observed in the E1 reactions of alcohols? Yes, they are. If a carbocation can undergo a 1,2-shift to a more stable carbocation, rearrangement products will be observed.

Rearrangements:

CH$_3$ OH   CH$_3$ CH$_3$
| | \   /
CH$_3$—C——CHCH$_3$ $\xrightarrow[95°]{H_2SO_4}$ C=C
| /   \
CH$_3$ CH$_3$ CH$_3$

3,3-dimethyl-2-butanol   2,3-dimethyl-2-butene

CH$_3$ H   CH$_3$ CH$_3$
| | \   /
CH$_3$CH——C—CH$_2$OH $\xrightarrow[140°]{H_2SO_4}$ C=C
| /   \
CH$_3$ CH$_3$ CH$_3$

2,3-dimethyl-1-butanol   2,3-dimethyl-2-butene

Because rearrangements can occur in the dehydration of alcohols and because the dehydrations of primary alcohols are slow, the dehydration of an alcohol is not always the best method for preparing an alkene. In many cases, it is preferable to convert the alcohol to the alkyl halide and then subject the alkyl halide to an E2 reaction.

SAMPLE PROBLEM

Predict the major products of dehydration of   (a) 3-pentanol and (b) neopentyl alcohol (2,2-dimethyl-1-propanol). (*Hint*: Neopentyl alcohol undergoes rearrangement when treated with H$_2$SO$_4$.)

Solution:

OH
|
(a) CH$_3$CH$_2$CHCH$_2$CH$_3$ $\xrightarrow{-H_2O}$ CH$_3$CH$_2$CH=CHCH$_3$

trans-2-pentene

more stable alkene

CH$_3$ ⌈   CH$_3$ ⌉
| | | |
(b) (CH$_3$)$_2$C—CH$_2$—ŌH$_2$ $\xrightarrow{-H_2O}$ ⌊(CH$_3$)$_2$C̈—CH$_2$⌋ $\xrightarrow{-H^+}$ (CH$_3$)$_2$C=CHCH$_3$

rearranged   2-methyl-2-butene

more stable alkene

A.   Pinacol rearrangement

When treated with strong acids, 1,2-diols do not yield alkenes or dienes, as we might expect. Instead, a rearrangement called a **pinacol rearrangement** occurs. **Pinacol** is the trivial name for 2,3-dimethyl-2,3-butanediol. (The name comes from the Greek *pinox*, "plate," which is descriptive of the appearance of crystalline pinacol.) When pinacol is treated with an acid, a ketone called **pinacolone** results.

$$
\underset{\substack{\text{2,3-dimethyl-2,3-butanediol} \\ \text{(pinacol)}}}{(CH_3)_2\overset{\overset{\displaystyle OH}{|}}{C}—\overset{\overset{\displaystyle OH}{|}}{C}(CH_3)_2} \quad \xrightarrow{H^+} \quad \underset{\substack{\text{3,3-dimethyl-2-butanone} \\ \text{(pinacolone)}}}{(CH_3)_3\overset{\overset{\displaystyle O}{||}}{C}CCH_3} \quad + \; H_2O
$$

This reaction is another example of a carbocation rearrangement.

$$
\underset{\text{pinacol}}{CH_3\overset{\overset{\displaystyle :\ddot{O}H}{|}}{\underset{\underset{\displaystyle CH_3}{|}}{C}}—\overset{\overset{\displaystyle OH}{|}}{\underset{\underset{\displaystyle CH_3}{|}}{C}CH_3}} \quad \underset{\rightleftharpoons}{\overset{H^+}{}} \quad \underset{\text{protonated}}{CH_3\overset{\overset{\displaystyle \overset{+}{O}H_2}{|}}{\underset{\underset{\displaystyle CH_3}{|}}{C}}—\overset{\overset{\displaystyle OH}{|}}{\underset{\underset{\displaystyle CH_3}{|}}{C}CH_3}} \quad \xrightarrow{-H_2\ddot{O}:}
$$

$$
\underset{\substack{\text{carbocation rearranging} \\ \text{(less stable)}}}{\left[CH_3\overset{+}{\underset{\underset{\displaystyle CH_3}{|}}{C}}—\overset{\overset{\displaystyle :\ddot{O}H}{|}}{\underset{\underset{\displaystyle CH_3}{|}}{C}CH_3} \right]} \quad \longrightarrow \quad \underset{\substack{\text{a protonated ketone} \\ \text{(more stable)}}}{\left[CH_3\overset{\overset{\displaystyle CH_3}{|}}{\underset{\underset{\displaystyle CH_3}{|}}{C}}—\overset{\overset{\displaystyle {}^+\ddot{O}H}{||}}{C}CH_3 \right]} \quad \underset{\rightleftharpoons}{\overset{-H^+}{}} \quad \underset{\text{pinacolone}}{(CH_3)_3\overset{\overset{\displaystyle \cdot\ddot{O}\cdot}{||}}{C}CCH_3}
$$

In this reaction, first one of the —OH groups is protonated to yield —OH$_2^+$, which leaves as water. The resulting carbocation undergoes a 1,2-shift of a methyl group to yield a resonance-stabilized cation (the protonated ketone) that loses a proton to yield a ketone.

$$
\underset{\substack{\text{a protonated ketone} \\ \text{(resonance-stabilized)}}}{\left[\underset{+}{R}—\overset{\overset{\displaystyle \cdot\ddot{O}\diagup^H}{|}}{C}—R \quad \longleftrightarrow \quad R—\overset{\overset{\displaystyle \overset{+}{\cdot\ddot{O}}\diagup^H}{||}}{C}—R \right]} \quad \underset{\rightleftharpoons}{} \quad \underset{\text{a ketone}}{R—\overset{\overset{\displaystyle \cdot\ddot{O}\cdot}{||}}{C}—R \; + \; H^+}
$$

The term "pinacol rearrangement" is used to describe similar rearrangements of any 1,2-diols, not just pinacol itself. The pinacol rearrangements of compounds containing a variety of groups have been studied to show which groups are most apt to migrate, or shift, in a rearrangement. We will consider one example. The first step in the pinacol rearrangement of 1-phenyl-1,2-propanediol is carbocation formation:

OH OH
| |
C₆H₅—CH—CHCH₃ $\xrightarrow[-H_2O]{H^+}$ [C₆H₅—$\overset{+}{C}$H—CHCH₃ OH]

1-phenyl-1,2-propanediol   *resonance-stabilized ·*

and not   [C₆H₅—CH—$\overset{+}{C}$HCH₃ OH]

not resonance-stabilized

Which migrates, the H or the CH_3? We can tell by identifying the product:

If H migrates:

$\overset{+}{C}$H—C—CH₃ (OH, H)   ⟶   CH₂CCH₃ (O)

1-phenyl-2-propanone

observed product

If CH₃ migrates:

$\overset{+}{C}$H—C—CH₃ (OH, H)   ⟶   CHCH (O), CH₃

2-phenylpropanal

not observed

The pinacol rearrangement of 1-phenyl-1,2-propanediol yields 1-phenyl-2-propanone and not 2-phenylpropanal. In this example, a hydrogen has migrated instead of a methyl group.

From such studies of a large number of pinacol rearrangements, it has been determined that (1) the more stable carbocation is formed initially (a benzylic carbocation in the preceding example), and (2) the migratory aptitude is Ar > R; that is, when either an alkyl or an aryl group can undergo a 1,2-shift, the aryl group shifts rather than the alkyl group. It has also been found that hydrogen atoms are unpredictable in their migratory aptitudes. With some 1,2-diols, a hydrogen atom migrates in preference to an Ar or R group; with other diols, an Ar or R group migrates in preference to H.

STUDY PROBLEM

7.10. Predict the product of the pinacol rearrangement of each of the following diols:

(a) $(C_6H_5)_2CCH_2OH$ (OH)

(b) $C_6H_5-C-C-C_6H_5$ (OH, OH, CH₃, CH₃)

SECTION 7.9.

Alcohols as Acids

An alcohol is similar to water in that it can act as a *base* and accept a proton (to yield a protonated alcohol, ROH_2^+). Like water, an alcohol can also act as an *acid* and lose a proton (to yield an alkoxide ion, RO^-). Like water, an alcohol is a very weak acid or base; for a pure alcohol or an aqueous alcohol, the equilibrium for the ionization reaction lies on the un-ionized side of the equation.

Water: $H\ddot{O} \! \stackrel{\frown}{-} \! H + H_2\ddot{O} \; \rightleftharpoons \; H\ddot{O}{:}^- + H_3O^+$

Methanol:

$$CH_3O{-}H + CH_3OH \; \rightleftharpoons \; CH_3O^- + CH_3\overset{+}{O}H_2$$

Methanol in H₂O:

$$CH_3O{-}H + H_2O \; \rightleftharpoons \; CH_3O^- + H_3O^+$$

In dilute aqueous solution, alcohols have approximately the same pK_a values as water (see Table 7.6). However, in their pure liquid state (no water), alcohols are much weaker acids than water. The pK_a of pure methanol is around 17, and other alcohols are even weaker acids. For comparison, the pK_a value for pure water is 15.7 (not 14, which is the pK_w).

TABLE 7.6. pK_a values for water and some alcohols in dilute aqueous solution

Compound	pK_a
H_2O	15.7
CH_3OH	15.5
CH_3CH_2OH	15.9
$(CH_3)_3COH$	~18

The lower acidity of nonaqueous alcohols is not due to any unique structural inability of RO^- to carry an ionic charge, but rather is a result of the low dielectric constants of alcohols compared to that of water (Table 7.4, page 258). Because they are less polar, alcohols are less able to support ions in solution than are water molecules.

SECTION 7.10.

Alkoxides and Phenoxides

An **alkoxide** is the salt of an alcohol. (The name is analogous to *hydroxide*.)

$CH_3O^- \; Na^+$ $CH_3CH_2O^- \; K^+$ $(CH_3)_2CHO^- \; Na^+$

sodium methoxide   potassium ethoxide   sodium isopropoxide

Alkoxides are *strong bases*, generally stronger than hydroxides. To prepare an alkoxide from an alcohol, a base stronger than the alkoxide itself is required. Sodamide ($NaNH_2$) and Grignard reagents are strong enough bases to abstract a hydrogen ion from an alcohol.

$$ROH + NaNH_2 \longrightarrow RO^- \ Na^+ + NH_3$$
$$ROH + R'MgX \longrightarrow RO^- \ {}^+MgX + R'H$$

The most convenient method for the preparation of alkoxides is the treatment of an alcohol with an alkali metal such as sodium or potassium. The reaction is not an acid–base reaction, but an oxidation–reduction reaction. The alkali metal is oxidized to a cation, and the hydrogens of the OH groups are reduced to hydrogen gas.

$$CH_3OH + Na \longrightarrow CH_3O^- \ Na^+ \ \ \ + \tfrac{1}{2}H_2\uparrow$$
$$CH_3CH_2OH + Na \longrightarrow CH_3CH_2O^- \ Na^+ + \tfrac{1}{2}H_2\uparrow$$
$$(CH_3)_3COH + K \longrightarrow (CH_3)_3CO^- \ K^+ \ \ + \tfrac{1}{2}H_2\uparrow$$

Methanol and ethanol undergo fairly vigorous reaction with sodium metal. As the size of the R group is increased, the vigor of the reaction decreases. Sodium and water react explosively; sodium and ethanol undergo reaction at a very controllable rate; and sodium and 1-butanol undergo a very sluggish reaction. With alcohols of four or more carbons, the more reactive potassium metal is generally used to prepare the alkoxide.

✳ H_2O CH_3OH CH_3CH_2OH $CH_3CH_2CH_2OH$
decreasing reactivity toward Na or K

A **phenoxide** is the salt of a phenol, a compound in which OH is attached directly to an aromatic ring.

phenol

sodium phenoxide

a phenoxide

4-methylphenol
(*p*-cresol)

potassium 4-methylphenoxide

a phenoxide

Phenols are much stronger acids than alcohols. The pK_a of phenol itself is 10. Thus, phenol is about halfway between ethanol and acetic acid ($pK_a = 4.75$) in acid strength. A phenoxide ion is a weaker base than OH^-; therefore, a phenoxide can be prepared by treating a phenol with aqueous NaOH. This reactivity is in direct contrast to that of alcohols.

$$CH_3CH_2OH \quad + \; NaOH \; \rightleftharpoons \; CH_3CH_2O^- \; Na^+ + H_2O$$

ethanol, $pK_a = 15.9$ sodium ethoxide

favored

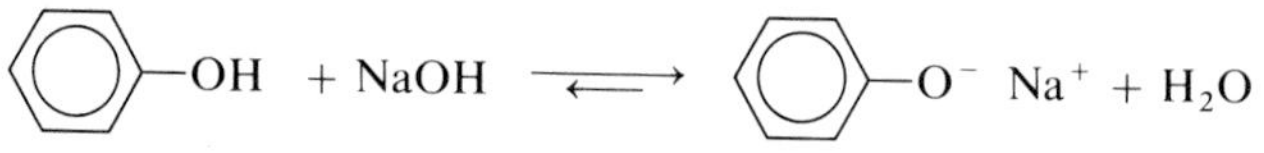

phenol, $pK_a = 10$ sodium phenoxide

favored

The degree of ionization of a weak acid is determined by the relative stabilities of the un-ionized compound and the anion:

$$HA \; \rightleftharpoons \; H^+ + A^-$$

If A^- is stabilized relative to HA, acidity is increased

The reason for the acidity of phenols is that the product anion is *resonance-stabilized*, with the negative charge delocalized by the aromatic ring.

Resonance structures for the phenoxide ion:

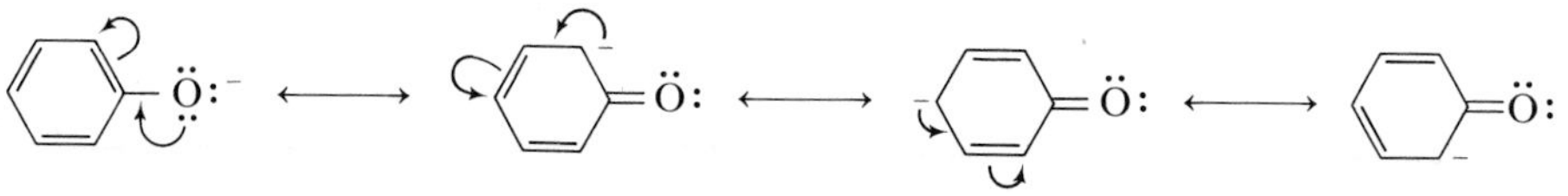

The negative charge in an alkoxide ion (RO^-) cannot be delocalized. Therefore, an alkoxide ion is of higher energy relative to the alcohol, and alcohols are not as strong acids as phenols.

Because of its acidity, phenol was originally called *carbolic acid*. In the 1800's, the British surgeon Joseph Lister urged that phenol be used as a hospital anti-septic. Prior to that time, no antiseptics were used because it was thought that odors, not microorganisms, were the cause of infection. As an antiseptic, phenol itself has been replaced by less irritating compounds. Interestingly, many modern anti-septics still contain phenolic groups.

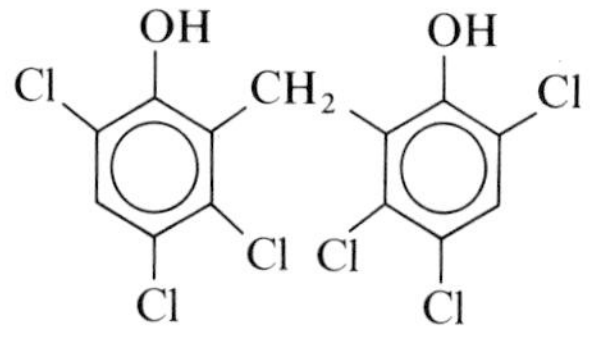

hexachlorophene

banned for most uses because it can be absorbed through the skin

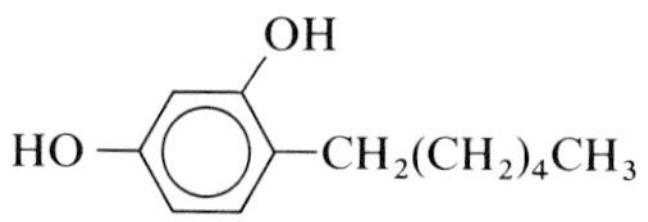

n-hexylresorcinol

Alkoxides and phenoxides are good nucleophiles. The use of these reagents for reaction with alkyl halides to yield ethers was mentioned in Chapter 5 and will also be discussed later in this chapter (Section 7.14).

STUDY PROBLEMS

7.11. What are the principal ions in solution when the following reagents are mixed?

(a) sodium ethoxide and phenol   (b) sodium phenoxide and ethanol

7.12. The pK_a of *p*-nitrophenol is 7. Use resonance structures to explain why this compound is more acidic than phenol.

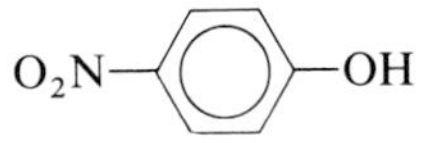

$$O_2N-\!\!\!\bigcirc\!\!\!-OH$$

p-nitrophenol

SECTION 7.11.

Esterification Reactions

Alcohols undergo reaction with carboxylic acids and carboxylic acid derivatives to yield *esters of carboxylic acids,* These reactions, called **esterification reactions,** and the product esters will be covered in detail in Chapters 12 and 13.

$$\underset{\substack{\text{acetic acid}\\\text{\textit{a carboxylic acid}}}}{CH_3\overset{O}{\overset{||}{C}}OH} + \underset{\text{ethanol}}{H\,OCH_2CH_3} \underset{}{\overset{H^+,\,heat}{\rightleftarrows}} \underset{\substack{\text{ethyl acetate}\\\text{\textit{an ester}}}}{CH_3\overset{O}{\overset{||}{C}}OCH_2CH_3} + H_2O$$

$$\underset{\substack{\text{benzoic acid}\\\text{\textit{a carboxylic acid}}}}{\bigcirc\!\!-\overset{O}{\overset{||}{C}}OH} + \underset{\text{1-propanol}}{H\,OCH_2CH_2CH_3} \overset{H^+,\,heat}{\rightleftarrows} \underset{\substack{\text{\textit{n}-propyl benzoate}\\\text{\textit{an ester}}}}{\bigcirc\!\!-\overset{O}{\overset{||}{C}}OCH_2CH_2CH_3} + H_2O$$

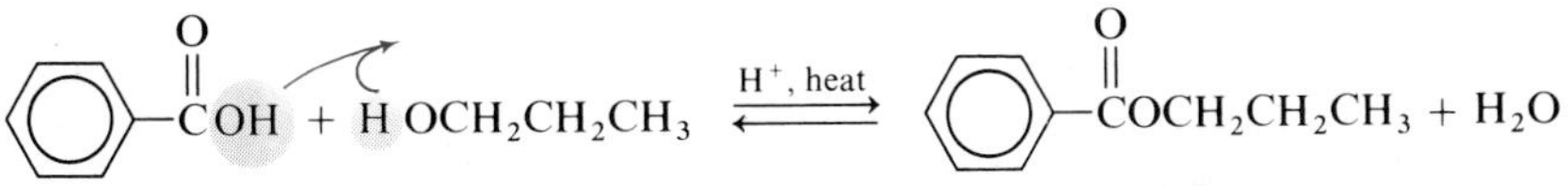

STUDY PROBLEM

7.13. Predict the esterification product of each of the following reactions:

(a) $CH_3CH_2CH_2CO_2H + CH_3OH \overset{H^+,\,heat}{\rightleftarrows}$

(b) 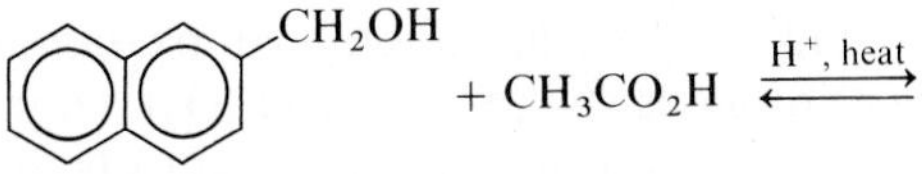 $\bigcirc\!\!\bigcirc\!\!-CH_2OH + CH_3CO_2H \overset{H^+,\,heat}{\rightleftarrows}$

SECTION 7.12.

Inorganic Esters of Alcohols

Inorganic esters of alcohols are compounds prepared by the reaction of alcohols and either mineral acids (such as HNO_3 or H_2SO_4) or acid halides of mineral acids (such as $SOCl_2$, Section 7.7).

$$
\begin{array}{c}
O \\
\parallel \\
CH_3OSOCH_3 \\
\parallel \\
O
\end{array}
\qquad
\begin{array}{c}
CH_2ONO_2 \\
| \\
CHONO_2 \\
| \\
CH_2ONO_2
\end{array}
$$

dimethyl sulfate   nitroglycerin

an inorganic ester   *an inorganic ester*

A. Nitrates

To see how an inorganic ester of an alcohol is formed, let us consider the formation of an alkyl nitrate ($RONO_2$). (Do not confuse the alkyl nitrates with nitroalkanes, RNO_2, in which carbon is attached to the N.) A nitrate esterification proceeds by (1) an ionization reaction of HNO_3 to yield a nitronium ion ($^+NO_2$), followed by (2) attack on $^+NO_2$ by the alcohol oxygen. This second reaction is a typical Lewis acid–base reaction. Loss of a proton from the intermediate adduct yields the nitrate ester.

1. Formation of $NO_2{}^+$:

$$
2\ H\ddot{O}{-}NO_2 \rightleftharpoons H_2\overset{+}{\ddot{O}}{-}NO_2 + NO_3{}^-
$$

$$
H_2\overset{+}{\ddot{O}}{-}NO_2 \rightleftharpoons H_2\ddot{O}\colon + \overset{+}{N}O_2
$$

2. Formation of ester:

$$
\begin{array}{c}
CH_3\ddot{O}\colon \overset{\frown}{+}\ \overset{+}{N}O_2 \\
| \\
H
\end{array}
\rightleftharpoons
\begin{array}{c}
CH_3\overset{+}{\ddot{O}}{-}NO_2 \\
| \\
H
\end{array}
\xrightarrow{-H^+}
\begin{array}{c}
CH_3\ddot{O}NO_2 \\
\text{methyl nitrate}
\end{array}
$$

methanol   *protonated*
          nitrate

STUDY PROBLEM

7.14. Give the steps in the formation of ethyl nitrite (CH_3CH_2ONO) from nitrous acid (HONO) and ethanol.

Nitric acid is a strong oxidizing agent, and oxidation of the alcohol (sometimes explosively) can accompany the formation of nitrate esters. The nitrate esters themselves (for example, nitroglycerin and PETN) are explosives. When detonated, these compounds undergo fast intramolecular oxidation–reduction reactions to yield large volumes of gases (N_2, CO_2, H_2O, O_2). Organic nitrates (such as nitroglycerin) and nitrites are also used as *vasodilators* (substances that dilate blood vessels) in the treatment of certain types of heart disease.

$$
\begin{array}{c}
CH_2ONO_2 \\
| \\
O_2NOCH_2CCH_2ONO_2 \\
| \\
CH_2ONO_2
\end{array}
$$

pentaerythritol nitrate (PETN)

an explosive

B.   Sulfates

The reaction of concentrated sulfuric acid with alcohols can lead to monoalkyl or dialkyl sulfate esters. The monoesters are named as **alkyl hydrogen sulfates, alkylsulfuric acids**, or **alkyl bisulfates**; the three terms are synonymous. The names of the diesters are straightforward; the alkyl groups are named and the word **sulfate** is added. Alkyl hydrogen sulfates are strong acids, but dialkyl sulfates are not acidic.

$$
\begin{array}{ccc}
\underset{\substack{\| \\ O}}{\overset{\substack{O \\ \|}}{CH_3OSOH}} \;\; \overset{\textit{acidic}}{\searrow} & \underset{\substack{\| \\ O}}{\overset{\substack{O \\ \|}}{CH_3OSOCH_3}} & \underset{\substack{\| \\ O}}{\overset{\substack{O \\ \|}}{CH_3OSOCH_2CH_3}} \\
\text{methyl hydrogen} & \text{dimethyl sulfate} & \text{methyl ethyl sulfate} \\
\text{sulfate} & &
\end{array}
$$

Earlier in this chapter, we mentioned that the action of sulfuric acid on alcohols results in alkenes. In Section 7.14, we will discuss the fact that diethyl ether is prepared by the reaction of ethanol and sulfuric acid. What actually happens when an alcohol is treated with concentrated sulfuric acid?

When an alcohol is mixed with H_2SO_4, a series of reversible reactions occurs. (The scheme that follows is simplified; alkenes can also go to sulfates, sulfates can go to ethers, etc.) Which reaction product predominates depends on the structure of the alcohol, the relative concentrations of reactants, and the temperature of the reaction mixture. In general, *primary alcohols* give sulfate esters at low temperatures, ethers at moderate temperatures, and alkenes at high temperatures. (In all cases, mixtures would be expected.) *Tertiary alcohols* and, to a large extent, *secondary alcohols* yield alkene products.

$$
1°: \quad ROH + H_2SO_4 \;\; \overset{0°}{\nearrow} \quad ROSO_2OH + ROSO_2OR + H_2O
$$

$$
\underset{170°}{\overset{140°}{\rightleftharpoons}} \quad ROR + H_2O
$$

$$
\qquad\qquad\qquad\qquad\qquad\qquad\qquad alkenes + H_2O
$$

$$
2° \text{ and } 3°: \quad ROH + H_2SO_4 \;\; \rightleftharpoons \quad alkenes + H_2O
$$

Dimethyl sulfate and diethyl sulfate are volatile and therefore can easily be isolated from a reaction mixture by distillation. The other alkyl sulfates are not so volatile; the high temperatures needed for distillation are sufficient to crack the sulfates and yield alkenes.

A more general method for preparing sulfates is by using one of the chlorides of sulfuric acid. For example:

$$
2\,CH_3OH + \underset{\substack{\| \\ O}}{\overset{\substack{O \\ \|}}{Cl-S-Cl}} \longrightarrow \underset{\substack{\| \\ O}}{\overset{\substack{O \\ \|}}{CH_3O-S-OCH_3}} + 2\,HCl
$$

$$
\qquad\quad \text{sulfuryl chloride} \qquad\qquad\qquad \text{dimethyl sulfate}
$$

An <u>alkyl sulfate group is a good leaving group</u>. Dimethyl sulfate, which is commercially available, is often used as an alkylating agent. The following example shows its use in preparing a methyl ester.

$$CH_3C\overset{O}{\overset{\|}{C}}\ddot{O}{:}^- +CH_3 {-}\ddot{O}SO_2OCH_3 \longrightarrow CH_3C\overset{O}{\overset{\|}{C}}OCH_3 + {:}\ddot{O}SO_2OCH_3$$

acetate ion   dimethyl sulfate   methyl acetate

an ester

C. Sulfonates

A **sulfonate** is an <u>inorganic ester</u> with the <u>general formula</u> $\boxed{RSO_2OR.}$ (Do not confuse the sulfonate structure with the sulfate structure. A sulfonate has an alkyl or aryl group attached *directly to the sulfur atom.*)

benzenesulfonic acid   methyl benzenesulfonate

a sulfonic acid   *a sulfonate*
(a strong acid)

Sulfonates are the most widely used of the various sulfur derivatives of alcohols. They are <u>often solids</u>, a fact that simplifies their laboratory purification. In addition, a <u>sulfonate group ($RSO_2O{-}$) is an excellent leaving group</u> and can be <u>displaced by a variety of nucleophiles.</u> Conversion of an alcohol to a sulfonate, followed by nucleophilic displacement, provides an excellent synthetic route to a variety of products.

We will concentrate our attention on only one class of sulfonates—the *p*-toluenesulfonates (4-methylbenzenesulfonates), usually called **tosylates** and abbreviated ROTs. The tosylates are <u>prepared</u> by the <u>reaction of an alcohol with</u> *p*-toluenesulfonyl chloride (tosyl chloride). An amine, such as <u>pyridine</u> (page 753), is often <u>added to the reaction mixture</u> to "trap" the HCl as it is formed ($R_3N{:} + HCl \rightarrow R_3NH^+ \ Cl^-$).

ROH + ClS $-\langle\bigcirc\rangle-$ CH$_3$ $\longrightarrow$ ROS $-\langle\bigcirc\rangle-$ CH$_3$ + HCl
an alcohol

tosyl chloride   *an alkyl tosylate*
(TsCl)   (ROTs)

In the formation of the tosylate, the <u>carbon–oxygen bond of the alcohol is</u> <u>not broken.</u> If a tosylate is <u>prepared using a single enantiomer of a chiral alcohol,</u> the tosylate <u>retains the configuration of the starting alcohol.</u>

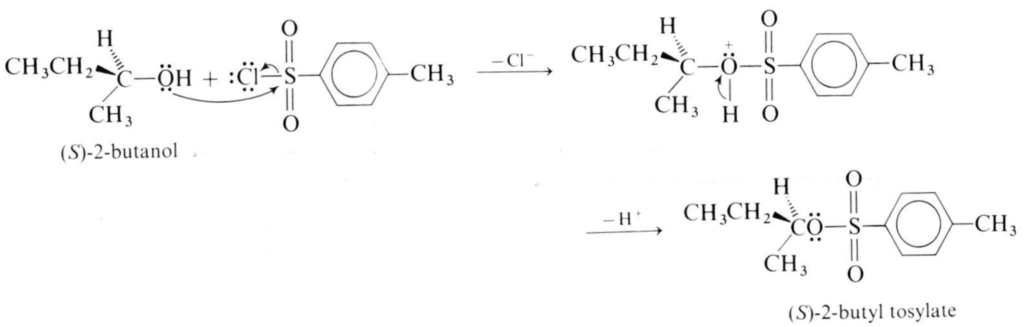

(S)-2-butanol

(S)-2-butyl tosylate

STUDY PROBLEM

7.15. How would you prepare   (a) *n*-propyl tosylate and   (b) (*R*)-2-hexyl tosylate?

The tosylate anion (as well as other sulfonate anions) is resonance-stabilized and is a very weak base.

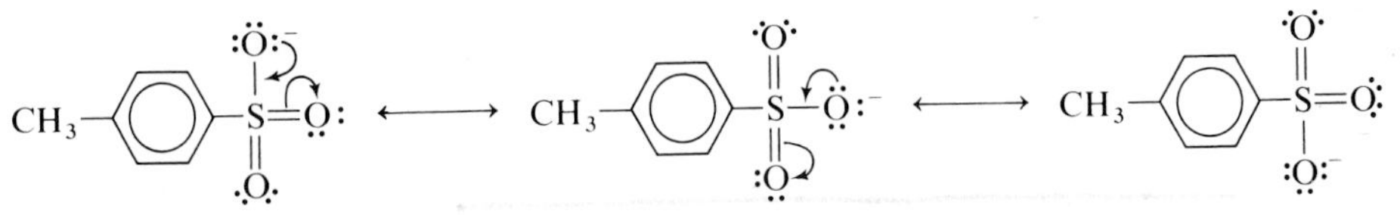

For this reason, the tosylate group is a far better leaving group than an OH group. The tosylate group may be displaced in S_N2 reactions by such weak nucleophiles as halide ions or alcohols. (No acidic catalyst is necessary.)

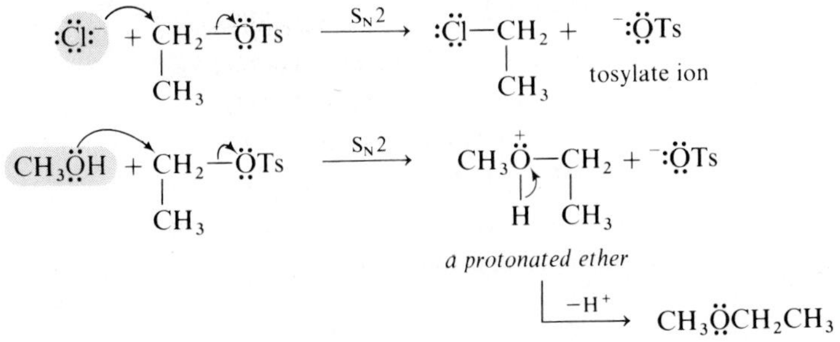

STUDY PROBLEMS

7.16. Write resonance structures for the methylsulfonate ion.

7.17. Predict the S_N2 product of the reaction of water with (*R*)-2-octyl tosylate.

SECTION 7.13.

Oxidation of Alcohols

In inorganic chemistry, **oxidation** is defined as the *loss of electrons* by an atom, while **reduction** is the *gain of electrons* by an atom.

Oxidation:

$$Na^0 \xrightarrow{-e^-} Na^+$$

$$Mg^0 \xrightarrow{-2e^-} Mg^{2+}$$

Reduction:

$$Fe^{3+} \xrightarrow{+e^-} Fe^{2+}$$

$$Cu^{2+} \xrightarrow{+2e^-} Cu^0$$

In organic reactions, it is not always easy to determine whether a carbon atom "gains" or "loses" electrons. However, oxidation and reduction of organic compounds are common reactions. Good rules of thumb to determine if an organic compound has been oxidized or reduced follow.

If a molecule gains oxygen or loses hydrogen, it is oxidized:

$$CH_3CH_2OH \xrightarrow{[O]} CH_3CO_2H$$

$$\underset{\displaystyle CH_3\overset{\displaystyle |}{\underset{}{C}}HCH_3}{\overset{\displaystyle OH}{}} \xrightarrow{[O]} \underset{\displaystyle CH_3\overset{\displaystyle \|}{\underset{}{C}}CH_3}{\overset{\displaystyle O}{}}$$

If a molecule loses oxygen or gains hydrogen, it is reduced:

$$CH_3CO_2H \xrightarrow{[H]} CH_3CH_2OH$$

$$\underset{\displaystyle CH_3\overset{\displaystyle \|}{\underset{}{C}}CH_3}{\overset{\displaystyle O}{}} \xrightarrow{[H]} \underset{\displaystyle CH_3\overset{\displaystyle |}{\underset{}{C}}HCH_3}{\overset{\displaystyle OH}{}}$$

We may list a series of compounds according to the increasing oxidation state of carbon:

CH_3CH_3	$CH_2{=}CH_2$	$CH{\equiv}CH$	$\overset{\displaystyle O}{\underset{}{\overset{\|}{CH_3COH}}}$	CO_2
	CH_3CH_2OH	CH_3CHO		
	CH_3CH_2Cl	CH_3CHCl_2	and its derivatives	

increasing oxidation state of C

Note that $CH_2{=}CH_2$ and CH_3CH_2OH are at the same oxidation level. This is not surprising because the difference between the two molecules is only a molecule of water. No oxidation–reduction reaction takes place in the interconversion of ethylene and ethanol.

$$CH_2{=}CH_2 \underset{-H_2O}{\overset{H_2O}{\rightleftarrows}} CH_3CH_2OH$$

STUDY PROBLEMS

7.18. List the following compounds in order of increasing oxidation state.

(a) $HO_2C(CH_2)_4CO_2H$ (b)  —OH   (c) =O

7.19. List the following carboxylic acids in order of increasing oxidation state:

(a) $\underset{\text{hydroxyacetic acid}}{HOCH_2\overset{\displaystyle O}{\overset{\|}{C}}OH}$ (b) $\underset{\text{acetic acid}}{CH_3\overset{\displaystyle O}{\overset{\|}{C}}OH}$ (c) $\underset{\text{oxalic acid}}{HO\overset{\displaystyle O}{\overset{\|}{C}}-\overset{\displaystyle O}{\overset{\|}{C}}OH}$

Alcohols may be oxidized to ketones, aldehydes, or carboxylic acids. These oxidations are widely used in the laboratory and in industry, and they also occur in biological systems.

$$\underset{\underset{\text{a 1° alcohol}}{}}{RCH_2OH} \overset{[O]}{\longrightarrow} \underset{\text{an aldehyde}}{\overset{\displaystyle O}{\overset{\|}{RCH}}}$$

$$\overset{[O]}{\longrightarrow} \underset{\text{a carboxylic acid}}{RCO_2H}$$

$$\underset{\underset{\text{a 2° alcohol}}{}}{\overset{\displaystyle OH}{\overset{|}{RCHR}}} \overset{[O]}{\longrightarrow} \underset{\text{a ketone}}{\overset{\displaystyle O}{\overset{\|}{RCR}}}$$

Theoretically, alcohols can also be reduced to hydrocarbons; however, such reactions are not common. Instead, indirect methods of reduction are used; for example, dehydration of an alcohol leads to an alkene, which is readily reduced with H_2 to the alkane (Section 9.13).

$$\underset{\text{cyclohexanol}}{\bigcirc\text{—OH}} \underset{\text{heat}}{\overset{H_2SO_4}{\longrightarrow}} \underset{\text{cyclohexene}}{\bigcirc} \underset{\text{Pt catalyst}}{\overset{H_2}{\longrightarrow}} \underset{\text{cyclohexane}}{\bigcirc}$$

A. Combustion of ethanol

Alcohols, like other organic compounds, can undergo combustion.

$$\underset{\text{ethanol}}{CH_3CH_2OH} + 3\,O_2 \longrightarrow 2\,CO_2 + 3\,H_2O + \text{energy}$$

The burning of ethanol has an interesting history. In the days of pirates and sailing ships, the alcoholic content of rum or whiskey was determined by pouring it on a small pile of gunpowder and igniting the vapor. If the flames died down and the gunpowder did not burn, the conclusion was that the rum had been watered down. If the gunpowder did burn, this was proof that the rum had not been diluted.

The term "proof" derives from this custom of testing alcoholic beverages. Proof is twice the percent of alcohol; 100-proof spirits are 50 percent ethanol.

B.   Biological oxidation of ethanol

In a mammalian system, ingested ethanol is oxidized primarily in the liver with the aid of an enzyme called *alcohol dehydrogenase*. The product of this dehydrogenation is acetaldehyde, CH_3CHO. (The biological oxidation of methanol leads to formaldehyde, HCHO, which is toxic.) The acetaldehyde from ethanol is further oxidized enzymatically to the acetate ion, $CH_3CO_2^-$, which undergoes esterification with the thiol **coenzyme A** (often abbreviated HSCoA). The product of the esterification is **acetylcoenzyme A.** (The complete structure of acetylcoenzyme A is shown on page 638.) The acetyl group (CH_3CO-) in acetylcoenzyme A can be converted to CO_2, H_2O, and energy, or it can be converted to other compounds, such as fat.

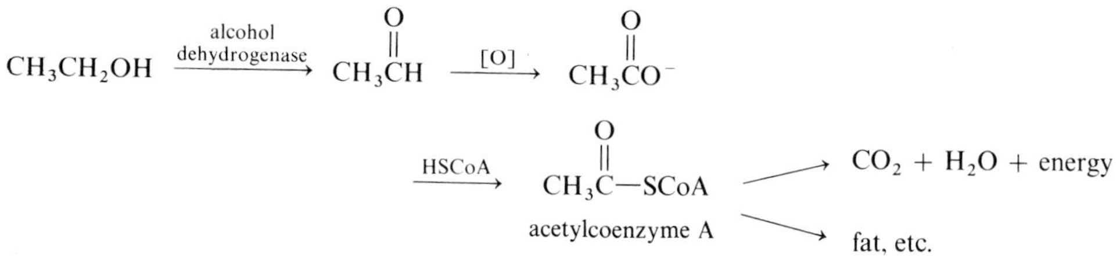

$$CH_3CH_2OH \xrightarrow[\text{dehydrogenase}]{\text{alcohol}} CH_3\overset{\overset{\displaystyle O}{\|}}{C}H \xrightarrow{[O]} CH_3\overset{\overset{\displaystyle O}{\|}}{C}O^-$$

$$\xrightarrow{\text{HSCoA}} CH_3\overset{\overset{\displaystyle O}{\|}}{C}-SCoA$$

acetylcoenzyme A

$$\longrightarrow CO_2 + H_2O + \text{energy}$$
$$\longrightarrow \text{fat, etc.}$$

C.   Laboratory oxidation of alcohols

In general, laboratory oxidizing agents oxidize primary alcohols to carboxylic acids and secondary alcohols to ketones.

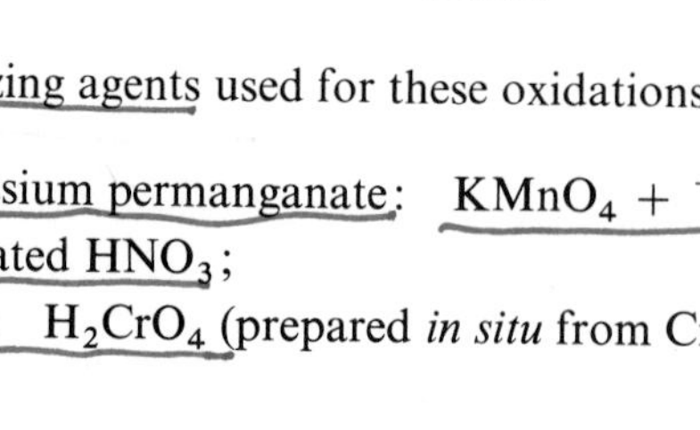

$$RCH_2OH \xrightarrow{[O]} R\overset{\overset{\displaystyle O}{\|}}{C}OH$$

a 1° alcohol   *a carboxylic acid*

$$R\overset{\overset{\displaystyle OH}{|}}{C}HR \xrightarrow{[O]} R\overset{\overset{\displaystyle O}{\|}}{C}R$$

a 2° alcohol   *a ketone*

Some typical oxidizing agents used for these oxidations are:

1.   alkaline potassium permanganate: $KMnO_4 + {}^-OH$;
2.   hot, concentrated HNO_3;
3.   chromic acid: H_2CrO_4 (prepared *in situ* from CrO_3 or $Na_2Cr_2O_7$ with H_2SO_4).

Primary alcohols are oxidized first to aldehydes. Aldehydes are more easily oxidized than alcohols; therefore, the oxidation usually continues until the carboxylic acid (or, in alkaline solution, its anion) is formed.

$$CH_3CH_2CH_2OH + H_2CrO_4 \xrightarrow{H^+} \left[CH_3CH_2\overset{\overset{\displaystyle O}{\|}}{C}H \right] \longrightarrow CH_3CH_2CO_2H + Cr^{3+}$$

1-propanol   propanoic acid

cyclohexylmethanol

cyclohexane-
carboxylic acid

If the intermediate aldehyde has a low boiling point, it can be distilled from the reaction mixture before it is oxidized to the carboxylic acid. Yields of aldehydes by this method are usually low; therefore, this technique is of limited synthetic value. A better reagent for oxidizing a primary alcohol to an aldehyde is a chromic oxide–pyridine complex, a reagent that does not oxidize an aldehyde to a carboxylic acid.

1-propanol

pyridine

propanal

Secondary alcohols are oxidized to ketones in excellent yields by standard oxidizing agents. (Acidic conditions are usually used because ketones can be oxidized further in alkaline solution.)

$$\underset{\text{2-octanol}}{CH_3(CH_2)_5\overset{\overset{\displaystyle OH}{|}}{C}HCH_3} \xrightarrow[H^+]{H_2CrO_4} \underset{\substack{\text{2-octanone}\\96\%}}{CH_3(CH_2)_5\overset{\overset{\displaystyle O}{\|}}{C}CH_3}$$

cyclohexanol

cyclohexanone
95%

Tertiary alcohols are not oxidized under alkaline conditions. If the oxidation is attempted in acidic solution, the tertiary alcohol undergoes dehydration and then the alkene is oxidized. Alkene oxidation will be discussed in Chapter 9.

$$R_3COH \underset{H^+}{\overset{\underset{\displaystyle OH^-}{[O]}}{\nearrow\searrow}} \begin{array}{l} \text{no reaction} \\[1em] \text{alkenes} \xrightarrow{[O]} \text{alkene oxidation products} \end{array}$$

a 3° alcohol

The mechanisms of many oxidation reactions are not completely understood. Because of the possible variations of reduced forms of Mn(VII) from MnO_4^-, or

Cr(VI) from $CrO_4{}^{2-}$, the mechanisms have the potential for being quite complicated. In some reactions, the oxidizing agent probably forms an inorganic ester
with the alcohol and, by appropriate shifts of electrons and protons, the oxidized
product results.

$$R_2CHOH + HCrO_4{}^- \xrightarrow{-H_2O} \left[\begin{array}{c} \text{H} \quad\quad \ddot{\text{O}}: \\ | \quad\quad || \\ R_2C-\ddot{\text{O}}-CrO^- \\ || \\ O \end{array} \right] \longrightarrow R_2C=\ddot{O} + \begin{array}{c} H\ddot{O}: \\ | \\ CrO^- \\ || \\ O \end{array}$$

$\quad\quad$ Cr(VI) $\quad\quad\quad\quad\quad\quad$ *an inorganic ester* $\quad\quad\quad\quad\quad\quad\quad\quad$ Cr(IV)

STUDY PROBLEM

7.20. Predict the organic products of H_2CrO_4 oxidation of: (a) cyclopentanol, and
(b) benzyl alcohol. (*Hint*: The benzene ring is not affected by H_2CrO_4.)

D.   Oxidation of 1,2-diols

The **periodic acid oxidation** is a test for 1,2-diols and for 1,2- or α-hydroxy aldehydes
and ketones. A compound containing such a grouping is oxidized and cleaved by
periodic acid (HIO_4). In the case of a simple 1,2-diol, the products are two aldehydes or ketones.

— cleavage

$$\begin{array}{c} OH \vdots OH \\ | \quad\vdots\quad | \\ RCH \!\!\vdots\!\! CHR \end{array} \xrightarrow{HIO_4} \begin{array}{c} O \quad\quad O \\ || \quad\quad || \\ RCH + HCR + HIO_3 \end{array}$$

$$\begin{array}{c} O \quad OH \\ || \quad\quad | \\ RCCH_2CHR \end{array} \xrightarrow{HIO_4} \text{no reaction}$$

$$\begin{array}{c} CH_3O \quad OH \\ | \quad\quad\quad | \\ RCH-CHR \end{array} \xrightarrow{HIO_4} \text{no reaction}$$

The periodic acid reaction goes through a cyclic intermediate, a fact that
explains why isolated hydroxyl groups are not oxidized.

$$\begin{array}{c} R_2C-OH \\ | \\ R_2C-OH \end{array} + HIO_4 \xrightarrow{-H_2O} \left[\begin{array}{c} R_2C-O \quad\quad OH \\ \diagup \quad\quad\quad I=O \\ R_2C-O \quad\quad O \end{array} \right] \longrightarrow \begin{array}{c} R_2C=O \\ R_2C=O \end{array} + HIO_3$$

STUDY PROBLEM

7.21. Predict the products of the periodic acid oxidation of the following compound:

$$\begin{array}{c} OH \quad OH \quad OCH_3 \\ | \quad\quad | \quad\quad | \\ CH_3CH-CH-CH_2 \end{array}$$

In the case of an α-hydroxy aldehyde or ketone, the carbonyl group is oxidized to a carboxyl group, while the hydroxyl group is again oxidized to an aldehyde or ketone.

$$
\begin{array}{c}
to -CO_2H \qquad to -CHO \\[4pt]
\underset{RC\!-\!CHR}{\overset{O \;\; OH}{\underset{\|\;\;\;\;|}{}}} \xrightarrow{\;HIO_4\;} \underset{RCOH}{\overset{O}{\|}} + \underset{HCR}{\overset{O}{\|}} + HIO_3
\end{array}
$$

$$
\underset{HC\!-\!CHR}{\overset{O \;\; OH}{\underset{\|\;\;\;\;|}{}}} \xrightarrow{\;HIO_4\;} \underset{HCOH}{\overset{O}{\|}} + \underset{HCR}{\overset{O}{\|}} + HIO_3
$$

The periodic acid oxidation is used in carbohydrate chemistry for the analysis of sugars. Most sugars contain an aldehyde group plus hydroxyl groups on the other carbon atoms. In these cases, the oxidation of the interior hydroxyl groups proceeds further than it does with a simple 1,2-diol. For example, the products of the periodic acid oxidation of the sugar erythrose (2,3,4-trihydroxybutanal) are formaldehyde and formic acid in the molar ratio of 1:3.

$$
\underset{CH_2-CH-CH-CH}{\overset{OH \;\; OH \;\; OH \;\; O}{}} \xrightarrow{\;HIO_4\;} \underset{\text{formaldehyde}}{\overset{O}{HCH}} + \underset{\text{formic acid}}{3\,\overset{O}{HCOH}}
$$

oxidized to HCHO   *oxidized to*
                           HCO₂H

This oxidation of the CHOH groups in erythrose to formic acid can be rationalized by considering the reaction to be stepwise.

Step 1:

$$
\underset{CH_2-CH-CH-CH}{\overset{OH \;\; OH \;\; \overset{oxidized}{\overbrace{OH \;\; O}}}{}} \xrightarrow{\;HIO_4\;} \underset{CH_2-CH-CH}{\overset{OH \;\; OH \;\; O}{}} + \underset{\text{formic acid}}{HCOH}
$$

Step 2:

$$
\underset{CH_2-CH-CH}{\overset{OH \;\; \overset{oxidized}{\overbrace{OH \;\; O}}}{}} \xrightarrow{\;HIO_4\;} \underset{CH_2-CH}{\overset{OH \;\; O}{}} + \underset{\text{formic acid}}{HCOH}
$$

Step 3:

$$
\underset{CH_2-CH}{\overset{OH \;\; O}{}} \xrightarrow{\;HIO_4\;} \underset{\text{formaldehyde}}{\overset{O}{HCH}} + \underset{\text{formic acid}}{\overset{O}{HCOH}}
$$

STUDY PROBLEM

7.22.   Predict the periodic acid oxidation products of glucose (page 807).

SECTION 7.14.

Preparation of Ethers

A.   Diethyl ether

Under the proper conditions, the reaction of sulfuric acid with ethanol produces diethyl ether by way of the intermediate ethyl hydrogen sulfate. (This reaction was first reported in the 1500's! Until 1800, it was thought that diethyl ether contained sulfur in its structure when, in fact, sulfur was an impurity arising from the sulfuric acid.)

$$CH_3CH_2OH \xrightarrow{\text{H}_2\text{SO}_4} CH_3CH_2OSO_3H \xrightarrow{\text{CH}_3\text{CH}_2\text{OH}} CH_3CH_2OCH_2CH_3$$

ethanol   ethyl hydrogen sulfate   diethyl ether

     Diethyl ether is undoubtedly the most popular organic laboratory solvent. Of historical interest is its introduction in the 1800's, along with chloroform and nitrous oxide (N_2O, laughing gas), as a general anesthetic. Diethyl ether and nitrous oxide are still used as anesthetics. (Chloroform, however, has a narrow margin of safety and leads to liver damage, as do many of the chlorinated hydrocarbons.) Diethyl ether is volatile, its vapors are explosive, and it has a tendency to cause nausea. Despite these drawbacks, it is physiologically a relatively safe anesthetic. Other ethers that are used as anesthetics are methyl propyl ether ($CH_3OCH_2CH_2CH_3$) and ethyl vinyl ether ($CH_3CH_2OCH=CH_2$).

B.   Williamson ether synthesis

The **Williamson ether synthesis** is the most versatile laboratory procedure for synthesizing ethers. The Williamson synthesis is the S_N2 reaction of an alkyl halide with an alkoxide or phenoxide, a reaction we discussed in detail in Chapter 5.

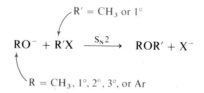

$$RO^- + R'X \xrightarrow{\text{S}_N2} ROR' + X^-$$

R' = CH$_3$ or 1°

R = CH$_3$, 1°, 2°, 3°, or Ar

Best yields are obtained when the alkyl halide is methyl or primary. (Secondary and tertiary alkyl halides lead to alkenes, while aryl and vinyl halides do not undergo S_N2 reactions.) The alkoxide that may be used in a Williamson synthesis has fewer limitations. It may be methyl, primary, secondary, tertiary, or allylic. Usually either the sodium or potassium alkoxide or phenoxide is used.

Synthesis of dialkyl ethers:

$$CH_3O^- + CH_3CH_2CH_2{-}Cl \xrightarrow{S_N2} CH_3OCH_2CH_2CH_3 + Cl^-$$

methoxide ion   1-chloropropane   methyl propyl ether

$$CH_3CH_2CH_2O^- + CH_3I \xrightarrow{S_N2} CH_3OCH_2CH_2CH_3 + I^-$$

propoxide ion   iodomethane   methyl propyl ether

$$(CH_3)_3CO^- + CH_3I \xrightarrow{S_N2} (CH_3)_3COCH_3 + I^-$$

t-butoxide ion   *t*-butyl methyl ether

but not $CH_3O^- + (CH_3)_3CI$ (Why not?)

Synthesis of an alkyl aryl ether:

$$\langle\!\!\bigcirc\!\!\rangle{-}O^- + CH_3CH_2Br \xrightarrow{S_N2} CH_3CH_2O{-}\langle\!\!\bigcirc\!\!\rangle + Br^-$$

phenoxide ion   bromoethane   ethyl phenyl ether

but not $CH_3CH_2O^- + \langle\!\!\bigcirc\!\!\rangle{-}Br$ (Why not?)

STUDY PROBLEM

7.23.   Show by equations the best method for preparing each of the following ethers by a Williamson synthesis:

(a) [naphthalene ring]–OCH$_3$ (b) $(CH_3)_2CHOCH_2CH_3$

SECTION 7.15.

Substitution Reactions of Ethers

Ethers are quite unreactive and behave more like alkanes than like organic compounds containing functional groups. Ethers undergo auto-oxidation (Section 6.6C) and combustion (which occurs readily), but they are not oxidized by laboratory reagents; nor do ethers undergo reduction, elimination, or reactions with bases. When they are heated with strong acids, ethers do undergo substitution reactions. For example, when heated with HI or HBr, an ether undergoes a substitution reaction to yield an alcohol and an alkyl halide. (Under these conditions, the alcohol can undergo further reaction with the HI or HBr to yield additional alkyl iodide or bromide.)

$$CH_3CH_2OCH_2CH_3 + HI \xrightarrow{heat} CH_3CH_2I + HOCH_2CH_3$$

diethyl ether   iodoethane   ethanol

$$\xrightarrow{HI} CH_3CH_2I$$

Ether cleavage with HI or HBr proceeds by almost the same path as the reaction of an alcohol with HX: protonation of the oxygen, followed by S_N1 or S_N2 reaction. (Protonation is necessary because RO^- is a poor leaving group, while ROH, like H_2O, is easily displaced.)

$$CH_3CH_2-\overset{..}{\underset{..}{O}}-CH_2CH_3 \xrightarrow{\text{ } H^+ \text{ }} \underset{protonated}{CH_3CH_2-\overset{\overset{H}{|}}{\underset{..}{O}}\overset{+}{-}CH_2CH_3} \xrightarrow{\text{ } I^- \text{ }}$$

$$\left[\overset{..}{\underset{..}{I}}\overset{\delta-}{---}CH_2\overset{\overset{CH_3}{|}}{---}\overset{\overset{H}{|}}{\underset{\delta+}{O}}-CH_2CH_3 \right] \longrightarrow CH_3CH_2\overset{..}{\underset{..}{I}} + H\overset{..}{\underset{..}{O}}CH_2CH_3$$

$$\underset{S_N2 \text{ transition state}}{}$$

An alkyl phenyl ether, such as anisole, yields the alkyl iodide and phenol (not iodobenzene) because the bond from the sp^2 carbon is stronger than the bond from the sp^3 carbon (Section 2.4F).

methyl phenyl ether   phenol   iodomethane
(anisole)

SAMPLE PROBLEM

Give the steps for the cleavage of diisopropyl ether by HI (an S_N1 reaction).

Solution:

$$(CH_3)_2CH-\overset{..}{\underset{..}{O}}-CH(CH_3)_2 \xrightarrow{\text{ } H^+ \text{ }} \underset{protonated}{(CH_3)_2CH\overset{\overset{H}{|}}{\underset{}{O}}\overset{+}{-}CH(CH_3)_2}$$

$$\xrightleftharpoons{-HOCH(CH_3)_2} [(CH_3)_2\overset{+}{C}H] \xrightarrow{\text{ } I^- \text{ }} (CH_3)_2CHI$$

$$\underset{a \text{ carbocation}}{}$$

STUDY PROBLEM

7.24.   The Zeisel procedure for estimating the number of methoxyl (CH_3O-) or ethoxyl (CH_3CH_2O-) groups in an alkyl aryl ether consists of ether cleavage with excess HI, followed by distillation of the volatile methyl or ethyl iodide from the reaction mixture. The amount of volatile iodide is determined by reaction with aqueous silver nitrate. Write the equations for the reactions that would occur in the determination of the methoxyl groups in 1,2,3-trimethoxybenzene:

SECTION 7.16.

Substitution Reactions of Epoxides

Epoxides are synthesized by the reaction of alkenes with peroxybenzoic acid ($C_6H_5CO_3H$). This reaction will be discussed in Section 9.14B.

Before we discuss the reactions of epoxides, let us consider a few aspects of the structure of an epoxide. An epoxide ring, like a cyclopropane ring, cannot have normal sp^3 bond angles of 109°; instead, the internuclear angles are 60°, a geometric requirement of the three-membered ring. The orbitals forming the ring bonds are incapable of maximum overlap; therefore, epoxide rings are strained. The polarity of the C—O bonds, along with this ring strain, contributes to the high reactivity of epoxides compared to the reactivity of other ethers.

polar and strained

Because an epoxide is cyclic, a substituted epoxide may be capable of geometric isomerism.

cis   *trans*

Epoxide rings may be part of fused-ring systems; in these cases, the epoxide must be *cis* on the other ring. (The required bond angles for the three-membered ring make the *trans*-configuration impossible.)

cyclopentene oxide

a fused-ring system

STUDY PROBLEM

7.25. Which of the following epoxides can exist as a pair of geometric isomers?

(a)   (b) CH_3CH—CH_2 (c) CH_3CH—$CHCH_3$

Opening of the strained three-membered ring results in a lower-energy, more stable product. The characteristic reaction of epoxides is ring opening, which can occur under either alkaline or acidic reaction conditions. These reactions of epoxides are referred to as **base-catalyzed** or **acid-catalyzed cleavage reactions,**

In base:

$$\underset{\substack{\text{oxirane}\\\text{(ethylene oxide)}}}{\overset{O}{CH_2{-}CH_2}} + H_2O \xrightarrow{OH^-} \underset{\substack{\text{1,2-ethanediol}\\\text{(ethylene glycol)}}}{\overset{OH\quad OH}{CH_2{-}CH_2}}$$

$$\overset{O}{CH_2{-}CH_2} + CH_3OH \xrightarrow{^-OCH_3} \underset{\substack{\text{2-methoxy-1-ethanol}}}{\overset{OH\quad OCH_3}{CH_2{-}CH_2}}$$

In acid:

$$\overset{O}{CH_2{-}CH_2} + H_2O \xrightarrow{H^+} \overset{OH\quad OH}{CH_2{-}CH_2}$$

$$\overset{O}{CH_2{-}CH_2} + HCl \longrightarrow \underset{\substack{\text{2-chloro-1-ethanol}\\\text{(ethylene chlorohydrin)}}}{\overset{OH\quad Cl}{CH_2{-}CH_2}}$$

A. Base-catalyzed cleavage

Epoxides undergo S_N2 attack by nucleophiles such as the hydroxide ion or alkoxides. The steps in the reactions of ethylene oxide with hydroxide ion (NaOH or KOH in water) and with methoxide ion (NaOCH$_3$ in methanol) follow:

$$\overset{:\ddot{O}:}{CH_2{-}CH_2} + :\ddot{O}H^- \xrightarrow{S_N2} \underset{\substack{\text{abstracting a}\\\text{proton from } H_2O}}{\overset{:\ddot{O}:^-}{CH_2{-}CH_2OH}} \longrightarrow H{-}\ddot{O}H$$

$$\overset{:\ddot{O}H}{CH_2CH_2OH} + :\ddot{O}H^-$$
$$\underset{\text{1,2-ethanediol}}{}$$

$$\overset{O}{CH_2{-}CH_2} + {^-}OCH_3 \xrightarrow{S_N2} \overset{O^-}{CH_2{-}CH_2OCH_3} \underset{\substack{\text{excess}\\CH_3OH}}{\rightleftharpoons} \underset{\text{2-methoxy-1-ethanol}}{\overset{OH}{CH_2CH_2OCH_3}} + CH_3O^-$$

In base-catalyzed cleavage, the nucleophile attacks the *less hindered carbon*, just as we would expect from an S_N2 attack ($1° > 2° > 3°$).

$$\underset{\substack{CH_3}}{\overset{O}{CH_3{-}C{-}CH_2}} \qquad \text{\textit{in base, attack at}}\\ \text{\textit{less hindered carbon}}$$
$$\overset{\downarrow}{\underset{CH_3\; {-}Nu^-}{}}$$

$$\underset{\text{2,2-dimethyloxirane}}{\overset{O}{(CH_3)_2C{-}CH_2}} + CH_3OH \xrightarrow{^-OCH_3} \underset{\text{1-methoxy-2-methyl-2-propanol}}{\overset{OH}{(CH_3)_2CCH_2OCH_3}}$$

A Grignard reagent contains a partially negative carbon atom and attacks an epoxide ring in the same manner as other nucleophiles. The product is the magnesium salt of an alcohol; the alcohol may be obtained by hydrolysis. The reaction of a Grignard reagent with ethylene oxide is a method by which the hydrocarbon chain of the Grignard reagent may be *extended by two carbons*.

From methylmagnesium iodide to 1-propanol:

$$\overset{\delta-}{CH_3}-MgI + CH_2-CH_2 \longrightarrow CH_3CH_2CH_2\overset{O^- \ ^+MgI}{|} \xrightarrow{H_2O, H^+} CH_3CH_2CH_2OH$$

From cyclohexylmagnesium bromide to 2-cyclohexyl-1-ethanol:

MgBr + CH_2-CH_2 (epoxide) $\longrightarrow$ —$CH_2CH_2\overset{O^- \ ^+MgBr}{|}$ $\xrightarrow{H_2O, H^+}$

—CH_2CH_2OH

SAMPLE PROBLEM

Suggest a synthesis for 2-phenyl-1-ethanol starting with bromobenzene.

Solution:

1. Convert bromobenzene to a Grignard reagent.

$$C_6H_5Br + Mg \xrightarrow{ether} C_6H_5MgBr$$

2. Treat the Grignard reagent with ethylene oxide.

$$C_6H_5MgBr + CH_2-CH_2 \overset{O}{\triangle} \longrightarrow C_6H_5CH_2CH_2OMgBr$$

3. Hydrolyze the magnesium alkoxide.

$$C_6H_5CH_2CH_2OMgBr + H^+ \xrightarrow{H_2O} C_6H_5CH_2CH_2OH + Mg^{2+} + Br^-$$

B.   Acid-catalyzed cleavage

In acidic solution, the oxygen of an epoxide is protonated. A protonated epoxide can be attacked by weak nucleophiles such as water, alcohols, or halide ions.

General:

$$CH_2-CH_2 \overset{\cdot\ddot{O}\cdot}{\triangle} \underset{}{\overset{H^+}{\rightleftharpoons}} CH_2-CH_2 \overset{\cdot\overset{H}{\ddot{O}^+}}{\triangle} \underset{-Nu^-}{\longrightarrow} CH_2-CH_2 \overset{:\ddot{O}H}{|} \ \underset{Nu}{|}$$

protonated

$$CH_2-CH_2 \overset{O}{\triangle} \overset{H^+}{\rightleftharpoons} CH_2-CH_2 \overset{\overset{H}{O^+}}{\triangle}$$

$$\xrightarrow{H_2O} \underset{\overset{+}{O}-H}{\overset{OH}{\underset{|}{CH_2-CH_2}}} \overset{-H^+}{\rightleftharpoons} \overset{OH}{\underset{}{CH_2CH_2OH}}$$
1,2-ethanediol

$$\xrightarrow{CH_3OH} \underset{HOCH_3}{\overset{OH}{\underset{|}{CH_2-CH_2}}} \overset{-H^+}{\rightleftharpoons} \overset{OH}{\underset{}{CH_2CH_2OCH_3}}$$
2-methoxy-1-ethanol

$$\xrightarrow{Cl^-} \overset{OH}{\underset{}{CH_2CH_2Cl}}$$
2-chloro-1-ethanol

As contrasted with base-catalyzed cleavage, attack in acid occurs at the more hindered carbon.

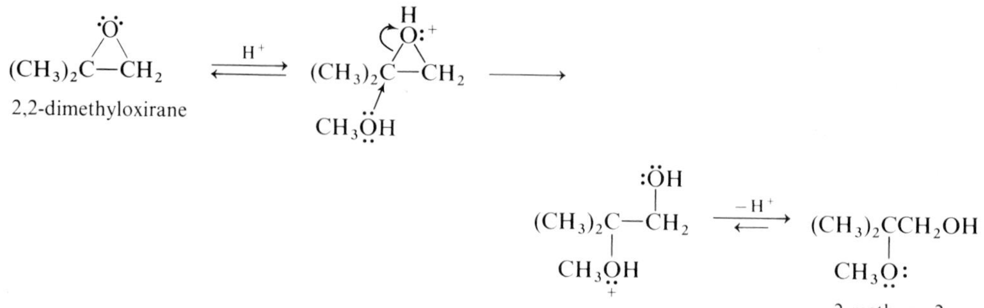

$(CH_3)_2C\overset{\ddot{\ddot{O}}}{\triangle}CH_2$ 2,2-dimethyloxirane $\overset{H^+}{\rightleftharpoons}$ $(CH_3)_2\underset{CH_3\overset{..}{O}H}{C}\overset{\overset{H}{\overset{|}{C}\ddot{O}:^+}}{CH_2}$ $\longrightarrow$

$(CH_3)_2\underset{CH_3\overset{+}{\underset{..}{O}}H}{C}\overset{:\ddot{O}H}{\underset{|}{CH_2}}$ $\overset{-H^+}{\longleftarrow}$ $(CH_3)_2\underset{CH_3\overset{..}{\underset{..}{O}:}}{CCH_2OH}$

2-methoxy-2-
methyl-1-propanol

We must conclude that the protonated epoxide has a fair amount of carbo-cation character. If this is the case, there is a partial positive charge on the carbon with the greater number of alkyl groups (carbocation stability: $3° > 2° > 1°$). The subsequent nucleophilic attack is favored at the more positive carbon even though this carbon is more hindered.

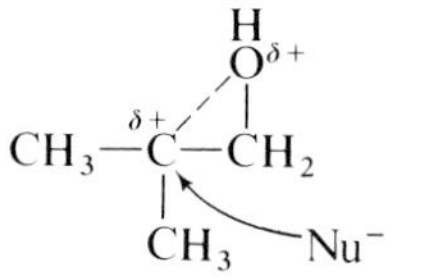

$CH_3\overset{\delta+}{\underset{\underset{Nu^-}{\overset{|}{CH_3}}}{C}}\overset{\overset{H}{\overset{|}{O}^{\delta+}}}{CH_2}$

In acid, attack occurs at the more hindered carbon because it has a greater + charge.

How do we know that a true carbocation is not formed? When the product of an epoxide-cleavage reaction is capable of geometric isomerism, only the *trans*-product is observed. If the reaction went through a true carbocation intermediate, we would observe both *cis*- and *trans*-products.

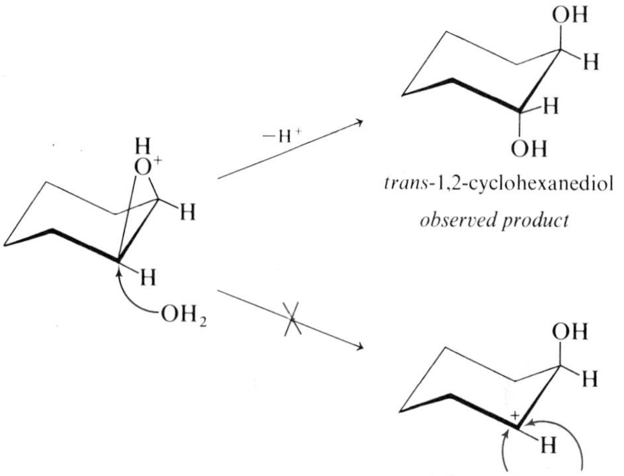

trans-1,2-cyclohexanediol

observed product

attack from either side would give both cis and trans, which does not occur

SAMPLE PROBLEM

Predict the products:

(a) $\underset{\displaystyle CH_3\overset{\displaystyle O}{\overset{\displaystyle \triangle}{CH}}-CH_2}{} + CH_3OH \xrightarrow{CH_3O^-}$

(b) $\underset{\displaystyle CH_3\overset{\displaystyle O}{\overset{\displaystyle \triangle}{CH}}-CH_2}{} + CH_3OH \xrightarrow{H^+}$

Solution:

(a) In base, attack occurs at the less hindered carbon: $CH_3\overset{\displaystyle OH}{\underset{\displaystyle |}{CH}}CH_2OCH_3.$

(b) In acid, attack occurs at the more hindered carbon: $CH_3\underset{\displaystyle \underset{\displaystyle OCH_3}{|}}{CH}CH_2OH.$

STUDY PROBLEM

7.26. Predict the products of reaction of the following epoxide with   (a) sodium methoxide in methanol, and   (b) aqueous HCl:

SECTION 7.17.

Thiols and Sulfides

Sulfur is just below oxygen in the periodic table. Many organic compounds containing oxygen have sulfur analogs. The sulfur analog of an alcohol is called an **alkanethiol,** or simply **thiol,** or by its older name **mercaptan.** The —SH group is called a **thiol group** or a **sulfhydryl group.**

$$CH_3SH \qquad CH_3CH_2\overset{\displaystyle SH}{\underset{\displaystyle |}{C}}HCH_3$$

methanethiol   2-butanethiol

The most characteristic property of a thiol is its odor! The human nose is very sensitive to these compounds and can detect their presence at levels of about 0.02 parts thiol to one billion parts air. The odor of a skunk's spray is due primarily to a few simple thiols.

We may deduce some of the properties of thiols as compared to alcohols by comparing hydrogen sulfide and water. For example, H_2S ($pK_a = 7.04$) is a stronger acid than water ($pK_a = 15.7$). Thiols ($pK_a = \sim 8$) are also substantially stronger acids than alcohols ($pK_a = \sim 16$).

$$CH_3CH_2SH + \quad OH^- \quad \rightleftharpoons \quad CH_3CH_2S^- + H_2O$$

stronger acid   *stronger base*
than H₂O   *than RS⁻*

Sulfur is less electronegative than oxygen, and its outer electrons are more diffuse; therefore, sulfur atoms form weaker hydrogen bonds than oxygen atoms. For this reason, H_2S (bp $-61°$) is more volatile than water (bp 100°), and thiols are more volatile than their analogous alcohols.

Treatment of an alkyl halide with the hydrogen sulfide ion (HS^-) leads to thiols. Good yields are obtained only if an excess of inorganic hydrogen sulfide is used because the resulting thiol (which is acidic) can ionize to form the RS^- ion, also a good nucleophile. The subsequent reaction of RS^- with the alkyl halide yields the **sulfide**, R_2S.

$$CH_3I + SH^- \longrightarrow CH_3SH + I^-$$

methanethiol

$$CH_3SH \underset{}{\overset{-H^+}{\rightleftharpoons}} CH_3S^- \xrightarrow[-I^-]{CH_3I} CH_3SCH_3$$

dimethyl sulfide

When a thiol is treated with a mild oxidizing agent (such as I_2), it undergoes coupling to form a **disulfide**, a compound containing the S—S linkage. This reaction can be reversed by treatment of the disulfide with a reducing agent (such as lithium metal in liquid NH_3).

$$2\ CH_3CH_2SH \underset{[H]}{\overset{[O]}{\rightleftharpoons}} CH_3CH_2S{-}SCH_2CH_3$$

ethanethiol   diethyl disulfide

This disulfide link is an important structural feature of some proteins (Section 19.1B). The disulfide bond helps hold protein chains together in their proper shapes. The locations of the disulfide bonds determine, for example, whether hair (a protein) is curly or straight.

A sulfide can be oxidized to a **sulfoxide** or a **sulfone**, depending upon the reaction conditions. For example, 30% hydrogen peroxide in the presence of an acid oxidizes a sulfide to a sulfoxide at 25° or to a sulfone at 100°.

$$CH_3SCH_3 + H_2O_2$$

dimethyl sulfide

$H^+, 25°$ →

$$\overset{\displaystyle O}{\overset{\displaystyle \|}{CH_3SCH_3}}$$

dimethyl sulfoxide

$H^+, 100°$ →

$$\overset{\displaystyle O}{\overset{\displaystyle \|}{\underset{\displaystyle \|}{\underset{\displaystyle O}{CH_3SCH_3}}}}$$

dimethyl sulfone

Dimethyl sulfoxide (DMSO) is prepared industrially by the air oxidation of dimethyl sulfide; it is also a by-product of the paper industry. DMSO is a unique and versatile solvent. It has a high dielectric constant (49 D), but does not form hydrogen bonds in the pure state. (Why not?) It is a powerful solvent for both inorganic ions and organic compounds. Reactants often have enhanced reactivities in DMSO, compared to alcohol solvents. DMSO readily penetrates the skin and has been used to promote the dermal absorption of drugs; however, DMSO can also cause the absorption of dirt and poisons. A common complaint of people working with DMSO is that, when it is spilled on their hands, they can taste it!

STUDY PROBLEM

7.27.   Explain   (a) why dimethyl sulfoxide is miscible with water, and   (b) why this compound can act as a weak acid.

SECTION 7.18.

Use of Alcohols and Ethers in Synthesis

Except for epoxides, ethers are not very useful for the synthesis of other organic compounds. Alcohols, however, are versatile starting materials for the preparation of alkyl halides, alkenes, carbonyl compounds, and ethers. The types of compounds that can be obtained from alcohols and epoxides are shown in Tables 7.7 and 7.8. From the reactions presented in this chapter, together with those presented previously, a large number of types of compounds can be prepared from a variety of starting materials.

TABLE 7.7. Types of compounds that can be obtained from alcohols

Reaction	Principal product		Section reference
Substitution:			
$ROH + HX$	$\longrightarrow$ RX	**alkyl halide**	7.6
$ROH + PX_3$ or $SOCl_2$	$\longrightarrow$ RX	**alkyl halide**	7.7
Elimination:			
$\underset{R_2\overset{\displaystyle |}{\text{C}}CHR_2}{\overset{\displaystyle OH}{}} + H_2SO_4$	$\xrightarrow{\text{heat}}$ $R_2C{=}CR_2$	**alkene**	7.8
Rearrangement:			
$\underset{R_2\overset{\displaystyle |}{\text{C}}{-}\overset{\displaystyle |}{\text{C}}R'_2}{\overset{\displaystyle OH\,OH}{}} + H_2SO_4$	$\longrightarrow$ $R_2\overset{\displaystyle O}{\overset{\displaystyle \|}{C}}{-}CR'$ ketone with R' below	**ketone**	7.8

(*continued*)

TABLE 7.7. (continued)

Reaction	Principal product		Section reference
Alkoxide formation:[a]			
$ROH + Na$ ⟶	$RO^-\ Na^+$	**alkoxide**	7.10
Esterification:			
$ROH + R'CO_2H \xrightarrow{H^+}$	$R'CO_2R$	**ester**	7.11, 12.9, 13.5B
$ROH + TsCl$ ⟶	$ROTs$	**inorganic ester**[b]	7.12
Oxidation:			
$RCH_2OH + CrO_3 \cdot 2\,pyridine$ ⟶	$RCHO$	**aldehyde**	7.13
$RCH_2OH + [O]$ ⟶	RCO_2H	**carboxylic acid**	7.13
$R_2CHOH + [O]$ ⟶	$R_2C{=}O$	**ketone**	7.13
$\overset{\displaystyle OH\quad OH}{\underset{\displaystyle\quad\mid\quad\ \ \mid}{RCH-CHR'}} + HIO_4$ ⟶	$RCHO + R'CHO$	**aldehyde**[c]	7.13

[a] Phenoxides may be prepared from phenols and aqueous NaOH (Section 7.10). Both RO^- and ArO^- are used (with methyl and 1° RX) to synthesize ethers.

[b] For the synthesis of other inorganic esters, see Section 7.12.

[c] Other compounds may also be obtained (see Section 7.13D).

TABLE 7.8. Types of compounds that can be obtained from epoxides

Reaction	Principal product		Section reference
$R_2C\overset{O}{-}CR_2 + H_2O \xrightarrow{H^+\ or\ OH^-}$	$R_2C\overset{OH}{\underset{OH}{-}}CR_2$	**1,2-diol**	7.16
$R_2C\overset{O}{-}CR_2 + R'OH \xrightarrow{H^+\ or\ R'O^-}$	$R_2C\overset{OH}{\underset{OR'}{-}}CR_2$	**1,2-alkoxy alcohol**	7.16
$R_2C\overset{O}{-}CR_2 + HX$ ⟶	$R_2C\overset{OH}{\underset{X}{-}}CR_2$	**1,2-halohydrin**	7.16
$CH_2\overset{O}{-}CH_2 \xrightarrow[(2)\ H_2O,\ H^+]{(1)\ RMgX}$	RCH_2CH_2OH	**1° alcohol**	7.16

SAMPLE PROBLEM

Suggest a synthesis for 3-methyl-3-hexanol from alcohols of four carbons or less.

Solution:

1.   Write the stucture of 3-methyl-3-hexanol:

$$\underset{\displaystyle \underset{CH_3}{|}}{\overset{\displaystyle \overset{OH}{|}}{CH_3CH_2CH_2CCH_2CH_3}}$$

2.   Decide on the reactants. (Remember to approach a synthesis problem backwards.) This is a 3° alcohol, so it may be prepared by a Grignard reaction.

$$\overset{\displaystyle \overset{O}{\|}}{CH_3CCH_2CH_3} \xrightarrow[\text{(2) } H_2O,\, H^+]{\text{(1) } CH_3CH_2CH_2MgBr} \quad \text{product}$$

3.   The organic reactants in the preceding step may be obtained from alcohols.

$$\overset{\displaystyle \overset{OH}{|}}{CH_3CHCH_2CH_3} \xrightarrow{H_2CrO_4} \overset{\displaystyle \overset{O}{\|}}{CH_3CCH_2CH_3}$$

$$CH_3CH_2CH_2OH \xrightarrow{HBr} CH_3CH_2CH_2Br \xrightarrow[\text{diethyl ether}]{Mg} CH_3CH_2CH_2MgBr$$

4.   Write the entire synthetic sequence.

$$CH_3CH_2CH_2OH \xrightarrow{HBr} CH_3CH_2CH_2Br \xrightarrow[\text{diethyl ether}]{Mg} CH_3CH_2CH_2MgBr$$

$$\overset{\displaystyle \overset{OH}{|}}{CH_3CHCH_2CH_3} \xrightarrow{H_2CrO_4} \overset{\displaystyle \overset{O}{\|}}{CH_3CCH_2CH_3} \xrightarrow[\text{(2) } H_2O,\, H^+]{\text{(1) } CH_3CH_2CH_2MgBr} \quad \text{product}$$

STUDY PROBLEMS

7.28.   Suggest a method for preparing ethyl acetate ($CH_3CO_2CH_2CH_3$) from ethanol and *no other organic reagent.*

7.29.   Suggest a method of converting cyclopentanol to 1-cyclopentyl-1-ethanol. (Other organic reagents may be used.)

7.30.   Show how you would prepare 4-methylpentanoic acid from alcohols or epoxides of four carbons or less.

Summary

Alcohols and **ethers** each contain an sp^3 oxygen with two filled orbitals. These compounds are *polar*. Alcohols can undergo *hydrogen bonding* with themselves, and either type of compound can undergo hydrogen bonding with water or any other compound containing NH or OH.

Alcohols may be prepared in the laboratory by the S_N2 reaction of a primary alkyl halide with OH^-; by the reaction of a Grignard reagent with a carbonyl compound or with an epoxide; or by the hydration of alkenes. These reactions are summarized in Table 7.5, page 264.

Alcohols undergo *substitution reactions* with HX ($1°$ alcohols, S_N2; $2°$ and $3°$ alcohols, S_N1). Alcohols undergo *elimination reactions* with H_2SO_4 or other strong acids. In either case, the order of reactivity of alcohols is $3° > 2° > 1°$. Alkyl halides may be prepared from alcohols without rearrangement with $SOCl_2$ or PCl_3.

Substitution: $ROH + HX \longrightarrow RX + H_2O$

Elimination: $R_2CH\overset{\overset{\displaystyle OH}{|}}{-}CR_2 \xrightarrow[\text{heat}]{H_2SO_4} R_2C{=}CR_2 + H_2O$

An **alkoxide** (RO^-) can be prepared from an alcohol and a strong base or an alkali metal ($RMgX$, NH_2^-, Na, K). A **phenoxide** (ArO^-) can be formed by the reaction of a phenol and an alkali metal hydroxide (NaOH, KOH).

$$ROH + Na \longrightarrow RO^- \; Na^+ + \tfrac{1}{2} H_2$$

$$ArOH + NaOH \rightleftharpoons ArO^- \; Na^+ + H_2O$$

An alcohol can undergo reaction with an acid or an acid derivative to yield an **ester of a carboxylic acid** (RCO_2R) or an **inorganic ester**, such as $RONO_2$ or $ROSO_3H$. (These reactions are shown in Table 7.7, page 301.) Most inorganic ester groups are good leaving groups.

The **oxidation** of a primary alcohol results in a carboxylic acid (or aldehyde), while the oxidation of a secondary alcohol yields a ketone (see Table 7.7).

Ethers may be prepared by the reaction of an alkoxide or phenoxide (good nucleophiles) with a methyl or primary alkyl halide.

$$RO^- + R'X \longrightarrow ROR' + X^-$$

Ethers do not undergo elimination reactions, but can undergo *substitution reactions* when heated with HBr or HI.

$$ROR + HI \xrightarrow{\text{heat}} RI + ROH \xrightarrow{HI} RI$$

Epoxides are more reactive than other ethers and undergo S_N2 ring opening with nucleophiles (in either alkaline or acidic solution) or with Grignard reagents. These reactions are summarized in Table 7.8, page 302.

Thiols may be used to prepare **sulfides** or **disulfides**. **Sulfides** may be oxidized to **sulfoxides** or **sulfones**.

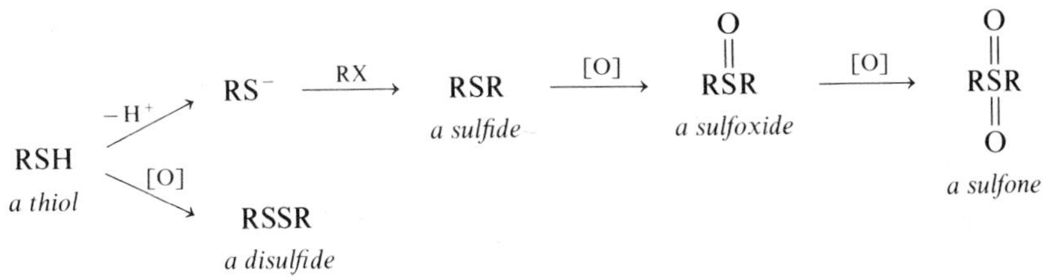

RSH
a thiol
−H⁺
RS⁻
RX
RSR
a sulfide
[O]
RSR
a sulfoxide
[O]
RSR
a sulfone
[O]
RSSR
a disulfide

STUDY PROBLEMS

7.31. Name all the oxygen-containing functional groups in the following compounds:

(a) the antihistamine *Benadryl*, $(CH_3)_2NCH_2CH_2OCH(C_6H_5)_2$
(b) the drug morphine (p. 763)
(c) tetrahydrocannabinol (the principal active ingredient in marijuana):

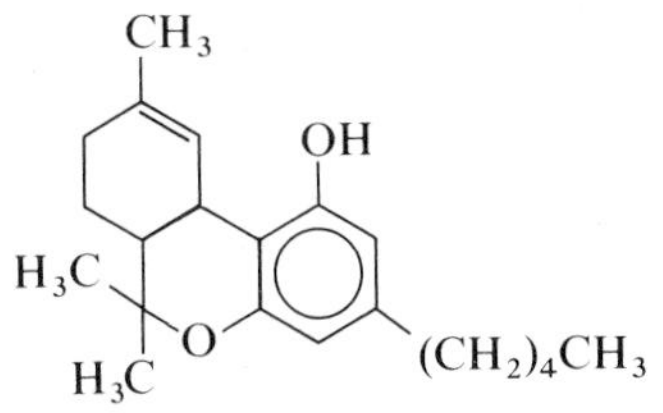

CH3
OH
H3C
O
H3C
(CH2)4CH3

7.32. Which compound in each group would be most water-soluble? Explain.

(a) *n*-butyl alcohol, isobutyl alcohol, *t*-butyl alcohol
(b) diethyl ether, tetrahydrofuran
(c) 1-bromooctane, 1-octene, 1-octanol
(d) pentane, 1-pentanol, 3-pentanol, 1,5-pentanediol

7.33. Write an acceptable name for each of the following alcohols, and classify each as 1°, 2°, or 3°.

(a) $(CH_3)_2CHCH_2CH_2OH$

(b) 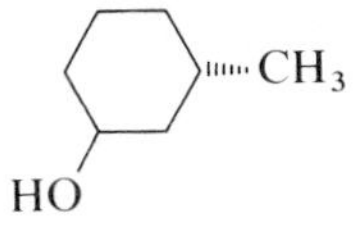

CH3
HO

(c) 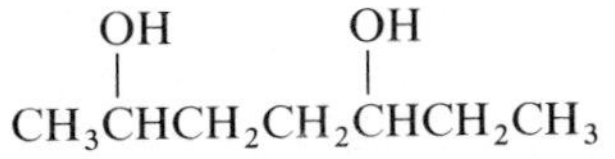

OH
OH
CH3CHCH2CH2CHCH2CH3

(d) 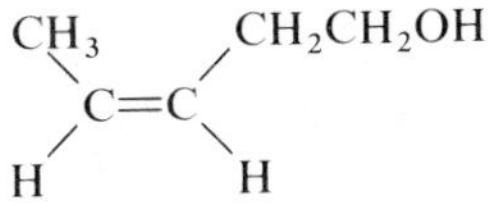

CH3
CH2CH2OH
C=C
H
H

7.34. Provide suitable names for the following structures:

(a) $CH_3OCH_2CH_2OCH_3$

(b) $(CH_3)_2CHOCH_2CH_2CH_3$

(c) 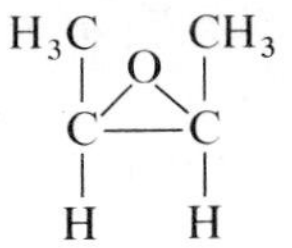

H3C
CH3
O
C
C
H
H

(d) 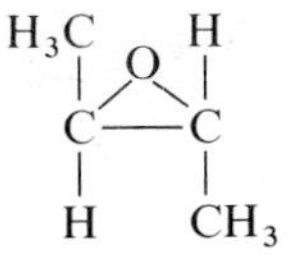

H3C
H
O
C
C
H
CH3

7.35. Complete the following equations, giving the major organic products:

(a) $(CH_3)_2CHCH_2CH_2Cl + OH^- \longrightarrow$

(b) $(CH_3)_3CI + OH^- \longrightarrow$

(c) 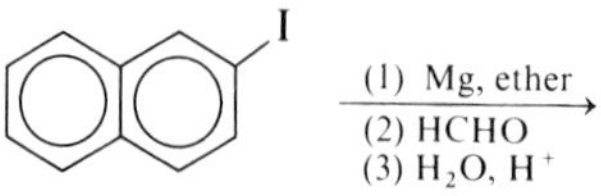 $\xrightarrow[\text{(3) H}_2\text{O, H}^+]{\begin{array}{l}\text{(1) Mg, ether}\\\text{(2) HCHO}\end{array}}$

(d) 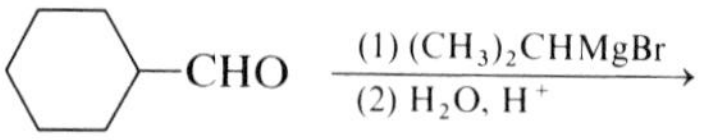 —CHO $\xrightarrow[\text{(2) H}_2\text{O, H}^+]{\text{(1) (CH}_3)_2\text{CHMgBr}}$

7.36. Write equations that show how each of the following alcohols can be prepared (1) by the reduction of a carbonyl compound, and (2) by a Grignard reaction:

(a) 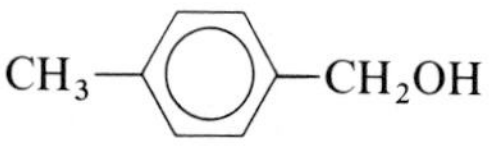 CH_3——CH_2OH

(b) $(CH_3)_2CHCH_2\overset{\displaystyle OH}{\overset{\displaystyle |}{C}}HCH_3$

(c) 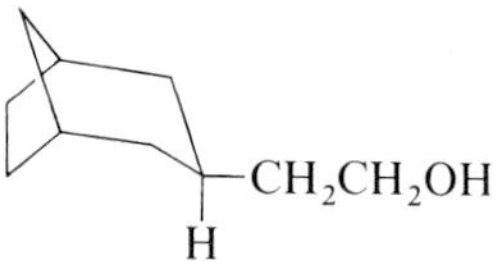 —CH_2CH_2OH

7.37. Write the equation for the reaction that occurs when each of the following alcohols is treated with HI. (Show the mechanisms.)

(a) 2-propanol   (b) cyclohexanol   (c) 1-butanol

7.38. Complete the following equations:

(a) $(CH_3)_2CHCH_2CH_2OH + HBr \longrightarrow$

(b)  —$OH + HCl \xrightarrow{\text{ZnCl}_2}$

(c) 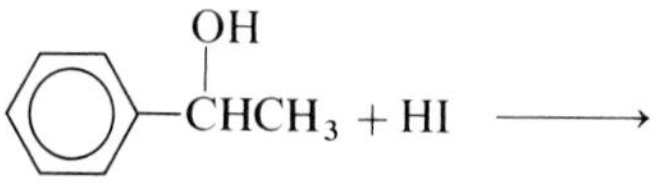 —$\overset{\displaystyle OH}{\overset{\displaystyle |}{C}}HCH_3 + HI \longrightarrow$

(d) 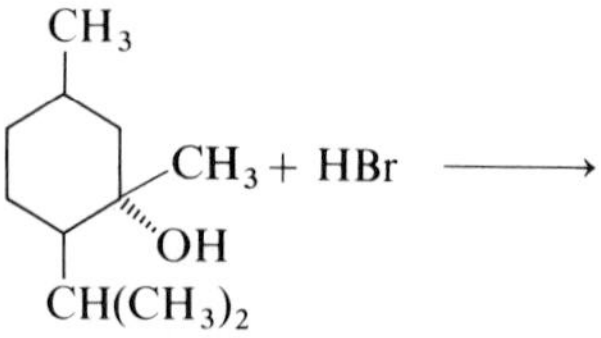 $CH_3 + HBr \longrightarrow$

7.39. Complete the following equations for substitution reactions. Which reaction of the three has the fastest rate of reaction? Which has the slowest rate? Explain.

(a) 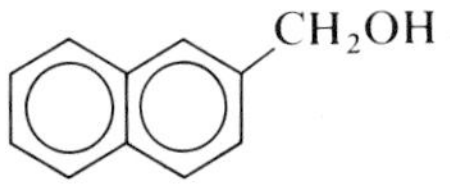   [structure: naphthalene-type ring with CH$_2$OH substituent]   + HBr ⟶

(b) $CH_3CH_2CH_2OH$ + HBr ⟶

(c) 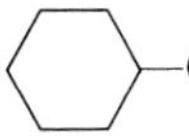   [cyclohexane ring]—OH + HBr ⟶

7.40. Sketch energy diagrams for the reactions in Problems 7.39 (b) and (c). (Use the *protonated* alcohol as the organic reactant.)

7.41. Give the structure of the expected rearranged halide from each of the following reactions:

(a)   3,3-dimethyl-2-butanol + HCl $\xrightarrow{\text{ZnCl}_2}$
(b)   2,2-diphenyl-1-ethanol + HI ⟶

7.42. In Problem 7.41, what alkenes would you expect as by-products in each case?

7.43. What would be the product when (*S*)-2-hexanol is treated with: (a) PCl$_3$;   (b) SOCl$_2$ in ether;   (c) HCl + ZnCl$_2$;   (d) SOCl$_2$ in pyridine?

7.44. Predict the major organic products of dehydration reactions of the following alcohols. Include the stereochemistry of the product if applicable.   (a) 2-hexanol; (b) 1-phenyl-2-propanol;   (c) 1-butanol;   (d) 2-butanol;   (e) 4-methyl-1,4-pentanediol (elimination of 1.0 mole H$_2$O only).

7.45. Upon dehydration, 2,2-dimethyl-1-cyclohexanol yields two alkenes, both the result of rearrangements. One contains a five-membered ring. What are these alkenes?

7.46. Predict the major products when the following diols are treated with sulfuric acid:

(a) 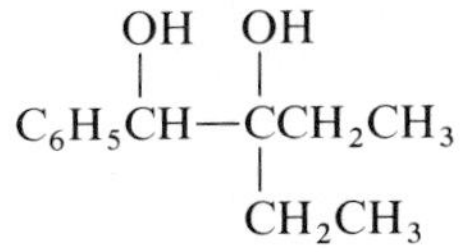 $C_6H_5\overset{\displaystyle OH}{\underset{\displaystyle CH_2CH_3}{CH}}-\overset{\displaystyle OH}{C}CH_2CH_3$

(b) 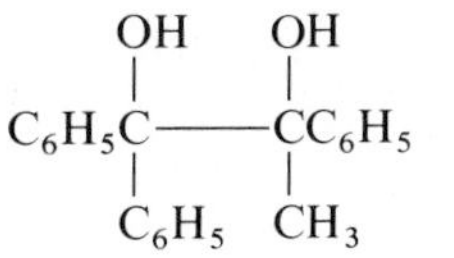 $C_6H_5\overset{\displaystyle OH}{\underset{\displaystyle C_6H_5}{C}}-\overset{\displaystyle OH}{\underset{\displaystyle CH_3}{C}}C_6H_5$

7.47. Complete the following equations for acid–base reactions:

(a) $CH_3CH_2OH + H^+$ ⇌

(b) 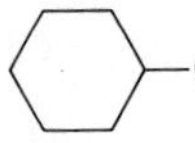   [cyclohexane ring]—OH + OH$^-$ ⇌

(c) 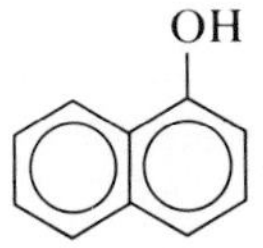   [naphthalene-type ring with OH substituent] + OH$^-$ ⇌

(d) $(CH_3)_3CO^- + H_2O$ ⇌

7.48. What will be the products of reaction (if any) of the following compounds with sodium ethoxide?   (a) 2-bromopropane;   (b) water;   (c) acetic acid;   (d) phenol.

7.49. If 1-butanol is added to each of the following reagents, what would be the expected major products? (a) methylmagnesium iodide; (b) phenyllithium; (c) sodium phenoxide; (d) sodium acetate; (e) HBr; (f) potassium metal.

7.50. Which of the following compounds are resonance-stabilized? Write the important resonance structures.

(a) 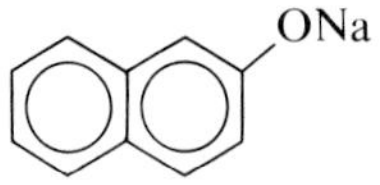 [structure with ONa group on fused bicyclic aromatic]   (b) CH_2=$CHCH_2OK$ (c) $(CH_3)_3COK$

7.51. Give the major organic product of the reaction (if any) of (R)-2-heptanol with each of the following reagents: (a) H_2CrO_4; (b) HI; (c) Li metal; (d) hot, conc. H_2SO_4; (e) CH_3MgI; (f) aqueous NaCl; (g) aqueous NaOH; (h) $SOCl_2$ in ether.

7.52. Write equations for the following reactions: (a) (S)-2-pentanol and tosyl chloride; (b) (R)-2-butanol and H_2SO_4 at 180°; (c) (R)-2-butanol and $ClSO_3H$; (d) $C_6H_5CO_2^-$ Na^+ and diethyl sulfate; (e) (R)-2-butyl tosylate and ethanol under S_N1 conditions.

7.53. Suggest a method for the preparation of each of the following compounds from an alcohol, a diol, or a phenol:

(a) $CH_2(CH_2ONO_2)_2$

(b) 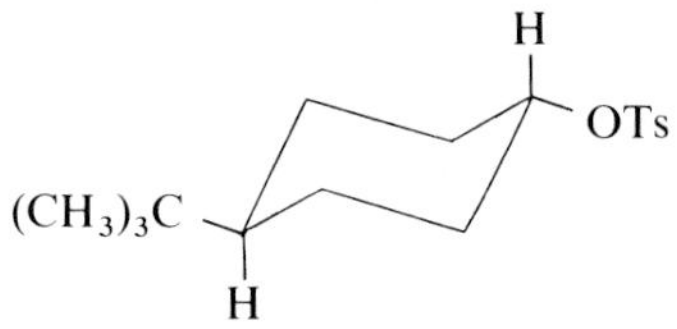 [cyclohexane ring structure with $(CH_3)_3C$, H, and OTs substituents]

(c) 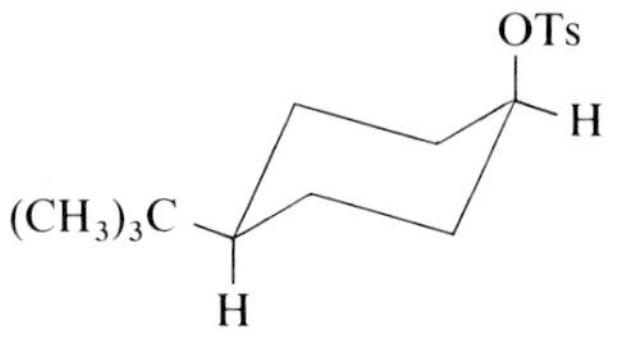 [cyclohexane ring structure with OTs, H, $(CH_3)_3C$, and H substituents]

(d) 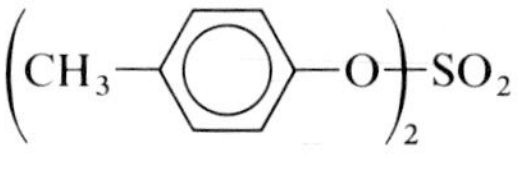 $\left(CH_3-\bigcirc-O \right)_2 SO_2$

7.54. In which one of each of the following pairs of compounds is carbon in the higher oxidation state?

(a) $CH_3CH_2C{\equiv}CH$ or $CH_3CH_2CO_2H$

(b) CH_3CH_2CHO or $CH_3CH_2CH_2OH$

(c) 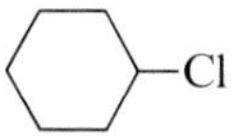 [cyclohexane ring]—Cl   or   [cyclohexane ring with two Cl substituents]

7.55. Suggest an alcohol and an oxidizing agent for preparing: (a) 3-methyl-1-cyclohexanone; (b) 2-butanone; (c) butanal ($CH_3CH_2CH_2CHO$); (d) butanoic acid.

7.56. Three compounds (A, B, and C) were subjected to oxidation with HIO_4. The following products were obtained from each reaction, respectively. What are the structures of A, B, and C?

(a) $CH_3COH + HCCH_2CH_3$ (b) $CH_3CCH_2CH_2CH_2CH$

(c) $HCH + CH_3CCH_3$

7.57. (a) There are two ways to prepare $C_6H_5CH_2OCH_2CH_3$ by a Williamson synthesis. Write the equations.

(b) Write an equation for the preparation of $C_6H_5OCH_2CH_2CH_3$.

7.58. Predict the principal organic products:

(a) 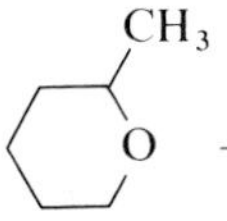 [structure] + one equivalent HI $\xrightarrow{heat}$

(b) 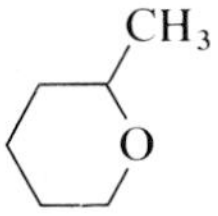 [structure] + excess HI $\xrightarrow{heat}$

7.59. Predict the major organic products when propylene oxide (methyloxirane) is treated with the following reagents: (a) NH_3; (b) 1-pentanol and HCl; (c) $CH_2=CHCH_2MgBr$, then dilute HCl; (d) phenyllithium, then dilute HCl; (e) a solution of phenol and NaOH.

7.60. What would be the product of the following reaction? (Do not forget stereochemistry.)

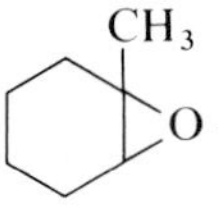

[structure] O + HBr $\longrightarrow$

7.61. Complete the following equations, giving only the major organic products:

(a) $CH_3CH_2SCH_2CH_3 + H_2O_2 \xrightarrow[heat]{H^+}$

(b) 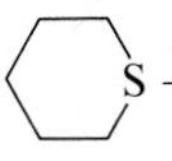 [structure] S + $H_2O_2 \xrightarrow[room\ temperature]{H^+}$

(c) 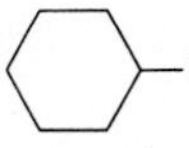 [structure] —SH + NaOH $\longrightarrow$

7.62. A chemist had at his disposal a mixture of HCl + $ZnCl_2$, called the **Lucas reagent**, and a solution of methylmagnesium iodide. He found that a compound of unknown structure underwent rapid reaction with the Grignard reagent and slow reaction when heated with the Lucas reagent. Which compound was it?

(a) $(CH_3CH_2)_3COH$ (b) ⬡—OCH_3 (c) $(CH_3)_3CCl$

(d) ⬡—CH_2CH_2OH (e) ⬡

7.63. Each of the following hydroxy compounds can undergo intramolecular hydrogen bonding (that is, hydrogen bonding between two groups in the same molecule). Rewrite each structure, showing the intramolecular hydrogen bond.

(a) [structure: benzene ring with NO_2 and OH]   (b) $\overset{OH}{\underset{|}{CH_3}}CHCH_2\overset{O}{\overset{||}{C}}CH_3$ (c) [structure: cyclohexane ring with OH and O]

7.64. How would you make the following conversions?

   (a) bromobenzene to benzoic acid ($C_6H_5CO_2H$)
   (b) benzyl bromide to stilbene (*trans*-1,2-diphenylethene)
   (c) (R)-2-butanol to (S)-2-butanol

7.65. Vinyl alcohol (CH_2=$CHOH$) is unstable and spontaneously forms acetaldehyde (CH_3CH=O). Suggest a mechanism for this reaction.

7.66. When *trans*-2-chloro-1-cyclohexanol is treated with base, cyclohexene oxide is the product; however, when *cis*-2-chloro-1-cyclohexanol is treated with base, the product is cyclohexanone.

   (a) Why does the *cis*-isomer not yield the oxide?
   (b) Write a mechanism for each reaction.

7.67. What products could be obtained when *cis*-1,2-cyclohexanediol is treated with sulfuric acid?

7.68. Propose a reaction sequence for each of the following conversions. Use any inorganic or organic reagents you wish.

   (a) $C_6H_5CH_3$ to $C_6H_5CH_2\overset{OH}{\underset{|}{C}}HCH_3$

   (b) [cyclohexene oxide structure]   to   *cis*- [cyclohexane with two Cl]

   (c) [cyclohexene oxide structure]   to   *trans*- [cyclohexane with two Cl] (racemic)

   (d) CH_2—CH_2 (epoxide)   to $C_6H_5OCH_2CH_2OC_6H_5$

   (e) $(CH_3CH_2)_2C$—$C(CH_2CH_3)_2$ (epoxide)   to $(CH_3CH_2)_3C\overset{O}{\overset{||}{C}}CH_2CH_3$

7.69. When treated with $SOCl_2$, Compound A yields Compound B. Reaction of B with magnesium, followed by reaction with acetaldehyde (CH_3CHO) and then aqueous acid, results in 5-methyl-2-heptanol. What are the structures of Compounds A and B?

7.70. Allyl disulfide (CH_2=$CHCH_2SSCH_2CH$=CH_2) is a contributor to the odor of garlic. How would you prepare this compound from allyl alcohol?

7.71. Suggest a synthesis for each of the following compounds from an alcohol of four carbons or less and any other reagents needed:

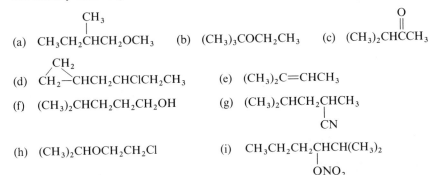

(a) $CH_3CH_2\overset{\underset{\displaystyle |}{CH_3}}{C}HCH_2OCH_3$ (b) $(CH_3)_3COCH_2CH_3$ (c) $(CH_3)_2CHC\overset{\underset{\displaystyle }{O}}{C}CH_3$

(d) $CH_2\overset{CH_2}{\diagdown}CHCH_2CHClCH_2CH_3$ (e) $(CH_3)_2C$=$CHCH_3$

(f) $(CH_3)_2CHCH_2CH_2CH_2OH$ (g) $(CH_3)_2CHCH_2\overset{\underset{\displaystyle |}{}}{C}HCH_3$
                                                         $\underset{\displaystyle CN}{}$

(h) $(CH_3)_2CHOCH_2CH_2Cl$ (i) $CH_3CH_2CH_2\overset{\underset{\displaystyle |}{}}{C}HCH(CH_3)_2$
                                                         $\underset{\displaystyle ONO_2}{}$

7.72. Suggest a synthesis for each of the following compounds, starting with any organic compounds of three or fewer carbon atoms:

(a) 1-bromo-2-butene   (b) 3-ethyl-3-pentanol

(c) 2-butyl acetate, $CH_3CH_2\overset{\underset{\displaystyle |}{}}{C}H O\overset{\underset{\displaystyle }{O}}{C}CH_3$
                                  $\underset{\displaystyle CH_3}{}$

(d) $CH_3\overset{\underset{\displaystyle |}{SCH_3}}{C}HCH$=$CH_2$ (e) octane

7.73. When Compound A ($C_7H_{14}O_3$) is treated with periodic acid, formaldehyde and Compound B ($C_6H_{10}O_3$) result. Oxidation of B with $KMnO_4$ solution yields adipic acid, $HO_2C(CH_2)_4CO_2H$. What are the structures of A and B?

CHAPTER 8

Spectroscopy I: Infrared and Nuclear Magnetic Resonance

Spectroscopy is the study of the interactions between radiant energy and matter. The colors that we see and the fact that we can see at all are consequences of energy absorption by organic and inorganic compounds. The capture of the sun's energy by plants in the process of photosynthesis is another aspect of the interaction of organic compounds with radiant energy. Of primary interest to the organic chemist is the fact that the wavelengths at which an organic compound absorbs radiant energy are *dependent upon the structure of the compound*. Therefore, spectroscopic techniques may be used to determine the structures of unknown compounds and to study the bonding characteristics of known compounds.

In this chapter, our emphasis will be on **infrared spectroscopy** and **nuclear magnetic resonance spectroscopy**, both of which are used extensively in organic chemistry. In Chapter 21, we will broaden our discussion to include some other types of spectroscopy.

SECTION 8.1.

Electromagnetic Radiation

Electromagnetic radiation is energy that is transmitted through space in the form of waves. Each type of electromagnetic radiation (radio waves, ultraviolet, infrared, visible, and so forth) is characterized by its **wavelength** (λ), the distance from the crest of one wave to the crest of the next wave (see Figure 8.1).

The entire spectrum of electromagnetic radiation is represented in Figure 8.2. The wavelengths that lead to vision range from 400 nm to 750 nm (1 nm = 10^{-9} m or 10^{-7} cm); however, the visible region is a very small part of the entire

wavelength, λ
λ
λ
λ
longer wavelength
shorter wavelength

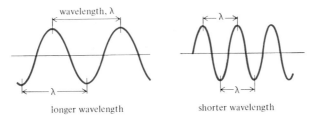

FIGURE 8.1. Wavelength of electromagnetic radiation.

cosmic rays and γ rays
x-rays
ultra-violet
visible
infrared
microwaves and radio waves
wavelength, cm: 10⁻⁶
10⁻⁵
10⁻⁴
10⁻³
10⁻²
10⁻¹
increasing wavelength, decreasing energy

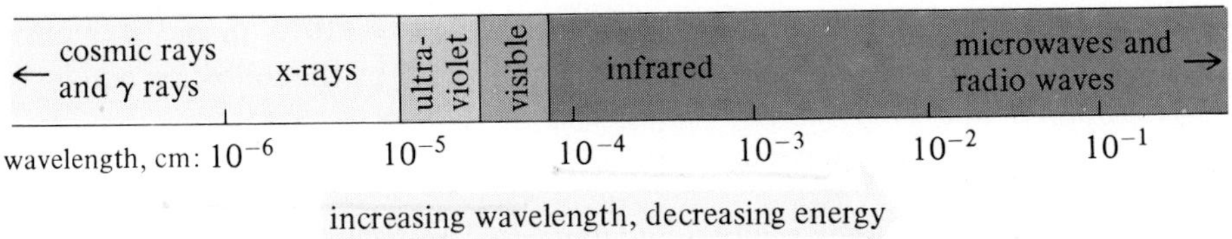

FIGURE 8.2. The electromagnetic spectrum.

electromagnetic spectrum. Wavelengths slightly shorter than those of the visible region fall into the ultraviolet region, while slightly longer wavelengths fall into the infrared region.

In addition to being characterized by its wavelength, radiation may be characterized by its **frequency** (ν), which is defined as the number of complete cycles per second (cps), also called *Hertz* (Hz). (See Figure 8.3.) Radiation of a higher frequency contains more waves per second; therefore, the wavelength must be shorter. By their definitions, wavelength and frequency are *inversely proportional*. This relationship may be expressed mathematically:

$$\nu = \frac{c}{\lambda}$$

where ν = frequency in Hz,
$c = 3 \times 10^{10}$ cm/sec (the speed of light), and
λ = wavelength in cm.

In infrared spectroscopy, frequency is expressed as **wavenumbers**: the number of cycles per centimeter. Wavenumbers have units of *reciprocal centimeters* (1/cm, or cm^{-1}). The unit used for wavelength in infrared spectroscopy is the

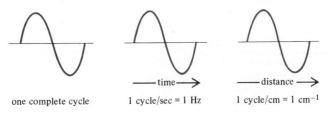

time
distance
one complete cycle
1 cycle/sec = 1 Hz
1 cycle/cm = 1 cm⁻¹

FIGURE 8.3. Frequency of electromagnetic radiation.

TABLE 8.1. Symbols commonly encountered in spectroscopy

Symbol	Definition
v	frequency in Hz (cycles per second)
λ	wavelength
μm	micrometer, same as micron (μ), 10^{-6} m
nm	nanometer, same as millimicron (mμ), 10^{-9} m
Å	Angstrom, 10^{-10} m or 10^{-1} nm
cm^{-1}	wavenumber: frequency in reciprocal cm, or $1/\lambda$

micrometer, μm (or *micron*, μ), where 1.0 μm $= 10^{-6}$ m or 10^{-4} cm. Wavenumbers and wavelengths can be interconverted by the following equation:

$$\text{wavenumber in cm}^{-1} = \frac{1}{\lambda \text{ in cm}} = \frac{1}{\lambda \text{ in } \mu\text{m}} \times 10^4$$

A few common symbols and units used in spectroscopy are listed in Table 8.1.

Electromagnetic radiation is transmitted in particle-like packets of energy called **photons** or **quanta.** The energy of a photon is *inversely proportional to the wavelength.* (Mathematically, $E = hc/\lambda$, where $h =$ Planck's constant.) Radiation of shorter wavelength has a higher energy; therefore, a photon of ultraviolet light has more energy than a photon of visible light and has substantially more energy than a photon of radio waves.

Conversely, the energy of a photon of radiation is *directly proportional to the frequency* (more waves per unit time mean higher energy). The relationship is expressed in the equation $E = hv$. (The symbol hv is often used in chemical equations to represent electromagnetic radiation.)

ultraviolet   visible   infrared   radio

increasing λ (or decreasing v) means decreased energy

Molecules absorb only specific wavelengths of electromagnetic radiation. Absorption of ultraviolet light (high-energy radiation) results in the promotion of an electron to a higher-energy orbital. Infrared radiation does not contain enough energy to promote an electron; absorption of infrared radiation results in increased amplitudes of vibration of bonded atoms. When a sample absorbs photons of radiation, the number of photons being transmitted through the sample decreases. This absorption is observed as a decrease in **intensity**, or quantity, of radiation. It is this change in intensity that is used as a measurement in spectroscopy.

STUDY PROBLEMS

8.1. Which has the higher energy:

(a) Infrared radiation of 1500 cm^{-1} or of 1600 cm^{-1}?
(b) Ultraviolet radiation of 200 nm or of 300 nm?
(c) Radio waves of 60,000 Hz or of 60,004 Hz?

8.2. Make the following conversions:

(a) 6.000 μm to cm^{-1} (b) 800 cm^{-1} to μm   (c) 1.5 μ to μm

SECTION 8.2.

Features of a Spectrum

An infrared, visible, or ultraviolet spectrum of a compound is a graph of either *wavelength* or *frequency*, continuously changing over a small portion of the electromagnetic spectrum, versus either *percent transmission* (%T) or *absorbance* (A).

$$\% T = \frac{\text{intensity}}{\text{original intensity}} \times 100 \qquad A = \log\left(\frac{\text{original intensity}}{\text{intensity}}\right)$$

Most infrared spectra record wavelength or frequency versus %T. The absence of absorption by a compound at a particular wavelength is recorded as 100 %T (ideally). When a compound absorbs radiation at a particular wavelength, the intensity of radiation being transmitted decreases. This results in a *decrease* in %T and appears in the spectrum as a dip, called an **absorption peak**, or **absorption band**. The portion of the spectrum where %T measures 100 (or near 100) is called the **base line**, which is recorded at the top of an infrared spectrum.

Visible and ultraviolet spectra (Chapter 21) are usually presented as graphs of *A* versus wavelength. In these cases, the base line (zero absorbance) runs along the bottom of the spectrum, and absorption is recorded as an *increase* of the signal. The general appearance of spectra using %T and A is shown in Figure 8.4.

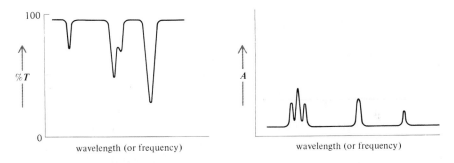

FIGURE 8.4. Spectra are graphs of percent transmission (% T) or absorbance (A) by a sample versus wavelength or frequency of radiation.

Although the physical appearance of a nuclear magnetic resonance spectrum is similar to that of a visible or ultraviolet spectrum (base line at the bottom), the physical principles of nuclear magnetic resonance spectroscopy are different from those of other types of spectroscopy. Therefore, we will defer a discussion of these spectra until Section 8.6.

SECTION 8.3.

Absorption of Infrared Radiation

Nuclei of atoms bonded by covalent bonds undergo vibrations, or oscillations, in a manner similar to two balls attached by a spring. When molecules absorb infrared radiation, the absorbed energy causes an increase in the amplitude of the vibrations of the bonded atoms. The molecule is then in an **excited vibrational**

state. (The absorbed energy is subsequently dissipated as heat when the molecule returns to the ground state.) The exact wavelength of absorption by a given type of bond depends upon the type of vibration of that bond. Therefore, different types of bonds (C—H, C—C, O—H and so forth) absorb infrared radiation at different wavelengths.

A bond within a molecule may undergo different types of oscillations; therefore, a particular bond may absorb energy at more than one wavelength. For example, an O—H bond absorbs energy at about 3330 cm^{-1} (3.0 μm); energy of this wavelength causes increased **stretching vibrations** of the O—H bond. An O—H bond also absorbs at about 1250 cm^{-1} (8.0 μm); energy of this wavelength causes increased **bending vibrations**. These different types of vibrations are called different **fundamental modes of vibration**.

$$
\begin{array}{cc}
\text{stretching} & \text{bending}
\end{array}
$$

The relative amounts of energy absorbed also vary from bond to bond. This is partly due to changes in bond moment when energy is absorbed. Nonpolar bonds (such as C—H or C—C bonds) give rise to weak absorption, while polar bonds (such as C=O) exhibit much stronger absorption.

SECTION 8.4.

The Infrared Spectrum

The instrument used to measure absorption of infrared radiation is called an **infrared spectrophotometer**. A diagram of a typical instrument is shown in Figure 8.5. At one end of the instrument is the light source, which emits all wavelengths of infrared radiation. The light from this source is split by mirrors (not shown) into two beams, the reference beam and the sample beam. After passing through the reference cell (which contains solvent, if used in the sample, or nothing if the sample is pure) and the sample cell, the two beams are combined in the chopper (another mirror system) into one beam that alternates from reference beam to sample beam. This alternating beam is diffracted by a grating that separates the beam into its different wavelengths. The detector measures the difference in intensities of the two segments of the beam at each wavelength and passes this information on to the recorder, which produces the spectrum.

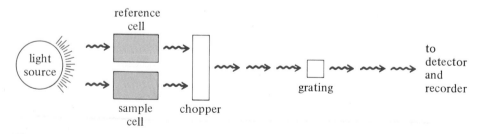

FIGURE 8.5. The infrared spectrophotometer.

Figure 8.6 shows two infrared spectra of 1-hexanol. The top spectrum is a typical example of the type of spectrum obtained directly from a spectrophotometer. The lower spectrum is an artist's rendition of the real spectrum. In the artist-rendered spectrum, many small extraneous peaks arising from impurities and electronic "noise" have been deleted. For the sake of clarity, we will use this second type of spectrum in this text.

The scales at the bottoms of these spectra are in wavenumbers, decreasing from 4000 cm^{-1} to about 670 cm^{-1} or lower. The corresponding wavelengths in μm (or μ) are shown at the top. The wavelength or frequency of the *minimum point* of an absorption band is used to identify each band. This point is more reproducible than the range of a wide band, which may vary with concentration or with the sensitivity of the instrument.

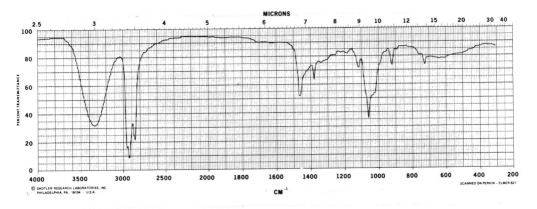

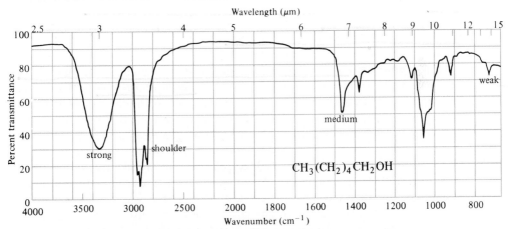

FIGURE 8.6. Infrared spectra of 1-hexanol. The upper spectrum is a reproduction of an actual spectrum, while the lower one is an artist's rendition. Upper spectrum © Sadtler Research Laboratories, Division of Bio-Rad Laboratories, Inc., 1982.

The infrared bands in a spectrum may be classified by intensity: *strong (s)*, *medium (m)*, and *weak (w)*. A weaker band overlapping a stronger band is called a *shoulder (sh)*. These terms are relative, and the assignment of any given band as *s, m, w,* or *sh* is qualitative at best. In Figure 8.6, a few bands in the lower spectrum are labeled according to this classification method.

The number of identical groups in a molecule alters the relative strengths of the absorption bands in a spectrum. For example, a single OH group in a molecule produces a relatively strong absorption, while a single CH absorption is relatively weak. However, if a compound has many CH bonds, the collective effect of the CH absorption gives a peak that is medium or even strong.

SECTION 8.5.

Interpretation of Infrared Spectra

Chemists have studied thousands of infrared spectra and have determined the wavelengths of absorption for each of the functional groups. **Correlation charts** provide summaries of this information.

A typical correlation chart for the stretching and bending frequencies of various groups is shown inside the front cover of this book. From the chart we see that OH and NH stretching bands are found between 3000–3700 cm^{-1} (2.7–3.3 μm). (Figure 8.6, the infrared spectrum of 1-hexanol, shows absorption at this position.) If the infrared spectrum of a compound of unknown structure shows absorption in this region, then we suspect that the compound contains either an OH or an NH group in its structure. If this region does not contain an absorption band, we conclude that the structure probably does not have an OH group or an NH group.

The region from 1400–4000 cm^{-1} (2.5 μm to about 7.1 μm), to the left in the infrared spectrum, is especially useful for identification of the various functional groups. This region shows absorption arising from stretching modes. The region to the right of 1400 cm^{-1} is often quite complex because both stretching and bending modes give rise to absorption here. In this region, correlation of an individual band with a specific functional group usually cannot be made with accuracy; however, each organic compound has its own unique absorption here. This part of the spectrum is therefore called the **fingerprint region**. Although the left-hand portion of a spectrum may appear the same for similar compounds, the fingerprint region must also match for two spectra to represent the same compound.

Figure 8.7 shows the infrared spectra of two alkanes of the formula C_8H_{18}: *n*-octane and 2-methylheptane. Note that the two spectra are practically identical from 1400–4000 cm^{-1}, but that the fingerprint regions are slightly different.

In the following sections, we will discuss the characteristic infrared absorption of compounds containing aliphatic C—C and C—H bonds and a few functional groups. Our intent is to develop familiarity with the features of typical infrared spectra. As we discuss the various functional groups in future chapters, we will also include discussions of their infrared spectral characteristics.

A.   Carbon–carbon and carbon–hydrogen

Bonds between sp^3 carbon atoms (C—C single bonds) give rise to weak absorption bands in the infrared spectrum. In general, these absorption bands are not very useful for structure identification. Bonds between sp^2 carbons (C=C) often exhibit characteristic absorption (variable in strength) around 1600–1700 cm^{-1}

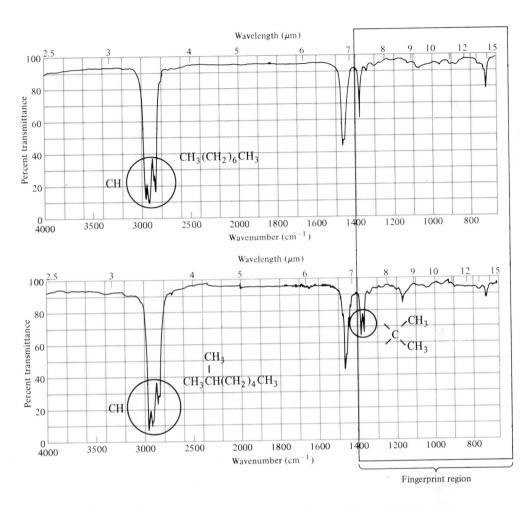

FIGURE 8.7.   Infrared spectra of *n*-octane and 2-methylheptane, showing slight differences in the fingerprint regions. Also note the CH absorption peaks.

(5.8–6.2 μm). Aryl carbon–carbon bonds show absorption at slightly lower frequencies (to the right in the spectrum). Bonds between *sp* carbons (C≡C) show weak, but extremely characteristic, absorption at 2100–2250 cm⁻¹ (4.4–4.8 μm), a region of the spectrum where most other groups show no absorption.

$$sp^2 \text{ C=C:}\quad 1600\text{–}1700 \text{ cm}^{-1} \ (5.8\text{–}6.2 \ \mu m)$$
$$sp^2 \text{ C—C (aryl):}\quad 1450\text{–}1600 \text{ cm}^{-1} \ (6.25\text{–}6.9 \ \mu m)$$
$$sp \text{ C≡C:}\quad 2100\text{–}2250 \text{ cm}^{-1} \ (4.4\text{–}4.8 \ \mu m)$$

Almost all organic compounds contain CH bonds. Absorption arising from CH stretching is seen at about 2800–3300 cm⁻¹ (3.1–3.75 μm). The CH stretching peaks are often useful in determining the hybridization of the carbon atom. In

Figure 8.7, the CH absorption is indicated. Spectra showing C—C and CH absorption by alkenes, alkynes, and benzene compounds will be presented in Chapters 9 and 10.

sp^3 C—H (alkanes or alkyl groups):   2800–3000 cm^{-1} (3.3–3.6 μm)
sp^2 C—H (=CH—):   3000–3300 cm^{-1} (3.0–3.3 μm)
sp C—H (≡CH):   ~3300 cm^{-1} (3.0 μm)

The **geminal-** or **gem-dimethyl** grouping (two methyl groups on the same carbon) often exhibits a double CH bending peak in the 1360–1385 cm^{-1} (7.22–7.35 μm) region (see Figure 8.7). Unfortunately, the two peaks are not visible in all spectra; sometimes only a single peak is observed.

C(CH$_3$)$_2$:   1360–1385 cm^{-1} (7.22–7.35 μm)   (two peaks)

B.   Haloalkanes

The stretching absorption of the CX bond of a haloalkane falls in the fingerprint region of the infrared spectrum, from 500–1430 cm^{-1} (7–20 μm) (see Figure 8.8). Without additional information, the presence or absence of a band in this region cannot be used for verifying the presence of a halogen in an organic compound.

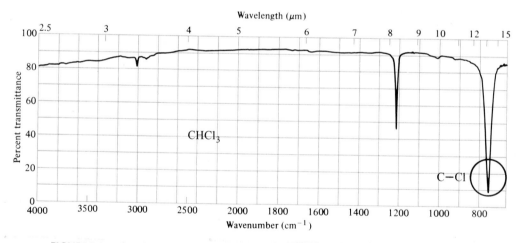

FIGURE 8.8. Infrared spectrum of chloroform, a trihaloalkane.

C.   Alcohols and amines

Alcohols and amines exhibit conspicuous OH or NH stretching absorption at 3000–3700 cm^{-1} (2.7–3.3 μm), to the *left* of CH absorption. If there are two hydrogens on an amine nitrogen (—NH$_2$), the NH absorption appears as a double peak. If there is only one H on the N, then only one peak is observed. Of course, if there is no NH (as in the case of a tertiary amine, R$_3$N), then there is no absorption in this region. Figure 8.9 shows the spectrum of an alcohol, while Figure 8.10 (page 322) shows spectra of the three types of amines.

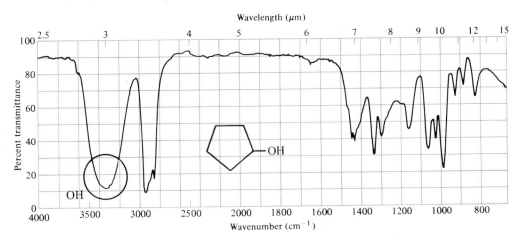

FIGURE 8.9. Infrared spectrum of cyclopentanol, an alcohol.

Alcohols and amines also exhibit C—O and C—N absorption in the finger-print region. These bands are not always easy to identify because this region of the spectrum often contains a large number of peaks.

OH or NH: 3000–3700 cm^{-1} (2.7–3.3 μm)

C—O or C—N: 900–1300 cm^{-1} (8–11 μm)

Hydrogen bonding changes the position and appearance of an infrared absorption band. The spectra in Figures 8.9 and 8.10 are those of pure liquids in which hydrogen bonding is extensive. Note that the OH absorption in Figure 8.9 appears as a wide band at about 3330 cm^{-1} (3.0 μm). When hydrogen bonding is less extensive, a sharper, less intense OH peak is observed. Figure 8.11 (page 323) shows two partial spectra of an alcohol. One spectrum is of a pure liquid (hydrogen bonded); the other spectrum is of the alcohol in the vapor phase (not hydrogen bonded). The differences in OH absorption are apparent in this figure.

Absorption by NH bonds is less intense than OH absorption, partly because of weaker hydrogen bonds in amines and partly because NH bonds are less polar.

STUDY PROBLEM

8.3. Figure 8.12 (page 323) gives the infrared spectra of two compounds, I and II, both of which appear in the following list. Which is I and which is II?

(a) $CH_3(CH_2)_6CH_3$ (b) $CH_3(CH_2)_6CH_2OH$
(c) $CH_3(CH_2)_5N(CH_3)_2$ (d) $CH_3CH_2CH_2CH_2NH_2$
(e) $CH_3(CH_2)_6CH_2I$ (f) $CH_3CH_2CH_2NHCH_3$

D.   Ethers

Ethers have a C—O stretching band that falls in the fingerprint region at 1050–1260 cm^{-1} (7.9–9.5 μm). Because oxygen is electronegative, the stretching causes a large change in bond moment; therefore, the C—O absorption is usually strong

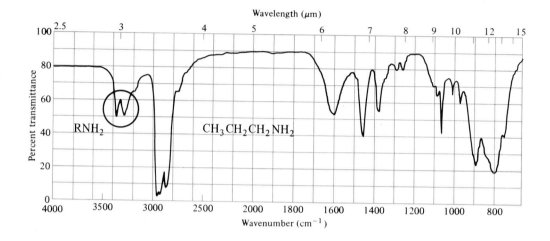

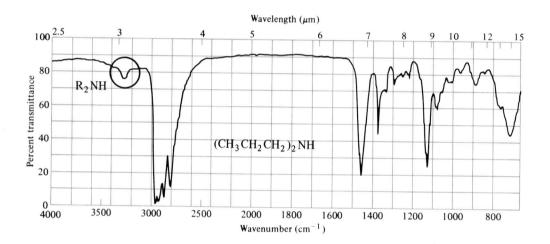

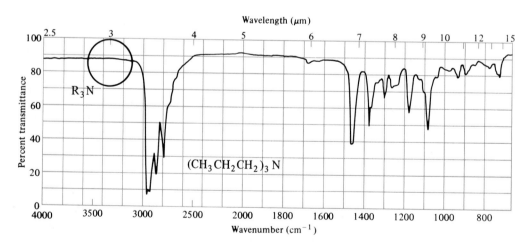

FIGURE 8.10. Infrared spectra of a primary amine, *n*-propylamine (top); a secondary amine, dipropylamine (center); and a tertiary amine, tripropylamine (bottom).

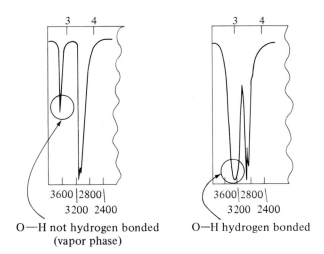

O—H not hydrogen bonded   O—H hydrogen bonded
(vapor phase)

FIGURE 8.11. Partial infrared spectra of an alcohol in the vapor and liquid phases, showing hydrogen-bonded and nonhydrogen-bonded OH absorption.

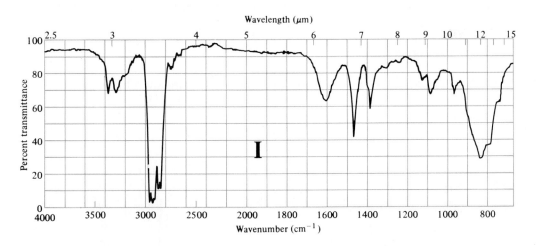

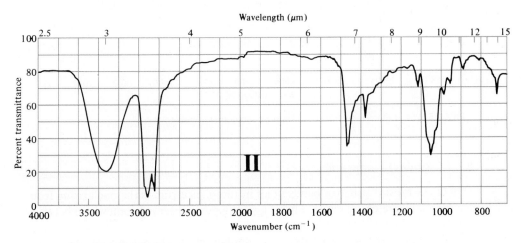

FIGURE 8.12. Infrared spectra for Problem 8.3.

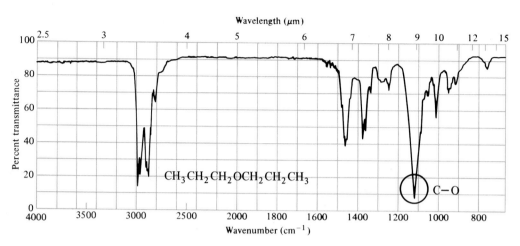

FIGURE 8.13. Infrared spectrum of di-*n*-propyl ether, an ether.

(see Figure 8.13). Alcohols, esters, and other compounds containing C—O single bonds also show absorption here.

STUDY PROBLEM

8.4. A compound yielded only iodoethane as a product when heated with HI. The infrared spectrum of the original compound is shown in Figure 8.14. What is its structure?

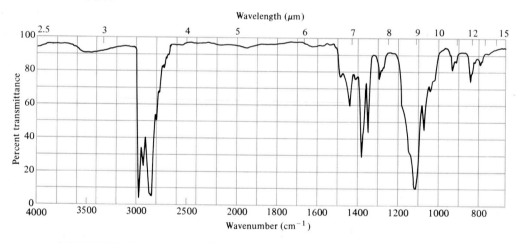

FIGURE 8.14. Infrared spectrum of unknown compound for Problem 8.4.

E. Carbonyl compounds

One of the most distinctive bands in an infrared spectrum is the one arising from the carbonyl stretching mode. This is a strong peak observed somewhere between 1640 and 1820 cm^{-1} (5.5–6.1 μm).

The carbonyl group is part of a number of functional groups. The exact position of the carbonyl absorption, the positions of other absorption bands in the infrared spectrum, and other spectral techniques (particularly nmr) may be needed to identify the functional group. The positions of $C{=}O$ absorption for aldehydes, ketones, carboxylic acids, and esters are listed in Table 8.2.

TABLE 8.2. Stretching vibrations for some carbonyl compounds[a]

	Position of absorption	
Type of compound	*cm^{-1}*	*μm*
aldehyde, $\overset{\displaystyle O}{\overset{\|}{RCH}}$	1720–1740	5.75–5.80
ketone, $\overset{\displaystyle O}{\overset{\|}{RCR}}$	1705–1750	5.70–5.87
carboxylic acid, $\overset{\displaystyle O}{\overset{\|}{RCOH}}$	1700–1725	5.80–5.88
ester, $\overset{\displaystyle O}{\overset{\|}{RCOR}}$	1735–1750	5.71–5.76

[a] In each case, R is saturated and aliphatic.

Ketones give the simplest spectra of the carbonyl compounds. If a compound is an aliphatic ketone, all strong stretching infrared absorption arises from either $C{=}O$ or CH. Other functionality may increase the complexity of the spectrum, of course. The infrared spectrum of a ketone is shown in Figure 8.15.

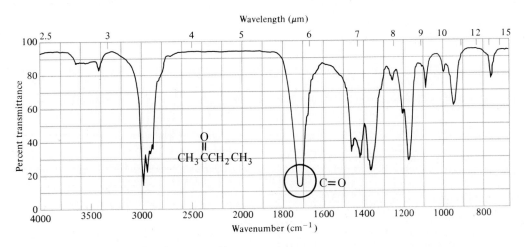

FIGURE 8.15. Infrared spectrum of butanone, a ketone.

Aldehydes give infrared spectra very similar to those of ketones. The important difference between an aldehyde and a ketone is that an aldehyde has an H bonded to the carbonyl carbon. This particular CH bond shows two characteristic stretching bands (just to the right of the aliphatic CH band) at 2820–2900 cm^{-1} (3.45–3.55 μm) and 2700–2780 cm^{-1} (3.60–3.70 μm). Both these CH peaks are sharp, but weak, and the peak at 2900 cm^{-1} (3.45 μm) may be obscured by overlapping absorption of other CH bonds (see Figure 8.16). The aldehyde CH also has a very characteristic nmr absorption (see Section 8.7). If the infrared spectrum of a compound suggests that the structure is an aldehyde, the nmr spectrum should be checked.

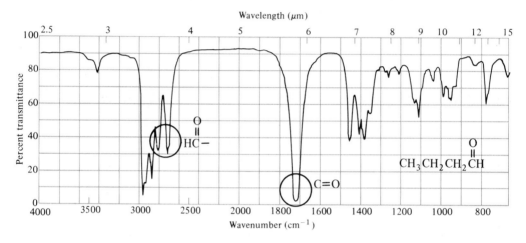

FIGURE 8.16.   Infrared spectrum of butanal, an aldehyde.

Carboxylic acids exhibit typical C=O absorption and also show a very distinctive O—H band, which begins at about 3330 cm^{-1} (3.0 μm) and slopes into the aliphatic CH absorption band (Figure 8.17). The reason that a carboxyl OH gives

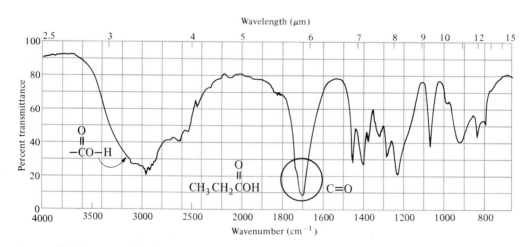

FIGURE 8.17.   Infrared spectrum of propanoic acid, a carboxylic acid.

a different-looking spectrum from that of an alcohol OH is that carboxylic acids form hydrogen-bonded *dimers*:

$$R-C \overset{\overset{\ddot{O}:\cdots H-\ddot{O}:}{\|}}{\underset{\underset{\ddot{O}:-H\cdots:\ddot{O}}{}}{}} C-R$$

Esters exhibit both a typical carbonyl band and a C—O band. The C—O band, like that in ethers, is observed in the fingerprint region, 1110–1300 cm^{-1} (7.7–9.0 μm) and is sometimes difficult to assign. However, this C—O band is strong and, in some cases, may be used to distinguish between esters and ketones. See Figure 8.18 for the infrared spectrum of a typical ester.

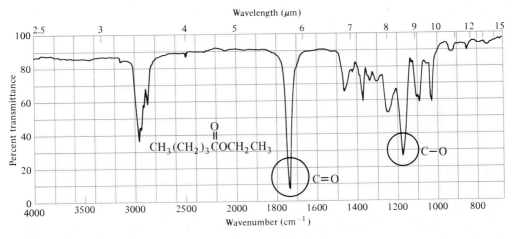

FIGURE 8.18. Infrared spectrum of ethyl pentanoate, an ester.

STUDY PROBLEM

8.5. One of the two spectra in Figure 8.19 is that of a ketone; the other is of an ester. Which is which?

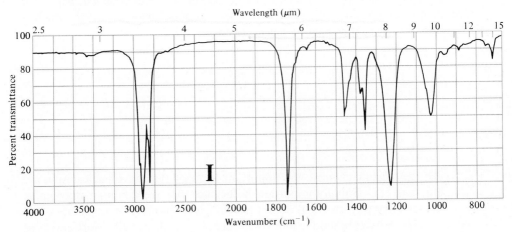

FIGURE 8.19. Infrared spectra of unknown compounds for Problem 8.5.   *(continued)*

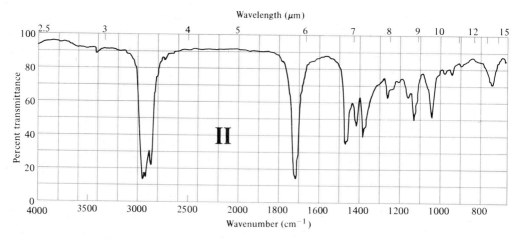

FIGURE 8.19. (continued) Infrared spectra of unknown compounds for Problem 8.5.

SECTION 8.6.

Nuclear Magnetic Resonance Spectroscopy

The infrared spectrum of a compound gives a picture of the different functional groups in an organic molecule, but gives only meager clues about the hydrocarbon portion of the molecule. **Nuclear magnetic resonance (nmr) spectroscopy** fills this gap by providing a picture of the hydrogen atoms in the molecule.

Nmr spectroscopy is based upon the absorption of radio waves by certain nuclei in organic molecules when they are in a strong magnetic field. Before proceeding with a discussion of nmr spectra and their use in organic chemistry, let us first consider the physical principles that give rise to the nmr phenomenon.

A.   Origin of the nmr phenomenon

The nuclei of atoms of all elements can be classified as either *having spin* or *not having spin*. A nucleus with spin gives rise to a small magnetic field, which is described by a **nuclear magnetic moment,** a vector.

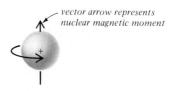

To the organic chemist, the important isotopes that have nuclear spin are ^1_1H, $^{13}_6C$, and $^{19}_9F$. Equally important is the fact that the common isotopes of carbon ($^{12}_6C$) and oxygen ($^{16}_8O$) do not have nuclear spin. While all isotopes that have nuclear spin can be used in nmr spectroscopy, they do not all absorb energy at the same radiofrequency. The most common isotope studied by nmr methods is ^1_1H, the proton; consequently, we will concentrate our discussion on this one isotope. Then, in Section 8.12, we will mention ^{13}C spectroscopy.

In nmr spectroscopy, an **external magnetic field** is generated by a permanent horseshoe magnet or an electromagnet. The strength of this external field is symbolized by H_0, and its direction is represented by an arrow.

$$\uparrow$$

Symbol representing the external magnetic field: H_0

A spinning proton with its nuclear magnetic moment is similar, in many respects, to a tiny bar magnet. When molecules containing hydrogen atoms are placed in an external magnetic field, the magnetic moment of each hydrogen nucleus, or proton, aligns itself in one of two different orientations with respect to the external magnetic field. (Keep in mind that it is only the magnetic moments of *hydrogen nuclei*, not molecules, that become aligned.) The two orientations that the nuclear magnetic moment may assume are **parallel** and **antiparallel** to the external field. In the parallel state, the magnetic moment of the proton points in the *same direction* as that of the external field. In the antiparallel state, the magnetic moment of the proton *opposes* the external field. At any given time, approximately half the protons in a sample are in the parallel state and half are in the antiparallel state.

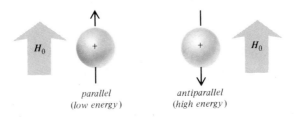

parallel   *antiparallel*
(low energy)   *(high energy)*

The parallel state of a proton is slightly more stable (lower energy) than the antiparallel state. When exposed to the proper frequency of radio waves, the magnetic moments of a small fraction of the parallel protons absorb energy and turn around, or *flip*, to the higher-energy antiparallel state. The amount of energy required to flip the magnetic moment of a proton from parallel to antiparallel depends, in part, upon the strength of H_0. If H_0 is increased, the energy difference between the parallel and antiparallel states increases. Thus, if H_0 is increased, the nucleus is more resistant to being flipped and higher-energy, higher-frequency radiation is required.

When the particular combination of external magnetic field strength and radiofrequency causes a proton to flip from parallel to antiparallel, the proton is said to be in **resonance** (a different type of resonance from that of "resonance" structures of benzene). The term **nuclear magnetic resonance** means "nuclei in resonance in a magnetic field."

It would seem that all protons should come into resonance at the same combination of H_0 and radiofrequency; however, this is not the case. The magnetic field actually observed by a proton in a particular molecule is a combination of two fields: (1) the applied external magnetic field (H_0) and (2) an **induced molecular magnetic field**, a small magnetic field induced in the molecule by H_0. The magnetic field observed by a proton is also modified by the spin states of nearby protons, a topic we will discuss later in this chapter. The protons of a molecule flip at different combinations of H_0 and radiofrequency because they are in different molecular (and magnetic) environments. Because of the differences in the energy absorption by protons, we are able to obtain a spectrum of the different types of protons.

B.   The nmr spectrum

A diagram of an nmr spectrometer is shown in Figure 8.20. The sample is placed between the poles of a magnet and is irradiated with radio waves. When the protons flip from the parallel to the antiparallel state, the absorption of energy is detected by a power indicator.

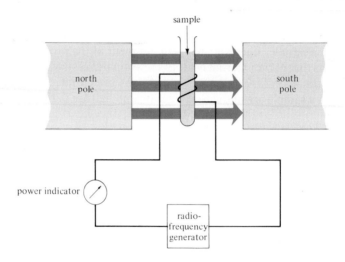

FIGURE 8.20. Schematic diagram of an nmr spectrometer.

In one type of nmr spectrometer, the radiofrequency is held constant at 60 MHz (60 megaHertz, or 60×10^6 Hz), H_0 is varied over a small range, and the frequency of energy absorption is recorded at the various values for H_0. Thus, the nmr spectrum is a graph of the amount of energy absorbed (I, or intensity) versus magnetic field strength.

Figure 8.21 shows two nmr spectra of methanol: a reproduction of an actual spectrum and an artist's copy. On the lower spectrum, we have indicated the direction of field sweep from weak H_0 to strong H_0. You can see two principal peaks in these spectra. The left-hand peak arises from the OH proton; the right-hand peak, from the CH_3 protons. The protons that flip more easily absorb energy at a lower H_0; they give rise to an absorption peak **downfield** (farther to the left). The protons that flip with greater difficulty absorb energy at a higher H_0 and give a peak that is **upfield** (to the right).

Protons in different molecular environments flip at different strengths of the applied field because the induced molecular magnetic field can either aid or oppose the external magnetic field. (Technically, all induced molecular fields oppose the external field; therefore, we are speaking of molecular fields relative to other molecular fields.) If the two fields (H_0 and the induced molecular field) oppose one

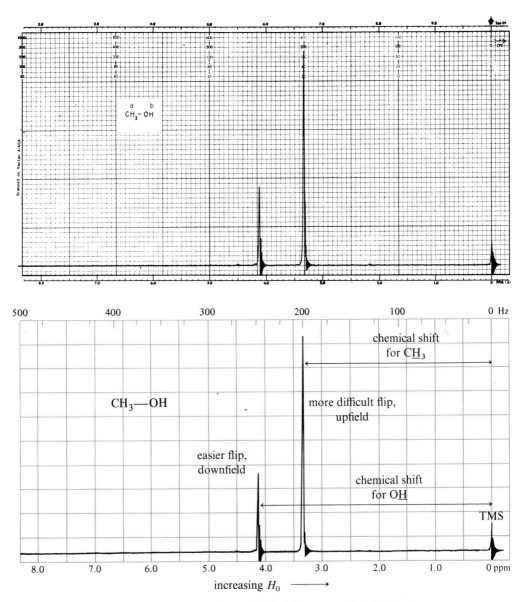

FIGURE 8.21. Nmr spectra of methanol (solvent, CCl_4). The upper spectrum is a reproduction of an actual spectrum, while the lower one is an artist's rendition. The lower spectrum shows the field sweep and chemical shifts. (The fluctuations to the right of each signal, called "ringing," arise from distortion of the magnetic field when the sample is scanned rapidly. Ringing is an indication that the instrument is properly adjusted.) Upper spectrum © Sadtlér Research Laboratories, Division of Bio-Rad Laboratories, Inc., 1982.

another, then a greater applied H_0 is needed to bring a proton into resonance. In this case, the proton is said to be **shielded**, and we observe its absorption *upfield* in a spectrum. If the two fields add, then less applied H_0 is needed to bring the proton into resonance. This proton is **deshielded**, and the absorption appears *downfield* (see Figure 8.22).

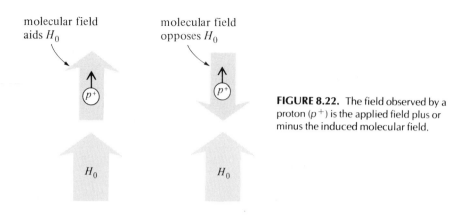

FIGURE 8.22. The field observed by a proton (p^+) is the applied field plus or minus the induced molecular field.

Shielding and deshielding are relative terms. In order to obtain quantitative measurements, we need a reference point. The compound that has been chosen for this reference point is tetramethylsilane (TMS), $(CH_3)_4Si$, the protons of which absorb to the far right in the nmr spectrum. The absorption for most other protons is observed downfield from that of TMS. In practice, a small amount of TMS is added directly to the sample, and the peak for TMS is observed on the spectrum along with any absorption peaks from the sample compound. The difference between the position of absorption of TMS and that of a particular proton is called the **chemical shift**. You can see the TMS peak in the spectra in Figure 8.21, where we have also indicated the chemical shifts.

Chemical shifts are reported in **δ values**, which are expressed as *parts per million (ppm) of the applied radiofrequency.* At 60 MHz, 1.0 ppm is 60 Hz; therefore, a δ value of 1.0 ppm is 60 Hz downfield from the position of absorption of TMS, which is set at 0 ppm. As may be seen in Figure 8.21, the two types of protons in CH_3OH have δ values of 3.4 ppm and 4.15 ppm, respectively. (Note that the lower scale on an nmr spectrum shows δ values in ppm. The use of the Hz scale at the top of an nmr spectrum will be discussed in Section 8.10.)

SECTION 8.7.

Types of Induced Molecular Magnetic Fields

A.   Fields induced by sigma electrons

Any hydrogen atom in an organic compound is bonded to carbon, oxygen, or some other atom by a sigma bond. The external magnetic field causes these sigma-bond electrons to circulate; the result is a small molecular magnetic field that *opposes* H_0 (Figure 8.23).

Because the induced field opposes the external field, the sigma-bonded proton is shielded. It takes a slightly higher external field strength to overcome the effect of the induced field in order to bring the proton into resonance; therefore, the proton absorbs upfield compared to a hypothetical naked proton. The strength of the induced field depends upon the electron density near the hydrogen atom in the sigma bond. The higher the electron density, the greater is the induced field (and the farther upfield is the observed absorption).

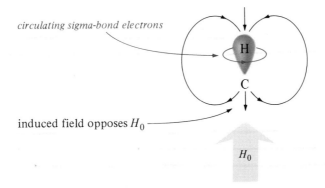

circulating sigma-bond electrons

induced field opposes H_0

H_0

FIGURE 8.23. The induced field from circulating sigma-bond electrons opposes H_0 in the
vicinity of the proton. (For simplicity, the figure shows only one CH bond of an
organic molecule.)

The electron density of a carbon–hydrogen covalent bond is affected by the
electronegativity of the other atoms bonded to the carbon. Let us consider a
specific example. The CF bond in CH_3F is polar: the fluorine atom carries a
partial negative charge and the carbon atom carries a partial positive charge.
Because the carbon has a partial positive charge, the electrons in each C—H
sigma bond are drawn toward the carbon and away from the hydrogen atom.
Recall from our discussion of carbocation stability that the polarization of bonds
by positive or negative centers is called the **inductive effect**. This shift in electron
density toward an electronegative element (F) is another example of the inductive
effect. In this case, the result of the electron-withdrawing effect of F is that there is a
greater electron density around F and a lesser electron density around each hydrogen atom. The protons of CH_3F are deshielded and absorb downfield compared
to the protons of CH_4.

$$H\rightarrow\overset{\displaystyle\underset{\uparrow}{H}}{C}\rightarrow F$$

F causes a decrease in
e^- *density around each H*

The following series of methane and the methyl halides shows increased
deshielding of the hydrogen nuclei with increasing electronegativity of the atom
attached to —CH_3.

H_3C—H H_3C—I H_3C—Br H_3C—Cl H_3C—F

increased deshielding of H

SAMPLE PROBLEM

Which of the circled protons is the most shielded? The most deshielded? What
are the relative positions of nmr absorption?

(a) CH_3CH_2Cl (b) CH_3CH_3 (c) CH_3CH_2I

Solution: Chlorine is more electronegative than iodine, which in turn is more
electronegative than hydrogen. Therefore, (a) is the most deshielded and absorbs
downfield; (b) is the most shielded and absorbs upfield, closer to TMS; and
(c) is intermediate.

In a single molecule, a proton that is attached to the same carbon as an electronegative atom is more deshielded than protons on other carbons. Figure 8.31 (page 340) shows the spectrum of chloroethane. The peaks at $\delta = 1.5$ ppm arise from the CH_3 protons, while the deshielded CH_2Cl protons absorb farther downfield, at $\delta = 3.55$ ppm. (The complexity of these peaks will be discussed later in this chapter.)

The inductive effect of an electronegative atom falls off rapidly when it is passed through a number of sigma bonds. In nmr spectra, the inductive effect is negligible three carbons away from the electronegative atom.

effect of X is
of little importance
in decreasing
e^- density
around this proton

$$H-\overset{\overset{\displaystyle H}{|}}{\underset{\underset{\displaystyle H}{|}}{C}}-\overset{\overset{\displaystyle H}{|}}{\underset{\underset{\displaystyle H}{|}}{C}}-\overset{\overset{\displaystyle H}{|}}{\underset{\underset{\displaystyle H}{|}}{C}}-X$$

effect of X is important
in decreasing
e^- density around
this proton

Most elements encountered in organic compounds are more electronegative than carbon. Their inductive effect is one of electron-withdrawal, and protons affected by them are deshielded. However, silicon is *less* electronegative than carbon. The silicon–carbon bond is polarized such that carbon carries a partial negative charge. The electrons in the CH bond of an $SiCH_3$ group are repelled from the negative carbon and pushed toward the hydrogen atoms. In this case, the inductive effect is *electron-releasing*. The protons of an $SiCH_3$ group are highly shielded because of the increased electron density around them. This is the reason that the protons of TMS absorb upfield and that TMS provides a good reference peak in an nmr spectrum.

greater e^- density on H; highly shielded

$$H_3C \leftarrow \underset{\underset{\displaystyle CH_3}{\downarrow}}{\overset{\overset{\displaystyle CH_3}{\uparrow}}{Si}} \rightarrow CH_3$$

B.   Fields induced by pi electrons

The magnetic fields induced by *pi electrons* are *directional* (that is, unsymmetrical). A measurement that varies depending on the direction in which the measurement is taken is said to be **anisotropic**. Because the effects of molecular fields induced by pi electrons are direction-dependent, thay are called **anisotropic effects**. (These effects are contrasted to inductive effects, which are symmetrical around the proton.) Anisotropic effects occur in addition to the ever-present molecular fields induced by sigma-bond electrons.

In benzene, the pi electrons are delocalized around the ring. Under the influence of an external magnetic field, these pi electrons *circulate around the ring*. This circulation, called **ring current**, induces a molecular magnetic field with the geometry shown in Figure 8.24. (All the benzene rings, of course, do not line up as we have represented the ring in Figure 8.24; Figure 8.24 shows the net effect of the vector forces.)

The result of this induced field is that H_0 is augmented (relatively speaking) in the vicinity of the benzene protons. Less applied field is required to bring aryl

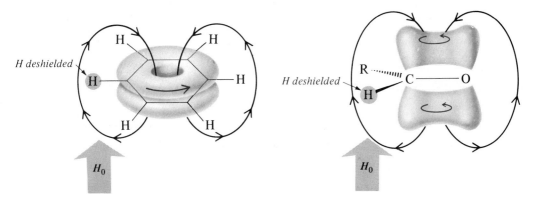

FIGURE 8.24. Circulating pi electrons in benzene or aldehydes induce a magnetic field that deshields the adjacent protons. (All six benzene protons are deshielded.)

protons into resonance than is necessary for alkyl protons. Therefore, aryl protons are deshielded and absorb farther downfield than alkyl protons.

An analogous situation is observed in the case of a vinylic hydrogen or an aldehyde hydrogen. In either case, the pi electrons are set in motion and induce a field that adds to the applied field in the vicinity of the $=CH$ proton (see Figure 8.24). A proton attached to an sp^2 carbon of $C=O$ or $C=C$ absorbs downfield from an alkyl proton. Figure 8.25 shows the nmr spectrum of a compound with aryl protons, vinylic protons, and a CH_3 group. In this spectrum, it is evident that the vinylic protons absorb downfield from CH_3 protons and that aryl protons absorb even farther downfield.

Figure 8.26 shows the spectrum of an aldehyde. The aldehyde proton is shifted downfield both by anisotropic effects and by electron-withdrawal by the

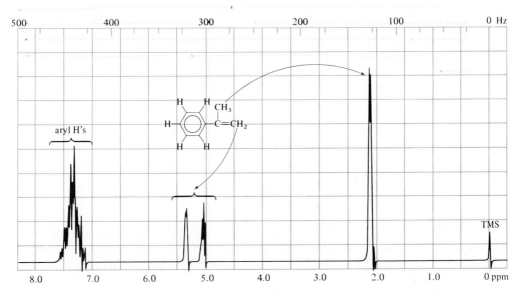

FIGURE 8.25. Nmr spectrum of 2-phenylpropene, showing absorption by aryl, vinylic, and methyl protons.

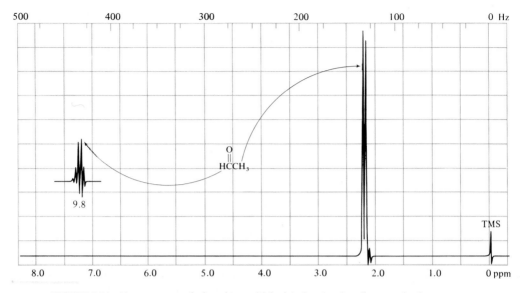

FIGURE 8.26. Nmr spectrum of ethanal (acetaldehyde), showing the offset scan for the aldehyde proton.

carbonyl oxygen. The combination of effects results in absorption that is far downfield (9–10 ppm), <u>outside the normal scan for an nmr spectrometer</u>. Instrument design allows <u>scanning at field strengths that are lower than normal</u>; the scan of this region (8–20 ppm), traced above the standard nmr scan, is called an **offset scan**.

C.   Summary of induced-field effects

The <u>presence of an electronegative atom causes a decrease in electron density around a proton by the **inductive effect**</u>. Such a <u>proton is deshielded and absorbs downfield. In aromatic compounds, alkenes, and aldehydes, a proton attached to the sp^2 carbon is deshielded by **anisotropic effects** and absorbs even farther downfield.</u> The absorption positions are summarized in Figure 8.27. A chart inside the front cover lists the δ values for a number of types of protons.

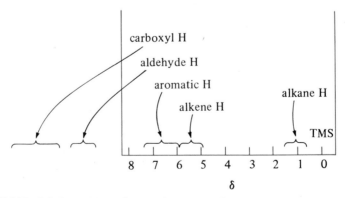

FIGURE 8.27. Relative positions of proton absorption in the nmr spectrum.

SECTION 8.8.

Counting the Protons

A.   Equivalent and nonequivalent protons

Protons that are in the same magnetic environment in a molecule have the same chemical shift in an nmr spectrum. Such protons are said to be **magnetically equivalent protons.** Protons that are in different magnetic environments have different chemical shifts and are said to be **nonequivalent protons**.

Magnetically equivalent protons in nmr spectroscopy are generally the same as *chemically equivalent protons.* In chloroethane, the three methyl protons are magnetically equivalent and are also chemically equivalent. To see that they are chemically equivalent, imagine a chemical reaction in which one of these three protons is replaced by another atom, such as bromine. If only one product can result, regardless of which proton is replaced, then the protons are chemically equivalent. In the following example, note that replacement of any of the methyl protons by Br leads to the same compound, 1-bromo-2-chloroethane.

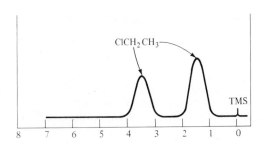

The two protons of the CH_2Cl group are also magnetically and chemically equivalent to each other. However, the three protons of the CH_3 group are not equivalent to the two CH_2Cl protons.

three equivalent protons   *two equivalent protons*
(but nonequivalent to CH_2)   *(but nonequivalent to CH_3)*

$$CH_3CH_2Cl$$

The three CH_3 protons have the same chemical shift and absorb at the same position in the nmr spectrum. The two CH_2 protons are deshielded compared to the methyl protons and have a greater chemical shift. A low-resolution spectrum of chloroethane would look like the stylized spectrum shown in Figure 8.28.

ClCH₂CH₃

TMS

8   7   6   5   4   3   2   1   0

FIGURE 8.28. Stylized low-resolution nmr spectrum of CH_3CH_2Cl.

In chloroethene, the proton *cis* to the Cl atom is in a different environment from that of the *trans*-proton. Both of these protons are in different environments from that of the proton on the C—Cl carbon. In chloroethene, all three protons are nonequivalent.

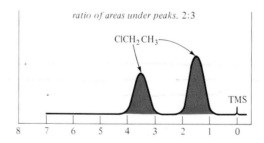

chloroethene
(vinyl chloride)

three nonequivalent protons

SAMPLE PROBLEM

Which protons in *p*-chlorotoluene (1-chloro-4-methylbenzene) are chemically equivalent, and which are nonequivalent?

Solution:

the three methyl protons are equivalent;
the two H_a's are equivalent;
the two H_b's are equivalent;
the methyl protons, H_a, and H_b, are nonequivalent

B. Areas under the peaks

If we <u>measure the areas under the peaks in an nmr spectrum,</u> we <u>find that</u> *the areas are in the same ratio as the number of protons that give rise to each signal.* In the case of chloroethane (Figure 8.29), the ratio is 2:3. (Note that the *height* of the peak is not the important feature, but rather the *area* <u>under the peak.</u>)

ratio of areas under peaks, 2:3

$ClCH_2CH_3$

TMS

8 7 6 5 4 3 2 1 0

FIGURE 8.29. Stylized low-resolution nmr spectrum of CH_3CH_2Cl, showing areas under the peaks.

SAMPLE PROBLEM

How many types of equivalent protons are there in each of the following structures? What would be the relative areas under the nmr absorption bands?

(a) $(CH_3)_2CHCl$ (b) $CH_3CH_2OCH_2CH_3$ (c) Cl—⬡—OCH_3

Solution: (a) two, 6:1; (b) two, 3:2; (c) three, 3:2:2 (3 for CH_3, 2 for aryl protons adjacent to the oxygen, 2 for aryl protons adjacent to Cl).

Most nmr spectrometers are equipped with **integrators**, which give a signal that shows the relative areas under the peaks in a spectrum. The integration appears as a series of steps superimposed upon the nmr spectrum; the height of the step over each absorption peak is proportional to the area under that peak. From the relative heights of the steps on the integration curve, the relative areas under the peaks may be determined. In Figure 8.30, the heights of the steps of the integration curve were measured with a ruler and found to be 33 mm, 100 mm, and 50 mm. For determining the relative numbers of equivalent protons, these values are converted to ratios of small whole numbers, 2:6:3. We can compare these values with the numbers of protons in the known structure of 1-bromo-2,4,6-trimethylbenzene and, indeed, the values agree.

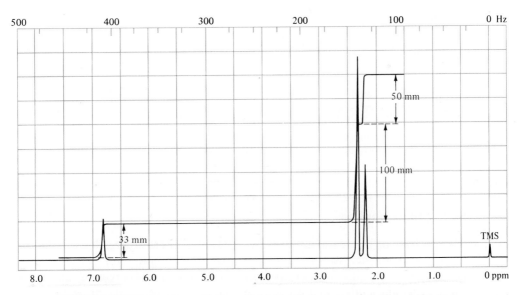

FIGURE 8.30. Nmr spectrum of 1-bromo-2,4,6-trimethylbenzene, showing an integration curve.

In this text, we will not show the integration lines in the spectra, but will indicate relative areas as numbers directly above the proton absorption peaks, as shown in Figure 8.31.

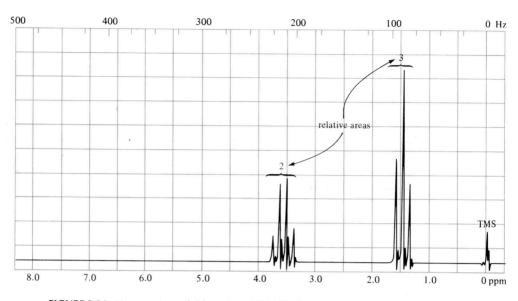

FIGURE 8.31. Nmr spectrum of chloroethane, CH_3CH_2Cl.

SAMPLE PROBLEM

The heights of four steps of an integration curve were measured to be 90 mm, 36 mm, 37 mm, and 54 mm. Calculate the ratio of the different types of protons in the sample compound.

Solution:

1. Divide each height by the smallest one (36 mm in this case):

$$\frac{90 \text{ mm}}{36 \text{ mm}} = 2.5 \qquad \frac{36 \text{ mm}}{36 \text{ mm}} = 1.0 \qquad \frac{37 \text{ mm}}{36 \text{ mm}} = 1.0 \qquad \frac{54 \text{ mm}}{36 \text{ mm}} = 1.5$$

2. Multiply the quotients by the integer that will convert them to the smallest whole numbers possible (rounding off, if necessary). In this case, multiplying by 2 gives the ratio of $5:2:2:3$. Therefore, the ratio of the numbers of nonequivalent protons is $5:2:2:3$.

SECTION 8.9.

Spin–Spin Coupling

The spectrum of chloroethane in Figures 8.28 and 8.29 is a stylized low-resolution spectrum. If we increase the resolution (that is, the sensitivity), the peaks are resolved, or split, into groups of peaks (see Figure 8.31). This type of splitting is called **spin–spin splitting** and is caused by the presence of *vicinal, or neighboring, protons (protons on an adjacent carbon) that are nonequivalent to the proton in question.* Protons that split the signals of each other are said to have undergone **spin–spin coupling**.

*these protons split
the signal for ——
the CH₂ protons*   *these protons split
                          — the signal for
                          the CH₃ protons*

CH_3CH_2Cl

Why do protons undergo spin–spin coupling? The splitting of the signal arises from the two spin states (parallel and antiparallel) of the neighboring protons.

Signal for H_a split by the two spin states of H_b:

parallel   antiparallel

The spin of a proton generates a magnetic moment. If the spin of the neighboring proton (H_b in the preceding formula) is parallel, its magnetic moment *adds* to the applied magnetic field; consequently, the first proton (H_a in the preceding formula) sees a slightly stronger field and comes into resonance at a slightly lower applied field strength. If the neighboring proton is in the antiparallel state, its magnetic moment *decreases* the magnetic field around the first proton. In this case, it takes slightly more applied H_0 for that proton to come into resonance. Because approximately half of the H_b nuclei are parallel and half are antiparallel at any given time, we might say that there are two types of H_a in the sample: those with a neighboring H_b in a parallel spin state and those with a neighboring H_b in an anti-parallel spin state. Consequently, we observe two peaks for H_a instead of just one.

For many compounds, we can predict the number of spin–spin splitting peaks in the nmr absorption of a particular proton (or a group of equivalent protons) by counting *the number (n) of neighboring protons nonequivalent to the proton in question and adding 1.* This is called the **$n + 1$ rule**.   ✳ ✳

| These three equivalent protons see *two* neighboring, nonequivalent protons. Their nmr band is split into 2 + 1, or 3, peaks. | These two equivalent protons see *three* neighboring, nonequivalent protons. Their nmr band is split into 3 + 1, or 4, peaks. |

$$CH_3CH_2Cl$$

Protons that have the same chemical shift do not split the signals of each other. Only neighboring protons that have different chemical shifts cause splitting. Some examples follow:

no neighboring H: not split

one neighboring, nonequivalent H: H_a split into two

one neighboring, nonequivalent H: H_b split into two

twelve equivalent H's: no splitting

six equivalent H's: no splitting

In some cases, the magnetic environments of chemically nonequivalent protons are so similar that the protons exhibit identical chemical shifts. In this case, no splitting is observed. For example, toluene has four groups of chemically nonequivalent protons, yet the nmr spectrum shows only two absorption peaks (one for the CH_3 protons and one for the ring protons).

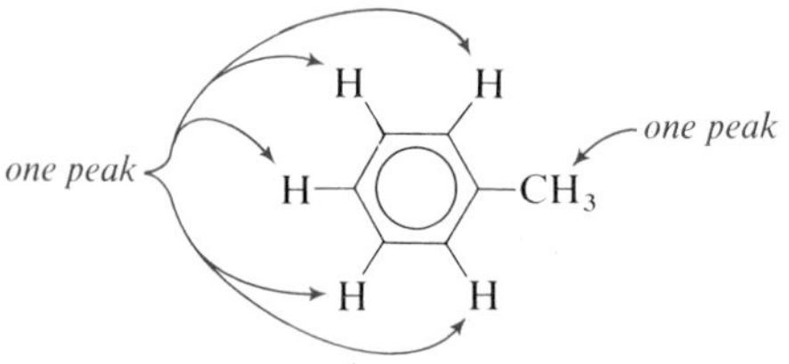

In summary, the **chemical shift** for a particular proton is determined by the molecular magnetic field surrounding it. The **area** under an absorption band is determined by the number of equivalent protons giving rise to the particular signal. **Spin–spin splitting** of a signal is dependent on the number of neighboring protons nonequivalent to the proton giving the signal.

SAMPLE PROBLEM

Predict the spin–spin splitting patterns for the protons in 2-chloropropane.

Solution:

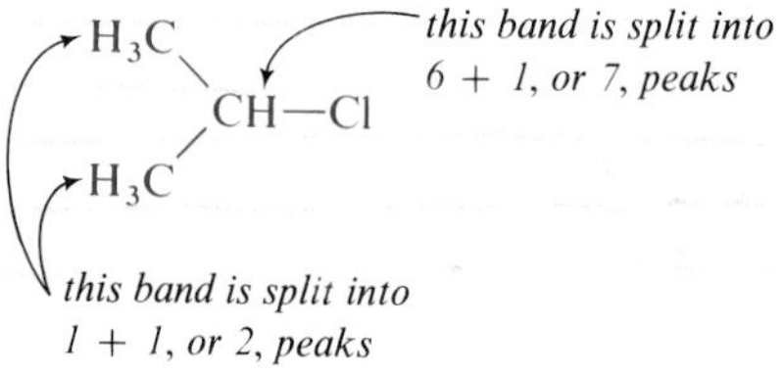

STUDY PROBLEM

8.6.   Using the $n + 1$ rule, predict the number of nmr peaks for each of the indicated protons:

(a) 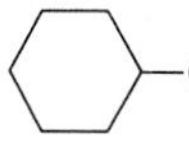   (b) $CH_3C\underline{H}_2CH_3$ (c) $(CH_3)_3C\underline{H}$ (d) $(C\underline{H}_3)_3CH$

Splitting Patterns

A.   The singlet

A proton with no neighboring protons magnetically nonequivalent to it shows a single peak, called a **singlet**, in the nmr spectrum. Figure 8.32, the nmr spectrum of *p*-methoxybenzaldehyde, shows two singlets: one for the $CH_3O—$ protons and one for the $—CHO$ proton.

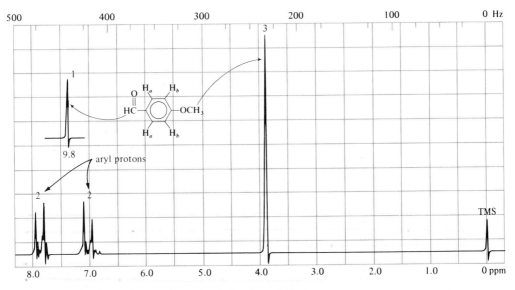

FIGURE 8.32. Nmr spectrum of *p*-methoxybenzaldehyde.

STUDY PROBLEMS

8.7. Which of the following compounds would show at least one singlet in the nmr spectrum?

(a) CH_3CH_3 (b) $(CH_3)_2CHCH(CH_3)_2$ (c) $(CH_3)_3CCl$
(d) Cl_3CCHCl_2 (e) $ClSi(CH_3)_3$

8.8. How many peaks would you expect to observe in the nmr spectra of
(a) cyclohexane, and   (b) benzene?

B.   The doublet

A proton with one neighboring, nonequivalent proton gives a signal that is split into a double peak, or **doublet**. In the following example, a pair of doublets is produced, one for each proton.

H_a *gives a doublet* ⟶   ⟵ H_b *also gives a doublet*

$$\begin{array}{cc} H_a & H_b \\ | & | \\ -C & -C- \\ | & | \end{array}$$

In the nmr spectrum of the preceding hypothetical structure, the δ value for each proton is the value at the *center* of the doublet (see Figure 8.33). The relative areas under the entire doublets in this case are $1:1$, reflecting the fact that each doublet arises from the absorption by one proton. (The two peaks *within* any doublet also have an area ratio that is ideally $1:1$, but may be slightly different, as may be seen for the aryl protons in Figure 8.32.)

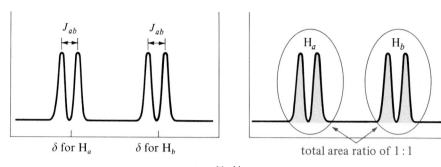

FIGURE 8.33. Spin–spin splitting pattern for $-\overset{\underset{\textstyle |}{H_a}}{\underset{\textstyle |}{C}}-\overset{\underset{\textstyle |}{H_b}}{\underset{\textstyle |}{C}}-$

The separation between the two peaks of a doublet is called the **coupling constant J**, and varies with the environment of the protons and their geometric relationship to each other. (Because splitting patterns are caused by internal forces, coupling constants are independent of the strength of H_0.) The symbol J_{ab} means the coupling constant for H_a split by H_b or for H_b split by H_a. For any pair of coupled protons, the J value is the same in each of the two doublets. By convention, J values are reported in Hz; therefore, the Hz scale at the top of an nmr spectrum is used to determine coupling constants. For a pair of neighboring, nonequivalent protons attached to freely rotating carbons, the J value is about 7 Hz.

Figure 8.32, the nmr spectrum of *p*-methoxybenzaldehyde, shows a pair of doublets in the aromatic region. The two protons labeled H_a are equivalent, as are the two H_b protons. H_a and H_b are nonequivalent to each other and are neighboring; their signals are split into a pair of doublets. Note that the peaks within the doublets are not perfectly symmetrical; the inside peaks of a doublet are taller. This phenomenon is called **leaning**.

STUDY PROBLEM

8.9.   Each of the following indicated protons or groups of protons gives rise to a doublet in the nmr spectrum. Tell how many peaks arise from the signal for the *other, single proton*. (Use the $n + 1$ rule.)

(a) $-\underline{C}H-CH-$ (b) $-\underline{C}H_2-CH-$ (c) $\underline{C}H_3-CH-$

SAMPLE PROBLEMS

Which of the following compounds would show a doublet (as well as other signals) in its nmr spectrum?

(a) CH_3CHCl_2 (b) $CH_3CH_2CH_2Cl$ (c) $CH_3CHClCH_3$

Solution: Underlined protons would appear as doublets: (a) $C\underline{H}_3CHCl_2$; (b) no doublet;   (c) $C\underline{H}_3CHCl\underline{C}H_3$ as one doublet and not as a pair of doublets. (Why?)

Give the total relative areas under all of the absorption bands for the compounds in the preceding problem.

Solution: (a) 3:1   (b) 3:2:2   (c) 6:1

Tell how many singlets and how many doublets would appear in the nmr spectrum of each of the following substituted benzenes:

(a) CH_3—⟨○⟩—OCH_3 (b) Cl—⟨○⟩—NH_2

Solution: (a) two singlets (CH_3 and OCH_3) and two doublets for the aryl protons; (b) one singlet (NH_2) and two doublets. (A pair of doublets in the aryl region is typical of 1,4-disubstituted benzenes.)

STUDY PROBLEM

$$\overset{O}{\overset{\|}{}}$$

8.10. The nmr spectrum of an aryl ketone (ArCR), C_8H_7ClO, is shown in Figure 8.34. What is the structure of this compound?

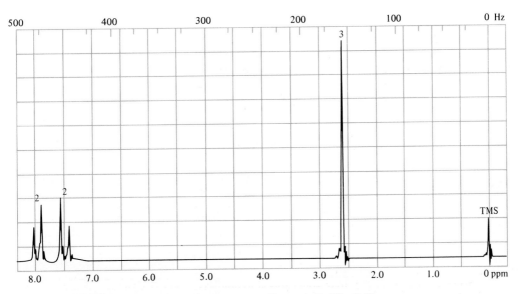

FIGURE 8.34. Nmr spectrum for unknown aryl ketone in Problem 8.10.

C. The triplet

If a proton (H_a) sees two neighboring protons equivalent to each other, but not equivalent to itself, the nmr signal of H_a is a **triplet** ($2 + 1 = 3$). If the two protons labeled H_b are equivalent, they give a signal that is split into a doublet by H_a.

H_a *sees two neighboring H's,*
signal is a triplet

H_b *sees one neighboring H,*
signal is a doublet

$$\begin{array}{cc} H_a & H_b \\ | & | \\ -C-C-H_b \\ | & | \end{array}$$

The nmr absorption pattern for all three protons in the preceding partial structure consists of a doublet and a triplet. The peaks in the triplet each are separated by the *same J value as that for the doublet.* The total width of the triplet (from side peak to side peak) is therefore $2J$ (see Figure 8.35). The areas under the entire triplet and the entire doublet in our example are in the ratio of 1 for H_a to 2 for H_b. (The relative areas *within* the triplet, however, are in a ratio of $1:2:1$. The reasons for this will be discussed in Section 8.11.)

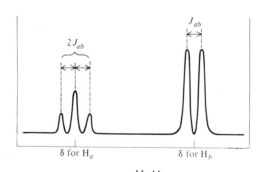

FIGURE 8.35. Spin–spin splitting pattern for $\begin{array}{cc} H_a & H_b \\ | & | \\ -C-C-H_b \\ | & | \end{array}$.

The nmr spectrum of 1,1,2-trichloroethane, which has a doublet and a triplet, is found in Figure 8.36.

$$\begin{array}{c} Cl \\ | \\ Cl-C-CH_2-Cl \\ | \\ H \end{array}$$

more shielded: upfield
sees one H: doublet
relative total area: 2

less shielded: downfield
sees two H's: triplet
relative total area: 1

STUDY PROBLEM

8.11. Suggest a reason for the fact that the proton in the Cl_2CH group in 1,1,2-trichloroethane is less shielded than the protons in the CH_2Cl group.

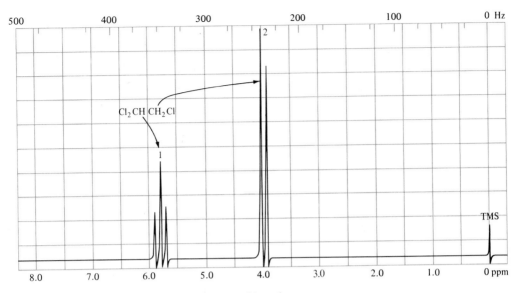

FIGURE 8.36. Nmr spectrum of 1,1,2-trichloroethane.

SAMPLE PROBLEM

Which of the following compounds shows a triplet (among other signals) in the nmr? How many triplets will each compound show?

$$\begin{array}{cc} Cl & OCH_3 \\ | & | \\ CH_2CH_2 \end{array}$$

(a) CH_2CH_2 (b) $CH_3CH_2-\bigcirc$ (c) $Cl_2CHCH_2CHCl_2$

Solution: (a) two triplets, one for each CH_2;   (b) one triplet for CH_3;   (c) two triplets, H_a split by two H_b's and H_b split by two H_a's.

$$\begin{array}{ccc} H_a & H_b & H_a \\ | & | & | \\ Cl_2C-C-CCl_2 \\ & | \\ & H_b \end{array}$$

STUDY PROBLEM

8.12. Figure 8.37 (page 348) shows the nmr spectrum of 2-phenylethyl acetate. Assign each signal to the proper protons

D.   The quartet

Consider a compound with a methyl group and one nonequivalent proton on the adjacent carbon.

quartet for H_a H_a H_b ⟵ *doublet for $CH_3(H_b)$*

$$\begin{array}{cc} H_a & H_b \\ | & | \\ -C-C-H_b \\ | & | \\ H_b \end{array}$$

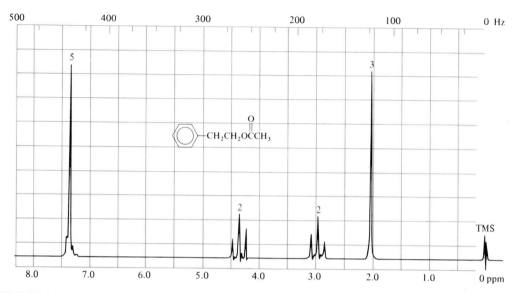

FIGURE 8.37. Nmr spectrum of 2-phenylethyl acetate.

The three equivalent methyl protons (H_b) see one neighboring proton and appear in the spectrum as a doublet, the total relative area of which is 3 (for three protons).

The signal arising from H_a is observed as a **quartet** $(3 + 1)$ because it sees three neighboring protons. The J values between each pair of peaks in the quartet is the same as the J value between the peaks in the doublet. In our example, the total area under the quartet for H_a is 1. (The area ratio *within* a quartet is $1:3:3:1$; see Figure 8.38.)

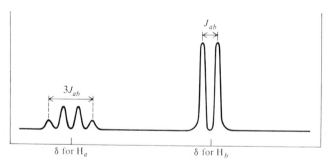

FIGURE 8.38. Spin–spin splitting pattern for

$$\begin{matrix} & H_a & H_b \\ & | & | \\ -C & -C & -H_b \\ & | & | \\ & & H_b \end{matrix}.$$

The ethyl group (CH_3CH_2-), which is very common in organic compounds, exhibits a characteristic nmr pattern—a triplet and a quartet.

CH_3 sees two neighboring H's and is split into a triplet

$$CH_3CH_2-$$

CH_2 sees three neighboring H's and is split into a quartet

The chemical shifts of an ethyl group are also characteristic. The CH_2 is often bonded to an electronegative atom, such as oxygen, which deshields the CH_2 protons. The quartet is thus observed downfield, while the triplet for the more shielded CH_3 group is observed upfield. The nmr spectrum of chloroethane (Figure 8.31, page 340) shows typical ethyl absorption. Figure 8.39 contains another example of an nmr spectrum that shows an ethyl pattern: an upfield triplet and a downfield quartet.

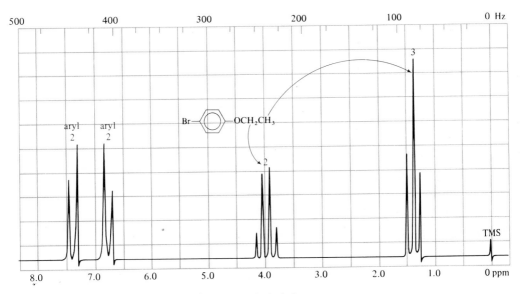

FIGURE 8.39. Nmr spectrum showing a typical ethyl pattern.

STUDY PROBLEM

8.13. Match each nmr spectrum in Figure 8.40 (pages 350–351) with a compound in the following list:

(a) $CH_3CO_2CH_2CH_3$ (b) CH_3CH_2—⟨◯⟩—I   (c) $CH_3CO_2CH(CH_3)_2$

(d) $CH_3CH_2CH_2NO_2$ (e) CH_3CH_2I (f) $(CH_3)_2CHNO_2$

E.   Chemical exchange and hydrogen bonding

On the basis of the preceding discussions, one would expect the nmr spectrum of methanol (Figure 8.21, page 331) to show a doublet (for the CH_3 protons) and a quartet (for the OH proton). If the spectrum of methanol is run at a very low temperature ($-40°$), this is exactly what is observed. However, if the spectrum is run at room temperature, only two singlets are observed.

The reason for this behavior of methanol is that alcohol molecules undergo rapid reaction with each other at room temperature in the presence of a trace of acid, exchanging OH protons in a process called **chemical exchange**. This exchange is so rapid that neighboring protons cannot distinguish any differences in spin

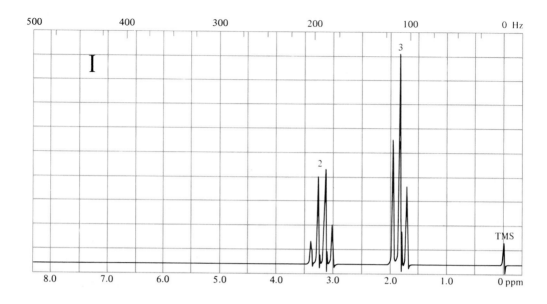

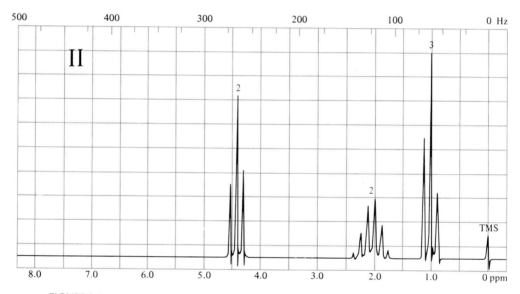

FIGURE 8.40. Nmr spectra for Problem 8.13.   *(continued)*

states and thus see only the average value of zero $[(+\frac{1}{2}) + (-\frac{1}{2}) = 0]$. Amines ($RNH_2$ and R_2NH) also undergo chemical exchange.

$$CH_3\overset{..}{O}:\overset{\frown}{+}H'^{+} \rightleftharpoons CH_3\overset{+}{\underset{\underset{H}{|}}{O}}H' \rightleftharpoons CH_3\overset{..}{\underset{..}{O}}H' + H^{+}$$

Of more practical importance to most organic chemists is the fact that the chemical shifts of OH and NH protons are *solvent- and concentration-dependent* because of hydrogen bonding. In a nonhydrogen-bonding solvent (such as CCl_4) and at low concentrations (1 % or less), OH proton absorption is observed at

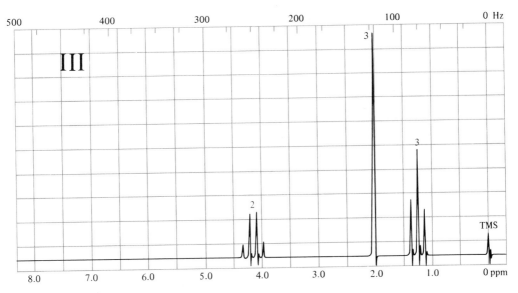

FIGURE 8.40. **(continued)** Nmr spectra for Problem 8.13.

a δ value of about 0.5 ppm. At the more usual, higher concentrations, the absorption is observed in the 4–5.5 ppm region because of hydrogen bonding between CH₃OH molecules. In hydrogen-bonding solvents, the OH proton absorption can be shifted even farther downfield.

F.   Other factors that affect splitting patterns

The spin–spin splitting patterns we have been discussing in this chapter are idealized cases. Most nmr spectra are more complex than those we have been using as examples. The complexity arises from a number of factors, of which we will mention only two. One factor that adds complexity to an nmr spectrum is the *nonequivalence of neighboring protons*:

$$\text{ClCH}_2\overset{\displaystyle O}{\underset{\displaystyle \underset{CHCl_2}{|}}{\overset{\displaystyle \|}{C}}} H\, CH \qquad \textit{signal is split by three types of nonequivalent, neighboring H's}$$

Another factor that adds complexity is the *magnitude of the chemical shift*. The *n* + 1 splitting patterns are truly apparent in an nmr spectrum only if the signals for coupled protons are separated from one another by a fairly large chemical shift. When the chemical shifts are close, the inside peaks of a multiplet increase in size, while the outside peaks decrease in size. This is the phenomenon of leaning, mentioned previously. When chemical shifts become very close, the peaks can coalesce into a singlet.

Figure 8.41 is the nmr spectrum of a typical alkane, *n*-octane. The chemical shifts of the CH₂ groups are all about the same (δ = 1.27 ppm); the CH₂ signal is a single, rather broad peak. The two methyl groups, which are equivalent to each other, are shielded compared to the methylene protons and absorb at a slightly

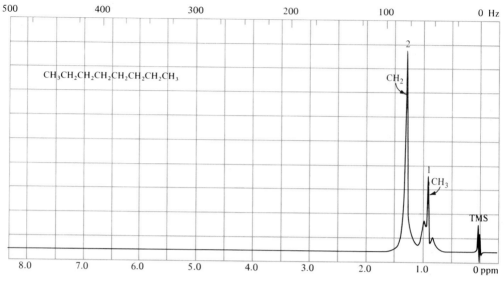

FIGURE 8.41. Nmr spectrum of *n*-octane.

higher field strength ($\delta = 0.83$ ppm). According to spin–spin splitting rules, the methyl absorption should be a triplet. With a little imagination, you can see the triplet leaning toward the methylene absorption peak.

SECTION 8.11.

Spin–Spin Splitting Diagrams

A **spin–spin splitting diagram**, also called a **tree diagram**, is a convenient technique for the analysis of splitting patterns. Let us consider H_a in the simple partial structure $>CH_a—CH_b<$. The splitting of the signal for H_a into a doublet by H_b may be symbolized by the following tree diagram, which we "read" starting at the top and moving toward the bottom.

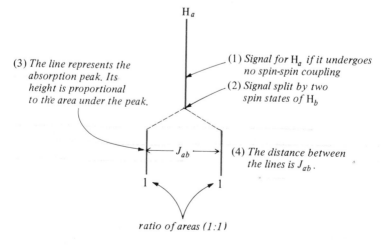

The splitting of the signal for H_b by the spin states of H_a may be represented by a similar tree diagram. These two diagrams may then be superimposed on an nmr spectrum (see Figure 8.42).

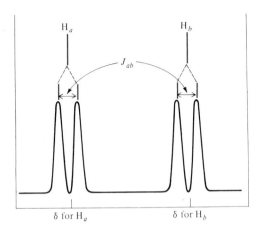

FIGURE 8.42. Spin–spin splitting pattern of two nonequivalent neighboring H's.

The tree diagram describing a triplet is a direct extension of that for a doublet. Consider the absorption pattern for H_a in the following grouping:

$$\begin{array}{cc} H_a & H_b \\ | & | \\ -C-C & -H_b \\ | & | \end{array}$$

In this case, H_a sees two neighboring protons and is split into a triplet. The coupling constant is J_{ab}. The triplet comes about because the peak for H_a is split twice, once for each H_b. Using a tree diagram, we can see the result of each of the two splits. H_a is first split into a pair of doublets, then the resulting two peaks are split again. We observe a triplet because the center two peaks absorb at the same position in the nmr. Consequently, the area of the center peak of the triplet is twice that of the two outside peaks.

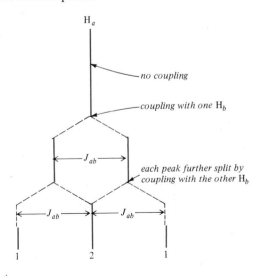

SAMPLE PROBLEMS

Draw a spin–spin splitting diagram for H_m of the following system, where $J_{am} = 10$ Hz and $J_{mx} = 5$ Hz:

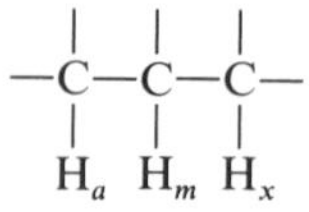

$$-\overset{|}{C}-\overset{|}{C}-\overset{|}{C}- \\ \; H_a\;H_m\;H_x$$

Solution:

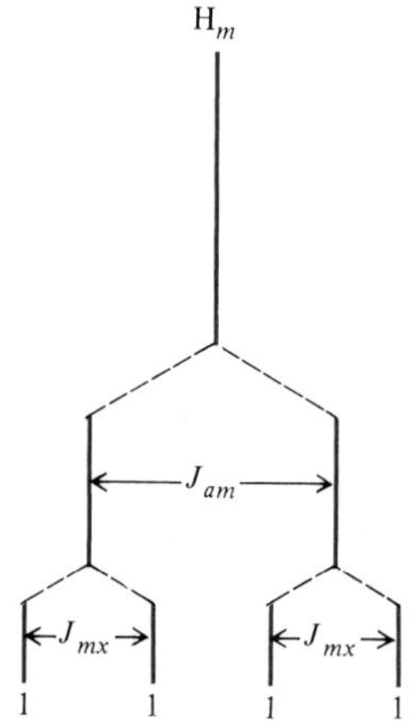

Note that the absorption pattern is not a triplet, as would be predicted by the $n + 1$ rule. Instead, four distinct lines of equal heights are observed because the coupling constants J_{am} and J_{mx} are not equal.

Draw a splitting diagram for the absorption of H_a in the partial structure

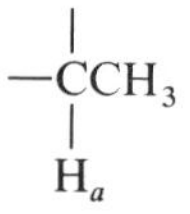

$$-\overset{|}{C}CH_3 \\ \; H_a$$

Solution:

all J values $= J_{ax}$

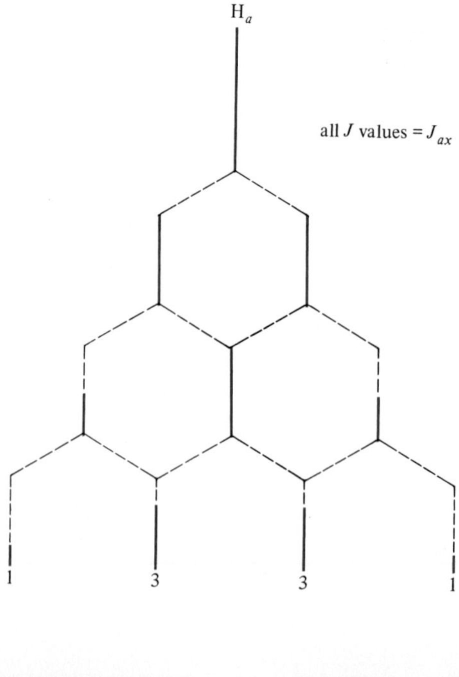

The original signal is split three times. The $1:3:3:1$ ratio of the areas arises from the fact that all the protons have the same coupling constant and consequently have superimposed absorption positions.

SECTION 8.12.

Carbon-13 Nmr Spectroscopy

Proton, or ^{1}H, nmr spectroscopy provides a picture of the hydrogen atoms in an organic molecule. Carbon-13, or ^{13}C, nmr spectrometers produce a picture of the *carbons* in an organic molecule. Carbon-13 spectra are not as widely used as proton spectra today, partly because ^{1}H nmr was developed first. We can expect a greater emphasis upon this new tool in the years ahead.

In ^{1}H nmr spectroscopy, we are dealing with the common natural isotope of hydrogen; 99.985% of natural hydrogen atoms are $^1_1$H. However, 98.9% of the carbon atoms in nature are $^{12}_6$C, an isotope with nuclei that have no spin. Carbon-13 constitutes only 1.1% of naturally occurring carbon atoms. Also, the transition of a ^{13}C nucleus from parallel to antiparallel is a low-energy transition. A ^{13}C nmr spectrometer must be extremely sensitive (about 6000 times more sensitive than a standard ^{1}H nmr instrument) to detect flips that are both few in number and low in energy absorption. However, recent advances in instrumentation allow the normally weak ^{13}C signals to be rapidly differentiated from background "noise."

While the low abundance of ^{13}C nuclei complicates instrument design, it also reduces the complexity of ^{13}C spectra compared to ^{1}H spectra. Although adjacent ^{13}C nuclei in a molecule will split each other's signals, the chances of finding adjacent ^{13}C nuclei are very small. For this reason, no ^{13}C–^{13}C splitting patterns are observed in ^{13}C spectra. On the negative side, the low abundance of ^{13}C nuclei means that the areas under the peaks are not necessarily proportional to the actual number of carbons in the sample. Area analysis is not used in ^{13}C nmr as it is in ^{1}H nmr.

There are two principal types of ^{13}C spectra: those that show ^{13}C–^{1}H spin–spin splitting patterns and those that do not. These two types of spectra are frequently used in conjunction with each other. In both types of spectra, TMS is used as an internal standard, and chemical shifts are measured downfield from the TMS peak. The chemical shifts in ^{13}C nmr are far greater than those observed in ^{1}H nmr. Most protons in ^{1}H nmr spectra show absorption between δ values of 0–10 ppm downfield from TMS; only a few, such as aldehyde or carboxyl protons, show peaks outside this range. Carbon-13 absorption is observed over a range of 0–200 ppm downfield from TMS. This large range of chemical shifts is another factor that simplifies ^{13}C spectra compared to ^{1}H spectra: in ^{13}C spectra we are less likely to observe overlapping absorption.

The relative shifts in ^{13}C nmr spectroscopy are roughly parallel to those in ^{1}H nmr spectroscopy. TMS absorbs upfield, while aldehyde and carboxyl carbons absorb far downfield. Figure 8.43 shows the general locations of absorption by different types of carbon atoms.

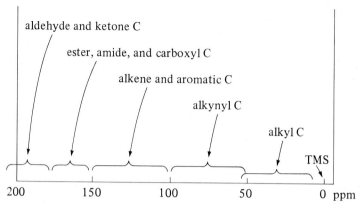

FIGURE 8.43. Relative positions of ^{13}C nmr absorption.

A.   Proton-decoupled spectra

A ^{13}C **proton-decoupled spectrum** is, as the name implies, a spectrum in which ^{13}C is not coupled with ^1H and thus shows no spin–spin splitting. Decoupling is accomplished electronically by the application of a second radiofrequency to the sample. This extra energy causes rapid interconversions between the parallel and antiparallel spin states of the protons. As a result, a ^{13}C nucleus sees only an average of the two spin states of the protons, and its signal is not split.

Because there is no splitting in a proton-decoupled spectrum, the signal for each group of magnetically equivalent carbon atoms appears as a *singlet*. By simply counting the number of peaks in the spectrum, we can determine the number of different types of carbon atoms in a molecule of the sample.

Figure 8.44 shows the proton-decoupled ^{13}C nmr spectrum of *n*-octane, a compound containing four types of nonequivalent carbons (carbons 1 and 8, carbons 2 and 7, carbons 3 and 6, carbons 4 and 5). The spectrum shows four singlets. Compare this spectrum with the ^1H spectrum of *n*-octane in Figure 8.41 (page 352). In the ^1H spectrum, the proton signals are crowded together in the 0–1.5 ppm area, and analysis of the peaks is impossible. In the ^{13}C spectrum, the signals are farther apart, in the 0–35 ppm range, and are easily distinguished.

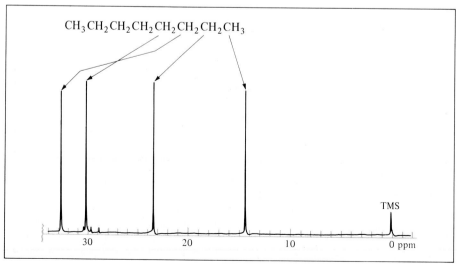

FIGURE 8.44. The proton-decoupled ^{13}C nmr spectrum of *n*-octane.

B. Proton-coupled spectra

The second common type of ^{13}C nmr spectra is the type in which the $^{13}C-^{1}H$ splitting is not suppressed. In this case, the signal for each carbon is split by the *protons bonded directly to it*. As in ^{1}H nmr spectroscopy, the $n + 1$ rule is followed. In ^{13}C nmr, n is the number of hydrogen atoms bonded to the carbon in question.

$$-\overset{|}{\underset{|}{C}}- \quad no\ H's:\ singlet \qquad\qquad -CH \quad one\ H:\ doublet$$

$$\overset{}{\underset{}{CH_2}} \quad two\ H's:\ triplet \qquad\qquad -CH_3 \quad three\ H's:\ quartet$$

An impressive amount of information can be deduced about the structure of a compound when the proton-coupled and the proton-decoupled spectra of the compound are compared. Consider the two spectra of vinyl acetate shown in Figure 8.45. Inspection of the spectra reveals that the carbonyl carbon at $\delta = 169$ ppm is a singlet in both spectra; therefore, this carbon is not bonded to any hydrogens. The signal for the $CH_3CO_2\underline{C}H=CH_2$ carbon is split into a doublet in the coupled spectrum, indicating one bonded H. Similarly, the $\underline{C}H_2$ signal at $\delta = 98$ ppm is a triplet, and the $\underline{C}H_3$ signal at $\delta = 20$ ppm is a quartet.

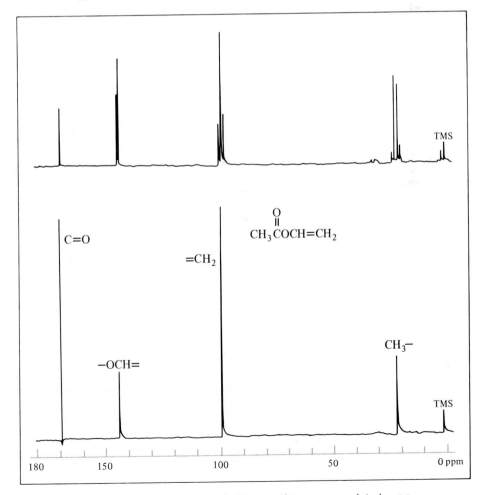

FIGURE 8.45. Coupled (top) and decoupled (bottom) ^{13}C nmr spectra of vinyl acetate.

STUDY PROBLEM

8.14. Figure 8.46 shows the ^1H and ^{13}C nmr spectra of a compound with the molecular formula $C_6H_{11}O_2Br$. What is the structure of this compound?

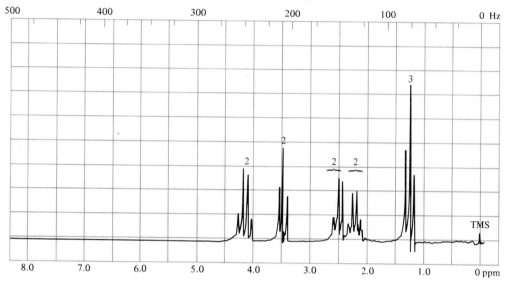

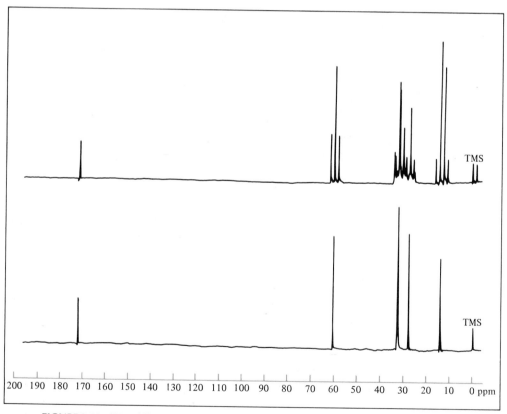

FIGURE 8.46. ^1H and ^{13}C nmr spectra for Problem 8.14.

SECTION 8.13.

Using Infrared and Nmr Spectra for the Identification of Organic Structures

Because ^{13}C nmr spectrometers are not yet widely used (especially by students), we will limit our discussion in this section to ^{1}H nmr spectroscopy. Also, throughout the rest of the book, when we speak of nmr spectroscopy, we are referring only to ^{1}H nmr.

From an infrared spectrum, we can deduce the identities of functional groups. From a proton nmr spectrum, we can often deduce the structure of the hydrocarbon portion of a molecule. Sometimes it is possible to deduce the complete structure of a compound from just the infrared and nmr spectra. More commonly, additional information is needed (such as chemical reactivity, elemental analysis, or other spectra). In this text, we provide additional information generally in the form of a molecular formula. From the molecular formula, we can calculate the number of rings or double bonds. (*Example*: C_6H_{12} is C_nH_{2n}; the structure contains either one double bond or a ring.) We can determine fragments of the structure from the spectra. Then, we try to match the fragments with the molecular formula.

SAMPLE PROBLEM

A compound has a molecular formula of C_3H_6O. Its infrared and nmr spectra are given in Figure 8.47. What is the structure of this compound?

Solution: From the molecular formula, we know that the compound has one oxygen; therefore, the compound must be an alcohol, an ether, an aldehyde, or a ketone. Because the molecular formula is of the type $C_nH_{2n}O$, we know that the structure contains either one double bond (C=C or C=O) or one ring. To distinguish between the possible functional groups, we use the infrared spectrum. We see strong absorption in the C=O region of 1750 cm^{-1} (5.8 μm). We conclude that the oxygen is not in a hydroxyl group or ether group, but rather is in a carbonyl group (aldehyde or ketone). This means, then, that no ring is present.

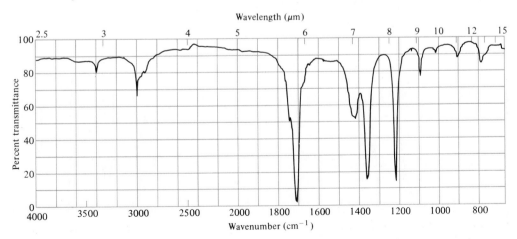

FIGURE 8.47. Spectra for C_3H_6O, Sample Problem.   (*continued*)

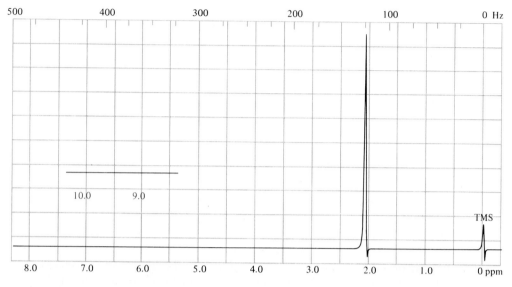

FIGURE 8.47. (continued) Spectra for C_3H_6O, Sample Problem.

In the nmr spectrum, we see no downfield, offset absorption; we conclude that the compound is a ketone, rather than an aldehyde. The rest of the nmr spectrum shows only a singlet; therefore all six hydrogens are equivalent and have no neighboring, nonequivalent hydrogens. The compound must be propanone (acetone): $(CH_3)_2C=O$.

STUDY PROBLEMS

8.15.   A compound has the formula C_3H_8O. Its infrared and nmr spectra are given in Figure 8.48. What is the structure of this compound?

8.16.   Figure 8.49 shows the infrared and nmr spectra for a compound with the molecular formula C_7H_8O. What is the structure of this compound?

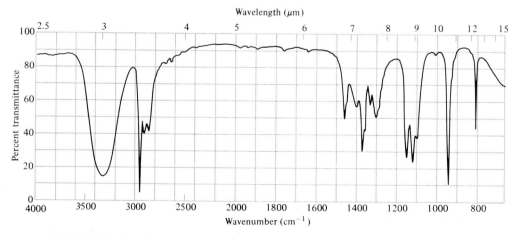

FIGURE 8.48. Spectra for C_3H_8O, Problem 8.15.   (continued)

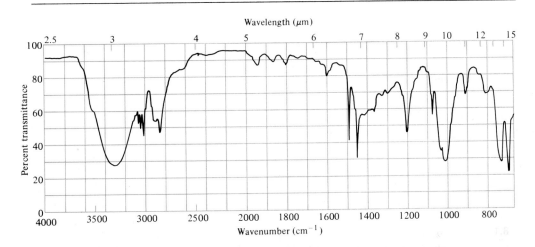

FIGURE 8.48. (continued) Spectra for C₃H₈O, Problem 8.15.

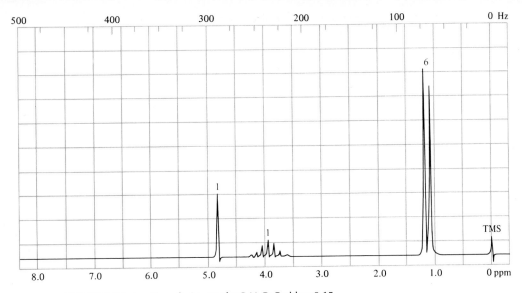

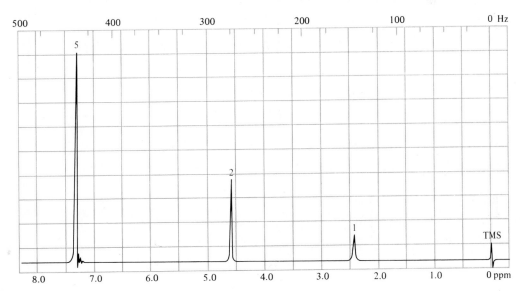

FIGURE 8.49. Spectra for C₇H₈O, Problem 8.16.

8.17.   Figure 8.50 shows nmr and infrared spectra for a compound with the molecular
formula $C_4H_7ClO_2$. What is the structure of the compound?

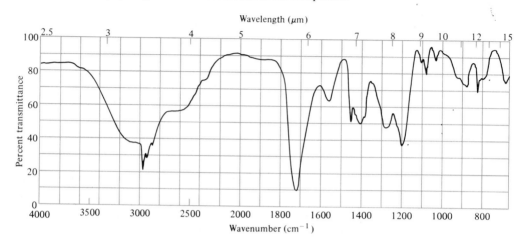

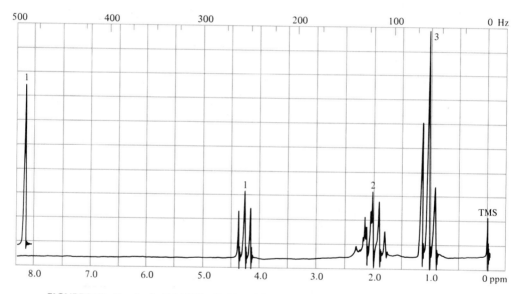

FIGURE 8.50. Spectra for $C_4H_7ClO_2$, Problem 8.17.

Summary

Organic compounds can absorb **electromagnetic radiation** of various wavelengths. Absorption in the **infrared region** results in **vibrational excitations** of bonds. Different types of bonds require differing amounts of energy for vibrational excitations. In an infrared spectrum, the region of 1400–4000 cm^{-1} (2.5–7.1 μm) is useful for determination of functional groups, while the region beyond is the **fingerprint region**.

 Nuclear magnetic resonance is the result of protons in a magnetic field (H_0) absorbing electromagnetic radiation in the radiofrequency region and flipping from the *parallel* to the *antiparallel spin state*. An **induced molecular magnetic field** can *shield* protons (oppose H_0) or *deshield* protons ("augment" H_0) and cause a *chemical shift* (δ) of the absorption band. The induced field is a result of **anisotropic effects** and **inductive effects**. A shielded proton absorbs *upfield*, close to the reference TMS, while a deshielded proton absorbs *downfield*.

 Spin–spin splitting of an absorption band results from the spin states of neighboring nonequivalent protons. The signal of a particular proton (or group of equivalent protons) is split into $n + 1$ peaks, where n is the number of neighboring protons equivalent to each other, but nonequivalent to the proton in question.

 The distance (in Hz) between any two peaks in a split band is the **coupling constant J**. For protons that are *coupled* (splitting the signals of each other), the J values are the same.

 The *area* under an entire absorption band is proportional to the relative number of protons giving rise to that signal.

 Carbon-13 nmr spectra show the absorption frequencies of ^{13}C atoms instead of protons. Spectra may show ^{13}C–^{1}H coupling or they may be **proton-decoupled**, in which case no splitting is observed. In ^{13}C nmr spectra the areas are not necessarily proportional to the numbers of C atoms giving rise to the signals.

STUDY PROBLEMS

8.18.   Make the following conversions: (a) 3000 cm^{-1} to μm;   (b) 1760 cm^{-1} to μm;
(c) 5.60 μm to cm^{-1};   (d) 8.20 μm to cm^{-1};   (e) 30 Hz to MHz.

8.19.   In otherwise similar compounds, which one of each of the following pairs of partial
structures would give stronger infrared absorption, and why?

(a) C=O   or   C=C   (b) C=C—Cl   or   C=C—H
(c) O—H   or   N—H

8.20.   Tell how you could distinguish each of the following pairs of compounds by their infrared
spectra:

(a) $CH_3CH_2CH_2N(CH_3)_2$ and $CH_3CH_2CH_2NH_2$
(b) $CH_3CH_2CH_2CO_2H$ and $CH_3CH_2CO_2CH_3$

$$\begin{array}{c} O \\ \| \end{array}$$

(c) $CH_3CH_2CCH_3$ and $CH_3CH_2CO_2CH_3$

8.21.   A chemist is oxidizing cyclohexanol to cyclohexanone. How can infrared spectroscopy be
used to determine when the reaction is completed?

8.22.   A compound with the molecular formula $C_5H_{10}O$ contains no double bonds and undergoes
reaction when heated with HI to yield $C_5H_{10}I_2$. The infrared spectrum of the original
compound shows absorption at about 2850 cm^{-1} (3.5 μm) and 1110 cm^{-1} (9 μm), but none
near 3330 cm^{-1} (3 μm) nor 1720 cm^{-1} (5.8 μm). What are the possible structures of the
compound and the diiodide?

8.23.   When an induced molecular magnetic field *opposes* H_0, a proton affected by this field:
(a) is *shielded* or *deshielded*?   (b) absorbs *upfield* or *downfield*?   (c) is to the *right* or *left* in
a typical nmr spectrum?

8.24.   At 60 MHz, how many Hz downfield from TMS is a chemical shift of $\delta = 7.5$ ppm?

8.25.   Indicate which underlined proton in each of the following groups of compounds will absorb
farther upfield:

(a) $CH_3CH_2C\underline{H}_2Cl$, $CH_3CHClC\underline{H}_3$, $C\underline{H}_2ClCH_2CH_3$

(b) $CH_3C\underline{H}_2Cl$, $CH_3C\underline{H}Cl_2$ (c) 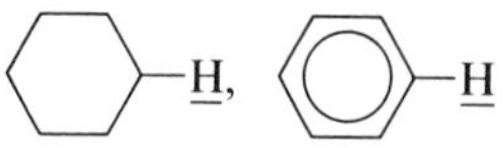 —$\underline{H}$, —$\underline{H}$

(d) 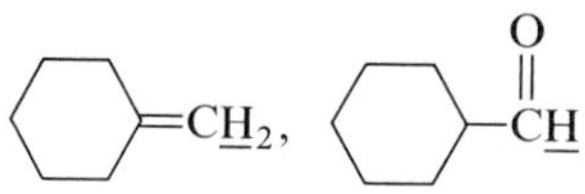 =$C\underline{H}_2$, —$C\underline{H}$

8.26.   Why would tetraethylsilane, $(CH_3CH_2)_4Si$, not be as good an internal standard as TMS in
nmr spectra?

8.27.   How many different groups of chemically equivalent protons are present in each of the
following structures? If more than one, indicate each group.

(a) $CH_3CH_2CH_2CH_3$ (b) $CH_3CH_2OCH_2CH_3$
(c) $CH_3CH=CH_2$ (d) *trans*-CHCl=CHCl

(e) *cis*-CHCl=CHCl
(g) (*R*)-2-chlorobutane
(i) $(CH_3)_2CHN(CH_3)CH(CH_3)_2$

(f) $CH_3CH_2CO_2CH_2CH_3$
(h) (*S*)-2-chlorobutane
(j) $BrCH_2CH_2CH(CH_3)_2$

(k) 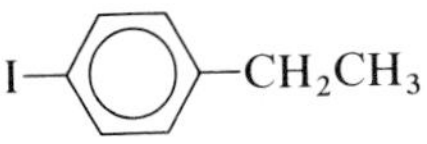 I—⟨O⟩—CH_2CH_3

(l) 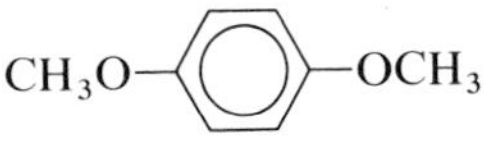 CH_3O—⟨O⟩—OCH_3

(m) 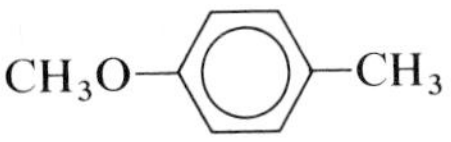 CH_3O—⟨O⟩—CH_3

8.28. In Problem 8.27, tell how many principal signals would probably be observed in the proton nmr spectrum of each compound, and predict the relative areas under the signals.

8.29. An nmr spectrum shows two principal signals with an area ratio of 3:1. Based on this information only, which of the following structures are possibilities?

(a) $CH_2=CHCH_3$
(c) $CH_3CH_2CH_2CH_2CH_2CH_3$
(e) $CH_2=C(CH_3)_2$

(b) $CH_2=CHCO_2CH_3$
(d) $CH_3CH_2CH_3$
(f) $CH_3CO_2CH_3$

8.30. Calculate the ratios of the different types of hydrogen atoms in a sample when the steps of the integration curve measure 81.5, 28, 55, and 80 mm.

8.31. For each of the following compounds, predict the *multiplicity* (number of peaks arising from spin–spin coupling) and the *total relative area* under the signal of each set of equivalent protons:

(a) $CH_3CH_2CO_2CH_3$ (b) $CH_3OCH_2CH_2OCH_3$ (c) $CH_3\overset{\displaystyle O}{\overset{\|}{C}}CH_2CH_3$

(d) 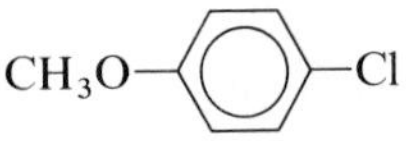 CH_3O—⟨O⟩—Cl   (e) Cl_2CHCH_2Br

8.32. A chemist has two isomeric chloropropanes, A and B. The nmr spectrum of A shows a doublet and a septet (seven peaks), while that of B shows two triplets and a sextet (six peaks). Identify the structures of A and B.

8.33. How would you distinguish between each of the following pairs of compounds by nmr spectroscopy?

(a) $CH_3CH_2CH_2CH_2CH=CH_2$ and $(CH_3)_2C=C(CH_3)_2$

(b) CH_3CH_2CHO and $CH_3\overset{\displaystyle O}{\overset{\|}{C}}CH_3$

(c) $CH_3\overset{\displaystyle O}{\overset{\|}{C}}CH_3$ and $CH_3\overset{\displaystyle O}{\overset{\|}{C}}OCH_3$

(d) 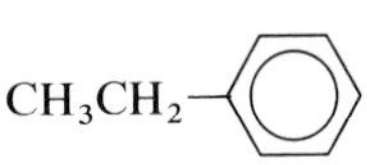 CH_3CH_2—⟨O⟩ and CH_3—⟨O⟩—CH_3

8.34. How would you use either infrared or nmr spectroscopy to distinguish between:

(a) 1-propanol and propylene oxide?

(b) diisopropyl ether and di-*n*-propyl ether?

(c) ethanol and 1,2-ethanediol?

(d) ⬡NCH$_3$ and ⬡NH?

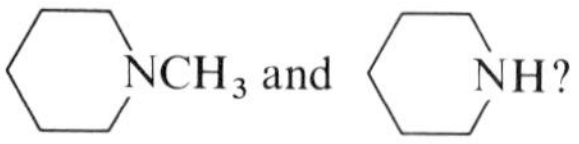

(e) ethanol and chloroethane?

(f) acetic acid and acetone?

8.35. Construct a tree diagram for each of the indicated protons:

(a) ClCH$_2$C$\underline{\text{H}}$$_2CH_2$Cl   (b) Cl$_2CHC\underline{\text{H}}$$_3$ (c) CH$_3$C$\underline{\text{H}}$$_2OCH_3$

8.36. Draw a tree diagram for H$_m$ in the following partial structure, where J_{am} = 5 Hz and J_{mx} = 11 Hz.

$$\begin{array}{c} \text{H}_a \\ | \\ \text{H}_a - \text{C} - \text{C} - \text{C} - \\ \quad | \quad | \quad | \\ \text{H}_a \ \text{H}_m \ \text{H}_x \end{array}$$

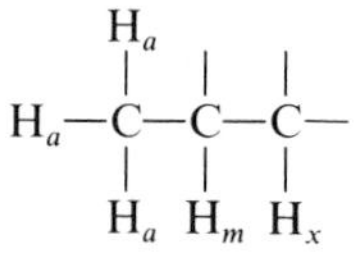

8.37. Predict (a) the number of principal peaks in the nmr spectrum of the local anesthetic *xylocaine* (following); (b) the splitting patterns of these peaks; (c) the characteristic absorption peaks in the infrared spectrum.

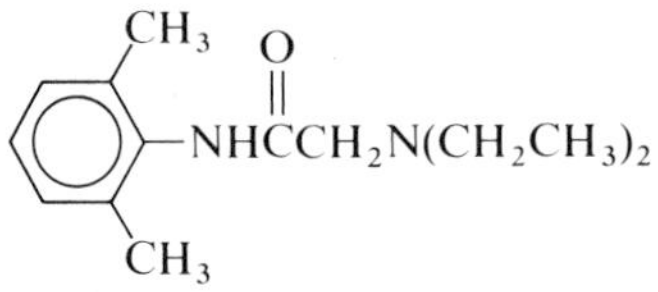

xylocaine

8.38. Sketch the expected nmr spectrum for 1,1-dichloroethane. Be sure to include (*qualitatively*) anticipated chemical shifts, splitting patterns, and appropriate areas.

8.39. Sketch the expected nmr spectrum for each of the following compounds:

(a) CH$_3$CHClCHClCH$_3$ (b) CH$_3$CH$_2$O—⬡—CH$_2$OCCH$_3$

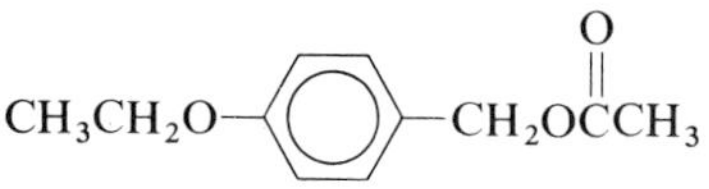

8.40. Under the influence of H_0, [18]annulene has an induced ring current not too different from that of benzene. Predict the shielding and deshielding of the protons of this ring system.

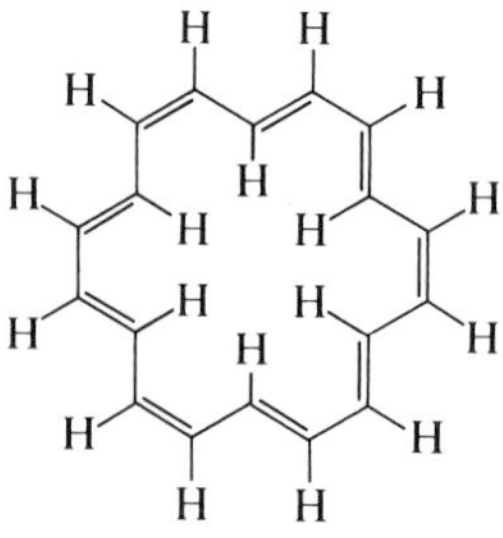

[18]annulene

8.41. The methyl protons of 15,16-dimethylpyrene do not have the same chemical shift as those of $C_6H_5CH_3$. Instead, their absorption is *upfield* from TMS ($\delta = -4.2$ ppm). Why?

15,16-dimethylpyrene

8.42. Predict the multiplicity of each absorption band in the proton-coupled ^{13}C nmr spectra of the following compounds:

$$\overset{\displaystyle OH}{\underset{\displaystyle |}{}}$$

(a) $(CH_3)_2CC\equiv CH$ 　　(b) $-COCH_2CH_3$

8.43. The ^1H nmr spectrum of an alcohol ($C_5H_{12}O$) shows the following absorption: one singlet (relative area 1); two doublets (areas 3 and 6); and two multiplets (areas both 1). When treated with HBr, the alcohol yields an alkyl bromide ($C_5H_{11}Br$). Its nmr spectrum shows only a singlet (area 6); a triplet (area 3); and a quartet (area 2). What are the structures of the alcohol and the alkyl bromide?

8.44. Figures 8.51 through 8.57 each gives a molecular formula, an infrared spectrum, and a ^1H nmr spectrum for an unknown compound. What is the structure of each compound?

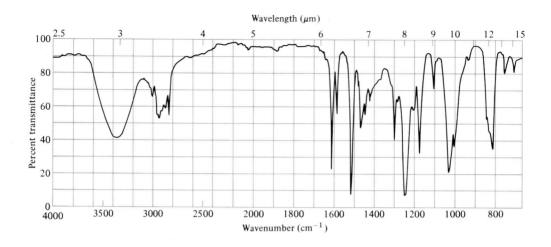

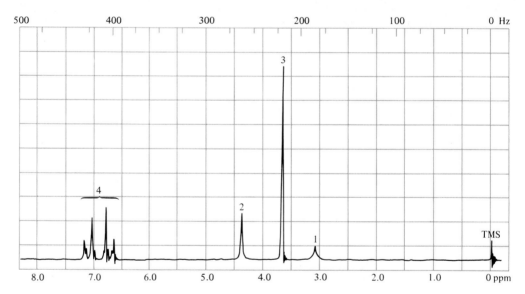

FIGURE 8.51. Spectra for Problem 8.44 (a): $C_8H_{10}O_2$.

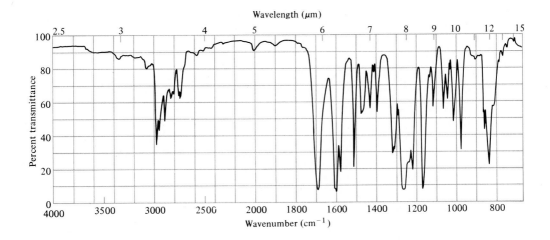

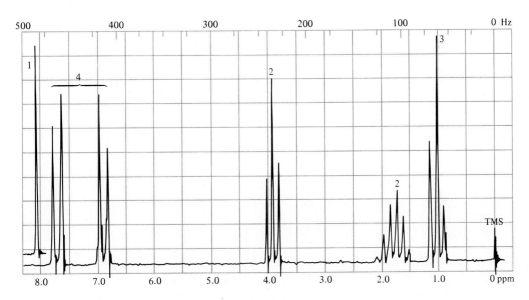

FIGURE 8.52. Spectra for Problem 8.44 (b): $C_{10}H_{12}O_2$.

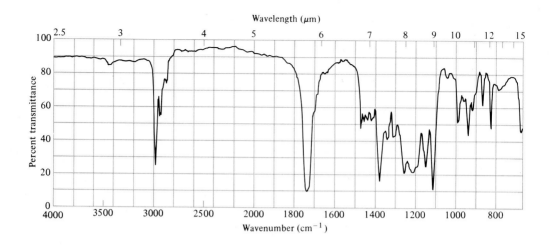

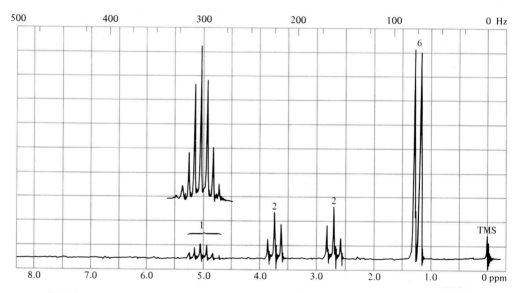

FIGURE 8.53. Spectra for Problem 8.44 (c): $C_6H_{11}O_2Cl$.

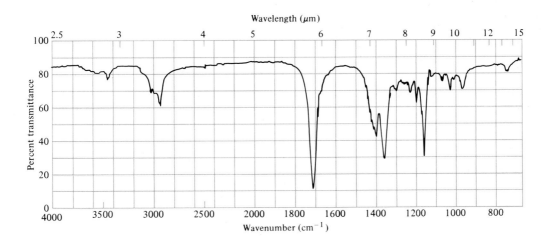

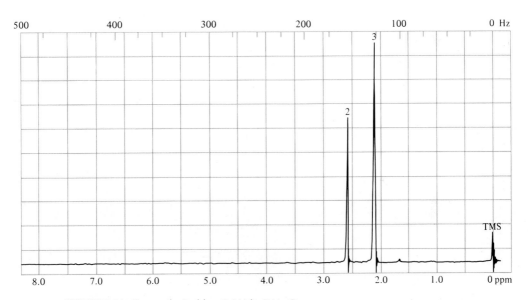

FIGURE 8.54. Spectra for Problem 8.44 (d): $C_6H_{10}O_2$.

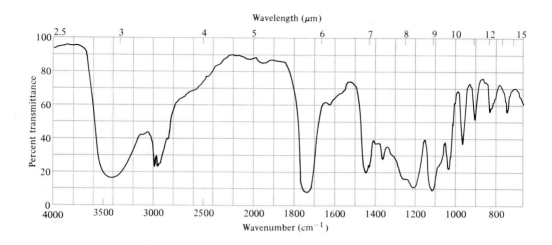

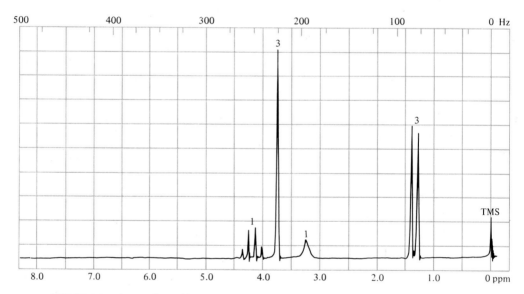

FIGURE 8.55. Spectra for Problem 8.44 (e): $C_4H_8O_3$.

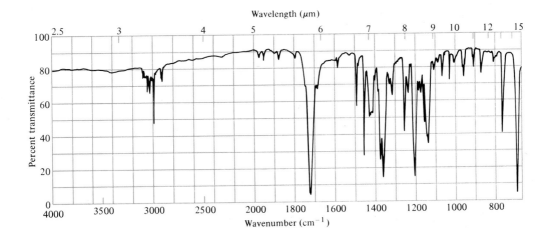

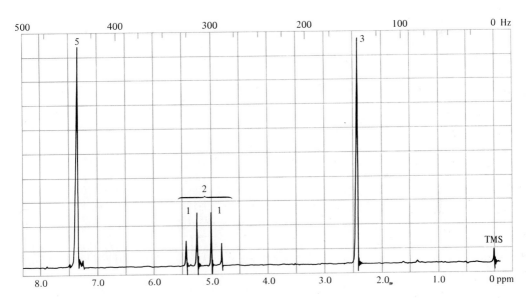

FIGURE 8.56. Spectra for Problem 8.44 (f): $C_{10}H_{10}Br_2O$.

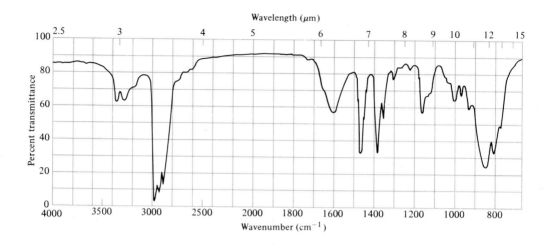

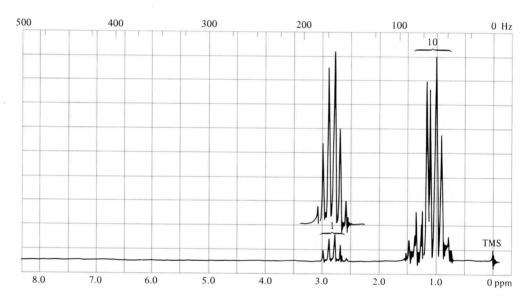

FIGURE 8.57. Spectra for Problem 8.44 (g): $C_4H_{11}N$.

CHAPTER 9

Alkenes and Alkynes

n **alkene** is a hydrocarbon with one double bond. Alkenes are sometimes called **olefins,** from *olefiant gas* ("oil-forming gas"), an old name for ethylene (CH_2=CH_2). An **alkyne** is a hydrocarbon with one triple bond; acetylene (CH≡CH) is the simplest alkyne.

A carbon–carbon double bond is a common functional group in natural products. Most frequently, the double bond is found in conjunction with other functional groups. However, alkenes with no other functionality are not at all rare and are often found as plant products and in petroleum. Two interesting examples of naturally occurring compounds containing carbon–carbon double bonds follow.

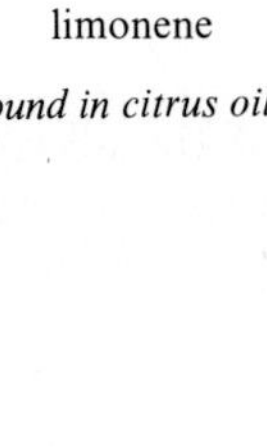

limonene

found in citrus oils

$$(CH_3)_2C=CHCH_2CH_2\overset{\overset{\displaystyle CH_3}{|}}{C}=CHCH_2CH_2\overset{\overset{\displaystyle CH_2}{||}}{C}CH=CH_2$$

3-methylene-7,11-dimethyl-1,6,10-dodecatriene

a compound secreted by aphids that signals "danger" to other aphids

SECTION 9.1.

Bonding in Alkenes and Alkynes; Acidity of Alkynes

The bonding in ethylene and acetylene was discussed in detail in Chapter 2. Recall that the two carbon atoms in ethylene are in the sp^2-hybrid state. The three sp^2 bonds from each carbon atom lie in the same plane with bond angles of approximately 120°. The pi bond joining the two sp^2 carbons lies above and below the plane of the sigma bonds (see Figure 2.14, page 52).

All the atoms in the ethylene molecule lie in the same plane; however, in a molecule that also has sp^3 carbons, only those atoms bonded to the double-bond carbons lie in the same plane.

these atoms lie in a plane

The electronic structure of the triple bond of an alkyne is very similar to that of the double bond in an alkene. Acetylene has two sp carbons with linear sigma bonds and *two* pi bonds joining the sp carbons (Figure 2.18, page 57).

$$\overset{180°}{H-C\equiv C-H}$$

acetylene

A triple-bond carbon is in the sp-hybrid state. The sp orbital is one-half s, while an sp^2 orbital is one-third s and an sp^3 orbital is only one-fourth s. Because an sp orbital has more s character, the electrons in this orbital are closer to the carbon nucleus than electrons in an sp^2 or sp^3 orbital. (See Section 2.4F.) In an alkyne, the sp carbon is therefore *more electronegative* than most other carbon atoms. Thus, an alkynyl CH bond is *more polar* than an alkane CH bond or an alkene CH bond.

$$\overset{\delta+}{R}-\overset{\delta-}{C}\equiv\overset{\delta-}{C}-\overset{\delta+}{H}$$

sp carbons electron-withdrawing

One of the most important results of the polarity of the alkynyl carbon–hydrogen bond is that RC≡CH can lose a hydrogen ion to a strong base. The resulting anion (RC≡C⁻) is called an **acetylide ion**. With a pK_a of 26, alkynes are not strong acids. They are weaker acids than water (p$K_a \sim 15$), but stronger acids than ammonia (p$K_a \sim 35$). Alkynes undergo reaction with a strong base like sodamide ($NaNH_2$) or a Grignard reagent or with sodium metal. Alkanes and alkenes do not react under these conditions.

$$CH_3C{\equiv}C{-}H + :\ddot{N}H_2^- \xrightarrow{\text{liq. } NH_3} CH_3C{\equiv}C{:}^- + :NH_3$$

propyne   an acetylide ion

$$CH_3C{\equiv}CH + CH_3MgI \longrightarrow CH_3C{\equiv}CMgI + CH_4$$

$$CH_3C{\equiv}CH + Na \longrightarrow CH_3C{\equiv}C^- + Na^+ + \tfrac{1}{2}H_2$$

SECTION 9.2.

Nomenclature of Alkenes and Alkynes

In the IUPAC system, the continuous-chain alkenes are named after their alkane parents, but with the **-ane** ending changed to **-ene**. For example, CH_3CH_3 is ethane and $CH_2{=}CH_2$ is ethene (trivial name, ethylene).

$$CH_2{=}CH_2 \qquad CH_3CH{=}CH_2$$

IUPAC: ethene   propene

cyclohexene

A hydrocarbon with two double bonds is called a **diene**, while one with three double bonds is called a **triene**. The following examples illustrate diene and triene nomenclature:

$$CH_2{=}CHCH{=}CH_2 \qquad \overset{\overset{\textstyle CH_3}{|}}{CH_2{=}CCH{=}CH_2} \qquad CH_2{=}CHCH{=}CHCH{=}CH_2$$

1,3-butadiene   2-methyl-1,3-butadiene   1,3,5-hexatriene.

a diene   *a diene*   *a triene*

In the names of most alkenes, we need a prefix number to show the position of the double bond. Unless there is functionality of higher nomenclature priority, the chain is numbered from the end that gives the lowest number to the double bond. The prefix number specifies the carbon atom in the chain where the double bond begins.

$$\overset{\overset{\textstyle CH_3}{|}}{CH_3C}{=}CHCH_2CH_3 \qquad CH_2{=}CHCH_2CH_2OH \qquad \overset{\overset{\textstyle O}{\|}}{CH_3CH{=}CHCOH}$$

2-methyl-2-pentene   3-buten-1-ol   2-butenoic acid

SAMPLE PROBLEM

Write the IUPAC name for $\underset{\overset{\|}{CH_2}}{CH_3CH_2CH_2CCH_2CH_2CH_3}$.

Solution: While indeed the molecule does have a seven-carbon chain, the longest continuous chain that *contains the double bond* has only five carbons. Consequently, the structure is named as a pentene. The IUPAC name is 2-propyl-1-pentene.

Some alkenes and alkenyl groups have trivial names that are in common use. A few of these are summarized in Table 9.1.

A pi bond prevents free rotation of groups around a double bond; consequently, alkenes may exhibit geometric isomerism. This topic and the nomenclature of geometric isomers was covered in Section 4.1. (Because alkynes are linear molecules, they do not exhibit geometric isomerism.)

$$
\begin{array}{cc}
\underset{H_3C}{}\diagup C=C\diagup \underset{H}{}CH_3 & \underset{H_3C}{}\diagup C=C\diagup \underset{CH_3}{}H
\end{array}
$$

cis-2-butene   *trans*-2-butene

or (*Z*)-2-butene   or (*E*)-2-butene

The IUPAC nomenclature of alkynes is directly analogous to that of the alkenes. The suffix for an alkyne is **-yne**, and a position number is used to signify the position of the triple bond in the parent hydrocarbon chain. Unless there is functionality of higher nomenclature priority in the molecule, the chain is numbered to give the triple bond the lowest number.

In an older, trivial system of nomenclature for the simple alkynes, acetylene ($CH{\equiv}CH$) is considered the parent. Groups attached to the *sp* carbons are named

---

TABLE 9.1. Trivial names of some alkenes and alkenyl groups

Structure	*Name*	*Example*
alkenes:		
$CH_2{=}CH_2$	ethylene	—
$CH_3CH{=}CH_2$	propylene	—
$CH_3\overset{\displaystyle CH_3}{\underset{\displaystyle \vert}{C}}{=}CH_2$	isobutylene	—
$CH_2{=}\overset{\displaystyle CH_3}{\underset{\displaystyle \vert}{C}}CH{=}CH_2$	isoprene	—
$CH_2{=}C{=}CH_2$	allene	—
alkenyl groups:		
$CH_2{=}$	methylene[a]	$\bigcirc{=}CH_2$
		methylenecyclohexane
$CH_2{=}CH{-}$	vinyl	$CH_2{=}CHCl$
		vinyl chloride
$CH_2{=}CHCH_2{-}$	allyl	$CH_2{=}CHCH_2Br$
		allyl bromide

[a] The term *methylene* is also used to refer to a disubstituted *sp*³ carbon ($-CH_2-$); for example, CH_2Cl_2 is called methylene chloride.

as substituents on acetylene. In this text we will use the IUPAC nomenclature system for the alkynes except for acetylene itself.

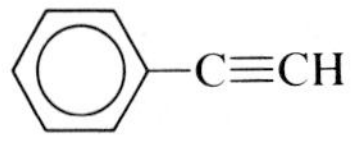

$\langle\bigcirc\rangle$—C≡CH	$CH_3C≡CCH_2CH_3$
IUPAC: phenylethyne	2-pentyne
trivial: phenylacetylene	ethylmethylacetylene

STUDY PROBLEMS

9.1.   Give the structures of   (a) (*E*)-1-chloro-3,4-dimethyl-3-hexene;
(b) *cis*-1,3-pentadiene;   (c) cyclohexylethyne;   (d) diphenylethyne.

9.2.   Name the following compounds:

$$CH_3$$
(a) $CH_2{=}CCH_2CH{=}CH_2$ (b) $CH≡CCH_2OH$

SECTION 9.3.

Physical Properties of Alkenes and Alkynes

The physical properties of alkenes (but not the chemical properties) are practically identical to those of their alkane parents. Table 9.2 lists the boiling points of a few alkenes and alkynes. The boiling points of an homologous series of alkenes

TABLE 9.2. Physical properties of some alkenes and alkynes

Name	*Structure*	*Bp,* °*C*
alkenes:		
ethene (ethylene)	$CH_2{=}CH_2$	−102
propene (propylene)	$CH_3CH{=}CH_2$	−48
methylpropene (isobutylene)	$(CH_3)_2C{=}CH_2$	−7
1-butene	$CH_3CH_2CH{=}CH_2$	−6
1-pentene	$CH_3CH_2CH_2CH{=}CH_2$	30
alkynes:		
ethyne (acetylene)	$CH≡CH$	−75
propyne	$CH_3C≡CH$	−23
1-butyne	$CH_3CH_2C≡CH$	8.1
2-butyne	$CH_3C≡CCH_3$	27

increase about 30° per CH_2 group. This is the same increase observed for an homologous series of alkanes. As with alkanes, branching in an alkene lowers the boiling point slightly.

Although alkenes are considered to be nonpolar, they are slightly more soluble in water than the corresponding alkanes because the pi electrons, which are somewhat exposed, are attracted to the partially positive hydrogen of water.

SECTION 9.4.

Spectra of Alkenes and Alkynes

A. Infrared spectra

Alkenes.   Stretching of the C=C double bond gives rise to absorption at 1600–1700 cm^{-1} (5.8–6.2 μm) in the infrared spectrum. Because the double bond is nonpolar, the stretching results in only a small change in bond moment; consequently, the absorption is weak, ten to 100 times less intense than that of a carbonyl group. The absorption due to the stretching of the alkenyl, or vinylic, carbon–hydrogen bond (=C—H) at about 3000–3100 cm^{-1} (3.2–3.6 μm) is also weak. Alkenyl carbon–hydrogen bonds exhibit bending absorption in the fingerprint region of the infrared spectrum (see Table 9.3). Figure 9.1 shows the spectra of heptane and 1-heptene; the differences between an alkane and an alkene are evident in these spectra.

Alkynes.   The C≡C stretching frequency of alkynes is at 2100–2250 cm^{-1} (4.4–4.8 μm). This absorption is quite weak and can easily be lost in the background noise of the spectrum. However, with the exception of C≡N and Si—H, no other groups show absorption in this region. The ≡C—H stretching frequency is found at about 3300 cm^{-1} (3.0 μm) as a sharp peak (see Figure 9.2).

TABLE 9.3. Infrared absorption characteristic of alkenes and alkynes

	Position of absorption	
Type of vibration	*cm^{-1}*	*μm*
alkenes:		
=C—H stretching	3000–3100	3.2–3.3
=C—H bending	800–1000	10.0–12.5
=CH$_2$ bending	855–885	11.2–11.3
C=C stretching	1600–1700	5.8–6.2
alkynes:		
≡C—H stretching	~3300	~3.0
C≡C stretching	2100–2250	4.4–4.8

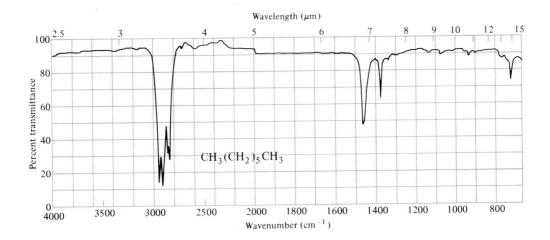

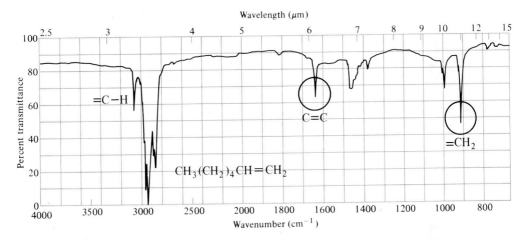

FIGURE 9.1. Infrared spectra of heptane and 1-heptene.

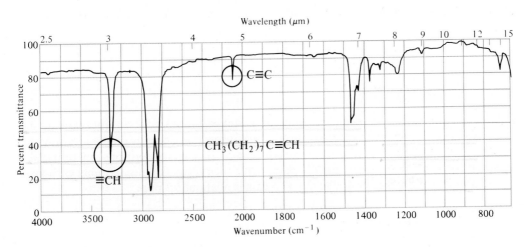

FIGURE 9.2. Infrared spectrum of 1-decyne.

STUDY PROBLEM

9.3.   Figure 9.3 is the spectra of two alcohols. One compound contains a carbon–carbon double bond, and one contains a carbon–carbon triple bond. Identify each.

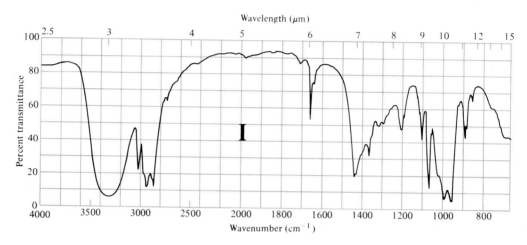

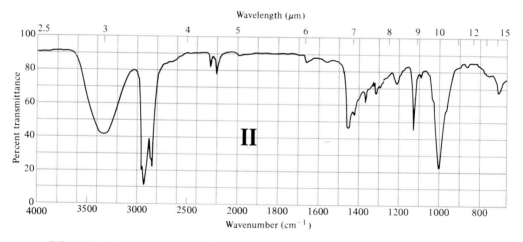

FIGURE 9.3. Infrared spectra for Problem 9.3.

B.   Nmr spectra

Alkenes. The chemical shift for a vinylic proton is at an approximate δ value of 5.0 ppm; the exact position of the absorption depends on the location of the double bond in the hydrocarbon chain. In general, protons on terminal alkenyl carbons absorb near 4.7 ppm, while the protons on nonterminal carbons absorb slightly farther downfield, at δ values of about 5.3 ppm.

$$CH_3CH_2CH{=}CH_2$$

5.3 ppm   4.7 ppm

The splitting patterns for vinylic protons are more complex than those for alkyl protons. The complexity arises from the lack of rotation around the double bond. Let us look at a general example:

$$\underset{R}{\overset{H_x}{\diagdown}}C=C\underset{H_b}{\overset{H_a}{\diagup}}$$

In this example, all three vinylic protons (H_a, H_b, and H_x) are nonequivalent and thus have different chemical shifts and give rise to three separate signals. In addition, the coupling constants between any two of the protons (J_{ax}, J_{bx}, and J_{ab}) are different. Each of the three signals is therefore split into four peaks. (For example, the signal for H_x is split into two by H_b and again into two by H_a.) A total of twelve peaks is observed in the nmr spectrum for these three protons.

This pattern of twelve peaks can be seen in the nmr spectrum of *p*-chlorostyrene in Figure 9.4. Figure 9.5 shows tree diagrams for the splitting patterns of the double-bond protons in *p*-chlorostyrene. With the aid of the tree diagram, we can see that none of the absorption peaks are superimposed because of the differences in *J* values.

In the spectrum of *p*-chlorostyrene (Figure 9.4), note that the chemical shift for H_a is at $\delta = 5.3$ ppm, while the chemical shift for H_b, which is *cis* (and closer) to the benzene ring, is at $\delta = 5.7$ ppm. The signal for H_b is downfield because H_b is somewhat deshielded by the induced field of the benzene ring. The signal for H_x is even farther downfield because H_x is more deshielded by the induced field of the ring.

Although any terminal vinylic group of the type $RCH=CH_2$ should give a spectrum with twelve vinyl-proton peaks, the twelve peaks are not always evident. For example, in the spectrum of *p*-chlorostyrene, the four peaks for H_a almost

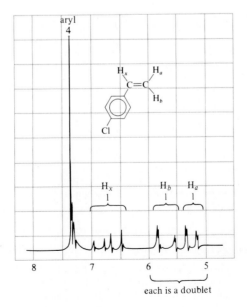

FIGURE 9.4. Partial nmr spectrum of *p*-chlorostyrene.

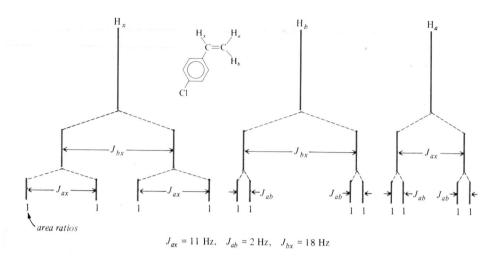

$$J_{ax} = 11 \text{ Hz}, \quad J_{ab} = 2 \text{ Hz}, \quad J_{bx} = 18 \text{ Hz}$$

FIGURE 9.5. Tree diagrams for the splitting patterns of the three alkenyl protons in
p-chlorostyrene.

look like two peaks because J_{ab} is small. Figure 9.6, the nmr spectrum of styrene,
shows almost-superimposed vinyl-proton signals.

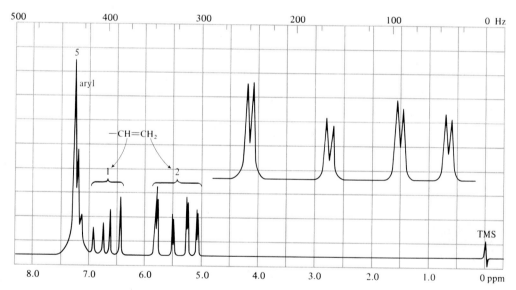

FIGURE 9.6. Nmr spectrum of styrene, $C_6H_5CH=CH_2$. (The superimposed line is a
higher-resolution, expanded spectrum of the eight peaks with δ values between
5.0 and 6.0 ppm.)

STUDY PROBLEMS

9.4. Figure 9.7 is the nmr spectrum of methyl vinyl sulfone, $CH_2=CHSO_2CH_3$. The
spectrum contains eight peaks (not 12) in the alkene region. Construct a tree
diagram that would explain the "missing" peaks.

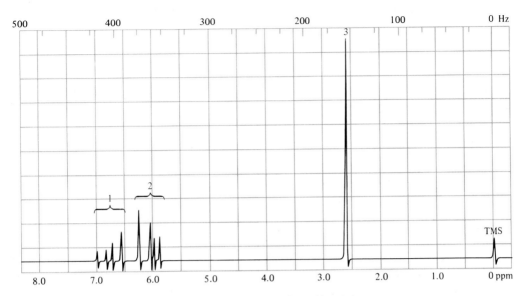

FIGURE 9.7. Nmr spectrum of $CH_2{=}CHSO_2CH_3$ for Problem 9.4.

9.5. In the nmr spectrum for *p*-chlorostyrene, the absorption for the aryl protons is observed as a *singlet*. In styrene ($C_6H_5CH{=}CH_2$), the absorption for the aryl protons is observed as a *multiplet*. Suggest a reason for these observations.

Alkynes. An alkyne of the type $RC{\equiv}CR$ has no acetylenic protons; therefore, a disubstituted alkyne has no characteristic nmr absorption. (The rest of the molecule may give rise to absorption, however.) A monosubstituted alkyne, $RC{\equiv}CH$, shows absorption for the alkynyl proton at a δ value of about 3 ppm. This absorption is not nearly as far downfield as that for a vinylic or aryl proton because the alkynyl proton is *shielded* by the induced field of the triple bond. Figure 9.8

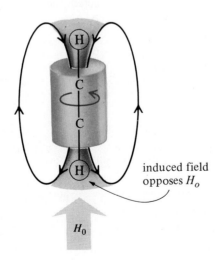

FIGURE 9.8. An alkynyl proton is shielded by the induced magnetic field of the triple bond.

shows how the circulation of the pi electrons results in this field. Note the difference between this anisotropic effect and the effect for a vinylic proton, $=CHR$ (Section 8.7B). In the case of an alkyne, the induced field opposes, rather than augments, H_0.

SECTION 9.5.

Preparation of Alkenes and Alkynes

Alkenes may be prepared by the elimination reactions of alcohols (in strong acid) or alkyl halides (in base). Table 9.4 (page 388) lists the sections where these reactions are discussed in detail.

Primary alcohols undergo elimination reactions slowly. In hot, concentrated H_2SO_4, the product alkene may also undergo isomerization and other reactions; therefore, primary alcohols are not usually useful in alkene preparation. *Primary alkyl halides* also undergo elimination reactions slowly by an E2 path. However, if a bulky base such as the *t*-butoxide ion is used, an alkene may be obtained in good yield (along with some S_N2 product).

1° RX (E2 and S_N2):

$$CH_3CH_2CH_2CH_2CH_2Br \xrightarrow[\text{warm}]{K^+ \ ^-OC(CH_3)_3}$$

1-bromopentane

$$\begin{cases} CH_3CH_2CH_2CH=CH_2 \\ \text{1-pentene} \\ 85\% \\ \\ + \\ \\ CH_3CH_2CH_2CH_2CH_2OC(CH_3)_3 \\ \text{t-butyl pentyl ether} \\ 12\% \end{cases}$$

Secondary alcohols undergo elimination by an E1 path when heated with a strong acid, and rearrangement of the intermediate carbocation may occur. Except in simple cases, secondary alcohols are therefore not useful intermediates for the preparation of alkenes. *Secondary alkyl halides* can undergo E2 reactions. Although product mixtures may be expected, the predominant product is usually the more highly substituted *trans*-alkene.

2° RX (E2):

$$\underset{\text{2-bromopentane}}{CH_3CH_2CH_2\overset{\overset{\displaystyle Br}{|}}{C}HCH_3} \xrightarrow[\text{warm}]{Na^+ \ ^-OCH_2CH_3}$$

$$\begin{cases} \underset{H}{\overset{CH_3CH_2}{\diagup}}C{=}C\underset{CH_3}{\overset{H}{\diagup}} \\ \text{\textit{trans}-2-pentene} \\ 51\% \\ \\ + \\ \\ \underset{H}{\overset{CH_3CH_2}{\diagup}}C{=}C\underset{H}{\overset{CH_3}{\diagup}} \\ \text{\textit{cis}-2-pentene} \\ 18\% \\ \\ + \\ \\ CH_3CH_2CH_2CH=CH_2 \\ \text{1-pentene} \\ 31\% \end{cases}$$

Tertiary alcohols undergo elimination readily through carbocations (E1) when treated with a strong acid. *Tertiary alkyl halides* undergo elimination with base principally by the E2 reaction. In both cases, excellent yields can be obtained if all three R groups of R_3CX or R_3COH are the same; otherwise, mixtures might be obtained.

STUDY PROBLEM

9.6.   Each of the following reactions yields one alkene in a yield of over 80 percent. Give the structure of each alkene product.

(a) 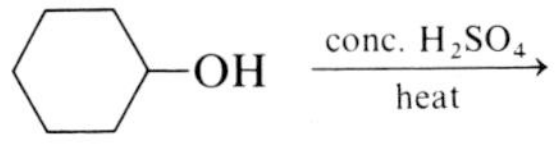 (cyclohexyl)—OH $\xrightarrow[\text{heat}]{\text{conc. } H_2SO_4}$

(b) $(CH_3)_3CBr$ $\xrightarrow[\text{CH}_3\text{CH}_2\text{OH}]{\text{Na}^+ \ ^-OCH_2CH_3}$

(c) $CH_3CH_2\overset{\overset{\displaystyle OH}{|}}{C}(CH_3)_2$ $\xrightarrow[\text{heat}]{50\% \ H_2SO_4}$

Alkynes may also be prepared by elimination reactions. In the following examples, note that a stronger base than OH^- is used for dehydrohalogenation of the vinylic halide. The reason for this is that the sp^2 bonds of a vinylic halide are stronger than the sp^3 bonds of an alkyl halide. (Why?)

$$\underset{\text{1,2-dibromopropane}}{CH_3\overset{\overset{\displaystyle Br}{|}}{CH}-\overset{\overset{\displaystyle Br}{|}}{CH_2}} \xrightarrow[-\text{HBr}]{OH^-} \underset{\text{1-bromopropene}}{CH_3CH=\overset{\overset{\displaystyle Br}{|}}{CH}} \xrightarrow[\substack{-\text{HBr} \\ -\text{H}^+}]{NH_2^-} \underset{\text{an acetylide}}{CH_3C\equiv C^-} \xrightarrow{H_2O} \underset{\text{propyne}}{CH_3C\equiv CH}$$

In Section 9.1, we mentioned that treatment of an alkyne with a strong base yields an acetylide. An acetylide ion may be used as a nucleophile in S_N2 reactions with primary alkyl halides. (Secondary and tertiary alkyl halides are more likely to give elimination products.) This reaction provides a synthetic route for obtaining substituted or more complex alkynes from simpler ones.

Preparation of acetylide:

$$CH_3C\equiv CH + NaNH_2 \xrightarrow[-30°]{\text{liq. } NH_3} CH_3C\equiv C^- \ Na^+ + NH_3$$

Reaction with an alkyl halide:

$$CH_3C\equiv C:^- + \underset{\text{1-chloropropane}}{CH_3CH_2CH_2-Cl} \xrightarrow{S_N2} \underset{\text{2-hexyne}}{CH_3CH_2CH_2C\equiv CCH_3} + Cl^-$$

As we also mentioned in Section 9.1, alkynyl Grignard reagents may be prepared by the reaction of a Grignard reagent and a 1-alkyne. In this reaction, the Grignard reagent acts as a base while the alkyne acts as an acid.

Preparation of RC≡CMgX:

$$CH_3C\equiv C\overset{\delta+}{H} + \overset{\delta-}{C}H_3MgI \longrightarrow CH_3C\equiv CMgI + CH_4$$

propyne

As with other Grignard reagents, the nucleophilic carbon of an alkynyl Grignard reagent attacks partially positive centers, such as the carbon of a carbonyl group. The advantage of this type of Grignard synthesis is that more-complex alkynes can be prepared more easily this way than by S_N2 reactions.

Reaction with a ketone:

$$CH_3C\overset{\delta-}{\equiv}\overset{\delta+}{C}MgI + CH_3\overset{\delta+}{C}CH_3 \longrightarrow \underset{\underset{C\equiv CCH_3}{|}}{\overset{\overset{:\ddot{O}MgI}{|}}{CH_3CCH_3}} \xrightarrow[H_2O]{H^+} \underset{\underset{C\equiv CCH_3}{|}}{\overset{\overset{:\ddot{O}H}{|}}{CH_3CCH_3}}$$

2-methyl-3-pentyn-2-ol

Table 9.4 summarizes the methods of synthesizing alkenes and alkynes.

TABLE 9.4. Summary of laboratory syntheses of alkenes and alkynes.

Reaction	Section reference
alkenes:	
$\underset{R_2CCHR_2 + OH^-}{\overset{\overset{X}{\mid}}{}} \xrightarrow{heat} R_2C{=}CR_2$	5.9–5.10
$\underset{R_2CCHR_2 + H_2SO_4}{\overset{\overset{OH}{\mid}}{}} \xrightarrow{heat} R_2C{=}CR_2$	7.8
$R_2C{=}O + (C_6H_5)_3P{=}CR'_2 \longrightarrow R_2C{=}CR'_2$	11.12
$RC{\equiv}CR + H_2 \xrightarrow{catalyst} RCH{=}CHR$	9.13
alkynes:	
$\underset{RCHCH_2}{\overset{\overset{X\ \ X}{\mid\ \ \mid}}{}} \xrightarrow[(2)\ H_2O]{(1)\ NaNH_2,\ liq.\ NH_3} RC{\equiv}CH$	9.5
$RC{\equiv}CH \xrightarrow[liq.\ NH_3]{NaNH_2} RC{\equiv}C^-\ Na^+ \xrightarrow{R'X} RC{\equiv}CR'$	9.5
$RC{\equiv}CH \xrightarrow{CH_3MgI} RC{\equiv}CMgI \xrightarrow[(2)\ H_2O,\ H^+]{(1)\ R'_2C{=}O} \underset{RC{\equiv}CCR'_2}{\overset{\overset{OH}{\mid}}{}}$ 9.5	

STUDY PROBLEMS

9.7.   Suggest synthetic routes, starting with propyne, to: (a) 2-pentyne, and
(b) 4-phenyl-2-pentyne.

9.8.   Show how you would synthesize 1-pentyn-3-ol from acetylene, propanal
(CH_3CH_2CHO), and ethylmagnesium bromide.

SECTION 9.6.

Preview of Addition Reactions

Three typical reactions of alkenes are the reactions with hydrogen, with chlorine,
and with a hydrogen halide:

$$CH_2{=}CH_2 \xrightarrow[\text{Pt catalyst}]{H_2} CH_3CH_3 \quad \text{ethane}$$

$$CH_2{=}CH_2 \xrightarrow{Cl_2} \underset{\text{1,2-dichloroethane}}{\overset{\displaystyle Cl \quad Cl}{\underset{|\quad\;\;|}{CH_2{-}CH_2}}}$$

$$CH_2{=}CH_2 \xrightarrow{HCl} CH_3CH_2Cl \quad \text{chloroethane}$$

ethylene

Each of these reactions is an **addition reaction**. In each case, a reagent has
added to the alkene without the loss of any other atoms. We will find that the
principal characteristic of unsaturated compounds is the *addition of reagents
to pi bonds*.

In an addition reaction of an alkene, the pi bond is broken and its pair of
electrons is used in the formation of two new sigma bonds. In each case, the sp^2
carbon atoms are rehybridized to sp^3. Compounds containing pi bonds are usually
of higher energy than comparable compounds containing only sigma bonds;
consequently, an addition reaction is usually exothermic.

$$\overset{sp^2}{\diagdown C{=}C\diagup} \longrightarrow \overset{sp^3}{-\overset{|}{C}-\overset{|}{C}-}$$

In general, carbon–carbon double bonds are not attacked by nucleophiles
because there is no partially positive carbon atom to attract a nucleophile. However,
the exposed pi electrons in the carbon–carbon double bond are attractive to
electrophiles (E^+), such as H^+. Many reactions of alkenes and alkynes are therefore
initiated by an *electrophilic attack*, a reaction step that results in a carbocation.
The carbocation can then be attacked by a nucleophile to yield the product. We

will discuss this type of addition reaction first, then proceed to other types of alkene reactions.

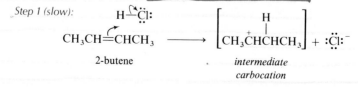

two electrons in the pi bond

an alkene

a carbocation

an addition product

SECTION 9.7.

Addition of Hydrogen Halides to Alkenes and Alkynes

Hydrogen halides add to the pi bonds of alkenes to yield alkyl halides. Alkynes react in an analogous manner and yield either vinylic halides or 1,1-dihaloalkanes, depending on the amount of HX used. We will not stress alkynes in our discussions, however, because alkenes are more important both in the laboratory and in nature.

$$CH_2{=}CH_2 + HX \longrightarrow CH_3CH_2X$$

ethylene   *an ethyl halide*

$$CH{\equiv}CH \xrightarrow{HX} CH_2{=}CHX \xrightarrow{HX} CH_3CHX_2$$

acetylene   *a vinyl halide*   *a 1,1-dihaloethane*

The addition of hydrogen halides to alkenes to prepare alkyl halides is often used as a synthetic reaction. Usually, the gaseous HX is bubbled through a solution of the alkene. (Concentrated aqueous solutions of hydrogen halides give mixtures of products because water can also add to double bonds.) The relative reactivity of HX in this reaction is HI > HBr > HCl > HF. The strongest acid (HI) is the most reactive toward alkenes, while the weakest acid (HF) is the least reactive.

A hydrogen halide contains a highly polar H—X bond and can easily lose H^+ to the pi bond of an alkene. The result of the attack of H^+ is an intermediate carbocation, which quickly undergoes reaction with a negative halide ion to yield an alkyl halide. Because the initial attack is by an electrophile, the addition of HX to an alkene is called an **electrophilic addition reaction.**

Step 1 (slow):

$$CH_3CH{=}CHCH_3 \longrightarrow \left[CH_3\overset{+}{C}HCHCH_3 \right] + :\ddot{Cl}:^-$$

H—$\ddot{C}l$:

2-butene   *intermediate carbocation*

Step 2 *(fast):*

$$\left[CH_3\overset{+}{C}HCH_2CH_3\right] + :\overset{..}{\underset{..}{C}l}:^- \longrightarrow CH_3\overset{\overset{\overset{..}{\underset{..}{Cl}:}}{|}}{C}HCH_2CH_3$$

2-chlorobutane

A.   Markovnikov's rule

If an alkene is *unsymmetrical* (that is, the groups attached to the two sp^2 carbons differ), there is the possibility of two different products from the addition of HX:

$$CH_3CH{=}CHCH_3 \qquad \qquad CH_3CH{=}CH_2$$

symmetrical alkenes   *unsymmetrical alkenes*

$$CH_3CH{=}CHCH_3 \xrightarrow{\ HCl\ } CH_3\overset{\overset{H}{|}}{C}H{-}\overset{\overset{Cl}{|}}{C}HCH_3$$

2-butene   2-chlorobutane

symmetrical   *only one possible product*

$$CH_3CH{=}CH_2 \xrightarrow{\ HCl\ } \begin{cases} CH_3CH_2{-}CH_2Cl & \text{1-chloropropane} \\ CH_3\overset{\overset{Cl}{|}}{C}H{-}CH_3 & \text{2-chloropropane} \end{cases}$$

two possible products

propene

unsymmetrical

In an electrophilic addition that can lead to two products, one product usually predominates over the other. In 1869, the Russian chemist Vladimir Markovnikov formulated the following empirical rule: *In additions of HX to unsymmetrical alkenes, the H^+ of HX goes to the double-bonded carbon that already has the greatest number of hydrogens.* By Markovnikov's rule, we would predict that the reaction of HCl with propene yields 2-chloropropane (and not the 1-chloro isomer). Examples of reactions that obey Markovnikov's rule follow:

H goes here

$$CH_3CH{=}CH_2 \xrightarrow{\ HCl\ } CH_3\overset{\overset{Cl}{|}}{C}H{-}CH_3$$

propene   2-chloropropane

H goes here

$$(CH_3)_2C{=}CHCH_3 \xrightarrow{\ HBr\ } (CH_3)_2\overset{\overset{Br}{|}}{C}{-}CH_2CH_3$$

2-methyl-2-butene   2-bromo-2-methylbutane

H goes here

1-methyl-1-cyclohexene   1-iodo-1-methylcyclohexane

B.   The reason for Markovnikov's rule

Markovnikov formulated his rule because of experimental observations. Why is this empirical rule followed? To answer this question, let us return to the mechanism of HX addition. *Step 1* is the formation of a carbocation. For propene, two possible carbocations could be formed:

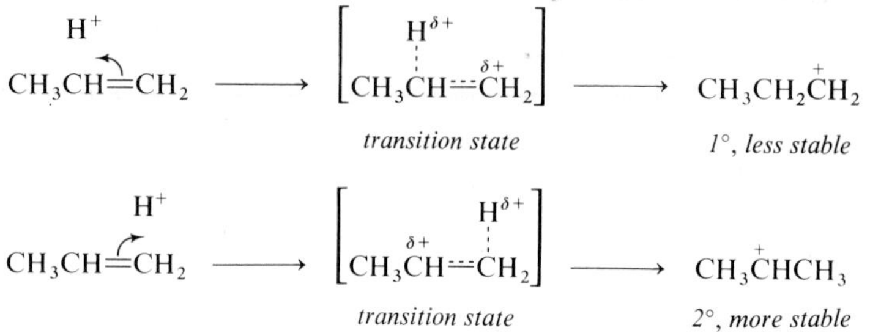

$$CH_3CH{\overset{H^+}{=\!=}}CH_2 \longrightarrow \left[\begin{array}{c} H^{\delta+} \\ \vdots \\ CH_3CH\text{---}\overset{\delta+}{CH_2} \end{array} \right] \longrightarrow CH_3CH_2\overset{+}{CH_2}$$

transition state   *1°, less stable*

$$CH_3CH{\overset{H^+}{=\!=}}CH_2 \longrightarrow \left[\begin{array}{c} H^{\delta+} \\ \vdots \\ \overset{\delta+}{CH_3CH}\text{---}CH_2 \end{array} \right] \longrightarrow CH_3\overset{+}{CH}CH_3$$

transition state   *2°, more stable*

The order of stability of carbocations is 3° > 2° > 1°. For propene, the two positions of H^+ addition lead to (1) a high-energy, unstable, primary carbocation, or (2) a lower-energy, more-stable, secondary carbocation. The transition states leading to these intermediates have carbocation character. Therefore, the secondary carbocation has a lower-energy transition state and a faster rate of formation (see Figure 9.9).

Addition of a reagent to an unsymmetrical alkene proceeds by way of the more stable carbocation. This is the reason that Markovnikov's rule is followed.

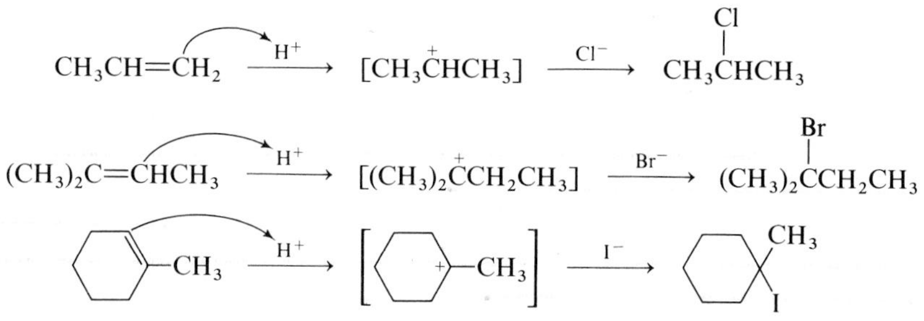

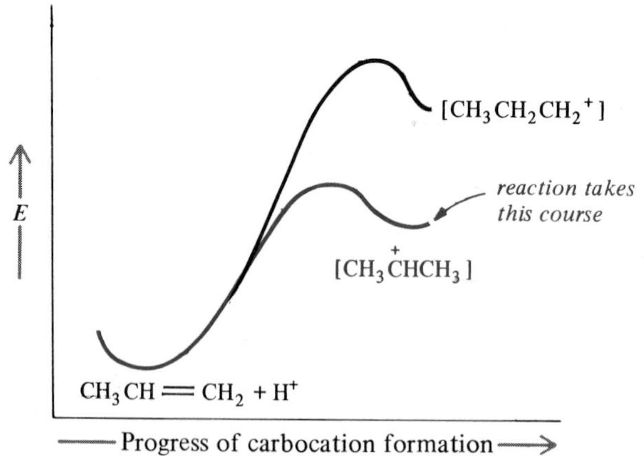

$$CH_3CH{=}CH_2 \xrightarrow{H^+} [CH_3\overset{+}{CH}CH_3] \xrightarrow{Cl^-} CH_3\overset{\overset{\displaystyle Cl}{|}}{CH}CH_3$$

$$(CH_3)_2C{=}CHCH_3 \xrightarrow{H^+} [(CH_3)_2\overset{+}{C}CH_2CH_3] \xrightarrow{Br^-} (CH_3)_2\overset{\overset{\displaystyle Br}{|}}{C}CH_2CH_3$$

FIGURE 9.9. Energy diagram for the protonation of propene.

SAMPLE PROBLEM

Predict the relative rates of reaction of the following alkenes toward HBr (lowest rate first):

(a) $CH_3CH_2CH{=}CH_2$ (b) $CH_2{=}CH_2$ (c) $(CH_3)_2C{=}CHCH_3$

Solution: The alkene that can form the most stable carbocation has the lowest E_{act} and the fastest rate. Therefore,

$$CH_2{=}CH_2 \qquad CH_3CH_2CH{=}CH_2 \qquad (CH_3)_2C{=}CHCH_3$$

increasing rate of reaction

STUDY PROBLEMS

9.9.   For the alkenes in (a) and (c) in the preceding sample problem, give the structures of the carbocation intermediate and the major product of the reaction with HBr.

9.10.   When propene is treated with HCl in ethanol, one of the products is ethyl isopropyl ether. Suggest a mechanism for its formation.

C.   Rearrangements

In Section 5.6F, we discussed carbocation rearrangements in S_N1 reactions. When a carbocation can form a more stable carbocation by a 1,2-shift of H, R, or Ar, rearrangement occurs. Is rearrangement observed in HX addition reactions? Yes, indeed. The intermediate carbocations in these HX additions are no different from those in S_N1 or E1 reactions.

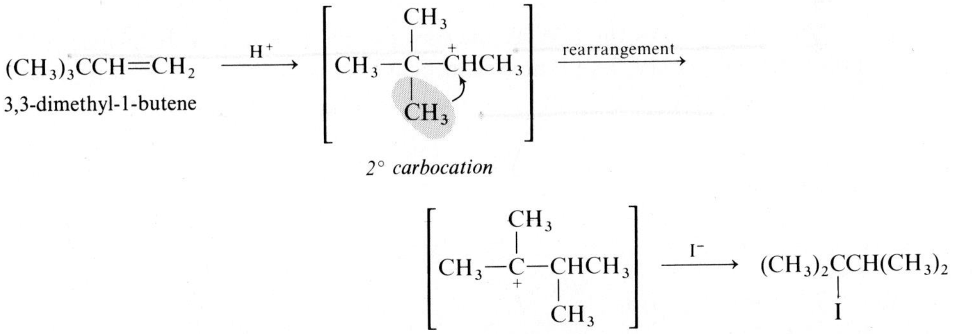

$$(CH_3)_3CCH{=}CH_2 \xrightarrow{H^+} \left[CH_3{-}\overset{\displaystyle CH_3}{\underset{\displaystyle CH_3}{\overset{|}{\underset{|}{C}}}}{-}\overset{+}{C}HCH_3 \right] \xrightarrow{\text{rearrangement}}$$

3,3-dimethyl-1-butene

2° carbocation

$$\left[CH_3{-}\overset{\displaystyle CH_3}{\underset{\displaystyle CH_3}{\overset{|}{\underset{|}{\overset{+}{C}}}}}{-}CHCH_3 \right] \xrightarrow{I^-} (CH_3)_2CCH(CH_3)_2$$

3° carbocation

2-iodo-2,3-dimethylbutane
rearranged product

STUDY PROBLEM

9.11.   Predict the rearrangement product of each of the following reactions:

(a)   3-methyl-1-butene + HCl $\longrightarrow$

(b)   4,4-dimethyl-2-pentene + HBr $\longrightarrow$

D.   Anti-Markovnikov addition of HBr

The addition of HBr to alkenes sometimes proceeds by Markovnikov's rule, but sometimes it does not. (This effect is not observed with HCl or HI.)

$$CH_3CH{=}CH_2 \xrightarrow{\;HBr\;} \underset{\text{2-bromopropane}}{CH_3\overset{\displaystyle Br}{\overset{|}{C}}HCH_3} \qquad \text{but sometimes} \quad \underset{\substack{\text{1-bromopropane}\\ \textit{anti-Markovnikov product}}}{CH_3CH_2CH_2Br}$$

propene

It has been observed that the primary alkyl bromide is obtained only when a peroxide or O_2 is present in the reaction mixture. Oxygen is a stable diradical (see Section 6.6C), and peroxides (ROOR) are easily cleaved into free radicals. When O_2 or peroxides are present, HBr addition proceeds through a *free-radical* mechanism instead of an ionic one.

Formation of Br·: $ROOR \longrightarrow 2\,RO\cdot$

$$RO\cdot + HBr \longrightarrow ROH + Br\cdot$$

Addition of Br· to alkene:

$$CH_3\overset{\frown}{CH}{=}CH_2 + Br\cdot \longrightarrow \underset{\substack{Br\\ 2°,\ more\ stable}}{CH_3\overset{\cdot}{C}HCH_2} \quad \text{and not} \quad \underset{\substack{Br\\ 1°,\ less\ stable}}{CH_3CH\overset{\cdot}{C}H_2}$$

Formation of product:

$$CH_3\overset{\cdot}{C}HCH_2Br + H{\frown}Br \longrightarrow CH_3CH_2CH_2Br + Br\cdot$$

When Br· attacks the alkene, the more stable free radical is formed. (The stability of free radicals, like that of carbocations, is in the order $3° > 2° > 1°$.) The result of the free-radical addition in our example is 1-bromopropane.

STUDY PROBLEM

9.12. Predict the products:

(a) $(CH_3)_2C{=}CH_2 + HBr \xrightarrow{\;ROOR\;}$

(b) $(CH_3)_2C{=}CH_2 + HBr \xrightarrow{\;\text{no peroxides}\;}$

SECTION 9.8.

Addition of H_2SO_4 and H_2O to Alkenes and Alkynes

Sulfuric acid undergoes addition to an alkene just as a hydrogen halide does. The product is an alkyl hydrogen sulfate, which can be used to synthesize alcohols or ethers (see Section 7.12B).

$$CH_3CH{=}CH_2 + H{-}OSO_3H \longrightarrow CH_3\overset{\overset{\displaystyle OSO_3H}{|}}{C}H{-}CH_3$$

propene   2-propyl hydrogen sulfate

In strongly acidic solution (such as aqueous sulfuric acid), water adds to a double bond to yield an alcohol. This reaction is called the **hydration of an alkene.**

$$CH_3CH{=}CH_2 + H_2O \xrightarrow{\;H^+\;} CH_3\overset{\overset{\displaystyle OH}{|}}{C}H{-}CH_3$$

propene   2-propanol
                                           60%

Both reactions occur in two steps, just like the addition of a hydrogen halide. The first step is the protonation of the alkene to yield a carbocation. The second step is the addition of a nucleophile to the carbocation. Because initially a carbocation is formed, both reactions follow Markovnikov's rule. Rearrangements are to be expected if the carbocation can undergo a 1,2-shift of H or R to yield a more stable carbocation.

Step 1: $R_2C{=}CHR + H^+ \longrightarrow [R_2\overset{+}{C}{-}CH_2R]$

Step 2:

$$[R_2\overset{+}{C}{-}CH_2R] + H_2\ddot{O}: \longrightarrow R_2\overset{\overset{\displaystyle H{-}\overset{+}{\ddot{O}}H}{|}}{C}{-}CH_2R \rightleftharpoons R_2\overset{\overset{\displaystyle :\ddot{O}H}{|}}{C}{-}CH_2R + H^+$$

a protonated alcohol   an alcohol

STUDY PROBLEMS

9.13. Show the mechanism for the addition of H_2SO_4 to 1-butene to yield a butyl hydrogen sulfate.

9.14. Predict the major products:

(a) $CH_3CH_2CH{=}CH_2 + H_2O \xrightarrow{\;H^+\;}$

(b) $(CH_3)_3CCH{=}CH_2 + H_2O \xrightarrow{\;H^+\;}$

(c) $(CH_3)_2CHCH{=}CH_2 + H_2SO_4 \longrightarrow$

Alkynes also undergo hydration, but the initial product is a vinylic alcohol, or **enol.** An enol is in equilibrium with an aldehyde or ketone (see Section 11.17). The equilibrium favors the carbonyl compound; therefore, hydration of an alkyne actually results in an aldehyde or ketone. (The hydration of alkynes proceeds more smoothly when a mercuric salt is added to catalyze the reaction.)

$$CH_3C{\equiv}CH + H_2O \xrightarrow[\;Hg^{2+}\;]{H^+} \left[CH_3C{=}CH_2 \atop \overset{\displaystyle :\ddot{O}{-}H}{} \right] \rightleftharpoons CH_3\overset{\overset{\displaystyle \ddot{O}:}{\|}}{C}CH_3$$

propyne   propanone
                                                            (acetone)

Acid-catalyzed hydration of an alkene or alkyne is seldom used in the laboratory because of the relatively low yields and possibilities of rearrangements and polymerization (Section 9.17). However, many simple alcohols, such as ethanol and 2-propanol, are synthesized industrially by this technique.

SECTION 9.9.

Hydration Using Mercuric Acetate

Mercuric acetate, $Hg(O_2CCH_3)_2$, and water add to alkenes in a reaction called **oxymercuration**. Unlike the addition reactions we have discussed so far, oxymercuration proceeds *without rearrangement*. The product of oxymercuration is usually reduced with sodium borohydride ($NaBH_4$) in a subsequent reaction called **demercuration** to yield an alcohol, the same alcohol that would be formed if water had been added across the double bond. Oxymercuration–demercuration reactions usually give better yields of alcohols than the addition of water with H_2SO_4.

Oxymercuration:

$$CH_3CH_2CH_2CH{=}CH_2 \xrightarrow[H_2O]{Hg(OCCH_3)_2} \underset{\underset{HgO_2CCH_3}{|}}{\overset{\overset{OH}{|}}{CH_3CH_2CH_2CHCH_2}}$$

1-pentene

Demercuration:

$$\underset{}{\overset{\overset{OH}{|}}{CH_3CH_2CH_2CHCH_2}}{-}HgO_2CCH_3 \xrightarrow{NaBH_4} \overset{\overset{OH}{|}}{CH_3CH_2CH_2CHCH_3} + Hg$$

2-pentanol
90% overall

Like the addition of other reagents to alkenes, oxymercuration is a two-step process. The addition proceeds by electrophilic attack of $^+HgO_2CCH_3$ followed by nucleophilic attack of H_2O. Because rearrangements do not occur, the intermediate formed by electrophilic attack cannot be a true carbocation. On the other hand, since Markovnikov's rule is followed, the intermediate must have some carbocation character. Both these facts are explained by postulating a **bridged ion**, or **cyclic ion**, as the intermediate.

Dissociation of mercuric acetate:

$$Hg(O_2CCH_3)_2 \rightleftharpoons {}^+HgO_2CCH_3 + {}^-O_2CCH_3$$

Electrophilic attack:

$$R_2C{=}CHR \atop \searrow \atop {}^+HgO_2CCH_3 \longrightarrow \left[\underset{HgO_2CCH_3}{\overset{+}{R_2C}{-}CHR} \right]$$

a bridged intermediate

Attack of H₂O and proton loss:

$$\left[\begin{array}{c} R_2\overset{+}{C}\text{---CHR} \\ \quad | \\ \text{HgO}_2\text{CCH}_3 \end{array}\right] \xrightarrow{\text{H}_2\text{O}} \left[\begin{array}{c} {}^+\text{OH}_2 \\ | \\ R_2\text{C---CHR} \\ \quad\quad | \\ \text{HgO}_2\text{CCH}_3 \end{array}\right] \xrightarrow{-\text{H}^+} \begin{array}{c} \text{OH} \\ | \\ R_2\text{C---CHR} \\ \quad\quad | \\ \text{HgO}_2\text{CCH}_3 \end{array}$$

The formation of a bridged intermediate is not very different from the formation of a carbocation. The reaction of this intermediate is also very similar to that of a carbocation. The difference between this bridged intermediate and a true carbocation is that Hg is partially bonded to *each* double-bond carbon and rearrangements cannot occur. The more positive carbon in the bridged intermediate (the carbon attacked by H₂O) may be predicted by knowledge of carbocation stabilities (3° > 2° > 1°). We can compare the reaction of this type of bridged ion with the one formed in acid-catalyzed substitution of an epoxide (Section 7.16B).

more positive carbon;
H₂O attacks here

$$\underset{\text{bridged intermediate}}{\begin{array}{c} \downarrow \\ R_2\overset{+}{C}\text{---CHR} \\ \quad\searrow\quad | \\ \text{HgO}_2\text{CCH}_3 \end{array}} \qquad \text{similar to} \qquad \underset{\text{true carbocation}}{\begin{array}{c} R_2\overset{+}{C}\text{---CHR} \\ \quad\quad | \\ \text{HgO}_2\text{CCH}_3 \end{array}}$$

The reducing agent in the demercuration reaction, sodium borohydride, is an important reducing agent in organic chemistry. It forms stable solutions in aqueous base, but decomposes and releases H₂ in acidic solution. We will encounter this reagent again as a reducing agent for aldehydes and ketones (Section 11.14).

$$\text{Na}^+ \; \begin{array}{c} \text{H} \\ | \\ \text{H---B}\overset{=}{\text{---}}\text{H} \\ | \\ \text{H} \end{array}$$

sodium borohydride

STUDY PROBLEMS

9.15. (a) Write the steps in the oxymercuration–demercuration of 3,3-dimethyl-1-butene.

(b) Compare the product of this reaction sequence to the product from the reaction of 3,3-dimethyl-1-butene with dilute, aqueous HCl.

9.16. Oxymercuration–demercuration results in alcohols as products. If the reaction is run in an alcohol instead of in water (a sequence called **solvomercuration–demercuration**), an *ether* is obtained as a product. Write equations to show how you would prepare 2-methoxy-2-methylbutane from 2-methyl-1-butene by this technique.

SECTION 9.10.

Addition of Borane to Alkenes

Diborane (B_2H_6) is a toxic gas prepared by the reaction of sodium borohydride and boron trifluoride ($3\,NaBH_4 + 4\,BF_3 \rightarrow 2\,B_2H_6 + 3\,NaBF_4$). In diethyl ether solution, diborane dissociates into borane (BH_3) solvated by an ether molecule: $(CH_3CH_2)_2\ddot{O}\!:\!\cdots\!BH_3$. Borane undergoes rapid and quantitative reaction with alkenes to form **organoboranes** (R_3B). The overall reaction is the result of three separate reaction steps. In each step, one alkyl group is added to borane until all three hydrogen atoms have been replaced by alkyl groups. This sequence of reactions is called **hydroboration**.

Step 1:

$$CH_2{=}CH_2 + \overset{\displaystyle H}{\underset{\displaystyle H}{B{-}H}} \longrightarrow CH_3{-}\overset{\displaystyle BH_2}{CH_2}$$

Step 2: $CH_2{=}CH_2 + CH_3CH_2BH_2 \longrightarrow (CH_3CH_2)_2BH$

Step 3: $CH_2{=}CH_2 + (CH_3CH_2)_2BH \longrightarrow (CH_3CH_2)_3B$

triethylborane

an organoborane

Organoboranes were discovered in the 1950's by Herbert C. Brown at Purdue University, who was awarded a Nobel Prize in 1979 for his work with organoboron compounds. The value of these compounds arises from the variety of other compounds that can be synthesized from them. Let us first consider the addition of BH_3 to alkenes, and then look at some of the products that can be obtained from the resulting organoboranes.

Borane is different from the other addition reagents we have mentioned because **H** is the *electronegative* portion of the molecule instead of the electropositive portion, as it is in HCl or H_2O. When borane adds to a double bond, the hydrogen (as a hydride ion, H^-) becomes bonded to the *more substituted carbon*. The result is what appears to be anti-Markovnikov addition.

$$CH_3CH{=}CH_2 \longrightarrow CH_3CH{-}CH_2$$

$$\underset{\delta-\quad\delta+}{H{-}BH_2} \qquad\qquad \underset{H\quad BH_2}{}$$

H on more substituted carbon

Steric hindrance also plays a role in the course of this reaction. Best yields of the anti-Markovnikov organoborane are obtained when one carbon of the double bond is substantially more hindered than the other.

less hindered

more hindered

$$CH_3CH_2CH{=}CH_2 \qquad CH_3CH_2\overset{\displaystyle CH_3}{C}{=}CH_2$$

93% yield of anti-
Markovnikov product

99% yield of anti-
Markovnikov product

Organoboranes are easily oxidized to alcohols by alkaline hydrogen peroxide. The final result of borane addition, followed by H_2O_2 oxidation, appears as if water had been added to the double bond in an anti-Markovnikov manner. Overall yields are often 95–100%.

$$3 \ CH_3CH{=}CH_2 \xrightarrow{\ BH_3\ } (CH_3CH_2CH_2)_3B \xrightarrow{\ H_2O_2, \ OH^-\ } 3 \ CH_3CH_2CH_2OH + BO_3{}^{3-}$$

propene   tripropylborane   1-propanol   borate ion

$$\text{1-methyl-1-cyclohexene} \xrightarrow[\text{(2) } H_2O_2, \ OH^-]{\text{(1) } BH_3} \text{2-methyl-1-cyclohexanol} + BO_3{}^{3-}$$

1-methyl-1-cyclohexene   2-methyl-1-cyclohexanol

Now, we have discussed three routes to alcohols from alkenes. These are summarized in the following flow diagram:

$$R_2C{=}CHR$$

$\xrightarrow{H_2O, \ H^+}$	$R_2\overset{\text{OH}}{\underset{|}{C}}CH_2R$	*low yields, possible rearrangement*
$\xrightarrow[\text{(2) NaBH}_4]{\text{(1) Hg(O}_2CCH_3)_2, \ H_2O}$	$R_2\overset{\text{OH}}{\underset{|}{C}}CH_2R$	*excellent yields, Markovnikov product, no rearrangement*
$\xrightarrow[\text{(2) } H_2O_2, \ OH^-]{\text{(1) } BH_3}$	R_2CHCHR with OH	*excellent yields, anti-Markovnikov, no rearrangement*

Besides being oxidized to alcohols, organoboranes can be converted to alkanes, alkyl halides, or other products. In each case, the newly introduced atom or group becomes bonded to the less-substituted carbon of the double bond.

$$3 \ CH_3CH{=}CH_2 \xrightarrow{\ BH_3\ } (CH_3CH_2CH_2)_3B$$

$\xrightarrow{3 CH_3CO_2D}$ $3 \ CH_3CH_2CH_2D$

1-deuteriopropane

a deuteriated alkane

$\xrightarrow{3 Br_2, \ OH^-}$ $3 \ CH_3CH_2CH_2Br$

1-bromopropane

an alkyl halide

A.   Stereochemistry of hydroboration

When borane adds to a double bond, the boron atom and the hydride ion become bonded to the two carbon atoms of the double bond *simultaneously*. The result is that the **B** and **H** must be added to the same side of the double bond. An addition reaction in which two species add to the same side is called a *cis*-**addition**, or **syn-addition**. (Syn, like *cis*, means "on the same side, or face.")

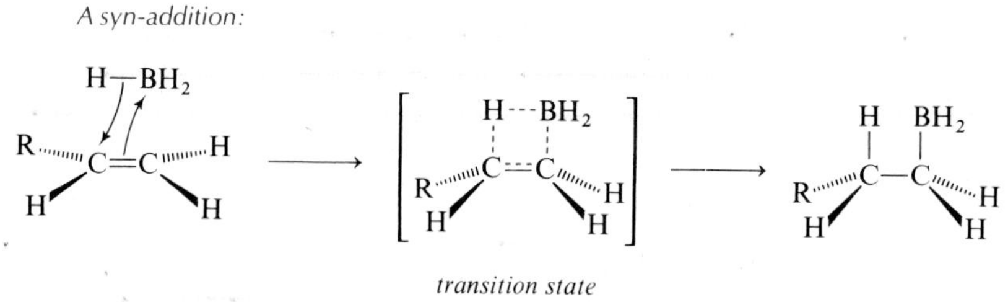

A syn-addition:

transition state

If the addition product is capable of geometric isomerism, such as the addition product of 1-methyl-1-cyclohexene, the B and the H are thus *cis* to each other in the product.

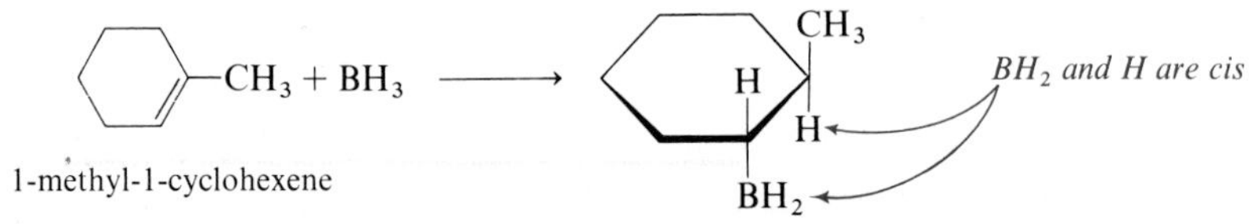

1-methyl-1-cyclohexene

BH₂ and H are cis

When an organoborane is subsequently oxidized to an alcohol, the hydroxyl group ends up in the *same position* as the boron atom that it replaced—that is, with retention of configuration at that carbon.

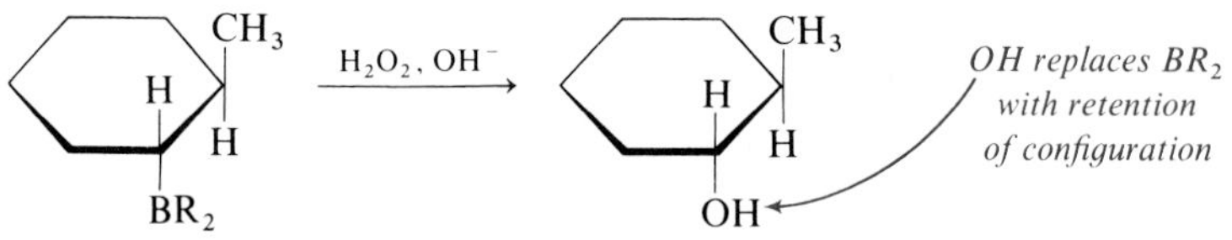

trans-2-methyl-1-cyclohexanol

OH replaces BR₂ with retention of configuration

The reason that the configuration is retained is that the oxidation proceeds by a 1,2-shift (similar in some respects to a carbocation rearrangement), followed by hydrolysis of the BO bond to yield the alcohol. The RO bond is not affected in this hydrolysis.

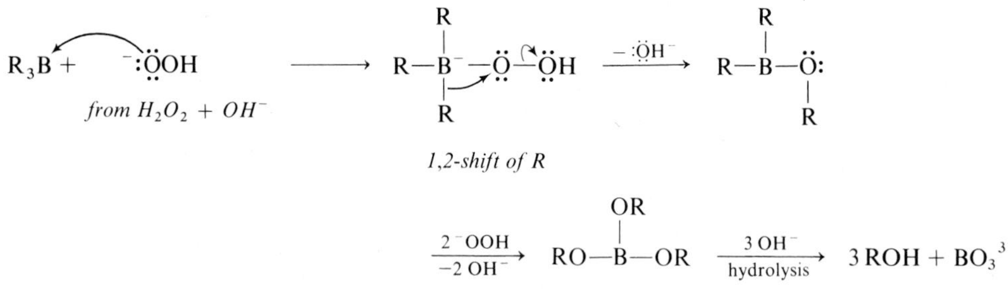

from H₂O₂ + OH⁻

1,2-shift of R

STUDY PROBLEMS

9.17. Use a transition-state structure to show why a *trans*-2-methyl-1-cyclohexylborane cannot yield *cis*-2-methyl-1-cyclohexanol.

9.18. Predict the major organic products (including stereochemistry) of the hydroboration and oxidation of   (a) 1-ethyl-1-cyclopentene; (b) (*Z*)-3-methyl-2-pentene.

SECTION 9.11.

Addition of Halogens to Alkenes and Alkynes

Like acids, chlorine and bromine add to carbon–carbon double bonds and triple bonds. A common laboratory test for the presence of a double or triple bond in a compound of unknown structure is the treatment of the compound with a dilute solution of bromine in CCl_4. The test reagent has the reddish-brown color of Br_2; disappearance of this color is a positive test. The decolorization of a Br_2/CCl_4 solution by an unknown is suggestive, but not definitive proof, that a double or triple bond is present. A few other types of compounds, such as aldehydes, ketones, and phenols, also decolorize Br_2/CCl_4 solution.

$$CH_3CH{=}CHCH_3 + Br_2 \longrightarrow \overset{\displaystyle Br}{\underset{\displaystyle |}{C}}H_3CH{-}CHCH_3$$

CH

$$\underset{\text{2-butene}}{CH_3CH{=}CHCH_3} + \underset{\text{red}}{Br_2} \longrightarrow \underset{\substack{\text{2,3-dibromobutane} \\ \textit{colorless}}}{\overset{Br\ \ \ Br}{\underset{|\ \ \ \ |}{CH_3CH{-}CHCH_3}}}$$

$$\underset{\text{2-butyne}}{CH_3C{\equiv}CCH_3} + \underset{\text{red}}{2\,Br_2} \longrightarrow \underset{\substack{\text{2,2,3,3-tetrabromobutane} \\ \textit{colorless}}}{\overset{Br\ \ Br}{\underset{Br\ \ Br}{CH_3C{-}CCH_3}}}$$

Neither F_2 nor I_2 is a useful reagent in alkene addition reactions. Fluorine undergoes explosive reaction with organic compounds. Iodine adds to a double bond, but the 1,2-diiodo product is unstable and loses I_2 to re-form the alkene.

$$R_2CI{-}CIR_2 \rightleftharpoons R_2C{=}CR_2 + I_2$$

Therefore, we can say this addition reaction is general only for *chlorine* and *bromine*. A more substituted alkene is more reactive toward X_2 than a less substituted alkene (see Table 9.5). This is the same order of reactivity as that toward HX.

$$CH_2{=}CH_2 \quad RCH{=}CH_2 \quad R_2C{=}CH_2 \quad R_2C{=}CHR \quad R_2C{=}CR_2$$

increasing reactivity toward X_2 or HX addition

TABLE 9.5. Relative reactivities of some alkenes toward Br_2 in methanol

Compound	Relative rate
$CH_2{=}CH_2$	1
$CH_3CH_2CH{=}CH_2$	10^3
cis-$CH_3CH_2CH{=}CHCH_3$	10^5
$(CH_3)_2C{=}C(CH_3)_2$	10^7

A. Electrophilic attack of X_2

The reaction of X_2 with alkenes is similar to that of HX. But what is the source of the electrophile in X_2? When X_2 approaches the pi-bond electrons, polarity is induced in the X_2 molecule by repulsion between the pi electrons and the electrons in the X_2 molecule.

$$\delta+ \quad \delta-$$
$$X \longrightarrow X$$

polarized by pi electrons

As the X—X bond becomes more polarized, it becomes progressively weaker until it finally breaks. The result is a halide ion and a positively charged organohalogen ion, called a **halonium ion**. There is evidence that the halonium ion is not a simple carbocation, but is bridged, similar to the intermediate in oxymercuration. In the case of addition of X_2 to ethylene or some other symmetrical alkene, the bridged halonium ion is symmetrical, with X equally bonded to each carbon.

$$CH_2{=}CH_2 \xrightarrow{-Br^-} \left[\begin{array}{c} \overset{+}{:\!\ddot{B}r\!:} \\ CH_2{-}CH_2 \end{array} \right] \quad \text{instead of} \quad \left[\begin{array}{c} :\!\ddot{B}r\!: \\ \overset{+}{CH_2}{-}CH_2 \end{array} \right]$$

a bridged bromonium ion

If the alkene is unsymmetrical, most of the positive charge is carried on the more substituted carbon. Carbocation stability is followed.

$$CH_3CH{=}CH_2 \xrightarrow{-Br^-} \left[CH_3\overset{\delta+}{CH}{-}\overset{\delta+}{CH_2} \right]$$

propene

more positive carbon

B. *anti*-Attack of X^-

The bridged intermediate ion is positively charged and of high energy. Like a carbocation, it exists only momentarily in solution; reaction is completed by attack of a nucleophile (in this case, Br^-). A negative Br^- cannot attack a carbon of the bridged intermediate from the top (as we have shown the structure); that path is blocked by the Br bridge. Therefore, Br^- attacks from the opposite side of the intermediate. The result is ***anti*-addition** of Br_2 to the double bond. (Contrast this mode of addition to the syn-addition of BH_3, Section 9.10A.)

1,2-dibromopropane

General mechanism:

Step 1 (slow):

$$R_2C{=}CHR + X_2 \longrightarrow \left[R_2\overset{\delta+}{C}\overset{\overset{\overset{\delta+}{X}}{|}}{\cdots}CHR \right] + X^-$$

Step 2 (fast):

$$\left[R_2\overset{\delta+}{C}\overset{\overset{\overset{\delta+}{X}}{|}}{\cdots}CHR \right] + X^- \longrightarrow R_2\overset{\overset{X}{|}}{C}{-}\overset{}{C}HR \\ \overset{|}{X}$$

STUDY PROBLEM

9.19. Predict the products of Br_2 addition to:

(a) $CH_3CH{=}CHCH_3$ (b) $(CH_3)_2C{=}CH_2$ (c) $(CH_3)_2C{=}CHCH_3$

C. Evidence for *anti*-addition

Two pieces of evidence point to a bridged ion as the intermediate in halogen addition and to *anti*-addition as the mechanism. Both pieces of evidence are based upon the fact that only one stereoisomeric product is observed in reactions where two or more products would be expected from a simple carbocation intermediate.

The first piece of evidence is that *trans*-dihalides (and not *cis*-dihalides) are formed when the product of halogen addition is capable of stereoisomerism. If the intermediate were a simple carbocation, both the *cis*- and *trans*-isomers would be formed.

cyclohexene

trans-1,2-dibromo-cyclohexane

Additional evidence for *anti*-addition is encountered in the reactions of geometric isomers of open-chain alkenes. When either *cis*- or *trans*-2-butene is treated with Br_2, two chiral carbons are generated.

$$CH_3CH{=}CHCH_3 \longrightarrow CH_3\overset{*}{C}\overset{\overset{Br}{|}}{\underset{\underset{H}{|}}{}}{-}\overset{*}{C}\overset{\overset{H}{|}}{\underset{\underset{Br}{|}}{}}CH_3$$

2-butene
(*cis* or *trans*)

2,3-dibromobutane

The product of this addition, 2,3-dibromobutane, can exist in three stereo-isomeric forms: a pair of enantiomers and a *meso* form. The addition of bromine to *cis*-2-butene yields *only the enantiomeric pair*. No *meso* form is produced in this reaction.

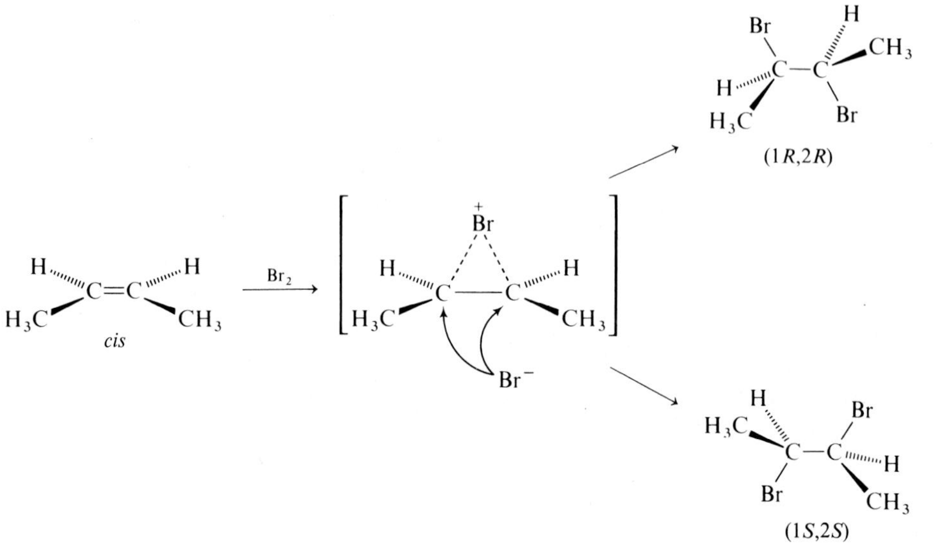

(1*R*,2*R*)

cis

Br₂

Br⁻

(1*S*,2*S*)

On the other hand, the *trans*-isomer yields only the *meso* form and not the enantiomeric pair.

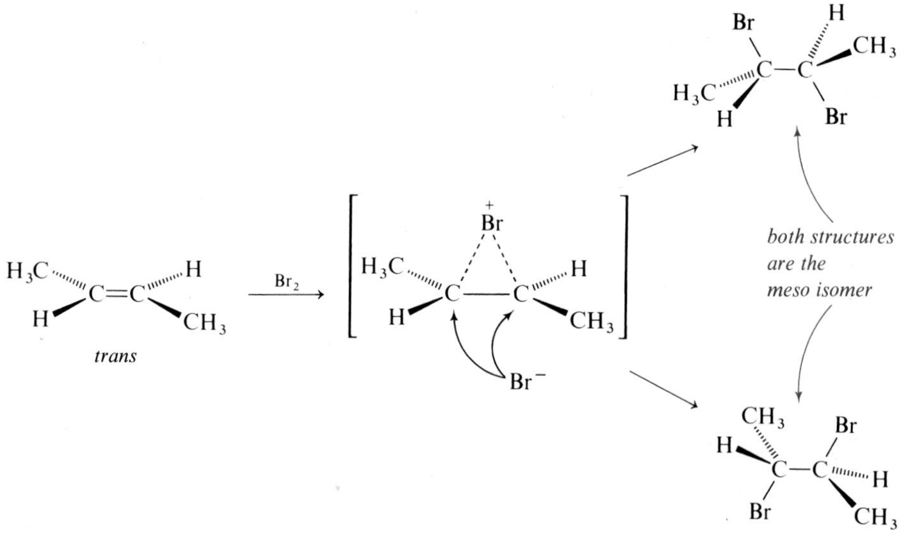

both structures are the meso isomer

$trans$

Br₂

Br⁻

The conclusion is that the intermediate is not an open carbocation. If it were, then rotation around the carbon 2–carbon 3 sigma bond would allow all three stereoisomers to be formed regardless of the geometry of the starting alkene.

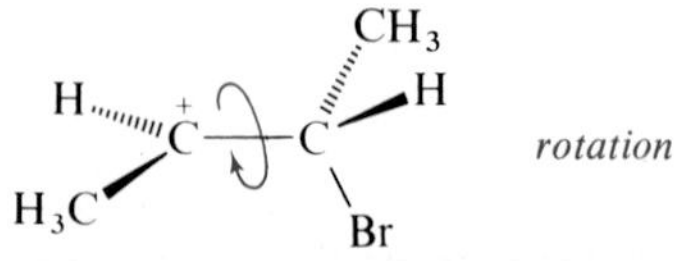

rotation

STUDY PROBLEM

9.20.   What would be the stereochemistry of the reaction of bromine with
(a) (E)-1,2-dideuterioethene;   (b) (Z)-1,2-dideuterioethene?

D.   Mixed addition

Bromination reactions of alkenes proceed by way of a bromonium-ion intermediate, followed by attack by a bromide ion to yield the dibromide. Is the
second step limited to attack only by bromide ion? Can other nucleophiles
compete with bromide ion in the second step to yield other products? Consider the
case of a bromination reaction carried out with Br_2 in a solution containing Cl^-
(from, say, NaCl). In this reaction, two nucleophiles (Br^- and Cl^-) are present.
In such a case, *mixed dihalide products* are observed—along with the dibromo-
alkane, we find some bromochloroalkane.

$$CH_2{=}CH_2 + Br_2 \longrightarrow \left[\overset{+}{\underset{CH_2-CH_2}{Br}} \right] \overset{Br^-}{\nearrow} \quad \underset{1,2\text{-dibromoethane}}{\overset{Br}{\underset{|}{CH_2{-}CH_2Br}}}$$

$$\underset{Cl^-}{\searrow} \quad \underset{1\text{-bromo-2-chloroethane}}{\overset{Br}{\underset{|}{CH_2{-}CH_2Cl}}}$$

STUDY PROBLEM

9.21.   Would you expect to find 1,2-dichloroethane as a product in the preceding
example? Explain.

SAMPLE PROBLEM

When propene is treated with $Br_2 + Cl^-$, only one bromochloropropane is
isolated as a product. What is its structure? Show, by equations, its formation.

Solution:

Step 1:

$$CH_3CH{=}CH_2 + Br_2 \xrightarrow{-Br^-} \left[CH_3\underset{\delta+}{CH}{-}CH_2 \overset{\delta+}{\overset{.Br}{\underset{|}{}}} \right], \quad not \quad \left[CH_3CH{-}\underset{\delta+}{CH_2} \overset{\delta+}{\overset{Br}{\underset{|}{}}} \right]$$

Step 2:

$$\left[CH_3\underset{\delta+}{CH}{-}CH_2 \overset{\delta+}{\overset{.Br}{\underset{|}{}}} \right] + Cl^- \longrightarrow CH_3\overset{Br}{\underset{|}{CH}}CH_2, \quad not \quad CH_3\overset{Br}{\underset{|}{CH}}CH_2Cl$$

$$\underset{|}{Cl}$$

1-bromo-2-
chloropropane

E. Addition of halogens and water

When an <u>alkene is treated with a mixture of Cl$_2$ or Br$_2$ in water</u>, a **1,2-halohydrin** (a <u>compound with X and OH on adjacent carbon atoms) is formed</u>.

General:

$$R_2C{=}CH_2 \xrightarrow{X_2,\ H_2O} \underset{\underset{\underline{OH}}{|}}{\overset{\overset{\overline{X}}{|}}{R_2C{-}CH_2}}$$

an alkene   a 1,2-halohydrin

The path is similar to that for mixed halogen addition:

Step 1:

$$CH_3CH{=}CH_2 + Cl_2 \longrightarrow \left[\underset{\delta+}{CH_3\overset{\overset{\overset{\delta+}{Cl}}{|}}{CH}{-}CH_2} \right] + Cl^-$$

Step 2:

$$\left[\underset{\delta+}{CH_3\overset{\overset{\overset{\delta+}{Cl}}{|}}{CH}{-}CH_2} \right] + H_2\ddot{O}: \longrightarrow \underset{\overset{|}{{}^+\underset{}{O}H_2}}{\overset{\overset{Cl}{|}}{CH_3CHCH_2}}$$

$$\xrightleftharpoons{-H^+} \underset{\overset{|}{:\ddot{O}H}}{\overset{\overset{Cl}{|}}{CH_3CHCH_2}}$$

1-chloro-2-propanol
(propylene chlorohydrin)

STUDY PROBLEM

9.22. What would be the product of the reaction of cyclopentene with aqueous Cl$_2$? Write equations for the steps in this reaction (complete with stereochemistry).

SECTION 9.12.

Addition of Carbenes to Alkenes

If a student were asked whether a compound with the structure <u>CH$_2$</u> exists, the reply might be, "No, because the carbon only has two bonds." However, such a species <u>does have a fleeting existence. It is called *methylene* and belongs to a class</u> of <u>highly reactive intermediates called **carbenes** (R$_2$C:)</u>.

The existence of :CH$_2$ was established spectroscopically in 1959. It has been shown that there are two different methylenes, each with a carbon containing six bonding electrons rather than the usual eight. **Singlet methylene (:CH₂) has an** *sp²*-hybridized carbon and a pair of unshared electrons. **Triplet methylene (HĊH)** contains an *sp*-hybridized carbon and two unpaired electrons.* The orbital pictures of these two structures are presented in Figure 9.10.

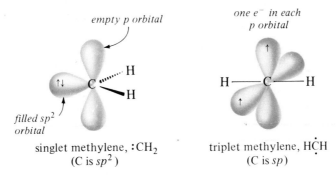

empty p orbital

one e⁻ in each p orbital

*filled sp²
orbital*

singlet methylene, :CH₂
(C is *sp²*)

triplet methylene, HĊH
(C is *sp*)

FIGURE 9.10. Orbital pictures of the two methylenes.

Singlet methylene is formed by the **photolysis** (cleavage by light) of diazo-methane (CH₂N₂) or ketene (CH₂=C=O), both of which are unusual and reactive compounds themselves.

$$\left[:\overset{\frown}{CH_2}-\overset{\frown}{\ddot{N}}\overset{+}{=}\ddot{N}: \quad \longleftrightarrow \quad \overset{\frown}{CH_2}\overset{+}{=}\overset{+}{N}=\ddot{N}:^- \quad \longleftrightarrow \quad :\overset{-}{C}H_2\overset{\frown}{-}\overset{+}{N}\equiv N: \right]$$

resonance structures for diazomethane

$$\xrightarrow{\;hv\;} \quad :CH_2 + :N\equiv N:$$

$$\overset{\frown}{CH_2}\overset{\frown}{=}C=O \quad \xrightarrow{\;hv\;} \quad :CH_2 + CO$$

ketene

Triplet methylene cannot be prepared directly. However, if singlet methylene is dissolved in an inert gas, it undergoes a slow transformation to triplet methylene, which is the more stable of the two methylenes.

Both methylenes are electron-deficient and electrophilic. Their most im-portant reaction is addition to alkenes to yield substituted cyclopropanes. Singlet methylene gives a stereospecific syn-addition. This type of addition suggests a concerted, or one-step, reaction.

---

* The terms *singlet* and *triplet* arise from the multiplicity of electronic states of the two methylenes. The spin number of an electron may be denoted as $+\frac{1}{2}$ or $-\frac{1}{2}$, depending on the direction of spin. The magnitude of the vector sum of the spin numbers is called S. If all electrons are paired (of opposite spin numbers), $S = 0$ $[(+\frac{1}{2}) + (-\frac{1}{2}) = 0]$. If two electrons are unpaired, $S = +1$ $[(+\frac{1}{2}) + (+\frac{1}{2}) = +1$ or $(-\frac{1}{2}) + (-\frac{1}{2}) = -1]$. *Multiplicity* is defined as $2S + 1$. For :CH₂ (paired electrons), $S = 0$. Its multiplicity is therefore $2(0) + 1$, or $+1$. "Singlet" refers to $+1$. For HĊH (two unpaired electrons), $S = 1$. Its multiplicity is $2(1) + 1$, or $+3$. "Triplet" refers to $+3$.

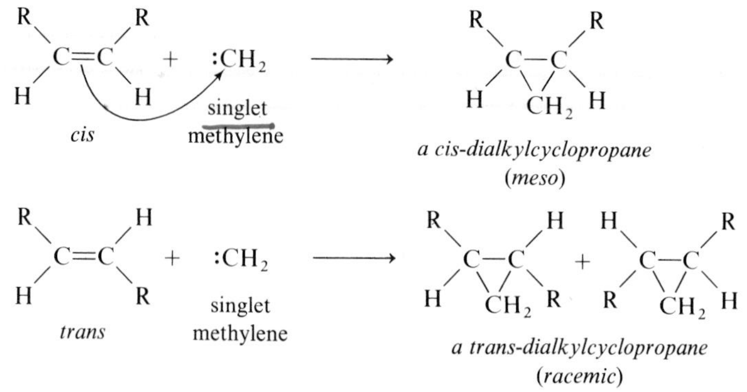

a cis-dialkylcyclopropane
(*meso*)

a trans-dialkylcyclopropane
(*racemic*)

Triplet methylene, on the other hand, gives nonstereospecific addition. Its addition follows a free-radical path, in which rotation can occur in the diradical intermediate. The result is a mixture of stereoisomeric products.

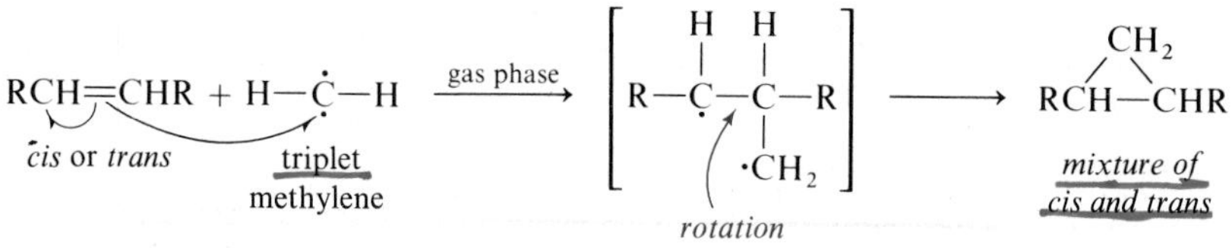

A major side reaction in the formation of cyclopropane rings with singlet methylene is an **insertion reaction**, in which :CH_2 inserts itself into a C—H bond. Because of the high reactivity of :CH_2, insertion reactions are unselective and yield mixtures.

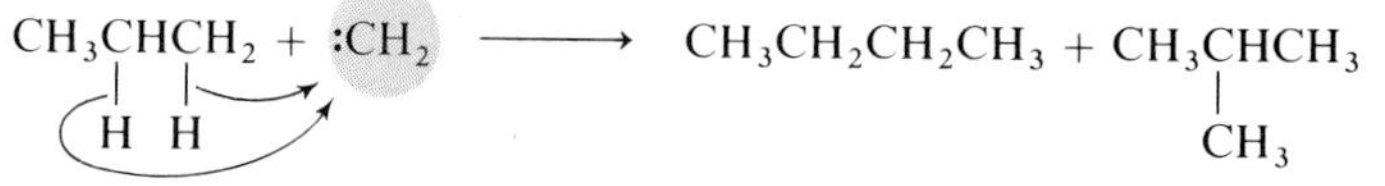

Another carbene, dichlorocarbene (Cl_2C:), is formed by the reaction of strong base and chloroform. This reaction is related to elimination reactions of alkyl halides in that the base removes the elements of HCl from the molecule. The elimination reaction of $CHCl_3$ is an **α elimination** instead of a β elimination. (A molecule that has a β hydrogen will lose the β hydrogen in preference; however, $CHCl_3$ has no β hydrogen.)

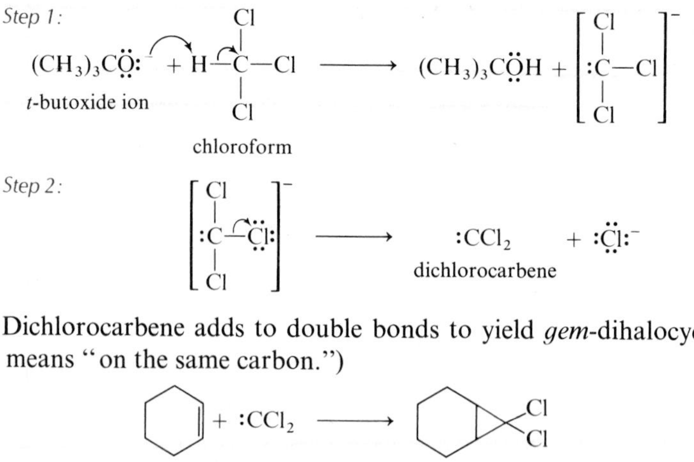

Step 1:

Step 2:

Dichlorocarbene adds to double bonds to yield *gem*-dihalocyclopropanes. (*Gem* means "on the same carbon.")

STUDY PROBLEMS

9.23. Predict the structures and stereochemistry of the products of the following addition reactions:

(a) *cis*-3-hexene + singlet methylene $\longrightarrow$

(b) *cis*-3-hexene + triplet methylene $\longrightarrow$

(c) *trans*-3-hexene + singlet methylene $\longrightarrow$

(d) *trans*-3-hexene + triplet methylene $\longrightarrow$

9.24. Give equations for the carbene reactions that would yield the following products:

(a) 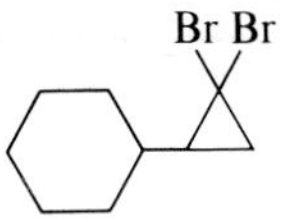   (b)

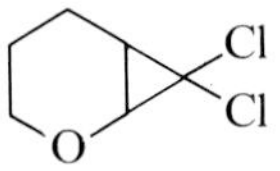

SECTION 9.13.

Catalytic Hydrogenation

The catalytic addition of hydrogen gas to an alkene or alkyne is a reduction of the pi-bonded compound. The reaction is general for alkenes, alkynes, and other compounds with pi bonds.

Alkenes and alkynes:

$$CH_3CH{=}CH_2 + H_2 \xrightarrow{\text{Pt}} CH_3CH_2CH_3$$

propene   propane

$$CH_3C{\equiv}CH + 2 H_2 \xrightarrow{\text{Pt}} CH_3CH_2CH_3$$

propyne   propane

Other pi systems:

$$\underset{\text{acetone}}{CH_3\overset{\overset{\text{O}}{\|}}{C}CH_3} + H_2 \xrightarrow[\text{heat, pressure}]{\text{Pt}} \underset{\text{2-propanol}}{CH_3\overset{\overset{\text{OH}}{|}}{C}HCH_3}$$

$$\underset{\substack{\text{ethanenitrile}\\\text{(acetonitrile)}}}{CH_3C{\equiv}N} + 2 H_2 \xrightarrow{\text{Pt}} \underset{\text{ethylamine}}{CH_3CH_2NH_2}$$

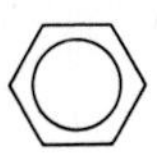

$$\underset{\text{benzene}}{⬡} + 3 H_2 \xrightarrow[\text{heat, pressure}]{\text{Pt}} \underset{\text{cyclohexane}}{⬡}$$

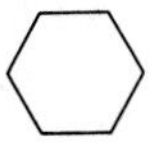

A.   Action of the catalyst

Hydrogenation reactions are exothermic, but they do not proceed spontaneously because the energies of activation are extremely high. Heating cannot
supply the energy needed to get the molecules to the transition state; however,
reaction proceeds smoothly when a catalyst is added.

A finely divided metal or a metal adsorbed onto an insoluble, inert carrier
(such as elemental carbon or barium carbonate) is often used as a hydrogenation
catalyst. The metal chosen depends on the compound to be reduced and the
conditions of the hydrogenation. For example, platinum, palladium, nickel,
rhenium, and copper are all suitable for the reduction of alkenes. For esters, which
are far more difficult to reduce, a copper–chromium catalyst (plus heat and
pressure) is usually used.

A **poisoned catalyst** (one that is partly deactivated) is used for hydrogenation
of an alkyne to an alkene instead of to an alkane. Palladium that has been treated
with quinoline (page 757) is a typical poisoned catalyst.

$$CH_3C\equiv CH + H_2 \xrightarrow{\text{deactivated Pd}} CH_3CH=CH_2$$

$$\text{propyne} \qquad\qquad\qquad\qquad \text{propene}$$

How does a hydrogenation catalyst ease the course of a hydrogenation
reaction? Experimental evidence supports the theory that first the hydrogen
molecules are adsorbed onto the metallic surface, then the H_2 sigma bonds are
broken, and metal–H bonds are formed. The alkene is also adsorbed onto the
metallic surface, with its pi bond interacting with the empty orbitals of the metal.
The alkene molecule moves around on the surface until it collides with a metal-
bonded hydrogen atom, undergoes reaction, and then leaves as the hydrogenated
product (see Figure 9.11).

The overall effect of the catalyst is to provide a surface on which the reaction
can occur and to weaken the bonds of both H_2 and the alkene. The result is a
lowering of the energy of activation for the reaction. Figure 9.12 shows energy
diagrams for a hydrogenation reaction. Note that *the catalyst does not affect the
energies of reactants or products.* The ΔH for the reaction is not changed by catalytic
action; only the E_{act} is changed.

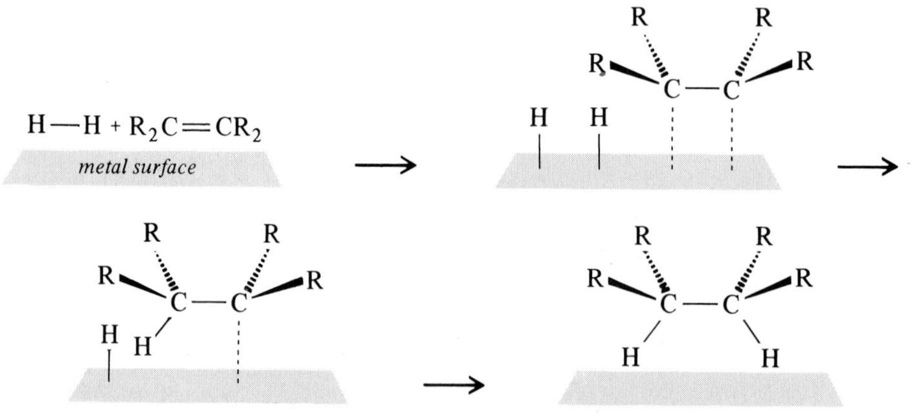

FIGURE 9.11. Hydrogenation of an alkene.

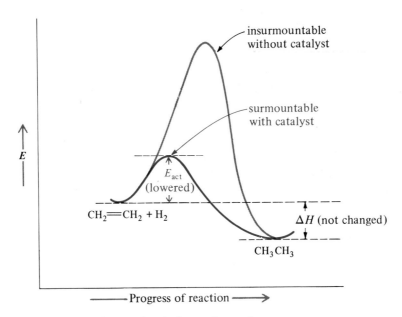

FIGURE 9.12. Energy diagrams for a hydrogenation reaction.

In recent years, soluble catalysts have been developed so that hydrogenation can occur in a homogeneous solution, rather than on a surface. These catalysts are organic metal complexes, such as $[(C_6H_5)_3P]_3RhCl$. If the organic portion of the catalyst is chiral, it is sometimes possible to reduce an achiral compound to a single enantiomer instead of the usual racemic mixture! This type of synthesis is analogous to the action of enzymes in living systems. Unfortunately, these soluble catalysts are difficult to use and are not practical in all situations.

B. Stereochemistry of hydrogenation

Evidence shows that the two hydrogen atoms add to the same side of, or syn to, the double bond when a solid catalyst is used. The syn-addition arises from the fact that the reaction occurs on a surface; access by H_2 to only one side of the pi bond is more favorable than access to both sides. If the hydrogenation products are capable of geometric isomerism, the *cis*-product is usually observed as the predominant product. (In some cases, however, isomerization to the more stable *trans*-product occurs.)

C.   How heats of hydrogenation show alkene stability

The **heat of hydrogenation** of an alkene is the energy difference between the starting alkene and the product alkane. It is calculated from the amount of heat released in a hydrogenation reaction. Table 9.6 lists the heats of hydrogenation of a few alkenes.

TABLE 9.6. Heats of hydrogenation for some alkenes and dienes

Name	Structure	$-\Delta H$, kcal/mole
ethene (ethylene)	$CH_2=CH_2$	32.8
propene (propylene)	$CH_3CH=CH_2$	30.1
1-butene	$CH_3CH_2CH=CH_2$	30.3
cis-2-butene	cis-$CH_3CH=CHCH_3$	28.6
trans-2-butene	trans-$CH_3CH=CHCH_3$	27.6
2-methyl-2-butene	$CH_3\overset{\underset{\displaystyle CH_3}{\mid}}{C}=CHCH_3$	26.9
3-methyl-1-butene	$CH_3\overset{\underset{\displaystyle CH_3}{\mid}}{C}HCH=CH_2$	30.3
1,3-butadiene	$CH_2=CHCH=CH_2$	57.1
1,4-pentadiene	$CH_2=CHCH_2CH=CH_2$	60.8

Let us consider the three alkenes that can be reduced to butane:

$$CH_3CH_2CH=CH_2$$
$$cis\text{-}CH_3CH=CHCH_3 \xrightarrow{\underset{H_2,\,Pt}{H_2,\,Pt}} CH_3CH_2CH_2CH_3$$
$$trans\text{-}CH_3CH=CHCH_3 \qquad \text{butane}$$

The product butane has the same energy regardless of the starting alkene. Any differences in ΔH for the three reactions reflect *differences in the energies of the starting alkenes*. The greater the value of the ΔH of hydrogenation, the higher is the energy of the starting alkene (see Figure 9.13).

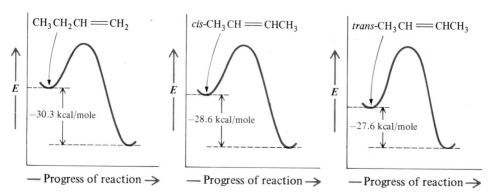

FIGURE 9.13. Comparison of the heats of hydrogenation of the three butenes that yield butane upon reduction.

From the differences in ΔH, we can see that 1-butene contains 1.7 kcal/mole more energy than *cis*-2-butene. *cis*-2-Butene, in turn, contains 1.0 kcal/mole more energy than *trans*-2-butene. The relative heats of hydrogenation of these compounds show that *trans*-2-butene is the most stable of the three butenes and that 1-butene is the least stable.

From just such ΔH comparisons, the relative stabilities of a large number of alkenes have been determined. The following statements summarize what we have learned about alkene stabilities:

1. Alkenes with more alkyl groups on the pi-bond carbons are more stable (probably because of the inductive effect of alkyl groups, which release electron density toward the sp^2 carbons).

$$CH_2{=}CH_2 \quad RCH{=}CH_2 \quad RCH{=}CHR \quad R_2C{=}CH_2 \quad R_2C{=}CHR \quad R_2C{=}CR_2$$

increasing stability →

2. Conjugated dienes are more stable than dienes with isolated double bonds (because of delocalization of the pi-electron density).

3. *trans*-Alkenes are more stable than *cis*-alkenes (because there are fewer steric repulsions in *trans*-isomers).

repulsion

less stable   *more stable*

STUDY PROBLEM

9.25. Which of each of the following pairs of alkenes would you expect to show the greater difference in energy between the *cis*- and *trans*-isomers? Why?

(a) $(CH_3)_3CCH{=}CHCH_2CH_3$ and $CH_3CH{=}CHCH_3$
(b) $ClCH{=}CHCl$ and $CH_3CH{=}CHCl$

D. Hydrogenation of fats and oils

The molecules of animal fats and vegetable oils contain long hydrocarbon chains. In vegetable oils, these chains are **polyunsaturated** (have several double bonds). Solid fats, on the other hand, usually contain few, if any, double bonds.

A vegetable oil may be converted to a substance of more solid consistency by partial hydrogenation of the double bonds. The process of converting liquid oils to solid fats by this technique is called **hardening**. Although the polyunsaturates may be more healthful, the hydrogenated products are generally more palatable. Partially hydrogenated peanut oil is used to make peanut butter, and partially hydrogenated corn oil or safflower oil is used in margarine. Note that the carbonyl groups in the vegetable oil do not undergo hydrogenation under these conditions because they are more difficult to hydrogenate. Also note that any or all of the carbon–carbon double bonds could be hydrogenated; therefore, a mixture of partially hydrogenated products would result.

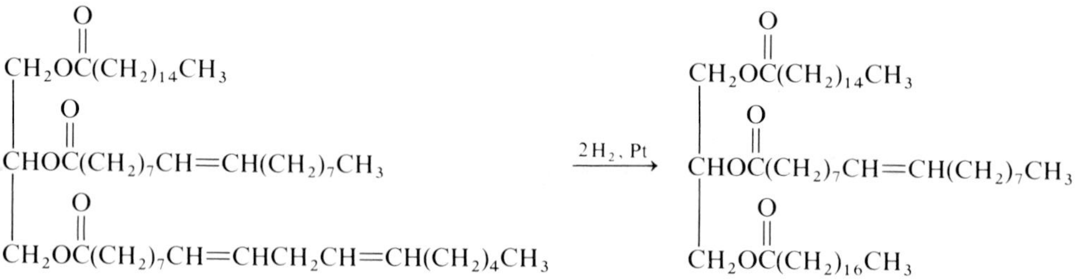

$$
\begin{array}{c}
\text{O} \\
\| \\
\mathrm{CH_2OC(CH_2)_{14}CH_3} \\
| \\
\text{O} \\
\| \\
\mathrm{CHOC(CH_2)_7CH{=}CH(CH_2)_7CH_3} \\
| \\
\text{O} \\
\| \\
\mathrm{CH_2OC(CH_2)_7CH{=}CHCH_2CH{=}CH(CH_2)_4CH_3}
\end{array}
\qquad \xrightarrow{2\,\mathrm{H_2},\ \mathrm{Pt}} \qquad
\begin{array}{c}
\text{O} \\
\| \\
\mathrm{CH_2OC(CH_2)_{14}CH_3} \\
| \\
\text{O} \\
\| \\
\mathrm{CHOC(CH_2)_7CH{=}CH(CH_2)_7CH_3} \\
| \\
\text{O} \\
\| \\
\mathrm{CH_2OC(CH_2)_{16}CH_3}
\end{array}
$$

a typical vegetable oil   a typical fat

SAMPLE PROBLEM

Dehydrogenation, the reverse reaction of hydrogenation, is carried out by heating a compound in the presence of the same type of catalyst as used for hydrogenation.

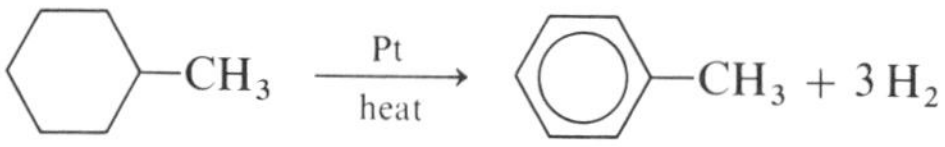

$$
\langle\ \rangle\!-\!\mathrm{CH_3} \xrightarrow[\text{heat}]{\mathrm{Pt}} \langle\!\bigcirc\!\rangle\!-\!\mathrm{CH_3} + 3\,\mathrm{H_2}
$$

Predict the dehydrogenation products of the following reactions:

(a) 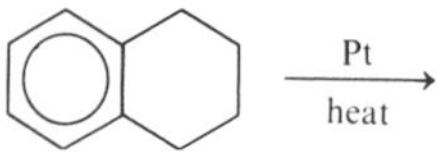 [bicyclic structure] $\xrightarrow[\text{heat}]{\mathrm{Pt}}$

(b) 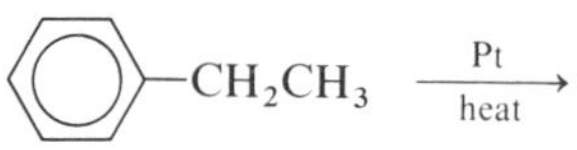 $\langle\!\bigcirc\!\rangle\!-\!\mathrm{CH_2CH_3} \xrightarrow[\text{heat}]{\mathrm{Pt}}$

Solution: (a) 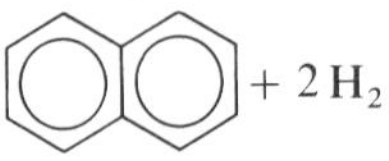 [naphthalene structure] $+ 2\,\mathrm{H_2}$

(b) 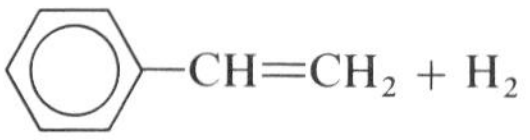 $\langle\!\bigcirc\!\rangle\!-\!\mathrm{CH{=}CH_2} + \mathrm{H_2}$

SECTION 9.14.

Oxidation of Alkenes

Alkenes can be oxidized to a variety of products, depending on the reagent used. Reactions involving oxidation of a carbon–carbon double bond may be classified into two general groups: (1) oxidation of the pi bond *without cleavage of the sigma bond*, and (2) oxidation of the pi bond *with cleavage of the sigma bond*. The products of oxidation without cleavage are either 1,2-diols or epoxides.

Without cleavage:

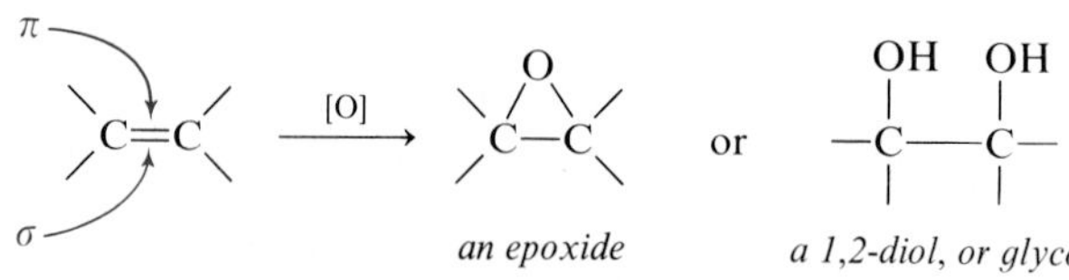

$$
\begin{array}{c}
\pi \\
\diagdown \quad \diagup \\
\mathrm{C}{=}\mathrm{C} \\
\diagup \quad \diagdown \\
\sigma
\end{array}
\xrightarrow{[\mathrm{O}]}
\begin{array}{c}
\text{O} \\
\diagup \ \diagdown \\
\diagdown\mathrm{C}{-}\mathrm{C}\diagup \\
\diagup \quad \diagdown
\end{array}
\quad \text{or} \quad
\begin{array}{c}
\text{OH}\quad\text{OH} \\
|\qquad| \\
-\mathrm{C}{-}\mathrm{C}{-} \\
|\qquad|
\end{array}
$$

an epoxide   a 1,2-diol, or glycol

When both the sigma bond and the pi bond of an alkene are cleaved in an oxidation, the products are ketones, aldehydes, or carboxylic acids.

With cleavage:

$$\ce{\underset{\diagdown}{\overset{\diagup}{C}}=\underset{\diagup}{\overset{\diagdown}{C}}} \xrightarrow{[O]} \underset{ketones}{\ce{-C-}} \quad or \quad \underset{aldehydes}{\ce{-CH}} \quad or \quad \underset{carboxylic\ acids}{\ce{-COH}}$$

A variety of reagents are used to oxidize alkenes. Some of the more common ones are listed in Table 9.7.

TABLE 9.7. Common reagents for oxidation of alkenes

Reagent	Products
Oxidation without cleavage:	
$KMnO_4$ with OH^- (cold)	1,2-diols
OsO_4 followed by Na_2SO_3	1,2-diols
$C_6H_5CO_3H$	epoxides
Oxidation with cleavage:	
$KMnO_4$ (hot)	carboxylic acids and ketones
O_3 followed by H_2O_2 with H^+	carboxylic acids and ketones
O_3 followed by Zn with H^+	aldehydes and ketones

A. Diol formation

The most popular reagent used to convert an alkene to a 1,2-diol is a cold, alkaline, aqueous solution of potassium permanganate (even though this reagent usually gives low yields). Osmium tetroxide (OsO_4) gives better yields of diols, but the use of this reagent is limited because it is both expensive and toxic. Both the permanganate and the OsO_4 oxidations proceed by way of cyclic inorganic esters, which yield *cis*-diols if the product is capable of geometric isomerism.

cis-1,2-cyclohexanediol

General:

$$\underset{an\ alkene}{R_2C=CR_2} \xrightarrow[\text{syn-addition}]{\text{cold } MnO_4^- \text{ or } OsO_4} \underset{a\ 1,2\text{-}diol}{R_2\overset{\overset{\displaystyle OH}{|}}{C}-\overset{\overset{\displaystyle OH}{|}}{C}R_2}$$

The reaction with cold permanganate solution constitutes the **Baeyer test** for unsaturation in compounds of unknown structure. The test solution ($KMnO_4$) is purple. As the reaction proceeds, the purple color disappears and a brown precipitate of MnO_2 is observed. The Baeyer test for double bonds, while widely used, has a serious limitation: any easily oxidized group (aldehyde, alkene, alkyne) gives a positive test result.

STUDY PROBLEM

9.26. The following compounds are treated with OsO_4, followed by Na_2SO_3. What products would you expect? (Indicate any stereoisomerism.)

(a) $CH_3CH{=}C(CH_3)_2$ (b)

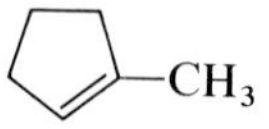

B.   Epoxide formation

Treatment of an alkene with peroxybenzoic acid in an inert solvent, such as $CHCl_3$ or CCl_4, yields an *epoxide, or oxirane*.

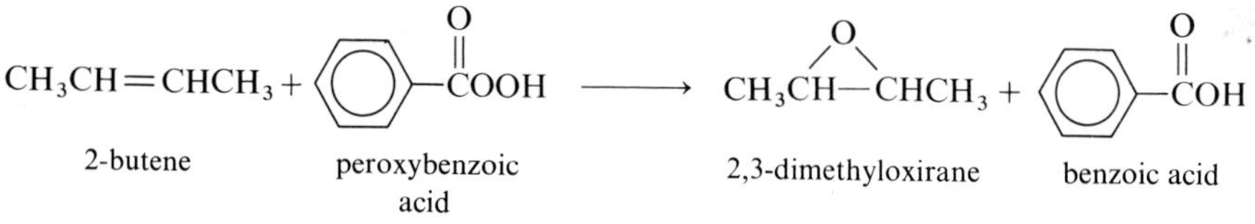

2-butene	peroxybenzoic acid	2,3-dimethyloxirane	benzoic acid

The reaction path involves transfer of an oxygen from the peroxyacid directly to the alkene.

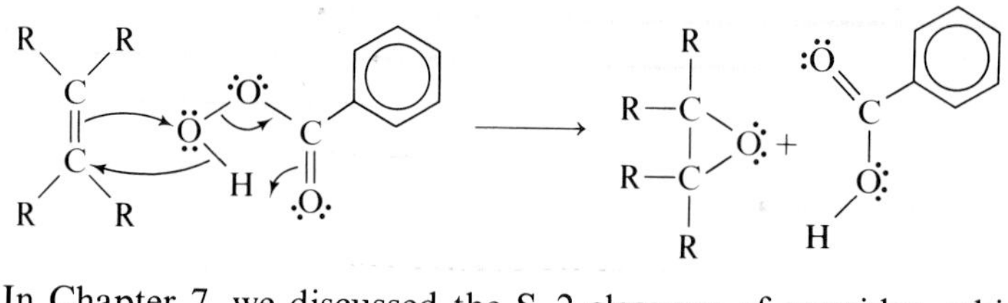

In Chapter 7, we discussed the S_N2 cleavage of epoxides, which results in *trans*-1,2-diols. By contrast, oxidation of an alkene with OsO_4 or cold $KMnO_4$ yields *cis*-1,2-diols. Thus, either type of diol may be prepared from an alkene, depending on the choice of reagents.

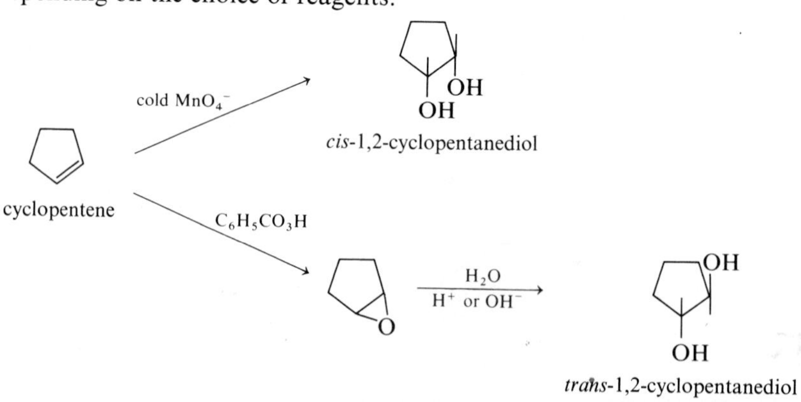

C. Oxidation with cleavage

The products of oxidation with cleavage depend both upon the oxidizing conditions and upon the structure of the alkene. Let us consider first the structure of the alkene.

The structural feature of the alkene that determines the products of oxidative cleavage is the *presence or absence of hydrogen atoms on the alkene carbons*. If each alkene carbon is not bonded to a hydrogen atom (that is, each alkene carbon is disubstituted), then oxidative cleavage results in a pair of *ketone* molecules.

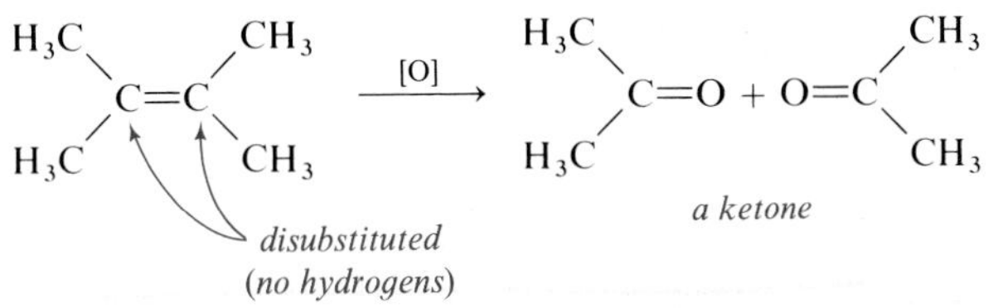

disubstituted
(no hydrogens)

a ketone

If, on the other hand, each alkene carbon has a hydrogen attached to it, then the products of oxidative cleavage are either *aldehydes* or *carboxylic acids*, depending upon the reaction conditions.

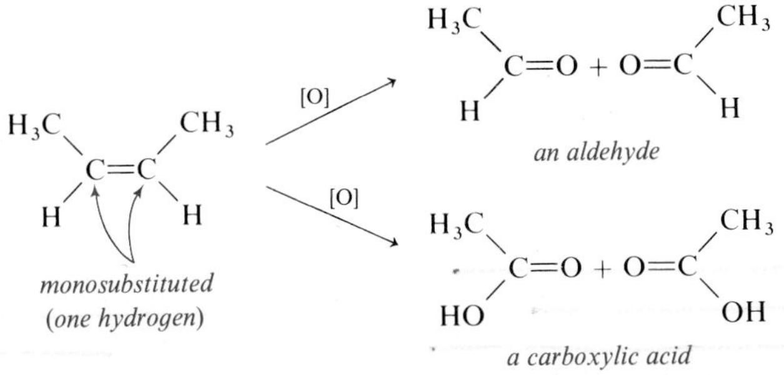

an aldehyde

monosubstituted
(one hydrogen)

a carboxylic acid

If one side of the double bond is disubstituted while the other side is monosubstituted, then oxidative cleavage results in a ketone from the disubstituted side, and an aldehyde or carboxylic acid from the monosubstituted side.

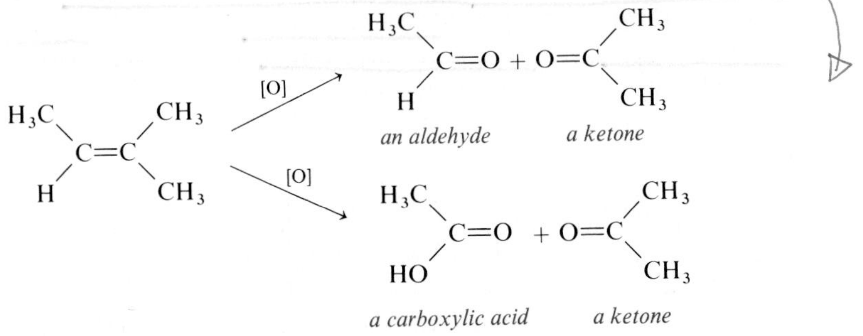

an aldehyde   *a ketone*

a carboxylic acid   *a ketone*

D. Cleavage with KMnO₄

A hot solution of $KMnO_4$ is a vigorous oxidizing agent that leads to only ketones and carboxylic acids. (Aldehydes cannot be isolated from $KMnO_4$ solutions; they are oxidized promptly to carboxylic acids.)

$$\text{2-methyl-2-butene} \xrightarrow[\text{heat}]{\text{MnO}_4^-} \underset{\text{acetic acid}}{CH_3\overset{O}{\overset{\|}{C}}OH} + \underset{\text{acetone}}{CH_3\overset{O}{\overset{\|}{C}}CH_3}$$

Under these vigorous oxidizing conditions, the carbon of a terminal double bond is oxidized to CO_2.

to CO_2

methylene-cyclopentane $\xrightarrow[\text{heat}]{\text{MnO}_4^-}$ cyclopentanone $= O + CO_2 + H_2O$

The reason for CO_2 formation is that the methylene group is first oxidized to formic acid, which is further oxidized to carbonic acid. The latter undergoes spontaneous decomposition to CO_2 and H_2O.

$$\underset{R}{\overset{R}{>}}C=CH_2 \xrightarrow{[O]} \underset{\text{formic acid}}{\left[H O\overset{O}{\overset{\|}{C}}H \right]} \xrightarrow{[O]} \underset{\text{carbonic acid}}{\left[HO\overset{O}{\overset{\|}{C}}OH \right]} \longrightarrow H_2O + CO_2$$

E. Ozonolysis

Ozonolysis (cleavage by ozone) has been used for determining the structures of unsaturated compounds because it results in the degradation of large molecules into smaller, identifiable fragments. Ozonolysis consists of two separate reactions: (1) oxidation of the alkene by ozone to an **ozonide**, and (2) either oxidation or reduction of the ozonide to the final products.

The initial oxidation is usually carried out by bubbling ozone through a solution of the alkene in an inert solvent such as CCl_4. Ozone attacks the pi bond to yield an unstable intermediate called a 1,2,3-trioxolane. This intermediate then goes through a series of transformations in which the carbon–carbon sigma bond is cleaved. The product is an ozonide (a 1,2,4-trioxolane), which is rarely isolated, but is carried on to the second step.

2-methyl-2-butene   ozone   *a 1,2,3-trioxolane*   many steps →

*an ozonide
(a 1,2,4-trioxolane)*

The second reaction in ozonolysis is either the oxidation or the reduction of the ozonide. If the ozonide is subjected to a *reductive work-up,* a monosubstituted carbon of the original alkene yields an aldehyde. If an *oxidative work-up* is used, then a monosubstituted carbon yields a carboxylic acid. In either case, a disubstituted carbon of the alkene yields a ketone.

Reductive work-up to aldehydes and ketones:

$$\underset{\text{to aldehyde}}{\underset{H}{\overset{H_3C}{}}} \underset{}{C} \underset{O-O}{} \underset{O}{C} \underset{\text{to ketone}}{\overset{CH_3}{\underset{CH_3}{}}} \xrightarrow[\text{H}^+,\,H_2O]{\text{Zn}} \underset{\text{acetaldehyde}}{CH_3\overset{\overset{\displaystyle O}{\|}}{C}H} + \underset{\text{acetone}}{CH_3\overset{\overset{\displaystyle O}{\|}}{C}CH_3}$$

Oxidative work-up to carboxylic acids and ketones:

$$\underset{\text{to carboxylic acid}}{\underset{H}{\overset{H_3C}{}}} \underset{}{C} \underset{O-O}{} \underset{O}{C} \underset{\text{to ketone}}{\overset{CH_3}{\underset{CH_3}{}}} \xrightarrow[]{H_2O_2,\,H^+} \underset{\text{acetic acid}}{CH_3\overset{\overset{\displaystyle O}{\|}}{C}OH} + \underset{\text{acetone}}{CH_3\overset{\overset{\displaystyle O}{\|}}{C}CH_3}$$

SAMPLE PROBLEM

Predict the products of reductive ozonolysis of γ-terpinene, a compound found in coriander oil:

$$\text{(structure of γ-terpinene with } CH_3 \text{ at top and } CH(CH_3)_2 \text{ at bottom)}$$

Solution:

$$\begin{array}{c} HC\overset{\displaystyle O}{\diagup} \\ | \\ CH_2 \\ \overset{|}{C}\diagup O \\ | \\ CH(CH_3)_2 \end{array} \quad + \quad \begin{array}{c} CH_3 \\ | \\ O\diagdown \overset{\displaystyle C}{\diagup} \\ \diagdown CH_2 \\ | \\ CH \\ \diagup O \end{array}$$

STUDY PROBLEM

9.27. What would be the products of both reductive and oxidative ozonolyses of each of the following alkenes or dienes?

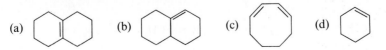

(a)   (b)   (c)   (d)

SECTION 9.15.

1,2-Addition and 1,4-Addition to Conjugated Dienes

Many of the reactions of conjugated dienes are identical to those of compounds with isolated double bonds. Acidic reagents and the halogens can add across one or both of the pi bonds. In conjugated-diene systems, these simple addition reactions are called **1,2-additions**, a term that refers to addition to the first and second carbons of a four-carbon, conjugated-diene system, and not necessarily to nomenclature numbers.

1,2-Addition reactions:

$$CH_2=CHCH=CH_2 \xrightarrow{\text{HBr}} \underset{\substack{| \quad\quad | \\ \text{3-bromo-1-butene}}}{\overset{\text{H} \quad \text{Br}}{CH_2-CHCH=CH_2}}$$

1,3-butadiene

$$CH_3CH=CHCH=CHCH_3 \xrightarrow{\text{Cl}_2} \underset{\substack{| \quad\quad | \\ \text{4,5-dichloro-2-hexene}}}{\overset{\text{Cl} \quad \text{Cl}}{CH_3CH-CHCH=CHCH_3}}$$

2,4-hexadiene

STUDY PROBLEM

9.28. Predict the product of 1,2-addition to the second double bond in each of the two preceding examples.

Along with 1,2-addition, conjugated dienes can also undergo **1,4-addition**. In these reactions, one equivalent of the reagent adds to the two end carbons (carbons 1 and 4) of the diene system; the remaining double bond ends up in the center of the original diene system.

1,4-Addition reactions:

$$CH_2=CHCH=CH_2 \xrightarrow{\text{HBr}} \underset{\substack{| \quad\quad\quad\quad\quad | \\ \text{1-bromo-2-butene}}}{\overset{\text{H} \quad\quad\quad\quad\quad \text{Br}}{CH_2-CH=CH-CH_2}}$$

1,3-butadiene

$$CH_3CH=CHCH=CHCH_3 \xrightarrow{\text{Cl}_2} \underset{\substack{| \quad\quad\quad\quad\quad | \\ \text{2,5-dichloro-3-hexene}}}{\overset{\text{Cl} \quad\quad\quad\quad\quad \text{Cl}}{CH_3CH-CH=CH-CHCH_3}}$$

2,4-hexadiene

Let us look at the mechanism of each type of addition. The mechanism for 1,2-addition is the same as that for addition to an isolated double bond. (The reaction of 1,3-butadiene goes through the more stable secondary carbocation and not through the less stable $^+CH_2CH_2CH=CH_2$.)

1,2-Addition:

$$CH_2=CHCH=CH_2 \xrightarrow{\text{H}^+} \left[\underset{\substack{| \\ \text{a } 2° \text{ carbocation} \\ \text{more stable}}}{\overset{\text{H}}{CH_2-\overset{+}{C}HCH=CH_2}} \right] \xrightarrow{\text{Br}^-} \underset{\substack{| \quad\quad | \\ \text{3-bromo-1-butene}}}{\overset{\text{H} \quad \text{Br}}{CH_2-CHCH=CH_2}}$$

The mechanism for 1,4-addition is a direct extension of that for 1,2-addition. The carbocation in the preceding example is an *allylic cation* (Section 5.7A) and is resonance-stabilized. Because of the resonance-stabilization of the allylic cation, there is a partial positive charge on carbon 4 of the diene system as well as on carbon 2. Attack at carbon 4 leads to the 1,4-addition product.

1,4-Addition:

$$CH_2{=}CHCH{=}CH_2 \xrightarrow{H^+} \left[\overset{\overset{\displaystyle H}{|}}{CH_2}{-}\overset{+}{CH}{\frown}CH{=}CH_2 \longleftrightarrow \overset{\overset{\displaystyle H}{|}}{CH_2}{-}CH{=}CH{-}\overset{+}{CH_2} \right]$$

$$\xrightarrow{Br^-} \overset{\overset{\displaystyle H}{|}}{CH_2}{-}CH{=}CH{-}\overset{\overset{\displaystyle Br}{|}}{CH_2}$$

1-bromo-2-butene

If just one equivalent of reagent is added to 1,3-butadiene, a mixture of two products results: 3-bromo-1-butene from 1,2-addition, and 1-bromo-2-butene from 1,4-addition.

SAMPLE PROBLEM

(a) Write the structures of *all possible* carbocation intermediates in the addition of one equivalent of HI to 2,4-hexadiene.

(b) Which carbocation would you expect to be formed at the faster rate?

Solution:

(a) $CH_3CH{=}CH{-}CH{=}CHCH_3 \xrightarrow{H^+}$

$[CH_3CH_2{-}\overset{+}{CH}{-}CH{=}CHCH_3]$ or $[CH_3\overset{+}{CH}{-}CH_2{-}CH{=}CHCH_3]$

(Addition of H^+ to the other double bond gives identical intermediates.)

(b) The first carbocation shown would be formed at the faster rate because it is a resonance-stabilized, allylic carbocation.

STUDY PROBLEMS

9.29. Write the *resonance structures* for the principal intermediate in the preceding sample problem, and give the structures of the principal products.

9.30. Predict the products of the addition of one equivalent of bromine to the following dienes:

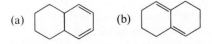

(a)   (b)

In the reaction of 1,3-butadiene with one equivalent of HBr, the ratio of 1,2-addition to 1,4-addition varies with the temperature at which the reaction is carried out. At $-80°$, the 1,2-addition product predominates. At $40°$, the 1,4-addition product predominates.

$$
\underset{\text{1,3-butadiene}}{CH_2{=}CHCH{=}CH_2} + HBr
$$

$-80°$ →

$$
\underset{\substack{\text{1,2-product} \\ 80\%}}{CH_3\overset{\displaystyle Br}{\overset{\displaystyle |}{C}}HCH{=}CH_2} + \underset{\substack{\text{1,4-product} \\ 20\%}}{CH_3CH{=}CHCH_2Br}
$$

$40°$ →

$$
\underset{20\%}{CH_3\overset{\displaystyle Br}{\overset{\displaystyle |}{C}}HCH{=}CH_2} + \underset{80\%}{CH_3CH{=}CHCH_2Br}
$$

It has also been observed that warming 3-bromo-1-butene (the 1,2-addition product) to $40°$ with a trace of acid results in an equilibrium mixture that is predominantly 1-bromo-2-butene (the 1,4-addition product).

$$
\underset{20\%}{CH_3\overset{\displaystyle Br}{\overset{\displaystyle |}{C}}HCH{=}CH_2} \underset{}{\overset{H^+,\ 40°}{\rightleftharpoons}} \underset{80\%}{CH_3CH{=}CHCH_2Br}
$$

How can we explain these observations? At low temperature, the reaction yields predominantly the 1,2-addition product because the 1,2-addition has the lower E_{act} (because the 2° carbon carries more + charge than the 1° carbon) and thus the faster rate. The relative rates of the reactions control the product ratio. We say that the reaction is under **kinetic control** at low temperatures. Figure 9.14 shows the energy diagram for the competing reactions.

At higher temperatures, a greater percentage of molecules can reach the higher-energy transition state, and the two products are in *equilibrium*. The more stable 1,4-product (which is the more substituted alkene) predominates. At the higher temperatures, the relative stabilities of the products control the product ratios, and the reaction is under **equilibrium control**, or **thermodynamic control**. Figure 9.15 shows the energy diagram for the equilibrium.

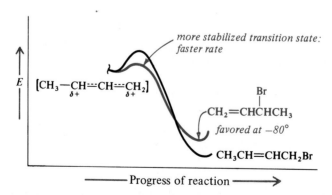

FIGURE 9.14. Partial energy diagram for the reaction of 1,3-butadiene with HBr. At $-80°$, the 1,2-addition product predominates.

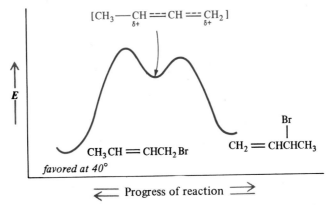

FIGURE 9.15. Energy diagram for the equilibrium between the 1,2- and 1,4-addition products. At 40°, the 1,4-addition product predominates.

STUDY PROBLEMS

9.31. Which *geometric isomer* of 1-bromo-2-butene would predominate at 40°?

9.32. 1,3-Butadiene is treated with one equivalent of Br_2 at $-15°$. Two structural isomers are obtained: 46% A and 54% B. When the reaction is carried out at 60°, the product mixture contains 90% A. What are the structures of A and B?

SECTION 9.16.

The Diels–Alder Reaction

One very important type of 1,4-addition is represented by the **Diels–Alder reaction,** which is a route to cyclic compounds from acyclic ones. The reaction is named after the German chemists Otto Diels and Kurt Alder, who jointly received the 1950 Nobel prize for their work in this area. In a Diels–Alder reaction, a diene is heated with a second unsaturated compound, called the **dienophile** ("diene-lover") to yield a product containing a six-membered ring.

1,3-butadiene	propenal	3-cyclohexene-1-carboxaldehyde
the diene	*the dienophile*	100%

The Diels–Alder reaction is but one example of a larger class of reactions called **pericyclic reactions.** In this section, we will limit our discussion to the Diels–Alder reaction. In Chapter 17, we will discuss the theory of pericyclic reactions.

A.   Line formulas and conformations

Because the Diels–Alder reaction converts open-chain compounds to cyclic compounds, the use of **line formulas** is very convenient for representing the open-chain compounds in this reaction. These line formulas are directly analogous to polygon formulas that represent rings.

|| or ═   means $CH_2{=}CH_2$

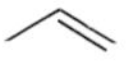

⌒⌒ or ⋎   means $CH_3CH{=}CH_2$

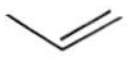

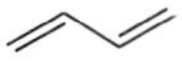

⌒⌒⌒ or //⟍   means $CH_2{=}CHCH{=}CH_2$

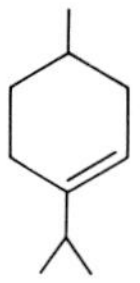

means

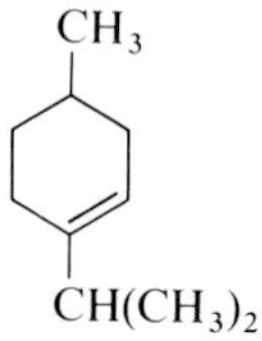

Another convention is the use of the terms **s-cis** and **s-trans** to describe the conformations of conjugated dienes. (The letter "s" is used because it is the geometry around the central *single bond* that determines the conformation.) The following compounds illustrate the use of the terms. For open-chain compounds, these formulas do not represent true isomers, but conformers, because sigma-bond rotation (requiring about 5 kcal/mole for 1,3-butadiene) is all that is needed for conversion from one to the other.

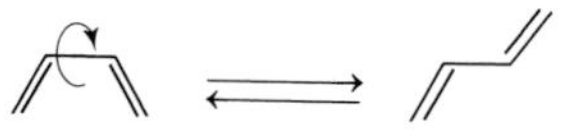

⇌

s-cis   *s-trans*

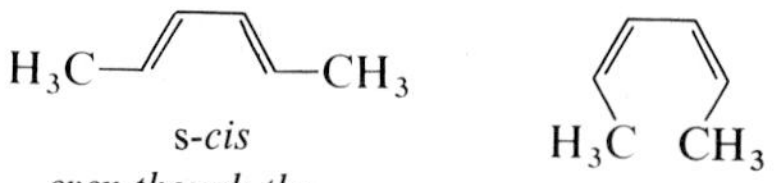

H_3C⌒⟍CH_3
s-cis
even though the
stereochemistry at
each double bond is
(E), or trans

H_3C CH_3
s-cis
and (Z), or
cis, at each
double bond

When the diene function is part of a cyclic system, the s-cis and s-trans structures represent different compounds; interconversion cannot occur without bonds being broken.

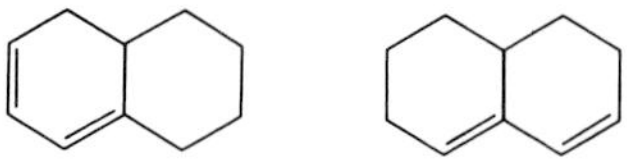

s-cis   *s-trans*

STUDY PROBLEM

9.33.   Classify each of the following dienes as s-*cis* or s-*trans*. Indicate which is interconvertible with the other form.

(a) 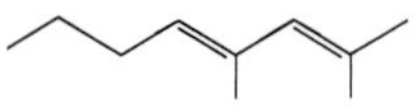 ⌒⌒⋎⌒⋎   (b)  ⬡   (c)

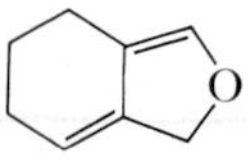

In a Diels–Alder reaction, the diene must have the s-*cis*, not the s-*trans*, conformation. In some structures (such as 1,3-butadiene), the s-*cis* and s-*trans* conformers are readily interconvertible. In other diene systems (such as in ring systems), the s-*trans* isomer does not undergo reaction.

Some s-cis dienes that can be used in a Diels–Alder reaction:

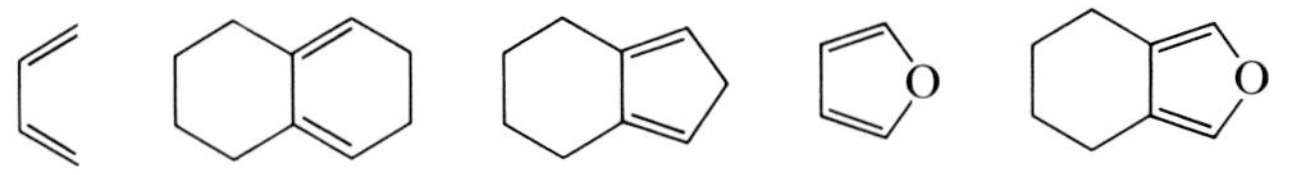

B.   Examples of Diels–Alder reactions

Using the line-formula symbolism, we can show the Diels–Alder reaction that was presented on page 423 as:

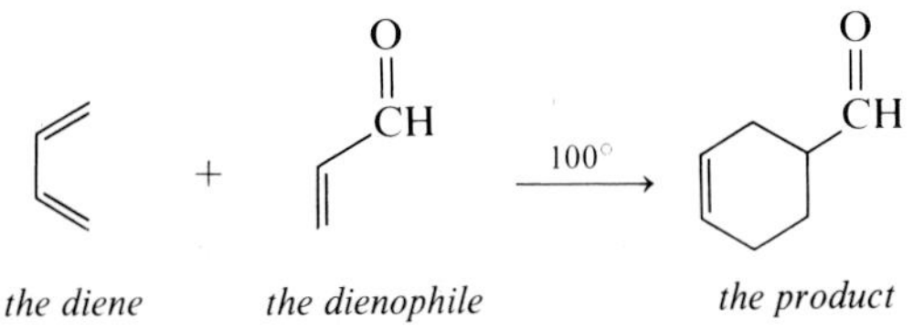

The dienophile usually contains other unsaturation (an aldehyde group in the preceding example) that does not participate directly in the addition reaction. This group does, however, enhance the reactivity of the dienophile's carbon–carbon double bond (the site of reaction) by electron-withdrawal. (Remember, the carbon in C=O carries a partial positive charge.)

A few examples of the types of dienes and dienophiles used in Diels–Alder reactions follow. You can see from these examples the versatility of this reaction in the synthesis of cyclic compounds.

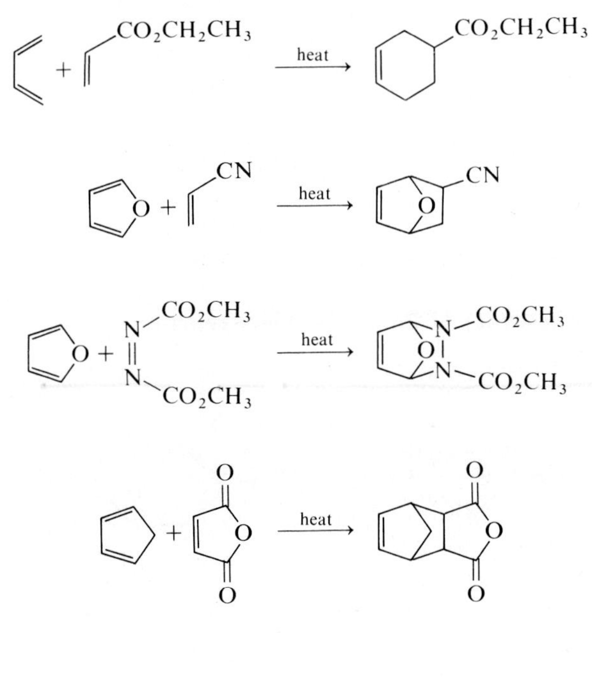

Industrially, a number of insecticides are prepared by Diels–Alder reactions.

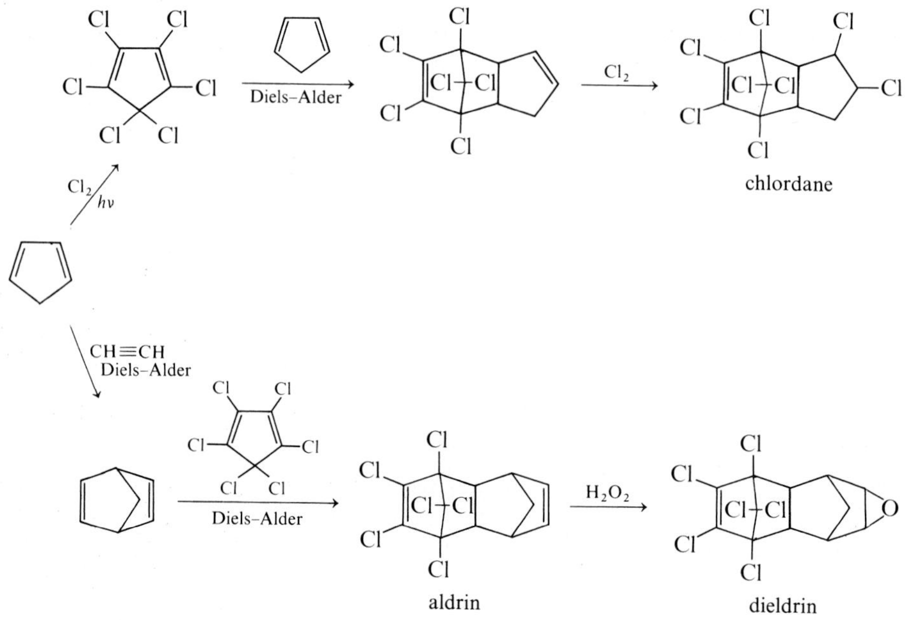

chlordane

aldrin   dieldrin

STUDY PROBLEMS

9.34. Predict the Diels–Alder products:

(a) 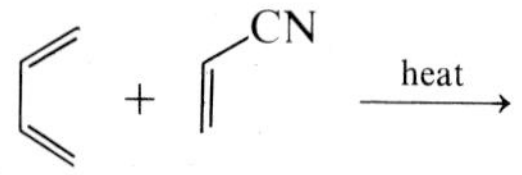   (b)

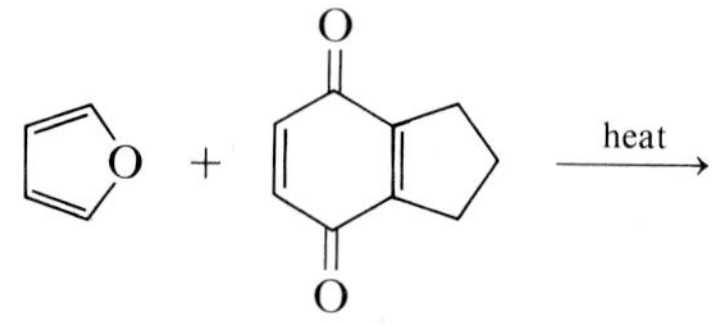

9.35. Suggest a synthesis for the following compound. (*Hint*: An alkyne may be used as a dienophile.)

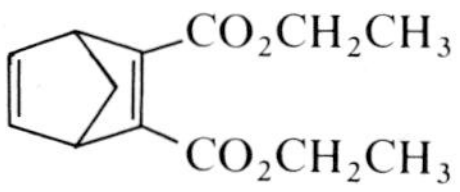

C. Stereochemistry of the Diels–Alder reaction

A Diels–Alder reaction is a concerted, *cis-*, or syn-addition and is thus stereo-specific.

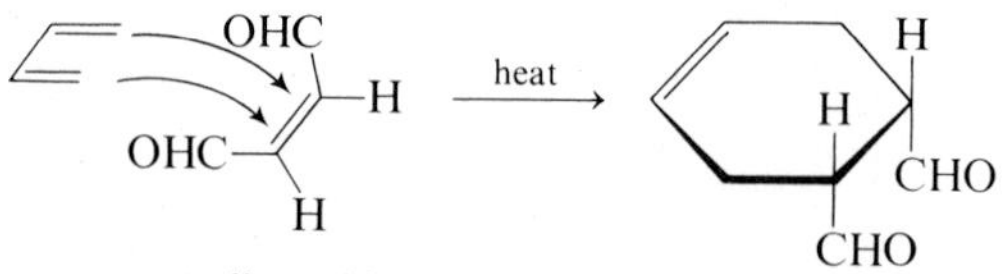

a *cis*-dienophile   a *cis*-product

There are two ways a diene can approach a dienophile. In some cases, two possible products can be formed. The two modes of addition are called *endo* (inside) and *exo* (outside), as the following examples show. The *endo* addition is usually favored, probably because of favorable interactions of the pi orbitals of the developing double bond and the pi orbitals in the unsaturated group. In the *endo* product of our example, the carbonyl group of the dienophile is *trans* to the bridge. In the *exo* product, the dienophile C=O is *cis* to the bridge.

STUDY PROBLEMS

9.36. Predict the product (including stereochemistry):

9.37. The following Diels–Alder reaction can yield two isomeric cyclohexenes. What are their structures?

$$CH_2=CHC=CH_2 + CH_2=CHCHO \xrightarrow{heat}$$
with CH_3 substituent

SECTION 9.17.

Polymers

Polymers are giant molecules, or **macromolecules.** Natural polymers include proteins (such as silk, muscle fibers, and enzymes), polysaccharides (starch and cellulose), rubber, and nucleic acids. Man-made polymers are almost as diverse as nature's polymers. We wear polyester clothes, sit on vinyl chairs, and write on Formica table tops. Our rugs may be made of polyester, polyacrylic, or polypropylene. Sky divers use nylon parachutes. We paint walls with latex paint and protect wood floors with polyurethane. Automobiles may have synthetic rubber tires and vinyl upholstery. Dishes may be melamine. Other common polymeric products include food wrap, Teflon coating for frying pans, hairbrushes, toothbrushes, epoxy glue, electrical insulators, plastic jugs, heart valves, airplane windshields—the list could continue! The technology of macromolecules has become a giant in the world of industry.

Polymers fall into three general classifications: *elastomers*, those polymers with elastic properties, like rubber; *fibers*, the threadlike polymers, such as cotton, silk, or nylon; and *plastics*, which can be thin sheets (kitchen wrap), hard and moldable solids (piping, children's toys), or coatings (car finishes, varnishes). The multiplicity of properties depends on the variety of structures that are possible in polymers.

Many useful polymers come from alkenes, and we will discuss some of these here. Other types of polymers will be covered in appropriate sections elsewhere in this text.

A polymer (Greek, "many parts") is made up of thousands of repeating units of small parts, the **monomers** (Greek, "one part"). In a polymerization reaction, the first products are **dimers** ("two parts"), then **trimers**, **tetramers**, and finally, after a series of reaction steps, the polymer molecules. The polymers that we will discuss here are called **addition polymers** because they are formed by the addition of monomers to each other without the loss of atoms or groups.

A synthetic polymer is usually named from the name of its monomer prefixed with **poly-**. For example, ethylene forms the simple polymer *polyethylene*, which is used for things like cleaners' bags and plastic piping.

$$CH_2{=}CH_2 \xrightarrow[\text{or catalyst}]{O_2,\ \text{heat, pressure}} {-}CH_2CH_2{-}CH_2CH_2{-}CH_2CH_2{-}$$

ethylene   *repeating monomeric units*   polyethylene

the monomer   *the polymer*

The equations for polymerization are conveniently represented in the following format, where x is used to mean "a large number."

$$x\ CH_2{=}CH_2 \xrightarrow{\text{catalyst}} {\{}CH_2CH_2{\}}_x$$

Frequently, the end groups of macromolecules are unknown—they may arise from impurities in the reaction mixture. In some cases, the end groups can be controlled. The properties of a polymer are governed almost entirely by the bulk of the polymer molecule rather than by the end groups. To emphasize the basic structure of the polymer, it is customary not to include the end groups in the formula unless they are specifically known.

A.   Free-radical polymers

A common mode of polymerization of alkenes is by a free-radical path. The polymerization is started by a catalyst or an initiator such as O_2 or a peroxide. The resulting polymer is formed by a chain-propagation process. Let us use the polymerization of propylene as our example.

Overall reaction:

$$x \ CH_2=\overset{\overset{\displaystyle CH_3}{|}}{CH} \xrightarrow[\text{initiator}]{\text{free-radical}} \left(\overset{\overset{\displaystyle CH_3}{|}}{CH_2CH}\right)_x$$

propylene   polypropylene

used for rugs and upholstery

Initiation: $ROOR \longrightarrow 2\,RO\cdot$

Propagation:

$$RO\cdot \ \overparen{CH_2=\underset{\underset{\displaystyle CH_3}{|}}{CH}} \longrightarrow RO-CH_2\underset{\underset{\displaystyle CH_3}{|}}{\dot{C}H}$$

$$ROCH_2\underset{\underset{\displaystyle CH_3}{|}}{\dot{C}H} \ \overparen{CH_2=\underset{\underset{\displaystyle CH_3}{|}}{CH}} \longrightarrow ROCH_2\underset{\underset{\displaystyle CH_3}{|}}{CH}-CH_2\underset{\underset{\displaystyle CH_3}{|}}{\dot{C}H} \ \text{etc.}$$

Theoretically, the chain growth could go on indefinitely, which of course does not happen. The termination steps for polymerization are typical free-radical termination steps. Two radicals may meet and join, or two radicals might undergo disproportionation. (The squiggles in the following formulas are used to indicate that a much larger molecule than shown is involved in the reaction.)

Coupling:

$$\text{⊱}CH_2\underset{\underset{\displaystyle CH_3}{|}}{\dot{C}H} \ + \ \underset{\underset{\displaystyle CH_3}{|}}{\dot{C}H}CH_2\text{⊰} \longrightarrow \text{⊱}CH_2\underset{\underset{\displaystyle CH_3}{|}}{CH}-\underset{\underset{\displaystyle CH_3}{|}}{CH}CH_2\text{⊰}$$

terminated

Disproportionation:

$$\text{⊱}CH_2\underset{\underset{\displaystyle CH_3}{|}}{\dot{C}H} \ + \ \underset{\underset{\displaystyle CH_3}{|}}{\dot{C}H}CH_2\text{⊰} \longrightarrow \text{⊱}CH=CH \ + \ \underset{\underset{\displaystyle CH_3}{|}}{CH_2}\underset{\underset{\displaystyle CH_3}{|}}{CH_2}\text{⊰}$$

terminated

There are two ways in which propylene molecules could join together to form polypropylene: (1) **head-to-tail**, or (2) **head-to-head and tail-to-tail**.

$$\textit{tail}\nearrow \qquad \nwarrow\textit{head}$$

$$CH_2=\underset{\underset{\displaystyle CH_3}{|}}{CH}$$

head-to-tail   *head-to-head*   *tail-to-tail*

$$\text{⊱}CH_2\underset{\underset{\displaystyle CH_3}{|}}{CH}-CH_2\underset{\underset{\displaystyle CH_3}{|}}{CH}-CH_2\underset{\underset{\displaystyle CH_3}{|}}{CH}\text{⊰} \qquad \text{⊱}CH_2\underset{\underset{\displaystyle CH_3}{|}}{CH}-\underset{\underset{\displaystyle CH_3}{|}}{CH}CH_2-CH_2\underset{\underset{\displaystyle CH_3}{|}}{CH}\text{⊰}$$

Polypropylene is an example of a *head-to-tail polymer*. Let us discuss the reason for this orientation of the monomers. A more stable free-radical intermediate means a lower-energy transition state and a faster rate of reaction. Of the two possible modes of free-radical attack, one leads to a less stable *primary* free radical, while the other leads to the more stable *secondary* free radical. The repetitive formation of secondary free radicals leads to a head-to-tail joining of propylene monomers.

$$R\cdot + CH_2{=}\overset{\displaystyle |}{\underset{\displaystyle CH_3}{CH}} \longrightarrow RCH_2{-}\overset{\displaystyle |}{\underset{\displaystyle CH_3}{\dot{C}H}} \quad \text{and not} \quad RCH{-}\overset{\displaystyle |}{\underset{\displaystyle CH_3}{\dot{C}H_2}}$$

$$2°, \textit{more stable} \qquad\qquad 1°, \textit{less stable}$$

$$\xrightarrow{\;CH_2{=}CHCH_3\;} RCH_2CH{-}CH_2\dot{C}H \quad \text{etc.}$$
$$\overset{\displaystyle |}{CH_3} \quad \overset{\displaystyle |}{CH_3}$$

A large number of polymers may be synthesized from alkenes in analogous reactions to those of ethylene and propylene.

$$x\; CH_2{=}\overset{\displaystyle \overset{Cl}{|}}{CH} \xrightarrow{\text{catalyst}} \left(CH_2{-}\overset{\displaystyle \overset{Cl}{|}}{CH}\right)_x$$

vinyl chloride   polyvinyl chloride (PVC)

used for flooring, piping, siding,
phonograph records, and garbage bags

$$x\; CH_2{=}\overset{\displaystyle \overset{CH_3}{|}}{\underset{\displaystyle \underset{CO_2CH_3}{|}}{C}} \xrightarrow{\text{catalyst}} \left(CH_2{-}\overset{\displaystyle \overset{CH_3}{|}}{\underset{\displaystyle \underset{CO_2CH_3}{|}}{C}}\right)_x$$

methyl methacrylate   polymethyl methacrylate

Plexiglas and Lucite

$$x\; CH_2{=}CH\!\!-\!\!\bigcirc \xrightarrow{\text{catalyst}} \left(CH_2{-}CH\!\!-\!\!\bigcirc\right)_x$$

styrene   polystyrene

used in toothbrush handles and
in the manufacture of styrofoam

A chemist is not limited to using a single monomer in the production of a polymer. To achieve desired properties, he might use a mixture of two, three, or even more monomers. A mixture of two different monomers results in a **copolymer**, such as Saran (used in kitchen wrap).

$$x\; CH_2{=}CHCl + x\; CH_2{=}CCl_2 \xrightarrow{\text{catalyst}} (CH_2CHCl{-}CH_2CCl_2)_x$$

vinyl chloride   1,1-dichloroethene   Saran
                  (vinylidene chloride)
                                                a copolymer
                                        (*not necessarily alternating*)

STUDY PROBLEMS

9.38. The monomer for *Teflon* is $CF_2=CF_2$. What is the structure of Teflon?

9.39. *Orlon* has the formula $\left(CH_2\overset{\displaystyle CN}{\underset{|}{CH}}\right)_x$. What is the structure of its monomer?

Until 1955, most addition polymers were made by free-radical paths. In that year, however, Karl Ziegler and Giulio Natta introduced a new technique for polymerization. These two chemists received the 1963 Nobel prize for their discovery, a type of catalyst that permits control of the stereochemistry of a polymer during its formation. (A commonly used Ziegler–Natta catalyst is $(CH_3CH_2)_3Al$ complexed with $TiCl_4$.) Ziegler–Natta catalysts function by undergoing reaction with the monomeric alkene; then new monomers are *inserted* between the catalyst and the growing polymer.

$$\boxed{catalyst}\!\!-CH_2CH_3 + CH_2\!\!=\!\!CH_2 \longrightarrow \boxed{catalyst}\!\!-CH_2CH_2-CH_2CH_3 \quad etc.$$

B.   1,4-Addition polymers

Conjugated dienes can be polymerized by **1,4-addition**. The product still contains unsaturation; therefore, the polymer could contain all *cis*-units, all *trans*-units, or a mixture of *cis*- and *trans*-units. The following equation shows the free-radical, 1,4-polymerization of isoprene.

$$RO\cdot + CH_2\!\!=\!\!\overset{\displaystyle CH_3}{\underset{|}{C}}\!\!-CH\!\!=\!\!CH_2 \xrightarrow{\text{1,4-addition}} ROCH_2-\overset{\displaystyle CH_3}{\underset{|}{C}}\!\!=\!\!CH-\overset{\displaystyle \cdot}{C}H_2 \quad etc. \longrightarrow$$

initiator   2-methyl-1,3-butadiene
(isoprene)

all-*cis*-polyisoprene
(natural rubber)

or

all-*trans*-polyisoprene
(gutta percha)

Natural rubber is polyisoprene with *cis* double bonds. The *trans*-polymer, called **gutta percha**, is a hard polymer used as a golf-ball covering and in temporary dental fillings. The reason for the difference in properties of these two polymers will be discussed in Section 9.17D. Neither one of these polymers is synthesized in nature from isoprene itself, as was once believed. Instead, their precursor is mevalonic acid (Section 20.5).

STUDY PROBLEM

9.40. *Neoprene*, developed in 1932, was the first synthetic rubber. It is used in washers, tubing, and the like. Neoprene is the all-*trans*, head-to-tail polymer of 2-chloro-1,3-butadiene. What is the structure of this polymer?

C.   Ionic addition polymers

Besides free-radical addition, polymers may be formed through **cationic addition**, a reaction that proceeds through a carbocation intermediate. An acid such as H_2SO_4 or $AlCl_3$ may be used to form the initial carbocation. (This procedure is not used for the synthesis of polyethylene because of the difficulty in forming a primary carbocation.)

$$H^+ \quad CH_2{=}\overset{\overset{\displaystyle CH_3}{|}}{C}{-}CH_3 \longrightarrow CH_3{-}\overset{\overset{\displaystyle CH_3}{|}}{\underset{+}{C}}{-}CH_3$$

methylpropene
(isobutylene)

$$CH_3\overset{+}{\underset{\overset{|}{CH_3}}{\overset{|}{C}}} \quad CH_2{=}\overset{\overset{\displaystyle CH_3}{|}}{\underset{\overset{|}{CH_3}}{C}} \longrightarrow CH_3\overset{\overset{\displaystyle CH_3}{|}}{\underset{\overset{|}{CH_3}}{C}}{-}CH_2\overset{+}{\underset{\overset{|}{CH_3}}{\overset{|}{C}}} \xrightarrow{\text{many steps}} \left(CH_2\overset{\overset{\displaystyle CH_3}{|}}{\underset{\overset{|}{CH_3}}{C}}\right)_x$$

first unit   second unit   polyisobutylene

SAMPLE PROBLEM

Suggest a termination step in the carbocation polymerization of isobutylene.

Solution: Tertiary carbocations often undergo elimination in the presence of H_2SO_4 or other strong acid:

$$RCH_2\overset{+}{\underset{\overset{|}{CH_3}}{\overset{|}{C}}} \longrightarrow RCH{=}C(CH_3)_2 + H^+$$
terminated

D.   Structure and stereochemistry of polymers

Polymers, like any other organic compound, may have functional groups and chiral carbons. They can undergo hydrogen bonding and dipole–dipole inter-actions. The chemical composition of a polymer chain is referred to as its **primary structure**. How the chain is arranged in relation to itself and to other chains is called the **secondary structure**. This secondary structure may be as important to the properties of a polymer as its chemical composition.

A polymer may be a tangled mass of continuous chains or branched chains. The result is a soft amorphous solid such as soft rubber. On the other hand, a polymer may be composed of continuous chains held together by hydrogen bonds or by other dipole–dipole attractions. This type of polymer structure lends itself to fibers or hard, moldable plastics. A more-ordered polymer is said to have a higher degree of **crystallinity** than the amorphous, or noncrystalline, polymer.

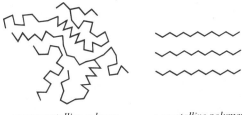

a noncrystalline polymer   *a crystalline polymer*

The differences in properties between crystalline and noncrystalline polymers are beautifully demonstrated by gutta percha and natural rubber. Gutta percha is a highly crystalline polymer, while natural rubber has a molecular shape that does not lend itself to an ordered, crystalline arrangement.

gutta percha (all *trans*)

natural rubber (all *cis*)

Let us reconsider the polymerization of propylene. There are three types of products that could result from the head-to-tail polymerization of propylene. (1) The methyl groups at the newly formed chiral carbons could be protruding from the chain in a random fashion; this is an **atactic polymer** (a soft, amorphous product). (2) The methyl groups could alternate from one side of the chain to the other; this is a **syndiotactic polymer**. (3) The methyl groups could all be on the same side; then the polymer is said to be **isotactic** (see Figure 9.16). Because of their orderly arrangements, the chains of the latter two polymers can lie closer together and the polymers are more crystalline.

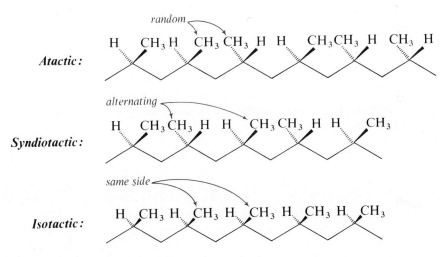

FIGURE 9.16. The three types of polypropylene molecules.

SECTION 9.18.

Use of Alkenes and Alkynes in Synthesis

From Table 9.8, you can see that alkenes are valuable starting materials for synthesizing other organic compounds. Alkynes are not widely used in synthesis (nor are they as readily available).

The addition reactions that use H^+ as a catalyst proceed through carbocations and yield Markovnikov products (and possible rearrangement products). These reactions include those with HX and with H_2O (H^+).

Markovnikov products:

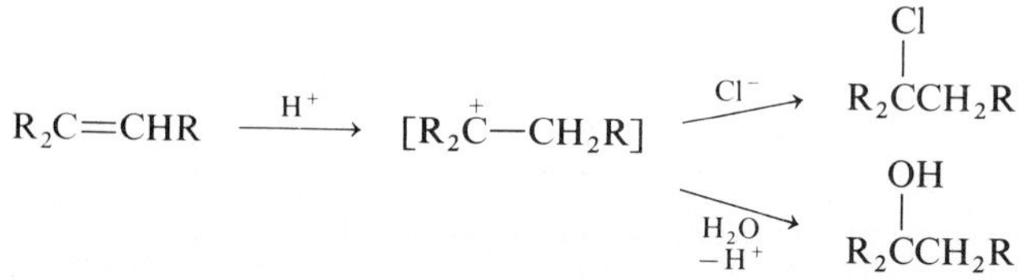

Alcohols (or ethers) can be prepared *without rearrangement*, in excellent yield, by oxymercuration–demercuration reactions, a sequence that also leads to Markovnikov products.

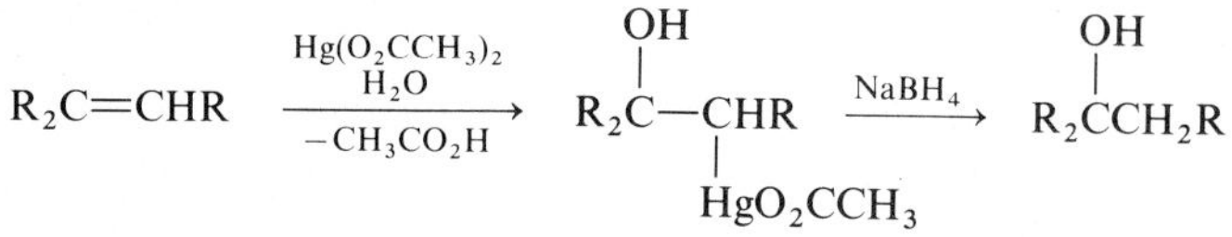

"Anti-Markovnikov" products, with the functional group added to the less substituted carbon of the alkene, may be prepared by HBr (with O_2 or a peroxide catalyst) or by hydroboration.

"anti-Markovnikov" products:

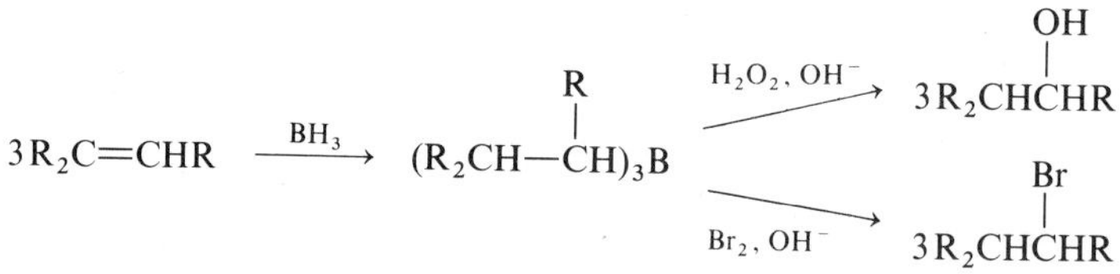

Difunctional compounds arise from the addition of X_2 (or $X_2 + H_2O$) to a double bond; by 1,2- or 1,4-addition to a diene; or by oxidation of an alkene to a 1,2-diol. For example,

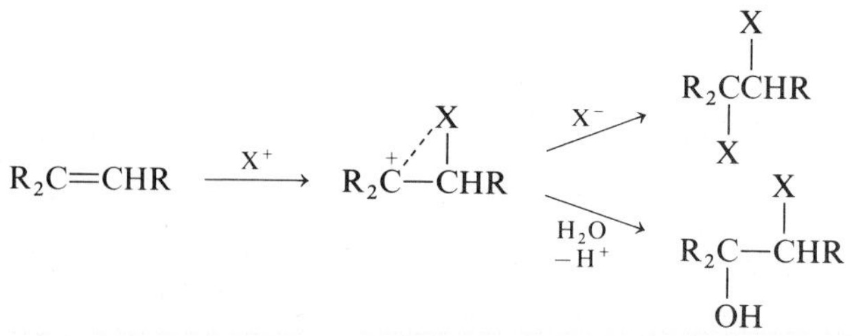

TABLE 9.8. Types of compounds that can be obtained from alkenes and dienes.

Reaction		*Product*	*Section reference*

Markovnikov addition:

$$\underset{}{R_2C{=}CHR} + HX \longrightarrow R_2\overset{\overset{\displaystyle X}{|}}{C}CH_2R$$

                                    alkyl halide   9.7A–C

$$+ H_2O \xrightarrow{H^+} R_2\overset{\overset{\displaystyle OH}{|}}{C}CH_2R$$

    alcohol   9.8

$$\xrightarrow[\text{(2) NaBH}_4]{\text{(1) Hg(O}_2\text{CCH}_3)_2,\ H_2O} R_2\overset{\overset{\displaystyle OH}{|}}{C}CH_2R$$

    alcohol   9.9

$$\xrightarrow[\text{(2) NaBH}_4]{\text{(1) Hg(O}_2\text{CCH}_3)_2,\ R'OH} R_2\overset{\overset{\displaystyle OR'}{|}}{C}CH_2R$$

    ether   9.9

$$+ X_2 \longrightarrow R_2\overset{\overset{\displaystyle X}{|}}{C}{-}\overset{\overset{\displaystyle X}{|}}{C}HR$$

    dihaloalkane   9.11A–C

$$+ X_2 \xrightarrow{H_2O} R_2\overset{\overset{\displaystyle OH}{|}}{\underset{\underset{\displaystyle X}{|}}{C}}CHR$$

    1, 2-halohydrin   9.11D

"Anti-Markovnikov" addition:

$$R_2C{=}CHR \xrightarrow[\text{(2) H}_2\text{O}_2,\ \text{OH}^-]{\text{(1) BH}_3} R_2CH\overset{\overset{\displaystyle OH}{|}}{C}HR$$

    alcohol   9.10

$$\xrightarrow[\text{(2) RCO}_2\text{H}]{\text{(1) BH}_3} R_2CHCH_2R$$

    alkane   9.10

$$\xrightarrow[\text{(2) Br}_2,\ \text{OH}^-]{\text{(1) BH}_3} R_2CH\overset{\overset{\displaystyle Br}{|}}{C}HR$$

    alkyl bromide   9.10

$$+ HBr \xrightarrow[\text{peroxide}]{\text{O}_2\text{ or}} R_2CH\overset{\overset{\displaystyle Br}{|}}{C}HR$$

    alkyl bromide   9.7D

1,4-Addition:[a]

$$R_2C{=}CHCH{=}CR_2 + HX \longrightarrow R_2CHCH{=}\overset{\overset{\displaystyle X}{|}}{C}HCR_2$$

    allylic halide   9.15

$$R_2C{=}CHCH{=}CR_2 + X_2 \longrightarrow R_2\overset{\overset{\displaystyle X}{|}}{C}CH{=}\overset{\overset{\displaystyle X}{|}}{C}HCR_2$$

    dihaloalkene   9.15

                                                                         (continued)

TABLE 9.8. *(continued)*

Reaction	Product	Section reference
Addition leading to cyclic products:		
$R_2C{=}CHR + :CH_2 \longrightarrow R_2\overset{\displaystyle CH_2}{\overbrace{C{-}CHR}}$	**a cyclopropane**	9.12
where Y is usually $C{=}O$ or other unsaturated group	**a cyclohexene**	9.16
Reduction:		
$R_2C{=}CHR + H_2 \xrightarrow{\text{Pt}} R_2CHCH_2R$	**alkane**	9.13
Oxidation:		
$R_2C{=}CHR + MnO_4^- \xrightarrow{25^\circ} R_2\overset{\displaystyle OH}{\underset{\displaystyle OH}{C{-}CHR}}$	**1,2-diol**[b]	9.14A
$+ C_6H_5CO_3H \longrightarrow R_2\overset{\displaystyle O}{\overbrace{C{-}CHR}}$	**epoxide**[b]	9.14B
$+ MnO_4^- \xrightarrow{\text{heat}} R_2C{=}O + HO\overset{\displaystyle O}{\overset{\|}{C}}R$	**ketone, carboxylic acid**	9.14D
$\xrightarrow[\text{(2) Zn, H}^+\text{, H}_2\text{O}]{\text{(1) O}_3} R_2C{=}O + H\overset{\displaystyle O}{\overset{\|}{C}}R$	**ketone, aldehyde**[c]	9.14E

[a] 1,2-Addition may also occur.
[b] A *cis*-1,2-diol may be prepared from an alkene and cold MnO_4^-; the *trans*-1,2-diol may be prepared from hydrolysis of the epoxide.
[c] Ozonolysis with an oxidative work-up yields ketones and carboxylic acids.

Oxidation of alkenes is a way to introduce other functionality, such as an epoxide ring. Vigorous oxidizing conditions result in cleavage of the double bond and yield carbonyl compounds, as shown in Table 9.8.

A tremendous variety of *cyclic compounds* containing six-membered rings can be made by Diels–Alder reactions of conjugated dienes with dienophiles. Because many naturally occurring compounds, synthetic drugs, and so forth contain rings, the Diels–Alder reaction affords routes to some of these. Cyclopropane rings can be made by the reactions of carbenes with alkenes.

SAMPLE PROBLEM

Suggest a synthesis for 3-methyl-2-pentanone from 3-bromo-3-methylpentane.

Solution:

1. Write the structures:

$$\underset{\underset{CH_3}{|}}{CH_3CH_2\overset{\overset{Br}{|}}{C}CH_2CH_3} \longrightarrow \underset{\underset{CH_3}{|}}{CH_3\overset{\overset{O}{\|}}{C}CHCH_2CH_3}$$

2. There is no one-step reaction from **RX** to a ketone. Therefore, working backwards, ask yourself what reactions lead to ketones (without cleavage of the carbon skeleton). Oxidation of a 2° alcohol is a standard reaction leading to a ketone.

$$\underset{\underset{CH_3}{|}}{CH_3\overset{\overset{OH}{|}}{C}HCHCH_2CH_3} \xrightarrow{H_2CrO_4} \underset{\underset{CH_3}{|}}{CH_3\overset{\overset{O}{\|}}{C}CHCH_2CH_3}$$

3. The hydroxyl group in the alcohol is not in the same position as the Br in the starting material. Therefore, a simple substitution reaction (**RX** + OH⁻) would not be a route to the alcohol; however, an alcohol can be obtained from an alkene.

$$\underset{\underset{CH_3}{|}}{CH_3CH{=}CCH_2CH_3} \longrightarrow \underset{\underset{CH_3}{|}}{CH_3\overset{\overset{OH}{|}}{C}H{-}CHCH_2CH_3}$$

Because the alcohol is the *anti-Markovnikov product*, a reaction with $H_2O + H^+$ or oxymercuration–demercuration will not yield the proper alcohol, but hydroboration–oxidation will. The reagents for the above conversion are therefore (1) BH_3 (or, more correctly, B_2H_6); (2) H_2O_2, ⁻OH.

4. Can the preceding alkene be prepared from the starting alkyl bromide? Yes, this is an elimination reaction that proceeds by an E2 path.

$$\underset{\underset{CH_3}{|}}{CH_3CH_2\overset{\overset{Br}{|}}{C}CH_2CH_3} \xrightarrow[E2]{Na^+ \ ^-OCH_3} \underset{\underset{CH_3}{|}}{CH_3CH{=}CCH_2CH_3}$$

5. Now, the entire sequence can be written.

$$\underset{\underset{CH_3}{|}}{CH_3CH_2\overset{\overset{Br}{|}}{C}CH_2CH_3} \xrightarrow{Na^+ \ ^-OCH_3} \underset{\underset{CH_3}{|}}{CH_3CH{=}CCH_2CH_3} \xrightarrow[(2)\ H_2O_2,\ OH^-]{(1)\ BH_3}$$

$$\underset{\underset{CH_3}{|}}{CH_3\overset{\overset{OH}{|}}{C}HCHCH_2CH_3} \xrightarrow{H_2CrO_4} \underset{\underset{CH_3}{|}}{CH_3\overset{\overset{O}{\|}}{C}CHCH_2CH_3}$$

STUDY PROBLEM

9.41. Suggest syntheses for the following compounds:

(a) $(CH_3)_2C{=}\overset{\displaystyle CH_3}{\overset{|}{C}}CH_2CH_3$ from organic compounds containing four or fewer carbon atoms

(b) $CH_3CH_2C{\equiv}CH$ from an alkene

(c) from organic compounds containing six or fewer carbon atoms

Summary

The reactivities of alkenes and alkynes arise from the weakness and exposure of the pi bond, as well as from the exposure of the trigonal or linear carbon. Unlike other hydrocarbons, an alkyne with a $-C{\equiv}CH$ group is somewhat *acidic*.

Alkenes are susceptible to **electrophilic attack**. If the alkene and reagent are unsymmetrical, reaction goes through the most stable carbocation. (**Markovnikov's rule**: H^+ adds to the carbon that already has the most H's.) As reagents, hydrogen halides, H_2SO_4, $H_2O + H^+$, or BH_3 may be used. Reactions with halogens, halogens plus water, and $Hg(O_2CCH_3)_2$ proceed through bridged ions and thus are *stereospecific*. (These reactions are summarized in Table 9.8, page 435.)

$$R_2C{=}CHR \xrightarrow{\;H^+\;} \left[R_2\overset{+}{C}{-}\overset{\displaystyle H}{\overset{|}{C}}HR\right] \xrightarrow{\;Nu^-\;} R_2\overset{\displaystyle Nu}{\overset{|}{C}}{-}CH_2R$$

$$R_2C{=}CHR \xrightarrow{\;E^+\;} \left[R_2\overset{\delta+}{C}\cdots\overset{\overset{\displaystyle \overset{\delta+}{E}}{|}}{C}HR\right] \xrightarrow[\substack{\textit{anti}\text{-addition,}\\ \text{stereospecific}}]{\;Nu^-\;} R_2\overset{\displaystyle E}{\overset{|}{C}}{-}\overset{\displaystyle |}{C}HR{-}Nu$$

Some addition reactions proceed by concerted, **syn-additions** instead of by way of carbocations or bridged ions. Hydrogenation, hydroboration, carbene-addition, and Diels–Alder reactions are in this category. Syn-additions are usually

stereospecific and often lead to *cis* products. (Note that hydroboration also leads
to "anti-Markovnikov" products.)

$$\underset{R_2C \overset{}{=} CHR}{\overset{H \frown BH_2}{}} \quad \xrightarrow[\text{stereospecific}]{\text{syn-addition}} \quad \underset{R_2C - CHR}{\overset{H \;\;\; BH_2}{\mid \;\;\;\; \mid}}$$

Hydrogenation reactions may be used to determine relative *stabilities of
alkenes.* The most highly substituted alkenes are the most stable. Acyclic *trans*-
alkenes are usually more stable than *cis*-alkenes. Conjugated dienes are more
stable than nonconjugated dienes.

Oxidation of alkenes can lead to *cis*-1,2-diols, epoxides, ketones, aldehydes,
or carboxylic acids. These reactions are summarized in Table 9.8. Conjugated
dienes can undergo **1,2-addition** or **1,4-addition**, the latter by way of allylic carbocations.

$$CH_2 = CHCH = CH_2 \quad \begin{array}{l} \xrightarrow[1,2]{X_2} XCH_2CHXCH = CH_2 \\[2ex] \xrightarrow[1,4]{X_2} XCH_2CH = CHCH_2X \end{array}$$

Addition polymers are formed by free-radical or ionic addition reactions of
alkenes or by 1,4-addition of conjugated dienes.

$$x \; CH_2 = \overset{\overset{\displaystyle R}{\mid}}{CH} \quad \xrightarrow{\text{catalyst}} \quad \left(CH_2 \overset{\overset{\displaystyle R}{\mid}}{CH} \right)_x$$

$$x \; CH_2 = \overset{\overset{\displaystyle R}{\mid}}{C}CH = CH_2 \quad \xrightarrow{\text{catalyst}} \quad \left(CH_2 \overset{\overset{\displaystyle R}{\mid}}{C} = CHCH_2 \right)_x$$

STUDY PROBLEMS

9.42. Write the IUPAC name for each of the following compounds:

(a) $BrCH_2CH{=}CH_2$

(b) $(CH_3)_2CHC{\equiv}CH$

(c) $CH_3CH{=}CHCHCH_2OH$
　　　　　　　　　$|$
　　　　　　　　CH_3

(d) $CH_2{=}CHCH{=}CH_2$

(e)

(f) $CH_2{=}CHCO_2H$

9.43. Give the structure for each of the following compounds:

(a) 2-methyl-1-butene

(b) 2-pentyne

(c) 1,3-hexadiene

(d) *cis*-2-hexene

(e) *trans*-1,2-diphenylethene

(f) *cis*-1,2-dibromoethene

(g) (2Z,4Z)-hexadiene

(h) (*E*)-3-phenyl-2-butenoic acid

9.44. Tell whether each of the following alkenes is (*E*), (*Z*), or neither:

(a)

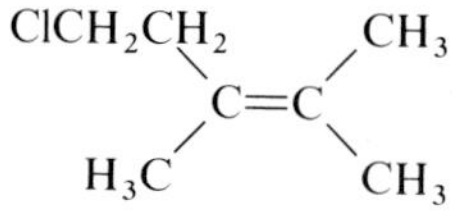

(b)

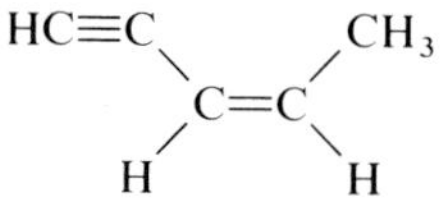

(c)

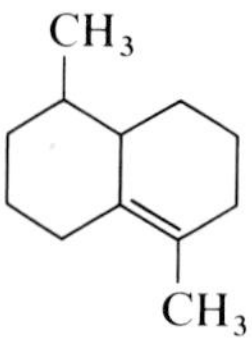

(d)

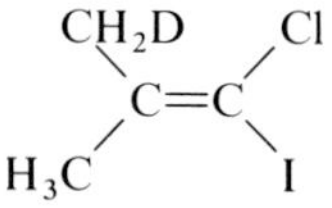

9.45. Tell which would be the reaction of choice to prepare 1-methyl-1-cyclohexene. (Each reaction is carried out in ethanol.) Explain your answer.

(a)

(b)

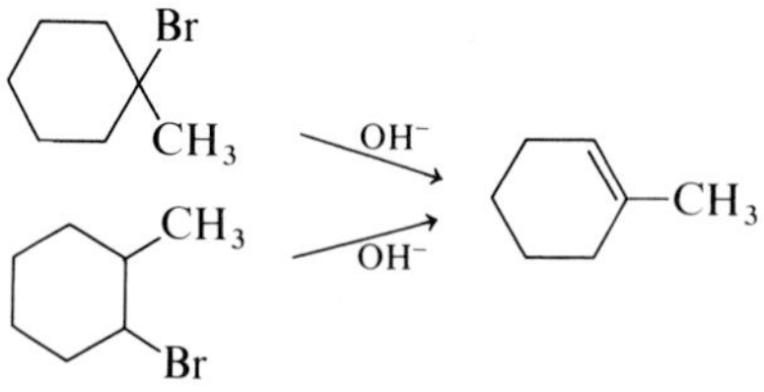

9.46. Starting with propyne, how would you prepare the following compounds?

(a) 2-butyne

(b)

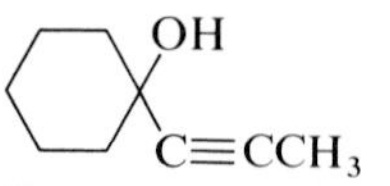

(c) $CH_3C{\equiv}CD$

9.47. What would be the product if 1-pentyne were treated with each of the following reagents? (*Hint*: Alkynes also follow Markovnikov's rule.) (a) 1 equivalent Cl_2; (b) 2 equivalents Cl_2; (c) 2 equivalents HCl; (d) phenylmagnesium bromide; (e) $NaNH_2$ and iodomethane.

9.48. One equivalent of HI is added to each of the following alkenes. Give the structure of the likely products in each case. (a) 1-pentene; (b) 1,3-pentadiene; (c) 2-methyl-1,3-butadiene; (d) 2,2-dimethyl-3-heptene.

9.49. Which one of each pair would be more reactive toward addition of HCl?

(a) $ClCH$=$CHCl$ or CH_2=$CHCl$
(b) CH_3CH_2CH=$CHCH_3$ or CH_3CH=$C(CH_3)_2$

9.50. (1) Arrange the following compounds in order of increasing reactivity toward H_2SO_4 (least reactive first). (2) Write formulas for the intermediate and the major product in each case.

(a) propene   (b) 2-methylpropene   (c) 2-butene

9.51. Predict the major organic products of the following reactions:

(a) 3-methyl-2-pentene with aqueous H_2SO_4
(b) 2-methylpropene with H_2SO_4 in ethanol
(c) 2,2-dimethyl-3-hexene with aqueous H_2SO_4
(d) 1-butene with $0.1M$ aqueous HI
(e) methylenecyclohexane (page 378) with the strong acid trifluoroacetic acid (CF_3CO_2H).

9.52. Predict the major organic products:

(a) $(CH_3)_3CCH$=$CHCH_3$ $\xrightarrow[\text{(2) NaBH}_4]{\text{(1) Hg(O}_2\text{CCH}_3)_2,\ \text{H}_2\text{O}}$

(b) $(CH_3)_2CHCH$=CH_2 $\xrightarrow[\text{(2) NaBH}_4]{\text{(1) Hg(O}_2\text{CCH}_3)_2,\ \text{CH}_3\text{CH}_2\text{OH}}$

(c) $(CH_3)_2CHCH$=CH_2 $\xrightarrow[\text{(2) H}_2\text{O}_2,\ \text{OH}^-]{\text{(1) BH}_3}$

(d) ⬠$-CH_3$ $\xrightarrow[\text{(2) Br}_2,\ \text{OH}^-]{\text{(1) BH}_3}$

(e) ⬠=$CHCH_3$ + HBr $\xrightarrow{\text{no peroxides}}$

(f) ⬠=$CHCH_3$ + HBr $\xrightarrow{\text{C}_6\text{H}_5\overset{\text{O}}{\overset{\|}{\text{C}}}\text{OOC}\overset{\text{O}}{\overset{\|}{\text{C}}}\text{C}_6\text{H}_5}$

9.53. Predict the likely organic products of the following reactions: (a) *cis*-2-pentene with Cl_2; (b) 1-methyl-1-cyclohexene with Br_2; (c) 1-methyl-1-cyclohexene with aqueous Br_2; (d) myrcene, which is found in oil of bayberry, with one equivalent of Cl_2.

myrcene

9.54.   (a)   What products would probably be observed when 1-butene is treated with bromine water that contains sodium chloride?
   (b)   Write mechanisms that explain the formation of each product.

9.55.   Suggest reagents for the conversion of cyclopentene to   (a) cyclopentane;
   (b) *trans*-2-bromo-1-cyclopentanol.

9.56.   When 2-methyl-1,3-butadiene is treated with one equivalent of Br_2, four possible dibromoalkenes could result. However, only two dibromoalkenes are actually formed in this reaction.   (a) Which two are formed?   (b) Why are the other ones not formed?

9.57.   The following reaction proceeds by a carbene-type mechanism. What is the product?

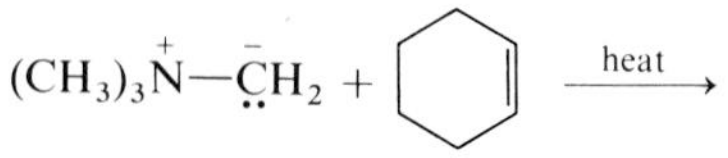

$$(CH_3)_3\overset{+}{N}-\overset{-}{\underset{\cdot\cdot}{C}}H_2 + \bigcirc \xrightarrow{\ heat\ }$$

9.58.   Write an equation or equations for the preparation of each of the following compounds from methylenecyclohexane (page 378):

(a)  $\bigcirc \begin{smallmatrix} CH_3 \\ OH \end{smallmatrix}$ (b) 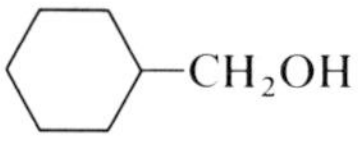 $\bigcirc-CH_2OH$ (c) 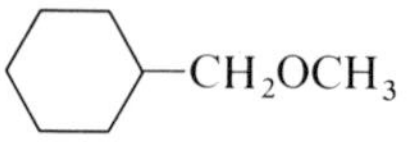 $\bigcirc-CH_2OCH_3$

9.59.   Show how you could synthesize the following compounds from propyne:

(a) $CH_3\overset{O}{\overset{||}{C}}CH_3$ (b) $CH_3\overset{O}{\overset{||}{C}}CH_2CH_3$ (c) $CH_3\overset{OCH_3}{\overset{|}{C}H}C\equiv CCH_3$

9.60.   Predict the principal products (including stereochemistry) of the hydroboration–oxidation of the following cycloalkenes:

(a) 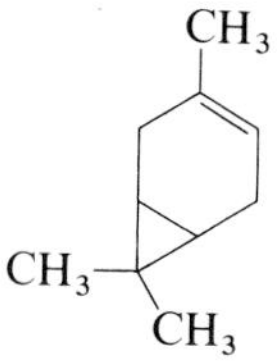   (b)

Δ^3-carene

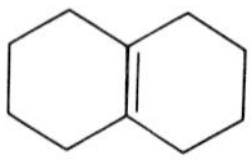

9.61.   Predict the major organic products (showing stereochemistry):

(a) 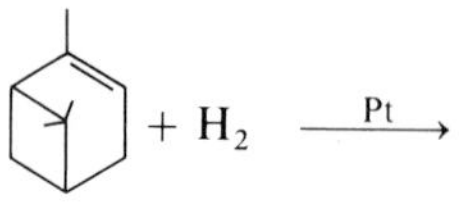 $\bigcirc + H_2 \xrightarrow{\ Pt\ }$

(b) 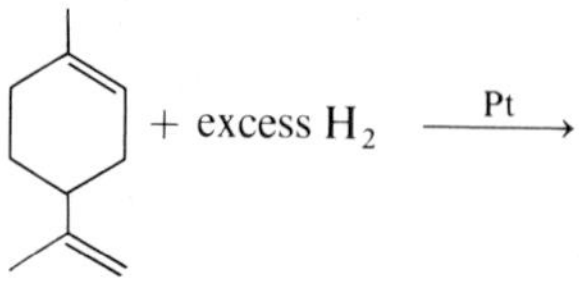 $\bigcirc + excess\ H_2 \xrightarrow{\ Pt\ }$

(c) 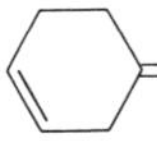 =O + H$_2$ $\xrightarrow{\text{Pt}}$

(d) CH$_3$CH$_2$C≡CCH(CH$_3$)$_2$ + H$_2$ $\xrightarrow{\text{deactivated Pd}}$

9.62. Of each of the following pairs of compounds, tell which one would be the more stable.

(a) 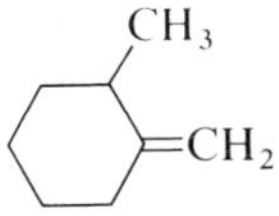 =CH$_2$ or =CHCH$_3$

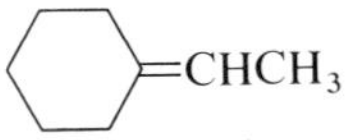

(b) *cis-* or *trans*-CH$_3$CH$_2$CH=CHCH$_3$

(c) (*E*)- or (*Z*)-CH$_3$COCH=CHCOCH$_3$

(d) CH$_2$=CHCH$_2$COCH$_3$ or CH$_2$=CHCOCH$_2$CH$_3$

9.63. Predict the products:

(a) 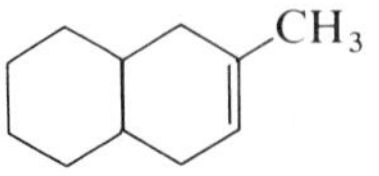 + C$_6$H$_5$CO$_3$H $\longrightarrow$

(b) 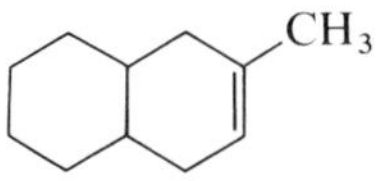 $\xrightarrow[\text{(2) Na}_2\text{SO}_3]{\text{(1) OsO}_4}$

(c) 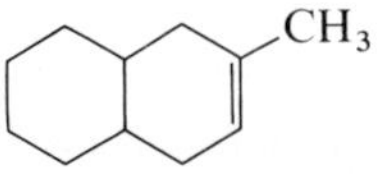 + hot KMnO$_4$ solution $\longrightarrow$

9.64. Predict the ozonolysis products when each of the following alkenes is subjected to (1) reaction with O$_3$, followed by (2) reaction with H$_2$O$_2$ and H$^+$:

(a) 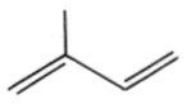 (b) 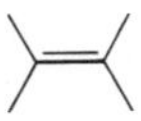 (c) 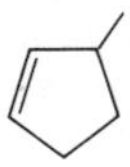 (d)

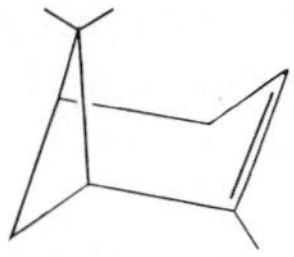

α-pinene

found in turpentine

9.65. Predict the organic product of the reaction of *trans*-2-butene with each of the following reagents. (Show the stereochemistry where appropriate.) (a) HI; (b) OsO$_4$, followed by treatment with Na$_2$SO$_3$; (c) C$_6$H$_5$CO$_3$H followed by H$_2$O + HCl; (d) H$_2$SO$_4$; (e) singlet :CH$_2$

9.66. Give the structure of the product when 1-methyl-1-cyclopentene is treated with each of the following reagents: (a) H_2 (Pt catalyst); (b) Cl_2 in H_2O; (c) Cl_2 in CCl_4; (d) cold alkaline $KMnO_4$; (e) hot $KMnO_4$ solution; (f) dilute solution of Br_2 in CH_3OH; (g) BH_3 followed by H_2O_2–NaOH solution; (h) HCl; (i) HBr with H_2O_2; (j) mercuric acetate in water followed by $NaBH_4$.

9.67. Each of a series of unknown alkenes is treated with (1) O_3, and (2) Zn, H^+, H_2O. The following products are obtained. Give the structure (or structures, if there is more than one possibility) of each unknown alkene.

$$\text{(a)} \quad \overset{O}{\overset{\|}{HC}}CH_2CH_2CH_2\overset{O}{\overset{\|}{CH}} \text{ only} \qquad \text{(b)} \quad CH_3CH_2\overset{O}{\overset{\|}{CH}} + CH_3\overset{O}{\overset{\|}{C}}CH_3$$

$$\text{(c)} \quad CH_3\overset{O}{\overset{\|}{CH}} \text{ only} \qquad\qquad \text{(d)} \quad H\overset{O}{\overset{\|}{CH}} + \text{(cyclohexylidene)}{=}O$$

$$\text{(e)} \quad CH_3\overset{O}{\overset{\|}{C}}{-}\overset{O}{\overset{\|}{CH}} + 2\,CH_3\overset{O}{\overset{\|}{C}}CH_3 \qquad \text{(f)} \quad CH_3\overset{O}{\overset{\|}{CH}} + H\overset{O}{\overset{\|}{CH}} + H\overset{O}{\overset{\|}{C}}CH_2\overset{O}{\overset{\|}{CH}}$$

9.68. A chemist has obtained the following information about an alkene of unknown structure. What is the structure of the alkene?

cyclohexanecarboxylic acid $(C_6H_{11}CO_2H)$

hot $KMnO_4$

unknown

H_2, Pt

ethylcyclohexane

9.69. When subjected to catalytic hydrogenation, a hydrocarbon takes up 1.0 equivalent of H_2. When oxidized with a hot solution of $KMnO_4$, only one compound is obtained, $HO_2CCHCH_2CHCO_2H$. What is the structure of the hydrocarbon?
$$\quad\quad\quad\; \underset{CH_3}{|} \quad \underset{CH_3}{|}$$

9.70. Each of the following dienes is shown in the *s-trans* form. Which of these compounds are interconvertible with their *s-cis* forms? Show the *s-cis* conformations in these cases.

(a)   (b)   (c)

9.71. Predict the products of the following Diels–Alder reactions, showing stereochemistry where appropriate:

(a)

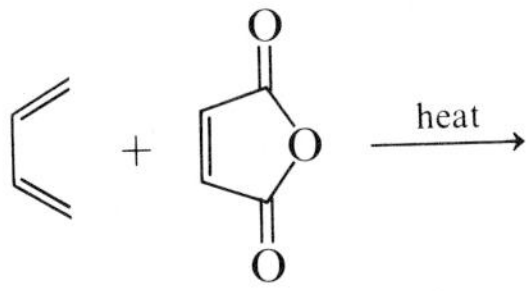

(b)

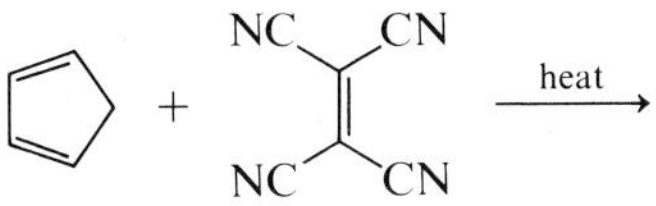

(c)

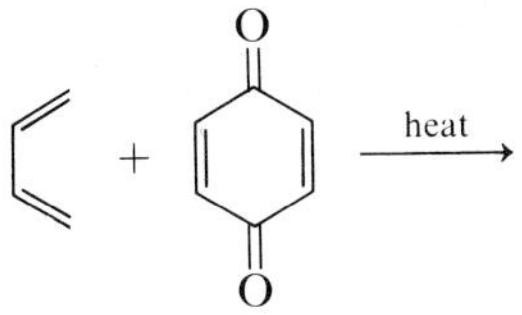

(d)

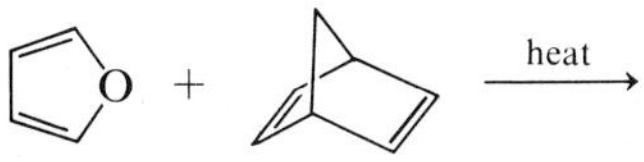

(e) 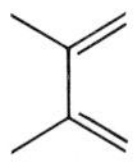   + $CH_3CH_2O_2CC\equiv CCO_2CH_2CH_3$ $\xrightarrow{heat}$

9.72. Suggest a Diels–Alder reaction that would lead to each of the following compounds. Show the stereochemistry of the product where applicable.

(a)

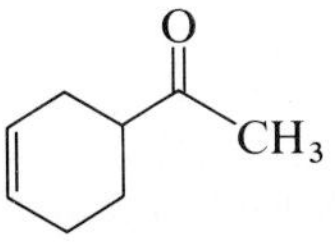

(b)

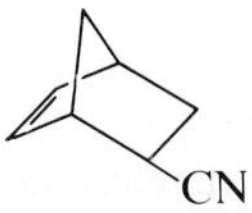

(c)

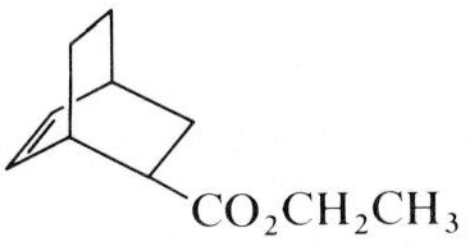

(d)

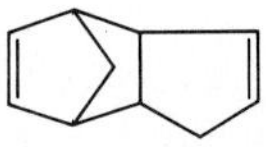

9.73. Styrene (page 430) may be copolymerized with maleic anhydride. Give an equation for this reaction.

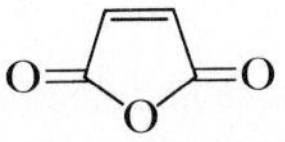

maleic anhydride

9.74. Dimers, as well as polymers, may be made by a cationic route. Give the mechanisms for the acid-catalyzed dimerizations of   (a) propene;   (b) 2-methyl-1-butene.

9.75. Suggest synthetic paths for the following conversions:

(a) 1,2-dibromopropane from 2-bromopropane
(b) 2-bromopropane from 1-propanol

(c) 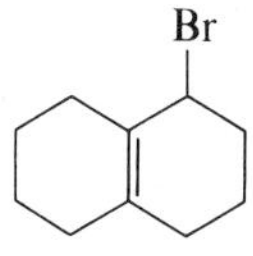 [structure: bicyclic ring with Br] from [structure: bicyclic ring]

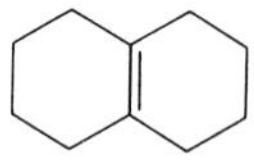

(d) $\overset{O}{\overset{||}{HC}}CH_2CH_2CH_2\overset{O}{\overset{||}{CH}}$ from cyclopentene

(e) 1,2-dideuteriopropane from propane

(f) acetic acid from 1-bromopropane

(g) *trans*-1,2-cyclohexanediol from cyclohexane

(h) *trans*-2-chloro-1-cyclopentanol from chlorocyclopentane

9.76. Because of its geometry, *trans*-cyclooctene exists as a pair of enantiomers.

(a) Give the structure of the enantiomer of the structure shown.

(b) Would *cis*-cyclooctene also exist as a pair of enantiomers? Explain.

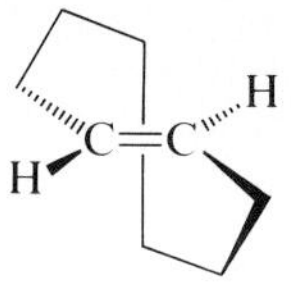

[structure of trans-cyclooctene]

trans-cyclooctene

one enantiomer

9.77. Reaction of 3-hexene with $Cl_2 + H_2O$ yields 3,4-dichlorohexane and the 3,4-chlorohydrin. However, the chlorohydrin obtained from *cis*-3-hexene is different from the chlorohydrin obtained from *trans*-3-hexene. Explain.

9.78. The toxic components of poison ivy and poison oak are a group of 1,2-hydroquinones (dihydroxybenzenes) called *urushiols*. One of these urushiols (A), $C_{21}H_{34}O_2$, is treated with dilute NaOH, followed by iodomethane, and yields a dimethylated compound (B), $C_{23}H_{38}O_2$. When B is treated with ozone, followed by treatment with Zn dust and water, the products are:

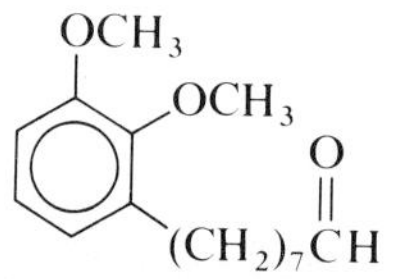

[structure: benzene ring with OCH$_3$, OCH$_3$, and $(CH_2)_7\overset{O}{\overset{||}{CH}}$]   and $HC\overset{O}{\overset{||}{}}(CH_2)_5CH_3$

What are the structures of A and B?

9.79. Starting with only acetylene, ethylene, iodomethane, and any appropriate inorganic reagents, show how you would prepare the following compounds: (a) *cis*-4-octene; (b) acetone; (c) 1-butene; (d) 1-chloro-2-butanol; (e) 2-methyl-2-butanol; (f) *meso*-2,3-butanediol; (g) racemic 2,3-butanediol.

9.80. Suggest a mechanism for each of the following reactions:

(a) $(CH_3)_2C=CH(CH_2)_2CH=C(CH_3)_2 \xrightarrow{H^+}$ [structure: cyclopentane ring with CH$_3$, CH$_3$, C(CH$_3$)$_2$ substituents]

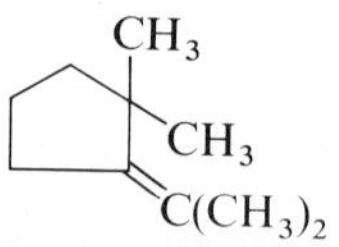

(b) 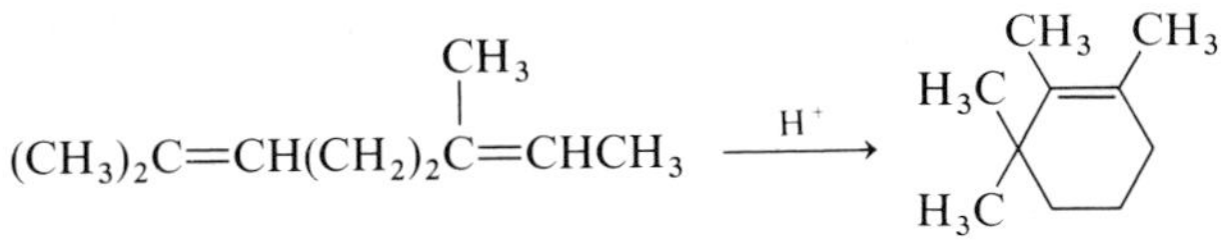 $(CH_3)_2C=CH(CH_2)_2C=CHCH_3$ $\xrightarrow{H^+}$

9.81. A chemist treated (*R*)-3-butene-1,2-diol with OsO_4, obtaining two tetraols. What were they? (Specify the stereochemistry.)

9.82. How would you prepare the following compounds from nondeuteriated hydrocarbons? Show the stereochemistry of the products where applicable.

(a) 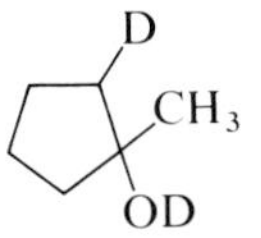   (b) 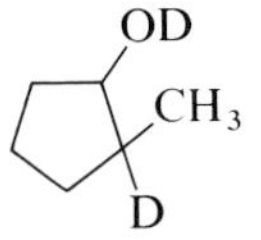   (c)

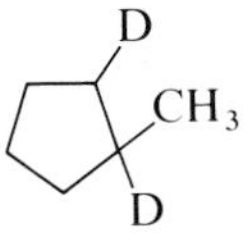

(d) 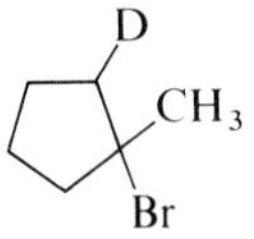   (e) 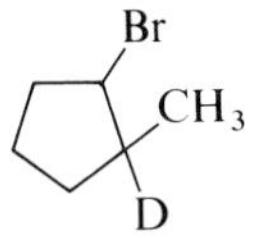   (f)

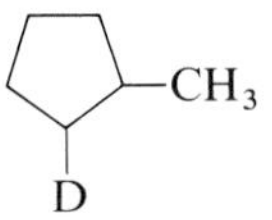

9.83. The sex attractant of the housefly is a hydrocarbon with the formula $C_{23}H_{46}$. Catalytic hydrogenation of this compound yields $C_{23}H_{48}$, while oxidation with a hot solution of $KMnO_4$ yields $CH_3(CH_2)_{12}CO_2H$ and $CH_3(CH_2)_7CO_2H$. Addition of bromine to the hydrocarbon yields one pair of enantiomeric dibromides ($C_{23}H_{46}Br_2$). What is the structure of the housefly sex attractant?

9.84. A chemist wants to devise a catalytic hydrogenation experiment that will determine if *endocyclic double bonds* are more or less stable than *exocyclic double bonds*. (An *endocyclic* double bond is one that joins two ring carbons, while an *exocyclic* double bond is one that goes from a ring carbon to a nonring carbon.) Of the following pairs of compounds, which pair would give the best hydrogenation data for comparison purposes? Why?

(a) 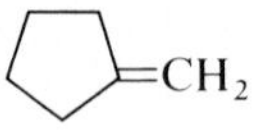 $=CH_2$ and $-CH_3$

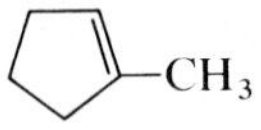

     exocyclic   *endocyclic*

(b) 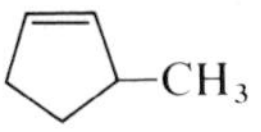 $-CH_3$ and $-CH_3$

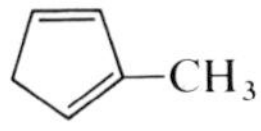

(c) 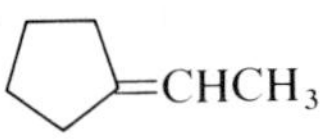 $=CHCH_3$ and $-CH_2CH_3$

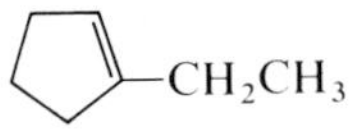

9.85. Suggest syntheses for the following compounds from alkenes or dienes containing six or fewer carbons and other appropriate reagents:

(a)   1-cyclohexyl-1-propanone   (b)   cyclohexylmethanol
(c)   *trans*-2-methyl-1-cyclohexanol   (d)   *cis*-1,2-dibromocyclohexane

(e)   racemic HO_2C⸺ ⸺CO_2H with ⸺$\overset{\overset{\textstyle O}{\|}}{C}CH_3$

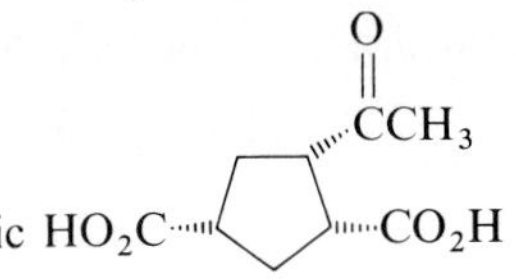

9.86. The two nmr spectra in Figure 9.17 were obtained from samples of two hydrocarbons, each with the formula $C_{14}H_{12}$. Both samples yielded only benzoic acid upon treatment with hot $KMnO_4$ solution. What are the structures of these two hydrocarbons?

9.87. A six-carbon compound, A, can be converted to B by hydrogenation. The nmr spectrum of A and the infrared spectrum of B are shown in Figure 9.18. What are the structures of A and B?

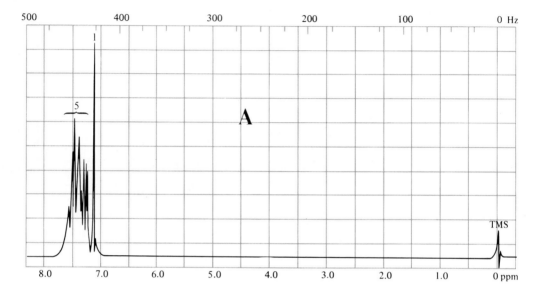

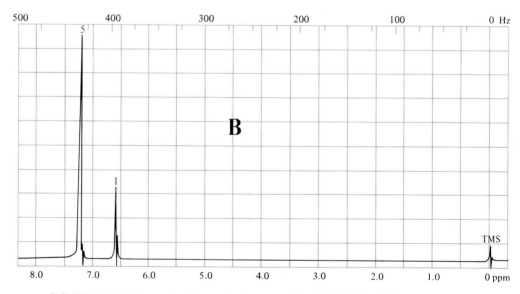

FIGURE 9.17. Nmr spectra for unknown hydrocarbons ($C_{14}H_{12}$) in Problem 9.86.

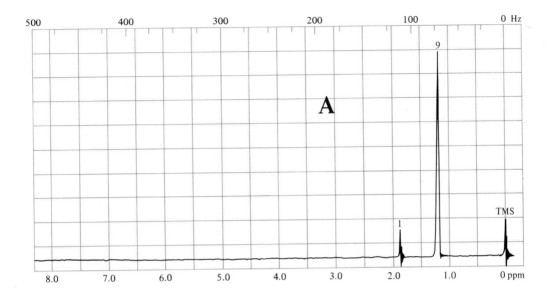

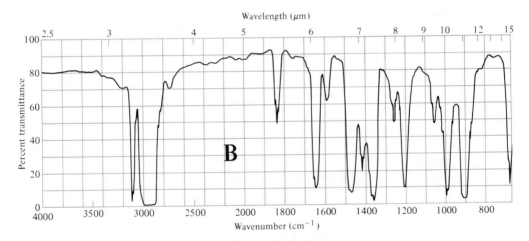

FIGURE 9.18. Spectra for unknown compounds in Problem 9.87.

CHAPTER 10

Aromaticity, Benzene, and Substituted Benzenes

Benzene is the simplest of the aromatic compounds and one that we have already encountered many times. In this chapter, we will formalize the definition of aromaticity, and we will discuss the properties and reactions of benzene and substituted benzenes. In Chapter 16, we will discuss the chemistry of polycyclic and heterocyclic aromatic compounds. (*Heterocyclic* aromatic compounds have at least one ring atom different from carbon.)

Benzene was first isolated in 1825 by Michael Faraday from oily residues that had accumulated in London gas mains. Today the main source of benzene, substituted benzenes, and other aromatic compounds is petroleum. Until the 1940's, coal tar was the principal source. The types of aromatic compounds obtained from these sources are hydrocarbons, phenols, and aromatic heterocycles.

Aromatic hydrocarbons:

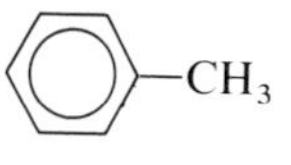

toluene 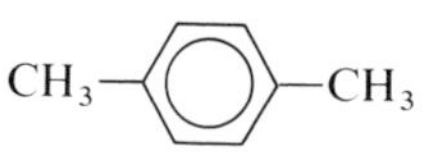   *p*-xylene   naphthalene   phenanthrene

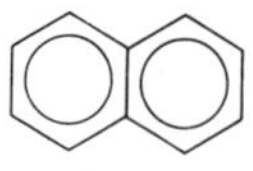

substituted benzenes   *polycyclic compounds*

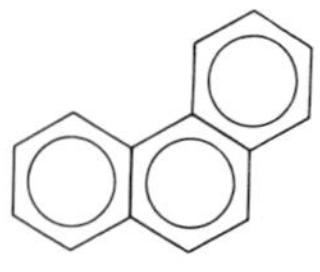

Aromatic nitrogen heterocycles:

pyridine 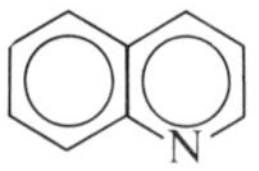   quinoline

Compounds containing benzene rings and aromatic heterocyclic rings are also exceedingly common in biological systems.

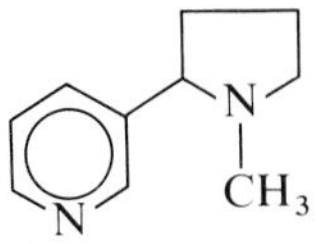

nicotine

in tobacco

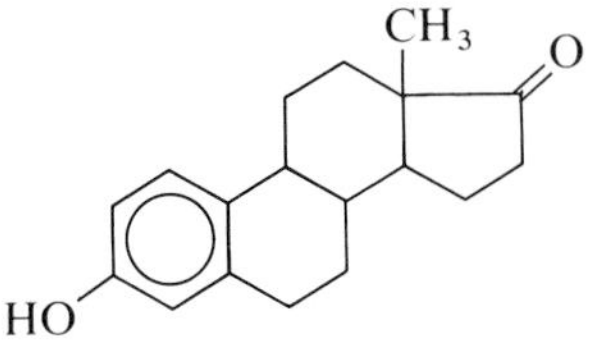

estrone

an estrogen, or female hormone

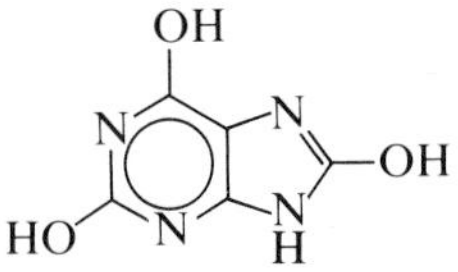

uric acid

associated with gout

SECTION 10.1.

Nomenclature of Substituted Benzenes

The naming of monosubstituted benzenes was mentioned in Section 3.3M. Many common benzene compounds have their own names, names that are not necessarily systematic. Some of these more-commonly used names are listed in Table 10.1.

Disubstituted benzenes are named with the prefixes *ortho*, *meta*, and *para* rather than with position numbers. The prefix *ortho* signifies that two substituents are 1,2 to each other on a benzene ring; *meta* signifies a 1,3-relationship; and

TABLE 10.1. Structures and names of some common benzene compounds

Structure	Name	Structure	Name
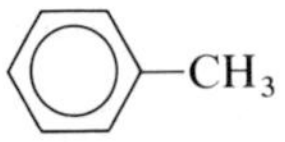 $\bigcirc$—CH$_3$	toluene	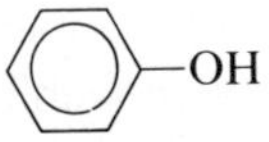 $\bigcirc$—OH	phenol
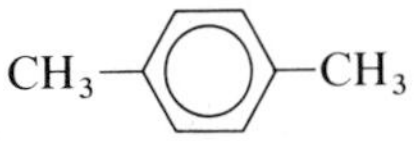 CH$_3$—$\bigcirc$—CH$_3$	*p*-xylene	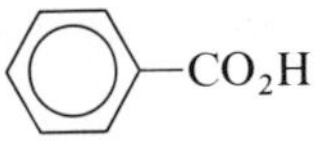 $\bigcirc$—CO$_2$H	benzoic acid
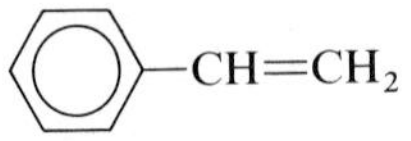 $\bigcirc$—CH=CH$_2$	styrene	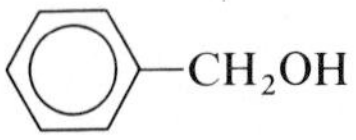 $\bigcirc$—CH$_2$OH	benzyl alcohol
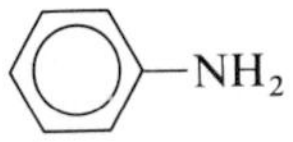 $\bigcirc$—NH$_2$	aniline	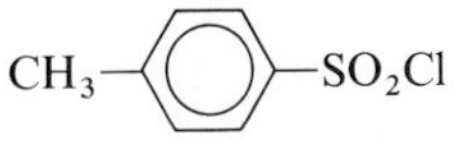 CH$_3$—$\bigcirc$—SO$_2$Cl	*p*-toluenesulfonyl chloride (tosyl chloride)
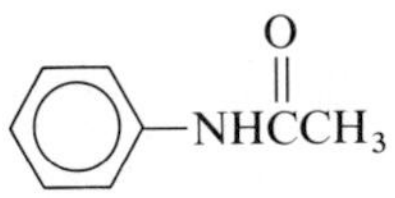 $\bigcirc$—NHCCH$_3$	acetanilide	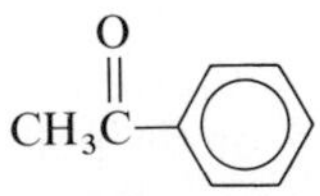 CH$_3$C—$\bigcirc$	acetophenone
		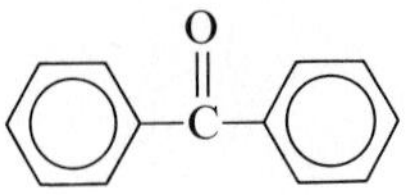 $\bigcirc$—C—$\bigcirc$	benzophenone

para means a 1,4-relationship. The use of *ortho*, *meta*, and *para* in lieu of position numbers is reserved exclusively for disubstituted benzenes—the system is never used with cyclohexanes or other ring systems.

ortho-, or o-   meta-, or m-   para, or p-

The use of these prefixes in the naming of some disubstituted benzenes follows:

o-dibromobenzene   m-chloroaniline   p-chlorophenol

In reactions of benzene compounds, we will speak of *ortho*-substitution (or *meta*- or *para*-substitution). Note that a monosubstituted benzene has two *ortho* and two *meta* positions, but only one *para* position.

ortho   meta   para

If there are three or more substituents on a benzene ring, the *o-*, *m-*, *p*-system is no longer applicable. In this case, numbers must be used. As in numbering any compound, we number the benzene ring in such a way as to keep the prefix numbers as low as possible and give preference to the group of highest nomenclature priority. If a substituted benzene, such as aniline or toluene, is used as the parent, that substituent is understood to be at position 1 on the ring.

1,2,4-tribromobenzene   2-chloro-4-nitroaniline   2,4,6-trinitrotoluene
                                                            (TNT)

Benzene as a substituent is called a **phenyl group**. How a *toluene* substituent is named depends on the point of attachment.

phenyl   benzyl   p-tolyl   o-tolyl

SAMPLE PROBLEM

Name the following substituted benzenes:

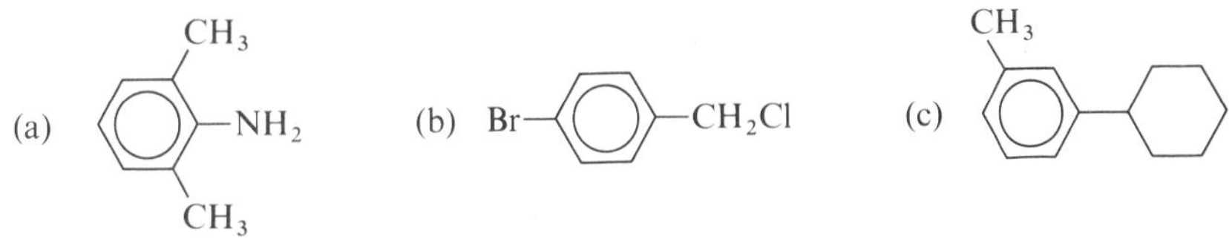

(a) [structure: benzene ring with CH₃ at top, —NH₂ at right, CH₃ at bottom]   (b) Br—[benzene ring]—CH₂Cl   (c) [benzene ring with CH₃ at top, connected to cyclohexane ring]

Solution: (a) 2,6-dimethylaniline; (b) *p*-bromobenzyl chloride;
(c) *m*-tolylcyclohexane (or *m*-cyclohexyltoluene).

STUDY PROBLEM

10.1. Write structures for: (a) ethylbenzene; (b) 2,4,6-tribromoaniline;
(c) *p*-ethylphenol; (d) 2-phenyl-1-ethanol; (e) benzyl bromide.

SECTION 10.2.

Physical Properties of Aromatic Hydrocarbons

Like aliphatic and alicyclic hydrocarbons, benzene and other aromatic hydro-carbons are nonpolar. They are insoluble in water, but are soluble in organic solvents such as diethyl ether, carbon tetrachloride, or hexane. Benzene itself has been widely used as a solvent. It has the useful property of forming an azeotrope with water. (The azeotrope, a mixture that distills with a constant composition, consists of 91 % benzene–9 % H_2O and boils at 69.4°C.) Compounds dissolved in benzene are easily dried by distillation of the azeotrope.

Although the boiling points and melting points of aromatic hydrocarbons (listed in Table 10.2) are typical of those of nonpolar organic compounds, note that *p*-xylene has a higher melting point than *o*- or *m*-xylene. A higher melting point is typical of *p*-substituted benzenes; a *p*-isomer is more symmetrical and can form a more orderly and stronger crystal lattice in the solid state than the less symmetrical *o*- or *m*-isomers.

It is interesting to note that many of the compounds found in coal tar (and cigarette tar) that contain four or more fused benzene rings are carcinogenic (cancer-causing). Benzene itself is toxic and somewhat carcinogenic; therefore, it should be used in the laboratory only when necessary. (In many cases, toluene may be used as a substitute.)

Two highly carcinogenic hydrocarbons:

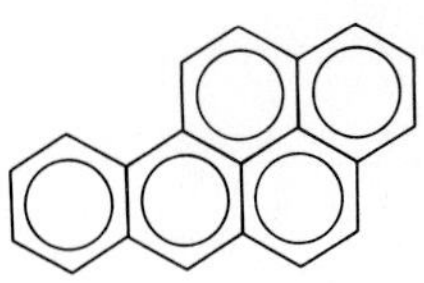

benzo[a]pyrene 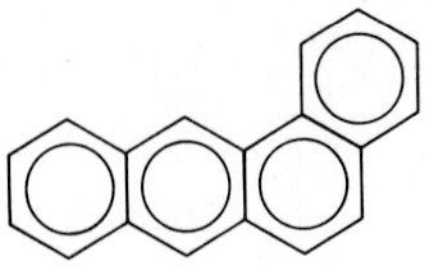   benzanthracene

TABLE 10.2. Melting points and boiling points of some aromatic hydrocarbons

Name	Structure	Mp, °C	Bp, °C
benzene		5.5	80
toluene	—CH$_3$	−95	111
o-xylene	CH$_3$ / —CH$_3$	−25	144
m-xylene	CH$_3$ / —CH$_3$	−48	139
p-xylene	CH$_3$— —CH$_3$	13	138

SECTION 10.3.

Spectra of Substituted Benzenes

Both infrared and nmr spectra provide data that are useful in the structure determination of substituted benzenes. The nmr spectrum provides the most clear-cut answer to absence or presence of aromatic protons (and thus an aromatic ring).

A. Infrared spectra

The infrared absorption bands of substituted benzenes are summarized in Table 10.3. The presence of a benzene ring in a compound of unknown structure can often be determined by inspection of two regions of the infrared spectrum. The absorption for aryl CH stretching, which is generally weak, falls near 3030 cm^{-1} (3.3 μm), just to the left of aliphatic CH absorption. Absorption for aryl C—C vibrations gives a series of four peaks, generally between 1450 and 1600 cm^{-1} (about 6–7 μm); however, all four peaks are not always apparent. In the spectrum of chlorobenzene (Figure 10.1), the first peak is visible, but the second peak is a barely

TABLE 10.3. Infrared absorption characteristics of benzene compounds

Type of vibration	Position of absorption	
	cm^{-1}	μm
aryl C—H	3000–3300	3.0–3.3
aryl C—C (four peaks)	1450–1600	6.25–6.9

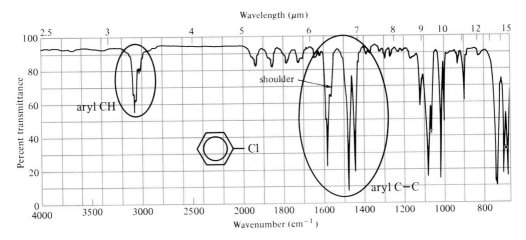

FIGURE 10.1. Infrared spectrum of chlorobenzene.

visible shoulder. In this spectrum, the third and fourth peaks are quite evident; however, in some spectra, the fourth peak at 1450 cm^{-1} (6.90 μm) is obscured by aliphatic CH$_2$ bending absorption.

The positions of substitution on a benzene ring may sometimes be determined by examination of the infrared spectrum. Differently substituted benzene rings often give characteristic absorption at about 680–900 cm^{-1} (11–15 μm). The patterns observed are summarized in Table 10.4. Figure 10.2 (page 456) shows infrared spectra of the three isomeric chlorotoluenes. By comparison of these spectra with Table 10.4, you can see how the absorption in the fingerprint region may be used. Also compare these spectra with that of chlorobenzene in Figure 10.1. Unfortunately, the absorption in this region is not always so clear-cut.

TABLE 10.4. The C—H bending absorption of substituted benzenes

		Position of absorption	
Substitution	*Appearance*	*cm^{-1}*	*μm*
monosubstituted,	two peaks	730–770	12.9–13.7
		690–710	14.0–14.5
o-disubstituted,	one peak^a	735–770	12.9–13.6
m-disubstituted,	three peaks	860–900	11.1–11.6
		750–810	12.3–13.3
		680–725	13.7–14.7
p-disubstituted,	one peak	800–860	11.6–12.5

^{a} An additional band is often observed at about 680 cm^{-1} (14.7 μm).

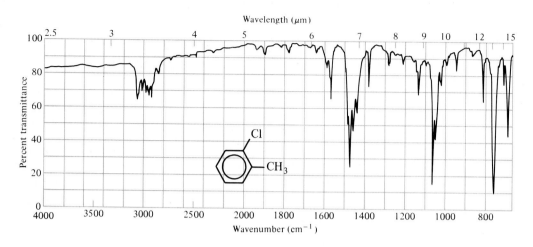

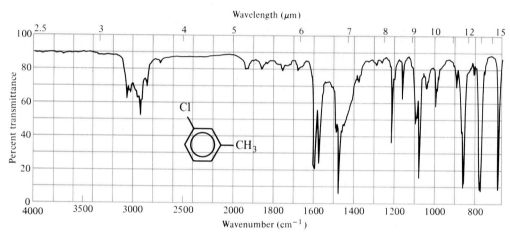

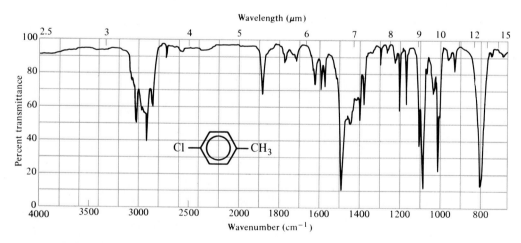

FIGURE 10.2. Infrared spectra for *o*-, *m*-, and *p*-chlorotoluene.

B.   Nmr spectra

The nmr spectra of aromatic compounds are distinctive. Protons on an aromatic ring absorb downfield, with δ values between 6.5 ppm and 8 ppm. This downfield absorption is due to the ring current, which gives rise to a molecular magnetic field that deshields protons attached to the ring. The chemical shift for the protons on benzene itself is at $\delta = 7.27$ ppm. Electronegative substituents on the ring shift the absorption of adjacent protons farther *downfield*, while electron-releasing groups shift absorption *upfield* from that of unsubstituted benzene. Simple splitting patterns of aryl protons are sometimes observed (Figure 10.3); in many cases, however, the splitting patterns are very complex.

Benzylic protons are not as affected by the aromatic ring current as are ring protons; their absorption is observed farther upfield, in the region of 2.3 ppm. (See the nmr spectrum of toluene in Figure 10.3.)

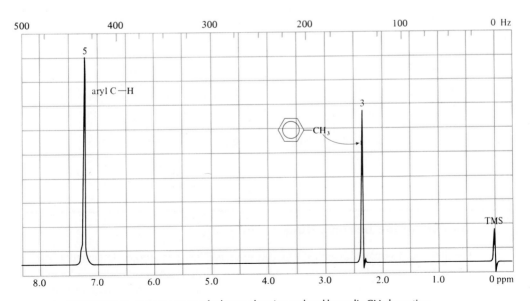

STUDY PROBLEM

10.2.   Figure 10.4 shows two nmr spectra (without the relative areas under the peaks). One spectrum is that of *p*-chloroaniline, while the other is that of *p*-iodoanisole (p-$IC_6H_4OCH_3$). On the basis of the chemical shifts (compared to the chemical shift of benzene protons), assign structures to the two spectra.

FIGURE 10.3. Nmr spectrum of toluene, showing aryl and benzylic CH absorption.

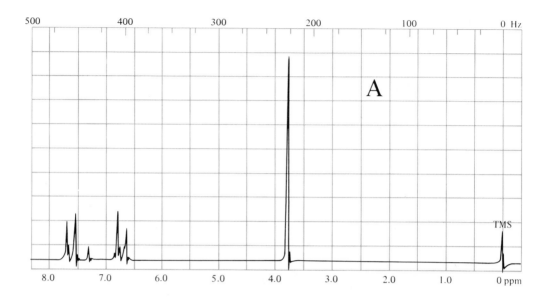

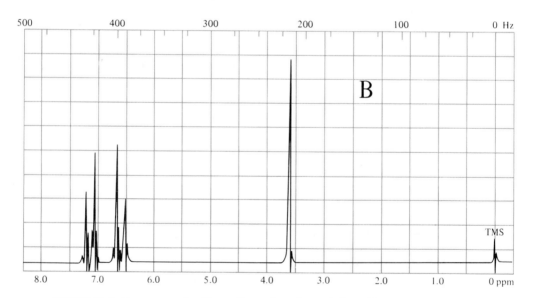

FIGURE 10.4. Nmr spectra of *p*-chloroaniline and *p*-iodoanisole for Problem 10.2.

SECTION 10.4.

Stability of the Benzene Ring

The heat of hydrogenation of cyclohexene is 28.6 kcal/mole. If benzene contained three alternate single and double bonds without any pi-electron delocalization, we would expect its heat of hydrogenation to be 3 × 28.6, or 85.8, kcal/mole. However, benzene liberates only 49.8 kcal/mole of energy when it is hydrogenated.

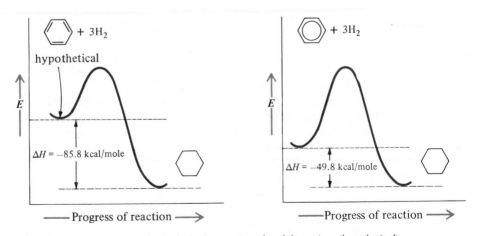

cyclohexene + H_2 $\xrightarrow{\text{Pt}}$ cyclohexane + 28.6 kcal/mole

benzene + 3 H_2 $\xrightarrow[\substack{225°,\\35\ \text{atm}}]{\text{Pt}}$ cyclohexane + 49.8 kcal/mole

The hydrogenation of benzene liberates 36 kcal/mole less energy than would be liberated by the hydrogenation of the hypothetical cyclohexatriene. Therefore, benzene, with delocalized pi electrons, contains 36 kcal/mole less energy than it would contain if the pi electrons were localized in three isolated double bonds. This difference in energy between benzene and the imaginary cyclohexatriene is called the **resonance energy** of benzene. The resonance energy is the energy lost (stability gained) by the complete delocalization of electrons in the pi system. It is a measure of the added stability of the aromatic system compared to that of the localized system. Figure 10.5 shows the energy diagrams for these hydrogenations.

What does the resonance energy of benzene mean in terms of chemical reactivity? It means that more energy is required for a reaction in which the aromatic character of the ring is lost. Hydrogenation is one example of such a reaction. Whereas an alkene can be hydrogenated at room temperature under atmospheric pressure, benzene requires high temperature and high pressure. Also, benzene does not undergo most reactions that are typical of alkenes. For example, benzene does not undergo addition of HX or X_2, nor does it undergo oxidation with $KMnO_4$ solution.

benzene $\xrightarrow{\text{HX}}$ no addition reaction

$\xrightarrow{MnO_4^-}$ no reaction

FIGURE 10.5. Energy diagrams for the hydrogenation of cyclohexatriene (hypothetical) and benzene.

SECTION 10.5.

The Bonding in Benzene

Although the molecular formula of benzene (C_6H_6) was determined shortly after its discovery in 1825, 40 years elapsed before Kekulé proposed a hexagonal structure for benzene. This first proposed structure contained no double bonds (because benzene does not undergo reactions characteristic of alkenes). To be consistent with the tetravalence of carbon, Kekulé proposed in 1872 that benzene contains three alternate single and double bonds. To explain the existence of only three (not five) isomeric disubstituted benzenes, Kekulé suggested that the benzene ring is in rapid equilibrium with the structure in which the double bonds are in the alternative positions. This idea survived for almost 50 years before it was replaced by the theories of resonance and molecular orbitals.

benzene in 1865   benzene in 1872   benzene in 1940

The Kekulé formulas for benzene, showing three double bonds instead of a circle in the ring, do not explain the unique stability of the benzene ring. However, these formulas have one advantage—they allow us to count the number of pi electrons at a glance. In conjunction with resonance theory, these formulas are quite useful; therefore, we will use Kekulé formulas when we discuss the reactions of benzene.

Let us now consider the π molecular orbitals of benzene. Benzene has six sp^2 carbons in a ring. The ring is planar, and each carbon atom has a p orbital perpendicular to this plane. Figure 2.23 (page 67) shows a representation of the p orbitals of benzene and how they overlap in the lowest-energy bonding molecular orbital.

Overlap of six atomic p orbitals leads to the formation of six π molecular orbitals. When we look at all six possible π molecular orbitals of benzene (Figure 10.6), we see that our representation of the aromatic pi cloud as a "double donut" represents only one, π_1, of the six molecular orbitals. In the π_1 orbital, all six p orbitals of benzene are in phase and overlap equally; this orbital is of lowest energy because it has no nodes between carbon nuclei.

The π_2 orbital and the π_3 orbital each has one node between the carbon nuclei. These two bonding orbitals are degenerate (equal in energy) and of higher energy than the π_1 molecular orbital. Benzene, with six p electrons, has the π_1, π_2, and π_3 orbitals each filled with a pair of electrons. These are the bonding orbitals of benzene.

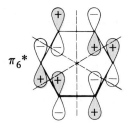

(3 nodes; all six p orbitals out of phase)

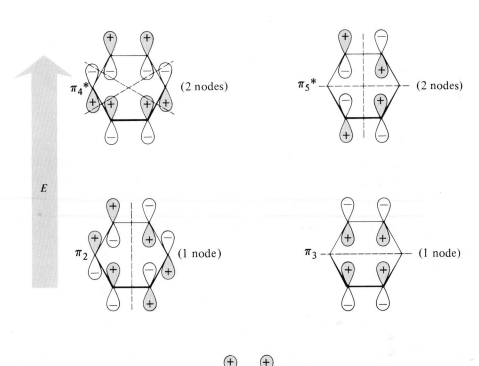

(2 nodes)

(2 nodes)

(1 node)

(1 node)

(no nodes; all six p orbitals in phase)

FIGURE 10.6. The π orbitals in benzene. Nodes are represented by dashed lines; the "missing p orbitals" in π_3 and $\pi_5{}^*$ are a result of nodes at these positions. (The + and − signs are mathematical signs of phase, not electrical charges.)

Along with the three bonding orbitals in benzene, there are three antibonding orbitals. Two of these antibonding orbitals ($\pi_4{}^*$ and $\pi_5{}^*$) have two nodes each, and the highest-energy orbital ($\pi_6{}^*$) has three nodes. Recall that a node is a region of very low electron density. A molecular orbital with a node between nuclei is of higher energy than a molecular orbital without a node between nuclei. Note that as we progress from π_1 to $\pi_6{}^*$, the number of nodes increases; this is the reason that the energy associated with these orbitals increases.

What Is an Aromatic Compound?

Benzene is one member of the large class of aromatic compounds, compounds that are *substantially stabilized by pi-electron delocalization.* The resonance energy of an aromatic compound is a measure of its gain in stability. (The structural features that give rise to aromaticity will be discussed shortly.)

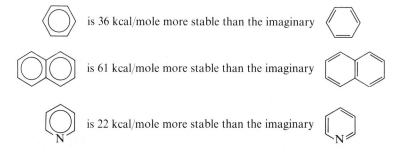

The most convenient way to determine if a compound is aromatic is by the position of absorption in the nmr spectrum by protons attached to ring atoms. Protons attached to the outside of an aromatic ring are highly deshielded and absorb far downfield from most other protons, usually beyond 7 ppm.

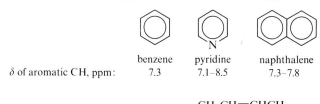

	benzene	pyridine	naphthalene
δ of aromatic C H, ppm:	7.3	7.1–8.5	7.3–7.8

$$CH_3CH{=}CHCH_3$$

δ of nonaromatic sp^2 C H, ppm: 5.3

STUDY PROBLEM

10.3.   The nmr spectrum of cyclooctatetraene shows only a singlet at $\delta = 5.7$ ppm. On the basis of this chemical shift, would you say that cyclooctatetraene is aromatic or not?

cyclooctatetraene

Requirements for Aromaticity

What structural features are necessary for a molecule to be aromatic? The first two criteria are that the molecule must be *cyclic* and *planar*. Third, each atom of the ring or rings must have a *p orbital perpendicular to the plane of the ring.*

If a system does not fit these criteria, there cannot be complete delocalization of the pi electrons. Whether or not these three criteria are met may often be deduced from inspection of the formula of an organic compound. The valence-bond formula of an aromatic compound usually shows a ring with alternate single and double bonds. There are cases, however, of cyclic organic compounds with alternate single and double bonds that are *not* aromatic. Cyclooctatetraene is such a compound. Cyclooctatetraene undergoes addition reactions with the hydrogen halides and with the halogens. These reactions are typical of alkenes, but are not typical of benzene and other aromatic compounds. Cyclooctatetraene is not planar, but has been shown to be shaped like a tub.

cyclooctatetraene

a cyclic tetraene: not aromatic

Why is cyclooctatetraene not aromatic? To answer this question, we must proceed to a fourth criterion for aromaticity, a criterion usually called **the Hückel rule**.

A.   The Hückel rule

In 1931, the German chemist Erich Hückel proposed that, to be aromatic, a monocyclic (one ring), planar compound must have **($4n + 2$) pi electrons**, where *n* is an integer. According to the Hückel rule, a ring with 2, 6, 10, or 14 pi electrons may be aromatic, but a ring with 8 or 12 pi electrons may not. Cyclooctatetraene (with 8 pi electrons) does not fit the Hückel rule for aromaticity.

$six\ \pi\ e^-$	$ten\ \pi\ e^-$	$eight\ \pi e^-$
($4n + 2$)	($4n + 2$)	$4n$
$n = 1$	$n = 2$	$n = 2$
aromatic	*aromatic*	*not aromatic*

Why can a monocyclic compound with six or ten pi electrons be aromatic, but not a compound with eight pi electrons? The answer is found in the number of pi electrons versus the number of pi orbitals available. To be aromatic, a molecule must have *all pi electrons paired.* This system provides the maximum and complete overlap required for aromatic stabilization. If some pi orbitals are *not* filled (that is, there are unpaired pi electrons), overlap is not maximized, and the compound is not aromatic. Benzene has six pi electrons and three bonding pi orbitals. The three bonding pi orbitals are filled to capacity, all pi electrons are paired, and benzene is aromatic.

Let us look at the π molecular orbitals for cyclooctatetraene. This compound has eight p orbitals on the ring. Overlap of eight p orbitals would result in eight π molecular orbitals.

If cyclooctatetraene were planar and had a pi system similar to that in benzene, the π_1, π_2, and π_3 orbitals would be filled with six of the pi electrons. The remaining two electrons would be found, one each, in the degenerate π_4 and π_5 orbitals. The pi electrons of cyclooctatetraene would not all be paired, and overlap would not be maximal. Thus, cyclooctatetraene cannot be aromatic.

B.   The ions of cyclopentadiene

Cyclopentadiene is a conjugated diene and is not aromatic. The principal reason that cyclopentadiene is not aromatic is that one of the carbon atoms is sp^3, not sp^2. This sp^3 carbon has no p orbital available for bonding; however, removal of a hydrogen ion from cyclopentadiene changes the hybridization of that carbon so that it is sp^2 and has a p orbital containing a pair of electrons.

cyclopentadiene   cyclopentadienyl
                            anion

The cyclopentadiene cation also has all its carbon atoms in the sp^2 state. (The orbital pictures for the ions of cyclopentadiene are shown in Figure 10.7.)

cyclopentadienyl
cation

Would either or both of these ions be aromatic? Either ion would have five π molecular orbitals (from five p orbitals, one per carbon). The cyclopentadienyl anion, with six pi electrons ($4n + 2$), has three pi orbitals filled and all pi electrons paired. *The anion is aromatic.* The cation, however, would have only four p

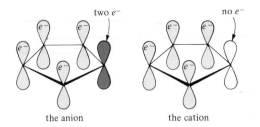

FIGURE 10.7. Orbital pictures of the anion and the cation of cyclopentadiene. (Each carbon is *sp²*-hybridized and is also attached to one H.)

electrons (4*n*) in the three orbitals. The pi electrons would *not* all be paired, and there would *not* be maximum overlap. *The cation is not aromatic.*

Cyclopentadiene is a weak acid because its anion is aromatic and thus is resonance-stabilized. Although cyclopentadiene is not nearly as strong an acid as a carboxylic acid, it is a proton donor in the presence of a strong base. (The pK_a of cyclopentadiene is 16, similar to that of an alcohol. By contrast, the pK_a of cyclopentane is about 50.)

STUDY PROBLEM

10.4. Which of the following compounds or ions could be aromatic?

(a)   (b)   (c)

Electrophilic Aromatic Substitution

The aromaticity of benzene confers a unique stability to the pi system, and benzene does not undergo most of the reactions typical of alkenes. However, benzene is far from inert. Under the proper conditions, benzene readily undergoes **electrophilic**

aromatic substitution reactions: reactions in which an electrophile is substituted for one of the hydrogen atoms on the aromatic ring. Two examples of this type of substitution reaction follow. Note that the aromaticity of the ring is retained in each product.

Halogenation:

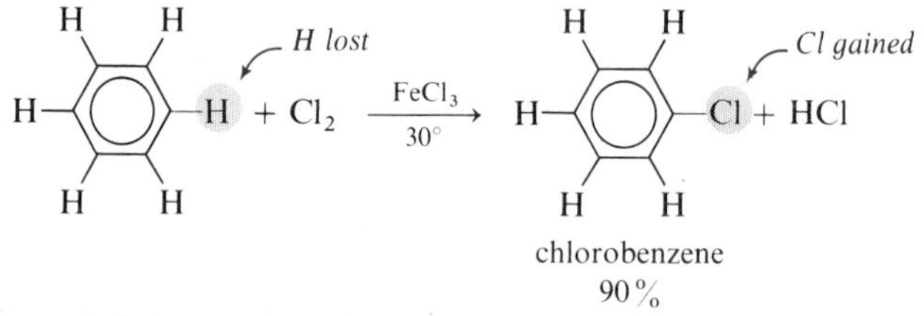

chlorobenzene
90%

Nitration:

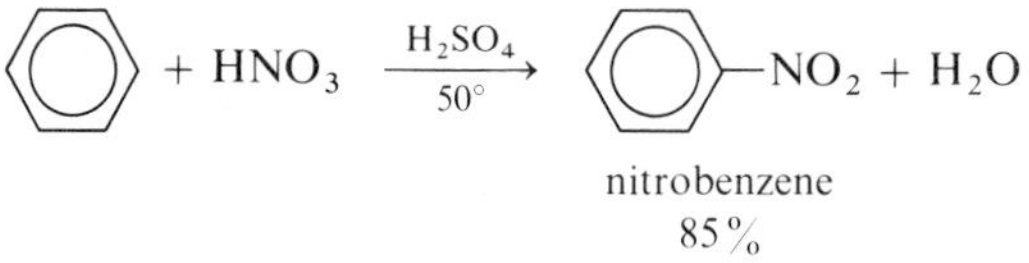

nitrobenzene
85%

The preceding examples show **monosubstitution** of the benzene ring. Further substitution is possible:

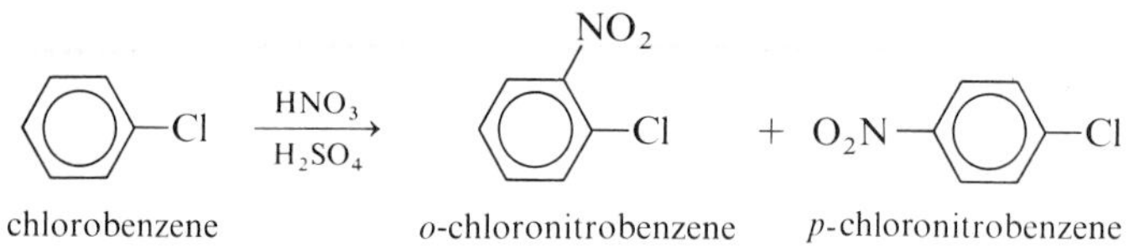

chlorobenzene   o-chloronitrobenzene   p-chloronitrobenzene

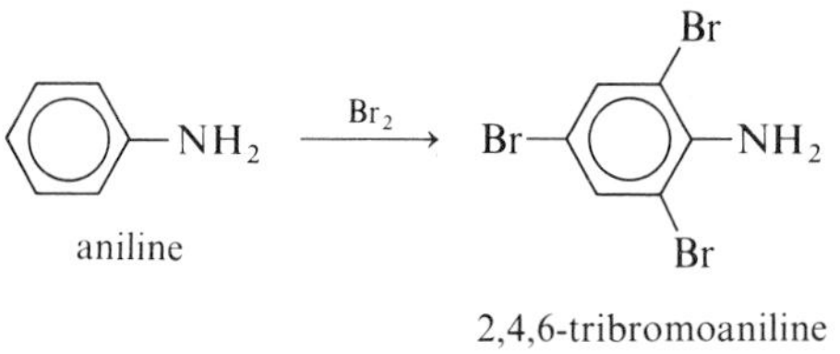

aniline   2,4,6-tribromoaniline

We will first consider the mechanism of monosubstitution (that is, the first substitution), and then the mechanism of further substitution leading to disubstituted benzenes.

SECTION 10.9.

The First Substitution

In each of the two monosubstitution reactions shown, a Lewis acid is used as a catalyst. The Lewis acid undergoes reaction with the reagent (such as X_2 or HNO_3) to generate an electrophile, which is the actual agent of substitution. For example, H_2SO_4 (a very strong acid) can remove a hydroxyl group from nitric acid to yield the nitronium ion, $^+NO_2$.

Formation of an electrophile by a Lewis acid:

$$\ddot{HO}-NO_2 + H_2SO_4 \underset{-HSO_4^-}{\rightleftharpoons} H_2\overset{+}{\ddot{O}} \frown NO_2 \rightleftharpoons H_2\ddot{O}: + \quad ^+NO_2$$

an electrophile

An electrophile can attack the pi electrons of a benzene ring to yield a type of resonance-stabilized carbocation called a **benzenonium ion.** Like other carbocations, a benzenonium ion reacts further. In this case, a hydrogen ion is removed from the intermediate (by HSO_4^-, for example) to yield the substitution product. In showing these structures, we use Kekulé formulas, which allow us to keep track of the number of pi electrons. In the following equation, we also show the hydrogen atoms attached to the ring so you can see how they are affected (or not) in the reaction.

benzene   intermediate   product
                    benzenonium ion

As we discuss the various types of electrophilic aromatic substitution reactions, you will see that the mechanisms are all simply variations of this general one.

A.   Halogenation

Aromatic halogenation is typified by the bromination of benzene. The catalyst in aromatic bromination is $FeBr_3$ (often generated *in situ* from Fe and Br_2). The function of the catalyst is to generate the electrophile Br^+. This may occur by direct reaction and fission of the Br—Br bond. More likely, Br_2 is not completely cleaved upon reaction with the $FeBr_3$ catalyst, but is polarized. For the sake of simplicity, we will show Br^+ as the electrophile.

electrophilic

$$:\ddot{Br}-\ddot{Br}: + FeBr_3 \rightleftharpoons \overset{\delta+}{:\ddot{Br}}-\overset{\delta-}{\ddot{Br}}:---FeBr_3 \rightleftharpoons :\ddot{Br}^+ + FeBr_4^-$$

polarized   *cleaved*

When an electrophile such as Br^+ collides with the electrons of the aromatic pi cloud, a pair of the pi electrons forms a sigma bond with the electrophile. This step is the slow step, and therefore the rate-determining step, in the reaction.

Step 1 (slow):

The benzenonium ion loses a proton to a base in the reaction mixture. The product is bromobenzene, a product in which the aromatic character of the ring has been recaptured.

Step 2 (fast):

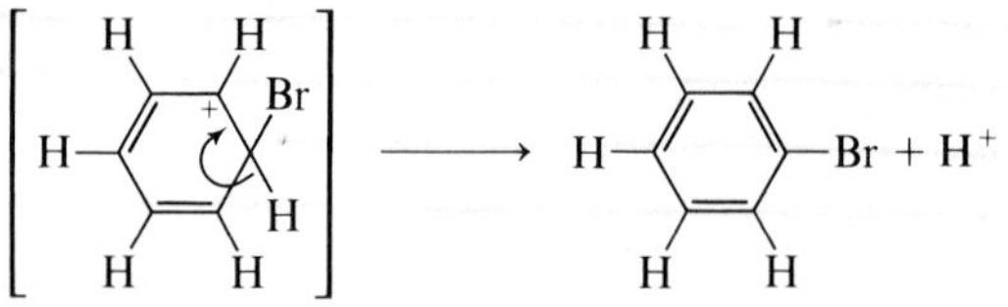

bromobenzene

The third step in the reaction mechanism (which is concurrent with the loss of H^+) is the regeneration of the Lewis acid catalyst. The proton released in Step 2 undergoes reaction with the $FeBr_4^-$ ion to yield HBr and $FeBr_3$.

Step 3 (fast):

$$H^+ + FeBr_4^- \rightleftharpoons FeBr_3 + HBr$$

Ignoring the function of the catalyst, we may write an equation for the overall reaction of the aromatic bromination of benzene:

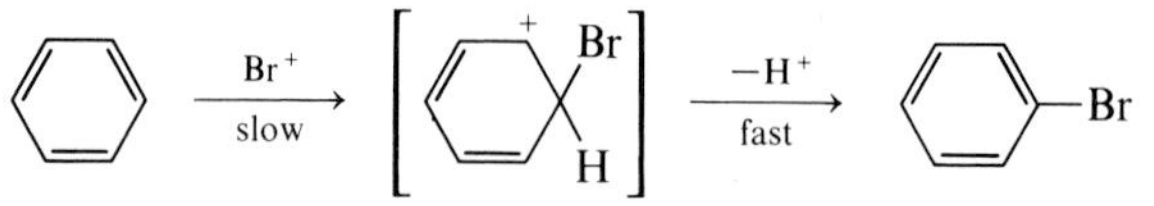

benzene   *intermediate*   bromobenzene
                          benzenonium ion

SAMPLE PROBLEM

Write resonance structures for the intermediate benzenonium ion in the aromatic bromination of benzene.

Solution:

The intermediate is often represented as

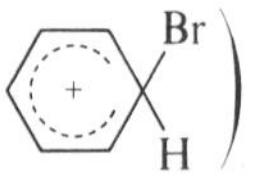

Note the similarity between an electrophilic aromatic substitution reaction and an E1 reaction. In an E1 reaction, an intermediate alkyl carbocation eliminates a proton to form an alkene.

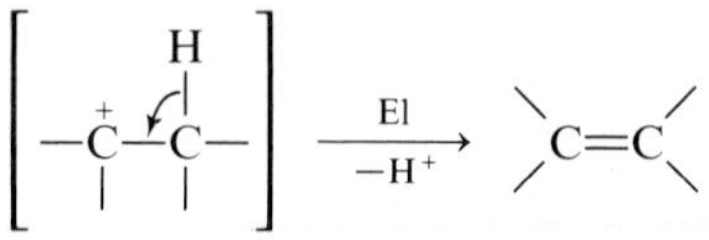

An alkyl carbocation also can undergo reaction with a nucleophile in an S_N1 reaction. However, the intermediate benzenonium ion does *not* undergo this reaction with a nucleophile. The addition of the nucleophile would destroy the aromatic stabilization of the benzene ring.

B. Isotope effect

Recall from Section 5.9A that a CD bond is stronger than a CH bond. If the breaking of a CH bond is part of the rate-determining step of a reaction, the rate of reaction for a CD compound is slower than the rate for the corresponding CH compound.

If, as we have said, the rate-determining step of electrophilic aromatic substitution is the formation of the benzenonium ion, then reaction of deuteriated benzene would proceed at the *same rate* as the reaction of normal benzene. Experiments have shown this to be true; benzene and perdeuteriobenzene undergo electrophilic bromination at the same rate, and no kinetic isotope effect is observed.

Step 2 in the reaction mechanism, loss of H^+ or D^+, does involve breaking of the CH or CD bond. Undoubtedly, the rate of elimination of D^+ is slower than that of H^+, but in either case the second step is so fast compared to Step 1 that no change in overall rate of reaction is observed. This fact may be illustrated with an energy diagram (see Figure 10.8).

C. Nitration

Benzene undergoes nitration when treated with concentrated HNO_3. The Lewis acid catalyst in this reaction is concentrated H_2SO_4. Like halogenation, aromatic nitration is a two-step reaction. The first step (the slow step) is electrophilic attack.

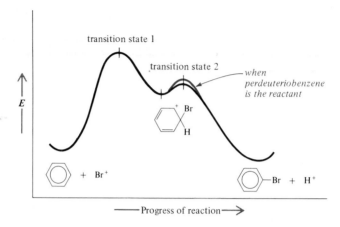

FIGURE 10.8. Energy diagram for the bromination of benzene and perdeuteriobenzene.

In nitration, the electrophile is $^+NO_2$; the equation for its formation is shown on page 467. The result of the attack is a benzenonium ion, which undergoes rapid loss of H^+ in the second step. This H^+ combines with HSO_4^- to regenerate the catalyst, H_2SO_4.

$$\underset{\text{benzene}}{\bigcirc} \xrightarrow[\text{slow}]{^+NO_2} \left[\underset{H}{\bigcirc \!\!\! \times \!\!\! ^+NO_2} \right] \xrightarrow[\text{fast}]{-H^+} \underset{\text{nitrobenzene}}{\bigcirc \!\!\! - NO_2}$$

a benzenonium ion

D. Alkylation

The alkylation of benzene is the substitution of an alkyl group for a hydrogen on the ring. Alkylation with an alkyl halide and a trace of $AlCl_3$ as catalyst is often referred to as a **Friedel–Crafts alkylation,** after Charles Friedel, a French chemist, and James Crafts, an American chemist, who developed this reaction in 1877. The reaction of 2-chloropropane with benzene in the presence of $AlCl_3$ is a typical Friedel–Crafts alkylation reaction.

$$\bigcirc + (CH_3)_2CHCl \xrightarrow[30°]{AlCl_3} \bigcirc\!\!-CH(CH_3)_2 + HCl$$

2-chloropropane   isopropylbenzene
(isopropyl chloride)   (cumene)

The first step in alkylation is the generation of the electrophile: a carbocation.

$$R-\ddot{\underset{..}{Cl}}:+AlCl_3 \rightleftharpoons R^+ + AlCl_4^-$$

The second step is electrophilic attack on benzene, while the third step is elimination of a hydrogen ion. The product is an alkylbenzene.

$$\bigcirc \xrightarrow[\text{slow}]{R^+} \left[\underset{H}{\bigcirc \!\!\! \times \!\!\! ^+R} \right] \xrightarrow[\text{fast}]{-H^+} \bigcirc\!\!-R$$

an alkylbenzene

One problem in Friedel–Crafts alkylations is that the substitution of an alkyl group on the benzene ring activates the ring so that a *second* substitution may also occur. (We will discuss ring activation and second substitutions later in this chapter.) To suppress this second reaction, an excess of the aromatic compound is commonly used.

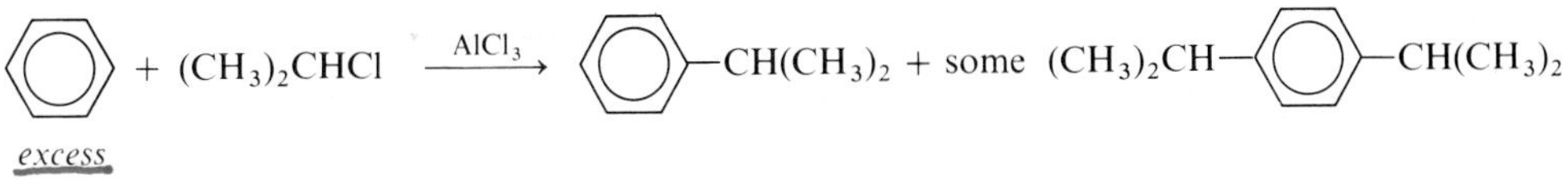

excess

Another problem in Friedel–Crafts alkylations is that the attacking electrophile may undergo rearrangement by 1,2-shifts of H or R.

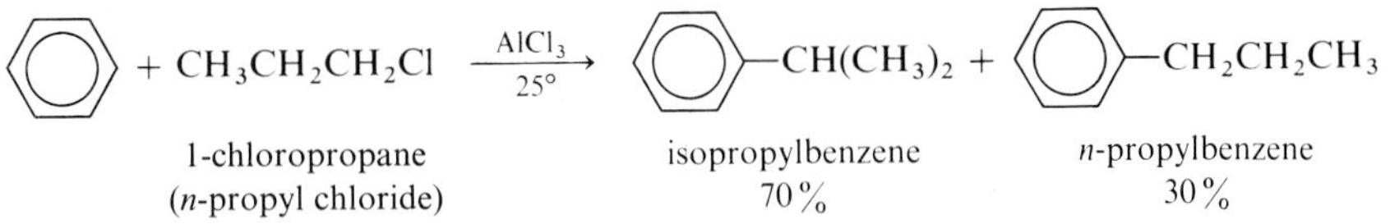

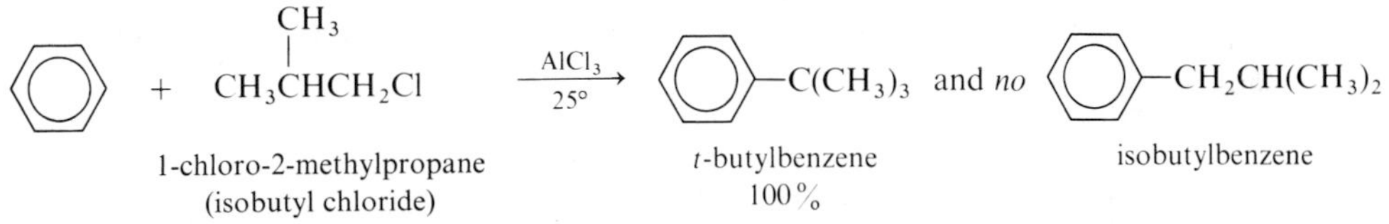

1-chloropropane   isopropylbenzene   *n*-propylbenzene
(*n*-propyl chloride)   70%   30%

1-chloro-2-methylpropane   *t*-butylbenzene   isobutylbenzene
(isobutyl chloride)   100%

The rearrangements shown are of primary alkyl halides, which do not readily form carbocations. In these cases, reaction probably proceeds through an RX—AlCl$_3$ complex.

$$CH_3CH_2CH_2-Cl + AlCl_3 \rightleftharpoons CH_3CH_2\overset{\delta+}{C}H_2-\overset{\delta-}{C}l ---AlCl_3$$

a complex

This complex may (1) undergo reaction with benzene to yield the nonrearranged product, or (2) undergo rearrangement to a secondary or tertiary carbocation, which leads to the rearranged product.

No rearrangement:

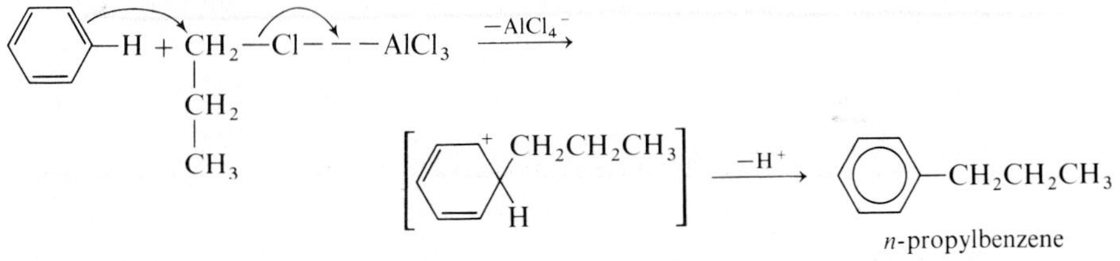

n-propylbenzene

Rearrangement:

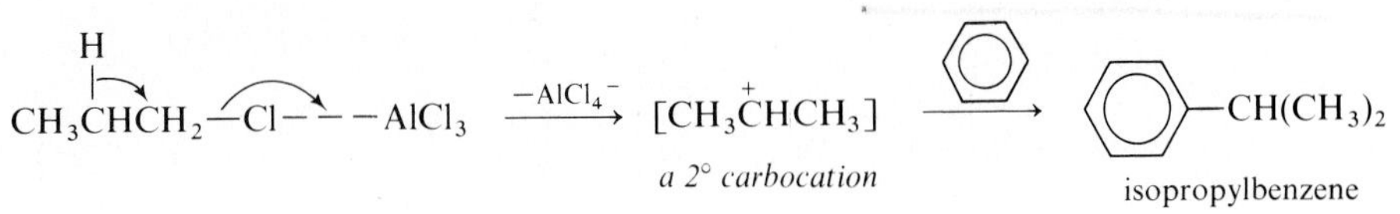

a 2° carbocation   isopropylbenzene

Alkylations may also be accomplished with alkenes in the presence of HCl and AlCl₃. The mechanism is similar to alkylation with an alkyl halide and proceeds by way of the more stable carbocation.

$$CH_3CH=CH_2 + HCl + AlCl_3 \longrightarrow [CH_3\overset{+}{C}HCH_3] + AlCl_4^-$$

propene

more stable
2° carbocation

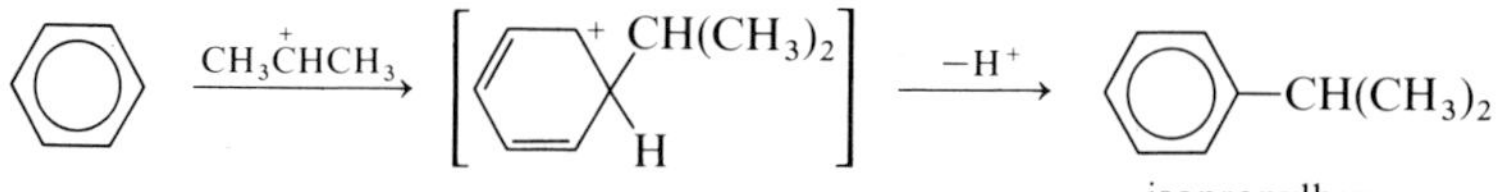

isopropylbenzene
85%

STUDY PROBLEM

10.5. Predict the *major* organic product of the reaction of benzene in the presence of AlCl₃ (and HCl in the case of an alkene) with: (a) 1-chlorobutane; (b) methylpropene; (c) neopentyl chloride (1-chloro-2,2-dimethylpropane); (d) dichloromethane (using an excess of benzene).

E. Acylation

The RC— group or the ArC— group is called an **acyl group**. Substitution of an acyl group on an aromatic ring by reaction with an acid halide is called an **aromatic acylation reaction**, or a **Friedel–Crafts acylation**.

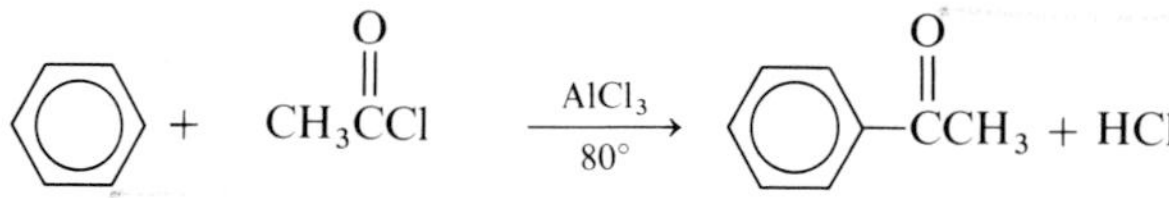

acetyl chloride

an acid halide

acetophenone
97%

This reaction is often the method of choice for preparing an aryl ketone. The carbonyl group of the aryl ketone may be reduced to a CH₂ group (Section 11.14C). By the combination of a Friedel–Crafts acylation and reduction, an alkylbenzene may be prepared without rearrangement of the alkyl group.

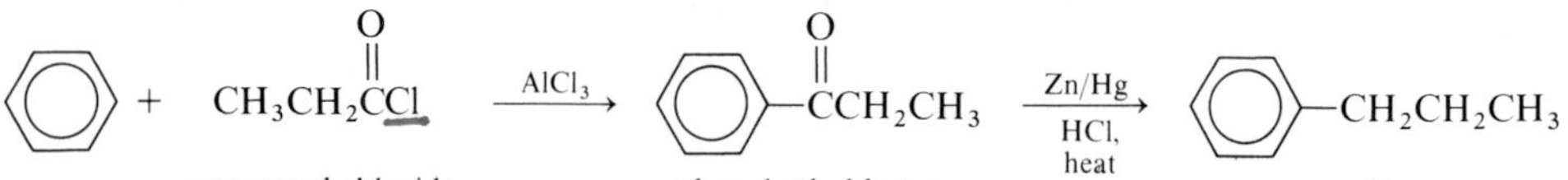

propanoyl chloride

phenyl ethyl ketone
90%

n-propylbenzene
80%

The mechanism of the Friedel–Crafts acylation reaction is similar to that of the other electrophilic aromatic substitution reactions. The attacking nucleophile is an **acylium ion** (R$\overset{+}{C}$=O).

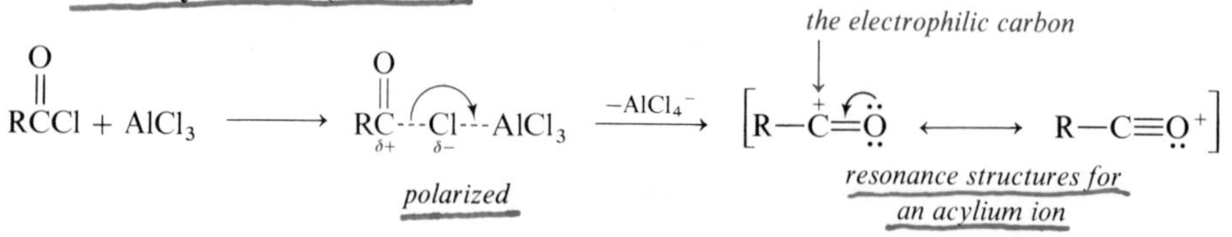

the electrophilic carbon

polarized

resonance structures for
an acylium ion

STUDY PROBLEMS

10.6. Write a mechanism for the reaction of an acylium ion with benzene to yield a ketone.

10.7. Predict the major organic products of the reaction of benzene with each of the following compounds and $AlCl_3$ as the catalyst:

$$
\begin{array}{cc}
\text{CH}_3 & \text{CH}_3 \ \text{O} \\
| & | \ \ || \\
\text{(a)} \ \ \text{CH}_3\text{CH}_2\text{CHCH}_2\text{Cl} & \text{(b)} \ \ \text{CH}_3\text{CH}_2\text{CH}-\text{CCl,} \quad \text{followed by Zn/Hg, HCl}
\end{array}
$$

F.   Sulfonation

The sulfonation of benzene with fuming sulfuric acid ($H_2SO_4 + SO_3$) yields benzenesulfonic acid.

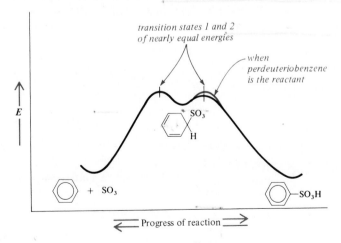

benzenesulfonic acid
50%

Unlike the other electrophilic substitution reactions of benzene, sulfonation is a readily reversible reaction and shows a moderate kinetic isotope effect. Perdeuteriobenzene undergoes sulfonation at about half the rate of ordinary benzene. From these data, we conclude that the intermediate benzenonium ion in sulfonation can revert to benzene or go on to benzenesulfonic acid with almost equal ease (or the reaction would not be readily reversible). Also, the rates of reaction of Steps 1 and 2 must be more nearly equal than for other electrophilic aromatic substitution reactions (or the reaction would not show a kinetic isotope effect). Figure 10.9 shows an energy diagram for the sulfonation of benzene. Note that the energies of the transition states of Steps 1 and 2 are roughly the same.

The sulfonic acid group is easily displaced by a variety of other groups; therefore, arylsulfonic acids are useful synthetic intermediates. We will introduce a few of these reactions of arylsulfonic acids in Section 10.16.

FIGURE 10.9. Energy diagram for the sulfonation of benzene.

SECTION 10.10.

The Second Substitution

A substituted benzene may undergo substitution of a second group. Some substituted benzenes undergo reaction *more easily* than benzene itself, while other substituted benzenes undergo reaction *less easily*. For example, aniline undergoes electrophilic substitution a million times faster than benzene. Nitrobenzene, on the other hand, undergoes reaction at approximately one-millionth the rate of benzene! (Rather than allowing a reaction to proceed over a much longer period of time, a chemist commonly uses stronger reagents and higher temperatures for a less reactive compound.)

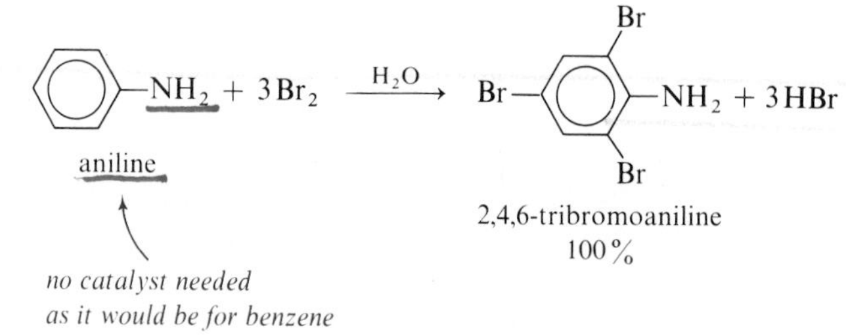

aniline

no catalyst needed
as it would be for benzene

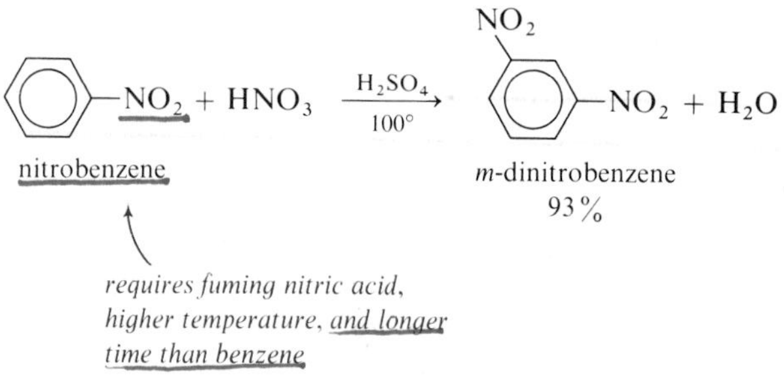

2,4,6-tribromoaniline
100%

nitrobenzene

requires fuming nitric acid,
higher temperature, and longer
time than benzene

m-dinitrobenzene
93%

In these examples, we would say that NH_2 is an **activating group**: its presence causes the ring to be *more* susceptible to further substitution. On the other hand, the NO_2 group is a **deactivating group**: its presence causes the ring to be *less* susceptible than benzene to substitution.

Besides the differences in reaction rates of substituted benzenes, the *position* of the second attack varies:

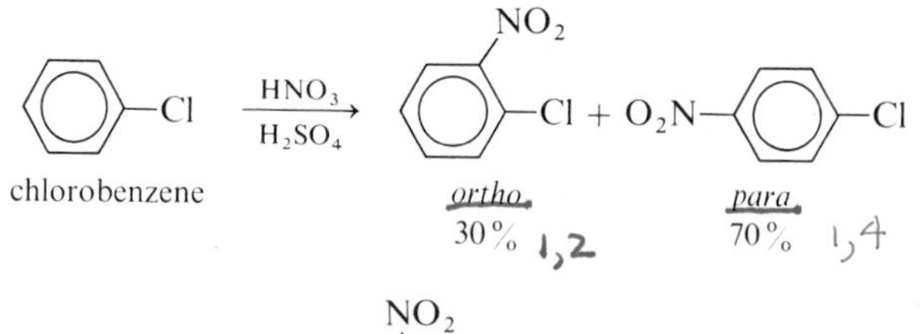

chlorobenzene

ortho.
30% 1,2

para.
70% 1,4

but *no meta*

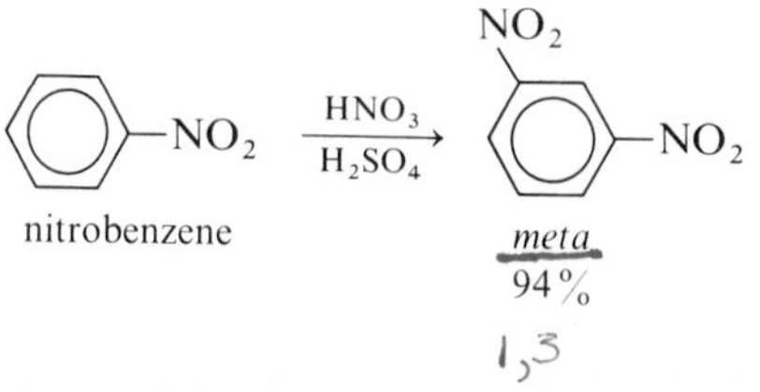

nitrobenzene

meta
94%

1,3

and very little *ortho* or *para*

STUDY PROBLEM

10.8. If chlorobenzene were nitrated at equal rates at each possible position of
substitution, what would be the ratio of *o*-, *m*-, and *p*-products?

Chlorobenzene is nitrated in the *ortho*- and *para*-positions, but not in the
meta-position. However, nitrobenzene undergoes a second nitration in the
meta-position; very little substitution at the *ortho*- or *para*-positions occurs. These
examples show that the nature of the incoming group has no effect on its own
positioning on the ring. The position of second substitution is determined by the
group that is already on the ring.

To differentiate between these two types of substituent, Cl is called an
ortho, *para*-**director**, while NO_2 is called a **meta-director**. Any substituent on a
benzene ring is either an *o,p*-director or a *m*-director, although to varying degrees,
as is shown in Table 10.5.

TABLE 10.5. Orientation of the nitro group in aromatic nitration

Reactant	Approximate percents of products		
	o	*p*	*m*
C_6H_5OH	50	50	—
$C_6H_5CH_3$	60	40	—
C_6H_5Cl	30	70	—
C_6H_5Br	40	60	—
$C_6H_5NO_2$	6	—	94
$C_6H_5CO_2H$	20	—	80

Table 10.6 contains a summary of commonly encountered benzene sub-
stituents classified as *activating* or *deactivating* and as *o,p-directors* or *m-directors*.
Note that all *o,p*-directors except the halogens are also activating groups. All *m*-
directors are deactivating. Also note that all *o,p*-directors except aryl and alkyl
groups have an unshared pair of electrons on the atom attached to the ring. None
of the *m*-directors has an unshared pair of electrons on the atom attached to the
ring.

- unshared e⁻

─ÖH

o, *p*-director,
activating

- no unshared e⁻

─N⁺=O, O⁻

m-director,
deactivating

TABLE 10.6. Effect of the first substituent on the second substitution

o,p-Directors	m-Directors (all deactivating)
$-\ddot{N}H_2, -\ddot{N}HR, -\ddot{N}R_2$	$-\overset{\overset{O}{\|\|}}{C}R$
$-\ddot{O}H$	$-CO_2R$
$-\ddot{O}R$	$-SO_3H$
$-\overset{\overset{O}{\|\|}}{\ddot{N}}HCR$	$-CHO$
	$-CO_2H$
$-C_6H_5$ (aryl)	$-CN$
$-R$ (alkyl)	$-NO_2$
$-\ddot{\underset{\cdot\cdot}{X}}:$ (deactivating)	$-NR_3{}^+$

increasing activation (arrow up, left side)

increasing deactivation (arrow down, right side)

STUDY PROBLEM

10.9. What would be the major organic products of second substitution in each of the following reactions?

(a) $C_6H_5Cl \xrightarrow{HNO_3, H_2SO_4}$

(b) $C_6H_5NO_2 \xrightarrow{Br_2, FeBr_3}$

(c) $C_6H_5OH \xrightarrow{SO_3, H_2SO_4}$

(d) $C_6H_5CH_3 \xrightarrow[AlCl_3, HCl]{CH_2=CH_2}$

A. Mechanism of the second substitution with an o, p-director

Why are most o,p-directors activating groups? Why do they direct incoming groups to the o- and p-positions? To answer these questions, let us consider aniline, a compound with the o,p-directing NH$_2$ group on the ring. The resonance structures for aniline show that the NH$_2$ group is *electron-releasing by resonance*, even though N is an electronegative atom.

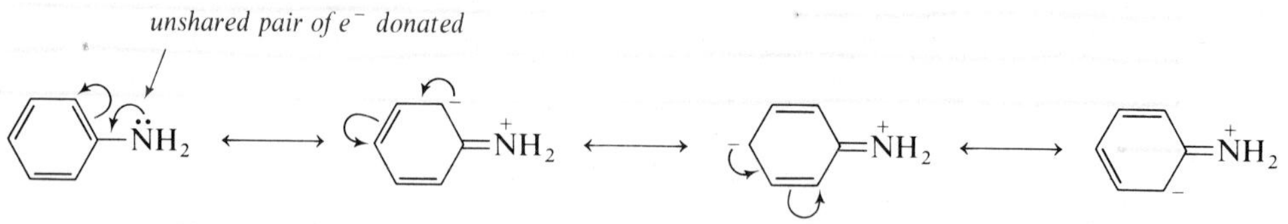

unshared pair of e⁻ donated

The result of the resonance-stabilization of aniline is that the ring is partially negative and is highly attractive to an incoming electrophile. All positions (o-, m-, and p-) on the aniline ring are activated to electrophilic substitution; however, the o- and p-positions are more highly activated than the m-position. The resonance structures above show that the o- and p-positions carry partial negative charges, while the m-position does not.

The amino group in aniline activates the benzene ring toward substitution to such an extent that (1) no Lewis acid catalyst is needed, and (2) it is very difficult to obtain a monobromoaniline. Aniline quickly undergoes reaction to form the 2,4,6-tribromoaniline (both *o*-positions and the *p*-position brominated).

The mechanism of the bromination of aniline is similar to the mechanism of the bromination of benzene itself.

aniline   a benzenonium ion   *p*-bromoaniline

The difference between the mechanism of the bromination of aniline and that of benzene lies in the stabilization of the intermediate benzenonium ion. A substituted benzenonium ion is resonance-stabilized, just as is the unsubstituted benzenonium ion, but in the case of aniline, the amino group can increase the stabilization. An increased stabilization of the intermediate means a transition state of lower energy in Step 1 (see Figure 10.10) and therefore a faster rate of reaction.

Resonance structures for the p-intermediate:

added stabilization

In the intermediate for *m*-substitution, the nitrogen of the amino group cannot help share the positive charge. (Draw the Kekulé structures for the *m*-substituted intermediate and verify this statement for yourself.) Therefore, the intermediate for *m*-substitution is of higher energy than are the intermediates leading to either *o*- or *p*-substituted products. Because this intermediate is of higher energy, its transition state also has a higher energy, and the rate of reaction at the *m*-position is lower.

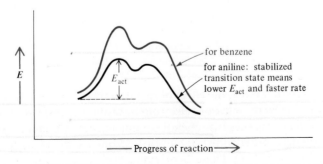

FIGURE 10.10. Energy diagrams for the bromination of aniline and benzene.

The amino group, along with —OH, —OR, —NHCOR, and phenyl groups, activates the benzene ring toward electrophilic substitution by donating a pair of electrons to the ring through resonance. Substitution occurs at the *o*- and *p*-positions because the group helps share the positive charge in these intermediates.

Although the amino group is an *o*,*p*-director and a ring activator, its character changes in a reaction mixture containing a Lewis acid such as H_2SO_4, HNO_3, or $AlCl_3$. The reason for this is that the amino group reacts with a Lewis acid to yield a *meta*-directing, deactivating ammonium ion group.

o,p-director, activating   *m-director, deactivating*

$$\text{—}\ddot{N}H_2 + H_2SO_4 \longrightarrow \text{—}\overset{+}{N}H_3 + HSO_4^-$$

STUDY PROBLEMS

10.10. Draw the resonance structures for the intermediate in the nitration of phenol to yield *o*-nitrophenol.

10.11. The benzene ring in acetanilide (Table 10.1, page 451) is *less* reactive toward electrophilic substitution than the ring in aniline. Suggest a reason for this lesser activation by the —NHCCH$_3$ group.
                                        ‖
                                        O

Halogens are different from the other *o*,*p*-directors. They direct an incoming group *ortho* or *para*, but they *deactivate* the ring to electrophilic substitution. A halogen substituent on the benzene ring directs an incoming group to the *o*- or *p*-position for the same reason the amino or hydroxyl group does. The halogen can donate electrons and help share the positive charge in the intermediate.

But why does a halogen deactivate the ring? A halogen, oxygen, or nitrogen withdraws electronic charge from the ring by the inductive effect. We would expect that any electronegative group would decrease the electron density of the ring and would make the ring less attractive to an incoming electrophile.

e⁻ withdrawal should deactivate the ring (and does deactivate the ring when X is Cl, Br, I)

In phenol or aniline, the effect of ring deactivation by electron withdrawal is counterbalanced by the release of electrons by resonance. Why is the same effect not observed in halobenzenes? In phenol or aniline, the resonance structures of the intermediate that confer added stability arise from overlap of 2*p* orbitals of carbon and 2*p* orbitals of N or O. These 2*p* orbitals are about the same size, and overlap is maximal.

In chlorobenzene, bromobenzene, or iodobenzene, the overlap in the intermediate is $2p$–$3p$, $2p$–$4p$, or $2p$–$5p$, respectively. The overlap is between orbitals of different size and is not so effective. The intermediate is less stabilized, the transition-state energy is higher, and the rate of reaction is lower. (Fluorobenzene, with $2p$–$2p$ overlap in the intermediate, is more reactive than the other halobenzenes toward electrophilic substitution.)

In summary, OH, NH_2, or a halogen determines the orientation of an incoming group by *donating electrons* by resonance and by adding resonance-stabilization to *o*- and *p*-intermediates. For OH and NH_2, electron release by resonance activates the ring toward electrophilic substitution. Electron release by resonance is less effective for Cl, Br, or I than it is for OH or NH_2. Chlorobenzene, bromobenzene, and iodobenzene contain deactivated rings because the electron-withdrawal by these substituents is relatively more effective.

An *alkyl group* does not have unshared electrons to donate for resonance-stabilization. However, an alkyl group is electron-releasing by the inductive effect, a topic we discussed in Section 5.6F. Because an alkyl group releases electrons to the benzene ring, the ring gains electron density and becomes more attractive toward an incoming electrophile.

R groups activate because they are e^--releasing

To see why alkyl groups direct incoming electrophiles to *o*- and *p*-positions, we must again look at the resonance structures of the intermediates.

added stabilizations; + next to e^--releasing R

The intermediates for *o*- or *p*-substitution both have resonance structures in which the positive charge is adjacent to the R group. These structures are especially important contributors to resonance-stabilization because the R group can help delocalize the positive charge by electron release and lower the energy of the transition state leading to these intermediates. The situation is directly analogous to that of an R group stabilizing a carbocation. The resonance structures for the intermediate in *m*-substitution have no such contributor. The *m*-intermediate is of higher energy. Attack on an alkylbenzene occurs *ortho* and *para* at a rate that is much faster than attack at a *meta* position.

SAMPLE PROBLEM

Which compound would you expect to undergo aromatic nitration more readily, $C_6H_5CH_3$ or $C_6H_5CCl_3$?

Solution: While the CH_3 group is electron-releasing and activates the ring, the CCl_3 group is strongly *electron-withdrawing* because of the influence of the electronegative chlorines. $C_6H_5CH_3$ has an activated ring; $C_6H_5CCl_3$ has a *deactivated* ring and undergoes substitution more slowly.

STUDY PROBLEM

10.12. Give the major organic products of the reaction of each of the following compounds with 2-chloropropane and $AlCl_3$. Indicate the relative order of the reaction rates.

(a) bromobenzene   (b) phenol   (c) toluene   (d) benzene

B.   Mechanism of the second substitution with a *meta*-director

In benzene substituted with a *meta*-director (such as NO_2 or CO_2H), the atom attached to the benzene ring has no unshared pair of electrons and carries a positive or a partial positive charge. It is easy to see why the *m*-directors are de-activating. Each one is *electron-withdrawing* and cannot donate electrons by resonance. Each one decreases the electron density of the ring and makes it less attractive to an incoming electrophile. The energy of the Step-1 transition state is higher than it would be for unsubstituted benzene.

A *meta*-director does not activate the *m*-position toward electrophilic sub-stitution. A *meta*-director *deactivates all positions in the ring*, but it deactivates the *m*-position less than the other positions. The resonance structures of the inter-mediates resulting from attack at the various positions show that the *o*- and *p*-intermediates are destabilized by the nearness of two positive charges. The *m*-intermediate has no such destabilizing resonance structure. The following re-sonance structures are for the intermediate in the bromination of nitrobenzene:

favored: no adjacent + charges

C.   Summary of substituent effects

1.   A substituent that is electron-releasing activates a benzene ring and is an *o,p*-director.

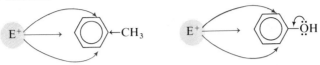

2.   A halogen is an *o, p*-director because it releases electrons by resonance, but it deactivates the ring by its electron-withdrawing inductive effect.

3.   A *meta*-director deactivates all positions on the ring by electron withdrawal and deactivates the *o*- and *p*-positions especially.

SECTION 10.11.

The Third Substitution

What if a benzene ring has two substituents already? Where does a third substituent go? A few general rules will cover the majority of cases.

1.   If two substituents direct an incoming group to the same position, that will be the principal position of the third substitution.

o to CH₃ and m to NO₂

p to CH₃ and m to NO₂

o-nitrotoluene   2,6-dinitrotoluene   2,4-dinitrotoluene

2.   If two groups conflict in their directive effects, the more powerful activator (Table 10.6) will exert the predominant directive effect.

more powerful o,p-director

CH_3—⟨⟩—OH $\xrightarrow{\text{dil. } HNO_3}$ CH_3—⟨⟩—OH

p-methylphenol
(*p*-cresol)

4-methyl-2-nitrophenol

3.   If two deactivating groups are on the ring, regardless of their position, it may be difficult to effect a third substitution.

4.   If two groups on a ring are *meta* to each other, the ring does not usually undergo substitution at the position between them, even if the ring is activated. The lack of reactivity at this position is probably due to steric hindrance.

m-hydroxybenzaldehyde

6-bromo-3-hydroxybenzaldehyde

STUDY PROBLEM

10.13. Predict the products of the next substitution:

(a)

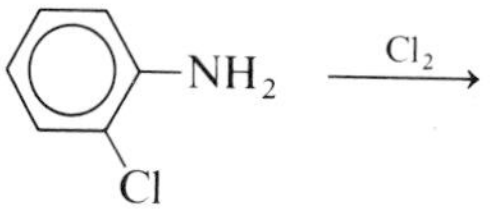

(b)

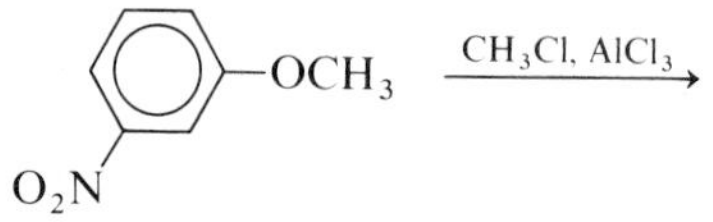

(c)

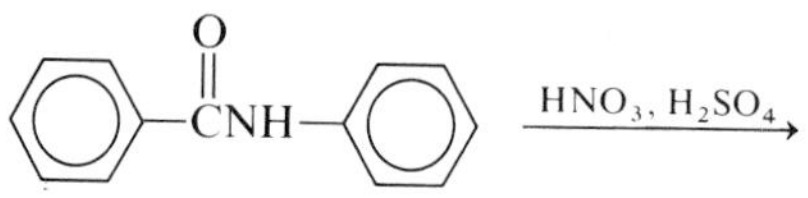

SECTION 10.12.

Alkylbenzenes

The benzene ring often has a large effect on the chemical properties of its substituents. For example, the alkyl groups attached to a benzene ring are no different from other alkyl groups, with one important exception: the carbon adjacent to the benzene ring is a benzylic carbon.

Benzylic carbons:

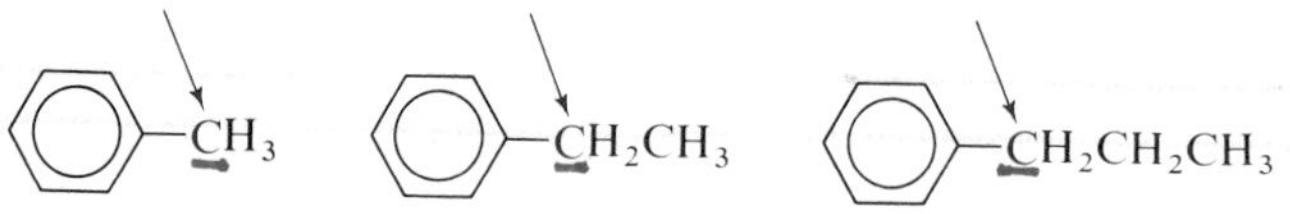

A benzyl cation, a benzyl free radical, and a benzyl carbanion are all resonance-stabilized by the benzene ring. Consequently, the benzylic position is a site of attack in many reactions.

STUDY PROBLEM

10.14. Draw the resonance structures of the carbocation intermediate in the S_N1 reaction of benzyl bromide with water.

Although an alkene is readily oxidized by such reagents as hot KMnO$_4$ solution, a benzene ring is not oxidized under these laboratory conditions. However, the *alkyl group* of an alkylbenzene may be oxidized. Because of the reactivity of the benzylic position, alkylbenzenes all yield the same product, benzoic acid, upon oxidation.

toluene $\xrightarrow{\text{hot MnO}_4{}^-}$

ethylbenzene $\xrightarrow{\text{hot MnO}_4{}^-}$ benzoic acid $-CO_2H$

$\xrightarrow{\text{hot MnO}_4{}^-}$

isopropylbenzene $-CH(CH_3)_2$

·STUDY PROBLEM

10.15. What would be the oxidation product of tetralin with hot KMnO$_4$ solution?

tetralin

Free-radical halogenation is another reaction that takes place preferentially at the benzylic position (Section 6.5B). (Note that the conditions of free-radical halogenation are different from those for electrophilic aromatic halogenation, which is an ionic reaction, not a free-radical reaction.)

$-CH_2CH_3$ $\xrightarrow[hv]{Br_2}$ $-\overset{\overset{\displaystyle Br}{|}}{C}HCH_3$

ethylbenzene   1-bromo-1-phenylethane

SECTION 10.13.

Phenols

A phenol (ArOH) is a compound with an OH group attached to an aromatic ring. As we mentioned in Section 10.10, the OH group is a powerful activator in electrophilic aromatic substitution reactions.

Because a bond from a sp^2 carbon is stronger than a bond from an sp^3 carbon (Section 2.4F), the C—O bond of a phenol is not easily broken. Phenols do not undergo S_N1 or S_N2 reactions or elimination reactions as alcohols do.

$$R-OH + HBr \xrightarrow{\text{S_N1 or S_N2}} RBr + H_2O$$

an alcohol

$$Ar-OH + HBr \longrightarrow \text{no reaction}$$

a phenol

Although the C—O bond of a phenol is not easily broken, the OH bond is readily broken. Phenol, with a pK_a of 10, is a stronger acid than an alcohol or water. (The acidity of phenol was discussed in Section 7.10.) A phenoxide ion (ArO^-) is easily prepared by treatment of a phenol with aqueous NaOH. Recall that phenoxides are useful in the preparation of aryl alkyl ethers.

$$ArO^- + RX \longrightarrow ArOR + X^-$$

a phenoxide	*a 1° alkyl*	*an ether*
ion	*halide*	

A. Esterification of phenols

Esterification of phenol does not involve cleavage of the strong C—O bond of the phenol, but depends on the cleavage of the OH bond. (We will discuss the mechanism of esterification reactions in Chapters 12 and 13.) Therefore, esters of phenols can be synthesized by the same reactions that lead to alkyl esters.

Although a carboxylic acid could be used in the esterification of a phenol, yields are usually low. Better results are obtained if a more reactive derivative of a carboxylic acid is used instead of the acid itself. Acetic anhydride is a reactive derivative that leads to acetate esters (Section 13.4C).

acetic anhydride   phenol   phenyl acetate

an ester

STUDY PROBLEM

10.16. Suggest syntheses for the following compounds:

(a)   (b)

B. The Kolbe reaction

Because the OH group is such a powerful ring activator in electrophilic substitution reactions, phenols and phenoxides undergo some rather interesting reactions. One of these is the **Kolbe reaction**, the reaction of sodium phenoxide with CO_2 to yield sodium salicylate, which yields salicylic acid when acidified.

The reaction mechanism diagram showing the conversion to salicylate ion and salicylic acid.

Salicylic acid is used to synthesize acetylsalicylic acid, commonly called *aspirin*.

The reaction of salicylic acid with acetic anhydride:

$$\text{salicylic acid} + (CH_3C)_2O \longrightarrow \text{acetylsalicylic acid (aspirin)} + CH_3COH$$

acetic anhydride

acetylsalicylic acid
(aspirin)

C. The Reimer–Tiemann reaction

Another interesting reaction of phenol is its reaction with chloroform in aqueous base, followed by treatment with aqueous acid. The product is salicylaldehyde. This reaction is called the **Reimer–Tiemann reaction**.

$$\text{phenol} \xrightarrow[\text{(2) } H_2O, H^+]{\text{(1) } CHCl_3, OH^-, 70°} \text{salicylaldehyde}$$

salicylaldehyde

The first step in this reaction is the formation of the very reactive intermediate dichlorocarbene from the action of the base on chloroform. (This type of reaction has been discussed in Section 9.12.)

$$HO:^- + H-CCl_3 \xrightarrow{-H_2O} {}^-:CCl_3 \xrightarrow{-Cl^-} :CCl_2$$

dichlorocarbene

The carbon atom in $:CCl_2$ contains only six valence electrons; thus, $:CCl_2$ can act as an electrophile in aromatic substitution.

$$\text{phenoxide} + :CCl_2 \longrightarrow [\text{intermediate}] \longrightarrow o\text{-(dichloromethyl)phenoxide ion}$$

o-(dichloromethyl)phenoxide ion

The (dichloromethyl)phenoxide ion contains two chlorides in a benzylic position. Recall that benzylic halides undergo rapid S_N1 and S_N2 reactions with OH^-. The dichloro compound undergoes just such a substitution to yield an unstable 1,1-chlorohydrin that subsequently loses H^+ (to the base) and Cl^-.

$$\text{CHCl}_2 \text{ compound} \xrightarrow[-Cl^-]{^-OH} \text{a 1,1-chlorohydrin} \xrightarrow[-Cl^-]{^-OH, -H_2O} \text{salicylaldehyde}$$

a 1,1-chlorohydrin

Salicylaldehyde is obtained by acidification of the alkaline solution.

salicylaldehyde

D. Oxidation of phenols

Phenol itself resists oxidation because formation of a carbonyl group would lead to loss of aromatic stabilization. However, 1,2- and 1,4-dihydroxybenzenes, called **hydroquinones**, can be oxidized to **quinones**. The oxidation proceeds with very mild oxidizing agents, such as Ag^+ or Fe^{3+}, and is readily reversible. (The biological reduction of the quinone ubiquinone was mentioned in Section 6.6B.) Although simple hydroquinones are colorless, quinones are colored. The quinone ring system is found in many dyes (Section 21.5).

hydroquinone   1,4-benzoquinone
(quinone)

catechol   1,2-benzoquinone

The ability of hydroquinone to reduce silver ions to silver metal is the chemical basis of photography. Silver ions in a silver halide crystal that has been exposed to light are more easily reduced than the silver ions of an unexposed crystal. Hydroquinone in the developer fluid reduces these light-activated silver ions at a faster rate than the nonexposed silver ions. In the fixing process, unreacted silver halide is converted to a water-soluble silver complex by sodium thiosulfate, $Na_2S_2O_3$ (called *hypo*), and washed from the film. The result is the familiar photographic negative.

Bombardier beetles rely on the easy oxidation of hydroquinones as a defensive mechanism against spiders, mice, and frogs. These beetles contain glands that store H_2O_2 and other glands that store hydroquinones. When the beetle is threatened, secretions from the two types of glands are mixed together, along with enzymes that catalyze the oxidation of the hydroquinones. The products are irritating quinones and O_2 (which acts as a propellant) that are squirted at the predator.

SECTION 10.14.

Benzenediazonium Salts

A. Preparation and reactivity of benzenediazonium chloride

Arylamines such as aniline can be prepared in the laboratory by the nitration of aromatic compounds followed by reduction of the nitro group. A mixture of iron filings and concentrated HCl is a common reducing agent for this reaction. Because the reaction is run in acid, a *protonated* amine is the product. Subsequent treatment with base generates the amine itself.

Aniline and other arylamines undergo reaction with cold nitrous acid (HONO) in HCl solution to yield **aryldiazonium chlorides** (ArN_2^+ Cl^-). The nitrous acid is usually generated *in situ* by the reaction of sodium nitrite (Na^+ ^-ONO) with HCl. Diazonium salts are very reactive; therefore, the reaction mixture must be chilled. (Alkyldiazonium salts, RN_2^+ X^-, are not stable even in the cold; see Section 15.9.)

aniline   benzenediazonium
chloride

The high reactivity of diazonium salts arises from the excellent leaving ability of nitrogen gas, N_2. Because of this leaving ability, the diazonium group may be displaced by a variety of nucleophiles, such as I^-. Some of the substitution reactions that we will present are thought to proceed by a free-radical mechanism. In other cases, the substitution reaction proceeds through an aryl cation by a mechanism similar to that of an S_N1 reaction.

In these displacement reactions, the diazonium salt is generally prepared (but not isolated), the nucleophilic reagent is added, and the mixture is allowed to warm or is heated. Yields of substitution products are generally good to excellent: 70–95 % from the starting arylamine.

B. Reactions of benzenediazonium salts

Displacement by halide ions and cyanide ions.   The diazonium group is readily displaced as N_2 by halide ions (F^-, Cl^-, Br^-, or I^-) or the cyanide ion (^-CN). These reactions provide synthetic routes to aryl fluorides, iodides, and nitriles (ArCN), none of which can be obtained by direct electrophilic substitution. Although aryl bromides and chlorides can be synthesized by electrophilic substitution reactions, the products are often contaminated with disubstituted by-products. Diazonium displacement yields pure monochloro or monobromo compounds, uncontaminated by disubstituted products.

For the substitution of $-Cl$, $-Br$, or $-CN$, a copper(I) salt is used as the source of the nucleophile, and the reaction mixture is warmed to 50–100°. (The Cu^+ ion acts as a catalyst in these reactions.) This reaction, using the copper(I) salt, is called the **Sandmeyer reaction**.

Aryl iodides and fluorides are prepared with KI (no catalyst needed) or fluoroboric acid (HBF_4), respectively.

One use of these diazonium reactions is the preparation of compounds that cannot be readily synthesized by other routes. For example, consider the synthesis of *p*-toluic acid (*p*-methylbenzoic acid). The usual route to an aryl carboxylic acid is the oxidation of an alkylarene (R—Ar). Oxidation cannot be used for the preparation of *p*-toluic acid from *p*-xylene (*p*-dimethylbenzene) because *both* methyl groups would be oxidized to carboxyl groups. However, *p*-toluic acid can be obtained by way of a diazonium intermediate, followed by treatment with CuCN–KCN and subsequent hydrolysis of the —CN group to —CO_2H, a reaction that will be discussed in Section 13.12.

Displacement by —OH.   Phenols may be prepared from diazonium salts by reaction with hot aqueous acid. This reaction provides one of the few laboratory routes to phenols.

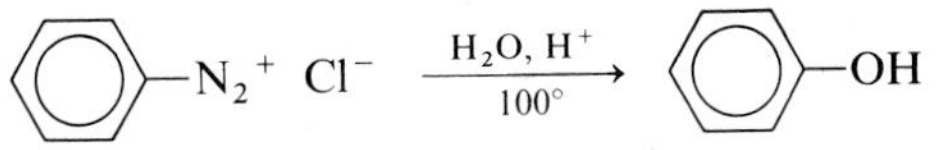

$$\text{C}_6\text{H}_5-\text{N}_2{}^+ \text{ Cl}^- \xrightarrow[100°]{\text{H}_2\text{O, H}^+} \text{C}_6\text{H}_5-\text{OH}$$

phenol

Displacement by —H.   The —$\text{N}_2{}^+$ group may be replaced with —H by treatment of the diazonium salt with hypophosphorous acid, H_3PO_2. This sequence provides a method of *removal of an* NH_2 *group from an aromatic ring.*

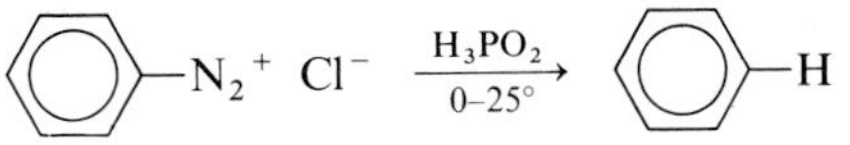

$$\text{C}_6\text{H}_5-\text{N}_2{}^+ \text{ Cl}^- \xrightarrow[0-25°]{\text{H}_3\text{PO}_2} \text{C}_6\text{H}_5-\text{H}$$

benzene

SAMPLE PROBLEM

Show how you would synthesize 1,3,5-tribromobenzene from aniline. (Remember that bromine is an *o,p*-director in electrophilic substitution reactions.)

Solution: Recall from page 477 that aniline undergoes tribromination readily. Therefore, the following reaction sequence will yield the desired product:

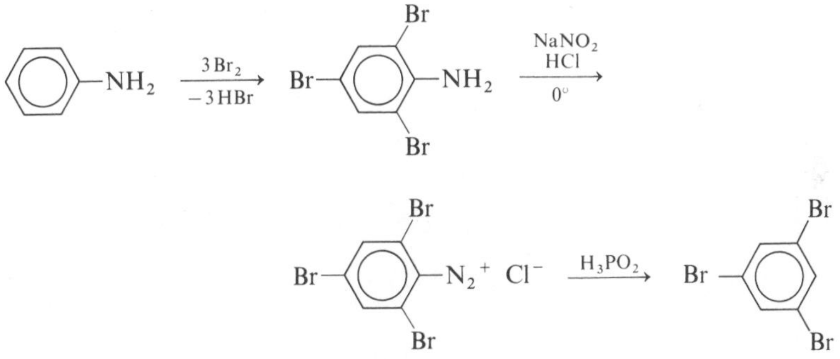

1,3,5-tribromobenzene

Coupling reactions.   Coupling reactions of aryldiazonium salts are used to prepare dyes from aniline and from substituted anilines. In these reactions, the diazonium ion acts as an *electrophile*. Resonance structures for the diazonium ion show that both nitrogens carry a partial positive charge.

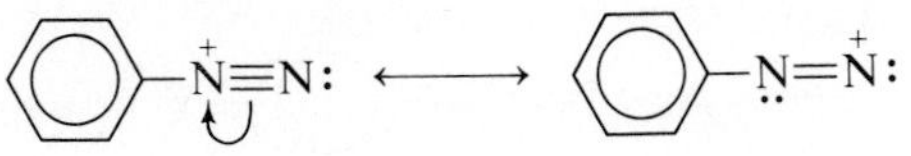

The terminal nitrogen attacks the *para*-position of an activated benzene ring (one substituted with an electron-releasing group like NH_2 or OH). The coupling product contains an **azo group** (—N=N—) and is generally referred to as an **azo compound.** Many azo compounds are used as dyes (see Section 21.5B).

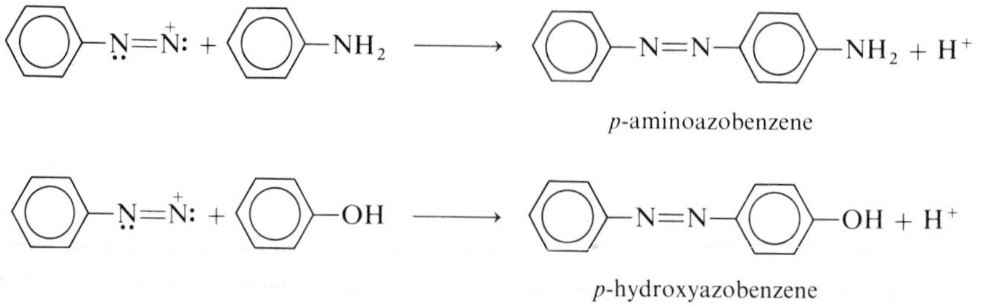

p-aminoazobenzene

p-hydroxyazobenzene

The reactions of aryldiazonium salts are summarized in Figure 10.11.

STUDY PROBLEM

10.17. How would you make the following conversions?

(a)

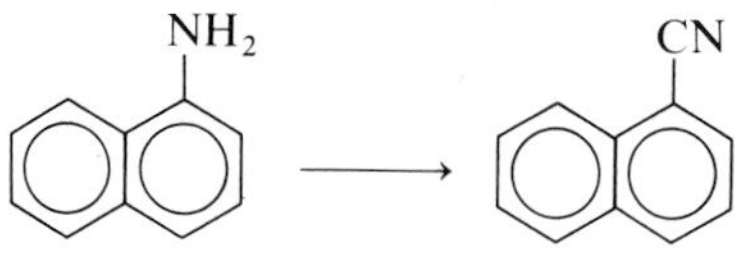

(b) CH_3—⬡—NO_2 ⟶ CH_3—⬡—Br

(c) ⬡ ⟶ ⬡—N=N—⬡—OCH_3

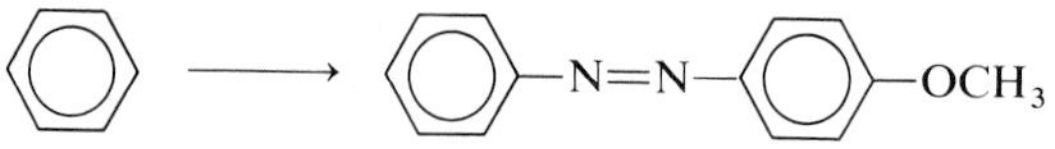

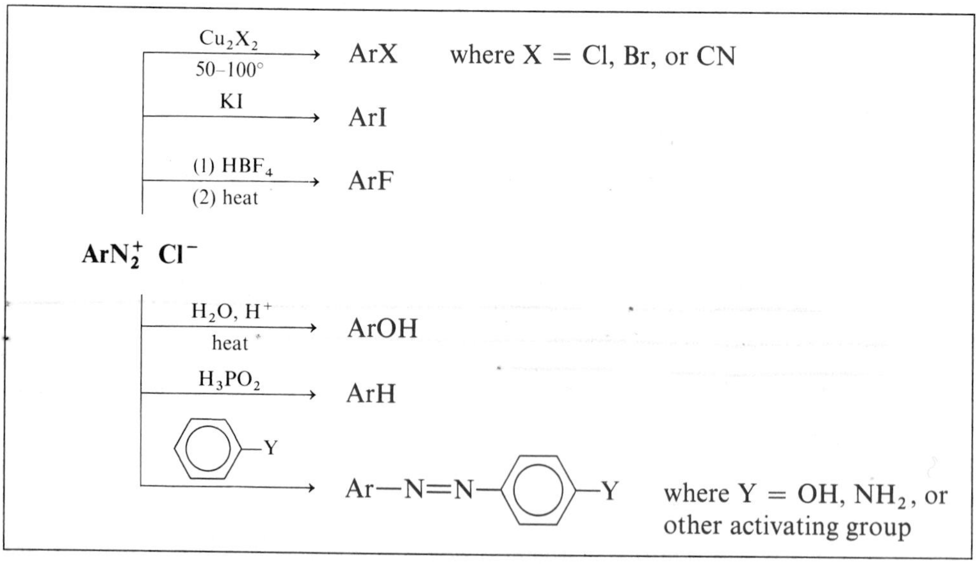

FIGURE 10.11. Reactions of an aryldiazonium chloride.

SECTION 10.15.

Halobenzenes and Nucleophilic Aromatic Substitution

In Chapter 5, we mentioned that <u>aryl halides do not undergo the substitution</u> <u>and elimination reactions characteristic of alkyl halides</u> because of the extra strength of a bond from an sp^2 carbon.

$$\text{\LARGE \bigcirc}\!-X + Nu^- \longrightarrow \text{no } S_N1 \text{ or } S_N2 \text{ reaction}$$

an aryl halide

Under certain circumstances, however, an aryl halide can undergo a **nucleophilic aromatic substitution reaction.**

$$\text{\LARGE \bigcirc}\!-X + Nu^- \longrightarrow \text{\LARGE \bigcirc}\!-Nu + X^-$$

Although this reaction appears to be similar to an S_N1 or S_N2 reaction, it is quite different. It is also different from *electrophilic* substitution, which is initiated by E^+, not Nu^-.

The <u>reactivity of aryl halides toward nucleophiles is *enhanced by the presence*</u> *of electron-withdrawing substituents on the ring.* This is the opposite of a substituent's effect in electrophilic aromatic substitution. Here, an electron-withdrawing substituent makes the ring less rich in electrons and more attractive to an attacking nucleophile.

increasing reactivity toward Nu⁻

chlorobenzene → phenol

$$\xrightarrow[350°]{10\% \text{ NaOH}} \quad 300 \text{ atm}$$

p-chloronitrobenzene → *p*-nitrophenol

$$\xrightarrow[160°]{15\% \text{ NaOH}}$$

1-chloro-2,4,6-trinitrobenzene → 2,4,6-trinitrophenol (picric acid) *an explosive*

$$\xrightarrow[\text{warm}]{H_2O}$$

Other nucleophiles besides OH^- and H_2O can also undergo reaction with activated aryl chlorides.

1-chloro-2,4-dinitrobenzene $\xrightarrow{NH_3}$ 2,4-dinitroaniline

The mechanism of nucleophilic aromatic substitution is thought to proceed by one of two routes. The first is through a carbanion intermediate. If the ring is activated toward nucleophilic substitution by an electron-withdrawing group, the reaction proceeds by a two-step mechanism: (1) addition of the nucleophile to form a carbanion (stabilized by the electron-withdrawing group), and (2) subsequent loss of the halide ion.

carbanion intermediate

Although the carbanion intermediate is unstable and reactive, it is stabilized to some extent by resonance and by dispersal of the negative charge by the electron-withdrawing group. Resonance structures of this intermediate show that an electron-withdrawing *o*- or *p*-substituent lends more stability to the carbanion intermediate than does a *meta*-substituent. When the substituent is *o*- or *p*-, the negative charge is adjacent to the electron-withdrawing group in one resonance structure. In the case of the nitro group, added stability is gained because the nitro group helps disperse the negative charge by resonance.

Resonance structures for p-substituted intermediate:

added stabilization

For m-substituted intermediate (no added stabilization):

STUDY PROBLEM

10.18. Give the resonance structures for the carbanion intermediate in the reaction of sodium hydroxide and *o*-chloronitrobenzene.

If there is no electron-withdrawing substituent on the ring, nucleophilic aromatic substitution is very difficult and proceeds by the second mechanistic sequence. In these cases, the mechanism is thought to proceed through a **benzyne** intermediate.

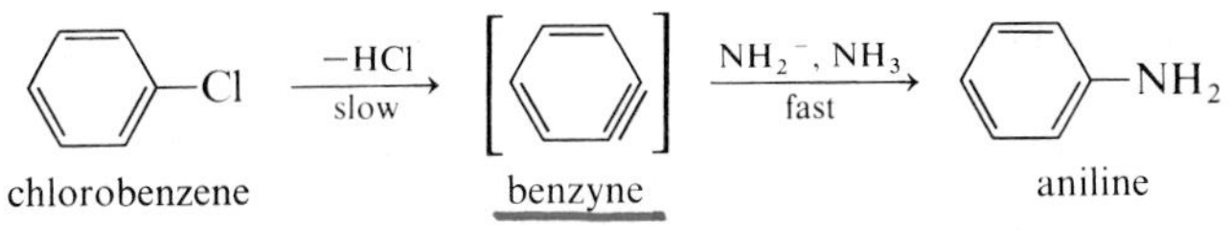

chlorobenzene   benzyne   aniline

The structure of benzyne is not like that of an alkyne. The two ring carbons that have a triple bond between them are joined by an sp^2–sp^2 sigma bond and p–p overlap (the aromatic pi cloud). These two bonds are the same as in benzene. The third bond is the side-to-side overlap of two sp^2 orbitals, the ones that originally were used in the bonds to H and X. Figure 10.12 shows this overlap. Because of the rigid geometry of the ring and the unfavorable angles of normal sp^2 orbitals, this overlap cannot be very good. The new bond is very weak, and benzyne is a highly reactive intermediate.

The formation of the benzyne intermediate in the reaction of chlorobenzene with $NaNH_2$ in liquid NH_3 is thought to occur by the following route:

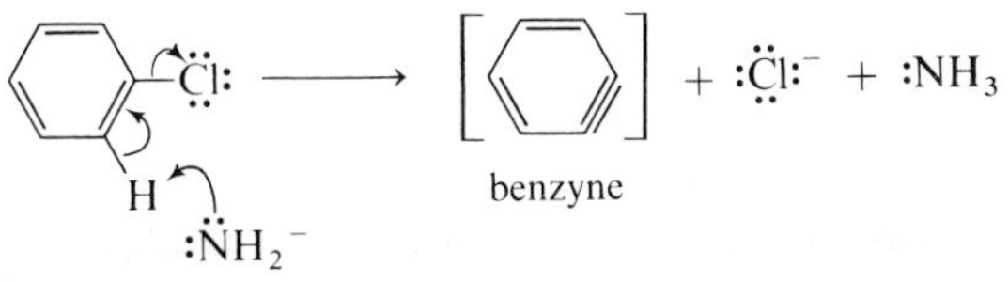

Benzyne undergoes a rapid addition of a nucleophile to yield a carbanion that can abstract a proton from NH_3 to yield the product (aniline, here) and another NH_2^- ion.

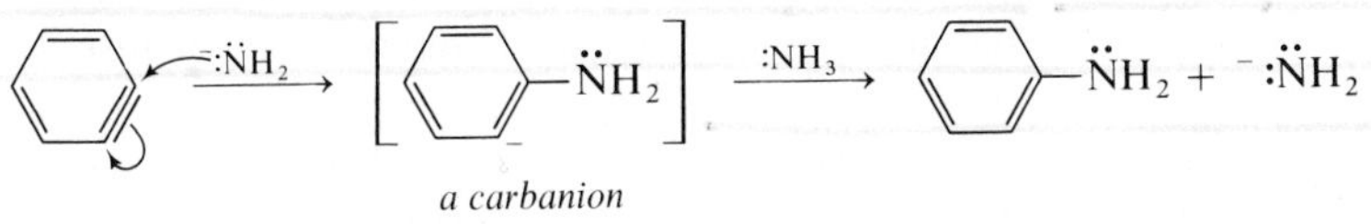

a carbanion

One interesting reaction of a benzyne intermediate is its Diels–Alder reaction with anthracene (its structure is over the arrow in the following equation) to yield an unusual compound called triptycene.

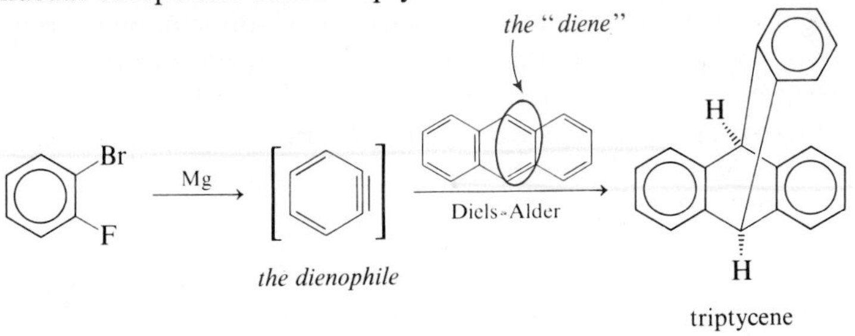

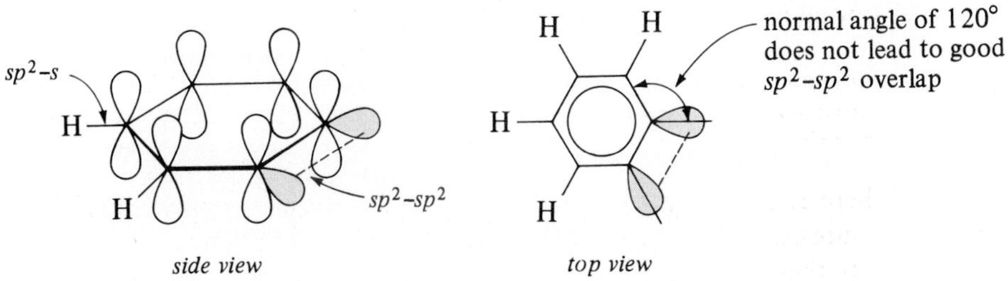

the dienophile   triptycene

side view   top view

FIGURE 10.12. The bonding in the benzyne intermediate.

STUDY PROBLEM

10.19. In an experiment designed to test the existence of benzyne as an intermediate in nucleophilic aromatic substitution, chlorobenzene was treated with sodamide (NaNH$_2$) in liquid ammonia. The chlorobenzene was labeled with carbon-14 at the C—Cl position.

(a) If this reaction proceeds in accordance with the benzyne mechanism, what would be the distribution of carbon-14 in the product aniline—that is, at what positions and in what percentages would carbon-14 be found?

(b) What would be the distribution of carbon-14 if the reaction proceeds by simple nucleophilic displacement?

SECTION 10.16.

Syntheses Using Benzene Compounds

Table 10.7 (page 495) shows the synthesis possibilities of benzene and its derivatives. In the laboratory, electrophilic aromatic substitutions are more widely used as synthetic reactions than nucleophilic aromatic substitution because there are fewer limitations on the starting material. (To effect nucleophilic aromatic substitution, a chemist must either use a ring containing electron-withdrawing groups, or else use "forcing" conditions, as shown in the equations on page 491.)

When using electrophilic aromatic substitution reactions to prepare substituted benzene compounds, a chemist may need to use ingenuity. For example, if we are preparing *m*-chloronitrobenzene, chlorination would be a poor choice as the first step because this reaction places an *o,p*-director on the ring. Subsequent nitration would give *o*- and *p*-chloronitrobenzene but not the desired *m*-chloronitrobenzene. It would be best to start with nitration because the nitro group is a *meta*-director. In synthetic work with substituted benzenes, the *order of substitution reactions* is of importance.

$$\text{benzene} \xrightarrow[\text{FeCl}_3]{\text{Cl}_2} \text{C}_6\text{H}_5\text{—Cl} \xrightarrow[\text{H}_2\text{SO}_4]{\text{HNO}_3}$$

o-chloro-
nitrobenzene   *p*-chloro-
nitrobenzene

$$\text{benzene} \xrightarrow[\text{H}_2\text{SO}_4]{\text{HNO}_3} \text{C}_6\text{H}_5\text{—NO}_2 \xrightarrow[\text{FeCl}_3]{\text{Cl}_2}$$

m-chloronitrobenzene

TABLE 10.7. Compounds that can be obtained from benzene and its derivatives

Reaction		Product	Section reference
Electrophilic substitution:			
$C_6H_6 + X_2$ $\xrightarrow{FeX_3}$	C_6H_5X	halobenzene	10.9A
$+ HNO_3$ $\xrightarrow{H_2SO_4}$	$C_6H_5NO_2$	nitrobenzene[a]	10.9C
$+ RX$ $\xrightarrow{AlX_3}$	C_6H_5R	alkylbenzene	10.9D
$+ R_2C{=}CHR$ $\xrightarrow[HX]{AlX_3}$	$C_6H_5CR_2CH_2R$	alkylbenzene	10.9D
$+ RCX$ (RC=O) $\xrightarrow{AlX_3}$	C_6H_5CR (C_6H_5C=O)	aryl ketone[b]	10.9E
$+ SO_3$ $\xrightarrow{H_2SO_4}$	$C_6H_5SO_3H$	benzenesulfonic acid[c]	10.9F
$C_6H_5-\overset{..}{Y} + E^+ \longrightarrow$ where Y is OH, NH_2, X, R, etc.		o- and p-products	10.10
$C_6H_5-\overset{\delta+}{Z} + E^+ \longrightarrow$ where Z is NO_2, CO_2H, etc.		m-products	10.10
Nucleophilic substitution:			
$C_6H_5X + OH^- \longrightarrow$	C_6H_5OH	phenol[d]	10.15
$C_6H_5X + NH_3 \longrightarrow$	$C_6H_5NH_2$	aniline[d]	10.15
Substituent reactions:			
$C_6H_5R + MnO_4^- \xrightarrow{heat}$	$C_6H_5CO_2H$	carboxylic acid	10.12
$C_6H_5CH_2R + Br_2$ or NBS $\xrightarrow{catalyst}$	C_6H_5CHBrR	benzylic bromide	6.5B
$C_6H_5OH + OH^- \longrightarrow$	$C_6H_5O^-$	phenoxide	7.10
$C_6H_5OH + RCX$ (RC=O) $\longrightarrow$	C_6H_5OCR (C_6H_5OC=O)	ester	10.13A, 13.3C, 13.4C
C_6H_5OH $\xrightarrow[(2)\ H_2O,\ H^+]{(1)\ CHCl_3,\ OH^-}$	(o-CHO–C₆H₄–OH)	o-hydroxy aryl aldehyde	10.13C
$HO{-}C_6H_4{-}OH + [O] \longrightarrow$	$O{=}C_6H_4{=}O$	quinone	10.13D
Reactions of aryldiazonium salts:[e]			
$C_6H_5N_2^+\ Cl^- + Nu^- \longrightarrow$	C_6H_5Nu	substituted (or unsubstituted) benzene	10.14B
$C_6H_5N_2^+\ Cl^- + C_6H_5Y \longrightarrow$ where Y is OH, NH_2, etc.	$C_6H_5N{=}NC_6H_4Y$	azo compound	10.14B

[a] May be reduced to aniline (Section 10.14).

[b] May be reduced to an alkylbenzene (Section 11.14C).

[c] May be converted to C_6H_6 or C_6H_5OH (Section 10.16, page 497).

[d] Practical in the laboratory only if one or more other electron-withdrawing groups is present on the ring.

[e] The preparation of aryldiazonium salts is discussed in Section 10.14A. Some specific reactions of these salts are summarized in Figure 10.11, page 490.

Conversion of one group to another may also be necessary. For instance, re-duction of the nitro group to an amino group gives a route to *m*-substituted anilines. Benzene may first be nitrated, then subjected to *m*-substitution, and finally the nitro group may be reduced.

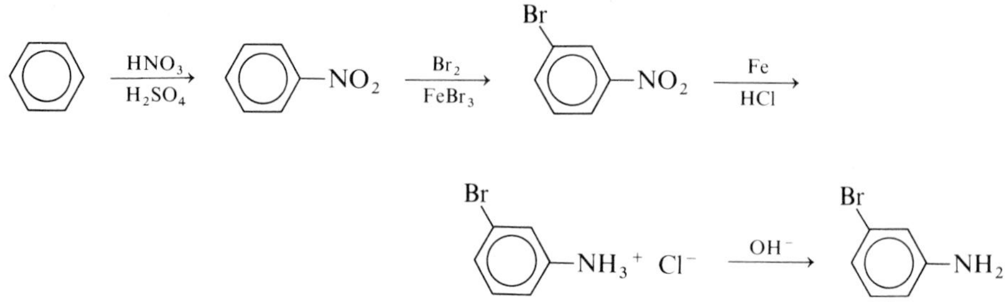

an amine salt   *m-bromoaniline*

Arylamines can also be converted to aryldiazonium salts, making possible a diversity of substitution products, as was shown in Figure 10.11, page 490. In viewing aromatic compounds from a synthetic standpoint, every nitro group should be viewed as a potential diazonium group.

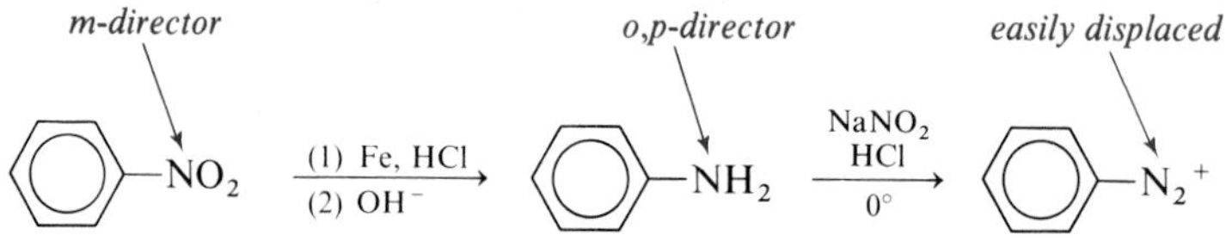

Knowledge of individual chemical characteristics of aromatic compounds is often required for successful solutions of synthetic problems. For example, aniline does not undergo Friedel–Crafts reactions because an amino group (a basic group) undergoes reaction with Lewis acids.

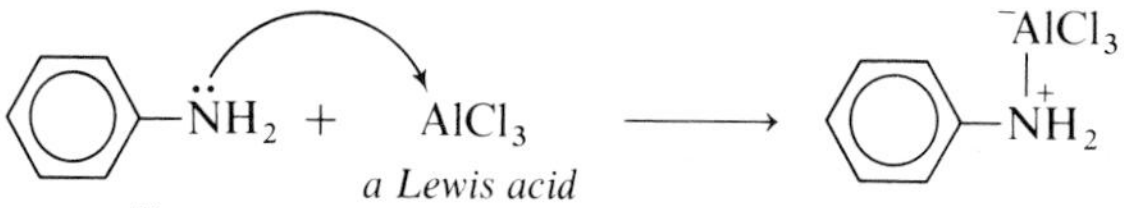

aniline

a base

Nitrobenzene also does not undergo Friedel–Crafts reactions; in this case, lack of reaction is due to deactivation of the ring by the electron-withdrawing nitro group.

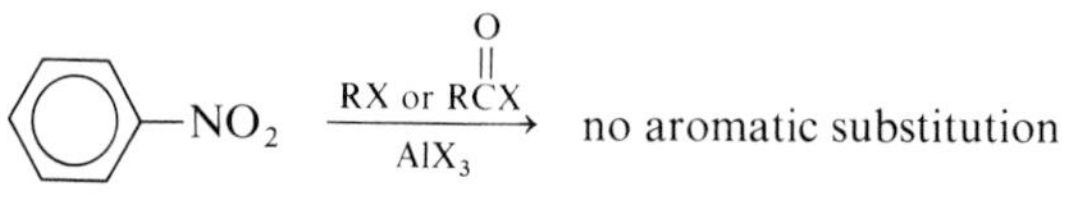

nitrobenzene

Like the diazonium group, the sulfonic acid group is easily removed from an aromatic ring and therefore may be displaced by a variety of reagents. Phenols, for example, can be prepared from arylsulfonic acids as well as from diazonium salts.

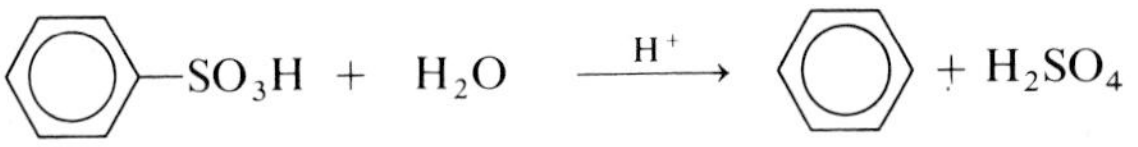

benzenesulfonic acid + H_2O —H^+→ benzene + H_2SO_4

benzenesulfonic   *as steam*   benzene
acid

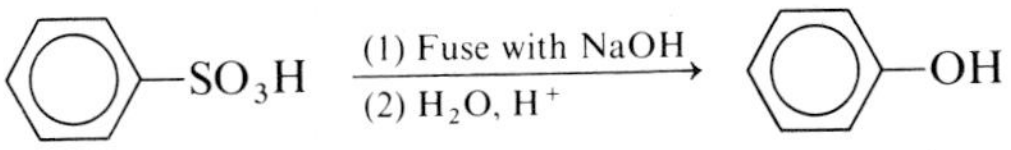

—SO_3H —(1) Fuse with NaOH / (2) H_2O, H^+→ —OH

phenol

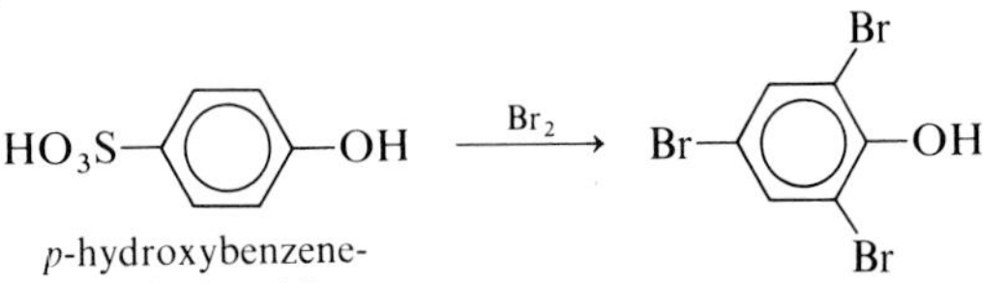

HO_3S——OH —Br_2→ Br——OH (with Br at positions 2,4,6)

p-hydroxybenzene-
sulfonic acid

2,4,6-tribromophenol

STUDY PROBLEM

10.20. Show how you would synthesize the following compounds from benzene:
(a) *m*-chloroaniline;   (b) *n*-pentylbenzene;   (c) 1,2-diphenylethane;
(d) triphenylmethane.

Summary

An **aromatic compound** is a type of compound that gains substantial stabilization by pi-electron delocalization. To be aromatic, a compound must be cyclic and
planar. Each ring atom must have a *p* orbital perpendicular to the plane of the ring,
and the *p* orbitals must contain $(4n + 2)$ pi electrons **(Hückel rule)**.

Benzene and other aromatics undergo **electrophilic aromatic substitution
reactions**. These reactions are summarized in Table 10.7, page 495.

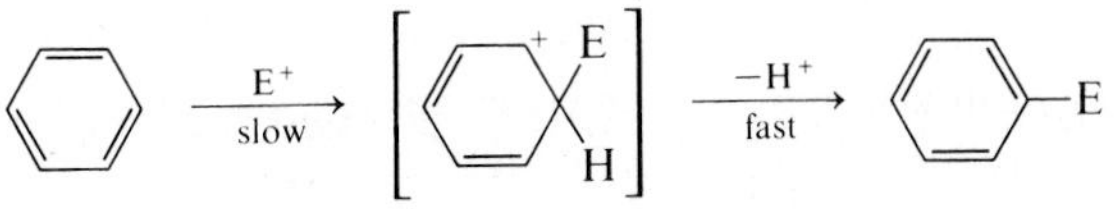

benzene —E^+, slow→ [intermediate with $+$, E, H] —$-H^+$, fast→ —E

A second substitution results in *o*- and *p*-isomers or else *m*-isomers, depending
on the first substituent (see Table 10.6, page 476). The *o,p*-directors (except R) have
electrons that can be donated to the ring by resonance.

Electron release by resonance:

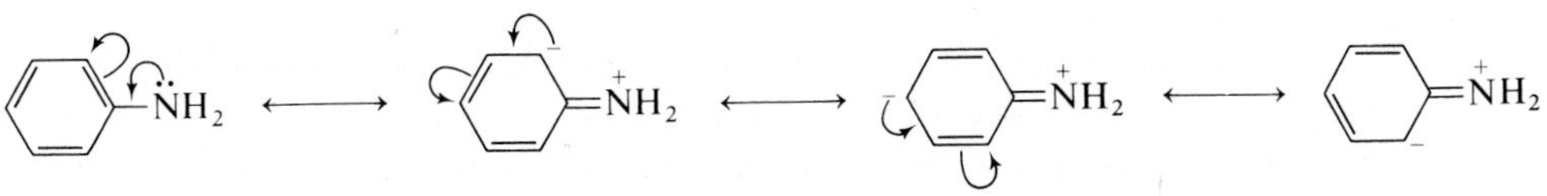

—$\ddot{N}H_2$ ⟷ =$\overset{+}{N}H_2$ ⟷ =$\overset{+}{N}H_2$ ⟷ =$\overset{+}{N}H_2$

Electron release by inductive effect:

$$\langle\!\langle \delta- \rangle\!\rangle\!\!-\!\!R^{\delta+}$$

All *o,p*-directors except X activate the entire ring toward electrophilic substitution. The *o*- and *p*-positions are the preferred positions of substitution because of added resonance-stabilization of their intermediates.

All *m*-directors and X *deactivate* the ring toward further electrophilic substitution by electron withdrawal.

$$\langle\!\langle \delta+ \rangle\!\rangle\!\!\rightarrow\!\!X^{\delta-} \qquad \langle\!\langle \delta+ \rangle\!\rangle\!\!\rightarrow\!\!NO_2{}^{\delta-}$$

deactivated toward E⁺

Alkylbenzenes contain a benzylic position that is active toward many reagents:

$$\langle\!\langle \rangle\!\rangle\!\!-\!\!CH_2CH_3$$

hot $MnO_4{}^-$ → $\langle\!\langle \rangle\!\rangle\!\!-\!\!CO_2H$

$X_2, h\nu$ → $\langle\!\langle \rangle\!\rangle\!\!-\!\!\overset{X}{\underset{|}{C}}HCH_3$

Phenols are more acidic than alcohols and contain a ring highly activated toward electrophilic aromatic substitution. They may be esterified with an acid anhydride. **Hydroquinones** (1,2- and 1,4-dihydroxybenzenes) undergo reversible oxidation to **quinones**.

Arylamines are converted to **aryldiazonium salts** by reaction with HONO. These salts are stable at 0°, but are highly reactive toward a variety of nucleophiles (see Figure 10.11, page 490).

$$ArNH_2 \xrightarrow[\substack{HCl \\ 0°}]{NaNO_2} ArN_2{}^+ \; Cl^- \xrightarrow{Nu^-} ArNu + N_2 + Cl^-$$

Halobenzenes do not undergo S_N1 or S_N2 reactions, but X^- can be displaced in **nucleophilic aromatic substitution reactions**, especially if the ring is activated by electron-withdrawing groups such as NO_2.

$$O_2N\!-\!\langle\!\langle \delta+ \rangle\!\rangle\!\!-\!\!Cl + Nu^- \longrightarrow O_2N\!-\!\langle\!\langle \rangle\!\rangle\!\!-\!\!Nu + Cl^-$$

(with NO_2 substituents on each ring)

STUDY PROBLEMS

10.21. Draw the structure of each of the following compounds: (a) *o*-dideuteriobenzene;
(b) 1,3,5-trichlorobenzene;   (c) *m*-bromotoluene;   (d) *p*-bromonitrobenzene;
(e) 4-bromo-2,3-dinitrotoluene;   (f) *m*-chlorobenzoic acid;   (g) isopropylbenzene.

10.22. Name each of the following compounds:

(a) $C_6H_5CH=CH_2$ (b) 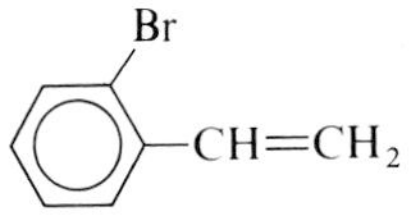 [structure: benzene ring with Br and $-CH=CH_2$]   (c) 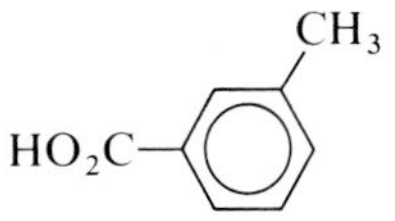 HO_2C-[structure: benzene ring with CH_3]

(d) 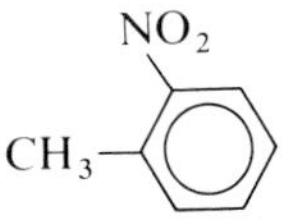 CH_3-[structure: benzene ring with NO_2]   (e) 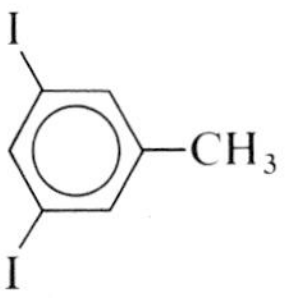 [structure: benzene ring with two I and $-CH_3$]   (f) $C_6H_5CH=CHCH=CHC_6H_5$

10.23. Give the structures and the names of all the isomeric: (a) monobromoanilines;
(b) monochlorophenols;   (c) dinitrotoluenes.

10.24. Which of the following compounds could be aromatic? (*Hint*: In many cases, the Hückel
rule may be extrapolated to structures with more than one ring.)

(a) 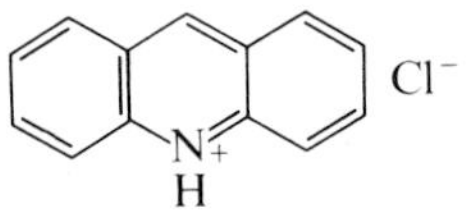 [structure: acridinium, N^+-H] Cl^- (b) 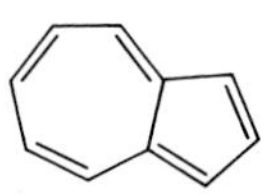 [structure: azulene]   (c)  [structure: cyclooctatetraene dianion]

azulene

cyclooctatetraene
dianion

10.25. The heat of hydrogenation of benzene is -49.8 kcal/mole, while that of 1,4-cyclohexadiene
is -59.3 kcal/mole. Calculate ΔH for the conversion of benzene to 1,4-cyclohexadiene.

10.26. From the following heats of hydrogenation, calculate the energy of the added stabilization
by conjugation of the double bond in styrene ($C_6H_5CH=CH_2$): styrene, -76.9 kcal/mole;
benzene, -49.8 kcal/mole; propene, -28.6 kcal/mole.

10.27. Naphthalene undergoes electrophilic aromatic substitution at the α-position:

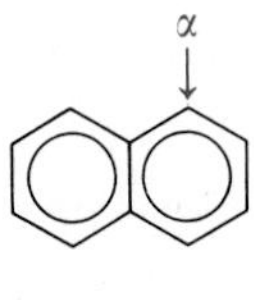

[structure: naphthalene with α-position indicated]

Predict the organic products of the monosubstitution reactions of naphthalene with each of
the following sets of reagents:

(a) Br_2, $FeBr_3$ (b) HNO_3, H_2SO_4
(c) $ClCH_2CH_3$, $AlCl_3$ (d) $CH_2=CH_2$, $AlCl_3$, HCl

(e) $CH_3\overset{O}{\underset{||}{C}}Cl$, $AlCl_3$

10.28. Nitrobenzene is sometimes used as a solvent for Friedel–Crafts alkylations. Why does the reaction of nitrobenzene not interfere with the desired reaction?

10.29. Friedel–Crafts acylation reactions may be accomplished with acid anhydrides as well as with acid halides:

$$\text{benzene} + CH_3\overset{O}{\overset{||}{C}}-O\overset{O}{\overset{||}{C}}CH_3 \xrightarrow{AlCl_3} \text{benzene}-\overset{O}{\overset{||}{C}}CH_3 + HO\overset{O}{\overset{||}{C}}CH_3$$

an acid anhydride

Predict the product of the AlCl$_3$-catalyzed reaction of benzene with:

(a)   (b)

10.30. Which one of each of the following pairs is more reactive toward aromatic bromination?

(a)   acetanilide ($C_6H_5NH\overset{O}{\overset{||}{C}}CH_3$) or benzene
(b)   bromobenzene or toluene
(c)   *p*-xylene (*p*-dimethylbenzene) or *p*-toluic acid (*p*-methylbenzoic acid)
(d)   *m*-dinitrobenzene or *m*-nitrotoluene
(e)   chlorobenzene or *m*-dichlorobenzene

10.31. Label each of the following groups as an *o,p*-director or as a *m*-director:

(a) $-N\langle$ (b) $\overset{+}{N}H-$ (c) $-OC_6H_5$

(d) $-\overset{O}{\overset{||}{C}}NHCH_3$ (e) $-\overset{O}{\overset{||}{C}}C_6H_5$ (f)

(g) $-O\overset{O}{\overset{||}{C}}CH_3$ (h) $-\overset{O}{\overset{||}{C}}OCH_3$ (i)

10.32. In acidic solution, nitrous acid (HONO) reacts with some aromatic compounds in a manner similar to nitric acid.

(a)   Write an equation showing the formation of the electrophile.
(b)   Write the equation showing the electrophilic substitution of phenol by this species. (Include the structure of the intermediate.)

10.33. Predict the major organic products :

(a)   ethylbenzene + Cl$_2$ $\xrightarrow{FeCl_3}$

(b)   ethylbenzene + Br$_2$ $\xrightarrow{hv}$

(c)   styrene + MnO$_4^-$ $\xrightarrow{heat}$

(d)   toluene + 1-chloropropane $\xrightarrow{\text{AlCl}_3}$

(e)   toluene + fuming H_2SO_4 $\longrightarrow$

(f)   styrene + excess H_2 $\xrightarrow[200°, 100 \text{ atm}]{\text{Ni}}$

(g)   1-propylbenzene + cyclohexene $\xrightarrow[\text{(a Lewis acid)}]{\text{HF}}$

10.34. Predict the major organic products:

(a) $C_6H_5CH_2CH_2CH_2\overset{\overset{\displaystyle O}{\|}}{C}Cl$ $\xrightarrow{\text{AlCl}_3}$

(b)   toluene + 1-chloro-2-methylpropane $\xrightarrow{\text{AlCl}_3}$

(c)   benzene + 1-chloro-2-butene $\xrightarrow{\text{AlCl}_3}$

10.35. Explain the following observations: (a) Treatment of benzene with an excess of fuming nitric acid and sulfuric acid yields *m*-dinitrobenzene, but not the trinitrated product. (b) Under the same conditions, toluene yields 2,4,6-trinitrotoluene (TNT).

10.36. Ring monobromination of a diethylbenzene yields three isomeric bromodiethylbenzenes (two formed in minor amounts). Give the structures of the diethylbenzene and the brominated products.

10.37. Predict the major products of aromatic monochlorination of: (a) chlorobenzene; (b) *o*-dichlorobenzene;   (c) *m*-bromochlorobenzene;   (d) *m*-xylene;   (e) acetophenone (see Table 10.1);   (f) (trichloromethyl)benzene ($C_6H_5CCl_3$).

10.38. Predict the major products of aromatic mononitration of the following compounds:

(a)

(b)

(c)

(d)

(e)

(f)

10.39. How would you synthesize each of the following compounds from benzene (assuming that you can separate *o*-, *m*-, and *p*-isomers)?

(a)   anisole ($C_6H_5OCH_3$)
(b)   terephthalic acid (*p*-$HO_2C—C_6H_4—CO_2H$)
(c)   *m*-dibromobenzene
(d)   *m*-nitrophenol
(e)   2-phenyl-2-propanol
(f)   1,1-diphenylethane
(g)   benzyl methyl ether

10.40. Suggest syntheses for the three nitrobenzoic acids from toluene.

10.41. What ions are formed when triphenylmethanol is treated with concentrated H_2SO_4? Suggest a reason for the formation of these ions.

10.42. Predict the major organic products:

    (a) *p*-ethylaniline + $NaNO_2$ + HCl $\xrightarrow{0°}$

    (b) *β*-naphthylamine (page 712) + $NaNO_2$ + HCl $\xrightarrow{0°}$

    (c) [the product from (a)] + CuCN $\xrightarrow[\text{KCN}]{100°}$

    (d) [the product from (b)] + phenol $\xrightarrow{100°}$

10.43. Complete the following equations:

    (a) O_2N-⟨◯⟩$-Cl + NH_3 \xrightarrow{\text{heat}}$

    (b) O_2N-⟨◯⟩$-Cl + {}^-OCH_3 \xrightarrow{\text{heat}}$ (with NO_2 substituent)

    (c) O_2N-⟨◯⟩$-Cl + H_2NNH_2 \xrightarrow{\text{heat}}$ (with NO_2 substituent)

10.44. Predict the major organic product of the reaction of benzene with ICl (with iron filings added as catalyst).

10.45. (a) Suggest a reason for the fact that aniline (with a pK_b of 9.4) is about one-millionth as basic as cyclohexylamine, $C_6H_{11}NH_2$ ($pK_b = 3.3$).

     (b) Would you expect *p*-nitroaniline to be more or less basic than aniline? Why?

10.46. (a) The herbicide (2,4,5-trichlorophenoxy)acetic acid (called 2,4,5-T) is synthesized by heating a tetrachlorinated benzene with methanolic NaOH, followed by reaction with $ClCH_2CO_2Na$ and acidification. Write equations for these reactions.

     (b) A by-product of this synthesis is 2,3,7,8-tetrachlorodibenzodioxin (often called 2,3,7,8-TCDD, or simply *dioxin*), which is highly toxic and not readily biodegradable. Suggest a mechanism for the formation of 2,3,7,8-TCDD.

                 2,4,5-T   2,3,7,8-TCDD

10.47. 2,4-Dinitrofluorobenzene is used in structure determinations of polypeptides (small proteins). This compound undergoes reaction with the free amino group at one end of a peptide chain. Write the equation for its reaction with glycylglycine.

$$H_2NCH_2\overset{\overset{\textstyle O}{\|}}{C}NHCH_2CO_2H$$

                glycylglycine

10.48. Suggest syntheses for the following compounds, starting with benzene, toluene, or phenol and other appropriate reagents. (Assume that you can separate *o-*, *m*, and *p*-isomers.)

(a) 2,4,6-trichlorophenyl acetate

(b) bromodiphenylmethane

(c) *p*-nitroacetophenone

(d) 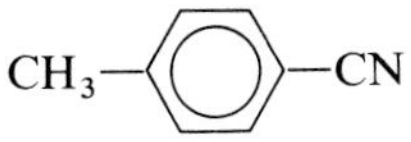 CH_3—⟨benzene ring⟩—CN

(e) *p*-fluorotoluene

(f) 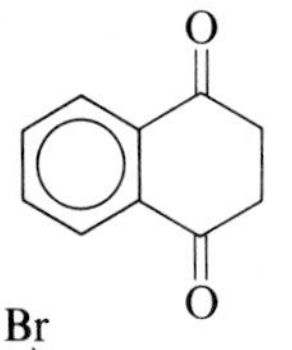 ⟨bicyclic structure with two ketone oxygens⟩

(g) 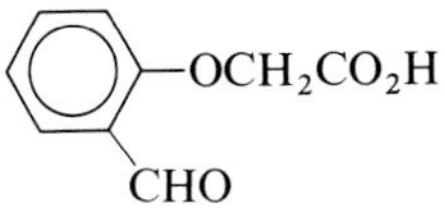 ⟨benzene ring⟩—OCH_2CO_2H with CHO

(h) 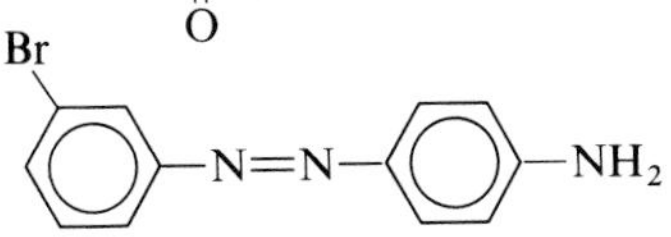 ⟨benzene ring with Br⟩—N=N—⟨benzene ring⟩—NH_2

(i) 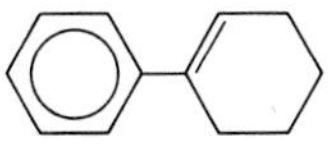 ⟨phenyl-cyclohexene structure⟩

10.49. Suggest a structure for a compound, C_7H_6BrCl, that yields *m*-chlorobenzoic acid upon oxidation.

10.50. There are three dibromobenzenes, which melt at 87°, 6°, and −7°, respectively. The dibromo isomer that melts at 87° yields only one mononitrodibromobenzene. The dibromo isomer that melts at 6° yields two mononitrodibromobenzenes. The dibromo isomer that melts at −7° yields three mononitrodibromobenzenes. Using these data, assign the structures of the three dibromobenzenes.

10.51. Michler's ketone is used as an intermediate in the dye industry. It is synthesized from phosgene ($Cl_2C=O$) and an excess of *N,N*-dimethylaniline with a $ZnCl_2$ catalyst. Suggest a structure for Michler's ketone.

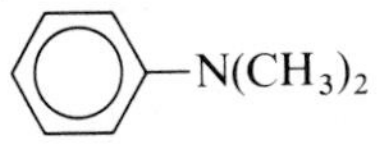

⟨benzene ring⟩—$N(CH_3)_2$

N,N-dimethylaniline

10.52. When bromobenzene is chlorinated, two isomeric compounds (A and B, C_6H_4BrCl) can be isolated. Bromination of A gives a number of isomeric products of the composition $C_6H_3Br_2Cl$, while bromination of B yields two isomers (C and D) of the composition $C_6H_3Br_2Cl$. Compound C is identical with one of the compounds obtained from the bromination of A; however, D is different from any of the isomeric compounds obtained from the bromination of A. Give the structures of A, B, C, and D.

10.53. The following reaction is called a **Claisen rearrangement**. Suggest a mechanism for this reaction.

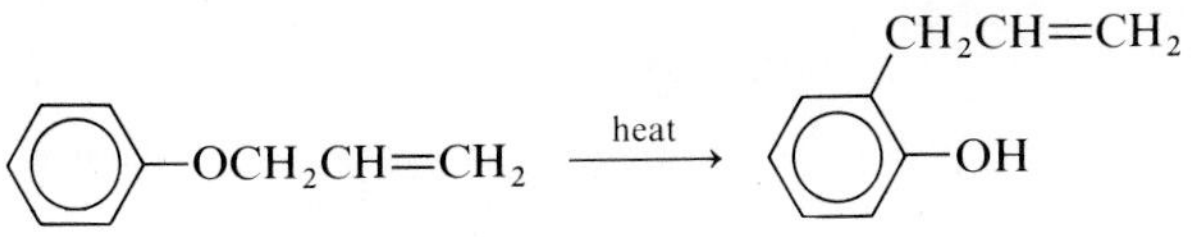

⟨benzene ring⟩—$OCH_2CH=CH_2$ $\xrightarrow{\text{heat}}$ ⟨benzene ring with $CH_2CH=CH_2$⟩—OH

73%

10.54. Suggest mechanisms for the following reactions:

(a) benzene + $\overset{O}{\overset{\|}{FCH}}$ $\xrightarrow{BF_3}$ benzaldehyde $\left(\overset{O}{\overset{\|}{C_6H_5CH}}\right)$

(b) *t*-butylbenzene + Br_2 $\xrightarrow{AlBr_3}$ bromobenzene + propene

(c)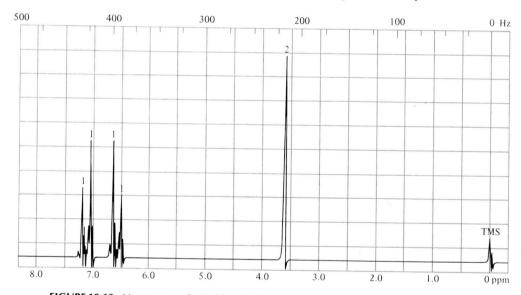

10.55. Reaction of *o*-chlorotoluene with KNH_2 in liquid NH_3 yields a mixture of *o*- and *m*-$CH_3C_6H_4NH_2$. The *p*-isomer is not observed. A similar reaction with *p*-chlorotoluene gives *m*- and *p*-$CH_3C_6H_4NH_2$, but not the *o*-isomer. Finally, *m*-chlorotoluene gives all three isomers. Write equations for these reactions (with intermediates).

10.56. A compound C_6H_5Cl was treated with HNO_3 and H_2SO_4. The products of this reaction were then treated with Fe and HCl, followed by neutralization. The nmr spectrum of one of the products of this reaction is given in Figure 10.13. What is the structure of this component? What would be the structures of the other components of the product mixture?

FIGURE 10.13. Nmr spectrum for Problem 10.56.

10.57. A compound with the formula $C_{15}H_{14}O$ gives the nmr spectrum shown in Figure 10.14. Its infrared spectrum shows strong absorption at about 1750 cm^{-1} (5.75 μm). What is the structure of the compound?

10.58. One of the isomeric cresols (methylphenols) was treated with NaOH and iodomethane. The nmr spectrum of the product is shown in Figure 10.15. What are the structures of the cresol and the product?

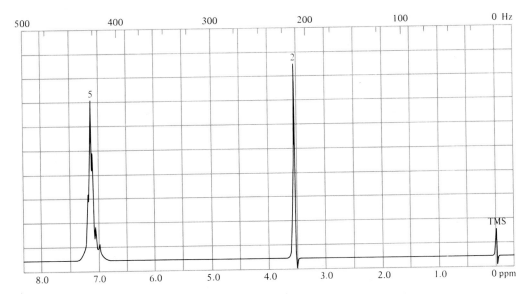

FIGURE 10.14. Nmr spectrum for Problem 10.57.

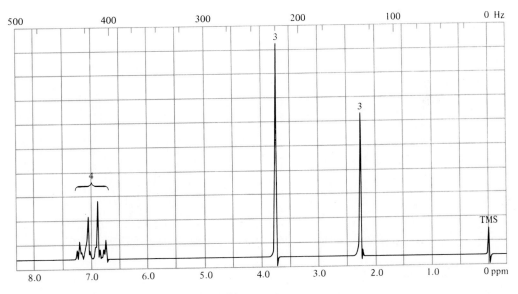

FIGURE 10.15. Nmr spectrum for Problem 10.58.

10.59. Upon treatment with HNO_3, a monosubstituted aromatic compound yielded two isomeric products, A and B. Treatment of A with NaOH followed by CH_3I gave C. In an identical manner, compound B yielded compound D. The infrared spectra of A, B, C, and D are given in Figure 10.16. Reaction of C with Fe and HCl, followed by treatment with base, yielded E. Following this same procedure, compound D yielded compound F. The nmr spectra of E and F are given in Figure 10.17. What are the structures of the original compound, A, B, C, D, E, and F?

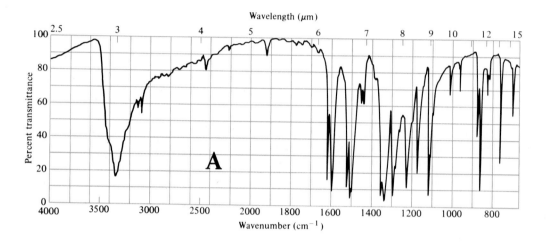

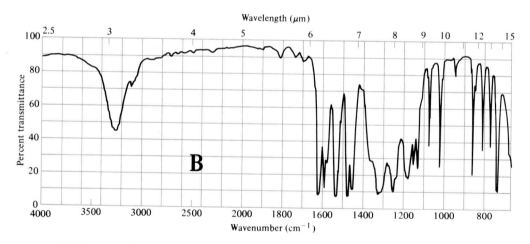

FIGURE 10.16. Infrared spectra for Problem 10.59.

(continued)

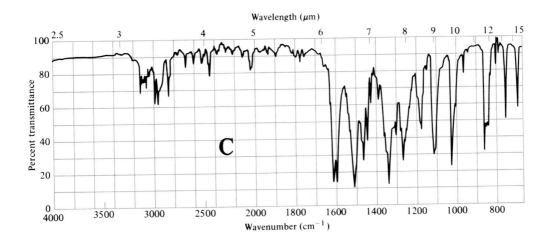

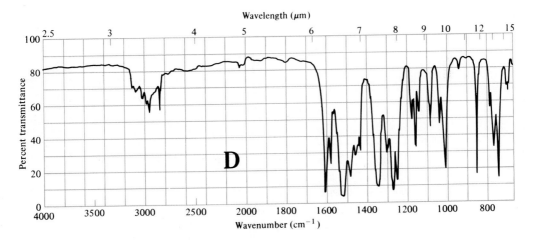

FIGURE 10.16 (continued). Infrared spectra for Problem 10.59.

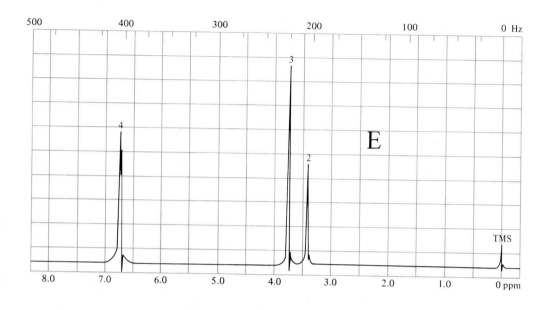

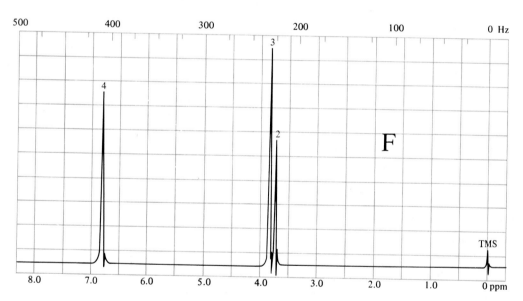

FIGURE 10.17. Nmr spectra for Problem 10.59.

CHAPTER 11

Aldehydes and Ketones

Aldehydes and ketones are but two of many classes of organic compounds that contain carbonyl groups. A **ketone** has *two alkyl (or aryl) groups attached to the carbonyl carbon*, while an **aldehyde** has *one alkyl (or aryl) group and one hydrogen* attached to the carbonyl carbon.

$$\underset{\text{an aldehyde}}{R-\overset{\displaystyle O}{\overset{\|}{C}}-H \quad \text{or} \quad RCHO} \qquad \underset{\text{a ketone}}{R-\overset{\displaystyle O}{\overset{\|}{C}}-R \quad \text{or} \quad RCOR}$$

Other carbonyl compounds, such as carboxylic acids or esters, have electronegative groups connected to the carbonyl carbon; consequently, their chemistry is somewhat different from that of aldehydes and ketones. These other carbonyl compounds will be covered in later chapters.

SECTION 11.1.

Nomenclature of Aldehydes and Ketones

In the IUPAC system, the name of an aldehyde is derived from the name of the parent alkane by changing the final **-e** to **-al**. No number is needed; the —CHO group always contains carbon 1.

$$\underset{\text{ethanal}}{\text{CH}_3\overset{\displaystyle O}{\overset{\|}{\text{CH}}}} \qquad \underset{\text{2-chloropropanal}}{\text{CH}_3\underset{\displaystyle \text{Cl}}{\text{CH}}\overset{\displaystyle O}{\overset{\|}{\text{CH}}}} \qquad \underset{\text{2-butenal}}{\text{CH}_3\text{CH}=\text{CH}\overset{\displaystyle O}{\overset{\|}{\text{CH}}}}$$

IUPAC: ethanal   2-chloropropanal   2-butenal

Ketones are named by changing the -e of the alkane name to -one. A number is used where necessary:

| IUPAC: | cyclohexanone | 2-pentanone | 2,4-pentanedione |

Trivial names for the common aldehydes and ketones are widely used. Aldehydes are named after the parent carboxylic acids with the -*oic acid* or -*ic acid* ending changed to -*aldehyde*. Table 11.1 lists a few examples.

Propanone is usually called *acetone*, while the other simple ketones are sometimes named by a functional-group name. The alkyl or aryl groups attached to the carbonyl group are named, then the word *ketone* is added:

$$CH_3CCH_3 \qquad CH_3CCH_2CH_3 \qquad (CH_3)_2CHCC(CH_3)_3$$

IUPAC:	propanone	butanone	2,2,4-trimethyl-3-pentanone
trivial:	acetone	methyl ethyl ketone	isopropyl *t*-butyl ketone

Other positions in a molecule in relation to the carbonyl group may be referred to by Greek letters. The carbon adjacent to the C=O is called the **alpha** (α) **carbon.** The next carbon is **beta** (β), then **gamma** (γ), **delta** (δ), and so forth, according to the Greek alphabet. Occasionally, **omega** (ω), the last letter in the Greek alphabet, is used to designate the terminal carbon of a long chain, regardless of the actual length of the chain. The groups (or atoms) attached to an α carbon are called α groups; those attached to the β carbon are called β groups.

TABLE 11.1. Trivial names for some carboxylic acids and aldehydes

Carboxylic acid		Aldehyde	
HCOH	formic acid	HCH	formaldehyde
CH_3COH	acetic acid	CH_3CH	acetaldehyde
CH_3CH_2COH	propionic acid	CH_3CH_2CH	propionaldehyde
$CH_3CH_2CH_2COH$	butyric acid	$CH_3CH_2CH_2CH$	butyraldehyde
—COH	benzoic acid	—CH	benzaldehyde

$$\underset{\substack{\nearrow \quad \nwarrow \\ \alpha\ carbon \\ \beta\ carbon}}{CH_3CH_2CH_2CH}{\overset{\overset{\displaystyle O}{\|}}{}}$$

$$\underset{a\ \beta\text{-}diketone}{CH_3\overset{\overset{\displaystyle O}{\|}}{C}CH_2\overset{\overset{\displaystyle O}{\|}}{C}CH_3}$$

$$\underset{an\ \alpha\text{-}bromoaldehyde}{CH_3\overset{\overset{\displaystyle Br}{|}}{CH}CHO}$$

$$BrCH_2CH_2CH_2CH_2\overset{\overset{\displaystyle O}{\|}}{C}C_6H_5 \qquad Br(CH_2)_{10}\overset{\overset{\displaystyle O}{\|}}{C}CH_3 \qquad Br(CH_2)_{25}\overset{\overset{\displaystyle O}{\|}}{C}CH_2CH_3$$

all are ω-bromoketones

The Greek-letter designations may be used in the trivial names of carbonyl compounds, but *not in the IUPAC names* because this latter practice would be mixing two systems of nomenclature.

IUPAC:	3-phenylpropanal	2-bromopropanal
trivial:	β-phenylpropionaldehyde	α-bromopropionaldehyde

SECTION 11.2.

Preparation of Aldehydes and Ketones

In the laboratory, the most common way of synthesizing a simple aldehyde or ketone is by the *oxidation of an alcohol*. Aryl ketones can be prepared by *Friedel–Crafts acylation reactions*. General equations for these reactions are shown in Table 11.2.

$$\underset{\text{1-heptanol}}{CH_3(CH_2)_5CH_2OH} + CrO_3\cdot2N\bigcirc \longrightarrow \underset{\substack{\text{heptanal}\\75\%}}{CH_3(CH_2)_5\overset{\overset{\displaystyle O}{\|}}{C}H}$$

$$\underset{\text{menthol}}{} + H_2CrO_4 \xrightarrow{60°} \underset{\substack{\text{menthone}\\85\%}}{}$$

$$\underset{p\text{-isopropyltoluene}}{(CH_3)_2CH-\bigcirc-CH_3} + \underset{\text{acetyl chloride}}{CH_3\overset{\overset{\displaystyle O}{\|}}{C}Cl} \xrightarrow{AlCl_3} \underset{\substack{\text{5-isopropyl-2-methylacetophenone}\\55\%}}{(CH_3)_2CH-\bigcirc-CH_3}$$

TABLE 11.2. Summary of laboratory syntheses of aldehydes and ketones[a]

Reaction	Section reference
aldehydes:	
$RCH_2OH \xrightarrow[\text{pyridine}]{CrO_3} RCHO$ *1° alcohol*	7.13C
$\underset{\text{acid chloride}}{\overset{\overset{\text{O}}{\|}}{RCCl}} \xrightarrow[(2)\ H_2O]{(1)\ LiAlH[OC(CH_3)_3]_3} RCHO$	13.3C
ketones:	
$R_2CHOH \xrightarrow{H_2CrO_4} R_2C{=}O$ *2° alcohol*	7.13C
$\underset{\text{acid chloride}}{\overset{\overset{\text{O}}{\|}}{RCCl}} \xrightarrow{R'_2Cd} \overset{\overset{\text{O}}{\|}}{RCR'}$	13.3C
aryl ketones:	
	10.9E

[a] Polyfunctional aldehydes and ketones can be prepared by condensation and alkylation reactions, which will be discussed in Chapter 14.

One of the more important aldehydes, *formaldehyde*, which is used as a reagent, as an olefactory desensitizer in wick deodorizers, and as a preservative for biological specimens, is a gas. However, it is conveniently shipped or stored in water solution (formalin = 37% formaldehyde and 7–15% methanol in H_2O) or as a solid polymer or trimer. Heating any one of these preparations yields the gaseous formaldehyde.

Acetaldehyde, with a boiling point near room temperature (20°), is also conveniently stored or shipped in a cyclic trimer or tetramer form. Acetaldehyde is used as an intermediate in industrial syntheses of acetic acid, acetic anhydride (Section 13.4A), and other compounds.

$$CH_3CH \xrightarrow{\text{heat}} \begin{array}{c} O \\ \parallel \\ CH_3CH \end{array}$$

paraldehyde

a sedative and hypnotic
bp 125°C

metaldehyde

used as snail bait
mp 246°C

ethanal
(acetaldehyde)

SECTION 11.3.

The Carbonyl Group

The carbonyl group consists of an sp^2 carbon atom joined to an oxygen atom by a sigma bond and by a pi bond. (See Figure 2.21, page 64, for the orbital picture.) The sigma bonds of the carbonyl group lie in a *plane* with bond angles of approximately 120° around the sp^2 carbon. The pi bond joining the C and the O lies above and below the plane of these sigma bonds. The carbonyl group is *polar*, the electrons in the sigma bond, and especially those in the pi bond, being drawn toward the electronegative oxygen. The oxygen of the carbonyl group has *two pair of unshared valence electrons*. All these structural features—the flatness, the pi bond, the polarity, and the unshared electrons—contribute to the reactivity of the carbonyl group.

unshared electrons

$\mu = 2.85$ D

Isolated carbon–carbon double bonds are nonpolar. For reaction, an electrophile is generally needed to attack the pi-bond electrons. However, the carbon–oxygen double bond is polar even without electrophilic attack. A carbonyl compound may be attacked either by a nucleophile or by an electrophile.

Many reactions of carbonyl groups involve an initial protonation of the oxygen. This protonation enhances the positive charge of the carbonyl carbon so that this carbon is more easily attacked by weaker nucleophiles.

*resonance structures for
a protonated carbonyl group*

SECTION 11.4.

Physical Properties of Aldehydes and Ketones

The unique features of the carbonyl group influence the physical properties of the aldehydes and ketones. Because they are polar, and therefore undergo intermolecular dipole–dipole attractions, aldehydes and ketones have higher boiling points than nonpolar compounds of similar molecular weight (see Table 11.3). To a limited extent, aldehydes and ketones can solvate ions (for example, NaI is soluble in acetone).

$$
\begin{array}{ccc}
\overset{\displaystyle CH_3}{\underset{|}{\vphantom{x}}} & \overset{\displaystyle O}{\underset{\parallel}{\vphantom{x}}} & \overset{\displaystyle OH}{\underset{|}{\vphantom{x}}} \\
CH_3CHCH_3 & CH_3CCH_3 & CH_3CHCH_3 \\
bp\ -12° & bp\ 56° & bp\ 82.5°
\end{array}
$$

Because of the unshared electrons on the oxygen, a carbonyl compound can undergo hydrogen bonding (but not with another carbonyl compound unless it has an acidic hydrogen available for hydrogen bonding).

$$
\begin{array}{cc}
:\!O\!:\!-\!-\!-H\!-\!O & \\
\parallel\quad\quad\ | & \\
R\!-\!C\!-\!R\quad\ H &
\end{array}
$$

The result of this ability to form hydrogen bonds is that the aldehydes and ketones of low molecular weight, like alcohols, are soluble in water (Table 11.3).

TABLE 11.3. Physical properties of some aldehydes and ketones

Trivial name	Structure	Bp, °C	Solubility in H_2O
aldehydes:			
formaldehyde	HCHO	−21	∞
acetaldehyde	CH_3CHO	20	∞
propionaldehyde	CH_3CH_2CHO	49	16 g/100 mL
butyraldehyde	$CH_3CH_2CH_2CHO$	76	7 g/100 mL
benzaldehyde	C_6H_5CHO	178	slightly
ketones:			
acetone	$CH_3\overset{O}{\overset{\parallel}{C}}CH_3$	56	∞
methyl ethyl ketone	$CH_3\overset{O}{\overset{\parallel}{C}}CH_2CH_3$	80	26 g/100 mL
acetophenone	$C_6H_5\overset{O}{\overset{\parallel}{C}}CH_3$	202	insoluble
benzophenone	$C_6H_5\overset{O}{\overset{\parallel}{C}}C_6H_5$	306	insoluble

However, since they cannot undergo hydrogen bonding with themselves, the boiling points are substantially lower than those of corresponding alcohols.

SECTION 11.5.

Spectral Properties of Aldehydes and Ketones

A.   Infrared spectra

The infrared spectrum is useful in the detection of a carbonyl group in a ketone or an aldehyde. (Characteristic absorption bands are listed in Table 11.4.) However, carbonyl groups are also found in other compounds (carboxylic acids, esters, and so forth). For this reason, the fact that a carbonyl group is present does not mean that an unknown is necessarily an aldehyde or a ketone.

TABLE 11.4. Characteristic infrared absorption of aldehydes and ketones

	Position of absorption	
Type of vibration	*cm^{-1}*	*μm*
aldehydes:		
C—H stretching of —CHO	2700–2900	3.45–3.7
C=O stretching	1700–1740	5.7–5.9
ketones:		
C=O stretching	1660–1750	5.7–6.0

For aldehydes, corroborating evidence may be found in both the infrared and nmr spectra because of the unique absorption of the aldehyde hydrogen. Ketones, unfortunately, cannot be positively identified by spectral methods. The usual procedure is to eliminate the other carbonyl compounds as possibilities. If a carbonyl compound is not an aldehyde, carboxylic acid, ester, amide, etc., it is probably a ketone.

The C=O absorption of both aldehydes and ketones appears somewhere near 1700 cm^{-1} (about 5.8 μm). If the carbonyl group is in conjugation with a double bond or benzene ring, the position of absorption is shifted to a slightly lower frequency (about 1675 cm^{-1}, or 6 μm, for ketones). Figure 11.1 shows the infrared spectra of cyclohexanone (nonconjugated) and 2-cyclohexen-1-one (conjugated).

The CH stretching of the aldehyde group, which shows absorption just to the right of aliphatic CH absorption, is characteristic of an aldehyde. Usually two peaks are found in this region. The two CH peaks of the aldehyde are clearly evident in the spectrum of butanal in Figure 11.2, but the peak closer to the aliphatic CH absorption is often obscured by that absorption.

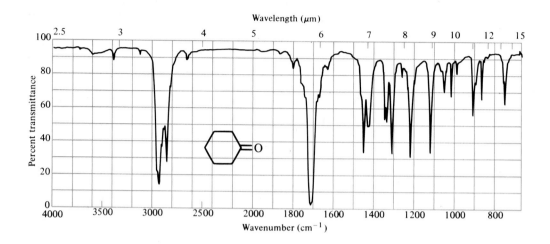

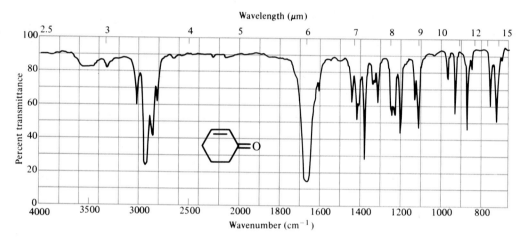

FIGURE 11.1. Infrared spectra of cyclohexanone and 2-cyclohexen-1-one, illustrating the slight shift of C=O absorption to lower frequency (longer wavelength) by conjugation.

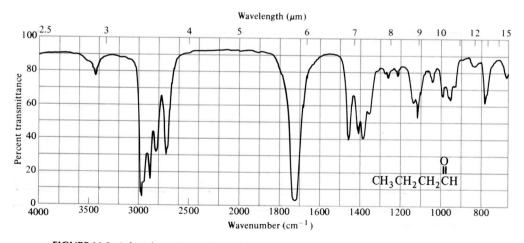

FIGURE 11.2. Infrared spectrum of butanal.

STUDY PROBLEM

11.1. Match the infrared spectrum in Figure 11.3 to one of the following structures:

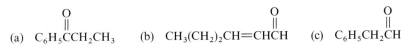

(a) $C_6H_5\overset{\displaystyle O}{\overset{\displaystyle \|}{C}}CH_2CH_3$ (b) $CH_3(CH_2)_2CH{=}CHCH\overset{\displaystyle O}{\overset{\displaystyle \|}{}}$ (c) $C_6H_5CH_2C\overset{\displaystyle O}{\overset{\displaystyle \|}{}}H$

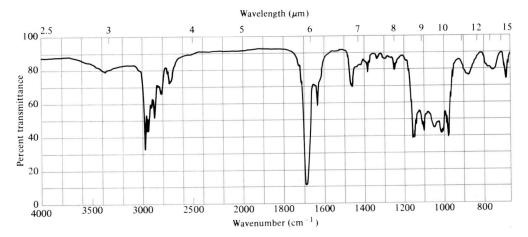

FIGURE 11.3. Infrared spectrum for Problem 11.1.

B. Nmr spectra

The electrons in a carbonyl group, like those in a double bond or an aromatic pi cloud, are set in motion by an external magnetic field. The resulting induced molecular field has a profound effect upon the nmr absorption of the aldehyde proton. You will recall from Section 8.7B that nmr absorption for an aldehyde proton is shifted far downfield ($\delta = 9$–10 ppm, offset from the usual spectral range). This large shift arises from the additive effects of both anisotropic deshielding by the pi electrons and inductive deshielding by the electropositive carbon of the carbonyl group.

   The α hydrogens of either aldehydes or ketones are not affected to such a large extent by the carbonyl group. The nmr absorption for the α protons ($\delta = 2.1$–2.6 ppm) appears slightly downfield from that of other CH absorption (about 1.5 ppm) because of electron withdrawal by the electronegative oxygen atom. The effects of this deshielding by the inductive effect are evident in the nmr spectra of butanal (Figure 11.4) and 1-phenyl-2-propanone (Figure 11.5). In an aldehyde, the splitting of the aldehyde proton may sometimes be used to determine the number of α hydrogens. The spectrum of butanal shows a triplet for the —CHO proton, an indication of two α hydrogens.

STUDY PROBLEM

11.2. A compound with the molecular formula C_7H_6O has the nmr spectrum shown in Figure 11.6. What is the structure of this compound?

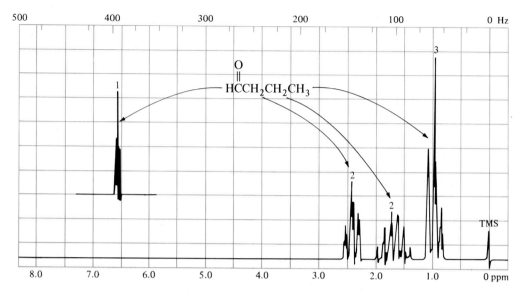

FIGURE 11.4. Nmr spectrum of butanal, showing the relative chemical shifts of α, β, and γ protons and the aldehyde proton

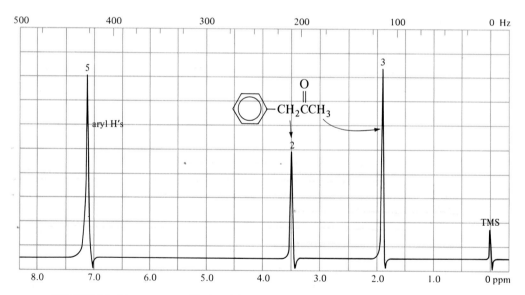

FIGURE 11.5. Nmr spectrum of 1-phenyl-2-propanone.

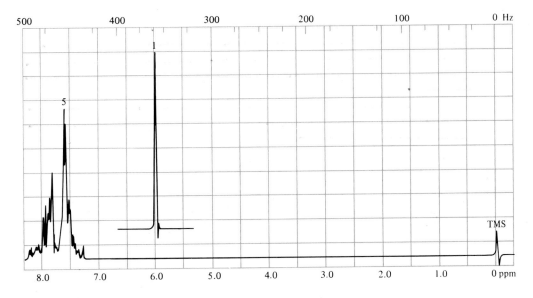

FIGURE 11.6. Nmr spectrum for Problem 11.2.

SECTION 11.6.

Addition of Reagents to the Carbonyl Group

The focal point of reactivity in aldehydes and ketones is the pi bond of the carbonyl group. Like alkenes, aldehydes and ketones undergo addition of reagents to the pi bond.

✳   *General:*   R—C—R + H—Nu $\rightleftharpoons$ R—C—R

(with O double-bonded to C on the left, OH and Nu on the right)

$$CH_3-\overset{\overset{\delta-}{O}}{\underset{\delta+}{\overset{\|}{C}}}-CH_3 + H-OH \underset{}{\overset{H^+}{\rightleftharpoons}} CH_3-\overset{OH}{\underset{OH}{\overset{|}{C}}}-CH_3$$

a hydrate

$$CH_3-\overset{O}{\overset{\|}{C}}-CH_3 + H-OCH_3 \overset{H^+}{\rightleftharpoons} CH_3-\overset{OH}{\underset{OCH_3}{\overset{|}{C}}}-CH_3$$

a hemiketal

$$CH_3-\overset{O}{\overset{\|}{C}}-CH_3 + H-CN \rightleftharpoons CH_3-\overset{OH}{\underset{CN}{\overset{|}{C}}}-CH_3$$

a cyanohydrin

The relative reactivities of aldehydes and ketones in addition reactions may be attributed partly to the *amount of positive charge on the carbonyl carbon.* A greater positive charge means a higher reactivity. If this partial positive charge is dispersed throughout the molecule, then the carbonyl compound is more stable and less reactive.

The carbonyl group is stabilized by adjacent alkyl groups, which are electron-releasing. A ketone, with two R groups, is about 7 kcal/mole more stable than an aldehyde, with only one R group. Formaldehyde, with no alkyl groups, is the most reactive of the aldehydes and ketones.

$$
\begin{array}{ccc}
\overset{\displaystyle O}{\underset{\displaystyle R-C-R}{\|}} & \overset{\displaystyle O}{\underset{\displaystyle R-C-H}{\|}} & \overset{\displaystyle O}{\underset{\displaystyle H-C-H}{\|}} \\
\textit{a ketone} & \textit{an aldehyde} & \textit{formaldehyde}
\end{array}
$$

increasing reactivity →

SAMPLE PROBLEM

List the following aldehydes in terms of increasing reactivity: CH_3CHO, $ClCH_2CHO$, Cl_2CHCHO, Cl_3CCHO.

Solution: Cl is *electron-withdrawing.* The carbonyl carbon becomes *increasingly positive* and *increasingly reactive* as more Cl atoms are added to the α carbon. Therefore, the order of reactivity is as already shown, with CH_3CHO being the least reactive and Cl_3CCHO being the most reactive.

$$
CH_3 \rightarrow \overset{\delta+}{C}HO \qquad\qquad Cl_3C \leftarrow \overset{\delta+}{C}HO
$$
$$
\textit{stabilized} \qquad\qquad\quad \textit{destabilized}
$$

Steric factors also play a role in the relative reactivities of aldehydes and ketones. An addition reaction of the carbonyl group leads to an increase in steric hindrance around the carbonyl carbon.

$$
\underset{R\ \ R}{\overset{sp^2,\ less\ hindered}{\underset{\displaystyle \diagdown}{\overset{\displaystyle O}{\underset{\displaystyle C}{\|}}}}} + CH_3CH_2OH \underset{H^+}{\overset{}{\rightleftharpoons}} \underset{\displaystyle OCH_2CH_3}{\overset{\displaystyle OH}{\underset{\displaystyle R-C-R}{|}}} \quad sp^3,\ more\ hindered
$$

Bulky groups around the carbonyl group lead to more steric hindrance in the product (and in the transition state). The product is of higher energy because of steric repulsions. A more hindered ketone is therefore less reactive than an aldehyde or a less hindered ketone. The lack of steric hindrance is another reason that formaldehyde is more reactive than other aldehydes.

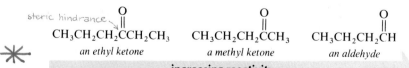

$$
\underset{\textit{an ethyl ketone}}{\overset{\textit{steric hindrance}\,\diagdown\,\overset{\displaystyle O}{\|}}{CH_3CH_2CH_2CCH_2CH_3}} \qquad \underset{\textit{a methyl ketone}}{\overset{\displaystyle O}{\overset{\|}{CH_3CH_2CH_2CCH_3}}} \qquad \underset{\textit{an aldehyde}}{\overset{\displaystyle O}{\overset{\|}{CH_3CH_2CH_2CH}}}
$$

increasing reactivity →

STUDY PROBLEM

11.3. An aldehyde or ketone undergoes a reversible acid-catalyzed reaction with ethanol, as shown in the preceding equation. A series of compounds is treated with ethanol and a trace of H_2SO_4, and the reaction mixtures are allowed to come to equilibrium. Of each of the following pairs of compounds, which compound yields the greater percentage of addition product in the equilibrium mixture?

   (a) 3-pentanone   or   cyclopentanone
   (b) 2-pentanone   or   3-pentanone
   (c) 2-pentanone   or   pentanal
   (d) 2-chlorocyclopentanone   or   2-methylcyclopentanone

SECTION 11.7.

Reaction with Water

Water can add to a carbonyl group to form a 1,1-diol, called a *gem*-diol, or **hydrate**. The reaction is reversible, and the equilibrium generally lies on the carbonyl side.

$$CH_3-\overset{\overset{\displaystyle O}{||}}{C}-CH_3 + H_2O \;\rightleftharpoons\; CH_3-\overset{\overset{\displaystyle OH}{|}}{\underset{\underset{\displaystyle OH}{|}}{C}}-CH_3$$

acetone

a hydrate

Stable hydrates are known, but they are the exception rather than the rule. *Chloral hydrate* (a hypnotic and chief ingredient of the "Mickey Finn") is an example of a stable hydrate. *Formalin* also contains a stable hydrate of formaldehyde.

$$Cl_3C\overset{\overset{\displaystyle O}{||}}{C}H + H_2O \;\rightleftharpoons\; Cl_3CCH(OH)_2$$

chloral   chloral hydrate

$$H\overset{\overset{\displaystyle O}{||}}{C}H + H_2O \;\rightleftharpoons\; HCH(OH)_2$$

formaldehyde   *in formalin*

Both formaldehyde and chloral are more reactive than most other aldehydes or ketones because the carbonyl carbon in each has a fairly large amount of positive charge. In formaldehyde, there are no alkyl groups to help disperse the positive charge. In chloral, the strongly electron-withdrawing Cl_3C- group enhances the positive charge by removing electron density.

    Let us compare the equilibrium constants for the hydration reactions of chloral (with a more positive carbonyl carbon) and acetone (with the positive charge more dispersed). The equilibrium constants differ by a factor of 10^6!

$$Cl_3CCH\ (O) + H_2O \rightleftharpoons Cl_3CCH(OH)_2 \qquad K = \dfrac{[Cl_3CCH(OH)_2]}{[Cl_3CCHO][H_2O]} = 3000$$

$$CH_3CCH_3\ (O) + H_2O \rightleftharpoons CH_3C(OH)_2CH_3 \qquad K = \dfrac{[CH_3C(OH)_2CH_3]}{[CH_3COCH_3][H_2O]} = 0.002$$

STUDY PROBLEM

11.4.   Which of the following compounds would you predict to form stable hydrates? Explain.

(a) Cl_3CCCCl_3 (with C=O)   (b)   (cyclohexanone ring)=O   (c) C_6H_5CHO

SECTION 11.8.

Reaction with Alcohols

Like water, an <u>alcohol can add to a carbonyl group</u>. In most cases, the <u>equilibrium lies on the aldehyde or ketone side of the equation</u>, just as in the reaction with water.

The product of addition of *one* molecule of an alcohol to an aldehyde is called a **hemiacetal**, while the product of addition of *two* molecules of alcohol (with the loss of H_2O) is called an **acetal**. (**Hemiketal** and **ketal** are the corresponding terms used for ketone products.) All these reactions are <u>catalyzed by a trace of strong acid.</u>

General:

$$RCH\ (O) \underset{H^+}{\overset{R'OH,}{\rightleftharpoons}} RCH(OR')(OH) \underset{H^+}{\overset{R'OH,}{\rightleftharpoons}} RCH(OR')(OR') + H_2O$$

an aldehyde   *a hemiacetal*   *an acetal*
                                (*OH and OR on C*)   (*two OR's on C*)

$$CH_3CH\ (O) \underset{H^+}{\overset{CH_3CH_2OH,}{\rightleftharpoons}} CH_3CH(OCH_2CH_3)(OH) \underset{H^+}{\overset{CH_3CH_2OH,}{\rightleftharpoons}} CH_3CH(OCH_2CH_3)(OCH_2CH_3) + H_2O$$

acetaldehyde   *a hemiacetal*   *an acetal*

SAMPLE PROBLEM

Give the structures of the organic compounds present in a methanol solution of cyclohexanone that contains a trace of HCl.

Solution:

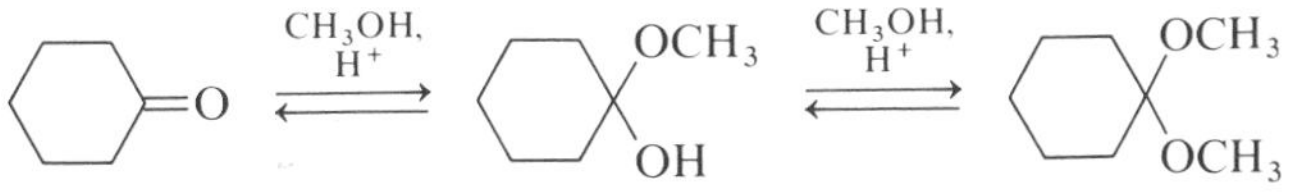

hemiketal   ketal

STUDY PROBLEMS

11.5. Which of the following structures contains a hemiacetal or hemiketal group, and which contains an acetal or ketal group? Circle and identify each group.

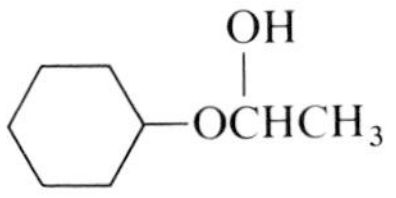

11.6. Give the structures of the alcohol and the aldehyde or ketone that are needed to prepare each of the compounds shown in the preceding problem.

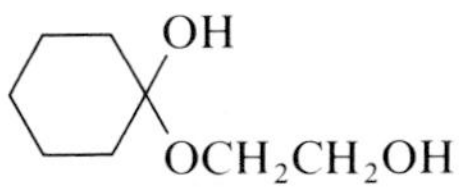

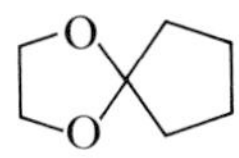

The mechanism for the reversible reactions of aldehydes or ketones with alcohols is typical of the mechanisms for many acid-catalyzed addition reactions of carbonyl compounds: a series of protonations and deprotonations of oxygen-containing groups.

Protonation:

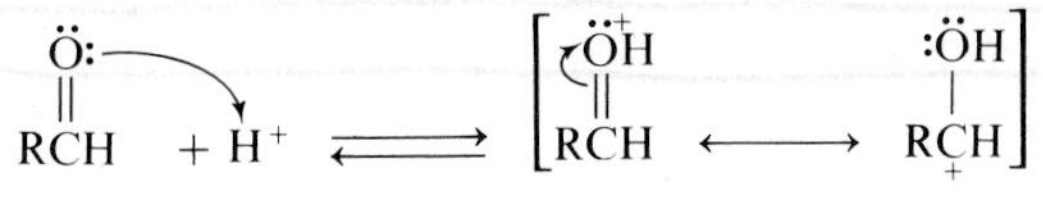

an aldehyde

resonance structures for a protonated aldehyde

Attack of R'OH:

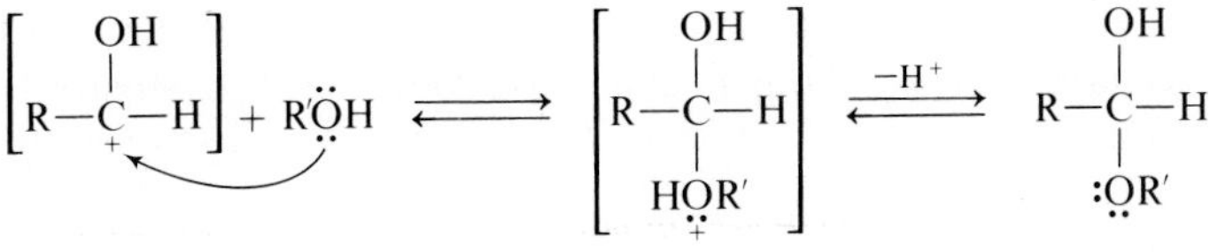

a protonated hemiacetal

a hemiacetal

In the mechanism for acetal formation from the hemiacetal, again protonation and deprotonation, along with loss of water, are the major reaction steps. Acetal formation from a hemiacetal is therefore a two-step *substitution* of an OR group for an OH group.

Protonation and loss of water:

$$R-\underset{\underset{OR'}{|}}{\overset{\overset{:\ddot{O}H}{|}}{C}}-H \quad \underset{H^+}{\rightleftarrows} \quad \left[R-\underset{\underset{OR}{|}}{\overset{\overset{\overset{+}{\ddot{O}H_2}}{|}}{C}}-H \right] \quad \underset{-H_2O}{\rightleftarrows} \quad \left[R-\underset{\underset{OR'}{|}}{\overset{+}{C}}-H \right]$$

a hemiacetal   *a carbocation*

Attack of R'OH:

$$\left[R-\underset{\underset{OR'}{|}}{\overset{+}{C}}-H \right] \quad \underset{R'OH}{\rightleftarrows} \quad \left[R-\underset{\underset{OR'}{|}}{\overset{\overset{\overset{+}{H\ddot{O}R'}}{|}}{C}}-H \right] \quad \underset{-H^+}{\rightleftarrows} \quad R-\underset{\underset{OR'}{|}}{\overset{\overset{:\ddot{O}R'}{|}}{C}}-H$$

an acetal

STUDY PROBLEM

11.7. An intermediate in the formation of an acetal is a carbocation. Would you expect this hemiacetal carbocation to be more or less stabilized than a comparable alkyl carbocation? Why?

$$R-\underset{\underset{:\underset{\cdot\cdot}{O}R'}{|}}{\overset{+}{C}}-H \quad\quad versus \quad\quad R-\underset{\underset{R'}{|}}{\overset{+}{C}}-H$$

    In the equilibrium between an aldehyde, a hemiacetal, and an acetal, the aldehyde is generally favored. In an equilibrium mixture, we would usually find a large amount of aldehyde and only small amounts of hemiacetal and acetal. There is one important exception to this generality. A molecule that has an OH group γ or δ (1,4 or 1,5) to an aldehyde or ketone carbonyl group undergoes an intramolecular reaction to form a five- or six-membered hemiacetal ring. These cyclic hemiacetals are favored over the open-chain aldehyde forms.

the hemiacetal carbon has OH and OR

favored

    The reason that cyclic hemiacetals are important is that glucose and other sugars contain hydroxyl groups γ and δ to carbonyl groups; sugars, therefore, form cyclic hemiacetals in water solution. This topic will be discussed in Section 18.4.

the hemiacetal carbon

glucose

a sugar

favored

STUDY PROBLEM

11.8. Predict the cyclic hemiacetal or hemiketal products in water solutions of
(a) 5-hydroxy-2-hexanone, and (b) 1,3,4,5,6-pentahydroxy-2-hexanone. [One
stereoisomer of (b) is *fructose*, or *grape sugar*.]

In most cases, a hemiacetal cannot be isolated. Acetals, however, are stable
in nonacidic solution and can be isolated. (In acidic solution, of course, they are in
equilibrium with their aldehydes.) If an acetal is the desired product from reaction
of an aldehyde and an alcohol, an *excess of alcohol* is used to drive the series of
reaction steps to that product. *Removing water* as it is formed also helps drive the
reversible reactions to the acetal side. Best results in this type of reaction are obtained when the acetal is *cyclic*:

$$CH_3\overset{\overset{O}{\|}}{C}H \ + \ HOCH_2CH_2OH \ \underset{}{\overset{H^+}{\rightleftharpoons}} \ CH_3CH\overset{O-CH_2}{\underset{O-CH_2}{|}} \ + \ H_2O$$

acetaldehyde   1,2-ethanediol   *a cyclic acetal*
                  (ethylene glycol)

A **blocking group** is a group used to prevent a functional group from reacting
while a reaction is carried out on another part of the molecule. A blocking group
must be inert to the desired reaction, yet must be easy to remove when the desired
reaction is completed. If a desired reaction can be carried out under alkaline conditions, acetals and ketals are effective blocking groups for aldehydes and ketones.
For example, by blocking an aldehyde group as an acetal, we can oxidize a double
bond in the same molecule without oxidizing the aldehyde to a carboxylic acid.

Blocking:

$$CH_2{=}CHCH \overset{\overset{O}{\|}}{} + \ \overset{HOCH_2}{\underset{HOCH_2}{|}} \ \underset{}{\overset{H^+}{\rightleftharpoons}} \ CH_2{=}CHCH\overset{O-CH_2}{\underset{O-CH_2}{|}} \ + \ H_2O$$

propenal   *an acetal*
(acrolein)

Oxidation of double bond and regeneration of aldehyde:

$$CH_2{=}CHCH\overset{O}{\underset{O}{\big\rangle}} \ \xrightarrow[\text{cold}]{MnO_4{}^-,\ OH^-} \ \overset{OH\ OH}{\underset{}{CH_2CH-CH}}\overset{}{\underset{O}{\big\rangle}} \ \underset{}{\overset{H_2O,\ H^+}{\rightleftharpoons}} \ \overset{OH\ OH\ O}{\underset{}{CH_2CH-CH}}\overset{}{\|}$$

                                                              2,3-dihydroxypropanal
                                                              (glyceraldehyde)

SECTION 11.9.

Reaction with Hydrogen Cyanide

Hydrogen cyanide (bp 26°) can be considered to be either a gas or a low-boiling
liquid. In normal laboratory operations, it is used as a gas, but by use of a special
apparatus it can be used as a liquid (and, in some cases, even as a solvent). Often,
HCN is generated directly in a reaction mixture from KCN or NaCN and a

strong acid. Hydrogen cyanide is toxic and is particularly insidious because the human nose can detect its odor only at levels that may be lethal.

Like water and alcohols, hydrogen cyanide can add to the carbonyl group of an aldehyde or a ketone. The product in either case is referred to as a **cyanohydrin**.

$$
\underset{\substack{\text{an aldehyde}\\\text{or ketone}}}{R-\overset{\displaystyle O}{\overset{\|}{C}}-R} + HCN \;\rightleftharpoons\; \underset{\text{a cyanohydrin}}{R-\overset{\displaystyle OH}{\underset{\displaystyle CN}{\overset{|}{\underset{|}{C}}}}-R}
$$

General:

$$
\underset{\text{acetaldehyde}}{CH_3\overset{\displaystyle \ddot{O}:}{\overset{\|}{C}}H} + H\;CN \;\rightleftharpoons\; \underset{\substack{\text{acetaldehyde cyanohydrin}\\75\%}}{CH_3\overset{\displaystyle :\ddot{O}H}{\underset{}{\overset{|}{C}}}H-CN}
$$

Hydrogen cyanide (pK_a = 9.3) is too weak an acid to add directly to a carbonyl group. Successful addition requires slightly alkaline reaction conditions. In this way, the concentration of cyanide ion is increased, and addition proceeds by nucleophilic attack of CN$^-$ upon the carbonyl group. Although weak nucleophiles (such as H$_2$O and ROH) require acid catalysis to add to the carbonyl group, the strongly nucleophilic CN$^-$ does not require a catalyst.

$$
HCN: + :\ddot{O}H^- \;\rightleftharpoons\; H_2O + :CN:^-
$$

$$
R-\overset{\displaystyle \ddot{O}:}{\underset{\displaystyle :CN:}{\overset{\|}{C}}}-H \;\rightleftharpoons\; R-\overset{\displaystyle :\ddot{O}:^-}{\underset{\displaystyle CN:}{\overset{|}{C}}}-H \;\overset{HCN}{\rightleftharpoons}\; R-\overset{\displaystyle :\ddot{O}H}{\underset{\displaystyle CN:}{\overset{|}{C}}}-H + CN^-
$$

Cyanohydrins are useful synthetic intermediates. For example, the CN group can be hydrolyzed to a carboxyl group, or it can be reduced to a CH$_2$NH$_2$ group. (Reactions of the CN group will be discussed in Chapter 13.)

$$
\underset{\substack{\text{acetone cyanohydrin}}}{(CH_3)_2\overset{\displaystyle OH}{\overset{|}{C}}CN}
\;\overset{H_2O,\,H^+}{\underset{[H]}{\diagdown}}\;
$$

$$
\underset{\substack{\text{2-hydroxy-2-methylpropanoic acid}\\ \textit{an } \alpha\text{-hydroxy acid}}}{(CH_3)_2\overset{\displaystyle OH}{\overset{|}{C}}CO_2H}
$$

$$
\underset{\substack{\text{1-amino-2-methyl-2-propanol}\\ \textit{a } \beta\text{-amino alcohol}}}{(CH_3)_2\overset{\displaystyle OH}{\overset{|}{C}}CH_2NH_2}
$$

The millipede (*Apheloria corrigata*) carries its own poison-gas generator in the form of *mandelonitrile*, a cyanohydrin stored in its defensive glands. When the millipede is attacked, the cyanohydrin is mixed with an enzyme that causes a rapid dissociation to a mixture of benzaldehyde and HCN, which is squirted on the predator to ward off the attack. A single millipede can emit enough HCN to kill a mouse! It is interesting to note that mandelonitrile, benzaldehyde, and HCN all have the odor of bitter almonds, despite their disparity in structure.

$$
\underset{\text{mandelonitrile}}{\overset{\displaystyle\text{OH}}{\underset{\displaystyle}{\overset{\displaystyle|}{C_6H_5CHCN}}}} \xrightarrow{\text{enzyme}} \underset{\text{benzaldehyde}}{\overset{\displaystyle\overset{\text{O}}{\|}}{C_6H_5CH}} + HCN
$$

In plants of the genus *Prunus* (which includes plums, apricots, cherries, and peaches), cyanohydrins are biosynthesized and stored as sugar derivatives in the kernels of the pits. Amygdalin and laetrile are the best known of these cyanohydrins. (These two compounds are closely related structurally; indeed, amygdalin is often sold as laetrile.) Because these cyanohydrins can be hydrolyzed enzymatically to HCN, the pits of cherries and other *Prunus* species should not be eaten in quantity.

amygdalin

laetrile

("*laevorotatory glycosidic nitrile*")

STUDY PROBLEMS

11.9. Suggest a synthesis for lactic acid, $CH_3CH(OH)CO_2H$, from ethanol and HCN.

11.10. Aldehydes and nonhindered ketones (primarily methyl ketones) yield water-soluble **bisulfite addition products** when treated with concentrated, aqueous sodium bisulfite.

$$
\underset{}{\overset{\displaystyle\overset{\text{O}}{\|}}{RCH}} + Na^+ \ {}^-SO_3H \longrightarrow \underset{\textit{bisulfite addition product}}{\overset{\displaystyle\overset{\text{OH}}{|}}{RCH}\!-\!SO_3{}^- \ Na^+}
$$

This reaction is sometimes used to separate an aldehyde or ketone from water-insoluble organic compounds. (The aldehyde or ketone can be regenerated by treatment of the aqueous bisulfite solution with acid or base.)

(a) Suggest a mechanism for the formation of the bisulfite addition product of acetaldehyde.

(b) Explain how the following compounds (in a diethyl ether solution) could be separated from one another by a series of extractions: heptanoic acid, 4-heptanone, and heptanal.

SECTION 11.10.

Reaction with Ammonia and Amines

A.   Imines

Ammonia is a nucleophile and, like other nucleophiles, can attack the carbon of the carbonyl group of either an aldehyde or a ketone. The reaction is catalyzed by a trace of acid. For the moment, we may view the first step in the reaction as a simple addition of ammonia across the carbonyl group. The product of the addition is unstable and eliminates water to form an **imine**, a compound that contains the C=N grouping.

$$RCH + H-NH_2 \underset{}{\overset{H^+}{\rightleftharpoons}} \left[\begin{array}{c} OH \\ | \\ RCH-NH_2 \end{array} \right] \overset{-H_2O}{\rightleftharpoons} RCH=NH$$

an imine

Unsubstituted imines formed from NH_3 are unstable and polymerize on standing. However, if a *primary amine* (RNH_2) is used instead of ammonia, a more stable, substituted imine (sometimes called a **Schiff base**) is formed. Aromatic aldehydes (such as benzaldehyde) or arylamines (such as aniline) give the most stable imines, but other aldehydes, ketones, or primary amines may be used.

$$CH_3CH + H_2N-\bigcirc \underset{}{\overset{H^+}{\rightleftharpoons}} \left[\begin{array}{c} OH \\ | \\ CH_3CH \\ | \\ NH-\bigcirc \end{array} \right] \overset{-H_2O}{\rightleftharpoons} CH_3CH=N-\bigcirc$$

acetaldehyde   aniline   *an imine*

$$\bigcirc=O + H_2NCH(CH_3)_2 \overset{-H_2O}{\rightleftharpoons} \bigcirc=NCH(CH_3)_2$$

cyclopentanone   isopropylamine   *an imine*

The mechanism for imine formation is essentially a two-step process. The first step is the *addition* of the nucleophile amine to the partially positive carbonyl carbon, followed by the loss of a proton from the nitrogen and the gain of a proton by the oxygen.

Step 1, addition:

$$RCR + R'NH_2 \overset{fast}{\rightleftharpoons} \begin{array}{c} :\ddot{O}:^- \\ | \\ RCR \\ | \\ R'\overset{+}{N}H_2 \end{array} \overset{fast}{\rightleftharpoons} \begin{array}{c} :\ddot{O}H \\ | \\ RCR \\ | \\ R'\ddot{N}H \end{array}$$

Step 2 is the protonation of the OH group, which then can be lost as water in an *elimination* reaction.

Step 2, elimination:

$$R_2CNHR' \overset{H^+}{\underset{fast}{\rightleftharpoons}} R_2C-NHR' \overset{-H_2O}{\underset{slow}{\rightleftharpoons}} R_2C=\overset{+}{N}HR' \overset{-H^+}{\underset{fast}{\rightleftharpoons}} R_2C=\ddot{N}R'$$

the imine

This type of two-step reaction is often referred to as an **addition–elimination reaction**.

Imine formation is a reaction that is *pH-dependent.* Why? Consider the two steps in the mechanism. The first step is the addition of the free, nonprotonated amine to the carbonyl group. If the solution is too acidic, the concentration of the amine becomes negligible. If this happens, the usually fast addition step becomes slow and actually becomes the rate-determining step in the sequence.

In acid:

— *not nucleophilic*

$$R\ddot{N}H_2 + H^+ \rightleftharpoons RNH_3^+$$

The second step in the reaction is the elimination of the protonated OH group as water. Unlike the first step (amine addition), the rate of the second step increases with increasing acid concentration. (Remember, OH^- is a strong base and a poor leaving group, while $-OH_2^+$ can leave as the weak base and good leaving group H_2O.) Consequently, an increase in acidity causes Step 2 to go faster, but Step 1 to go slower. Conversely, decreasing acidity causes Step 1 to go faster, but Step 2 to go slower.

In between these two extremes is the optimum pH (about pH 3–4), at which the rate of the overall reaction is greatest. At this pH, some of the amine is protonated, but some is free to initiate the nucleophilic addition. At this pH, too, enough acid is present so that elimination can proceed at a reasonable rate.

STUDY PROBLEMS

11.11. Why does an arylamine yield a more stable imine than isopropylamine?

11.12. What is the geometry of $CH_3CH=NCH_3$? Would you expect this compound to have any stereoisomeric forms?

11.13. How would you prepare each of the following imines from a carbonyl compound?

(a) $-CH=NCH_2CH_3$

(b) $CH_3-$$-CH=N-$

(c) $(CH_3)_2C=N-$$-CH_3$

(d)

B.   Biological transamination

Imines are important intermediates in the biosynthesis of α-amino acids, $RCH(NH_2)CO_2H$, which are used by an organism in the synthesis of proteins. If a diet does not contain the required proportions of necessary amino acids, an organism can, in some cases, convert an unneeded amino acid to a desired amino acid in a **transamination reaction.** The process involves the transfer of an amino group from the unneeded amino acid to a keto acid.

Transamination:

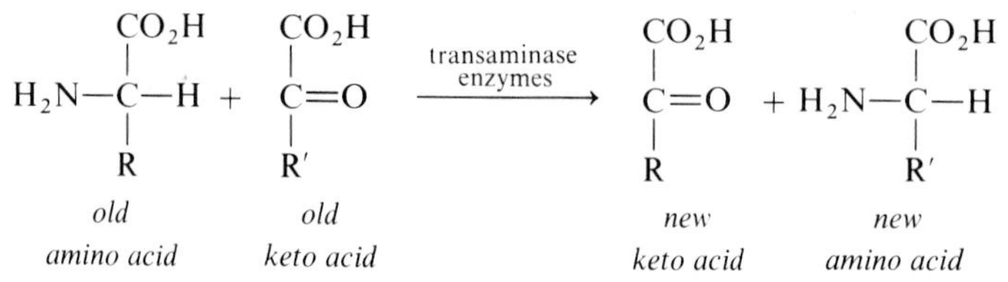

$$\underset{\substack{old \\ amino\ acid}}{H_2N-\overset{\displaystyle CO_2H}{\underset{\displaystyle R}{\overset{|}{\underset{|}{C}}}}-H} \;+\; \underset{\substack{old \\ keto\ acid}}{\overset{\displaystyle CO_2H}{\underset{\displaystyle R'}{\overset{|}{\underset{|}{C}}}}=O} \quad\xrightarrow[\text{enzymes}]{\text{transaminase}}\quad \underset{\substack{new \\ keto\ acid}}{\overset{\displaystyle CO_2H}{\underset{\displaystyle R}{\overset{|}{\underset{|}{C}}}}=O} \;+\; \underset{\substack{new \\ amino\ acid}}{H_2N-\overset{\displaystyle CO_2H}{\underset{\displaystyle R'}{\overset{|}{\underset{|}{C}}}}-H}$$

The reaction is thought to proceed through a series of imine intermediates:

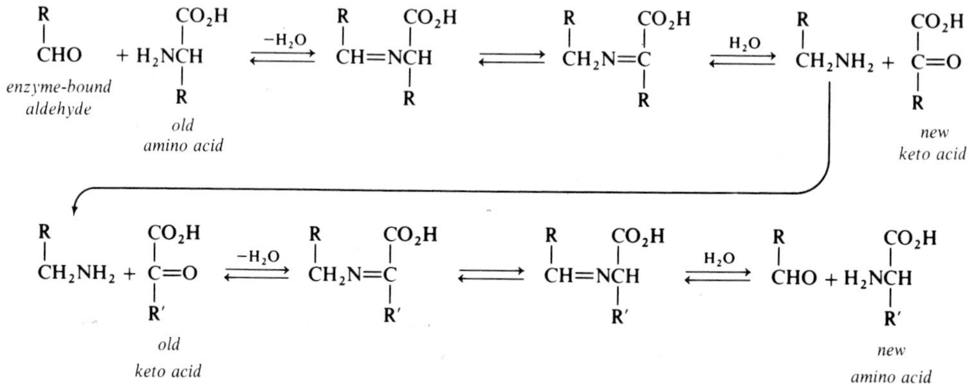

C. Enamines

With primary amines, aldehydes and ketones yield imines. With *secondary amines* (R_2NH), aldehydes and ketones yield **iminium ions**, which undergo further reaction to **enamines** (vinylamines). The enamine is formed by loss of a proton from a carbon atom β to the nitrogen, which results in a double bond between the α and β carbon atoms. Enamines are useful synthetic intermediates. You will encounter them again in Chapter 14.

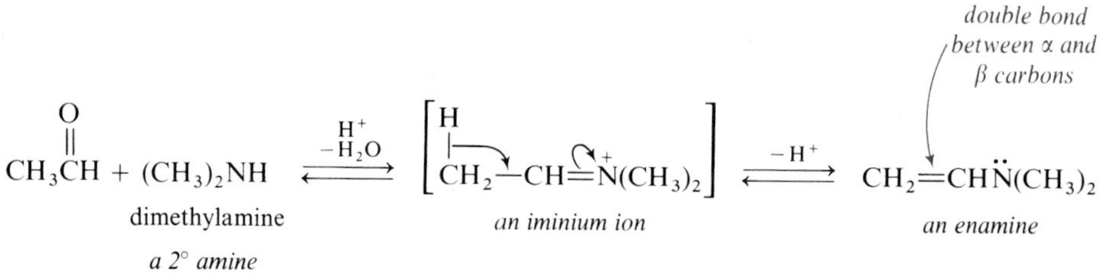

STUDY PROBLEMS

11.14. Predict the product of the reaction of cyclohexanone with:

(a) CH_3NH_2 (b) $(CH_3)_2NH$ (c) 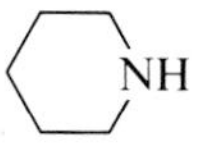 NH

11.15. Predict the enamine product of the following reaction:

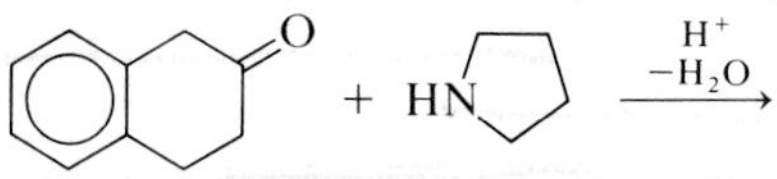

SECTION 11.11.

Reaction with Hydrazine and Related Compounds

Imines are easily hydrolyzed (cleaved by water). The initial step of hydrolysis is protonation of the imine nitrogen. If an *electronegative group* is attached to the imine nitrogen, the basicity of the nitrogen is reduced and the hydrolysis is suppressed.

$$
\underset{an\ imine}{\overset{H_3C}{\underset{H_3C}{>}}C=\ddot{N}-H} \;+\; H^+ \;\rightleftharpoons\; \overset{H_3C}{\underset{H_3C}{>}}C=\overset{+}{N}\overset{H}{\underset{H}{<}} \quad\underset{steps}{\overset{many}{\rightleftharpoons}}\quad \overset{H_3C}{\underset{H_3C}{>}}C=O \;+\; NH_3
$$

$$H_2O$$

$$
\underset{electron\text{-}withdrawing}{\overset{H_3C}{\underset{H_3C}{>}}C=\ddot{N}-NH_2} \;+\; H^+ \;\rightleftharpoons\; \underset{not\ favored}{\overset{H_3C}{\underset{H_3C}{>}}C=\overset{+}{N}\overset{H}{\underset{NH_2}{<}}}
$$

Imine-type products formed from aldehydes or ketones and a nitrogen compound of the type H_2N-NH_2 or H_2N-OH (reagents with an electronegative group attached to the N) are quite stable. Table 11.5 lists the variety of nitrogen

TABLE 11.5. Some nitrogen compounds that form stable substitution products with aldehydes and ketones

Name	Structure	Product with RCHO
hydroxylamine	$HONH_2$	$RCH=NOH$ *an oxime*
hydrazine	H_2NNH_2	$RCH=NNH_2$ *a hydrazone*
phenylhydrazine	⟨C₆H₅⟩—$NHNH_2$	$RCH=NNHC_6H_5$ *a phenylhydrazone*
2,4-dinitro-phenylhydrazine	O_2N—⟨ring, NO_2⟩—$NHNH_2$	$RCH=NNH$—⟨ring, NO_2⟩—NO_2 *a 2,4-dinitrophenylhydrazone*
semicarbazide	$H_2NNH\overset{O}{\overset{\|}{C}}NH_2$	$RCH=NNH\overset{O}{\overset{\|}{C}}NH_2$ *a semicarbazone*

compounds that undergo reaction with aldehydes and most ketones to form stable imine-type products.

$$
\begin{array}{c}
H_3C \\
 \\
H_3C
\end{array}
C=O + H_2NNH_2 \quad \underset{}{\overset{H^+}{\rightleftarrows}} \quad
\begin{array}{c}
H_3C \\
 \\
H_3C
\end{array}
C=NNH_2 + H_2O
$$

acetone   hydrazine   acetone hydrazone

$$
\text{cyclopentanone} \quad + H_2NNHC_6H_5 \quad \overset{H^+}{\rightleftarrows} \quad =NNHC_6H_5 + H_2O
$$

cyclopentanone   phenylhydrazine   cyclopentanone
                                              phenylhydrazone

The hydrazones and other products listed in Table 11.5, especially the high-molecular-weight 2,4-dinitrophenylhydrazones, or DNP's, are generally solids. Before the wide use of spectrometers, these derivatives were used extensively for identification purposes. A liquid ketone of unknown structure could be converted to the solid DNP, purified by crystallization, and its melting point compared to those of DNP's of known structure.

STUDY PROBLEM

11.16. Predict the product of the reaction of: (a) butanone with semicarbazide:
(b) cyclohexanone with 2,4-dinitrophenylhydrazine; and   (c) acetophenone
($C_6H_5COCH_3$) with hydrazine.

SECTION 11.12.

The Wittig Reaction

In 1954, George Wittig reported a general synthesis of alkenes from carbonyl compounds using *phosphonium ylids*. This synthesis is called the **Wittig reaction**.

$$
\begin{array}{c}
R \\
 \\
R
\end{array}
C=O + (C_6H_5)_3P=C
\begin{array}{c}
R' \\
 \\
R'
\end{array}
\longrightarrow
\begin{array}{c}
R \\
 \\
R
\end{array}
C=C
\begin{array}{c}
R' \\
 \\
R'
\end{array}
+ (C_6H_5)_3\overset{+}{P}-\overset{..}{\underset{..}{O}}{:}^-
$$

an aldehyde   a phosphonium ylid   *an alkene*   triphenylphosphine
or ketone   oxide

An **ylid**, or **ylide**, is a molecule with *adjacent + and − charges* (see the following resonance structures). An ylid is formed by removal of a proton from the carbon adjacent to a positively charged heteroatom (such as P^+, S^+, or N^+). The phosphonium ylid for a Wittig reaction is prepared by (1) nucleophilic substitution (S_N2) of an alkyl halide with a tertiary phosphine, such as triphenylphosphine (a good nucleophile, a weak base), and (2) treatment with base, a reaction in which the intermediate phosphonium ion eliminates a proton to form the ylid.

$$\underset{\text{triphenylphosphine}}{(C_6H_5)_3P:} + \underset{R'}{\overset{R'}{\underset{|}{CH}}}\!\!-X \xrightarrow[S_N2]{-X^-} \underset{\text{a phosphonium ion}}{(C_6H_5)_3\overset{+}{P}\!\!-\!\!\overset{\overset{H}{|}}{CR'_2}} \xrightarrow{\ ^-OCH_2CH_3\ }$$

$$CH_3CH_2OH + (C_6H_5)_3\overset{+}{P}\!\!-\!\!\ddot{C}R'_2 \quad\longleftrightarrow\quad (C_6H_5)_3P{=}CR'_2$$

resonance structures for the ylid

The Wittig reaction is versatile. The alkyl halide used to prepare the ylid may be methyl, primary, or secondary, but not tertiary (why not?). The halide may also contain other functionality, such as double bonds or alkoxyl groups. The product of the Wittig reaction is an alkene with the double bond in the desired position, even if it is not the most stable alkene. Yields are generally good (about 70%). Unfortunately, it is sometimes difficult to predict whether the *cis* or the *trans* product will predominate in a particular reaction.

$$\underset{\text{cyclohexanone}}{\bigcirc{=}O} + (C_6H_5)_3P{=}C(CH_3)_2 \longrightarrow \underset{\substack{\text{isopropylidene-}\\\text{cyclohexane}}}{\bigcirc{=}C(CH_3)_2} + (C_6H_5)_3\overset{+}{P}\!\!-\!\!O^-$$

$$\underset{\text{benzaldehyde}}{\bigcirc\!\!-\!\!\overset{\overset{O}{\|}}{CH}} + (C_6H_5)_3P{=}CH_2 \longrightarrow \underset{\text{styrene}}{\bigcirc\!\!-\!\!CH{=}CH_2} + (C_6H_5)_3\overset{+}{P}\!\!-\!\!O^-$$

$$\underset{\text{acetone}}{\overset{\overset{O}{\|}}{CH_3CCH_3}} + (C_6H_5)_3P{=}CHCH{=}CH_2 \longrightarrow \underset{\text{4-methyl-1,3-pentadiene}}{(CH_3)_2C{=}CHCH{=}CH_2} + (C_6H_5)_3\overset{+}{P}\!\!-\!\!O^-$$

The mechanism of the Wittig reaction is still being investigated. A currently accepted theory is that the carbonyl group undergoes nucleophilic attack by the negative carbon of the ylid.

nucleophilic carbon

$$(C_6H_5)_3P{=}CR'_2 \quad\longleftrightarrow\quad (C_6H_5)_3\overset{+}{P}\!\!-\!\!\ddot{C}R'_2$$

resonance structures

Addition to carbonyl:

$$\underset{\ddot{O}:}{\overset{\overset{R}{|}}{\underset{\|}{R\!\!-\!\!C}}} + \underset{^+P(C_6H_5)_3}{:CR'_2} \longrightarrow \underset{\substack{:\ddot{O}:\quad ^+P(C_6H_5)_3\\ \text{a betaine}}}{\overset{\overset{R}{|}}{R\!\!-\!\!C\!\!-\!\!CR'_2}}$$

The addition product of the ylid and an aldehyde or ketone is a **betaine** (a molecule having *nonadjacent* opposite charges). The betaine undergoes elimination of triphenylphosphine oxide to form the alkene.

Elimination to alkene:

$$\underset{:\ddot{O}:\quad P(C_6H_5)_3}{R_2C\!\!-\!\!CR'_2} \longrightarrow R_2C{=}CR'_2 + :\ddot{O}\!\!-\!\!\overset{+}{P}(C_6H_5)_3$$

STUDY PROBLEMS

11.17. (a) Suggest a Wittig reaction for the synthesis of methylenecyclohexane.
  (b) What product would you obtain if you tried to make this compound by the dehydration of 1-methyl-1-cyclohexanol?

$$\text{(structure)} = CH_2$$

methylenecyclohexane

11.18. What organic halide and what carbonyl compound would you use to prepare each of the following compounds by a Wittig reaction?

  (a) $C_6H_5CH=CHCH_2CH_3$ (b) (structure)   (c) (structure)

SECTION 11.13.

Reaction with Grignard Reagents

The reaction of a Grignard reagent with a carbonyl compound is another example of a nucleophilic addition to the positive carbon of a carbonyl group.

$$\begin{array}{ccc}
\overset{\ddot{O}:}{\underset{|}{\|}} & & :\overset{..}{\overset{..}{O}}:^- \ \overset{+}{MgI} \\
-\overset{|}{C}- + \underset{\delta-}{CH_3}\underset{\delta+}{MgI} & \longrightarrow & -\overset{|}{\underset{|}{C}}- \\
& & CH_3
\end{array}$$

Reaction of a Grignard reagent with an aldehyde or a ketone provides an excellent method for the synthesis of alcohols and was discussed in that context earlier (Sections 6.9 and 7.4B). The reaction sequence consists of two separate steps: (1) the reaction of the Grignard reagent with the carbonyl compound, and (2) hydrolysis of the resulting magnesium alkoxide to yield the alcohol. Recall that the Grignard reaction of formaldehyde yields a *primary alcohol*; other aldehydes yield *secondary alcohols*; and ketones yield *tertiary alcohols*.

$$
RMgX \begin{cases}
\xrightarrow{HCHO} & \underset{\underset{R}{|}}{\overset{\overset{OMgX}{|}}{HCH}} \xrightarrow{H_2O, H^+} RCH_2OH \quad \textit{1° alcohol} \\[2em]
\xrightarrow{R'CHO} & \underset{\underset{R}{|}}{\overset{\overset{OMgX}{|}}{R'CH}} \xrightarrow{H_2O, H^+} \underset{}{\overset{\overset{OH}{|}}{R'CHR}} \quad \textit{2° alcohol} \\[2em]
\xrightarrow{\overset{O}{\overset{\|}{R'CR''}}} & \underset{\underset{R}{|}}{\overset{\overset{OMgX}{|}}{R'CR''}} \xrightarrow{H_2O, H^+} \underset{\underset{R}{|}}{\overset{\overset{OH}{|}}{R'CR''}} \quad \textit{3° alcohol}
\end{cases}
$$

SAMPLE PROBLEMS

Suggest *two* synthetic routes to 2-butanol from an aldehyde or ketone and a Grignard reagent.

Solution: In the synthesis of 2-butanol, either (1) CH_3MgX and CH_3CH_2CHO may be used, or (2) CH_3CH_2MgX and CH_3CHO may be used. In the laboratory, the choice would depend on a number of factors, including availability and cost of the appropriate alkyl halides and aldehydes.

$$\overset{\displaystyle OH}{\underset{\displaystyle |}{CH_3-CHCH_2CH_3}}$$

from CH_3MgI from $HCCH_2CH_3$ ($\overset{O}{\overset{||}{}}$)

$$\overset{\displaystyle OH}{\underset{\displaystyle |}{CH_3CH-CH_2CH_3}}$$

from CH_3CH ($\overset{O}{\overset{||}{}}$)   from $BrMgCH_2CH_3$

The two sequences to 2-butanol:

$$CH_3CH_2CHO \xrightarrow[\text{(2) } H_2O, H^+]{\text{(1) } CH_3MgI}$$

$$\overset{\displaystyle OH}{\underset{\displaystyle |}{CH_3CHCH_2CH_3}}$$

$$CH_3CHO \xrightarrow[\text{(2) } H_2O, H^+]{\text{(1) } CH_3CH_2MgBr}$$

Suggest three different Grignard reactions leading to 2-phenyl-2-butanol.

Solution:

(1) C_6H_5MgBr $\xrightarrow[\text{(2) } H_2O, H^+]{\text{(1) } CH_3\overset{O}{\overset{||}{C}}CH_2CH_3}$

(2) CH_3MgI $\xrightarrow[\text{(2) } H_2O, H^+]{\text{(1) } C_6H_5\overset{O}{\overset{||}{C}}CH_2CH_3}$ $C_6H_5-\overset{\displaystyle OH}{\underset{\displaystyle CH_3}{\overset{\displaystyle |}{\underset{\displaystyle |}{C}}}}-CH_2CH_3$

(3) CH_3CH_2MgBr $\xrightarrow[\text{(2) } H_2O, H^+]{\text{(1) } C_6H_5\overset{O}{\overset{||}{C}}CH_3}$

STUDY PROBLEMS

11.19. What Grignard reagent would you use to effect the following conversions?
(a) formaldehyde to benzyl alcohol;   (b) cyclohexanone to
1-propyl-1-cyclohexanol.

11.20. Which of the following compounds could *not* be used as a carbonyl starting
material in a Grignard synthesis? (*Hint*: See Section 6.11.)

(a)
$$
\underset{\text{OH}}{\overset{}{\underset{|}{\text{CH}_3\text{CHCH}_2}}}\overset{\text{O}}{\overset{||}{\text{CH}}}
$$

(b)
$$
\overset{\text{O}\quad\text{O}}{\overset{||\ \ ||}{\text{CH}_3\text{CCH}_2\text{COH}}}
$$

(c)
$$
\overset{\text{O}}{\overset{||}{\text{C}_6\text{H}_5\text{CH}_2\text{CH}_2\text{CH}}}
$$

(d) $\text{H}_2\text{N}\!-\!\!\left\langle \bigcirc \right\rangle\!\!-\!\overset{\text{O}}{\overset{||}{\text{CH}}}$

(e) $\overset{\text{O}}{\overset{||}{\text{HCCH}_2\text{CH}_2\text{CH}_2\text{CH}}}$

(f) $\left\langle \bigcirc \right\rangle\!\!=\!\!\text{O}$

SECTION 11.14.

Reduction of Aldehydes and Ketones

An aldehyde or a ketone can be reduced to an alcohol, a hydrocarbon, or an
amine. The product of the reduction depends on the reducing agent and on the
structure of the carbonyl compound.

A.   Hydrogenation

The pi bond of a carbonyl group can undergo catalytic hydrogenation, just as
can the pi bond in an alkene. Alkenes can be hydrogenated at low pressure and
at room temperature; for the hydrogenation of a carbonyl group, heat and
pressure are usually required. A ketone is reduced to a secondary alcohol by
catalytic hydrogenation, while an aldehyde yields a primary alcohol. Yields are
excellent (90–100%).

STUDY PROBLEM

11.21. 2-Heptanol can be obtained by two Grignard reactions as well as by a hydrogenation reaction. Write the equations for the three sets of reactions that would lead to this alcohol.

If both a double bond and a carbonyl group are present in a structure, the double bond may be hydrogenated, leaving the carbonyl intact, or both may be hydrogenated. However, the carbonyl group cannot be hydrogenated independently of the double bond. If it is desired to reduce a carbonyl group while leaving a double bond intact, a metal hydride reduction is the method of choice.

C=C reduced (but not C=O):

$$CH_3CH{=}CHCH_2\overset{\displaystyle O}{\overset{\|}{C}}H + H_2 \xrightarrow[25°]{Ni} CH_3CH_2CH_2CH_2\overset{\displaystyle O}{\overset{\|}{C}}H$$

3-pentenal   pentanal

C=C and C=O reduced:

$$CH_3CH{=}CHCH_2\overset{\displaystyle O}{\overset{\|}{C}}H + 2 H_2 \xrightarrow[\text{heat, pressure}]{Ni} CH_3CH_2CH_2CH_2CH_2OH$$

3-pentenal   1-pentanol

B. Metal hydrides

Hydrogen gas is inexpensive on a molar basis; however, a hydrogenation reaction is rather inconvenient. The apparatus usually consists of gas tanks and a metal pressure vessel. An alternative reduction procedure involves the use of a metal hydride. Two valuable reducing agents are *lithium aluminum hydride* (often abbreviated LAH) and *sodium borohydride*, both of which reduce aldehydes and ketones to alcohols.

$$Li^+ \quad H{-}\overset{\displaystyle H}{\underset{\displaystyle H}{Al}}{-}H \qquad Na^+ \quad H{-}\overset{\displaystyle H}{\underset{\displaystyle H}{B}}{-}H$$

lithium aluminum hydride   sodium borohydride

$$CH_3CH_2\overset{\displaystyle O}{\overset{\|}{C}}CH_3$$

butanone

$$\xrightarrow[\text{(2) H}_2\text{O}]{\text{(1) LiAlH}_4} \quad CH_3CH_2\overset{\displaystyle OH}{\overset{|}{C}}HCH_3$$

2-butanol
80%

$$\xrightarrow[\text{(2) H}_2\text{O, H}^+]{\text{(1) NaBH}_4} \quad CH_3CH_2\overset{\displaystyle OH}{\overset{|}{C}}HCH_3$$

87%

These two metal hydrides are quite different in their reactivities. LAH is a powerful reducing agent that reduces not only aldehydes and ketones, but also carboxylic acids, esters, amides, and nitriles. LAH undergoes violent reaction with water; reductions are usually carried out in a solvent such as anhydrous ether.

Sodium borohydride is a milder reducing agent than LAH. Its reactions can be carried out in water or aqueous alcohol as the solvent. For the reduction of an aldehyde or ketone, $NaBH_4$ is the preferred reagent; it is certainly more convenient to use because of its lack of reactivity toward water. While $NaBH_4$ reduces aldehydes and ketones rapidly, it reduces esters very slowly. Therefore, an aldehyde or ketone carbonyl group can be reduced without the simultaneous reduction of an ester group in the same molecule. This selectivity is not possible with LAH.

$NaBH_4$
· selectivity possible

ester not reduced

$$\underset{\text{HCCH}_2\text{CH}_2\text{COCH}_2\text{CH}_3}{\overset{\text{O} \quad\quad \text{O}}{\| \quad\quad \|}} \xrightarrow[\text{(2) H}_2\text{O, H}^+]{\text{(1) NaBH}_4} \underset{\text{CH}_2\text{CH}_2\text{CH}_2\text{COCH}_2\text{CH}_3}{\overset{\text{OH} \quad\quad\quad \text{O}}{| \quad\quad\quad \|}}$$

Neither $NaBH_4$ nor LAH reduces isolated carbon–carbon double bonds, but C=C in conjugation with a carbonyl group is sometimes attacked. Consequently, a structure that contains both a double bond and a carbonyl group can often be reduced selectively at the carbonyl position. In this respect, the metal hydrides are complementary to hydrogen gas as reducing agents.

C=O reduced (but not C=C):

$$\underset{\text{3-pentenal}}{\text{CH}_3\text{CH}=\text{CHCH}_2\overset{\overset{\text{O}}{\|}}{\text{CH}}} \xrightarrow[\text{(2) H}_2\text{O, H}^+]{\text{(1) NaBH}_4} \underset{\text{3-penten-1-ol}}{\text{CH}_3\text{CH}=\text{CHCH}_2\text{CH}_2\text{OH}}$$

Diisobutylaluminum hydride (DBAH), $[(CH_3)_2CHCH_2]_2AlH$, is a newer, but popular, metal hydride reducing agent that is similar to LAH in reducing power. Besides reducing aldehydes or ketones to alcohols, DBAH reduces carboxylic acids and esters to aldehydes or alcohols (depending on the reaction conditions). DBAH can also reduce isolated carbon–carbon double bonds. Other hydrides that have specialized reducing activity are also available; for example, see Section 13.3C.

STUDY PROBLEMS

11.22. Show how each of the following alcohols could be prepared by the $NaBH_4$ reduction of an aldehyde or ketone:

(a) $CH_3CH_2CH_2OH$ (b) ⬡—OH   (c) ⬡—CHCH₃ with OH above

11.23. When glucose (page 807) is treated with $NaBH_4$, then with aqueous acid, an artificial sweetener called *sorbitol* results. What is the structure of sorbitol?

Metal hydrides react by transferring a negative hydride ion to the positive carbon of a carbonyl group, just as a Grignard reagent transfers R to the carbonyl group.

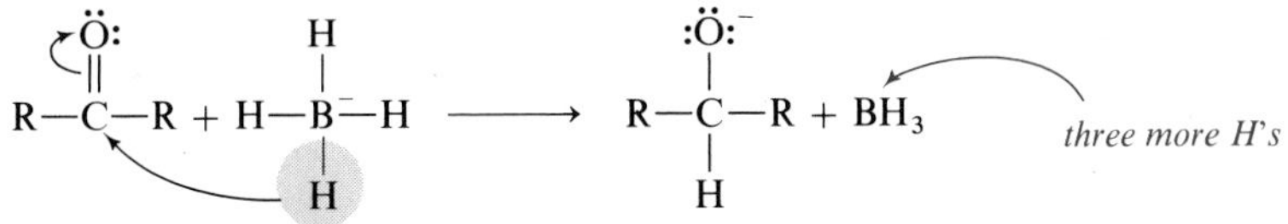

Each hydride ion can reduce one carbonyl group. Therefore, one mole of LAH or $NaBH_4$ can reduce *four* moles of aldehyde or ketone, theoretically. After the reaction is completed, treatment with water or aqueous acid liberates the alcohol from its salt. (Of course, if water, methanol, or ethanol is used as a solvent for a borohydride reduction, this step occurs spontaneously.) In the hydrolysis, the boron portion of the organoborate is converted to boric acid, H_3BO_3.

Step 1: $4\ RCR + NaBH_4 \longrightarrow Na^+\ ^-B(OCHR)_4$

Step 2: $RCHR + H^+ \longrightarrow RCHR$

Camphor is a bridged, cyclic compound with a ketone group. Reduction of camphor with LAH leads to 90% of the isomer in which the OH group is *cis* to the bridge. Why is this so? Let us look at the structure of camphor. Note that the bridge provides substantial steric hindrance on one side of the carbonyl group—on top, as it is shown here.

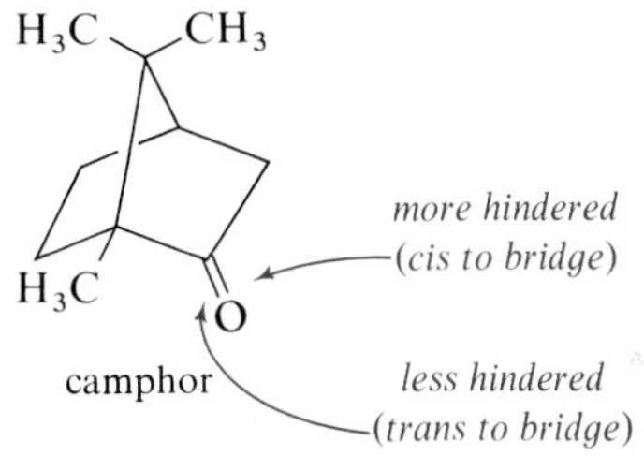

When a ketone is reduced by LAH, it is not just a tiny hydride ion attacking; it is the relatively bulky AlH_4^- ion or an alkoxyaluminum hydride ion, such as $^-AlH_2(OR)_2$. There is evidence that the electropositive, metallic portion of the metal hydride ion forms a complex with the carbonyl oxygen while the hydride ion is transferred to the carbonyl carbon. A look at the likely structure of the transition state shows how AlH_4^- attacks the *less hindered* side of the camphor structure—that is, AlH_4^- attacks the carbonyl group from the side that is *trans* to the bridge. The resultant OH group is then formed *cis* to the bridge.

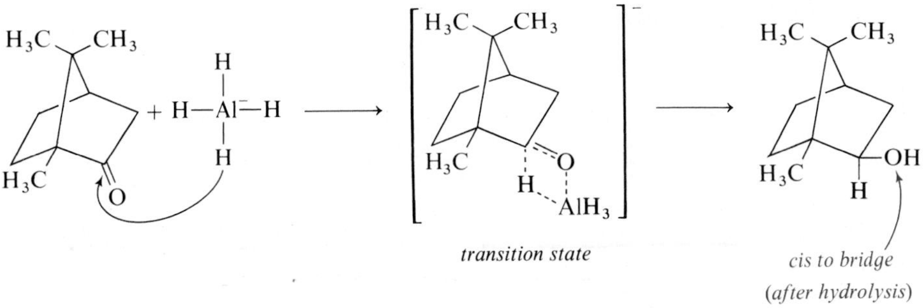

transition state

cis to bridge
(after hydrolysis)

SAMPLE PROBLEM

When a *cis*-3-alkyl-4-*t*-butylcyclohexanone is subjected to a reaction with LAH, the predominant product is that in which the OH is *cis* to the *t*-butyl group and the alkyl group. Suggest a reason.

Solution: The equatorial *t*-butyl group forces the R group into the axial position (see Section 4.5). The transition state leading to an equatorial OH shows more steric hindrance (1,3-diaxial interactions) than that leading to the axial OH. Therefore, the transition state leading to an axial OH is of lower energy and is favored, and the *cis*-product results.

To equatorial:

To axial:

favored

C.   Wolff–Kishner and Clemmensen reductions

The Clemmensen reduction and the Wolff–Kishner reduction are primarily used to reduce aryl ketones obtained from Friedel–Crafts reactions (Section 10.9E), but may sometimes be used to reduce other aldehydes and ketones. Both these methods of reduction result in the conversion of a C=O group to a CH₂ group.
    In the Wolff–Kishner reduction, the aldehyde or ketone is first converted to a hydrazone by reaction with hydrazine. The hydrazone is then treated with a strong base, such as potassium *t*-butoxide in dimethyl sulfoxide as solvent. The reaction is therefore limited to carbonyl compounds that are stable in base.

Wolff–Kishner reduction:

acetophenone   ethylbenzene

In the Clemmensen reduction, on the other hand, a zinc amalgam (an alloy of zinc and mercury) and concentrated HCl are used; these reagents would be the reagents of choice for a compound unstable in base but stable in acid.

Clemmensen reduction:

$$\text{C}_6\text{H}_5-\overset{\overset{\displaystyle O}{\|}}{\text{C}}\text{CH}_3 \xrightarrow[\text{HCl}]{\text{Zn/Hg}} \text{C}_6\text{H}_5-\text{CH}_2\text{CH}_3$$

D. Reductive amination

If an amine is the desired reduction product, the carbonyl compound is treated with ammonia or a primary amine to form an imine in the presence of hydrogen and a catalyst. The imine $C{=}N$ group then undergoes catalytic hydrogenation in the same way that a $C{=}C$ or a $C{=}O$ group does.

$$\overset{\overset{\displaystyle O}{\|}}{\text{C}_6\text{H}_5-\text{CH}} \xrightarrow[-\text{H}_2\text{O}]{\text{NH}_3} \left[\text{C}_6\text{H}_5-\text{CH}{=}\text{NH} \right] \xrightarrow{\text{H}_2,\ \text{Ni}} \text{C}_6\text{H}_5-\text{CH}_2\text{NH}_2$$

benzaldehyde   *an imine*   benzylamine
                                                                85%

$$\text{CH}_3\text{CH}_2\overset{\overset{\displaystyle O}{\|}}{\text{C}}\text{CH}_3 \xrightarrow[-\text{H}_2\text{O}]{\text{CH}_3\text{NH}_2} \left[\text{CH}_3\text{CH}_2\overset{\overset{\displaystyle NCH_3}{\|}}{\text{C}}\text{CH}_3 \right] \xrightarrow{\text{H}_2,\ \text{Ni}} \text{CH}_3\text{CH}_2\overset{\overset{\displaystyle NHCH_3}{|}}{\text{C}}\text{HCH}_3$$

butanone   *an imine*   *N*-methyl-2-butylamine

Reductive amination is a good method for the preparation of an amine with a secondary alkyl group: $R_2\text{CHNH}_2$. (Treating the secondary alkyl halide $R_2\text{CHX}$ with NH_3 in an S_N2 reaction may result in elimination or in dialkylamines, a reaction we will discuss in Chapter 15.)

$$\text{C}_6\text{H}_{11}{-}\text{Br} \ + \ \text{NH}_3 \xrightarrow{\text{E2}} \text{cyclohexene} \qquad \text{instead of} \qquad \text{C}_6\text{H}_{11}{-}\text{NH}_2$$

bromocyclohexane   cyclohexene

$$\text{C}_6\text{H}_{10}{=}\text{O} \xrightarrow[\text{H}_2,\ \text{Ni}]{\text{NH}_3} \text{C}_6\text{H}_{11}{-}\text{NH}_2$$

cyclohexanone   cyclohexylamine

Oxidation of Aldehydes and Ketones

Ketones are not easily oxidized (see Sections 11.17 and 11.18 for exceptions), but aldehydes are very easily oxidized to carboxylic acids. Almost any reagent that oxidizes an alcohol also oxidizes an aldehyde (see Section 7.13C). Permanganate or dichromate salts are the most popular oxidizing agents, but are by no means the only reagents that can be used.

$$\underset{\substack{\text{propanal} \\ \textit{an aldehyde}}}{CH_3CH_2\overset{\displaystyle O}{\overset{\|}{C}}H} \xrightarrow{KMnO_4} \underset{\substack{\text{propanoic acid} \\ \textit{a carboxylic acid}}}{CH_3CH_2\overset{\displaystyle O}{\overset{\|}{C}}OH}$$

$$\underset{\substack{\text{acetone} \\ \textit{a ketone}}}{CH_3\overset{\displaystyle O}{\overset{\|}{C}}CH_3} \xrightarrow{KMnO_4} \text{no reaction}$$

In addition to oxidation by permanganate or dichromate, aldehydes are oxidized by very mild oxidizing agents, such as Ag^+ or Cu^{2+}. **Tollens reagent** (an alkaline solution of the silver–ammonia complex ion) is used as a test for aldehydes. The aldehyde is oxidized to a carboxylate anion; the Ag^+ in the Tollens reagent is reduced to Ag metal. A positive test is indicated by the formation of a silver mirror on the wall of the test tube. With the widespread use of spectroscopy, the Tollens test is no longer the test of choice for an aldehyde, but mirrors are sometimes still made this way.

$$\underset{\substack{ \\ \textit{from Tollens reagent}}}{\overset{\displaystyle O}{\overset{\|}{R}C}H + \quad Ag(NH_3)_2{}^+} \xrightarrow{OH^-} \underset{\substack{ \\ \textit{mirror}}}{\overset{\displaystyle O}{\overset{\|}{R}C}O^- + \quad Ag}$$

SECTION 11.16.

Reactivity of the Alpha Hydrogens

A carbon–hydrogen bond is usually stable, nonpolar, and certainly not acidic. But the presence of a carbonyl group results in an *acidic alpha hydrogen*. If a hydrogen is alpha to *two* carbonyl groups, it is acidic enough that salts can be formed by treatment with an alkoxide. The pK_a of ethyl acetoacetate $(CH_3COCH_2CO_2CH_2CH_3)$ is 11; it is more acidic than ethanol $(pK_a = 16)$ or water $(pK_a = 15)$. Treatment of this β-dicarbonyl compound with sodium ethoxide (or any other strong base, such as NaH or $NaNH_2$) yields the sodium salt. (Sodium hydroxide is usually not used with keto esters because the ester groups undergo hydrolysis in NaOH and water. This reaction will be discussed in Section 13.5C.)

$$\underset{\substack{\text{acetone} \\ pK_a = 20}}{\overset{\displaystyle O}{\overset{\|}{CH_3C}}CH_3} \quad \overset{\alpha \text{ to one } C=O}{\frown} \qquad\qquad \underset{\substack{\text{ethyl acetoacetate} \\ pK_a = 11}}{\overset{\displaystyle O \quad\; O}{\overset{\| \quad \|}{CH_3CCH_2COCH_2CH_3}}} \overset{\alpha \text{ to two } C=O \text{ groups}}{\frown}$$

$$\underset{\substack{| \\ H}}{CH_2\overset{\displaystyle O}{\overset{\|}{C}}CH_3} + Na^+ \ {}^-OCH_2CH_3 \quad \rightleftharpoons \quad Na^+ \ {}^-CH_2\overset{\displaystyle O}{\overset{\|}{C}}CH_3 + CH_3CH_2OH$$
$$\underset{\textit{not favored}}{}$$

$$\underset{H}{\overset{\overset{O}{\|}\overset{O}{\|}}{CH_3CCHCOCH_2CH_3}} + Na^+\ {}^-OCH_2CH_3 \rightleftharpoons \underset{Na^+}{\overset{\overset{O}{\|}\overset{O}{\|}}{CH_3CCHCOCH_2CH_3}} + CH_3CH_2OH$$

favored

Why is a hydrogen alpha to a carbonyl group acidic? The answer is twofold. First, the alpha carbon is adjacent to one (or two) partially positive carbon atoms. The alpha carbon, too, partakes of some of this positive charge (inductive effect by electron-withdrawal), and C—H bonds are consequently weakened.

$$\overset{O}{\overset{\|}{-C}}{\leftarrow}\overset{\delta+}{CH_2}{\rightarrow}\overset{O}{\overset{\|}{C-}}$$

Second, and more important, is the resonance-stabilization of the **enolate ion**, the anion formed when the proton is lost. From the resonance structures, we can see that the negative charge is carried by the carbonyl oxygens as well as by the α carbon. This delocalization of the charge stabilizes the enolate ion and favors its formation.

Adjacent to one carbonyl group:

$$\left[-\overset{..}{C}H-\overset{\overset{..}{\ddot O}}{\overset{\|}{C}}- \longleftrightarrow -CH=\overset{:\ddot O:^-}{C}- \right] \quad \text{or} \quad -CH{=\!=}\overset{O^{\delta-}}{\overset{\|}{C}}-$$

Adjacent to two carbonyl groups:

$$\left[-\overset{:\ddot O}{\overset{\|}{C}}-\overset{\overset{:\ddot O:^-}{}}{CH{=}C}- \longleftrightarrow -\overset{:\ddot O}{\overset{\|}{C}}-\overset{..}{C}H-\overset{\overset{\ddot O:}{\|}}{C}- \longleftrightarrow -\overset{:\ddot O:^-}{C}=CH-\overset{\overset{\ddot O:}{\|}}{C}- \right]$$

$$\text{or} \quad -\overset{O^{\delta-}}{\overset{\|}{C}}{=\!=}\overset{}{CH}{=\!=}\overset{O^{\delta-}}{\overset{\|}{C}}-$$

STUDY PROBLEM

11.24. Give the resonance structures of the enolate ions formed when the following diones are treated with sodium ethoxide:

(a) $\langle\!\bigcirc\!\rangle\!-\!\overset{\overset{O}{\|}}{C}CH_2\overset{\overset{O}{\|}}{C}CH_3$ (b) $CH_3\overset{\overset{O}{\|}}{C}CH_2\overset{\overset{O}{\|}}{C}CH_3$

SECTION 11.17.

Tautomerism

Even if a strong base is not present, the acidity of the alpha hydrogen may be evident. A carbonyl compound with an acidic alpha hydrogen may exist in two forms called **tautomers**: a **keto** tautomer and an **enol** tautomer. The keto tautomer of a carbonyl compound has the expected <u>carbonyl structure</u>. The <u>enol tautomer</u> <u>(from *-ene* + *-ol*), which is a vinylic alcohol, is formed by transfer of an acidic</u> <u>hydrogen from the α carbon to the carbonyl oxygen</u>. Because a hydrogen atom is in different positions, the two tautomeric forms are not resonance structures, but are two different structures in equilibrium. (Remember that resonance structures vary only in the positions of *electrons*.)

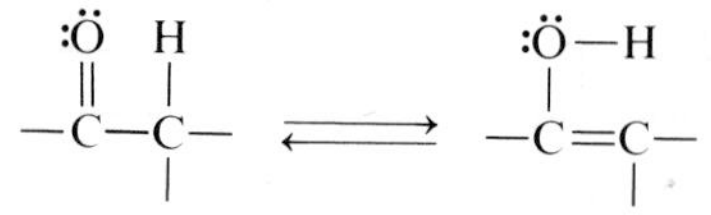

keto form   enol form

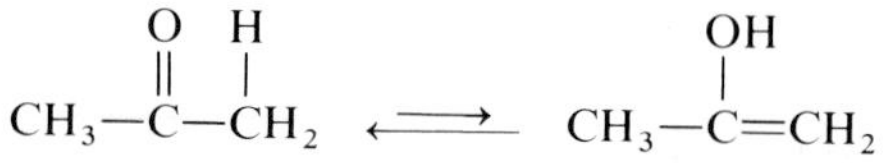

keto form   enol form
of acetone   of acetone

STUDY PROBLEM

11.25 Which of each of the following pairs are tautomers and which are resonance structures?

(a) 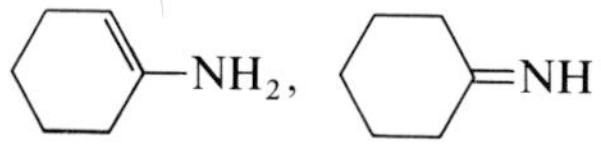 [cyclohexene]—NH₂,   [cyclohexene]=NH

(b) 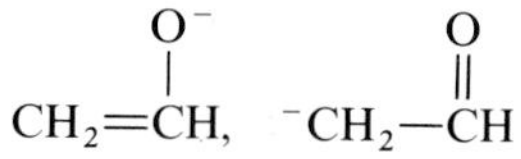 $CH_2=CH-O^-$, $^-CH_2-CH=O$

(c) 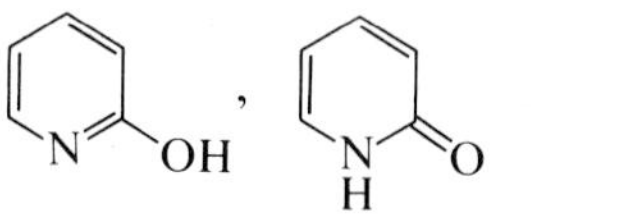 [pyridine-OH], [pyridinone-NH]

(d) 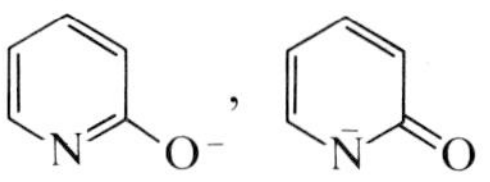 [pyridine-O⁻], [pyridinone]

The relative quantities of enol versus keto in a pure liquid may be estimated by infrared or nmr spectroscopy. Acetone exists primarily in the keto form (99.99%, as determined by a specialized titration procedure). Most other simple aldehydes and ketones also exist primarily in their keto forms; however, 2,4-pentanedione exists as 80% enol! How can this tremendous difference be explained? Let us consider the structures of the 2,4-pentanedione tautomers:

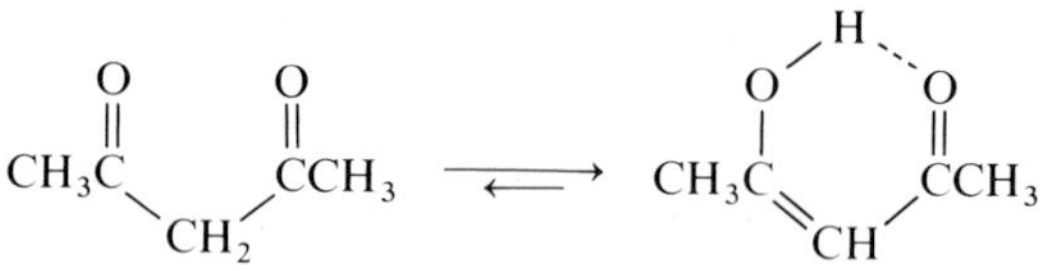

keto form   enol form
20%   80%

The enol form not only has conjugated double bonds, which add a small amount of stability, but it is also structurally arranged for internal hydrogen bonding, which helps stabilize this tautomer.

SAMPLE PROBLEM

Suggest reasons why 1,2-cyclohexanedione exists 100% in an enol form.

Solution:

dipole–dipole
repulsions

repulsions relieved;
stabilized by hydrogen
bonding

STUDY PROBLEM

11.26. Write the formulas for the principal tautomers of the following compounds:

(a) CH_3CCH_2CH (b) $CH_3CCHCCH_3$ (c) CH_3C-
                              |
                             CH_3

Tautomerism can affect the reactivity of a compound. An exception to the generality that ketones are not easily oxidized is the oxidation of a ketone with at least one α hydrogen. A ketone that can undergo tautomerism can be oxidized by a strong oxidizing agent at the carbon–carbon double bond of the enol tautomer. Yields in this reaction are poor because, under these conditions, other C=C bonds may be cleaved. This reaction is not used in synthetic work, but is used often in structure determinations.

$$CH_3CH_2CH_2-\overset{O}{\overset{\|}{C}}-CH_2CH_3 \ \rightleftharpoons$$

3-hexanone

$$\begin{cases} \underset{CH_3CH_2CH=CCH_2CH_3}{\overset{OH}{|}} \xrightarrow[\text{heat}]{\text{conc. HNO}_3} 2\ CH_3CH_2CO_2H \\ \qquad\qquad + \qquad\qquad\qquad\qquad\quad \text{propanoic acid} \\ \underset{CH_3CH_2CH_2C=CHCH_3}{\overset{OH}{|}} \xrightarrow[\text{heat}]{\text{conc. HNO}_3} \begin{cases} CH_3CH_2CH_2CO_2H \\ \text{butanoic acid} \\ + \\ CH_3CO_2H \\ \text{acetic acid} \end{cases} \end{cases}$$

A. Tautomerism in carbohydrate metabolism

The first step in the metabolism of carbohydrates (starches and sugars) is their breakdown to glucose in the digestive tract. This breakdown is the hydrolysis of the acetal bonds. In the cells of an organism, glucose is eventually converted to CO_2 and H_2O. The first step in this cellular sequence of reactions is the formation of <u>glucose 6-phosphate</u>, followed by <u>isomerism to fructose 6-phosphate</u>. The isomerism reaction is simply an enzyme-catalyzed tautomerism that proceeds by way of an *enediol intermediate*, an intermediate that can lead to two carbonyl products.

$$
\begin{array}{ccc}
\text{HC=O} & \left[\begin{array}{c}\text{HCOH}\\ \|\| \\ \text{COH}\end{array}\right] & \text{H}_2\text{COH} \\
| & & | \\
\text{H—C—OH} & & \text{C=O} \\
| & & | \\
\textit{an aldehyde} & \textit{an enediol} & \textit{a ketone}
\end{array}
$$

$$
\begin{array}{ccc}
\text{O} & & \text{O} \\
\| & & \| \\
\text{CH} & & \text{CH} \\
| & & | \\
\text{H—C—OH} & \overset{\text{enzyme-}}{\underset{\text{phosphate}}{\rightleftharpoons}} & \text{H—C—OH} \\
| & & | \\
\text{HO—C—H} & & \text{HO—C—H} \qquad \rightleftharpoons \\
| & & | \\
\text{H—C—OH} & & \text{H—C—OH} \\
| & & | \\
\text{H—C—OH} & & \text{H—C—OH} \\
| & & | \\
\text{CH}_2\text{OH} & & \text{CH}_2\text{OPO}_3\text{H}^- \\
\text{glucose} & & \text{glucose 6-phosphate}
\end{array}
$$

$$
\begin{array}{ccc}
\boxed{\begin{array}{c}\text{CHOH}\\ \|\| \\ \text{COH}\end{array}} & & \text{CH}_2\text{OH} \\
| & & | \\
\text{HO—C—H} & & \text{C=O} \\
| & & | \\
\text{H—C—OH} & \rightleftharpoons & \text{HO—C—H} \\
| & & | \\
\text{H—C—OH} & & \text{H—C—OH} \\
| & & | \\
\text{CH}_2\text{OPO}_3\text{H}^- & & \text{H—C—OH} \\
\textit{an enediol} & & | \\
& & \text{CH}_2\text{OPO}_3\text{H}^- \\
& & \text{fructose 6-phosphate}
\end{array}
$$

STUDY PROBLEM

11.27. After fructose 6-phosphate is formed, it is converted enzymatically to the 1,6-diphosphate, which in turn is cleaved into two three-carbon compounds:

$$
\begin{array}{cc}
\text{CH}_2\text{OPO}_3\text{H}^- & \text{CHO} \\
| & \vdots \\
\text{C=O} & \text{H—C—OH} \\
| & \vdots \\
\text{CH}_2\text{OH} & \text{CH}_2\text{OPO}_3\text{H}^- \\
\text{dihydroxyacetone} & \text{glyceraldehyde} \\
\text{phosphate} & \text{3-phosphate}
\end{array}
$$

Before being carried on to the next step, the dihydroxyacetone phosphate is also converted to glyceraldehyde 3-phosphate. Suggest an intermediate in this conversion.

SECTION 11.18.

Alpha Halogenation

Ketones are readily halogenated at the α carbon. The reaction requires either alkaline conditions or an acidic catalyst. (Note that base is a *reactant*, whereas acid is a *catalyst*.)

In base:

$$CH_3\overset{O}{\overset{\|}{C}}CH_3 + Br_2 + OH^- \longrightarrow BrCH_2\overset{O}{\overset{\|}{C}}CH_3 + Br^- + H_2O$$

acetone   bromoacetone

cyclohexanone $+ Br_2 + OH^- \longrightarrow$ 2-bromo-cyclohexanone $+ Br^- + H_2O$

In acid:

$$CH_3\overset{O}{\overset{\|}{C}}CH_3 + Br_2 \xrightarrow{H^+} BrCH_2\overset{O}{\overset{\|}{C}}CH_3 + HBr$$

cyclohexanone $+ Br_2 \xrightarrow{H^+}$ 2-bromocyclohexanone $+ HBr$

The first step (the slow step) in the reaction under alkaline conditions is the formation of the enolate ion. The anion of a ketone with only one carbonyl group is a much stronger base than the hydroxide ion. Therefore, the acid–base equilibrium favors the hydroxide ion rather than the enolate ion. Nonetheless, a *few* enolate ions exist in alkaline solution. As these few anions are used up, more are generated to go on to Step 2. In Step 2, the enolate ion quickly undergoes reaction with halogen to yield the α-halogenated ketone and a halide ion.

In base:

Step 1 (slow):

$$CH_3\overset{O}{\overset{\|}{C}}CH_3 + OH^- \underset{-H_2O}{\rightleftharpoons} \left[CH_3\overset{\ddot{O}:}{\overset{\|}{C}}\overset{..}{C}H_2 \longleftrightarrow CH_3\overset{:\ddot{O}:^-}{\overset{|}{C}}=CH_2 \right]$$

resonance structures for the enolate ion

Step 2 (fast):

$$CH_3\overset{O}{\overset{\|}{C}}\overset{..}{C}H_2 + :\overset{..}{\underset{..}{Br}}-\overset{..}{\underset{..}{Br}}: \longrightarrow CH_3\overset{O}{\overset{\|}{C}}CH_2\overset{..}{\underset{..}{Br}}: + :\overset{..}{\underset{..}{Br}}:^-$$

STUDY PROBLEMS

11.28. Upon which species does the rate of base-promoted α halogenation depend? Would the reaction follow first-order or second-order kinetics?

11.29. One of the disadvantages of base-promoted α halogenation of a ketone is that a second halogen atom is introduced more easily than the first.

$$\underset{\text{O}}{\overset{\text{O}}{\underset{\|}{CH_3\overset{\|}{C}CH_2Br}}} + Br_2 + OH^- \longrightarrow CH_3\overset{\|}{C}CHBr_2 + Br^- + H_2O$$

Write the steps in the mechanism of the second halogenation and suggest a reason why this reaction has a faster rate than that of the first halogenation.

Alpha halogenation in acid usually gives higher yields than the reaction in base. The acid-catalyzed reaction proceeds by way of the enol, the formation of which is the rate-determining step. The carbon–carbon double bond of the enol undergoes electrophilic addition, just like any carbon–carbon double bond, to form the more stable carbocation. In this case, the more stable carbocation is the one in which the positive charge is on the carbon of the carbonyl group (because this intermediate is resonance-stabilized). This carbocation intermediate quickly loses a proton and forms the ketone, which is now halogenated in the alpha position.

In acid:

Step 1 (fast):

$$CH_3\overset{\overset{\cdot\cdot}{O}\cdot}{\underset{\|}{C}}CH_3 + H^+ \rightleftharpoons \left[CH_3\overset{\overset{+}{\overset{\cdot\cdot}{O}}H}{\underset{\|}{C}}CH_3\right]$$

Step 2 (slow):

$$\left[CH_3\overset{\overset{+}{\overset{\cdot\cdot}{O}}H}{\underset{\|}{C}}CH_2\;H\right] \rightleftharpoons \left[CH_3\overset{\overset{\cdot\cdot}{O}H}{\underset{|}{C}}=CH_2\right] + H^+$$

the enol

Step 3 (fast):

$$\left[CH_3\overset{OH}{\underset{|}{C}}=CH_2\right] + Br-Br \longrightarrow \left[CH_3\overset{\overset{\cdot\cdot}{O}H}{\underset{|}{C}}-CH_2Br\right] \longleftrightarrow \left[CH_3\overset{\overset{+}{\overset{\cdot\cdot}{O}}H}{\underset{\|}{C}}-CH_2Br\right] + Br^-$$

Step 4 (fast):

$$\left[CH_3\overset{\overset{+}{\overset{\cdot\cdot}{O}}H}{\underset{\|}{C}}CH_2Br\right] \rightleftharpoons CH_3\overset{\overset{\cdot\cdot}{O}\cdot}{\underset{\|}{C}}CH_2Br + H^+$$

STUDY PROBLEM

11.30. (a) What species are involved in the rate-determining step of acid-catalyzed α halogenation of acetone?
(b) What would be the relative rates of α bromination and α iodination?

A. Haloform reaction

Alpha halogenation is the basis of a chemical test, called the **iodoform test**, for methyl ketones. The methyl group of a methyl ketone is iodinated stepwise until the yellow solid iodoform (CHI_3) is formed.

Iodoform test:

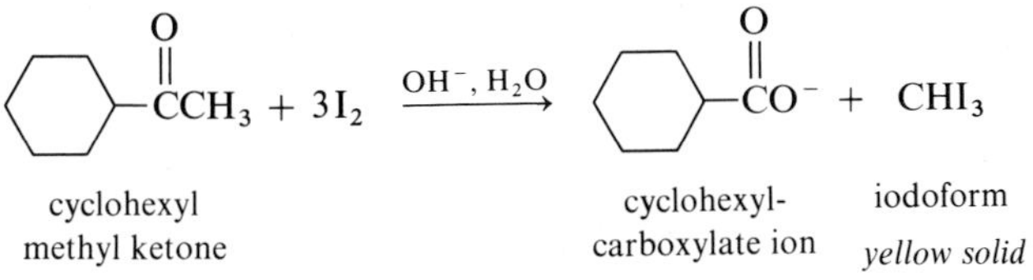

cyclohexyl
methyl ketone

cyclohexyl-
carboxylate ion

iodoform
yellow solid

Steps in the reaction:

(1) $RCCH_3$ $\underset{}{\overset{OH^-}{\rightleftharpoons}}$ $RCCH_2^-$ $\overset{I_2}{\longrightarrow}$ $RCCH_2I + I^-$

(2) $RCCH_2I$ $\underset{}{\overset{OH^-}{\rightleftharpoons}}$ $RC\overset{-}{C}HI$ $\overset{I_2}{\longrightarrow}$ $RCCHI_2 + I^-$

(3) $RCCHI_2$ $\underset{}{\overset{OH^-}{\rightleftharpoons}}$ $RC\overset{-}{C}I_2$ $\overset{I_2}{\longrightarrow}$ $RCCI_3 + I^-$

(4)

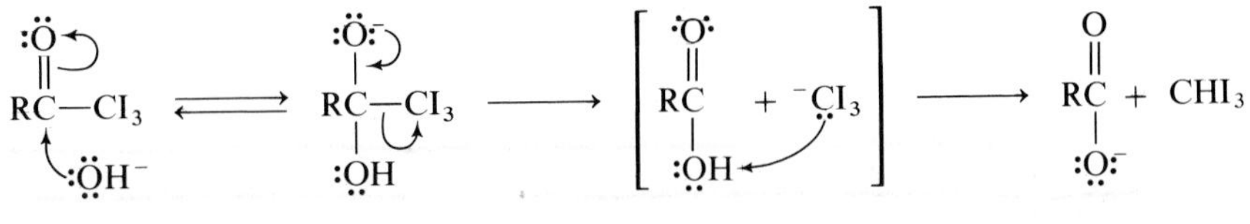

The test is not uniquely specific for methyl ketones. Iodine is a mild oxidizing agent, and any compound that can be oxidized to a methyl carbonyl compound also gives a positive test.

CH_3CH_2OH $\overset{I_2}{\longrightarrow}$ CH_3CH $\underset{OH^-}{\overset{I_2}{\longrightarrow}}$ $HCO^- + CHI_3$

ethanol

acetaldehyde

formate ion

$CH_3\overset{OH}{\underset{}{C}}HCH_3$ $\overset{I_2}{\longrightarrow}$ CH_3CCH_3 $\underset{OH^-}{\overset{I_2}{\longrightarrow}}$ $CH_3CO^- + CHI_3$

2-propanol

acetone

acetate ion

Bromine and chlorine also undergo reaction with methyl ketones to yield bromoform ($CHBr_3$) and chloroform ($CHCl_3$), respectively. "Haloform" is the general term used to describe CHX_3; hence this reaction is often referred to as the **haloform reaction**. Because bromoform and chloroform are nondistinctive liquids, their formations are not useful for test purposes. However, the reaction of a methyl ketone with any of these halogens provides a method for the conversion of these compounds to carboxylic acids.

STUDY PROBLEMS

11.31. Which of the following compounds gives a positive iodoform test?

(a) ICH_2CH (with =O above CH)
(b) CH_3CH_2CH (with =O above CH)
(c) $CH_3CH_2CHCH_3$ (with OH above CH)

11.32. What methyl ketones can be used to prepare the following carboxylic acids by haloform reactions?

(a) [bicyclic decalin structure with CO_2H substituent]

(b) $(CH_3)_2CHCO_2H$

(c) HO_2C—[benzene ring]—CO_2H

SECTION 11.19.

1,4-Addition to α,β-Unsaturated Carbonyl Compounds

A.   Electrophilic 1,4-addition

When an alkene undergoes reaction with HCl, the reaction proceeds by electrophilic attack of H^+ to yield the more stable carbocation, followed by attack of Cl^-.

$$CH_3CH{=}CH_2 \xrightarrow{\;H^+\;} [CH_3\overset{+}{C}HCH_3] \xrightarrow{\;Cl^-\;} CH_3\overset{\underset{|}{Cl}}{C}HCH_3$$

propene   2-chloropropane

An *α,β-unsaturated* aldehyde or ketone has a carbon–carbon double bond in conjugation with a carbonyl group. The carbon–carbon double bond in an alkene is nonpolar. However, a carbon–carbon double bond in conjugation with a carbonyl group is *polar*, as the following resonance structures indicate.

$$CH_2{=}CH{-}CH{\overset{\overset{\ddot{O}:}{\|}}{}} \longleftrightarrow CH_2{=}CH{-}\overset{+}{C}H{\overset{:\ddot{O}:^-}{}} \longleftrightarrow {}^+CH_2{-}CH{=}CH{\overset{:\ddot{O}:^-}{}}$$

The resonance structures show that the *β* carbon, as well as the carbonyl carbon, carries a partial positive charge, while the carbonyl oxygen carries a partial negative charge.

$$CH_2{=}CH{-}CH{\overset{\overset{O^{\delta-}}{\|}}{}}$$

β carbon is δ+   *carbonyl carbon is δ+*

Because the $C{=}C$ grouping in an α,β-unsaturated carbonyl compound is polarized, the mechanism for electrophilic addition is somewhat different from that for electrophilic addition to an isolated, nonpolar, alkene double bond. Let us consider two examples of electrophilic addition reactions of α,β-unsaturated carbonyl compounds; then we will discuss the mechanism.

$$CH_2=CHCH\overset{\overset{O}{\|}}{} + HCl \longrightarrow CH_2CH_2CH\overset{\overset{Cl}{|}\ \overset{O}{\|}}{}$$

propenal   3-chloropropanal

$$CH_3CH=CHCCH_3\overset{\overset{O}{\|}}{} + H_2O \xrightarrow{H^+} CH_3CHCH_2CCH_3\overset{\overset{OH}{|}\ \ \overset{O}{\|}}{}$$

3-penten-2-one   4-hydroxy-2-pentanone

Note that, in each reaction, the nucleophilic part of the reagent (not H^+) becomes bonded to the β carbon. The reason for this is that the β carbon has a partial positive charge. The initial attack by H^+ occurs not at this positive carbon, but at the partially negative oxygen of the carbonyl group.

$$\overset{\delta-\ddot{O}:\quad H^+}{\underset{\delta+}{CH_2}=CHCH}$$

The protonated intermediate is resonance-stablized. In this intermediate, the β carbon still carries a partial positive charge and can be attacked by a nucleophile.

Protonation:

$$CH_2=CHCH\overset{\overset{\ddot{O}:}{\|}}{} \ \underset{}{\overset{H^+}{\rightleftharpoons}}\ \left[CH_2=CHCH\overset{\overset{+\ddot{O}H}{\|}}{} \longleftrightarrow CH_2=CH-CH\overset{\overset{:\ddot{O}H}{|}}{}_+ \longleftrightarrow \overset{+}{CH_2}CH=CH\overset{\overset{:\ddot{O}H}{|}}{} \right]$$

resonance-stabilized

Attack of Nu⁻:

$$\left[{}^+CH_2CH=CH\overset{\overset{OH}{|}}{} \right] + Cl^- \longrightarrow \left[ClCH_2CH=CH\overset{\overset{OH}{|}}{} \right] \rightleftharpoons ClCH_2CH_2CH\overset{\overset{O}{\|}}{}$$

an enol of an aldehyde   *an aldehyde*

Note that this reaction is a **1,4-addition**, the same type of addition observed with conjugated dienes. The difference is that the initial addition product is an enol, which undergoes tautomerism to the final keto form of the aldehyde.

You might wonder why the nucleophile cannot also attack the carbonyl carbon (which also carries a partial positive charge in the intermediate). This attack on the carbonyl carbon can occur, but the product is unstable and reverts to starting material. This is a concurrent, but nonproductive, side reaction.

$$\left[CH_2=CHCH\underset{+}{} \right] + Cl^- \rightleftharpoons \left[CH_2=CHCH-Cl\overset{\overset{O\ H}{|}}{} \right] \rightleftharpoons CH_2=CHCH\overset{\overset{O}{\|}}{} + HCl$$

B. Nucleophilic 1,4-addition

The pi bond of an alkene is not normally attacked by a nucleophile unless there has been prior attack by an electrophile. However, a double bond in conjugation with a carbonyl group is polarized. In this case, nucleophilic addition can occur *either at the C=C double bond or at the C=O double bond* (at either of the two partially positive carbons).

$$CH_2=CHCCH_3 + {}^-CN \xrightarrow{HCN} CH_2=CHCCH_3 \quad \text{or} \quad CH_2-CH_2CCH_3$$
(with O double bond, OH, and CN substituents)

Let us look at the mechanism of each reaction. First, we will consider the attack of the cyanide ion (from HCN and dilute base) on the carbonyl group. In this case, the nucleophilic CN^- attacks the partially positive carbon of the carbonyl group. The reaction is no different from cyanohydrin formation by an ordinary ketone.

Attack of CN⁻ on the carbonyl carbon:

$$CH_2=CHCCH_3 + {:}CN \longrightarrow \left[CH_2=CHCCH_3 \atop CN \right] \underset{HCN}{\rightleftharpoons} CH_2=CHCCH_3 + CN^-$$

Now we will consider attack of the nucleophilic CN^- on the β carbon. This reaction is a 1,4-addition of CN^- and H^+ to the conjugated system. The product of the 1,4-addition is an enol, which forms the product ketone.

Attack of CN⁻ on the β carbon:

$$CH_2=CH-CCH_3 \rightleftharpoons \left[CH_2-CH-CCH_3 \atop CN \longleftrightarrow CH_2-CH=CCH_3 \atop CN \right] \xrightarrow{HCN}$$

$$CN^- + \left[CH_2-CH=CCH_3 \atop CN \right] \rightleftharpoons CH_2CH_2CCH_3 \atop CN$$

an enol

Which of the two addition reactions occurs? Sometimes both do, and a mixture of products results. In most cases, however, one product or the other predominates. Steric hindrance around the double bond or the carbonyl group may lead to preferred attack at the nonhindered position. Aldehydes, less hindered than ketones, usually undergo carbonyl attack.

$$(CH_3CH_2)_2C=CHCH \qquad CH_2=CCH_2CH_3 \atop CH_3$$

Nu⁻ attacks here

Nu⁻ attacks here

A highly basic nucleophile (such as RMgX or LiAlH$_4$) attacks preferentially at the carbonyl group, while a weaker base (such as CN$^-$ or R$_2$NH) usually attacks at the carbon–carbon double bond.

Stronger bases attack 1,2 (at C=O):

$$CH_2=CHCCH_3 + AlH_4^- \xrightarrow{\quad H_2O \quad} CH_2=CHCHCH_3$$

$$CH_2=CHCCH_3 + CH_3MgI \xrightarrow[H_2O]{H^+} CH_2=CHCCH_3$$

Weaker bases attack 1,4 (at C=C):

$$CH_2=CHCCH_3 + CH_3NH_2 \longrightarrow CH_2CH_2CCH_3$$

$$CH_2=CHCCH_3 + CN^- \xrightarrow{\quad HCN \quad} CH_2CH_2CCH_3$$

STUDY PROBLEM

11.33. Predict the major organic product of the reaction of each of the following reagents with 2-cyclohexen-1-one: (a) CH$_3$MgI (followed by H$^+$, H$_2$O); (b) 1 equivalent of H$_2$ with Ni catalyst (25°); (c) NaBH$_4$ (followed by H$^+$, H$_2$O); (d) NH$_3$.

SECTION 11.20.

Use of Aldehydes and Ketones in Synthesis

Aldehydes and ketones are readily available by the oxidation of alcohols and may be converted to a variety of other types of compounds, as may be seen in Table 11.6. When viewed from a synthesis standpoint, the reactions in this table can be grouped into three major categories:

1. reactions in which the carbonyl group is retained (α halogenation and 1,4-addition);

2. reactions in which the carbonyl group is converted to another functional group (for example, reduction or conversion to a hemiacetal);

3. reactions in which extension of the carbon skeleton occurs at the carbonyl group (Grignard and Wittig reactions).

In designing synthetic sequences, you should keep all three classes of reactions in mind.

TABLE 11.6. Types of compounds that can be obtained from aldehydes and ketones[a]

Reaction	Product	Section reference
Addition:[b]		

$$RCHO + 2R'OH \xrightarrow{H^+} \underset{\overset{|}{RCHOR'}}{\overset{OR'}{}}$$ acetal 11.8

$$RCHO + HCN \xrightarrow{CN^-} \underset{\overset{|}{RCHCN}}{\overset{OH}{}}$$ cyanohydrin 11.9

Addition–elimination:[b]

$RCHO + R'NH_2 \longrightarrow RCH{=}NR'$	imine	11.10A
$RCH_2CHO + R'_2NH \longrightarrow RCH{=}CHNR'_2$	enamine	11.10C, 14.5
$RCHO + R'NHNH_2 \longrightarrow RCH{=}NNHR'$	hydrazone	11.11

Wittig reaction:

$$R_2C{=}O + (C_6H_5)_3P{=}CR'_2 \longrightarrow R_2C{=}CR'_2 \quad \text{alkene} \quad 11.12$$

Grignard reaction:

$$\underset{RCR}{\overset{O}{\overset{\|}{}}} \xrightarrow[\text{(2) } H_2O, H^+]{\text{(1) } R'MgX} \underset{\underset{R'}{\overset{|}{}}{\overset{|}{RCR}}}{\overset{OH}{\overset{|}{}}}$$ alcohol 6.9, 7.4B, 11.13

Reduction:

$$R_2C{=}O \xrightarrow[\text{or metal hydride}]{H_2,\ \text{catalyst}} R_2CHOH$$ alcohol 11.14A and B

$$R_2C{=}O \xrightarrow[\text{or Zn/Hg, HCl}]{\text{(1) } NH_2NH_2,\ \text{(2) } {}^-OC(CH_3)_3} R_2CH_2$$ alkane or alkylbenzene 11.14C

$$R_2C{=}O \xrightarrow[H_2,\ Ni]{R'_2NH} R_2CHNR'_2$$ amine 11.14D, 15.5B

α Halogenation:

$$\underset{RCH_2CR}{\overset{O}{\overset{\|}{}}} + X_2 \xrightarrow{H^+} \underset{\underset{X}{\overset{|}{}}{\overset{\|}{RCHCR}}}{\overset{O}{}}$$ α-halo carbonyl 11.18

$$\underset{RCCH_3}{\overset{O}{\overset{\|}{}}} + 3X_2 \xrightarrow[\text{(2) } H^+]{\text{(1) } OH^-} RCO_2H$$ carboxylic acid 11.18A

1,4-Addition:

$$\underset{RCH{=}CHCR}{\overset{O}{\overset{\|}{}}} + Nu^- \text{ or } HNu \longrightarrow \underset{RCHCH_2CR}{\overset{Nu\ \ \ \ O}{\overset{|\ \ \ \ \ \|}{}}}$$ β-substituted carbonyl 11.19

---

[a] The condensation and alkylation reactions of aldehydes and ketones will be discussed in Chapter 14.

[b] Nonhindered ketones, such as methyl ketones, can also be used.

SAMPLE PROBLEMS

How would you make the following conversion?

$$CH_3\overset{\overset{\displaystyle O}{\|}}{C}CH_3 \longrightarrow CH_3\overset{\overset{\displaystyle O}{\|}}{C}CH_2OH$$

Solution: In this conversion, the carbonyl group is retained, but a functional group is inserted α to the C=O group. Alpha halogenation is a way to insert an α functional group.

$$CH_3\overset{\overset{\displaystyle O}{\|}}{C}CH_3 \xrightarrow{Cl_2,\ H^+} CH_3\overset{\overset{\displaystyle O}{\|}}{C}CH_2Cl$$

The alcohol can then be obtained by treatment of the α-chloroketone with aqueous NaOH (S_N2 reaction).

$$CH_3\overset{\overset{\displaystyle O}{\|}}{C}CH_2Cl \xrightarrow{^-OH} CH_3\overset{\overset{\displaystyle O}{\|}}{C}CH_2OH$$

How would you make the following conversion?

$$\langle\!\bigcirc\!\rangle\!-\!\overset{\overset{\displaystyle O}{\|}}{C}H \longrightarrow \langle\!\bigcirc\!\rangle\!-\!CH\!=\!CHCH_2CH_3$$

Solution: This conversion involves an extension of the carbon skeleton at the carbonyl group. The Wittig reaction could be used for this conversion.

$$C_6H_5\overset{\overset{\displaystyle O}{\|}}{C}H \xrightarrow{(C_6H_5)_3P=CHCH_2CH_3} \text{product}$$

The product could also be obtained from the alcohol $C_6H_5CH(OH)CH_2CH_2CH_3$ by dehydration (spontaneous in this case because the product C=C is in conjugation with a benzene ring). The alcohol, in turn, can be obtained by a Grignard reaction, another reaction that allows us to build up a carbon skeleton.

$$C_6H_5\overset{\overset{\displaystyle O}{\|}}{C}H \xrightarrow[\text{(2) } H_2O,\ H^+]{\text{(1) } CH_3CH_2CH_2MgBr} C_6H_5\overset{\overset{\displaystyle OH}{|}}{C}HCH_2CH_2CH_3 \xrightarrow{-H_2O} \text{product}$$

Suggest a synthesis for *N*-ethyl-3-hexylamine from organic compounds containing three or fewer carbon atoms.

Solution:

1. Write the structure: $CH_3CH_2\overset{\overset{\displaystyle NHCH_2CH_3}{|}}{C}HCH_2CH_2CH_3$

2. It is apparent that more than one step will be needed to synthesize this compound. We must both build up a carbon skeleton and also insert the ethylamino group. Let us consider the amino group first. It could be placed in this structure by reductive amination of a ketone.

$$CH_3CH_2\overset{\overset{\displaystyle O}{\|}}{C}CH_2CH_2CH_3 \xrightarrow[H_2,\ Ni]{CH_3CH_2NH_2} \text{product}$$

3.   Is there a one-step reaction to 3-hexanone from starting materials containing three or fewer carbons? No, but a Grignard reaction can provide us with an alcohol, which is readily oxidized to a ketone. Thus, we can write the initial series of reactions.

$$\underset{\substack{\| \\ \text{O}}}{\text{CH}_3\text{CH}_2\text{CH}} \xrightarrow[\text{(2) H}_2\text{O, H}^+]{\text{(1) CH}_3\text{CH}_2\text{CH}_2\text{MgBr}} \underset{\substack{| \\ \text{OH}}}{\text{CH}_3\text{CH}_2\text{CHCH}_2\text{CH}_2\text{CH}_3} \xrightarrow{\text{H}_2\text{CrO}_4}$$

$$\underset{\substack{\| \\ \text{O}}}{\text{CH}_3\text{CH}_2\text{CCH}_2\text{CH}_2\text{CH}_3}$$

The Grignard reagent can be prepared by the reaction of RX with Mg.

$$\text{CH}_3\text{CH}_2\text{CH}_2\text{Br} \xrightarrow[\text{diethyl ether}]{\text{Mg}} \text{CH}_3\text{CH}_2\text{CH}_2\text{MgBr}$$

The proposed synthesis is complete, and the series of reactions may be written forward.

STUDY PROBLEM

11.34. Suggest synthetic routes to the following compounds from aldehydes or ketones of six or fewer carbon atoms and other appropriate reagents:

(a) 2-phenylpropene   (b) —CO$_2$H   (c) 2-cyclopentyl-1-heptene

Summary

The **carbonyl group** is *planar* and *polar*, and the oxygen has *two filled orbitals*. The carbonyl group may undergo *electrophilic* or *nucleophilic* attack.

$$\overset{\delta+\quad\delta-}{\underset{\underset{E^+}{}}{\text{C}=\ddot{\text{O}}:}} \longrightarrow \text{C}=\overset{+}{\ddot{\text{O}}}\text{E} \qquad \underset{\underset{Nu^-}{}}{\text{C}=\overset{\frown}{\ddot{\text{O}}}:} \longrightarrow \underset{\underset{Nu}{}}{-\overset{|}{\underset{|}{\text{C}}}-\ddot{\text{O}}:^-}$$

Because of (1) *inductive stabilization* of the partial positive charge on the carbonyl carbon, and (2) *steric hindrance*, ketones are less reactive than aldehydes:

$$\underset{\substack{\| \\ \text{O}}}{\text{RCR}} < \underset{\substack{\| \\ \text{O}}}{\text{RCH}} < \underset{\substack{\| \\ \text{O}}}{\text{HCH}}$$

Many reactions of aldehydes and ketones are simple **addition reactions** at the C=O pi bond. These reactions can lead to hydrates, hemiacetals (or hemiketals) and cyanohydrins.

$$
\underset{\text{R}-\overset{\overset{\text{O}}{\|}}{\text{C}}-\text{H}}{} + \text{H}-\text{Nu} \ \rightleftharpoons \ \text{R}-\underset{\underset{\text{Nu}}{|}}{\overset{\overset{\text{OH}}{|}}{\text{C}}}-\text{H}
$$

where HNu = H_2O, ROH, or HCN

Other addition reactions are **reduction** and **Grignard reactions**.

$$
\underset{\text{RCR}'}{\overset{\overset{\text{O}}{\|}}{}} + \text{H}-\text{H} \ \xrightarrow[\text{heat, pressure}]{\text{catalyst}} \ \underset{\text{RCHR}'}{\overset{\overset{\text{OH}}{|}}{}}
$$

$$
\underset{\text{RCR}}{\overset{\overset{\text{O}}{\|}}{}} + \text{R}'\text{MgX} \ \xrightarrow{\text{diethyl ether}} \ \underset{\underset{\text{R}'}{|}}{\underset{\text{RCR}}{\overset{\overset{\text{OMgX}}{|}}{}}}
$$

Substitution reactions of aldehydes and ketones are the result of initial addition reactions followed by elimination, or **addition–elimination reactions**. The formation of imines, enamines, hydrazones, and alkenes (by the Wittig reaction) all fall into this category. These reactions are summarized in Table 11.6.

$$
\underset{\text{RCR}}{\overset{\overset{\text{O}}{\|}}{}} \ \xrightarrow[\text{addition}]{\text{R}'\text{NH}_2} \ \left[\underset{\underset{\text{NHR}'}{|}}{\underset{\text{RCR}}{\overset{\overset{\text{OH}}{|}}{}}} \right] \ \xrightarrow[\text{elimination}]{-\text{H}_2\text{O}} \ \underset{\text{RCR}}{\overset{\overset{\text{NR}'}{\|}}{}}
$$

Because the carbonyl group is polar and because its pi electrons can participate in resonance-stabilization, an α *hydrogen is acidic*, especially if it is α to two carbonyl groups. This acidity can give rise to **tautomerism**. Because of tautomerism, ketones can undergo **α halogenation**, as shown in Table 11.6, or **oxidative cleavage** between the carbonyl carbon and the α carbon.

$$
\underset{\text{CH}_3\text{CH}}{\overset{\overset{\text{O}}{\|}}{}} + \text{OH}^- \ \rightleftharpoons \ {}^-\underset{\text{CH}_2\text{CH}}{\overset{\overset{\text{O}}{\|}}{}} + \text{H}_2\text{O}
$$

acidic

$$
\underset{\text{CH}_3\text{CCH}_2\text{CCH}_3}{\overset{\overset{\text{O}\quad\text{O}}{\|\quad\|}}{}} + \text{OH}^- \ \rightleftharpoons \ \underset{\text{CH}_3\text{CCHCCH}_3}{\overset{\overset{\text{O}\quad\text{O}}{\|\ -\ \|}}{}} + \text{H}_2\text{O}
$$

$$
\underset{\text{CH}_3\text{C}-\text{CH}-\text{CCH}_3}{\overset{\overset{\text{O}\ \ \text{H}\ \ \text{O}}{\|\ \ |\ \ \|}}{}} \ \rightleftharpoons \ \underset{\text{CH}_3\text{C}=\text{CH}-\text{CCH}_3}{\overset{\overset{\text{OH}\qquad\text{O}}{|\qquad\|}}{}}
$$

$$
\quad\quad\quad\quad \textit{keto tautomer} \quad\quad\quad\quad\quad\quad \textit{enol tautomer}
$$

When a C=C double bond is in conjugation with C=O, addition reactions may occur 1,2 or 1,4.

Electrophilic:

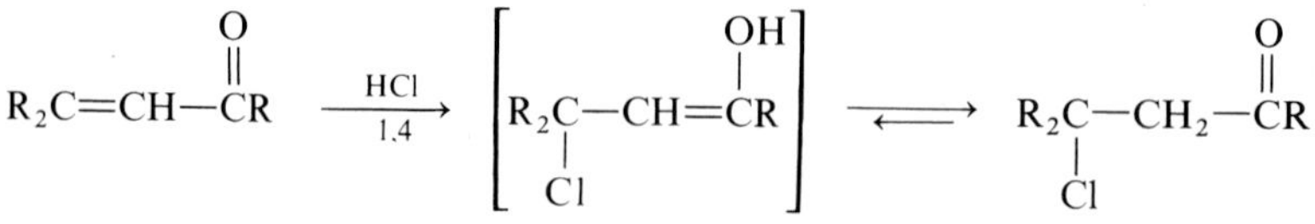

$$R_2C=CH-\overset{\overset{\displaystyle O}{\|}}{C}R \xrightarrow[1,4]{HCl} \left[R_2\underset{\underset{\displaystyle Cl}{|}}{C}-CH=\overset{\overset{\displaystyle OH}{|}}{C}R \right] \rightleftharpoons R_2\underset{\underset{\displaystyle Cl}{|}}{C}-CH_2-\overset{\overset{\displaystyle O}{\|}}{C}R$$

Nucleophilic:

$$\underset{\substack{nonhindered\ C=O}}{R_2C=CH\overset{\overset{\displaystyle O}{\|}}{C}H} + \underset{\substack{strong\ base}}{R'MgX} \xrightarrow{1,2} R_2C=CH\underset{\underset{\displaystyle R'}{|}}{\overset{\overset{\displaystyle OMgX}{|}}{C}}H$$

$$\underset{\substack{hindered\ C=O}}{CH_2=CH\overset{\overset{\displaystyle O}{\|}}{C}R} + \underset{\substack{weak\ base}}{NH_3} \xrightarrow{1,4} \left[CH_2-CH=\overset{\overset{\displaystyle OH}{|}}{\underset{\underset{\displaystyle NH_2}{|}}{C}}R \right] \longrightarrow CH_2CH_2\overset{\overset{\displaystyle O}{\|}}{C}R$$
$$\underset{\underset{\displaystyle NH_2}{|}}{}$$

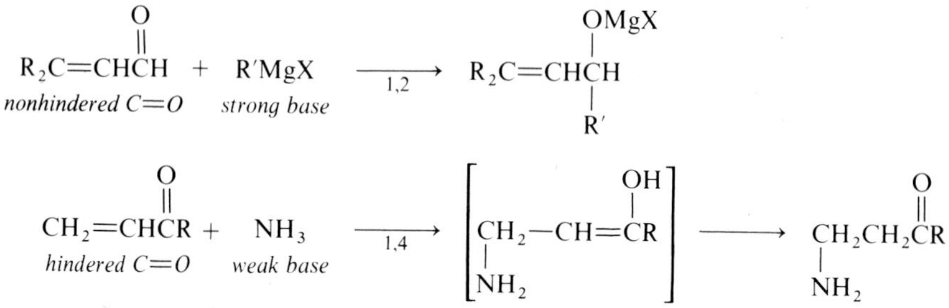

STUDY PROBLEMS

11.35. Write the structures for: (a) 2-methyl-1-cyclopentanone; (b) *sec*-butyl isopropyl ketone; (c) bromoacetone; (d) 2-iodopentanal; (e) 2-methyl-3-heptanone.

11.36. Write an acceptable name for each of the following structures:

(a) 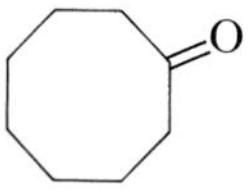   (b)

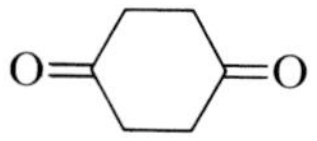

(c) $C_6H_5CH_2CHO$ (d) $(CH_3)_2CH\overset{\overset{\displaystyle O}{\|}}{C}CH_2CH_2CH_3$

(e) $BrCH_2CH_2CH_2CHO$ (f) $(CH_3)_2CHCHO$

11.37. Give formulas for: (a) a β-ketoaldehyde; (b) an α,β-unsaturated ketone; (c) an α-bromoaldehyde; (d) a β-hydroxyketone.

11.38. Tell whether each of the following equations represents an *electrophilic* attack or a *nucleophilic* attack on the carbonyl group:

(a) 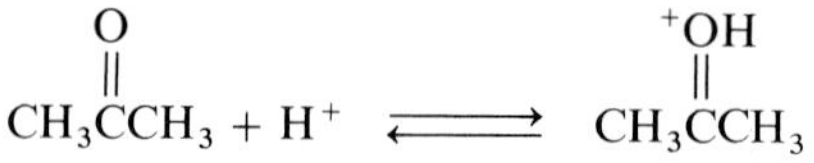 $CH_3\overset{\overset{\displaystyle O}{\|}}{C}CH_3 + H^+ \rightleftharpoons CH_3\overset{\overset{\displaystyle ^+OH}{\|}}{C}CH_3$

(b) 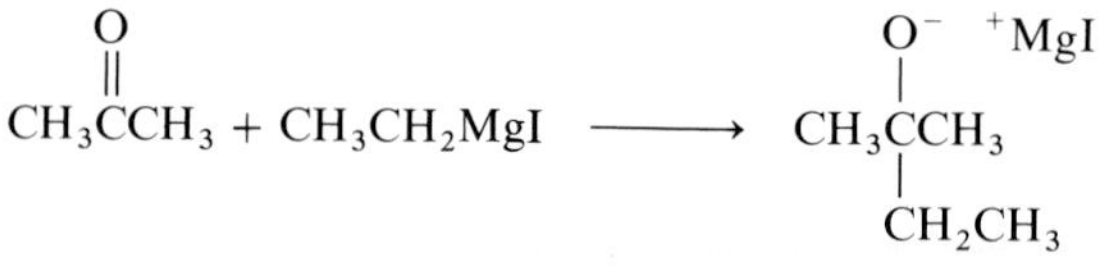 $CH_3\overset{\overset{\displaystyle O}{\|}}{C}CH_3 + CH_3CH_2MgI \longrightarrow CH_3\underset{\underset{\displaystyle CH_2CH_3}{|}}{\overset{\overset{\displaystyle O^-\ ^+MgI}{|}}{C}}CH_3$

(c) $CH_3CH_2CHO + Li^+ \ AlH_4^-$ $\longrightarrow$ $\left[\begin{array}{c} CH_3CH_2CH \text{---} O \\ \vdots \vdots \\ H \text{-----} AlH_3 \end{array} \right]^-$ Li^+

11.39. Which of the following aldehydes would form the most stable hydrate? the least stable? Why?

(a) $BrCH_2CHO$ (b) Br_2CHCHO (c) Br_3CCHO

11.40. Write the equation for hydrate formation by Br_2CHCHO.

11.41. Write equations for the reactions leading to the hemiacetal and acetal of: (a) propanal and methanol; (b) acetone and 1,2-ethanediol; (c) 5-hydroxy-2-hexanone and methanol.

11.42. Each of the following compounds is dissolved in water to which a trace of HCl has been added. Give the structures of any other compounds (besides HCl, water, and the compound in question) that would be found in each solution.

(a)   (b) CH_3 (c)

11.43. Write equations for the reactions of   (a) acetone with 2,4-dinitrophenylhydrazine; (b) cyclohexanone with $NaCN + H_2SO_4$;   (c) benzaldehyde with aniline;   (d) propanal with $(CH_3)_2NH$.

11.44. Show how you would prepare the following compounds from compounds containing six or fewer carbon atoms:

(a) $CH_3CH_2CH{=}CHN(CH_2CH_3)_2$ (b)

(c) $CH_3CH_2CH{=}NC_6H_5$

11.45. Predict the major organic products:

(a) $\xrightarrow[\text{warm}]{H_2SO_4}$

(b) $+ \ (C_6H_5)_3P{=}CH_2$ $\longrightarrow$

(c) $\xrightarrow[\text{(2) } H_2O, \ H^+]{\text{(1) } NaC{\equiv}CCH_3}$

(d) $+ \ HSCH_2CH_2SH$ $\underset{}{\overset{H^+}{\rightleftharpoons}}$

(e) $+ \ H_2NOH$ $\underset{}{\overset{H^+}{\rightleftharpoons}}$

(f) 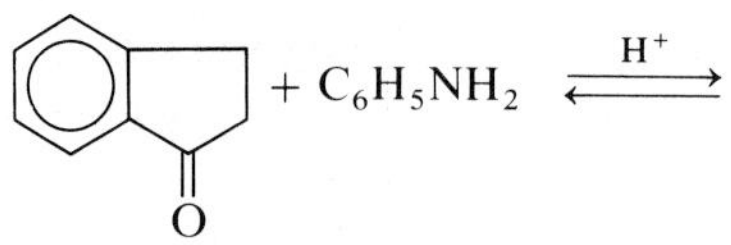 $+ C_6H_5NH_2 \xrightleftharpoons{H^+}$

(g) 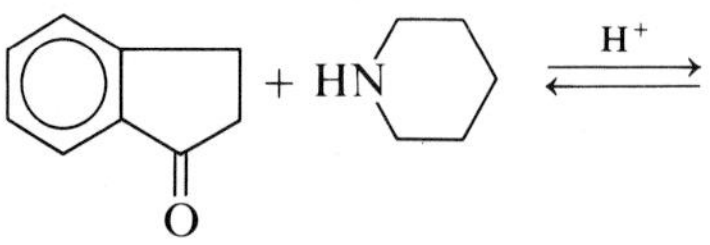 $+ HN \bigcirc \xrightleftharpoons{H^+}$

11.46. Give the product of the reaction of cyclopentanone with:

   (a) Br_2 in acetic acid
   (b) $NaBH_4$, followed by H^+, H_2O
   (c) phenylhydrazine with H^+
   (d) $(CH_3)_2CHMgBr$, followed by H^+, H_2O
   (e) $(C_6H_5)_3P\!=\!C(CH_3)_2$
   (f) hydrazine, followed by heating in K^+ $^-OC(CH_3)_3$ solution

11.47. List the following compounds in terms of reactivity toward 2,4-dinitrophenylhydrazine (least reactive first):   (a) 2-pentanone;   (b) 3-pentanone;   (c) pentanal.

11.48. How would you effect the following conversions by Wittig reactions? (Start with alkyl halides.)

   (a)

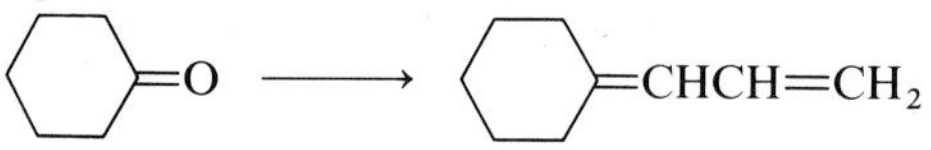

   (b)

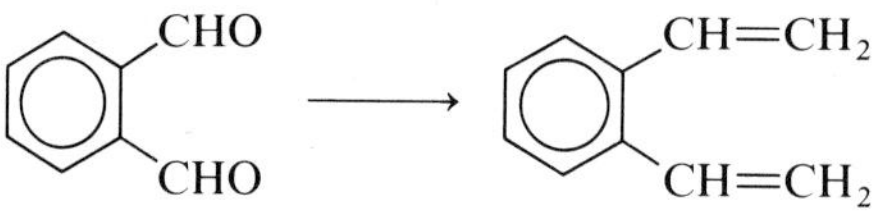

   (c) 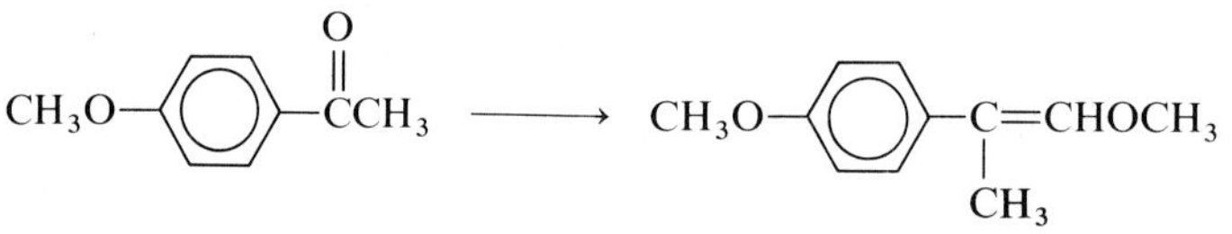 $CH_3O-$$\overset{\overset{\textstyle O}{\|}}{-C}CH_3 \longrightarrow CH_3O-$$-\underset{\underset{\textstyle CH_3}{|}}{C}\!=\!CHOCH_3$

11.49. Predict the major organic products:

   (a) 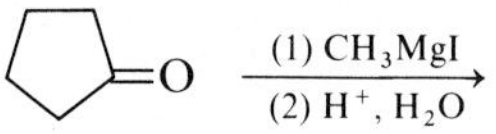 $\xrightarrow[\text{(2) } H^+, H_2O]{\text{(1) } CH_3MgI}$

   (b) 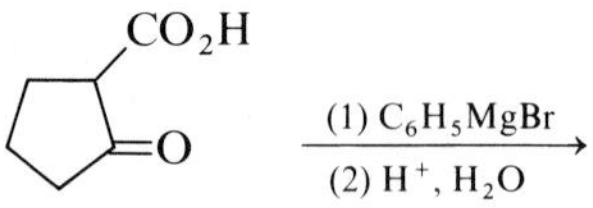 $\xrightarrow[\text{(2) } H^+, H_2O]{\text{(1) } C_6H_5MgBr}$

   (c) HCHO $\xrightarrow[\text{(2) } H^+, H_2O]{\text{(1)}}$

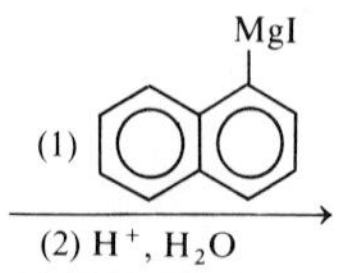

11.50. Suggest a synthesis for each of the following compounds, starting with iodomethane and other appropriate reagents: (a) 3-methyl-3-pentanol; (b) ethanol; (c) 2-pentanol.

11.51. Predict the organic products:

(a) $OHCCH_2CH_2CHO$ $\xrightarrow[\text{heat, pressure}]{\text{excess } H_2, \text{ Pt}}$

(b) ⬡$=O$ + MnO_4^- $\xrightarrow[25°]{CH_3CO_2H}$

(c) ⬡$-CHO$ + MnO_4^- $\xrightarrow[25°]{OH^-}$

(d) $CH_3\overset{O}{\overset{\|}{C}}CH_3$ + NH_3 $\xrightarrow{H_2, \text{ Pt}}$

(e) $CH_3\overset{O}{\overset{\|}{C}}CH_3$ + $HOCH_2CH_2NH_2$ $\xrightarrow{H_2, \text{ Pt}}$

(f) (cyclopentanone with $CH_2CH_2CH_2NH_2$ substituent) $\xrightarrow{H_2, \text{ Pt}}$

(g) ⬡$-\overset{O}{\overset{\|}{C}}CH_3$ $\xrightarrow[OH^-]{Cl_2, H_2O}$

11.52. Which of the following compounds would give a positive Tollens test?

(a) CH_3CHO (b) $CH_3\overset{O}{\overset{\|}{C}}CH_3$ (c) (ring)$-OH$ (d) (ring)$-OCH_3$

11.53. Suggest a synthetic scheme to convert 3-hydroxypropanal to HO_2CCH_2CHO.

11.54. Show how the following compounds could be prepared from four-carbon aldehydes or ketones: (a) 3-chloropropanoic acid; (b) 4-octene; (c) 3-methyl-3-heptanol; (d) di-*n*-butyl ether; (e) *n*-butylamine; (f) 3-butenoic acid; (g) 3-buten-1-ol.

11.55. Suggest syntheses for the following compounds from starting materials containing six or fewer carbon atoms and appropriate inorganic reagents. (Triphenylphosphine may also be used.)

(a) $(CH_3CH_2)_3COH$ (b) ⬡$=CHCH_2CH=$⬡

(c) (spiro dioxolane structure)   (d) ⬡$-CO_2H$

11.56. Rank the following compounds in terms of increasing acidity (least acidic first):

(a) 2,4-pentanedione   (b) butanal   (c) water

11.57. Complete the following equations for acid–base reactions:

(a) $CH_3\overset{\overset{\textstyle O}{\|}}{C}H + OH^-$ ⇌

(b) [structure: cyclopentane-1,3-dione with two C=O] $+ CO_3{}^{2-}$ ⇌

(c) $(C_6H_5)_2CH\overset{\overset{\textstyle O}{\|}}{C}OCH_2CH_3 + Na^+ \ ^-OCH_2CH_3$ ⇌

(d) $CH_3CH_2\overset{\overset{\textstyle O}{\|}}{C}CH_3 + \text{excess } D_2O \xrightarrow{\ ^-OD\ }$

(e) $(S)\text{-}CH_3CH_2\overset{\overset{\textstyle CH_3}{|}}{C}HCH_2CHO + OH^-$ ⇌

(f) $(S)\text{-}CH_3CH_2CH_2\overset{\overset{\textstyle CH_3}{|}}{C}HCHO + OH^-$ ⇌

11.58. Each of the following compounds is about as acidic as a β-diketone. Show the resonance structures for the anions of each that would account for this acidity.

(a) CH_3NO_2 (b) $CH_3\overset{\overset{\textstyle O}{\|}}{C}CH_2CN$ (c) $C_6H_5CH_2\overset{\overset{\textstyle O}{\|}}{C}H$

11.59. Write equations that illustrate tautomerism of the following compounds:

(a) [cyclopentanone-2-one structure]   (b) [phenyl-CCH_3 with C=O]   (c) [cyclic HN–C=O ... NH structure]   (d) [cyclohexyl-N=O structure]

11.60. Predict the organic products:

(a) [phenyl-$\overset{\overset{\textstyle O}{\|}}{C}CH_3$] $+ Cl_2 \xrightarrow{H^+}$

(b) $O={}$[cyclohexadiene ring] $+ Cl_2 \xrightarrow{CCl_4}$

(c) [cyclohexane-1,3-dione structure] $+ Br_2 \xrightarrow{H^+}$

11.61. What would be observed if each of the following compounds were placed in a test tube and a solution of I_2 in dilute aqueous NaOH were added?

(a) cyclohexyl–$\overset{\overset{\displaystyle O}{||}}{C}CH_3$
(b) cyclohexyl–$CH_2\overset{\overset{\displaystyle O}{||}}{C}OCH_3$
(c) cyclohexyl=O

(d) cyclohexyl–$\overset{\overset{\displaystyle OH}{|}}{C}HCH_3$
(e) $CH_3CH_2CH_2CH_2\overset{\overset{\displaystyle O}{||}}{C}CH_3$
(f) cyclohexyl–$\overset{\overset{\displaystyle O}{||}}{C}H$

11.62. How would you differentiate the following pairs of compounds by simple chemical tests?

(a) cyclohexanone and cyclohexanol
(b) 2-pentanone and 3-pentanone
(c) pentanal and 2-pentanone
(d) 2-pentene and 2-pentanone

11.63. Complete the following equations:

(a) $CH_2{=}CHCH\overset{\overset{\displaystyle O}{||}}{C}CH_3 + HBr \longrightarrow$

(b) $CH_2{=}CHCH_2CH_2\overset{\overset{\displaystyle O}{||}}{C}CH_3 + HCl \longrightarrow$

11.64. Show how you could synthesize the following compounds from organic starting materials containing only C, H, and O:

(a) $CH_3CH_2NHCH_2CH_2\overset{\overset{\displaystyle O}{||}}{C}CH_2CH_3$
(b) cyclohexane with CN and $-\overset{\overset{\displaystyle }{|}}{\underset{\underset{\displaystyle O}{||}}{C}}CH_3$ substituents

11.65. Suggest a technique by which ^{18}O-labeled aldehydes and ketones might be used to determine the relative rates of hydrate formation.

11.66. In the reductive amination of RCHO with ammonia, a secondary amine is a common by-product. Suggest a general structure for this secondary amine and explain how it is formed.

11.67. When quinone (page 486) is treated with HCl, a chlorohydroquinone is formed. Write the equation for this reaction, showing the intermediates.

11.68. When 3,3,5-trimethyl-1-cyclohexanone is treated with $LiAlH_4$, followed by hydrolysis, an 83% yield of one diastereomer of an alcohol is obtained. Write the formula for this diastereomer, and explain why it is the principal product.

11.69. Explain why an α-alkylcyclohexanone tends to form the less-substituted enamine. (*Hint:* Consider the resonance structures of the enamine.)

11.70. The LAH reduction of a stereoisomeric 3-phenyl-2-pentanone yielded 75% of the (2R,3S) alcohol and 25% of the (2S,3S) alcohol.
(a) What was the stereochemistry of the starting ketone?
(b) Why was the (2R,3S) alcohol the predominant product? (*Hint:* Use models.)

11.71. Write flow equations for syntheses of the following compounds:

(a) 4-methoxy-2-phenyl-2-butene from compounds containing six or fewer carbons
(b) 4-methyl-1,3-pentadiene from acetone
(c) 3,5-heptanediol from compounds containing four or fewer carbons
(d) 1,3-cyclohexanedione from cyclohexanone

(e) from cyclopentanol

(f) 2,2-dimethylpropanoic acid from 2,3-dimethyl-2,3-butanediol

11.72. Suggest syntheses for: (a) *biacetyl* (2,3-butanedione), which is used in margarine as a butter flavoring, from compounds containing only one or two carbons; (b) 2-chloro-1-phenylethanone (α-chloroacetophenone), the lachrymator used in Mace, from benzene.

11.73. (a) 3-Phenyl-2-butanone can form two enols, but one is preferred. Which one?
(b) If 3-phenyl-2-butanone is dissolved in D_2O with an acidic or basic catalyst, then recovered, it is found to contain deuterium in its structure at primarily one location. Which location?
(c) If (R)-3-phenyl-2-butanone is dissolved in aqueous acid or base, what would you expect to happen to it?

11.74. Predict the major organic product:

$$CH_2{=}\underset{\underset{CH_3}{|}}{C}CCH_3 + {}^-CH(COCH_2CH_3)_2 \xrightarrow{CH_3CH_2OH} \xrightarrow{\text{cold, dil. HCl}}$$

11.75. The formation of a hemiacetal from an aldehyde and an alcohol is catalyzed by base, as well as by acid. The mechanism for the acid-catalyzed reaction was presented in Section 11.8. Suggest a mechanism for the base-catalyzed reaction of acetaldehyde and methanol.

11.76. An acetal can be formed in acidic solution, but not in base. Why not?

11.77. Write equations for the mechanism of base-promoted chlorination of (R)-1-phenyl-2-methyl-1-butanone.

11.78. *Glutaraldehyde* forms a cyclic hydrate that contains one equivalent of water. What is its structure, and how is it formed? (*Hint*: If *one* aldehyde group is hydrated, what reaction might the molecule undergo?)

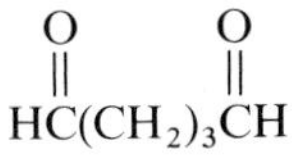

$$\underset{\text{glutaraldehyde}}{HC(CH_2)_3CH}$$

glutaraldehyde
used as an antiseptic

11.79. Benzene undergoes reaction with formaldehyde and HCl in the presence of zinc chloride to yield benzyl chloride. Write a mechanism for this reaction.

11.80. Suggest a mechanism for the following reaction:

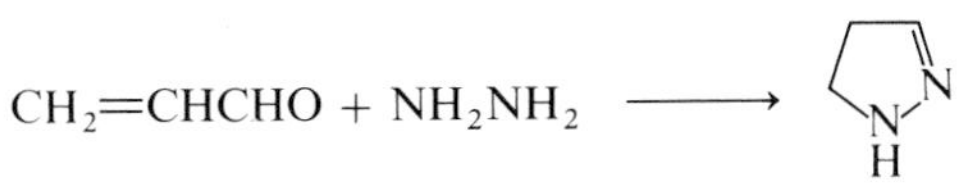

$$CH_2{=}CHCHO + NH_2NH_2 \longrightarrow$$

11.81. Explain why compound **A** forms two enols, while compound **B** forms only one. (Include the enol structures in your answer.)

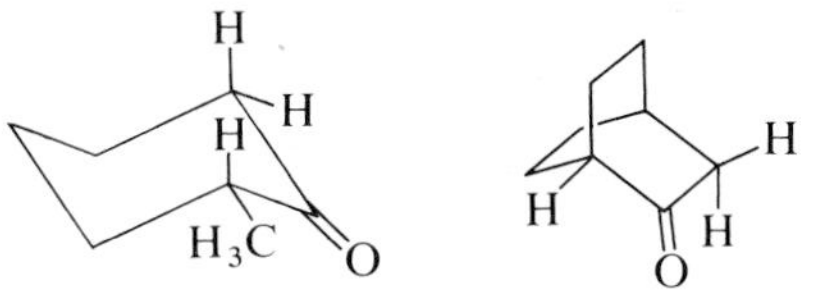

A   **B**

11.82. The following structure is that of *puberulic acid*, an antibiotic found in *Penicillium puberulum*. When treated with dilute acid, it readily forms a cation. Give the structure of the cation (and its resonance structures).

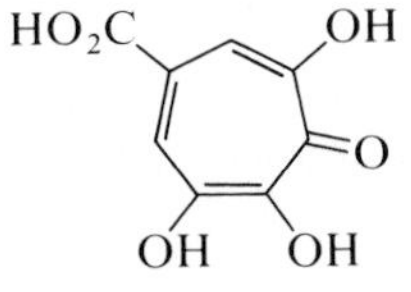

puberulic acid

11.83. *Civetone* is an active ingredient of *civet*, a mixture isolated from the scent glands of the African civet cat and used in perfumes. Civetone, which has the formula $C_{17}H_{30}O$, shows strong absorption in the infrared spectrum at 1700 cm^{-1} (5.8 μm) and shows no offset downfield absorption in the nmr spectrum. Treatment of civetone with Br_2 in CCl_4 yields a single dibromide A, $C_{17}H_{30}Br_2O$. Oxidation of civetone with $KMnO_4$ solution yields a diacid B, $C_{17}H_{30}O_5$. Oxidation of civetone with hot, concentrated HNO_3 yields principally $HO_2C(CH_2)_7CO_2H$ and $HO_2C(CH_2)_6CO_2H$. Hydrogenation of civetone (Pd catalyst, no heat or pressure), followed by oxidation with hot HNO_3, yields a diacid $HO_2C(CH_2)_{15}CO_2H$. What are the structures of civetone, A, and B?

11.84. *Muscone* ($C_{16}H_{30}O$) is the active ingredient of *musk*, which is obtained from the scent glands of the male musk deer and is also used in perfumes. Muscone is oxidized by hot HNO_3 to a mixture of diacids, two of which are

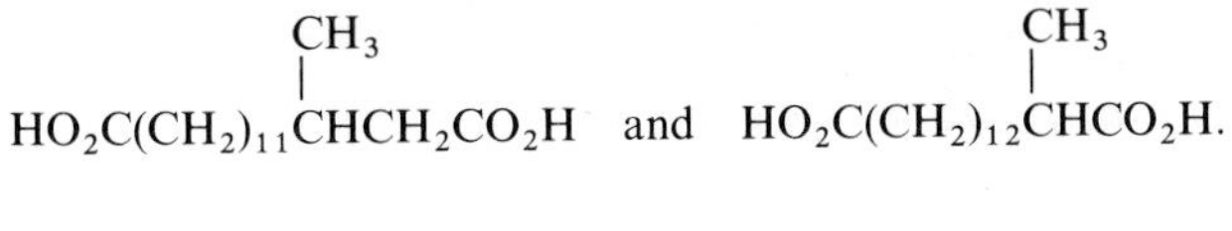

$$\underset{}{HO_2C(CH_2)_{11}\overset{\overset{\displaystyle CH_3}{|}}{C}HCH_2CO_2H} \quad \text{and} \quad \underset{}{HO_2C(CH_2)_{12}\overset{\overset{\displaystyle CH_3}{|}}{C}HCO_2H.}$$

Reduction of muscone with a zinc–mercury amalgam and HCl yields methylcyclopentadecane:

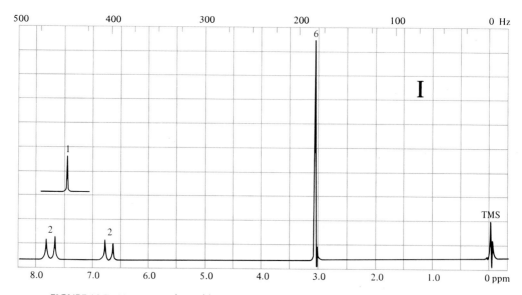

What is the structure of muscone?

11.85. Queen bees secrete a compound A, $C_{10}H_{16}O_3$, that attracts worker bees. Compound A is insoluble in water, but soluble in alkaline solution. It gives a positive iodoform test. When A is subjected to ozonolysis (oxidative work-up), Compound B, $C_8H_{14}O_3$, is obtained. Compound B is also soluble in aqueous base and also gives a positive iodoform test. Oxidation of A with hot HNO_3 yields, among other products, hexanedioic acid and heptanedioic acid. What are the structures of A and B?

11.86. Match each of the spectra in Figure 11.7 with one of the following structures:

(a) $(CH_3)_3CCHO$

(b) (furan)$C=C$ with H, H, CHO

(c) $NCCH_2CH_2CCHO$ with CH_3, CH_3

(d) $(CH_3)_2N-$(benzene ring)$-CHO$

(e) $(CH_3)_2N-$(benzene ring)$-CH_2CHO$

FIGURE 11.7. Nmr spectra for Problem 11.86.   *(continued)*

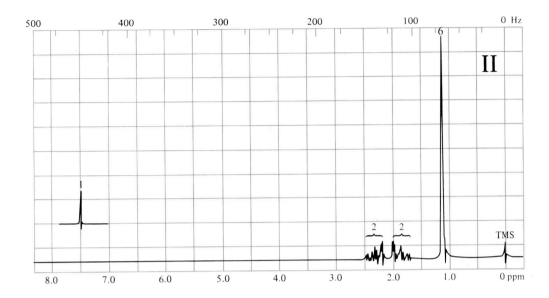

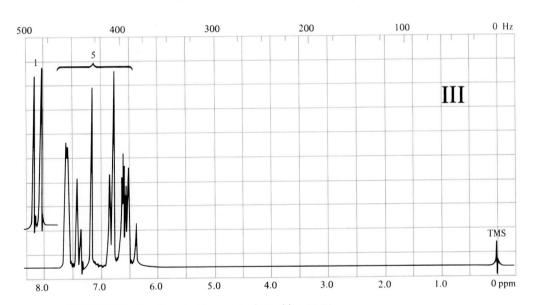

FIGURE 11.7. (continued) Nmr spectra for Problem 11.86.

11.87. Compound A has a molecular weight of 132. Treatment of A with NaBH$_4$ in aqueous methanol yields Compound B. The nmr spectrum of A and the infrared spectrum of B are shown in Figure 11.8. What are the structures of A and B?

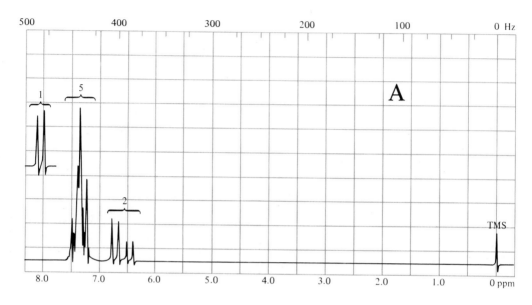

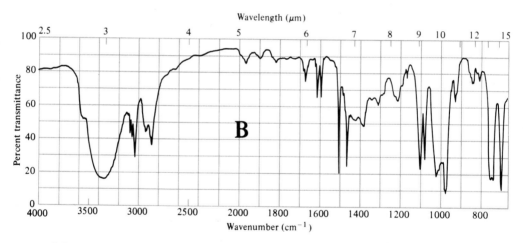

FIGURE 11.8. Spectra for Problem 11.87.

CHAPTER 12

Carboxylic Acids

A **carboxylic acid** is an organic compound containing the **carboxyl group**, —CO$_2$H. The carboxyl group contains a carbonyl group and a hydroxyl group; the interactions of these two groups lead to a chemical reactivity that is unique to carboxylic acids.

planar *polar* *unshared electrons*

Because the carboxyl group is polar and nonhindered, its reactions are not affected to a great extent by the rest of the molecule. All of the following carboxylic acids undergo similar reactions:

CH$_3$COH	CH$_3$CH$_2$COH	(CH$_3$)$_2$CHCOH	—COH
IUPAC: ethanoic acid	propanoic acid	2-methylpropanoic acid	benzoic acid
trivial: acetic acid	propionic acid	isobutyric acid	

The most notable chemical property of carboxylic acids is their acidity. Ranked against the mineral acids, such as HCl or HNO$_3$ (pK_a values about 1 or smaller), the carboxylic acids are weak acids (pK_a values typically about 5). However, the carboxylic acids are more acidic than alcohols or phenols, primarily because of <u>resonance-stabilization of the carboxylate anion, RCO$_2^-$</u>. A *p*-orbital picture of the carboxylate ion is shown in Figure 12.1.

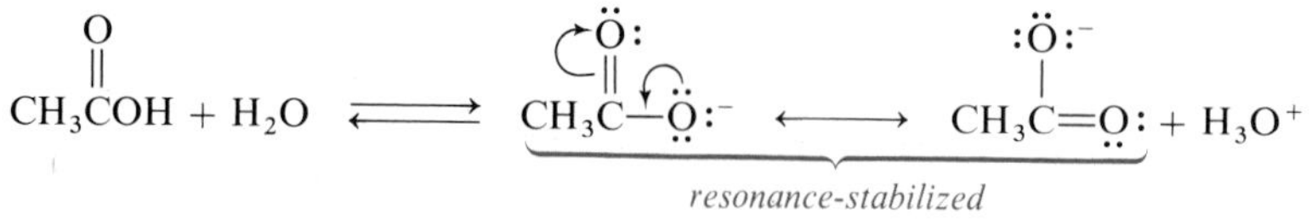

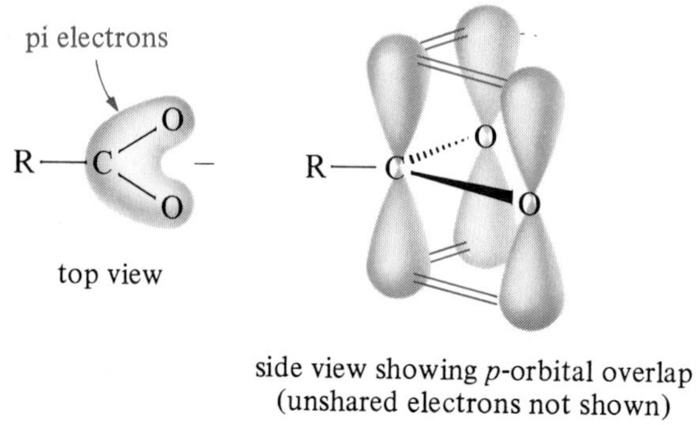

$$CH_3\overset{\overset{\textstyle O}{\|}}{C}OH + H_2O \rightleftharpoons CH_3\overset{\overset{\textstyle \ddot{O}:}{\|}}{C}-\ddot{\underset{..}{O}}:^- \longleftrightarrow CH_3\overset{\overset{\textstyle :\ddot{O}:^-}{|}}{C}=\ddot{\underset{..}{O}}: + H_3O^+$$

resonance-stabilized

FIGURE 12.1. Bonding in the carboxylate ion, RCO$_2^-$.

SECTION 12.1.

Nomenclature of Carboxylic Acids

The IUPAC name of an aliphatic <u>carboxylic acid</u> is that of the alkane parent with the <u>-e changed to -oic acid</u>. The carboxyl carbon is carbon 1, just as the aldehyde carbon is.

	Cl	
HCO$_2$H	CH$_3$CH$_2$CHCO$_2$H	HO$_2$CCH$_2$CO$_2$H
IUPAC: methanoic acid	2-chlorobutanoic acid	propanedioic acid

For the first four carboxylic acids, the trivial names are used more often than the IUPAC names. (See Table 12.1.) The name **formic acid** comes from *formica* (Latin for "ants"): in medieval times, alchemists obtained formic acid by distilling red ants! **Acetic acid** is from the Latin *acetum*, "vinegar." In its pure form, it is called *glacial* acetic acid. The term "glacial" arises from the fact that pure acetic acid is a viscous liquid that solidifies into an icy-looking solid. The name for **propionic acid** literally means "first fat." <u>Propionic acid is the first carboxylic acid (the one of lowest molecular weight) to exhibit some of the properties of fatty acids</u>, which are carboxylic acids obtained from the hydrolysis of fats (discussed in Section 20.1). **Butyric acid** is found in rancid butter (Latin, *butyrum*). Some other commonly encountered carboxylic acids and names follow:

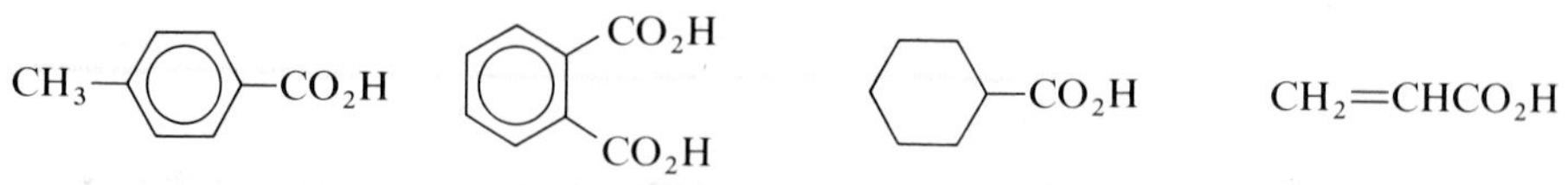

CH$_3$—⟨benzene⟩—CO$_2$H   ⟨benzene with CO$_2$H and CO$_2$H ortho⟩   ⟨cyclohexane⟩—CO$_2$H   CH$_2$=CHCO$_2$H

p-toluic acid   *o*-phthalic acid   cyclohexanecarboxylic acid   acrylic acid

TABLE 12.1. Trivial names of the first ten carboxylic acids

Number of carbons	Structure	Trivial name	Occurrence and derivation of name
1	HCO_2H	formic	ants (L. *formica*)
2	CH_3CO_2H	acetic	vinegar (L. *acetum*)
3	$CH_3CH_2CO_2H$	propionic	milk, butter, and cheese (Gr. *protos*, first; *pion*, fat)
4	$CH_3(CH_2)_2CO_2H$	butyric	butter (L. *butyrum*)
5	$CH_3(CH_2)_3CO_2H$	valeric	valerian root (L. *valere*, to be strong)
6	$CH_3(CH_2)_4CO_2H$	caproic	goat (L. *caper*)
7	$CH_3(CH_2)_5CO_2H$	enanthic	(Gr. *œnanthe*, vine blossom)
8	$CH_3(CH_2)_6CO_2H$	caprylic	goat
9	$CH_3(CH_2)_7CO_2H$	pelargonic	Its ester is found in *Pelargonium roseum*, a geranium.
10	$CH_3(CH_2)_8CO_2H$	capric	goat

As with aldehydes and ketones, Greek letters may be used in the trivial names of carboxylic acids to refer to a position in the molecule relative to the carboxyl group.

$$CH_3CH_2CO_2H \qquad CH_3CH_2\overset{\overset{\displaystyle Br}{|}}{C}HCO_2H$$

β carbon *α carbon*   α-bromobutyric acid
or 2-bromobutanoic acid

It is sometimes convenient to refer to the RCO— group as an **acyl group** and to $RCO_2{}^-$ as an **acyloxy group**. For example, the *acylation* of benzene is the substitution of a RCO— group on the aromatic ring.

$$\overset{\overset{\displaystyle O}{\|}}{R}C— \qquad \overset{\overset{\displaystyle O}{\|}}{R}CO—$$

an acyl group   *an acyloxy group*

$$\overset{\overset{\displaystyle O}{\|}}{CH_3}C— \qquad \underset{}{\bigcirc}—\overset{\overset{\displaystyle O}{\|}}{C}— \qquad \overset{\overset{\displaystyle O}{\|}}{CH_3}CO—$$

acetyl group   benzoyl group   acetoxy group
(Ac —)   (Bz —)   (AcO—)

STUDY PROBLEMS

12.1. Write IUPAC names for the following carboxylic acids:

(a) $CH_2{=}CHCO_2H$ (b) $HO_2CCH_2CH_2CH_2CO_2H$ (c) $BrCHCO_2H$
$\qquad\qquad\qquad\qquad\qquad\qquad\qquad\qquad\qquad\qquad\qquad\qquad\quad |$
$\qquad\qquad\qquad\qquad\qquad\qquad\qquad\qquad\qquad\qquad\qquad\qquad\; CH_3$

12.2.   Write trivial names for

(a) $CH_3CHBrCO_2H$. and   (b) $HOCH_2CH_2CO_2H$.

SECTION 12.2.

Physical Properties of Carboxylic Acids

The carboxyl group is ideally structured for forming two hydrogen bonds between a pair of molecules. A pair of hydrogen-bonded carboxylic acid molecules is often referred to as a **carboxylic acid dimer**. Because of the strength of these hydrogen bonds (a total of about 10 kcal/mole for the two hydrogen bonds), carboxylic acids are found as dimers to a limited extent even in the vapor phase.

a carboxylic acid dimer   hydrogen bonds with H_2O

STUDY PROBLEM

12.3.   Show the principal hydrogen bonds between molecules of the following compounds:

The physical properties of carboxylic acids reflect the strong hydrogen bonding between carboxylic acid molecules. The melting points and boiling points are relatively high. The infrared spectra of carboxylic acids also show the effects of hydrogen bonding (see Section 12.3A). The lower-molecular-weight acids are water-soluble, as well as soluble in organic solvents. The melting points, boiling points, and water-solubilities of some carboxylic acids are listed in Table 12.2.

A notable property (not a physical property, but a physiological one) of the lower-molecular-weight carboxylic acids is their odor. Formic and acetic acids have pungent odors. Propionic acid has a pungent odor reminiscent of rancid fat. The odor of rancid butter arises in part from butyric acid. Caproic acid smells like a goat. (Goat sweat, incidentally, contains caproic acid.) Valeric acid (from the Latin *valere*, "to be strong") is not a strong acid, but it does have a strong odor somewhere in between that of rancid butter and goat sweat. (Interestingly, valeric acid is the sex attractant of the sugar beet wire worm.) Dogs can differentiate the odors of individual humans because of differing proportions of carboxylic acids in human sweat. The odors of the aliphatic carboxylic acids of ten and more carbons diminish, probably because of their lack of volatility.

TABLE 12.2. Physical properties of some carboxylic acids

Name	Structure	Mp, °C	Bp, °C	Solubility in H_2O at 20°C
formic	HCO_2H	8	100.5	∞
acetic	CH_3CO_2H	16.6	118	∞
propionic	$CH_3CH_2CO_2H$	−22	141	∞
butyric	$CH_3(CH_2)_2CO_2H$	−6	164	∞
valeric	$CH_3(CH_2)_3CO_2H$	−34	187	3.7 g/100 mL
caproic	$CH_3(CH_2)_4CO_2H$	−3	205	1.0 g/100 mL
cyclohexane-carboxylic	$C_6H_{11}CO_2H$	31	233	0.2 g/100 mL
benzoic	$C_6H_5CO_2H$	122	250	0.3 g/100 mL

SECTION 12.3.

Spectral Properties of Carboxylic Acids

A.   Infrared spectra

Carboxylic acids, either as pure liquids or in solution at concentrations in excess of about 0.01M, exist primarily as hydrogen-bonded dimers rather than as discrete monomers. The infrared spectrum of a carboxylic acid is therefore the spectrum of the dimer. Because of the hydrogen bonding, the OH stretching absorption of carboxylic acids is very broad and very intense. This OH absorption starts around 3300 cm^{-1} (3.0 μm) and slopes into the region of aliphatic carbon–hydrogen absorption (see Figure 12.2). The broadness of the carboxylic acid OH band can often obscure both aliphatic and aromatic CH absorption, as well as any other OH or NH absorption in the spectrum.

The carbonyl absorption is observed at about 1700–1725 cm^{-1} (5.8–5.88 μm) and is of moderately strong intensity. Conjugation shifts this absorption to lower frequencies: 1680–1700 cm^{-1} (5.9–5.95 μm).

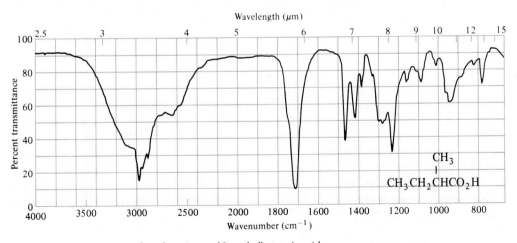

FIGURE 12.2. Infrared spectrum of 2-methylbutanoic acid.

The fingerprint region in an infrared spectrum of a carboxylic acid often shows C—O stretching and OH bending (see Table 12.3). Another OH bending vibration of the dimer results in a broad absorption near 925 cm^{-1} (10.8 μm).

TABLE 12.3. Characteristic infrared absorption for carboxylic acids

Type of vibration	Position of absorption	
	cm^{-1}	μm
O—H stretching	2860–3300	3.0–3.5
C=O stretching	1700–1725	5.8–5.88
C—O stretching	1210–1330	7.5–8.26
O—H bending	1300–1440	6.94–7.71
O—H bending (dimer)	∼925	∼10.8

B. Nmr spectra

In the nmr spectrum, the absorption of the acidic proton of a carboxylic acid is seen as a singlet far downfield ($\delta = 10$–13 ppm), offset from the usual spectral range. The alpha protons are only slightly affected by the C=O group; their absorption is slightly downfield (about 2.2 ppm) because of the inductive effect of the partially positive carbonyl carbon. There is no unique splitting pattern associated with the carboxylic acid group because the carboxyl proton has no neighboring protons. (See Figure 12.3.)

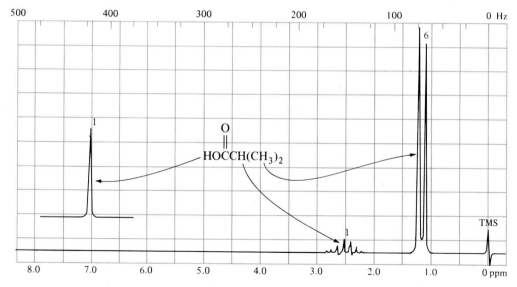

FIGURE 12.3. Nmr spectrum of 2-methylpropanoic acid.

SECTION 12.4.

Preparation of Carboxylic Acids

The numerous synthetic paths that lead to carboxylic acids may be grouped into three types of reactions: (1) *hydrolysis of the derivatives of the carboxylic acids*; (2) *oxidation reactions*; and (3) *Grignard reactions*. These reactions are summarized in Table 12.4.

Hydrolysis of carboxylic acid derivatives results from the attack of water or OH^- on the carbonyl carbon (or the $-CN$ carbon of a nitrile) of the derivatives.

TABLE 12.4. Summary of laboratory syntheses of carboxylic acids

Reaction			Section reference
Hydrolysis:[a]			
ester:	$\overset{O}{\overset{\|}{RC}}-OR' + H_2O \xrightarrow{H^+ \text{ or } OH^-}$	$RCO_2H + HOR'$	13.5C
amide:	$\overset{O}{\overset{\|}{RC}}-NR'_2 + H_2O \xrightarrow{H^+ \text{ or } OH^-}$	$RCO_2H + HNR'_2$	13.9C
anhydride:	$\overset{O}{\overset{\|}{RC}}-\overset{O}{\overset{\|}{O}}CR' + H_2O \xrightarrow{H^+ \text{ or } OH^-}$	$RCO_2H + HO_2CR'$	13.4C
acid halide:	$\overset{O}{\overset{\|}{RC}}-X + H_2O \xrightarrow{H^+ \text{ or } OH^-}$	$RCO_2H + X^-$	13.3C
nitrile:	$RC{\equiv}N + H_2O \xrightarrow{H^+ \text{ or } OH^-}$	$RCO_2H + NH_3$	13.12D
Oxidation:			
1° alcohol:	$RCH_2OH + [O]^b \longrightarrow$	RCO_2H	7.13
aldehyde:	$\overset{O}{\overset{\|}{RCH}} + [O] \longrightarrow$	RCO_2H	11.15
alkene:	$RCH{=}CR_2 + [O] \longrightarrow$	$RCO_2H + R_2C{=}O$	9.14
alkylarene:	$Ar-R + [O] \longrightarrow$	$ArCO_2H$	10.12
methyl ketone:[a]	$\overset{O}{\overset{\|}{RCCH_3}} + X_2 \xrightarrow{OH^-}$	$RCO_2H + CHX_3$	11.18A
Grignard reaction:			
	$RX \xrightarrow[\substack{(2)\ CO_2 \\ (3)\ H_2O,\ H^+}]{(1)\ Mg,\ ether} RCO_2H$		6.9A

[a] In alkaline solution, the carboxylate is obtained. The acid can be generated by
   acidification: $RCO_2^- + H^+ \rightarrow RCO_2H$.
[b] Typical oxidizing agents are $KMnO_4$ or H_2CrO_4 solutions.

The hydrolysis of an ester to yield a carboxylic acid and an alcohol is typical of this class of reactions. We will discuss these reactions in detail in Chapter 13.

$$CH_3\overset{\overset{\displaystyle O}{\|}}{C}OCH_2CH_3 + H_2O \underset{\longleftarrow}{\overset{H^+}{\rightleftharpoons}} CH_3\overset{\overset{\displaystyle O}{\|}}{C}OH + HOCH_2CH_3$$

ethyl acetate   acetic acid   ethanol

an ester   *a carboxylic acid*   *an alcohol*

Oxidation of primary alcohols and aldehydes to yield carboxylic acids was discussed in Sections 7.13C and 11.15, respectively. The principal limitation of alcohol-oxidation is that the necessary strength of the oxidizing agent precludes the presence of another oxidizable functional group in the molecule (unless it is protected by a blocking group, such as the dibromide of an alkene or an acetal of an aldehyde). Even with this limitation, the oxidation of primary alcohols is the most common oxidative procedure for obtaining carboxylic acids because alcohols are often available.

$$RCH_2OH \xrightarrow[\text{heat}]{H_2CrO_4} RCO_2H$$

no double bond,
aldehyde, benzyl
group, or other
—OH group

The oxidation of aldehydes proceeds with mild oxidizing agents (such as Ag^+) that do not oxidize other groups; however, aldehydes are not as readily available as primary alcohols.

Oxidation of alkenes is used primarily as an analytical tool, but also can be employed in synthesis of carboxylic acids. Like alcohols, alkenes require vigorous oxidizing agents.

$$\text{cyclohexene} \xrightarrow[\text{heat}]{MnO_4^-} HO_2CCH_2CH_2CH_2CH_2CO_2H$$

cyclohexene   hexanedioic acid
                              (adipic acid)

Oxidation of substituted alkylbenzenes is an excellent route to the substituted benzoic acids. A carboxyl group is a *meta*-director, but an alkyl group is an *o,p*-director. Electrophilic substitution of an alkylbenzene, followed by oxidation, yields *o*- and *p*-substituted benzoic acids.

$$O_2N\text{—}\underset{\underset{\displaystyle}{\bigcirc}}{\overset{\overset{\displaystyle NO_2}{|}}{}}\text{—}CH_3 \xrightarrow[\text{heat}]{MnO_4^-} O_2N\text{—}\underset{}{\overset{\overset{\displaystyle NO_2}{|}}{\bigcirc}}\text{—}CO_2H$$

2,4-dinitrotoluene   2,4-dinitrobenzoic acid

A Grignard reaction between a Grignard reagent (1°, 2°, 3°, vinylic, or aryl) and carbon dioxide (as a gas or as dry ice) is often the method of choice for preparing a carboxylic acid.

$$\text{C}_6\text{H}_5\text{—MgBr} \xrightarrow[\text{(2) H}_2\text{O, H}^+]{\text{(1) CO}_2} \text{C}_6\text{H}_5\text{—CO}_2\text{H}$$

phenylmagnesium   benzoic acid
bromide   85%

Let us summarize the common synthetic routes to carboxylic acids from a different viewpoint—that of what happens to the molecule as a whole. Synthesis from an alkyl halide by way of the nitrile or Grignard reagent leads to a carboxylic acid with *one more carbon* than in the alkyl halide.

Chain-lengthening:

$$1° \text{ RX} \xrightarrow[\text{S}_\text{N}2]{\text{CN}^-} \text{RCN} \xrightarrow{\text{H}_2\text{O, H}^+} \text{RCO}_2\text{H}$$

$$\text{RX} \xrightarrow[\text{ether}]{\text{Mg}} \text{RMgX} \xrightarrow[\text{(2) H}_2\text{O, H}^+]{\text{(1) CO}_2} \text{RCO}_2\text{H}$$

Oxidation of a primary alcohol (or aldehyde) does not affect the length of the carbon chain, nor does hydrolysis of a carboxylic acid derivative.

Same number of carbons:

$$\text{RCH}_2\text{OH} \xrightarrow[\text{heat}]{\text{MnO}_4^-} \text{RCO}_2\text{H}$$

$$\underset{\text{RCOR}'}{\overset{\text{O}}{\underset{\|}{}}} \xrightarrow{\text{H}_2\text{O, H}^+} \underset{\text{RCOH}}{\overset{\text{O}}{\underset{\|}{}}} + \text{R}'\text{OH}$$

Oxidation of an alkene (unless the alkene is cyclic) causes fragmentation of the parent chain.

Fragmentation:

$$\text{RCH=CR}_2 \xrightarrow[\text{heat}]{\text{MnO}_4^-} \text{RCO}_2\text{H} + \text{O=CR}_2$$

STUDY PROBLEM

12.4. Considering only nitrile formation and Grignard reactions, propose a feasible route for each of the following conversions:

(a) $\text{HOCH}_2\text{CH}_2\text{Cl} \longrightarrow \text{HOCH}_2\text{CH}_2\text{CO}_2\text{H}$

(b) $(\text{CH}_3)_3\text{CCH}_2\text{Cl} \longrightarrow (\text{CH}_3)_3\text{CCH}_2\text{CO}_2\text{H}$

(c) naphthalene—I $\longrightarrow$ naphthalene—CO$_2$H

(d) $\text{C}_6\text{H}_5\text{—CH}_2\text{Br} \longrightarrow \text{C}_6\text{H}_5\text{—CH}_2\text{CO}_2\text{H}$

SECTION 12.5.

Acidity of Carboxylic Acids

Carboxylic acids, sulfonic acids (RSO_3H), and alkyl hydrogen sulfates ($ROSO_3H$) are the only classes of organic compounds more acidic than carbonic acid (H_2CO_3). Of these three classes, carboxylic acids are by far the most common.

	RCH_3	RNH_2	$RC{\equiv}CH$	ROH	H_2O	$ArOH$	H_2CO_3	RCO_2H
approx. pK_a:	45	35	25	18	15	10	6.4	5

increasing acid strength →

Because it is more acidic than carbonic acid, a carboxylic acid undergoes an acid–base reaction with sodium bicarbonate as well as with stronger bases such as NaOH.

$$CH_3\overset{\displaystyle O}{\overset{\|}{C}}OH + OH^- \longrightarrow CH_3\overset{\displaystyle O}{\overset{\|}{C}}O^- + H_2O$$

acetic acid   acetate ion

$$CH_3\overset{\displaystyle O}{\overset{\|}{C}}OH + HCO_3^- \longrightarrow CH_3\overset{\displaystyle O}{\overset{\|}{C}}O^- + [H_2CO_3] \longrightarrow H_2O + CO_2$$

While carboxylic acids undergo reaction with sodium bicarbonate, phenols require the stronger base NaOH, and alcohols require a yet stronger base, such as $NaNH_2$.

In NaHCO₃: $RCO_2H \xrightarrow{\;HCO_3^-\;} RCO_2^- + CO_2 + H_2O$

$ArOH$ and $ROH \xrightarrow{\;HCO_3^-\;}$ no appreciable reaction

In NaOH: $RCO_2H \xrightarrow{\;OH^-\;} RCO_2^- + H_2O$

$ArOH \xrightarrow{\;OH^-\;} ArO^- + H_2O$

$ROH \xrightarrow{\;OH^-\;}$ no appreciable reaction

This difference between phenols and carboxylic acids in reactivity toward NaOH and $NaHCO_3$ is the basis of a simple classification and separation procedure. If a water-insoluble compound dissolves in NaOH solution, but not in $NaHCO_3$ solution, it is likely to be a phenol. On the other hand, if the compound dissolves in *both* NaOH and $NaHCO_3$ solution, it is probably a carboxylic acid.

A carboxylic acid can be extracted from a mixture of water-insoluble organic compounds with sodium bicarbonate solution. The acid forms a sodium salt and becomes water-soluble, while the other organic compounds remain insoluble. The free carboxylic acid is obtained from the water solution by acidification.

When used either as a test or as a separation procedure, the reaction with sodium bicarbonate has its limitations. If the hydrocarbon portion of the carboxylic acid is long, the compound fails to dissolve in $NaHCO_3$ solution and may even fail to dissolve in NaOH solution. In addition, some phenols, such as the nitrophenols, have acid strengths comparable to those of carboxylic acids; these phenols dissolve in $NaHCO_3$ as well as in NaOH solutions.

STUDY PROBLEMS

12.5. An aldehyde obtained from a storeroom is probably contaminated with some carboxylic acid (from air oxidation of the aldehyde). Describe in words how you would remove the unwanted acid.

12.6. You are confronted with an ether solution containing a mixture of heptanoic acid, 1-naphthol, and 1-octanol. How would you separate these three components?

1-naphthol

A.   Neutralization equivalent

The **equivalent weight** of an acid, or its **neutralization equivalent**, is the number of grams of the acid that undergoes reaction with 1.0 equivalent of OH^-. The neutralization equivalent is determined by titrating a known weight of the acid with a standardized solution of sodium hydroxide. The endpoint is often determined by the color change of an indicator such as phenolphthalein; however, a pH meter gives more reliable results.

$$RCO_2H + OH^- \longrightarrow RCO_2^- + H_2O$$

At the endpoint, the number of equivalents of base added—the normality of the base times its volume in liters, or (NV)—is equal to the number of equivalents of acid. From the weight of the sample of acid and its number of equivalents, the neutralization equivalent can be calculated.

$$NV = \text{no. of equivalents of } OH^- = \text{no. of equivalents of acid}$$

$$\text{neutralization equivalent of acid} = \frac{\text{wt. in g of acid}}{\text{no. of equivalents of acid}}$$

Example: A 0.528-g sample of a carboxylic acid requires 24.00 mL of 0.100*N* NaOH to be neutralized. What is its neutralization equivalent?

$$\text{equivalents acid} = \text{equivalents } OH^-$$
$$= NV$$
$$= (0.100 \text{ eq./liter})(0.0240 \text{ liter})$$
$$= 0.00240 \text{ eq.}$$

$$\text{neutralization equivalent} = \frac{\text{wt. in g}}{\text{no. of equivalents}}$$

$$= \frac{0.528 \text{ g}}{0.00240 \text{ eq.}}$$

$$= 220$$

If a carboxylic acid has only one acidic proton per molecule, the neutralization equivalent equals the molecular weight. If an acid has *two* acidic protons, the neutralization equivalent is only *half* the molecular weight.

$$\langle\bigcirc\rangle\text{—CO}_2\text{H} \qquad \text{HO}_2\text{C}\text{—}\langle\bigcirc\rangle\text{—CO}_2\text{H}$$

neutralization equivalent:	122.1	83.0
molecular weight:	122.1	166.1

STUDY PROBLEMS

12.7. A 0.200-g sample of an unknown acid requires 13.00 mL of 0.201 N sodium hydroxide solution to reach neutrality. What is the neutralization equivalent of the acid?

12.8. What are the equivalent weights of:

(a) succinic acid, $\text{HO}_2\text{CCH}_2\text{CH}_2\text{CO}_2\text{H}$ (b) citric acid,

$$\begin{array}{c}\text{CH}_2\text{CO}_2\text{H}\\|\\\text{HOCCO}_2\text{H}\\|\\\text{CH}_2\text{CO}_2\text{H}\end{array}$$

SECTION 12.6.

Salts of Carboxylic Acids

The reaction of a carboxylic acid with a base results in a *salt.* An organic salt has many of the physical properties of its inorganic counterparts. Like NaCl or KNO_3, an organic salt is high-melting, water-soluble, and odorless.

$$\text{HCO}_2\text{H} + \text{Na}^+\,\text{OH}^- \longrightarrow \text{HCO}_2^-\,\text{Na}^+ + \text{H}_2\text{O}$$
formic acid   sodium formate

$$\langle\bigcirc\rangle\text{—CO}_2\text{H} + \text{Na}^+\,\text{HCO}_3^- \longrightarrow \langle\bigcirc\rangle\text{—CO}_2^-\,\text{Na}^+ + \text{H}_2\text{O} + \text{CO}_2$$
cyclohexane-   sodium cyclohexane-
carboxylic acid   carboxylate

The carboxylate anion is named by changing the **-ic acid** ending of the carboxylic acid name to **-ate**. In the name of the salt, the name of the cation precedes the name of the anion as a separate word.

$$\text{CH}_3\text{CO}_2\text{H} \qquad \text{CH}_3\text{CO}_2^-\,\text{Na}^+ \qquad \langle\bigcirc\rangle\text{—CO}_2\text{H} \qquad \langle\bigcirc\rangle\text{—CO}_2^-\,\text{NH}_4^+$$
acetic acid   sodium acetate   benzoic acid   ammonium benzoate

$$\begin{array}{c}\text{NH}_2\\|\\\text{HO}_2\text{CCH}_2\text{CH}_2\text{CHCO}_2\text{H}\end{array} \qquad \begin{array}{c}\text{NH}_2\\|\\\text{HO}_2\text{CCH}_2\text{CH}_2\text{CHCO}_2^-\,\text{Na}^+\end{array}$$
glutamic acid   monosodium glutamate

an amino acid found in proteins   *a flavor enhancer (MSG)*

The carboxylate ion is a weak base and can act as a nucleophile. Esters, for example, may be prepared by the reaction of reactive alkyl halides and carboxylates (see Chapter 5).

$$CH_3CO_2^- \;+\; \overset{\displaystyle C_6H_5}{\underset{}{CH_2}}\!-\!Br \longrightarrow C_6H_5CH_2O_2CCH_3 + Br^-$$

acetate ion   benzyl bromide   benzyl acetate

an ester

SECTION 12.7.

How Structure Affects Acid Strength

Acid strength is a term describing the extent of ionization of an acid: the greater the amount of ionization, the more hydrogen ions are formed, and the stronger is the acid. The strength of an acid is expressed as its K_a or its pK_a (see Table 12.5). (You may want to review the discussion of K_a and pK_a in Section 1.10 before proceeding.) In this section, we will discuss the general structural features that affect the acid strength of an organic compound. Our emphasis will be on carboxylic acids, but we will not limit the discussion to just these compounds.

The reaction of a weak acid with water is reversible. The equilibrium lies on the lower-energy side of the equation. Any structural feature that *stabilizes the*

TABLE 12.5. pK_a values for some carboxylic acids

Trivial name	Structure	pK_a
formic	HCO_2H	3.75
acetic	CH_3CO_2H	4.75
propionic	$CH_3CH_2CO_2H$	4.87
butyric	$CH_3(CH_2)_2CO_2H$	4.81
trimethylacetic	$(CH_3)_3CCO_2H$	5.02
fluoroacetic	FCH_2CO_2H	2.66
chloroacetic	$ClCH_2CO_2H$	2.81
bromoacetic	$BrCH_2CO_2H$	2.87
iodoacetic	ICH_2CO_2H	3.13
dichloroacetic	Cl_2CHCO_2H	1.29
trichloroacetic	Cl_3CCO_2H	0.7
α-chloropropionic	$CH_3CHClCO_2H$	2.8
β-chloropropionic	$ClCH_2CH_2CO_2H$	4.1
lactic	$CH_3\overset{\displaystyle OH}{\underset{}{C}}HCO_2H$	3.87
vinylacetic	$CH_2{=}CHCH_2CO_2H$	4.35

anion with respect to its conjugate acid *increases the acid strength* by driving the equilibrium to the H_3O^+ and anion (A^-) side.

$$HA + H_2O \rightleftharpoons H_3O^+ + A^-$$

lower energy means stronger acid

The principal factors that affect the stability of A^-, and thus the acid strength of HA, are: (1) electronegativity of A^-; (2) size of A^-; (3) hybridization of A^-; (4) inductive effect of other atoms or groups attached to the negative atom in A^-; (5) resonance-stabilization of A^-; and (6) solvation of A^-. We will discuss each of these features in turn, but keep in mind that they do not work independently of one another.

A. Electronegativity

A more electronegative atom holds its bonding electrons more tightly than does a less electronegative atom. In comparisons of anions, the anion with a more electronegative atom carrying the negative ionic charge is generally the more stable anion. Therefore, as we proceed from left to right in the periodic table, we find that the elements form progressively more stable anions and that the conjugate acids are progressively stronger acids.

$$C \quad N \quad O \quad F$$

increasing electronegativity of elements

$$R_3C-H \quad R_2N-H \quad RO-H \quad F-H$$

increasing acid strength

Just the reverse is true when we consider the base strengths of the conjugate bases. The anion of a very weak acid is a very strong base, while the anion of a stronger acid is a weaker base.

$$R_3C^- \quad R_2N^- \quad RO^- \quad F^-$$

increasing base strength

Consider, for example, the ionization reactions of ethanol and HF in water. As an element, F is more electronegative than O; thus, a fluoride ion is better able to carry a negative charge than is the alkoxide ion. Although HF is a weak acid, it is a much stronger acid than ethanol. Conversely, the fluoride ion is a weaker base than the ethoxide ion.

$$CH_3CH_2OH + H_2O \rightleftharpoons CH_3CH_2O^- + H_3O^+$$

weaker acid · *stronger base*
$pK_a = 16$

$$HF + H_2O \rightleftharpoons F^- + H_3O^+$$

stronger acid · *weaker base*
$pK_a = 3.45$

B. Size

A larger atom is better able to disperse a negative charge than is a smaller atom. Dispersal of a charge results in stabilization. Thus, as the size of an atom attached to H increases through a series of compounds in any group of the periodic table, the stability of the anion increases and so does acid strength. Because of the small size of the fluorine atom, HF is a weaker acid than the other hydrogen halides, even though fluorine is more electronegative than the other halogens.

$$F^- \qquad Cl^- \qquad Br^- \qquad I^-$$

increasing ionic radii →

	HF	HCl	HBr	HI
pK_a:	3.45	−7	−9	−9.5

increasing acid strength →

C.   Hybridization

In Section 9.1, we discussed why an alkyne with a $\equiv$CH group is weakly acidic. The increasing s character of the hybrid orbitals of carbon in the series $sp^3 \rightarrow sp^2 \rightarrow sp$ means increasing electronegativity of the carbon, and thus increasing polarity of the CH bond and increasing acid strength. A greater electronegativity of the atom bonded to H also enhances anion stability and thus the acidity of the compound. For these reasons, an alkynyl proton is more acidic than an alkenyl proton, which, in turn, is more acidic than the proton of an alkane.

	CH_3CH_3	$CH_2{=}CH_2$	$CH{\equiv}CH$
approx. pK_a:	43	36	26

increasing acid strength →

Again, the strongest acid of the series yields the anion that is the weakest base,

$$CH_3CH_2^- \qquad CH_2{=}CH^- \qquad CH{\equiv}C^-$$

← **increasing base strength**

D.   Inductive effect

So far, we have discussed how the atom bonded directly to a hydrogen affects acid strength. However, other parts of a molecule can also affect acid strength. Compare the pK_a values for acetic acid and chloroacetic acid:

$$CH_3CO_2H \qquad\qquad ClCH_2CO_2H$$

acetic acid   chloroacetic acid
$pK_a = 4.75$ $pK_a = 2.81$

As an acid, chloroacetic acid is one hundred times stronger than acetic acid! This enhanced acidity arises from the inductive effect of the electronegative chlorine. In the un-ionized carboxylic acid, the electron-withdrawing Cl decreases the electron density of the α carbon. The result is a relatively high-energy structure with adjacent positive charges.

$$\overset{\delta^-}{Cl}{\leftarrow}\underset{\delta^+}{CH_2}{-}\overset{\displaystyle O \atop \displaystyle \|}{\underset{\delta^+}{C}}OH$$

adjacent $\delta+$ charges
destabilize acid

The presence of the <u>chlorine</u>, however, *reduces* <u>the energy of the anion</u>. In this case, the <u>negative charge of the carboxylate group</u> is <u>partially dispersed by the nearby $\delta+$ charge.</u>

$$
\underset{\underset{\delta-}{Cl}\leftarrow CH_2}{}\underset{\delta+}{\overset{\overset{\displaystyle O}{\|}}{\leftarrow C}}-O^-
$$

nearby $\delta+$ and $-$
charges stabilize anion

The effect of an <u>electronegative group near the carboxyl group is to strengthen the acid by destabilizing the acid and stabilizing the anion relative to each other.</u>

$$
\underset{\text{less stable}}{\overset{\overset{\displaystyle O}{\|}}{ClCH_2COH}} + H_2O \rightleftharpoons \underset{\text{more stable}}{\overset{\overset{\displaystyle O}{\|}}{ClCH_2CO^-}} + H_3O^+
$$

A list of groups in order of their electron-withdrawal power follows:

$$CH_3-\quad H-\quad CH_2{=}CH-\quad C_6H_5-\quad HO-\quad CH_3O-\quad I-\quad Br-\quad Cl-$$

increasing power of electron-withdrawal ⟶

The pK_a values of the following carboxylic acids reflect the differences in electron-withdrawal by groups attached to $-CH_2CO_2H$:

$CH_3CH_2CO_2H$	CH_3CO_2H	$CH_2{=}CHCH_2CO_2H$	$C_6H_5CH_2CO_2H$	$HOCH_2CO_2H$	$ClCH_2CO_2H$
pK_a: 4.87	4.75	4.35	4.31	3.87	2.81

increasing acid strength ⟶

Additional electron-withdrawing groups amplify the inductive effect. Dichloroacetic acid is a stronger acid than is chloroacetic acid, and trichloroacetic acid is the strongest of the three.

$ClCH_2CO_2H$	Cl_2CHCO_2H	Cl_3CCO_2H
chloroacetic acid	dichloroacetic acid	trichloroacetic acid
$pK_a = 2.81$	$pK_a = 1.29$	$pK_a = 0.7$

The <u>influence of the inductive effect upon acid strength diminishes with an increasing number of atoms between the carboxyl group and the electronegative group.</u> 2-Chlorobutanoic acid is a substantially stronger acid than butanoic acid itself; however, 4-chlorobutanoic acid has a pK_a value very close to that of the unsubstituted acid.

$CH_3CH_2CH_2CO_2H$	$\overset{\overset{\displaystyle Cl}{\|}}{CH_2}CH_2CH_2CO_2H$	$CH_3\overset{\overset{\displaystyle Cl}{\|}}{CH}CH_2CO_2H$	$CH_3CH_2\overset{\overset{\displaystyle Cl}{\|}}{CH}CO_2H$
4.8	4.5	4.0	2.9

pK_a:

increasing acid strength as distance between $-Cl$ and $-CO_2H$ decreases ⟶

STUDY PROBLEM

12.9. Which is the stronger acid: (a) phenylacetic acid or bromoacetic acid?
(b) dibromoacetic acid or bromoacetic acid? (c) 2-iodopropanoic acid or
3-iodopropanoic acid?

E.   Resonance-stabilization

Alcohols, phenols, and carboxylic acids all contain OH groups. Yet these classes
of compounds vary dramatically in acid strength. The differences may be attributed directly to the resonance-stabilization (or lack of it) of the anion with respect
to its conjugate acid.

	ROH	ArOH	RCO_2H
approx. pK_a:	18	10	5

In the case of *alcohols*, the anion is not resonance-stabilized. The negative
charge of an alkoxide ion resides entirely on the oxygen and is not delocalized.
At the opposite end of the scale are the *carboxylic acids*. The negative charge of the
carboxylate ion is equally shared by two electronegative oxygen atoms. *Phenols*
are intermediate between carboxylic acids and alcohols in acidity. The oxygen of a
phenoxide ion is adjacent to the aromatic ring and the negative charge is partially
delocalized by the aromatic pi cloud.

An alkoxide: $CH_3CH_2O^-$ *no resonance-stabilization*

A phenoxide:

major contributor

A carboxylate:

equal contributors

F.   Solvation

The solvation of the anion can play a major role in the acidity of a compound.
By associating with an anion, solvent molecules stabilize the anion by helping disperse the negative charge through dipole–dipole interactions. Any factor that increases the degree of solvation of the anion increases the acidity of that compound
in solution. For example, water has a greater ability to solvate ions than does
ethanol. A water solution of a carboxylic acid is more acidic than an ethanol
solution by a factor of about 10^5!

SECTION 12.8.

Acid Strengths of Substituted Benzoic Acids

You might expect that resonance-stabilization by the aromatic pi cloud
would play a large role in the relative acid strengths of benzoic acid and substituted benzoic acids—however, this is not the case. The negative charge of the

carboxylate ion is shared by the two carboxylate oxygen atoms but cannot be effectively delocalized by the aromatic ring. (The oxygens of the carboxylate anion are not directly adjacent to the aromatic ring; resonance structures in which the negative charge is delocalized by the ring cannot be drawn.)

Even though the negative charge of the benzoate ion is not delocalized by the benzene ring, benzoic acid is a stronger acid than phenol. In the benzoate ion, the negative charge is equally shared by two electronegative oxygen atoms. In the phenoxide ion, however, most of the negative charge resides on the single oxygen atom.

Because the benzene ring does not participate in resonance-stabilization of the carboxylate group, substituents on a benzene ring influence acidity primarily by the inductive effect. Regardless of the position of substitution, an *electron-withdrawing* group usually enhances the acidity of a benzoic acid (see Table 12.6).

TABLE 12.6. pK_a values for some benzoic acids

| Acid[a] | Position of substitution versus pK_a | | |
	ortho	meta	para
⬡—CO$_2$H	4.2	4.2	4.2
CH$_3$ ⬡—CO$_2$H	3.9	4.3	4.4
HO ⬡—CO$_2$H	3.0	4.1	4.5
CH$_3$O ⬡—CO$_2$H	4.1	4.1	4.5
Br ⬡—CO$_2$H	2.9	3.8	4.0
Cl ⬡—CO$_2$H	2.9	3.8	4.0
O$_2$N ⬡—CO$_2$H	2.2	3.5	3.4

[a] A bond to the center of a benzene ring denotes an unspecified position of substitution.

all are stronger acids than benzoic acid

The reasons that an electronegative substituent increases the acid strength are, again, destabilization of the acid and stabilization of the anion.

destabilized by   *stabilized by*
electron-withdrawal   *electron-withdrawal*

An *electron-releasing* alkyl substituent that is *m-* or *p-* to the carboxyl group decreases the acid strength of a benzoic acid. By releasing electrons, the substituent stabilizes the un-ionized acid and destabilizes the anion.

m- or p-alkyl group decreases
acid strength

Almost all *ortho*-substituents (whether electron-releasing or electron-withdrawing) increase the acid strength of a benzoic acid. The reasons for this *ortho*-effect, as it is called, are probably a combination of both steric and electronic factors.

o-substituents increase
acid strength

SAMPLE PROBLEM

In Table 12.6 you will notice that *p*-hydroxybenzoic acid is a weaker acid than benzoic acid, even though the hydroxyl group is electron-withdrawing. Suggest a reason for this. (*Hint*: Write resonance structures for the anion showing the delocalization of unshared electrons on the OH oxygen by the benzene ring.)

Solution: The lowering of the acid strength is an example of resonance-destabilization of the benzoate anion. The key is in the circled structure, in which the negative charge is *adjacent* to the $-CO_2^-$ group. This resonance structure lends high energy to the anion.

STUDY PROBLEMS

12.10. (a)   Would you expect *m*-isopropylbenzoic acid to be a stronger or a weaker acid than benzoic acid?

   (b)   Which would be the stronger acid, *m*-nitrobenzoic acid or 3,5-dinitrobenzoic acid?

12.11. The pK_a of *o*-methoxybenzoic acid is 4.1. It is a stronger acid than benzoic acid because of the *ortho*-effect. However, *o*-hydroxybenzoic acid (salicylic acid) is ten times stronger an acid than the methoxy acid—its pK_a is 3.0. Write the structures of the two anions and suggest why the hydroxyl group has the greater stabilizing effect.

12.12. Rank the following hydrocarbons in order of increasing acidity of the methyl hydrogen (weakest acid first):   (a) diphenylmethane;   (b) triphenylmethane;   (c) toluene;   (d) methane.

SECTION 12.9.

Esterification of Carboxylic Acids

An **ester of a carboxylic acid** is a compound that contains the —CO_2R group, where R may be alkyl or aryl. An ester may be formed by the direct reaction of a carboxylic acid with an alcohol, a reaction called an **esterification reaction**. Esterification is acid-catalyzed and reversible.

General:

$$\underset{\text{a carboxylic acid}}{RCOH} + \underset{\text{an alcohol}}{R'OH} \underset{\xrightarrow{\hspace{1cm}}}{\xleftarrow{\hspace{1cm}}} \overset{H^+, \text{ heat}}{} \underset{\text{an ester}}{RCOR'} + H_2O$$

$$\underset{\text{acetic acid}}{CH_3COH} + \underset{\text{ethanol}}{CH_3CH_2OH} \underset{\xrightarrow{\hspace{1cm}}}{\xleftarrow{\hspace{1cm}}} \overset{H^+, \text{ heat}}{} \underset{\text{ethyl acetate}}{CH_3COCH_2CH_3} + H_2O$$

benzoic acid   cyclohexanol   cyclohexyl benzoate

The rate at which a carboxylic acid is esterified depends primarily upon the steric hindrance in both the alcohol and the carboxylic acid. The acid strength of the carboxylic acid plays only a minor role in the rate at which the ester is formed.

Reactivity of alcohols toward esterification:

$$CH_3OH > 1° > 2° > 3°$$

Reactivity of carboxylic acids toward esterification:

$$HCO_2H > CH_3CO_2H > RCH_2CO_2H > R_2CHCO_2H > R_3CCO_2H$$

Like many reactions of aldehydes and ketones, esterification of a carboxylic acid proceeds through a series of protonation and deprotonation steps. The carbonyl oxygen is protonated, the nucleophilic alcohol attacks the positive carbon, and elimination of water yields the ester.

$$
RC-\ddot{O}H \;\overset{H^+}{\rightleftharpoons}\; \left[RC-\ddot{O}H \longleftrightarrow RC-\ddot{O}H \right] \;\overset{R'\ddot{O}H}{\rightleftharpoons}\; \left[\begin{array}{c} OH \\ | \\ RC-\ddot{O}H \\ | \\ R\overset{+}{O}-H \end{array} \right] \;\overset{-H^+}{\rightleftharpoons}\;
$$

$$
\left[\begin{array}{c} OH \\ | \\ RC-\ddot{O}H \\ | \\ R'\ddot{O}: \end{array} \right] \;\overset{H^+}{\rightleftharpoons}\; \left[\begin{array}{c} OH \\ | \\ RC-\overset{+}{O}H_2 \\ | \\ R'\ddot{O}: \end{array} \right] \;\overset{-H_2O}{\rightleftharpoons}\; \left[\begin{array}{c} \ddot{O}H \\ | \\ RC^+ \longleftrightarrow RC \\ | \\ R'\ddot{O}: \end{array} \begin{array}{c} \overset{+}{O}H \\ || \\ \\ | \\ R'\ddot{O}: \end{array} \right] \;\overset{-H^+}{\rightleftharpoons}\; RC\ddot{O}R'
$$

We may summarize this mechanism in the following way:

$$
\underset{\substack{a\ carboxylic \\ acid}}{RCOH} + R'OH \;\overset{H^+}{\rightleftharpoons}\; \left[\begin{array}{c} OH \\ | \\ R-C-OH \\ | \\ OR' \end{array} \right] \;\rightleftharpoons\; \underset{an\ ester}{RCOR'} + H_2O
$$

Note that in an esterification reaction, it is the C—O bond of the carboxylic acid that is broken and not the O—H bond of the acid nor the C—O bond of the alcohol. Evidence for the mechanism is the reaction of a labeled alcohol such as $CH_3{}^{18}OH$ with a carboxylic acid. In this case, the ^{18}O stays with the methyl group.

C—O bond not broken

$$
\bigcirc\!\!-COH + CH_3{}^{18}OH \;\overset{H^+}{\rightleftharpoons}\; \left[\bigcirc\!\!-\begin{array}{c} OH \\ | \\ COH \\ | \\ {}^{18}OCH_3 \end{array} \right] \;\rightleftharpoons\;
$$

$$
\bigcirc\!\!-\overset{O}{\overset{||}{C}}-{}^{18}OCH_3 + H_2O
$$

STUDY PROBLEM

12.13. Write the complete mechanism for the esterification of acetic acid with $CH_3{}^{18}OH$.

The esterification reaction is reversible. To obtain a high yield of an ester, we must shift the equilibrium to the ester side. One technique for accomplishing this is to use an excess of one of the reactants (the cheaper one). Another technique is to remove one of the products from the reaction mixture (for example, by the azeotropic distillation of water).

As the amount of steric hindrance in the intermediate increases, the rate of ester formation drops. The yield of ester also decreases. The reason is that esterification is a reversible reaction and the less hindered species (the reactants) are favored. If bulky esters are to be prepared, it is better to use another synthetic route, such as the reaction of an alcohol with an acid anhydride or an acid chloride, which are more reactive than the carboxylic acid (see Chapter 13) and undergo irreversible reactions with alcohols.

Phenyl esters ($RCO_2C_6H_5$) are not generally prepared directly from phenols and carboxylic acids because the equilibrium favors the acid–phenol side rather than the ester side. Phenyl esters, like bulky esters, can be obtained by using the more reactive acid derivatives.

STUDY PROBLEMS

12.14. Predict the esterification products of: (a) *p*-toluic acid (page 570) and 2-propanol;   (b) terephthalic acid (p-HO_2C—C_6H_4—CO_2H) and excess ethanol;   (c) acetic acid and (*R*)-2-butanol.

12.15. 4-Hydroxybutanoic acid spontaneously forms a cyclic ester, or *lactone*. What is the structure of this lactone?

SECTION 12.10.

Reduction of Carboxylic Acids

The carbonyl carbon of a carboxylic acid is at the highest oxidation state it can attain and still be part of an organic molecule. (The next higher oxidation state is in CO_2.) Other than combustion or oxidation by very strong reagents, such as hot H_2SO_4–CrO_3 (cleaning solution), the carboxylic acid group is inert toward further oxidative reaction.

Surprisingly, the carboxylic acid group is also inert toward most common reducing agents (such as hydrogen plus catalyst). This inertness made necessary the development of alternative reduction methods, such as conversion of the carboxylic acid to an ester and then reduction of the ester. However, the introduction of lithium aluminum hydride (LAH) in the late 1940's simplified the reduction because LAH reduces a carboxyl group directly to a —CH_2OH group. (Other carbonyl functionality in the molecule is, of course, reduced as well; see Section 11.14.)

$$CH_3CO_2H \xrightarrow[\text{(2) H}_2\text{O}]{\text{(1) LiAlH}_4} CH_3CH_2OH$$

acetic acid   ethanol

benzoic acid   benzyl alcohol

STUDY PROBLEM

12.16. Give the structures for the LAH reduction products of:

(a) (image)—CO_2H (b) O=(image)—CO_2H

SECTION 12.11.

Polyfunctional Carboxylic Acids

Dicarboxylic acids and carboxylic acids containing other functional groups often show unique chemical properties. In this section, we will consider a few of the more important of these polyfunctional carboxylic acids. Hydroxy acids will be mentioned in Chapter 13, and amino acids will be discussed in Chapter 19.

A.   Acidity of dibasic acids

A **dibasic acid** is one that undergoes reaction with *two equivalents of base*. The term **diprotic acid** (two acidic protons) would perhaps be a better term to describe such a compound. In general, dibasic carboxylic acids have a chemistry similar to that of the monocarboxylic acids, but let us examine a few of the differences.

With any dibasic acid (inorganic or organic), the first hydrogen ion is removed more easily than the second. Thus, K_1 (the acidity constant for the ionization of the first H^+) is larger than K_2 (that for the ionization of the second H^+), and the value for pK_1 is smaller than that for pK_2. The difference between pK_1 and pK_2 decreases with increasing distance between the carboxyl groups. (Why?) The first and second pK_a values for a few diacids are listed in Table 12.7.

$$HO_2CCH_2CO_2H + H_2O \rightleftharpoons HO_2CCH_2CO_2^- + H_3O^+$$

malonic acid
$$pK_1 = 2.83$$

$$HO_2CCH_2CO_2^- + H_2O \rightleftharpoons {}^-O_2CCH_2CO_2^- + H_3O^+$$

$$pK_2 = 5.69$$

TABLE 12.7. pK_a values for some diacids

Trivial name[a]	Structure	pK_1	pK_2
oxalic	$HO_2C—CO_2H$	1.2	4.2
malonic	$HO_2CCH_2CO_2H$	2.8	5.7
succinic	$HO_2C(CH_2)_2CO_2H$	4.2	5.6
glutaric	$HO_2C(CH_2)_3CO_2H$	4.3	5.4
adipic	$HO_2C(CH_2)_4CO_2H$	4.4	5.4
pimelic	$HO_2C(CH_2)_5CO_2H$	4.5	5.4
o-phthalic	⟨benzene ring with two CO_2H groups⟩	2.9	5.5

[a] To memorize the trivial names of the diacids with two to seven carbons, note that the first letters (o, m, s, g, a, p) fit the phrase: "Oh my, such good apple pie." (*o*-Phthalic acid is, of course, not part of the homologous series.)

B.   Anhydride formation by dibasic acids

An **anhydride of a carboxylic acid** has the structure of two carboxylic acid molecules joined together with the loss of water.

$$
\underset{\text{two carboxylic acids}}{RC\overset{O}{\overset{\|}{-}}O\,H \quad HO\overset{O}{\overset{\|}{-}}CR} \qquad \underset{\text{an anhydride}}{RC\overset{O}{\overset{\|}{-}}O\overset{O}{\overset{\|}{-}}CR}
$$

Although it would seem reasonable at first glance, heating most carboxylic acids to drive off water does *not* result in the anhydride. (The reactions that *do* lead to anhydrides will be discussed in Section 13.4.) Exceptions are dicarboxylic acids that can form five- or six-membered cyclic anhydrides. These diacids yield anhydrides when heated to 200°–300°.

succinic acid   succinic anhydride

glutaric acid   glutaric anhydride

maleic acid   maleic anhydride

A five- or six-membered cyclic anhydride may also be synthesized by heating the appropriate diacid with acetic anhydride. This reaction is the result of an equilibrium between a more stable, cyclic anhydride with a less stable, open-chain anhydride. Removing the acetic acid by distillation helps drive the reaction to completion.

85%

STUDY PROBLEMS

12.17. *o*-Phthalic acid (Table 12.7) forms a cyclic anhydride when heated at 200°. What is its structure?

12.18. When adipic acid (hexanedioic acid) is heated with a strong dehydrating agent such as P_2O_5, a cyclic anhydride is not formed. Instead, a polymeric product is obtained. What is the structure of this polymer?

C. Decarboxylation of β-keto acids and β-diacids

Simply heating most carboxylic acids does not result in any chemical reaction. However, a carboxylic acid with a β carbonyl group undergoes **decarboxylation** (loss of CO_2) when heated. (The temperature necessary varies with the individual compound.)

Decarboxylation takes place through a cyclic transition state:

transition state   *an enol*

Note that the cyclic transition state requires only a carbonyl group beta to the carboxyl group. This carbonyl group need not necessarily be a keto group. A β-diacid also undergoes decarboxylation when heated. Decarboxylation of substituted malonic acids is especially important in organic synthesis, and we will encounter this reaction again in Chapter 14.

STUDY PROBLEM

12.19. Write the mechanism for the decarboxylation of ethylmalonic acid.

A few α-carbonyl acids, such as oxalic acid, can also undergo decarboxylation. The decarboxylation of α-keto acids is common in biological systems in which enzymes catalyze the reaction.

$$\underset{\text{oxalic acid}}{HO\overset{\overset{O}{\|}}{C}-\overset{\overset{O}{\|}}{C}OH} \xrightarrow{150°} \underset{\text{formic acid}}{HCO_2H} + CO_2$$

$$\underset{\substack{\text{pyruvic acid}\\ \textit{formed in glucose}\\ \textit{metabolism}}}{CH_3\overset{\overset{O}{\|}}{C}-\overset{\overset{O}{\|}}{C}OH} \xrightarrow{\text{enzymes}} \underset{\text{acetaldehyde}}{CH_3\overset{\overset{O}{\|}}{C}H} + CO_2$$

STUDY PROBLEM

12.20. Predict the product (if any) when each of the following acids is heated:

(a) $CH_2(CO_2H)_2$

(b) $CH_3\overset{\overset{O}{\|}}{C}CHCO_2H$ with CH_3

(c) $CH_3CH_2\overset{\overset{O}{\|}}{C}CHCO_2H$ with CH_2CH_3

(d) $CH_3\overset{\overset{O}{\|}}{C}CH_2CH_2CH_2CO_2H$

D. α,β-Unsaturated carboxylic acids

An isolated double bond in an unsaturated carboxylic acid behaves independently of the carboxyl group. A *conjugated* double bond and carboxyl group, however, may undergo typical 1,4-addition reactions. These reactions take place in the same fashion as those of α,β-unsaturated aldehydes or ketones (Section 11.19).

$$\ddot{C}N^- + CH_2=CH-\overset{\overset{\ddot{O}}{\|}}{C}-OH \longrightarrow [NCCH_2CH=C\overset{:\ddot{O}:^-}{\underset{}{|}}OH] \rightleftharpoons NCCH_2CH_2\overset{\overset{\ddot{O}}{\|}}{C}\ddot{O}:^-$$

an enol

$$\overset{..}{N}H_3 + \underset{HO_2C}{\overset{H}{\underset{}{}}}C=C\underset{H}{\overset{CO_2H}{}} \longrightarrow \underset{\text{aspartic acid}}{^-O_2CCH\overset{^+NH_3}{\underset{}{|}}CH_2CO_2H}$$

fumaric acid / *an amino acid found in proteins*

STUDY PROBLEM

12.21. A double bond in conjugation with an aldehyde or ketone may undergo nucleophilic attack at the β position by a Grignard reagent (Section 11.19). What products would you expect from the reaction of 2-pentenoic acid with phenylmagnesium bromide? (Be careful; check Section 6.11.)

SECTION 12.12.

Use of Carboxylic Acids in Synthesis

The types of products that can be obtained from carboxylic acids are shown in Table 12.8. In many of these reactions, an acid halide or an acid anhydride gives better yields; esterification is such a reaction. However, the acids themselves are usually more readily available. Therefore, a chemist must balance time and expense versus yield.

TABLE 12.8. Compounds that can be obtained from carboxylic acids

Reaction		Product	Section reference
Neutralization:[a]			
$RCO_2H + Na^+ OH^- \xrightarrow{H_2O} RCO_2^- Na^+$		**carboxylate salt**	12.5, 12.6
Esterification:			
$RCO_2H + R'OH \xrightarrow{H^+} RCO_2R'$		**ester**	12.9
Reduction:			
$RCO_2H \xrightarrow[(2)\ H_2O]{(1)\ LiAlH_4} RCH_2OH$		**1° alcohol**	12.10
Anhydride formation:			
$HO_2C(CH_2)_nCO_2H \xrightarrow[\text{or heat}]{(CH_3C)_2O}$ where $n = 2$ or 3	$O=C\underset{(CH_2)_n}{\overset{O}{\diagup\diagdown}}C=O$	**cyclic anhydride**	12.11B, 13.4B
Decarboxylation:			
$R\overset{O}{\overset{\|}{C}}CR'_2CO_2H \xrightarrow{\text{heat}} R\overset{O}{\overset{\|}{C}}CHR'_2$		**ketone**	12.11C
$HO_2CCR_2CO_2H \xrightarrow{\text{heat}} R_2CHCO_2H$		**carboxylic acid**	12.11C
1,4-Addition:			
$R_2C=CHCO_2H + HNu \longrightarrow R_2\underset{\underset{Nu}{\|}}{C}CH_2CO_2H$		**β-substituted carboxylic acid**	12.11D

(continued)

TABLE 12.8. *(continued)*

Reaction	*Product*	*Section reference*
Acid halide formation:[b]		
$RCO_2H + SOCl_2$ or $PCl_3 \longrightarrow$ $\overset{\overset{\textstyle O}{\textstyle \|}}{\textstyle RCCl}$	**acid halide**	13.3B
α Halogenation:		
$RCH_2CO_2H + Cl_2 \xrightarrow{PCl_3} RCHClCO_2H$	**α-chloro acid**	13.3C

[a] Carboxylate salts can be used to make esters by S_N2 reactions with reactive halides:

$$RCO_2^- + R'X \longrightarrow RCO_2R' \quad \text{(See Section 5.11.)}$$

[b] Acid halides may be used to make a variety of compounds, including anhydrides and esters. These reactions are discussed in Chapter 13.

STUDY PROBLEMS

12.22. Show two routes to methyl butanoate from butanoic acid.

12.23. Show by flow equations how you would make the following conversions:

(a)   propanoic acid $\longrightarrow$ butanoic acid

(b)   3-chloropropanoic acid $\longrightarrow$ butanedioic acid

(c) $\longrightarrow$ —CH_2CO_2H

Summary

Carboxylic acids (RCO_2H) undergo hydrogen bonding to form **dimers**, a structural feature that has an effect on their physical and spectral properties.

Carboxylic acids may be synthesized by: (1) the *hydrolysis of their derivatives* (esters, amides, anhydrides, acid halides, or nitriles); (2) *oxidation of primary alcohols, aldehydes, alkenes, or alkylbenzenes*; and (3) *Grignard reactions* of $RMgX$ and CO_2.

Carboxylic acids are one of the few general classes of organic compounds that is more acidic than H_2CO_3; carboxylic acids undergo reaction with HCO_3^-. **Carboxylate salts** result from the reaction of a carboxylic acid with base.

$$\underset{\substack{\text{a carboxylic} \\ \text{acid}}}{RCO_2H} + \underset{\text{a base}}{NaOH} \longrightarrow \underset{\substack{\text{a sodium} \\ \text{carboxylate}}}{RCO_2^- \ Na^+} + H_2O$$

The strength of an acid is determined by the relative stabilities of the acid and its anion. Acid strength is affected by *electronegativity* (HF > ROH > R_2NH > RH); by *size* (HI > HBr > HCl > HF); and by *hybridization* ($\equiv$CH > =CH_2 > —CH_3). The *inductive effect* of electron-withdrawing groups causes an increase in acid strength ($ClCH_2CO_2H$ > CH_3CO_2H). *Resonance-stabilization* of the anion also strengthens an acid (RCO_2H > ArOH > ROH). The anion of an acid can be partially stabilized by *solvation*—increased solvation of the anion strengthens an acid.

The strength of a benzoic acid is determined largely by inductive effects because the —$CO_2{}^-$ group does not enter into resonance with the aromatic ring. Electron-withdrawing substituents strengthen the acid, while electron-releasing groups weaken the acid. Substitution at the *o*-position almost always increases acid strength.

$$O_2N \!\longleftarrow\!\!\!\bigcirc\!\!\!-CO_2H > \bigcirc\!\!\!-CO_2H > CH_3 \!\longrightarrow\!\!\!\bigcirc\!\!\!-CO_2H$$

The reactions of carboxylic acids are summarized in Table 12.8, pages 595–596.

STUDY PROBLEMS

12.24. Name the following acids and salts:

(a) $(CH_3)_3CCO_2H$

(b) $Cl\!-\!\bigcirc\!\!-CO_2H$

(c) $CH_3CH_2CHBrCHBrCO_2H$

(d) $(CH_3CH_2CO_2)_2Mg$

(e) $\begin{matrix} CO_2{}^- \\ | \\ CO_2{}^- \end{matrix}\ Ca^{2+}$

(f) $CH_3CHBrCO_2{}^-\ Na^+$

12.25. Give the structures for: (a) 4-iodobutanoic acid; (b) potassium formate; (c) disodium *o*-phthalate; (d) sodium benzoate; (e) *m*-methylbenzoic acid.

12.26. Give the structure of each of the following groups: (a) propionyl group; (b) butyryl group; (c) *m*-nitrobenzoyl group.

12.27. Show the principal types of hydrogen bonding (there may be more than one type) that occur in each of the following systems: (a) an aqueous solution of propanoic acid; (b) an aqueous solution of lactic acid, $CH_3CH(OH)CO_2H$.

12.28. Show how you could synthesize butanoic acid from each of the following compounds:

(a) 1-bromopropane

(b) 1-butanol

(c) butanal

(d) 4-octene

(e) $CH_3CH_2CH_2CO_2CH_2CH_3$

(f) $CH_3CH_2CH_2CO_2\!-\!\bigcirc$

12.29. Suggest reagents for the following conversions:

(a) $H_2NCH_2CH_2CH_2Br \longrightarrow {}^+H_3NCH_2CH_2CH_2CO_2{}^-$

(b) $(CH_3)_3CBr \longrightarrow (CH_3)_3CCO_2H$

(c) $CH_3CH_2OH \longrightarrow CH_3\underset{\underset{\displaystyle OH}{|}}{CH}CO_2H$

(d) $BrCH_2CH_2CH_2Br \longrightarrow HO_2CCH_2CH_2CH_2CO_2H$

(e) ⬠$=CH_2 \longrightarrow$ ⬠$-CH_2CO_2H$

(f) $C_6H_5Br \longrightarrow C_6H_5CO_2H$

(g) (cyclohexane ring with $CO_2CH_2CH_3$ and $\underset{\underset{\displaystyle O}{||}}{C}CH_3$ substituents) $\longrightarrow$ (cyclohexane ring with $\overset{\overset{\displaystyle O}{||}}{C}CH_3$ substituent)

(h) $CH_3CH_2CH_2Br \longrightarrow CH_3(CH_2)_3CO_2H$

12.30. Predict the major organic products:

(a) (cyclohexane ring with CO_2H and CO_2H substituents) + excess NaOH $\longrightarrow$

(b) $C_6H_5CH_2Cl + CH_3CH_2\overset{\overset{\displaystyle O}{||}}{C}O^-\ Na^+ \longrightarrow$

(c) $CH_3CO_2H + CH_3O^- \longrightarrow$

(d) $CH_3CO_2H + CH_3NH_2 \longrightarrow$

(e) $CH_3CO_2{}^- + ClCH_2CO_2H \longrightarrow$

(f) $HO_2CCH_2CO_2H + 1$ equivalent $NaHCO_3 \longrightarrow$

(g) (naphthalene ring with OH and HO substituents) + 2 equivalents $NaHCO_3 \longrightarrow$

(h) $C_6H_5CO_2H + CH_3I + CO_3{}^{2-} \longrightarrow$

(i) $C_6H_5CO_2H + LiOH \longrightarrow$

(j) $C_6H_5OH + LiOH \longrightarrow$

12.31. If a $0.200M$ solution of pentanoic acid has a hydrogen-ion concentration of $0.00184M$, what is the dissociation constant (K_a) of the acid? (*Hint*: See Section 1.10.)

12.32. A mixture contains *p*-ethylphenol, benzoic acid, and benzaldehyde. The mixture is dissolved in ether and washed with an aqueous solution of $NaHCO_3$. The bicarbonate solution is Solution A. The mixture is then washed with an aqueous solution of NaOH (Solution B) and finally with water (Solution C). The remaining ether solution is Solution D. Identify the main organic components in Solutions A, B, C, and D.

12.33. How would you separate each of the following pairs of compounds?

(a) octanoic acid and ethyl octanoate

(b) phenol and phenyl propanoate

(c) phenol and cyclohexanol

12.34. What is the neutralization equivalent of each of the following acids?

(a) $\underset{\text{lactic acid}}{CH_3\overset{\displaystyle OH}{\overset{|}{C}}HCO_2H}$

(b) $\underset{\text{oxaloacetic acid}}{\overset{\displaystyle O}{\overset{||}{C}}CO_2H \atop CH_2CO_2H}$

12.35. A 0.250-g sample of an unknown acid is titrated with a standardized NaOH solution that is $0.307M$. The volume of the NaOH solution required to neutralize the acid is 11.00 mL. Which of the following acids could the sample be?

(a) $CH_3CH_2CH_2CO_2H$
(b) $CH_3CH_2CO_2H$
(c) $HO_2CCH_2CO_2H$
(d) $HO_2CCH_2CH_2CH_2CH_2CO_2H$

(e) 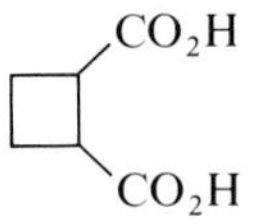   ![cyclobutane with two CO₂H groups] a cyclobutane ring bearing CO_2H and CO_2H

12.36. Calculate the pK_a for each of the following acids (see Section 1.10):

(a) $cis\text{-}C_6H_5CH{=}CHCO_2H$, $K_a = 1.3 \times 10^{-4}$
(b) $trans\text{-}C_6H_5CH{=}CHCO_2H$, $K_a = 3.65 \times 10^{-5}$

12.37. The pH of a $0.0100M$ solution of an acid was found to be 2.50. Calculate the K_a of this acid.

12.38. How do you account for the following acidities?

$$FCH_2CO_2H \qquad ClCH_2CO_2H \qquad BrCH_2CO_2H$$
$$pK_a = 2.66 \qquad pK_a = 2.81 \qquad pK_a = 2.87$$

12.39. List the following compounds in order of increasing acid strength (least acidic first);

(a) $CH_3CH_2CHBrCO_2H$
(b) $CH_3CHBrCH_2CO_2H$
(c) $CH_3CH_2CH_2CO_2H$
(d) $CH_3CH_2CH_2CH_2OH$
(e) C_6H_5OH
(f) H_2CO_3
(g) Br_3CCO_2H
(h) H_2O

12.40. Which one of each of the following pairs of carboxylic acids is the stronger acid? (a) benzoic acid or *p*-bromobenzoic acid;   (b) benzoic acid or *m*-bromobenzoic acid; (c) *m*-bromobenzoic acid or 3,5-dibromobenzoic acid.   Explain.

12.41. Arrange the following compounds in order of increasing acidity (weakest acid first). Give a reason for your answer.

$$p\text{-nitrophenol} \qquad p\text{-methylphenol} \qquad \text{phenol}$$

12.42. Which of the following acids is most acidic? least acidic?

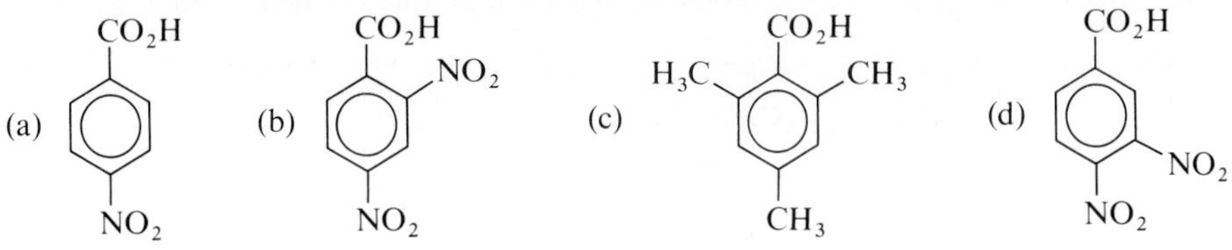

(a) benzene ring with CO_2H para to NO_2

(b) benzene ring with CO_2H and NO_2 (ortho), and NO_2 para

(c) benzene ring with CO_2H, H_3C and CH_3 (meta), CH_3 para

(d) benzene ring with CO_2H, and NO_2, NO_2 substituents

12.43. In each pair, which is the stronger base?

(a) $CH_3CH{=}CH^-$ or $CH_3C{\equiv}C^-$
(b) Cl^- or $CH_3CO_2^-$
(c) $ClCH_2CO_2^-$ or $Cl_2CHCO_2^-$
(d) $(CH_3)_3CO^-$ or $(CH_3)_3CCO_2^-$
(e) $CH_3CHClCO_2^-$ or $ClCH_2CH_2CO_2^-$

12.44. Suggest reasons why *o*-phthalic acid has a pK_1 of 2.9 and a pK_2 of 5.5, while terephthalic acid has a pK_1 of 3.5 and a pK_2 of 4.8.

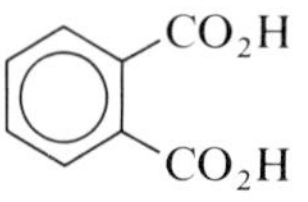
CO2H
CO2H
o-phthalic acid

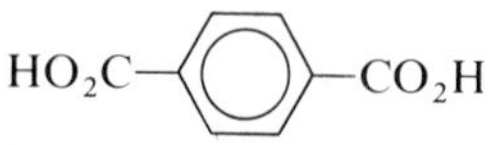
HO2C
CO2H
terephthalic acid

12.45. Complete the following equations, giving the organic products:

(a) 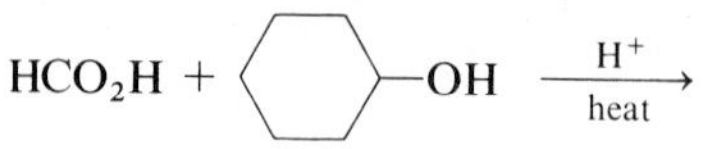
HCO2H +
OH
H+
heat

(b) 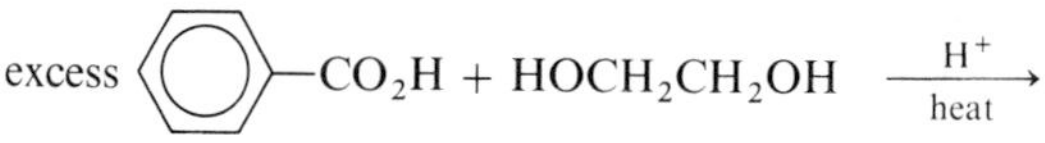
excess
CO2H + HOCH2CH2OH
H+
heat

(c) 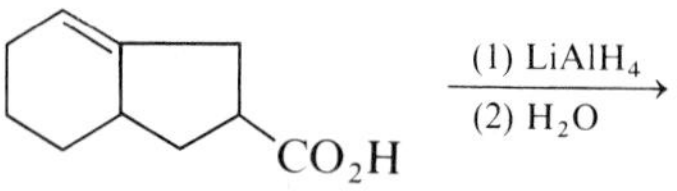
CO2H
(1) LiAlH4
(2) H2O

(d) 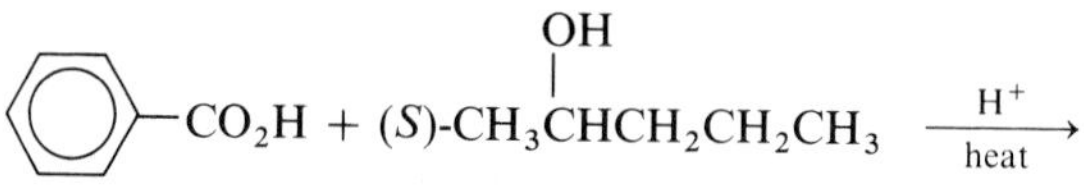
OH
CO2H + (S)-CH3CHCH2CH2CH3
H+
heat

(e) 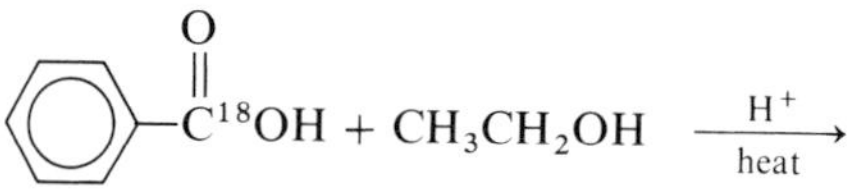
O
C18OH + CH3CH2OH
H+
heat

12.46. List the following acids and alcohols in order of increasing rates of reaction leading to the ester (slowest first). Explain your answer.

(a) acetic acid with methanol and HCl
(b) cyclohexanecarboxylic acid with *t*-butyl alcohol and HCl
(c) cyclohexanecarboxylic acid with ethanol and HCl

12.47. Racemic 2-bromopropanoic acid undergoes reaction with (*R*)-2-butanol to produce an ester mixture. (a) What are the structures (stereochemistry included) of the esters? (b) Are these esters enantiomers or diastereomers?

12.48. What are the major organic products when the following compounds are heated?

(a) 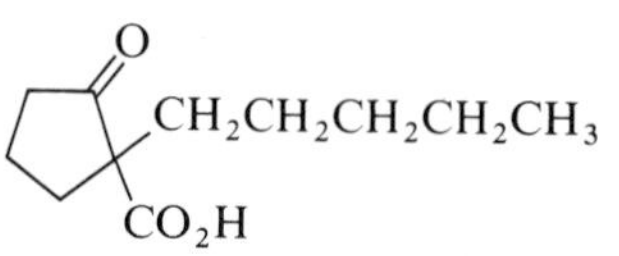
O
CH2CH2CH2CH2CH3
CO2H

(b) 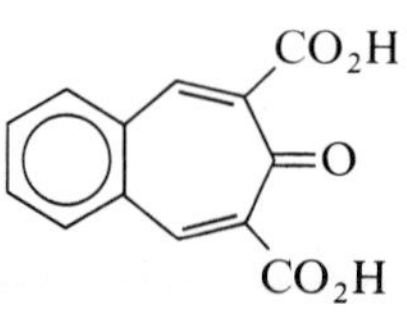
CO2H
O
CO2H

12.49. One of the steps in the biological oxidation of glucose to CO_2 and H_2O is the β decarboxylation of oxalosuccinic acid. What is the product of the decarboxylation?

$$\underset{\text{oxalosuccinic acid}}{HO_2CCH_2\overset{\displaystyle CO_2H}{\underset{\displaystyle \underset{O}{\|}}{CHCCO_2H}}}$$

12.50. Explain why the following compound does not undergo decarboxylation when heated:

12.51. Predict the major organic products:

(a) a concentrated solution (inert solvent) of $HOCH_2(CH_2)_8CO_2H$ $\xrightarrow[\text{heat}]{H^+}$

(b) a dilute solution (inert solvent) of $HOCH_2(CH_2)_8CO_2H$ $\xrightarrow[\text{heat}]{H^+}$

(c) $+$ excess $C_6H_5\overset{O}{\overset{\|}{C}}O\overset{O}{\overset{\|}{C}}C_6H_5$ $\xrightarrow{\text{heat}}$

(d) $CH_3\overset{O}{\overset{\|}{C}}CH_2CO_2CH_2CH_3$ $\xrightarrow[\text{heat}]{H_2O, H^+}$

(e) $-CO_2H + HCl \longrightarrow$

(f) $-CO_2H + H_2O \xrightarrow{H^+}$

12.52. *Nicotinic acid* (also called *niacin*), a B vitamin, can be obtained by the vigorous oxidation of nicotine (page 763). What is the structure of nicotinic acid?

12.53. *Dacron* is a synthetic polymer that can be made by the reaction of terephthalic acid (Problem 12.44) and 1,2-ethanediol. What is the structure of Dacron?

12.54. Compound A, $C_4H_6O_4$, yields compound B, $C_3H_6O_2$, when heated. The neutralization equivalent of A is 60 $\pm$ 1, while that of B is 75 $\pm$ 1. What are the structures of A and B?

12.55. With $^{14}CH_3I$ as the only source of ^{14}C, show how you would prepare the following labeled compounds:

(a) $CH_3CH_2CO_2{}^{14}CH_3$ (b) $^{14}CH_3CO_2H$ (c) $CH_3{}^{14}CO_2H$

12.56. Write equations to show how benzoic acid can be converted to the following compounds:

(a) $-CH_2OH$ (b) $-CH_2CN$ (c) $-CO_2CH_3$

12.57. Propose mechanisms for the following reactions:

(a) $CH_3\overset{O}{\overset{\|}{C}}OH + H_2{}^{18}O \xrightleftharpoons{H^+} CH_3\overset{O}{\overset{\|}{C}}{-}{}^{18}OH + CH_3\overset{{}^{18}O}{\overset{\|}{C}}OH + H_2O$

(b) 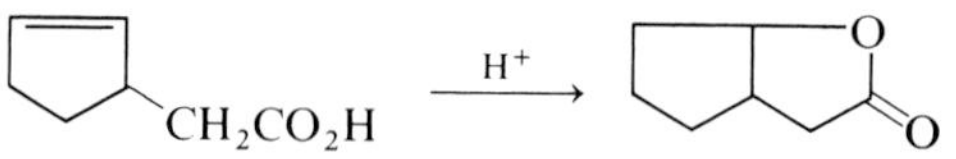   [cyclopentane ring with CH_2CO_2H substituent] $\xrightarrow{H^+}$ [bicyclic lactone]

(c) $BrCH_2(CH_2)_4CH_2CO_2H$ $\xrightarrow[\text{(2) H}^+,\ \text{heat}]{\text{(1) OH}^-}$ [cyclic lactone]

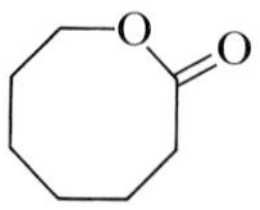

(d) $C_6H_5CCl_3$ $\xrightarrow[\text{(2) H}_2O,\ H^+]{\text{(1) OH}^-,\ \text{heat}}$ $C_6H_5CO_2H$

12.58. When tartaric acid (page 148) is heated, it is converted to pyruvic acid. Write a flow equation to show the steps in this reaction.

$$CH_3\overset{O}{\overset{\|}{C}}CO_2H$$

pyruvic acid

12.59. The salt of a carboxylic acid does not show carbonyl absorption at 1660–2000 cm^{-1} (5–6 μm) in the infrared spectrum. Explain.

12.60. Compound A, $C_4H_6O_4$, yielded Compound B, $C_4H_4O_3$, when heated. When A was treated with an excess of methanol and a trace of sulfuric acid, Compound C, $C_6H_{10}O_4$, was obtained. Upon treatment with LiAlH$_4$ followed by hydrolysis, Compound A yielded Compound D, $C_4H_{10}O_2$. What are the structures of Compounds A, B, C, and D?

12.61. Compound A ($C_5H_8O_2$) dissolves in water to yield a neutral solution. Treatment of A with HIO$_4$ gives no reaction, but treatment with I$_2$ in aqueous NaOH, followed by acidification, yields B ($C_3H_4O_4$). An aqueous solution of B is acidic. What are the structures of A and B?

12.62. Suggest syntheses for the following compounds from compounds containing six or fewer carbons:

(a) $CH_3(CH_2)_7CO_2H$

(b) $CH_3(CH_2)_2\overset{\overset{\displaystyle CO_2CH_3}{\displaystyle |}}{C}HCH_2CO_2CH_3$

(c) 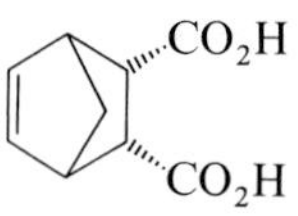   [bicyclic structure with CO_2H and CO_2H substituents]

(d) 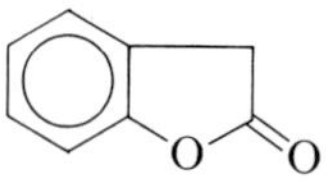   [benzofuranone-type bicyclic structure]

12.63. Suggest synthetic schemes for the following conversions:

(a)   ethyl propanoate from bromoethane as the only organic starting material

(b) 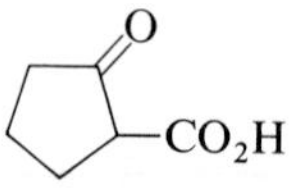   [cyclopentanone ring]$-CO_2H$ from cyclopentanol (c)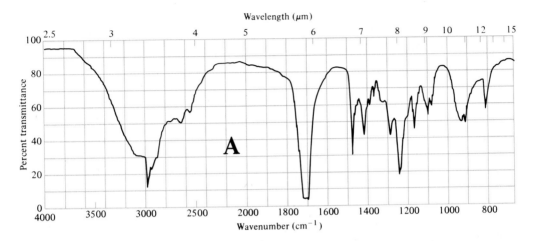

—CO$_2$H   from cyclopentanol

12.64. Compound A (C$_4$H$_8$O$_2$) was treated with LiAlH$_4$. After hydrolysis, Compound B was
isolated. The infrared spectrum of A and the nmr spectrum of B are shown in Figure 12.4.
What are the structures of A and B?

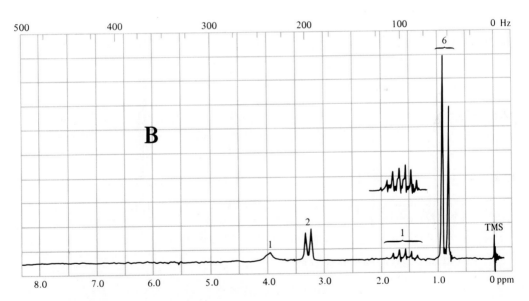

FIGURE 12.4. Spectra for Problem 12.64.

12.65. A chemist treated Compound A ($C_4H_7O_2Br$) with potassium t-butoxide in t-butyl alcohol. After acidification of the product mixture, the chemist isolated two isomeric products: B and C. The infrared spectrum of B and the nmr spectrum of C are shown in Figure 12.5. What are the possible structures of A, B, and C?

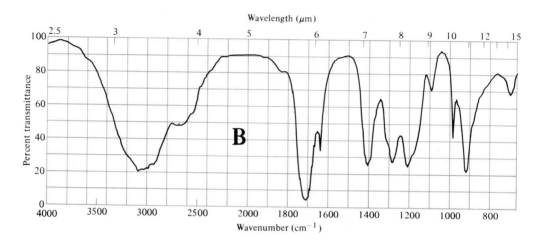

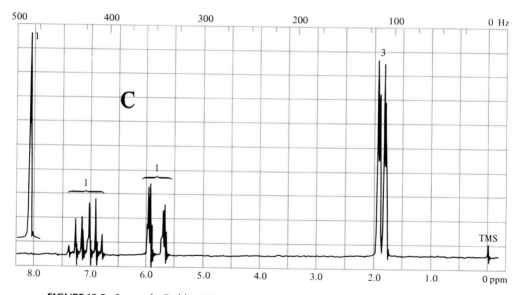

FIGURE 12.5. Spectra for Problem 12.65.

CHAPTER 13

Derivatives of Carboxylic Acids

A derivative of a carboxylic acid is a compound that yields a carboxylic acid upon reaction with water.

$$\underset{\text{ethyl acetate}}{CH_3\overset{\overset{\displaystyle O}{\|}}{C}OCH_2CH_3} + H_2O \underset{}{\overset{H^+,\text{heat}}{\rightleftharpoons}} \underset{\text{acetic acid}}{CH_3\overset{\overset{\displaystyle O}{\|}}{C}OH} + HOCH_2CH_3$$

$$\underset{\text{acetamide}}{CH_3\overset{\overset{\displaystyle O}{\|}}{C}NH_2} + H_2O + H^+ \xrightarrow{\text{heat}} CH_3\overset{\overset{\displaystyle O}{\|}}{C}OH + NH_4{}^+$$

In this chapter, we will discuss acid halides, acid anhydrides, esters, amides, and nitriles. Table 13.1 shows some representative examples of these compounds. Note that all the derivatives except nitriles contain **acyl groups, RCO—**. In each case, an electronegative atom is attached to the carbonyl carbon of the acyl group. For this reason, the chemistry of each of these classes of compounds is similar.

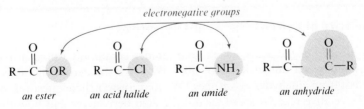

Carboxylic acids and some of their derivatives are found in nature. Fats are triesters, waxes are monoesters, and proteins are polyamides, to name just a few.

TABLE 13.1. Some derivatives of carboxylic acids

Class	Structure	Trivial name
	Examples	
acid halide	CH_3COCl	acetyl chloride
	⟨◯⟩—COCl	benzoyl chloride
acid anhydride	$(CH_3CO)_2O$	acetic anhydride
	(⟨◯⟩—CO)$_2$O	benzoic anhydride
ester	$CH_3CO_2CH_2CH_3$	ethyl acetate
	⟨◯⟩—CO_2CH_3	methyl benzoate
amide	CH_3CONH_2	acetamide
	⟨◯⟩—$CONH_2$	benzamide
nitrile	CH_3CN	acetonitrile
	⟨◯⟩—CN	benzonitrile

Acid halides are never found in nature, and anhydrides are found only rarely. One example of a naturally occurring anhydride is *cantharidin*, a cyclic anhydride found in Spanish flies. Cantharidin is an irritant of the urinary tract. The dried flies were used by the ancient Greeks and Romans as an aphrodisiac, and it is also reputed to remove warts.

cantharidin

SECTION 13.1.

Reactivity of Carboxylic Acid Derivatives

Why are carboxylic acids, esters, and amides commonly found in nature, while acid halides and anhydrides are not? Why are the carboxylic acid derivatives different from aldehydes or ketones? We may answer these questions by considering the relative reactivities of carboxylic acid derivatives and how they undergo reaction.

The derivatives of carboxylic acids contain *leaving groups* attached to the acyl carbons, whereas aldehydes and ketones do not. Reagents generally *add* to the carbonyl group of ketones and aldehydes, but *substitute* for the leaving groups of acid derivatives:

Addition:

$$CH_3CCH_3 \underset{}{\overset{HCN}{\rightleftharpoons}} CH_3-\underset{CN}{\overset{OH}{\underset{|}{\overset{|}{C}}}}-CH_3$$

Substitution:

$$CH_3C-Cl \xrightarrow{H_2O} CH_3C-OH + HCl$$

good leaving group

In Chapter 5, we mentioned that a good leaving group is a weak base. Therefore, Cl⁻ is a good leaving group, but ⁻OH and ⁻OR are poor leaving groups. The reactivity of carbonyl compounds toward substitution at the carbonyl carbon may be directly attributed to the basicity of the leaving group:

$$^-NH_2 \quad ^-OR \quad ^-OCR \quad X^-$$

decreasing basicity
(increasing ease of displacement)

$$R-\overset{O}{\underset{}{\overset{||}{C}}}-R \quad R\overset{O}{\underset{}{\overset{||}{C}}}-NH_2 \quad R\overset{O}{\underset{}{\overset{||}{C}}}-OR' \quad R\overset{O}{\underset{}{\overset{||}{C}}}-O\overset{O}{\underset{}{\overset{||}{C}}}R \quad R\overset{O}{\underset{}{\overset{||}{C}}}-Cl$$

increasing reactivity

Acid chlorides and acid anhydrides, with good leaving groups, are readily attacked by water. Therefore, we would not expect to find these compounds in the cells of plants or animals. Because of their high reactivity, however, these acid derivatives are invaluable in the synthesis of other organic compounds. A relatively nonreactive carboxylic acid may be converted to one of these more reactive derivatives and then converted to a ketone, an ester, or an amide (Figure 13.1).

Esters and amides are relatively stable toward water. In the laboratory, these compounds require an acid or a base and, usually, heating to undergo reaction. In nature, enzymes can perform the functions of acid or base and heat.

STUDY PROBLEM

13.1. Both acetyl chloride and vinyl chloride (CH_2=CHCl) have a chlorine atom attached to an sp^2 carbon. Suggest a reason why acetyl chloride is quite reactive, while vinyl chloride is almost inert to substitution reactions.

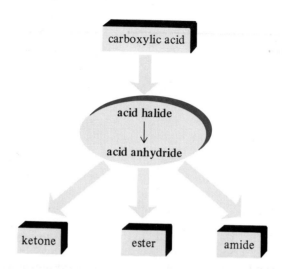

FIGURE 13.1. The synthetic relationship of acid halides and anhydrides to the other carboxylic acid derivatives.

SECTION 13.2.

Spectral Properties of Carboxylic Acid Derivatives

The nmr spectra of carboxylic acid derivatives provide little information about the functionality in these compounds. The signals for α hydrogens of these carbonyl compounds are shifted slightly downfield from the signals for ordinary aliphatic hydrogens because of deshielding by the partially positive carbonyl carbon atom. Note that the α hydrogens of an acid chloride exhibit a greater chemical shift than those of the other acid derivatives. This large chemical shift arises from the greater ability of the Cl (compared to O or N) to withdraw electron density from nearby bonds.

$$
\begin{array}{ccccc}
& \overset{O}{\underset{\parallel}{}} & \overset{O}{\underset{\parallel}{}} & \overset{O}{\underset{\parallel}{}} & \overset{O}{\underset{\parallel}{}} \\
\underline{CH}_3CN & \underline{CH}_3COCH_3 & \underline{CH}_3CNH_2 & \underline{CH}_3COH & \underline{CH}_3CCl
\end{array}
$$

δ in ppm: 2.00   2.03   2.08   2.10   2.67

increasing chemical shift for α hydrogens

The infrared spectra of acid derivatives provide more information about the type of functional group than do nmr spectra. With the exception of the nitriles, the principal distinguishing feature of the infrared spectra of all the carboxylic acid derivatives is the carbonyl absorption found at about 1630–1840 cm^{-1} (5.4–6 μm). The positions of carbonyl absorption for the various acid derivatives are summarized in Table 13.2. Anhydrides and esters also show C—O absorption in the region of 1050–1250 cm^{-1} (8–9.5 μm).

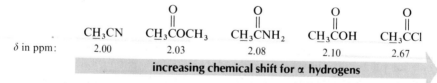

$$
\begin{array}{cccc}
\overset{O}{\underset{\parallel}{}} & \overset{O}{\underset{\parallel}{}} & \overset{O}{\underset{\parallel}{}} & \overset{O}{\underset{\parallel}{}}\;\overset{O}{\underset{\parallel}{}} \\
RCCl & RCNH_2 & RC\!-\!OR & RC\!-\!O\!-\!CR
\end{array}
$$

show C=O   *also show C—O*

TABLE 13.2. Infrared carbonyl absorption for carboxylic acid derivatives

		Position of absorption	
Class	Structure	cm^{-1}	μm
acid chloride	$\overset{\displaystyle O}{\overset{\displaystyle \|}{RCCl}}$	1785–1815	5.51–5.60
acid anhydride	$\overset{\displaystyle O\ \ O}{\overset{\displaystyle \|\ \ \|}{RCOCR}}$	1740–1840 (usually two peaks)	5.45–5.75
ester	$\overset{\displaystyle O}{\overset{\displaystyle \|}{RCOR}}$	1740	5.75
amide	$\overset{\displaystyle O}{\overset{\displaystyle \|}{RCNH_2}}$	1630–1700	5.9–6.0

A. Acid chlorides

The carbonyl infrared absorption of acid chlorides is observed at slightly higher frequencies than that of other acid derivatives. There is no other distinguishing feature in the infrared spectrum that signifies "This is an acid chloride." See Figure 13.2 for the infrared spectrum of a typical acid chloride.

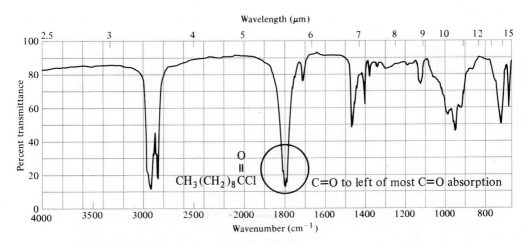

FIGURE 13.2. Infrared spectrum of decanoyl chloride.

B. Anhydrides

A carboxylic acid anhydride, which has two C=O groups, generally exhibits a *double carbonyl peak* in the infrared spectrum. Anhydrides also exhibit a C—O stretching band around $1100 \, cm^{-1}$ ($9 \, \mu m$). Figure 13.3 shows the infrared spectrum of a typical aliphatic anhydride.

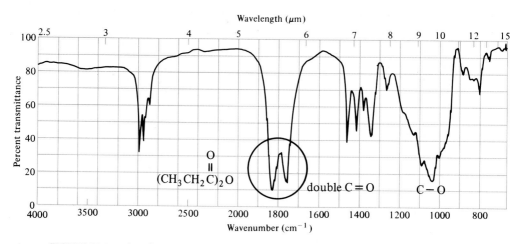

FIGURE 13.3. Infrared spectrum of propanoic anhydride.

C.   Esters

The carbonyl infrared absorption of aliphatic esters is observed at about 1740 cm^{-1} (5.75 μm). However, conjugated esters (either α,β-unsaturated esters or α-aryl esters) absorb at slightly lower frequencies, about 1725 cm^{-1} (5.8 μm). Esters also exhibit C—O stretching absorption in the fingerprint region. See Figure 13.4 for the spectrum of a typical ester.

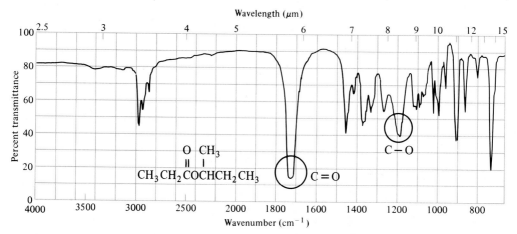

FIGURE 13.4. Infrared spectrum of *sec*-butyl propanoate. (ester)

D.   Amides

The position of absorption of the carbonyl group of an amide is variable and depends upon the extent of hydrogen bonding between molecules. The infrared spectrum of a pure liquid amide (maximum hydrogen bonding) shows a carbonyl peak called the **amide I bond** around 1650 cm^{-1} (6.0 μm). (The spectra in Figure 13.5 show this C=O peak.) As the sample is diluted with a nonhydrogen-bonding solvent, the extent of the hydrogen bonding diminishes, and the C=O absorption is shifted to a higher frequency (1700 cm^{-1}; 5.88 μm).

The **amide II band** appears between 1515–1670 cm^{-1} (6.0–6.6 μm), just to the

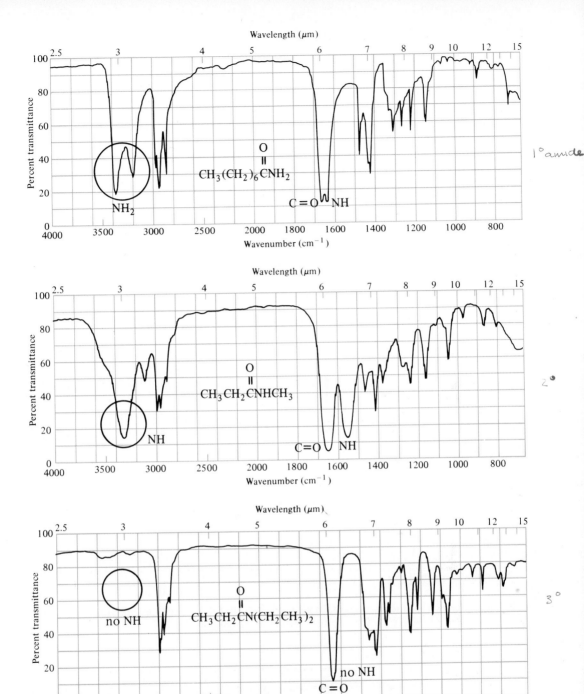

1° amide

2°

3°

FIGURE 13.5. Infrared spectra of a 1° amide (top), a 2° amide (center), and a 3° amide (bottom).

right of the C=O absorption. This absorption arises from NH bending. Therefore, a disubstituted, or tertiary, amide does not show an amide II band.

The NH stretching vibrations give rise to absorption to the left of aliphatic CH absorption at 3125–3570 cm^{-1} (2.8–3.2 μm). (This is about the same region where the NH of amines and OH absorb.) *Primary amides* (RCONH$_2$) show a double peak in this region. *Secondary amides* (RCONHR), with only one NH bond, show a single peak. *Tertiary amides* (RCONR$_2$), with no NH, show no absorption in this region. Figure 13.5 shows the infrared spectra of the three types of amides; compare the NH stretching and bending absorptions of these three compounds.

E.   Nitriles

The C≡N absorption is found in the triple-bond region of the infrared spectrum (2200–2300 cm^{-1}; 4.3–4.5 μm) and is medium to weak in intensity (see Figure 13.6).

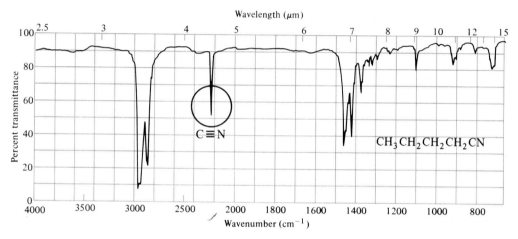

FIGURE 13.6. Infrared spectrum of pentanenitrile.

STUDY PROBLEM

13.2.   An unknown has the molecular formula C$_5$H$_7$O$_2$N. Its infrared and nmr spectra are given in Figure 13.7. What is the structure of the unknown?

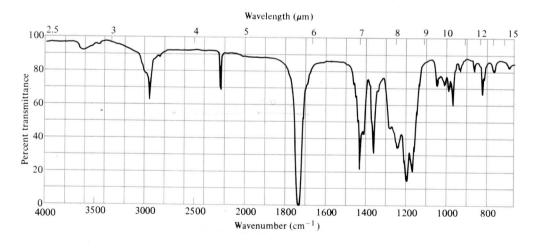

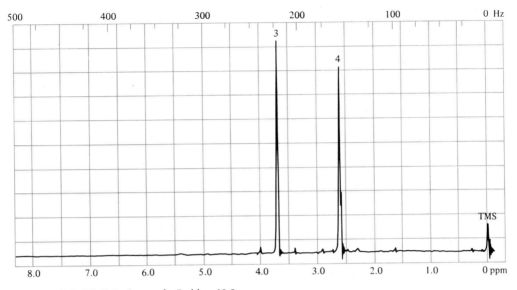

FIGURE 13.7. Spectra for Problem 13.2.

SECTION 13.3.

Acid Halides

Acid fluorides, chlorides, bromides, and iodides <u>all undergo similar reactions.</u> Because acid chlorides are the most popular of the acid halides, we will limit our discussion to just these compounds.

A.   Nomenclature of acid chlorides

Acid chlorides are named after the <u>parent carboxylic acid with the **-ic acid** ending changed to **-yl chloride**.</u>

$$\underset{\text{CH}_3\text{CCl}}{\overset{\overset{\displaystyle O}{\displaystyle \|}}{}} \qquad \underset{\text{CH}_3\text{CH}_2\text{CCl}}{\overset{\overset{\displaystyle O}{\displaystyle \|}}{}} \qquad \underset{\text{CH}_3\text{CH}_2\text{CH}_2\text{CCl}}{\overset{\overset{\displaystyle O}{\displaystyle \|}}{}}$$

IUPAC: ethanoyl chloride   propanoyl chloride   butanoyl chloride
trivial: acetyl chloride   propionyl chloride   butyryl chloride

B.   Preparation of acid chlorides

Acid chlorides may be <u>obtained directly from their parent carboxylic acids</u> by reaction with <u>thionyl chloride (SOCl$_2$) or some other active halogenating agent,</u> such as <u>phosphorus trichloride (PCl$_3$):</u>

$$\underset{\substack{\text{a carboxylic} \\ \text{acid}}}{\overset{\overset{\displaystyle O}{\displaystyle \|}}{\text{RCOH}}} + \text{SOCl}_2 \longrightarrow \underset{\substack{\text{an acid} \\ \text{chloride}}}{\overset{\overset{\displaystyle O}{\displaystyle \|}}{\text{RCCl}}} + \text{SO}_2 + \text{HCl}$$

$$3\ \overset{\overset{\displaystyle O}{\displaystyle \|}}{\text{RCOH}} + \text{PCl}_3 \longrightarrow \overset{\overset{\displaystyle O}{\displaystyle \|}}{\text{RCCl}} + \text{H}_3\text{PO}_3$$

Note the similarity between these reactions and the corresponding reactions of alcohols (Section 7.7).

$$CH_3CH_2OH + SOCl_2 \longrightarrow CH_3CH_2Cl + SO_2 + HCl$$

$$\underset{\displaystyle \parallel}{\overset{\displaystyle O}{CH_3COH}} + SOCl_2 \longrightarrow \underset{\displaystyle \parallel}{\overset{\displaystyle O}{CH_3CCl}} + SO_2 + HCl$$

STUDY PROBLEM

13.3.   Write an equation for the preparation of each of the following acid chlorides:

(a) $\underset{\displaystyle \parallel}{\overset{\displaystyle O}{CH_3CH_2CCl}}$ (b) $\underset{\displaystyle \parallel}{\overset{\displaystyle O}{ClC}}-\underset{\displaystyle \parallel}{\overset{\displaystyle O}{CCl}}$ (c) $\underset{\displaystyle \parallel}{\overset{\displaystyle O}{C_6H_5CH_2CCl}}$

C.   Reactions of acid chlorides

The acid halides are the most reactive of all the derivatives of carboxylic acids. The halide ion is a good leaving group. Attached to the positive carbon of a carbonyl group, it is displaced even more easily than when it is attached to an alkyl carbon. In the following general mechanism for the reaction of an acid chloride with a nucleophile, note that displacement of Cl⁻ is not a simple displacement like an S_N2 reaction. Rather, the reaction consists of two steps: (1) *addition of the nucleophile to the carbonyl group,* followed by (2) *elimination of the chloride ion.* The result of this reaction is a **nucleophilic acyl substitution**, which means "nucleophilic substitution on an acyl (RCO—) carbon."

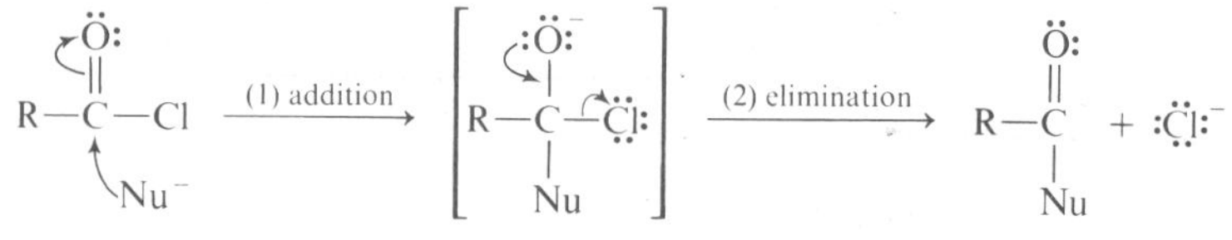

intermediate

Hydrolysis.   Cleavage by water, called **hydrolysis**, is a typical reaction of an acid chloride with a nucleophile.

Nucleophilic attack and elimination of Cl⁻:

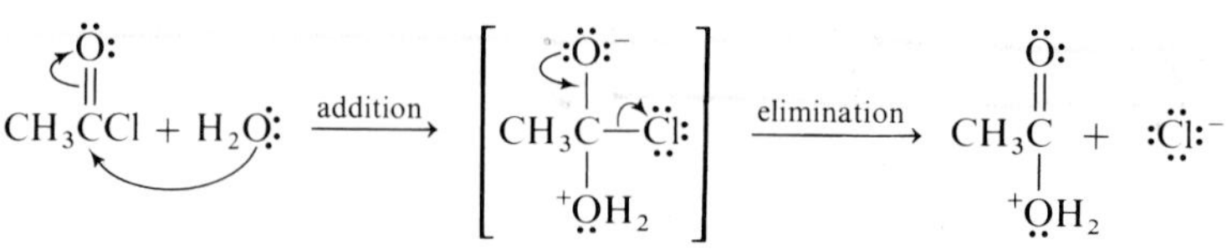

Loss of a proton:

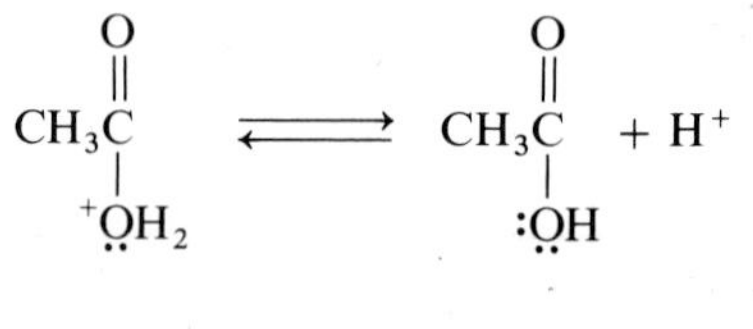

Overall reaction:

$$\underset{\text{acetyl chloride}}{CH_3\overset{\overset{\displaystyle O}{\|}}{C}Cl} + H_2O \longrightarrow \underset{\text{acetic acid}}{CH_3\overset{\overset{\displaystyle O}{\|}}{C}OH} + HCl$$

Although all acid chlorides undergo acidic and alkaline hydrolysis to yield carboxylic acids, their rates of reaction vary. An acid chloride that has a bulky alkyl group attached to the carbonyl group undergoes reaction more slowly than an acid chloride with a small alkyl group. For example, acetyl chloride reacts almost explosively with water, but butanoyl chloride requires gentle reflux.

The effect of the size of the alkyl group on the reaction rate is one of *solubility in water* rather than steric hindrance. An acid chloride with a small alkyl group is more soluble and undergoes reaction faster. An increase in the size of the alkyl portion of the molecule renders the acid chloride less water-soluble; the reaction is slower. If hydrolysis of a variety of acid chlorides is carried out in an inert solvent that dissolves both the acid chloride and water, the rates of hydrolysis are similar.

$$CH_3CH_2CH_2\overset{\overset{\displaystyle O}{\|}}{C}Cl \qquad CH_3CH_2\overset{\overset{\displaystyle O}{\|}}{C}Cl \qquad CH_3\overset{\overset{\displaystyle O}{\|}}{C}Cl$$

increasing rate of hydrolysis in pure H₂O ➡

SAMPLE PROBLEM

Suggest a reason why benzoyl chloride is less reactive toward nucleophilic attack than most aliphatic acid chlorides.

Solution: Conjugation of the carbonyl group can slow the rate of hydrolysis by dispersal of the positive charge on the carbonyl carbon. (A less positive carbon is less attractive to a nucleophile.)

+ charge delocalized

STUDY PROBLEMS

13.4. Suggest a mechanism for the hydrolysis of butanoyl chloride in dilute, aqueous NaOH.

13.5. Complete the following equation and predict the position(s) of ^{18}O in the product. Explain your answer.

$$\overset{\overset{\displaystyle O}{\|}}{C}Cl + H_2{}^{18}O \longrightarrow$$

Reaction with alcohols. Acid chlorides undergo reaction with alcohols to yield esters and HCl in a reaction that is directly analogous to hydrolysis. The reaction of an organic compound with an alcohol is referred to as **alcoholysis**. Alcoholysis of acid chlorides is valuable for the synthesis of hindered esters or phenyl esters.

$$\underset{\text{acetyl chloride}}{CH_3\overset{\displaystyle O}{\overset{\displaystyle \|}{C}}Cl} + \underset{\text{methanol}}{CH_3OH} \longrightarrow \underset{\text{methyl acetate}}{CH_3\overset{\displaystyle O}{\overset{\displaystyle \|}{C}}OCH_3} + HCl$$

It is usually wise to remove HCl from the reaction mixture as it is formed. The reason is that HCl can undergo reaction with the alcohol and produce alkyl chlorides or alkenes and water. A tertiary amine or pyridine is usually added as a scavenger for HCl.

2,4,6-trimethyl-
benzoyl chloride   *t*-butyl alcohol   pyridine

t-butyl 2,4,6-
trimethylbenzoate
79%

pyridine
hydrochloride

STUDY PROBLEM

13.6.   Write the mechanism for the reaction of butanoyl chloride with phenol in the presence of pyridine.

Reaction with ammonia and amines. Ammonia and amines are good nucleophiles. Like other nucleophiles, they undergo reaction with acid chlorides. The organic product of the reaction is an *amide*. As protons are lost in the deprotonation step, they react with the basic NH_3 or amine. For this reason, at least two equivalents of NH_3 or amine must be used.

$$\underset{\text{ammonia}}{CH_3\overset{\displaystyle O}{\overset{\displaystyle \|}{C}}Cl + NH_3} \longrightarrow \underset{\text{a 1° amide}}{CH_3\overset{\displaystyle O}{\overset{\displaystyle \|}{C}}NH_2} + HCl \xrightarrow{NH_3} NH_4^+ \; Cl^-$$

$$\underset{\substack{\text{methylamine} \\ \text{a 1° amine}}}{CH_3\overset{\displaystyle O}{\overset{\displaystyle \|}{C}}Cl + 2\,CH_3NH_2} \longrightarrow \underset{\text{a 2° amide}}{CH_3\overset{\displaystyle O}{\overset{\displaystyle \|}{C}}NHCH_3} + CH_3NH_3^+ \; Cl^-$$

$$CH_3CCl + 2 (CH_3)_2NH \longrightarrow CH_3CN(CH_3)_2 + (CH_3)_2NH_2{}^+ \ Cl^-$$

dimethylamine   a 3° amide

a 2° amine

If an amine is expensive, a chemist may not want to use an excess in the reaction with acid chloride. Only one mole of amine is needed to undergo reaction with acid chloride; a second mole is wasted when it is acting merely as a scavenger for HCl. In this case, the chemist will use another base to remove the HCl. For example, an inexpensive tertiary amine might be used. (A tertiary amine is reactive toward HCl, but cannot form an amide with an acid chloride. Why not?)

If an acid halide (such as benzoyl chloride) is not very reactive toward water, aqueous NaOH may be added to remove HX. The reactants and the aqueous NaOH form two layers. As HX is formed, it undergoes reaction with NaOH in the water layer. This reaction of an acid chloride and an amine in the presence of NaOH solution is called the **Schotten–Baumann reaction.**

benzoyl   piperidine   an amide
chloride   80%

STUDY PROBLEM

13.7. The Schotten–Baumann reaction is not applicable if the acid chloride and amine are water-soluble. What products would you expect from a mixture of acetyl chloride, methylamine, NaOH, and H_2O?

Conversion to anhydrides. Carboxylate ions are nucleophiles, and carboxylate salts (RCO_2Na) can be used for displacement of the chloride of acid chlorides. The product of the reaction is an acid anhydride.

$$CH_3(CH_2)_5CCl + CH_3(CH_2)_5CO^- \longrightarrow CH_3(CH_2)_5COC(CH_2)_5CH_3 + Cl^-$$

heptanoyl chloride   heptanoate ion   heptanoic anhydride
                                                60%

STUDY PROBLEM

13.8. Write the mechanism for the reaction of sodium acetate and benzoyl chloride.

Conversion to aryl ketones. Acid halides are usually the reagents of choice for Friedel–Crafts acylation reactions (Section 10.9E). This reaction is a route to aryl alkyl ketones without rearrangement of the alkyl side chain. (Recall that similar *alkylations* go through alkyl carbocations and that rearrangements are common.)

$$
\underset{\text{benzene}}{\bigcirc} + \underset{\substack{\text{propanoyl} \\ \text{chloride}}}{CH_3CH_2\overset{\displaystyle O}{\overset{\|}{C}}Cl} \xrightarrow{AlCl_3} \underset{\substack{\text{1-phenyl-1-propanone} \\ 90\%}}{\bigcirc -\overset{\displaystyle O}{\overset{\|}{C}}CH_2CH_3} + HCl
$$

STUDY PROBLEM

13.9. Without referring to Chapter 10, tell how you would prepare isobutylbenzene from benzene.

__Reaction with organometallic compounds.__ An acid chloride undergoes reaction with a variety of nucleophiles, including organometallic compounds. Reaction of an acid chloride with a Grignard reagent first yields a ketone, which then undergoes further reaction with the Grignard reagent to yield a *tertiary alcohol* after hydrolysis. (If an excess of the acid halide is used and temperatures of about $-25°$ are maintained, the intermediate ketone may be isolated.)

$$
\underset{\text{an acid chloride}}{R-\overset{\displaystyle O}{\overset{\|}{C}}-Cl} \xrightarrow{R'MgX} \underset{\text{a ketone}}{R-\overset{\displaystyle O}{\overset{\|}{C}}-R'} \xrightarrow{R'MgX} \underset{}{R-\overset{\displaystyle OMgX}{\underset{\displaystyle R'}{\overset{\|}{C}}}-R'} \xrightarrow{H_2O,\ H^+} \underset{\substack{\\ \\ \text{a } 3° \text{ alcohol}}}{R-\overset{\displaystyle OH}{\underset{\displaystyle R'}{\overset{\|}{C}}}-R'}
$$

$$
\underset{\text{benzoyl chloride}}{\bigcirc -\overset{\displaystyle O}{\overset{\|}{C}}Cl} \xrightarrow[\text{(2) H}_2\text{O, H}^+]{\text{(1) 2 CH}_3\text{MgI}} \underset{\substack{\text{2-phenyl-2-propanol} \\ \text{a } 3° \text{ alcohol}}}{\bigcirc -\overset{\displaystyle OH}{\overset{\|}{C}}(CH_3)_2}
$$

A suitable organometallic reagent for preparing a ketone from an acid chloride is the **cadmium reagent,** an organocadmium compound prepared from a Grignard reagent and cadmium chloride.

Formation of a cadmium reagent:

$$
2\,RMgCl + CdCl_2 \longrightarrow \underset{\substack{\text{a cadmium} \\ \text{reagent}}}{R_2Cd} + 2\,MgCl_2
$$

Cadmium is less electropositive than magnesium; therefore, the C—Cd bond is less polar than the C—Mg bond. For this reason, cadmium reagents are less reactive than Grignard reagents.

$$
\overset{\delta-}{R}\!-\!\overset{\delta+}{MgX} \qquad \overset{\delta-}{R}\!-\!\overset{\delta+}{Cd}\!-\!\overset{\delta-}{R}
$$

more negative   *less negative*

Cadmium reagents do not undergo reaction with ketones, but they do undergo reaction with acid halides and offer an excellent method of ketone synthesis.

Reaction with an acid halide:

$$R_2Cd + R'\overset{\overset{\displaystyle O}{\|}}{C}Cl \longrightarrow R'-\overset{\overset{\displaystyle O}{\|}}{C}-R + RCdCl$$

a ketone

An example of a ketone synthesis with a cadmium reagent follows. Note that the acid halide in this example contains an ester group. This synthesis would not have been successful if a Grignard reagent had been used because the ester group would also react with the Grignard reagent.

$$CH_3CH_2CH_2MgBr \xrightarrow{CdCl_2} (CH_3CH_2CH_2)_2Cd \xrightarrow{CH_3O\overset{\overset{\displaystyle O}{\|}}{C}CH_2CH_2\overset{\overset{\displaystyle O}{\|}}{C}Cl}$$

$$CH_3O\overset{\overset{\displaystyle O}{\|}}{C}CH_2CH_2\overset{\overset{\displaystyle O}{\|}}{C}CH_2CH_2CH_3$$

a keto ester
75%

Secondary and tertiary alkylcadmium reagents are unstable; therefore, an organocadmium reagent is useful only when R is methyl, primary, or phenyl. A more general route to ketones is the use of **lithium dialkylcuprates**. (In Section 6.10, we described the use of these agents and alkyl halides in alkane synthesis.) Like cadmium reagents, cuprates undergo reaction with acid halides to yield ketones. These reactions are usually carried out in an ether solvent at low temperatures (0 to −78°).

$$CH_3I \xrightarrow{Li} CH_3Li \xrightarrow{CuI} LiCu(CH_3)_2$$

a lithium dialkylcuprate

$$LiCu(\overset{\delta-}{CH_3})_2 + (CH_3)_3\overset{\overset{\displaystyle O}{\|}}{\underset{\delta+}{C}}-Cl \longrightarrow (CH_3)_3C\overset{\overset{\displaystyle O}{\|}}{C}-CH_3$$

3,3-dimethyl-2-butanone
(t-butyl methyl ketone)

STUDY PROBLEM

13.10. Show by equations how you would prepare the following ketones from alkyl halides and acid chlorides (1) by way of a cadmium reagent, and (2) by way of a cuprate:

(a) $\bigcirc\!\!\!\!\!\bigcirc-\overset{\overset{\displaystyle O}{\|}}{C}CH(CH_3)_2$ (b) $\bigcirc-\overset{\overset{\displaystyle O}{\|}}{C}CH_3$

Reduction. Reduction of acid chlorides with lithium aluminum hydride yields primary alcohols. Since primary alcohols also can be obtained by LAH reduction of the parent acids, this reaction finds little synthetic use.

benzoic acid — SOCl₂ → (benzoyl chloride) — (1) LiAlH₄ (2) H₂O → benzyl alcohol —CH₂OH

(1) LiAlH₄   (2) H₂O

The <u>partial reduction of an acid chloride to an aldehyde</u> can be effected, and this reaction is very useful. (A carboxylic acid itself is not readily reduced to an aldehyde.) A milder reducing agent than LAH is required to <u>reduce RCOCl to RCHO instead of to RCH₂OH</u>. A <u>suitable reagent is **lithium tri-*t*-butoxyaluminum hydride**</u>, which is <u>obtained from *t*-butyl alcohol and LAH</u>. This <u>reducing reagent is less reactive than LAH</u> because of <u>both steric hindrance and electron-withdrawal by the oxygen atoms</u>.

Preparation of reducing agent:

$$3 \ (CH_3)_3COH + LiAlH_4 \longrightarrow Li^+ \ H{-}\overset{\overset{\displaystyle OC(CH_3)_3}{|}}{\underset{\underset{\displaystyle OC(CH_3)_3}{|}}{Al}}{=}OC(CH_3)_3 + 3 \ H_2$$

t-butyl alcohol

lithium tri-*t*-butoxy-
aluminum hydride

Reaction with RCOCl:

benzoyl chloride —CCl — (1) LiAlH[OC(CH₃)₃]₃ (2) H₂O → benzaldehyde —CH

Alpha halogenation.

Ketones can be halogenated in the α position by treatment with X₂ and H⁺ or OH⁻. This reaction, which was discussed in Section 11.18, proceeds by way of the enol. <u>Acid halides also undergo tautomerism and therefore undergo α halogenation.</u>

$$CH_3\overset{\overset{\displaystyle O}{\|}}{C}Cl \ \rightleftharpoons \ CH_2{=}\overset{\overset{\displaystyle OH}{|}}{C}Cl \ \xrightarrow{Cl_2} \ \left[CH_2{-}\overset{\overset{\displaystyle OH}{|}}{\underset{\underset{\displaystyle Cl}{|}}{C}}Cl \ \underset{Cl}{} \right] \ \xrightarrow{-HCl} \ CH_2{-}\overset{\overset{\displaystyle O}{\|}}{\underset{\underset{\displaystyle Cl}{|}}{C}}Cl$$

"keto" form   *"enol" form*

chloroacetyl chloride

an α-halo acid halide

<u>Carboxylic acids do not tautomerize readily</u> and thus <u>do not undergo α halogenation</u>. However, the <u>halogenation of acid halides</u> provides a technique by which we can obtain α-halocarboxylic acids. If a <u>catalytic amount of PCl₃</u> is added to a <u>carboxylic acid along with the halogenating agent (usually Cl₂)</u>, the <u>PCl₃ converts a small proportion of the acid to the acid chloride</u>, which <u>undergoes α halogenation</u>.

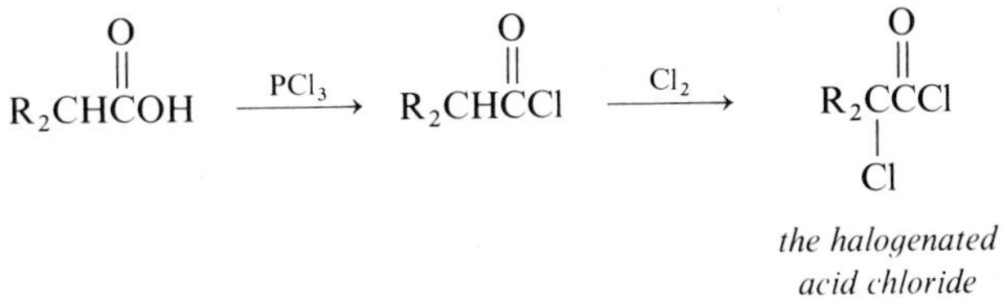

$$\underset{\text{R}_2\text{CHCOH}}{\overset{\overset{\text{O}}{\|}}{}} \xrightarrow{\text{PCl}_3} \underset{\text{R}_2\text{CHCCl}}{\overset{\overset{\text{O}}{\|}}{}} \xrightarrow{\text{Cl}_2} \underset{\underset{\text{Cl}}{|}}{\underset{\text{R}_2\text{CCCl}}{\overset{\overset{\text{O}}{\|}}{}}}$$

the halogenated
acid chloride

In the reaction mixture, the acid halide is in equilibrium with the carboxylic acid itself. Because the acid is present in excess, the α-halo acid chloride is converted to the α-halo acid.

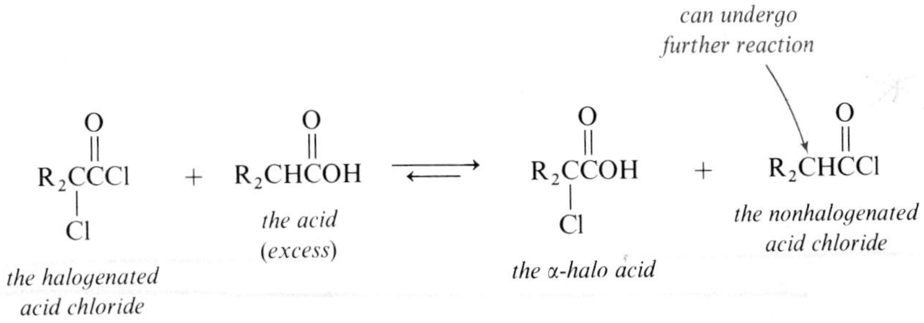

can undergo
further reaction

$$\underset{\underset{\text{Cl}}{|}}{\underset{\text{R}_2\text{CCCl}}{\overset{\overset{\text{O}}{\|}}{}}} + \underset{\text{R}_2\text{CHCOH}}{\overset{\overset{\text{O}}{\|}}{}} \; \rightleftharpoons \; \underset{\underset{\text{Cl}}{|}}{\underset{\text{R}_2\text{CCOH}}{\overset{\overset{\text{O}}{\|}}{}}} + \underset{\text{R}_2\text{CHCCl}}{\overset{\overset{\text{O}}{\|}}{}}$$

the halogenated   *the acid*   *the α-halo acid*   *the nonhalogenated*
acid chloride   *(excess)*   *acid chloride*

In the process, more nonhalogenated acid chloride is formed. This acid chloride undergoes α halogenation, and the reaction sequence is repeated. The overall result of this sequence is, in effect, the α halogenation of a carboxylic acid. This reaction sequence is called the **Hell–Volhard–Zelinsky reaction** after the chemists who developed the technique.

Overall reaction:

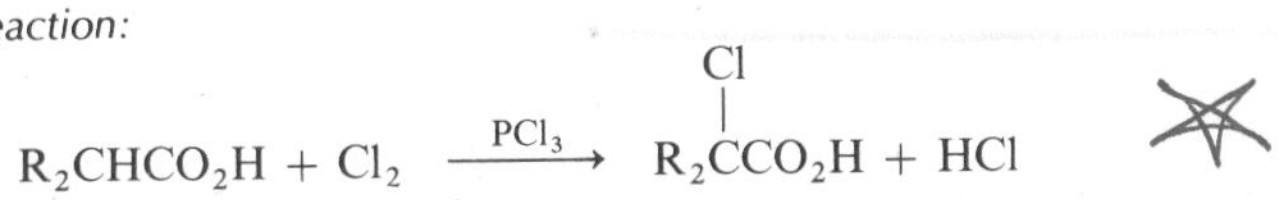

$$\text{R}_2\text{CHCO}_2\text{H} + \text{Cl}_2 \xrightarrow{\text{PCl}_3} \underset{\underset{\text{Cl}}{|}}{\text{R}_2\text{CCO}_2\text{H}} + \text{HCl}$$

Some specific examples of α halogenation of carboxylic acids (by way of acid halides) follow:

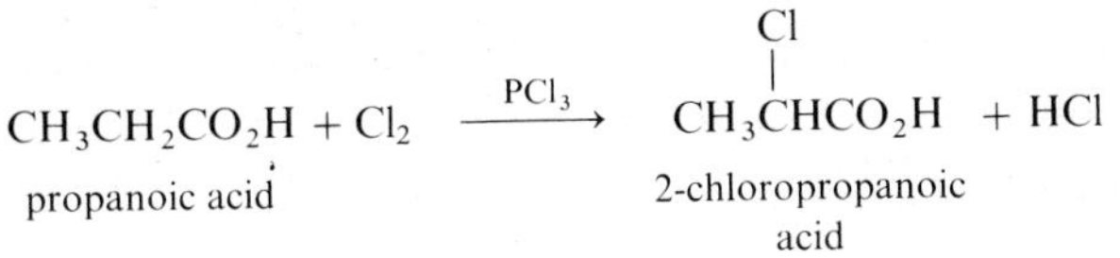

$$\text{CH}_3\text{CH}_2\text{CO}_2\text{H} + \text{Cl}_2 \xrightarrow{\text{PCl}_3} \underset{\underset{\text{Cl}}{|}}{\text{CH}_3\text{CHCO}_2\text{H}} + \text{HCl}$$

propanoic acid   2-chloropropanoic
                                     acid

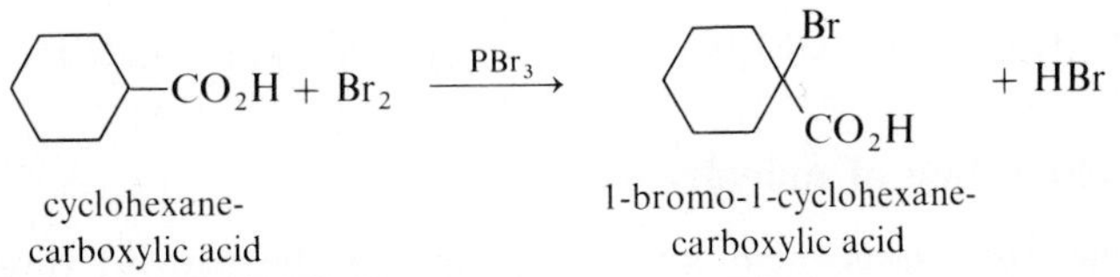

cyclohexane-
carboxylic acid   1-bromo-1-cyclohexane-
                       carboxylic acid

STUDY PROBLEMS

13.11. Give the organic product of each of the following reactions:

(a) $(\text{CH}_3)_2\text{CHCO}_2\text{H} + \text{Cl}_2 \xrightarrow{\text{PCl}_3}$

(b) 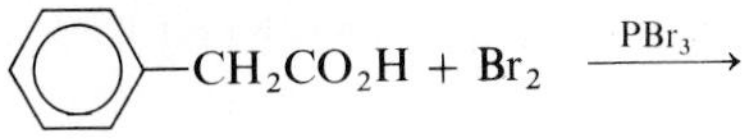 ⬡—$\text{CH}_2\text{CO}_2\text{H} + \text{Br}_2 \xrightarrow{\text{PBr}_3}$

13.12. Write the mechanism for the reaction of chloroacetyl chloride with acetic acid to yield chloroacetic acid.

SECTION 13.4.

Anhydrides of Carboxylic Acids

An **anhydride of a carboxylic acid** has the structure of two carboxylic acid molecules with a molecule of water removed. (*Anhydride* means "without water.")

$$
\overset{\text{✗}}{\underset{}{\text{CH}_3\overset{\text{O}}{\overset{\|}{\text{C}}}\text{O H}}} \quad \underset{}{\text{HO}\overset{\text{O}}{\overset{\|}{\text{C}}}\text{CH}_3} \quad \xrightarrow{-\text{H}_2\text{O}} \quad \underset{\textit{an anhydride}}{\text{CH}_3\overset{\text{O}}{\overset{\|}{\text{C}}}\text{O}\overset{\text{O}}{\overset{\|}{\text{C}}}\text{CH}_3}
$$

A.   Nomenclature of anhydrides

Symmetrical anhydrides are those in which the two acyl groups are the same. They are named after the parent carboxylic acid followed by the word **anhydride**.

$$
\underset{}{\text{CH}_3\overset{\text{O O}}{\overset{\|\ \|}{\text{C}}\text{O}}\text{CCH}_3} \qquad \text{CH}_3\text{CH}_2\overset{\text{O O}}{\overset{\|\ \|}{\text{C}}\text{O}}\text{CCH}_2\text{CH}_3
$$

IUPAC:	ethanoic anhydride	propanoic anhydride
trivial:	acetic anhydride	propionic anhydride

Unsymmetrical anhydrides are named by the word **anhydride** preceded by the names of the *two* parent acids.

$$
\text{CH}_3\overset{\text{O O}}{\overset{\|\ \|}{\text{C}}\text{O}}\text{CCH}_2\text{CH}_3
$$

IUPAC:	ethanoic propanoic anhydride
trivial:	acetic propionic anhydride

STUDY PROBLEM

13.13. Write a suitable name for each of the following anhydrides:

(a) $\text{CH}_3\text{CH}_2\text{CH}_2\text{CH}_2\overset{\text{O O}}{\overset{\|\ \|}{\text{C}}\text{O}}\text{CCH}_2\text{CH}_2\text{CH}_2\text{CH}_3$ (b) $\text{C}_6\text{H}_5\overset{\text{O O}}{\overset{\|\ \|}{\text{C}}\text{O}}\text{CCH}_3$

B.   Preparation of anhydrides

With few exceptions, acid anhydrides cannot be formed directly from their parent carboxylic acids, but must be prepared from the more reactive derivatives of carboxylic acids. One route to an anhydride is from an acid chloride with a carboxylate, a reaction mentioned on page 617.

$$
\underset{\textit{an acid chloride}}{\text{R}\overset{\text{O}}{\overset{\|}{\text{C}}}-\text{Cl}} \quad + \quad \underset{\textit{a carboxylate ion}}{^-\text{O}\overset{\text{O}}{\overset{\|}{\text{C}}}\text{R}'} \quad \longrightarrow \quad \underset{\textit{an anhydride}}{\text{R}\overset{\text{O}}{\overset{\|}{\text{C}}}-\text{O}\overset{\text{O}}{\overset{\|}{\text{C}}}\text{R}'} + \text{Cl}^-
$$

Another route to an anhydride is by treatment of the carboxylic acid with acetic anhydride. A reversible reaction occurs between a carboxylic acid and an anhydride. The equilibrium can be shifted to the right by distilling the acetic acid as it is formed. This reaction has also been mentioned previously (Section 12.11B).

$$2 \; C_6H_5{-}CO_2H + CH_3\overset{O}{\overset{\|}{C}}O\overset{O}{\overset{\|}{C}}CH_3 \; \rightleftharpoons \; C_6H_5{-}\overset{O}{\overset{\|}{C}}O\overset{O}{\overset{\|}{C}}{-}C_6H_5 + 2\;CH_3CO_2H\uparrow$$

benzoic acid   acetic anhydride   benzoic anhydride   acetic acid

A five- or six-membered cyclic anhydride can be formed by heating the appropriate diacid (see Section 12.11B).

$$\begin{array}{l} CH_2CO_2H \\ | \\ CH_2CO_2H \end{array} \xrightarrow{\;heat\;} \text{(succinic anhydride)} + H_2O$$

succinic acid   succinic anhydride

a diacid   *a cyclic anhydride*

C. Reactions of anhydrides

Like acid halides, acid anhydrides are more reactive than carboxylic acids and may be used to synthesize ketones, esters, or amides. Acid anhydrides undergo reactions with the same nucleophiles that the acid chlorides react with; however, the rates of reaction are slower. (A carboxylate ion is not quite as good a leaving group as a halide ion.) Note that the other product in these reactions is a carboxylic acid or, when the reaction mixture is alkaline, its anion.

$$RC{-}OCR + Nu^- \xrightarrow{addition} \left[RC{-}\overset{\cdot\cdot}{\underset{Nu}{O}}CR \right] \xrightarrow{elimination} RC{-}Nu \; + \; {}^-:OCR$$

intermediate

Hydrolysis. Anhydrides undergo reaction with water to yield carboxylic acids. The rate of reaction, like the rate of hydrolysis of an acid chloride, depends on the solubility of the anhydride in water.

$$CH_3\overset{O}{\overset{\|}{C}}O\overset{O}{\overset{\|}{C}}CH_3 + H_2O \longrightarrow 2\;CH_3\overset{O}{\overset{\|}{C}}OH$$

acetic anhydride   acetic acid

General:
$$RC\overset{O}{\overset{\|}{}}OCR' + H_2O \longrightarrow RCO_2H + R'CO_2H$$

an anhydride   *carboxylic acids*

Reaction with alcohols and phenols.   The reaction of an anhydride with either an alcohol or a phenol yields an ester. The reaction is particularly useful with the commercially available acetic anhydride, which results in acetates. A tertiary amine or pyridine may be added to the reaction mixture to remove the acid (CH_3CO_2H in our examples) as it is formed.

$$
\underset{\substack{\text{acetic} \\ \text{anhydride}}}{CH_3\overset{O}{\overset{\|}{C}}O\overset{O}{\overset{\|}{C}}CH_3} + \underset{\text{ethanol}}{CH_3CH_2OH} + \underset{\text{pyridine}}{\bigcirc\!\!N\!\!:} \longrightarrow \underset{\text{ethyl acetate}}{CH_3\overset{O}{\overset{\|}{C}}OCH_2CH_3} + CH_3CO_2^- \; H\overset{+}{N}\bigcirc
$$

$$
CH_3\overset{O}{\overset{\|}{C}}O\overset{O}{\overset{\|}{C}}CH_3 + \underset{\text{phenol}}{\bigcirc\!-\!OH} + \bigcirc\!\!N\!\!: \longrightarrow \underset{\text{phenyl acetate}}{CH_3\overset{O}{\overset{\|}{C}}O\!-\!\bigcirc} + CH_3CO_2^- \; H\overset{+}{N}\bigcirc
$$

STUDY PROBLEM

13.14. Complete the following equation:

$$
\text{(phthalic anhydride)} + CH_3CH_2OH + \bigcirc\!\!N \longrightarrow
$$

Reaction with ammonia and amines.   Ammonia, primary amines, and secondary amines undergo reaction with anhydrides to yield *amides*. Again, acetic anhydride is the most popular anhydride used in this reaction. Ammonia and acetic anhydride yield acetamide, while amines and acetic anhydride give sub-stituted acetamides. One mole of amine is consumed in neutralization of the acetic acid formed in the reaction.

$$
CH_3\overset{O}{\overset{\|}{C}}O\overset{O}{\overset{\|}{C}}CH_3 + \underset{\text{ammonia}}{2\,NH_3} \longrightarrow \underset{\text{acetamide}}{CH_3\overset{O}{\overset{\|}{C}}NH_2} + CH_3CO_2^- \; NH_4^+
$$

$$
CH_3\overset{O}{\overset{\|}{C}}O\overset{O}{\overset{\|}{C}}CH_3 + \underset{\substack{\text{a 1° amine}}}{2\,RNH_2} \longrightarrow \underset{\substack{\text{an N-alkyl-} \\ \text{acetamide}}}{CH_3\overset{O}{\overset{\|}{C}}NHR} + CH_3CO_2^- \; RNH_3^+
$$

$$
CH_3\overset{O}{\overset{\|}{C}}O\overset{O}{\overset{\|}{C}}CH_3 + \underset{\substack{\text{a 2° amine}}}{2\,R_2NH} \longrightarrow \underset{\substack{\text{an N,N-dialkyl-} \\ \text{acetamide}}}{CH_3\overset{O}{\overset{\|}{C}}NR_2} + CH_3CO_2^- \; R_2NH^+
$$

SAMPLE PROBLEM

Predict the product of reaction of one equivalent of succinic anhydride with two equivalents of ammonia.

Solution:

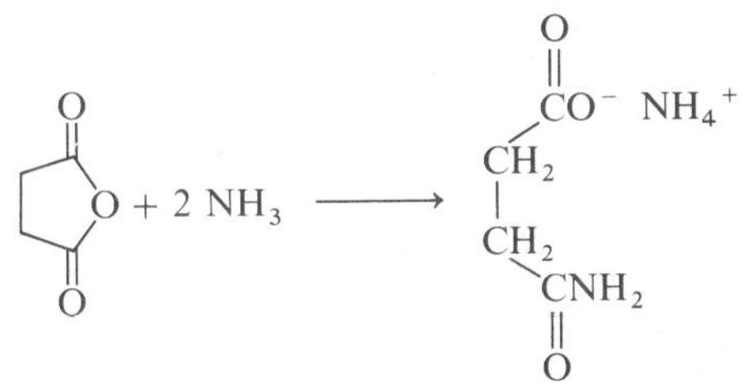

STUDY PROBLEM

13.15. Give the structures of all principal organic products:

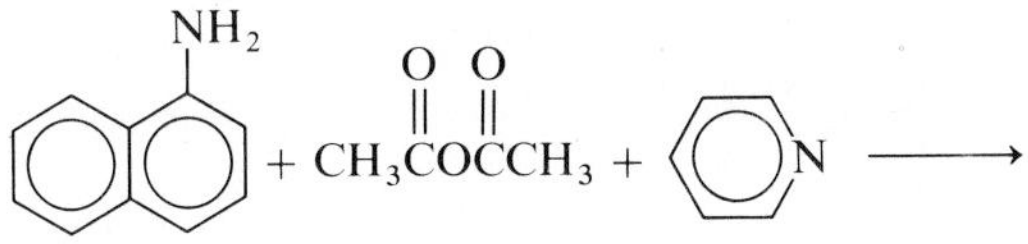

SECTION 13.5.

Esters of Carboxylic Acids

Esters, one of the most useful classes of organic compounds, may be converted to a variety of other compounds (see Figure 13.8). Esters are common in nature. Volatile esters lend pleasant aromas to many fruits and perfumes. Fats and waxes are esters (Section 20.1). Esters are also used for synthetic polymers: for example, Dacron is a polyester. Table 13.3 lists some representative esters. You might find it interesting to compare the odors of some of these esters to the odors of carboxylic acids, Section 12.2.

A.   Nomenclature of esters

The name of an ester consists of two words. The first word is the name of the alkyl group attached to the ester oxygen. The second word is derived from the carboxylic acid name with **-ic acid** changed to **-ate**. Note the similarity between the name of an ester and that of a carboxylate salt.

$$CH_3CH_2\overset{\overset{\displaystyle O}{\|}}{C}O—H \qquad CH_3CH_2\overset{\overset{\displaystyle O}{\|}}{C}O^-\ Na^+ \qquad CH_3CH_2\overset{\overset{\displaystyle O}{\|}}{C}O—CH_3$$

an acid	*a salt*	*an ester*

IUPAC:	propanoic acid	sodium propanoate	methyl propanoate
trivial:	propionic acid	sodium propionate	methyl propionate

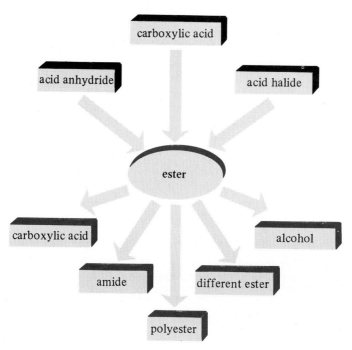

FIGURE 13.8. The synthetic relationship of esters to other compounds.

TABLE 13.3. Names, odors, and boiling points of selected esters

Trivial name	Structure	Odor	Bp, °C
methyl acetate	$CH_3CO_2CH_3$	pleasant	57.5
ethyl acetate	$CH_3CO_2CH_2CH_3$	pleasant	77
propyl acetate	$CH_3CO_2CH_2CH_2CH_3$	like pears	102
ethyl butyrate	$CH_3(CH_2)_2CO_2CH_2CH_3$	like pineapple	121
isoamyl acetate	$CH_3CO_2(CH_2)_2CH(CH_3)_2$	like pears	142
isobutyl propionate	$CH_3CH_2CO_2CH_2CH(CH_3)_2$	like rum	137
methyl salicylate		like wintergreen	220

B.   Preparation of esters

Most methods for ester synthesis have been covered elsewhere in this text. In this section, we will provide a summary of these methods. One additional reaction, that of a carboxylic acid with diazomethane, will be discussed here.

From carboxylic acids and alcohols (Section 12.9):

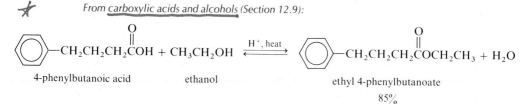

From <u>acid halides and alcohols</u> (for hindered systems and for phenols) (Section 13.3):

$$\underset{\text{malonyl chloride}}{ClCCH_2CCl} \; + \; \underset{\text{t-butyl alcohol}}{2 \; (CH_3)_3COH} \quad \xrightarrow{-2 \; HCl} \quad \underset{\substack{\text{di-t-butyl malonate} \\ 80\%}}{(CH_3)_3COCCH_2COC(CH_3)_3}$$

From an <u>anhydride and an alcohol or phenol</u> (Section 13.4):

phthalic anhydride $+ \; CH_3CH_2CHCH_3$ (2-butanol) $\longrightarrow$ 2-butyl hydrogen phthalate 97%

From a <u>carboxylate and a reactive alkyl halide</u> (Section 12.6):

$$\underset{\text{acetate ion}}{CH_3CO^-} + \underset{\text{benzyl chloride}}{\text{—}CH_2Cl} \longrightarrow \underset{\substack{\text{benzyl acetate} \\ 93\%}}{\text{—}CH_2OCCH_3} + Cl^-$$

From a <u>carboxylic acid and diazomethane</u>:

$$\underset{\substack{\text{cyclohexane-} \\ \text{carboxylic acid}}}{\text{—COH}} + \underset{\text{diazomethane}}{CH_2N_2} \xrightarrow{\text{ether}} \underset{\substack{\text{methyl cyclohexane-} \\ \text{carboxylate} \\ 100\%}}{\text{—COCH}_3} + N_2$$

With the exception of the diazomethane reaction, the preceding esterification reactions usually give good, but not quantitative, yields. The reaction of a carboxylic acid with diazomethane usually gives a quantitative yield (eliminating the need for purification of the product ester). Diazomethane (CH_2N_2), which has been mentioned earlier (Section 9.12), is a toxic, explosive gas; it is almost always prepared just prior to the time it is needed. Despite these disadvantages, diazomethane is often the reagent of choice. For example, if a chemist wishes to esterify a few milligrams of a carboxylic acid for spectral analysis or if, for any reason, a near-quantitative yield of an ester is desired, diazomethane may be the reagent of choice.

How does esterification with diazomethane occur? The carbon of diazomethane carries a partial negative charge, and the acidic proton of the carboxylic acid is removed by this carbanion-like structure. Then nitrogen (N_2), one of the best leaving groups known, is displaced by the carboxylate anion to yield the methyl ester.

Resonance structures of CH_2N_2:

$$[^-:CH_2\text{—}\overset{..}{N}\!\!=\!\!\overset{+}{N}: \quad \longleftrightarrow \quad CH_2\!\!=\!\!\overset{+}{N}\!\!=\!\!\overset{..}{N}:^- \quad \longleftrightarrow \quad {}^-:CH_2\text{—}\overset{+}{N}\!\!\equiv\!\!N:]$$

Reaction with RCO₂H:

$$RC\overset{O}{\overset{\|}{\underset{..}{O}}}H \quad \bar{:}CH_2-N_2{}^+ \longrightarrow \left[RC\overset{O}{\overset{\|}{\underset{..}{O}}}{}^{\bar{}} \quad CH_3 \overset{+}{\underset{\frown}{N_2}} \right] \longrightarrow RC\overset{O}{\overset{\|}{\underset{..}{O}}}CH_3 + N_2$$

a methyl ester

STUDY PROBLEM

13.16. How would you prepare aspirin (page 485) from methyl salicylate (Table 13.3)?

C.   Reactions of esters

In *acidic solution*, the carbonyl oxygen of an ester may be protonated. The partially positive carbon can then be attacked by a weak nucleophile such as water.

Protonation:

$$R-\overset{\overset{..}{\overset{..}{O}}}{\overset{\|}{C}}-OR \xrightarrow{H^+} \left[R-\overset{\overset{+}{\overset{..}{O}}H}{\overset{\|}{C}}-OR \longleftrightarrow R-\overset{\overset{..}{O}H}{\overset{|}{\underset{+}{C}}}-OR \right] Nu^-$$

resonance structures
for a protonated ester

The carbonyl oxygen is protonated rather than the alkoxyl oxygen because the carbonyl oxygen of an ester is the more basic of the two. From the resonance structures of the protonated ester, you can see that this resulting cation is resonance-stabilized. (The cation formed by protonation of the alkoxyl oxygen is not resonance-stabilized.)

In an *alkaline solution*, the carbonyl carbon of an ester may be attacked by a good nucleophile without prior protonation. This is the same addition–elimination path as for nucleophilic attack on acid chlorides or anhydrides.

$$R-\overset{\overset{..}{O}:^{\delta-}}{\overset{\|}{\underset{\delta+}{C}}}-OR + Nu^- \xrightarrow{\text{addition}} \left[R-\overset{\overset{..}{O}:^{\bar{}}}{\underset{\underset{Nu}{|}}{\overset{|}{C}}}\overset{..}{O}R \right] \xrightarrow{\text{elimination}} R-\overset{\overset{..}{O}\cdot}{\underset{\underset{Nu}{|}}{\overset{\|}{C}}} + {}^-\overset{..}{O}R$$

intermediate

Acid hydrolysis.   The esterification of a carboxylic acid with an alcohol (Section 12.9) is a reversible reaction. When a carboxylic acid is esterified, an excess of the alcohol is used. To cause the reverse reaction—that is, *acid-catalyzed hydrolysis of an ester to a carboxylic acid*—an excess of water is used. The excess of water shifts the equilibrium to the carboxylic acid side of the equation.

Esterification:

$$\bigcirc\!\!-\overset{O}{\overset{\|}{C}}OH + CH_3OH \underset{\xleftarrow{\hspace{1cm}}}{\overset{H^+,\text{heat}}{\xrightarrow{\hspace{1cm}}}} \bigcirc\!\!-\overset{O}{\overset{\|}{C}}OCH_3 + H_2O$$

benzoic acid   methanol   methyl benzoate
(*excess*)

Hydrolysis:

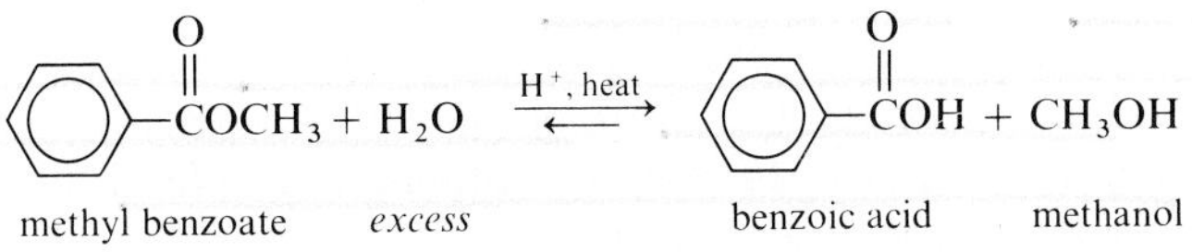

methyl benzoate   *excess*   benzoic acid   methanol

If water labeled with oxygen-18 is used in the hydrolysis, the labeled oxygen ends up in the carboxylic acid.

The reason for this is that the water attacks the *carbonyl group*. The RO bond is not broken in the hydrolysis.

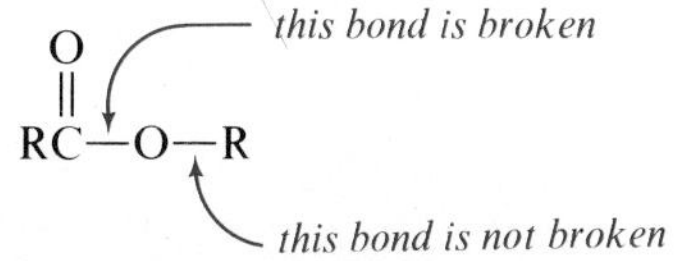

The following mechanism accounts for this observation. Note that the first step is *protonation* (only one resonance structure is shown), followed by *addition of H_2O*, then *elimination of R'OH*, followed by *deprotonation*,

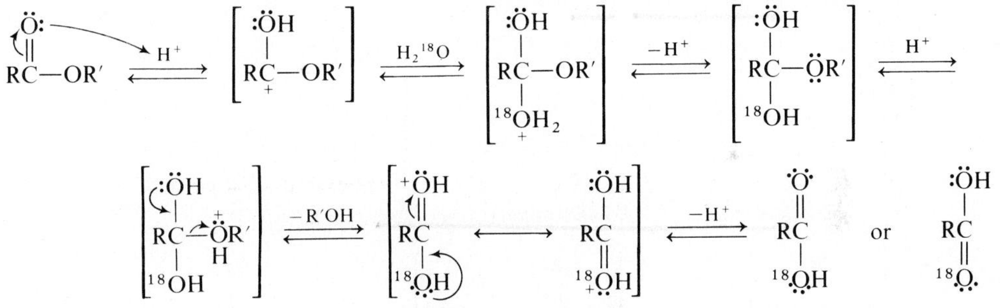

resonance structures for protonated acid

A simplified mechanism for ester hydrolysis may be written:

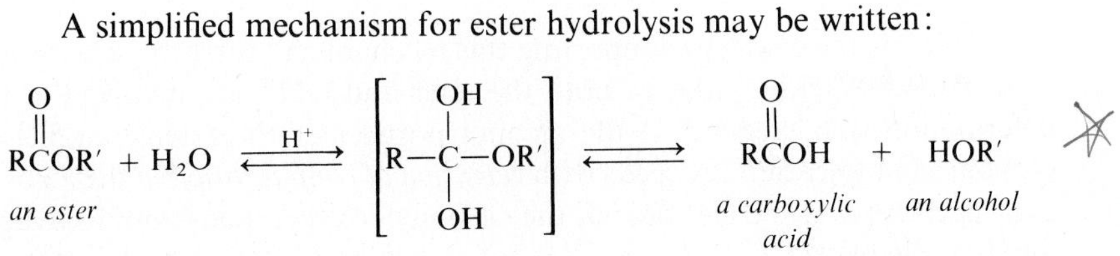

an ester   *a carboxylic* *an alcohol* *acid*

STUDY PROBLEM

13.17. What would you expect as acid-hydrolysis products of the following labeled ester?

$$CH_3C^{18}OCH_2CH_3$$
$$\overset{O}{\overset{\|}{}}$$

Alkaline hydrolysis (saponification). Hydrolysis of an ester in base, or **saponification,** is an *irreversible reaction*. Because it is irreversible, saponification often gives better yields of carboxylic acid and alcohol than does acid hydrolysis. Because the reaction occurs in base, the product of saponification is the carboxylate salt. The free acid is generated when the solution is acidified. Note that OH$^-$ is a reactant, not a catalyst, in this reaction.

Saponification:

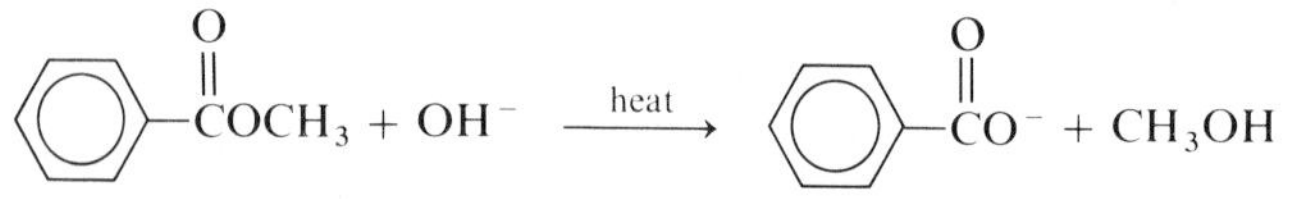

methyl benzoate   benzoate ion

Acidification:

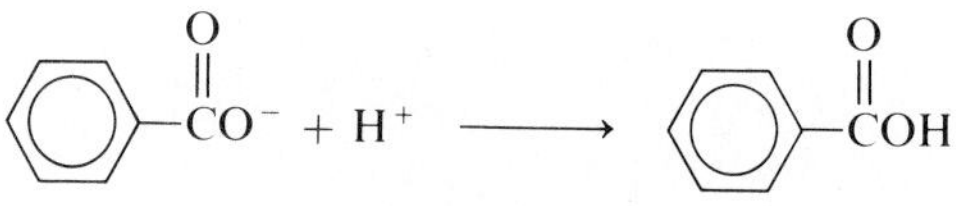

benzoic acid

A large body of evidence has been accumulated to support the following mechanistic scheme, which is typical of nucleophilic attack on a carboxylic acid derivative.

Step 1 (addition of OH$^-$) (slow):

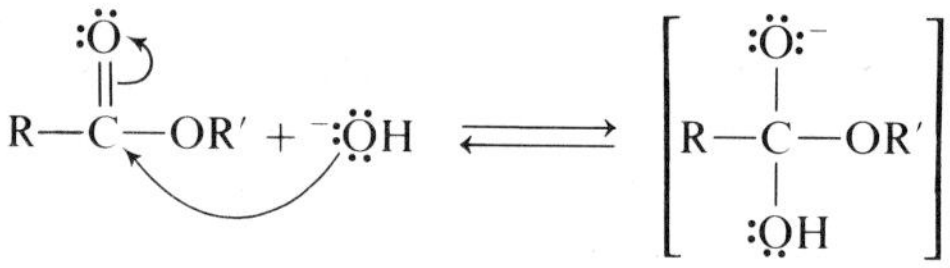

Step 2 (elimination of $^-$OR' and proton transfer) (fast):

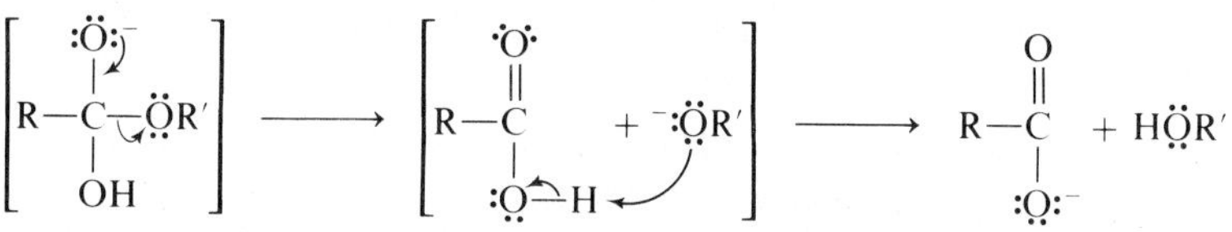

What is the evidence supporting this mechanism? First, the reaction follows *second-order kinetics*—that is, both the ester and OH$^-$ are involved in the rate-determining step. Second, if the alcohol portion of the ester contains a chiral carbon, saponification proceeds with *retention of configuration* in the alcohol. This evidence supports the cleavage of the carbonyl–oxygen bond, not cleavage of the alkyl–oxygen bond.

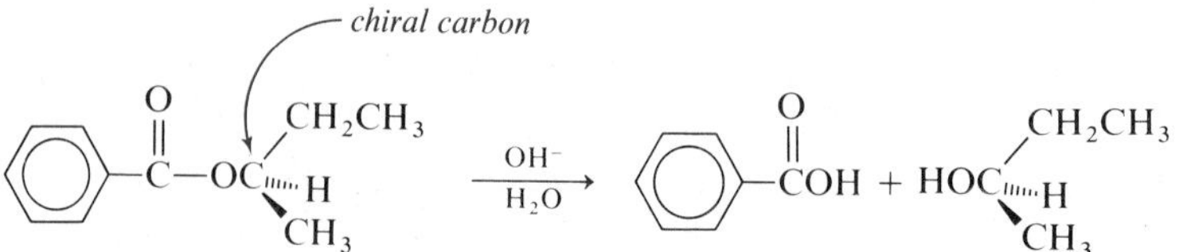

(*R*)-2-butyl benzoate   benzoic acid   (*R*)-2-butanol

*if this bond were cleaved, we
would expect racemization or inversion*

$$\underset{\underset{\text{if this bond is cleaved}}{\text{no racemization or inversion observed;}}}{\langle\text{Ph}\rangle-\overset{\overset{\text{O}}{\|}}{C}-O-\overset{\overset{\text{CH}_2\text{CH}_3}{|}}{\underset{\underset{\text{CH}_3}{|}}{C_{\prime\prime\prime\prime}\text{H}}}}$$

STUDY PROBLEM

13.18. Predict the products of saponification of $CH_3\overset{\overset{O}{\|}}{C}{}^{18}OCH_2CH_3$.

The word *saponification* comes from the word "soap." Soaps, which are synthesized by the saponification of fats, will be discussed at greater length in Chapter 20.

$$\underset{\underset{\underset{\textit{a fat, or triglyceride}}{\text{tripalmitin}}}{}}{\underset{\text{CH}_2\text{OC(CH}_2)_{14}\text{CH}_3}{\overset{\overset{\text{O}}{\|}}{\underset{\underset{\text{CH}_2\text{OC(CH}_2)_{14}\text{CH}_3}{\overset{\overset{\text{O}}{\|}}{|}}}{\underset{\overset{\overset{\text{O}}{\|}}{\text{CHOC(CH}_2)_{14}\text{CH}_3}}{|}}}} + 3\ \text{NaOH} \longrightarrow \underset{\underset{\text{glycerol}}{\text{CH}_2\text{OH}}}{\underset{\overset{\text{CH}_2\text{OH}}{|}}{\overset{\text{CH}_2\text{OH}}{\underset{|}{\text{CHOH}}}}} + \underset{\underset{\textit{a soap}}{\text{sodium palmitate}}}{3\ \text{CH}_3(\text{CH}_2)_{14}\overset{\overset{\text{O}}{\|}}{C}\text{O}^-\ \text{Na}^+}$$

SAMPLE PROBLEM

Which has a faster rate of saponification: (a) ethyl benzoate, or   (b) ethyl *p*-nitrobenzoate?

Solution: (b) has the faster rate because it has an electron-withdrawing nitro group. The transition state leading to the intermediate in Step 1 is more stabilized by dispersal of the negative charge:

$$O_2N{\leftarrow}\langle\bigcirc\rangle{\leftarrow}\overset{\overset{\text{O}^-}{|}}{\underset{\underset{\text{OH}}{|}}{C}}-OCH_2CH_3$$

STUDY PROBLEM

13.19. Write equations for the saponification of the following esters with aqueous NaOH:

(a) $\langle\bigcirc\rangle-\text{CO}_2\text{CH}_2\text{CH}_3$ (b)   (c)

Transesterification. Exchange of the alcohol portion of an ester can be accomplished in acidic or basic solution by a reversible reaction between the ester and an alcohol. These **transesterification reactions** are directly analogous to hydrolysis in acid or base. Because the reactions are reversible, an excess of the initial alcohol is generally used.

$$\underset{\substack{\text{methyl benzoate}}}{C_6H_5-\overset{\overset{\displaystyle O}{\|}}{C}-OCH_3} + \underset{\substack{\text{excess}}}{CH_3CH_2OH} \xrightleftharpoons{H^+ \text{ or } CH_3CH_2O^-} \underset{\substack{\text{ethyl benzoate}}}{C_6H_5-\overset{\overset{\displaystyle O}{\|}}{C}-OCH_2CH_3} + CH_3OH$$

STUDY PROBLEM

13.20. Suggest mechanisms for the transesterification reactions of ethyl acetate with
(a) methanol and HCl, and   (b) methanol and sodium methoxide.

Reaction with ammonia. Esters undergo reaction with aqueous ammonia to yield amides. The reaction is slow compared to the reactions of acid halides or anhydrides with ammonia. This slowness of the ester reaction can be an advantage because the reaction of an acid chloride with an amine can sometimes be violent. The ester route to amides is also the reaction of choice when a chemist desires an amide with another functional group that would not be stable toward an acid chloride. Such a case is illustrated in our second example.

$$\underset{\substack{\text{ethyl chloroacetate}}}{ClCH_2\overset{\overset{\displaystyle O}{\|}}{C}OCH_2CH_3} + NH_3 \xrightarrow[\text{1 hr}]{0°} \underset{\substack{\text{chloroacetamide} \\ 80\%}}{ClCH_2\overset{\overset{\displaystyle O}{\|}}{C}NH_2} + CH_3CH_2OH$$

$$\underset{\substack{\text{ethyl 2-hydroxypropanoate} \\ \text{(ethyl lactate)}}}{\underset{\underset{\displaystyle OH}{|}}{CH_3CHCOCH_2CH_3}\overset{\overset{\displaystyle O}{\|}}{}} + NH_3 \xrightarrow[\text{24 hr}]{25°} \underset{\substack{\text{2-hydroxypropanamide} \\ 70\%}}{\underset{\underset{\displaystyle OH}{|}}{CH_3CHCNH_2}\overset{\overset{\displaystyle O}{\|}}{}} + CH_3CH_2OH$$

STUDY PROBLEM

13.21. What products would arise from the reaction of 2-hydroxypropanoic acid (lactic acid) with $SOCl_2$?

Reduction.   Esters can be reduced by catalytic hydrogenation, a reaction sometimes called **hydrogenolysis of esters**, or by lithium aluminum hydride. An older technique is the reaction of the ester with sodium metal in ethanol. Regardless of the reducing agent, a pair of alcohols (with at least one being primary) results from the reduction of an ester.

General:

$$RC\overset{\overset{\displaystyle O}{\|}}{-}OR' \xrightarrow{\text{[H]}} RCH_2OH + HOR'$$

to a primary alcohol   *to the other alcohol*

ethyl benzoate $\xrightarrow[\text{(2) H}_2\text{O}]{\text{(1) LiAlH}_4}$ benzyl alcohol 90% + $HOCH_2CH_3$ ethanol

$$CH_3(CH_2)_8\overset{\overset{\displaystyle O}{\|}}{C}OCH_2CH_3 \xrightarrow[\text{CH}_3\text{CH}_2\text{OH}]{\text{Na}} CH_3(CH_2)_8CH_2OH + HOCH_2CH_3$$

ethyl decanoate   1-decanol 70%   ethanol

$$CH_3CH_2O\overset{\overset{\displaystyle O}{\|}}{C}(CH_2)_4\overset{\overset{\displaystyle O}{\|}}{C}OCH_2CH_3 \xrightarrow[255°, 200 \text{ atm}]{\underset{\text{CuCr}_2\text{O}_4 \text{ catalyst}}{\text{H}_2}} HOCH_2(CH_2)_4CH_2OH + HOCH_2CH_3$$

diethyl adipate   1,6-hexanediol 90%   ethanol

STUDY PROBLEM

13.22. Show by equations how you would prepare $CH_3(CH_2)_{14}CH_2OH$ from tripalmitin (page 631).

Reaction with Grignard reagents.   The reaction of esters with Grignard reagents is an excellent technique for the preparation of *tertiary alcohols with two identical R groups.*

General:

$$RCOR' \xrightarrow[\text{(2) H}_2\text{O, H}^+]{\text{(1) 2 R''MgX}} R\overset{\overset{\displaystyle OH}{|}}{\underset{\underset{\displaystyle R''}{|}}{C}}R''$$

an ester   *two R'' groups the same*

a 3° alcohol

ethyl cyclohexanecarboxylate $\xrightarrow[\text{(2) H}_2\text{O, H}^+]{\text{(1) 2 CH}_3\text{MgI}}$ 2-cyclohexyl-2-propanol

If a formate ester is subjected to a Grignard reaction, a *secondary alcohol with two identical R groups* is obtained. Formates are a special case because the carbonyl carbon is attached to an H atom, not an alkyl or aryl group.

$$\underset{a\ formate\ ester}{\underset{\text{HCOR}}{\overset{\overset{\textstyle O}{\|}}{}}} \xrightarrow[\text{(2) H}_2\text{O, H}^+]{\text{(1) 2 R'MgX}} \underset{a\ 2°\ alcohol}{\underset{\underset{\text{R}'}{|}}{\overset{\overset{\textstyle OH}{|}}{\text{H—C—R'}}}} \quad \substack{\text{two R' groups} \\ \text{the same}}$$

The mechanism of the reaction of a Grignard reagent with an ester is similar to that of the reaction of a Grignard reagent with an aldehyde or ketone—that is, the nucleophilic carbon of the Grignard reagent attacks the positive carbon of the carbonyl group. In the case of an ester, as in the case of an acid halide (page 618), *two* equivalents of Grignard reagent attack the carbonyl carbon atom. To see why this is so, let us consider the reaction stepwise. First, the negative carbon of the Grignard reagent attacks the carbon of the carbonyl group. The product of this step has a hemiketal-like structure that loses an alkoxyl group to yield a ketone.

Initial attack:

$$\underset{}{\overset{\overset{\textstyle \ddot{O}:}{\|}}{\text{RC—OR'}}} + \overset{\delta-\quad\delta+}{\text{CH}_3\text{MgX}} \longrightarrow \left[\underset{\substack{| \\ \text{CH}_3}}{\overset{\overset{\textstyle :\ddot{O}:^- \ ^+\text{MgX}}{|}}{\text{RC}\overset{}{\ddot{O}\text{R'}}}}\right] \longrightarrow \left[\underset{\substack{| \\ \text{CH}_3}}{\overset{\overset{\textstyle \ddot{O}:}{\|}}{\text{RC}}}\right] + \text{R}\ddot{O}:^- \ ^+\text{MgX}$$

a type of hemiketal   a ketone

The ketone then undergoes a *second* reaction with the Grignard reagent. The rate of the second reaction is greater than that of the first; hence the ketone cannot be isolated.

Second attack and hydrolysis:

$$\left[\underset{\text{RCCH}_3}{\overset{\overset{\textstyle \ddot{O}:}{\|}}{}}\right] + \text{CH}_3\text{MgX} \longrightarrow \underset{\substack{| \\ \text{CH}_3}}{\overset{\overset{\textstyle :\ddot{O}\text{MgX}}{|}}{\text{RCCH}_3}} \xrightarrow{\text{H}_2\text{O, H}^+} \underset{\substack{| \\ \text{CH}_3}}{\overset{\overset{\textstyle OH}{|}}{\text{RCCH}_3}} + \text{Mg}^{2+} + \text{X}^- + \text{H}_2\text{O}$$

a 3° alcohol

STUDY PROBLEMS

13.23. Show by equations how you would prepare the following alcohols from methyl propanoate: (a) 2-methyl-2-butanol; (b) 3-ethyl-3-pentanol.

13.24. Predict the product of the reaction of ethyl formate with ethylmagnesium bromide, followed by work-up with aqueous acid.

SECTION 13.6.

Lactones

A **lactone** is a cyclic ester. Lactones are fairly common in natural sources. For example, vitamin C and nepetalactone, the cat attractant in catnip, are both lactones. It is interesting to note the close structural relationship of nepetalactone to iridomyrmecin, an odorous compound found in the *Iridomyrmex* species of ants.

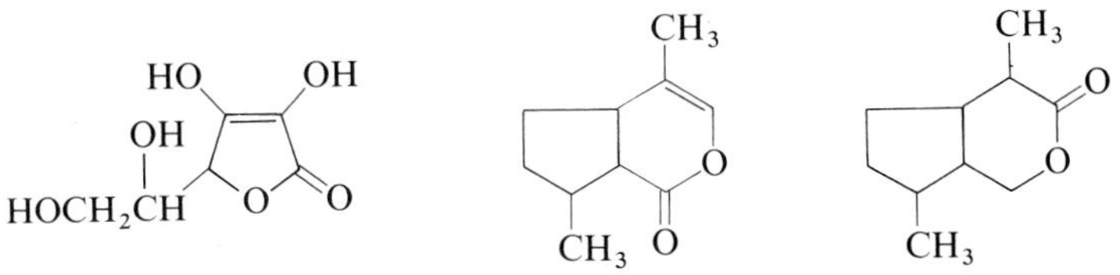

vitamin C	nepetalactone	iridomyrmecin
(ascorbic acid)	*cat attractant in catnip*	*in ants*

Lactones are formed from molecules that contain a carboxyl group and a hydroxyl group. These molecules can undergo an intramolecular esterification.

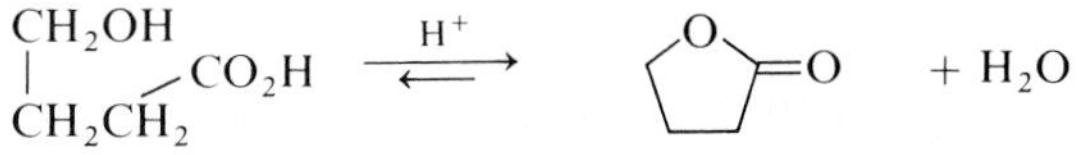

4-hydroxybutanoic acid — a γ-hydroxy acid → 4-hydroxybutanoic acid lactone (γ-butyrolactone) + H_2O

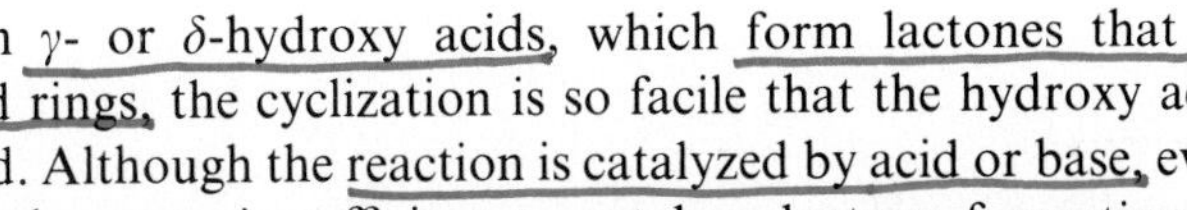

5-hydroxypentanoic acid — a δ-hydroxy acid → 5-hydroxypentanoic acid lactone (δ-valerolactone) + H_2O

With γ- or δ-hydroxy acids, which form lactones that are five- or six-membered rings, the cyclization is so facile that the hydroxy acids often cannot be isolated. Although the reaction is catalyzed by acid or base, even a trace of acid from the glassware is sufficient to catalyze lactone formation if a five- or six-membered ring is the product. The percentages of hydroxy acid and lactone in some equilibrium mixtures in aqueous acid are shown in Table 13.4.

STUDY PROBLEM

13.25. Lactic acid (2-hydroxypropanoic acid) does not form a lactone. However, when it is heated, it yields a dimeric cyclic ester called a **lactide**. What is the structure of this lactide?

TABLE 13.4. Composition of equilibrium mixtures of hydroxy acid versus lactone

Hydroxy acid	Lactone	% Acid	% Lactone
$HOCH_2CH_2CO_2H$		100	0
$HOCH_2CH_2CH_2CO_2H$		27	73
$HOCH_2(CH_2)_2CH_2CO_2H$		9	91
$HOCH_2(CH_2)_3CH_2CO_2H$		100	0

Carboxylic acids with hydroxyl groups in positions other than the γ- or δ-positions are stable and do not cyclize spontaneously. (Compare the percents of lactone in the equilibrium mixtures of the hydroxy acids in Table 13.4.) However, the lactones of these hydroxy acids may be synthesized under the usual conditions for esterification. In these cases, a *dilute solution* of hydroxy acid in an inert solvent is used. An intramolecular reaction is favored by dilute solution because collisions between molecules are less apt to occur. If the solution is *concentrated*, then the hydroxy acid molecules undergo reaction with each other to yield a **polyester**. In either case, a solvent such as benzene allows the product water to be distilled as an azeotrope and drives the reaction toward the lactone (or polyester).

SECTION 13.7.

Polyesters

The synthetic fiber **Dacron** is a polyester made by a transesterification reaction of dimethyl terephthalate and ethylene glycol. The reason that polymer-formation can occur is that the reactants are *bifunctional*, and thus each reactant can undergo reaction with two other molecules.

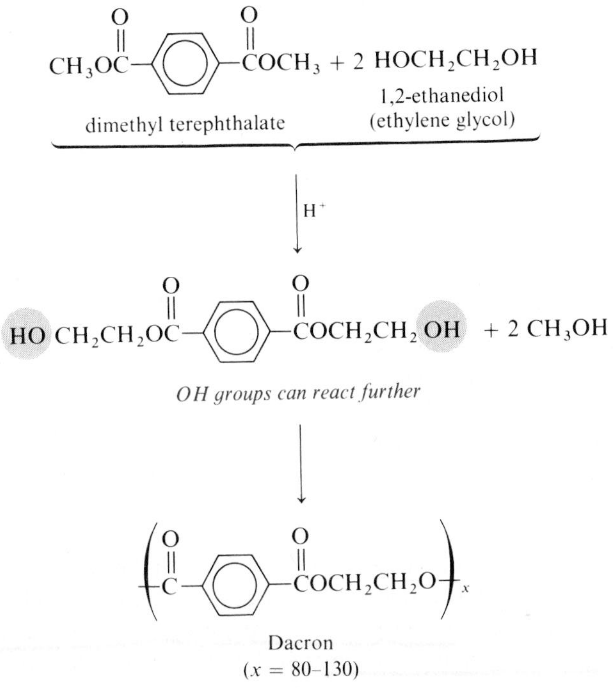

$$CH_3OC-\langle\bigcirc\rangle-COCH_3 + 2\ HOCH_2CH_2OH$$

dimethyl terephthalate   1,2-ethanediol
(ethylene glycol)

$$\downarrow H^+$$

$$HO\ CH_2CH_2OC-\langle\bigcirc\rangle-COCH_2CH_2\ OH\ + 2\ CH_3OH$$

OH groups can react further

$$\left(\begin{matrix} O \\ \| \\ C \end{matrix}-\langle\bigcirc\rangle-COCH_2CH_2O\right)_x$$

Dacron
($x = 80$–130)

When the monomers are bifunctional, such as dimethyl terephthalate and ethylene glycol, polymer growth must occur in a *linear* fashion. Linear polymers often make excellent textile fibers. If more than two reactive sites are present in one of the monomers, then the polymer can grow into a cross-linked network. **Glyptal** (a polymer of glycerol and phthalic anhydride) is an example of a cross-linked polyester.

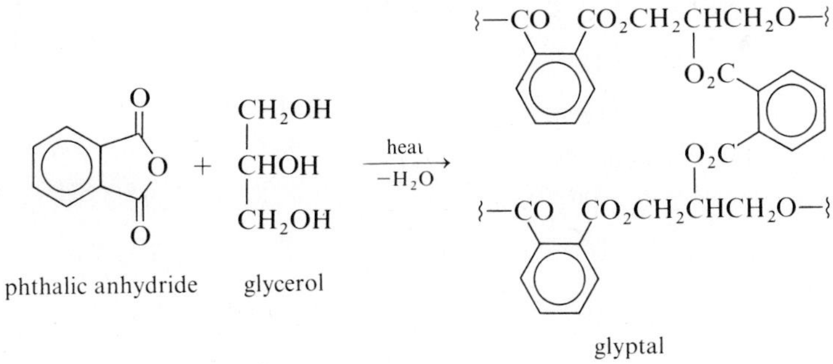

$$\begin{matrix} CH_2OH \\ | \\ CHOH \\ | \\ CH_2OH \end{matrix}$$

phthalic anhydride   glycerol

$$\xrightarrow[-H_2O]{heat}$$

{—CO   CO_2CH_2CHCH_2O—}
                    |
                   O_2C
        O_2C
                    |
{—CO   CO_2CH_2CHCH_2O—}

glyptal

SECTION 13.8.

Thioesters

Esters with the $RCS-$ unit are called **thioesters**. Thioesters undergo reactions, such as hydrolysis, similar to those of ordinary esters. **Acetylcoenzyme A**, which is important in biological reactions, is a thioester.

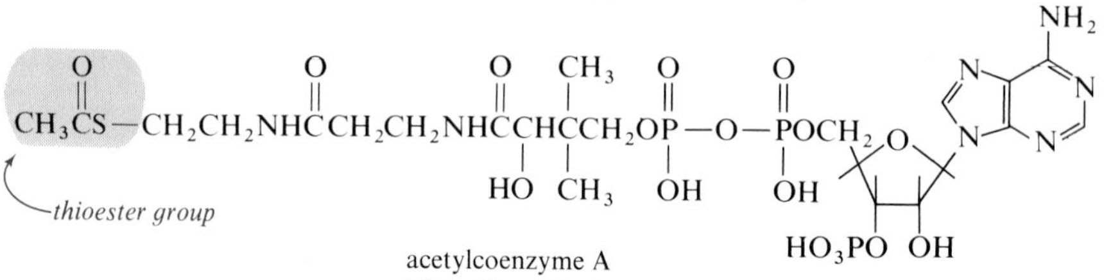

acetylcoenzyme A
$(CH_3CO-SCoA)$

In an organism, acetylcoenzyme A has two principal functions. The first function is that of an **acylating agent** (a reagent that places an acyl group, $RCO-$, in a molecule). The following example shows the transfer of an acetyl group from acetylcoenzyme A to a phosphate group. In this reaction, acetylcoenzyme A is hydrolyzed to the thiol **coenzyme A**.

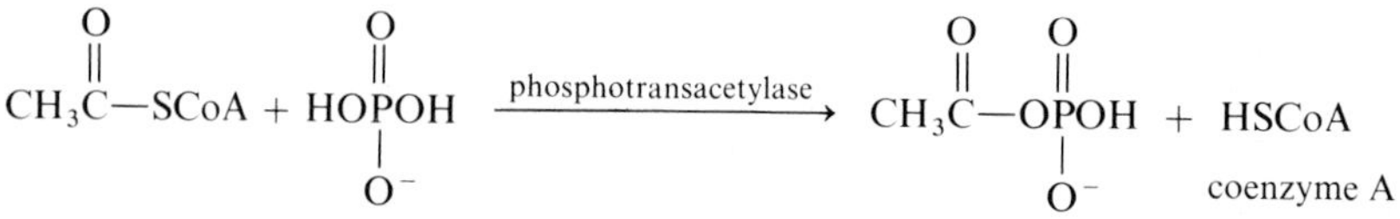

$$CH_3\overset{O}{\overset{\|}{C}}-SCoA + HO\overset{O}{\overset{\|}{P}}OH \xrightarrow{\text{phosphotransacetylase}} CH_3\overset{O}{\overset{\|}{C}}-O\overset{O}{\overset{\|}{P}}OH + HSCoA$$

coenzyme A

The second function of acetylcoenzyme A is that of an **alkylating agent**. The α hydrogen of a thioester is acidic and can be removed rather easily by the appropriate enzyme. Therefore, the acetyl carbon in acetylcoenzyme A can act as a nucleophile and can attack a carbonyl group. (This type of reaction will be discussed in Chapter 14.)

Loss of proton:

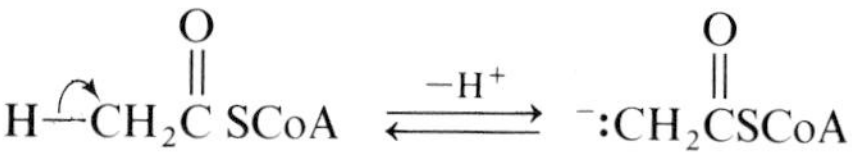

$$H-CH_2\overset{O}{\overset{\|}{C}}SCoA \underset{}{\overset{-H^+}{\rightleftharpoons}} {}^-:CH_2\overset{O}{\overset{\|}{C}}SCoA$$

Nucleophilic attack and thioester hydrolysis:

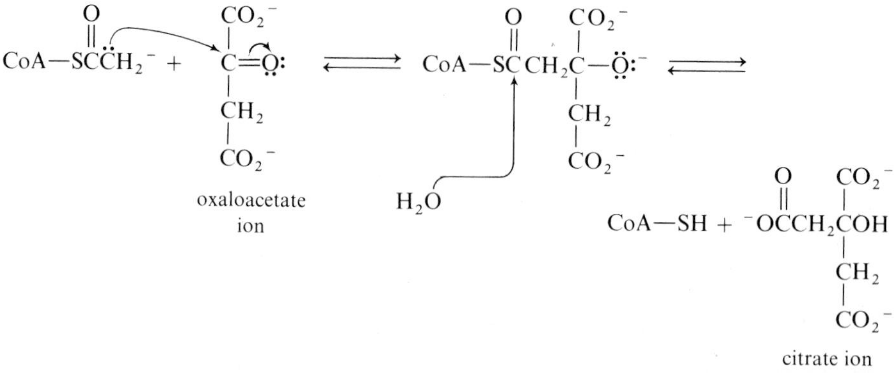

oxaloacetate ion

H_2O

citrate ion

SECTION 13.9.

Amides

A.   Nomenclature of amides

An **amide** is a compound that has a trivalent nitrogen attached to a carbonyl group. An amide is named from the parent carboxylic acid with the **-oic (or -ic) acid** ending changed to **-amide**.

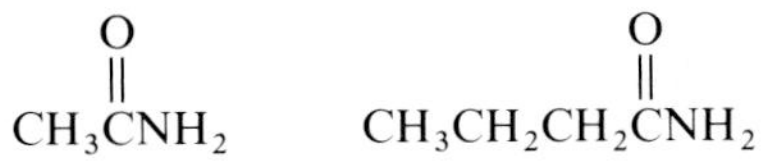

$$\underset{\substack{\text{IUPAC:} \quad \text{ethanamide} \\ \text{trivial:} \quad \text{acetamide}}}{CH_3\overset{\displaystyle O}{\overset{\|}{C}}NH_2} \qquad \underset{\substack{\text{butanamide} \\ \text{butyramide}}}{CH_3CH_2CH_2\overset{\displaystyle O}{\overset{\|}{C}}NH_2}$$

Amides with alkyl substituents on the nitrogen have their names preceded by *N*-alkyl, where *N* refers to the nitrogen atom.

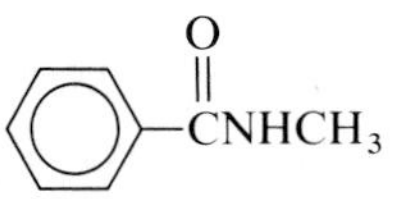

$$\underset{N\text{-methylbenzamide}}{\text{C}_6\text{H}_5\text{—}\overset{\displaystyle O}{\overset{\|}{C}}NHCH_3} \qquad \underset{N,N\text{-dimethylformamide}}{H\overset{\displaystyle O}{\overset{\|}{C}}N(CH_3)_2}$$

A few amides of interest follow; the amide groups are circled:

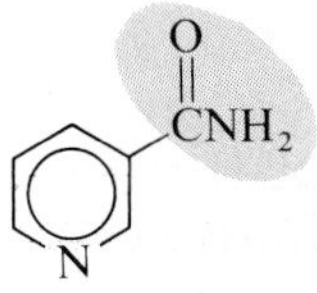

nicotinamide, or niacinamide

a B vitamin

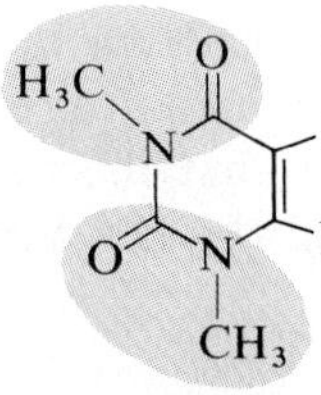

caffeine

a lactam, or cyclic amide

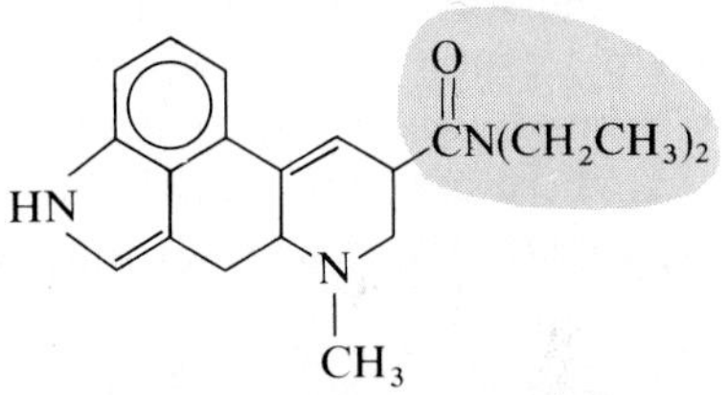

lysergic acid diethylamide (LSD)

B.   Preparation of amides

Amides are synthesized from derivatives of carboxylic acids and ammonia or the appropriate amine. These reactions have been discussed previously in this chapter.

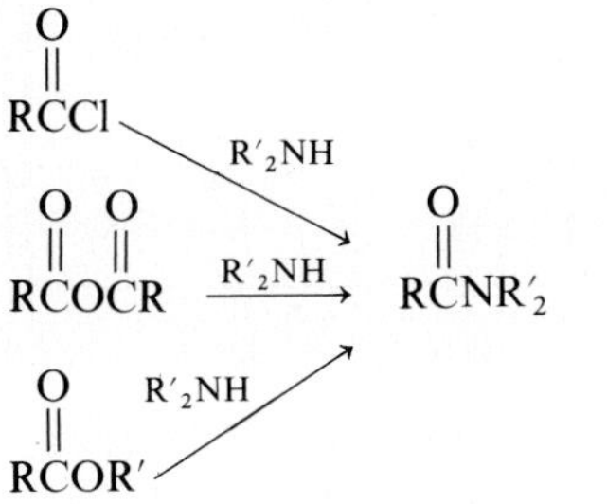

$$\begin{array}{c}
R\overset{\displaystyle O}{\overset{\|}{C}}Cl \\[4pt]
R\overset{\displaystyle O}{\overset{\|}{C}}O\overset{\displaystyle O}{\overset{\|}{C}}R \\[4pt]
R\overset{\displaystyle O}{\overset{\|}{C}}OR'
\end{array}
\quad \xrightarrow{R'_2NH} \quad
R\overset{\displaystyle O}{\overset{\|}{C}}NR'_2$$

C.   Reactions of amides

An amide contains a nitrogen that has a pair of unshared electrons in a filled orbital. It would be reasonable to expect amides to undergo reaction with acids, as do amines; however, they do not. Amides are *very* weak bases with pK_b values of 15–16. (By contrast, methylamine has a pK_b of 3.34.) The resonance structures for an amide show why the nitrogen of an amide is neither particularly basic nor nucleophilic.

$$CH_3\ddot{N}H_2 \ + \ dilute \ HCl \ \longrightarrow \ CH_3NH_3{}^+ \ Cl^-$$

methylamine   methylammonium
                                                                chloride

$$CH_3\overset{\overset{\displaystyle O}{\|}}{C}\ddot{N}H_2 + dilute \ HCl \ \longrightarrow \ no \ appreciable \ salt \ formation$$

acetamide

Resonance structures for an amide:

$$R-\overset{\overset{\displaystyle \ddot{O}:}{\|}}{C}-\ddot{N}H_2 \ \longleftrightarrow \ R-\overset{\overset{\displaystyle :\ddot{O}:^-}{|}}{C}=\overset{+}{N}H_2 \qquad \textit{less basic than an amine nitrogen}$$

The effect of the partial double-bond character of the bond between the carbonyl carbon and the nitrogen of an amide is evident in the nmr spectrum of *N,N*-dimethylformamide (Figure 13.9). The spectrum shows one peak for *each* methyl group. If the two methyl groups underwent free rotation around the CN bond, they would be equivalent and would give rise to one singlet. The fact that there are two methyl singlets shows that the two methyl groups are *not equivalent*. The restricted rotation around the CN bond results in two methyl

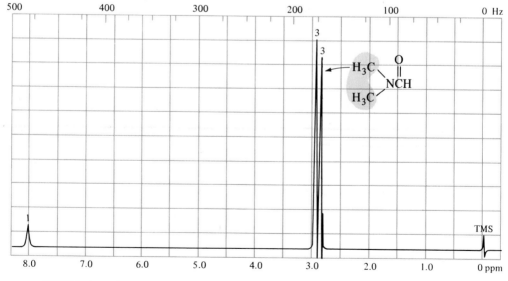

FIGURE 13.9. The nmr spectrum of *N,N*-dimethylformamide, showing a pair of peaks for the *N*-methyl groups.

groups, each in a different magnetic environment. (The energy barrier for rotation around the CN bond in an amide has been found to be 18 kcal/mole.)

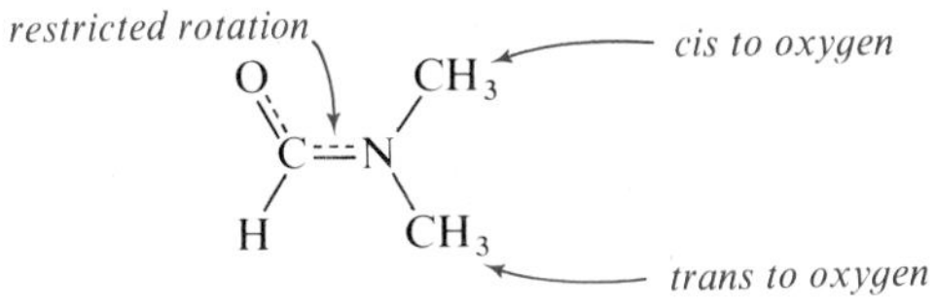

Hydrolysis. Like esters, amides may be hydrolyzed in either acidic or alkaline solution. In either case, the acid or base is a reactant, not a catalyst, and must be used in a 1:1 molar ratio or in excess. Neither type of hydrolysis reaction is reversible.

In acid:

$$CH_3CH_2\overset{\overset{\displaystyle O}{||}}{C}N(CH_3)_2 \ + H_2O + H^+ \longrightarrow CH_3CH_2CO_2H + H_2\overset{+}{N}(CH_3)_2$$

N,N-dimethylpropanamide   propanoic   dimethyl-
                                                acid   ammonium ion

In base:

$$CH_3CH_2\overset{\overset{\displaystyle O}{||}}{C}N(CH_3)_2 + OH^- \longrightarrow CH_3CH_2\overset{\overset{\displaystyle O}{||}}{C}O^- + HN(CH_3)_2$$

                                             propanoate ion   dimethylamine

Hydrolysis of an amide in acidic solution proceeds in a fashion similar to hydrolysis of an ester. The carbonyl oxygen is protonated, the carbonyl carbon is attacked by H_2O, and an amine is expelled. This amine then undergoes reaction with H^+ to yield the amine salt. The formation of the amine salt explains (1) why H^+ is a reactant, not a catalyst, and (2) why the reverse reaction does not proceed. (Although R_3N is a nucleophile, R_3NH^+ is not, and it cannot attack the carbonyl group.)

In acid:

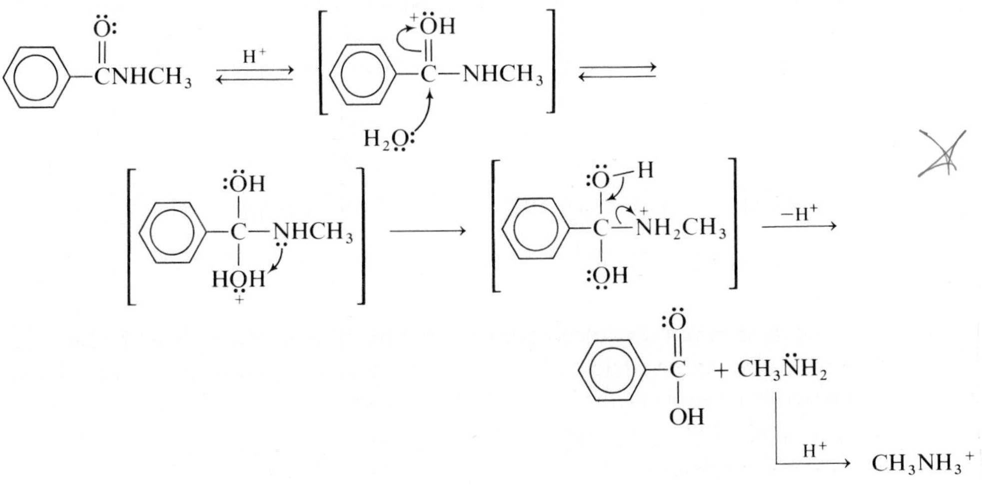

Alkaline hydrolysis of an amide is similar to saponification of an ester. The products are the carboxylate salt of the acid and a free amine or ammonia.

In base:

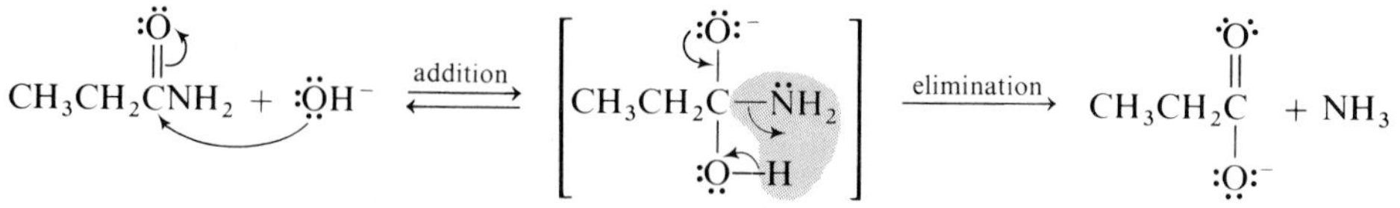

$$
CH_3CH_2\overset{\overset{\displaystyle :\ddot O}{\|}}{C}NH_2 + :\ddot OH^- \underset{}{\overset{\text{addition}}{\rightleftharpoons}} \left[CH_3CH_2\overset{\overset{\displaystyle :\ddot O:^-}{|}}{\underset{\underset{\displaystyle :\ddot O\!-\!H}{|}}{C}}\!-\!\ddot NH_2 \right] \xrightarrow{\text{elimination}} CH_3CH_2\overset{\overset{\displaystyle :\ddot O}{\|}}{\underset{\underset{\displaystyle :\ddot O:^-}{|}}{C}} + NH_3
$$

STUDY PROBLEMS

13.26. The following structure represents a portion of a polyamide molecule, similar in structure to a protein. What would be the alkaline hydrolysis products? What would be the acid hydrolysis products?

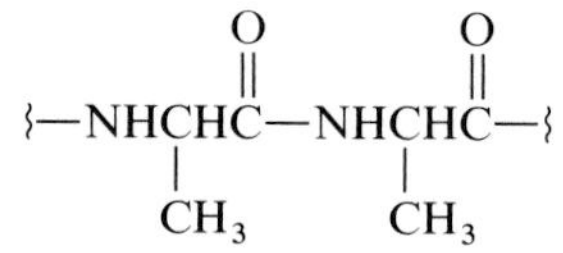

$$
\text{\}-NHCHC\overset{\overset{\displaystyle O}{\|}}{}-NHCHC\overset{\overset{\displaystyle O}{\|}}{}-\{}
$$
$$
\qquad\quad|\qquad\qquad\ |
$$
$$
\qquad\ \ CH_3\qquad\quad CH_3
$$

13.27. What would be the hydrolysis products of   (a) nicotinamide (page 639) in acid?   (b) LSD (page 639) in base?   (c) the following lactam in base?

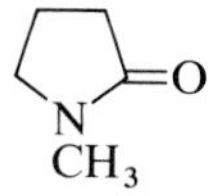

$$
\underset{\underset{\displaystyle CH_3}{N}}{\boxed{}}\!=\!O
$$

SECTION 13.10.

Polyamides

There can be no question that the most important polyamides are the *proteins.* Chapter 19 is devoted to this subject. The most notable example of a man-made polyamide is the synthetic polyamide **nylon 66**, which is prepared from adipic acid (a diacid) and hexamethylenediamine (a diamine). As in the synthesis of the polyester Dacron, the result of the reaction of two types of bifunctional molecules is a linear polymer.

$$
x\ HO_2C(CH_2)_4CO_2H + x\ H_2N(CH_2)_6NH_2 \xrightarrow[-H_2O]{\text{heat}} \left[\overset{\overset{\displaystyle O}{\|}}{C}(CH_2)_4\overset{\overset{\displaystyle O}{\|}}{C}-NH(CH_2)_6NH\right]_x
$$

hexanedioic acid   1,6-hexanediamine   nylon 66
(adipic acid)   (hexamethylenediamine)

Nylon 66 is but one member of the family of synthetic nylons. Nylon 66 is made from a *six-carbon* diacid and *six-carbon* diamine. **Nylon 6**, on the other hand, is prepared from caprolactam, a monomer that contains the acid and amine in the same molecule (with *six carbons*). In this reaction, caprolactam undergoes ring opening with water; then, in the polymerization, water is eliminated.

$$x \underset{\text{caprolactam}}{\text{(caprolactam)}} \xrightarrow[\text{H}_2\text{O}]{} x\ \text{H}_2\text{N(CH}_2)_5\overset{\text{O}}{\overset{\|}{\text{C}}}\text{OH} \xrightarrow[-\text{H}_2\text{O}]{250°} \underset{\text{nylon 6}}{\left[\text{NH(CH}_2)_5\overset{\text{O}}{\overset{\|}{\text{C}}} \right]_x}$$

SECTION 13.11.

Compounds Related to Amides

Some types of compounds that are related to amides are shown in Table 13.5. **Urea** is one of the most important amide relatives. Excess nitrogen from the metabolism of proteins is excreted by the higher animals as urea. Some lower animals excrete ammonia, while reptiles and birds excrete **guanidine**. Both guanidine and

TABLE 13.5. Some types of compounds related to amides

Partial structure	Class of compound	Example
$-\overset{\text{O}}{\overset{\|}{\text{C}}}\text{N}\underset{\diagdown}{\diagup}$	amide	$\text{CH}_3\overset{\text{O}}{\overset{\|}{\text{C}}}\text{NH}_2$
$-\overset{\text{O}}{\overset{\|}{\text{C}}}\text{N}\diagup$ in ring	lactam	(six-membered lactam ring with NH)
$-\overset{\text{O O}}{\overset{\|\ \|}{\text{C}}}\text{N}\text{C}-$	imide	(six-membered imide ring with NH)
$\diagdown\text{N}\overset{\text{O}}{\overset{\|}{\text{C}}}\text{N}\diagup$	urea	$\text{H}_2\text{N}\overset{\text{O}}{\overset{\|}{\text{C}}}\text{NH}_2$
$\diagdown\text{N}\overset{\text{O}}{\overset{\|}{\text{C}}}\text{O}-$	carbamate, or urethane	$\text{H}_2\text{N}\overset{\text{O}}{\overset{\|}{\text{C}}}\text{OCH}_3$
$-\overset{\text{O}}{\underset{\text{O}}{\overset{\|}{\underset{\|}{\text{S}}}}}\text{N}\diagup$	sulfonamide	$\text{C}_6\text{H}_5-\overset{\text{O}}{\underset{\text{O}}{\overset{\|}{\underset{\|}{\text{S}}}}}\text{NH}_2$

urea, as well as ammonia, are widely used as nitrogen fertilizers and as starting materials for synthetic polymers and drugs.

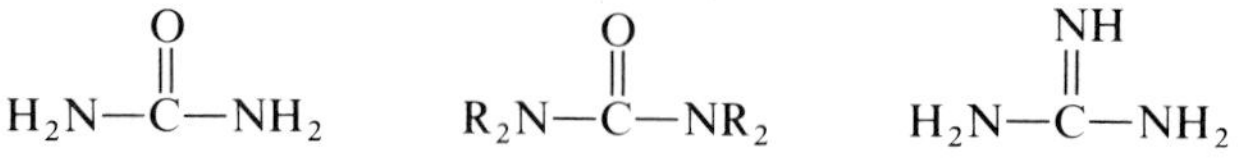

$$H_2N-\underset{\underset{\text{urea}}{}}{\overset{\overset{O}{\|}}{C}}-NH_2 \qquad R_2N-\underset{\underset{\text{a substituted urea}}{}}{\overset{\overset{O}{\|}}{C}}-NR_2 \qquad H_2N-\underset{\underset{\text{guanidine}}{}}{\overset{\overset{NH}{\|}}{C}}-NH_2$$

Urea is used for the synthesis of barbiturates (used as sedatives) by reaction with α-substituted diethyl malonates. This reaction is similar to the reaction of an ester with an amine to yield an amide.

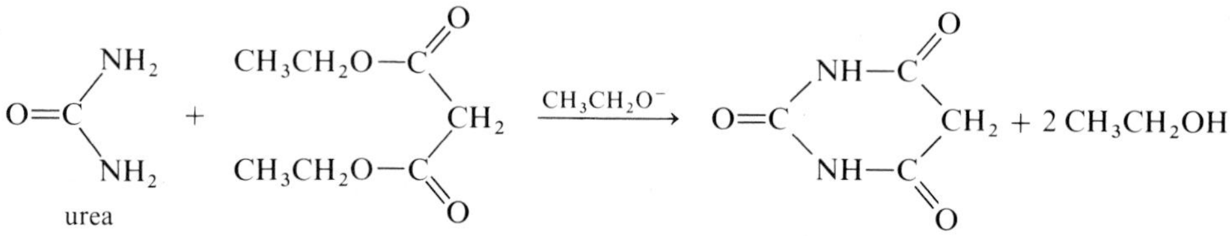

urea   +   diethyl malonate $\xrightarrow{CH_3CH_2O^-}$ barbituric acid   + 2 CH₃CH₂OH

not substituted

STUDY PROBLEM

13.28. What two reactants would you use to synthesize each of the following barbiturates?

(a)

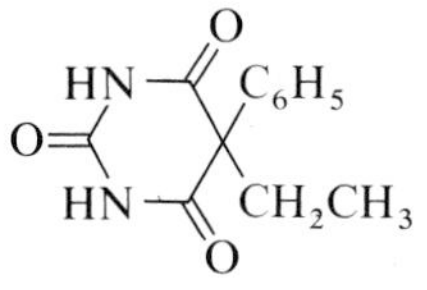

phenobarbital

(b)

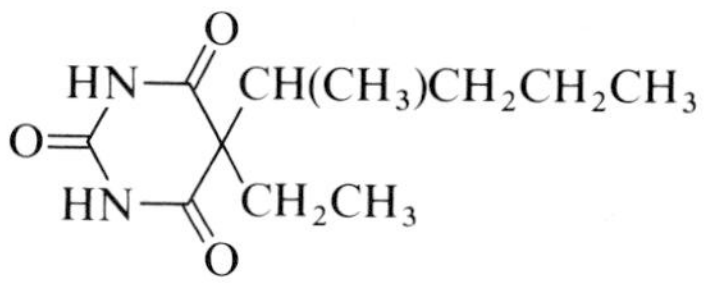

pentobarbital
(Nembutal)

An **imide**, a compound with the $-\overset{\overset{O}{\|}}{C}NH\overset{\overset{O}{\|}}{C}-$ group, is the nitrogen analog of an acid anhydride. Like amides, an imide can be made from ammonia and an acid anhydride.

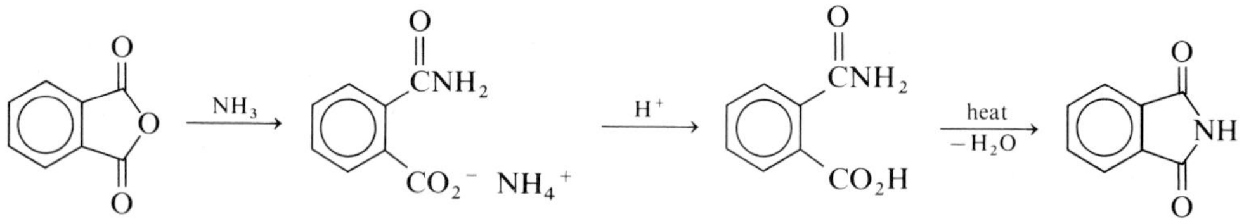

phthalic
anhydride

$\xrightarrow{NH_3}$ $\xrightarrow{H^+}$ $\xrightarrow[-H_2O]{heat}$ phthalimide
a cyclic imide

A **carbamate**, or **urethane**, is a compound in which the $-NH_2$, $-NHR$, or $-NR_2$ group is attached to an ester carbonyl group. A carbamate is related to a carbonate structure, with one O replaced by N.

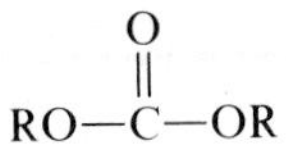

$$
\underset{\text{a carbonate}}{RO-\overset{\overset{\displaystyle O}{\|}}{C}-OR}
\qquad\qquad
\underset{\text{a carbamate}}{H_2N-\overset{\overset{\displaystyle O}{\|}}{C}-OR}
$$

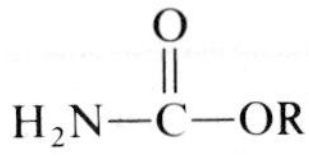

$$
\underset{\substack{\text{meprobamate}}}{H_2NCOCH_2\overset{\overset{\displaystyle CH_3}{|}}{\underset{\underset{\displaystyle CH_2CH_2CH_3}{|}}{C}}CH_2OCNH_2}
$$

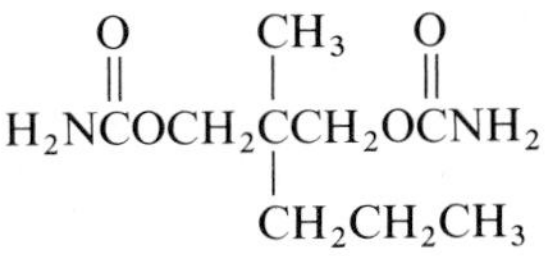

meprobamate

*a dicarbamate used as
a tranquilizer (Miltown, Equanil)*

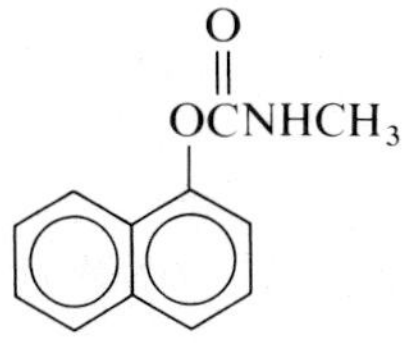

1-naphthyl-*N*-methylcarbamate
(Sevin)

a biodegradable insecticide

One way in which a carbamate may be prepared is by the action of an alcohol or phenol on an **isocyanate**, a compound containing the —N=C=O group.

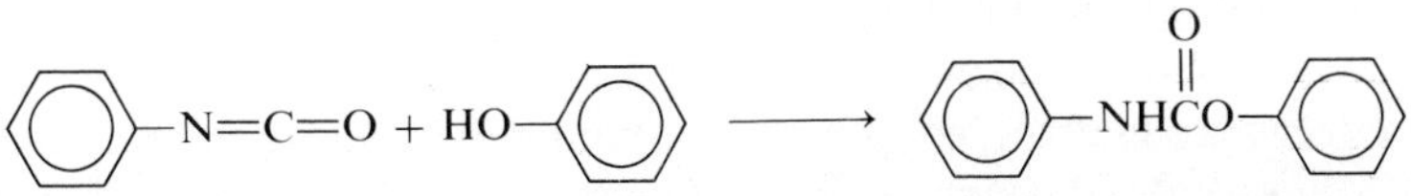

phenyl isocyanate   phenol   phenyl *N*-phenylcarbamate

An analogous reaction is used to make **polyurethanes** (used for polyurethane foam insulation, for example). As in the formation of other polymers mentioned in this chapter, bifunctional starting materials must be used. (The foaming effect in polyurethane foam is achieved by adding a low-boiling liquid, such as dichloromethane, that vaporizes during the polymerization.)

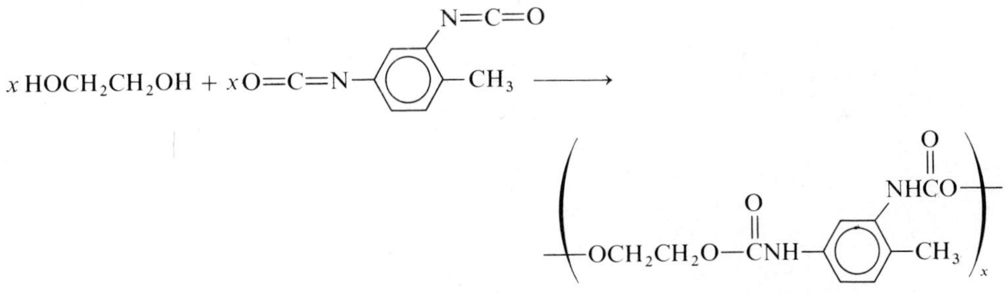

$$
x\,HOCH_2CH_2OH + x\,O{=}C{=}N-\underset{}{\bigcirc}-CH_3 \longrightarrow
$$

a polyurethane

Sulfa drugs are **sulfonamides**, compounds in which the nitrogen is attached to a sulfonyl group rather than to an acyl group. A sulfonamide is prepared by the action of an arylsulfonyl chloride on ammonia or on a primary or secondary amine.

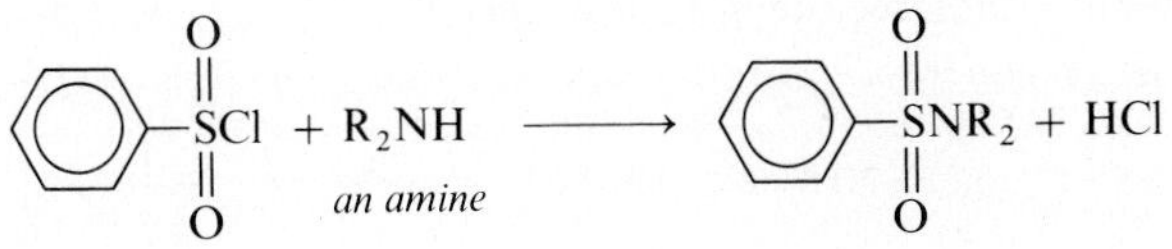

benzenesulfonyl
chloride

an amine

a sulfonamide

Many of the *p*-aminosulfonamides are effective bacteriostatic agents. Although the bacteriological properties of one of the sulfonamides were observed in 1909, these compounds were not widely used against infections in humans until around 1940.

*essential part of structure
for drug activity*

H_2N—⟨O⟩—SO_2NH_2

sulfanilamide

H_2N—⟨O⟩—SO_2NH—⟨S≡N⟩

sulfathiazole

H_2N—⟨O⟩—SO_2NH—⟨N=N / O⟩—OCH_3

sulfamethoxypyridazine

H_2N—⟨O⟩—SO_2NH—⟨N=N⟩ CH_3

sulfamerazine

Sulfa drugs inhibit the growth and multiplication of some types of bacteria that require *p*-aminobenzoic acid (PABA) for the biosynthesis of folic acids. The sulfa drugs are hydrolyzed *in vivo* to sulfanilamide, which is mistaken for PABA by certain bacterial enzymes. When the enzyme is associated with sulfanilamide, it is inhibited from catalyzing the incorporation of PABA into a folic acid molecule. This process is called **competitive inhibition of an enzyme-catalyzed reaction**. Sulfa drugs are not effective against all bacteria because some strains do not require PABA and some strains can synthesize their own PABA, which successfully competes against the sulfonamide for a position on the enzyme.

H_2N—⟨O⟩—CO_2H

p-aminobenzoic acid
(PABA)

from PABA

$$\text{OH}$$
$$\underset{H_2N}{\overset{N}{\diagdown}}\text{...}CH_2\text{—NH}\text{—}⟨O⟩\text{—}\overset{\overset{O}{\|}}{C}\text{—NHCHCH}_2\text{CH}_2\text{CO}_2\text{H}$$
$$\overset{|}{\text{CO}_2\text{H}}$$

folic acid

STUDY PROBLEM

13.29. Sodium cyclohexylsulfamate (a **cyclamate**) is an artificial sweetener that is thirty times sweeter than cane sugar. This compound may be prepared from cyclohexylamine and chlorosulfonic acid, followed by treatment with sodium hydroxide. What is the structure of this cyclamate?

$$⟨\hexagon⟩\text{—NH}_2 \;+\; \overset{\overset{O}{\|}}{\underset{\underset{O}{\|}}{\text{HOSCl}}} \xrightarrow{\hspace{1cm}} \xrightarrow{\text{OH}^-}$$

cyclohexylamine   chlorosulfonic
acid

SECTION 13.12.

Nitriles

A.   Nomenclature of nitriles

Nitriles are organic compounds containing the C≡N group. They are also sometimes called *cyano compounds* or *cyanides*. In the IUPAC system, the number of carbon atoms, including that in the CN group, determines the alkane parent. The alkane name is suffixed with **-nitrile**. Some nitriles are named after the trivial names of their carboxylic acid parent with the **-ic acid** changed to **-nitrile**, or to **-onitrile** if the parent name lacks an *o-*.

$$CH_3C{\equiv}N$$

IUPAC: ethanenitrile
trivial: acetonitrile   benzonitrile

STUDY PROBLEM

13.30. Write formulas for   (a) propanenitrile;   (b) butyronitrile.

B.   Bonding in nitriles

The cyano group contains a triple bond—one sigma bond and two pi bonds (Figure 13.10). Although the nitrogen has a pair of unshared electrons, a nitrile is a very weak base. The pK_b of a nitrile is about 24, while the pK_b of NH_3 is about 4.5 (about 20 powers of ten difference). The lack of basicity of a —CN: group results from the unshared electrons being in an *sp* orbital. The greater amount of *s* character in an *sp* orbital (compared to that in an *sp²* or *sp³* orbital) means that these *sp* electrons are more tightly held and less available for bonding to a proton.

C.   Preparation of nitriles

The CN⁻ ion (from NaCN, for example) is a good nucleophile for S_N2 displacement of a halide ion from an alkyl halide. This reaction is the principal route to

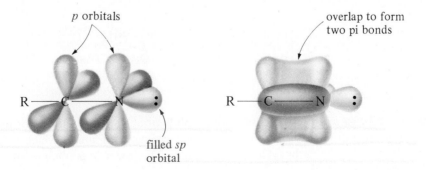

FIGURE 13.10. The bonding in a nitrile, RC≡N:.

nitriles; however, because of elimination reactions, high yields are obtained only with *primary alkyl halides* and, to a lesser extent, *secondary alkyl halides.*

$$CH_3CH_2CH_2CH_2Br + CN^- \xrightarrow{\ S_N2\ } CH_3CH_2CH_2CH_2CN + Br^-$$

1-bromobutane
a 1° alkyl halide

pentanenitrile
90%

Aryl nitriles are best obtained through the **diazonium salts**, compounds that were discussed in Section 10.14.

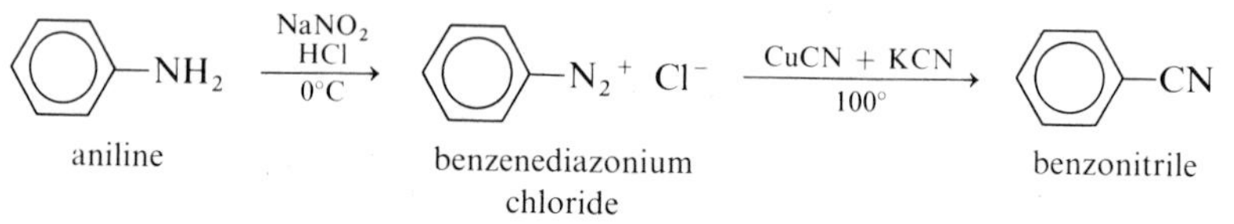

aniline

benzenediazonium
chloride

benzonitrile

D.   Reactions of nitriles

Hydrolysis.   Nitriles are included as carboxylic acid derivatives because their hydrolysis yields carboxylic acids. The hydrolysis of a nitrile may be carried out by heating with either dilute acid or base.

$$CH_3CH_2CH_2CH_2CN + 2\,H_2O + H^+ \xrightarrow{\ heat\ } CH_3CH_2CH_2CH_2CO_2H + NH_4^+$$

pentanenitrile

pentanoic acid
85%

In *acidic hydrolysis,* the weakly basic nitrogen is protonated and then water attacks the electropositive carbon atom. The reaction goes through an amide, which is further hydrolyzed to the carboxylic acid and ammonia. Because the ammonia undergoes reaction with hydrogen ions, an excess of acid must be used.

In acid:

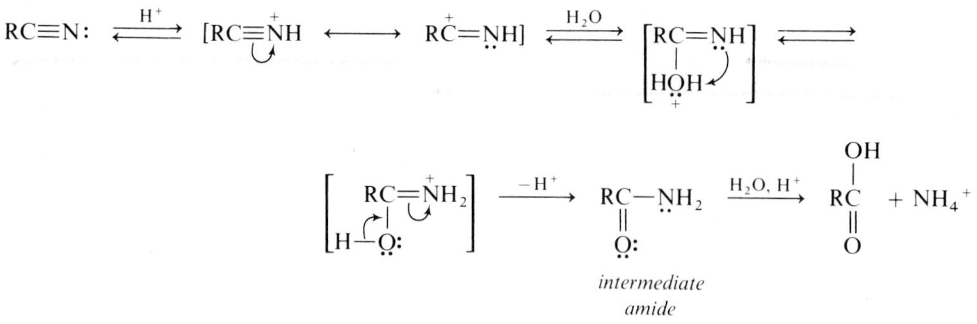

intermediate
amide

Alkaline hydrolysis occurs by nucleophilic attack on the partially positive carbon of the nitrile group. The reaction again results in an amide, which is further hydrolyzed to the carboxylate and ammonia. The free acid is obtained when the solution is acidified.

In base:

$$RC{\equiv}N: + \ddot{O}H^- \overset{heat}{\rightleftharpoons} \left[RC{=}\ddot{N}:^- \atop \overset{|}{:}\underset{}{\ddot{O}H} \right] \xrightarrow[-OH^-]{H_2O} RC{=}\ddot{N}H \rightleftharpoons \checkmark$$

$$\underset{\overset{|}{:\ddot{O}}}{RC}{-}\ddot{N}H_2 \quad \text{(intermediate amide)}$$

$$RC{-}\ddot{N}H_2 \xrightarrow{OH^-} \underset{\overset{||}{O}}{RC}^{O^-} + NH_3$$

intermediate
amide

$$\xrightarrow{H^+} RCO_2H$$

Reduction. Nitriles can be reduced to *primary amines* of the type RCH_2NH_2 either by catalytic hydrogenation or by lithium aluminum hydride.

General: $RC{\equiv}N \xrightarrow{[H]} RCH_2NH_2$

$$C_6H_5{-}CH_2C{\equiv}N \xrightarrow[140°]{2H_2,\ Ni} \left[C_6H_5{-}CH_2CH{=}NH \right] \longrightarrow C_6H_5{-}CH_2CH_2NH_2$$

phenylacetonitrile   (2-phenylethyl)amine
                                                                    70%

$$CH_3CH_2CH_2C{\equiv}N \xrightarrow[(2)\ H_2O]{(1)\ LiAlH_4} CH_3CH_2CH_2CH_2NH_2$$

butanenitrile   *n*-butylamine
                                                     85%

SECTION 13.13.

Use of Carboxylic Acid Derivatives in Synthesis

Carboxylic acids and their derivatives are all synthetically interconvertible. However, of the carboxylic acid derivatives, the acid halides and anhydrides are probably the most versatile because they are more reactive than other carbonyl compounds. Either of these two reactants can be used to synthesize hindered esters or phenyl esters, which cannot be prepared in good yield by heating RCO_2H and $R'OH$ with an acidic catalyst because of an unfavorable equilibrium. (Suggest a reason why the reactions of acid halides or anhydrides with $R'OH$ are essentially *irreversible*.)

$$\left. \begin{array}{c} \overset{O}{\overset{||}{RC}}{-}Cl \\ \text{or} \\ \overset{O\quad O}{\overset{||\quad ||}{RC}}{-}OCR \end{array} \right\} + R'OH \longrightarrow \overset{O}{\overset{||}{RC}}OR' \quad \text{where R and R' can be hindered or aryl}$$

These two derivatives are also the most useful reagents for making *N*-substituted amides.

$$
\left.
\begin{array}{c}
\overset{\displaystyle O}{\overset{\displaystyle \|}{RC}}-Cl \\[2mm]
\text{or} \\[2mm]
\overset{\displaystyle O}{\overset{\displaystyle \|}{RC}}\ \overset{\displaystyle O}{\overset{\displaystyle \|}{-OCR}}
\end{array}
\right\}
+ R'_2NH \longrightarrow \overset{\displaystyle O}{\overset{\displaystyle \|}{RCNR'_2}}
$$

In addition, the reduction of an acid chloride with $LiAlH(OR)_3$ affords one of the few routes to aldehydes. (Can you think of one other?)

Although esters are not as reactive as acid chlorides or anhydrides, they are useful in the synthesis of alcohols (by reduction or by Grignard reactions) and are valuable starting materials in the synthesis of complex molecules. We will discuss these reactions of esters in Chapter 14.

The synthesis of nitriles affords one of the most convenient techniques for extending an aliphatic carbon chain by one or for introducing a carboxyl group or an NH_2 group.

$$
RX \xrightarrow{\ CN^-\ } RCN
\begin{array}{l}
\xrightarrow{\ H_2O,\ H^+\ } RCO_2H \\[3mm]
\xrightarrow{\ [H]\ } RCH_2NH_2
\end{array}
$$

As we have mentioned, the reaction of RX and CN^- gives best yields with *primary* alkyl halides. Secondary alkyl halides can also be used, but give lower yields. (What other products can be expected?)

The preparations and reactions of the carboxylic acid derivatives are summarized in Tables 13.6 and 13.7.

STUDY PROBLEM

13.31. Suggest synthetic routes to the following compounds:

(a) ⬡—$\overset{\displaystyle O}{\overset{\displaystyle \|}{C}}N(CH_3)_2$ from a carboxylic acid

(b) 2-methyl-4-propyl-4-heptanol   from compounds containing six or fewer carbon atoms

(c) $CH_3CH_2\overset{\displaystyle O}{\overset{\displaystyle \|}{C}}NH(CH_2)_3CH_3$ from 1-propanol and *no other organic reagents*

TABLE 13.6. Summary of laboratory syntheses of carboxylic acid derivatives

Reaction	Section reference

Acid chlorides:

$$RCO_2H + SOCl_2 \quad \text{or} \quad PCl_3 \longrightarrow \underset{\text{O}}{\overset{\overset{\text{O}}{\|}}{R\ddot{C}Cl}} \qquad 13.3B$$

$$RCO_2H + SOCl_2 \quad \text{or} \quad PCl_3 \longrightarrow \overset{O}{\overset{\|}{RCCl}} \qquad \qquad 13.3B$$

Acid anhydrides:

$$\overset{O}{\overset{\|}{RCCl}} + \overset{O}{\overset{\|}{{}^-OCR'}} \longrightarrow \overset{O\ O}{\overset{\|\ \|}{RCOCR'}} \qquad 13.4B$$

$$RCO_2H + \text{excess}\ \overset{O}{\overset{\|}{(CH_3C)_2O}} \longrightarrow \overset{O\ O}{\overset{\|\ \|}{RCOCR}} \qquad 13.4B$$

$$\underset{n\ =\ 2\ \text{or}\ 3}{HO_2C(CH_2)_nCO_2H} \xrightarrow{\text{heat}} \overset{O=C \diagdown O \diagup C=O}{\underset{(CH_2)_n}{}} \qquad 12.11B$$

Esters:[a]

$$RCO_2H + R'OH \xrightarrow{H^+} RCO_2R' \qquad 12.9$$

$$\overset{O}{\overset{\|}{RCCl}} + R'OH \longrightarrow RCO_2R' \qquad 13.3C$$

$$\overset{O}{\overset{\|}{(RCO)_2O}} + R'OH \longrightarrow RCO_2R' \qquad 13.4C$$

$$RCO_2^- + R'X \longrightarrow RCO_2R'[b] \qquad 12.6$$

$$RCO_2H + CH_2N_2 \longrightarrow RCO_2CH_3 \qquad 13.5B$$

$$RCO_2R' + R''OH \xrightarrow{H^+ \text{ or } {}^-OR''} RCO_2R'' \qquad 13.5C$$

Amides:

$$\overset{O}{\overset{\|}{RCCl}} + HNR'_2 \longrightarrow \overset{O}{\overset{\|}{RCNR'_2}} \qquad 13.3C$$

$$\overset{O}{\overset{\|}{(RCO)_2O}} + HNR'_2 \longrightarrow \overset{O}{\overset{\|}{RCNR'_2}} \qquad 13.4C$$

$$RCO_2R' + NH_3 \longrightarrow \overset{O}{\overset{\|}{RCNH_2}} \qquad 13.5C$$

Nitriles:

$$RX + CN^- \longrightarrow RCN \qquad 13.12C$$

$$ArNH_2 \xrightarrow[0°]{\overset{NaNO_2}{HCl}} ArN_2^+\ Cl^- \xrightarrow[\text{heat}]{CuCN + KCN} ArCN \qquad 10.14,\ 13.12C$$

[a] The syntheses of some complex esters will be discussed in Chapter 14.

[b] For this reaction to be successful, a reactive halide must be used.

TABLE 13.7. Types of compounds that can be obtained from carboxylic acid derivatives

Reaction		Product	Section reference
Acid chlorides:[a]			
$\underset{\overset{\|}{O}}{\overset{\|\|}{R\ddot{C}Cl}} + H_2O \longrightarrow$	RCO_2H	carboxylic acid	13.3C
$\overset{\overset{\|\|}{O}}{RCCl} + R'OH \longrightarrow$	RCO_2R'	ester	13.3C
$\overset{\overset{\|\|}{O}}{RCCl} + R'_2NH \longrightarrow$	$\overset{\overset{\|\|}{O}}{RCNR'_2}$	amide	13.3C
$\overset{\overset{\|\|}{O}}{RCCl} + R'CO_2^- \longrightarrow$	$\overset{\overset{\|\| \;\; \|\|}{O \;\; O}}{RCOCR'}$	anhydride	13.3C
$\overset{\overset{\|\|}{O}}{RCCl} + C_6H_6 \xrightarrow{AlCl_3}$	$\overset{\overset{\|\|}{O}}{RCC_6H_5}$	aryl ketone	10.9E, 13.3C
$\overset{\overset{\|\|}{O}}{RCCl} \xrightarrow[\text{(2) } H_2O,\ H^+]{\text{(1) } 2\,R'MgX}$	$\overset{\overset{\|}{OH}}{RCR'_2}$	3° alcohol	13.3C
$\overset{\overset{\|\|}{O}}{RCCl} + R'_2Cd \;$ or $\; LiCuR'_2 \longrightarrow$	$\overset{\overset{\|\|}{O}}{RCR'}$	ketone	13.3C
$\overset{\overset{\|\|}{O}}{RCCl} \xrightarrow[\text{(2) } H_2O]{\text{(1) } LiAlH(OR')_3}$	$\overset{\overset{\|\|}{O}}{RCH}$	aldehyde	13.3C
$\overset{\overset{\|\|}{O}}{RCH_2CCl} + Cl_2 \longrightarrow$	$\overset{\overset{\|\|}{O}}{RCHClCCl}$	α-chloro acid chloride	13.3C
Acid anhydrides:			
$\overset{\overset{\|\| \;\; \|\|}{O \;\; O}}{RCOCR} + H_2O \longrightarrow$	$2\,RCO_2H$	carboxylic acid	13.4C
$(RCO)_2O + R'OH \longrightarrow$	RCO_2R'	ester	13.4C
$(RCO)_2O + R'_2NH \longrightarrow$	$\overset{\overset{\|\|}{O}}{RCNR'_2}$	amide	13.4C
Esters:[a]			
$RCO_2R' + H_2O \xrightarrow{H^+\text{ or }OH^-}$	RCO_2H	carboxylic acid	13.5C
$RCO_2R' + R''OH \xrightarrow{H^+\text{ or }^-OR''}$	RCO_2R''	ester	13.5C
$RCO_2R' + NH_3 \longrightarrow$	$\overset{\overset{\|\|}{O}}{RCNH_2}$	amide	13.5C

TABLE 13.7. *(continued)*

Reaction	Product	Section reference
$RCO_2R' + [H] \longrightarrow RCH_2OH + HOR'$	**alcohols**	13.5C
$RCO_2R' \xrightarrow[(2)\ H_2O,\ H^+]{(1)\ 2\ R''MgX} \overset{\overset{\textstyle OH}{\textstyle |}}{RCR''_2}$	**3° alcohol**	13.5C

Amides:

Reaction	Product	Section reference
$\overset{\overset{\textstyle O}{\textstyle \|}}{RCNR'_2} + H_2O \xrightarrow{H^+\ or\ OH^-} RCO_2H + HNR'_2$	**carboxylic acid and amine**	13.9C
$\overset{\overset{\textstyle O}{\textstyle \|}}{RCNR'_2} + [H] \longrightarrow RCH_2NR'_2$	**amine**	15.5C
$\overset{\overset{\textstyle O}{\textstyle \|}}{RCNH_2} + Br_2 + {}^-OH \longrightarrow RNH_2$	**amine**	15.5C

Nitriles:[a]

Reaction	Product	Section reference
$RCN + H_2O \xrightarrow{H^+\ or\ OH^-} RCO_2H$	**carboxylic acid**	13.12D
$RCN + [H] \longrightarrow RCH_2NH_2$	**amine**	13.12D

---

[a] Acid chlorides, esters, and nitriles may be used to synthesize more-complex compounds. Some of these reactions will be discussed in Chapter 14.

Summary

The derivatives of carboxylic acids are usually prepared from the carboxylic acids themselves or from other more reactive derivatives, as shown in Table 13.6.

The reactions of the various carboxylic acid derivatives with nucleophiles are similar to one another. Differences arise from differences in reactivity of the various derivatives.

$$
\left.
\begin{array}{l}
\overset{\overset{\displaystyle O}{\|}}{RC}-Cl \\[2ex]
\overset{\overset{\displaystyle O}{\|}}{RC}-O_2CR \\[2ex]
\overset{\overset{\displaystyle O}{\|}}{RC}-OR \\[2ex]
\overset{\overset{\displaystyle O}{\|}}{RC}-NH_2 \\[2ex]
RCN
\end{array}
\right\}
\xrightarrow[\text{H}^+ \text{ or OH}^-]{\text{H}_2\text{O}}
\quad \overset{\overset{\displaystyle O}{\|}}{RC}-OH \quad \textit{a carboxylic acid}
$$

increasing reactivity (leftward arrow)

$$
\left.
\begin{array}{l}
RCOCl \\
(RCO)_2O \\
RCO_2R
\end{array}
\right\}
\xrightarrow{\text{NH}_3}
\quad \overset{\overset{\displaystyle O}{\|}}{RC}-NH_2 \quad \textit{an amide}
$$

increasing reactivity

$$
\left.
\begin{array}{l}
RCOCl \\
(RCO)_2O \\
RCO_2R
\end{array}
\right\}
\xrightarrow{\text{R}'\text{OH}}
\quad \overset{\overset{\displaystyle O}{\|}}{RC}-OR' \quad \textit{an ester}
$$

increasing reactivity

In addition to these reactions, the more reactive acid halides undergo Friedel–Crafts reactions with aromatic compounds and also undergo reaction with cadmium reagents or lithium dialkylcuprates to yield ketones.

$$
\bigcirc + \overset{\overset{\displaystyle O}{\|}}{RC}Cl \xrightarrow{\text{AlCl}_3} \bigcirc\!-\overset{\overset{\displaystyle O}{\|}}{C}R \quad \textit{an aryl ketone}
$$

$$
\overset{\overset{\displaystyle O}{\|}}{RC}Cl \xrightarrow[\text{(2) H}_2\text{O, H}^+]{\text{(1) R}_2'\text{Cd}} \overset{\overset{\displaystyle O}{\|}}{RC}R' \quad \textit{a ketone}
$$

Esters undergo reaction with Grignard reagents to yield tertiary alcohols.

$$
RCO_2R \xrightarrow[\text{(2) H}_2\text{O, H}^+]{\text{(1) R}'\text{MgX}} \overset{\overset{\displaystyle OH}{|}}{RC}R'_2 \quad \textit{a 3}°\textit{ alcohol}
$$

All the derivatives may be reduced by catalytic hydrogenation or by LAH; the reductions of esters and nitriles are shown:

$$RCO_2R' \xrightarrow{[H]} RCH_2OH + HOR' \quad \textit{alcohols}$$

$$RCN \xrightarrow{[H]} RCH_2NH_2 \quad \textit{a 1° amine}$$

STUDY PROBLEMS

13.32. Name the following compounds:

(a) 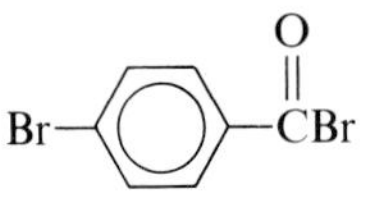 Br—⟨C₆H₄⟩—C(=O)Br

(b) $CH_3CH_2COCCH_2CH_2CH_2CH_3$ (with two C=O groups)

(c) $C_6H_5CN(CH_2CH_3)_2$ (with C=O)

(d) $CH_3CH_2CH_2CH_2CH_2CN$

(e) 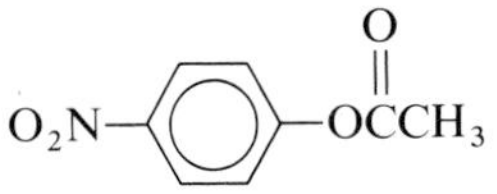 O_2N—⟨C₆H₄⟩—$OCCH_3$ (with C=O)

(f) 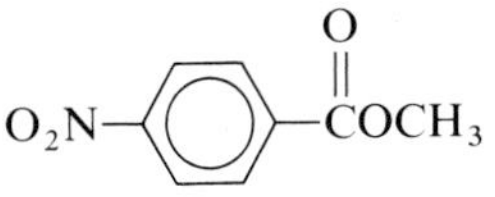 O_2N—⟨C₆H₄⟩—$COCH_3$ (with C=O)

13.33. Give structures for the following compounds:

(a) 5-hydroxy-2-octenoic acid lactone
(b) *N,N*-diethylpropanamide
(c) *N,N'*-dimethylurea   (the *N* and *N'* refer to two different nitrogens)
(d) 2-methylpentanenitrile
(e) ethyl 2-aminopropanoate
(f) β-chlorobutyronitrile
(g) benzoic formic anhydride

13.34. List the following compounds in order of increasing bond moment of the indicated bond:

(a) $CH_3C(=O){-}OCH_3$ (b) $CH_3C(=O){-}NH_2$ (c) $CH_3C(=O){-}Cl$

13.35. What would be the products of the reaction of each of the following reagents with acetic anhydride?
(a) cyclohexanol
(b) *p*-bromophenol
(c) piperidine (page 617)
(d) sodium ethoxide in ethanol
(e) excess aqueous NaOH

13.36. Give the organic product of the reaction with benzoyl chloride of each of the reagents in Problem 13.35.

13.37. Predict the organic products and the relative rates of reaction for each of the following acid derivatives toward alkaline hydrolysis in $1N$ NaOH:

$$\text{(a) } C_6H_5CO_2CH_2CH_3 \quad \text{(b) } C_6H_5\overset{\overset{\text{O}}{||}}{C}Cl \quad \text{(c) } C_6H_5\overset{\overset{\text{O}}{||}}{C}O\overset{\overset{\text{O}}{||}}{C}CH_2CH_3$$

13.38. Predict the products of acidic hydrolysis of *acetaminophen*, the active analgesic (pain reliever) and antipyretic (fever reducer) in some popular headache remedies.

$$HO-\overset{}{\underset{}{\bigcirc}}-NH\overset{\overset{\text{O}}{||}}{C}CH_3$$

acetaminophen

13.39. Complete the following equations:

(a) $CH_3CH_2\overset{\overset{\text{O}}{||}}{C}HCBr$ with CH_3 $+ \bigcirc -CO_2^- \longrightarrow$

(b) $\bigcirc -CO_2H + PBr_3 \longrightarrow$

(c) $\bigcirc -CO_2H + \bigcirc -OH \xrightarrow[\text{h}\cdot\text{t}]{H_2SO_4}$

(d) $\bigcirc -CO_2CH_2CH_3 \xrightarrow[\text{(2) } H_2O]{\text{(1) LiAlH}_4}$

(e) $CH_3CH_2CH_2CO_2H + SOCl_2 \longrightarrow$

(f) $CH_3CH_2O\overset{\overset{\text{O}}{||}}{C}OCH_2CH_3 + D_2O \xrightarrow{D^+}$

(g) $CH_3CH_2\overset{\overset{\text{O}}{||}}{C}Cl + (CH_2=CHCH_2)_2Cd \longrightarrow$

(h) $CH_3CH_2\overset{\overset{\text{O}}{||}}{C}Cl + LiCu[CH(CH_3)_2]_2 \longrightarrow$

13.40. Predict the saponification products (aqueous $NaOH$) of the following compounds:

(a) [structure: phenyl-CHClOCCH$_3$ with C=O]

(b) [structure: phthalic anhydride with CO_2CH_3 substituent]

(c) nepetalactone (page 635)

13.41. Show by equations how you would convert ethyl acetate to: (a) acetic acid; (b) ethanol; (c) *t*-butyl alcohol; (d) acetophenone (page 452); (e) sodium acetate; (f) *N*-methylacetamide.

13.42. Would acid hydrolysis or saponification be the method of choice for the conversion of 2-butenyl acetate to acetic acid and 2-buten-1-ol? Why?

13.43. The following ketones can be prepared by the reaction of an acid chloride with either a cadmium reagent or a lithium dialkylcuprate. There may be more than one cadmium reagent or cuprate that could be used. Write the equations showing the various methods of preparing these ketones.

(a) $C_6H_5CH_2CCH_2CH_3$

(b) $(CH_3)_2CHCH_2C-CC_6H_5$ with CH_3 and H_3C substituents

13.44. *Phosgene*, $ClCCl$, is a toxic gas that was used as a war gas in World War I. Phosgene undergoes the usual reactions of an acid chloride, but can undergo these reactions twice. Predict the product of the reaction of phosgene with each of the following reagents: (a) excess H_2O; (b) excess ethanol; (c) 1.0 equivalent ethanol; (d) 1.0 equivalent ethanol, followed by 2.0 equivalents NH_3.

13.45. Predict the products of the reactions of (a) acetic anhydride with (*R*)-2-octylamine, and (b) benzoic anhydride with $(CH_3)_2CH^{18}OH$.

13.46. When 2,4,6-trimethylbenzoic acid is heated with ethanol and a trace of H_2SO_4, no ester is obtained. Suggest a reason.

13.47. *Kodel* is the name of a fiber formed by the following transesterification reaction:

$$CH_3OC-\bigcirc-COCH_3 + HOCH_2-\bigcirc-CH_2OH \longrightarrow Kodel$$

dimethyl terephthalate   1,4-di(hydroxymethyl)cyclohexane

(a) What is the structure of Kodel?
(b) How could the starting diol be prepared from dimethyl terephthalate?

13.48. Write the equation for the preparation of each of the following compounds from acetic anhydride and other appropriate reagents:

(a)  $CH_3\overset{\overset{\displaystyle O}{\displaystyle \|}}{C}NHCH(CH_3)_2$

(b)

CH_3CO_2 ⟍ ╱ O_2CCH_3

CH_3CO_2 ⟋ O ⟍ $CH_2O_2CCH_3$

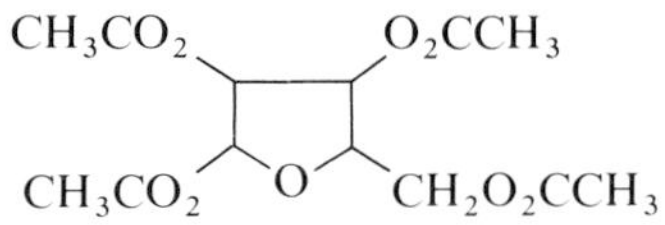

13.49. What is the structure of nylon 44? Suggest a laboratory synthesis.

13.50. An amide, like an aldehyde or ketone, is capable of tautomerism:

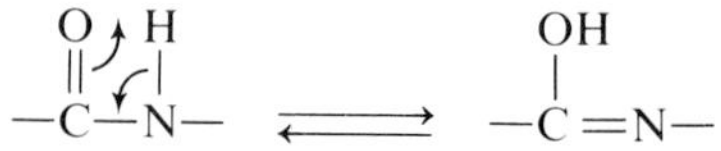

$$-\overset{\overset{\displaystyle O}{\displaystyle \|}}{C}-\overset{\overset{\displaystyle H}{\displaystyle |}}{N}- \quad \rightleftharpoons \quad -\overset{\overset{\displaystyle OH}{\displaystyle |}}{C}=N-$$

amide form   *"enol" form*

Show all tautomeric enol forms for each of the following cyclic structures, all of which are found in nucleic acids:

(a)

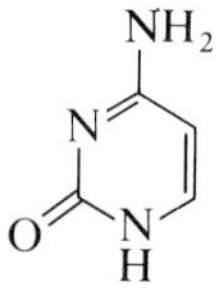

cytosine

(b)

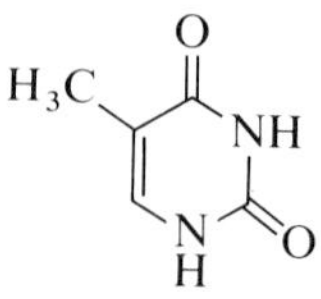

thymine

(c)

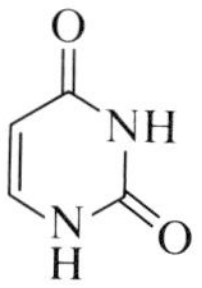

uracil

(d)

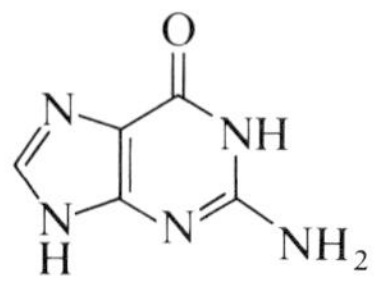

guanine

13.51. Suggest a practical method for the synthesis of:

(a)   hexanoic acid from 1-bromopentane
(b)   2-hydroxyhexanoic acid from pentanal
(c)   β-phenylethylamine ($C_6H_5CH_2CH_2NH_2$) from benzyl bromide
(d)   *p*-tolyl methyl ketone from toluene
(e)   *N*-cyclohexylacetamide from acetic acid

13.52. Propose chemical tests for distinguishing between the following pairs:

(a)   benzoic acid   and   methyl benzoate
(b)   ethyl benzoate   and   *N*-ethylbenzamide
(c)   benzoic acid   and   benzoyl chloride

13.53. Predict the products of catalytic hydrogenation (high temperature and pressure) of:

(a) 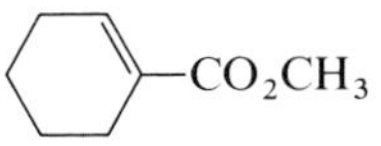 [benzene ring]—CO_2CH_3 (b) 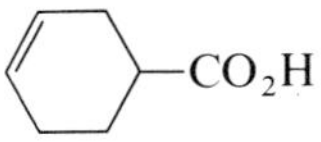 [cyclohexene ring]—CO_2H

13.54. Give the structure of the polyamide formed from the polymerization of methyl 2-aminoacetate.

13.55. Complete the following equations:

(a) (R)-$C_6H_5CO_2\overset{\underset{\displaystyle |}{CH_3}}{CH}CH_2CH_2CH_3 + OH^- \xrightarrow[\text{heat}]{H_2O}$

(b) $(CH_3)_3C\overset{\overset{\displaystyle O}{\|}}{C}NH_2 + D_2O \xrightarrow[\text{heat}]{OD^-}$

(c) 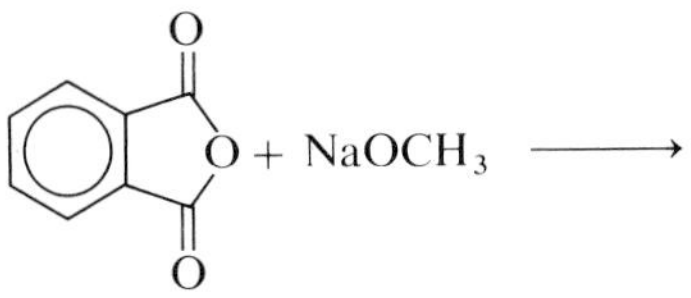   [phthalic anhydride structure] $+ NaOCH_3 \longrightarrow$

(d) $CH_3CH_2CH_2\overset{\overset{\displaystyle O}{\|}}{C}Cl + (CH_3CH_2CH_2)_2Cd \longrightarrow$

(e) 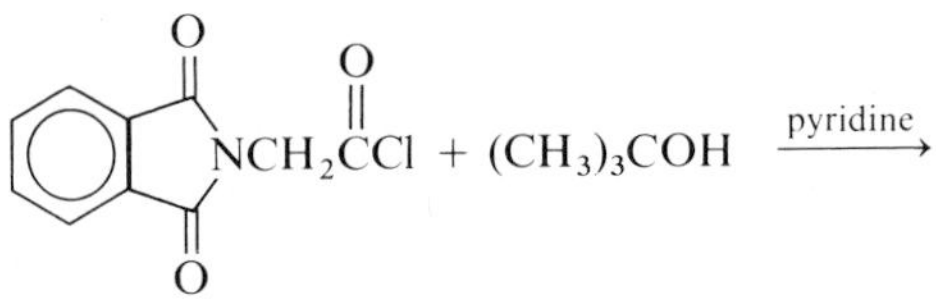   [phthalimide-N structure] $N CH_2\overset{\overset{\displaystyle O}{\|}}{C}Cl + (CH_3)_3COH \xrightarrow{\text{pyridine}}$

13.56. Show how the following compounds can be prepared from compounds containing four or fewer carbon atoms: (a) acetic butanoic anhydride;   (b) 2-hexanone;   (c) *t*-butyl pentanoate;   (d) 4-propyl-4-heptanol.

13.57. Give the products of each of the following reactions and the mechanisms for their formation:

(a) $CH_3\overset{\underset{\displaystyle |}{OH}}{CH}CO_2CH_3 + \text{dil. HCl} \xrightarrow{\text{heat}}$

(b) $(CH_3)_2CHCO_2CH_3 + \text{dil. NaOH} \xrightarrow{\text{heat}}$

13.58. The reaction of phosgene ($Cl_2C{=}O$) with 1,2-ethanediol can lead to two products: one is cyclic and one is polymeric. What are the structures of these products?

13.59. γ-Butyrolactone (page 635) is warmed with a dilute solution of HCl in $H_2{}^{18}O$. What lactones could be isolated from the product mixture?

13.60. Urea is hydrolyzed by aqueous HCl to yield NH_4Cl and CO_2. Suggest a mechanism.

13.61. Compound A (C_3H_5ON) gives a negative Tollens test, but decolorizes a solution of Br_2 in CCl_4. When A is heated with aqueous NaOH, the fumes have a strong ammoniacal odor.

(a) Suggest a structure for A.
(b) How would you confirm your structure assignment by infrared or nmr spectroscopy? (Tell what absorption you would look for.)

13.62. Suggest a mechanism for the following reaction:

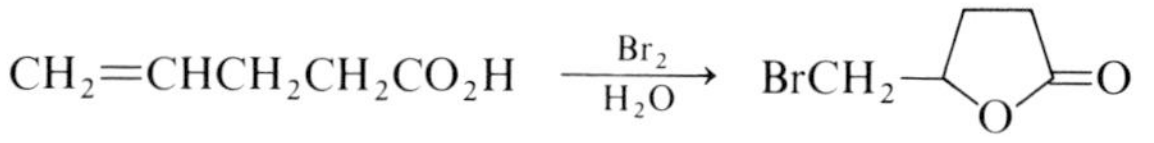

$$CH_2{=}CHCH_2CH_2CO_2H \xrightarrow[\text{H}_2\text{O}]{\text{Br}_2} BrCH_2-\overset{\displaystyle \diagup\!\!\diagdown}{\underset{O}{\bigcirc}}{=}O$$

13.63. *o*-Phthalic acid (page 591) does not give an acid chloride when treated with $SOCl_2$. Instead, the half-acid chloride undergoes an intramolecular reaction to yield another product. What is this product and how is it formed?

13.64. When Compound **A** is subjected to acid hydrolysis, no ^{18}O is found in the product acetic acid. However, when Compound **B** is subjected to acid hydrolysis, some acetic acid containing ^{18}O is isolated. Explain these observations.

$$\underset{\textstyle \text{A}}{\overset{\textstyle O}{\overset{\|}{CH_3C^{18}OCH_2CH_3}}} \qquad \underset{\textstyle \text{B}}{\overset{\textstyle O}{\overset{\|}{CH_3C^{18}OC(CH_3)_3}}}$$

13.65. The acid $\overset{O}{\overset{\|}{H}C}CH_2CH_2CO_2H$ was treated with HCN. The product has the molecular formula $C_5H_5O_2N$. What is this product? Suggest a route for its formation.

13.66. Suggest syntheses for the following compounds:

(a) butanal   from compounds containing three or fewer carbons
(b) $NCCH_2CO_2CH_2CH_3$ from acetic acid and other appropriate reagents
(c) the local anesthetic *benzocaine* (ethyl *p*-aminobenzoate)   from benzene

(d) 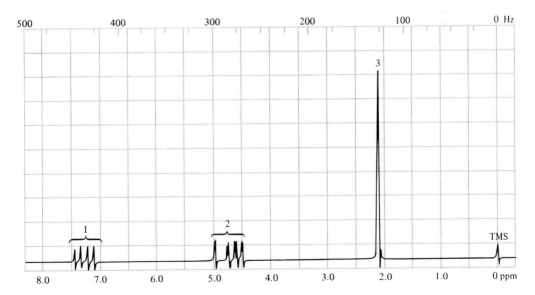 from maleic acid (page 592) and other appropriate reagents

(e) from cyclohexylmethanol and no other organic reactants

13.67. The nmr spectrum of Compound A is shown in Figure 13.11. When A is heated with aqueous acid, the products are acetic acid and acetaldehyde. What is the structure of A?

13.68. Compound A ($C_{10}H_{18}O_3$) was treated with dilute aqueous acid to yield the single compound B. When A was heated with ethanol and a trace of H_2SO_4, C was obtained as the only product. The infrared spectra of A and B and the nmr spectrum of C are shown in Figure 13.12. What are the structures of A, B, and C?

FIGURE 13.11. Nmr spectrum for Problem 13.67.

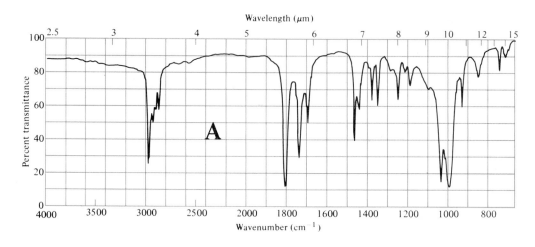

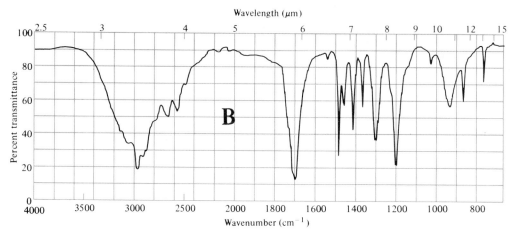

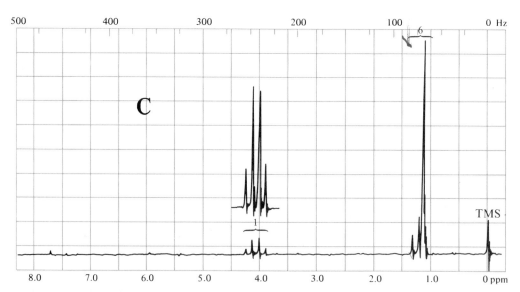

FIGURE 13.12. Spectra for Problem 13.68.

CHAPTER 14

Enolates and Carbanions: Building Blocks for Organic Synthesis

Nucleophilic reagents undergo reaction with compounds that contain partially positive carbon atoms.

$$\text{Nu}^- + \overset{\delta+}{\text{R}}-\overset{\delta-}{\text{X}} \longrightarrow \text{Nu}-\text{R} + \text{X}^-$$

$$\text{Nu}^- + \text{R}-\underset{\delta+}{\overset{\overset{\displaystyle O^{\delta-}}{\|}}{\text{C}}}-\text{R} \longrightarrow \text{R}-\underset{\underset{\displaystyle \text{Nu}}{|}}{\overset{\overset{\displaystyle O^-}{|}}{\text{C}}}-\text{R}$$

Reagents that contain **nucleophilic carbon atoms,** carbon atoms with carbanion character, also attack partially positive carbon atoms. As an example, a **Grignard reagent**, which has a partially negative carbon atom, attacks carbonyl groups.

$$\text{CH}_3\overset{\delta-}{\text{C}}\text{H}_2\overset{\delta+}{\text{M}}\text{gBr} + \text{CH}_3\overset{\overset{\displaystyle \ddot{\text{O}}:}{\|}}{\text{C}}\text{CH}_3 \longrightarrow \text{CH}_3\underset{\underset{\displaystyle \text{CH}_2\text{CH}_3}{|}}{\overset{\overset{\displaystyle :\ddot{\text{O}}:^- \ ^+\text{MgBr}}{|}}{\text{C}}}\text{CH}_3$$

The attack of one carbon upon another results in a new carbon–carbon bond. Reagents like Grignard reagents with nucleophilic carbon atoms allow a chemist to synthesize compounds with complex carbon skeletons from simple compounds.

Grignard reagents are but one of many reagents with nucleophilic carbons that are available to the organic chemist. Another versatile class of reagents for building complex molecules are the **enolates.** Recall from Section 11.16 that a hydrogen α to a carbonyl group is acidic and can be removed by a strong base. The resulting enolate anion contains a partially negative carbon atom.

$$
\begin{array}{c}
\text{CH}_3\text{CH}_2\text{OC} \\
\quad\quad\quad\text{CH}_2 \\
\text{CH}_3\text{CH}_2\text{OC}
\end{array}
+ \ \text{Na}^+ \ {}^-\text{OCH}_2\text{CH}_3 \ \rightleftharpoons \
\begin{array}{c}
\text{CH}_3\text{CH}_2\text{OC} \\
\quad\quad\quad\text{CH}^- \ \text{Na}^+ \\
\text{CH}_3\text{CH}_2\text{OC}
\end{array}
+ \ \text{CH}_3\text{CH}_2\text{OH}
$$

diethyl malonate   sodium ethoxide   diethyl sodium malonate   ethanol

an enolate

Enolates can enter into a number of organic reactions. One is S_N2 displacement of a halide ion from a methyl or primary alkyl halide. This reaction is called an **alkylation reaction** because an alkyl group becomes attached to a carbon of the enolate. A typical alkylation reaction is the reaction of the enolate of diethyl malonate with an alkyl halide:

$$
\begin{array}{c}
\text{CH}_3\text{CH}_2\text{OC} \\
\quad\quad\text{C}: \\
\text{CH}_3\text{CH}_2\text{OC} \quad \text{H}
\end{array}
+ \
\begin{array}{c}
\text{CH}_2\text{—Br:} \\
\text{CH}_2 \\
\text{CH}_2 \\
\text{CH}_3
\end{array}
\ \xrightarrow[\substack{\text{CH}_3\text{CH}_2\text{OH} \\ \text{as solvent}}]{S_N2} \
\begin{array}{c}
\text{CH}_3\text{CH}_2\text{OC} \\
\quad\quad\quad\text{CH(CH}_2)_3\text{CH}_3 \ + \ :\ddot{\text{Br}}:^- \\
\text{CH}_3\text{CH}_2\text{OC}
\end{array}
$$

the enolate ion   1-bromobutane   diethyl *n*-butylmalonate

85%

An alkylation reaction is just one technique for the synthesis of complex carbon skeletons using enolates as nucleophiles. However, before we begin a discussion of the synthetic utility of enolates, let us consider their formation.

SECTION 14.1.

Acidity of the Alpha Hydrogen

A hydrogen that is α to a carbonyl group is acidic primarily because of resonance-stabilization of the product anion.

$$
\text{CH}_3\ddot{\text{O}}:^- \ + \ \text{CH}_2\text{CCH}_3 \ \rightleftharpoons \ \text{CH}_3\text{OH} \ + \ \left[\ {}^-\ddot{\text{CH}}_2\text{—}\overset{\ddot{\text{O}}:}{\text{CCH}}_3 \ \longleftrightarrow \ \text{CH}_2\text{=}\overset{:\ddot{\text{O}}:^-}{\text{CCH}}_3 \ \right]
$$

acetone   *resonance structures for the enolate ion of acetone*

Because of the resonance-stabilization in the enolate ion of acetone, acetone is a far stronger acid than an alkane. (However, acetone is only one ten-thousandth as strong an acid as ethanol.)

$$
\begin{array}{ccc}
\overset{\text{H}}{\underset{|}{\text{CH}_3\text{CH}_2}} & \overset{\text{H} \quad \text{O}}{\underset{|}{\text{CH}_2\text{CCH}_3}} & \text{CH}_3\text{CH}_2\text{O—H} \\
\text{ethane} & \text{acetone} & \text{ethanol} \\
\text{p}K_a = 43 & \text{p}K_a = 20 & \text{p}K_a = 16
\end{array}
$$

The alpha hydrogen of an ester is less easily removed than that of an aldehyde or ketone because the carbonyl oxygen is already participating in delocalization. In an ester, the carbonyl oxygen already carries a partial negative charge from delocalization of the electrons on the alkoxyl oxygen. Therefore, the carbonyl group is somewhat less able to delocalize the anionic negative charge of the enolate. The resonance structures for a typical ester, ethyl acetate, follow:

$$CH_3-C-\overset{\ddot{O}:}{\underset{}{\overset{\|}{}}}-\overset{..}{\underset{..}{O}}CH_2CH_3 \longleftrightarrow CH_3-C=\overset{+}{O}CH_2CH_3$$

major contributor

Because the α hydrogen of an ester is less easily removed, a simple ester is less acidic than a ketone.

$$\overset{H}{\underset{}{\overset{|}{}}}\ \overset{O}{\underset{}{\overset{\|}{}}}$$
$$CH_2COCH_2CH_3$$

ethyl acetate
$pK_a = 25$

A hydrogen alpha to a single carbonyl group is less acidic than that of an alcohol; therefore, treatment of an aldehyde, ketone, or ester with an alkoxide results in a very low concentration of enolate ions. If we want a reasonably high concentration of the enolate, we must use a much stronger base, such as $NaNH_2$ or NaH.

$$CH_3COCH_2CH_3 + {}^-OCH_2CH_3 \longleftrightarrow {}^-CH_2COCH_2CH_3 + HOCH_2CH_3$$

ethyl acetate   *not favored*   ethanol
$pK_a = 25$ $pK_a = 16$

$$CH_3COCH_2CH_3 + {}^-NH_2 \longleftrightarrow {}^-CH_2COCH_2CH_3 + NH_3$$

$pK_a = 25$ *favored* $pK_a = 35$

If a hydrogen is alpha to two carbonyl groups, the negative charge on the anion can be delocalized by both C=O groups. Such a hydrogen is *more acidic than that of an alcohol.* A high concentration of enolate may be obtained by treatment of a β-dicarbonyl compound with an alkoxide. Table 14.1 (page 666) lists the pK_a values for some compounds with hydrogens alpha to one and two carbonyl groups.

$$CH_3CCH_2CCH_3 + {}^-OCH_3 \longleftrightarrow CH_3CCHCCH_3 + CH_3OH$$

2,4-pentanedione   *favored* $pK_a = 15.5$
$pK_a = 9$

Resonance structures of the enolate ion:

$$CH_3C-CHCCH_3 \longleftrightarrow CH_3C=CH-CCH_3 \longleftrightarrow CH_3CCH=CCH_3$$

TABLE 14.1. pK_a values for some carbonyl compounds

Structure	*Name*	*Approx.* pK_a
$\underset{\underset{H}{\mid}}{CH_3\overset{\overset{O}{\parallel}}{C}CH}\overset{\overset{O}{\parallel}}{C}CH_3$	2,4-pentanedione (acetoacetone)	9
$\underset{\underset{H}{\mid}}{CH_3\overset{\overset{O}{\parallel}}{C}CH}\overset{\overset{O}{\parallel}}{C}OCH_2CH_3$	ethyl acetoacetate (acetoacetic ester)	11
$\underset{R \quad H}{CH_3\overset{\overset{O}{\parallel}}{C}C}\overset{\overset{O}{\parallel}}{C}OCH_2CH_3$	an alkylacetoacetic ester	13
$\underset{\underset{H}{\mid}}{CH_3CH_2O\overset{\overset{O}{\parallel}}{C}CH}\overset{\overset{O}{\parallel}}{C}OCH_2CH_3$	diethyl malonate (malonic ester)	13
$\underset{H}{\overset{\mid}{CH_2}}\overset{\overset{O}{\parallel}}{C}CH_3$	acetone	20
$\underset{H}{\overset{\mid}{CH_2}}\overset{\overset{O}{\parallel}}{C}OCH_2CH_3$	ethyl acetate	25

Not only a carbonyl group, but <u>any strongly electron-withdrawing group, enhances the acidity of an alpha hydrogen.</u> Some other compounds that are more acidic than ethanol are:

$$\underset{H}{\overset{\mid}{CH_3CH_2O\overset{\overset{O}{\parallel}}{C}CHCN}} \qquad \underset{H}{\overset{\mid}{NCCHCN}} \qquad \underset{H}{\overset{\mid}{CH_2NO_2}}$$

STUDY PROBLEMS

14.1.   Which of the indicated hydrogens are acidic?

(a)   CH$_3$CH=CHCHO   (b)   C$_6$H$_5$CHO   (c)   CH$_3$CH$_2$CO$_2$CH$_2$CH$_3$

14.2.   Write an equation to show the acid–base reaction, if any, of each of the
following compounds with sodium ethoxide:

(a)   CH$_3$CH$_2$CHO

(b)   (CH$_3$)$_2$CHCO$_2$CH$_2$CH$_3$

(c)   (CH$_3$)$_3$CCO$_2$CH$_2$CH$_3$

(d)   O$_2$N—⟨NO$_2$ / ⟩—CH$_2$CH$_3$

(e)   C$_6$H$_5$CH$_2$NO$_2$

(f)   CH$_2$(CO$_2$H)$_2$

SECTION 14.2.

Alkylation of Malonic Ester

One of the more powerful tools at the disposal of the synthetic organic chemist
is the reaction of an enolate with an alkyl halide. In this section, we will emphasize
the *alkylation of malonic ester.* In general, the end products from alkylation of
malonic ester are *α-substituted acetic acids,* In the following example, the R group
comes from RX. (In the discussions that follow, we will use —C$_2$H$_5$ to represent
the ethyl group.)

General:

$$CH_2(CO_2C_2H_5)_2 \xrightarrow[\text{(2) RX}]{\text{(1) Na}^+ \ ^-OC_2H_5} RCH(CO_2C_2H_5)_2 \xrightarrow[\substack{\text{heat} \\ -CO_2}]{\text{H}_2\text{O, H}^+} RCH_2\overset{\displaystyle O}{\overset{\|}{C}}OH$$

diethyl malonate   *a diethyl*   *an α-substituted*
(malonic ester)   *alkylmalonate*   *acetic acid*

from RX

$$CH_2(CO_2C_2H_5)_2 \xrightarrow[\text{(2) CH}_3\text{CH}_2\text{Br}]{\text{(1) Na}^+ \ ^-OC_2H_5} CH_3CH_2CH(CO_2C_2H_5)_2 \xrightarrow[\substack{115° \\ -CO_2}]{\text{H}_2\text{O, H}^+} CH_3CH_2CH_2CO_2H$$

diethyl malonate   diethyl ethylmalonate   butanoic acid
                                                                  80%

A malonic ester alkylation consists of four separate reactions: (1) preparation of the enolate; (2) the actual alkylation; and (3) hydrolysis of the ester,
followed by (4) decarboxylation of the resulting β-dicarboxylic acid.

Preparation of the enolate:

$$CH_2(CO_2C_2H_5)_2 + Na^+ \ ^-OC_2H_5 \rightleftharpoons Na^+ \ ^-CH(CO_2C_2H_5)_2 + HOC_2H_5$$

Alkylation:

$$CH_3CH_2\!\!-\!\!Br + {}^-CH(CO_2C_2H_5)_2 \longrightarrow CH_3CH_2\!\!-\!\!CH(CO_2C_2H_5)_2 + Br^-$$

Hydrolysis and decarboxylation:

$$CH_3CH_2CH(CO_2C_2H_5)_2 \xrightarrow[\text{heat}]{H_2O,\ H^+} CH_3CH_2CH\!\!\begin{array}{c} {}^{CO_2H} \\ {}_{CO_2H} \end{array} \xrightarrow[\text{heat}]{-CO_2} CH_3CH_2CH_2CO_2H$$

While there are many chemical reactions involved in this sequence, the laboratory procedure is quite simple because all the reactions can occur in the same reaction vessel. One reactant is added after another, and the final product is isolated as the last step. (Sometimes the intermediate alkylated ester is purified prior to hydrolysis and decarboxylation in order to simplify the final purification.) Let us discuss each of the steps in this reaction in more detail.

A.   Formation of the enolate

The enolate of malonic ester is usually prepared by treatment of the ester with sodium ethoxide. Therefore, step 1 is the dissolving of sodium metal in anhydrous ethanol (not the common 95% ethanol). (Why not?) Excess ethanol serves as the solvent for the reaction. Step 2 is the addition of diethyl malonate. The ethoxide ion is a stronger base than the enolate ion; therefore, the acid–base equilibrium lies on the side of the resonance-stabilized enolate anion.

$$Na + C_2H_5OH \longrightarrow Na^+\ {}^-OC_2H_5 + \tfrac{1}{2}H_2\!\uparrow$$

$$\underset{}{C_2H_5O\overset{O}{\overset{\|}{C}}CH_2\overset{O}{\overset{\|}{C}}OC_2H_5} + {}^-OC_2H_5 \;\rightleftharpoons\; \underset{favored}{C_2H_5O\overset{O}{\overset{\|}{C}}\overset{-}{C}H\overset{O}{\overset{\|}{C}}OC_2H_5} + HOC_2H_5$$

STUDY PROBLEM

14.3.   Predict what would occur if a chemist added diethyl malonate to a solution of *sodium methoxide* in *methanol*.

B.   Alkylation

The alkylation reaction is a typical S_N2 displacement by a nucleophile. Methyl and primary alkyl halides give the best yields, with secondary alkyl halides giving lower yields because of competing elimination reactions. (Tertiary alkyl halides give exclusively elimination products, and aryl halides are nonreactive under S_N2 conditions.)

$$(C_2H_5O_2C)_2CH^- + \underset{\underset{CH_3}{|}}{CH_2}\!\!-\!\!Br \xrightarrow{S_N2} (C_2H_5O_2C)_2CHCH_2 + Br^- \atop \underset{CH_3}{|}$$

STUDY PROBLEM

14.4. Predict the products of the following reactions:

(a) $CH_2(CO_2C_2H_5)_2$ $\xrightarrow[\text{(2) } CH_3CH_2CH_2Br]{\text{(1) } Na^+ \ ^-OC_2H_5}$

(b) $CH_3CH(CO_2C_2H_5)_2$ $\xrightarrow[\text{(2) } CH_3CH_2I]{\text{(1) } Na^+ \ ^-OC_2H_5}$

The <u>product of the alkylation still contains an acidic hydrogen</u>:

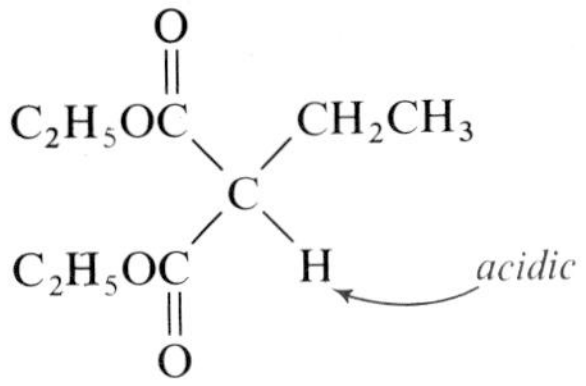

$$
\begin{array}{c}
\quad\quad\quad\quad O \\
\quad\quad\quad\quad \| \\
C_2H_5OC \quad CH_2CH_3 \\
\diagdown\;/ \\
C \\
/\;\diagdown \\
C_2H_5OC \quad H \quad \xleftarrow{\quad} acidic \\
\| \\
O
\end{array}
$$

This second hydrogen can be removed by base, and a *second* R group can be substituted on the malonic ester. This second R group may be the same as, or different from, the first.

$CH_3CH_2CH(CO_2C_2H_5)_2$ $\xrightarrow{Na^+ \ ^-OC_2H_5}$

diethyl ethylmalonate

$$CH_3CH_2\overset{-}{C}(CO_2C_2H_5)_2 \xrightarrow{CH_3I} \quad CH_3CH_2\overset{\overset{\displaystyle CH_3}{|}}{C}(CO_2C_2H_5)_2$$

an enolate   diethyl
                               ethylmethylmalonate

C.   Hydrolysis and decarboxylation

We have mentioned previously that a compound with a carboxyl group beta to a carbonyl group undergoes decarboxylation when heated. The mechanism for decarboxylation was shown in Section 12.11C. If <u>malonic ester</u> (substituted or not) is <u>hydrolyzed in hot acidic solution, a β-diacid is formed</u> and <u>may undergo decarboxylation</u>. (Sometimes decarboxylation does not occur until the diacid is distilled.)

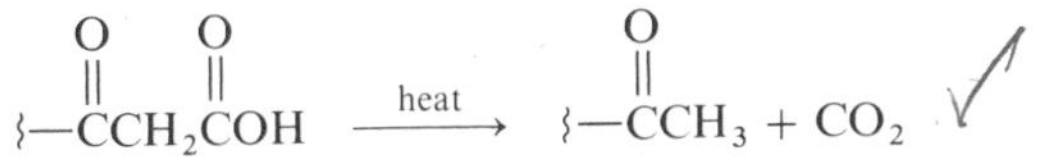

$$
\begin{array}{c}
O \quad O \\
\| \quad\; \| \\
\{-CCH_2COH \xrightarrow{heat} \quad \{-CCH_3 + CO_2 \quad \checkmark \\
\quad\quad\quad\quad\quad\quad\quad\quad\quad\quad \| \\
\quad\quad\quad\quad\quad\quad\quad\quad\quad\quad O
\end{array}
$$

Hydrolysis and decarboxylation:

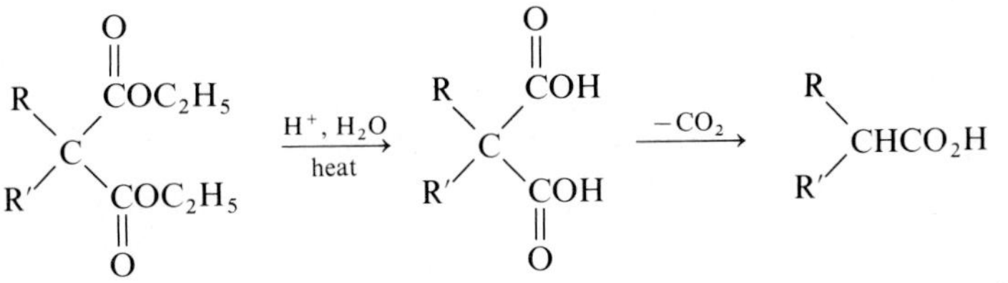

$$
\begin{array}{ccccc}
\quad\quad O & & & & \\
\quad\quad \| & & O & & \\
R \quad COC_2H_5 & & \| & & R \\
\diagdown\;/ & & R \quad COH & & \diagdown \\
C & \xrightarrow[heat]{H^+,\, H_2O} & \diagdown\;/ & \xrightarrow{-CO_2} & CHCO_2H \\
/\;\diagdown & & C & & / \\
R' \quad COC_2H_5 & & /\;\diagdown & & R' \\
\quad\quad \| & & R' \quad COH & & \\
\quad\quad O & & \| & & \\
& & O & &
\end{array}
$$

an α-disubstituted   *an α-disubstituted*
malonic ester   *acetic acid*

What if a chemist does not want a decarboxylation product, but wants a diacid? A diacid may be prepared by *saponification of the diester in base, followed by acidification.* This way, the carboxylic acid itself is not subjected to heat and is less likely to undergo decarboxylation.

Saponification and acidification:

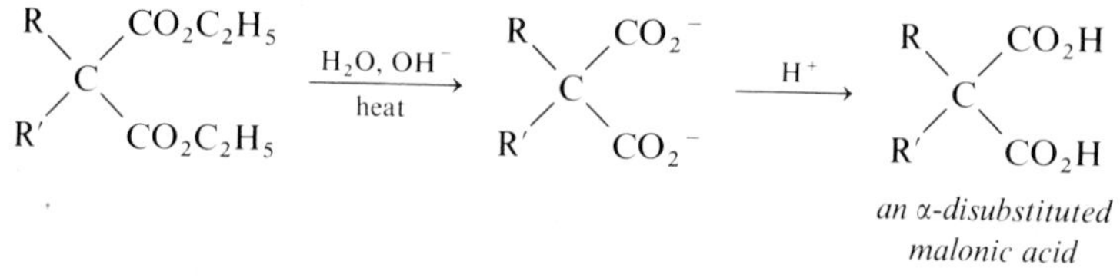

*an α-disubstituted
malonic acid*

STUDY PROBLEMS

14.5. Write equations for the following reactions:

(a) saponification of diethyl *n*-propylmalonate, followed by treatment with cold HCl
(b) acid hydrolysis of diethyl dimethylmalonate

14.6. Give the mechanism for the decarboxylation of methylmalonic acid.

SECTION 14.3.

Alkylation of Acetoacetic Ester

Alkylation reactions are not limited to the enolate of diethyl malonate. Other enolates also undergo S_N2 reaction with methyl or primary alkyl halides to yield alkylated products. Another commonly used enolate is that obtained from ethyl acetoacetate (acetoacetic ester). The end product of alkylation of acetoacetic ester is an *α-substituted acetone*.

General:

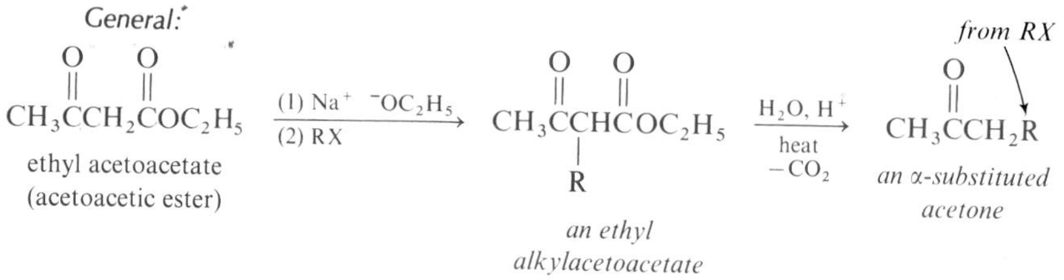

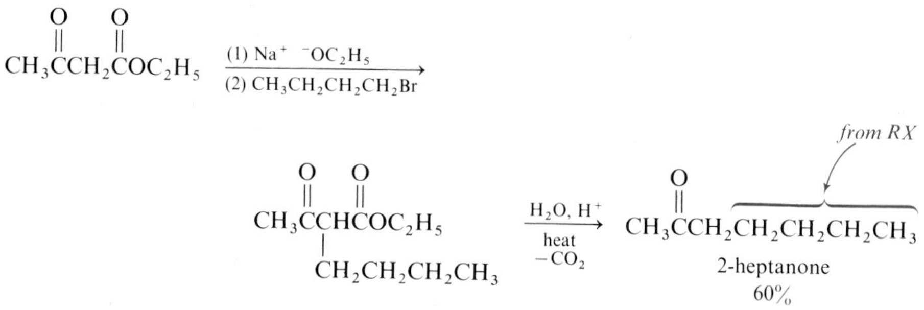

$$CH_3CCH_2COC_2H_5 \xrightarrow[\text{(2) } CH_3CH_2CH_2CH_2Br]{\text{(1) Na}^+ \ ^-OC_2H_5}$$

The steps in an acetoacetic ester synthesis are similar to those for a malonic ester synthesis.

Preparation of the enolate:

$$CH_3\overset{O}{\overset{\|}{C}}CH_2\overset{O}{\overset{\|}{C}}OC_2H_5 + {}^-OC_2H_5 \;\rightleftharpoons\; CH_3\overset{O}{\overset{\|}{C}}\overset{}{\underset{}{C}}H\overset{O}{\overset{\|}{C}}OC_2H_5 + HOC_2H_5$$

Alkylation:

$$CH_3\overset{O}{\overset{\|}{C}}\overset{}{C}H\overset{O}{\overset{\|}{C}}OC_2H_5 + R{-}X \;\xrightarrow{\;S_N2\;}\; CH_3\overset{O}{\overset{\|}{C}}\underset{R}{C}H\overset{O}{\overset{\|}{C}}OC_2H_5 + X^-$$

Hydrolysis and decarboxylation:

$$CH_3\overset{O}{\overset{\|}{C}}\underset{R}{C}H\overset{O}{\overset{\|}{C}}OC_2H_5 \;\xrightarrow[\text{heat}]{H_2O,\ H^+}\; CH_3\overset{O}{\overset{\|}{C}}\underset{R}{C}HC\overset{O}{\overset{\|}{}}OH \;\xrightarrow[\text{heat}]{-CO_2}\; CH_3\overset{O}{\overset{\|}{C}}CH_2R$$

STUDY PROBLEMS

14.7. Fill in the blanks:

$$CH_3\overset{O}{\overset{\|}{C}}CH_2CO_2C_2H_5 \;\xrightarrow{Na^+\ {}^-OC_2H_5}\; \underline{\hspace{3cm}} \quad \xrightarrow{\langle\rangle{-}CH_2CH_2Br} \quad \underline{\hspace{2cm}}$$

14.8. Show how the following compounds can be prepared from acetoacetic ester:

(a) $CH_2{=}CHCH_2\underset{CO_2C_2H_5}{C}H\overset{O}{\overset{\|}{C}}CH_3$ (b) $(CH_3CH_2)_2CH\overset{O}{\overset{\|}{C}}CH_3$

SECTION 14.4.

Syntheses Using Alkylation Reactions

In general, the products of alkylation reactions of malonic ester or acetoacetic ester are substituted acetic acids or substituted acetones.

From malonic ester:

$$R{-}CH_2CO_2H \quad \text{or} \quad \overset{R}{\underset{R'}{}}\!\!\diagdown\!\!\diagup\, CHCO_2H$$

From acetoacetic ester:

$$R{-}CH_2\overset{O}{\overset{\|}{C}}CH_3 \quad \text{or} \quad \overset{R}{\underset{R'}{}}\!\!\diagdown\!\!\diagup\, CH\overset{O}{\overset{\|}{C}}CH_3$$

However, we can also <u>obtain diacids, diesters, keto acids, and keto esters</u>. The assorted products that <u>may be obtained from alkylation of malonic ester or acetoacetic ester</u> are summarized in Figure 14.1.

It is comparatively easy to predict the products of a reaction when we are given the reactants. It is somewhat more difficult to decide upon specific reactants to use in a synthesis problem. Remember to work the problem backwards: if you

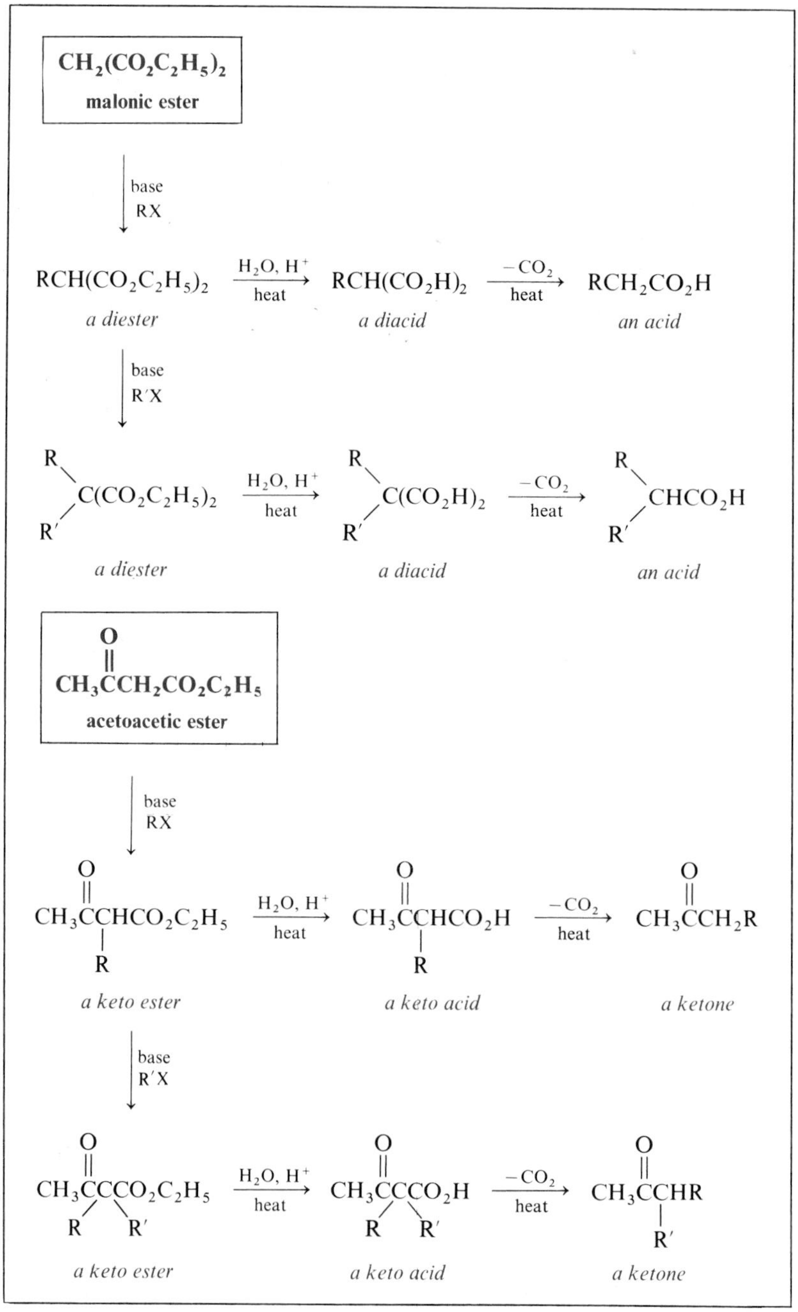

FIGURE 14.1. Products from the alkylations of malonic ester and acetoacetic ester.

are asked to synthesize a compound by an alkylation reaction, first decide what dicarbonyl compound you would need, then pick the alkyl halide.

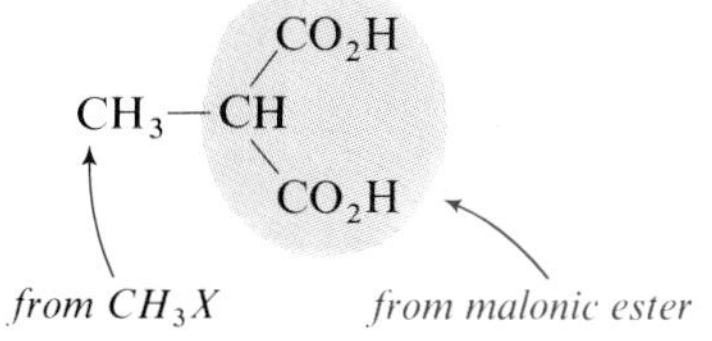

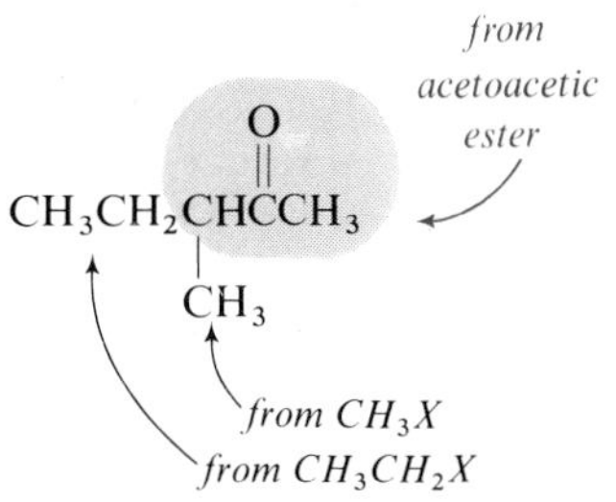

$$CH_3-CH\begin{array}{c} CO_2H \\ \\ CO_2H \end{array}$$

from CH$_3$X   *from malonic ester*

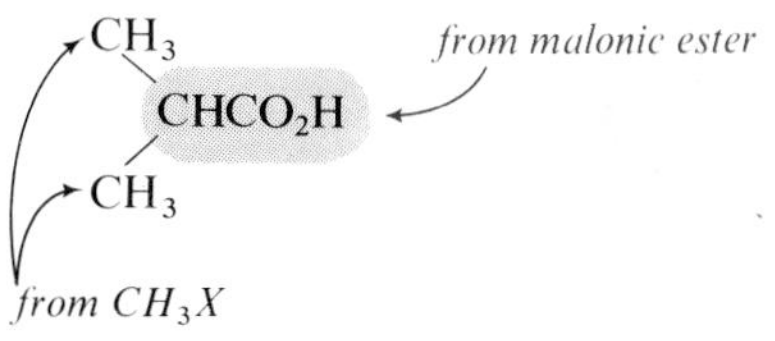

$$CH\begin{array}{c} CH_3 \\ \\ CO_2H \\ \\ CH_3 \end{array}$$ *from malonic ester*

from CH$_3$X

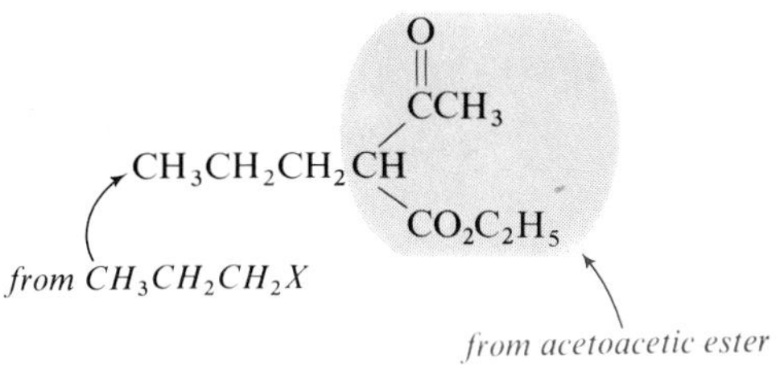

$$CH_3CH_2CH_2CH\begin{array}{c} \overset{O}{\overset{||}{C}}CH_3 \\ \\ CO_2C_2H_5 \end{array}$$

from CH$_3$CH$_2$CH$_2$X

from acetoacetic ester

$$CH_3CH_2\overset{}{\underset{CH_3}{CH}}\overset{O}{\overset{||}{C}}CH_3$$ *from acetoacetic ester*

from CH$_3$X
from CH$_3$CH$_2$X

Example: If you were asked to write the equations for the synthesis of 3-methyl-2-pentanone, you would:

1. *write the structure;*

2. *decide what β-dicarbonyl compound you would need; and*

3. *decide what alkyl halides would have to be used for the substitution.*

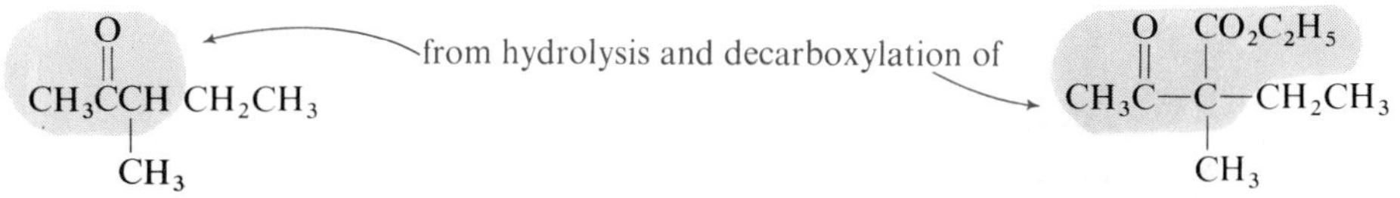

$$CH_3\overset{O}{\overset{||}{C}}\underset{CH_3}{CH}\ CH_2CH_3$$ from hydrolysis and decarboxylation of $$CH_3\overset{O}{\overset{||}{C}}-\underset{CH_3}{\overset{CO_2C_2H_5}{C}}-CH_2CH_3$$

It is evident that the β-dicarbonyl starting material is acetoacetic ester. The alkyl halides needed are CH$_3$X and CH$_3$CH$_2$X. Now, equations for the steps in the synthesis may be written:

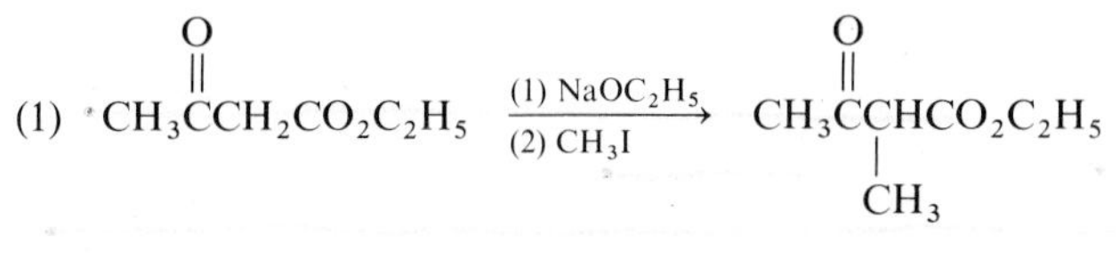

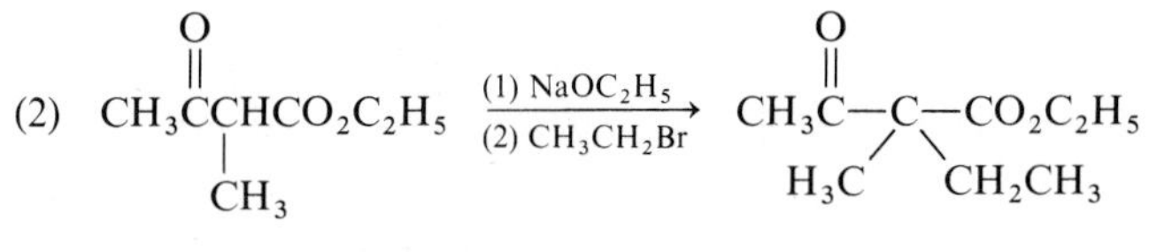

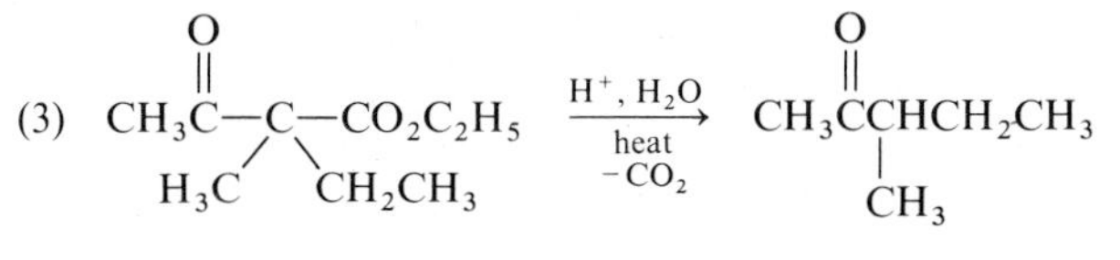

(1) $CH_3\overset{O}{\overset{||}{C}}CH_2CO_2C_2H_5 \xrightarrow[\text{(2) CH}_3\text{I}]{\text{(1) NaOC}_2\text{H}_5} CH_3\overset{O}{\overset{||}{C}}\underset{CH_3}{CHCO_2C_2H_5}$

(2) $CH_3\overset{O}{\overset{||}{C}}\underset{CH_3}{CHCO_2C_2H_5} \xrightarrow[\text{(2) CH}_3\text{CH}_2\text{Br}]{\text{(1) NaOC}_2\text{H}_5} CH_3\overset{O}{\overset{||}{C}}-\underset{H_3C\quad CH_2CH_3}{C}-CO_2C_2H_5$

(3) $CH_3\overset{O}{\overset{||}{C}}-\underset{H_3C\quad CH_2CH_3}{C}-CO_2C_2H_5 \xrightarrow[\substack{\text{heat} \\ -CO_2}]{H^+,\ H_2O} CH_3\overset{O}{\overset{||}{C}}\underset{CH_3}{CHCH_2CH_3}$

SAMPLE PROBLEM

Suggest a reaction sequence leading to 3-phenylpropanoic acid.

Solution:

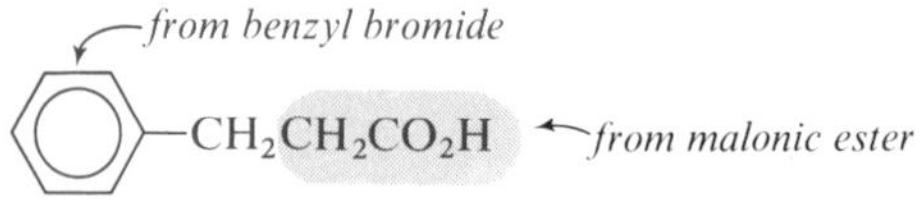

$$\text{C}_6\text{H}_5\text{—CH}_2\text{CH}_2\text{CO}_2\text{H}$$

— *from benzyl bromide*

← *from malonic ester*

(1) $\text{CH}_2(\text{CO}_2\text{C}_2\text{H}_5)_2 \xrightarrow{\text{Na}^+ \ ^-\text{OC}_2\text{H}_5} \ ^-\text{CH}(\text{CO}_2\text{C}_2\text{H}_5)_2$

(2) $\text{C}_6\text{H}_5\text{CH}_2\text{Br} + \ ^-\text{CH}(\text{CO}_2\text{C}_2\text{H}_5)_2 \xrightarrow{-\text{Br}^-} \text{C}_6\text{H}_5\text{CH}_2\text{CH}(\text{CO}_2\text{C}_2\text{H}_5)_2$

(3) $\text{C}_6\text{H}_5\text{CH}_2\text{CH}(\text{CO}_2\text{C}_2\text{H}_5)_2 \xrightarrow[\substack{\text{heat} \\ -\text{CO}_2}]{\text{H}_2\text{O, H}^+} \text{C}_6\text{H}_5\text{CH}_2\text{CH}_2\text{CO}_2\text{H}$

STUDY PROBLEM

14.9.   Show how you could synthesize the following compounds by alkylation reactions:

(a) $\text{CH}_3\text{CH}_2\text{CH}_2\text{CH}_2\text{CH}(\text{CO}_2\text{H})_2$ (b) $(\text{CH}_3)_2\text{CHCO}_2\text{H}$

(c) $\underset{\underset{\text{CH}_3}{|}}{\text{CH}_3\overset{\overset{\text{O}}{\|}}{\text{C}}\text{CHCH}_2\text{CH}_2\text{CH}_3}$ (d) $\underset{\text{CH}_2\text{CO}_2\text{C}_2\text{H}_5}{\overset{\text{CH}(\text{CO}_2\text{C}_2\text{H}_5)_2}{|}}$

(e)    (f)

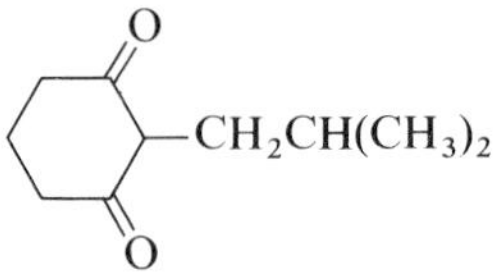

SECTION 14.5.

Alkylation and Acylation of Enamines

Another type of organic compound containing a nucleophilic carbon that can undergo alkylation reactions is an **enamine**. In Section 11.10C, we discussed the formation of enamines from secondary amines and aldehydes or ketones.

Formation of an enamine:

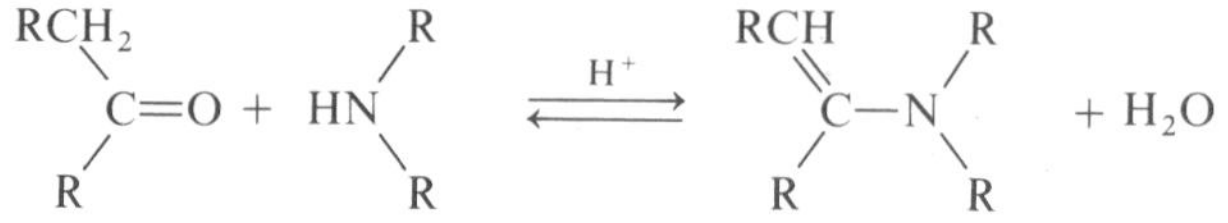

$$\underset{\text{a ketone}}{\underset{\underset{\text{R}}{|}}{\overset{\overset{\text{RCH}_2}{|}}{\text{C}}}=\text{O}} + \underset{\text{a 2° amine}}{\text{HN}\underset{\underset{\text{R}}{\diagup}}{\overset{\text{R}}{\diagdown}}} \ \underset{\text{H}^+}{\rightleftarrows} \ \underset{\text{an enamine}}{\overset{\overset{\text{RCH}}{\|}}{\text{C}}-\underset{\underset{\text{R}}{|}}{\overset{\overset{\text{R}}{|}}{\text{N}}}} + \text{H}_2\text{O}$$

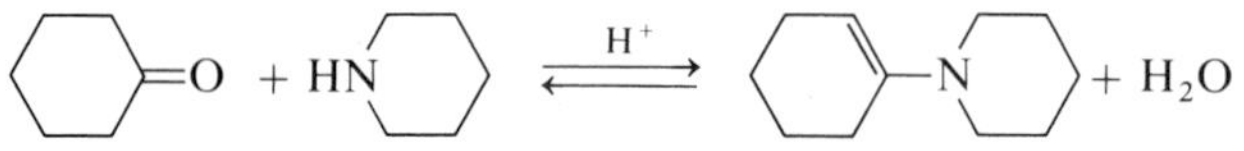

     cyclohexanone   *piperidine*   *an enamine*

The nitrogen of an enamine has an unshared pair of electrons. These electrons are, in a sense, in an *allylic position* and consequently are in conjugation with the double bond. Resonance structures for the enamine show that the carbon β to the nitrogen has a partial negative charge.

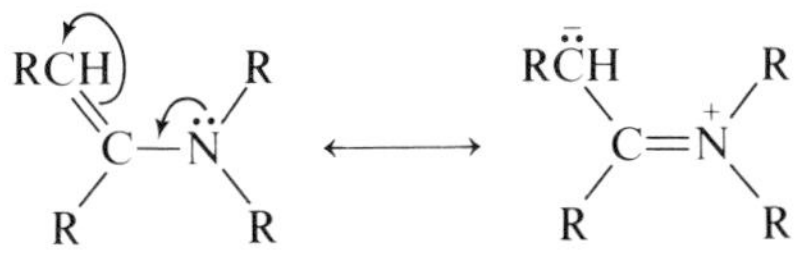

resonance structures for an enamine

This β carbon has carbanion character and can act as a nucleophile. For example, when an enamine is treated with an alkyl halide, such as CH_3I, the enamine displaces the halogen of the alkyl halide in an S_N2 reaction. The result is alkylation of the enamine at the position that is β to the nitrogen.

Alkylation:

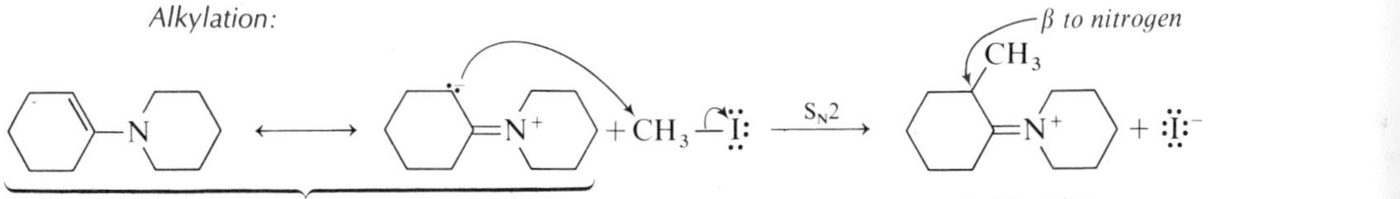

the enamine   *an iminium ion*

The product iminium ion is readily hydrolyzed to a ketone. The net result of the entire reaction sequence is alkylation of a ketone in the α position.

Hydrolysis:

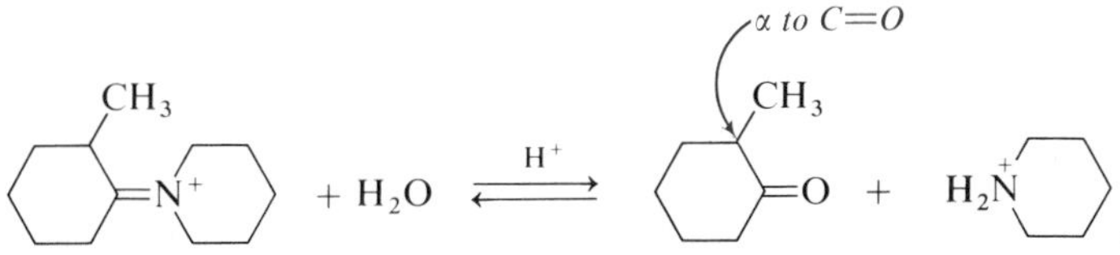

the iminium ion   2-methyl-1-cyclohexanone   piperidinium ion

a ketone

A generalized sequence for an enamine synthesis follows:

Preparation of enamine:

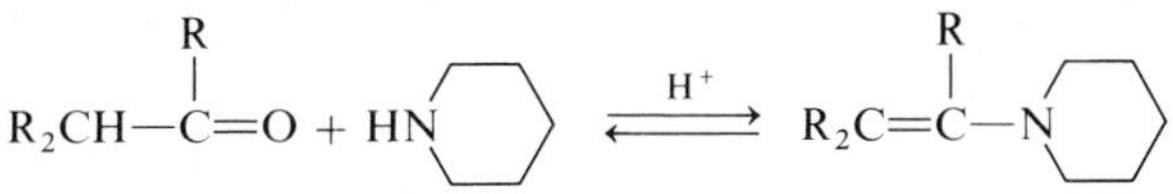

Substitution reaction:

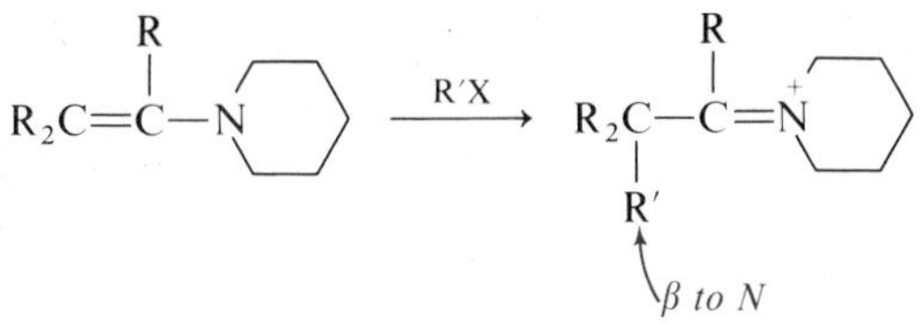

Hydrolysis:

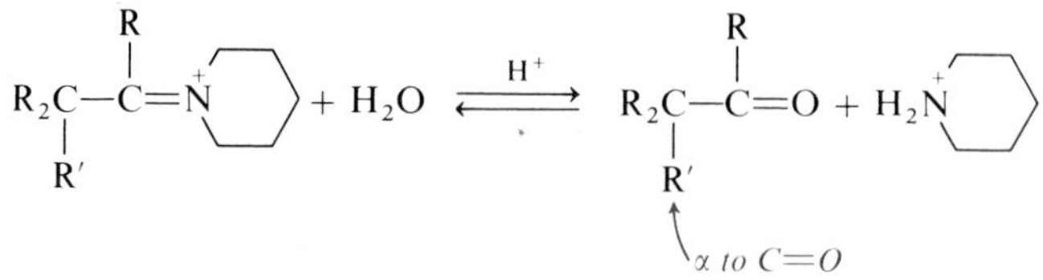

$$R_2C-\overset{\underset{\displaystyle R'}{|}}{\overset{\displaystyle R}{|}}C=\overset{+}{N}\bigcirc + H_2O \underset{}{\overset{H^+}{\rightleftharpoons}} R_2C-\overset{\underset{\displaystyle R'}{|}}{\overset{\displaystyle R}{|}}C=O + H_2\overset{+}{N}\bigcirc$$

α to C=O

SAMPLE PROBLEM

Give the steps in the preparation of the following ketone, using an enamine synthesis with piperidine as the amine.

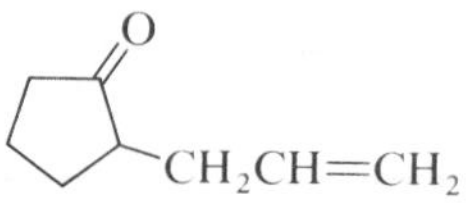

Solution:

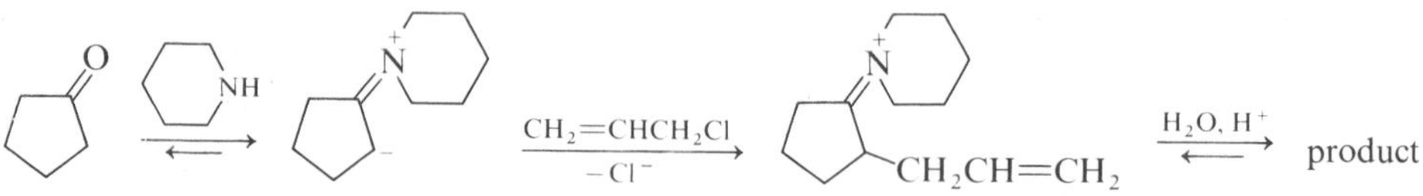

The alkylation step in an enamine synthesis is an S_N2 reaction with a rather weak nucleophile. (Why?) It is not surprising then that only the most reactive halogen compounds are suitable as alkylating agents. These compounds include allylic halides, benzylic halides, α-halocarbonyl compounds, and iodomethane. (Ordinary alkyl halides other than CH_3I are attacked by the enamine nitrogen, rather than by the carbon.)

Some reactive halides:

$\overset{O}{\overset{\|}{CH_3CCl}}$	$CH_2{=}CHCH_2Cl$	$C_6H_5CH_2Cl$	$\overset{O}{\overset{\|}{BrCH_2CCH_3}}$	CH_3I
acetyl chloride	allyl chloride	benzyl chloride	bromoacetone	iodomethane

The reactions of enamines with α-halocarbonyl compounds and acid halides follow similar paths to that of alkylation. In each case, the final product (after hydrolysis) is a ketone substituted at the α position.

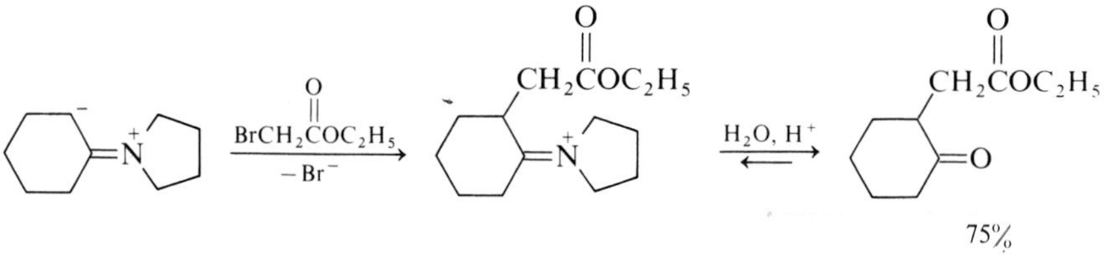

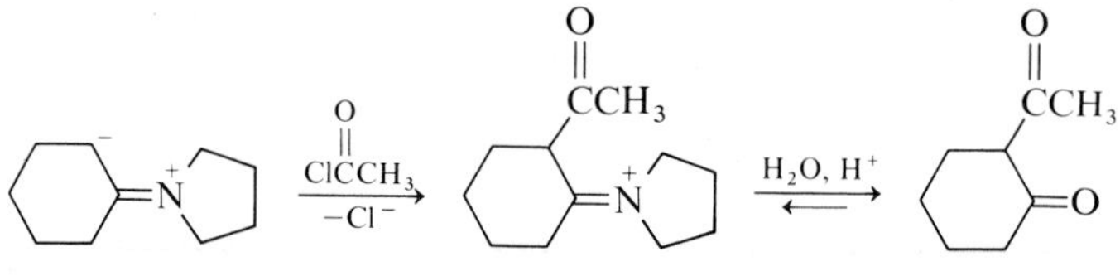

75%

SAMPLE PROBLEM

How would you prepare the following compound by an enamine synthesis?

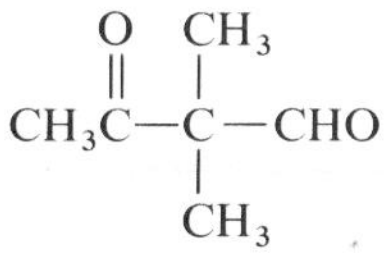

$$\underset{\underset{CH_3}{|}}{\overset{\overset{O}{\|}\quad\overset{CH_3}{|}}{CH_3C-C-CHO}}$$

Solution:

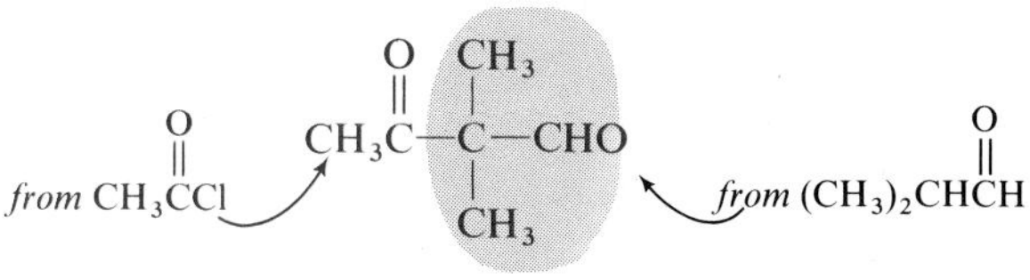

$$\text{from } CH_3CCl \nearrow \quad \underset{\underset{CH_3}{|}}{\overset{\overset{O}{\|}\quad\overset{CH_3}{|}}{CH_3C-C-CHO}} \quad \nwarrow \text{ from } (CH_3)_2CHCH\overset{O}{\overset{\|}{}}$$

Reaction sequence:

$$(CH_3)_2CHCH\overset{O}{\overset{\|}{}} \;\rightleftharpoons\; \left[(CH_3)_2C=CHN\bigcirc \;\longleftrightarrow\; (CH_3)_2\bar{C}CH=\overset{+}{N}\bigcirc \right]$$

$$\xrightarrow[-Cl^-]{CH_3CCl\overset{O}{\overset{\|}{}}} \quad (CH_3)_2\overset{\overset{\overset{O}{\|}}{CH_3C}}{\underset{}{C}}CH=\overset{+}{N}\bigcirc \quad \xrightarrow[]{H_2O,\,H^+} \quad \underset{\underset{CH_3}{|}}{\overset{\overset{O}{\|}\;\;\overset{CH_3}{|}}{CH_3C-C-CHO}}$$

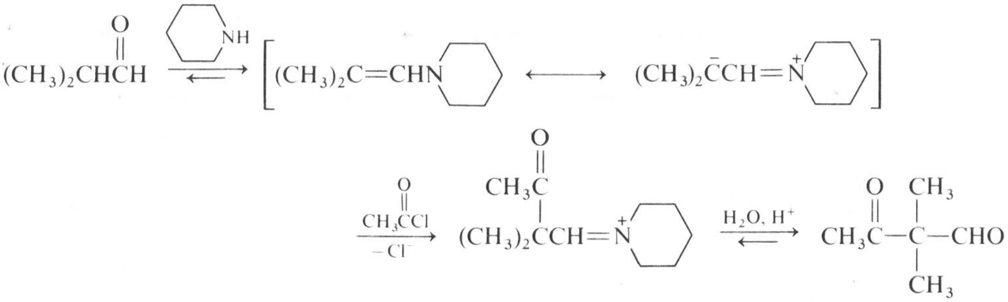

STUDY PROBLEMS

14.10. Suggest a mechanism for the hydrolysis of the iminium ion in the preceding sample problem.

14.11. Enamine syntheses are generally carried out using a cyclic amine as the secondary amine. Suggest a reason for this.

Three cyclic amines used in enamine syntheses:

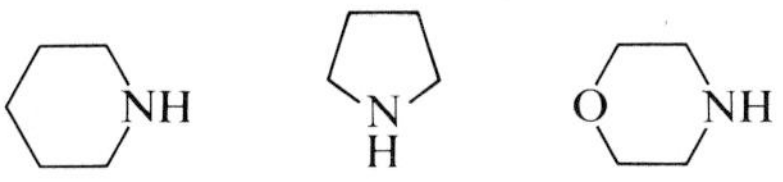

piperidine   pyrrolidine   morpholine

14.12. Predict the products:

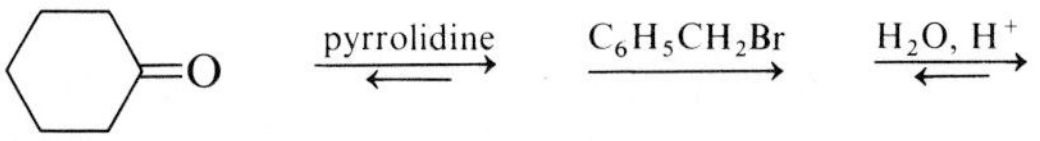

$$\bigcirc\!\!=\!\!O \;\xrightarrow[]{pyrrolidine}\; \xrightarrow{C_6H_5CH_2Br}\; \xrightarrow[]{H_2O,\,H^+}$$

14.13. Show how you would prepare the following compounds by enamine syntheses:

(a) 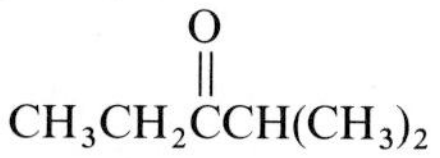 $CH_3CH_2\overset{O}{\overset{\|}{C}}CH(CH_3)_2$ (b)

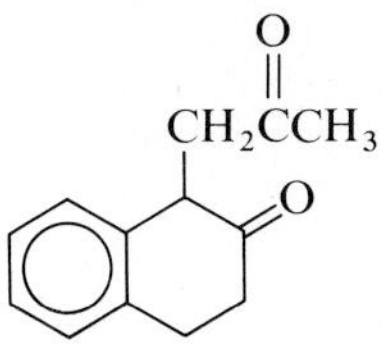

SECTION 14.6.

Aldol Condensations

So far, we have been discussing the displacement of halide ions by nucleophiles. A reagent with a nucleophilic carbon atom can also attack the partially positive carbon of a carbonyl group. The rest of this chapter will be devoted to the reactions of enolates and related anions with carbonyl compounds.

$$Nu^- + -\overset{\overset{\displaystyle \ddot{O}:}{\|}}{C}- \longrightarrow -\overset{\overset{\displaystyle :\ddot{O}:}{|}}{\underset{\underset{\displaystyle Nu}{|}}{C}}-$$

When an aldehyde is treated with a base such as aqueous NaOH, the resulting enolate ion can undergo reaction at the carbonyl group of another molecule of aldehyde. The result is the *addition of one molecule of aldehyde to another*.

$$2\ \overset{\overset{\displaystyle O}{\|}}{CH_3CH} \underset{\longleftarrow}{\overset{OH^-}{\longrightarrow}} \overset{\overset{\displaystyle OH}{|}}{CH_3CH}-\overset{\overset{\displaystyle O}{\|}}{CH_2CH}$$

from one aldehyde

acetaldehyde   3-hydroxybutanal
                            (acetaldol or aldol)
                            50%

This reaction is called an **aldol condensation reaction**. The word "aldol," derived from *ald*ehyde and alcoh*ol*, describes the product, which is a *β-hydroxy aldehyde*. A **condensation reaction** is one in which two or more molecules combine into a larger molecule with or without the loss of a small molecule (such as water). The aldol condensation is an addition reaction in which no small molecule is lost.

How does an aldol condensation proceed? If acetaldehyde is treated with dilute aqueous sodium hydroxide, a low concentration of enolate ions is formed. The reaction is reversible—as enolate ions undergo reaction, more are formed.

$$\overset{\overset{\displaystyle O}{\|}}{CH_3CH} + OH^- \Longleftrightarrow \left[{}^-\overset{\overset{\displaystyle \ddot{O}:}{\|}}{\ddot{C}H_2CH} \longleftrightarrow \overset{\overset{\displaystyle :\ddot{O}:^-}{|}}{CH_2{=}CH} \right] + H_2O$$

*resonance structures for
the enolate ion*

The enolate ion undergoes reaction with another acetaldehyde molecule by adding to the carbonyl carbon to form an alkoxide ion, which abstracts a proton from water to yield the product aldol.

$$\overset{\overset{\displaystyle \ddot{O}:}{\|}}{CH_3CH} + {}^-\overset{\overset{}{}}{CH_2}\overset{\overset{\displaystyle O}{\|}}{CH} \Longleftrightarrow \left[\overset{\overset{\displaystyle :\ddot{O}:^-}{|}}{CH_3CH}-CH_2\overset{\overset{\displaystyle O}{\|}}{CH} \right] \underset{\longleftarrow}{\overset{H_2O}{\longrightarrow}} \overset{\overset{\displaystyle OH}{|}}{CH_3CHCH_2}\overset{\overset{\displaystyle O}{\|}}{CH} + OH^-$$

α hydrogens

an alkoxide ion

Note that the <u>starting aldehyde in an aldol condensation must contain a hydrogen α to the carbonyl group</u> so that it can form an enolate ion in base. The aldol product still has a carbonyl group with α hydrogens. Can it undergo further reaction to form trimers? tetramers? polymers? Yes, these materials are by-products of the reaction. For simplicity, we will show only the dimer products and ignore the fact that other, higher-molecular-weight products may also be formed.

We have shown the aldol condensation for acetaldehyde. Other aldehydes also undergo this self-addition. <u>Ketones undergo aldol condensations,</u> but the <u>equilibrium does not favor the ketone-condensation product.</u> (Why not?) Although there are a number of laboratory procedures that can be used to induce ketone condensations of the aldol type, the reaction is not as useful with ketones as it is with aldehydes. Therefore, we will concentrate our present discussion on aldehydes. Two other examples of aldol condensations follow:

Other examples of aldol condensations:

$$\underset{\text{propanal}}{CH_3CH_2\overset{\overset{\displaystyle O}{\|}}{C}H} + CH_3CH_2\overset{\overset{\displaystyle O}{\|}}{C}H \;\underset{}{\overset{OH^-}{\rightleftharpoons}}\; CH_3CH_2\underset{\underset{\displaystyle CH_3}{|}}{\overset{\overset{\displaystyle OH}{|}}{C}H}-\overset{\overset{\displaystyle O}{\|}}{C}HCH$$

an aldol

cyclohexanecarboxaldehyde   *an aldol*

SAMPLE PROBLEM

Show how you could prepare 3-hydroxy-2,2,4-trimethylpentanal by an aldol condensation.

Solution: Write the structure and indicate the carbon–carbon bond formed in the condensation.

the new carbon–carbon bond

from $(CH_3)_2CHCH$

Write the equation.

$$2\,(CH_3)_2CH\overset{\overset{\displaystyle O}{\|}}{C}H \;\underset{}{\overset{OH^-}{\rightleftharpoons}}\; (CH_3)_2CH\underset{}{\overset{\overset{\displaystyle OH}{|}}{C}H}-C(CH_3)_2\overset{\overset{\displaystyle O}{\|}}{C}H$$

STUDY PROBLEMS

14.14. Which of the following aldehydes can undergo self-condensations? Explain.

(a) ⬡—CHO   (b) HCHO   (c) [cyclopentane with CH₃]—CHO

(d) $(CH_3)_3CCHO$ (e) $(CH_3CH_2)_2CHCHO$

14.15. Predict the product of the self-condensation of: (a) butanal;   (b) acetone;
(c) 3-methylbutanal.

A.   Dehydration of aldols

A β-hydroxy carbonyl compound, such as an aldol, undergoes dehydration readily
because the double bond in the product is in conjugation with the carbonyl group.
Therefore, an **α,β-unsaturated aldehyde** may be readily obtained as the product of an
aldol condensation.

$$\underset{\text{3-hydroxybutanal}}{\overset{\text{OH}\quad\;\;\text{O}}{CH_3\overset{|}{C}H-CH_2\overset{||}{C}H}} \xrightarrow[\text{warm}]{\text{dil. H}^+} \underset{\substack{\text{2-butenal}\\\text{(crotonaldehyde)}}}{CH_3CH=CH\overset{\overset{\text{O}}{||}}{C}H} + H_2O$$

$$\underset{}{\text{[cyclohexane ring]}\overset{\overset{\text{O}}{||}}{-CH}} \xrightarrow[\text{warm}]{\text{dil. H}^+} \text{[cyclohexene ring]}\overset{\overset{\text{O}}{||}}{-CH}$$
(with OH)

When dehydration leads to a double bond in conjugation with an aromatic
ring, dehydration is often spontaneous.

$$\underset{\substack{\text{3-hydroxy-3-}\\\text{phenylpropanal}}}{\text{[benzene ring]}\overset{\overset{\text{OH}\quad\;\;\text{O}}{}}{-\overset{|}{C}H-CH_2\overset{||}{C}H}} \xrightarrow{\text{spontaneous}} \underset{\substack{\text{3-phenylpropenal}\\\text{(cinnamaldehyde)}}}{\text{[benzene ring]}-CH=CH\overset{\overset{\text{O}}{||}}{C}H} + H_2O$$

B.   Crossed aldol condensations

An aldehyde with no α hydrogens cannot form an enolate ion and thus cannot
dimerize in an aldol condensation. However, if such an aldehyde is mixed with
an aldehyde that *does* have an alpha hydrogen, a condensation between the two
can occur. This reaction is called a **crossed aldol condensation**. A crossed aldol
condensation is most useful when only one of the carbonyl compounds has an
α hydrogen; otherwise, mixtures of products result. Methyl ketones may be used
successfully in crossed aldol condensations with aldehydes that contain no α
hydrogen, as in the second example following.

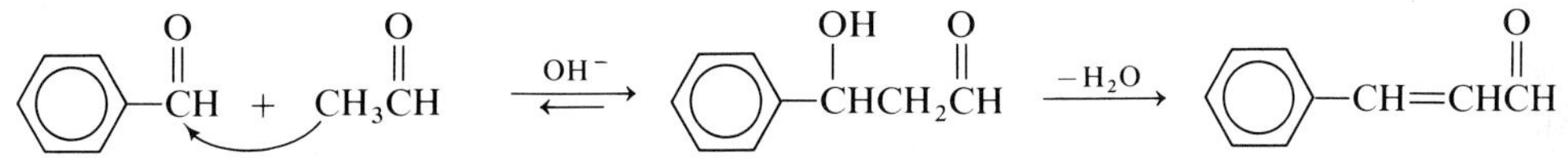

benzaldehyde   acetaldehyde

(*no α hydrogens*)

cinnamaldehyde

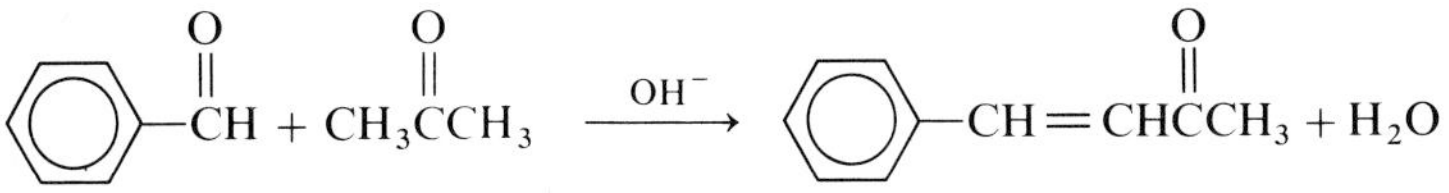

acetone   4-phenyl-3-buten-2-one
90%

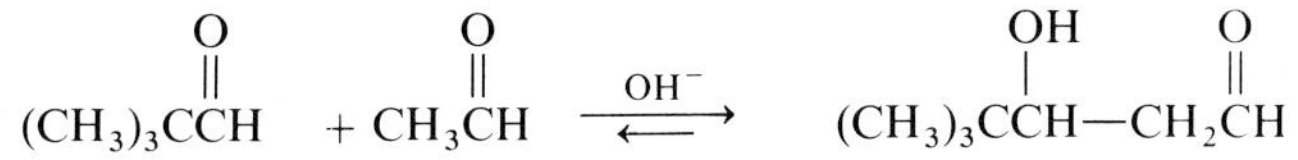

2,2-dimethylpropanal   3-hydroxy-4,4-dimethylpentanal

STUDY PROBLEM

14.16. Predict the major products:

(a) $C_6H_5CHO + CH_3CH_2CH_2CHO \xrightarrow{OH^-}$

(b) $C_6H_5CHO + CH_3CH_2\overset{O}{\overset{\|}{C}}CH_2CH_3 \xrightarrow{OH^-}$

C.   Syntheses using aldol condensations

In an aldol condensation, two types of product can result: (1) β-hydroxy aldehydes or ketones, and (2) α,β-unsaturated aldehydes or ketones (see Figure 14.2). In a synthesis problem, look for these functionalities and decide which aldehydes or ketones must be used for the starting materials.

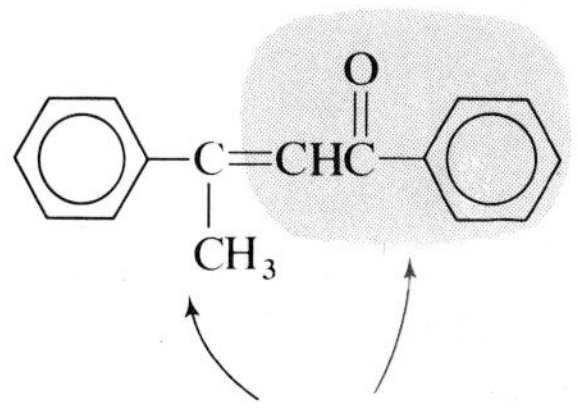

from acetophenone 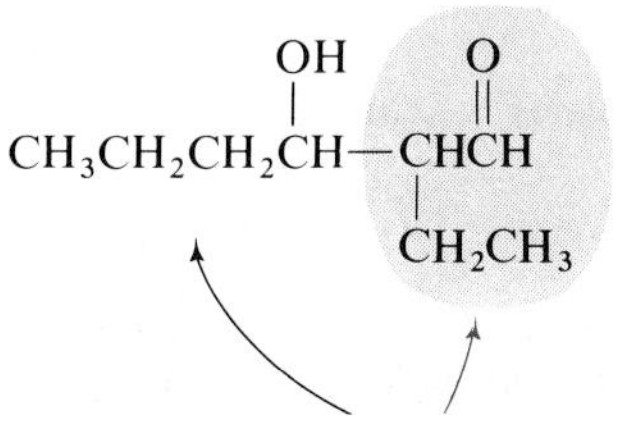   *from butanal*

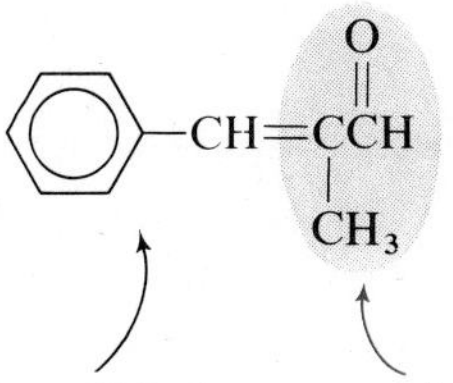

from benzaldehyde   *from propanal*

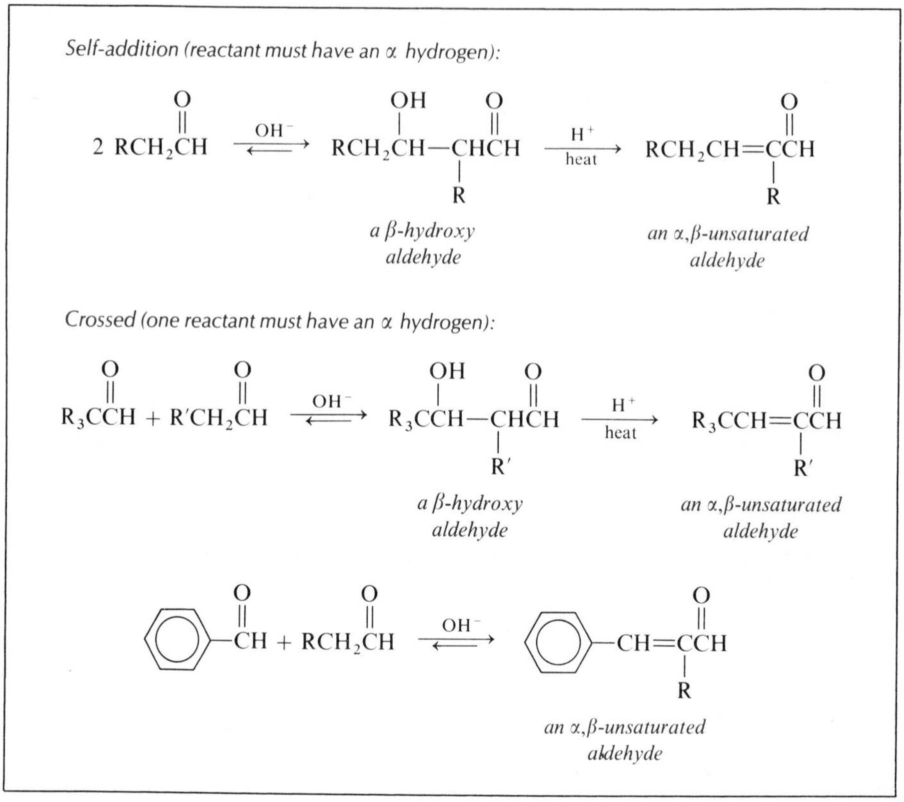

Self-addition (reactant must have an α hydrogen):

$$2\ \underset{\displaystyle}{RCH_2\overset{\displaystyle O}{\overset{\|}{C}}H} \ \underset{}{\overset{OH^-}{\rightleftharpoons}} \ RCH_2\underset{\underset{R}{|}}{\overset{OH}{\overset{|}{C}H}}-\overset{\displaystyle O}{\overset{\|}{C}}H \ \underset{heat}{\overset{H^+}{\longrightarrow}} \ RCH_2\underset{\underset{R}{|}}{CH}=\overset{\displaystyle O}{\overset{\|}{C}}H$$

a β-hydroxy
aldehyde   *an α,β-unsaturated*
aldehyde

Crossed (one reactant must have an α hydrogen):

$$R_3C\overset{\displaystyle O}{\overset{\|}{C}}H + R'CH_2\overset{\displaystyle O}{\overset{\|}{C}}H \ \underset{}{\overset{OH^-}{\rightleftharpoons}} \ R_3C\underset{\underset{R'}{|}}{\overset{OH}{\overset{|}{C}H}}-\overset{\displaystyle O}{\overset{\|}{C}}H \ \underset{heat}{\overset{H^+}{\longrightarrow}} \ R_3C\underset{\underset{R'}{|}}{CH}=\overset{\displaystyle O}{\overset{\|}{C}}H$$

a β-hydroxy
aldehyde   *an α,β-unsaturated*
aldehyde

$$C_6H_5-\overset{\displaystyle O}{\overset{\|}{C}}H + RCH_2\overset{\displaystyle O}{\overset{\|}{C}}H \ \underset{}{\overset{OH^-}{\rightleftharpoons}} \ C_6H_5-CH=\underset{\underset{R}{|}}{C}\overset{\displaystyle O}{\overset{\|}{C}}H$$

an α,β-unsaturated
aldehyde

FIGURE 14.2. Products from aldol condensations.

SAMPLE PROBLEMS

The following ketone can be prepared in 90% yield by a crossed aldol condensation:

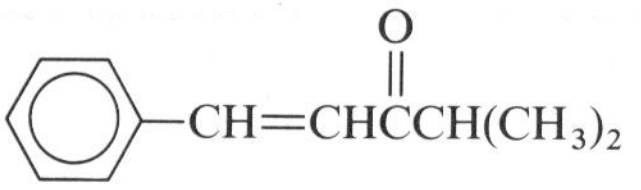

$$C_6H_5-CH=CHC\overset{\displaystyle O}{\overset{\|}{C}}CH(CH_3)_2$$

What organic reactants are needed?

Solution:

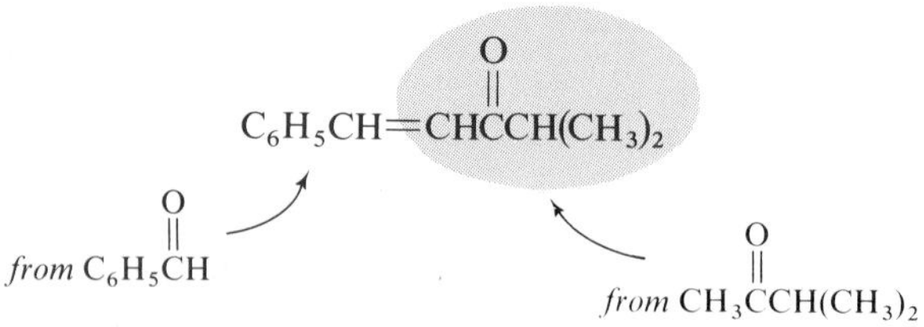

$$C_6H_5CH=CHC\overset{\displaystyle O}{\overset{\|}{C}}CH(CH_3)_2$$

from $C_6H_5\overset{\displaystyle O}{\overset{\|}{C}}H$ *from* $CH_3\overset{\displaystyle O}{\overset{\|}{C}}CH(CH_3)_2$

How would you make the following conversion?

$$CH_3\overset{\displaystyle O}{\overset{\|}{C}}CH_3 \ \longrightarrow \ CH_2=CHC\overset{\displaystyle O}{\overset{\|}{C}}CH_3$$

acetone   3-buten-2-one

Solution:

$$CH_2=CHCCH_3$$

$$\overset{\text{from HCHO}}{\nearrow} \quad \overset{\text{from } CH_3CCH_3}{\nwarrow}$$

$$HCH + CH_3CCH_3 \underset{\longleftarrow}{\overset{OH^-}{\longrightarrow}} \underset{CH_2-CH_2CCH_3}{\overset{OH}{|}} \overset{H^+}{\underset{\substack{\text{heat} \\ -H_2O}}{\longrightarrow}} CH_2=CHCCH_3$$

STUDY PROBLEM

14.17. Suggest syntheses for the following compounds from aldehydes or ketones:

(a) $C_6H_5CH=CCHO$
       $\underset{CH_3}{|}$

(b) [cyclopentene ring with CH₃ and —CCH₃ (C=O) substituents]

(c) [furan ring]—CH=CHCH (C=O)

(d) [cyclohexane ring]—CH (C=O) with OH

SECTION 14.7.

Reactions Related to the Aldol Condensation

We have shown aldol condensations and crossed aldol condensations; however, for this type of condensation to occur, all that is needed is one compound with a carbonyl group plus one compound with an acidic hydrogen. The **Knoevenagel condensation** is the reaction of an aldehyde with a compound that has a hydrogen α to *two* activating groups (such as C=O or C≡N), using ammonia or an amine as the catalyst. Under these conditions, malonic acid itself may be used as a reactant, as shown in the second example following.

Knoevenagel condensations:

$$(CH_3)_2CHCH_2CH + CH_2(CO_2C_2H_5)_2 \overset{\substack{\text{piperidine} \\ \text{benzene}}}{\underset{\text{heat}}{\longrightarrow}}$$

3-methylbutanal   diethyl malonate

$$(CH_3)_2CHCH_2CH=C(CO_2C_2H_5)_2 + H_2O$$
$$78\%$$

[benzene ring]—CH (C=O) + $CH_2(CO_2H)_2$ $\overset{NH_3}{\underset{\text{heat}}{\longrightarrow}}$ [benzene ring]—CH=CHCO_2H + H_2O + CO_2

benzaldehyde   malonic acid   3-phenylpropenoic acid
                                          (cinnamic acid)
                                          85%

A variation of the Knoevenagel reaction allows the less reactive ketones to undergo condensation with the more acidic ethyl cyanoacetate ($pK_a = 9$, compared to $pK_a = 11$ for diethyl malonate).

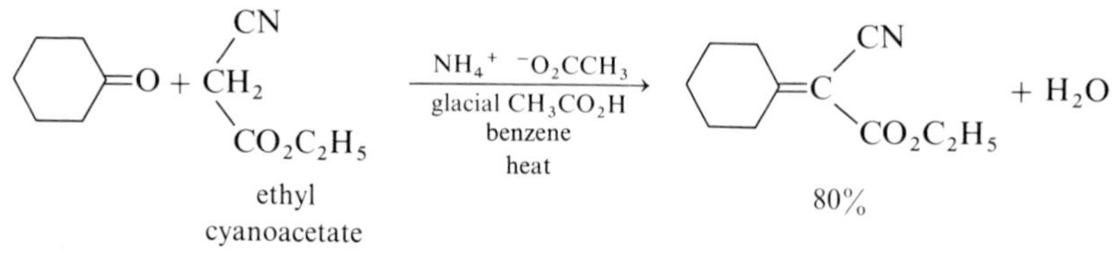

$$\underset{\substack{\text{ethyl}\\\text{cyanoacetate}}}{\text{(cyclohexanone)}=O + \underset{\substack{\text{CN}\\\text{CH}_2\\\text{CO}_2\text{C}_2\text{H}_5}}{}} \xrightarrow[\substack{\text{glacial CH}_3\text{CO}_2\text{H}\\\text{benzene}\\\text{heat}}]{\text{NH}_4^+ \ ^-\text{O}_2\text{CCH}_3} \underset{80\%}{=\text{C}\underset{\text{CO}_2\text{C}_2\text{H}_5}{\overset{\text{CN}}{}}} + \text{H}_2\text{O}$$

SAMPLE PROBLEMS

Suggest a technique for the preparation of $C_6H_5CH{=}CHNO_2$.

Solution:

$$C_6H_5CH{=}CHNO_2$$

from benzaldehyde   *from nitromethane*

$$\underset{}{C_6H_5\overset{O}{\overset{\|}{C}}H} + CH_3NO_2 \underset{}{\overset{OH^-}{\rightleftarrows}} C_6H_5CH{=}CHNO_2 + H_2O$$

How would you prepare the following compound?

$$\text{(phenyl)}{-}CH{=}\overset{\overset{\text{CN}}{|}}{C}{-}\text{(phenyl)}$$

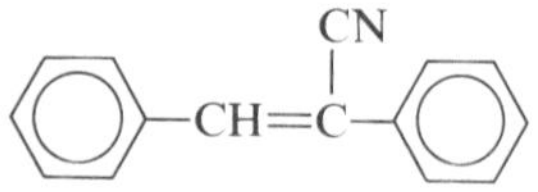

Solution:

$$C_6H_5CH{=}\overset{\overset{\text{CN}}{|}}{C}C_6H_5$$

from C_6H_5CHO   *from $C_6H_5CH_2CN$*

$$\underset{}{C_6H_5\overset{O}{\overset{\|}{C}}H} + \overset{\overset{\text{CN}}{|}}{C}H_2C_6H_5 \overset{OH^-}{\rightleftarrows} C_6H_5CH{=}\overset{\overset{\text{CN}}{|}}{C}C_6H_5 + H_2O$$

STUDY PROBLEM

14.18. Suggest synthetic routes to the following compounds:

(a) $(CH_3CH_2)_2C{=}\overset{\overset{O}{\|}}{C}\underset{\underset{\text{CN}}{|}}{}OC_2H_5$ (b) 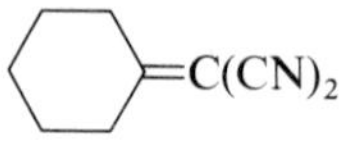 $\text{(cyclohexane)}{=}C(CN)_2$

SECTION 14.8.

Cannizzaro Reaction

An aldehyde with no α hydrogen cannot undergo self-addition to yield an aldol product.

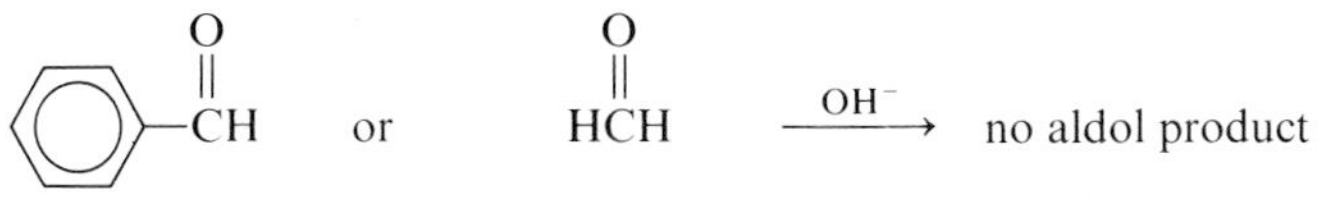

    benzaldehyde   formaldehyde

    If an aldehyde with no α hydrogen is heated with concentrated hydroxide solution, a disproportionation reaction occurs in which one half of the aldehyde molecules are oxidized to a carboxylic acid and one half are reduced to an alcohol. This reaction is known as the **Cannizzaro reaction**. Aldehydes with α hydrogens do not undergo this reaction; under these conditions, they undergo an aldol condensation.

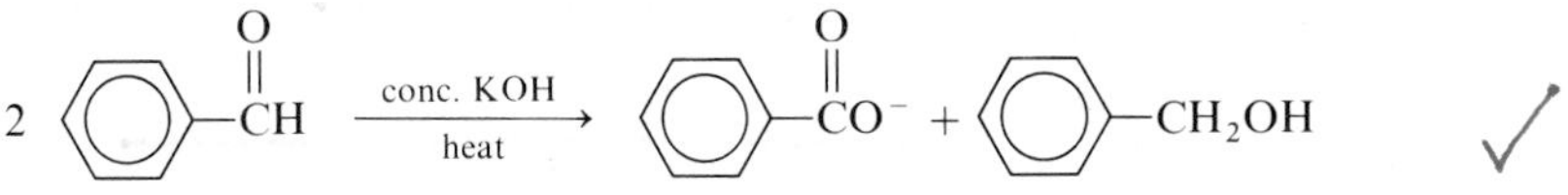

                benzoate ion   benzyl alcohol

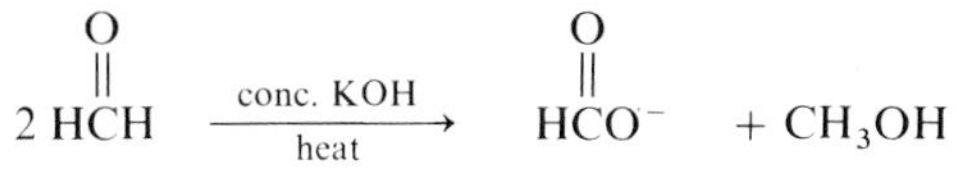

           formate ion   methanol

    The Cannizzaro reaction is initiated by attack of ⁻OH on the carbonyl carbon, followed by a hydride transfer.

Attack by ⁻OH:

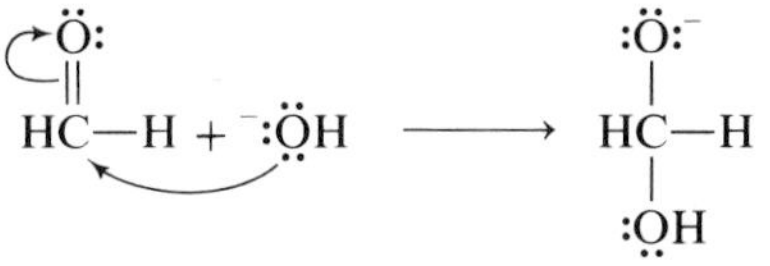

Hydride transfer:

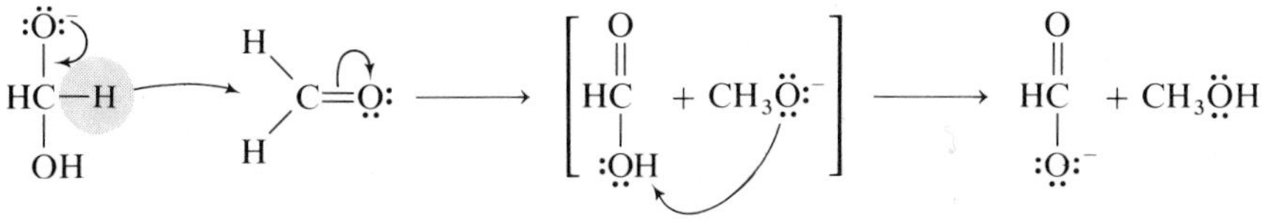

SECTION 14.9.

Ester Condensations

Esters with alpha hydrogens can undergo self-condensation reactions to yield β-keto esters. An ester condensation is similar to an aldol condensation; the difference is that the —OR group of an ester can act as a leaving group. The result is therefore *substitution* (whereas aldol condensations are *additions*). Simple ester condensations, such as the following examples, are called **Claisen condensations**. (Note the naming of the keto ester products.)

$$2 \ CH_3\overset{O}{\underset{||}{C}}OC_2H_5 \quad \xrightarrow{Na^+ \ ^-OC_2H_5} \quad CH_3\overset{O}{\underset{||}{C}}-CH_2\overset{O}{\underset{||}{C}}OC_2H_5 + C_2H_5OH$$

ethyl acetate   ethyl 3-oxobutanoate
(ethyl acetoacetate)
75%

$$2 \ CH_3CH_2\overset{O}{\underset{||}{C}}OC_2H_5 \quad \xrightarrow{Na^+ \ ^-OC_2H_5} \quad CH_3CH_2\overset{O}{\underset{||}{C}}-\underset{\underset{CH_3}{|}}{C}H\overset{O}{\underset{||}{C}}OC_2H_5 + C_2H_5OH$$

ethyl propanoate

ethyl 2-methyl-3-oxopentanoate
45%

Let us look at the stepwise reaction. First is the *formation of the enolate of the ester* by an acid–base reaction with the alkoxide ion. (An alkoxide, rather than a hydroxide, is used as the base to prevent saponification of the ester.) As in the aldol condensation, a low concentration of enolate is formed because the enolate (with only one carbonyl) is a stronger base than the alkoxide ion.

Enolate formation:

$$CH_3\overset{O}{\underset{||}{C}}OC_2H_5 + \ ^-\overset{..}{\underset{..}{O}}C_2H_5 \ \rightleftharpoons \ \left[\ ^-\overset{..}{C}H_2-\overset{:\overset{..}{O}:}{\underset{||}{C}}OC_2H_5 \ \longleftrightarrow \ CH_2=\overset{:\overset{..}{O}:^-}{\underset{|}{C}}OC_2H_5 \right] + C_2H_5\overset{..}{\underset{..}{O}}H$$

ethyl acetate   *resonance structures for the enolate ion*

The nucleophilic carbon then attacks the carbonyl group in a typical carbonyl *addition* reaction. This addition of the enolate is followed by *elimination* of ROH. The entire sequence is thus a typical nucleophilic acyl substitution reaction of a carbonyl compound, similar to those you encountered in Chapter 13.

Attack on carbonyl group:

$$CH_3\overset{\overset{..}{O}:}{\underset{||}{C}}OC_2H_5 + \ ^-CH_2CO_2C_2H_5 \ \rightleftharpoons \ \left[CH_3\underset{\underset{CH_2CO_2C_2H_5}{|}}{\overset{:\overset{..}{O}:^-}{\underset{|}{C}}}OC_2H_5 \right]$$

                  addition

Loss of ROH:

$$\left[CH_3\underset{\underset{CH_2CO_2C_2H_5}{|}}{\overset{:\overset{..}{O}:^-}{\underset{|}{C}}}\overset{..}{O}C_2H_5 \right] \ \underset{\text{elimination}}{\rightleftharpoons} \ \left[CH_3\underset{\underset{CH_2CO_2C_2H_5}{|}}{\overset{:\overset{..}{O}}{\underset{||}{C}}} \ + \ ^-:\overset{..}{\underset{..}{O}}C_2H_5 \right] \ \rightleftharpoons$$

$$CH_3\overset{O}{\underset{||}{C}}\overset{..}{C}HCO_2C_2H_5 + H\overset{..}{\underset{..}{O}}C_2H_5$$

*the enolate of
ethyl acetoacetate*

The product β-keto ester is more acidic than an alcohol because it has hydrogens that are α to two carbonyl groups. Therefore, the product of the condensation is the enolate salt of the β-keto ester. The β-keto ester itself is produced when the reaction mixture is acidified with cold, dilute mineral acid.

$$CH_3\overset{\overset{\displaystyle O}{||}}{C}\bar{C}HCO_2C_2H_5 \quad\xrightarrow{\text{H}^+}\quad CH_3\overset{\overset{\displaystyle O}{||}}{C}CH_2CO_2C_2H_5$$

ethyl acetoacetate

a β-keto ester

A β-keto ester may be hydrolyzed by heating in acidic solution, in which case decarboxylation may occur.

Hydrolysis and decarboxylation:

$$CH_3\overset{\overset{\displaystyle O}{||}}{C}CH_2CO_2C_2H_5 \quad\xrightarrow[-\text{C}_2\text{H}_5\text{OH}]{\text{H}^+,\ \text{H}_2\text{O, heat}}\quad CH_3\overset{\overset{\displaystyle O}{||}}{C}CH_2CO_2H \quad\xrightarrow[\text{heat}]{-\text{CO}_2}\quad CH_3\overset{\overset{\displaystyle O}{||}}{C}CH_3$$

a β-keto ester   *a ketone*

SAMPLE PROBLEM

Predict the product of the ester condensation of methyl butanoate with sodium methoxide as the base, followed by acidification.

Solution:

1. Write the structure of the starting ester and determine the structure of the enolate ion.

$$CH_3CH_2CH_2\overset{\overset{\displaystyle O}{||}}{C}OCH_3 + {}^-OCH_3 \;\rightleftharpoons\; CH_3CH_2\overset{-}{C}H\overset{\overset{\displaystyle O}{||}}{C}OCH_3 + CH_3OH$$

2. Write the equation for nucleophilic attack on the carbonyl group and loss of ROH.

$$CH_3CH_2CH_2\overset{\overset{\displaystyle O}{||}}{C}OCH_3 \longrightarrow CH_3CH_2CH_2\overset{\displaystyle O}{\underset{\underset{\displaystyle CH_2CH_3}{|}}{\underset{\displaystyle {}^-C-}{|}}}{C}\overset{\overset{\displaystyle O}{||}}{C}OCH_3 \quad + HOCH_3$$

$$\overset{\overset{\displaystyle O}{||}}{{}^-CHCOCH_3}$$
$$\underset{CH_2CH_3}{|}$$

3. Acidify:

$$\xrightarrow{\text{H}^+}\quad CH_3CH_2CH_2\overset{\overset{\displaystyle O}{||}}{C}\underset{\underset{\displaystyle CH_2CH_3}{|}}{C}H\overset{\overset{\displaystyle O}{||}}{C}OCH_3$$

4. Write the equation for the overall reaction:

$$2\ CH_3CH_2CH_2\overset{\overset{\displaystyle O}{||}}{C}OCH_3 \quad\xrightarrow[\text{(2) H}^+]{\text{(1) Na}^+\ {}^-\text{OCH}_3}\quad CH_3CH_2CH_2\overset{\overset{\displaystyle O}{||}}{C}\underset{\underset{\displaystyle CH_2CH_3}{|}}{C}H\overset{\overset{\displaystyle O}{||}}{C}OCH_3$$

STUDY PROBLEM

14.19. Predict the major organic product:

(a) $C_6H_5CH_2CO_2C_2H_5$ $\xrightarrow[\text{(2) } H^+]{\text{(1) } Na^+ \ \ ^-OC_2H_5}$

(b)   [the product from (a)] $\xrightarrow[\text{heat}]{H_2O, \ H^+}$

SAMPLE PROBLEM

A Claisen-like condensation, called a **Dieckmann ring closure**, is used to prepare the following cyclic ketone from a diester. What is the structure of the diester?

Solution: Since the keto group arises from attack of an alpha carbon, the ring must be closed at the following position:

Therefore, the starting diester must be diethyl adipate.

diethyl hexanedioate
(diethyl adipate)

STUDY PROBLEM

14.20. Give the equations for the preparation of the following cyclic compounds from open-chain starting materials:

A. Crossed Claisen condensations

Two different esters may be used in the Claisen condensation. Best results (that is, avoidance of mixtures) are obtained if only one of the esters has an alpha hydrogen.

$$\underset{\text{no } \alpha \text{ hydrogen}}{C_6H_5\!-\!\overset{\displaystyle O}{\overset{\|}{C}}\!-\!OCH_3} + CH_3CH_2CO_2CH_3 \xrightarrow[\text{(2) H}^+]{\text{(1) Na}^+ \ ^-\text{OCH}_3} \underset{\substack{\displaystyle CH_3 \\ 45\%}}{C_6H_5\!-\!\overset{\displaystyle O}{\overset{\|}{C}}\!-\!\overset{\displaystyle |}{C}H\!-\!CO_2CH_3} + CH_3OH$$

$$\underset{\text{no } \alpha \text{ hydrogen}}{H\!-\!\overset{\displaystyle O}{\overset{\|}{C}}\!-\!OC_2H_5} + CH_3CO_2C_2H_5 \xrightarrow[\text{(2) H}^+]{\text{(1) Na}^+ \ ^-\text{OC}_2\text{H}_5} H\!-\!\overset{\displaystyle O}{\overset{\|}{C}}\!-\!CH_2CO_2C_2H_5 + C_2H_5OH$$

Crossed Claisen condensations may be successfully carried out between ketones and esters whether the ester contains an α hydrogen or not. The α hydrogen of the ketone is removed preferentially because ketones are more acidic than esters. For this reason, the crossed Claisen is favored over a self-Claisen condensation of the ester.

$$\underset{\text{ethyl benzoate}}{C_6H_5\!-\!\overset{\displaystyle O}{\overset{\|}{C}}\!-\!OC_2H_5} + \underset{\text{acetone}}{CH_3\!-\!\overset{\displaystyle O}{\overset{\|}{C}}\!-\!CH_3} \xrightarrow[\text{(2) H}^+]{\text{(1) Na}^+ \ ^-\text{OC}_2\text{H}_5} \underset{\substack{\text{1-phenyl-1,3-butanedione} \\ 40\%}}{C_6H_5\!-\!\overset{\displaystyle O}{\overset{\|}{C}}\!-\!CH_2\!-\!\overset{\displaystyle O}{\overset{\|}{C}}\!-\!CH_3} + C_2H_5OH$$

SAMPLE PROBLEM

A mixture of acetone and diethyl oxalate $(C_2H_5O_2C\!-\!CO_2C_2H_5)$ is added to a mixture of sodium ethoxide in ethanol. After the reaction is completed, the mixture is treated with cold, dilute HCl. A condensation product is isolated from the mixture in 60% yield. What is the product?

Solution:

$$(1) \quad CH_3\overset{\displaystyle O}{\overset{\|}{C}}CH_3 + {}^-OC_2H_5 \rightleftharpoons {}^-CH_2\overset{\displaystyle O}{\overset{\|}{C}}CH_3 + HOC_2H_5$$

$$(2) \quad C_2H_5O\overset{\displaystyle O}{\overset{\|}{C}}\!-\!\overset{\displaystyle O}{\overset{\|}{C}}OC_2H_5 + {}^-CH_2\overset{\displaystyle O}{\overset{\|}{C}}CH_3 \longrightarrow \xrightarrow{\text{H}^+}$$

$$\underset{\text{from diethyl oxalate} \longrightarrow \quad \text{from acetone}}{C_2H_5O\overset{\displaystyle O}{\overset{\|}{C}}\!-\!\overset{\displaystyle O}{\overset{\|}{C}}\!-\!CH_2\overset{\displaystyle O}{\overset{\|}{C}}CH_3} + HOC_2H_5$$

STUDY PROBLEM

14.21. Predict the products:

(a) $C_2H_5O_2C-CO_2C_2H_5 + CH_3CO_2C_2H_5 \xrightarrow[\text{(2) H}^+]{\text{(1) Na}^+ \ ^-OC_2H_5}$

(b) [cyclopentanone structure]$=O + CH_3CO_2C_2H_5 \xrightarrow[\text{(2) H}^+]{\text{(1) Na}^+ \ ^-OC_2H_5}$

B.   Syntheses using ester condensations

Since the product of an ester condensation between two esters is a β-keto ester (or a ketone after hydrolysis and decarboxylation), the decision of which starting materials to use is not difficult. The <u>keto group comes from one starting ester;</u> the <u>ester group with its attachment comes from the other starting ester.</u> The different types of products obtained from ester condensations are summarized in Figure 14.3.

$$CH_3CH_2CH_2\overset{\displaystyle O}{\overset{\displaystyle \|}{C}}-CHCO_2C_2H_5$$
$$\underset{CH_2CH_3}{|}$$

from ethyl butanoate

[benzene ring]$-\overset{\displaystyle O}{\overset{\displaystyle \|}{C}}-\underset{\underset{CH_2CH_3}{|}}{CHCO_2H}$

from ethyl benzoate   *from ethyl butanoate*

[benzene ring]$-\overset{\displaystyle O}{\overset{\displaystyle \|}{C}}-CH_2CH_2CH_3$

from ethyl benzoate

from ethyl butanoate
(after decarboxylation)

SAMPLE PROBLEM

What reactants would you need to prepare the following compound by an ester condensation?

[furan ring]$-\overset{\displaystyle O}{\overset{\displaystyle \|}{C}}CHCO_2C_2H_5$
$$\underset{CH_3}{|}$$

Solution:

[furan ring]$-\overset{\displaystyle O}{\overset{\displaystyle \|}{C}}CHCO_2C_2H_5$
$$\underset{CH_3}{|}$$

from [furan ring]$-CO_2C_2H_5$

from $CH_3CH_2CO_2C_2H_5$

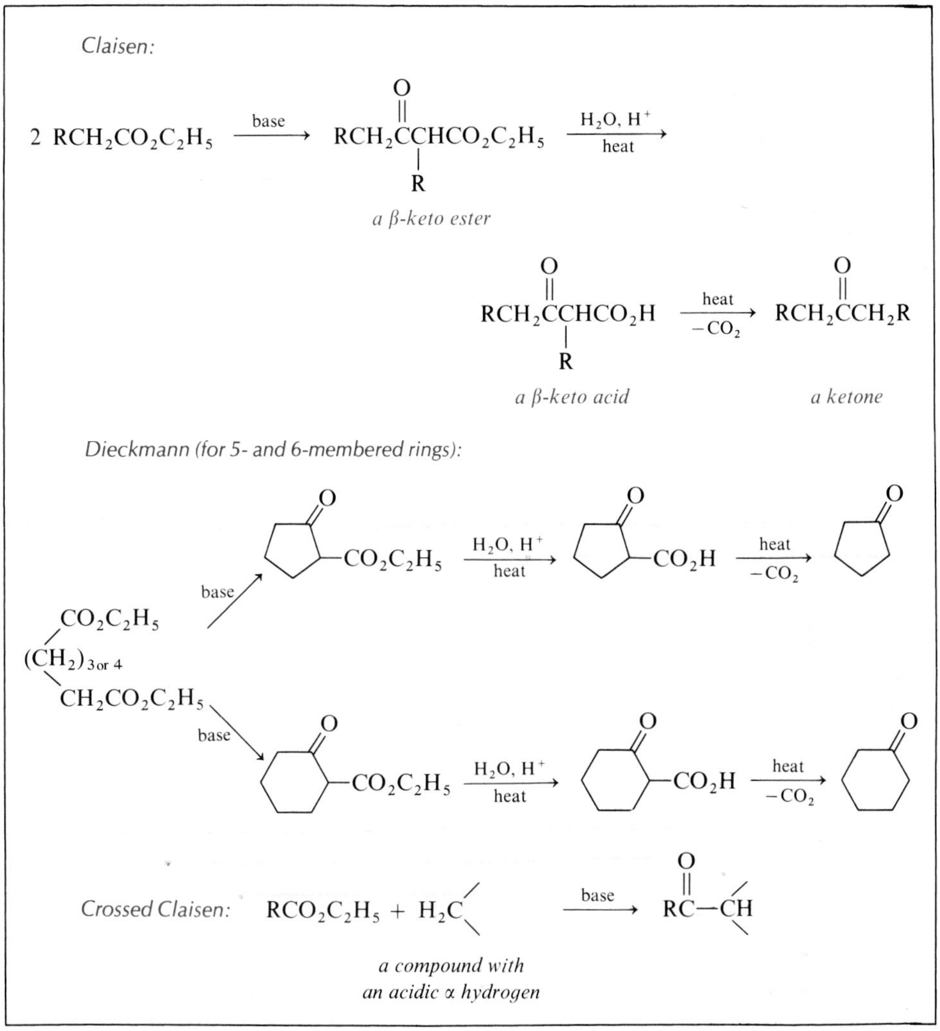

Claisen:

$$2 \ RCH_2CO_2C_2H_5 \xrightarrow{\text{base}} RCH_2\overset{\overset{\displaystyle O}{\|}}{C}\underset{\underset{\displaystyle R}{|}}{CH}CO_2C_2H_5 \xrightarrow[\text{heat}]{H_2O, \ H^+}$$

a β-keto ester

$$RCH_2\overset{\overset{\displaystyle O}{\|}}{C}\underset{\underset{\displaystyle R}{|}}{CH}CO_2H \xrightarrow[-CO_2]{\text{heat}} RCH_2\overset{\overset{\displaystyle O}{\|}}{C}CH_2R$$

a β-keto acid   *a ketone*

Dieckmann (for 5- and 6-membered rings):

$$\begin{array}{c} CO_2C_2H_5 \\ | \\ (CH_2)_{3 \text{ or } 4} \\ | \\ CH_2CO_2C_2H_5 \end{array}$$

Crossed Claisen: $\quad RCO_2C_2H_5 + H_2C\big\langle \xrightarrow{\text{base}} RC\overset{\overset{\displaystyle O}{\|}}{}\!-\!CH\big\langle$

a compound with
an acidic α hydrogen

FIGURE 14.3. Products of ester condensations.

STUDY PROBLEMS

14.22. Give equations for preparations of the following compounds:

(a)

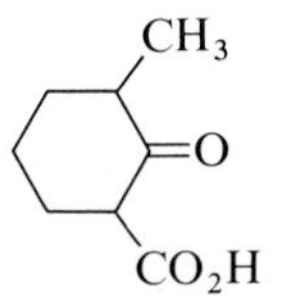

(b)

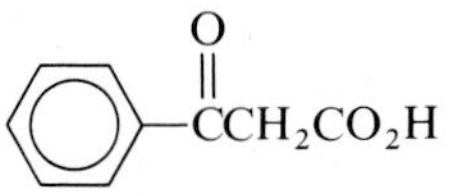

(c)

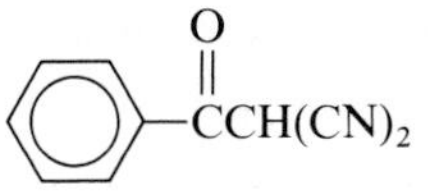

(d) $(CH_3)_3C\overset{\overset{\displaystyle O}{\|}}{C}CH_2\overset{\overset{\displaystyle O}{\|}}{C}CH_3$

14.23. Ester condensations are used for the synthesis of hydrocarbon chains in living organisms. Fatty acids, for example, are synthesized from acetyl groups by way of the thioester acetylcoenzyme A, the structure of which was shown in Section 13.8.

(a)   Indicate the acidic hydrogens in the abbreviated structure $CH_3\overset{\displaystyle O}{\overset{\|}{C}}SCoA$.

(b)   What is the product of an ester condensation between two molecules of acetylcoenzyme A? (Use the abbreviated structure.)

SECTION 14.10.

Nucleophilic Addition to α,β-Unsaturated Carbonyl Compounds

A double bond in conjugation with a carbonyl group is susceptible to nucleophilic attack in a 1,4-addition reaction (Section 11.19).

$$CH_2=CH-\overset{\displaystyle :\ddot{O}}{\overset{\|}{C}}CH_3 + :NH_3 \longrightarrow \left[\begin{array}{c} :\ddot{O}:^- \\ | \\ CH_2-CH=\overset{}{C}CH_3 \\ | \\ {}^+NH_3 \end{array}\right] \longrightarrow CH_2CH_2\overset{\displaystyle O}{\overset{\|}{C}}CH_3 \\ \qquad\qquad\qquad\qquad\qquad\quad NH_2$$

an enolate

If an α,β-unsaturated carbonyl compound can undergo nucleophilic attack, we might expect that an enolate ion could add to the double bond. Indeed, it does. This useful synthetic reaction is called a **Michael addition**.

$$CH_2=CH-\overset{\displaystyle :\ddot{O}}{\overset{\|}{C}}H + {}^-\overset{}{C}H(CO_2C_2H_5)_2 \rightleftharpoons \left[\begin{array}{c} :\ddot{O}:^- \\ | \\ CH_2-CH=CH \\ | \\ CH(CO_2C_2H_5)_2 \end{array}\right] \overset{H^+}{\rightleftharpoons} \begin{array}{c} O \\ \| \\ CH_2CH_2CH \\ | \\ CH(CO_2C_2H_5)_2 \end{array}$$

an enolate   50%

$$\text{\Large⬡}-CH=CH-\overset{\displaystyle O}{\overset{\|}{C}}OC_2H_5 + {}^-CH(CO_2C_2H_5)_2 \rightleftharpoons$$

$$\left[\text{\Large⬡}-\begin{array}{c} O^- \\ | \\ CH-CH=COC_2H_5 \\ | \\ CH(CO_2C_2H_5)_2 \end{array}\right] \overset{H^+}{\rightleftharpoons} \text{\Large⬡}-\begin{array}{c} O \\ \| \\ CHCH_2COC_2H_5 \\ | \\ CH(CO_2C_2H_5)_2 \end{array}$$

98%

The product of the last example shown is a triester. Saponification, followed by acidification, gives a triacid. Under the conditions of acid hydrolysis, however, decarboxylation may occur.

Acid hydrolysis and decarboxylation:

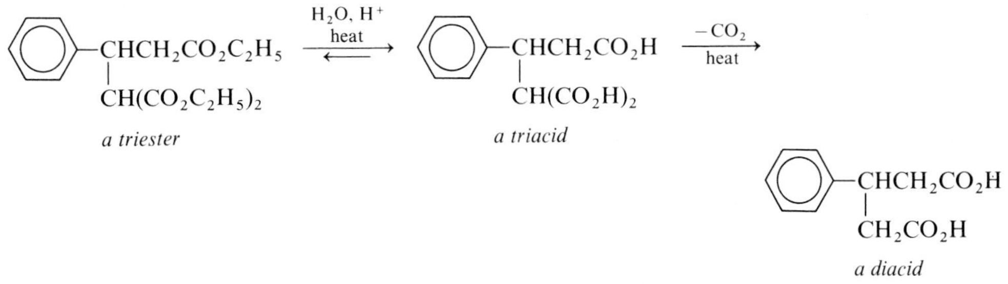

$$\underset{\substack{\big|\\ \text{CH(CO}_2\text{C}_2\text{H}_5)_2\\ \textit{a triester}}}{\text{C}_6\text{H}_5\text{—CHCH}_2\text{CO}_2\text{C}_2\text{H}_5} \quad \underset{\text{heat}}{\overset{\text{H}_2\text{O, H}^+}{\rightleftharpoons}} \quad \underset{\substack{\big|\\ \text{CH(CO}_2\text{H)}_2\\ \textit{a triacid}}}{\text{C}_6\text{H}_5\text{—CHCH}_2\text{CO}_2\text{H}} \quad \underset{\text{heat}}{\overset{-\text{CO}_2}{\longrightarrow}}$$

$$\underset{\substack{\big|\\ \text{CH}_2\text{CO}_2\text{H}\\ \textit{a diacid}}}{\text{C}_6\text{H}_5\text{—CHCH}_2\text{CO}_2\text{H}}$$

In the decarboxylation of the triacid, it is the malonic acid grouping that loses CO_2. (Why?) This group is circled in the following equation.

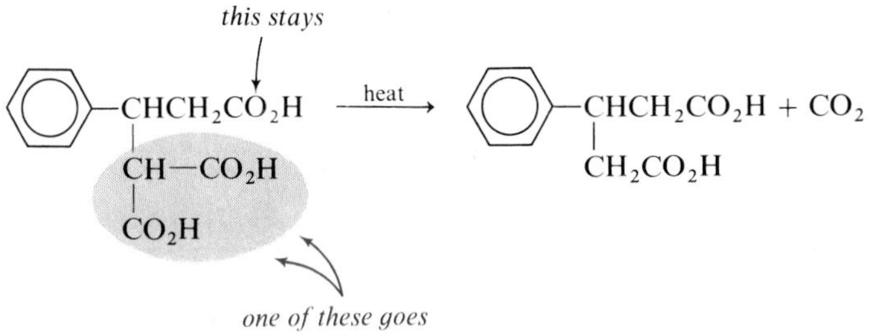

this stays

$$\underset{\substack{\big|\\ \text{CH—CO}_2\text{H}\\ \big|\\ \text{CO}_2\text{H}}}{\text{C}_6\text{H}_5\text{—CHCH}_2\overset{\downarrow}{\text{C}}\text{O}_2\text{H}} \quad \overset{\text{heat}}{\longrightarrow} \quad \underset{\substack{\big|\\ \text{CH}_2\text{CO}_2\text{H}}}{\text{C}_6\text{H}_5\text{—CHCH}_2\text{CO}_2\text{H}} + \text{CO}_2$$

one of these goes

The types of products that we can obtain from simple Michael additions are shown in Figure 14.4. Michael additions in combination with other condensations are exceedingly useful in laboratory syntheses of complex cyclic compounds, such

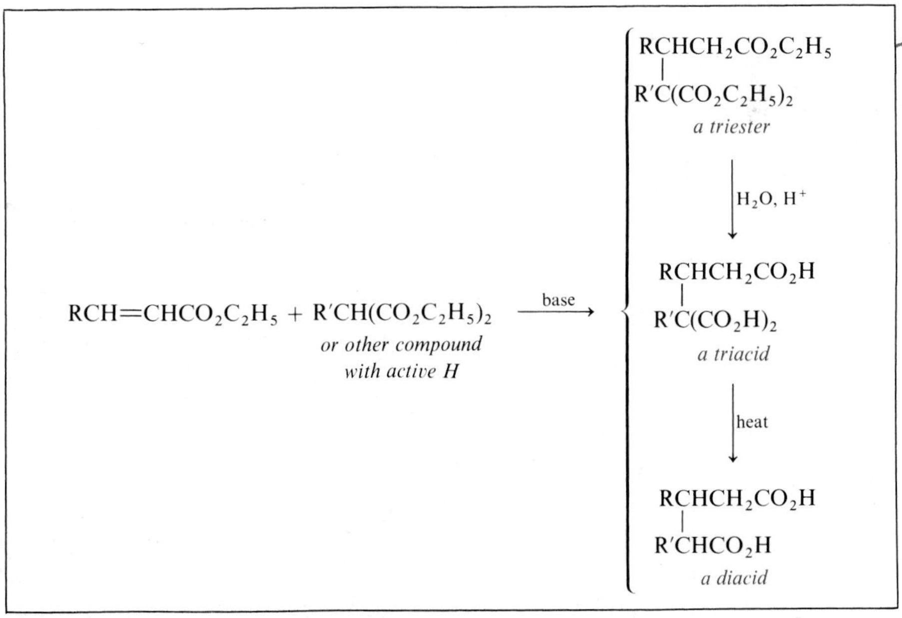

$$\text{RCH}{=}\text{CHCO}_2\text{C}_2\text{H}_5 + \text{R}'\text{CH(CO}_2\text{C}_2\text{H}_5)_2 \xrightarrow{\text{base}}$$

or other compound with active H

$$\left.\begin{array}{c} \underset{\substack{\big|\\ \text{R}'\text{C(CO}_2\text{C}_2\text{H}_5)_2\\ \textit{a triester}}}{\text{RCHCH}_2\text{CO}_2\text{C}_2\text{H}_5}\\[2em] \Big\downarrow {\scriptstyle \text{H}_2\text{O, H}^+}\\[2em] \underset{\substack{\big|\\ \text{R}'\text{C(CO}_2\text{H)}_2\\ \textit{a triacid}}}{\text{RCHCH}_2\text{CO}_2\text{H}}\\[2em] \Big\downarrow {\scriptstyle \text{heat}}\\[2em] \underset{\substack{\big|\\ \text{R}'\text{CHCO}_2\text{H}\\ \textit{a diacid}}}{\text{RCHCH}_2\text{CO}_2\text{H}} \end{array}\right\}$$

FIGURE 14.4. Products of Michael addition of a malonic ester with an α,β-unsaturated ester.

as steroids. A portion of one such synthesis is shown. This particular ring-forming sequence (Michael plus aldol) is called a **Robinson annelation**.

SAMPLE PROBLEMS

Show by equations how you would prepare the following keto acid by a Michael addition:

$$CH_3\overset{O}{\underset{||}{C}}CH_2CH_2CH_2CO_2H$$

Solution: Addition occurs β to the keto group in a 1,4-addition.

1. $CH_2(CO_2C_2H_5)_2 \xrightarrow{\text{NaOC}_2\text{H}_5} {}^-CH(CO_2C_2H_5)_2$

2.

3. $\xrightarrow[\text{heat}]{H_2O, H^+} CH_3\overset{O}{\underset{||}{C}}CH_2CH_2CH_2CO_2H$

A chemist decided to try the sequence of reactions in the preceding sample problem. After Step 2, he discovered that he had a *cyclic* product, along with the keto diester that he had predicted. What happened?

Solution: The initial product of Step 2 underwent a Dieckmann ring closure. (Note that this intermediate has a nucleophilic carbon that can attack a carbonyl group to yield a six-membered ring.)

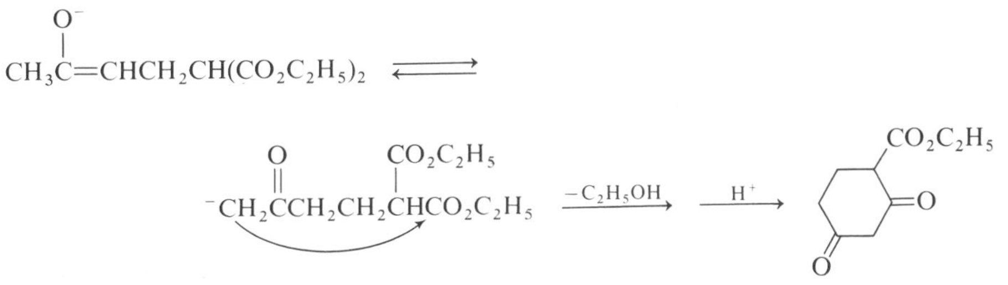

$$
\begin{array}{c}
\overset{\displaystyle O^-}{\underset{\displaystyle |}{}} \\
CH_3C{=}CHCH_2CH(CO_2C_2H_5)_2
\end{array}
\rightleftharpoons
$$

$$
\overset{O}{\overset{||}{}}\;\;\;\overset{CO_2C_2H_5}{\overset{|}{}}
$$
$$
{}^-CH_2CCH_2CH_2CHCO_2C_2H_5 \quad \xrightarrow{-C_2H_5OH} \quad \xrightarrow{H^+}
$$

STUDY PROBLEM

14.24. If the chemist had subjected the cyclic product in the preceding sample problem to acid hydrolysis and then distilled the organic material, what final product would have been observed?

SAMPLE PROBLEM

How would you prepare the following keto ester by a Michael addition?

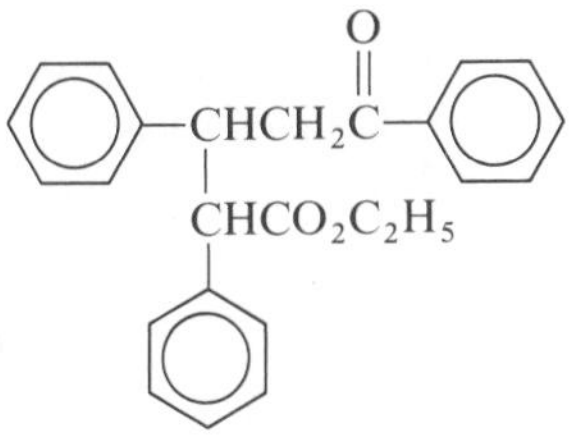

Solution: Addition occurs β to the ketone.

$$
\begin{array}{c}
\overset{\displaystyle O}{\overset{\displaystyle ||}{}} \\
C_6H_5CH{=}CHCC_6H_5 \;+\; CH_2CO_2C_2H_5 \\
\underset{\displaystyle |}{} \\
C_6H_5
\end{array}
\quad \xrightarrow{Na^+\ {}^-OC_2H_5} \quad \xrightarrow{H^+} \quad product
$$

STUDY PROBLEM

14.25. Show by equations how you would prepare the following carboxylic acids by Michael additions:

(a) 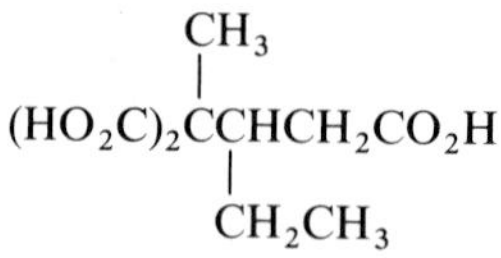 $(HO_2C)_2\overset{\overset{\displaystyle CH_3}{|}}{C}CHCH_2CO_2H$
     $\underset{\displaystyle CH_2CH_3}{|}$

(b) 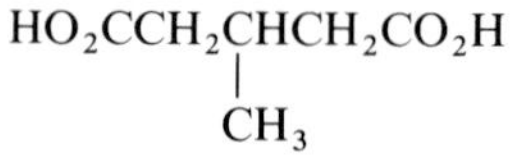 $HO_2CCH_2\overset{}{C}HCH_2CO_2H$
     $\underset{\displaystyle CH_3}{|}$

(c)

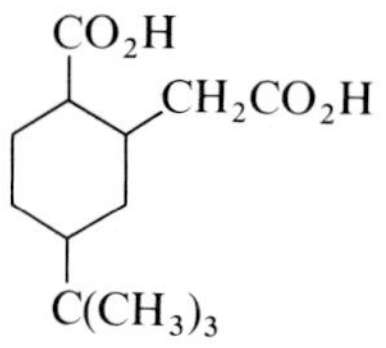

Summary

In this chapter, we have looked at a variety of ways to synthesize compounds with complex carbon skeletons. Each of these reactions is caused by a species with carbanion character.

Alkylations: $\diagdown\!CH^- + RX \longrightarrow \diagdown\!CHR + X^-$

Condensations:

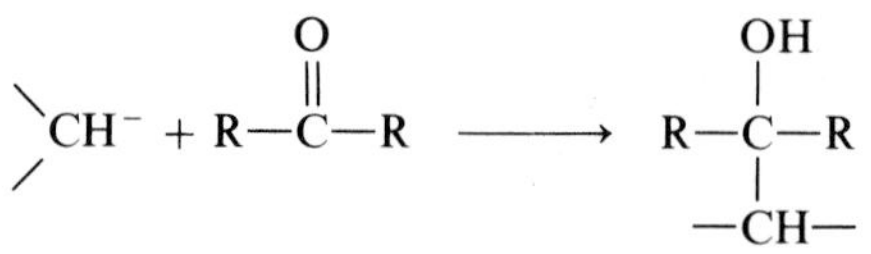

$$\diagdown\!CH^- + R\overset{\overset{\displaystyle O}{||}}{C}\!-\!R \longrightarrow R\!-\!\overset{\overset{\displaystyle OH}{|}}{\underset{\underset{\displaystyle -CH-}{|}}{C}}\!-\!R$$

1,4-Additions:

$$\diagdown\!CH^- + CH_2\!=\!CH\overset{\overset{\displaystyle O}{||}}{C}\!- \longrightarrow \underset{\underset{\displaystyle -CH-}{|}}{CH_2}\!-\!CH_2\overset{\overset{\displaystyle O}{||}}{C}\!-$$

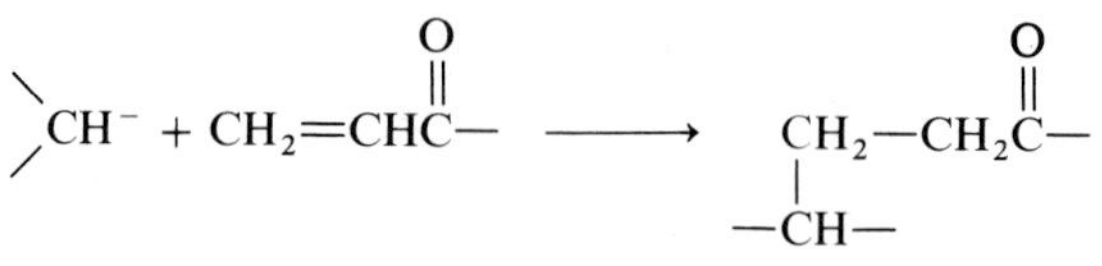

Table 14.2 gives an overview of the more important products available from these reactions.

TABLE 14.2. Major synthetic reactions involving enolates and carbanions

Malonic ester: $\quad CH_2(CO_2C_2H_5)_2 \quad \xrightarrow[\text{(2) }^-OC_2H_5,\ R'X]{\text{(1) }^-OC_2H_5,\ RX} \quad \overset{R}{\underset{R'}{\diagdown}} C(CO_2C_2H_5)_2$

Acetoacetic ester: $\quad CH_3\overset{\displaystyle O}{\overset{\|}{C}}CH_2CO_2C_2H_5 \quad \xrightarrow[\text{(2) }^-OC_2H_5,\ R'X]{\text{(1) }^-OC_2H_5,\ RX} \quad CH_3\overset{\displaystyle O}{\overset{\|}{C}}\underset{\underset{R\quad R'}{\diagup\diagdown}}{C}CO_2C_2H_5$

Enamine: $\quad R_2CH\overset{\displaystyle O}{\overset{\|}{C}}R \quad \xrightarrow[\text{(3) }H_2O,\ H^+]{\substack{\text{(1) }R_2NH,\ H^+ \\ \text{(2) }R'X}} \quad R_2\underset{\underset{R'}{|}}{C}\overset{\displaystyle O}{\overset{\|}{C}}R$

Aldol condensation: $\quad 2\ RCH_2CHO \quad \underset{\longleftarrow}{\overset{OH^-}{\longrightarrow}} \quad RCH_2\underset{\underset{R}{|}}{C}H\overset{\overset{\displaystyle OH}{|}}{C}HCHO$

Crossed aldol: $\quad RCHO + R'CH_2CHO \quad \underset{\longleftarrow}{\overset{OH^-}{\longrightarrow}} \quad RCH\overset{\overset{\displaystyle OH}{|}}{\underset{\underset{R'}{|}}{C}}HCHO$

Claisen condensation: $\quad 2\ RCH_2CO_2C_2H_5 \quad \xrightarrow[\text{(2) }H^+]{\text{(1) }^-OC_2H_5} \quad RCH_2\overset{\displaystyle O}{\overset{\|}{C}}\underset{\underset{R}{|}}{C}HCO_2C_2H_5$

Crossed Claisen: $\quad R\overset{\displaystyle O}{\overset{\|}{C}}OC_2H_5 + R'CH_2CO_2C_2H_5 \quad \xrightarrow[\text{(2) }H^+]{\text{(1) }^-OC_2H_5} \quad R\overset{\displaystyle O}{\overset{\|}{C}}\underset{\underset{R'}{|}}{C}HCO_2C_2H_5$

Michael addition:

$RCH{=}CHCO_2C_2H_5 + R'CH(CO_2C_2H_5)_2 \quad \xrightarrow[\text{(2) }H^+]{\text{(1) }^-OC_2H_5} \quad RCH{-}CH_2CO_2C_2H_5$
$\phantom{RCH{=}CHCO_2C_2H_5 + R'CH(CO_2C_2H_5)_2 \quad \xrightarrow[\text{(2) }H^+]{\text{(1) }^-OC_2H_5} \quad RCH{-}}\underset{\underset{R'C(CO_2C_2H_5)_2}{|}}{}$

STUDY PROBLEMS

14.26. In each structure, indicate the most acidic hydrogen:

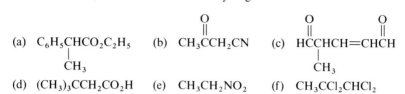

(a) $C_6H_5CHCO_2C_2H_5$
        $|$
        CH_3

(b) $CH_3\overset{O}{\overset{||}{C}}CH_2CN$

(c) $H\overset{O}{\overset{||}{C}}CHCH=CH\overset{O}{\overset{||}{C}}H$
        $|$
        CH_3

(d) $(CH_3)_3CCH_2CO_2H$

(e) $CH_3CH_2NO_2$

(f) $CH_3CCl_2CHCl_2$

14.27. Write an equation for the reversible acid–base reaction of each of the following compounds with sodium ethoxide in ethanol:

(a) $C_2H_5O\overset{O}{\overset{||}{C}}CH_2CN$

(b) (cyclohexane-1,3-dione structure)

(c) (cyclohexanone with —CN structure)

14.28. Write resonance structures that show delocalization of the negative charge in the following anions:

(a) $C_6H_5CH_2^-$

(b) $C_6H_5\overset{-}{C}H\overset{O}{\overset{||}{C}}CH_3$

(c) $C_6H_5CH_2\overset{O}{\overset{||}{C}}CH_2^-$

(d) $^-CH_2CH=CHCO_2C_2H_5$

(e) $CH_3CH_2CH_2\overset{O}{\overset{||}{C}}CH_2^-$

(f) $CH_3\overset{O}{\overset{||}{C}}\overset{-}{C}HCN$

14.29. Rank the following compounds in order of increasing acid strength (weakest acid first):

(a) (cyclohexane)—CHO

(b) (cyclohexane)—CO$_2$C$_2$H$_5$

(c) (cyclohexanone)—CHO

(d) (cyclohexanone)—CO$_2$C$_2$H$_5$

(e) C_2H_5OH

(f) (cyclohexane)—CO$_2$H

14.30. Complete the following equations for *acid–base reactions*, and indicate by arrow size if each equilibrium is to the right or to the left:

(a) $^-CH_2CO_2C_2H_5 + CH_3\overset{O}{\overset{||}{C}}CH_2CO_2C_2H_5 \rightleftharpoons$

(b) $CH_3\overset{O}{\overset{||}{C}}CH_2NO_2 + OH^- \rightleftharpoons$

(c) (cyclohexane)—$CO_2C_2H_5 + CH_3\overset{O}{\overset{||}{C}}\overset{-}{C}H\overset{O}{\overset{||}{C}}CH_3 \rightleftharpoons$

14.31. Predict the major organic product:

(a) O=⬠=O $\xrightarrow[\text{(2) NaOCH}_3, \text{CH}_3\text{I}]{\text{(1) NaOCH}_3, \text{CH}_3\text{I}}$

(b) $CH_3\overset{\overset{\displaystyle O}{\|}}{C}CH_2CO_2C_2H_5$ $\xrightarrow[\text{(2)} \ \bigcirc\text{—Br}]{\text{(1) NaOC}_2\text{H}_5}$

(c) [the product from (b)] $\xrightarrow[\text{heat}]{\text{H}_2\text{O, H}^+}$

(d) (decalin with CH$_3$ and OTs, H) $+ \ CH_2(CO_2C_2H_5)_2$ $\xrightarrow{\text{(CH}_3)_3\text{COK}}$

(*Hint*: See Section 7.12C.)

(e) $2 \ CH_2(CO_2C_2H_5)_2$ $\xrightarrow[\text{(2) CH}_2\text{I}_2]{\text{(1) 2 NaOC}_2\text{H}_5}$

(f) [the product from (e)] $\xrightarrow[\text{heat}]{\text{H}_2\text{O, H}^+}$

(g) $CH_3\overset{\overset{\displaystyle O}{\|}}{C}CH_2CO_2C_2H_5$ $\xrightarrow[\text{(2) BrCH}_2\text{CCH}_3]{\text{(1) NaOC}_2\text{H}_5}$ (with C=O on the side chain)

(h) [the product from (g)] $\xrightarrow[\text{heat}]{\text{H}_2\text{O, H}^+}$

(i) O_2N—(benzene ring with NO$_2$)—Cl $\xrightarrow[\text{(2) cold, dilute HCl}]{\text{(1) NC}\overline{\text{C}}\text{HCO}_2\text{C}_2\text{H}_5 \ \text{Na}^+}$

(*Hint :* See Section 10.15.)

14.32. How would you prepare each of the following compounds from diethyl malonate or ethyl acetoacetate?

(a) ⬠—CH$_2$CH$_2$CO$_2$H   (b) HO$_2$CCH$_2$CHCO$_2$H
                                                              |
                                                             CH$_3$

(c) $C_6H_5CH_2CH_2\overset{\overset{\displaystyle O}{\|}}{C}CH_3$ (d) (cyclopentanone)—CH$_2$CO$_2$H

14.33. Show the mechanism for the formation of the enamine of cyclopentanone and piperidine, followed by reaction with benzoyl chloride (C$_6$H$_5$COCl).

14.34. Show how you could synthesize the following ketones from cyclohexanone.

(a) (cyclohexanone)—CH$_2$—(benzene)—CH$_3$ (b) O_2N—(benzene with NO$_2$)—(cyclohexanone)

14.35. An aqueous NaOH solution is added to a mixture of acetone and formaldehyde.

(a)   What is the structure of the organic anion formed?

(b)   Would this anion undergo reaction at the faster rate with acetone or with formaldehyde?

(c)   Show each step in the mechanism for the predominant aldol condensation that would occur.

14.36. Predict the major organic product:

(a) $HC(CH_2)_4CH$ with O at each end $\xrightarrow[\text{(2) H}^+, \text{ heat}]{\text{(1) OH}^-, \text{ H}_2\text{O}}$

(b) $C_6H_5CHO +$ [cyclopentanone ylidene] $=O$ $\xrightarrow{\text{OH}^-}$

(c)   [cyclodecanedione structure] $\xrightarrow[\text{(2) H}^+, \text{ heat}]{\text{(1) OH}^-, \text{ H}_2\text{O}}$

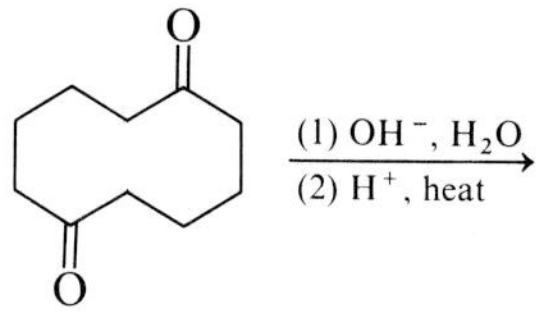

(d) 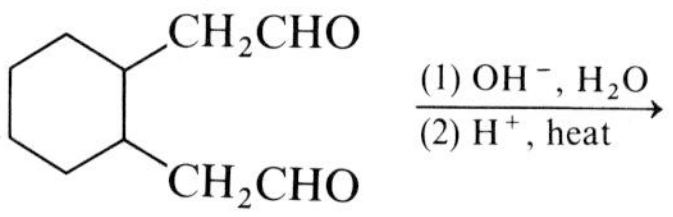   [cyclohexane with CH$_2$CHO and CH$_2$CHO substituents] $\xrightarrow[\text{(2) H}^+, \text{ heat}]{\text{(1) OH}^-, \text{ H}_2\text{O}}$

14.37. How would you make the following conversions?

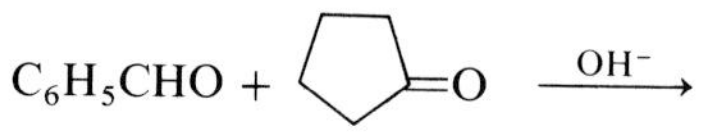

(a)   cyclopentanone   to   [cyclopentylidene]$=CC_6H_5$ with $CO_2C_2H_5$ below

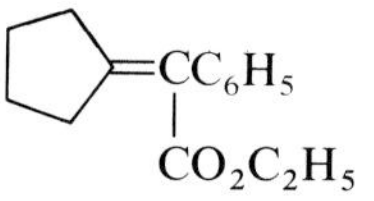

(b) 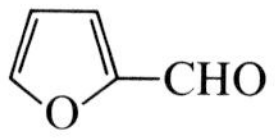   [furan]—CHO   to   [furan]—CH=CHNO$_2$

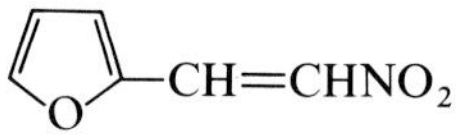

(c)   benzaldehyde   to $C_6H_5CH=CHCCH_3$ with O double bond

(d)   benzaldehyde   to $C_6H_5CH=CHCCH=CHC_6H_5$ with O double bond

14.38. Predict the major organic products:

(a) 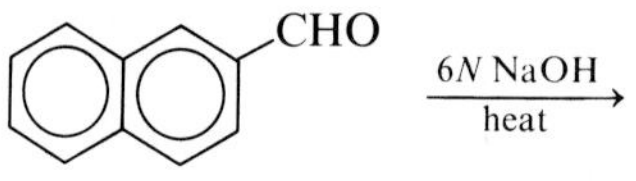   [naphthalene-CHO structure] $\xrightarrow[\text{heat}]{6N \text{ NaOH}}$

(b) $(CH_3)_3CCHO$ $\xrightarrow[\text{heat}]{6N \text{ NaOH}}$

(c) $2\,CH_3(CH_2)_3CO_2C_2H_5$ $\xrightarrow[\text{(2) H}_2\text{O, H}^+, \text{ heat}]{\text{(1) NaOC}_2\text{H}_5}$

(d) 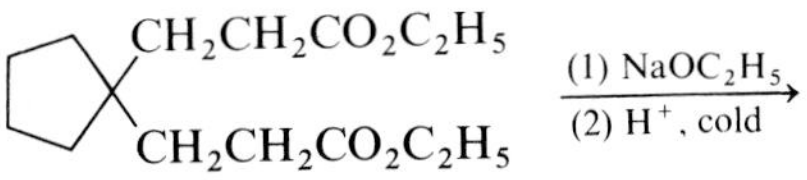 [cyclopentane ring with two CH₂CH₂CO₂C₂H₅ groups] $CH_2CH_2CO_2C_2H_5$ / $CH_2CH_2CO_2C_2H_5$ $\xrightarrow[\text{(2) H}^+, \text{ cold}]{\text{(1) NaOC}_2\text{H}_5}$

(e) $C_6H_5CH_2CO_2C_2H_5 + C_2H_5O\overset{O}{\overset{\|}{C}}OC_2H_5$ $\xrightarrow[\text{(2) H}^+, \text{ cold}]{\text{(1) NaOC}_2\text{H}_5}$
diethyl carbonate

(f) 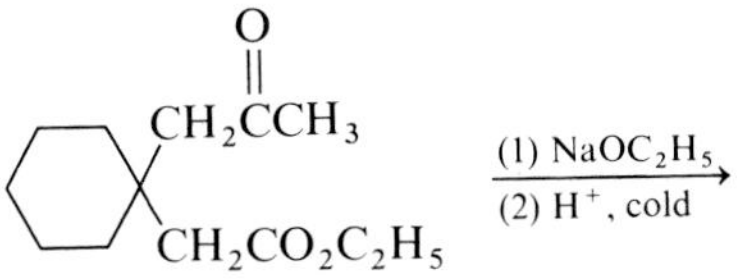 [cyclohexane ring with $CH_2\overset{O}{\overset{\|}{C}}CH_3$ and $CH_2CO_2C_2H_5$ groups] $\xrightarrow[\text{(2) H}^+, \text{ cold}]{\text{(1) NaOC}_2\text{H}_5}$

(g) $CH_3\overset{O}{\overset{\|}{C}}(CH_2)_4CO_2C_2H_5$ $\xrightarrow[\text{(2) H}^+, \text{ cold}]{\text{(1) NaOC}_2\text{H}_5}$

(h) 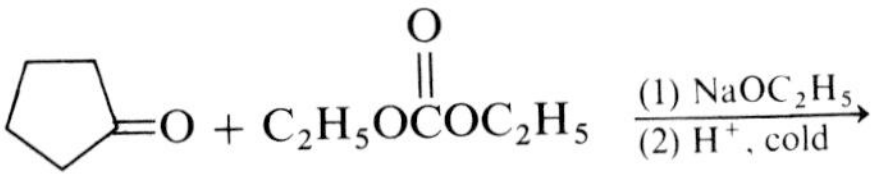 [cyclopentane ring]=O $+ C_2H_5O\overset{O}{\overset{\|}{C}}OC_2H_5$ $\xrightarrow[\text{(2) H}^+, \text{ cold}]{\text{(1) NaOC}_2\text{H}_5}$

14.39. Suggest synthetic paths leading to the following compounds:

(a) 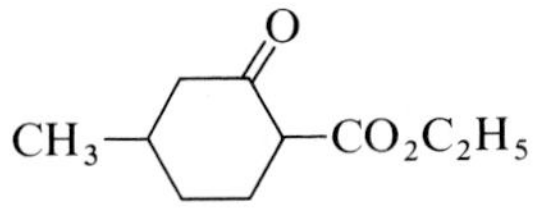 CH_3-[cyclohexanone ring]$-CO_2C_2H_5$

(b) $C_6H_5\overset{O}{\overset{\|}{C}}\underset{CN}{CH}CO_2C_2H_5$

(c) 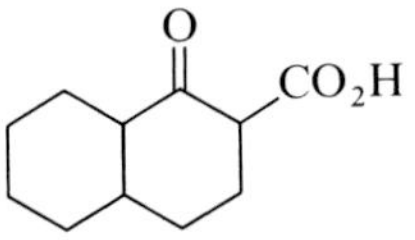 [decalone ring system with CO_2H]

(d) 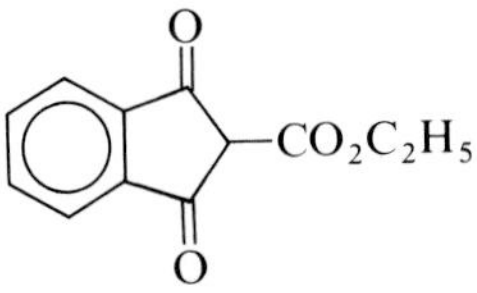 [indandione ring system with $-CO_2C_2H_5$]

(e) $(CH_3)_3C\overset{O}{\overset{\|}{C}}\underset{CH_3}{CH}CO_2C_2H_5$

(f) $C_6H_5\overset{O}{\overset{\|}{C}}\underset{C_6H_5}{CH}CO_2CH_3$

14.40. Predict the major organic product of each of the following Michael addition reactions:

(a) $1\ CH_2{=}CH\overset{O}{\overset{\|}{C}}CH_2CH_3 + CH_2(CO_2C_2H_5)_2$ $\xrightarrow[\text{(2) H}^+]{\text{(1) NaOC}_2\text{H}_5}$

(b) $2\ CH_2{=}CH\overset{O}{\overset{\|}{C}}CH_2CH_3 + CH_2(CO_2C_2H_5)_2$ $\xrightarrow[\text{(2) H}^+]{\text{(1) NaOC}_2\text{H}_5}$

(c) 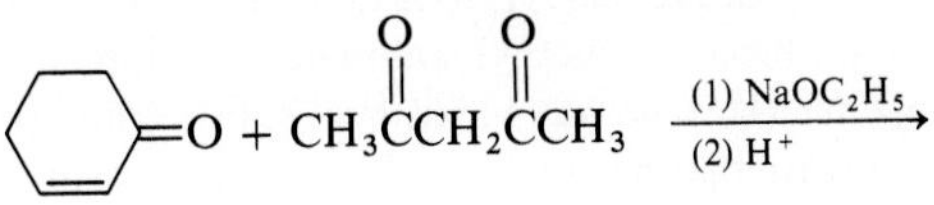 [cyclohexenone ring]=O $+ CH_3\overset{O}{\overset{\|}{C}}CH_2\overset{O}{\overset{\|}{C}}CH_3$ $\xrightarrow[\text{(2) H}^+]{\text{(1) NaOC}_2\text{H}_5}$

(d) $CH_3CH{=}CHCO_2C_2H_5 + C_6H_5CH_2CO_2C_2H_5$ $\xrightarrow[\text{(2) H}^+]{\text{(1) NaOC}_2\text{H}_5}$

14.41. How would you prepare the following compounds by Michael additions?

(a) $CH_3CCHCH_2CH_2CN$
      $\overset{O}{\overset{\|}{}}$
      $\underset{\overset{|}{CO_2C_2H_5}}{}$

(b)   [cyclohexane-1,3-dione with] $-CH_2CH_2CO_2C_2H_5$

14.42. Suggest syntheses for the following compounds from readily available starting materials:

(a) $C_6H_5CH{=}\overset{O}{\overset{\|}{C}}COC_2H_5$
      $\underset{C_6H_5}{}$

(b)   [cyclohexanone with] $-CH_2CH{=}CHCH_3$

(c) $(CH_3)_2CHCH(CO_2CH_3)_2$

(d) $C_6H_5CH{=}C(CN)_2$

(e) $C_6H_5CH{=}\overset{O}{\overset{\|}{C}}CCH_3$
      $\underset{CO_2C_2H_5}{}$

(f)   [cyclohexanone with] $-CHO$

(g) $C_2H_5O_2C\overset{O}{\overset{\|}{C}}CHCO_2C_2H_5$
      $\underset{CH_2CH_3}{}$

(h)   [cyclohexanone with] $-CH(CO_2C_2H_5)_2$

(i) $CH_3(CH_2)_3\overset{CHO}{\underset{|}{C}}HCH_2CH_3$

(j) $C_6H_5CH_2\overset{CH_3CHOH}{\underset{|}{C}}HCH_2-$ [cyclohexane]

14.43. (a)   How would you prepare the following compound from 2-methylcyclohexanone?

[structure: bicyclic compound with CH₃ groups, O]

(b)   From the dione in (a), show how you could prepare:

[structure: bicyclic compound with CH₃, O]

14.44. A chemist has on his shelf: dilute HCl, dilute NaOH, sodium metal, magnesium metal, anhydrous ethanol, anhydrous ether, bromine, diethyl malonate, a tank of anhydrous HCl, glacial acetic acid, PBr_3, and acetone. (He also has water, heat sources, and solvents for working up his products.) Suggest ways that he could synthesize the following compounds:
(a) ethyl acetate;   (b) ethyl acetoacetate;   (c) 2-pentanone;   (d) 2-ethylbutanoic acid;
(e) ethyl α-bromoacetate.

                         OH
                         |

14.45. It is observed that (R)-$C_6H_5CHCO_2C_2H_5$ undergoes racemization in alkaline solution. Explain.

14.46. Propose a mechanism for the following reaction:

$$\text{excess HCHO} + CH_3CHO \xrightarrow{\ CO_3{}^{2-}\ } HOCH_2\overset{\overset{\displaystyle CH_2OH}{|}}{\underset{\underset{\displaystyle CH_2OH}{|}}{C}}CHO$$

14.47. Show how the following compounds could be synthesized by Robinson annelations:

(a)

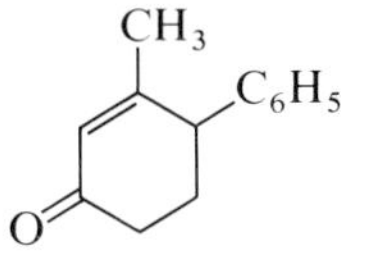

(b)

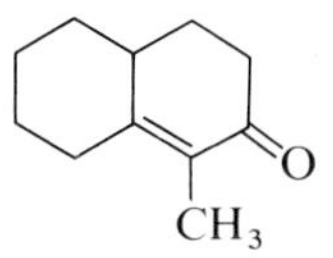

14.48. Suggest a mechanism for the following reaction:

$$\xrightarrow{\text{NaOC}_2H_5} \xrightarrow{H^+}$$

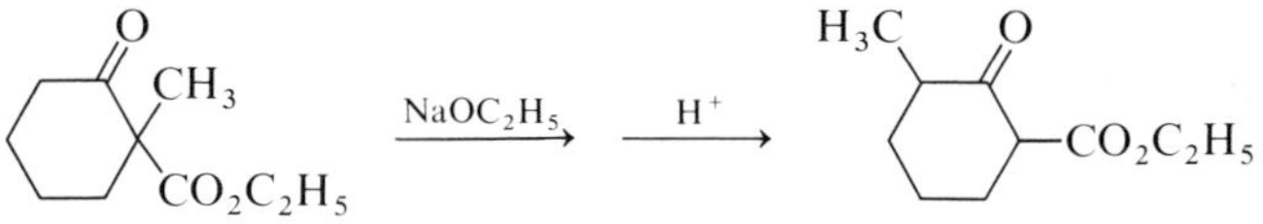

14.49. *Dimedone* can be obtained from *mesityl oxide* by a combination of a Michael condensation and a Dieckmann ring closure. Outline the synthesis.

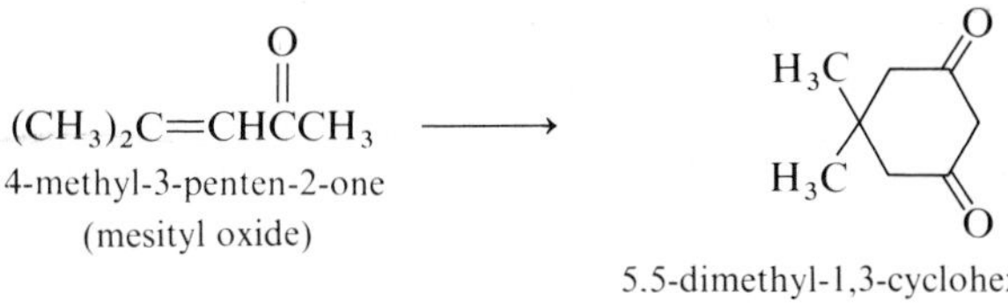

$$(CH_3)_2C{=}CHCCH_3 \longrightarrow$$

4-methyl-3-penten-2-one
(mesityl oxide)

5.5-dimethyl-1,3-cyclohexanedione
(dimedone)

14.50. Outline a general synthetic method for the preparation of cycloalkanecarboxylic acids, based upon the alkylation of malonic ester.

14.51. Cyclohexanone is treated with a strong base followed by iodomethane. Predict the monoalkylation and the dialkylation products. Which dialkylated product predominates? Why?

14.52. Devise a general method for the synthesis of γ-keto acids from ethyl α-chloroacetate.

14.53. How would you prepare the following carboxylic acids from the *same starting materials*?

(a) 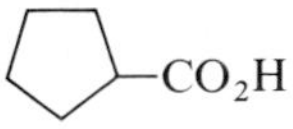 $\quad$ ⬠—CO_2H $\quad$ (b) $\ HO_2CCH_2(CH_2)_4CH_2CO_2H$

14.54. Give the structures of compounds **I** and **II**:

$$CH_3CH_2CH_2CO_2C_2H_5 + (CO_2C_2H_5)_2 \xrightarrow{NaOC_2H_5} \underset{\textbf{I}}{C_{10}H_{16}O_5} \xrightarrow[\text{heat}]{H_2O, H^+} \underset{\textbf{II}}{C_5H_8O_3}$$

14.55. Write each step in the mechanism of the following reaction sequence:

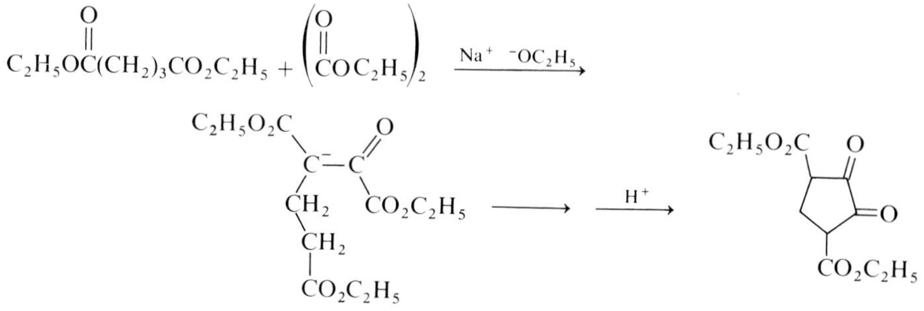

14.56. Give the structure of each indicated product:

excess malonic ester $\xrightarrow[\text{BrCH}_2\text{CH}_2\text{Br}]{\text{Na}^+ \ ^-OC_2H_5} \xrightarrow{H^+}$

$$\underset{\textbf{I}}{C_{16}H_{26}O_8} \xrightarrow[\text{BrCH}_2\text{CH}_2\text{Br}]{\text{Na}^+ \ ^-OC_2H_5} \xrightarrow{H^+} \underset{\textbf{II}}{C_{18}H_{28}O_8} \xrightarrow[\text{heat}]{H_2O, H^+} \underset{\textbf{III}}{C_8H_{12}O_4}$$

14.57. Show how the sedative *Seconal* can be prepared by a malonic ester synthesis.

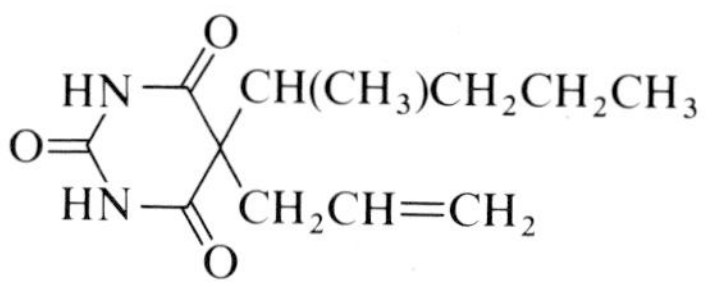

Seconal

14.58. *Pulegone* is a fragrant component of oil of pennyroyal. When this compound is heated in aqueous base, acetone is formed. What other organic product or products could be isolated from the reaction mixture? Explain your answer.

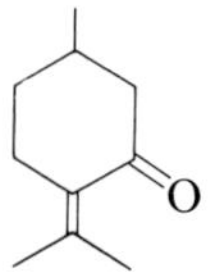

pulegone

14.59. Suggest appropriate reagents for the following conversions, and suggest a mechanism for each reaction.

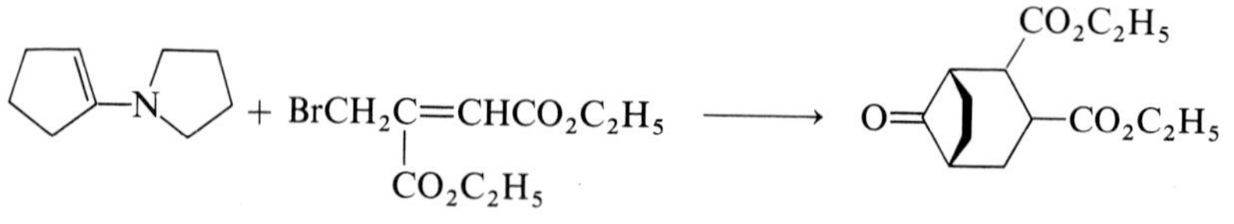

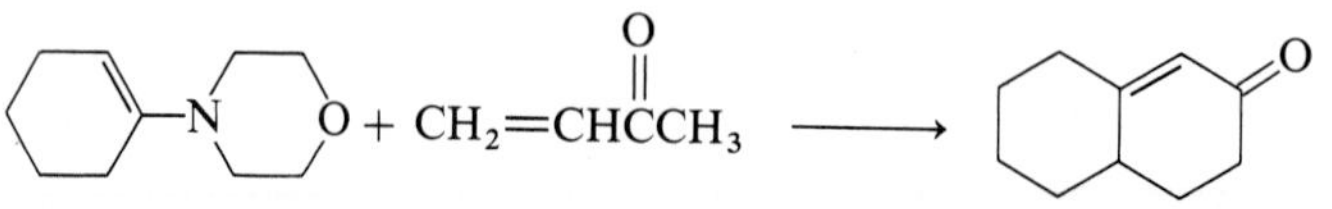

(c) $C_6H_5\overset{\overset{\displaystyle O}{\|}}{C}CH_3 + HCHO + (CH_3)_2NH \xrightarrow{H^+} C_6H_5\overset{\overset{\displaystyle O}{\|}}{C}CH_2CH_2N(CH_3)_2$

14.60. Suggest syntheses for the following compounds:

(a)   from compounds containing five or fewer carbons

(b)   from monocyclic or acyclic starting materials

(c)   from acyclic starting materials

(d)   from acyclic starting materials

(e)   from compounds containing six or fewer carbons

(f)   from compounds containing six or fewer carbons

CHAPTER 15

Amines

Carbon, hydrogen, and oxygen are the three most common elements in living systems. Nitrogen is fourth. Nitrogen is found in proteins and nucleic acids, as well as in many other naturally occurring compounds of both plant and animal origin. We have already discussed the nitrogen-containing amides and nitriles (Chapter 13) and later we will discuss some nitrogen-containing aromatic heterocycles, including nucleic acids (Chapter 16), and proteins (Chapter 19). In this chapter, we will discuss the **amines**, organic compounds containing trivalent nitrogen atoms bonded to one or more carbon atoms: RNH_2, R_2NH, or R_3N.

Amines are widely distributed in plants and animals, and many amines have physiological activity. For example, two of the body's natural stimulants of the sympathetic ("fight or flight") nervous system are norepinephrine and epinephrine (adrenaline).

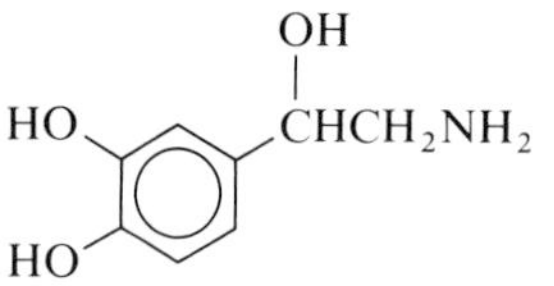

norepinephrine

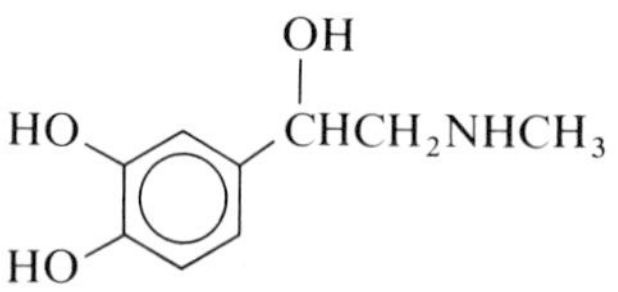

epinephrine
(adrenaline)

Both norepinephrine and epinephrine are β-phenylethylamines (2-phenylethylamines). A number of other β-phenylethylamines act upon the sympathetic receptors. These compounds are referred to as *sympathomimetic amines* because they "mimic," to an extent, the physiological action of norepinephrine and epinephrine.

Well before the birth of Christ, the drug *ephedrine* was extracted from the *ma-huang* plant in China and used as a drug. Today, it is the active decongestant in nose drops and cold remedies. Ephedrine causes shrinkage of swollen nasal membranes and inhibition of nasal secretions. (Overdoses cause nervousness and sleeplessness.) *Mescaline*, a hallucinogen isolated from the peyote cactus, has been used for centuries by the Indians of the southwestern U.S. and Mexico in religious ceremonies. *Amphetamine* is a synthetic stimulant that causes sleeplessness and nervousness. Amphetamine is sometimes prescribed for obesity, because it is also an appetite depressant. Like many other sympathomimetic amines, amphetamine contains a chiral carbon and has a pair of enantiomers. The more active enantiomer of amphetamine (the dextrorotatory one) is called *dexedrine*.

ephedrine	mescaline	amphetamine
a decongestant	*a hallucinogen*	*a stimulant*

Some other physiologically active amines, such as nicotine and morphine, are discussed in Section 16.10.

SECTION 15.1.

Classification and Nomenclature of Amines

Amines may be classified as **primary, secondary,** or **tertiary,** according to the number of alkyl or aryl substituents attached to the nitrogen.

one C attached	*two C's attached*	*three C's attached*
CH_3NH_2	—$NHCH_3$ $(CH_3CH_2)_3N$	
a 1° alkylamine	*a 2° arylalkylamine* *a 3° trialkylamine*	

Note that this classification is different from that of alkyl halides or alcohols. The classification of the latter is based upon the number of groups attached to the carbon that has the halide or hydroxyl group.

three C's attached to head C $CH_3{-}\underset{\underset{CH_3}{|}}{\overset{\overset{CH_3}{|}}{C}}{-}OH$ $CH_3{-}\underset{\underset{CH_3}{|}}{\overset{\overset{CH_3}{|}}{C}}{-}NH_2$ *one C attached to N*

t-butyl alcohol	*t*-butylamine
a 3° alcohol	*a 1° amine*

An amine nitrogen can have *four* groups or atoms bonded to it, in which case the nitrogen is part of a positive ion. These ionic compounds fall into two categories. If one or more of the attachments is H, the compound is an **amine salt.** If all four groups are alkyl or aryl (no H's on the N), the compound is a **quaternary ammonium salt.**

Amine salts:

$$(CH_3)_2NH_2{}^+ \ Cl^-$$

dimethylammonium chloride

salt of a 2° amine

N-methylpiperidinium bromide

salt of a 3° amine

Quaternary ammonium salts:

$$(CH_3)_4N^+ \ Cl^-$$

tetramethylammonium
chloride

$$CH_3CO_2CH_2CH_2\overset{+}{N}(CH_3)_3 \ Cl^-$$

acetylcholine chloride

*involved in the transmission
of nerve impulses*

STUDY PROBLEM

15.1.   Classify each of the following compounds as a 1°, 2°, or 3° amine; as a salt of a
1°, 2°, or 3° amine; or as a quaternary ammonium salt:

$$\underset{\text{(a)} \ \ \underset{\ }{CH_3CH_2}\overset{\displaystyle CH_3}{\underset{\displaystyle |}{C}}HNH_2}{}$$

(a) $CH_3CH_2\overset{\overset{\displaystyle CH_3}{\displaystyle |}}{C}HNH_2$

(b)   ⬡NH

(c) $(CH_3)_3NH^+ \ NO_3{}^-$

(d) $(CH_3CH_2)_3CNH_2$

(e)   ⬡$\overset{+}{N}(CH_3)_2 \ Cl^-$

(f)   ⬡NCH$_3$

Simple amines are usually named by the functional-group system. The
alkyl or aryl group is named, then the ending **-amine** is added.

$$CH_3CH_2CH_2NH_2 \qquad \qquad (CH_3CH_2)_2NH$$

propylamine   cyclohexylamine   diethylamine

Diamines are named from the name of the parent alkane (with appropriate
prefix numbers) followed by the ending **-diamine**.

$$H_2NCH_2CH_2CH_2NH_2$$

1,3-propanediamine

If more than one type of alkyl group is attached to the nitrogen, the largest
alkyl group is considered the parent. A subsidiary alkyl group is designated by an
N-alkyl- prefix.

$$\overset{\overset{\displaystyle CH_3}{\displaystyle |}}{CH_3CHNHCH_3} \qquad\qquad \overset{\overset{\displaystyle CH_3}{\displaystyle |}}{CH_3CHN(CH_3)_2}$$

N-methyl-2-propylamine   N,N-dimethyl-2-propylamine

If functionality of higher nomenclature priority is present, an **amino-** prefix is used.

$$H_2NCH_2CH_2OH$$

2-amino-1-ethanol

$$\overset{\displaystyle NHCH_3}{\underset{\displaystyle CH_3CHCO_2H}{|}}$$

2-(*N*-methylamino)propanoic acid

The chemistry of the **nonaromatic heterocyclic amines** is similar to that of their open-chain counterparts. Although we will not discuss the aromatic hetero-cycles until the next chapter, the chemistry of the nonaromatic nitrogen hetero-cycles will be included in this chapter. The commonly encountered compounds of this class generally have individual names.

pyrrolidine   piperidine   piperazine   morpholine

In the numbering of heterocyclic rings, the heteroatom is considered position 1. Oxygen has priority over nitrogen.

2-methylpyrrolidine   3,5-dimethylmorpholine

STUDY PROBLEM

15.2. Give each of the following compounds an acceptable name:

(a) $H_2NCH_2CH_2CH_2CH_2NH_2$ (commonly called *putrescine*, an odorous compound found in decaying flesh)

(b) $H_2NCH_2(CH_2)_4CH_2NH_2$ (*cadaverine*, origin similar to that of putrescine)

(c) $(CH_3)_2CHCH_2NH_2$ (d) $CH_3CH_2N(CH_3)_2$

(e) Cl—⟨○⟩—NHCH₃ (Name as a derivative of aniline.)

SECTION 15.2

Bonding in Amines

The bonding in an amine is directly analogous to that in ammonia: an sp^3 nitrogen atom bonded to three other atoms or groups (H or R) and with a pair of unshared electrons in the remaining sp^3 orbital (see Figure 2.19, page 62).

$$H-\overset{..}{\underset{|}{N}}-H \qquad CH_3-\overset{..}{\underset{|}{N}}-CH_3$$
$$\underset{H}{} \qquad\qquad \underset{CH_3}{}$$

ammonia   trimethylamine   piperidine

In an <u>amine salt</u> or a <u>quaternary ammonium salt</u>, the <u>unshared pair of elec-</u><u>trons forms the fourth sigma bond.</u> The cations are analogous to the ammonium ion.

$$H-\overset{\overset{\displaystyle H}{|}}{\underset{\underset{\displaystyle H}{|}}{\overset{+}{N}}}-H \ \ Cl^- \qquad CH_3-\overset{\overset{\displaystyle CH_3}{|}}{\underset{\underset{\displaystyle CH_3}{|}}{\overset{+}{N}}}-CH_3 \ \ Cl^-$$

ammonium   tetramethylammonium   *N*-methylpiperidinium
chloride   chloride   acetate

An amine molecule with three different groups attached to the nitrogen is chiral; however, enantiomers of most amines cannot be isolated because rapid inversion between mirror images occurs at room temperature. The inversion proceeds by way of a planar transition state (sp^2 nitrogen). The result is that the nitrogen pyramid flips inside out, much as an umbrella in a strong wind. The energy required for this inversion is about 6 kcal/mole, about twice the energy required for rotation around a carbon–carbon sigma bond.

The mirror images are interconvertible:

transition state with
two e$^-$ in p orbital

If an amine nitrogen has three different substituents and interconversion between the two mirror-image structures is restricted, then a pair of enantiomers may be isolated. **Tröger's base** is an example of such a molecule. The methylene bridge between the two nitrogens prevents interconversion between the mirror images, and Tröger's base can be separated into a pair of enantiomers.

Tröger's base

Another case in which the existence of isolable enantiomers is possible is that of the quaternary ammonium salts. These compounds are structurally similar to compounds containing sp^3 carbon atoms. If four different groups are attached to the nitrogen, the ion is chiral and the salt can be separated into enantiomers.

A pair of enantiomers:

STUDY PROBLEM

15.3. Which of the following structures could exist as isolable enantiomers?

(a) $[(CH_3)_2CH]_2\overset{+}{N}(CH_2CH_2Cl)_2$ Cl^-

(b) $CH_3NHCH_2CH_2Cl$

(c) 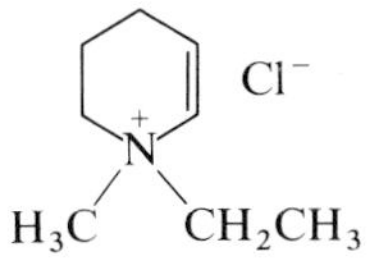 [structure: cyclic ring with $\overset{+}{N}$, bearing H_3C and CH_2CH_3 substituents] Cl^-

(d) 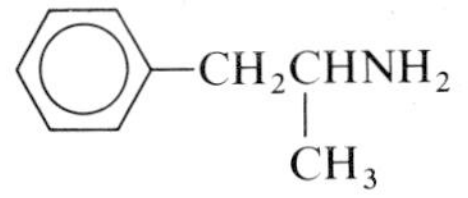 [benzene ring]—$CH_2\underset{\underset{CH_3}{|}}{C}HNH_2$

SECTION 15.3.

Physical Properties of Amines

As we mentioned in Chapter 1, amines undergo hydrogen bonding. The N———HN hydrogen bond is weaker then the O———HO hydrogen bond because N is less electronegative than O and therefore the NH bond is less polar. This weak hydrogen bonding between amine molecules results in boiling points that fall between those for nonhydrogen-bonded compounds (like alkanes or ethers) and those for strongly hydrogen-bonded compounds (like alcohols) of comparable molecular weight. (See Table 15.1.)

$$CH_3CH_2OCH_2CH_3 \qquad (CH_3CH_2)_2NH \qquad CH_3CH_2CH_2CH_2OH$$
$$\text{bp } 34.5° \qquad\qquad \text{bp } 56° \qquad\qquad\qquad \text{bp } 117°$$

Because they do not have an NH bond, tertiary amines in the pure liquid state cannot undergo hydrogen bonding. The boiling points of tertiary amines are lower than those for comparable primary or secondary amines, and are closer to the boiling points of alkanes of similar molecular weight.

no hydrogen bonding   *hydrogen bonding*

$$(CH_3)_3N \qquad (CH_3)_3CH \qquad CH_3CH_2CH_2NH_2$$
$$\text{bp } 3° \qquad\quad \text{bp } -10° \qquad\quad\quad \text{bp } 48°$$

Amines of low molecular weight are soluble in water because they can undergo hydrogen bonding with water. Tertiary amines, as well as primary and secondary

---

TABLE 15.1. Physical properties of some amines ·

Name	Structure	Bp, °C	Solubility in water
methylamine	CH_3NH_2	−7.5	∞
dimethylamine	$(CH_3)_2NH$	7.5	∞
trimethylamine	$(CH_3)_3N$	3	∞
ethylamine	$CH_3CH_2NH_2$	17	∞
benzylamine	$C_6H_5CH_2NH_2$	185	∞
aniline	$C_6H_5NH_2$	184	3.7 g/100 g

amines, can undergo this type of hydrogen bonding because they have <u>unshared</u> <u>pairs of electrons that can be used to form hydrogen bonds with water.</u>

$$(CH_3)_3N:----H-\overset{\overset{\displaystyle H}{|}}{O}$$

STUDY PROBLEM

15.4. Show all types of hydrogen bonding that could exist in   (a) pure dimethylamine and   (b) aqueous dimethylamine.

Volatile amines have very distinctive and usually offensive odors. Methylamine has an odor similar to that of ammonia; trimethylamine smells like dead salt-water salmon; and piperidine smells like dead fresh-water fish. Arylamines are not as unpleasant smelling as alkylamines; however, arylamines such as aniline are toxic and are especially insidious because they can be absorbed through the skin. Some, like β-naphthylamine, are carcinogenic.

β-naphthylamine

a carcinogenic arylamine

Amine salts and quaternary ammonium salts behave physically like inorganic salts—high-melting, water-soluble, and odorless.

SECTION 15.4

Spectral Properties of Amines

A.   Infrared spectra

The bonds that give rise to infrared absorption characteristic of amines are the <u>C—N</u> and N—H bonds (Table 15.2). <u>All aliphatic amines</u> show <u>C—N</u> <u>stretching in the fingerprint region</u>. However, only <u>primary and secondary amines</u> <u>show the distinctive NH stretching absorption,</u> which is observed to the left of CH absorption in the spectrum. This is the same region where OH absorption is observed; however, the two can often be differentiated because the <u>OH absorption</u> <u>is usually broader and stronger than NH absorption</u>. The stronger absorption by an OH bond is due to the greater polarity and hydrogen bonding of this group.

In Chapter 8, we mentioned that primary amines show two NH absorption peaks, secondary amines show one NH peak, and tertiary amines show no absorption in this region. Spectra of the three types of amines are shown in Figure 8.10 (page 322).

TABLE 15.2. Characteristic infrared absorption for amines

Type of absorption	Position of absorption	
	cm^{-1}	μm
1° amines:		
N—H stretching (pure liquid)	3250–3400 (2 peaks)	2.9–3.1
C—N stretching	1020–1250	8.0–9.8
2° amines:		
N—H stretching (pure liquid)	3330	3.0
C—N stretching	1020–1250	8.0–9.8
3° amines:		
C—N stretching	1020–1250	8.0–9.8

B.   Nmr spectra

The NH absorption in the nmr spectrum is generally a sharp singlet, not split by adjacent protons. In this respect, NH absorption is similar to OH absorption (Section 8.10E). Aliphatic amines show NH absorption at δ values of about 1.0–2.8 ppm, while arylamines absorb at around 2.6–4.7 ppm. (The exact position depends upon the solvent used.) The α protons are somewhat deshielded by the electronegative nitrogen; the chemical shift for these protons is from 2.2–2.8 ppm (see Figure 15.1).

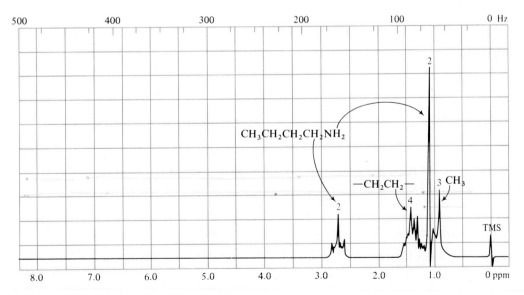

FIGURE 15.1. Nmr spectrum of *n*-butylamine.

SECTION 15.5.

Preparation of Amines

Techniques for the preparation of amines fall into three general categories. We will discuss each category in turn.

Nucleophilic substitution:

$$RX \quad + NH_3 \quad \xrightarrow{S_N2} \quad RNH_3^+ \ X^- \quad \xrightarrow{OH^-} \quad RNH_2$$

an alkyl halide

Reduction:
$$\overset{O}{\overset{||}{RCNH_2}} \quad \text{or} \quad RCN \quad \xrightarrow{[H]} \quad RCH_2NH_2$$

an amide or nitrile

Amide rearrangement:

$$\overset{O}{\overset{||}{RCNH_2}} \quad \xrightarrow{Br_2,\ OH^-} \quad RNH_2$$

an amide

A.   Synthesis by substitution reactions

Reaction of amines and alkyl halides.   Ammonia or an amine carries an unshared pair of electrons and can act as a nucleophile in a substitution reaction with an alkyl halide. The reaction of a nitrogen nucleophile is similar to the reaction of any other nucleophile with RX. The product of the reaction with ammonia or an amine is an amine salt. The free amine can be obtained by the treatment of this amine salt with a base such as NaOH.

S_N2 Reaction:

ammonia	bromoethane	ethylammonium bromide

an amine salt

Treatment with base:

$$CH_3CH_2NH_3^+ \ Br^- + OH^- \longrightarrow CH_3CH_2NH_2 + H_2O + Br^-$$

ethylamine

The order of reactivity of alkyl halides is typical for S_N2 reactions: $CH_3X > 1° > 2°$. Tertiary alkyl halides do not undergo substitution reactions with ammonia or amines; elimination products are obtained.

The principal disadvantage of this route to amines is that the product amine salt can exchange a proton with the starting ammonia or amine.

$$CH_3CH_2NH_3^+ \ Br^- + NH_3 \rightleftharpoons CH_3CH_2NH_2 + NH_4^+ \ Br^-$$

also a nucleophile

This proton exchange results in two or more nucleophiles competing in the reaction with the alkyl halide. For this reason, a mixture of mono-, di-, and trialkylamines and the quaternary ammonium salt is frequently obtained from the reaction of ammonia with an alkyl halide.

$$NH_3 \xrightarrow{RX} RNH_2 \xrightarrow{RX} R_2NH \xrightarrow{RX} R_3N \xrightarrow{RX} R_4N^+ \; X^-$$

Because all these products can be formed, the reaction of ammonia or an amine with an alkyl halide is not generally considered a useful synthetic reaction. If the amine is very inexpensive or if ammonia is used, a large excess can be used to favor monoalkylation. In this case, RX is more likely to collide with the molecules of the desired reactant and less likely to collide with those of the alkylated product. In the following example, an excess of ammonia favors the primary amine product.

$$CH_3CH_2CH_2CH_2Br + \text{excess } NH_3 \xrightarrow{\quad} \xrightarrow{OH^-} CH_3CH_2CH_2CH_2NH_2 + Br^-$$

1-bromobutane   *n*-butylamine
                                                         45%

If the quaternary ammonium salt is desired, the S_N2 reaction might also be useful. In this case, an excess of the alkyl halide would be used.

$$(CH_3CH_2)_2NH + \text{excess } CH_3CH_2I \xrightarrow{\quad} (CH_3CH_2)_4N^+ \; I^-$$

Gabriel phthalimide synthesis. A synthesis that gives primary amines without secondary and tertiary amines is the **Gabriel phthalimide synthesis**. The first step in the reaction sequence is an S_N2 reaction with the phthalimide anion as the nucleophile. The amine is then obtained by a hydrolysis reaction of the substituted phthalimide.

potassium phthalimide

bromoethane

N-ethylphthalimide

ethylamine

free of 2° and 3° amines

Phthalimide is prepared by heating phthalic anhydride with ammonia. The potassium salt is made by treating phthalimide with KOH. Normally, a proton cannot be removed from an amide nitrogen so easily. However, like other β-dicarbonyl compounds, imides are acidic because the anion is resonance-stabilized. Phthalimide has a pK_a of 9; it is ten times a stronger acid than phenol!

Preparation of phthalimide anion:

phthalic anhydride   phthalimide   *resonance-stabilized*

STUDY PROBLEM

15.5. Write resonance structures that show the delocalization of the negative charge in the phthalimide anion.

After the potassium phthalimide is prepared, it is treated with an alkyl halide. It is the nitrogen, not the oxygen, that attacks the carbon of the alkyl halide because the nitrogen is more nucleophilic than the oxygen.

Attack on RX:

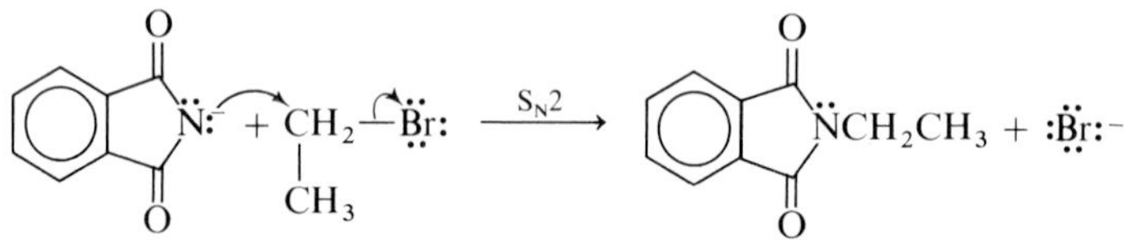

an N-alkylphthalimide

Finally, the alkylphthalimide is hydrolyzed. This reaction is simply the hydrolysis of an amide (Section 13.9C).

Hydrolysis:

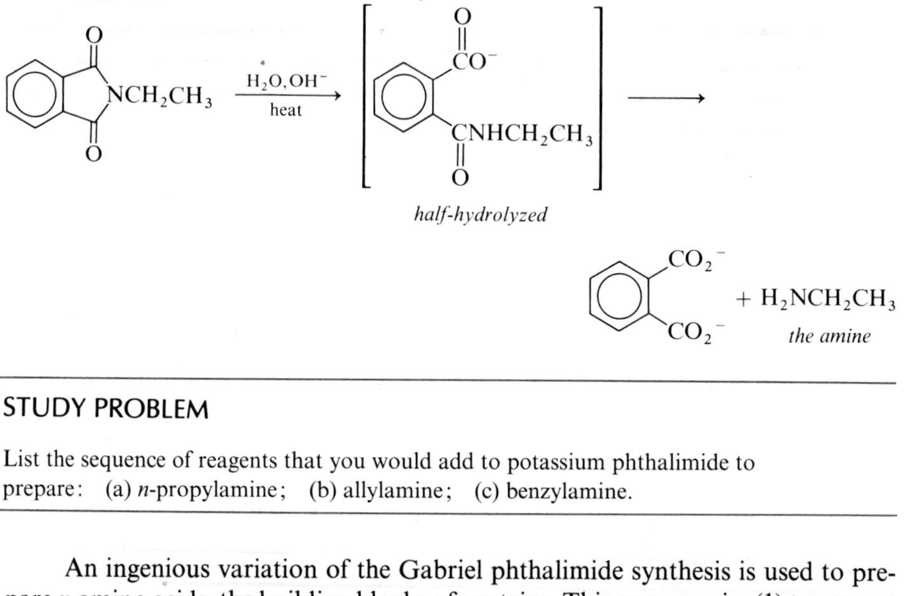

half-hydrolyzed

$$+ \; H_2NCH_2CH_3$$

the amine

STUDY PROBLEM

15.6. List the sequence of reagents that you would add to potassium phthalimide to prepare: (a) *n*-propylamine; (b) allylamine; (c) benzylamine.

An ingenious variation of the Gabriel phthalimide synthesis is used to prepare α-amino acids, the building blocks of proteins. This sequence is: (1) treatment of potassium phthalimide with diethyl bromomalonate; (2) treatment of the imide–malonate with base to remove the α hydrogen; and (3) treatment with RX, which gives a typical malonic ester alkylation reaction.

1. Reaction with bromomalonic ester:

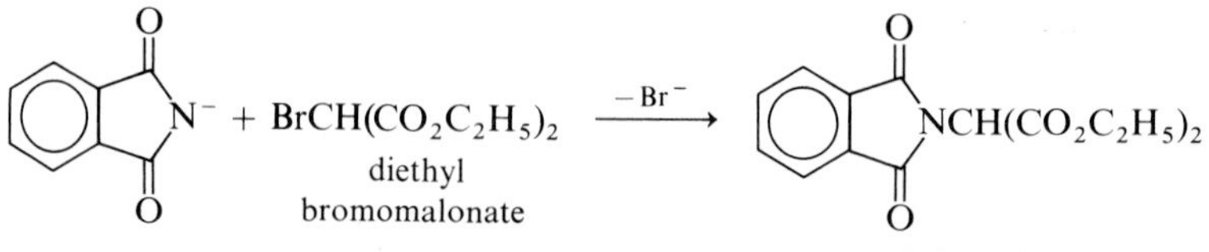

the imide–malonate

2. *Treatment with base:*

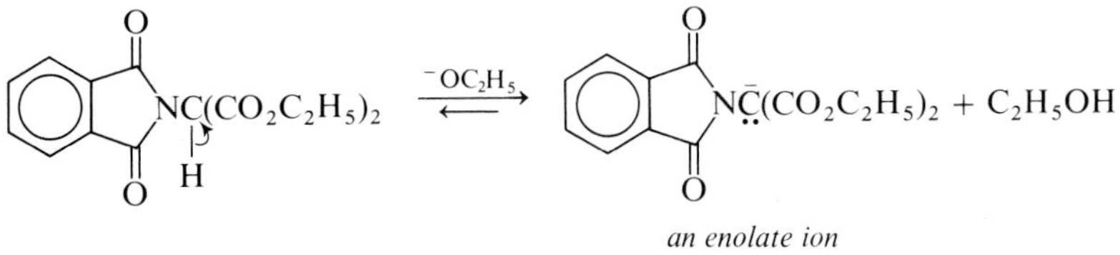

an enolate ion

3. *Reaction with RX:*

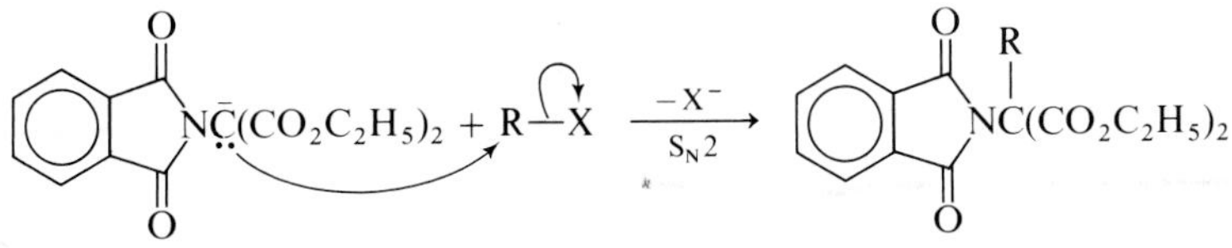

Acid hydrolysis results in hydrolysis of both imide and diester plus decarboxylation of the diacid. The product is the protonated amino acid.

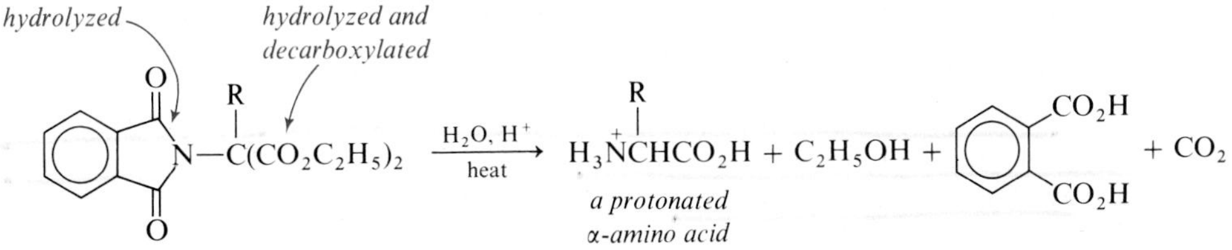

a protonated
α-amino acid

SAMPLE PROBLEM

How would you prepare phenylalanine, $C_6H_5CH_2CHCO_2H$, by a phthalimide synthesis?

$$\underset{NH_2}{C_6H_5CH_2CHCO_2H}$$

Solution:

1. *Determination of reagents needed:*

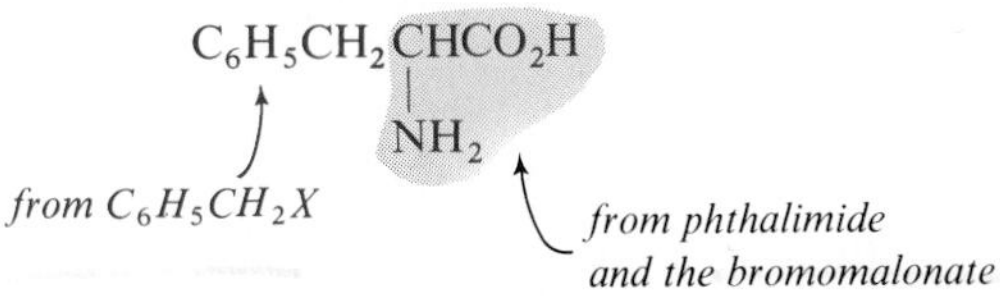

from $C_6H_5CH_2X$ *from phthalimide*
                          and the bromomalonate

2. *Steps in the synthesis:*

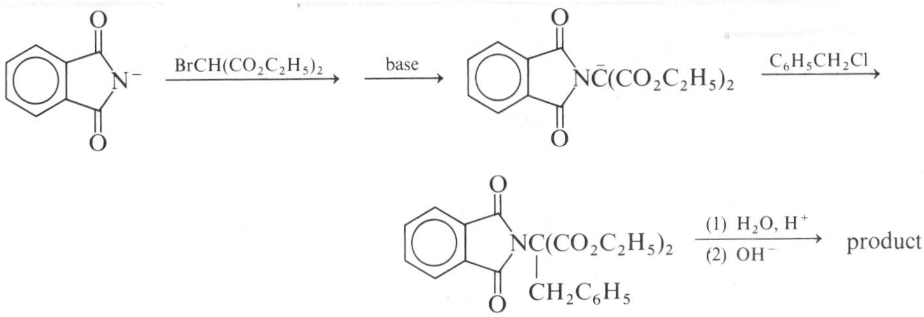

STUDY PROBLEM

15.7. How would you prepare the following amino acid by a phthalimide synthesis?

$$(CH_3)_2CHCH_2CHCO_2H$$
$$|$$
$$NH_2$$

leucine

B. Synthesis by reduction

Reduction reactions often provide convenient routes to amines. The reduction of *aromatic nitro compounds* to arylamines was discussed in Section 10.14.

2,4-dinitrotoluene	2,4-toluenediamine
	75%

Nitriles undergo catalytic hydrogenation or reduction with $LiAlH_4$ to yield primary amines of the type RCH_2NH_2 in yields of approximately 70%. Nitriles are available from alkyl halides; therefore, a nitrile synthesis is a technique for lengthening a carbon chain as well as for preparing an amine.

$$(CH_3)_2CHCH_2Br \xrightarrow[-Br^-]{CN^-} (CH_3)_2CHCH_2CN \xrightarrow[(2) H_2O]{(1) LiAlH_4} (CH_3)_2CHCH_2CH_2NH_2$$

1-bromo-2-methylpropane   3-methylbutanenitrile   (3-methyl-1-butyl)amine

a 1° alkyl halide

Amides also yield amines when treated with reducing agents:

$$\overset{O}{\overset{||}{CH_3(CH_2)_{10}CNHCH_3}} \xrightarrow[(2) H_2O]{(1) LiAlH_4} CH_3(CH_2)_{10}CH_2NHCH_3$$

N-methyldodecanamide   *N*-methyldodecylamine
                                   95%

Reductive amination, a reaction that converts ketones or aldehydes to primary amines, was discussed in Section 11.14D. This reaction is much better for synthesizing an amine of the type R_2CHNH_2 than is the reaction of R_2CHBr and NH_3 because the latter reaction may lead to elimination products. Secondary and tertiary amines may also be synthesized by reductive amination if a primary or secondary amine is used instead of ammonia.

benzaldehyde	*an imine*	benzylamine
		85%

$$HOCH_2CH_2NH_2 + CH_3\overset{\overset{\displaystyle O}{\displaystyle \|}}{C}CH_3 \longrightarrow [HOCH_2CH_2N{=}C(CH_3)_2]$$

2-aminoethanol   acetone   *an imine*
(ethanolamine)

$$\xrightarrow{\text{H}_2,\ \text{Pt}} HOCH_2CH_2NHCH(CH_3)_2$$

2-(*N*-isopropylamino)ethanol
95%

STUDY PROBLEMS

15.8. Show how you would carry out each of the following syntheses:

(a) cyclohexylamine from cyclohexanone
(b) $CH_2{=}CHCH_2CH_2CH_2NH_2$ from 4-bromo-1-butene
(c) *N,N*-dimethylbenzylamine from a carboxylic acid.

15.9. Suggest two techniques for preparing *sec*-butylamine free of higher alkylation products.

C. Amide rearrangement

When an <u>unsubstituted amide $(RCONH_2)$</u> is treated with an alkaline, <u>aqueous solution of bromine, it undergoes rearrangement to yield an amine</u>. This reaction is called the **Hofmann rearrangement**. Note that the <u>carbonyl group is lost as</u> CO_3^{2-}; therefore, the <u>amine contains one less carbon than the starting amide</u>.

$$CH_3(CH_2)_4\overset{\overset{\displaystyle O}{\displaystyle \|}}{C}NH_2 + 4\,OH^- + Br_2 \xrightarrow{\text{H}_2\text{O}} CH_3(CH_2)_4NH_2 + CO_3^{2-} + 2\,H_2O + 2\,Br^-$$

hexanamide   *n*-pentylamine
85%

Because the carbonyl group appears to be abstracted from the interior of the molecule, it is of interest to look at the mechanism of the Hofmann rearrangement. The reaction proceeds by a series of discrete steps. <u>Step 1 is bromination at the nitrogen</u>. *Step 2* is <u>loss of a proton from the nitrogen</u> and results in <u>an unstable anion</u>. The <u>rearrangement step is *Step 3*</u> of the sequence. Note that this rearrangement is a <u>*1,2-shift*</u> very similar to those we have encountered in carbocation rearrangements (Section 5.6F). The <u>product of the rearrangement is an isocyanate, stable under some conditions, but not in aqueous base</u>. In <u>aqueous base, the isocyanate undergoes hydrolysis (*Step 4*) to an amine and the carbonate ion</u>.

Step 1   (bromination of N):

$$R\overset{\overset{\displaystyle O}{\displaystyle \|}}{\underset{\underset{\displaystyle H}{\displaystyle |}}{C}}\ddot{N}H + OH^- \xrightarrow[\text{---}]{-\text{H}_2\text{O}} \left[R\overset{\overset{\displaystyle O}{\displaystyle \|}}{C}\ddot{\ddot{N}}H \right] \xrightarrow{Br_2} R\overset{\overset{\displaystyle O}{\displaystyle \|}}{\underset{\underset{\displaystyle Br}{\displaystyle |}}{C}}\ddot{N}H + Br^-$$

Step 2   (extraction of H$^+$ by OH$^-$):

$$\underset{Br}{\overset{O}{\underset{|}{RC\ddot{N}H}}} + \ ^-OH \ \rightleftharpoons \ \left[\underset{}{\overset{O}{\underset{}{RC\ddot{\underset{..}{N}}Br}}} \right] + H_2O$$

unstable

Step 3   (displacement of Br$^-$ by R—, a 1,2-shift):

$$\left[\overset{O}{\underset{R}{\overset{||}{C}}} \underset{\ddot{N}-Br}{} \right]^- \ \longrightarrow \ R-\ddot{N}=C=O + Br^-$$

an isocyanate

Step 4   (hydrolysis of the isocyanate):

$$RN=C=O \ \xrightarrow[H_2O]{OH^-} \ \left[\overset{O}{\underset{}{\overset{||}{RNHCO^-}}} \right] \ \xrightarrow[H_2O]{OH^-} \ RNH_2 + CO_3^{2-}$$

the amine

It has been found that the <u>Hofmann rearrangement proceeds with *retention of configuration*</u> at the α carbon of the amide. This evidence leads us to believe that the <u>rearrangement step (Step 3) has a bridged transition state.</u>

$$\underset{H}{\overset{CH_3}{\underset{|}{CH_3CH_2-C-C}}}\overset{O}{\underset{NH_2}{}} \ \xrightarrow{Br_2, \ OH^-} \ \underset{H}{\overset{CH_3}{\underset{|}{CH_3CH_2-C-NH_2}}}$$

(R)-2-methylbutanamide   (R)-2-butylamine

A bridged transition state in Step 3:

$$\underset{H}{\overset{CH_3}{\underset{|}{CH_3CH_2-C-C}}}\overset{O}{\underset{\underset{Br}{\ddot{N}}}{}} \ \longrightarrow \ \left[\underset{H}{\overset{CH_3}{\underset{|}{CH_3CH_2-C}}} \overset{O}{\underset{\underset{\ddot{N}^{\delta-}\cdots Br^{\delta-}}{}}{\overset{||}{C}}} \right]$$

$$\xrightarrow{-Br^-} \ \underset{H}{\overset{CH_3}{\underset{|}{CH_3CH_2-C-\ddot{N}=C=O}}}$$

The advantage of the Hofmann rearrangement is that <u>yields of pure primary amines are good.</u> This would be the best route to a primary amine containing a tertiary alkyl group, such as $(CH_3)_3CNH_2$. (The reaction of NH_3 with $(CH_3)_3CBr$ leads to the alkene and not the amine.)

STUDY PROBLEM

15.10. Predict the major organic products when the following compounds are treated with aqueous alkaline Br_2:

(a) (R)-$\langle$○$\rangle$—CH_2CHCNH_2 with $\overset{O}{\overset{||}{}}$ above and CH_3 below

(b) $H_2N\overset{O}{\overset{||}{C}}CH_2CH_2CH_2\overset{O}{\overset{||}{C}}NH_2$

D. Summary of amine syntheses

We have shown several routes to amines. By one or another of these routes, shown in the scheme on page 722, a chemist may synthesize:

1.   an amine with the same number of carbons as the starting material;

2.   an amine with one additional carbon; or

3.   an amine with one less carbon.

The specific reactions are summarized in Table 15.3.

TABLE 15.3. Summary of laboratory syntheses of amines

Reaction		*Section reference*
Primary amines:		
substitution	$1°\ RX$ $\xrightarrow[\text{(2) OH}^-]{\text{(1) excess NH}_3}$ RNH_2	15.5A
	RX $\xrightarrow[\substack{\text{(2) H}_2\text{O, H}^+ \\ \text{(3) OH}^-}]{\text{(1) K}^+\text{ phthalimide}}$ RNH_2	15.5A
reduction	$ArNO_2$ $\xrightarrow[\text{(2) OH}^-]{\text{(1) Fe, HCl}}$ $ArNH_2$	10.14
	RCN or $RCNH_2$ (with O above C) $\xrightarrow[\text{(2) H}_2\text{O}]{\text{(1) LiAlH}_4}$ RCH_2NH_2	15.5B
	$R_2C{=}O$ $\xrightarrow{\text{NH}_3,\ \text{H}_2,\ \text{Ni}}$ R_2CHNH_2	11.14D, 15.5B
rearrangement	$RCNH_2$ (with O above C) $\xrightarrow{X_2,\ \text{OH}^-}$ RNH_2	15.5C
Secondary and tertiary amines:		
reduction	$RCNR'_2$ (with O above C) $\xrightarrow[\text{(2) H}_2\text{O}]{\text{(1) LiAlH}_4}$ $RCH_2NR'_2$	15.5C
	$R_2C{=}O$ + R'_2NH $\xrightarrow{\text{H}_2,\ \text{Ni}}$ $R_2CHNR'_2$	11.14D, 15.5B

General routes to amines:

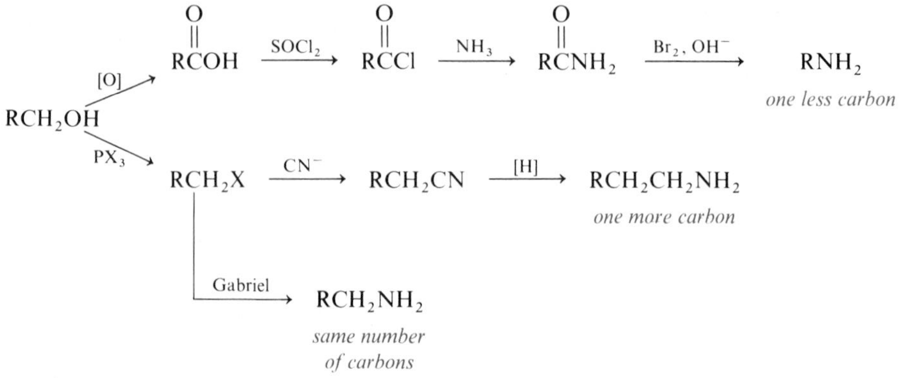

SECTION 15.6.

Basicity of Amines

The pair of electrons in the filled, nonbonded orbital of ammonia or an amine may be donated to an electron-deficient atom, ion, or molecule. In water solution, an amine is a *weak base* and accepts a proton from water in a reversible acid–base reaction.

$$(CH_3)_3N\text{:} + H{-}\ddot{O}H \rightleftharpoons (CH_3)_3\overset{+}{N}H + \text{:}\ddot{O}H^-$$

trimethylamine

The calculation of basicity constants and pK_b values for weak bases was discussed in Section 1.10. Table 15.4 lists a few amines along with their pK_b values. (Recall that decreasing values for pK_b indicate increasing base strength.)

TABLE 15.4. pK_b values for some amines

Structure	pK_b
NH_3	4.75
CH_3NH_2	3.34
$(CH_3)_2NH$	3.27
$(CH_3)_3N$	4.19
 ⬡NH	2.88
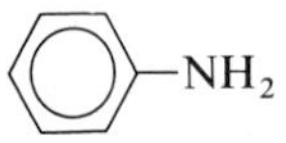 ⬡—NH_2	9.37

The same structural features that affect the relative acid strengths of carboxylic acids and phenols (Section 12.7) affect the relative base strengths of amines.

1.   *If the free amine is stabilized relative to the cation, the amine is less basic.*
2.   *If the cation is stabilized relative to the free amine, the amine is a stronger base.*

*if the free amine
is stabilized, R_3N
is a weaker base*

*if the cation is
stabilized, R_3N is
a stronger base*

$$R_3N: + H_2O \rightleftharpoons R_3NH^+ + OH^-$$

An *electron-releasing group*, such as an alkyl group, on the nitrogen increases basicity by dispersing the positive charge in the cation. (This dispersal of positive charge is analogous to that in carbocations, Section 5.6E.) By dispersal of the positive charge, the cation is stabilized relative to the free amine. Therefore, base strength increases in the series NH_3, CH_3NH_2, and $(CH_3)_2NH$.

NH_3	CH_3NH_2	CH_3NHCH_3
ammonia	methylamine	dimethylamine
$pK_b = 4.75$	$pK_b = 3.34$	$pK_b = 3.27$

$$CH_3 \rightarrow \ddot{N}H_2 + H_2O \rightleftharpoons CH_3 \rightarrow NH_3^+ + OH^-$$

*stabilized by
dispersal of
positive charge*

The cation is also stabilized by *increasing solvation*. In this case, the solvent helps disperse the positive charge. Dimethylamine ($pK_b = 3.27$) is a slightly stronger base than methylamine; however, trimethylamine ($pK_b = 4.19$) is a *weaker base* than dimethylamine. The reason is that trimethylamine is more hindered, and the cation is less stabilized by solvation. These arguments explain why the nonaromatic heterocyclic amines (with their alkyl groups "tied back" away from the unshared electrons of the nitrogen) are more basic than comparable open-chain secondary amines.

$$\begin{array}{c} CH_3 \\ | \\ CH_3-N-CH_3 \end{array} \quad \text{is a weaker base than} \quad \begin{array}{c} CH_3 \\ | \\ CH_3-N-H \end{array}$$

trimethylamine
$pK_b = 4.19$

dimethylamine
$pK_b = 3.27$

$$CH_3CH_2NHCH_2CH_3 \quad \text{is a weaker base than}$$

diethylamine
$pK_b = 3.01$

pyrrolidine
$pK_b = 2.73$

Hybridization of the nitrogen atom in a nitrogen compound also affects the base strength. An sp^2 orbital contains more *s* character than an sp^3 orbital. A molecule with an sp^2 nitrogen is less basic because its unshared electrons are more tightly held, and the free nitrogen compound is stabilized (rather than the cation).

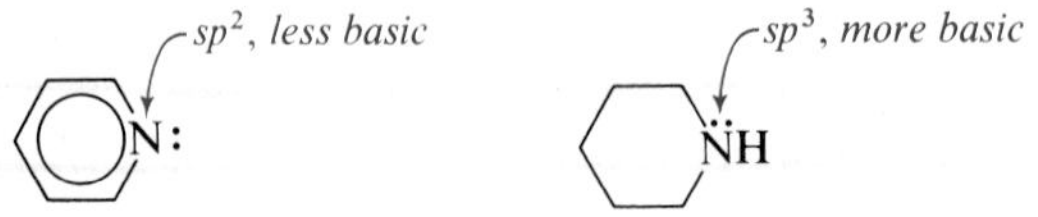

sp², less basic

pyridine
$pK_b = 8.75$

sp³, more basic

piperidine
$pK_b = 2.88$

Resonance also affects the base strength of an amine. Cyclohexylamine is a far stronger base than is aniline.

aniline
$pK_b = 9.37$

cyclohexylamine
$pK_b = 3.3$

The reason for the low basicity of aniline is that the positive charge of the anilinium ion cannot be delocalized by the aromatic pi cloud. However, the pair of electrons of the free amine are delocalized by the ring. The result is that the free amine is stabilized in comparison to the conjugate acid (the cation).

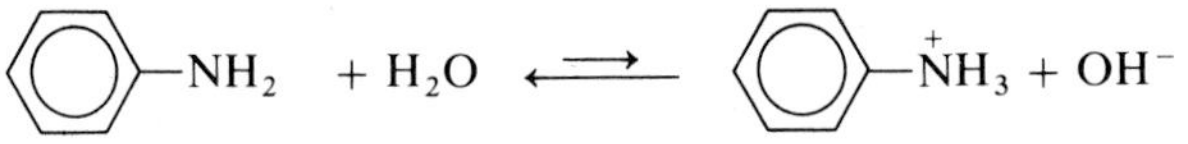

aniline

resonance-stabilized
and favored

anilinium ion

no resonance-stabilization
of positive charge

Resonance structures of aniline:

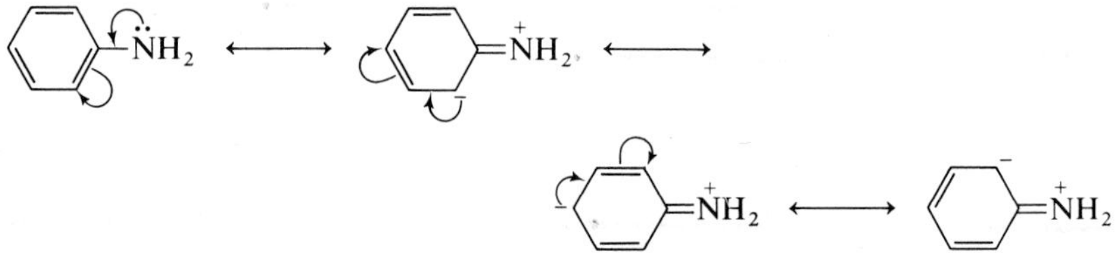

SAMPLE PROBLEM

Explain why piperidine is a stronger base than morpholine.

piperidine
$pK_b = 2.88$

morpholine
$pK_b = 5.67$

Solution: The oxygen atom in morpholine is *electron-withdrawing*, making the nitrogen more positive. The cation is not stabilized relative to the free morpholine, but is *destabilized*:

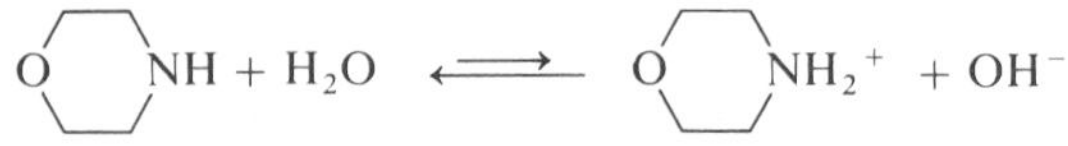

less stable
because of less
dispersal of + charge

Piperidine has no such destabilizing effect. Its cation is stabilized by electron release of the attached CH_2 groups:

$$\langle \ \rangle NH + H_2O \ \rightleftharpoons \ \langle \ \rangle NH_2^+ + OH^-$$

stabilized

STUDY PROBLEMS

15.11. Which would you expect to be more basic:

(a) piperazine (page 709) or piperidine? (b) piperazine or hexamethylenetetramine?

hexamethylenetetramine

15.12. Benzylamine ($C_6H_5CH_2NH_2$) has about the same base strength as an alkylamine, not that of an arylamine. How do you explain this fact?

15.13. Explain the following trend in pK_b values:

$$O_2N-\langle \bigcirc \rangle-NH_2 \qquad \langle \bigcirc \rangle-NH_2 \qquad CH_3-\langle \bigcirc \rangle-NH_2$$

$$pK_b = 13.0 \qquad\qquad pK_b = 9.37 \qquad\qquad pK_b = 8.9$$

SECTION 15.7.

Amine Salts

The reaction of an amine with a mineral acid (such as HCl) or a carboxylic acid (such as acetic acid) yields an **amine salt**. The salts are commonly named in one of two ways: as **substituted ammonium salts** or as **amine–acid complexes.**

$$(CH_3)_3N \ + HCl \ \longrightarrow \ (CH_3)_3NH^+ \ Cl^-$$

trimethylamine

trimethylammonium chloride
or
trimethylamine hydrochloride

$$CH_3CH_2NH_2 + CH_3CO_2H \ \longrightarrow \ CH_3CH_2\overset{+}{N}H_3 \ ^-O_2CCH_3$$

ethylamine   acetic acid

ethylammonium acetate
or
ethylamine acetate

Because of its ability to form salts, an amine that is insoluble in water may be made soluble by treatment with dilute acid. In this fashion, compounds containing amino groups may be separated from water- and acid-insoluble materials. Naturally occurring amines in plants, called **alkaloids** (Section 16.10), may be extracted from their sources, such as leaves or bark, by aqueous acid. Many compounds containing amino groups are used as drugs. These drugs are often administered as their water-soluble salts rather than as water-insoluble amines.

$$(CH_3CH_2)_2NCH_2CH_2OC{-}\!\!\langle\!\!\bigcirc\!\!\rangle\!\!-NH_2 \xrightarrow{\ HCl\ } (CH_3CH_2)_2\overset{+}{N}HCH_2CH_2OC{-}\!\!\langle\!\!\bigcirc\!\!\rangle\!\!-NH_2 \quad Cl^-$$

novocaine	novocaine hydrochloride
water-insoluble	*water-soluble*

STUDY PROBLEM

15.14. *Piperazine citrate* is a crystalline solid used in the treatment of pinworms and roundworms. Write an equation that shows the formation of the piperazine citrate formed from one molecule of each reactant.

$$\begin{matrix} CH_2CO_2H \\ | \\ HOCCO_2H \\ | \\ CH_2CO_2H \end{matrix} \quad + \quad HN\underset{\diagdown\!\!\_\!\!\diagup}{\overset{\diagup\!\!\overline{}\!\!\diagdown}{}}NH \quad \longrightarrow$$

citric acid	piperazine

A free amine may be regenerated from one of its salts by treatment with a strong base, usually NaOH. Quaternary ammonium salts, which have no acidic protons, do not undergo this reaction.

$$RNH_3^+ \; Cl^- \; + \; OH^- \; \longrightarrow \; RNH_2 \; + \; H_2O$$

an amine salt   *an amine*

$$R_4N^+ \; Cl^- \; + \; OH^- \; \longrightarrow \; \text{no reaction}$$

a quaternary
ammonium salt

Because of the ionic charge of a quaternary ammonium ion, quaternary ammonium salts have some interesting applications. For example, quaternary ammonium salts with long hydrocarbon chains are used as detergents. The combination of a long, hydrophobic, hydrocarbon tail with an ionic, hydrophilic head results in two types of interactions with other substances. One part of the molecule is soluble in nonpolar organic solvents, fats, and oils, while the other part is soluble in water. Soaps (Section 20.2) exhibit similar behavior.

Phospholipids, some of which are quaternary ammonium salts, are naturally occurring emulsifying agents. Phospholipids are also one of the principal components of cell membranes, where their unique structures are involved both in containing the cell and in the selective transport of ions and other substances through the membrane (see Section 20.3).

a phosphatidylcholine
(*a phospholipid*)

A fairly new application of quaternary ammonium salts is their use as **phase-transfer catalysts.** To show how these catalysts work, let us say that we wish to carry out a substitution reaction between an alkyl halide and CN^-. Unfortunately, alkyl halides are insoluble in water, and NaCN is insoluble in most organic solvents. When an aqueous NaCN solution is mixed with a solution of RX in a water-insoluble organic solvent such as benzene, two layers result. Reaction can occur only at the interface of these layers. A phase-transfer catalyst is used to transfer the CN^- ions to the organic solution so that reaction can occur in the organic solution as well as at the interface. An added advantage to reaction occurring in the organic solvent is that nucleophiles such as CN^- are more nucleophilic and more reactive when they are not solvated by water.

The catalytic action of $R_4N^+ \ X^-$ arises from the fact that it is water-soluble and also slightly soluble in organic solvents. If $R_4N^+ \ X^-$ is dissolved in the aqueous phase of the two-phase reaction mixture, some of the salt also becomes dissolved in the organic layer. However, if the aqueous layer contains an excess of CN^- ions, the salt that is transferred is mainly $R_4N^+ \ CN^-$, not $R_4N^+ \ X^-$.

Anion exchange in the aqueous phase:

$$(n\text{-}C_4H_9)_4N^+ \ Cl^- \ + CN^- \ \rightleftharpoons \ (n\text{-}C_4H_9)_4N^+ \ CN^- + Cl^-$$

tetrabutylammonium   *excess*   *migrates to organic*
     chloride   *phase*

S_N2 reaction in the organic phase:

$$CH_3(CH_2)_7Cl + (n\text{-}C_4H_9)_4N^+ \ CN^- \ \longrightarrow \ CH_3(CH_2)_7CN + (n\text{-}C_4H_9)_4N^+ \ Cl^-$$

1-chlorooctane   nonanenitrile   *returns to aqueous phase*
                                              90%   *for additional exchange*

SECTION 15.8.

Substitution Reactions with Amines

We have already mentioned a variety of substitution reactions with amines. The problems of the reaction of an amine with an alkyl halide were discussed earlier in this chapter (Section 15.5).

$$RNH_2 + R'Cl \xrightarrow{S_N2} R\overset{+}{N}H_2\ Cl^- \quad \text{and also} \quad R\overset{+}{N}HR'_2\ Cl^- \quad \text{and} \quad R\overset{+}{N}R'_3\ Cl^-$$
$$\underset{R'}{|}$$

In Chapter 13, we discussed the acylation of amines as a technique for the synthesis of amides. For example:

$$CH_3\overset{:\ddot{O}:}{\overset{\|}{C}}Cl + CH_3\ddot{N}H_2 \longrightarrow \left[CH_3\overset{:\ddot{O}:^-}{\underset{\underset{H}{|}}{\overset{|}{C}}}{\underset{+NHCH_3}{Cl}} \right] \xrightarrow{-HCl} CH_3\overset{:\ddot{O}}{\overset{\|}{C}}{\underset{:NHCH_3}{}}$$

$$\text{N-methylacetamide}$$

The utility of this reaction is that amines can be used to synthesize other amines by conversion to the amide, followed by reduction.

$$\underset{\text{an acid chloride}}{R\overset{O}{\overset{\|}{C}}Cl} + \underset{\text{an amine}}{R'_2NH} \longrightarrow \underset{\text{an amide}}{R\overset{O}{\overset{\|}{C}}NR'_2} \xrightarrow[\text{(2) H}_2\text{O}]{\text{(1) LiAlH}_4} \underset{\text{a new amine}}{RCH_2NR'_2}$$

Amides also undergo reactions with aldehydes and ketones to yield imines and enamines. (For the mechanisms, see Section 11.10.)

cyclohexanone

$$\xrightarrow[\text{H}^+]{\text{H}_2\text{NR (1°)}} \text{an imine} \quad =NR + H_2O$$

$$\xrightarrow[\text{H}^+]{\text{HNR}_2\ (2°)} \text{an enamine} \quad -NR_2 + H_2O$$

SECTION 15.9.

Reactions of Amines with Nitrous Acid

In Section 10.14, we discussed the formation of benzenediazonium chloride $(C_6H_5N_2^+\ Cl^-)$ by the treatment of aniline with cold aqueous nitrous acid, HNO_2 (prepared *in situ* from $NaNO_2$ and HCl). Recall that the aryldiazonium salts are stable at $0°$ and are useful synthetic intermediates because of the excellent leaving ability of N_2.

The treatment of a *primary alkylamine* with $NaNO_2$ and HCl also results in a diazonium salt, but an alkyldiazonium salt is unstable and decomposes to a mixture of alcohol and alkene products along with N_2. The decomposition proceeds by way of a carbocation.

$$(CH_3)_2CHNH_2 \xrightarrow[0°]{\substack{NaNO_2 \\ HCl}} (CH_3)_2CH-N_2^+ \ Cl^- \xrightarrow[-Cl^-]{-N_2}$$

isopropylamine   isopropyldiazonium

a 1 amine   chloride

$$[(CH_3)_2CH^+] \xrightarrow{H_2O} (CH_3)_2CHOH + CH_3CH=CH_2$$

When treated with $NaNO_2$ and HCl, *secondary amines* (alkyl or aryl) yield *N*-**nitrosoamines**, compounds containing the $N-N=O$ group. Many *N*-nitrosoamines are carcinogenic.

N-methylaniline   an *N-nitrosoamine*

a 2° amine

Tertiary amines are not entirely predictable in their reactions with nitrous acid. A tertiary arylamine usually undergoes ring substitution with $-NO$ because of the ring activation by the $-NR_2$ group. A tertiary alkylamine (and sometimes tertiary arylamines, too) may lose an R group and form an *N*-nitroso derivative of a secondary amine.

STUDY PROBLEMS

15.15. When *n*-butylamine is treated with a cold aqueous solution of HCl and $NaNO_2$, the following products are obtained: 1-chlorobutane, 2-chlorobutane, 1-butanol, 2-butanol, 1-butene, 2-butene, and nitrogen gas. Suggest a mechanism (or mechanisms) that accounts for each of these products.

15.16. Suggest a reason why benzenediazonium chloride is more stable than ethyldiazonium chloride.

SECTION 15.10.

Hofmann Elimination

Quaternary ammonium hydroxides ($R_4N^+ \ OH^-$) are amine derivatives that are used in structure-determination studies because they undergo elimination reactions to yield alkenes and amines. We will look briefly at how these compounds are prepared, then at their elimination reactions, and finally at their utility in structure studies.

A. Formation of quaternary ammonium hydroxides

When a quaternary ammonium halide is treated with aqueous silver oxide, the quaternary ammonium hydroxide is obtained:

$$2\,R_4N^+\,X^- \;+\; Ag_2O + H_2O \;\longrightarrow\; 2\,R_4N^+\,OH^- \;+\; 2\,AgX\downarrow$$

a quaternary ammonium halide   a quaternary ammonium hydroxide

N,N-dimethylpiperidinium chloride   N,N-dimethylpiperidinium hydroxide

A quaternary ammonium hydroxide cannot be obtained by the ionic reaction of $R_4N^+\ X^-$ with aqueous NaOH because the reactants and products are all water-soluble ionic compounds. If such a reaction were attempted, a mixture of $R_4N^+\ OH^-$ and $R_4N^+\ Cl^-$ (along with NaOH and NaCl) would result. However, silver hydroxide, which is formed *in situ* from moist silver oxide ($Ag_2O + H_2O \rightarrow 2\,AgOH$), removes the halide ion as an AgX precipitate. Removal of the AgX by filtration, followed by evaporation of the water, yields the pure quaternary ammonium hydroxide.

B. The elimination

When a quaternary ammonium hydroxide (as a solid) is heated, an elimination reaction called a **Hofmann elimination** occurs. This reaction is an E2 reaction in which the leaving group is an amine.

E2 transition state

This elimination generally yields the *Hofmann product*, the alkene with *fewer alkyl groups* on the pi-bonded carbons. The formation of the less substituted, less stable alkene may be attributed to steric hindrance in the transition state (Section 5.9E) due to the bulky $R_3N^{+}\!-$ group.

sec-butyltrimethylammonium hydroxide

$$(CH_3)_3N + CH_2{=}CHCH_2CH_3 \;+\; CH_3CH{=}CHCH_3 \;+\; H_2O$$

1-butene   2-butene
95%   5%

STUDY PROBLEMS

15.17. Predict the major organic products when the following compounds are heated:

(a) $CH_3CH_2CH_2CH_2\overset{+}{\underset{\underset{CH_3}{|}}{\overset{\overset{CH_3}{|}}{N}}}{-}CH_2CH_3\ OH^-$

(b) $C_6H_5CH_2CH_2\overset{+}{\underset{\underset{CH_3}{|}}{\overset{\overset{CH_3}{|}}{N}}}{-}CH_2CH_3\ OH^-$

C.   Exhaustive methylation

Many compounds in nature contain heterocyclic nitrogen rings. A quaternary ammonium hydroxide of a heterocyclic ring undergoes elimination in the same fashion as an open-chain amine. When the nitrogen atom is part of a ring, fragmentation does not occur. Instead, the product amino group and alkenyl group both remain in the same molecule.

N,N-dimethylpiperidinium hydroxide   *N,N*-dimethyl-4-penten-1-amine

In the determination of the structure of a compound, the goal is often to degrade the compound into small identifiable fragments. To show how the Hofmann elimination can do this, let us consider the simple ring system of piperidine. The quaternary ammonium hydroxide is prepared by treatment with CH_3I (S_N2 reaction) followed by reaction with Ag_2O. Heating results in elimination to yield an alkenylamine, as shown in the preceding equation.

Because the product of this elimination still contains an amino group, it can undergo another reaction with CH_3I and Ag_2O to yield a new quaternary ammonium hydroxide. Heating this product results in a new alkene, while the nitrogen is expelled, or "exhausted," from the molecule as trimethylamine. This series of reactions is called **exhaustive methylation**.

trimethylamine
"*exhausted*"

The beginning piperidine has gone through two *stages* of exhaustive methylation (two passes through the sequence CH_3I, Ag_2O, heat) before the nitrogen was lost from the parent compound. Two stages are typical for a nitrogen heterocycle. If the nitrogen had been attached to the ring instead of within the ring, one stage would have resulted in loss of the nitrogen.

STUDY PROBLEMS

15.18. Fill in the blanks:

(a) [pyrrolidine N–CH$_3$ structure] $\xrightarrow[\text{(2) Ag}_2\text{O}]{\text{(1) excess CH}_3\text{I}}$ _____ $\xrightarrow{\text{heat}}$ _____ $\xrightarrow[\text{(3) heat}]{\text{(1) CH}_3\text{I} \\ \text{(2) Ag}_2\text{O}}$ _____

(b) [cyclohexyl–NH$_2$ structure] $\xrightarrow[\text{(2) Ag}_2\text{O}]{\text{(1) excess CH}_3\text{I}}$ _____ $\xrightarrow{\text{heat}}$ _____

15.19. *Coniine*, $C_8H_{17}N$, is a toxic constituent of poison hemlock (*Conium maculatum*), the extract of which is believed to have killed Socrates. The nmr spectrum of coniine shows no doublets. Coniine undergoes reaction with two equivalents of CH_3I. Reaction with Ag_2O followed by pyrolysis yields an intermediate $(C_{10}H_{21}N)$ which, upon further methylation followed by conversion to the hydroxide and pyrolysis, yields trimethylamine, 1,5-octadiene, and 1,4-octadiene. What are the structures of coniine and the intermediate compound?

SECTION 15.11.

Use of Amines in Synthesis

The synthesis of nitrogen-containing compounds is of special interest to organic chemists involved in pharmacology and other biological sciences because many biomolecules contain nitrogen. Most of the reactions used to synthesize nitrogen compounds from amines have already been discussed in other chapters of this book.

Many of the reactions of amines are the result of nucleophilic attack by the unshared electrons of the amine nitrogen. The substitution reaction of an amine with an alkyl halide is one example of an amine acting as a nucleophile. Amines may also be used as nucleophiles in nucleophilic acyl substitution reactions. If a carboxylic acid derivative is the carbonyl reagent, the product is an amide. If an aldehyde or ketone is the carbonyl reactant, the product is an imine (from a primary amine, RNH_2) or an enamine (from a secondary amine, R_2NH). These and other reactions of amines are summarized in Table 15.5.

The conversion of an amino group to a good leaving group ($-N_2^+$ or $-NR_3^+$ OH^-) is another synthetic technique. Hofmann eliminations of quaternary ammonium hydroxides are more useful as an analytical tool than as a synthetic tool because mixtures of product alkenes can result. (Also, an elimination reaction of an alkyl halide is a more convenient route to an alkene in the laboratory.) Nmr spectroscopy has largely replaced Hofmann eliminations even as an aid in structure determinations. On the other hand, conversion of an arylamine to a diazonium salt and a subsequent substitution reaction are very useful in organic syntheses. Refer to Figure 10.11, page 490, to see the types of compounds that are readily obtainable from aryldiazonium salts.

Single enantiomers of chiral amines are common in plants. Because of their basicity, some of these amines find use in the resolution of racemic carboxylic acids. Two such amines are strychnine and brucine, both of which are isolated

TABLE 15.5. Some compounds that can be obtained from amines

Reaction		*Product*	*Section reference*
$R_3N + R'X \longrightarrow R_3\overset{+}{N}R'\ X^-$		amine salt or quaternary ammonium salt	15.5A
$R_2NH + R'\overset{\displaystyle O}{\overset{\displaystyle \|}{C}}Cl \longrightarrow R_2N\overset{\displaystyle O}{\overset{\displaystyle \|}{C}}R'$		amide[a]	13.3C, 15.8
$1° RNH_2 + R'_2C{=}O \xrightarrow{H^+} RN{=}CR'_2$		imine	11.10A
$2° R_2NH + R'_2CH\overset{\displaystyle O}{\overset{\displaystyle \|}{C}}R' \xrightarrow{H^+} R'_2C{=}\overset{\displaystyle NR_2}{\overset{\displaystyle \|}{C}}R'$		enamine[b]	11.10C
$ArNH_2 \xrightarrow[0°]{\underset{\text{HCl}}{\text{NaNO}_2}} ArN_2{}^+\ Cl^-$		aryldiazonium salt[c,d]	10.15, 15.9
$R_2CH\overset{\displaystyle NR'_2}{\overset{\displaystyle \|}{C}}R_2 \xrightarrow[\text{(3) heat}]{\underset{\text{(2) Ag}_2\text{O, H}_2\text{O}}{\text{(1) CH}_3\text{I}}} R_2C{=}CR_2$		alkene	15.10

[a] Other routes to amides are the similar reactions of amines with acid anhydrides (Section 13.4C) and with esters (Section 13.5C).

[b] Enamines may be converted to α-substituted aldehydes or ketones (see Section 14.5).

[c] Aryldiazonium salts may be converted to aryl halides, nitriles, etc. (see Section 10.15).

[d] Secondary amines (alkyl or aryl) yield *N*-nitrosoamines when treated with HNO_2.

from the seeds of the Asiatic tree *Strychnos nux-vomica*. (Both compounds are toxic stimulants of the central nervous system.) The steps in resolving a racemic carboxylic acid with a chiral amine were discussed in Section 4.10.

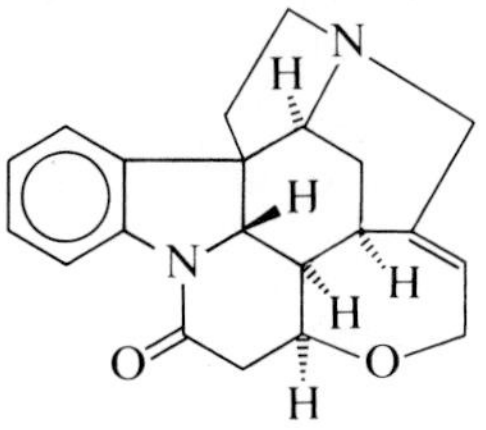

strychnine 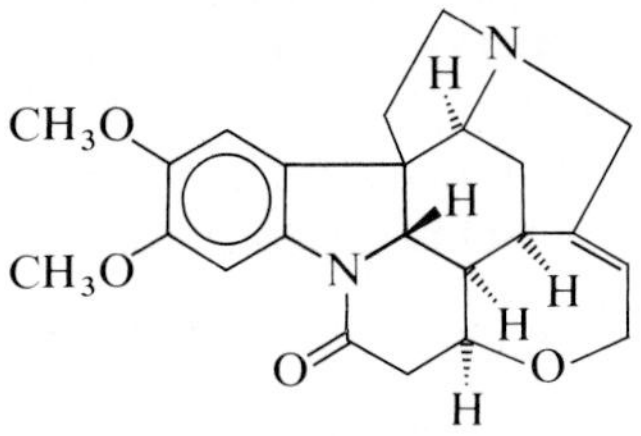   brucine

STUDY PROBLEM

15.20. Suggest synthetic routes to the following compounds:

   (a)   the pain reliever *phenacetin* (*p*-ethoxyacetanilide) from *p*-nitrophenol. (For the structure of acetanilide, see Table 10.1, page 451.)

   (b)   methyl orange (page 931) from substituted benzenes

   (c) 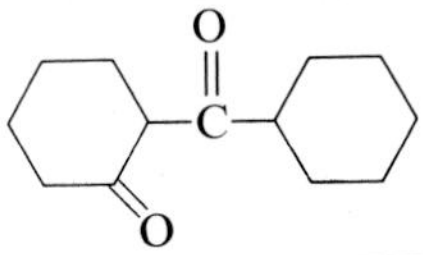   from compounds containing six or fewer carbon atoms

Summary

An **amine** is a compound that contains a trivalent nitrogen that has from one to three alkyl or aryl groups attached: RNH_2, R_2NH, or R_3N. A compound with four groups attached to the nitrogen is an **amine salt** (R_3NH^+ X^-) or a **quaternary ammonium salt** (R_4N^+ X^-).

Amines may be prepared by *substitution reactions*, by *reduction reactions*, or by *rearrangement*. These synthetic reactions are summarized in Table 15.3 (page 721).

Because the nitrogen of an amine has a pair of unshared electrons, amines are *weak bases*. The base strength is affected by *hybridization* ($sp^3 > sp^2 > sp$), by *electron-withdrawing groups* (base-weakening), by *electron-releasing groups* (base-strengthening), and by *conjugation* (base-weakening).

hybridization

$$CH_3{\rightarrow}NH_2 > NH_3$$

electron-release

$$CH_3NH_2 > Cl{\leftarrow}CH_2NH_2$$

electron-withdrawal

conjugation

Amines undergo reaction with acids to yield amine salts:

$$R_3N \xrightleftharpoons[OH^-]{HX} R_3NH^+ \ X^-$$

an amine salt

Most amines are nucleophiles and can displace good leaving groups or can attack carbonyl groups, reactions that are summarized in Section 15.11. When primary amines are treated with cold nitrous acid, diazonium salts are formed. Alkyldiazonium salts are unstable, but aryldiazonium salts may be used to prepare a variety of substituted aromatic compounds.

Quaternary ammonium hydroxides, when heated, undergo elimination of water and an amine. The least substituted alkene is usually formed. This reaction is called the **Hofmann elimination.**

$$CH_3CH_2\overset{+}{N}(CH_3)_2 \ OH^- \xrightarrow{\text{heat}} CH_2{=}CH_2 + N(CH_3)_2 + H_2O$$
$$\underset{CH_2CH_2CH_3}{|} \qquad\qquad\qquad \underset{CH_2CH_2CH_3}{|}$$

STUDY PROBLEMS

15.21. Classify each of the following compounds as a primary, secondary, or tertiary amine; as an amine salt of one of these; or as a quaternary ammonium salt:

(a) 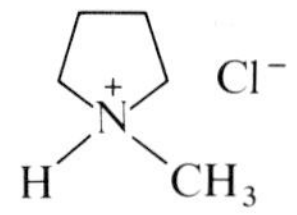 [structure: pyrrolidinium ring with N bearing H and CH_3] Cl^-

(b) 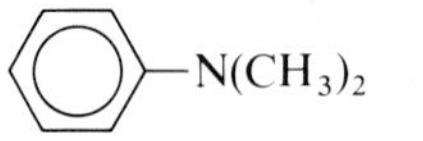 [structure: benzene ring]$-N(CH_3)_2$

(c) $(CH_3)_2CH\overset{+}{N}(CH_3)_3$ Br^-

(d) $(CH_3)_3CNH_3{}^+$ $HSO_4{}^-$

15.22. Write structures for the following compounds: (a) cyclopentylamine; (b) (2-methylcyclohexyl)amine; (c) *N,N*-diethyl-*p*-nitroaniline; (d) 2-(*N,N*-dimethylamino)hexanoic acid.

15.23. Name the following compounds:

(a) $C_6H_5CH_2NCH_3$
$\quad\quad\quad\quad\;\;|$
$\quad\quad\quad\;\;CH_2CH_3$

(b) 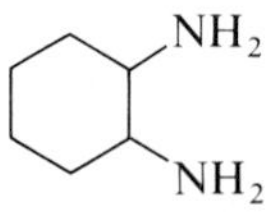 [structure: cyclohexane with two NH_2 groups]

(c) 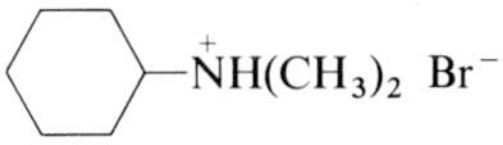 [structure: cyclohexane ring]$-\overset{+}{N}H(CH_3)_2$ Br^-

(d) $\underset{\quad\;|}{\overset{\quad\;NH_2}{CH_3CHCH_2CH_2CH}}\overset{\displaystyle O}{\overset{\|}{}}$

15.24. Which of the following structures has enantiomers, geometric isomers, both, or neither? [*Hint*: In (f), consider the hybridization of the N and the resultant geometry.]

(a) $C_6H_5CHNH_2$
$\quad\quad\quad\;|$
$\quad\quad\;CH_2CH_3$

(b) $CH_3NCH_2CH_3$
$\quad\quad\quad\quad|$
$\quad\quad\quad\;C_6H_5$

(c) 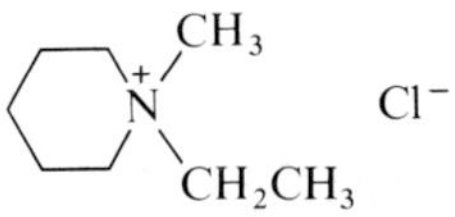 [structure: piperidinium ring with N bearing CH_3 and CH_2CH_3] Cl^-

(d) [structure: piperidinium ring with N bearing CH_3 and CH_2CH_3, with CH_3 on ring] Cl^-

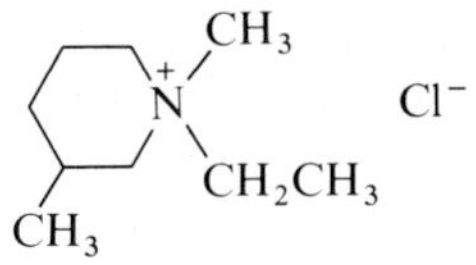

(e) $CH_3\overset{O^-}{\underset{|}{\overset{|}{\overset{+}{N}}}}CH_2CH_3$
$\quad\quad\;|$
$\quad\quad C_6H_5$

(f) $C_6H_5CH{=}NOH$

15.25. Which of the following species can act as a nucleophile?

(a) $(CH_3)_2NH$

(b) $(CH_3)_3N$

(c) H_2N-NH_2

(d) 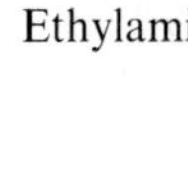 [structure: pyridine]

(e) 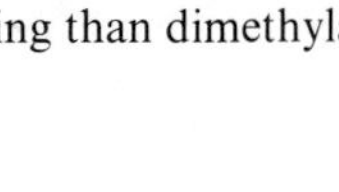 HN [ring] NCH_3

(f) [structure: benzene ring]$\overset{+}{N}-CH_3$

15.26. Explain the following observations:

(a) Cyclohexylamine is more water-soluble than cyclohexanol.
(b) Trimethylamine has a lower boiling point than dimethylamine.
(c) Ethylamine is higher boiling than dimethylamine.

15.27. Suggest syntheses for the following compounds from organohalogen compounds or alcohols:

(a) 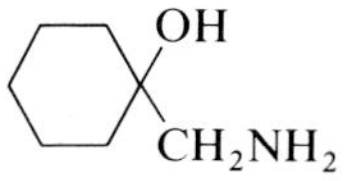 [cyclohexane with OH and CH$_2$NH$_2$ substituents]

(b) 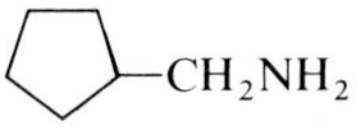 [cyclopentane]—CH$_2$NH$_2$

(c) HO$_2$CCH$_2$CHCO$_2$H
           |
          NH$_2$

(d) (CH$_3$)$_2$CHNH$_2$

(e)  [cyclohexane]—NHCH$_2$CH$_2$CH$_3$

(f) CH$_3$CH$_2$CH$_2$CH$_2$N(CH$_3$)$_2$

15.28. How would you convert 1-pentanol to: (a) *n*-pentylamine (free of higher alkylated amines)? (b) *n*-hexylamine? (c) *n*-butylamine?

15.29. Suggest a way to make each of the following conversions: (a) benzene to aniline; (b) benzamide to aniline; (c) aniline to acetanilide (C$_6$H$_5$NHCOCH$_3$); (d) glutaric anhydride (page 592) to 4-aminobutanoic acid; (e) (*R*)-2-butanol to (*S*)-2-butylamine; (f) toluene to benzylamine; (g) acetic acid to acetamide.

15.30. Which is more basic? (a) aniline or *p*-bromoaniline; (b) trimethylamine or tetramethylammonium hydroxide; (c) *p*-nitroaniline or 2,4-dinitroaniline; (d) ethylamine or ethanolamine (HOCH$_2$CH$_2$NH$_2$); (e) *p*-toluidine (*p*-methylaniline) or *p*-(trichloromethyl)aniline.

15.31. Complete the following equations:

(a) 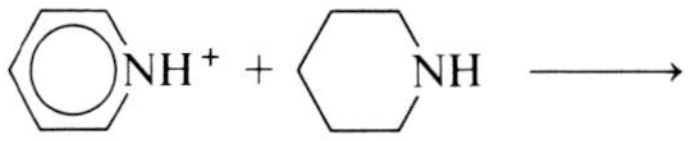 [pyridinium]NH$^+$ + [piperidine]NH $\longrightarrow$

(b) 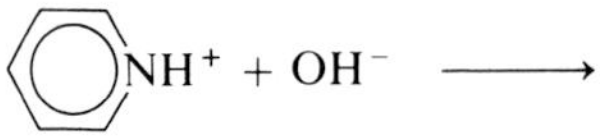 [pyridinium]NH$^+$ + OH$^-$ $\longrightarrow$

(c) 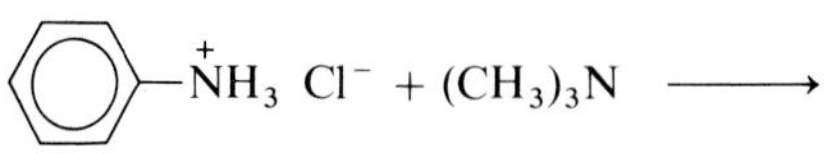 [benzene]—$\overset{+}{N}$H$_3$ Cl$^-$ + (CH$_3$)$_3$N $\longrightarrow$

(d) 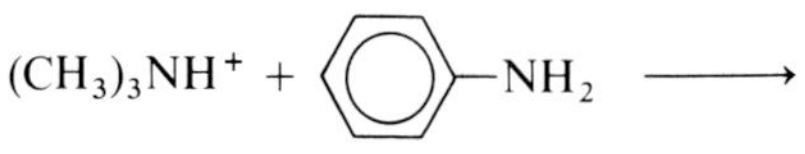 (CH$_3$)$_3$NH$^+$ + [benzene]—NH$_2$ $\longrightarrow$

(e) (CH$_3$)$_4$N$^+$ OH$^-$ + CH$_3$CO$_2$H $\longrightarrow$

(f)  [piperidine]NH + CH$_3$CO$_2$H $\longrightarrow$

(g) 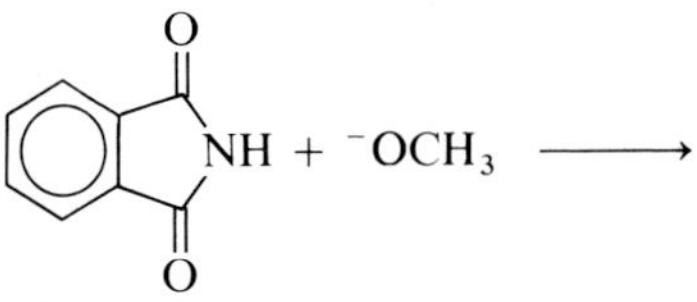 [phthalimide with two C=O groups]NH + $^-$OCH$_3$ $\longrightarrow$

15.32. List each of the following groups of cations in order of increasing acidity (weakest acid first):

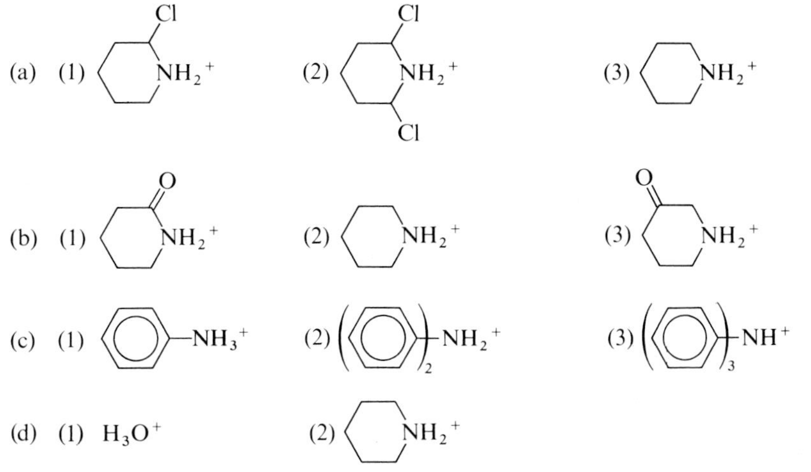

(a) (1) (2) (3)

(b) (1) (2) (3)

(c) (1) (2) (3)

(d) (1) H_3O^+ (2)

15.33. A chemist mixes a solution that is $0.0100M$ in NaOH and $0.00100M$ in methylamine ($pK_b = 3.34$).

(a) What concentration of methylammonium ion is present in the solution?
(b) At what pH would the concentrations of methylamine and methylammonium ion be equal?

15.34. Suggest techniques for separating the following mixtures:

(a) cyclohexanol, cyclohexylamine, and cyclohexanecarboxylic acid
(b) hexanamide and *n*-hexylamine

15.35. Which nitrogen in LSD is the most basic? Why?

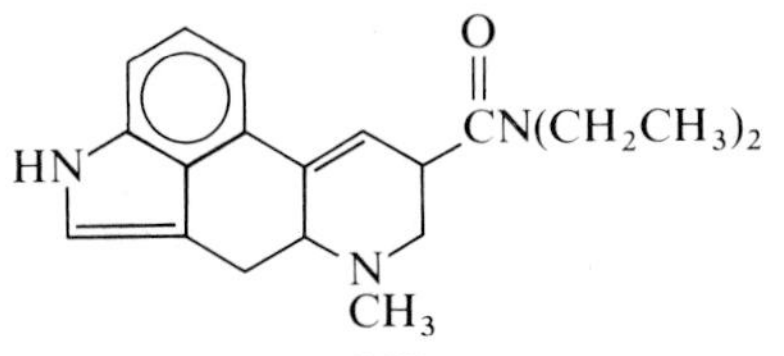

LSD

15.36. Predict the organic products when the local anesthetic novocaine (page 726) is treated with: (a) 1 equivalent cold dilute H_2SO_4; (b) excess dilute HCl and heat; (c) excess dilute NaOH and heat.

15.37. Predict the products when xylocaine (another local anesthetic) is treated with the reagents in Problem 15.36.

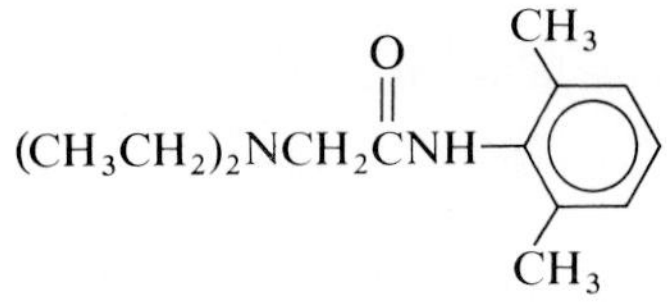

xylocaine

15.38. What would be an appropriate technique for resolving each of the following compounds?

(a) $C_6H_5CHCH_3$ with NH_2 group

(b) 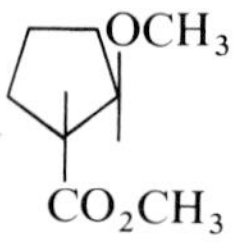 structure with OCH_3 and CO_2CH_3

15.39. Predict the major organic products:

(a) $CH_3I + NH_3$ (excess) $\longrightarrow$ (b) CH_3I (excess) $+ NH_3$ $\longrightarrow$

(c) 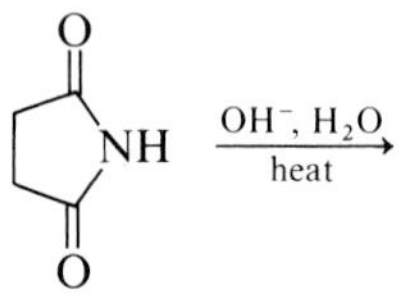 cyclic imide with NH, $\xrightarrow[\text{heat}]{OH^-, H_2O}$

(d) 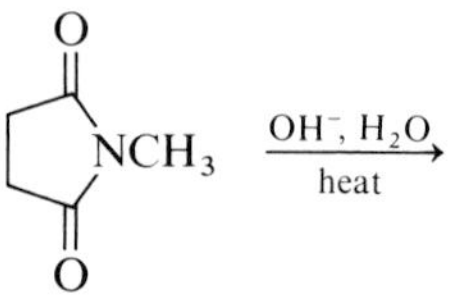 cyclic imide with NCH$_3$, $\xrightarrow[\text{heat}]{OH^-, H_2O}$

15.40. Predict the major product of the reaction of pyrrolidine (page 709) with   (a) benzoyl chloride;   (b) acetic anhydride;   (c) CH_3I (excess);   (d) phthalic anhydride; (e) benzenesulfonyl chloride ($C_6H_5SO_2Cl$);   (f) acetyl chloride, followed by LiAlH$_4$ (then hydrolysis);   (g) cold nitrous acid;   (h) dilute HCl;   (i) acetone + H$^+$.

15.41. Predict the major product of the reaction of *N*-methylpyrrolidine with each of the reagents [except (g)] in Problem 15.40.

15.42. What chemical reaction could you use to distinguish between:

(a)   aniline and *n*-hexylamine?
(b)   *n*-octylamine and octanamide?
(c)   triethylammonium chloride and tetraethylammonium chloride?

15.43. Predict the major organic products when each of the following compounds is heated:

(a) 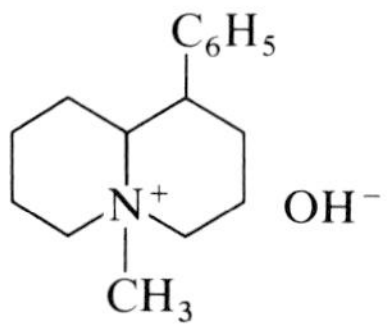 bicyclic ammonium structure with C_6H_5 and N^+, CH_3, OH$^-$

(b) 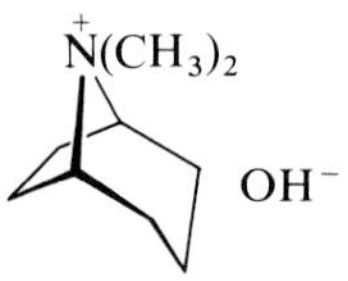 bicyclic structure with $\overset{+}{N}(CH_3)_2$, OH$^-$

15.44. Predict the products of the exhaustive methylation of the following heterocycle:

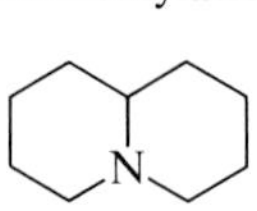

bicyclic heterocycle with N

15.45. Suggest synthetic routes to the following compounds:

(a)   isoleucine (2-amino-3-methylpentanoic acid)
(b)   β-phenylethylamine from toluene
(c)   2-methyl-1-phenyl-1,3-pentanedione from benzoic acid
(d)   aniline from benzoic acid

15.46. An amine ($C_8H_{15}N$) is subjected to two stages of exhaustive methylation. The final products are trimethylamine and the following diene:

What is a likely structure for the original amine?

15.47. The treatment of (cyclopentylmethyl)amine with nitrous acid resulted in a 76% yield of cyclohexanol. Another alcohol and three alkenes were also present in the product mixture.

   (a) Give a plausible mechanism for the formation of cyclohexanol.
   (b) What are the likely structures of the other products?

15.48. A chemist attempted to carry out a Hofmann rearrangement of butanamide with bromine and potassium hydroxide in methanol, rather than in water. Instead of *n*-propylamine, the chemist obtained a carbamate, $CH_3CH_2CH_2NHCO_2CH_3$. Explain how this product was formed.

15.49. When the following compound is subjected to catalytic hydrogenation, a compound with the formula $C_9H_{17}N$ is produced. What is the structure of this product?

15.50. A tertiary amine is oxidized by peroxides such as H_2O_2 to an **amine oxide**, a compound containing the —NO group. An amine oxide with a β hydrogen undergoes elimination when heated. (This reaction is called a **Cope elimination**.) Suggest a mechanism for the following elimination reaction of an amine oxide.

15.51. A chemist treated 1-bromobutane with ammonia and isolated products A and B. When A was treated with acetic anhydride, C was obtained. When B was treated with acetic anhydride, D was obtained. The infrared spectra of C and D are shown in Figure 15.2 (page 740). Identify A, B, C, and D.

15.52. The infrared spectrum for Compound A ($C_8H_{11}N$) is given in Figure 15.3 (page 740). A is soluble in dilute acid. Oxidation of A with hot $KMnO_4$ yields benzoic acid. What are the two possible structures for A? How could you distinguish these two possibilities by nmr spectroscopy?

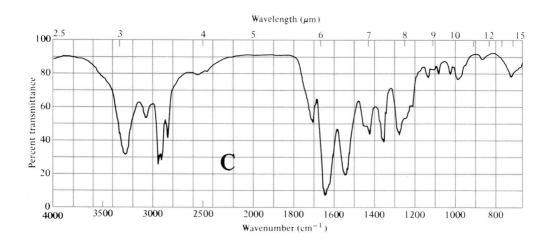

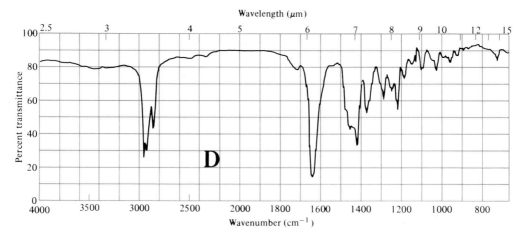

FIGURE 15.2. Infrared spectra for Problem 15.51.

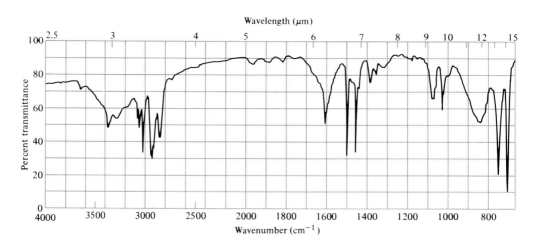

FIGURE 15.3 Infrared spectrum for Compound A in Problem 15.52.

15.53. A compound $C_{10}H_{13}NO$ (A) has the nmr spectrum shown in Figure 15.4. A is insoluble in dilute aqueous acid. Heating A with aqueous NaOH, followed by acidification, results in two compounds: acetic acid and an amine salt. The free amine (B) from this salt has the formula $C_8H_{11}N$.

Compound B was subjected to high-pressure hydrogenation to yield C ($C_8H_{17}N$). Compound C was subjected to one stage of exhaustive methylation (in which it took up 2 moles of CH_3I) and yielded trimethylamine and 3-methyl-1-cyclohexene. What are the structures of A, B, and C?

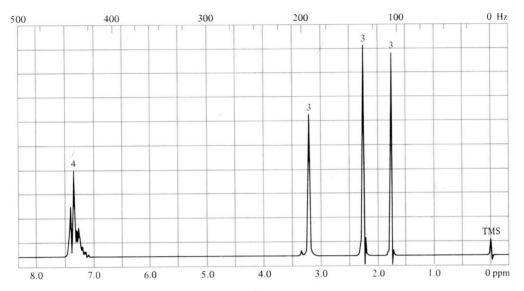

FIGURE 15.4. Nmr spectrum of compound A in Problem 15.53.

CHAPTER 16

Polycyclic and Heterocyclic Aromatic Compounds

In Chapter 10, we discussed benzene and substituted benzenes. However, benzene is but one member of a large number of aromatic compounds. Many other aromatic compounds may be grouped into two classes: **polycyclic and heterocyclic compounds**. The polycyclic aromatic compounds are also referred to as *polynuclear*, *fused-ring*, or *condensed-ring*, aromatic compounds. These aromatic compounds are characterized by rings that jointly share carbon atoms and by a common aromatic pi cloud.

Some polycyclic aromatic compounds:

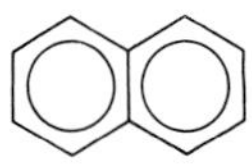

| naphthalene | anthracene | phenanthrene |

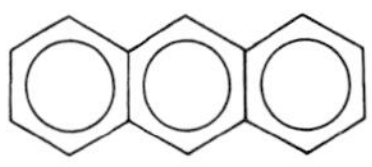

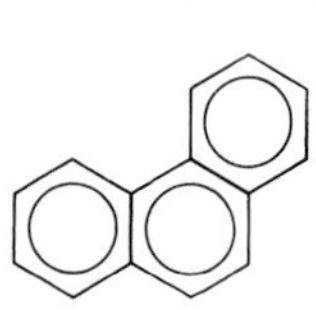

The polycyclic aromatic hydrocarbons and most of their derivatives are solids. Naphthalene has been used as mothballs and flakes, and derivatives of naphthalene are used in motor fuels and lubricants. The most extensive use made of the polycyclic aromatics is that of synthetic intermediates, for example, in the manufacture of dyes.

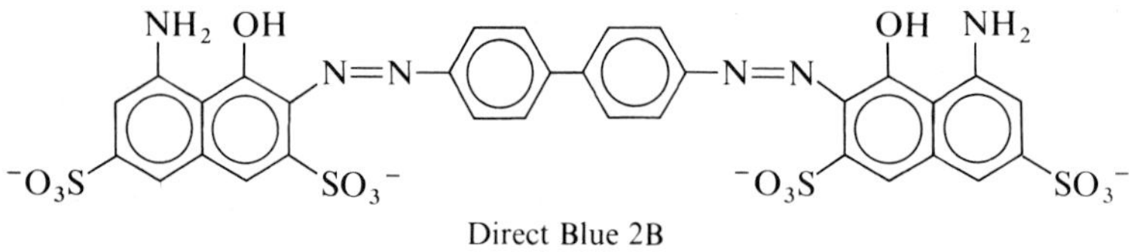

Direct Blue 2B

a dye

Graphite is one of the more interesting polycyclic compounds. The structure of graphite consists of planes of fused benzene rings (Figure 16.1). The distance (3.5Å) between any pair of planes is believed to be the thickness of the pi system of benzene. The "slipperiness" of graphite is due to the ability of these planes to slide across each other. Because of this property, graphite is a valuable lubricant that can be used even in outer space where ordinary oils and greases would solidify. Because of its mobile pi electrons, graphite can conduct electricity and finds use where an inert electrode is needed. Flashlight batteries, for example, contain graphite electrodes.

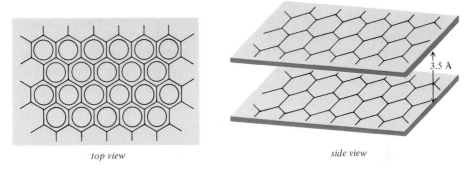

top view   *side view*

FIGURE 16.1. The structure of graphite.

A **heterocyclic compound** is a cyclic compound in which the ring atoms are of carbon and some other element. The atom of the other element (for example, N, S, or O) is called the **heteroatom**. Heterocyclic rings can be aromatic, just as carbon rings can. Approximately a third of the organic chemical literature deals with heterocyclic compounds. The importance of these compounds will become clear as we approach the end of the chapter and discuss some naturally occurring heterocycles—the **alkaloids**, such as morphine; the **nucleic acids**, the carriers of the genetic code; and a few other compounds of biological importance.

Some aromatic heterocyclic compounds:

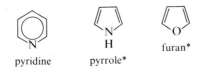

pyridine   pyrrole*   furan*

SECTION 16.1.

Nomenclature of Polycyclic Aromatic Compounds

The ring systems of common polycyclic aromatic compounds have individual names. Unlike the numbering of benzene or a cycloalkane ring, which starts at the

---

* A circle in the ring is not a proper representation of an aromatic pi cloud in a
       five-membered heterocycle. How the aromatic pi cloud is formed in these
       compounds will be discussed in Section 16.9.

position of a substituent, the numbering of a polycyclic ring is fixed by convention and does not change with the position of a substituent.

naphthalene   anthracene   phenanthrene

The position of a substituent in a monosubstituted naphthalene is often designated by a Greek letter. The positions adjacent to the ring-junction carbons are called α positions, while the next positions are β positions. By this system, 1-nitronaphthalene is called α-nitronaphthalene, while 2-nitronaphthalene is called β-nitronaphthalene. Naphthalene itself has four equivalent α positions and four equivalent β positions. (Only number designations are used in the anthracene and phenanthrene systems.)

1-nitronaphthalene   2-nitronaphthalene
(α-nitronaphthalene)   (β-nitronaphthalene)

SECTION 16.2.

Bonding in Polycyclic Aromatic Compounds

For a monocyclic ring system to be aromatic, it must meet three criteria:

1.   Each atom in the ring system must be in the sp^2 (or sp) hybrid state.

2.   The ring system must be planar.

3.   There must be $(4n + 2)$ pi electrons in the ring system (Hückel rule).

These criteria were discussed in Section 10.7. A polycyclic aromatic molecule must also contain only sp^2-hybridized atoms in the aromatic system, and the entire ring system must be planar. The Hückel rule, which was devised for monocyclic systems, also is applicable to polycyclic systems in which the sp^2 carbons are peripheral, or on the outside edge of the ring system. In polycyclic compounds, the number of pi electrons is easily counted when Kekulé formulas are used.

10 pi electrons   14 pi electrons   14 pi electrons
($n = 2$)   ($n = 3$)   ($n = 3$)

Like benzene, the polycyclic aromatic systems are more stable than the corresponding hypothetical polyenes with localized pi bonds. The energy differences between the hypothetical polyenes and the real compounds (that is, the *resonance energies*) have been calculated from heats of combustion and hydrogenation data.

resonance energy (kcal/mole):   36   61   84   92

Note that the resonance energy for a polycyclic aromatic compound is less than the sum of the resonance energies for a comparable number of benzene rings. Although the resonance energy for benzene is 36 kcal/mole, that for naphthalene is only 61 kcal/mole (about 30 kcal/mole for each ring).

In benzene, all carbon–carbon bond lengths are the same. This fact leads us to believe that there is an equal distribution of pi electrons around the benzene ring. In the polycyclic aromatic compounds, the carbon–carbon bond lengths are *not* all the same. For example, the distance between carbons 1 and 2 (1.36 Å) in naphthalene is smaller than the distance between carbons 2 and 3 (1.40 Å).

C—C in ethane: 1.54 Å

C=C in ethylene: 1.34 Å

C—C in benzene: 1.40 Å

From these measurements, we conclude that there is *not* an equal distribution of pi electrons around the naphthalene ring. From a comparison of bond lengths, we would say that the carbon 1–carbon 2 bond of naphthalene has more double-bond character than the carbon 2–carbon 3 bond. The resonance structures for naphthalene also indicate that the carbon 1–carbon 2 bond has more double–bond character.

two out of three resonance structures
show a carbon 1–carbon 2 double bond

Because all the carbon–carbon bonds in naphthalene are not the same, many chemists prefer to use Kekulé-type formulas for this compound instead of using circles to represent the pi cloud. We will use Kekulé-type formulas in our discussion of the reactions of naphthalene.

Phenanthrene shows similar differences between its bonds. The double-bond character of the 9,10-bond of phenanthrene is particularly evident in its chemical reactions. These positions of the phenanthrene ring system undergo addition reactions that are typical of alkenes but are not typical for benzene.

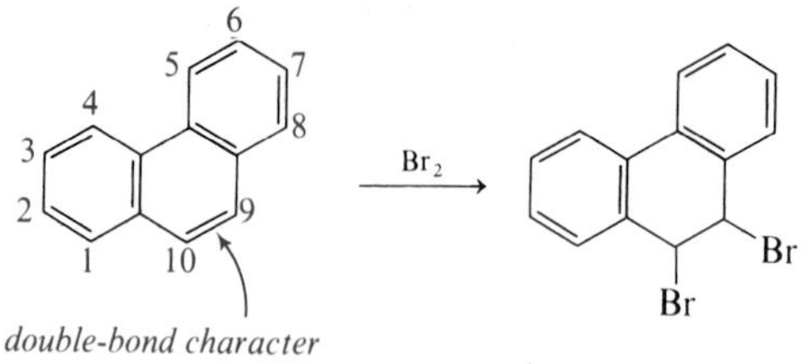

double-bond character

STUDY PROBLEM

16.1.   Write the Kekulé resonance structures for phenanthrene. On the basis of these structures, explain why the 9,10-bond has double-bond character.

SECTION 16.3.

Oxidation of Polycyclic Aromatic Compounds

The polycyclic aromatic compounds are more reactive toward oxidation, reduction, and electrophilic substitution than is benzene. The reason for the greater reactivity is that the polycyclic compounds can undergo reaction at one ring and still have one or more intact benzenoid rings in the intermediate and in the product. Less energy is required to overcome the aromatic character of a single ring of the polycyclic compounds than is required for benzene.

Benzene is not easily oxidized; however, naphthalene can be oxidized to products in which much of the aromaticity is retained. Phthalic anhydride is prepared commercially by the oxidation of naphthalene; this reaction probably proceeds by way of *o*-phthalic acid.

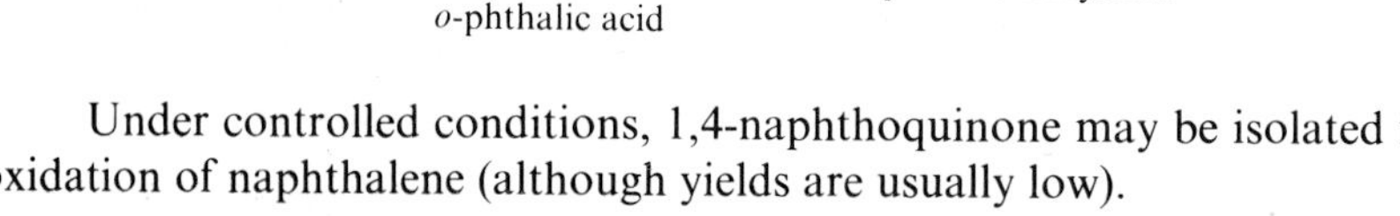

Under controlled conditions, 1,4-naphthoquinone may be isolated from an oxidation of naphthalene (although yields are usually low).

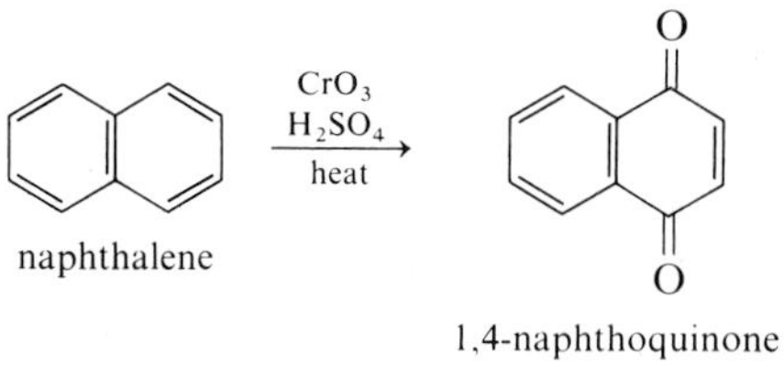

Anthracene and phenanthrene can also undergo oxidation to quinones:

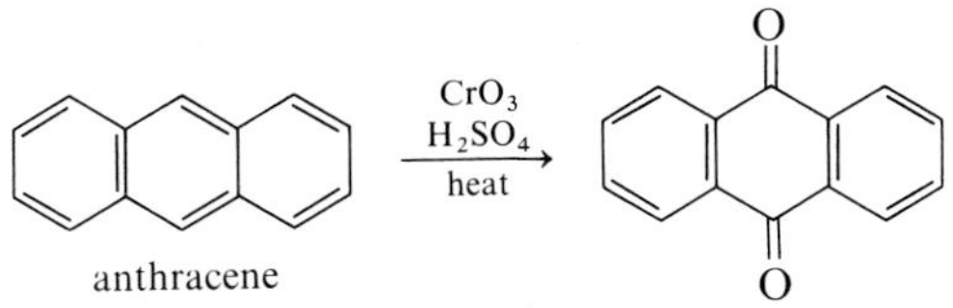

anthracene

$\xrightarrow[\text{heat}]{\begin{array}{c}\text{CrO}_3\\\text{H}_2\text{SO}_4\end{array}}$

9,10-anthraquinone

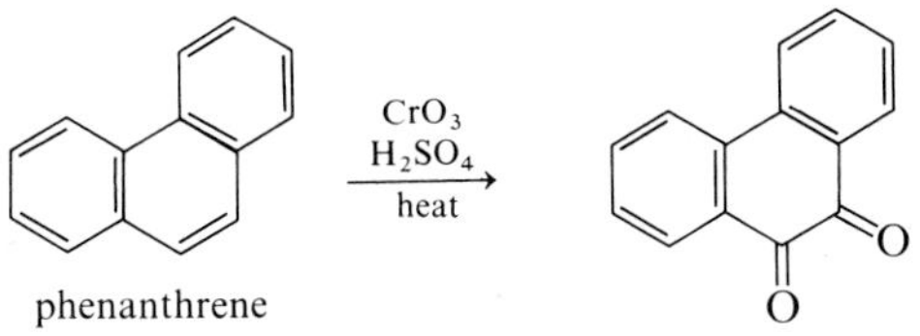

phenanthrene

$\xrightarrow[\text{heat}]{\begin{array}{c}\text{CrO}_3\\\text{H}_2\text{SO}_4\end{array}}$

9,10-phenanthraquinone

STUDY PROBLEM

16.2. From the following observations, tell whether chromate oxidation of naphthalene derivatives involves an initial *electrophilic attack* or *nucleophilic attack*.

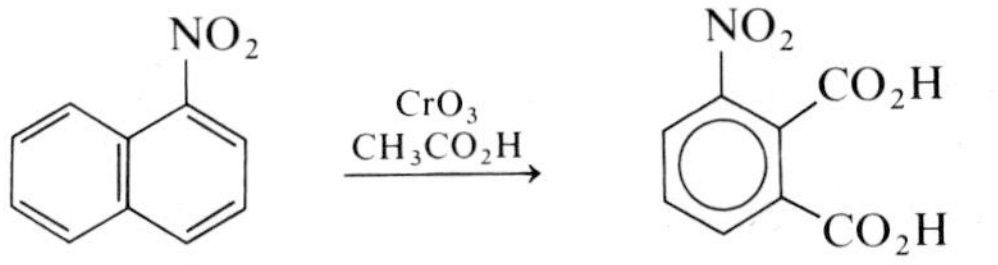

1-nitronaphthalene

$\xrightarrow[]{\begin{array}{c}\text{CrO}_3\\\text{CH}_3\text{CO}_2\text{H}\end{array}}$

3-nitro-1,2-phthalic acid

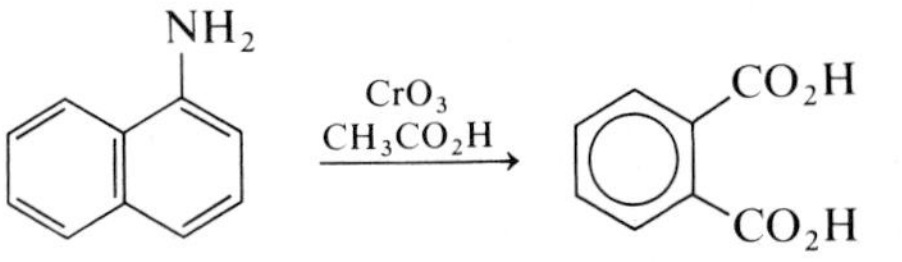

1-naphthylamine

$\xrightarrow[]{\begin{array}{c}\text{CrO}_3\\\text{CH}_3\text{CO}_2\text{H}\end{array}}$

o-phthalic acid

SECTION 16.4.

Reduction of Polycyclic Aromatic Compounds

Unlike benzene, the polycyclic compounds may be partially hydrogenated without heat and pressure, or they may be reduced with sodium and ethanol.

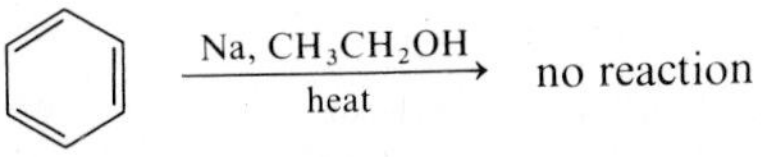

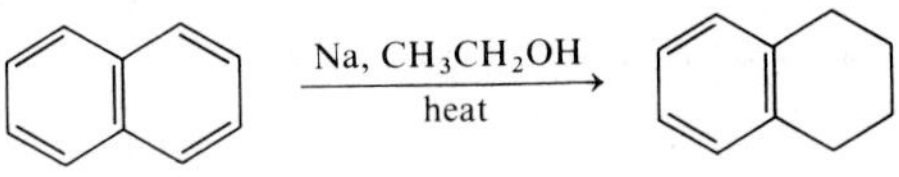

$\xrightarrow[\text{heat}]{\text{Na, CH}_3\text{CH}_2\text{OH}}$ no reaction

$\xrightarrow[\text{heat}]{\text{Na, CH}_3\text{CH}_2\text{OH}}$

tetralin

$\xrightarrow[\text{heat}]{\text{Na, CH}_3\text{CH}_2\text{OH}}$

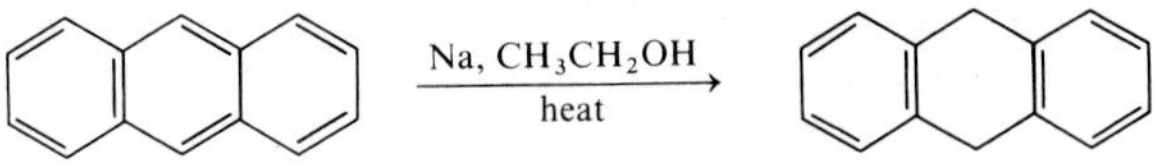

9,10-dihydroanthracene

Note that the partially reduced ring systems still contain one or more benzenoid rings. Most of the aromatic character of the original ring systems has been retained in these partially reduced products. To hydrogenate the polycyclic aromatics completely would, of course, require heat and pressure, just as it does for benzene.

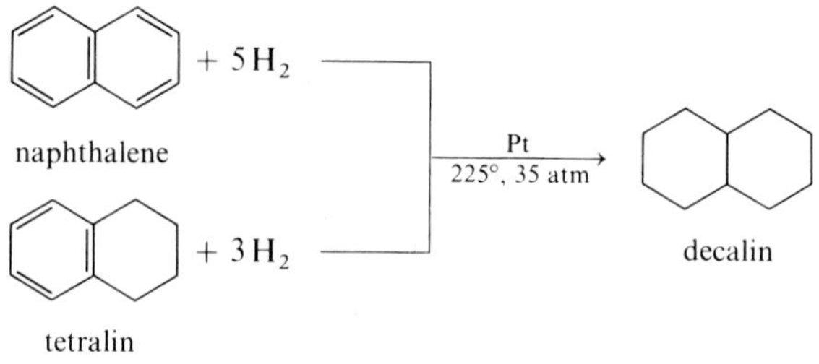

naphthalene   + 5 H$_2$

tetralin   + 3 H$_2$

$\xrightarrow[225°, 35 \text{ atm}]{\text{Pt}}$

decalin

STUDY PROBLEM

16.3.   Predict the product of the reaction of phenanthrene with sodium and ethanol.

SECTION 16.5.

Electrophilic Substitution Reactions of Naphthalene

The polycyclic aromatic ring systems are more reactive toward electrophilic attack than is benzene. Naphthalene undergoes electrophilic aromatic substitution reactions predominantly at the 1-position. The reasons for the enhanced reactivity and for this position of substitution will be discussed shortly.

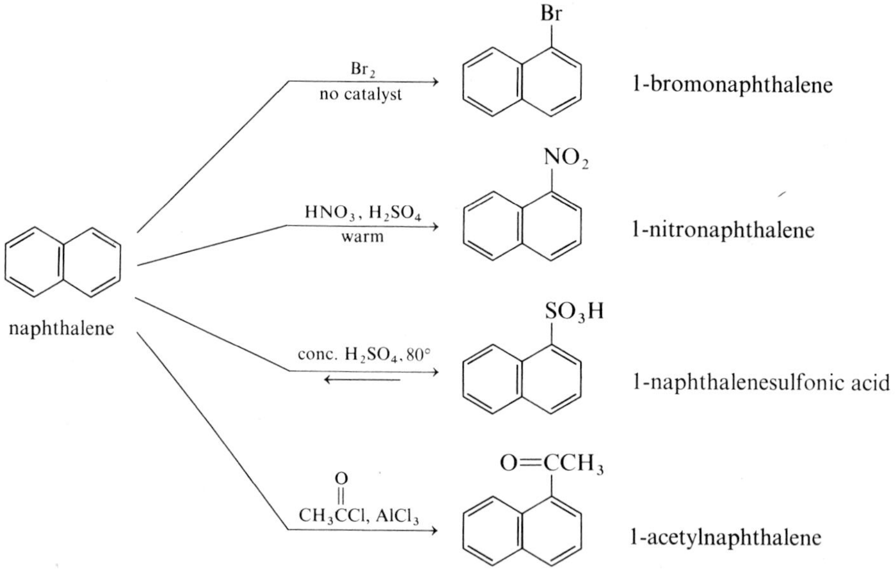

naphthalene

$\xrightarrow[\text{no catalyst}]{\text{Br}_2}$ 1-bromonaphthalene

$\xrightarrow[\text{warm}]{\text{HNO}_3, \text{H}_2\text{SO}_4}$ 1-nitronaphthalene

$\xrightleftharpoons{\text{conc. H}_2\text{SO}_4, 80°}$ 1-naphthalenesulfonic acid

$\xrightarrow{\underset{\|}{\overset{\text{O}}{\text{CH}_3\text{CCl}}}, \text{AlCl}_3}$ 1-acetylnaphthalene

Anthracene, phenanthrene, and larger fused-ring compounds are even more reactive than naphthalene toward electrophilic substitution. However, these reactions are not as important as those of naphthalene because mixtures of isomers (which are often difficult to separate) are obtained. Phenanthrene, for example, undergoes mononitration at each available position to yield five nitrophenanthrenes.

A. Position of substitution of naphthalene

The mechanism for naphthalene substitution is similar to that for benzene substitution. Let us look at the stepwise bromination reaction to see why substitution at the 1-position is favored and why this reaction occurs more readily than the bromination of benzene.

1-Substitution (favored):

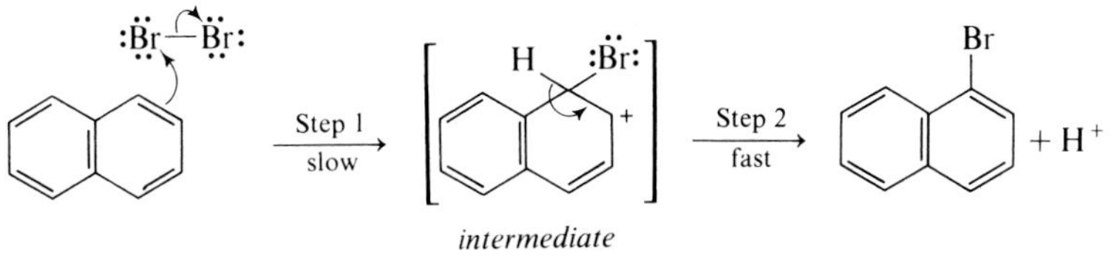

intermediate

Resonance structures for the 1-substitution intermediate:

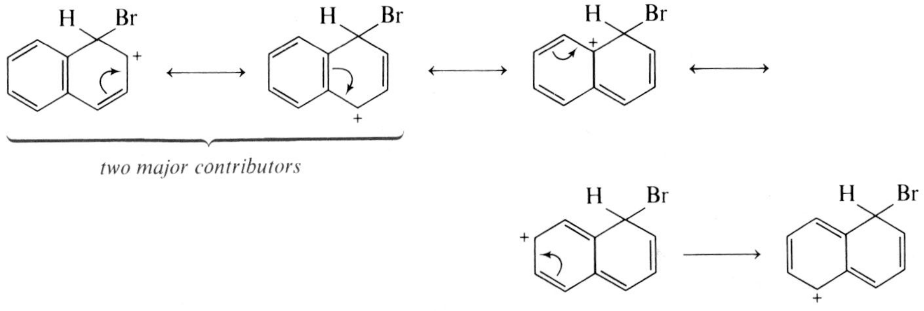

two major contributors

The resonance structures of the intermediate for substitution at the 1-position show two contributors in which the benzene ring is intact. Because of aromatic resonance-stabilization, these two structures are of lower energy than the other resonance structures and are major contributors to the real structure of the intermediate. This is the reason that naphthalene undergoes electrophilic substitution more readily than benzene. For benzene to go to a benzenonium ion requires the loss of aromaticity, about 36 kcal/mole. For naphthalene to go to its intermediate requires only partial loss of aromaticity, about 25 kcal/mole (the difference in resonance energy between naphthalene and benzene). Because of the lower E_{act} leading to the intermediate, the rate of bromination of naphthalene is faster than that of benzene.

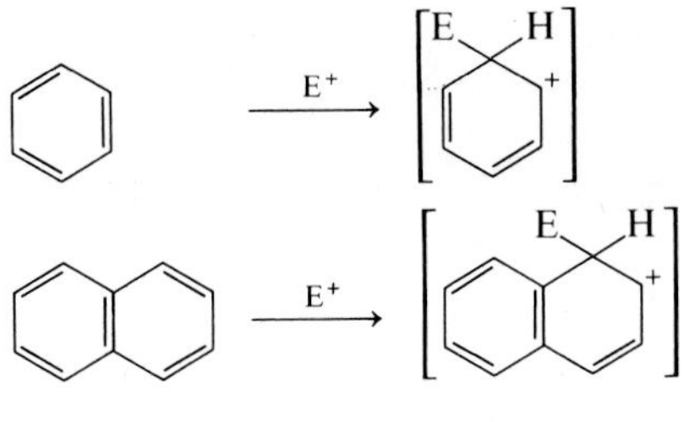

requires 36 kcal/mole to destroy aromaticity

requires only 25 kcal/mole

Why is 1-substitution favored over 2-substitution for naphthalene? Inspect the resonance structures for the intermediate leading to 2-substitution:

2-Substitution (not favored):

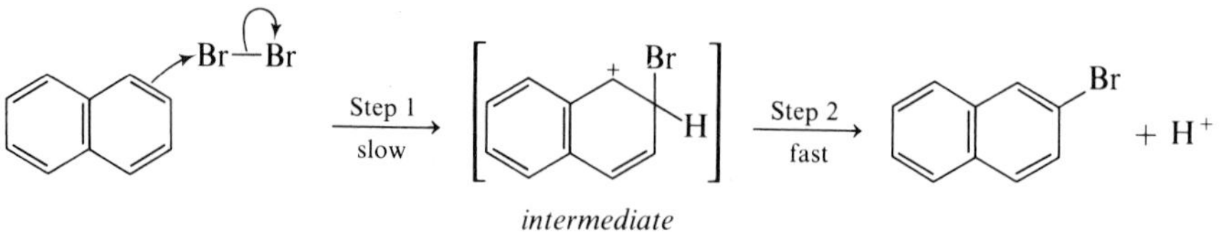

intermediate

Resonance structures for the 2-substitution intermediate:

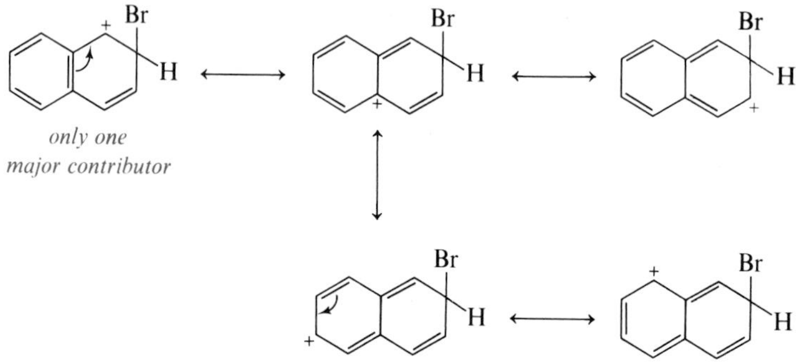

only one major contributor

The 2-intermediate has only one contributing resonance structure in which a benzenoid ring is intact, while the resonance structures for the 1-intermediate show two such structures. The 1-intermediate is more stabilized by resonance, and its transition state is of lower energy. For this reason, the E_{act} is lower and its rate of formation faster.

The sulfonation of naphthalene, which is a reversible reaction (Section 10.8F), is more complex than bromination. At 80°, the expected 1-naphthalenesulfonic acid is the product. However, at higher temperatures (160–180°), the product is 2-naphthalenesulfonic acid. At low temperatures, the reaction is under **kinetic control**—that is, the relative rates of reaction determine the product ratio. At high temperatures, the reaction is under **thermodynamic**, or **equilibrium**, **control**—the relative stabilities of the products determine the product ratio.

We have already seen why the rate of 1-substitution of naphthalene is faster than that of 2-substitution. At less than 80°, the rate of formation of either naphthalenesulfonic acid is relatively slow; the reaction proceeds through the lower-energy 1-intermediate just as it does for bromination.

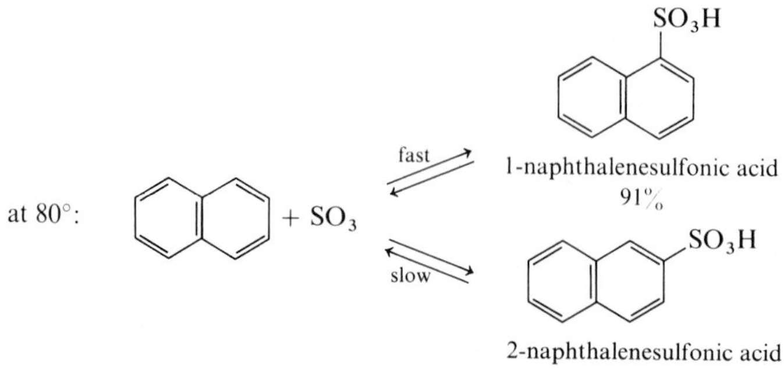

at 80°:

1-naphthalenesulfonic acid
91%

2-naphthalenesulfonic acid

Even though 1-naphthalenesulfonic acid is formed at low temperatures, this isomer is less stable than the 2-isomer because of repulsions between the —SO$_3$H group and the hydrogen at position 8.

repulsion   *less repulsion*

H⟋SO$_3$H   H H⟍

H⟍   ⟋H   H⟍   SO$_3$H

H⟍   ⟋H   H⟍   H

H   H   H   H

1-naphthalenesulfonic acid   2-naphthalenesulfonic acid

less stable   *more stable*

At a higher temperature, the rates of both forward reactions and the rates of both reverse reactions are all increased. Although the 1-product may be formed more readily, it can revert quickly to naphthalene. The 2-product is formed more slowly, but the rate of its reverse reaction is even slower because the 2-product is more stable and of lower energy. At higher temperatures, the 2-product accumulates in the reaction mixture and is the observed product (see Figure 16.2).

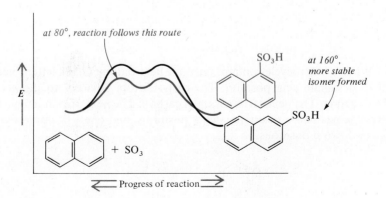

FIGURE 16.2. Energy diagram for the sulfonation of naphthalene.

STUDY PROBLEMS

16.4. On the basis of what you learned in Chapter 10 about activating and deactivating groups and the position(s) to which they direct an incoming substituent, predict the major organic products of aromatic nitration of the following compounds:

(a) [structure: naphthalene with CH₃ at 2-position]

(b) [structure: naphthalene with CH₃ at 1-position]

(c) [structure: naphthalene with SO₃H at 2-position]

(d) [structure: naphthalene with NO₂ at 1-position]

16.5. Write equations showing how you could prepare the following compounds from naphthalene.

(a) [structure: naphthalene with CN at 1-position]

(b) [structure: naphthalene with CH₂CH₂CH₃ at 1-position]

(c) [structure: naphthalene with Br at 1- and 3-positions]

SECTION 16.6.

Nomenclature of Aromatic Heterocyclic Compounds

Because of their widespread occurrence in nature, the aromatic heterocycles are of more general interest to chemists than are the polycyclic compounds containing only carbon atoms in their rings. Like the polycyclic aromatic compounds, the aromatic heterocycles generally have individual names. The names and structures of some of the more important members of this class of compounds are listed in Table 16.1.

The numbering of three representative heterocycles follows:

pyridine   thiazole   imidazole

When a heterocycle contains only one heteroatom, Greek letters may also be used to designate ring position. The carbon atom adjacent to the heteroatom is the α carbon. The next carbon is the β carbon. The next carbon in line, if any, is γ. Pyridine has two α positions, two β positions, and one γ position. Pyrrole has two α and two β positions.

pyridine   pyrrole

TABLE 16.1. Some important aromatic heterocyclic compounds

Structure	Name	Structure	Name
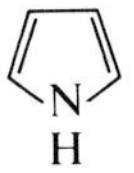	pyrrole	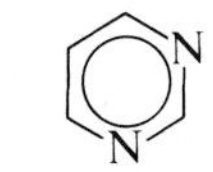	pyrimidine
	furan	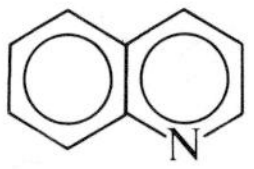	quinoline
	thiophene	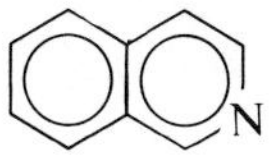	isoquinoline
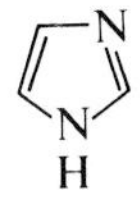	imidazole	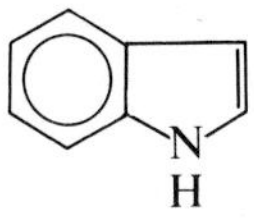	indole
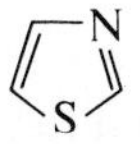	thiazole	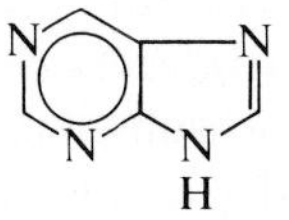	purine
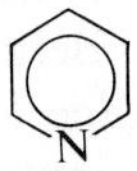	pyridine		

SECTION 16.7.

Pyridine, a Six-Membered Aromatic Heterocycle

Of the common six-membered heterocycles, only the nitrogen heterocycles are stable aromatic compounds.

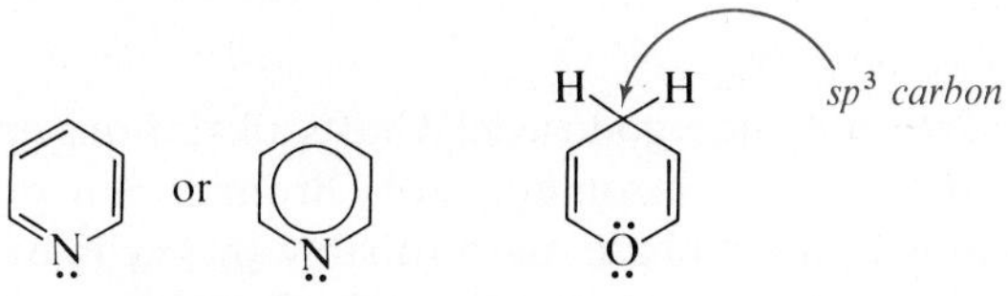

pyridine
aromatic

pyran
not aromatic

Pyridine has a structure similar to that of benzene. Pyridine contains a planar, six-membered ring consisting of five carbons and one nitrogen. Each of these ring atoms is sp^2-hybridized and has one electron in a p orbital that contributes to the aromatic pi cloud (six pi electrons). Figure 16.3 shows the lowest-energy π molecular orbital of pyridine.

Note the differences between benzene and pyridine. Benzene is symmetrical and nonpolar, but pyridine contains an electronegative nitrogen and therefore is *polar*.

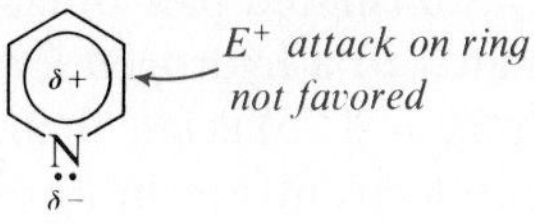

E^+ *attack on ring*
not favored

dipole moment $\mu = 2.26$ D

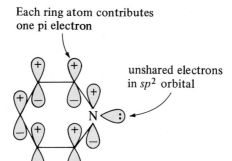

FIGURE 16.3. The lowest-energy π molecular orbital in pyridine.

Because the nitrogen is more electronegative than carbon, the rest of the pyridine ring is electron-deficient. An electron-deficient ring means that the carbon atoms in the ring carry a *partial positive charge*. A pyridine ring therefore has a low reactivity toward electrophilic substitution compared to benzene. Besides the electronegative nitrogen rendering the ring partially positive, pyridine forms a cation with many Lewis acids. Cation formation renders the ring even more electron-deficient.

Pyridine does not undergo Friedel–Crafts alkylations or acylations, nor does it undergo coupling with diazonium salts. Bromination proceeds only at high temperatures in the vapor phase and probably proceeds by a free-radical path. When substitution does occur, it occurs at the 3-position.

pyridine   3-bromopyridine   3,5-dibromopyridine
                                             37%   26%

Another major difference between pyridine and benzene is that the nitrogen in pyridine contains an unshared pair of electrons in an sp^2 orbital. This pair of electrons can be donated to a hydrogen ion. Like amines, pyridine is basic. The basicity of pyridine ($pK_b = 8.75$) is less than that of aliphatic amines ($pK_b = \sim 4$) because the unshared electrons are in an sp^2 orbital instead of an sp^3 orbital.

Nonetheless, pyridine undergoes many reactions typical of amines.

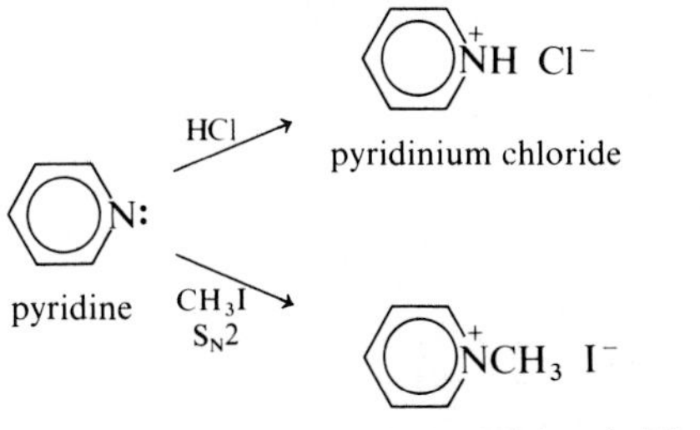

pyridinium chloride

pyridine

N-methylpyridinium iodide

Like that of benzene, the aromatic ring of pyridine is resistant to oxidation. Side chains can be oxidized to carboxyl groups under conditions that leave the ring intact.

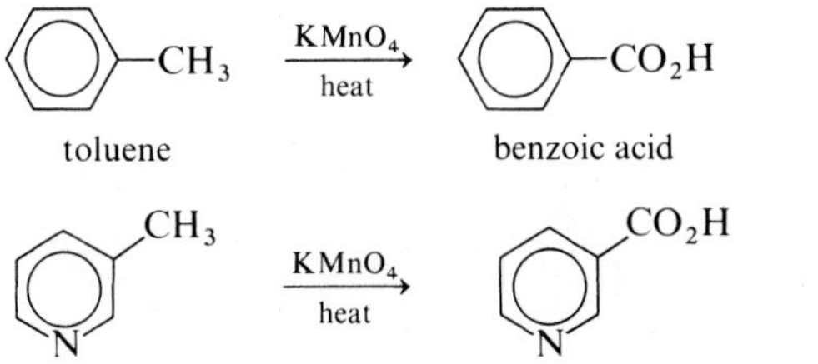

toluene   benzoic acid

3-methylpyridine   3-pyridinecarboxylic acid
                          (nicotinic acid)

a B vitamin

A.   Nucleophilic substitution on the pyridine ring

When a benzene ring is substituted with electron-withdrawing groups such as —NO_2, *aromatic nucleophilic substitutions* can take place (see Section 10.16).

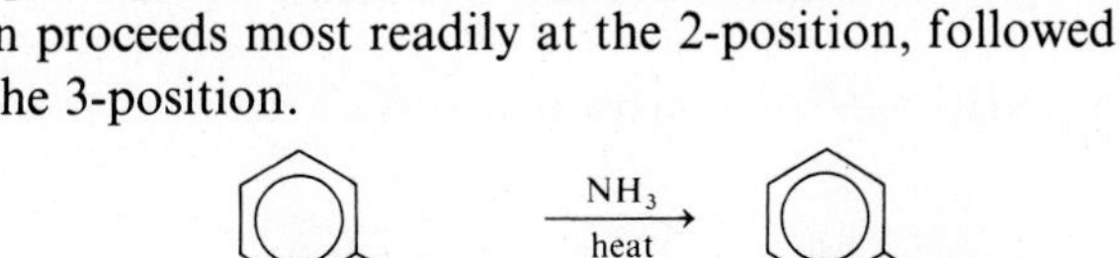

1-chloro-2,4-dinitrobenzene   2,4-dinitroaniline

The nitrogen in pyridine withdraws electron density from the rest of the ring. It is not surprising, then, that nucleophilic substitution also occurs with pyridine. Substitution proceeds most readily at the 2-position, followed by the 4-position, but not at the 3-position.

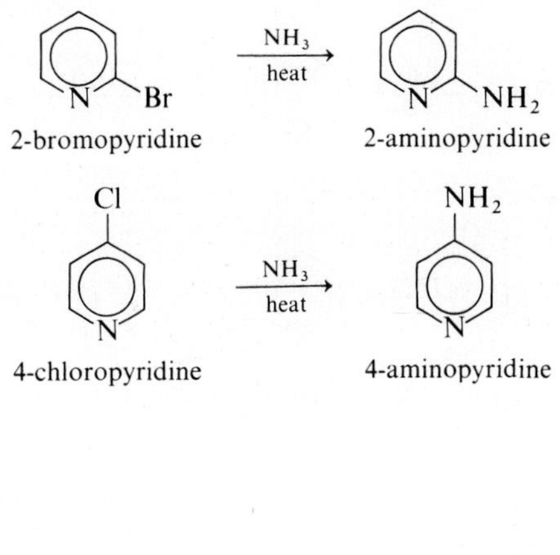

2-bromopyridine   2-aminopyridine

4-chloropyridine   4-aminopyridine

Let us look at the mechanisms for substitution at the 2- and 3-positions to see why the former reaction proceeds more readily.

2-Substitution (favored):

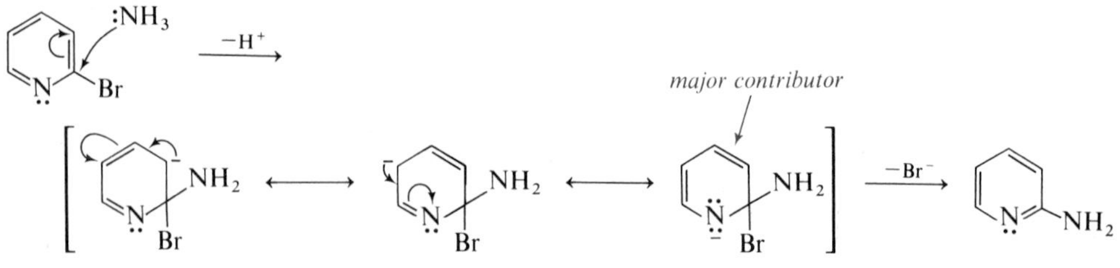

resonance structures for intermediate

3-Substitution (not favored):

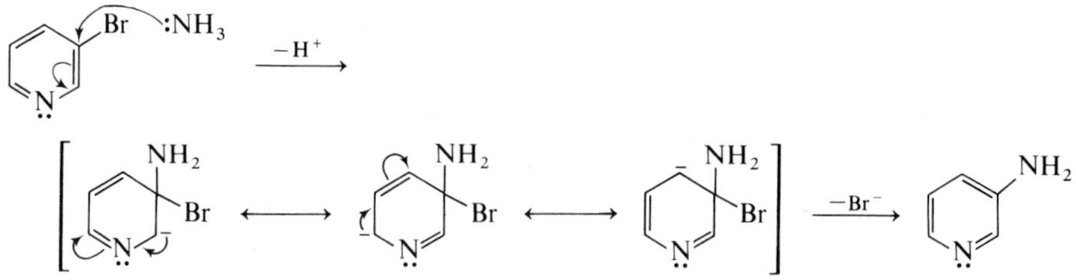

resonance structures for intermediate

The intermediate for 2-substitution is especially stabilized by the contribution of the resonance structure in which nitrogen carries the negative charge. Substitution at the 3-position goes through an intermediate in which the nitrogen cannot help stabilize the negative charge. The intermediate for 3-substitution is of higher energy; the rate of reaction going through this intermediate is slower.

STUDY PROBLEM

16.6. Give the resonance structures for the intermediate in the reaction of 4-chloropyridine with ammonia.

Benzene itself (with no substituents) does not undergo nucleophilic substitution. This reaction does occur with pyridine if an extremely strong base such as a lithium reagent or amide ion (NH_2^-) is used.

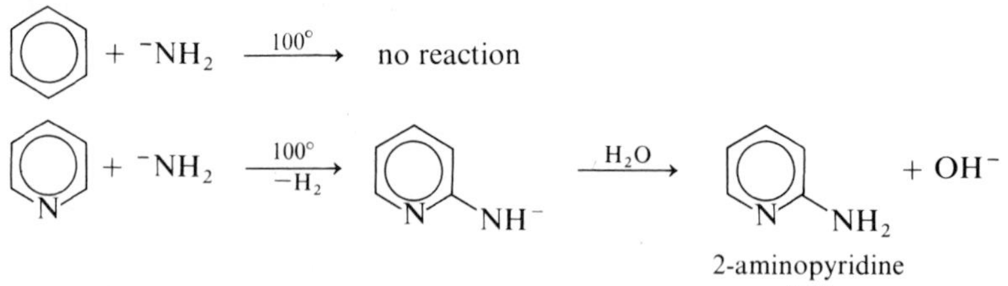

2-aminopyridine
70%

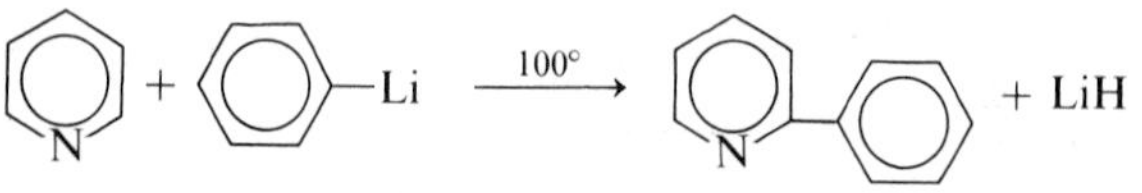

2-phenylpyridine
50%

In the reaction of pyridine with NH_2^-, the initial product is the anion of 2-aminopyridine. The free amine is obtained by treatment with water.

Step 1 (attack of NH$_2^-$ and loss of H$_2$):

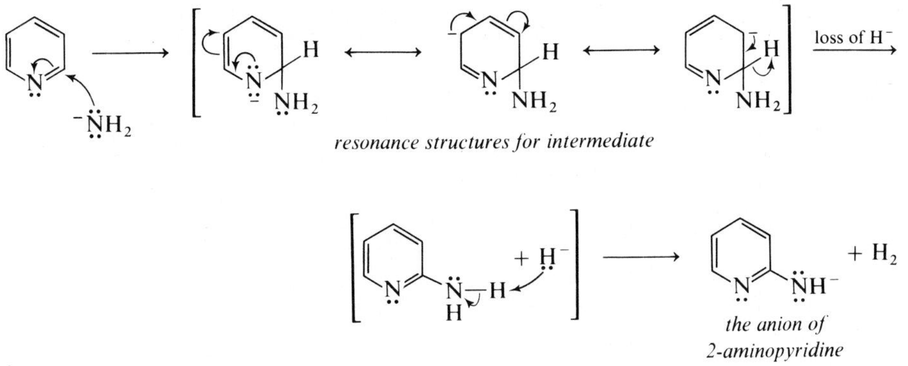

resonance structures for intermediate

the anion of
2-aminopyridine

Step 2 (treatment with H$_2$O):

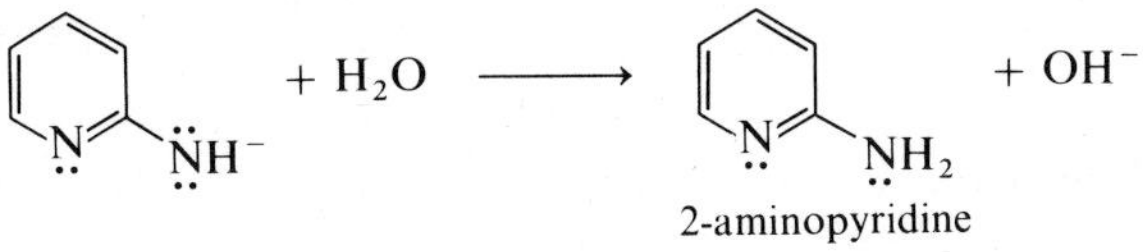

2-aminopyridine

SECTION 16.8.

Quinoline and Isoquinoline

Quinoline is a fused-ring heterocycle that is similar in structure to naphthalene, but with a nitrogen at position 1. Isoquinoline is the 2-isomer. (Note that the numbering of isoquinoline starts at a carbon, not at the nitrogen.)

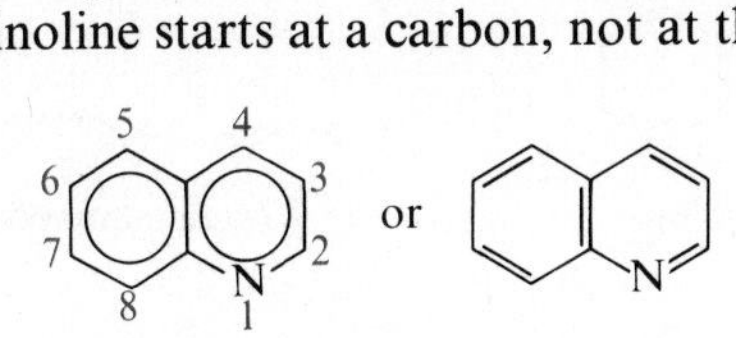

quinoline

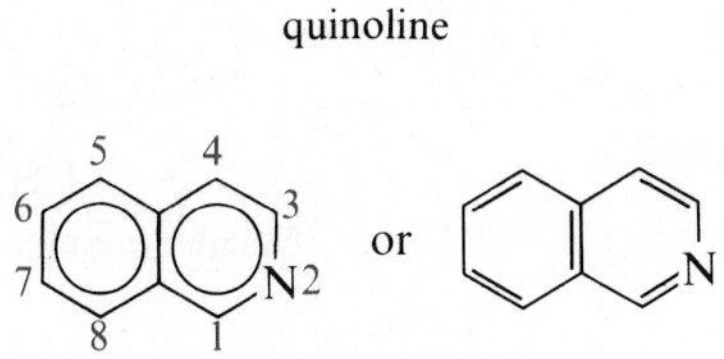

isoquinoline

Both quinoline and isoquinoline contain a pyridine ring fused to a benzene ring. The nitrogen ring in each of these two compounds behaves somewhat like the pyridine ring. Both quinoline and isoquinoline are weak bases ($pK_b = 9.1$ and 8.6, respectively). Both compounds undergo *electrophilic substitution* more easily than pyridine, but in positions 5 and 8 (on the benzenoid ring, not on the deactivated nitrogen ring). The positions of substitution are determined by intermediates similar to those in naphthalene substitution reactions.

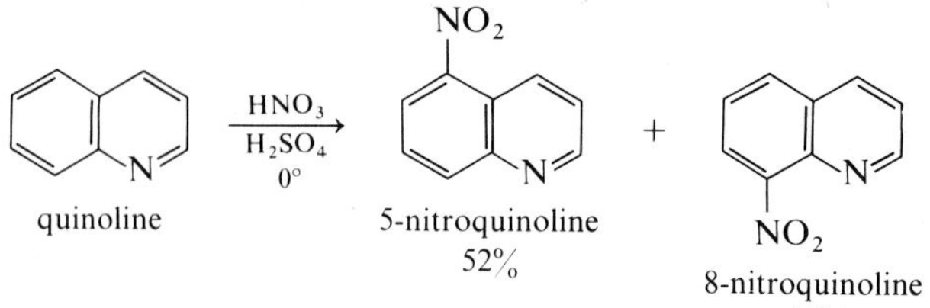

quinoline $\xrightarrow[\substack{H_2SO_4 \\ 0°}]{HNO_3}$ 5-nitroquinoline 52% + 8-nitroquinoline 48%

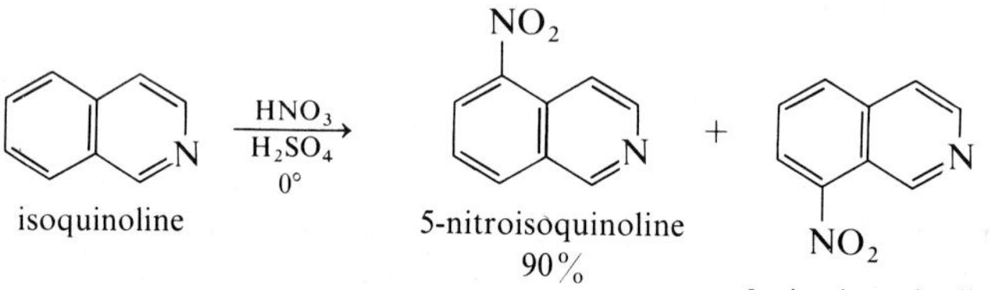

isoquinoline $\xrightarrow[\substack{H_2SO_4 \\ 0°}]{HNO_3}$ 5-nitroisoquinoline 90% + 8-nitroisoquinoline 10%

STUDY PROBLEM

16.7. Draw resonance structures of the intermediates for nitration at the 5- and 6-positions of quinoline to show why 5-nitroquinoline is formed preferentially.

Like pyridine, the nitrogen-containing ring of either quinoline or isoquinoline can undergo **nucleophilic substitution**. The position of attack is α to the nitrogen in either ring system, just as it is in pyridine.

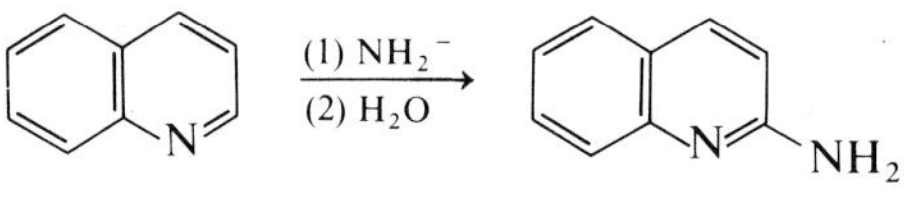

quinoline $\xrightarrow[\text{(2) } H_2O]{\text{(1) } NH_2^-}$ 2-aminoquinoline

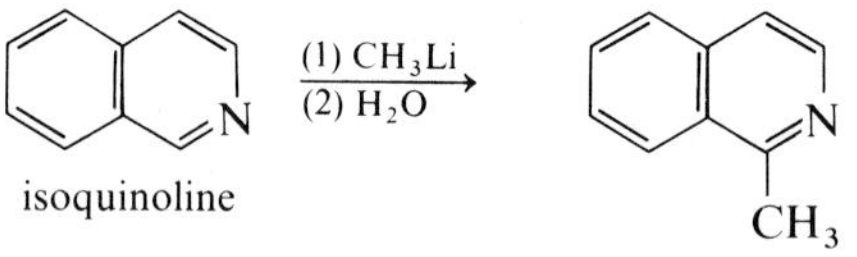

isoquinoline $\xrightarrow[\text{(2) } H_2O]{\text{(1) } CH_3Li}$ 1-methylisoquinoline

---

SECTION 16.9.

Pyrrole, a Five-Membered Aromatic Heterocycle

For a five-membered-ring heterocycle to be aromatic, the heteroatom must have *two electrons* to donate to the aromatic pi cloud. Pyrrole, furan, and thiophene all meet this criterion and therefore are aromatic. We will emphasize pyrrole in our discussion of five-membered aromatic heterocycles because it is typical in terms of both bonding and chemical reactivity.

pyrrole   furan   thiophene

Unlike pyridine and the amines, pyrrole ($pK_b = \sim 14$) is *not basic.*

$$+ \; H^+ \longrightarrow \text{no stable cation}$$

To see why pyrrole is not basic, we must consider the electronic structure of pyrrole. We know that pyrrole is aromatic because (1) its heat of combustion is about 25 kcal/mole less than that calculated for a diene structure; (2) pyrrole undergoes aromatic substitution reactions; and (3) the protons of pyrrole absorb in the aromatic region of the nmr spectrum (Figure 16.4). (Recall that aryl protons absorb downfield from most other protons because they are deshielded by the effects of the ring current; see Section 8.7B.)

In a five-membered ring, the minimum number of pi electrons needed for aromaticity is six ($4n + 2$, where $n = 1$). The four carbons of pyrrole each contribute one electron; therefore, the nitrogen atom of pyrrole must contribute *two* electrons (not just one as it does in pyridine). In addition to contributing two electrons to the pi molecular orbitals, the nitrogen in pyrrole shares three electrons in sigma bonds to two ring carbons and to a hydrogen. Consequently, all five bonding electrons of the nitrogen are used in bonding. The pyrrole nitrogen does not have unshared electrons and is not basic. Figure 16.5 shows the orbital picture for the lowest-energy π molecular orbital.

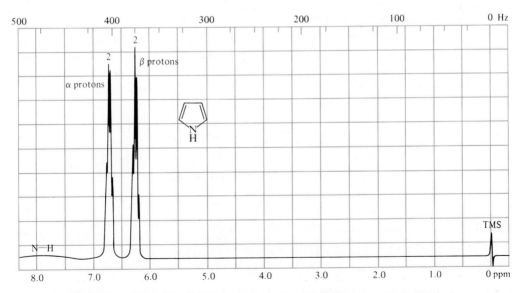

FIGURE 16.4. Nmr spectrum of pyrrole, C_4H_5N. (The NH absorption is a low, broad band near 8 ppm.)

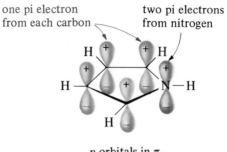

p orbitals in π_1

FIGURE 16.5. The bonding in pyrrole, C_4H_5N.

STUDY PROBLEM

16.8. If pyrrole did form a cation when treated with HCl, what would its structure be? (Show the *p* orbitals.) Would the pyrrole cation be aromatic or not?

Because the nitrogen atom in pyrrole contributes two electrons to the aromatic pi cloud, the nitrogen atom is electron-deficient and therefore not basic. The pyrrole ring, however, has six pi electrons for only five ring atoms. The ring is electron-rich and therefore partially negative. The dipole moments reflect this: nitrogen is usually the negative end of a dipole, but the nitrogen in pyrrole is the positive end of the molecule.

μ: 2.26 D   1.81 D

A.   Electrophilic substitution on the pyrrole ring

Because the ring carbons are the negative part of the pyrrole molecule, these carbons are *activated toward electrophilic attack, but deactivated to nucleophilic attack*. (This reactivity is opposite to that of pyridine.) The principal chemical characteristic of pyrrole and the other five-membered aromatic heterocycles is the ease with which they undergo electrophilic substitution.

$$\text{SO}_3, \text{ pyridine} \longrightarrow$$

2-pyrrolesulfonic acid
90%

pyrrole

$$\text{HNO}_3, (\text{CH}_3\text{CO})_2\text{O}, 5° \longrightarrow$$

2-nitropyrrole
80%

Electrophilic substitution occurs principally at the 2-position of the pyrrole ring. A look at the resonance structures for the intermediates of 2- and 3-nitration shows why. Three resonance structures can be drawn for the 2-intermediate, while only two can be drawn for the 3-intermediate. There is greater delocalization of the positive charge in the intermediate leading to 2-nitration than there is in that leading to 3-nitration.

2-Nitration (favored):

resonance structures for the intermediate

3-Nitration (not favored):

resonance structures for the intermediate

STUDY PROBLEM

16.9. Predict the major organic monosubstitution products:

    (a) furan + $(CH_3CO)_2O$ $\xrightarrow{BF_3}$

    (b) thiophene + H_2SO_4 $\xrightarrow{25°}$

    (c) pyrrole + $C_6H_5N_2{}^+$ Cl^- $\xrightarrow{25°}$

    (d) furan + Br_2 $\xrightarrow[25°]{dioxane}$

B.   Porphyrins

The **porphyrin ring system** is a biologically important unit found in *heme*, the oxygen-carrying component of hemoglobin; in *chlorophyll*, a plant pigment; and in the *cytochromes*, compounds involved in utilization of O_2 by animals. Structures of these compounds are shown in Figure 16.6.

    Note that the porphyrin ring system is composed of four pyrrole rings joined by =CH— groups. The entire ring system is aromatic.

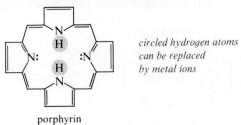

circled hydrogen atoms can be replaced by metal ions

porphyrin

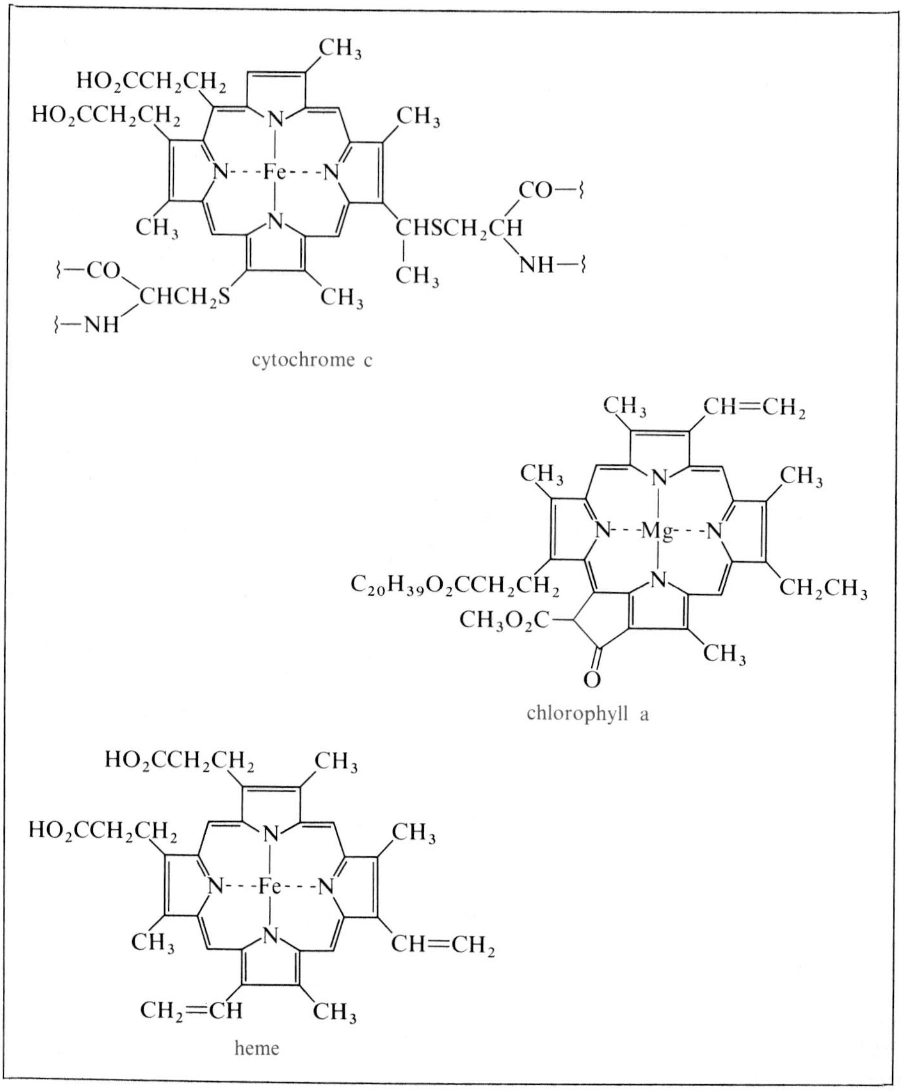

FIGURE 16.6. Some biologically important porphyrins.

The pyrrole hydrogens in the porphyrin ring system can be replaced by a variety of metal ions. The product is a **chelate** (Greek, *chele*, "crab's claw"), a compound or ion in which a metal ion is held by more than one bond from the original molecule. With porphyrin, the chelate is planar around the metal ion, and resonance results in four equivalent bonds from the nitrogen atoms to the metal.

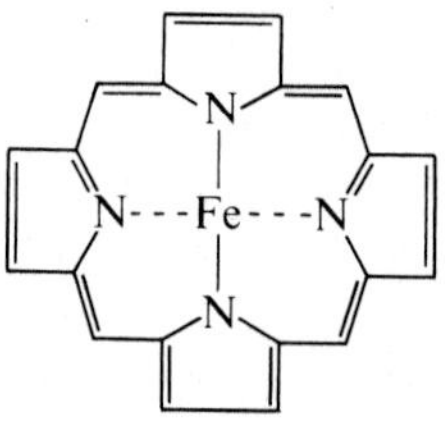

planar and resonance-stabilized

SECTION 16.10.

Alkaloids

A. Occurrence and structure

Primitive people have often used extracts of roots, bark, leaves, flowers, berries, and seeds as drugs. This use of plants for medicinal purposes was not necessarily based on superstition or wishful thinking. Many plants contain compounds that have profound physiological impact. The active agents in many of these plant substances have been isolated and have been found to be heterocyclic nitrogen compounds.

Many of the nitrogen compounds in plants contain basic nitrogen atoms and thus can be extracted from the bulk of the plant material by dilute acid. These compounds are called **alkaloids**, which means "like an alkali." After extraction, the free alkaloids can be regenerated by subsequent treatment with aqueous base.

Extraction: $\quad$ $R_3N: + HCl \xrightarrow{H_2O} R_3NH^+ \; Cl^-$

Regeneration: $R_3NH^+ \; Cl^- + OH^- \longrightarrow R_3N: + H_2O + Cl^-$

Alkaloids vary from simple to complex in their structures. One of the simplest in structure, but not in its physiological effects, is *nicotine*.

nicotine

In large doses, nicotine is toxic; nicotine sulfate is used as an insecticide. In small doses (such as a smoker obtains from cigarettes), nicotine acts by stimulating the autonomic (involuntary) nervous system. If small doses are continued, nicotine can depress this same nervous system into less than normal activity.

The first isolation of an alkaloid in the pure state was reported in 1805. This alkaloid was *morphine* (from the Greek *Morpheus*, the god of dreams), one of many alkaloids to come from the gum and seeds of the opium poppy, *Papaver somniferum.*

morphine $\qquad$ codeine $\qquad$ heroin

Codeine is the methoxy derivative of morphine (at the phenolic group), while *heroin* is the diacetyl derivative. Codeine, like morphine, is a powerful analgesic and occurs naturally in the seeds of the opium poppy. Codeine is also an excellent cough suppressant that is sometimes used in prescription cough medicines. In recent years, it has been largely replaced by *dextromethorphan,* a nonaddictive, synthetic drug that is an equally effective cough suppressant. (Note the similarities in structure.)

dextromethorphan

Heroin does not occur naturally, but may be synthesized from morphine in the laboratory. Heroin, like morphine and codeine, is a powerful analgesic. In some parts of the world, heroin is used to relieve pain in terminal cancer patients. Because it is even more addictive than morphine, its medicinal use is prohibited in the United States.

A large number of physiologically active alkaloids contain the **tropane ring system**:

tropane

One of the tropane alkaloids is *atropine*, found in *Atropa belladonna* and other members of the nightshade plant family. Atropine is used in eye drops to dilate the pupils. *Scopolamine* (a so-called "truth" serum) is used as a preoperative sedative; chemically, it is the epoxide of atropine. *Cocaine*, a habituating stimulant and pain reliever, also contains the tropane ring system.

atropine   scopolamine   cocaine

STUDY PROBLEM

16.10. When atropine is subjected to acid hydrolysis, two products can be isolated: *tropine*, which is not optically active, and *tropic acid*, which is obtained as a racemic mixture. What are the structures of these two compounds?

B. Synthesis of an alkaloid

The laboratory synthesis of an alkaloid can be a challenging problem. The goal of such a project is not only to synthesize the natural product, but to do so from simple molecules by a short, elegant pathway. Such syntheses often have practical importance because many alkaloids are potent and desirable drugs. Large amounts of these alkaloids are often difficult to obtain from natural sources. A simple synthesis thus can provide an alternative supply of such a drug.

Let us consider the synthesis of the simple alkaloid *arecaidine*, one of the alkaloids found in betel nuts (*Areca catechu*). "Betel," a mixture of these nuts with lime from shells and leaves from a type of pepper plant, has been chewed and consumed by natives of the East Indies as a euphoric (an agent that imparts a sense of well-being). Unfortunately for the chewers, betel stains the teeth black. Betel nuts have also been used in China for 1500 years in the treatment of intestinal worms.

The synthesis of arecaidine, summarized in Figure 16.7, was reported in 1946. Because this alkaloid has no stereoisomers, the synthesis was aimed only at converting an open-chain compound to a ring system that has the functional groups positioned correctly.

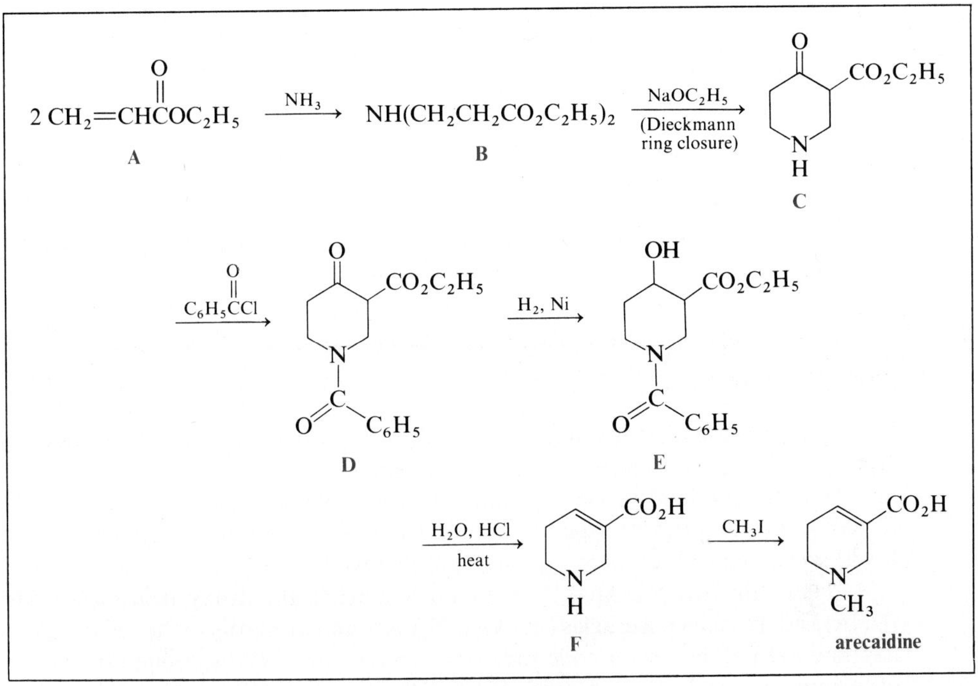

FIGURE 16.7. Synthesis of arecaidine, a betel-nut alkaloid.

The starting material for this synthesis is ethyl propenoate (**A** in Figure 16.7). Upon reaction with ammonia, **A** is converted by two stepwise 1,4-addition reactions (Section 11.19) to the secondary amine **B**. The product of the first 1,4-addition reaction is a β-amino ester, which then reacts with another molecule of the propenoate in the second reaction.

$$CH_2{=}CHCO_2C_2H_5 \xrightarrow{:NH_3} H_2\ddot{N}CH_2CH_2CO_2C_2H_5 \xrightarrow{CH_2=CHCO_2C_2H_5} \mathbf{B}$$

Cyclization is accomplished by a Dieckmann ring closure (page 688). The product is the β-keto ester **C**. This reaction places the nitrogen and the carboxyl group 1,3 to each other.

The remaining portion of the synthesis is aimed at reduction of the keto group to a hydroxyl group and dehydration to the desired alkenyl ring. For reduction, the original investigators used a catalytic hydrogenation. This reduction necessitates blocking the NH group because amines poison hydrogenation catalysts. Therefore, prior to hydrogenation, the amine **C** is converted to the amide **D** by reaction with benzoyl chloride (Section 13.3C). Compound **D** contains three carbonyl functional groups. Of these, the keto group is most reactive toward hydrogenation. By control of the reaction conditions, **D** can be converted to the alcohol **E**, leaving both the amide and the ester groups intact.

In the next step, the ester and amide groups are hydrolyzed by heating with aqueous HCl (Sections 13.5C and 13.9C). Under the hydrolysis conditions, the alcohol part of the molecule undergoes dehydration. (Why does dehydration occur so readily in this case, and why is only one isomeric alkenyl compound formed?) The synthesis is completed by the reaction of the amino group with CH_3I (an S_N2 reaction) to yield the tertiary amine (Section 15.5A).

STUDY PROBLEM

16.11. In 1946, when this synthesis of arecaidine was first carried out, metal hydride reducing agents were not readily available. How would you modify the synthesis of this alkaloid today?

SECTION 16.11.

Nucleic Acids

One of the most fascinating areas of modern research has been that of the **nucleic acids**, which are the carriers of the genetic codes in living systems. Because of information contained in the structure of the nucleic acids, an organism is able to biosynthesize different types of protein (hair, skin, muscles, enzymes, and so forth) and to reproduce more organisms of its own kind.

These are two principal types of nucleic acids. the **deoxyribonucleic acids (DNA)** and the **ribonucleic acids (RNA)**. DNA is found primarily in the cell nucleus; it is the carrier of the genetic code and can reproduce, or *replicate*, itself for purposes of forming new cells or for reproduction of the organism. In most organisms, the DNA of a cell directs the synthesis of RNA molecules. One type of RNA, **messenger RNA (mRNA)**, leaves the cell nucleus and directs the biosynthesis of the different types of proteins in the organism according to the DNA code.

DNA is a polymer. The recombination of DNA is a natural process by which bits of genetic material (fragments of the DNA polymer) are incorporated into another DNA molecule. The product DNA is referred to as **recombinant DNA**. In recent years, the artificial recombination of genes, often referred to as **gene-splicing** or **genetic engineering**, has become practical and of medical and economic importance. In 1978, it was reported that the genes that direct the synthesis of human insulin had been spliced onto the DNA of the bacteria *Escherichia coli*. The altered *E. coli* reproduces and becomes a virtual "insulin factory," producing human insulin right along with its own proteins. At the present time, most diabetics use animal insulin extracted from the pancreases of cattle and hogs. *Human growth hormone* (HGH) has also been produced through genetic engineering. The only previous source of this hormone (used to treat dwarfism) was pituitary glands taken from human cadavers! Another successful result of gene-splicing, reported in 1980, is the bacterial production of the antiviral agents called *human interferons*, which were previously obtainable only in minute amounts.

A.   The structure of DNA

The backbone of the DNA polymer is a long chain composed of molecules of the sugar deoxyribose (which will be discussed in Chapter 18) linked together by phosphate groups. The fundamental structure of a DNA molecule is shown in Figure 16.8. Note that one end of the chain has an OH group at carbon 5′, as the sugar is numbered, while the other end has an OH group at carbon 3′.

Each sugar molecule in DNA is also connected to a heterocyclic ring system, usually referred to as a *base*. Only four principal bases are found in DNA. Two of these are substituted pyrimidines and two are substituted purines.

pyrimidine

purine

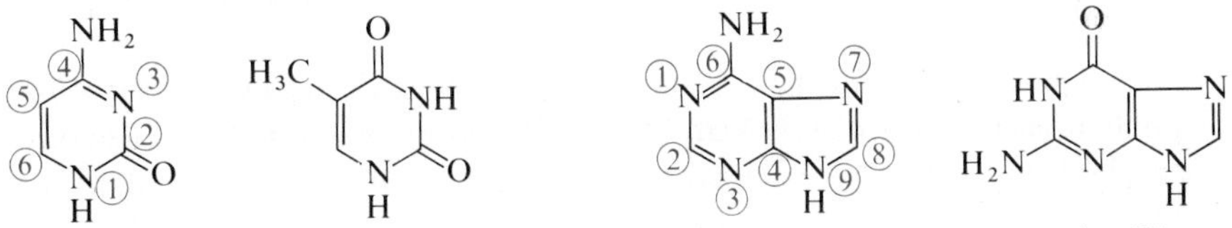

cytosine (C)   thymine (T)   adenine (A)   guanine (G)

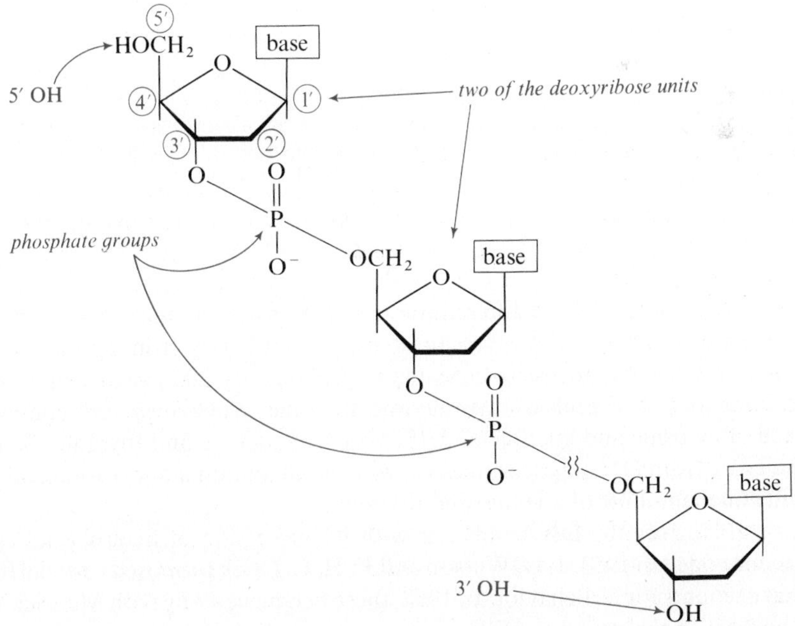

FIGURE 16.8.   The structure of a DNA molecule. Each base is one of the four heterocycles: cytosine, thymine, adenine, or guanine.

In DNA, the bases are attached to the deoxyribose at position 1 of the pyrimidines or position 9 of the purines.

Complete hydrolysis of DNA breaks it down into its smallest fragments: the sugar, the bases, and phosphate ions. Partial hydrolysis results in **nucleosides** (sugar bonded to base) and **nucleotides** (sugar bonded to base and phosphate).

The structures of the four nucleosides isolated after hydrolysis of the phosphate ester linkages of DNA are shown in Figure 16.9. The four nucleotides that can be isolated from partial hydrolysis have similar structures, but with a phosphate group attached to the sugar.

a nucleotide

DNA isolated from different tissues of the same organism contains the same proportions of bases; however, these proportions vary from species to species. For example, human tissue contains approximately 20% each of cytosine and guanine and 30% each of adenine and thymine. *Escherichia coli* contains 25% each of cytosine and guanine and 25% each of adenine and thymine. Notice that the bases from DNA appear in *pairs*, with equal amounts of cytosine and guanine and equal amounts of adenine and thymine.

How does the DNA polymer with its succession of heterocycles carry the genetic code? In 1953, J. D. Watson and F. H. C. Crick proposed a model for DNA that accounts for its behavior. In 1962, these two men, along with Maurice Wilkins, who verified the structure of the model by x-ray analysis, were awarded the Nobel Prize for their work. The Watson–Crick model of DNA is that of a *double helix of two long antiparallel DNA molecules held together by hydrogen bonds.* "Anti-

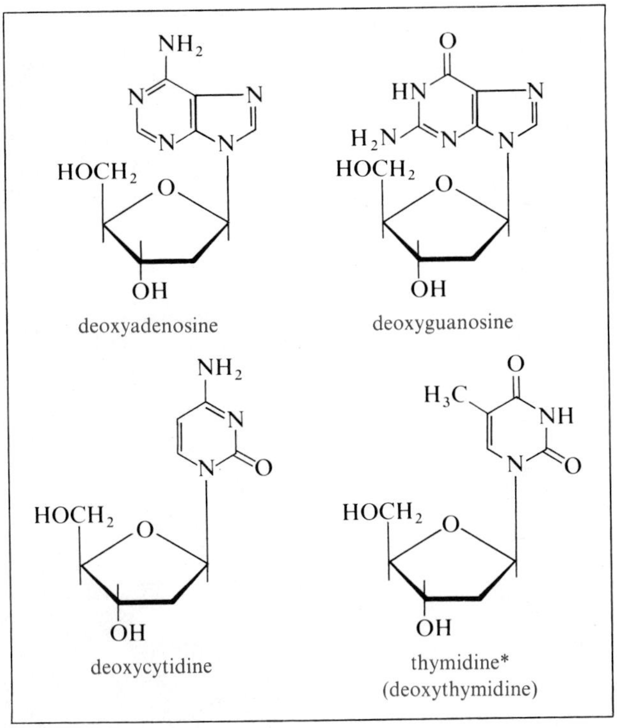

FIGURE 16.9. The nucleosides obtained from the hydrolysis of DNA

* Thymidine is the preferred name. The deoxy- prefix is unnecessary in this case
  because thymine is normally found only in DNA and not in other nucleic acids.
  The deoxy- prefix is necessary in the other names because the other bases are also
  found in RNA.

parallel" means that the two DNA molecules are parallel but aligned in opposite directions; each end of the double helix thus consists of a 3′ terminus (from one molecule) and a 5′ terminus (from the other molecule). Figure 16.10 shows three representations of double-stranded DNA.

The hydrogen bonds between DNA strands are not random, but are *specific between pairs of bases*: guanine is hydrogen-bonded to cytosine, and adenine to thymine. In Figure 16.10, initials are used to represent these bases (for example, **A** for adenine); the hydrogen bonds are represented by dashed lines. Why are these hydrogen bonds specific? Thymine and adenine can be joined by *two* hydrogen bonds (approximate strength, 10 kcal/mole). Cytosine and guanine can be joined by *three* hydrogen bonds (approximate strength, 17 kcal/mole). No other pairing of the four bases leads to such strong hydrogen bonding.

Thymine and adenine:

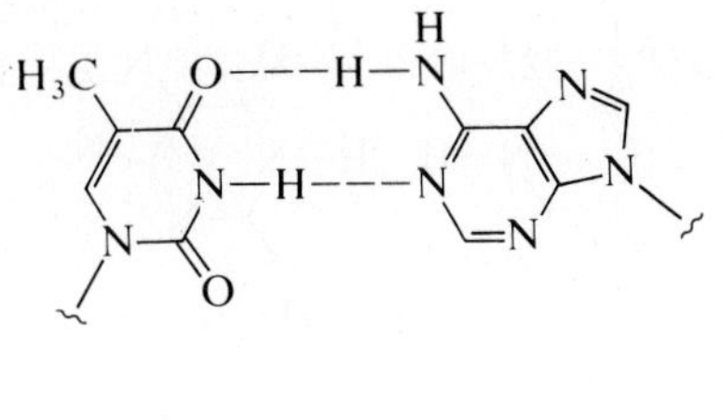

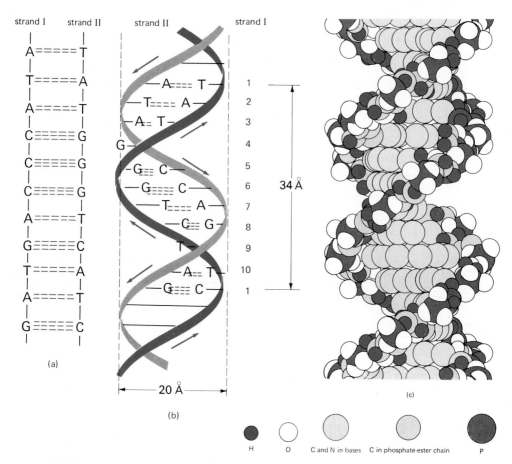

FIGURE 16.10. Three representations of the double-helical model of DNA.
(a) Two uncoiled DNA molecules are joined by hydrogen bonds between
complementary base pairs. (Not shown are the sugar and phosphate units.)
(b) The DNA strands are coiled in a double helix, with ten base pairs for every
complete turn of the helix. (c) A "space-filling" model of DNA. (Adapted from
William H. Brown and Judith A. McClarin, *Introduction to Organic and Biochemistry,
3rd ed.* (Willard Grant Press, Boston, 1981.)

Cytosine and guanine:

Not between thymine and guanine:

Now let us picture the double helix of DNA held together by series of particular hydrogen-bonded pairs, as shown in Figure 16.10 (a). Wherever an **A** appears in one strand, a **T** must appear opposite it in the other strand. The two strands are completely complementary in this respect. The pairing of bases explains why equal amounts of adenine and thymine and equal amounts of cytosine and guanine are found in DNA.

It is the *sequence of bases* that determines the genetic code; therefore, it is not surprising that different species contain differing amounts of the four bases. It is estimated that a single gene contains 1500 base pairs (a figure that is variable, depending on the gene). With this size gene, there can be 4^{1500} different possible combinations!

B.   Replication of DNA

The process of DNA replication in a typical cell begins with an enzyme-catalyzed unwinding of the double strand (see Figure 16.11). As the double strand unwinds, new nucleotides (triphosphates, in this case) become aligned along each strand. The nucleotides are incorporated, one by one, in an exactly complementary fashion: thymine across from adenine and cyostine across from guanine.

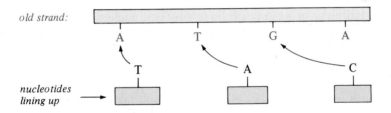

The polymerization of these nucleotides (and the simultaneous loss of diphosphate groups), catalyzed by the enzyme *DNA polymerase*, results in a pair of new strands. Each new strand is the complement of one of the old strands. The result is a pair of identical DNA helices where only one helix existed before.

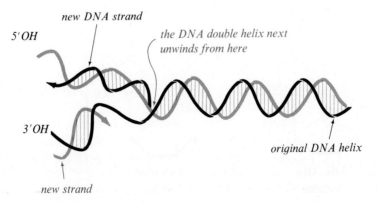

FIGURE 16.11. Replication of DNA involves the unwinding of the helix and the formation of new complementary strands.

C.   The structure of RNA

DNA holds the genetic code, but RNA "translates" the code into protein synthesis. The structure of RNA is similar to that of DNA: a series of sugar units (*ribose*, in this case) joined together by phosphate links, each sugar linked to a base. The principal bases in RNA are *adenine, guanine, cytosine,* and *uracil* (instead of thymine). Uracil forms the same favorable hydrogen bonds with adenine that thymine does, and is always paired with adenine in RNA synthesis.

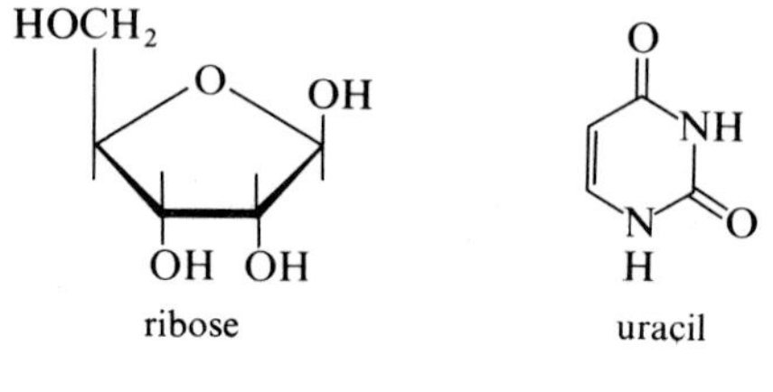

ribose   uracil

STUDY PROBLEM

16.12. Give the structures and show the hydrogen bonding between uracil and adenine.

Hydrolysis of RNA results in nucleotides, nucleosides, phosphate ions, and, finally, ribose and the bases. The structures of the nucleosides are shown in Figure 16.12.

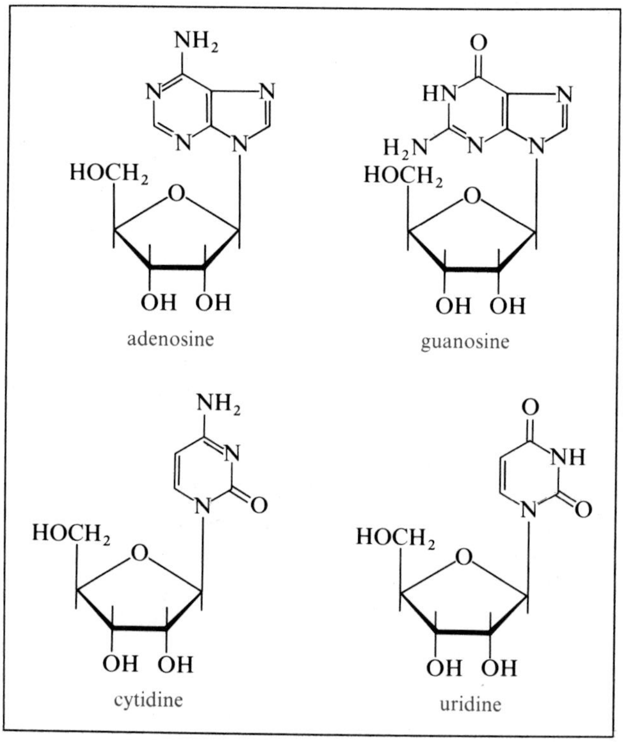

adenosine   guanosine

cytidine   uridine

FIGURE 16.12. The nucleosides obtained from the hydrolysis of RNA.

$$\text{RNA} \xrightarrow{\text{H}_2\text{O}} \underset{\textit{a nucleotide}}{\text{sugar}-\text{phosphate}} \xrightarrow{\text{H}_2\text{O}} \underset{\textit{a nucleoside}}{\overset{\overset{\text{base}}{|}}{\text{sugar}}} + \text{phosphate}$$

$$\xrightarrow{\text{H}_2\text{O}} \text{sugar} + \text{base}$$

Messenger RNA (mRNA) is synthesized under the direction of DNA. The mRNA molecules are smaller than DNA molecules. In the synthesis of an mRNA molecule, only a portion of the DNA helix unwinds; then complementary ribonucleotides line up and are polymerized. After polymerization, an mRNA molecule does not form a helix with the DNA molecule, but leaves the nucleus to aid in protein biosynthesis, a topic we will discuss in Section 19.9.

Summary

Three common **polycyclic aromatic compounds** are **naphthalene, anthracene,** and **phenanthrene**. These compounds exhibit less resonance energy per ring than benzene, and certain C—C bonds have more double-bond character than others. These compounds may be oxidized to quinones or may be partially hydrogenated, as shown in Table 16.2.

Naphthalene undergoes electrophilic substitution at the 1-position.

The six-membered aromatic heterocycle **pyridine** is a weak base and has a partially positive ring. Compared to benzene, pyridine is deactivated toward electrophilic substitution, but activated toward nucleophilic substitution.

Quinoline and **isoquinoline** undergo electrophilic substitution on the benzenoid ring, but nucleophilic substitution on the nitrogen ring.

Pyrrole is an aromatic five-membered nitrogen heterocycle. It is not basic. Its ring is partially negative and is activated toward electrophilic substitution, but deactivated toward nucleophilic substitution.

TABLE 16.2. Summary of the reactions of naphthalene, pyridine, and pyrrole

Reaction	Section reference
Naphthalene:	

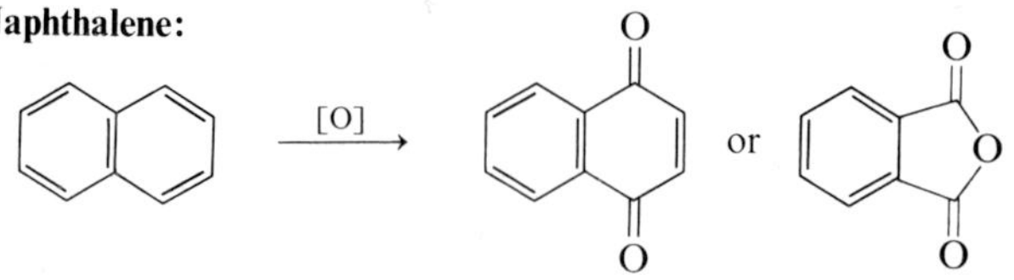	16.3
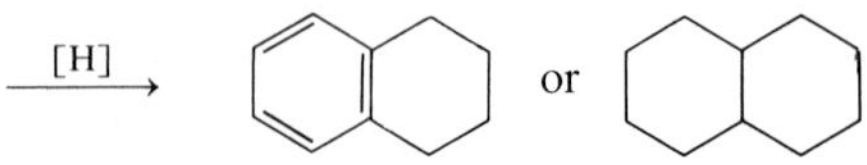	16.4
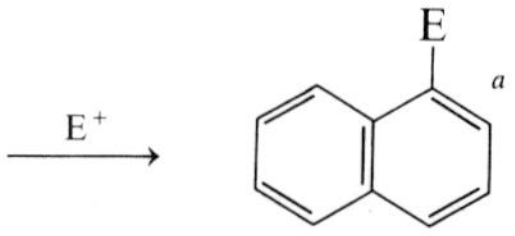	16.5

Pyridine:

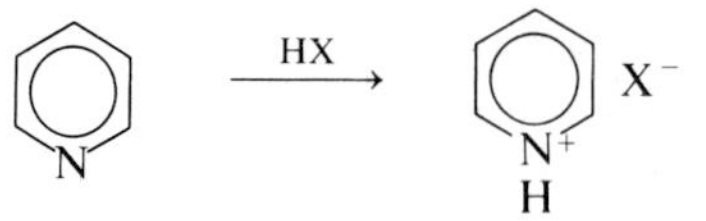	16.7
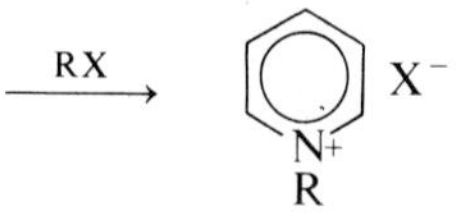	16.7
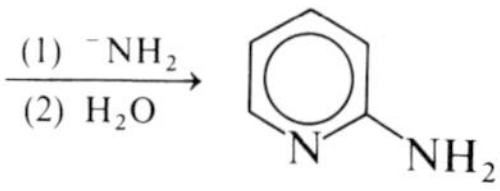	16.7A
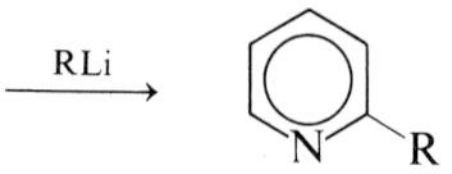	16.7A

Substituted pyridine:

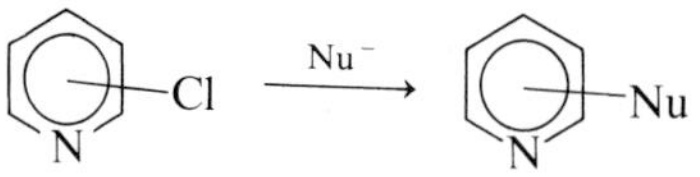

where Cl is 2- or 4-

| | 16.7A |

Quinoline:

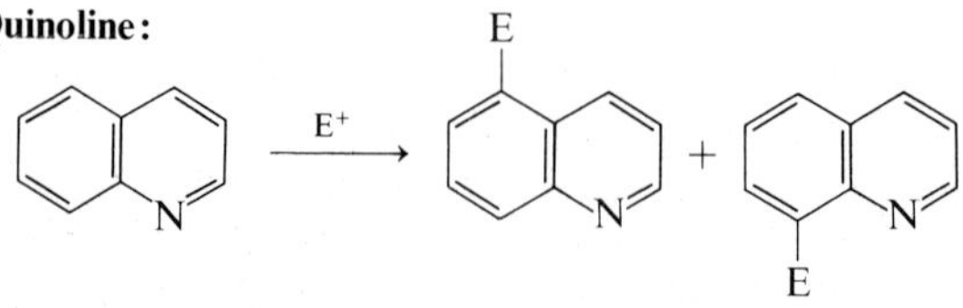

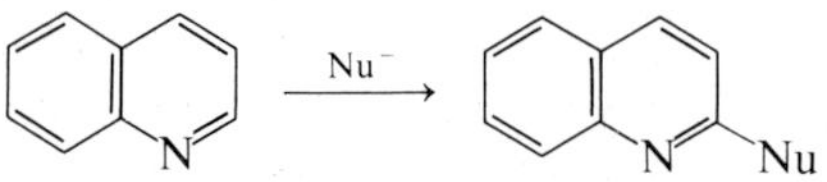

where Nu⁻ is ⁻NH$_2$ or R⁻ from RLi

| | 16.8 |
| | 16.8 |

TABLE 16.2 *(continued)*

Reaction	Section reference

Isoquinoline:

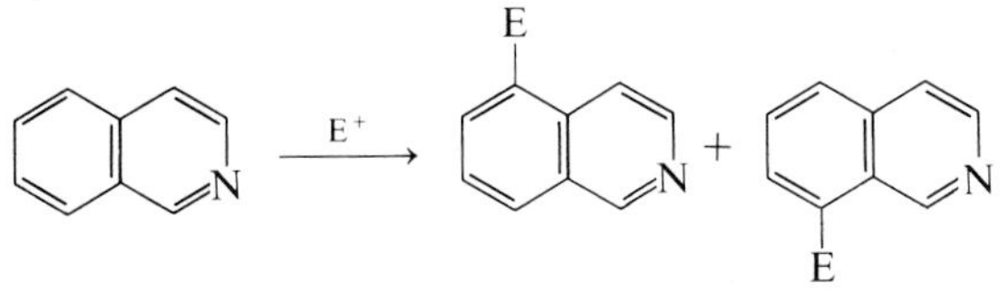

16.8

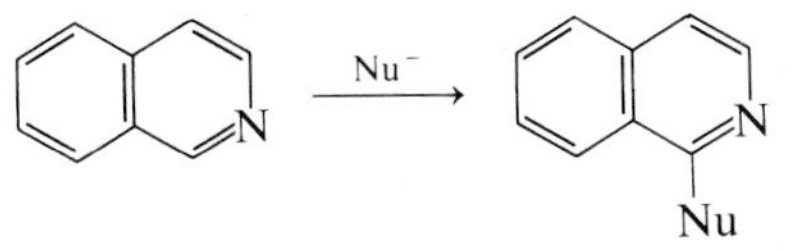

16.8

where Nu⁻ is ⁻NH₂ or R⁻ from R Li

Pyrrole:

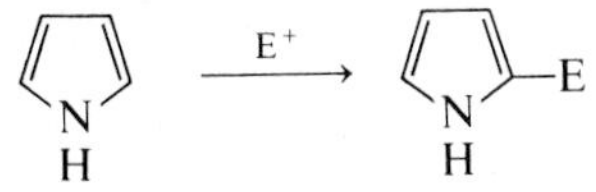

16.9A

---

^a At 160°, sulfonation yields 2-naphthalenesulfonic acid.

The order of reactivity of the heterocycles and benzene toward electrophilic and nucleophilic substitution follows:

Electrophilic substitution:

Nucleophilic substitution:

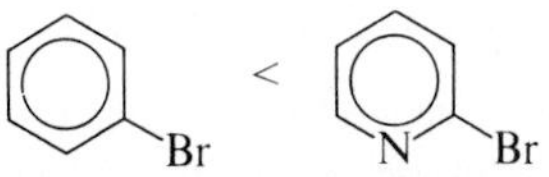

Alkaloids are acid-soluble, nitrogen-containing plant materials. Typical alkaloids are nicotine, morphine, codeine, and atropine.

The **nucleic acids** are polymers. The backbones of the nucleic acids **DNA** and **RNA** consist of sugar molecules (deoxyribose in DNA and ribose in RNA) linked together by phosphate units. Each sugar unit is also bonded to a heterocycle. Hydrogen bonding between specific heterocycles results in a pair of DNA molecules forming a **double helix**, and is one of the principal factors responsible for the genetic code. Nucleic acids may be hydrolyzed to **nucleotides** (phosphate–sugar–base) and to **nucleosides** (sugar–base).

STUDY PROBLEMS

16.13. Name the following compounds:

(a) 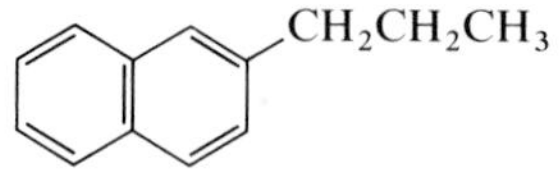 CH$_2$CH$_2$CH$_3$

(b) 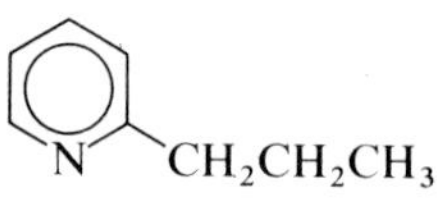 CH$_2$CH$_2$CH$_3$

(c) 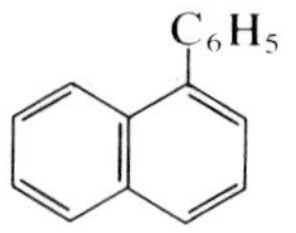 C$_6$H$_5$

(d) 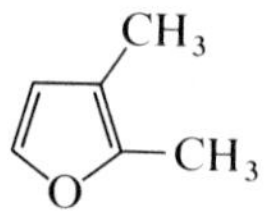 CH$_3$ —CH$_3$

(e) 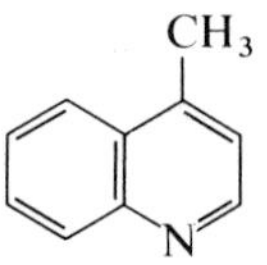 CH$_3$

(f) 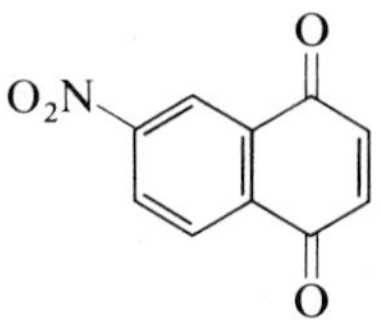 O$_2$N

16.14. Give structures for each of the following names: (a) 2-acetylnaphthalene; (b) 1-chloronaphthalene; (c) 2,4-dinitrothiophene; (d) *N*-phenylpyrrole.

16.15. How many pi electrons does each of the following structures have? Which structures are *fully aromatic*, the aromatic pi cloud involving the entire ring system?

(a) 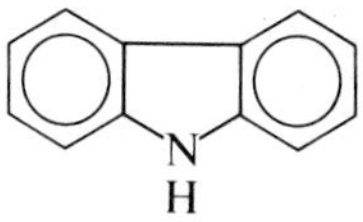 N H

(b) 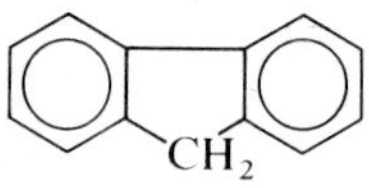 CH$_2$

(c) 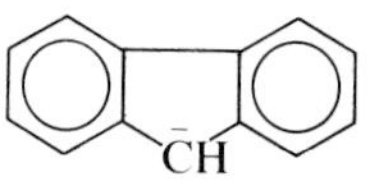 CH

16.16. Predict the major organic products. (If no reaction occurs, write *no reaction*.)

(a) 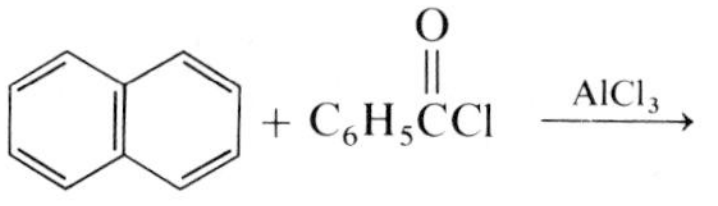 $\underset{\text{O}}{\overset{\text{O}}{\|}}$ + C$_6$H$_5$CCl $\xrightarrow{\text{AlCl}_3}$

(b) $\xrightarrow[\text{(2) H}_2\text{O, H}^+]{\text{(1) C}_6\text{H}_5\text{MgBr}}$

(c) OH + HNO$_3$ $\longrightarrow$

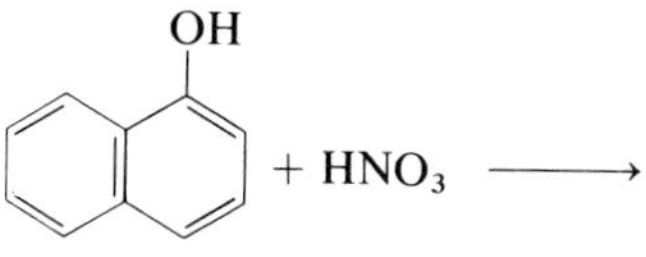

(d) 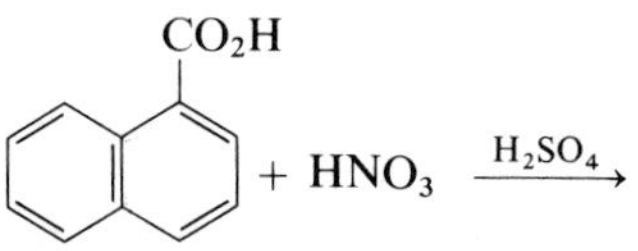 CO$_2$H + HNO$_3$ $\xrightarrow{\text{H}_2\text{SO}_4}$

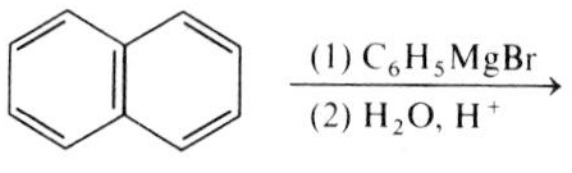

16.17. Show the stepwise mechanism for the following reaction. (Include resonance structures of the intermediate in your answer.)

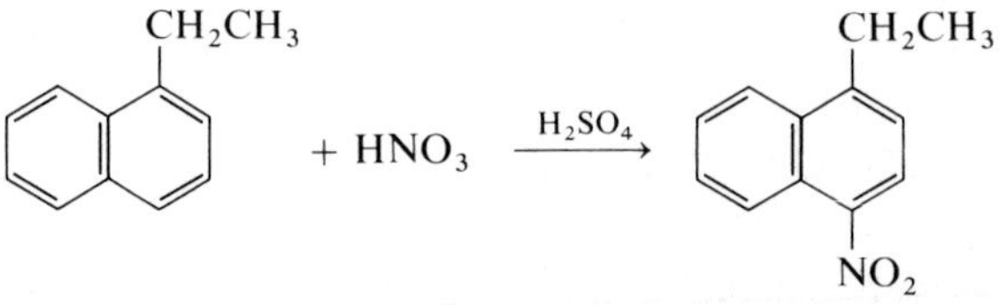

CH$_2$CH$_3$ + HNO$_3$ $\xrightarrow{\text{H}_2\text{SO}_4}$ CH$_2$CH$_3$ NO$_2$

16.18. Propose syntheses for the following compounds from naphthalene:

(a)

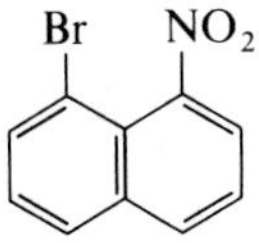

(b)

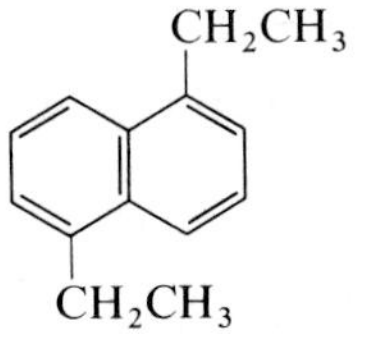

16.19. How would you explain the following observations? Aromatic nitration of 2-methylnaphthalene yields 75% 2-methyl-1-nitronaphthalene, but aromatic sulfonation of 2-methylnaphthalene yields 80% 6-methylnaphthalene-2-sulfonic acid.

16.20. Draw the *p*-orbital components in the lowest-energy pi molecular orbitals for the following compounds (see Figure 16.3). Indicate the number of *p* electrons contributed by each atom and any unshared electrons.

(a) thiazole   (b) pyrimidine   (c) purine   (d) thiophene   (e) pyran

16.21. Show the direction of the dipoles in   (a) 2-methylpyridine;   (b) 2-ethylpyrrole; (c) isoquinoline.

16.22. Complete the following equations. (If no reaction occurs, write *no reaction*.)

(a) 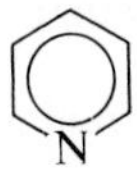 + HBr ⟶

(b)  + CH_3CH_2Br ⟶

(c) 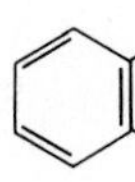 + $(CH_3)_3NH^+ \ Cl^-$ ⟶

(d) 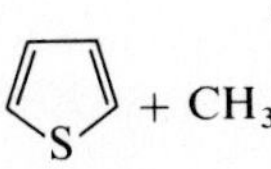 $\xrightarrow[\text{(2) } H_2O, H^+]{\text{(1) } C_6H_5MgBr}$

(e) 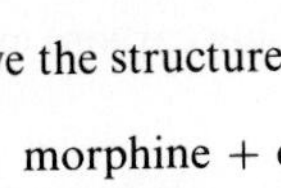 + Br_2 $\xrightarrow{FeBr_3}$

(f) 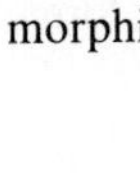 + $CH_3\overset{\displaystyle O}{\overset{\|}{C}}Cl$ $\xrightarrow{SnCl_4}$

16.23. Give the structures of the products:

    (a)   morphine + cold, dilute HCl ⟶

    (b)   morphine + cold, dilute NaOH ⟶

(c) codeine + cold, dilute NaOH ⟶

(d) psilocin + cold, dilute HCl

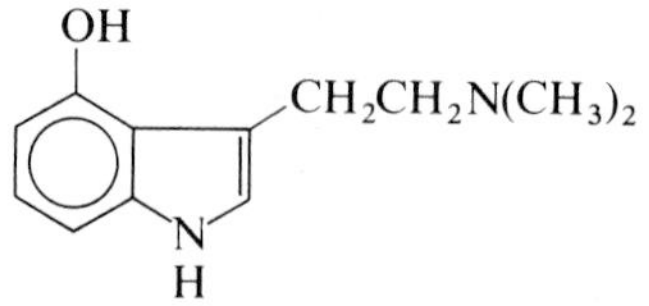

psilocin

a hallucinogen in
Psilocybe mushrooms

16.24. Which reaction would you expect to have the faster rate and why: (a) the reaction of pyridine and sodamide ($NaNH_2$) or   (b) the reaction of 2-chloropyridine and sodamide?

16.25. Suggest a reason why the rate of aromatic bromination of furan-2-carboxylic acid (2-furoic acid) is less than that of furan.

16.26. Although pyrrole is not basic, thiazole is. Explain.

16.27. Pyrrole ($pK_a = \sim 15$) is a weak acid: it can lose a proton and form an anion. Suggest a reason for the stability of this anion compared to that of $(CH_3CH_2)_2N^-$.

16.28. List the following compounds in order of increasing rates of reaction (slowest first) toward acetic anhydride with $AlCl_3$ catalyst:

(a) 3-nitrofuran   (b) 2,5-dinitrofuran   (c) 3,4-dinitrofuran   (d) 3-methylfuran

16.29. Suggest a reaction sequence for preparation of the following compound from naphthalene and furan:

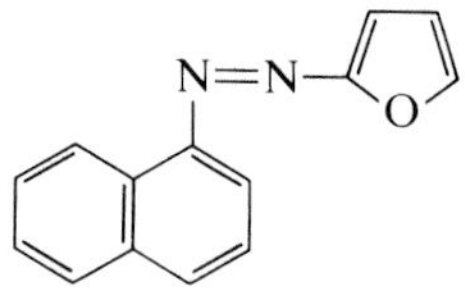

16.30. Using an unsubstituted porphyrin ring system with a chelated magnesium ion, show four resonance structures (there are more) that illustrate four equivalent bonds to the Mg.

16.31. Why are guanine and cytosine paired in the DNA helix, but not adenine and cytosine? (Use structures to show the hydrogen bonding.)

16.32. *Isoniazid* is used in the treatment of tuberculosis. Show how this compound may be prepared from 4-methylpyridine.

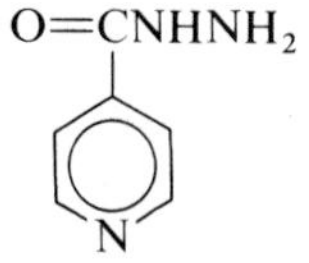

isoniazid

16.33. Only one of the three isomeric monohydroxypyridines (OH on a ring carbon) exhibits the chemical characteristics of a phenol. Explain.

16.34. When pyrrole is treated with strong acid, a polymer called *pyrrole red* is formed. Suggest a mechanism for the first step in the polymerization.

16.35. What is the most likely position of substitution by Br_2 on the indole ring system? Explain your answer by showing the mechanism.

indole

16.36. Predict the position of aromatic electrophilic and nucleophilic substitution of pyrimidine (Table 16.1). Would you expect pyrimidine to be more or less reactive than pyridine toward electrophilic substitution? Toward nucleophilic substitution?

16.37. *Imidazole* (Table 16.1) is found in proteins. The equilibrium between imidazole ($pK_b = 7$) and the imidazolium ion (imidazole—H^+) helps buffer proteins in biological systems. Use *p*-orbital pictures to explain why imidazole and the imidazolium ion are both aromatic.

16.38. Substituted naphthalene compounds can be prepared by Friedel–Crafts acylations of substituted benzenes with succinic anhydride. Tell what intermediate and final products would be obtained in the following sequence. [*Hints:* HF catalyzes aromatic acylation with a carboxylic acid, and Pd/heat causes catalytic dehydrogenation (loss of H_2) leading to aromatization of nonaromatic rings.]

toluene + succinic anhydride $\xrightarrow{\text{AlCl}_3}$ (a) $\xrightarrow[\text{HCl}]{\text{Zn(Hg)}}$ (b) $\xrightarrow{\text{HF}}$ (c) $\xrightarrow[\text{heat}]{\text{Pd}}$ (d)

16.39. The **Skraup synthesis** is the classical method for the synthesis of quinoline and substituted quinolines. In this reaction sequence, a mixture of glycerol and aniline is heated in the presence of concentrated H_2SO_4 and a mild oxidizing agent, such as nitrobenzene. The principal steps in the sequence are shown. Give mechanisms for steps 1 and 2 leading to 1,2-dihydroquinoline.

16.40. Predict the product of the reaction of aniline and 3-buten-2-one in the presence of a Lewis-acid catalyst.

16.41. Acid hydrolysis of the carbohydrates in oat hulls or corncobs gives Compound A, $C_5H_4O_2$, in almost 100% yield. Catalytic hydrogenation of A with $CuO—Cr_2O_3$ catalyst at 175° and 100 atm gives Compound B, $C_5H_6O_2$. The infrared spectrum of B and nmr spectra of A and B are shown in Figure 16.13. What are the structures of A and B?

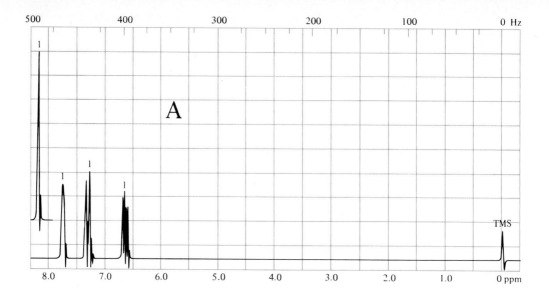

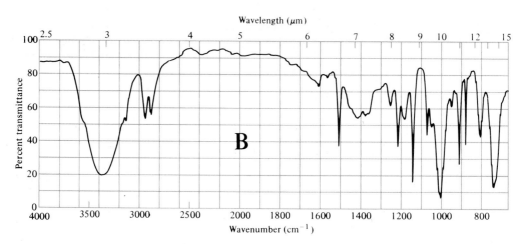

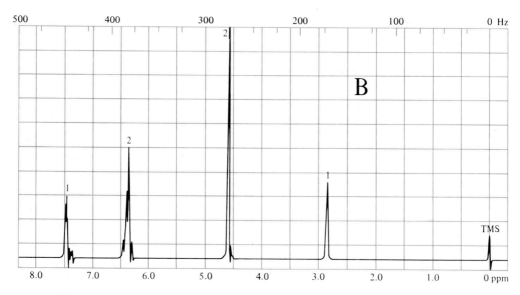

FIGURE 16.13. Spectra for Problem 16.41.

CHAPTER 17

Pericyclic Reactions

Except for S_N2 and E2 reactions, most of the organic reactions we have discussed so far proceed *stepwise* by way of intermediates such as carbocations or free radicals. A large number of reactions of conjugated polyenes, called **pericyclic reactions** (from *peri*, "around" or "about"), proceed by concerted (*single-step*) mechanisms just as an S_N2 reaction does—that is, old bonds are broken as new bonds are formed, all in one step. Pericyclic reactions are characterized by a cyclic transition state involving the pi bonds.

The energy of activation for pericyclic reactions is supplied by heat **(thermal induction)** or by ultraviolet light **(photo-induction)**. (Solvents and electrophilic or nucleophilic reagents have little or no effect on the course of a pericyclic reaction.) Pericyclic reactions are generally stereospecific, and it is not uncommon that the two modes of induction yield products of opposite stereochemistry. For example, a thermally induced pericyclic reaction might yield a *cis*-product, while the photo-induced reaction of the same reactant yields the *trans*-product.

There are three principal types of pericyclic reactions:

1. **Cycloaddition reactions,** in which two molecules combine to form a ring. In these reactions two pi bonds are converted to two sigma bonds. The best-known example of a cycloaddition reaction is the Diels–Alder reaction, discussed in Section 9.16. "Line" formulas, such as those in the following equation, were also discussed in that section. Recall that a reactant must be in the s-*cis* (not s-*trans*) form to undergo cycloaddition.

1,3-butadiene   ethylene   cyclohexene

2. **Electrocyclic reactions,** reversible reactions in which a compound with conjugated double bonds undergoes cyclization. In the cyclization, two pi electrons are used to form a sigma bond.

3. **Sigmatropic rearrangements,** concerted intramolecular rearrangements in which an atom or group of atoms shifts from one position to another.

For many years, a theoretical understanding of the mechanisms of pericyclic reactions eluded chemists. However, since 1960, several theories have been developed to rationalize these reactions. R. B. Woodward of Harvard University and R. Hoffmann of Cornell University have proposed explanations based upon the symmetry of the molecular orbitals of the reactants and products. These men received Nobel Prizes in 1965 and 1981, respectively, for their work. A similar treatment of pericyclic reactions has been developed by K. Fukui (Nobel Prize, 1981) of Kyoto University in Japan. In this text, we will emphasize Fukui's approach, which is called the **frontier orbital method** of analyzing pericyclic reactions.

Before we discuss the mechanisms of pericyclic reactions, we will introduce some features of the molecular orbitals of conjugated systems. We suggest that you first review Sections 2.1–2.3 to refresh your understanding of bonding and antibonding molecular orbitals.

SECTION 17.1.

Molecular Orbitals of Conjugated Polyenes

A conjugated polyene contains either $4n$ or $(4n + 2)$ pi electrons, where n is an integer, in its conjugated system. The simplest $4n$ system is represented by 1,3-butadiene, where $n = 1$. Any conjugated diene contains π molecular orbitals similar to those of 1,3-butadiene; therefore, we can use 1,3-butadiene as a model for all conjugated dienes.

In 1,3-butadiene, four p orbitals are used in the formation of the π molecular orbitals; thus, four π molecular orbitals result. In this system, π_1 and π_2 are the bonding orbitals and $\pi_3{}^*$ and $\pi_4{}^*$ are the antibonding orbitals. Figure 17.1 depicts these orbitals in terms of increasing energy. Note that the higher-energy molecular orbitals are those with a greater number of nodes between nuclei.

In the ground state, 1,3-butadiene has its four pi electrons in the two orbitals of lowest energy: π_1 and π_2. In this case, π_2 is the **Highest Occupied Molecular Orbital,** or **HOMO,** and $\pi_3{}^*$ is the **Lowest Unoccupied Molecular Orbital,** or **LUMO.** The HOMO and LUMO are referred to as **frontier orbitals** and are the orbitals used in the frontier orbital method of analyzing pericyclic reactions.

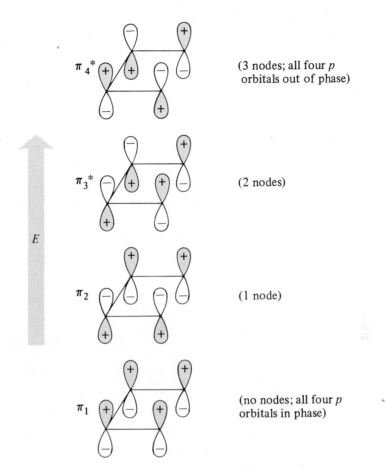

π_4^* (3 nodes; all four *p* orbitals out of phase)

π_3^* (2 nodes)

π_2 (1 node)

π_1 (no nodes; all four *p* orbitals in phase)

FIGURE 17.1. The bonding and antibonding π molecular orbitals of 1,3-butadiene, $CH_2{=}CHCH{=}CH_2$. The π_1 and π_2 orbitals are bonding orbitals; π_3^* and π_4^* are antibonding orbitals.

Ground state of 1,3-butadiene:

π_4^* ——
π_3^* —— ← the LUMO
π_2 ⇅ ← the HOMO
π_1 ⇅

When 1,3-butadiene absorbs a photon of the proper wavelength, an electron is promoted from the HOMO to the LUMO, which then becomes the new HOMO.

Excited state of 1,3-butadiene:

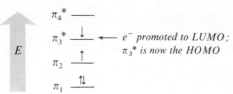

π_4^* ——
π_3^* ↓ ← *e⁻ promoted to LUMO;*
π_2 ↑ π_3^* *is now the HOMO*
π_1 ⇅

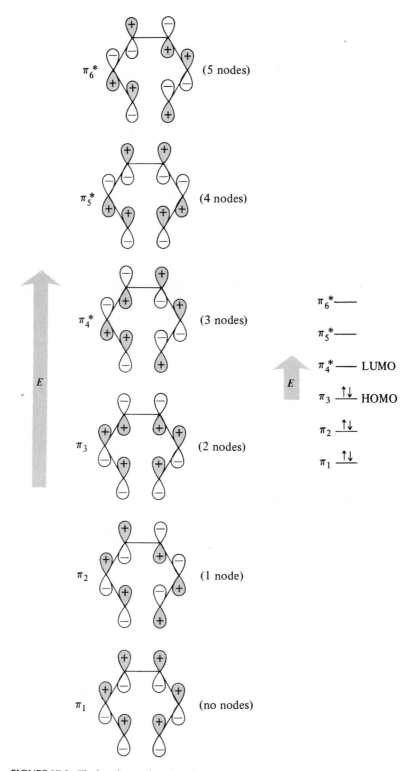

FIGURE 17.2. The bonding and antibonding π molecular orbitals of 1,3,5-hexatriene, $CH_2\!=\!CHCH\!=\!CHCH\!=\!CH_2$.

Aside from ethylene ($n = 0$), the simplest ($4n + 2$) system is represented by a conjugated triene ($n = 1$), such as 1,3,5-hexatriene. Because a triene contains a pi system formed from six p orbitals, a total of six π molecular orbitals results. These are depicted in Figure 17.2, along with the π orbital diagram of the ground state.

STUDY PROBLEMS

17.1. Draw the π orbital diagram for the lowest-energy *excited* state of 1,3,5-hexatriene.

17.2. Draw the π orbital diagram for the ground state of the nonconjugated diene 1,4-pentadiene. (Be careful!)

SECTION 17.2.

Cycloaddition Reactions

A **cycloaddition reaction** is a reaction in which two unsaturated molecules undergo an addition reaction to yield a cyclic product. For example,

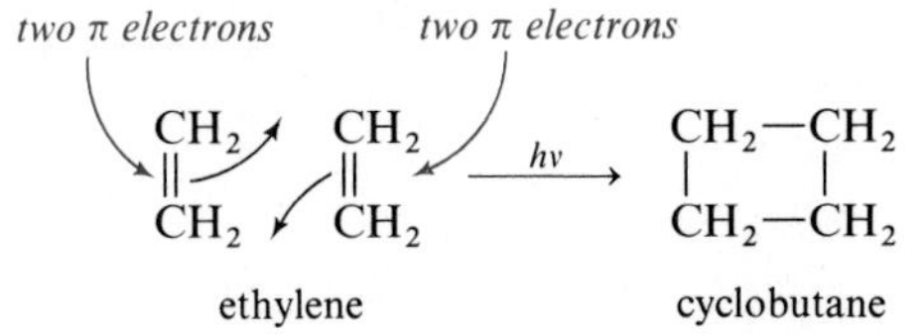

two π electrons   *two π electrons*

ethylene   cyclobutane

The cycloaddition of ethylene or any two simple alkenes is called a **[2 + 2] cycloaddition**, because *two pi electrons + two pi electrons* are involved. The Diels–Alder reaction (Section 9.16) is an example of a **[4 + 2] cycloaddition**. The diene contains four pi electrons that are used in the cycloaddition, while the dienophile contains two. (The carbonyl pi electrons in the following example are not used in bond formation in the reaction and therefore are not included in the number classification of this cycloaddition.)

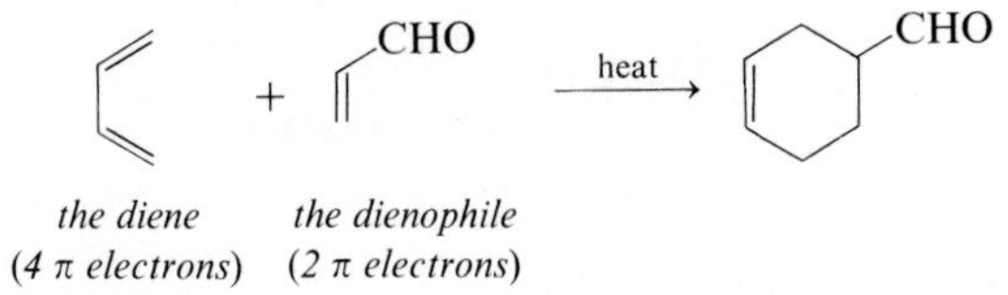

the diene   the dienophile
(4π *electrons*)   (2π *electrons*)

STUDY PROBLEM

17.3. Classify the following cycloaddition reaction by the number of pi electrons involved:

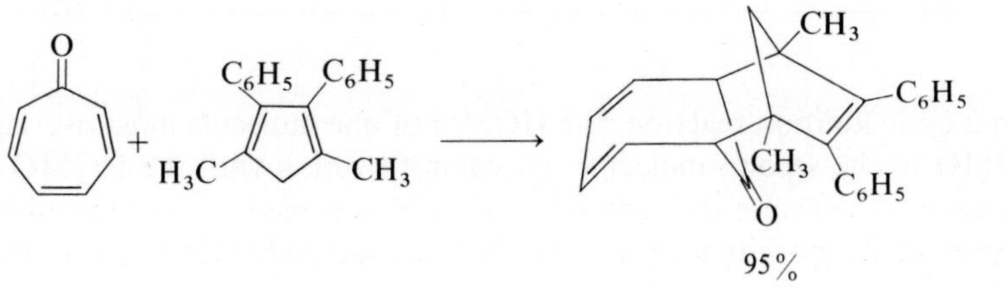

95%

Cycloaddition reactions are concerted, stereospecific reactions. (See Section 9.16 for a discussion of the stereochemistry of the Diels–Alder reaction.) Also, any particular cycloaddition reaction is either thermally induced or photo-induced, but not both.

A.   [2 + 2] Cycloadditions

Cycloaddition reactions of the [2 + 2] type proceed readily in the presence of light of the proper wavelength, but not when the reaction mixture is heated. The reason for this is readily explained by the frontier orbital theory: by assuming that electrons "flow" from the HOMO of one molecule to the LUMO of the other.

Let us consider the [2 + 2] cycloaddition of ethylene to yield cyclobutane. Ethylene has two π molecular orbitals: π_1 and $\pi_2{}^*$. In the ground state, π_1 is the bonding orbital and the HOMO, while $\pi_2{}^*$ is the antibonding orbital and the LUMO.

Ethylene in the ground state:

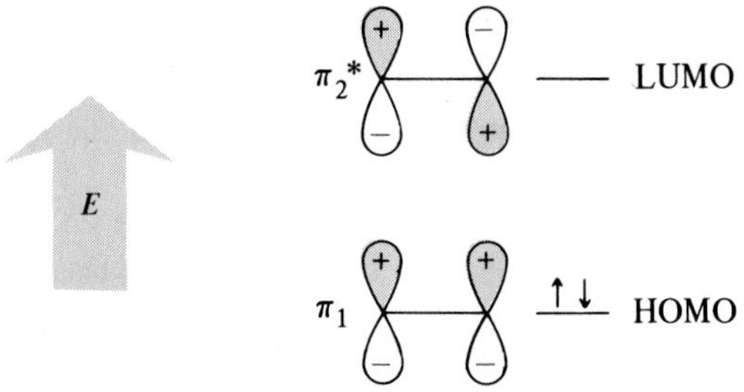

$\pi_2{}^*$ ———   LUMO

E

π_1 ↑↓   HOMO

SAMPLE PROBLEM

Draw the orbital diagram and the *p* orbitals for the lowest-energy *excited* state of ethylene, and indicate the HOMO.

Solution:

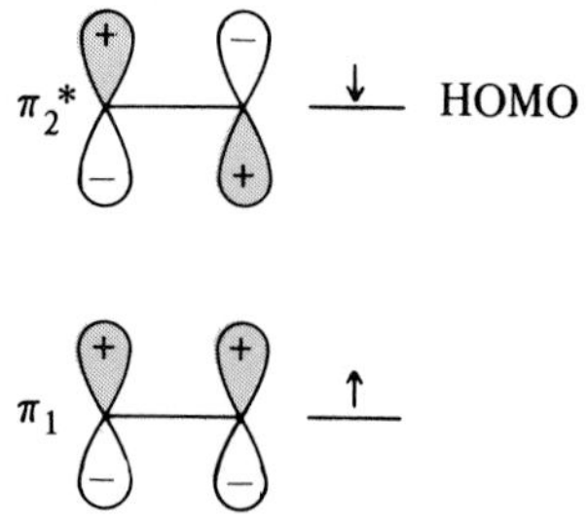

$\pi_2{}^*$ ↓   HOMO

π_1 ↑

In a cycloaddition reaction, the HOMO of one molecule must overlap with the LUMO of the second molecule. (It cannot overlap with the HOMO of the second molecule because that orbital is already occupied.) Simultaneously with the merging of the π orbitals, these orbitals also undergo hybridization to yield the new sp^3 sigma bonds.

When ethylene is heated, its π electrons are not promoted, but remain in the ground state, π_1. If we examine the phases of the ground-state HOMO of one ethylene molecule and the LUMO of another ethylene molecule, we can see why cyclization does not occur by thermal induction.

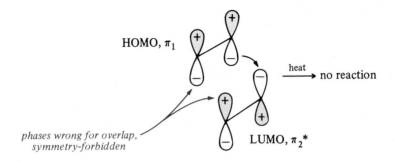

HOMO, π_1

$\xrightarrow{\text{heat}}$ no reaction

phases wrong for overlap,
symmetry-forbidden

LUMO, $\pi_2{}^*$

For bonding to occur, the phases of the overlapping orbitals must be the same. This is not the case for the ground-state HOMO and LUMO of two ethylene molecules or any other [2 + 2] system. Because the phases of the orbitals are incorrect for bonding, a thermally induced [2 + 2] cycloaddition is said to be a **symmetry-forbidden reaction**. A symmetry-forbidden reaction may occur under some circumstances, but the energy of activation would be very high—possibly so high that other reactions, such as free-radical reactions, would occur first.

When ethylene is irradiated with ultraviolet light, a pi electron is promoted from the π_1 to the $\pi_2{}^*$ orbital in some, but not all, of the molecules. The result is a mixture of the ground-state and excited-state ethylene molecules. If we examine the HOMO of an excited molecule ($\pi_2{}^*$) and the LUMO of a ground-state molecule (also $\pi_2{}^*$), we see that the phases are now correct for bonding. Such a reaction has a relatively low energy of activation, and is said to be **symmetry-allowed**.

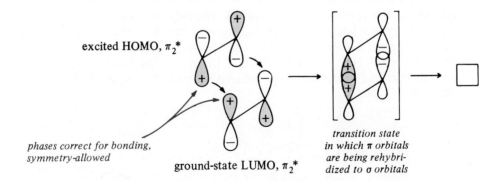

excited HOMO, $\pi_2{}^*$

phases correct for bonding,
symmetry-allowed

ground-state LUMO, $\pi_2{}^*$

transition state
in which π orbitals
are being rehybri-
dized to σ orbitals

Although the cycloaddition of ethylene itself proceeds in poor yield, other photo-induced [2 + 2] cycloadditions do find synthetic utility. Probably the widest use of this type of reaction is with intramolecular cyclizations, which can yield some very unusual "cage" structures.

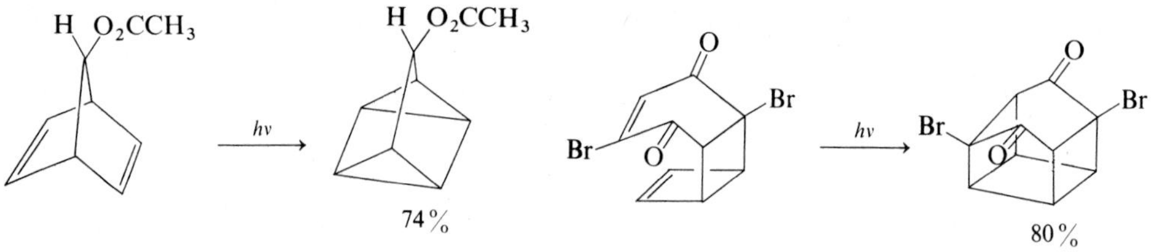

74%   80%

STUDY PROBLEM

17.4. Suggest synthetic routes to the following compounds from open-chain starting materials.

(a)

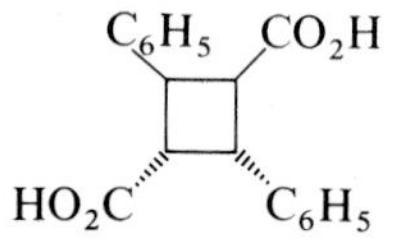

(b)

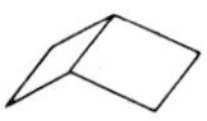

B. [4 + 2] Cycloadditions

As we have mentioned, the Diels–Alder reaction is the best-known [4 + 2] cyclo-addition. The examples on page 425 illustrate the versatility of this reaction. Note that the Diels–Alder reaction requires heat, not ultraviolet light, for its success. This experimental condition is different from that required for a [2 + 2] cyclo-addition. To see why this is so, we will examine the HOMO–LUMO interactions of *only the p-orbital components that will form the new sigma bonds* in a [4 + 2] cycloaddition. We will compare the HOMO–LUMO interactions for the ground state (for a thermally induced reaction) and those for the excited state (for an attempted photo-induced reaction). Based upon experimental observations, we would expect to find that the HOMO–LUMO interactions of the thermally induced reaction are symmetry–allowed and those of the photo-induced reaction are symmetry-forbidden.

   We will use the simplest [4 + 2] system: the cycloaddition of 1,3-butadiene (the diene) and ethylene (the dienophile). The frontier orbital pictures may, of course, be extrapolated to other [4 + 2] cycloadditions. In the thermally induced reaction, we can visualize the pi electrons "flowing" from the HOMO (π_2) of the diene (Figure 17.1) to the LUMO (π_2*) of the dienophile. Note the phases of the orbitals that lead to the thermally induced reaction. This reaction is symmetry-allowed.

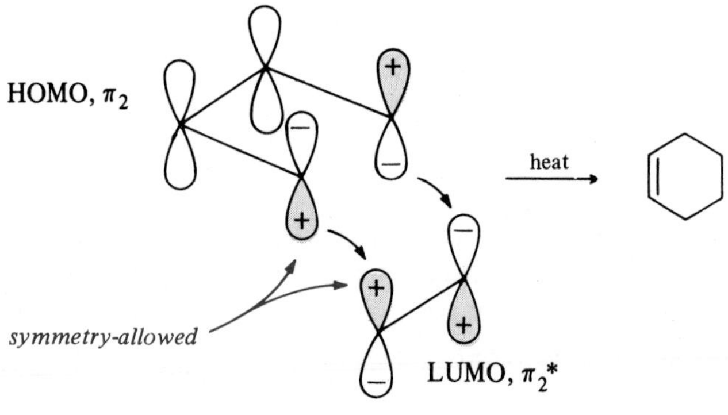

When a diene is excited by light, its HOMO becomes the $\pi_3{}^*$ orbital, and this molecular orbital cannot overlap with the LUMO of the dienophile. The photo-induced [4 + 2] cyclization is therefore symmetry-forbidden.

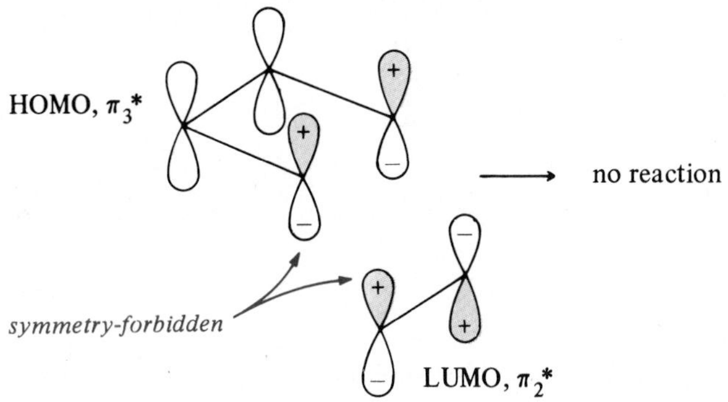

HOMO, $\pi_3{}^*$

symmetry-forbidden

→ no reaction

LUMO, $\pi_2{}^*$

STUDY PROBLEMS

17.5. Predict whether a [4 + 2] cycloaddition could be photo-induced if the dienophile, instead of the diene, were the excited reactant. Explain your answer.

17.6. Is the cycloaddition reaction in Problem 17.3 (page 785) a thermally induced or photo-induced reaction? Explain.

SECTION 17.3.

Electrocyclic Reactions

An **electrocyclic reaction** is the concerted interconversion of a conjugated polyene and a cycloalkene. We will discuss primarily the cyclization. The reverse reaction, ring opening, proceeds by the same mechanism, but in reverse.

Electrocyclic reactions are induced either thermally or photochemically:

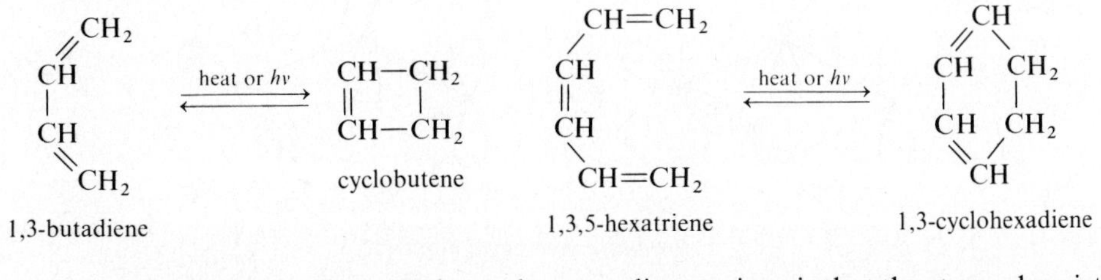

1,3-butadiene   heat or $h\nu$ cyclobutene   1,3,5-hexatriene   heat or $h\nu$ 1,3-cyclohexadiene

An intriguing feature about electrocyclic reactions is that the stereochemistry of the product is dependent on whether the reaction is thermally induced or photo-induced. For example, when (2*E*, 4*Z*)-hexadiene is heated, the *cis*-dimethylcyclobutene is the product. When the diene is irradiated with ultraviolet light, however, the *trans*-dimethylcyclobutene is formed.

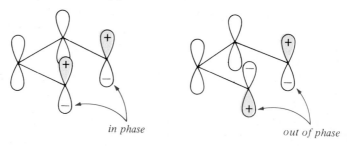

A.   Cyclization of 4n systems

A conjugated polyene yields a cycloalkene by the end-to-end overlap of its *p* orbitals and the simultaneous rehybridization of the carbon atoms involved in bond formation. 1,3-Butadiene, which has $4n$ pi electrons, is the simplest polyene; therefore, we will introduce the mechanism with this compound.

The two lobes of each *p* orbital that will form the new sigma bond in cycliza- tion are either *in phase* or *out of phase* with each other:

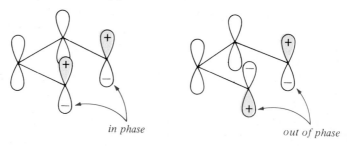

in phase   *out of phase*

To form a new sigma bond, the existing C—C sigma bonds must rotate so that the *p* orbitals can undergo end-to-end overlap. For this to occur, the existing pi bonds must be broken. The energy for the pi-bond breakage and the bond rotation is supplied by the heat or ultraviolet light. To form a sigma bond, the pair of overlapping lobes of the two *p* orbitals must be *in phase* after rotation.

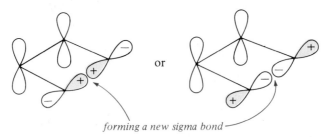

forming a new sigma bond

There are two different ways in which the existing C—C sigma bonds can rotate in order to position the *p* orbitals for overlap. (1) The two C—C sigma bonds can rotate in the *same direction* (either both clockwise or both counterclockwise). This type of rotation is referred to as **conrotatory motion**. (2) The two C—C sigma bonds can rotate in *different* directions, one clockwise and one counterclockwise.

This type of rotation is **disrotatory motion**.

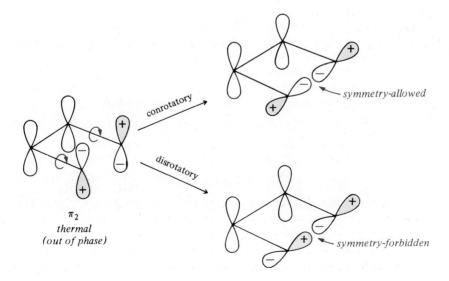

Note that in the two preceding equations, the phases of the *p* orbitals in the two starting dienes are different. Therefore, the direction of rotation for symmetry-allowed overlap depends upon the phases of the *p*-orbitals just prior to cyclization. If the *p* orbitals are out of phase before rotation, then conrotatory motion brings them into phase after rotation. If the *p* orbitals are in phase before rotation, then disrotatory motion is required. To determine which diene system is present just prior to reaction, we must consider the phases of the *p* orbitals in the ground and excited states of the diene.

When 1,3-butadiene is *heated*, reaction takes place from the ground state. The electrons that are used for sigma-bond formation are in the HOMO (π_2, in this case). In Figure 17.1 (page 783), it can be seen that the pertinent *p* orbitals in this HOMO are out of phase with each other. For the new sigma bond to form, rotation must be *conrotatory*. Only in this way are the in-phase lobes allowed to overlap. (Disrotatory motion would not place the in-phase lobes together.)

In *photo-induced* cyclization, the phases of the *p* orbitals of the HOMO (now $\pi_3{}^*$) are the reverse of that in thermal cyclization (see Figure 17.1); therefore, the symmetry-allowed rotation is *disrotatory* instead of conrotatory.

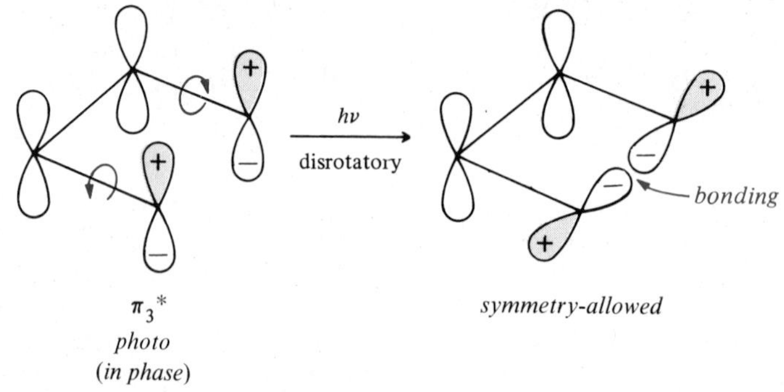

$\pi_3{}^*$
photo
(in phase)

symmetry-allowed

bonding

STUDY PROBLEM

17.7. Draw structures showing *conrotatory* motion of 1,3-butadiene in the excited state ($\pi_3{}^*$). Are the potential bonding *p* orbitals in a symmetry-allowed or symmetry-forbidden orientation?

B.   Stereochemistry of a 4*n* electrocyclization

Let us return to (2*E*,4*Z*)-hexadiene to see why the *cis*-dimethylcyclobutene results from thermal cyclization and the *trans*-isomer results from photocyclization.

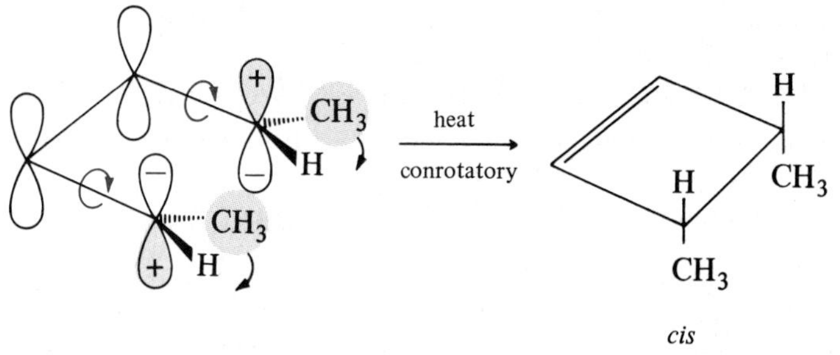

cis

In the case of thermal cyclization, conrotatory motion is required for sigma-bond formation. Both methyl groups rotate in the same direction; as a result they end up on the same side of the ring, or *cis*, in the product.

Just the reverse occurs in photochemical cyclization. In the disrotatory motion, one of the methyl groups rotates up and the other rotates down. The result is that the two methyl groups are *trans* in the product.

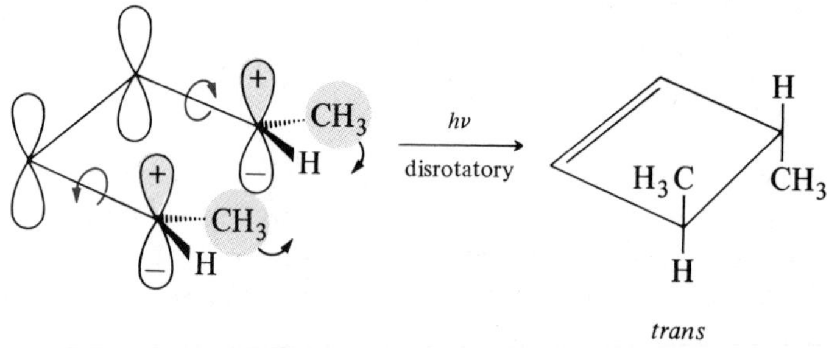

trans

SAMPLE PROBLEMS

Will the photochemical electrocyclic reaction of (2*E*, 4*E*)-hexadiene yield *cis*- or *trans*-3,4-dimethyl-1-cyclobutene?

Solution: 2,4-Hexadiene is a 4*n* polyene; therefore, the photochemical electrocyclic reaction takes place by disrotatory motion.

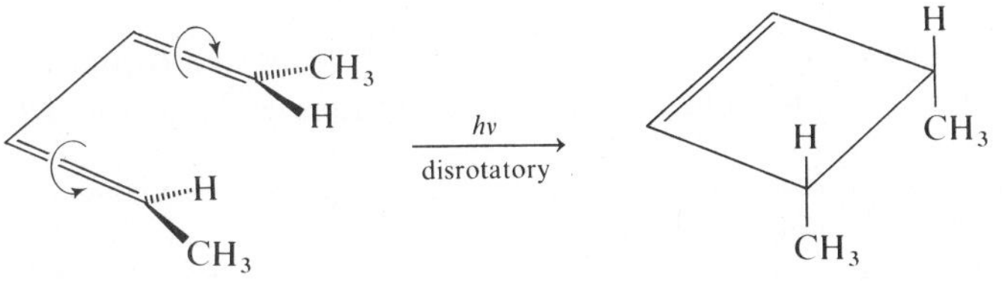

(2*E*,4*E*)-hexadiene   *cis*-3,4-dimethyl-1-cyclobutene

What is the structure and expected stereochemistry of the ring-opened product when *trans*-3,4-dimethyl-1-cyclobutene is heated?

Solution:

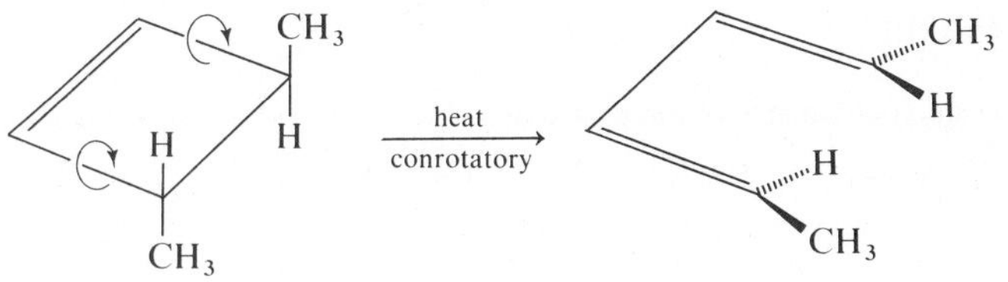

(2*E*,4*E*)-hexadiene

C.   Cyclization of (4*n* + 2) systems

Figure 17.2 (page 784) shows the π orbitals of 1,3,5-hexatriene, a (4*n* + 2) polyene. In the HOMO of the ground state (π_3), the *p* orbitals that form the sigma bond in the cyclization are in phase. Therefore, the thermal cyclization proceeds by *disrotatory motion*.

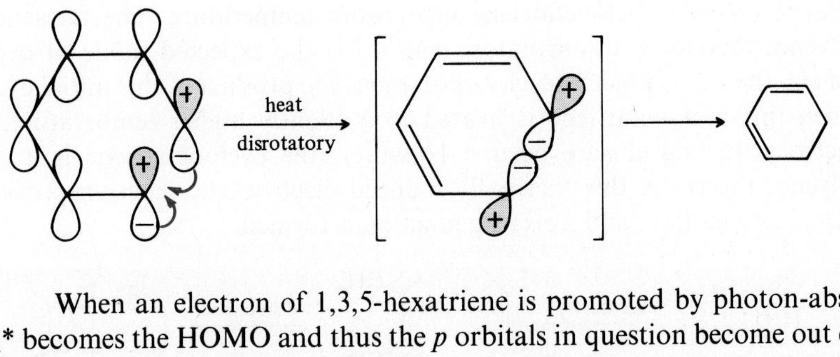

When an electron of 1,3,5-hexatriene is promoted by photon-absorption, π_4* becomes the HOMO and thus the *p* orbitals in question become out of phase. Therefore, photo-induced cyclization proceeds by *conrotatory motion*. The symmetry-allowed reactions of this (4*n* + 2) system are just the opposite of those for 1,3-butadiene, a 4*n* system.

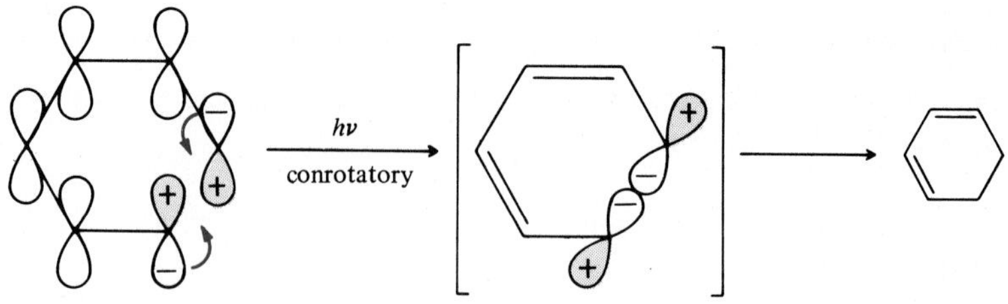

A summary of the types of motion to be expected from the different types of polyenes under the influence of heat and ultraviolet light is shown in Table 17.1.

TABLE 17.1. Types of electrocyclic reactions

Number of pi electrons	*Reaction*	*Motion*
$4n$	thermal	conrotatory
$4n$	photochemical	disrotatory
$(4n + 2)$	thermal	disrotatory
$(4n + 2)$	photochemical	conrotatory

STUDY PROBLEM

17.8. Predict the stereochemistry of the products.

(a)

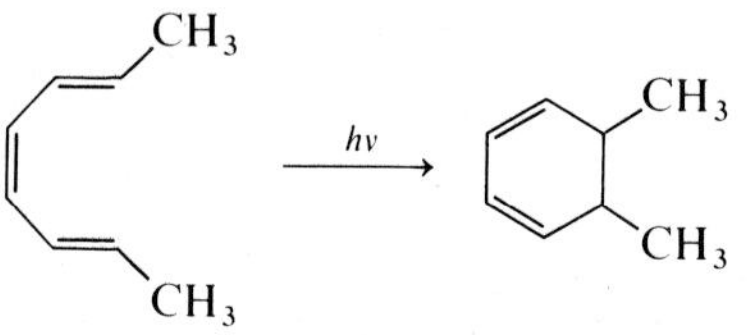

(b)

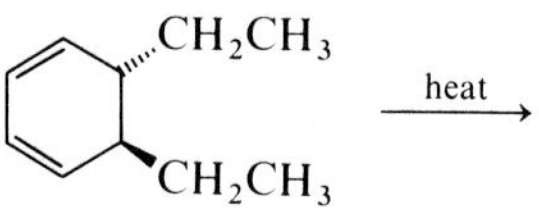

$CH_3CH_2CH{=}CHCH{=}CHCH{=}CHCH_2CH_3$

The thermally induced electrocyclic reactions of (2*E*,4*Z*,6*Z*,8*E*)-decatetraene provide elegant examples of electrocyclic reactions. The starting tetraene forms a cyclooctatriene near room temperature. The tetraene is a $4n$ polyene; therefore, a conrotatory motion is the expected mode of cyclization. Indeed, the *trans*-dimethylcyclooctatriene is the product of this initial cyclization. When this cyclooctatriene is heated to a slightly higher temperature, another electrocyclic ring closure occurs. However, the cyclooctatriene is a $(4n + 2)$ polyene; therefore, this thermally induced electrocyclic reaction proceeds with disrotatory motion, and a *cis* ring junction is formed.

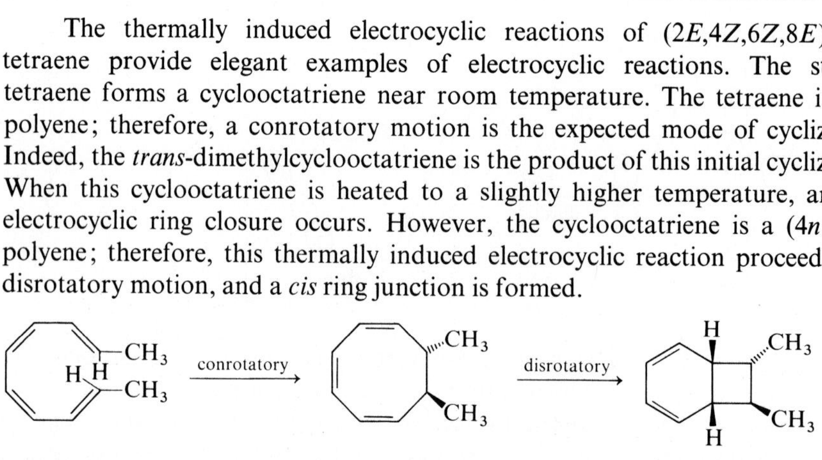

trans methyl groups *cis ring junction*

SECTION 17.4.

Sigmatropic Rearrangements

A **sigmatropic rearrangement** is a concerted intramolecular shift of an atom or a group of atoms. Two typical examples of sigmatropic rearrangements are:

Cope rearrangement:

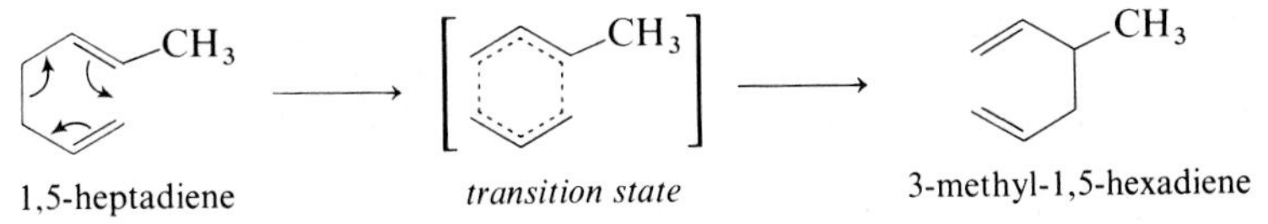

1,5-heptadiene   *transition state*   3-methyl-1,5-hexadiene

Claisen rearrangement:

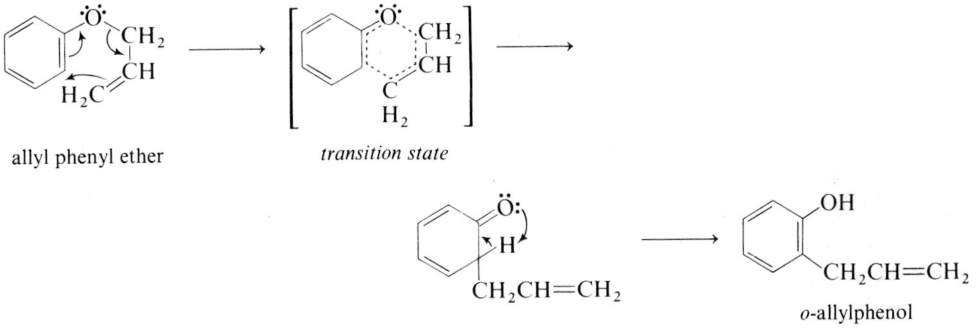

allyl phenyl ether   *transition state*

o-allylphenol

A. Classification of sigmatropic rearrangements

Sigmatropic rearrangements are classified by a double numbering system that refers to the relative positions of the atoms involved in the migration. This method of classification is different from those for cycloadditions or electrocyclic reactions, which are classified by the number of π electrons involved in the cyclic transition state.

The method used in classifying sigmatropic reactions is best explained by example. Consider the following rearrangement:

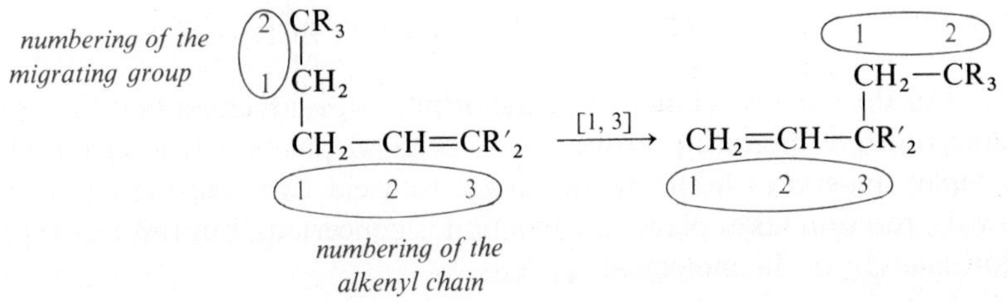

numbering of the migrating group

numbering of the alkenyl chain

Both the alkenyl chain and the migrating group are numbered *starting at the position of their original attachment.* (Note that these numbers are not related to nomenclature numbers.) In the example, atom 1 of the migrating group ends up on atom 3 of the alkenyl chain. Therefore, this sigmatropic rearrangement would be classified as a [1,3] sigmatropic rearrangement.

In a similar manner, the following reaction would be classified as a [1,7] sigmatropic shift. (In this example, there is no atom 2 in the migrating group.)

① H
|
CH₂CH=CHCH=CHCH=CD₂ $\xrightarrow{[1,7]}$ CH₂=CHCH=CHCH=CHCD₂
①   ⑦

H
|
CH₂=CHCH=CHCH=CHCD₂

It is not always the first atom of the migrating group that becomes bonded to the alkenyl chain in the rearrangement. Consider the following example. In this case, atom 3 of the migrating group becomes bonded to atom 3 of the alkenyl chain. This is an example of a [3,3] sigmatropic rearrangement.

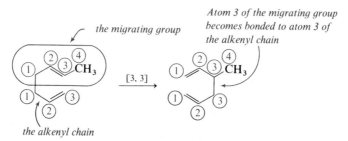

STUDY PROBLEM

17.9. Classify the following rearrangements by the preceding technique:

(a) the Claisen rearrangement

(b) $CH_3CH_2CH=CHCH=CDCH_3 \longrightarrow$

$CH_3CH=CHCH=CHCHDCH_3$

B. Mechanism of sigmatropic rearrangements

Sigmatropic rearrangements of the [1,3] type are relatively rare, while [1,5] sigmatropic rearrangements are fairly common. We can use the frontier orbital approach to analyze these reactions and see why this is so. Let us first consider the following thermally induced sigmatropic rearrangement, which is a [1,3] shift:

H   H
| |
CH₂CH=CD₂ $\xrightarrow{\text{difficult}}$ CH₂=CHCD₂

For the purpose of analyzing the orbitals, it is assumed that the sigma bond connecting the migrating group to its original position (the CH bond in our example) undergoes homolytic cleavage to yield two free radicals. This is *not* how the reaction takes place (the reaction is concerted), but this assumption does allow analysis of the molecular orbitals.

H   H·
|
CH₂CH=CD₂ $\xrightarrow[\text{homolytic cleavage}]{\text{hypothetical}}$ ·CH₂CH=CD₂

allyl radical

The products of the hypothetical cleavage are a hydrogen atom and an allyl radical, which contains three pi electrons and thus three π molecular orbitals. The π molecular orbitals of the allyl radical are shown in Figure 17.3.

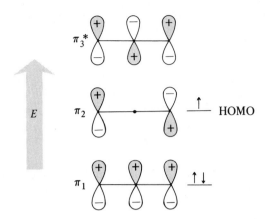

FIGURE 17.3. The three π molecular orbitals of an allyl radical. (Note that π_2 contains one node at carbon 2.)

The actual shift of the $H\cdot$ could take place in one of two directions. In the first case, the migrating group could remain on the same side of the π orbital system. Such a migration is termed a **suprafacial process**. As you can see, in this system a suprafacial migration is geometrically feasible but symmetry-forbidden.

Suprafacial migration:

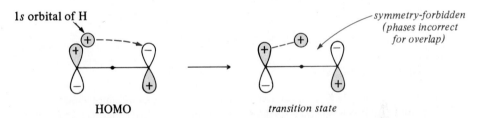

Let us consider the second mode of migration. For a symmetry-allowed [1,3] sigmatropic shift to occur, the migrating group ($H\cdot$ in our example) must shift by an **antarafacial process**—that is, it must migrate to the *opposite face* of the orbital system.

Antarafacial migration:

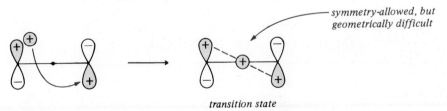

While symmetry-allowed, a [1,3] antarafacial sigmatropic rearrangement of H is not geometrically favorable. Our conclusion is that [1,3] sigmatropic shifts should not occur readily. This conclusion is in agreement with experimental facts; as we have mentioned, [1,3] sigmatropic rearrangements are rare.

By contrast, [1,5] sigmatropic shifts are quite common. A simple example follows:

$$\overset{\overset{\displaystyle H}{\displaystyle |}}{CH_2CH}=CHCH=CD_2 \xrightarrow{[1,\,5]} CH_2=CHCH=\overset{\overset{\displaystyle H}{\displaystyle |}}{CHCD_2}$$

If we again assume a homolytic bond cleavage for purposes of analysis, we must consider the π molecular orbitals of a pentadienyl radical, which contains five pi electrons. These orbitals are depicted in Figure 17.4.

$$\overset{\overset{\displaystyle H}{\displaystyle |}}{CH_2CH}=CHCH=CH_2 \xrightarrow[\text{homolytic cleavage}]{\text{hypothetical}} CH_2CH=CHCH=CH_2$$

pentadienyl radical

Considering the HOMO of this radical and the orbital symmetry, we can see that the **[1,5]** shift is both symmetry-allowed and suprafacial.

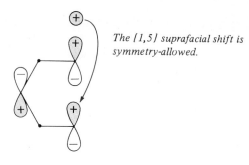

The [1,5] suprafacial shift is symmetry-allowed.

STUDY PROBLEM

17.10. Which of the following known sigmatropic rearrangements would proceed readily and which slowly? Explain your answers.

(a)

(b)

Pericyclic Reactions Leading to Vitamin D

Pericyclic reactions are not merely laboratory curiosities; they are also observed in natural processes. As one example, let us briefly consider some of the transformations that occur in the vitamin D group of compounds.

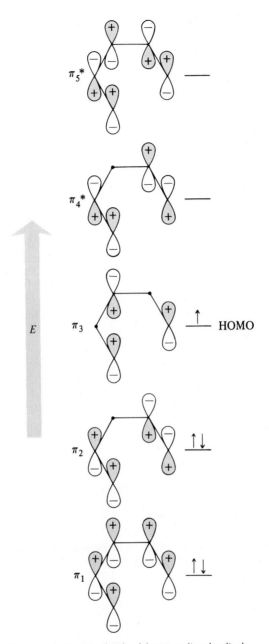

FIGURE 17.4 The five π molecular orbitals of the pentadienyl radical.

Vitamin D is essential for the proper growth of bones. Lack of vitamin D results in defective bone growth, a condition called *rickets*. Humans can obtain vitamin D by a number of routes. One is by the action of sunlight on a particular steroid, 7-dehydrocholesterol, which is found in the skin. (Steroids are discussed in Section 20.7.) This steroid undergoes a photo-induced electrocyclic ring opening to yield a triene. The triene, previtamin D, then undergoes a thermally induced [1,7] sigmatropic shift to yield vitamin D_3. (The subscript 3 is used to differentiate this vitamin D from other structurally similar compounds with vitamin D activity.)

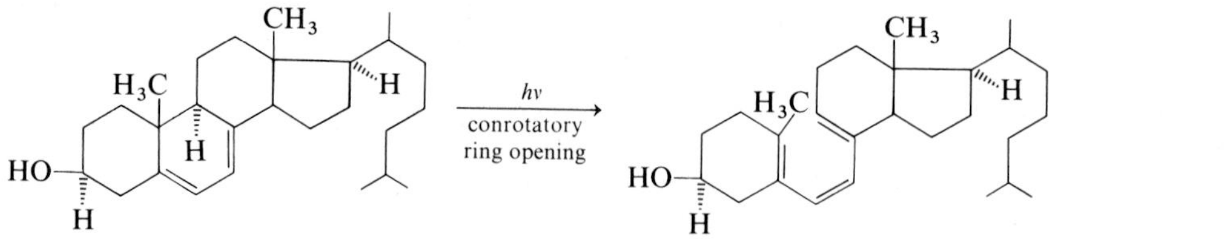

7-dehydrocholesterol

in the skin

previtamin D

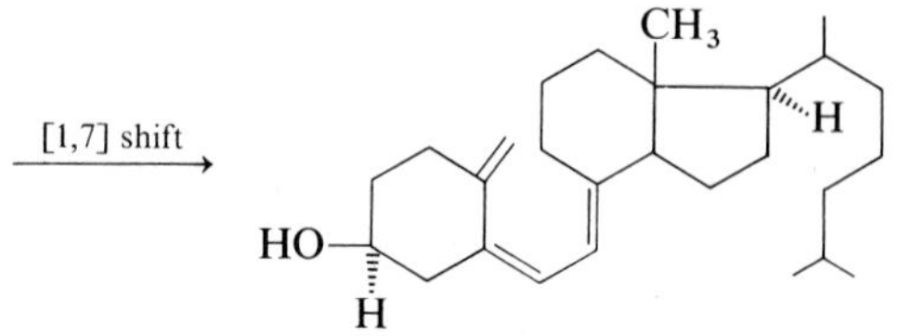

vitamin D$_3$

Another source of vitamin D is *irradiated ergosterol*, which is commonly added to milk. The conversion of ergosterol to vitamin D$_2$ proceeds by the same series of reaction steps as the conversion of 7-dehydrocholesterol.

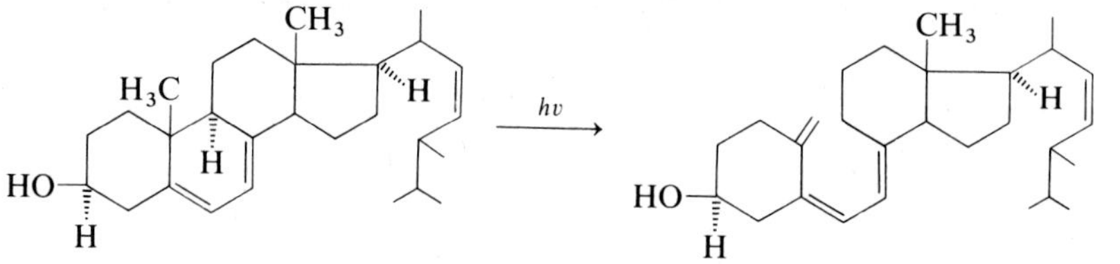

ergosterol

*in yeast, soybean oil,
and ergot (a fungus)*

vitamin D$_2$
(calciferol)

STUDY PROBLEM

17.11. Previtamin D can undergo both thermally induced and photo-induced electrocyclic ring-closure reactions. What will be the stereochemistry at positions 9 and 10 of the resulting products? (See page 905 for the numbering of the steroid ring.)

Summary

Pericyclic reactions are concerted, thermally induced or photo-induced reactions with cyclic transition states. Three types of pericyclic reactions are:

Cycloaddition:

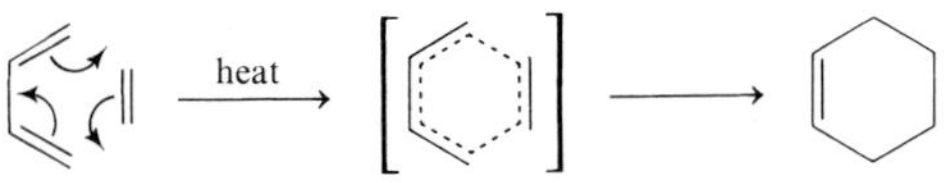

Electrocyclization:

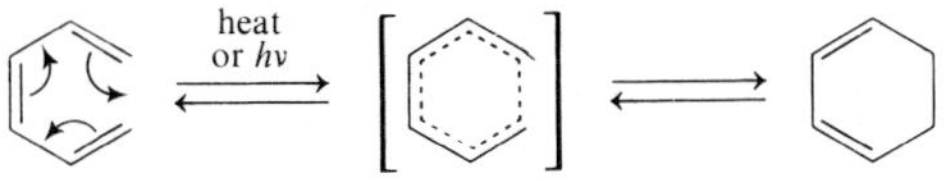

Sigmatropic rearrangement:

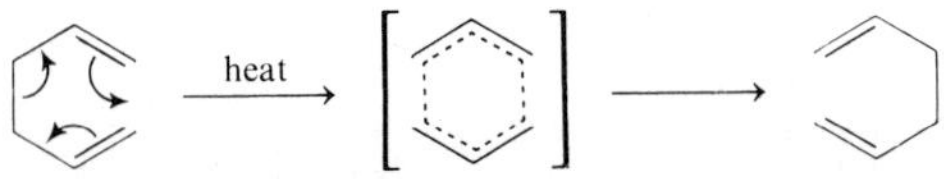

In the **frontier orbital method** of analyzing cycloaddition reactions, electrons are assumed to flow from the **HOMO** of one molecule to the **LUMO** of the other. If the phases of these orbitals are the same, the reaction is **symmetry-allowed**. If the orbital phases are opposite and show antibonding character, the reaction is **symmetry-forbidden**. Symmetry-allowed [2 + 2] cycloadditions are photo-induced, while [4 + 2] cycloadditions are thermally induced.

In electrocyclic reactions, *p*-orbital components of the HOMO undergo end-to-end overlap to form the new sigma bond. To do this, they must undergo **conrotatory** or **disrotatory** motion, which, in turn, determines the stereochemistry. A summary of the types of motion to be expected is shown in Table 17.1 (page 794).

Sigmatropic rearrangements occur **suprafacially** or **antarafacially**, depending on the phases of the interacting orbitals in the HOMO of a hypothetical radical system. The geometry of the transition state determines whether the rearrangement proceeds readily or not. The classification of sigmatropic rearrangements is discussed in Section 17.4A.

STUDY PROBLEMS

17.12. Identify each of the following pericyclic reactions as being (1) a cycloaddition; (2) an electrocyclic reaction; or (3) a sigmatropic rearrangement:

(a)

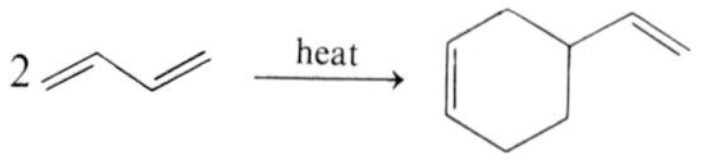

(b)

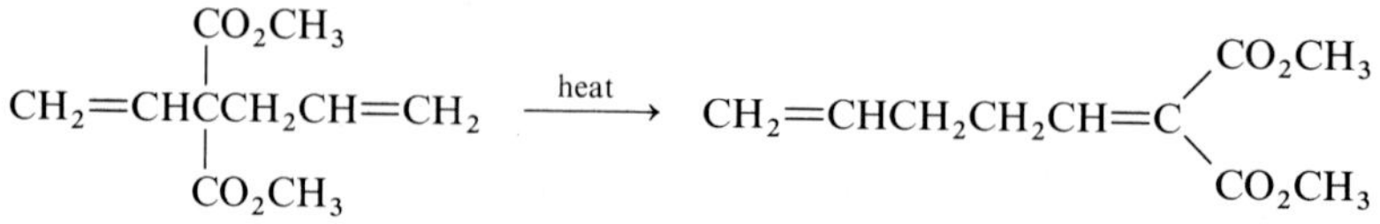

(c)

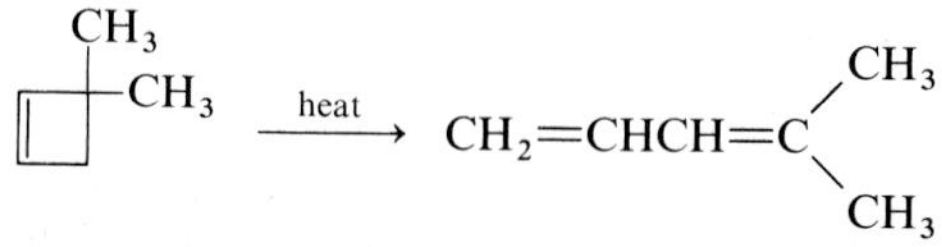

(d)

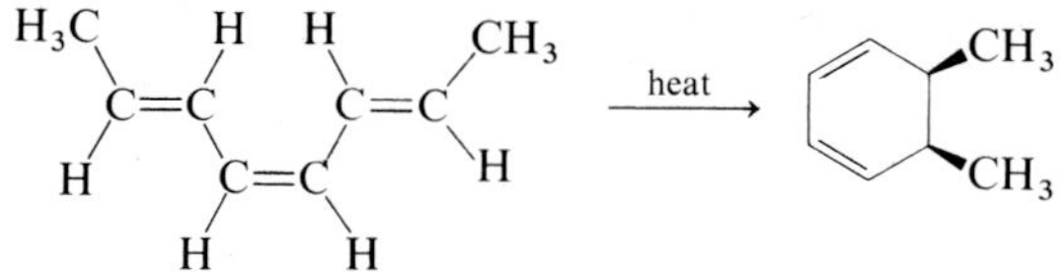

(e)

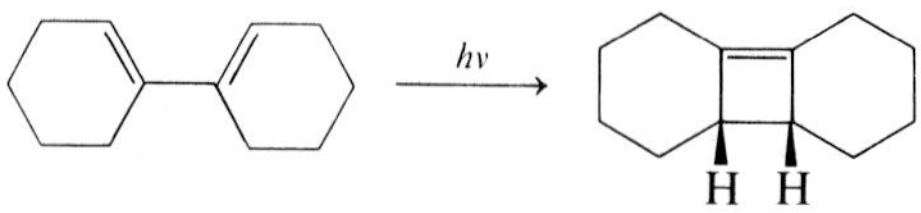

17.13. (1) Draw the *p*-orbital arrays in the π molecular orbitals of the following ions. (2) Draw diagrams showing the occupied orbitals of the ground states and indicate the HOMO's. (See Figure 17.2, page 784, for an example.)

(a) $CH_2{=}CH\overset{+}{C}HCH_3$ (b) $CH_2{=}CH\overset{-}{C}HCH_3$
(c) $CH_3CH{=}CHCH{=}CH\overset{-}{C}HCH_3$

17.14. Classify each of the following cycloadditions as $[2 + 2]$, $[4 + 2]$, etc.:

(a) 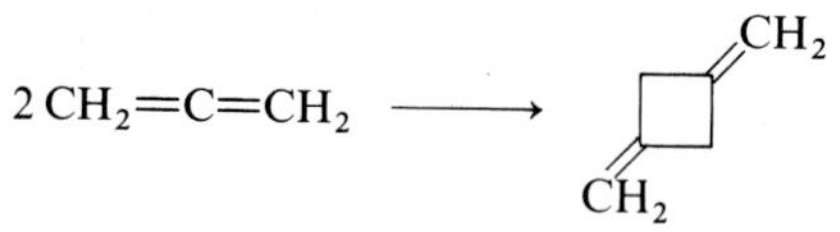 $2\,CH_2{=}C{=}CH_2 \longrightarrow$

(b)

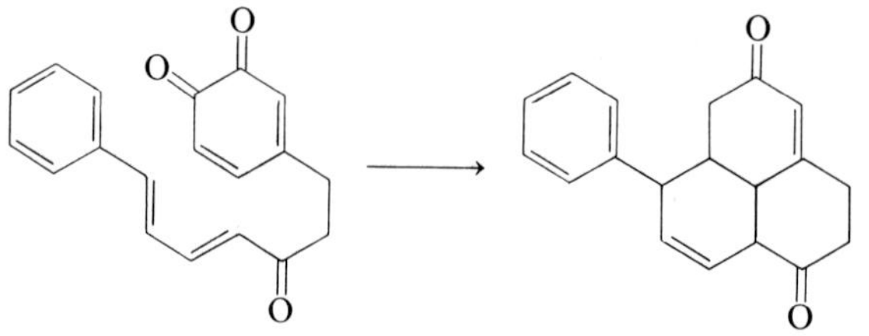

(c)

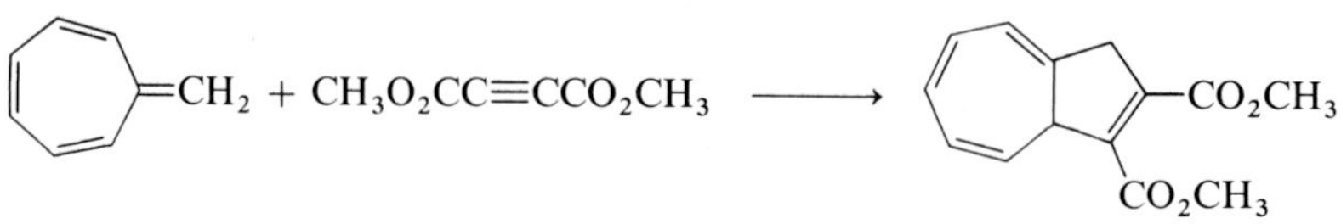

17.15. Which of the following types of cycloadditions would you predict to proceed easily upon heating?

(a) [6 + 2]   (b) [6 + 4]   (c) [8 + 2]   (d) [8 + 4]

17.16. 1,2-Diphenyl-1-cyclobutene undergoes a photochemical dimerization. What is the structure of the dimer? (Include the stereochemistry in your answer.)

17.17. Suggest synthetic routes to the following compounds from monocyclic or acyclic starting materials:

(a)

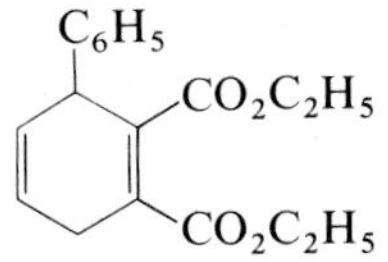

(b)

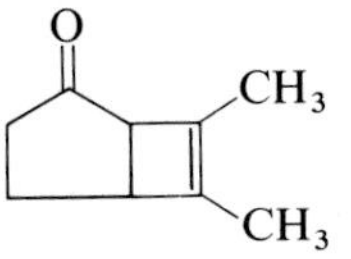

17.18. (1) Tell whether each of the following dienes or trienes would undergo conrotatory or disrotatory motion in a cyclization reaction. (2) What product would be observed? (Include stereochemistry where appropriate.)

(a)

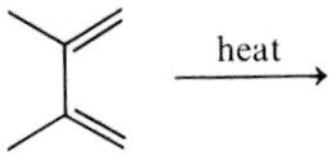

(b)

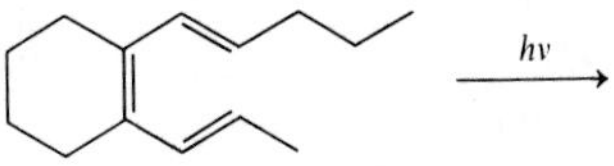

(c)

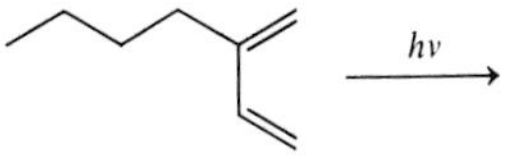

(d)

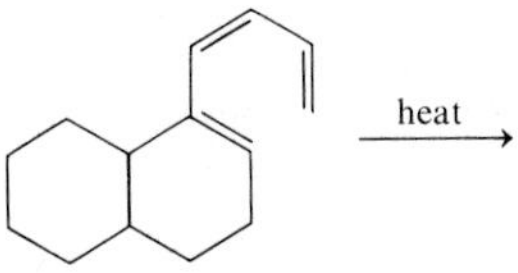

17.19. Predict the electrocyclic product and its stereochemistry:

(a)

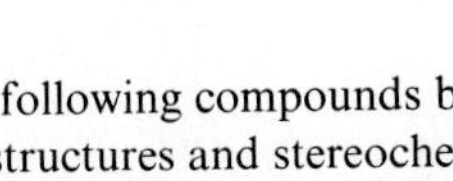

(b)

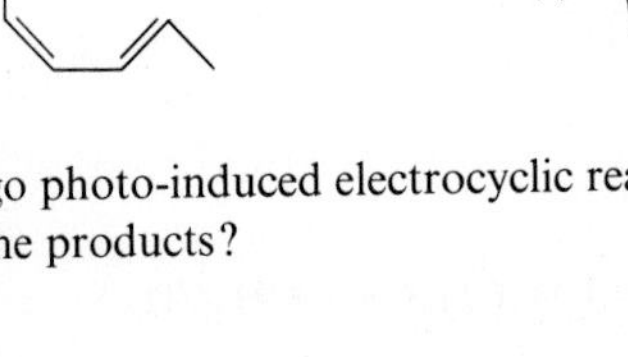

(c)

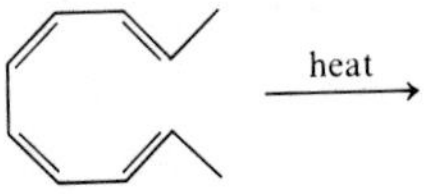

(d)

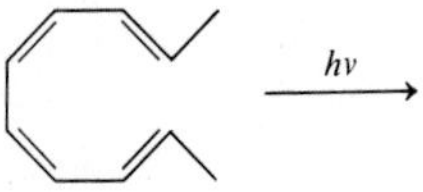

(e)

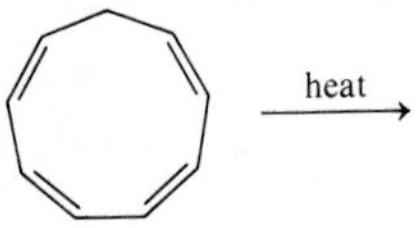

17.20. The following compounds both undergo photo-induced electrocyclic reactions. What are the structures and stereochemistry of the products?

(a)

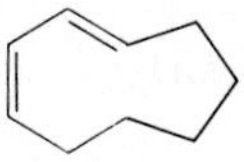

(b)

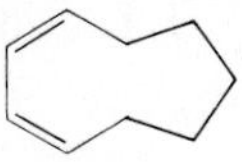

17.21. How would you carry out the following conversions?

(a)

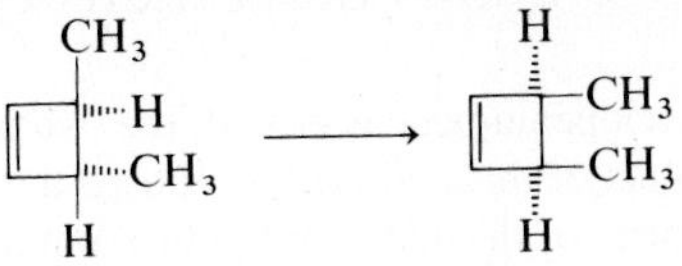

(b) 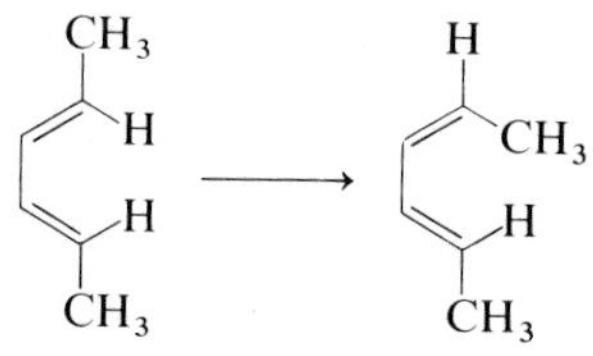

17.22. Classify each of the following sigmatropic rearrangements as [1,3], [3,3], etc.

(a) 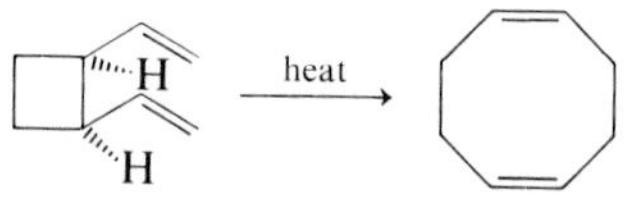

(b) 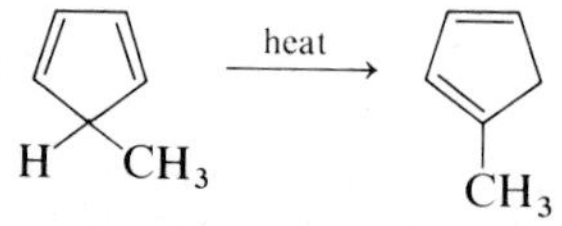

(c) 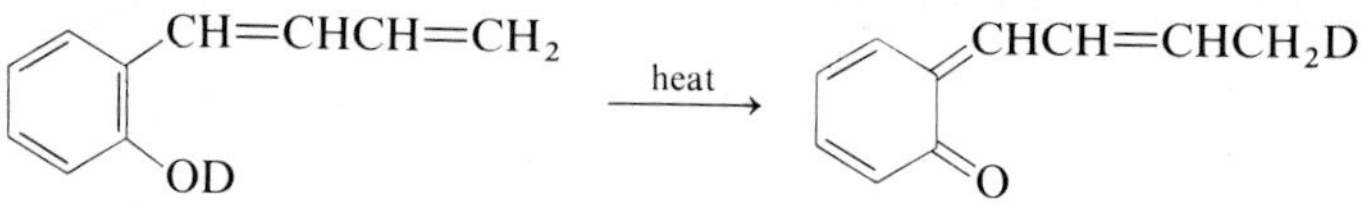

17.23. Predict the product or products of the sigmatropic rearrangement of 7,7-dideuterio-1,3,5-cycloheptatriene.

17.24. Explain the following observation:

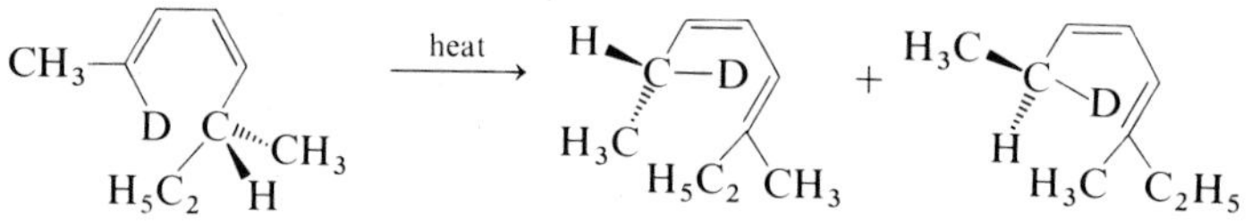

17.25. Should you use thermal induction or photo-induction to carry out the following reactions?

(a) 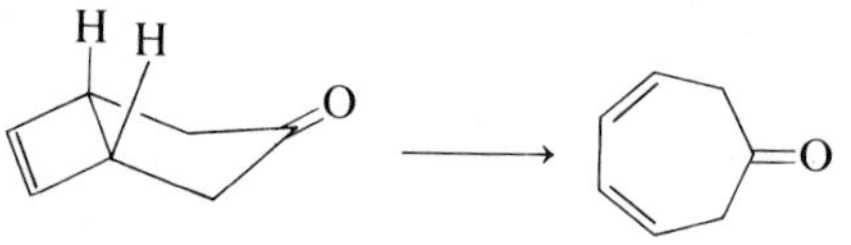

(b) 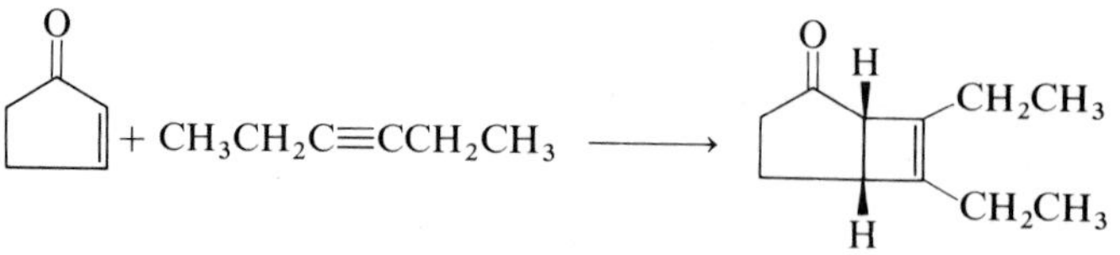

(c) $CH_2{=}CH\overset{CH_3}{\overset{|}{C}}HCH_2CH{=}CH_2 \longrightarrow CH_3CH{=}CHCH_2CH_2CH{=}CH_2$

17.26. When a mixture of quinone (page 486) and cyclopentadiene is heated, the two compounds undergo an addition reaction. When the product of this addition is exposed to sunlight, it undergoes isomerization. What are the structures of the addition product and its isomer?

17.27. Suggest mechanisms for the following reactions:

(a) 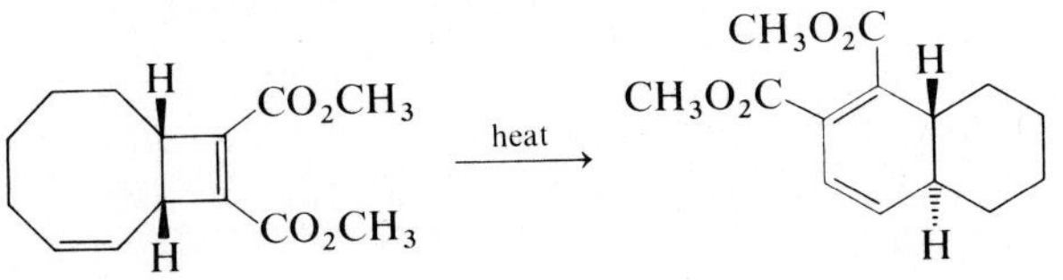
H
CO2CH3
CO2CH3
H
heat
CH3O2C
CH3O2C
H
H

(b) 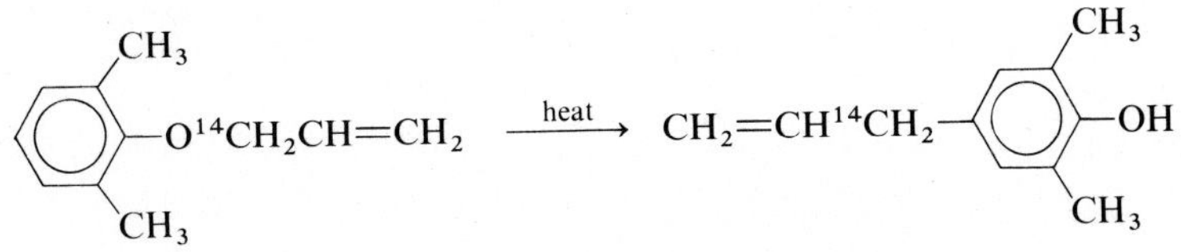
CH3
O14CH2CH=CH2
CH3
heat
CH2=CH14CH2
CH3
OH
CH3

17.28. Suggest synthetic routes to the following compounds from acyclic or monocyclic starting materials:

(a) 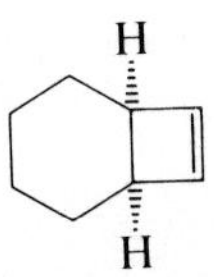
H
H

(b) 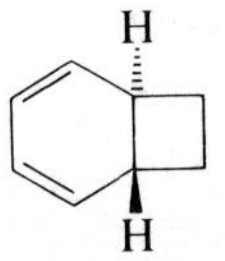
H
H

(c) 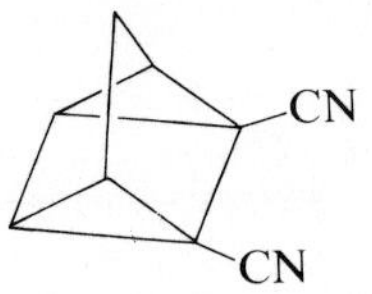
CN
CN

(d) 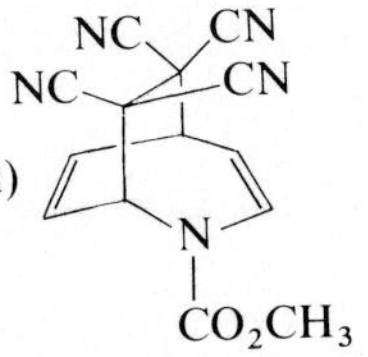
NC
CN
NC
CN
N
CO2CH3

17.29. Complete the following equations, showing the stereochemistry of the products:

(a) 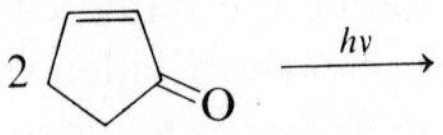
2
O
hν

(b) 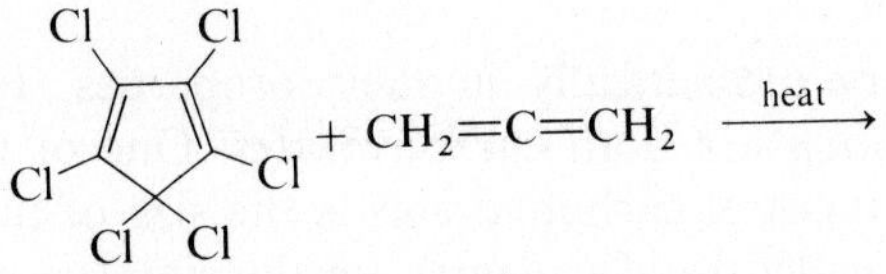
Cl Cl
Cl
Cl
Cl Cl
+ CH2=C=CH2
heat

(c) 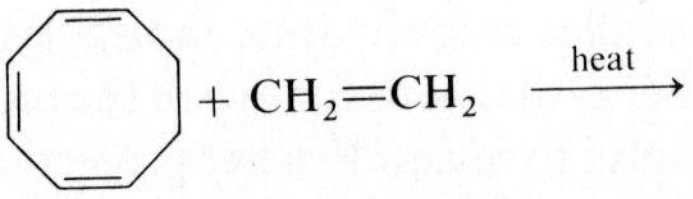
+ CH2=CH2
heat

(d) 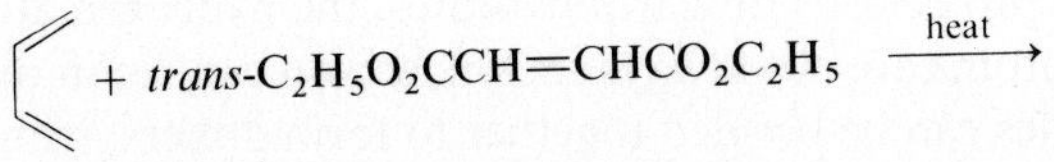
+ trans-C2H5O2CCH=CHCO2C2H5
heat

(e) 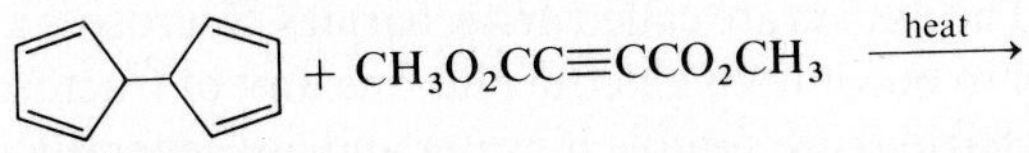
+ CH3O2CC≡CCO2CH3
heat

CHAPTER 18

Carbohydrates

Carbohydrates are naturally occurring compounds of carbon, hydrogen, and oxygen. Many carbohydrates have the empirical formula CH_2O; for example, the molecular formula for glucose is $C_6H_{12}O_6$ (six times CH_2O). These compounds were once thought to be "hydrates of carbon," hence the name carbohydrates. In the 1880's, it was recognized that the "hydrates of carbon" idea was a misconception and that carbohydrates are actually polyhydroxy aldehydes and ketones or their derivatives.

The carbohydrates vary dramatically in their properties. For example, *sucrose* (table sugar) and *cotton* are both carbohydrates. One of the principal differences between various types of carbohydrates is the size of the molecules. The **monosaccharides** (often called *simple sugars*) are the simplest carbohydrate units; they cannot be hydrolyzed to smaller carbohydrate molecules. Figure 18.1 shows dimensional formulas and Fischer projections for five of the most important monosaccharides. (You may find it helpful to review Fischer projections in Section 4.6C.) In the Fischer projections for carbohydrates, the hydrogen atoms attached to the chiral carbon atoms are not always shown, as you can see in the figure.

Monosaccharides can be bonded together to form dimers, trimers, etc. and, ultimately, polymers. The dimers are called **disaccharides**. Sucrose is a disaccharide that can be hydrolyzed to one unit of glucose plus one unit of fructose. The monosaccharides and disaccharides are soluble in water and are generally sweet-tasting.

$$1 \text{ sucrose} \quad \xrightarrow[\text{heat}]{H_2O, H^+} \quad 1 \text{ glucose} + 1 \text{ fructose}$$
a disaccharide

Carbohydrates composed of two to eight units of monosaccharide are referred to as **oligosaccharides** (Greek *oligo-*, "a few"). If more than eight units of

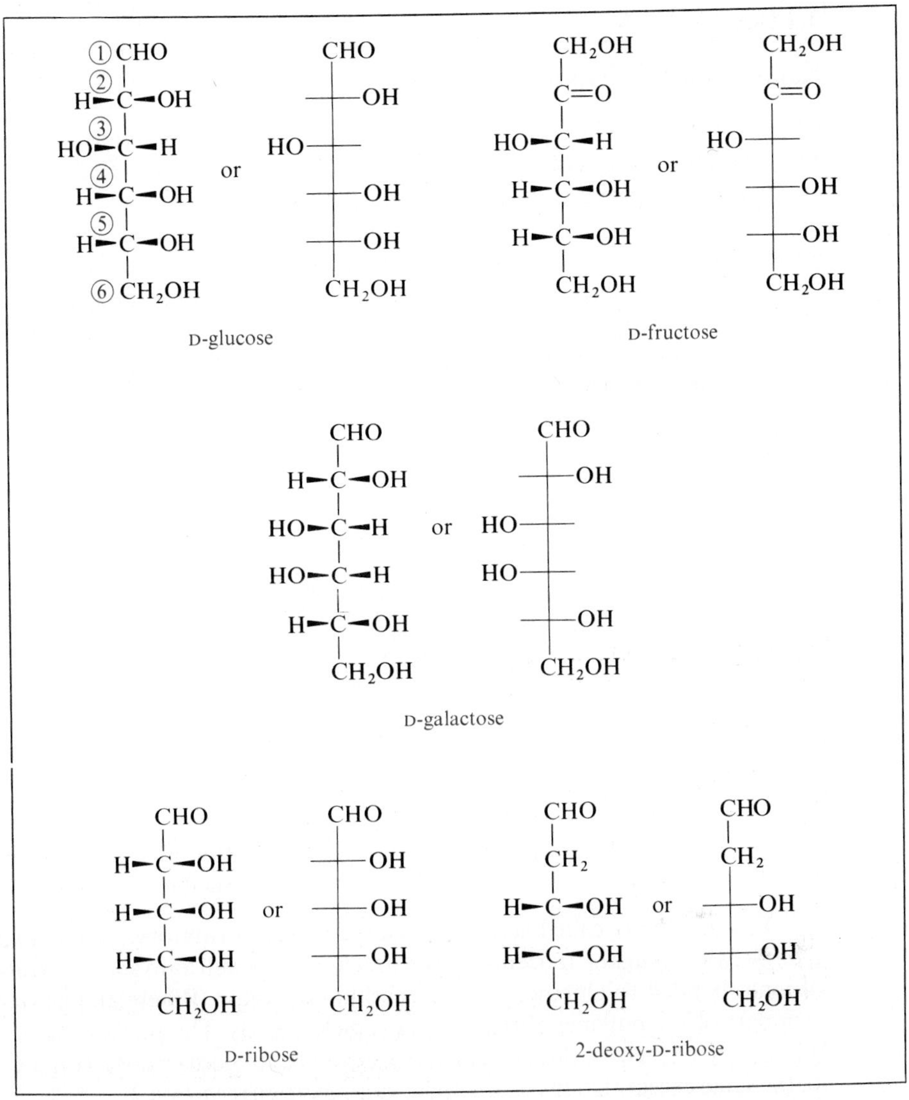

FIGURE 18.1. Some important monosaccharides.

monosaccharide result from hydrolysis, the carbohydrate is a **polysaccharide**. Examples of polysaccharides are *starch*, found in flour and cornstarch, and *cellulose*, a fibrous constituent of plants and the principal component of cotton.

$$\text{starch or cellulose} \xrightarrow[\text{heat}]{H_2O, \ H^+} \text{many units of glucose}$$

polysaccharides

In this chapter, we will consider first the monosaccharides and the conventions used by carbohydrate chemists. Then we will discuss some disaccharides, and finally, a few polysaccharides.

Some Common Monosaccharides

Glucose, the most important of the monosaccharides, is sometimes called *blood sugar* (because it is found in the blood), *grape sugar* (because it is found in grapes), or *dextrose* (because it is dextrorotatory). Mammals can convert sucrose, lactose (milk sugar), maltose, and starch to glucose, which is then used by the organism for energy or stored as *glycogen* (a polysaccharide). When the organism needs energy, the glycogen is again converted to glucose. Excess carbohydrates can be converted to fat; therefore a person can become obese on a fat-free diet. Carbohydrates can also be converted to steroids (such as cholesterol) and, to a limited extent, to protein. (A source of nitrogen is also needed in protein synthesis.) Conversely, an organism can convert proteins and fats to carbohydrates.

fats

$$CO_2 + H_2O + energy \longleftarrow \quad glucose \rightleftharpoons \quad CH_3\overset{\overset{O}{\|}}{C}- \rightleftharpoons \quad (proteins)$$

acetyl groups
in acetylcoenzyme A

cholesterol
and other steroids

Fructose, also called *levulose* because it is levorotatory, is the sweetest-tasting of all the sugars. It occurs in fruit and honey, as well as in sucrose. **Galactose** is found, bonded to glucose, in the disaccharide lactose. **Ribose** and **deoxyribose** form part of the polymeric backbones of nucleic acids. The prefix *deoxy-* means "minus an oxygen"; the structures of ribose and deoxyribose (Figure 18.1) are the same except that deoxyribose lacks an oxygen at carbon 2.

Classification of the Monosaccharides

The suffix **-ose** is used in systematic carbohydrate nomenclature to designate a **reducing sugar**, a sugar that contains an aldehyde group or an α-hydroxyketone grouping. Reducing sugars will be discussed in Section 18.6. Many of the oligo-saccharides and polysaccharides that are not reducing sugars have trivial names ending in *-ose* (for example, sucrose and cellulose). In this chapter, we will use both systematic and trivial nomenclature for carbohydrates; our emphasis will be on those names that are in common usage.

Monosaccharides that contain aldehyde groups are referred to as **aldoses** (*ald*ehyde plus *-ose*). Glucose, galactose, ribose, and deoxyribose are all aldoses.

Monosaccharides, such as fructose, with ketone groups are called **ketoses** (*ketone* plus *-ose*).

The number of carbon atoms in a monosaccharide (usually three to seven) may be designated by tri-, tetr-, etc. For example, a **triose** is a three-carbon monosaccharide, while a **hexose** is a six-carbon monosaccharide. Glucose is an example of a hexose. These terms may be combined. Glucose is an **aldohexose** (six-carbon aldose), while ribose is an **aldopentose** (five-carbon aldose). Ketoses are often given the ending -**ulose**. Fructose is an example of a **hexulose** (six-carbon ketose).

STUDY PROBLEM

18.1. Classify each of the following monosaccharides by the preceding system:

(a)
```
      CHO
    ——————OH
    ——————OH
     CH₂OH
```
erythrose

(b)
```
      CH₂OH
       |
      C═O
    ——————OH
    ——————OH
     CH₂OH
```
ribulose

(c)   galactose (Figure 18.1)

SECTION 18.3.

Configurations of the Monosaccharides

As may be seen from the five monosaccharides in Figure 18.1, monosaccharides are quite similar to one another in structure. Some monosaccharides are structurally different; for example, glucose is an aldehyde and fructose is a ketone. Other common monosaccharides are diastereomers (nonenantiomeric stereoisomers) of one another; for example, glucose and galactose are **epimers**, diastereomers that differ in configuration at only one of their chiral carbon atoms.

A.   The D and L system

In the late 19th century, it was determined that the configuration of the last chiral carbon in each of the naturally occurring monosaccharides is the same as that of (+)-glyceraldehyde. Today we call that configuration the (*R*)-configuration, but chemists at that time had no way to determine the absolute configuration around a chiral carbon atom. Instead, chemists devised the D and L system for designating relative configurations. (Do not confuse D and L with *d* and *l*, sometimes used to refer to the direction of rotation of the plane of polarization of plane-polarized light; see Section 4.7A.)

In the D and L system, (+)-glyceraldehyde was arbitrarily assigned the configuration with its OH on carbon 2 to the right in the Fischer projection (an assumption later shown to be correct). A monosaccharide is a member of the **D-series** if the hydroxyl group on the chiral carbon farthest from carbon 1 is also *on the right* in the Fischer projection. (Almost all naturally occurring carbo-hydrates are members of the D-series.) In addition, each monosaccharide was given its own name. For example, the following two diastereomeric aldopentoses are named D-lyxose and D-ribose.

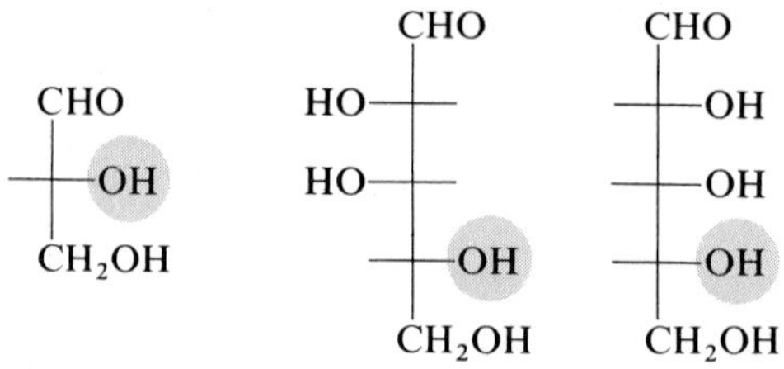

D-(+)-glyceraldehyde   D-lyxose   D-ribose

If the OH on the last chiral carbon is projected to the *left*, then the compound is a member of the L-series. The following two examples are the enantiomers of D-lyxose and D-ribose.

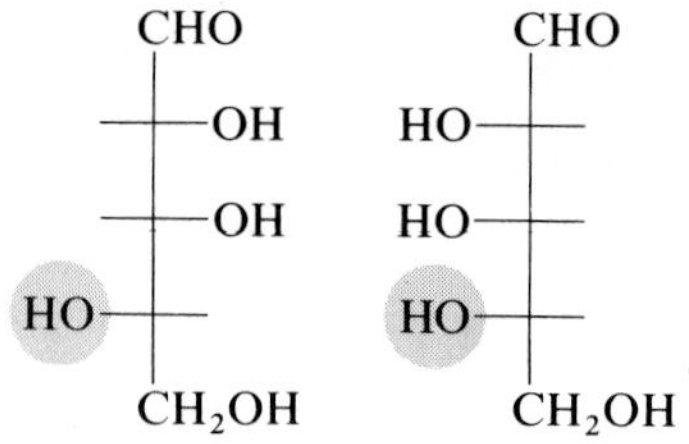

L-lyxose   L-ribose

STUDY PROBLEM

18.2. Assign each chiral carbon atom an (R) or (S) designation and tell whether each compound is D or L:

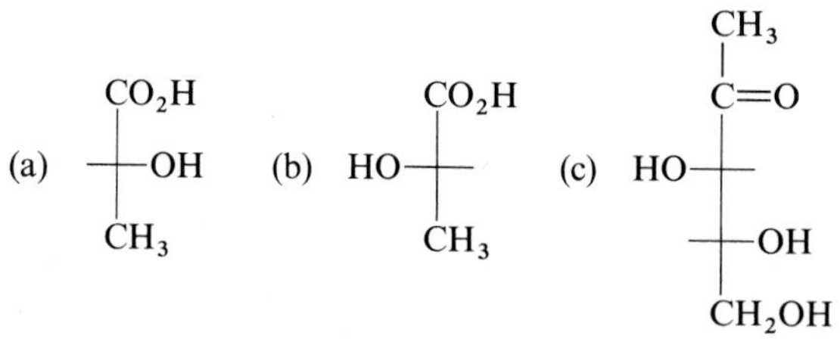

B. Relating configurations

We have mentioned that early chemists could not determine the absolute con-figurations of chiral carbons. Instead, configurations relative to that of (+)-glyceraldehyde were determined. How are other compounds related to glycer-aldehyde? One example of determining a relative configuration follows. If the aldehyde group of D-glyceraldehyde is oxidized to a carboxylic acid, the product, glyceric acid, necessarily has the same configuration around the chiral carbon as that in D-glyceraldehyde. The product, even though levorotatory, is still a member of the D-series.

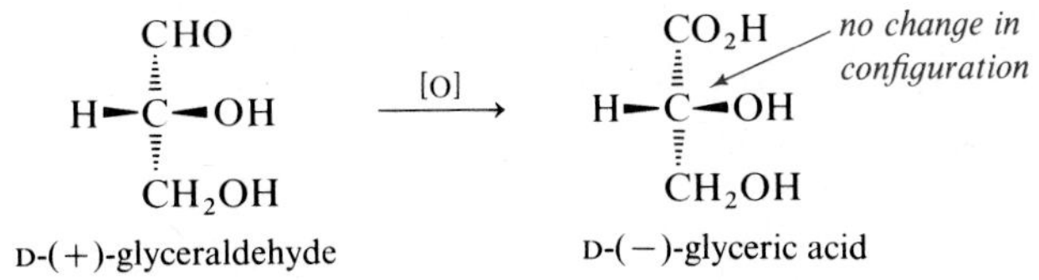

CHO
$\vdots$
H—C—OH
$\vdots$
CH$_2$OH

D-(+)-glyceraldehyde

$\xrightarrow{[O]}$

CO$_2$H
$\vdots$
H—C—OH
$\vdots$
CH$_2$OH

no change in configuration

D-(−)-glyceric acid

The configurations of the tartaric acids relative to D-glyceraldehyde were established in 1917 by the sequence shown in Figure 18.2, which produced two of the three isomers of tartaric acid.

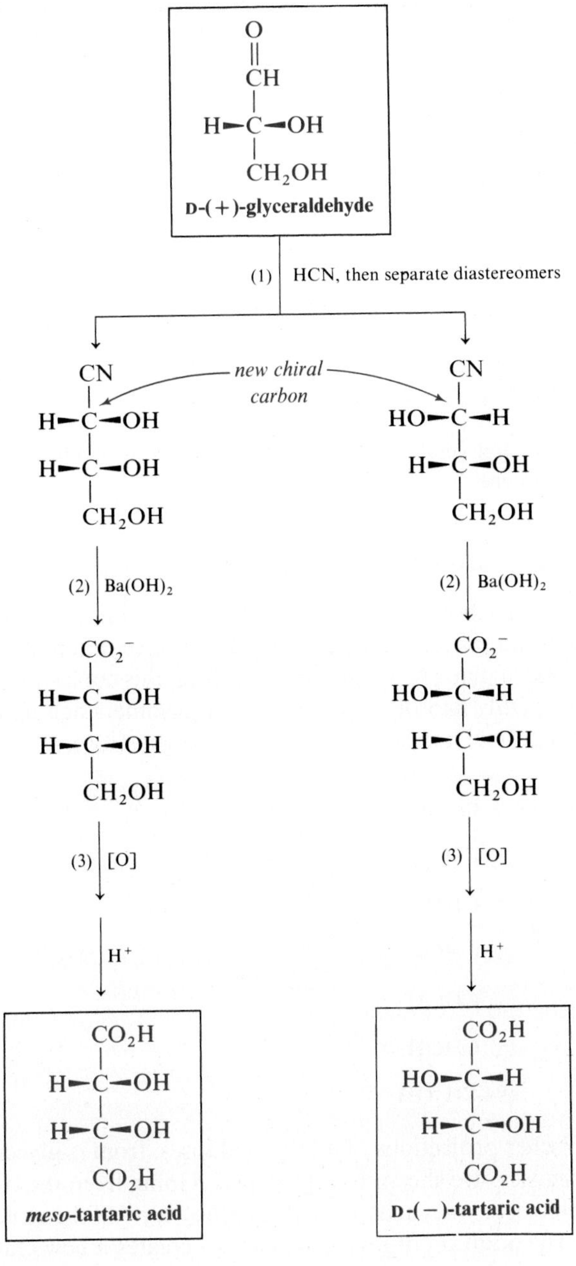

O
‖
CH
|
H—C—OH
|
CH$_2$OH

D-(+)-glyceraldehyde

(1)   HCN, then separate diastereomers

new chiral carbon

CN
|
H—C—OH
|
H—C—OH
|
CH$_2$OH

CN
|
HO—C—H
|
H—C—OH
|
CH$_2$OH

(2) Ba(OH)$_2$

(2) Ba(OH)$_2$

CO$_2$$^-$
|
H—C—OH
|
H—C—OH
|
CH$_2$OH

CO$_2$$^-$
|
HO—C—H
|
H—C—OH
|
CH$_2$OH

(3) [O]

(3) [O]

H$^+$

H$^+$

CO$_2$H
|
H—C—OH
|
H—C—OH
|
CO$_2$H

***meso*-tartaric acid**

CO$_2$H
|
HO—C—H
|
H—C—OH
|
CO$_2$H

D-(−)-tartaric acid

FIGURE 18.2. Determining the relative configurations of tartaric acids.

In the first step of the sequence, D-glyceraldehyde is treated with HCN to yield a mixture of cyanohydrins. A new site of chirality is introduced in this step, and both diastereomers are formed. The diastereomers are separated; then, in the second step, each diastereomeric cyanohydrin is hydrolyzed.

In the third step, the terminal CH_2OH group is oxidized to yield two tartaric acids. Carbon 3 of each of these tartaric acids has the same configuration as carbon 2 of D-glyceraldehyde because the series of reactions has not affected the configuration around that carbon. However, the configurations around carbon 2 in the tartaric acids are different. One of the tartaric acids obtained from this synthesis does not rotate plane-polarized light. This is the *meso* isomer, the one with an internal plane of symmetry. The other tartaric acid obtained from the synthesis rotates plane-polarized light to the left; it must have the second structure as follows:

meso-tartaric acid   D-(−)-tartaric acid

same configuration at carbon 3 as that of D-glyceraldehyde

STUDY PROBLEM

18.3. Starting with L-(−)-glyceraldehyde, what tartaric acid(s) would be produced in the preceding sequence?

C.   Configurations of the aldohexoses

Glucose has six carbon atoms, four of which are chiral (carbons 2, 3, 4, and 5). Because the terminal carbon atoms of glucose have different functionality, there can be no internal plane of symmetry; therefore, this compound has 2^4, or sixteen, stereoisomers. Only half of these sixteen stereoisomers belong to the D-series and are found in nature. Of these, only D-glucose, D-galactose, and D-mannose occur in abundance.

four chiral carbons: 16 stereoisomers

The Fischer projections of all the D-aldoses, from D-glyceraldehyde through the D-aldohexoses, are shown in Figure 18.3. Going from the triose, D-glyceraldehyde, to the tetroses, one carbon is added to the "top" of the molecule in the Fischer projection. The addition of one more carbon creates a new chiral carbon in each

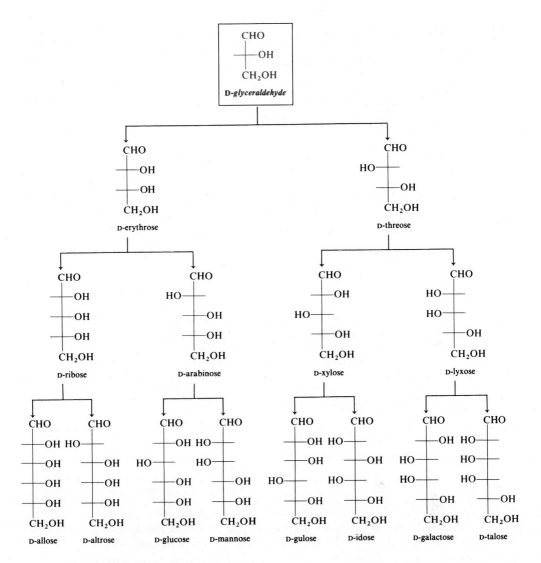

FIGURE 18.3. The D-aldoses.

step down in the figure. Therefore, D-glyceraldehyde leads to a pair of tetroses, each tetrose leads to a pair of pentoses, and each pentose leads to a pair of hexoses.

STUDY PROBLEMS

18.4. How many chiral carbon atoms does each of the following aldoses have?

(a) D-talose   (b) L-allose   (c) D-arabinose

18.5. There are eight D-aldohexoses and also eight L-aldohexoses. Write the Fischer projections of   (a) L-glucose,   and   (b) L-mannose.

18.6. What is the name of the aldohexose in which only the OH at carbon 5 has the opposite configuration from that of D-glucose?

SECTION 18.4.

Cyclization of the Monosaccharides

Glucose has an aldehyde group at carbon 1 and hydroxyl groups at carbons 4 and 5 (as well as at carbons 2, 3, and 6). A general reaction of alcohols and aldehydes is that of *hemiacetal formation* (see Section 11.8).

$$\underset{\text{RCH}}{\overset{\text{O}}{\|}} + \text{R'OH} \underset{\xleftarrow{\hspace{1cm}}}{\overset{\text{H}^+}{\xrightarrow{\hspace{1cm}}}} \underset{\text{RCH—OR'}}{\overset{\text{OH}}{|}}$$

a hemiacetal

In water solution, glucose can undergo an intramolecular reaction to yield *cyclic hemiacetals*. Either five-membered ring hemiacetals (using the hydroxyl group at carbon 4) or six-membered ring hemiacetals (using the hydroxyl group at carbon 5) can be formed.

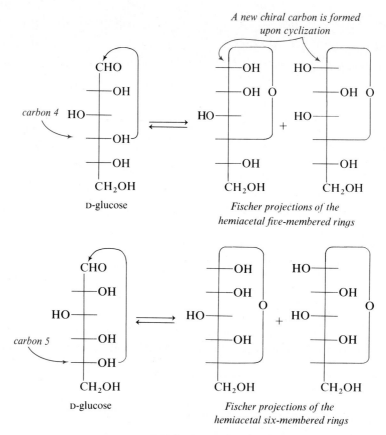

In the Fischer projections for the cyclic hemiacetals, note that carbon 1 (the aldehyde carbon), which is not chiral in the open-chain structure, becomes chiral in the cyclization. Therefore, a pair of diastereomers results from the cyclization.

Because all the hemiacetal structures are in equilibrium with the aldehyde in water solution, they are also in equilibrium with each other.

5-membered cyclic hemiacetal ⇌ glucose ⇌ 6-membered cyclic hemiacetal
  (two diastereomers)   (open-chain)   *(two diastereomers)*

STUDY PROBLEM

18.7. Write equations for the cyclization of 2-deoxy-D-ribose. Use Fischer projections, and show the formation of both the five- and six-membered cyclic hemiacetals.

A. Furanose and pyranose rings

A monosaccharide in the form of a five-membered ring hemiacetal is called a **furanose**. *Furan-* is from the name of the five-membered oxygen heterocycle *furan*. Similarly, the six-membered ring form is called a **pyranose** after *pyran*. The terms furanose and pyranose are often combined with the name of the mono-saccharide—for example, D-**glucopyranose** for the six-membered ring of D-glucose, or D-**fructofuranose** for the five-membered ring of fructose.

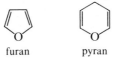

furan   pyran

Of the two ring systems for glucose, the six-membered cyclic hemiacetal, or glucopyranose, is favored; we will emphasize this ring size in our discussion. Part of the reason that glucose preferentially forms the six-membered ring in solution is that the bond angles and staggering of attached groups are favorable in the chair form of this ring. Even though the pyranose ring of a monosaccharide may pre-dominate in equilibrium in water, it may be the furanose ring that is incorporated enzymatically into natural products. For example, in ribonucleic acids, ribose is found as a furanose, and not as a pyranose.

B. Anomers

To yield a pyranose, the hydroxyl group at carbon 5 of glucose attacks the aldehyde carbon, carbon 1. A hemiacetal group is formed. Two of the most important consequences of this cyclization reaction are that a new chiral carbon is formed (carbon 1) and that a pair of diastereomers results. These diastereomers, mono-saccharides that differ only in the configuration at carbon 1, are called **anomers** of each other. The carbonyl carbon in any monosaccharide is the **anomeric carbon**. This is the carbon that becomes chiral in the cyclization reaction.

In the Fischer projection for one anomer of D-glucopyranose, the hydroxyl group at carbon 1 is projected to the *left*; this anomer is called **β-D-glucopyranose**, or simply **β-D-glucose**. The anomer in which the hydroxyl group at carbon 1 is pro-jected to the right is called **α-D-glucopyranose**, or **α-D-glucose**. (In the L-series, the *β*-anomer is the one with the OH at carbon 1 on the right; therefore, *β*-L-gluco-pyranose is the enantiomer of *β*-D-glucopyranose.)

β-D-glucopyranose   open-chain D-glucose   α-D-glucopyranose

STUDY PROBLEM

18.8. Write Fischer projections for the two D-ribofuranoses and label each as α or β.

C. Haworth and conformational formulas

In aqueous solution, only about 0.02 % of glucose exists in the open-chain aldehyde form; the rest exists as cyclic hemiacetals. While the Fischer projection is adequate to show the configurations about the chiral carbons of a carbohydrate in its open-chain form, it is a poor representation of the cyclic structures. To better represent the cyclic structures of the sugars, **Haworth perspective formulas** were developed. With a Haworth formula, it is clear that the hydroxyl groups (or other groups) at the chiral carbons are actually *cis* and *trans* to one another on the ring. Also, the Haworth formula eliminates the artificial, curved bonds to the ring oxygen that are necessary in a Fischer projection of a cyclic monosaccharide.

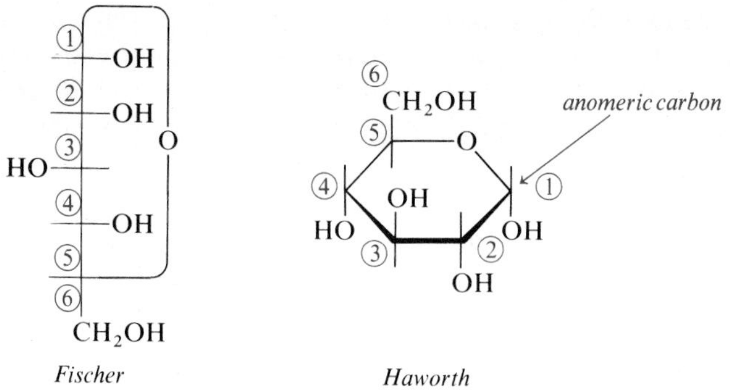

Fischer   *Haworth*

α-D-glucopyranose

By convention, a Haworth formula is drawn with the ring oxygen on the far side of the ring and the anomeric carbon on the right. The terminal CH_2OH group is positioned above the plane of the ring in the D-series and below the plane of the ring in the L-series. As in Fischer projections, hydrogen atoms are not usually shown.

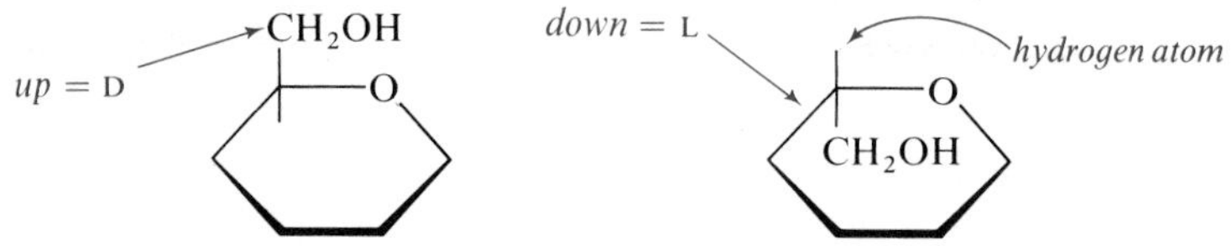

In a Haworth formula of a D-sugar, the structure in which the anomeric OH is projected *down* (*trans* to the terminal CH_2OH) is the α-anomer; the structure in which the anomeric OH is projected *up* is the β-anomer.

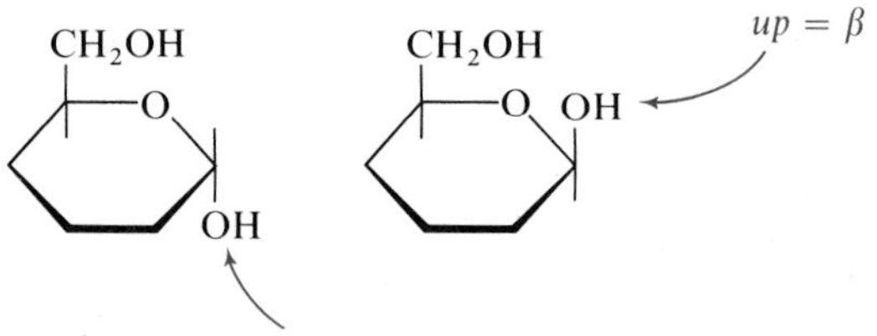

Note that any group that is to the right in the Fischer projection is down in the Haworth projection, and any group that is to the *left* in the Fischer projection is *up* in the Haworth formula.

The flat Haworth formula is not an entirely correct representation of a pyranose ring (although it is fairly correct for the more planar furanose ring). A pyranose, like cyclohexane, exists primarily in the chair form, as the following conformational formula shows. In this chapter, we will use both Haworth formulas and conformational formulas.

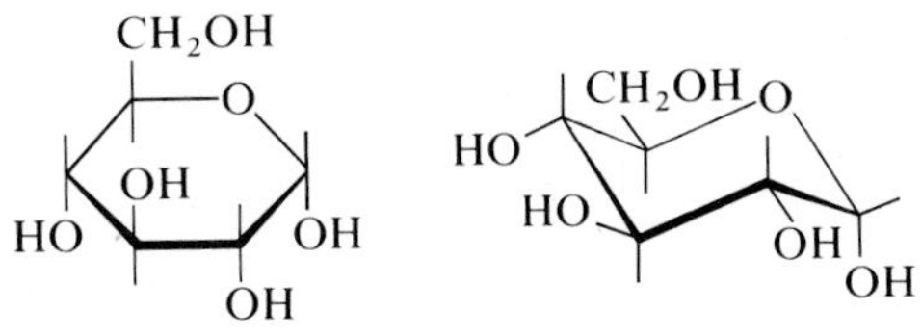

α-D-glucopyranose

If an OH is down in a Haworth formula, it is also down (below the plane of the ring) in the conformational formula. Similarly, if an OH is up in the Haworth formula, it is also up in the conformational formula. As for any substituted six-membered ring, the ring assumes the conformation in which the majority of the groups are equatorial.

STUDY PROBLEM

18.9. Draw a conformational formula of α-D-glucopyranose in which the CH_2OH is in an axial position.

SAMPLE PROBLEMS

Although fructose can form a six-membered cyclic hemiketal, in sucrose it is found in the furanose form. Draw Fischer projections and Haworth formulas for α- and β-D-fructofuranose. (*Note:* The CH_2OH group on the keto carbon of a sugar is treated in the formula just like the H on an aldehyde carbon.)

Solution:

Fischer:

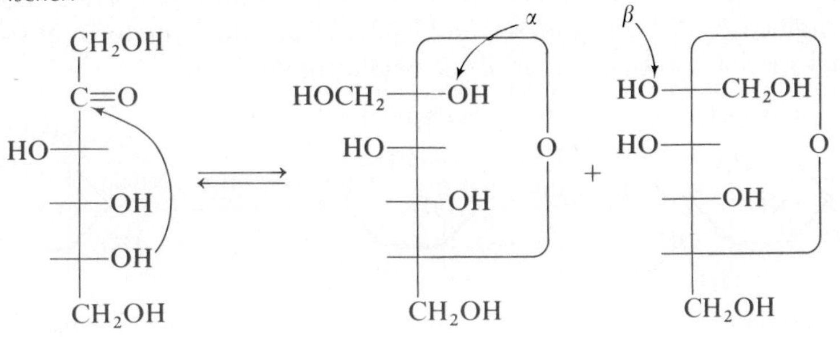

D-fructose

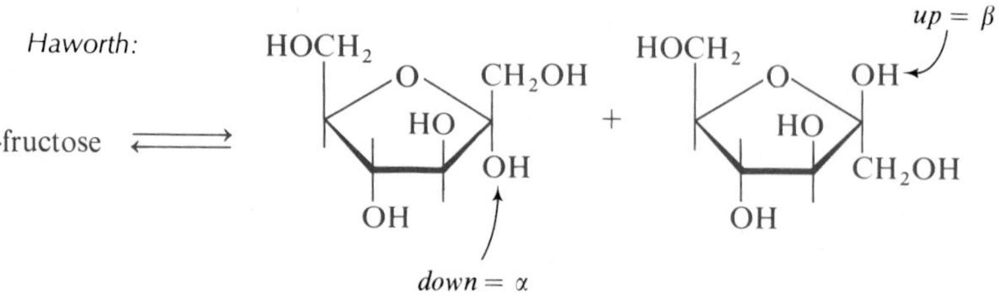

Haworth:

D-fructose $\rightleftharpoons$

up = β

down = α

Draw the Haworth formula and the Fischer projection for
2-deoxy-β-D-ribofuranose.

Solution:

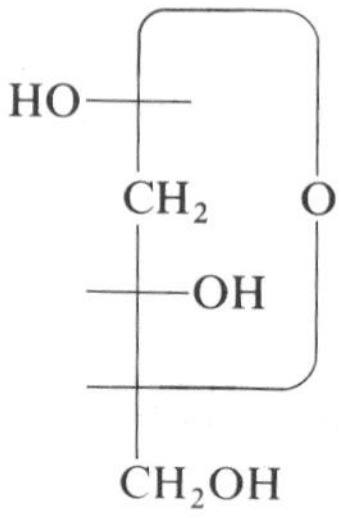

Fischer

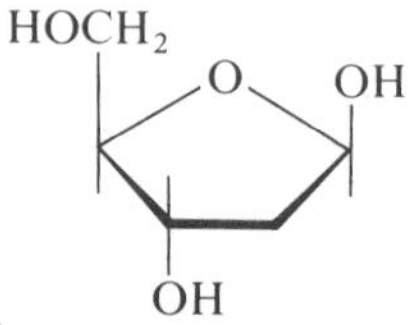

Haworth

STUDY PROBLEM

18.10. Draw the Fischer, Haworth, and conformational formulas for the anomers of
D-galactopyranose.

D.   Mutarotation

Pure glucose exists in two crystalline forms: α-D-glucose and β-D-glucose. Pure
α-D-glucose has a melting point of 146°C. The specific rotation of a freshly prepared solution is $+112°$. Pure β-D-glucose has a melting point of 150°C and a
specific rotation of $+18.7°$. The specific rotation of a solution of either α- or β-D-
glucose changes slowly until it reaches an equilibrium value of $+52.6°$. This slow
spontaneous change in optical rotation was first observed in 1846 and is called
mutarotation.

Mutarotation occurs because, in solution, either α- or β-D-glucose undergoes
a slow equilibration with the open-chain form and with the other anomer. Regardless of which anomer is dissolved, the result is an equilibrium mixture of 64%
β-D-glucose, 36% α-D-glucose, and 0.02% of the aldehyde form of D-glucose. The
final specific rotation is that of the equilibrium mixture.

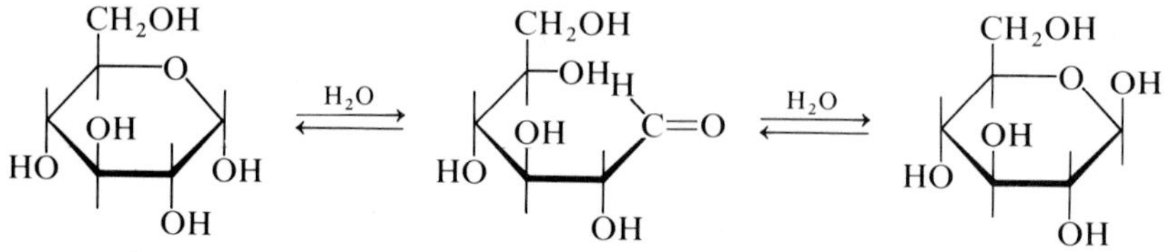

α-D-glucose
36%

D-glucose
0.02%

β-D-glucose
64%

Note that the equilibrium mixture of the anomers of D-glucose contains a greater percentage of the β-anomer than of the α-anomer. The reason is that the β-anomer is the more stable of the two. From our discussion of conformational analysis in Chapter 4, this is the expected result. The hydroxyl group at carbon 1 is *equatorial* in the β-anomer, but *axial* in the α-anomer.

α-D-glucose   β-D-glucose

Other monosaccharides also exhibit mutarotation. In water solution, the other aldoses (if they have a 5-hydroxyl group) also exist primarily in the pyranose forms. However, the percentages of the various species involved in the equilibrium may vary. For example, the equilibrium mixture of D-ribose in water is 56% β-pyranose, 20% α-pyranose, 18% β-furanose, and 6% α-furanose (plus a trace of the open-chain, aldehyde form).

β-D-ribopyranose   α-D-ribopyranose
56%   20%

β-D-ribofuranose   α-D-ribofuranose
18%   6%

Although the β-anomer of the pyranose rings is generally the more stable anomer, this is not always the case. For example, α-D-mannose is more stable than its β-anomer and predominates in an equilibrium mixture. This apparent anomaly, termed the **anomeric effect**, arises from interactions between the polar substituents on the ring.

Because of the facile conversion in water of the hemiacetal OH group between α and β, it often is not possible to specify the configuration at this carbon. For this reason, we will sometimes represent the hemiacetal OH bond with a squiggle, which means the structure may be α or β or a mixture.

α or β or a mixture

Glycosides

When a hemiacetal is treated with an alcohol, an acetal is formed (Section 11.8). The acetals of monosaccharides are called **glycosides** and have names ending in **-oside**.

$$\underset{\text{a hemiacetal}}{\overset{\overset{\displaystyle OH}{|}}{RCHOR}} + R'OH \xrightleftharpoons{H^+} \underset{\text{an acetal}}{\overset{\overset{\displaystyle OR'}{|}}{RCHOR}} + H_2O$$

β-D-glucopyranose   + CH₃OH ⇌   methyl β-D-glucopyranoside   + H₂O

a glycoside

The glycoside carbon (carbon 1 in an aldose) is easy to recognize because it has two OR groups attached.

an acetal   a glycoside

Although a hemiacetal of a monosaccharide is in equilibrium with the open-chain form and with its anomer in water solution, an acetal is stable in neutral or alkaline solution. Therefore, a glycoside is not in equilibrium with the aldehyde or with its anomer in water solution. However, glycosides may be hydrolyzed to the hemiacetal (and aldehyde) forms by treatment with aqueous acid. This reaction is simply the reverse of glycoside formation.

methyl β-D-glucopyranoside   + H₂O ⇌   D-glucopyranose

Disaccharides and polysaccharides are glycosides; we will discuss these compounds later in this chapter. Other types of glycosides are also common in plants and animals. *Amygdalin* and *laetrile* (Section 11.9) are glycosides found in the kernels of apricot pits and bitter almonds. *Vanillin* (used as vanilla flavoring) is another example of a structure found in nature as a glycoside, in this case, as a β-D-glucoside. In these types of glycosides, the nonsugar portion of the structure is called an **aglycone**. Vanillin is the aglycone in the following example.

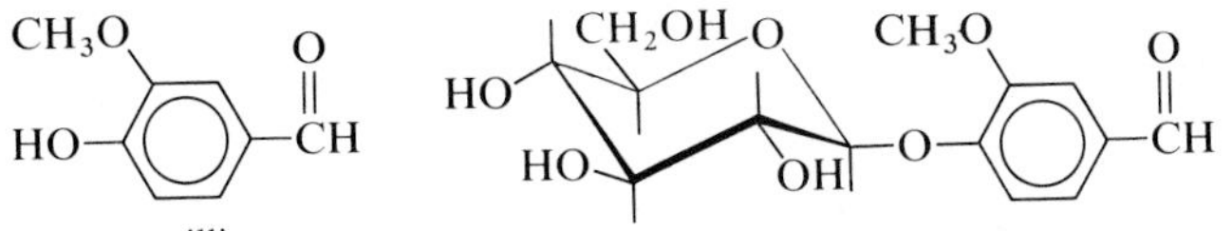

vanillin

vanillin β-D-glucoside
(glucovanillin)

SECTION 18.6.

Oxidation of Monosaccharides

An aldehyde group is very easily oxidized to a carboxyl group. Chemical tests for aldehydes depend upon this ease of oxidation (Section 11.15). Sugars that can be oxidized by such mild oxidizing agents as Tollens reagent, an alkaline solution of $Ag(NH_3)_2^+$, are called **reducing sugars** (because the inorganic oxidizing agent is *reduced* in the reaction). The cyclic hemiacetal forms of all aldoses are readily oxidized because they are in equilibrium with the open-chain aldehyde form.

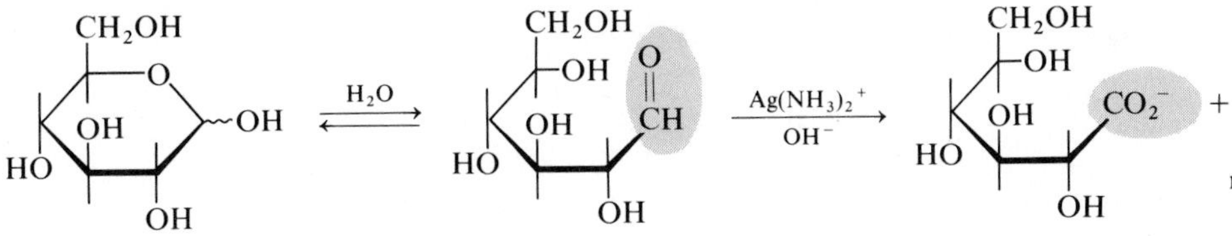

D-glucopyranose

a reducing sugar

Although fructose is a ketone, it is also a reducing sugar.

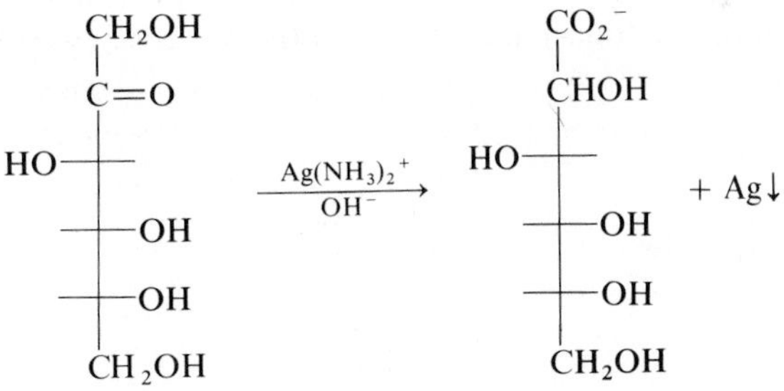

D-fructose

a reducing sugar

The reason that fructose can be oxidized so readily is that, in alkaline solution, fructose is in equilibrium with the aldehyde through an enediol tautomeric intermediate. (In Section 11.17A, a similar enzymatic isomerization was discussed.)

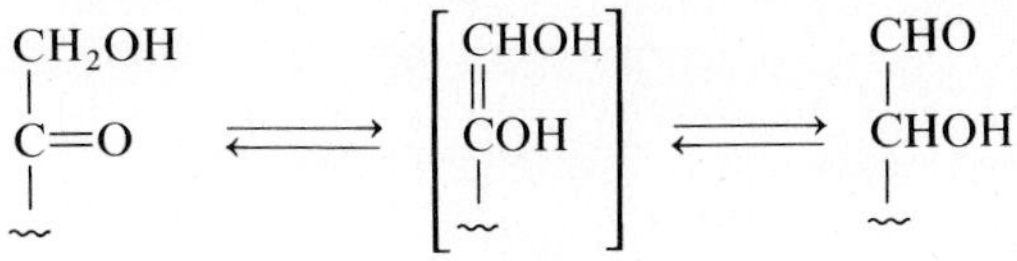

a ketose   *an enediol*   *an aldose*
                    intermediate

In glycosides, the carbonyl group is blocked. Glycosides are **nonreducing sugars**.

a glycoside

A.   Aldonic acids

The product of oxidation of the aldehyde group of an aldose is a polyhydroxy carboxylic acid called an **aldonic acid**. Although Tollens reagent can effect the conversion, a more convenient and less expensive reagent for the synthetic reaction is a buffered solution of bromine.

In alkaline solution, the aldonic acids exist as open-chain carboxylate ions. Upon acidification, they quickly form lactones (cyclic esters), just as any γ- or δ-hydroxy acid would (Section 13.6). Most aldonic acids have both γ and δ hydroxyl groups, and either a five- or a six-membered ring could be formed. The five-membered rings (γ-lactones) are favored in these cases.

Fischer projection of lactone

STUDY PROBLEM

18.11. Predict the product (if any) of bromine oxidation of each of the following compounds:

(a)

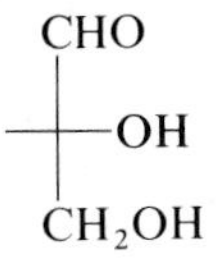

$$\underset{\text{CH}_2\text{OH}}{\overset{\text{CHO}}{\rule{0pt}{0pt}}}\!\!-\!\text{OH}$$

(b)

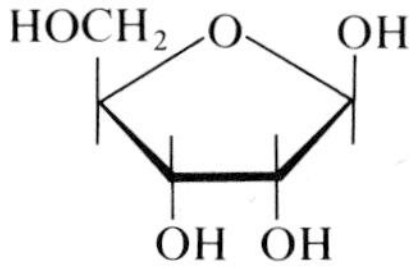

(c)

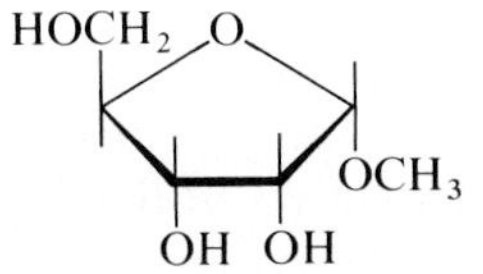

(d)

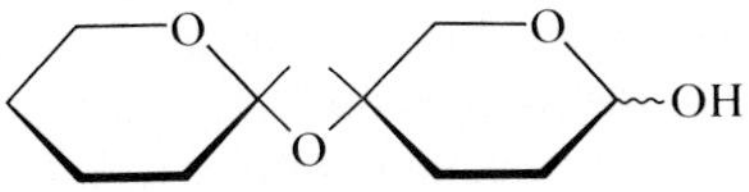

B.   Aldaric acids

Vigorous oxidizing agents oxidize the aldehyde group and also the terminal hydroxyl group (a primary alcohol) of a monosaccharide. The product is a polyhydroxy dicarboxylic acid called an **aldaric acid**.

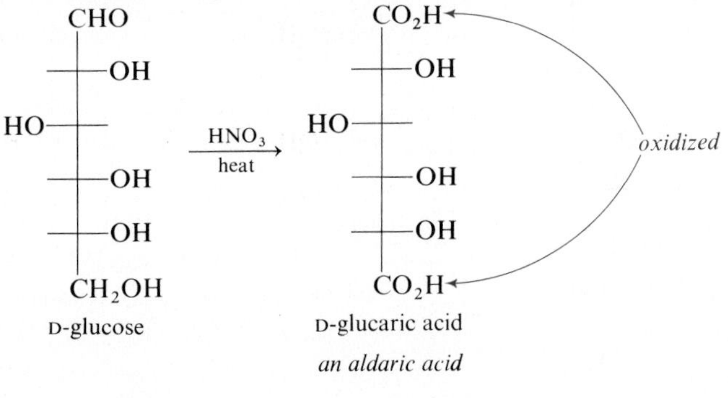

D-glucose $\xrightarrow[\text{heat}]{\text{HNO}_3}$ D-glucaric acid

D-glucose

D-glucaric acid
an aldaric acid

oxidized

The aldaric acids played an important role in the structure elucidation of the sugars (Section 18.9).

STUDY PROBLEM

18.12. Which of the aldohexoses form *meso*-aldaric acids upon oxidation with hot nitric acid?

C.   Uronic acids

Although it is not easy to do in the laboratory, in biological systems the terminal CH_2OH group may be oxidized enzymatically without oxidation of the aldehyde group. The product is called a **uronic acid**.

oxidized CH_2OH

D-glucose

$\xrightarrow[\text{enzymes}]{[O]}$

CO_2H

D-glucuronic acid

a uronic acid

or $CHOH$

CO_2H

Fischer projection

Glucuronic acid is important in animal systems because many toxic substances are excreted in the urine as **glucuronides**, derivatives of this acid. Also, in plant and animal systems, D-glucuronic acid may be converted to L-gulonic acid, which is used to biosynthesize L-ascorbic acid (vitamin C). (This last conversion does not take place in primates or guinea pigs, which require an outside source of vitamin C.) The fact that a compound of the D-series becomes a compound of the L-series is not due to a biochemical change in configuration; rather, the change arises from the change in the numbering of the carbons, as may be seen in the following equation.

carbon 1 reduced

$CHOH$

D-*series*

CO_2H

D-glucuronic acid

$\xrightarrow{[H]}$

new carbon 1

CH_2OH ← L-*series*

CO_2H

L-gulonic acid

$\xrightarrow{[O]}$

CH_2OH

$C=O$

L-ascorbic acid
(vitamin C)

SECTION 18.7.

Reduction of Monosaccharides

Both aldoses and ketoses can be reduced by carbonyl reducing agents, such as hydrogen and catalyst or a metal hydride, to polyalcohols called **alditols**. The suffix for the name of one of these polyalcohols is **-itol**. The product of reduction of D-glucose is called D-*glucitol*, or *sorbitol*.

D-glucose	D-glucitol (sorbitol)

Natural D-glucitol has been isolated from many fruits (for example, cherries, plums, apples, pears, and mountain ash berries) and from algae and seaweed. Synthetic D-glucitol is used as an artificial sweetener.

SECTION 18.8.

Reactions at the Hydroxyl Groups

The hydroxyl groups in carbohydrates behave in a manner similar to that of other alcohol groups. They may be esterified by either carboxylic acids or inorganic acids, and they may be subjected to ether formation. Carbohydrates may also act as diols and form cyclic acetals or ketals with aldehydes or ketones. These reactions were discussed in Chapters 7 and 11.

A.   Acetate formation

A common reagent for esterification of alcohols is acetic anhydride, with either sodium acetate or pyridine as an alkaline catalyst. If the reaction is carried out below 0°C, the acylation reaction is faster than the α–β anomeric interconversion. Under these conditions, either α- or β-D-glucose yields its corresponding pentacetate. At higher temperatures, a mixture of the α- and β-pentacetates is formed, with the β-pentacetate predominating.

β-D-glucopyranose	penta-O-acetyl-β-D-glucopyranose

B.   Ether formation

Treatment of an aldose, such as glucose, with methanol yields a methyl glycoside.

D-glucopyranose	methyl D-glucopyranoside

The other hydroxyl groups in a carbohydrate can be converted to methoxyl groups by reaction with dimethyl sulfate and NaOH.

methyl β-D-glucopyranoside   methyl tetra-O-methyl-β-D-glucopyranoside

In a typical Williamson ether synthesis ($RO^- + RX \rightarrow ROR + X^-$; Section 7.14B), the alkoxide must be prepared with a stronger base than NaOH. In the case of the carbohydrates, NaOH is a sufficiently strong base to yield alkoxide ions. (The inductive effect of the electronegative oxygens on adjacent carbons renders each hydroxyl group more acidic than a hydroxyl group in an ordinary alcohol.) Because the acetal bond is stable in base, the configuration at the anomeric carbon of a glycoside is not changed in this methylation reaction.

STUDY PROBLEM

18.13. Give the structure of the product of the treatment of methyl 2-deoxy-α-D-ribofuranoside with: (a) acetic anhydride;   (b) an alkaline solution of dimethyl sulfate.

C.   Cyclic acetal and ketal formation

Because carbohydrates contain numerous OH groups, it is sometimes desirable to block some of them so that selective reactions can be carried out on the other hydroxyl groups. Acetals and ketals are two common blocking groups (see Section 11.8). For example, an aldehyde, such as benzaldehyde, reacts with 1,3-diol groupings in sugar molecules. Other aldehydes and ketones can react preferentially at different diol groupings. (In some cases, the product is the furanose, rather than the pyranose, ring.) The different products arise because of subtle (generally unpredictable) steric and electronic effects.

In the commercial conversion of L-sorbose to vitamin C, acetone is used to block four hydroxyl groups so that a single CH_2OH group can be oxidized. This conversion is outlined in Figure 18.4.

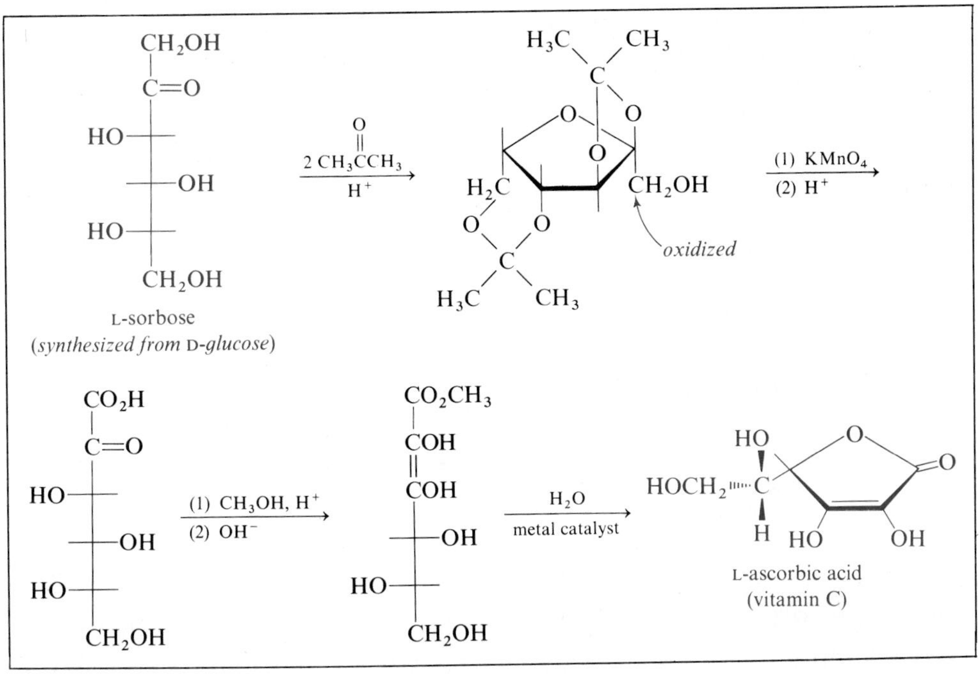

FIGURE 18.4. The conversion of L-sorbose to L-ascorbic acid (vitamin C).

SECTION 18.9.

The Structure Determination of Glucose

In 1888, it was known that glucose is an aldohexose. The question was, "Which of the 16 possible stereoisomeric aldohexoses is it?" In 1891, the German chemist Emil Fischer reported the structure of the open-chain aldehyde form of D-glucose; for this work, he received the Nobel Prize in 1902.

The determination of configuration of a compound with four chiral carbons might seem an overwhelming task, but Fischer accomplished this task in a series of simple reactions. From his data, it was possible to determine only the *relative configuration* of glucose, not the absolute configuration, which had to wait another 50 years for x-ray diffraction. Therefore, Fischer made the assumption that the OH on carbon 2 in D-(+)-glyceraldehyde and thus the OH on carbon 5 in D-(+)-glucose are projected to the right in the Fischer projections. (It turned out later that his assumption was correct; see Section 4.8.) This assumption narrowed the choices for the configuration of glucose to the eight D-aldohexoses shown in Figure 18.3, page 813.

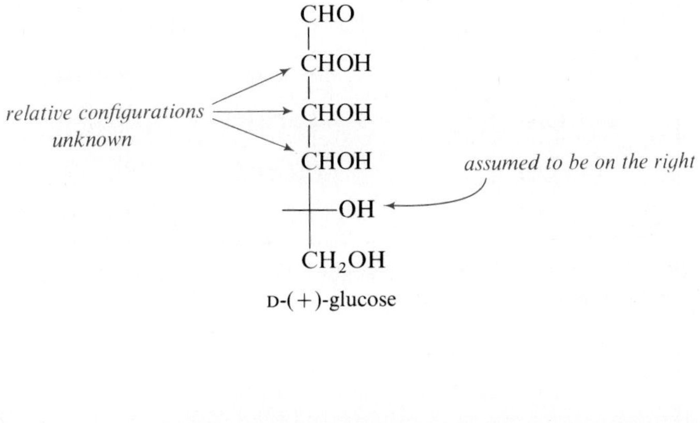

Fact 1. It was known that the aldopentose (−)-arabinose could be converted to the aldohexoses (+)-glucose and (+)-mannose. Heinrich Kiliani discovered the chain-lengthening step in 1886, and Fischer completed the synthesis by reduction of the resulting lactone to yield the aldohexoses in 1890. The following sequence is consequently known as the **Kiliani–Fischer synthesis**.

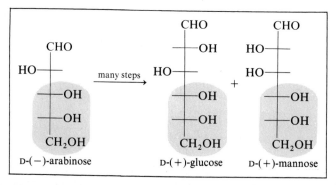

Because (−)-arabinose yields both (+)-glucose and (+)-mannose, all three of these sugars have *the same configuration at the last three chiral carbons* (carbons 3, 4, and 5 of glucose and mannose). (+)-Glucose and (+)-mannose must differ *only* by the configuration at carbon 2. These conclusions are pictured in Figure 18.5 with the now-known structures.

Fact 2: Fischer found that the oxidation of both end groups of (−)-arabinose yielded an optically active diacid, and not the *meso*-diacid.

FIGURE 18.5. D-(+)-Glucose and D-(+)-mannose have the same configuration at the last three chiral carbons.

$$
\begin{array}{c}
\text{CHO} \\
|\\
\text{CHOH} \\
|\\
\text{CHOH} \\
\underset{|}{\overset{|}{\text{—OH}}} \\
\text{CH}_2\text{OH}
\end{array}
\quad\xrightarrow[\text{heat}]{\text{HNO}_3}\quad
\begin{array}{c}
\text{CO}_2\text{H} \\
|\\
\text{CHOH} \\
|\\
\text{CHOH} \\
\underset{|}{\overset{|}{\text{—OH}}} \\
\text{CO}_2\text{H}
\end{array}
$$

(−)-arabinose *not meso*

Therefore, Fischer concluded that carbon 2 in (−)-arabinose must have OH on the *left*. If it were on the right, the *meso*-diacid would result.

left →

$$
\begin{array}{c}
\text{CO}_2\text{H} \\
\text{HO}\!-\!| \\
\text{CHOH} \\
|\!-\!\text{OH} \\
\text{CO}_2\text{H}
\end{array}
\qquad
\begin{array}{c}
\text{CO}_2\text{H} \\
|\!-\!\text{OH} \\
\text{CHOH} \\
|\!-\!\text{OH} \\
\text{CO}_2\text{H}
\end{array}
$$

this diacid would be meso, regardless of configuration at carbon 3

only possibility   *not this*

From the data presented so far, it is possible to write almost-complete structures for (−)-arabinose, (+)-glucose, and (+)-mannose. Figure 18.6 shows these data with the known structures.

$$
\begin{array}{c}
\text{CHO} \\
\text{HO}\!-\!| \\
\text{CHOH} \\
|\!-\!\text{OH} \\
\text{CH}_2\text{OH}
\end{array}
\qquad
\begin{array}{c}
\text{CHO} \\
|\!-\!\text{OH} \\
\text{HO}\!-\!| \\
\text{CHOH} \\
|\!-\!\text{OH} \\
\text{CH}_2\text{OH}
\end{array}
\qquad
\begin{array}{c}
\text{CHO} \\
\text{HO}\!-\!| \\
\text{HO}\!-\!| \\
\text{CHOH} \\
|\!-\!\text{OH} \\
\text{CH}_2\text{OH}
\end{array}
$$

D-(−)-arabinose

one is D-(+)-glucose;
one is D-(+)-mannose

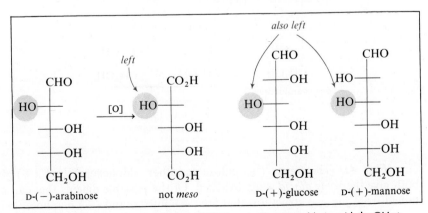

FIGURE 18.6. Because D-(−)-arabinose yields an optically active aldaric acid, the OH at carbon 2 must be projected on the left.

Fact 3: Fischer observed that both (+)-glucose and (+)-mannose are oxidized to optically active diacids. This means that the OH at carbon 4 of both of these monosaccharides is *on the right*. (If it were on the left, one of the two sugars would yield a *meso*-diacid.) Figure 18.7 shows these reactions with the known structures.

```
  CHO                                    CO2H
   |                                      |
  CHOH                                   CHOH
              right                                  not meso,
HO─┼─        ──╮      HNO3      HO─┼─               regardless of
              │      ─────→                   ←─    configuration
   ─┼─OH     ◀─        heat        ─┼─OH            at carbon 2
   ─┼─OH                            ─┼─OH
  CH2OH                             CO2H
```

If OH at carbon 4 were on the left:

```
  CHO                                    CO2H
   ─┼─OH                                  ─┼─OH
 HO─┼─                HNO3       HO─┼─
                     ─────→                   ------- plane of
 HO─┼─                heat       HO─┼─                symmetry
   ─┼─OH                            ─┼─OH
  CH2OH                             CO2H
                                    meso
```

Now a complete structure for (−)-arabinose may be written—all that is needed is to differentiate between (+)-glucose and (+)-mannose.

```
                              CHO            CHO
        CHO                    ─┼─OH      HO─┼─
     HO─┼─              HO─┼─           HO─┼─
        ─┼─OH              ─┼─OH           ─┼─OH
        ─┼─OH              ─┼─OH           ─┼─OH
       CH2OH             CH2OH            CH2OH
   D-(−)-arabinose   ╰─────────────┬─────────────╯
                          one is D-(+)-glucose;
                          one is D-(+)-mannose
```

Fact 4: The sugar (+)-gulose (another aldohexose) and (+)-glucose both give the *same diacid* when oxidized. Of the possible diacids that can be obtained from the two structures that are (+)-glucose and (+)-mannose, *only one* can come from two *different* sugars.

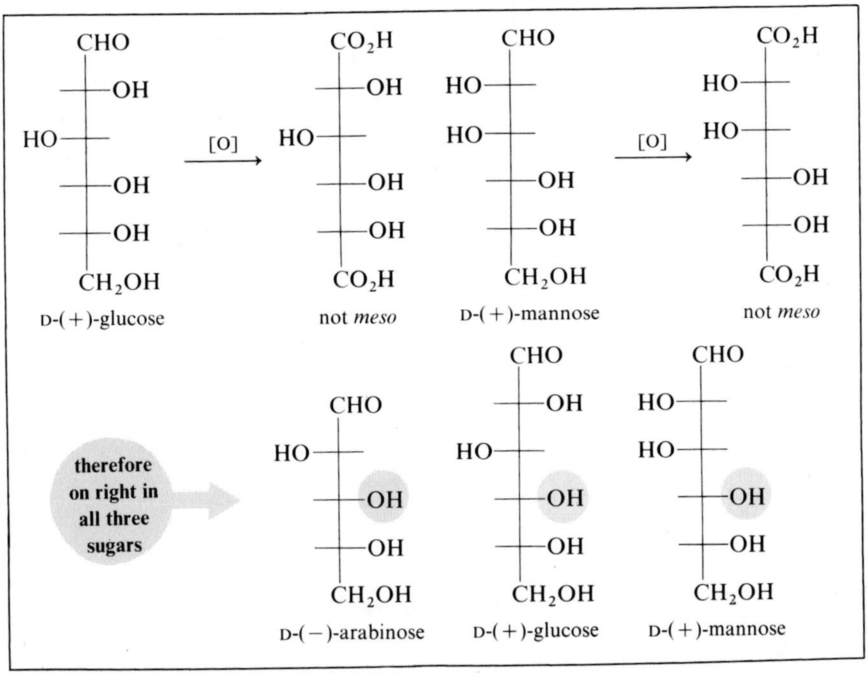

FIGURE 18.7. Because both D-(+)-glucose and D-(+)-mannose yield optically active aldaric acids, the 4-hydroxyl must be on the right.

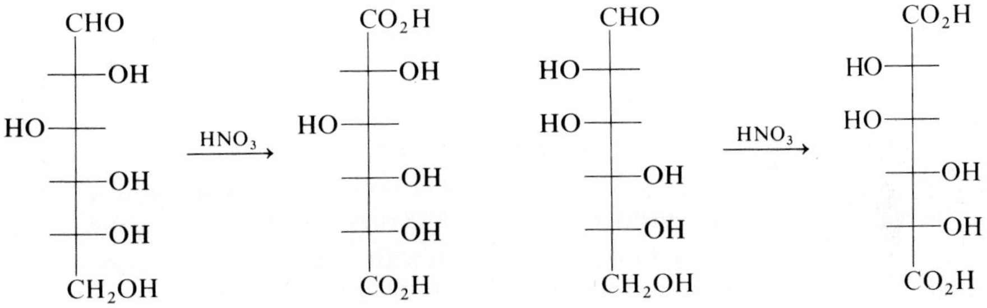

I

can arise from two aldohexoses

II

can arise from only one aldohexose

Let us look at diacid II first. The two potential aldohexoses that can lead to diacid II follow:

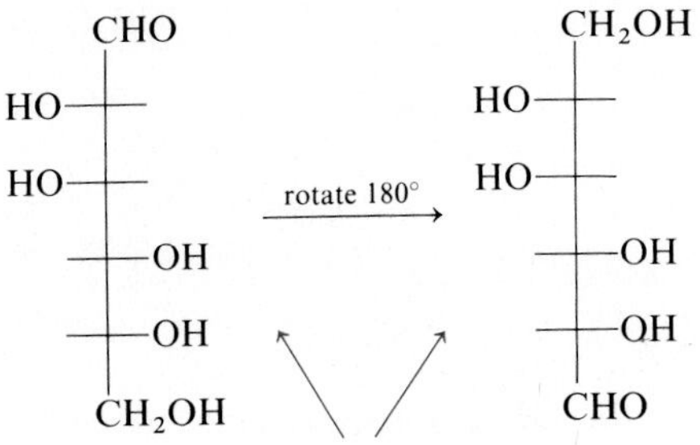

both are D-(+)-*mannose*

But these two aldohexoses are the *same*. If we rotate either projection 180° in the plane of the paper, the structure is the same as the other. Only this one aldohexose can yield diacid II. This aldohexose must be (+)-mannose.

Now let us look at diacid I. It can arise from two possible aldohexoses that are *not the same*: D-glucose and L-gulose. (Refer to Figure 18.3, page 813, for the structure of D-gulose.)

D-(+)-glucose   L-(+)-gulose

rotating one by 180°
does not give the other

If these two sugars give the same diacid (that is, diacid I), then the left-hand structure is (+)-glucose, the right-hand structure is (+)-gulose, and the aldohexose that leads to diacid II is (+)-mannose. See Figure 18.8 for these reactions with known structures.

A.   Determination of ring size

The cyclic structures for glucose were postulated in 1895, but it was not shown until 1926 that glucose forms six-membered cyclic hemiacetals and glycosides. The reactions that were used to determine the ring size are, again, not complex.

A methyl glycoside undergoes reaction with dimethyl sulfate to yield a completely methylated structure. In acidic solution, the methylated methyl glycoside can be hydrolyzed and the ring opened. The methoxyl groups, which are ethers, are not affected by this reaction. The hydrolyzed acetal therefore has only one hydroxyl group. Determination of the position of this hydroxyl group is the information needed to know the ring size of the original acetal.

The position of the OH group was determined by vigorous oxidation in which the —CHO group is oxidized to —CO$_2$H and the single OH group is

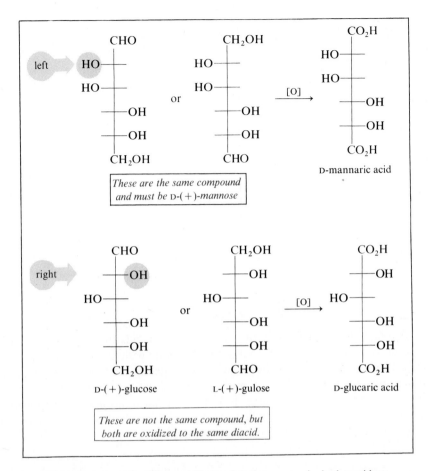

FIGURE 18.8. Because the aldaric acid from D-(+)-glucose can also be formed from L-(+)-gulose, D-(+)-glucose has the OH at carbon 2 on the right. Because the aldaric acid from D-(+)-mannose can come *only* from D-(+)-mannose, this sugar is the one with the OH at carbon 2 on the left.

oxidized to a ketone. Under these oxidizing conditions, cleavage of the molecule occurs adjacent to the ketone group (on either side) to yield two dicarboxylic acids. (See Section 11.17 for a discussion of this type of oxidation.)

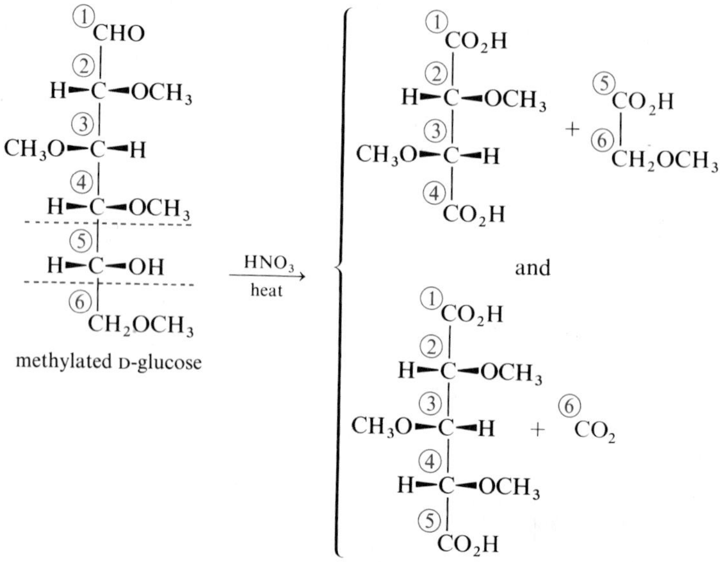

①CHO
②H—C—OCH₃
③CH₃O—C—H
④H—C—OCH₃
⑤H—C—OH
⑥CH₂OCH₃

methylated D-glucose

$\xrightarrow[\text{heat}]{\text{HNO}_3}$

①CO₂H
②H—C—OCH₃
③CH₃O—C—H
④CO₂H

+

⑤CO₂H
⑥CH₂OCH₃

and

①CO₂H
②H—C—OCH₃
③CH₃O—C—H
④H—C—OCH₃
⑤CO₂H

+ ⑥CO₂

From methylated glucose, the two methoxy diacids shown in the preceding equation were observed as principal products (along with methoxyacetic acid and CO_2). The structures of the two diacids were determined by comparing their physical properties with those of methoxy diacids of known structure. Because these product diacids are methoxy derivatives of *four-* and *five-carbon* diacids, it was concluded that carbon 5 holds the OH group in the hydrolyzed glycoside and that the original glycoside ring is a six-membered ring.

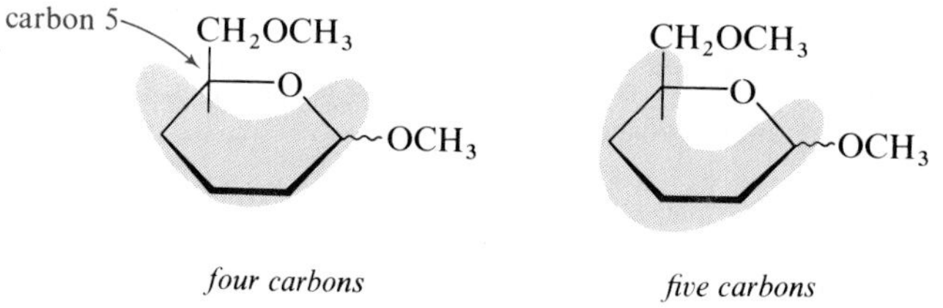

carbon 5

CH₂OCH₃   —O   ~OCH₃

four carbons

CH₂OCH₃   —O   ~OCH₃

five carbons

STUDY PROBLEMS

18.14. What products would be observed from methylation, hydrolysis, and oxidation of a methyl D-glucofuranoside?

18.15. Today, the ring size of a glycoside is determined by HIO_4 oxidation of the glycoside (Section 7.13D). Predict the products of the HIO_4 oxidation of methyl D-glucopyranoside and methyl D-glucofuranoside.

SECTION 18.10.

Disaccharides

A **disaccharide** is a carbohydrate composed of two units of monosaccharide joined together by a glycoside link from carbon 1 of one unit to an OH of the other unit. A common mode of attachment is an α or β glycoside link from the first unit to the 4-hydroxyl group of the second unit. This link is called a 1,4'-α or a 1,4'-β link, depending on the stereochemistry at the glycoside carbon.

a 1,4'-β link
(conformational)

a 1,4'-β link
(Haworth)

Let us look at the preceding structures more closely. Unit 1 (the left-hand unit in each structure) has a β-glycoside link to unit 2. In aqueous solution, this glycoside link is fixed. It is not in equilibrium with the anomer. However, unit 2 (the right-hand unit in each structure) contains a hemiacetal group. In aqueous solution, this particular group is in equilibrium with the open-chain aldehyde form and with the other anomer.

A. Maltose

The disaccharide **maltose** is used in baby foods and malted milk. It is the principal disaccharide obtained from the hydrolysis of starch. Starch is broken down into maltose in an apparently random fashion by an enzyme in saliva called *α-1,4-glucan 4-glucanohydrolase*. The enzyme *α-1,4-glucan maltohydrolase*, found in sprouted barley (*malt*), converts starch specifically into maltose units. In beer-making, malt is used for the conversion of starches from corn or other sources into maltose. An enzyme in yeast (*α-glucosidase*) catalyzes the hydrolysis of the maltose into D-glucose, which is acted upon by other enzymes from the yeast to yield ethanol. One molecule of maltose yields two molecules of D-glucose, regardless of whether the hydrolysis takes place in a laboratory flask, in an organism, or in a fermentation vat.

$$\text{starch} \xrightarrow[\text{H}^+ \text{ or enzymes}]{\text{H}_2\text{O}} \text{maltose} \xrightarrow[\text{H}^+ \text{ or enzymes}]{\text{H}_2\text{O}} \text{D-glucose} \xrightarrow[\text{enzymes}]{} \text{CH}_3\text{CH}_2\text{OH}$$
$$\text{ethanol}$$

A molecule of maltose contains two units of D-glucopyranose. The first unit (shown on the left) is in the form of an α-glycoside. This unit is attached to the oxygen at carbon 4′ in the second unit by a 1,4′-α link.

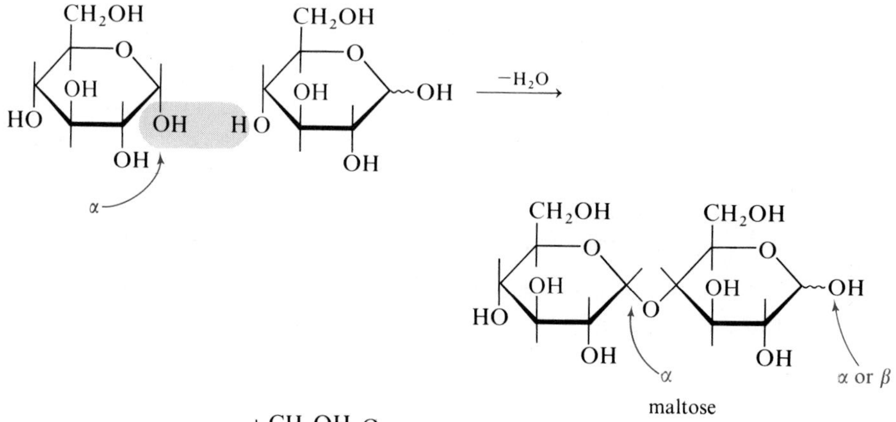

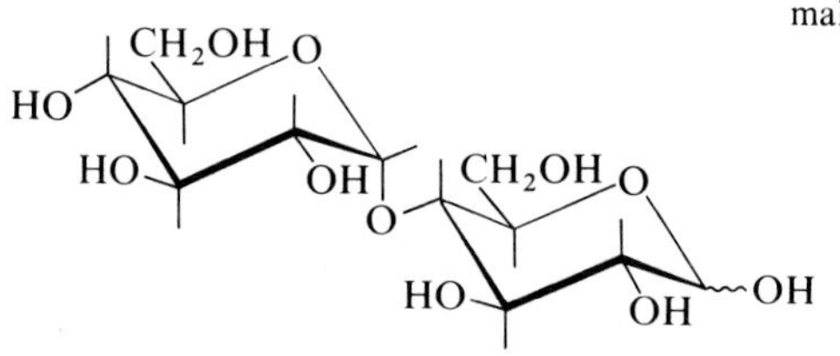

conformational formula for maltose
4-O-(α-D-glucopyranosyl)-D-glucopyranose

The anomeric carbon of the second unit of glucopyranose in maltose is part of a hemiacetal group. As a result, there are two forms of maltose (α- and β-maltose), which are in equilibrium with each other in solution. Maltose undergoes mutarotation, is a reducing sugar, and can be oxidized to the carboxylic acid **maltobionic acid** by a bromine–water solution.

STUDY PROBLEM

18.16. Give the structures of the products:

(a) α-maltose $\xrightarrow{\text{H}_2\text{O, H}^+}$ (b) β-maltose $\xrightarrow{\text{Br}_2, \text{H}_2\text{O}}$

B.   Cellobiose

The disaccharide obtained from the partial hydrolysis of cellulose is called **cellobiose**. Like maltose, cellobiose is composed of two glucopyranose units joined together by a 1,4′-link. Cellobiose differs from maltose in that the 1,4-linkage is β rather than α.

cellobiose
4-*O*-(β-D-glucopyranosyl)-D-glucopyranose

Chemical hydrolysis of cellobiose with aqueous acid yields a mixture of α- and β-D-glucose, the same products that are obtained from maltose. Cellobiose can also be hydrolyzed with the enzyme β-*glucosidase* (also called *emulsin*), but not by α-*glucosidase*, which is specific for the α link (that is, maltose).

STUDY PROBLEMS

18.17. Give the structures of the products:

(a) α-cellobiose $\xrightarrow{\text{H}_2\text{O, H}^+}$

(b) β-cellobiose $\xrightarrow{\text{H}_2\text{O, H}^+}$

(c) α-cellobiose $\xrightarrow{\text{Br}_2,\ \text{H}_2\text{O}}$

(d) α-cellobiose $\xrightarrow{\text{Tollens reagent}}$

(e) α-cellobiose $\xrightarrow{\text{β-glucosidase}}$

18.18. Which would you expect to be more stable, β-maltose or β-cellobiose? Why?

C.   Lactose

The disaccharide **lactose** (milk sugar) is different from maltose or cellobiose in that it is composed of two different monosaccharides, D-glucose and D-galactose.

lactose
4-*O*-(β-D-galactopyranosyl)-D-glucopyranose

Lactose is a naturally occurring disaccharide found only in mammals; cow's milk and human milk contain about 5% lactose. Lactose is obtained commercially as a by-product in the manufacture of cheese.

In normal human metabolism, lactose is hydrolyzed enzymatically to D-galactose and D-glucose; then the galactose is converted to glucose, which can undergo metabolism. A condition called **galactosemia** that affects some infants is caused by lack of the enzyme used to convert galactose to glucose. Galactosemia is characterized by high levels of galactose in the blood and urine. Symptoms range from vomiting to mental and physical retardation and sometimes death. Treatment consists of removing milk and milk products from the diet. (An artificial milk made from soybeans may be substituted.)

D.   Sucrose

The disaccharide **sucrose** is common table sugar. Sugar cane was grown domestically as early as 6000 B.C. in India. (The words "sugar" and "sucrose" come from the Sanskrit word *sarkara*.) Sucrose was encountered by the soldiers of Alexander the Great, who entered India in 325 B.C. In later centuries, the use of sucrose was spread by the Arabs and the Crusaders. Sugar cane was introduced into the New World by Columbus, who brought some to Santo Domingo in 1493. In the 1700's, it was discovered that certain beets also contain high levels of sucrose. The discovery meant that sugar could be obtained from plants grown in temperate climates as well as from sugar cane grown in the tropics. Today, more sucrose is produced than any other pure organic compound.

Whether it comes from beets or sugar cane, the chemical composition of sucrose is the same: one unit of fructose joined to one unit of glucose. The glycoside link joins carbon 1 of each monosaccharide and is β from fructose and α from glucose. Note the difference between sucrose and the other disaccharides we have discussed: in sucrose, *both* anomeric carbon atoms (not just one) are used in the glycoside link. In sucrose, neither fructose nor glucose has a hemiacetal group; therefore, sucrose in water is not in equilibrium with an aldehyde or keto form. Sucrose does not exhibit mutarotation and is not a reducing sugar.

sucrose
β-D-fructofuranosyl α-D-glucopyranoside

Invert sugar is a mixture of D-glucose and D-fructose obtained by the acidic or enzymatic hydrolysis of sucrose. The enzymes that catalyze the hydrolysis of sucrose, called *invertases*, are specific for the β-D-fructofuranoside link and are found in yeast and in bees. (Honey is primarily invert sugar.) Because of the presence of free fructose (the sweetest sugar), invert sugar is sweeter than sucrose. A synthetic invert sugar called *Isomerose* is prepared by the enzymatic isomerization of glucose in corn syrup. It has commercial use in the preparation of ice cream, soft drinks, and candy.

The name "invert sugar" is derived from inversion in the sign of the specific rotation when sucrose is hydrolyzed. Sucrose has a specific rotation of $+66.5°$, a *positive* rotation. The mixture of products (glucose, $[\alpha] = +52.7°$, and fructose, $[\alpha] = -92.4°$) has a net *negative* rotation.

SECTION 18.11.

Polysaccharides

A **polysaccharide** is a compound in which the molecules contain many units of monosaccharide joined together by glycoside links. Upon complete hydrolysis, a polysaccharide yields monosaccharides.

Polysaccharides serve three purposes in living systems: architectural, nutritional, and as specific agents. Typical architectural polysaccharides are *cellulose*, which gives strength to the stems and branches of plants, and *chitin*, the structural component of the exoskeletons of insects. Typical nutritional polysaccharides are *starch* (as is found in wheat and potatoes) and *glycogen*, an animal's internal store of readily available carbohydrate. *Heparin*, an example of a specific agent, is a polysaccharide that prevents blood coagulation.

heparin

Polysaccharides may also be bonded to other types of molecules, as in *glycoproteins* (polysaccharide–protein complexes; see Chapter 19), and *glycolipids* (polysaccharide–lipid complexes; see Chapter 20).

A.   Cellulose

Cellulose is the most abundant organic compound on earth. It has been estimated that about 10^{11} tons of cellulose are biosynthesized each year, and that cellulose accounts for about 50% of the bound carbon on earth! Dry leaves contain 10–20% cellulose; wood, 50%; and cotton, 90%. The most convenient laboratory source of pure cellulose is filter paper.

Cellulose forms the fibrous component of plant cell walls. The rigidity of cellulose arises from its overall structure. Cellulose molecules are chains, or microfibrils, of up to 14,000 units of D-glucose that occur in twisted rope-like bundles held together by hydrogen bonding.

A single molecule of cellulose is a linear polymer of 1,4'-β-D-glucose. Complete hydrolysis in 40% aqueous HCl yields only D-glucose. The disaccharide isolated from partially hydrolyzed cellulose is cellobiose, which can be further hydrolyzed to D-glucose with an acidic catalyst or with the enzyme emulsin.

Cellulose itself has no hemiacetal carbon—it cannot undergo mutarotation or be oxidized by such test reagents as Tollens reagent. (There may be a hemiacetal at one end of each cellulose molecule; however, this is but a small portion of the whole and does not lead to observable reaction.)

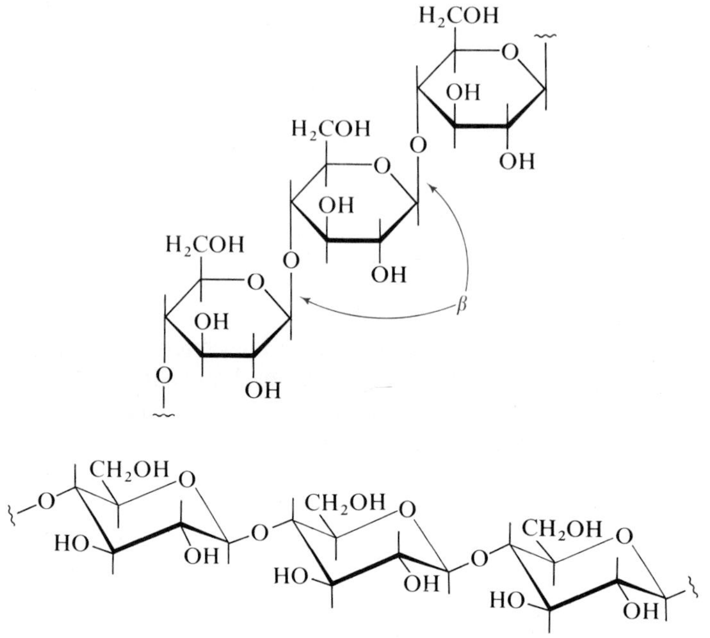

cellulose

STUDY PROBLEM

18.19. Predict the major product (and give a structure or partial structure) when cellulose is treated with: (a) an excess of hot aqueous H_2SO_4, then water; (b) hot water; (c) hot aqueous NaOH solution; (d) an excess of NaOH and dimethyl sulfate.

Although mammals do not produce the proper enzymes for breaking down cellulose into glucose, certain bacteria and protozoa do have these enzymes. Grazing animals are capable of using cellulose as food only indirectly. Their stomachs and intestines support colonies of microorganisms that live and reproduce on cellulose; the animal uses these microorganisms and their by-products as food.

B.   Starch

Starch is the second most abundant polysaccharide. Starch can be separated into two principal fractions based upon solubility when triturated (pulverized) with hot water: about 20% of starch is **amylose** (soluble) and the remaining 80% is **amylopectin** (insoluble).

Amylose. Complete hydrolysis of amylose yields only D-glucose; partial hydrolysis yields maltose as the only disaccharide. We conclude that amylose is a linear polymer of 1,4′-linked α-D-glucose. The difference between amylose and cellulose is the glycoside link: β in cellulose, α in amylose. This difference is responsible for the different properties of these two polysaccharides.

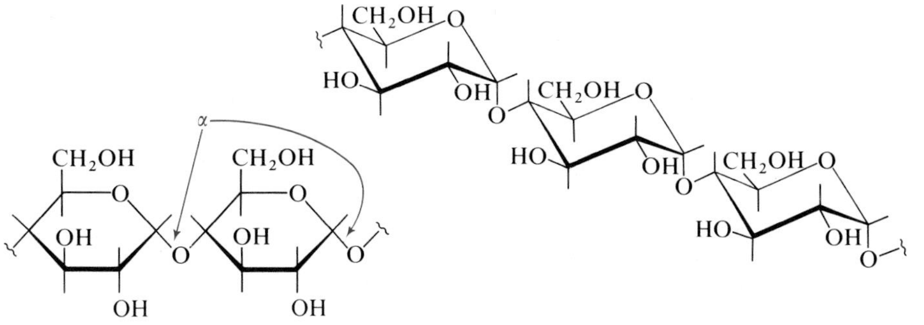

amylose

There are 250 or more glucose units per amylose molecule; the exact number depends upon the species of animal or plant. (Measurement of chain length is complicated by the fact that natural amylose degrades into smaller chains upon separation and purification.)

Amylose molecules form helices or coils around I_2 molecules; a deep blue color arises from electronic interactions between the two. This color is the basis of the **iodine test for starch**, in which a solution of iodine is added to an unknown as a test for the presence of starch.

Amylopectin, a much larger polysaccharide than amylose, contains 1000 or more glucose units per molecule. Like the chain in amylose, the main chain of amylopectin contains 1,4′-α-D-glucose. Unlike amylose, amylopectin is *branched* so that there is a terminal glucose about every 25 glucose units (Figure 18.9). The bonding at the branch point is a 1,6′-α-glycosidic bond.

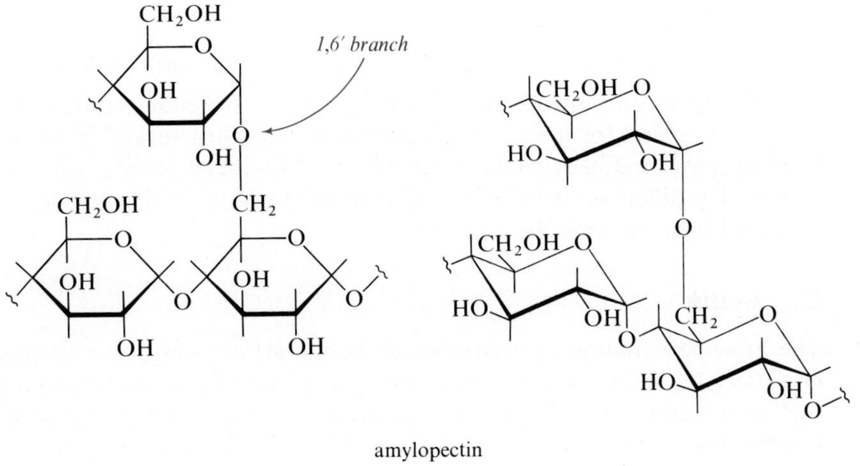

amylopectin

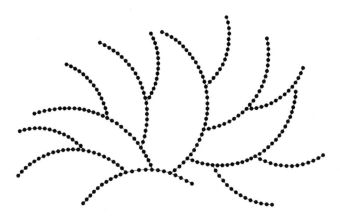

FIGURE 18.9. A representation of the branched structure of amylopectin. Each ——●— represents a glucose molecule.

Complete hydrolysis of amylopectin yields only D-glucose. However, incomplete hydrolysis yields a mixture of the disaccharides maltose and isomaltose, the latter arising from the 1,6'-branching. The oligosaccharide mixture obtained from the partial hydrolysis of amylopectin, which is referred to as **dextrins**, is used to make glue, paste, and fabric sizing.

$$\text{amylopectin} \xrightarrow{\text{H}_2\text{O}} \text{dextrins} \xrightarrow{\text{H}_2\text{O}} \text{maltose + isomaltose} \xrightarrow{\text{H}_2\text{O}} \text{D-glucose}$$

isomaltose
6-*O*-(α-D-glucopyranosyl)-D-glucopyranose

Glycogen is a polysaccharide that is used as a storehouse (primarily in the liver and muscles) for glucose in an animal system. Structurally, glycogen is related to amylopectin. It contains chains of 1,4'-α-linked glucose with branches (1,6'-α). The difference between glycogen and amylopectin is that glycogen is more branched than amylopectin.

C.   Chitin

The principal structural polysaccharide of the arthropods (for example, crabs and insects) is **chitin**. It has been estimated that 10^9 tons of chitin are biosynthesized each year! Chitin is a linear polysaccharide consisting of β-linked *N*-acetyl-D-glucosamine. Upon hydrolysis, chitin yields 2-amino-2-deoxy-D-glucose. (The acetyl group is lost in the hydrolysis step.) In nature, chitins are bonded to non-polysaccharide material (proteins and lipids).

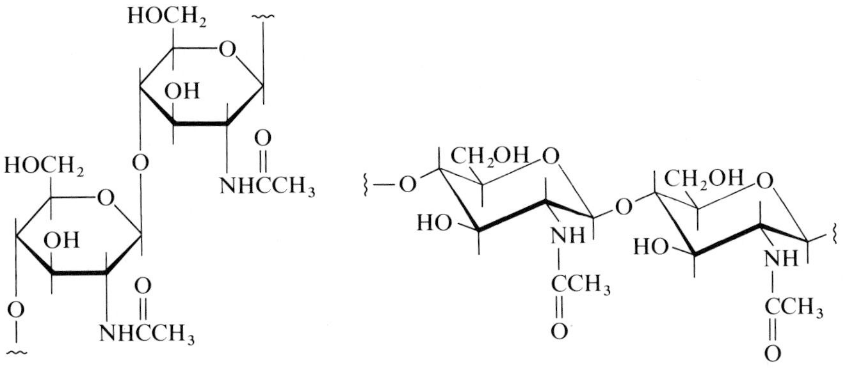

chitin

STUDY PROBLEM

18.20. Give the structures for the major organic products when chitin is treated with:
(a) hot dilute aqueous HCl;   (b) hot dilute aqueous NaOH.

Summary

Carbohydrates are polyhydroxy aldehydes and ketones or their derivatives. A **monosaccharide** is the smallest carbohydrate; it does not undergo hydrolysis to smaller units. Monosaccharides may be classified as to the number of carbons and to the principal functional group:

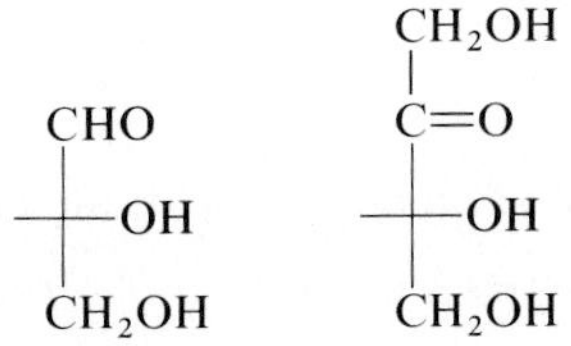

an aldotriose   *a tetrulose*

Natural monosaccharides generally belong to the D-**series**. Because of the presence of both hydroxyl and carbonyl groups, monosaccharides that can form **furanose** or **pyranose** hemiacetal or hemiketal rings undergo cyclization. The cyclization creates a new chiral carbon and therefore gives rise to a pair of diastereomers called α **and** β **anomers**. In solution, the anomers are in equilibrium with each other.

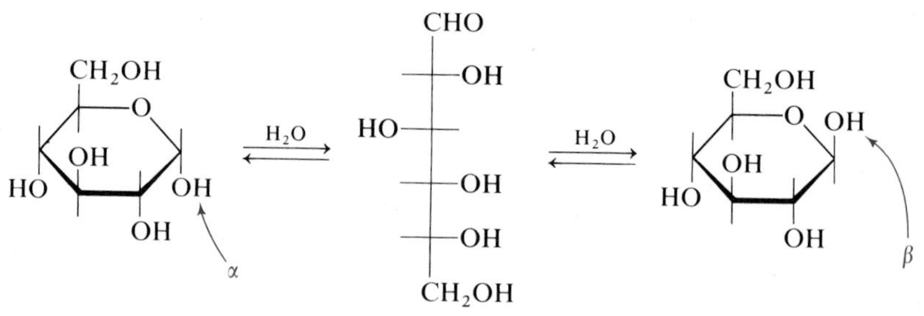

Because of the equilibrium, a monosaccharide undergoes reactions typical of aldehydes.

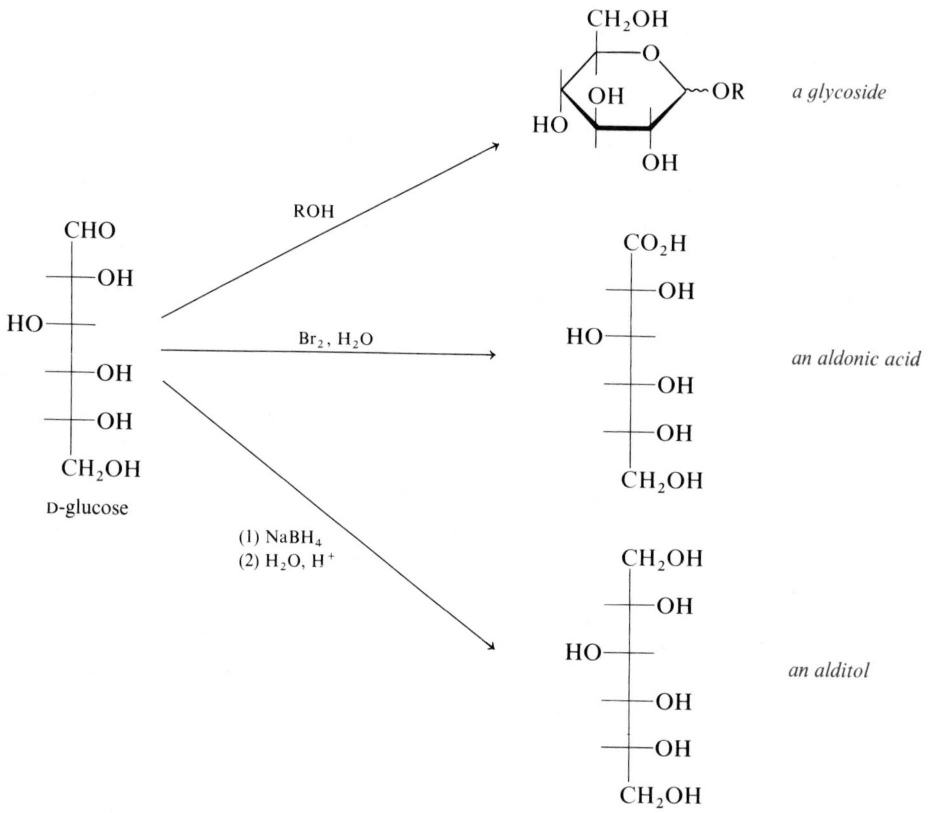

ROH → *a glycoside*

Br$_2$, H$_2$O → *an aldonic acid*

(1) NaBH$_4$
(2) H$_2$O, H$^+$ → *an alditol*

D-glucose

Vigorous oxidation of an aldose yields an **aldaric acid**.

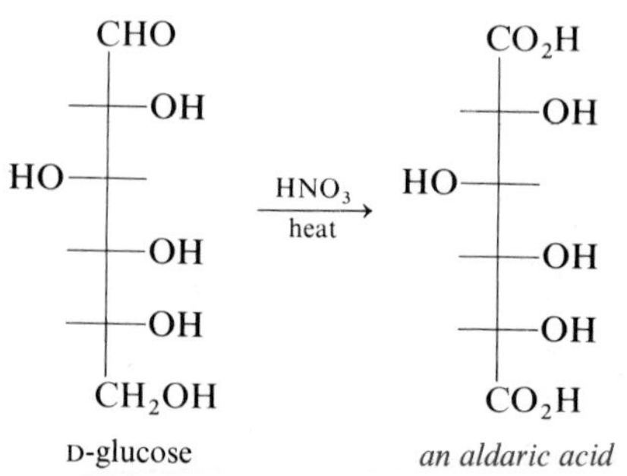

D-glucose $\xrightarrow[\text{heat}]{\text{HNO}_3}$ *an aldaric acid*

The hydroxyl groups of a glycoside may be acetylated, methylated, or converted to cyclic acetals or ketals.

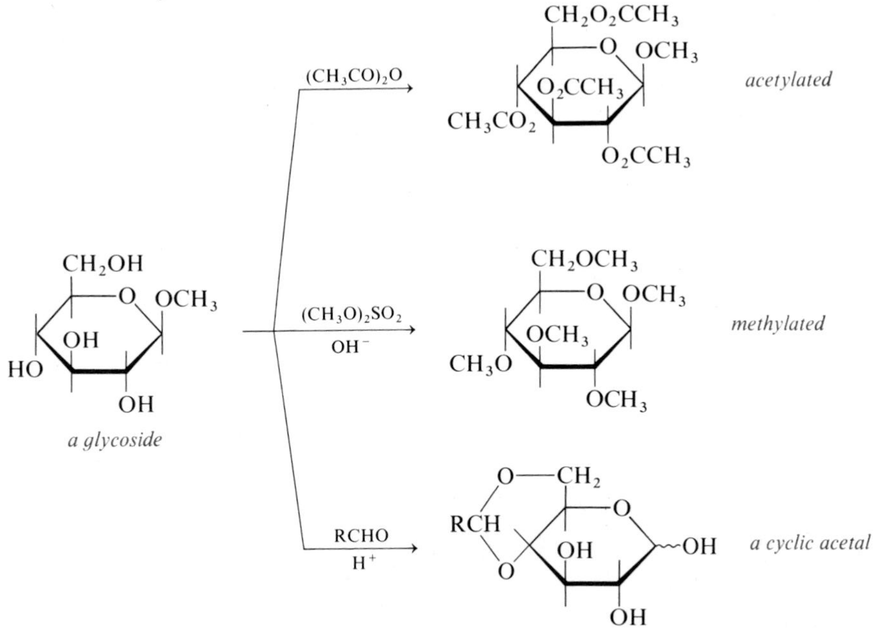

a glycoside

acetylated

methylated

a cyclic acetal

Disaccharides are composed of two units of monosaccharide joined by a glycoside link from one unit to an OH group of the second unit. Acid hydrolysis of a disaccharide yields the two monosaccharides. **Maltose** is composed of two D-glucopyranose units joined by a $1,4'$-α link. **Cellobiose** is composed of two D-glucopyranose units joined by a $1,4'$-β link. **Lactose** is composed of β-D-galactopyranose joined to the 4-position of D-glucopyranose. **Sucrose** is composed of α-D-glucopyranose and β-D-fructofuranose joined by a $1,1'$ link.

A **polysaccharide** is composed of many units of monosaccharide joined by glycosidic links:

cellulose:	$1,4'$-β-D-glucopyranose
amylose:	$1,4'$-α-D-glucopyranose
amylopectin:	$1,4'$-α-D-glucopyranose with $1,6'$-α-branching

STUDY PROBLEMS

18.21. Match each of the following classes of compound with a structure on the right:

(a)   a hexulose

(1)

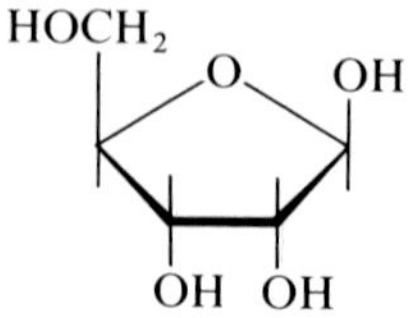

HOCH₂

(b)   a pentopyranose

(2)

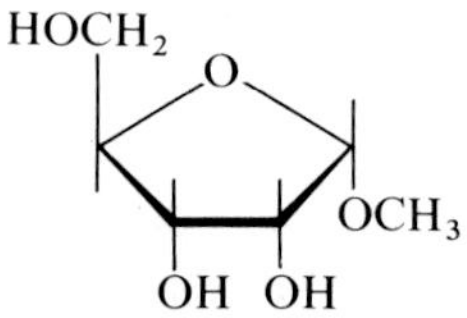

(c)   a pentofuranose

(3)

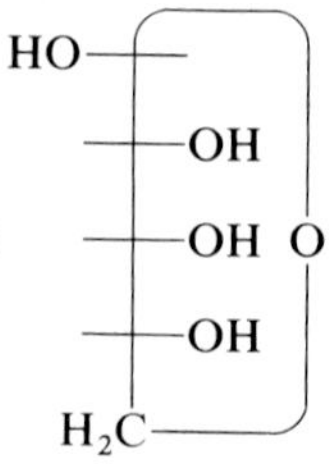

(d)   a pentofuranoside

(4)

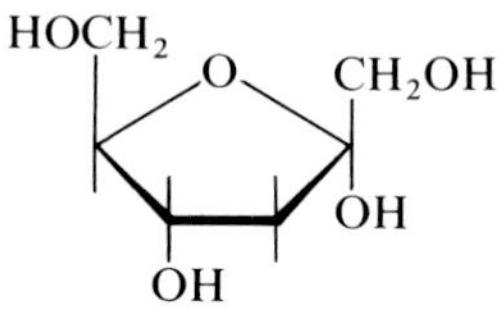

18.22. Label each of the structures in Problem 18.21 as D or L.

18.23. Give structures for (and label as α or β) the anomers of compounds (1) and (4) in Problem 18.21.

18.24. Draw the Fischer projections for the open-chain forms of all the isomeric D-hexuloses and indicate the chiral carbons in each.

18.25. (a)   When (R)-2-hydroxypropanal is treated with HCN, what products are obtained?
         (b)   Are those products optically active? Explain.

18.26. A chemist has a single stereoisomer of 1,2,3,4-butanetetraol of unknown configuration. He also has a flask of D-glyceraldehyde. Suggest a reaction sequence that could be used to relate the configuration of the tetraol to that of D-glyceraldehyde.

18.27. Draw Haworth projections for the following monosaccharides:

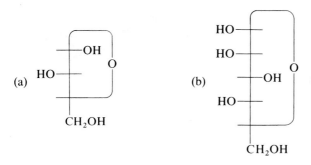

(a)

(b)

(c)

(d) β-D-altropyranose

18.28. Draw Fischer projections for the following monosaccharides:

(a)

(b)

(c)

(d) α-D-lyxofuranose

18.29. Write equations (using Haworth formulas) that illustrate:

(a) the mutarotation of pure β-D-arabinofuranose in water
(b) the conversion of β-D-fructofuranose to β-D-fructopyranose
(c) the mutarotation of β-maltose

18.30. Which of the following compounds would *not* undergo mutarotation? Why?

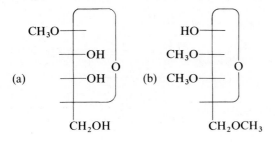

(a)

(b)

(c)

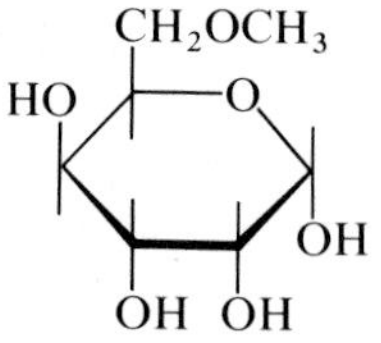

(d)

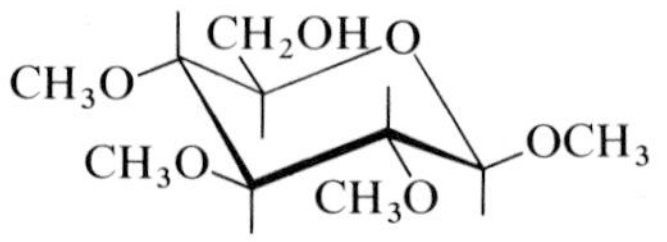

(e)

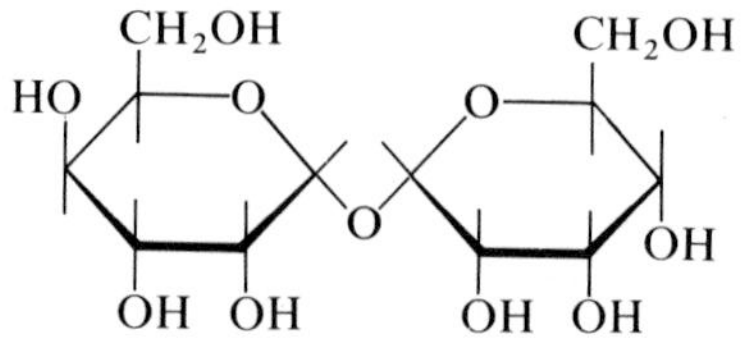

(f) 6-O-(α-D-galactopyranosyl)-β-D-glucopyranose

(g) α-D-glucopyranosyl α-D-glucopyranoside

18.31. Which of the compounds in Problem 18.30 are nonreducing sugars?

18.32. What would you expect to be the most stable conformation for: (a) β-D-galactopyranose, and   (b) α-D-idopyranose? Give structures and reasons for your answers.

18.33. Give the Haworth formulas for the major organic products:

(a)   D-glucose + $(CH_3)_2CHOH$ $\xrightarrow{\text{H}^+}$

(b)   D-galactose + CH_3CH_2OH $\xrightarrow{\text{H}^+}$

(c)   methyl α-D-ribofuranoside $\xrightarrow[\text{heat}]{\text{H}_2\text{O, H}^+}$

18.34. The oxidation of D-fructose with Tollens reagent yields a mixture of anions of D-mannonic acid and D-gluconic acid. Explain.

18.35. *Algin* is a polysaccharide obtained from seaweed. It is hydrolyzed to D-mannuronic acid. What is the structure of this acid? (Use a Fischer projection.)

18.36. Give the structure and the name of the organic product obtained when D-galactose is treated with: (a) aqueous Br_2;   (b) hot HNO_3;   (c) Tollens reagent.

18.37. Suggest a synthesis of 2,3-di-O-methylglucose from glucose. (Ignore the stereochemistry at the anomeric carbon.)

18.38. Fill in the blanks (give *all* possibilities):

(a) _____ $\xrightarrow{\text{hot HNO}_3}$ *meso*-tartaric acid
(a D-tetrose)

(b) _____ $\xrightarrow[\text{cold}]{\text{dil. HCl}}$

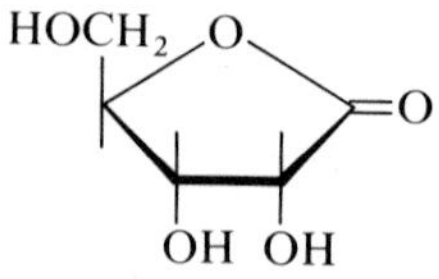

(c) _____ $\xrightarrow[\text{(2) H}_2\text{O, H}^+]{\text{(1) NaBH}_4}$ a *meso*-alditol
(a D-aldohexose)

18.39. Predict the major organic product for the reaction of D-mannose with: (a) $Br_2 + H_2O$; (b) HNO_3; (c) ethanol + H^+; (d) [the product from (c)] plus dimethyl sulfate and NaOH; (e) [the product from (c)] plus CH_3I and Ag_2O; (f) acetic anhydride; (g) acetyl chloride + pyridine; (h) $NaBH_4$; (i) HCN followed by H_2O and HCl; (j) $LiAlH_4$ followed by water; (k) H_2 and Ni catalyst; (l) [the product from (c)] plus H_2O, HCl.

18.40. Predict the major organic product of the treatment of amylose with dimethyl sulfate and NaOH, followed by hydrolysis in dilute HCl.

18.41. Methyl β-D-gulopyranoside is treated with: (1) dimethyl sulfate plus NaOH; (2) H_2O, H^+; and then (3) hot HNO_3. Write the equations that illustrate the steps in this reaction sequence.

18.42. How would you distinguish chemically between (a) maltose and sucrose? (b) D-lyxose and D-xylose?

18.43. How many possible disaccharides could be formed from just D-glucopyranose? (The α and β links form *different* disaccharides.)

18.44. If a polysaccharide were composed of 1,4'-α-linked D-glucopyranose with 1,3'-α-linked branches, what are the possible disaccharides that would be obtained upon partial hydrolysis?

18.45. *Trehalose* is a nonreducing sugar with the formula $C_{12}H_{22}O_{11}$. Upon hydrolysis, only D-glucose is obtained. What are the possible structures of trehalose?

18.46. A carbohydrate A ($C_{12}H_{22}O_{11}$) was treated with (1) CH_3OH, H^+, and (2) excess CH_3I and Ag_2O. The product B was hydrolyzed to 2,3,4,6-tetra-*O*-methyl-D-galactose and 2,3,6-tri-*O*-methyl-D-glucose. When A was treated with aqueous acid, the products were D-galactose and D-glucose in equal amounts. When A was treated with aqueous Br_2, a carboxylic acid C was isolated. Hydrolysis of C with aqueous HCl resulted in D-gluconic acid as the only acid. What are the structures of A, B, and C?

18.47. *Raffinose* is a trisaccharide found in beets. Complete hydrolysis of raffinose yields D-fructose, D-glucose, and D-galactose. Partial enzymatic hydrolysis of raffinose with invertase yields D-fructose and the disaccharide *melibiose*. Partial hydrolysis of raffinose with an α-glycosidase yields D-galactose and sucrose. Methylation of raffinose followed by hydrolysis yields 2,3,4,6-tetra-*O*-methylgalactose; 2,3,4-tri-*O*-methylglucose; and 1,3,4,6-tetra-*O*-methylfructose. What are the structures of raffinose and melibiose?

CHAPTER 19

Amino Acids and Proteins

Proteins are among the most important compounds in an animal organism. Appropriately, the word protein is derived from the Greek *proteios*, which means "first." Proteins are *polyamides*, and hydrolysis of a protein yields *amino acids*.

$$\{-NHCHC-NHCHC-\} \quad \xrightarrow[\text{heat}]{H_2O, H^+} \quad H_2NCHCO_2H + H_2NCHCO_2H \qquad \text{etc.}$$

with O (double bond) above each C, and R, R' below the left structure and R, R' below the right structures.

a protein   *amino acids*

Only twenty amino acids are commonly found in plant and animal proteins, yet these twenty amino acids can be combined in a variety of ways to form muscles, tendons, skin, fingernails, feathers, silk, hemoglobin, enzymes, antibodies, and many hormones. We will first consider the amino acids and then discuss how their combinations can lead to such diverse products.

SECTION 19.1.

The Structures of Amino Acids

The amino acids found in proteins are **α-aminocarboxylic acids**. Variation in the structures of these monomers occurs in the side chain.

$$\text{α amino group} \longrightarrow \underset{\underset{\displaystyle R}{|}}{\overset{\overset{\displaystyle CO_2H}{|}}{H_2N-C-H}} \quad \begin{array}{l}\textit{variation in structure}\\\textit{occurs in the side chain}\end{array}$$

The simplest amino acid is aminoacetic acid ($H_2NCH_2CO_2H$), called *glycine*, which has no side chain and consequently does not contain a chiral carbon. All other amino acids have side chains, and therefore their α carbons are chiral. Naturally occurring amino acids have an (*S*)-configuration at the α carbon and are said to belong to the L-series—that is, the groups around the α carbon have the same configuration as in L-glyceraldehyde. (It is interesting that *racemic* α-amino acids have been detected in certain carbonaceous meteorites.)

$$(S)\text{-}\textit{configuration}$$

$$\underset{\underset{\displaystyle CH_2OH}{|}}{\overset{\overset{\displaystyle CHO}{|}}{HO-C-H}} \qquad \underset{\underset{\displaystyle R}{|}}{\overset{\overset{\displaystyle CO_2H}{|}}{H_2N-C-H}}$$

L-glyceraldehyde   *an* L-*amino acid*

Table 19.1 contains a complete list of the twenty amino acids commonly found in proteins. Also included in this table are abbreviations for the amino acid names. The use of these abbreviations will be discussed later in this chapter.

A.   Essential amino acids

Most amino acids can be synthesized by an organism from its "pool" of organic compounds. One such mode of synthesis is the conversion of an amino acid that is present in excess to a desired amino acid by a **transamination reaction**. The mechanism for this reaction was presented in Section 11.10.

Circled groups exchanged:

$$\underset{\underset{\displaystyle R}{|}}{\overset{\overset{\displaystyle CO_2H}{|}}{H_2NCH}} \quad + \quad \underset{\underset{\displaystyle R'}{|}}{\overset{\overset{\displaystyle CO_2H}{|}}{C=O}} \quad \xrightarrow[\text{(many steps)}]{\overset{\text{transaminase}}{\text{enzymes}}} \quad \underset{\underset{\displaystyle R}{|}}{\overset{\overset{\displaystyle CO_2H}{|}}{C=O}} \quad + \quad \underset{\underset{\displaystyle R'}{|}}{\overset{\overset{\displaystyle CO_2H}{|}}{H_2NCH}}$$

old amino acid   old keto acid   *new keto acid   new amino acid*

Not all amino acids can be obtained by interconversions from other amino acids or by synthesis from other compounds in the animal system. Amino acids that are required for protein synthesis and cannot be synthesized by the organism must be present in the diet. Such compounds are referred to as **essential amino acids**. Which amino acids are essential depends upon the species of animal and even upon individual differences.

The amino acids usually considered essential for humans are starred in Table 19.1. Of these, tryptophan, phenylalanine, methionine, and histidine can be enzymatically converted from (*R*) to (*S*); thus, a racemic mixture of these amino acids can be totally utilized. The other essential amino acids must be provided in the diet in their (*S*)-configurations to be used in protein biosynthesis.

TABLE 19.1. Amino acids found in proteins

Name	Abbreviation	Structure
alanine	ala	CH_3CHCO_2H $\quad\ \ \ \vert$ $\quad\ \ NH_2$
arginine*	arg	$H_2NCNHCH_2CH_2CH_2CHCO_2H$ $\quad\ \ \Vert \qquad\qquad\qquad\quad \vert$ $\quad\ \ NH \qquad\qquad\qquad NH_2$
asparagine	asn	$\qquad\quad O$ $\qquad\quad \Vert$ $H_2NCCH_2CHCO_2H$ $\qquad\qquad\ \ \vert$ $\qquad\qquad\ NH_2$
aspartic acid	asp	$HO_2CCH_2CHCO_2H$ $\qquad\qquad\ \ \vert$ $\qquad\qquad\ NH_2$
cysteine	cyS	$HSCH_2CHCO_2H$ $\qquad\quad\ \vert$ $\qquad\ \ NH_2$
glutamic acid	glu	$HO_2CCH_2CH_2CHCO_2H$ $\qquad\qquad\qquad \vert$ $\qquad\qquad\quad NH_2$
glutamine	gln	$\qquad\qquad O$ $\qquad\qquad \Vert$ $H_2NCCH_2CH_2CHCO_2H$ $\qquad\qquad\qquad\ \ \vert$ $\qquad\qquad\qquad NH_2$
glycine	gly	CH_2CO_2H $\vert$ NH_2
histidine*	his	—CH_2CHCO_2H $\qquad\qquad\ \ \vert$ $\qquad\qquad\ NH_2$
isoleucine*	ile	$\qquad\quad CH_3$ $\qquad\qquad \vert$ $CH_3CH_2CHCHCO_2H$ $\qquad\qquad\qquad \vert$ $\qquad\qquad\quad NH_2$
leucine*	leu	$(CH_3)_2CHCH_2CHCO_2H$ $\qquad\qquad\qquad\ \ \vert$ $\qquad\qquad\qquad NH_2$
lysine*	lys	$H_2NCH_2CH_2CH_2CH_2CHCO_2H$ $\qquad\qquad\qquad\qquad\qquad \vert$ $\qquad\qquad\qquad\qquad\ NH_2$
methionine*	met	$CH_3SCH_2CH_2CHCO_2H$ $\qquad\qquad\qquad \vert$ $\qquad\qquad\quad NH_2$

TABLE 19.1. (*continued*)

Name	Abbreviation	Structure
phenylalanine*	phe	$\text{C}_6\text{H}_5-\text{CH}_2\text{CHCO}_2\text{H}$ with NH_2
proline	pro	pyrrolidine ring $-\text{CO}_2\text{H}$, N–H
serine	ser	$\text{HOCH}_2\text{CHCO}_2\text{H}$ with NH_2
threonine*	thr	$\text{CH}_3\text{CHCHCO}_2\text{H}$ with OH and NH_2
tryptophan*	try	indole$-\text{CH}_2\text{CHCO}_2\text{H}$ with NH_2
tyrosine	tyr	$\text{HO}-\text{C}_6\text{H}_4-\text{CH}_2\text{CHCO}_2\text{H}$ with NH_2
valine*	val	$(\text{CH}_3)_2\text{CHCHCO}_2\text{H}$ with NH_2

* Essential amino acid.

As an example of individual differences in requirements for amino acids, let us look at *tyrosine*. This amino acid is not considered an essential amino acid because, in most people, tyrosine can be synthesized from phenylalanine. A small percent of individuals who have inherited a condition known as **phenylketonuria (PKU)**, which means "phenyl ketones in the urine," do not have the enzyme necessary for this conversion. The diet of a person with PKU must contain some tyrosine and must be limited in its quantity of phenylalanine.

phenylalanine $\xrightarrow{\text{enzymes}}$ tyrosine

B.   Importance of side-chain structure

How can polymers composed of twenty similar amino acids have such a wide variety of properties? Part of the answer lies in the nature of the side chains in the amino acids. Note in Table 19.1 that some amino acids have side chains that contain carboxyl groups; these are classified as **acidic amino acids**. Amino acids containing side chains with amino groups are classified as **basic amino acids**. These acidic and basic side chains help determine the structure and reactivity of the proteins in which they are found. The rest of the amino acids are classified as **neutral amino acids**. The side chains of neutral amino acids are also important. For example, some of these side chains contain —OH, —SH, or other polar groups that can undergo hydrogen bonding, which we will find is an important feature in overall protein structure.

$$\begin{array}{cccc}
CO_2H & CO_2H & CO_2H & CO_2H \\
| & | & | & | \\
H_2NCH & H_2NCH & H_2NCH & H_2NCH \\
| & | & | & | \\
CH_2CH_2CO_2H & (CH_2)_4NH_2 & CH(CH_3)_2 & CH_2OH
\end{array}$$

glutamic acid	lysine	valine	serine
an acidic amino acid	*a basic amino acid*	*a neutral amino acid with a nonpolar side chain*	*a neutral amino acid with a polar side chain*

The characteristics of a protein are changed if a carboxylic acid group in a side chain is converted to an amide. Note the difference in the side chains of glutamine and glutamic acid.

$$\begin{array}{cc}
CO_2H \qquad \nearrow acidic & CO_2H \qquad \nearrow neutral \\
| & | \qquad O \\
H_2NCH & H_2NCH \qquad \| \\
| & | \\
CH_2CH_2CO_2H & CH_2CH_2CNH_2
\end{array}$$

glutamic acid   glutamine

The thiol side chain in cysteine plays a unique role in protein structure. Thiols can undergo oxidative coupling to yield disulfides (Section 7.17). This coupling between two units of cysteine yields a new amino acid called cystine and provides a means of cross-linking protein chains. Giving hair a "permanent wave" involves breaking some existing S—S cross-links by reduction and then reforming new S—S links in other positions of the protein chains.

first protein chain:

$$\left. \begin{array}{c}
O \\
\| \\
\{-NHCHC-\} \\
| \\
CH_2SH \\
\\
\underset{[H]}{\overset{[O]}{\rightleftarrows}} \\
\\
CH_2SH \\
| \\
\{-NHCHC-\} \\
\| \\
O
\end{array} \right\}$$

$$\begin{array}{c}
O \\
\| \\
\{-NHCHC-\} \\
| \\
CH_2 \\
| \\
S \\
| \\
S \\
| \\
CH_2 \\
| \\
\{-NHCHC-\} \\
\| \\
O
\end{array}$$

cross-link between chains

second protein chain:

The order in which amino acids are found in a protein molecule determines the relationship of the side chains to one another and consequently determines how the protein interacts with itself and with its environment. For example, a hormone or other water-soluble protein contains many amino acids with polar side chains, while an insoluble muscle protein contains a greater proportion of amino acids with nonpolar side chains.

The importance of the side chains of amino acids is illustrated by the condition known as **sickle-cell anemia**. The difference between normal hemoglobin and sickle-cell hemoglobin is that, in a protein molecule of 146 amino acid units, one single unit has been changed from glutamic acid (with an acidic side chain) to valine (with a nonpolar side chain). This one small error in the protein renders the affected hemoglobin less soluble and thus less able to perform its prescribed task of carrying oxygen to the cells of the body.

SECTION 19.2.

Amino Acids as Dipolar Ions

Amino acids do not always behave like organic compounds. For example, they have melting points of over $200°$, whereas most organic compounds of similar molecular weight are liquids at room temperature. Amino acids are soluble in water and other polar solvents, but insoluble in nonpolar solvents such as diethyl ether or benzene. Amino acids have large dipole moments. Also, they are less acidic than most carboxylic acids and less basic than most amines.

$$RCO_2H \qquad RNH_2 \qquad \begin{array}{c} CO_2H \\ | \\ H_2NCH \\ | \\ R \end{array}$$

$$pK_a = \sim 5 \qquad pK_b = \sim 4 \qquad \begin{array}{c} pK_a = \sim 10 \\ pK_b = \sim 12 \end{array}$$

Why do amino acids exhibit such unusual properties? The reason is that an amino acid contains a basic amino group and an acidic carboxyl group in the same molecule. An amino acid undergoes an internal acid–base reaction to yield a **dipolar ion**, also called a **zwitterion** (from German *zwitter*, "hybrid"). Because of the resultant ionic charges, an amino acid has many properties of a salt. Furthermore, the pK_a of an amino acid is not the pK_a of a $-CO_2H$ group, but that of an $-NH_3^+$ group. The pK_b is not that of a basic amino group, but that of the very weakly basic $-CO_2^-$ group.

$$\begin{array}{c} CO_2H \\ | \\ H_2\overset{..}{N}-C-H \\ | \\ R \end{array} \rightleftharpoons \begin{array}{c} CO_2^- \\ | \\ H_3\overset{+}{N}-C-H \\ | \\ R \end{array}$$

a dipolar ion

STUDY PROBLEMS

19.1. When each of the following amino acids is dissolved in water, would the solution be acidic, basic, or near-neutral? (Refer to Table 19.1 for the structures.)

(a) glutamic acid   (b) glutamine   (c) leucine   (d) lysine   (e) serine

19.2. Monosodium glutamate (MSG) is widely used as a condiment. What is the most likely structure for this compound? (*Hint:* Which carboxyl group in glutamic acid is more acidic?)

SECTION 19.3.

Amphoterism of Amino Acids

An amino acid contains both a carboxylate ion ($-CO_2^-$) and an ammonium ion ($-NH_3^+$) in the same molecule. Therefore, an amino acid is **amphoteric**: it can undergo reaction with either an acid or a base to yield a cation or an anion, respectively.

In acid:

$$H_3\overset{+}{N}-\underset{R}{\overset{CO_2^-}{\underset{|}{\overset{|}{C}}}}-H + H^+ \rightleftharpoons H_3\overset{+}{N}-\underset{R}{\overset{CO_2H}{\underset{|}{\overset{|}{C}}}}-H$$

a cation

In base:

$$H_2\overset{+}{N}-\underset{H \quad R}{\overset{CO_2^-}{\underset{|}{\overset{|}{C}}}}-H + OH^- \rightleftharpoons H_2\overset{..}{N}-\underset{R}{\overset{CO_2^-}{\underset{|}{\overset{|}{C}}}}-H + H_2O$$

an anion

STUDY PROBLEM

19.3. Predict the product of reaction of   (a) proline, and   (b) tyrosine with an excess of aqueous HCl and with an excess of aqueous NaOH.

You might think that an aqueous solution of a so-called neutral amino acid would be neutral. However, aqueous solutions of neutral amino acids are slightly *acidic* because the $-NH_3^+$ group is a stronger acid than $-CO_2^-$ is a base. The result of the difference in acidity and basicity is that an aqueous solution of alanine contains more amino acid anions than cations. We can say that alanine carries a *net negative charge* in aqueous solution.

At pH 7, alanine carries a net negative charge:

weaker base

$$H_3\overset{+}{N}-\underset{\underset{CH_3}{|}}{\overset{\overset{CO_2^-}{|}}{C}}-H + H_2O \rightleftharpoons H_2N-\underset{\underset{CH_3}{|}}{\overset{\overset{CO_2^-}{|}}{C}}-H + H_3O^+$$

stronger acid

no net charge   net negative charge

If a small amount of HCl or other acid is added to an alanine solution, the acid–base equilibrium is shifted so that the net charge on the alanine ions becomes zero. The pH at which an amino acid carries no net ionic charge is defined as the **isoelectric point** of that amino acid. The isoelectric point of alanine is 6.0.

At pH 6, alanine carries no net charge:

$$H_2N-\underset{\underset{CH_3}{|}}{\overset{\overset{CO_2^-}{|}}{C}}-H + H_3O^+ \rightleftharpoons H_3\overset{+}{N}-\underset{\underset{CH_3}{|}}{\overset{\overset{CO_2^-}{|}}{C}}-H$$

Isoelectric points can be determined by **electrophoresis**, a process of measuring the migration of ions in an electric field. This is accomplished by placing an aqueous solution of an amino acid on an adsorbent between a pair of electrodes. In this cell, anions migrate toward the positive electrode and cations migrate toward the negative electrode. If alanine or other neutral amino acid is dissolved in plain water, there is a net migration of amino acid ions toward the positive electrode.

$$H_2NCH \text{ with } CO_2^- \text{ and } CH_3$$

At its isoelectric point, an amino acid exhibits no net migration toward either electrode in an electrophoresis cell. The isoelectric point of a given amino acid is a physical constant. The value varies from amino acid to amino acid, but falls into one of three general ranges.

For a *neutral amino acid*, the isoelectric point, which depends primarily on the relative pK_a and pK_b of the $-NH_3^+$ and $-CO_2^-$ groups, is around 5.5–6.0.

The second carboxyl group in an *acidic amino acid* means that there is another group that can interact with water. A water solution of an acidic amino acid is definitely acidic, and the amino acid ion carries a net negative charge.

$$H_3\overset{+}{N}-\underset{\underset{CH_2CH_2CO_2H}{|}}{\overset{\overset{CO_2^-}{|}}{C}}-H + H_2O \rightleftharpoons H_3\overset{+}{N}-\underset{\underset{CH_2CH_2CO_2^-}{|}}{\overset{\overset{CO_2^-}{|}}{C}}-H + H_3O^+$$

acidic   net negative charge

A *greater concentration of H⁺* is required to bring an acidic amino acid to the iso-electric point than is needed for a neutral amino acid. The isoelectric points for acidic amino acids are near pH 3.

A *basic amino acid* has a second amino group that undergoes reaction with water to form a positive ion. Hydroxide ions are needed to neutralize a basic amino acid and bring it to its isoelectric point. For basic amino acids, we would expect the isoelectric points to be above pH 7, and that indeed is the case. These isoelectric points are in the range of 9–10. Isoelectric points of some representative amino acids are found in Table 19.2.

$$H_3\overset{+}{N}-\underset{\underset{\text{basic}}{(CH_2)_4\ NH_2}}{\overset{\overset{CO_2^-}{|}}{\underset{|}{C}}}-H \quad + H_2O \quad \rightleftharpoons \quad H_3\overset{+}{N}-\underset{(CH_2)_4\overset{+}{N}H_3}{\overset{\overset{CO_2^-}{|}}{\underset{|}{C}}}-H \quad + OH^-$$

net positive charge

TABLE 19.2. Isoelectric points for some amino acids

Name	Structure	Isoelectric point
Neutral:		
alanine	$CH_3\underset{\underset{NH_2}{|}}{C}HCO_2H$	6.00
glutamine	$H_2N\overset{\overset{O}{\|\|}}{C}CH_2CH_2\underset{\underset{NH_2}{|}}{C}HCO_2H$	5.65
Acidic:		
glutamic acid	$HO_2CCH_2CH_2\underset{\underset{NH_2}{|}}{C}HCO_2H$	3.22
aspartic acid	$HO_2CCH_2\underset{\underset{NH_2}{|}}{C}HCO_2H$	2.77
Basic:		
lysine	$H_2N(CH_2)_4\underset{\underset{NH_2}{|}}{C}HCO_2H$	9.74
arginine	$H_2N\overset{\overset{NH}{\|\|}}{C}NH(CH_2)_3\underset{\underset{NH_2}{|}}{C}HCO_2H$	10.76

STUDY PROBLEMS

19.4. Suggest a reason for the fact that the isoelectric point of lysine is 9.74, but that for tryptophan (Table 19.1) is only 5.89. (*Hint:* Think of a reason why the heterocyclic N in tryptophan is not basic.)

19.5. An amino acid undergoes reaction with two equivalents of the hydrate *ninhydrin* to yield *Ruhemann's purple*, a blue-violet product. This reaction is used as a test for the presence of amino acids in a sample of unknown composition. Draw the principal resonance structures that show the delocalization of the negative charge in Ruhemann's purple.

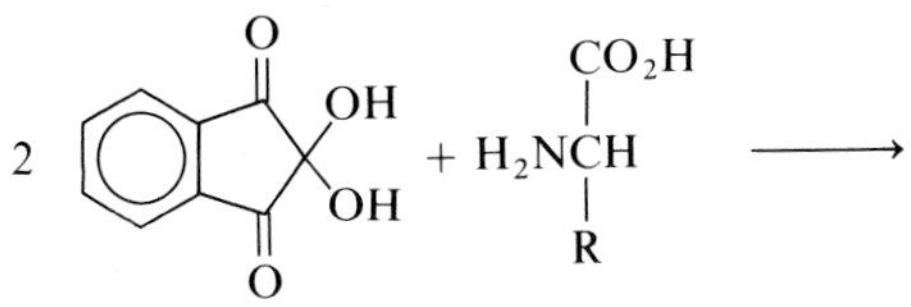

ninhydrin

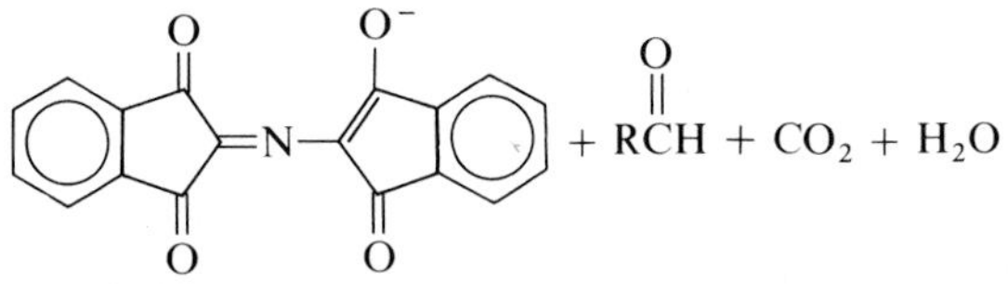

Ruhemann's purple

(*blue-violet*)

SECTION 19.4.

Synthesis of Amino Acids

The common amino acids are relatively simple compounds, and the synthesis of racemic mixtures of most of these amino acids may be accomplished by standard techniques. The racemic mixtures may then be resolved to yield the pure enantiomeric amino acids.

The **Strecker synthesis** of amino acids, developed in 1850, is a two-step sequence. The first step is the reaction of an aldehyde with a mixture of ammonia and HCN to yield an aminonitrile. Hydrolysis of the aminonitrile results in the amino acid.

Step 1:

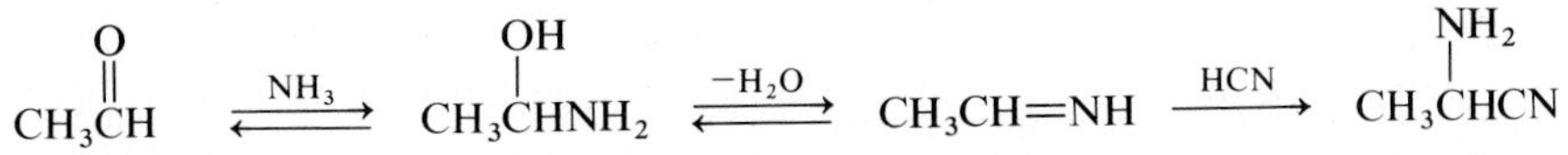

acetaldehyde

2-amino-
propanenitrile

Step 2:

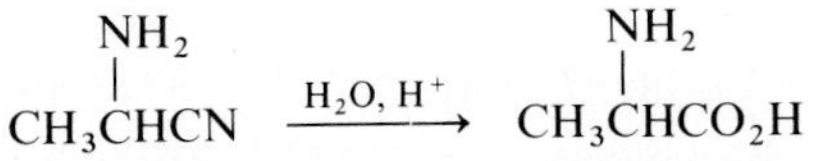

(*R*)(*S*)-alanine

Another synthetic route to amino acids is by the **amination of an α-halo acid** with an excess of ammonia. (An excess of NH_3 must be used to neutralize the acid and to minimize overalkylation. See Section 15.5A.)

Formation of an α-halo acid (Section 13.3C):

$$(CH_3)_2CHCH_2CH_2CO_2H \xrightarrow[PBr_3]{Br_2} (CH_3)_2CHCH_2\overset{\overset{\displaystyle Br}{|}}{C}HCO_2H$$

4-methylpentanoic acid   2-bromo-4-methyl-pentanoic acid

Amination:

$$(CH_3)_2CHCH_2\overset{\overset{\displaystyle }{|}}{C}HCO_2H \xrightarrow[-Br^-]{2\;\ddot{N}H_3} (CH_3)_2CHCH_2\overset{\overset{\displaystyle NH_2}{|}}{C}HCO_2^-\;NH_4^+ \xrightarrow{H^+}$$

$$(CH_3)_2CHCH_2\overset{\overset{\displaystyle NH_2}{|}}{C}HCO_2H$$

(*R*)(*S*)-leucine

The **Gabriel phthalimide synthesis** (Section 15.5A) is a more elegant route to amino acids. The advantage of this synthesis over direct amination is that over-alkylation cannot occur.

Reductive amination of an α-keto acid is another procedure used to obtain racemic amino acids. (Remember, carboxyl groups are not easily reduced.)

$$CH_3\overset{\overset{\displaystyle O}{||}}{C}CO_2H \xrightarrow{H_2,\;NH_3,\;Pd} CH_3\overset{\overset{\displaystyle NH_2}{|}}{C}HCO_2H$$

an α-keto acid   (*R*)(*S*)-alanine

The reactions leading to α-amino acids are summarized in Table 19.3.

TABLE 19.3. Summary of synthetic routes to α-amino acids

Reaction	Section reference
Substitution:	
$R\overset{\overset{\displaystyle X}{\|}}{C}HCO_2H \xrightarrow[(2)\;H^+]{(1)\;NH_3} R\overset{\overset{\displaystyle NH_2}{\|}}{C}HCO_2H$	15.5A
$XCH(CO_2C_2H_5)_2 \xrightarrow[\substack{(2)\;NaOC_2H_5,\;RX \\ (3)\;H_2O,\;H^+,\;heat}]{(1)\;K^+\;phthalimide} R\overset{\overset{\displaystyle NH_2}{\|}}{C}HCO_2H$	15.5A
Strecker synthesis:	
$R\overset{\overset{\displaystyle O}{\|\|}}{C}H \xrightarrow[\substack{(2)\;HCN \\ (3)\;H_2O,\;H^+}]{(1)\;NH_3,\;-H_2O} R\overset{\overset{\displaystyle NH_2}{\|}}{C}HCO_2H$	19.4
Reductive amination:	
$R\overset{\overset{\displaystyle O}{\|\|}}{C}CO_2H \xrightarrow{H_2,\;NH_3,\;Pd} R\overset{\overset{\displaystyle NH_2}{\|}}{C}HCO_2H$	11.14D

SECTION 19.5.

Peptides

A **peptide** is an amide formed from two or more amino acids. The amide link between an α-amino group of one amino acid and the carboxyl group of another amino acid is called a **peptide bond**. The following example of a peptide formed from alanine and glycine, called *alanylglycine*, illustrates the formation of a peptide bond.

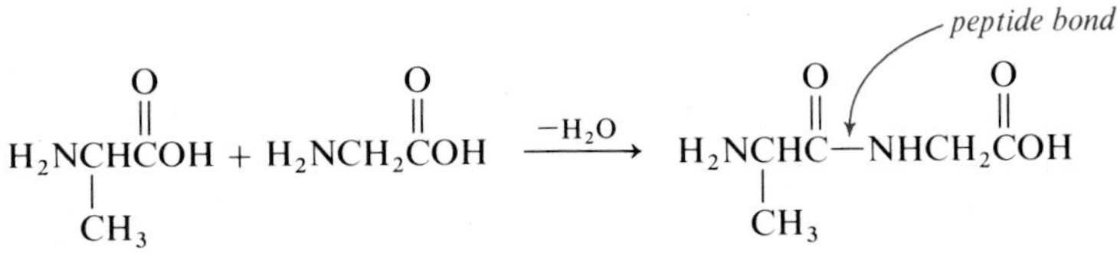

alanine glycine alanylglycine
a dipeptide

Each amino acid in a peptide molecule is called a **unit**, or a **residue**. Depending upon the number of amino acid units in the molecule, a peptide may be referred to as a **dipeptide** (two units), a **tripeptide** (three units), and so forth. A **polypeptide** is a peptide with a large number of amino acid residues. What is the difference between a polypeptide and a protein? None, really. Both are polyamides constructed from amino acids. By convention, a polyamide with fewer than 50 amino acid residues is classified as a peptide, while a larger polyamide is considered to be a protein.

In the dipeptide alanylglycine, the alanine residue has a free amino group and the glycine unit has a free carboxyl group. However, alanine and glycine could be joined another way to form *glycylalanine*, in which glycine has the free amino group and alanine has the free carboxyl group.

Two different dipeptides from alanine and glycine:

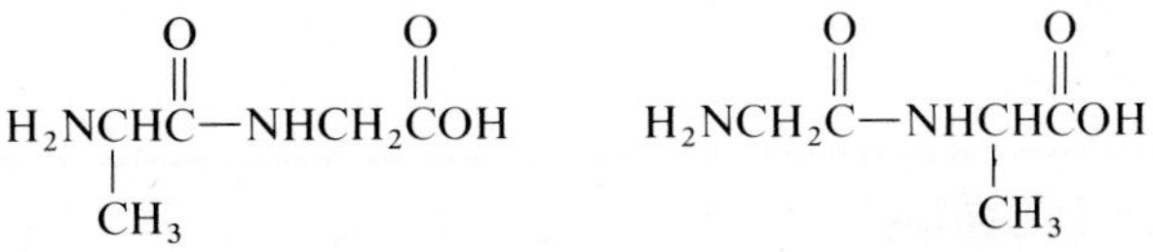

alanylglycine glycylalanine

The greater the number of amino acid residues in a peptide, the greater the number of structural possibilities. Glycine and alanine can be bonded together in two ways. In a tripeptide, three amino acids can be joined in six different ways. Ten different amino acids could lead to over four trillion decapeptides!

For purposes of discussion, it is necessary to represent peptides in a systematic manner. The amino acid with the free amino group is usually placed at the left end of the structure. This amino acid is called the ***N*-terminal amino acid**. The amino acid with the free carboxyl group is placed at the right and is called the ***C*-terminal amino acid**. The name of the peptide is constructed from the names of the amino acids as they appear left-to-right, starting with the *N*-terminal amino acid.

N-terminal on left

$$H_2NCHC\overset{\overset{\displaystyle O}{\|}}{}\!-\!NHCHC\overset{\overset{\displaystyle O}{\|}}{}\!-\!NHCH_2C\overset{\overset{\displaystyle O}{\|}}{}OH$$

C-terminal on right

CH₃   CH₂—⟨benzene⟩—OH

alanyltyrosylglycine

a tripeptide

For the sake of convenience and clarity, the names of the amino acids are often abbreviated. We have shown the abbreviations of the twenty common amino acids in Table 19.1. Using the abbreviated names, alanyltyrosylglycine becomes *ala–tyr–gly.*

SAMPLE PROBLEM

What is the structure of leu–lys–met?

Solution:

$$H_2NCHC\overset{\overset{\displaystyle O}{\|}}{}\!\!-\!\!-\!NHCHC\overset{\overset{\displaystyle O}{\|}}{}\!\!-\!\!-\!NHCHCO_2H$$

(CH₃)₂CHCH₂   (CH₂)₄NH₂   CH₂CH₂SCH₃

leu   lys   met

STUDY PROBLEM

19.6.   Give all possibilities for the structure of a tripeptide consisting of ala, gly, and phe. (Use abbreviated names rather than structural formulas.)

SECTION 19.6.

Bonding in Peptides

As was mentioned in Section 13.9C, an amide bond has some double-bond character due to partial overlap of the *p* orbitals of the carbonyl group with the unshared electrons of the nitrogen.

$$-\overset{\overset{\displaystyle \ddot{O}:}{\|}}{C}\!-\!\ddot{N}\!\!\big< \quad \longleftrightarrow \quad -\overset{\overset{\displaystyle :\ddot{O}:^-}{\|}}{C}\!=\!\overset{+}{N}\!\!\big<$$

Evidence for the double-bond character of a peptide bond is found in bond lengths. The bond length of the peptide bond is shorter than that of the usual C—N single bond: 1.32 Å in the peptide bond versus 1.47 Å for a typical C—N single bond in an amine.

Because of the double-bond character of the peptide bond, rotation of groups around this bond is somewhat restricted, and the atoms attached to the carbonyl group and to the N all lie in the same plane. X-ray analysis shows that the amino acid side chains around the plane of the peptide bond are in a *trans* type of relationship. This stereochemistry minimizes steric hindrance between side chains.

SECTION 19.7.

Determination of Peptide Structure

The determination of a peptide structure is not an easy task. Complete hydrolysis in acidic solution yields the individual amino acids. These may be separated and identified by such techniques as chromatography or electrophoresis. The molecular weight of the peptide may be determined by physical–chemical methods. With this information, the chemist can determine the number of amino acid residues, the identity of the amino acid residues, and the number of residues of each amino acid in the original peptide. But this information reveals nothing about the *sequence* of the amino acids in the peptide. Several techniques have been developed to determine this sequence. The first is **terminal-residue analysis**. There are many ways of determining the *N*-terminal and *C*-terminal amino acid residues; we will mention just one technique for each.

Analysis for the *N*-terminal residue can be accomplished by treating the peptide with phenyl isothiocyanate, followed by hydrolysis. The isothiocyanate undergoes reaction with the free amino group to form a thiourea derivative. Careful hydrolysis cleaves the *N*-terminal residue from the rest of the peptide and generates a cyclic structure called a *phenylthiohydantoin*. This modified end group can then be isolated from the reaction mixture and identified.

a phenylthiohydantoin

Why cannot the chemist continue to treat the peptide stepwise with phenyl isothiocyanate? In each step the *N*-terminal residue could be broken off until the entire peptide was degraded and its order of amino acids determined. To a certain extent, this can be done; however, each cycle of thiourea formation and hydrolysis results in some internal hydrolysis of the remaining peptide. After about 40 cycles, the hydrolysis of the peptide is sufficient to produce many smaller peptides, each of which has an *N*-terminal. The phenylthiohydantoin is thus no longer that of a single end group, but a mixture from a variety of end groups.

The *C*-terminal amino acid residue can be determined enzymatically. *Carboxypeptidase* is a pancreatic enzyme that specifically catalyzes hydrolysis of the *C*-terminal amino acid, but not that of other peptide bonds.

$$\{-\underset{\overset{|}{R}}{N}HCHCO_2H + H_2O \xrightarrow{\text{carboxypeptidase}} \text{rest of peptide} + \underset{\overset{|}{R}}{H_2N}CHCO_2H$$

C-terminal
amino acid

Partial hydrolysis of a peptide to dipeptides, tripeptides, tetrapeptides, and other fragments is one technique for determining the order of interior residues. The hydrolysis mixture is separated and the order of amino acid residues in each fragment determined (by end-group analysis, for example). The structures of the fragments are then pieced together like a jigsaw puzzle to give the structure of the entire peptide.

SAMPLE PROBLEM

What is the structure of a pentapeptide that yields the following tripeptides when partially hydrolyzed?

gly–glu–arg,   glu–arg–gly,   arg–gly–phe

Solution: Fit the pieces together:

gly–glu–arg
glu–arg–gly
arg–gly–phe

The pentapeptide is gly–glu–arg–gly–phe.

A new and indirect way of determining the sequence of amino acids in a protein molecule is by isolating the portion of a DNA molecule that is responsible for the biosynthesis of that protein and then determining the base sequence in the DNA fragment (see Section 16.11A).

SECTION 19.8.

Synthesis of Peptides

As a check on a proposed structure for a peptide, a chemist may synthesize the peptide from the individual amino acids. The first peptide was synthesized by

Emil Fischer, who in 1902 also put forth the idea that proteins were polyamides.

The synthesis of ordinary amides from acid chlorides and amines is a straight-forward reaction (Section 13.3C):

$$RCOCl + R'NH_2 \longrightarrow RCONHR'$$

However, the synthesis of peptides or proteins by this route is not straightforward. The principal problem is that there is more than one way in which the amino acids may join.

gly + ala $\longrightarrow$ gly–ala or ala–gly or gly–gly or ala–ala

gly–ala + phe $\longrightarrow$ phe–gly–ala or gly–ala–phe or

gly–ala–gly–ala or phe–phe

To prevent unwanted reactions, every other reactive group, including reactive groups in side chains, must be blocked. By leaving only the desired amino group and carboxyl group free, the chemist can control the positions of reaction.

all other reactive groups blocked

The criteria for a good blocking group are (1) that it be inert to the reaction conditions needed for forming the desired amide link, and (2) that it be readily removable when the synthesis is complete. One such blocking group is a *carbamate group*—inert to the amide-formation reaction, but easily removed in a later step without disturbing the rest of the molecule. This approach to peptide synthesis was developed in 1932.

Preparation of carbamate to protect an amino group:

from $Cl_2C{=}O + C_6H_5CH_2OH$ glycine   *carbamate group*

The amino-blocked glycine could be treated with $SOCl_2$ to form the acid chloride and then treated with a new amino acid to form an amide. However, acid chlorides are highly reactive and unwanted side reactions may occur despite blocking. To circumvent this problem, the amino-blocked glycine is usually treated with ethyl chloroformate to yield an **activated ester.**

amino-blocked   ethyl   *activated ester group*
glycine   chloroformate

Like an acid chloride, this activated ester can undergo reaction with an amino group of another amino acid to give the desired dipeptide.

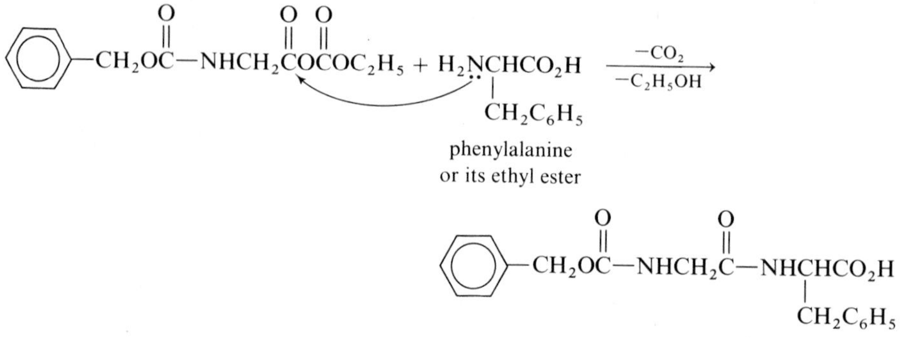

phenylalanine
or its ethyl ester

amino-blocked gly–phe

At this point, the sequence can be repeated to add a third amino acid. When the peptide synthesis is complete, the carbamate group is cleaved by reduction to yield the free peptide.

Removal of carbamate blocking group:

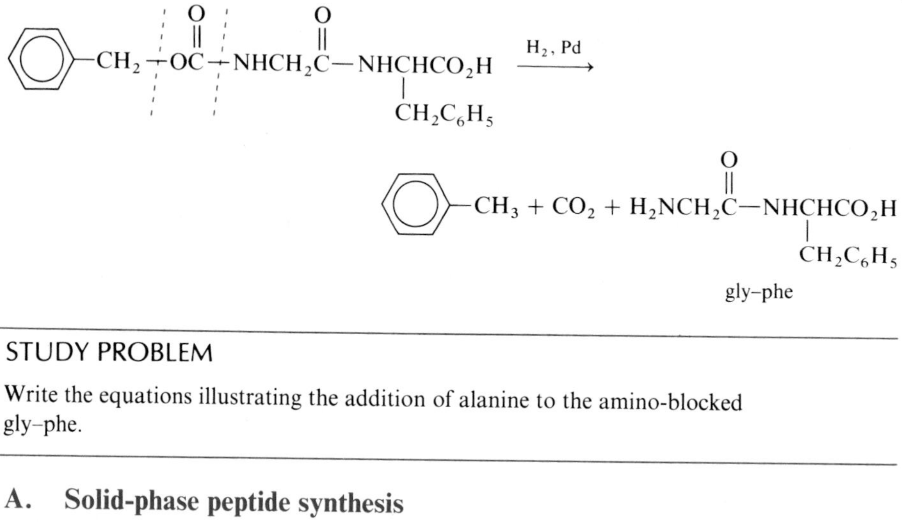

gly–phe

STUDY PROBLEM

19.7.   Write the equations illustrating the addition of alanine to the amino-blocked gly–phe.

A.   Solid-phase peptide synthesis

New and better methods of synthesizing peptides are always under investigation. One relatively new technique is called **solid-phase peptide synthesis,** or the **Merrifield peptide synthesis** (after Bruce Merrifield at Rockefeller University, who developed the technique). In this type of synthesis, resins hold the *C*-terminal amino acid by the carboxyl group as the peptide is being synthesized. The resin is a polystyrene that contains about 1% *p*-(chloromethyl)styrene units.

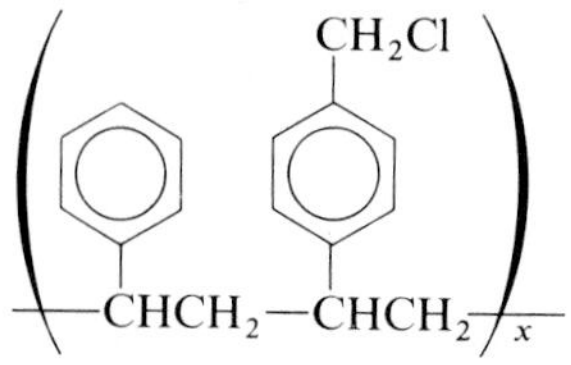

a polystyrene resin
containing *p*-(chloromethyl)styrene units

The amino group of the first amino acid is initially blocked, often as a *t*-butyl carbamate. This amino-blocked amino acid, as the carboxylate, reacts with the benzylic chloride groups of the resin to form ester groups (a typical substitution reaction between a carboxylate and a benzylic halide).

$$(CH_3)_3CO\overset{O}{\overset{\|}{C}}-NHCHCO_2^- \ + \ ClCH_2\text{-resin} \ \xrightarrow{-Cl^-} \ (CH_3)_3CO\overset{O}{\overset{\|}{C}}-NHCHCO_2-CH_2\text{-resin}$$

with R below the CH groups.

The amino-blocking group is removed by treatment with an anhydrous acid, such as HCl in acetic acid; then a second amino-blocked amino acid (with an activated carbonyl group) is added.

$$(CH_3)_3CO\overset{O}{\overset{\|}{C}}-NHCHC\overset{activated}{O_2H} \ + \ H_2NCHCO_2-CH_2\text{-resin} \ \xrightarrow{-H_2O}$$

$$(CH_3)_3CO\overset{O}{\overset{\|}{C}}-NHCH\overset{O}{\overset{\|}{C}}-NHCHCO_2-CH_2\text{-resin}$$

A common technique for activating the $-CO_2H$ group (so that it will undergo reaction with the amine) is by the addition of dicyclohexylcarbodiimide to the carboxylic acid. This compound reacts with the carboxylic acid to yield an intermediate with a leaving group that can be displaced by the amine in a typical nucleophilic acyl substitution reaction. The product is the amide.

$$R\overset{O}{\overset{\|}{C}}OH \ + \ \langle\rangle-N=C=N-\langle\rangle \ \xrightarrow{25°}$$

dicyclohexylcarbodiimide

$$\xrightarrow[\text{addition}]{R'NH_2} \ R\overset{:\ddot{O}:^-}{\underset{R'NH_2^+}{\overset{\|}{C}-\ddot{O}-\{}} \ \xrightarrow[\text{proton transfer}]{\text{elimination and}}$$

$$R\overset{:\ddot{O}}{\overset{\|}{C}}NHR' \ + \ \langle\rangle-N=\overset{OH}{\overset{|}{C}}-NH-\langle\rangle$$

the amide   the enol tautomer of dicyclohexylurea

In the solid-phase peptide synthesis, the product of this reaction is the amino-blocked dipeptide, still bonded to the resin at the carboxyl end. The amino-blocking group is removed, and the process is repeated until the peptide synthesis is complete. A controlled cleavage of the resin–ester bond with an acid such as anhydrous HF releases the peptide and removes the final amino-blocking group.

The classical type of peptide synthesis is tedious because intermediate peptides must be isolated and purified; however, in a solid-phase peptide synthesis, most impurities can simply be washed away from the resin after each step. This technique is successful enough that commercial **automated peptide synthesizers** have been developed.

Unfortunately, solid-phase peptide synthesis techniques are not practical for synthesizing large protein molecules in a pure state and in high yields. Poor-quality products may result because the resin tends to hold some impurities. However, recent advances in "gene-splicing" (page 766) have provided routes to high yields of pure peptides and proteins.

SECTION 19.9.

Biosynthesis of Peptides

The biosynthesis of peptides and proteins in a typical cell is accomplished by ribonucleic acids (RNA) and enzymes. The various types of RNA, such as **messenger RNA (mRNA)**, are synthesized under the direction of DNA in the nucleus of the cell (Section 16.11); then these RNA molecules leave the cell nucleus to perform their functions.

When mRNA leaves the nucleus, it becomes attached to **ribosomes**, granular bodies in the cytoplasm of the cell that are the sites of protein synthesis. A ribosome consists of approximately 60% **ribosomal RNA (rRNA)** and 40% protein. Each amino acid is brought to the mRNA and its ribosomes as an ester of one ribose unit in a **transfer RNA (tRNA)** molecule.

a ribose unit of tRNA

The mRNA is the template for the protein that is to be synthesized. The series of bases on a mRNA molecule is the code that determines the order of incorporation of amino acids into a growing protein molecule. (These bases were discussed in Section 16.11.) A series of *three bases in a row*, called a **codon**, directs the inclusion of one particular amino acid. For example, the presence of three uracil bases in a row **(U–U–U)** in a mRNA molecule is a codon for phenylalanine. The codon guanine–cytosine–cytosine **(G–C–C)** signals that alanine is to be incorporated.

The tRNA with the proper amino acid is able to recognize its proper site because tRNA carries an **anticodon**, a series of three bases that are complementary to the codon of mRNA.

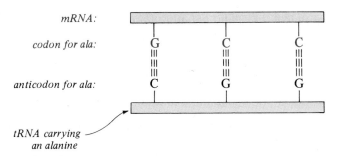

Protein biosynthesis may be likened to an assembly line. Figure 19.1 shows a ribosome moving along the mRNA chain (moving to the right in the figure). At

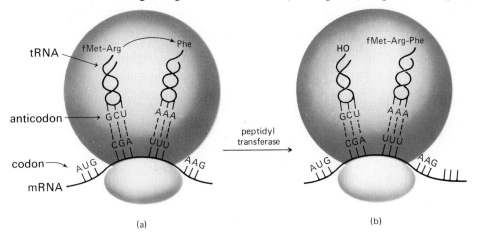

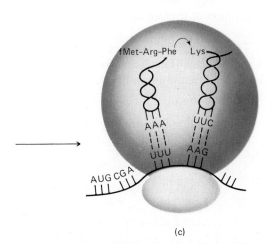

FIGURE 19.1. Biosynthesis of a peptide. (a) *fMet,* or *N*-formylmethionine (the initiating amino acid, with the codon AUG), has become bonded to *arg* (codon CGA). The next amino acid to be added will be *phe* (codon UUU). (b) The growing peptide chain has been transferred to phe. (c) The ribosome has moved along the mRNA chain (to the right) so that the peptide chain can be transferred to the next amino acid, *lys* (codon AAG). Adapted from William H. Brown and Judith A. McClarin, *Introduction to Organic and Biochemistry,* 3rd ed. (Willard Grant Press, Boston, 1981).

each codon, a new amino acid is added to the growing protein molecule. Other ribosomes follow the first one so that many molecules of the same protein are being synthesized simultaneously. When a completed protein molecule reaches the end of the mRNA chain, it leaves the site to carry out its own functions in the organism.

SECTION 19.10.

Some Interesting Peptides

Although most proteins are relatively large molecules, a peptide need not contain thousands of amino acid residues for biological activity. Consider the following tripeptide:

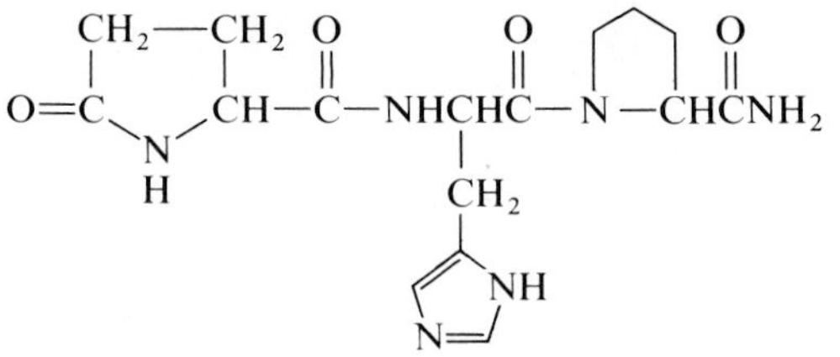

pyroglutamylhistidylprolinamide

This tripeptide is a **thyrotropic-hormone releasing factor** (TRF) that has been isolated from the hypothalamus glands of hogs and cattle. Administered to humans (intravenously or orally), it stimulates the secretion of thyrotropin, which in turn stimulates the secretion of thyroid hormones, the regulators of the body's metabolism.

The structure of TRF exemplifies some commonly encountered structural variations in peptides and proteins. The *N*-terminal amino acid is a derivative of glutamic acid: the side-chain carboxyl group has joined with the free amino group to form a lactam (a cyclic amide).

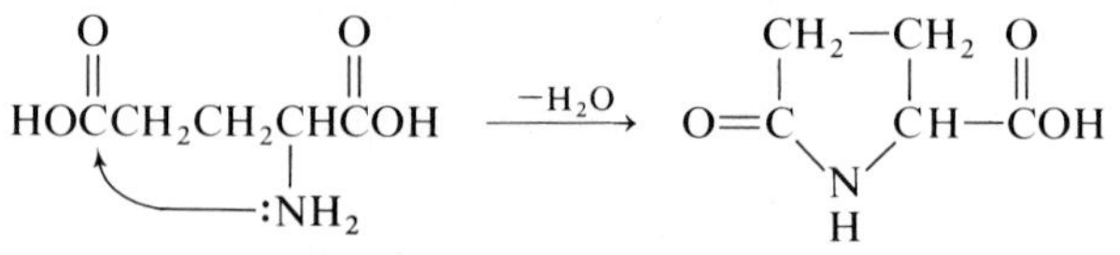

glutamic acid

pyroglutamic acid

a lactam

The *C*-terminal residue of TRF is an amide of proline. Amides of free carboxyl groups are not at all uncommon in natural protein structures. These are generally denoted in the abbreviated structures by adding NH_2 to the *C*-terminal: for example, *gly* means $-NHCH_2CO_2H$, but *gly*–NH_2 means $-NHCH_2CNH_2$.

STUDY PROBLEM

19.8.   Give the structural formula for pro–leu–ala–NH$_2$.

Oxytocin, a pituitary hormone that causes uterine contractions during parturition, is another important peptide. Note that oxytocin is a cyclic peptide joined by a S—S cystine link. Figure 19.2 shows the structural formula.

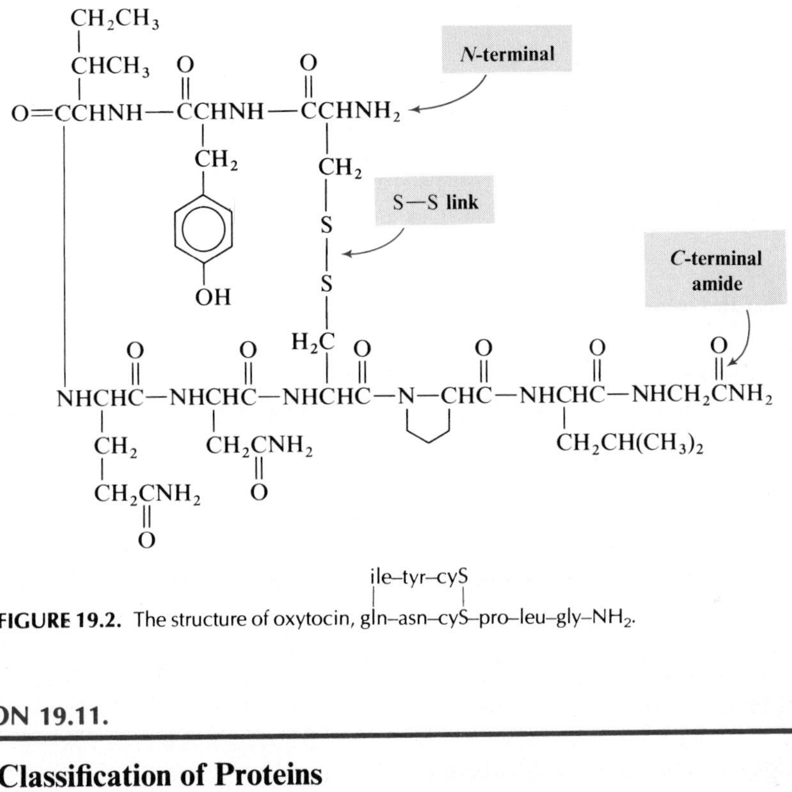

FIGURE 19.2.   The structure of oxytocin, gln–asn–cyS–pro–leu–gly–NH$_2$.

SECTION 19.11.

Classification of Proteins

Proteins may be roughly categorized by the type of function they perform. These classes are summarized in Table 19.4. **Fibrous proteins** (also called **structural proteins**), which form skin, muscles, the walls of arteries, and hair, are composed of long thread-like molecules that are tough and insoluble.

Another functional type of protein is the class of **globular proteins**. These are small proteins, somewhat spherical in shape because of folding of the protein chains upon themselves. Globular proteins are water-soluble and perform various functions in an organism. For example, *hemoglobin* transports oxygen to the cells; *insulin* aids in carbohydrate metabolism; *antibodies* render foreign protein inactive; *fibrinogen* (soluble) can form insoluble fibers that result in blood clots; and *hormones* carry messages throughout the body.

Conjugated proteins, proteins connected to a nonprotein moiety such as a sugar, perform various functions throughout the body. A common mode of linkage between the protein and nonprotein is by a functional side chain of the

TABLE 19.4. Classes of proteins

Class	Comments
Fibrous, or structural (insoluble):	
collagens	form connective tissue; comprise 30% of mammalian protein; lack cysteine and tryptophan; rich in hydroxyproline
elastins	form tendons and arteries
keratins	form hair, quills, hoofs, nails; rich in cysteine and cystine
Globular (soluble):	
albumins	examples are egg albumin and serum albumin
globulins	an example is serum globulin
histones	occur in glandular tissue and with nucleic acids; rich in lysine and arginine
protamines	associated with nucleic acids; contain no cysteine, methionine, tyrosine, or tryptophan; rich in arginine
Conjugated (combined with other substances):	
nucleoproteins	combined with nucleic acids
mucoproteins	combined with >4% carbohydrates
glycoproteins	combined with <4% carbohydrates
lipoproteins	combined with lipids, such as phosphoglycerides or cholesterol

protein. For example, an acidic side chain of the protein can form an ester with an —OH group of a sugar molecule.

SECTION 19.12.

Higher Structures of Proteins

The sequence of amino acids in a protein molecule is called the **primary structure** of the protein. However, there is much more to protein structure than just the primary structure. Many of the properties of a protein are due to the orientation of the molecule as a whole. The shape (such as a helix) into which a protein molecule arranges its backbone is called the **secondary structure**. Further interactions, such as folding of the backbone upon itself to form a sphere, are called the **tertiary structure**. Interactions between certain protein subunits, such as between the globins in hemoglobin (page 875), are called the **quaternary structure**. The secondary, tertiary, and quaternary structures are collectively referred to as the **higher structure** of the protein.

One protein that has been well-studied in terms of its secondary structure is **keratin**, found in fur and feathers. Each protein molecule in keratin has the shape of a spiral, called a *right-handed α-helix* (Figure 19.3). "Right-handed" refers to the direction of the turns in the helix; the mirror image is a left-handed helix. In the

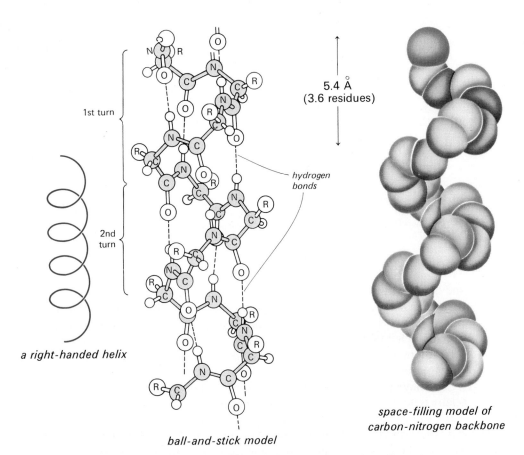

1st turn

2nd turn

a right-handed helix

5.4 Å
(3.6 residues)

hydrogen bonds

ball-and-stick model

space-filling model of carbon-nitrogen backbone

FIGURE 19.3. The protein chains in keratin form right-handed α-helices. The ball-and-stick model is adapted from C. B. Anfinsen, *The Molecular Basis of Evolution* (John Wiley and Sons, Inc., New York, 1964). The space-filling model is adapted from W. H. Brown and J. A. McClarin, *Introduction to Organic and Biochemistry, 3rd ed.* (Willard Grant Press, Boston, 1981).

mid-1930's, the term "α" was coined to differentiate the x-ray pattern of keratin from that of some other proteins.

In keratin, each turn of the helix contains 3.6 amino acid residues. The distance from one coil to the next is 5.4 Å. The helix is held in its shape primarily by hydrogen bonds between one amide–carbonyl group and an NH group that is 3.6 amino acid units away (Figure 19.4). The helical shape gives a strong, fibrous, flexible product.

Hydrogen bonding between an α-amino group and a carbonyl group is one contributor to the shape of a protein molecule. Other inter- and intramolecular interactions also contribute to the higher structure. Some of these interactions are hydrogen bonding between side chains, S—S cross-links, and *salt bridges* (ionic bonds such as RCO_2^- ^+H_3NR between side chains). The most stable higher structure is the one with the greatest number of stabilizing interactions. (Each hydrogen bond lends ∼ 5 kcal/mole of stability.) Given a particular primary structure, a protein naturally assumes its most stable higher structure.

Let us look at some other types of protein structure. **Collagen** is a general classification of a tough, strong protein that forms cartilage, tendons, ligaments,

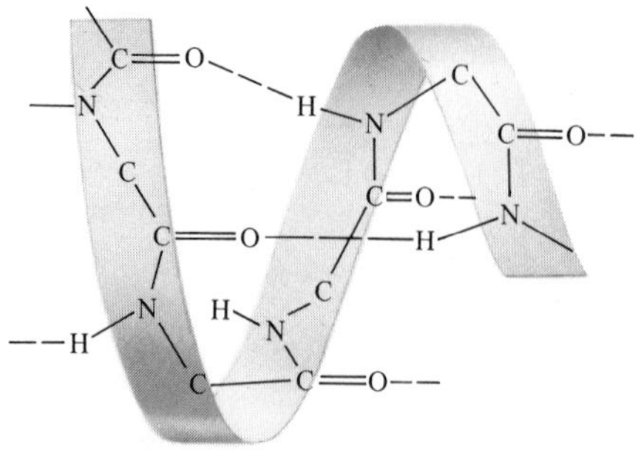

FIGURE 19.4. The helix in keratin is held in its shape by hydrogen bonds. (Nonparticipating R's and H's have been omitted.) From R. J. Fessenden and J. S. Fessenden, *Chemical Principles for the Life Sciences* (Allyn and Bacon, Inc., Boston, 1976).

and skin. Collagen derives its strength from its higher structure of " super-helices ": three right-handed, α-helical polypeptides entwined to form a triple, left-handed helical chain. The entwined molecules are collectively called a *tropocollagen molecule*. One of these tropocollagen molecules is about 15 Å in diameter and 2800 Å long. The tropocollagen triple helix, like a single helix, is held together by hydrogen bonding.

Gelatin is obtained by boiling collagen-containing animal material; however, gelatin is not the same type of protein as collagen. It has been found that the molecular weight of gelatin is only one-third that of collagen. Presumably, in gelatin-formation, the tropocollagen molecule is unraveled and the single strands form hydrogen bonds with water, resulting in the characteristic gel-formation.

Helical structures are not the only type of secondary structure of proteins. Another type of structure, referred to as a *β-sheet*, or *pleated sheet*, is found in silk fibroin. The pleated sheet is an arrangement in which single protein molecules are lined up side by side and held there by hydrogen bonds between the chains (Figure 19.5).

The protein chains in silk fibroin are not simply stretched-out zigzag chains. Analysis by x-ray diffraction shows repetitive units every 7.0 Å. This repetition probably arises from a puckering (or "pleating") in the chains that alleviates steric hindrance. It is interesting that silk fibroin contains 46% glycine (no side chain) and 38% of a mixture of alanine and serine (small side chains). The lack of bulky R groups in these amino acids allows the side-by-side arrangement of protein chains in the fibroin structure.

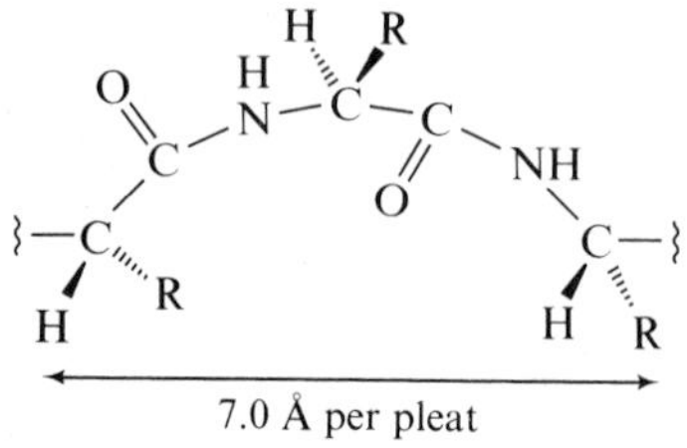

7.0 Å per pleat

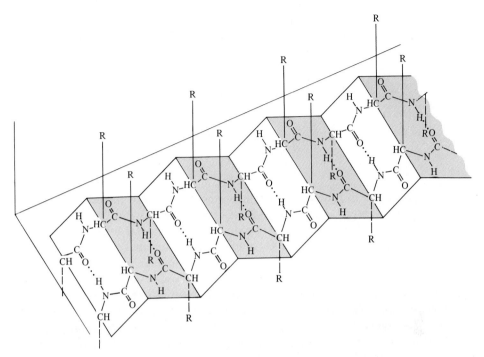

FIGURE 19.5. The pleated sheet structure of silk fibroin. From P. Karlson, *Introduction to Modern Biochemistry, 3rd ed.* (Georg Thieme Verlag, Stuttgart, 1968).

A *globular protein* depends on a tertiary structure to maintain its intricately folded, globular shape, which is necessary to maintain solubility. In a globular protein, polar, hydrophilic side chains are situated on the outside of the sphere (to increase water-solubility) and nonpolar, hydrophobic side chains are arranged on the interior surface, where they may be used to catalyze nonaqueous reactions. The unique surface of each globular protein enables it to "recognize" certain complementary organic molecules. This recognition allows enzymes to catalyze reactions of particular molecules, but not others.

Hemoglobin, the portion of erythrocytes (red blood cells) that is responsible for the transport of oxygen in the bloodstream, is a good example of a globular protein. One unit of hemoglobin, which has a "molecular weight" of about 65,000, contains four protein molecules called *globins*. Each globin is folded in such a way that (1) it fits perfectly with the three other globins to maintain the hemoglobin entity, and (2) it forms a molecular crevice of just the right size and shape to hold a unit of heme along with its O_2 molecule.

Heme (page 762) is a **prosthetic group**, a nonprotein organic molecule held firmly by a protein. It is composed of a porphyrin ring system with a chelated ferrous ion in the center. Each heme unit is bound to its globin by a coordinate bond from the ferrous ion to the nitrogen in an imidazole ring of histidine, one of the amino acids that forms the globin. The ferrous ion also can form a coordinate bond with an O_2 molecule.

For heme to function as a transporter of oxygen to the cells, the Fe^{2+} ion must release the oxygen readily. Such a release could not occur if Fe^{2+} were oxidized to Fe^{3+} and oxygen were reduced. The nonpolar hydrophobic environment of the protein around the O_2 binding site assures that electron-transfer from Fe^{2+} to O_2 does not occur. (Remember that nonpolar solvents cannot support ions in solution.) A similar phenomenon is observed in the laboratory when Fe^{2+} ions are embedded in polystyrene. Oxygen is attracted, but no oxidation–reduction reaction takes place as it would in an ionic medium.

Carbon monoxide poisoning is the result of CO molecules taking the place of O_2 molecules in hemoglobin. The CO molecules are firmly held by the iron and are not as readily released as oxygen molecules.

SECTION 19.13.

Denaturation of Proteins

Denaturation of a protein is the loss of its higher structural features by disruption of hydrogen bonding and other secondary forces that hold the molecule together. The result of denaturation is the loss of many of the biological properties of the protein.

One of the factors that can cause denaturation of a protein is a *change in temperature*. Cooking an egg white is an example of an irreversible denaturation. An egg white is a colorless liquid containing albumins, which are soluble globular proteins. Heating the egg whites causes the albumins to unfold and precipitate; the result is a white solid.

A change in pH can also cause denaturation. When milk sours, the change in pH arising from lactic acid formation causes *curdling*, or precipitation of soluble proteins. Other factors that can cause denaturation are detergents, radiation, oxidizing or reducing agents (which can alter S—S links), and changes in the type of solvent.

Some proteins (skin and the lining of the gastrointestinal tract, for example) are quite resistant to denaturation, while other proteins are more susceptible. Denaturation may be reversible if a protein has been subjected to only mild denaturing conditions, such as a slight change in pH. When this protein is replaced in its natural environment, it may resume its natural higher structure in a process called **renaturation**. Unfortunately, renaturation may be very slow or may not occur at all. One of the problems in protein research is how to study proteins without disrupting the higher structures.

SECTION 19.14.

Enzymes

The word **enzyme** means "in yeast." Even without any knowledge of their structures or functions, humans have used enzymes since prehistoric times in the production of wine, vinegar, and cheese. Pasteur thought that living yeast cells were necessary for fermentation processes. We now know that a living cell is not necessary; the proper enzymes, plus reaction conditions that do not cause denaturation, are all that are needed for enzymatic reactions.

An enzyme is a *biological catalyst*. A higher animal contains thousands of enzymes. Virtually every biochemical reaction is catalyzed by an enzyme. Even the equilibrium $CO_2 + H_2O \rightleftarrows H_2CO_3$ is enzyme-catalyzed because the rate of the uncatalyzed equilibration does not produce carbonic acid fast enough for an animal's needs.

Enzymes are more efficient catalysts than most laboratory or industrial catalysts (such as Pd in a hydrogenation reaction). Biological reactions in humans occur at 37°C and in aqueous media. High temperature, high pressure, or very reactive reagents (such as NaOH or $LiAlH_4$) are not available to an organism. Enzymes also allow a selectivity of reactants and a control over reaction rate that can be obtained with no other class of catalyst.

All enzymes are proteins. Some are relatively simple in structure; however, most are complex. The structures of many enzymes are still unknown. For biological activity, some enzymes require prosthetic groups, or **cofactors**. These cofactors are nonprotein portions of the enzyme. A cofactor may be a simple *metal ion*; for example, copper ion is the cofactor for the enzyme *ascorbic acid oxidase*. Other enzymes contain *nonprotein organic molecules* as cofactors. An organic prosthetic group is frequently referred to as a **coenzyme**. (One coenzyme, coenzyme A, was mentioned in Section 13.8.)

If an organism cannot synthesize a necessary cofactor, the cofactor must be present in small amounts in the diet. The active units of many cofactors are vitamins. Table 19.5 shows a few cofactors and the corresponding vitamins.

A. Naming enzymes

Most enzymes are named after the reactions that they catalyze. The ending for an enzyme name is usually **-ase**. The name may be *general* and refer to a class of enzymes that catalyze a general type of reaction. For example, a **polymerase** is any enzyme that catalyzes a polymerization reaction, and a **reductase** is any enzyme that catalyzes a reduction reaction. An enzyme name may also refer to a *specific* enzyme: **ascorbic acid oxidase** is the enzyme that catalyzes the oxidation of ascorbic acid, while **phosphoglucoisomerase** catalyzes the isomerization of glucose 6-phosphate to fructose 6-phosphate.

STUDY PROBLEM

19.9. Suggest the function of each of the following enzymes: (a) an acetyltransferase;
(b) phenylalanine hydroxylase;   (c) pyruvate dehydrogenase.

TABLE 19.5. Some cofactors that contain vitamins

Name of cofactor	Vitamin needed	Structure of vitamin
vitamin C (ascorbic acid)	vitamin C	
vitamin B$_1$ (thiamine)	vitamin B$_1$	
biotin	biotin	
coenzyme A	pantothenic acid	
NAD^{+a}	nicotinic acid (niacin) or nicotinamide (niacinamide)	
pyridoxyl phosphate	pyridoxine	

^{a} Nicotinamide adenine dinucleotide, a biological oxidizing agent.

B.   How enzymes work

Some enzymes have been studied in detail, yet there is still much to learn about even the well-known enzymes. It is believed that an enzyme fits itself around the substrate (the molecule to be acted upon) to form an **enzyme–substrate complex**. The bonds of the substrate may be strained by attractions between itself and the enzyme. Strained bonds are of higher energy and are more easily broken; therefore, the desired reaction proceeds easily and yields an **enzyme–product complex**.

In many cases, the product is not the same shape as the reacting substrate; thus the fit between the product and the enzyme is no longer perfect. The altered shape of the product causes a dissociation of the complex, and the enzyme surface is ready to accept another molecule of substrate. This theory of enzyme activity is called the **induced-fit theory**.

$$\text{E} \quad + \quad \text{S} \longrightarrow \qquad \text{E}-\text{S} \qquad \longrightarrow$$

enzyme   substrate   enzyme–substrate
                                 complex

$$\text{E}-\text{P} \qquad \longrightarrow \qquad \text{E} \quad + \quad \text{P}$$

        enzyme–product   enzyme   product
            complex

Enzymes have molecular weights of 12,000–120,000 and higher. Most substrates (for example, an amino acid or a unit of glucose) are much smaller molecules. The specific location of the large enzyme structure where reaction occurs is called the **active site**. This site is where the prosthetic group (if any) is located. Metallic prosthetic groups are thought to serve as electrophilic agents and, in this way, catalyze the desired reactions. In NAD^+, the active site is the nicotinamide end of the cofactor. NAD^+ is readily reduced and therefore catalyzes oxidation reactions.

$$R_2CHOH + \quad NAD^+ \longrightarrow R_2C{=}O + \quad NADH + H^+$$

oxidized

the reduced form of NAD^+

The rest of the enzyme molecule is not simply excess molecular weight! It is believed that this portion of the enzyme recognizes its substrate and holds it in place. It was suggested in the 1890's by Emil Fischer that enzymes are chiral molecules and that reactants must complement this chirality in order to undergo reaction. Fischer compared the fitting together of the substrate structure and the enzyme structure to a key fitting into a lock (see Figure 19.6).

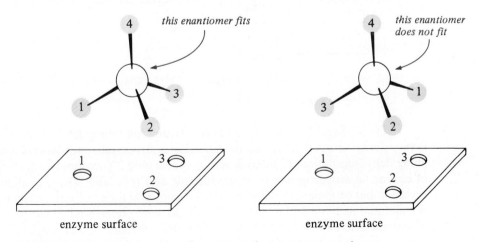

FIGURE 19.6. One enantiomer fits on the enzyme surface; its mirror image does not.

Recognition may occur by a series of dipole–dipole interactions, by hydrogen bonding, or by covalent bonding, in which the stereochemistry must be just right. In some cases, the rest of the enzyme molecule is folded to form a hydrophobic pocket that holds a nonpolar portion of the substrate in place. (We mentioned this type of structure for hemoglobin.) If the nonpolar end of a potential substrate does not fit in the pocket correctly, enzyme catalysis diminishes or is nonexistent. Therefore, the functional group to be acted upon must fit the active site on the enzyme, and the rest of the substrate molecule must fit together with other portions of the enzyme molecule for reaction to proceed. This dual type of recognition is the basis of the unique specificity of most enzymes.

Both the active site and the rest of an enzyme are important in enzyme activity. Let us look at one reaction in which the active site seems to be the more important factor in substrate recognition. The enzyme *succinate dehydrogenase* catalyzes the dehydrogenation of succinic acid to the *trans*-diacid fumaric acid. (The *cis*-isomer, maleic acid, is not produced in this reaction.) The oxidizing agent in this reaction is *flavin adenine dinucleotide* (FAD), which is reduced by a 1,4-addition of two hydrogen atoms (plus two electrons). (We show only the functional portion of FAD here.)

Other diacids, such as oxalic acid, malonic acid, and glutaric acid, inhibit the dehydrogenation of succinic acid.

Inhibitors of succinate dehydrogenase:

$$HO_2C-CO_2H \qquad HO_2CCH_2CO_2H \qquad HO_2CCH_2CH_2CH_2CO_2H$$

oxalic acid   malonic acid   glutaric acid

Of these diacids, malonic acid has the greatest inhibiting effect. Malonic acid is very similar in structure to succinic acid, but is structurally incapable of undergoing dehydrogenation. If succinic acid contains only 2% malonic acid, the rate of enzymatic dehydrogenation of the succinic acid is halved! The probability is excellent that malonic acid competes with succinic acid for a position on the active site and that malonic acid is attracted and held there preferentially. The presence of malonic acid on the active site thus blocks the approach of succinic acid.

Summary

A **protein** is a polyamide. Hydrolysis yields α-amino acids of (*S*)-configuration at the α carbon. Amino acids undergo an internal acid–base reaction to yield dipolar ions.

(S), or L

$$\{-NHCHC-\} \xrightarrow[\text{H}^+]{\text{H}_2\text{O}} \quad \text{H}_2\text{N}-\overset{\text{CO}_2\text{H}}{\underset{\text{R}}{\text{C}}}-\text{H} \quad \rightleftharpoons \quad \text{H}_3\overset{+}{\text{N}}-\overset{\text{CO}_2^-}{\underset{\text{R}}{\text{C}}}-\text{H}$$

an α-amino acid   *a dipolar ion*

Essential amino acids are those that cannot be synthesized by an organism and must be present in the diet. **Acidic amino acids** are those with a carboxyl group in the side chain (R in the preceding equation). **Basic amino acids** contain an amino group in the side chain. **Neutral amino acids** contain neither $-CO_2H$ nor $-NH_2$ in the side chain, but may contain OH, SH, or other polar group. **Cross-linking** in proteins may be provided by the SH group in cysteine, which can link with another SH in an oxidation reaction: $2\,RSH \rightarrow RSSR + 2\,H$.

The **isoelectric point** of an amino acid is the pH at which the dipolar ion is electrically neutral and does not migrate toward an anode or cathode. The isoelectric point depends on the acidity or basicity of the side chain.

in H₂O, a net positive charge   *at pH 9.74, no net charge*

Racemic amino acids may be synthesized by a variety of routes, which are summarized in Table 19.3, page 860.

A **peptide** is a polyamide of fewer than 50 amino acid residues. The *N*-terminal amino acid is the amino acid with a free α-amino group, while the *C*-terminal amino acid has a free carboxyl group at carbon 1. **End-group analysis** to determine the *C*- and *N*-terminals and **partial hydrolysis** to smaller peptides are two techniques for peptide structure-determination.

In the synthesis of a peptide, reactive groups (except for the groups desired to undergo reaction) must be blocked. A **carbamate group** may be used to protect an amino group. A **solid-phase peptide synthesis** provides a blocking group for the *C*-terminal carboxyl group. The biosynthesis of proteins is accomplished by RNA. The order of incorporation of amino acids is determined by the order of attachment of the bases (*N*-heterocycles) in mRNA.

Proteins are polyamides of more than 50 amino acid residues. The order of side chains in a protein determines its **higher structures**, which arise from internal and external hydrogen bonding, van der Waals forces, and other interactions between side chains. The higher structures of proteins give them a variety of physical and chemical properties so that they may perform a variety of functions.

Denaturation is the disruption of hydrogen bonds and thus the disruption of the higher structure of the protein.

Enzymes are proteins that catalyze biochemical reactions. Enzymes are efficient and specific in their catalytic action. The specificity is provided for by the unique shape and by the polar (or nonpolar) groups contained within the enzyme structure. Some enzymes work in conjunction with a nonprotein **cofactor**, which may be organic or inorganic.

STUDY PROBLEMS

19.10. Label each of the following amino acids as *acidic*, *basic*, or *neutral*:   (a) isoleucine;
(b) aspartic acid;   (c) asparagine;   (d) serine;   (e) histidine;   (f) glutamine.

19.11. Show the possible oxidative coupling products between 1.0 equivalent each of cysteine and the following compound:

$$\underset{\underset{CH_2SH}{|}}{\overset{\overset{O}{\overset{||}{}}}{HSCH_2CNHCHCO_2H}}$$

19.12. Predict the products of the reaction of alanine with: (a) dilute aqueous HCl;   (b) dilute aqueous KOH;   (c) methanol $+$ H_2SO_4 with heat;   (d) one equivalent of acetic anhydride.

19.13. Suggest a synthesis for valine from 3-methylbutanoic acid. Give the stereochemistry of the product.

19.14. Suggest a route to alanine from (*R*)-lactic acid (2-hydroxypropanoic acid) using a reductive amination. What is the stereochemistry of the product?

19.15. Explain the following observations:

(a)   Although saturated carboxylic acids absorb at about 1720 cm^{-1} (5.81 μm) in the infrared region, amino acids do not absorb at this position.
(b)   If a neutral solution of an amino acid is acidified, the infrared spectrum then shows absorption at 1720 cm^{-1}.

19.16. Outline a Strecker synthesis for: (a) phenylalanine,   and   (b) valine.
(c) What would be the configuration of the chiral carbons in the products?

19.17. Predict the major organic products:

(a) $CH_3CH_2CO_2H$ $\xrightarrow[\text{(2) excess NH}_3]{\text{(1) Br}_2,\ \text{PBr}_3}$

(b) $\xrightarrow[\text{(2) HCl, H}_2\text{O, heat}]{\text{(1) ClCH}_2\text{CO}_2\text{C}_2\text{H}_5}$

(c) $\xrightarrow[\substack{\text{(2) C}_6\text{H}_5\text{CH}_2\text{Cl}\\ \text{(3) HCl, H}_2\text{O, heat}}]{\text{(1) Na}^+\ ^-\text{OC}_2\text{H}_5}$

(d) $(CH_3)_2CHCH_2\overset{\overset{\displaystyle O}{\|}}{C}H + NH_3 + HCN \longrightarrow$

(e) $C_6H_5CH_2\underset{\underset{\displaystyle NH_2}{|}}{C}HCO_2H + CH_3OH + HCl \xrightarrow{\text{heat}}$

19.18. Without referring to the text, match the amino acid with its isoelectric point.

(a) cysteine   (1) 10.76
(b) aspartic acid   (2)   6.30
(c) proline   (3)   5.07
(d) arginine   (4)   2.77

19.19. Without referring to the text, predict the approximate isoelectric point of each of the following amino acids: (a) serine;   (b) histidine;   (c) glutamic acid;   (d) glutamine;   (e) lysine.

19.20. Predict the major products:

(a) $H_2NCHCO_2^-$ + 1.0 equivalent HCl $\longrightarrow$
      $\qquad |$
      $\quad CH(CH_3)_2$

(b) $H_2NCHCO_2^-$ + 1.0 equivalent HCl $\longrightarrow$
      $\qquad |$
      $\quad (CH_2)_4NH_2$

(c) $^+H_3NCHCO_2H$ + 1.0 equivalent NaOH $\longrightarrow$
      $\qquad |$
      $\quad CH(CH_3)_2$

(d) $^+H_3NCHCO_2H$ + 1.0 equivalent NaOH $\longrightarrow$
      $\qquad |$
      $\quad (CH_2)_4NH_3^+$

19.21. Write equations to represent: (a) dipolar-ion formation of histidine;   (b) the equilibria between histidine and water;   (c) the reaction of histidine with dilute aqueous HCl; (d) the reaction of histidine with dilute aqueous NaOH.

19.22. In the following structure, label: (a) the peptide bond(s);   (b) the *N*-terminal amino acid;
(c) the *C*-terminal amino acid.
(d)   Is this structure a dipeptide, tripeptide, or tetrapeptide?
(e)   Would this peptide be considered acidic, basic, or neutral?

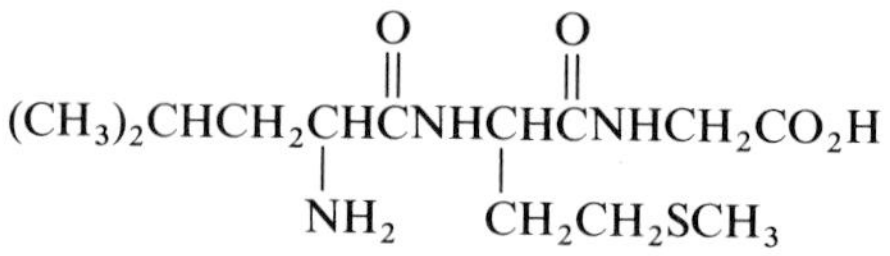

$$(CH_3)_2CHCH_2CHCNHCHCNHCH_2CO_2H$$

with carbonyl groups (O) above, and substituents NH_2 and $CH_2CH_2SCH_3$ below.

19.23. Write the structures for   (a) glycylglycine, and   (b) alanylleucylmethionine.

19.24. Write the full name and the abbreviated name for each of the following peptides:

(a) 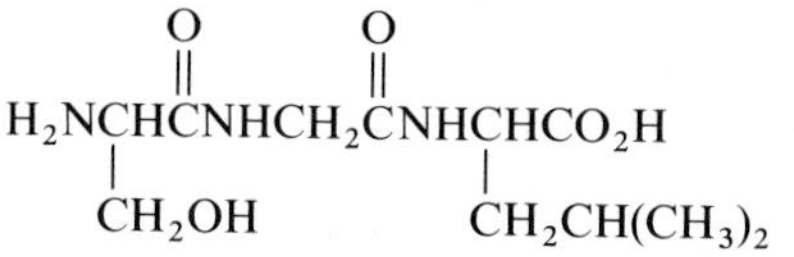 $H_2NCHCNHCH_2CNHCHCO_2H$

with substituents CH_2OH and $CH_2CH(CH_3)_2$ below.

(b) 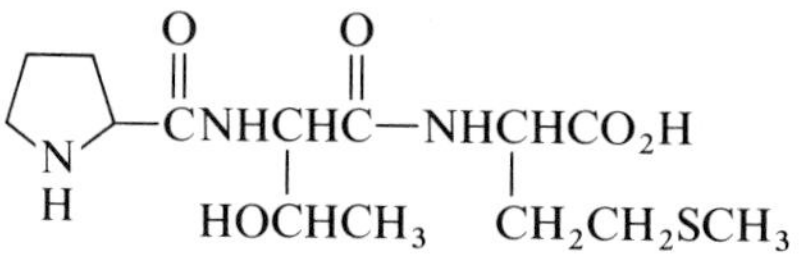   (pyrrolidine ring with N–H)—$CNHCHC—NHCHCO_2H$

with substituents $HOCHCH_3$ and $CH_2CH_2SCH_3$ below.

19.25. Give the structures of the principal organic ionic species in the following solutions:

(a)   glycyllysine in dilute aqueous HCl
(b)   glycylglutamic acid in dilute aqueous NaOH
(c)   glycyltyrosine in dilute aqueous NaOH

19.26. A globular protein is hydrolyzed by aqueous acid. When the reaction mixture is neutralized, ammonia is given off. What can you deduce about the structure of the protein from this observation?

19.27. Each of the following peptides is subjected to reaction with an alkaline solution of phenyl isothiocyanate, followed by acid hydrolysis. Give the structures of the products.

(a)   gly–ala   (b)   ala–gly   (c)   ser–phe–met

19.28. Each of the peptides in Problem 19.27 is treated with the enzyme carboxypeptidase. What products are formed?

19.29. *Bradykinin* is a pain-causing nonapeptide that is released by globulins in blood plasma as a response to toxins in wasp stings. Partial hydrolysis of bradykinin results in the following tripeptides:

ser–pro–phe   gly–phe–ser   pro–phe–arg   arg–pro–pro

pro–gly–phe   pro–pro–gly   phe–ser–pro

What is the amino acid sequence in bradykinin?

19.30. A chemist decided to prepare the dipeptide val–ala by the following sequence of reactions. Using structural formulas, rewrite each set of reactants and tell what possible products were obtained in each case.

 (a) val + $SOCl_2$ (b) [product(s) from (a)] + ala

19.31. Starting with monomeric amino acids, show how to prepare ala–gly and phe–val and then show how these dipeptides can be joined to yield ala–gly–phe–val. (Do not use a solid-phase synthesis.)

19.32. Three different codons of mRNA result in the incorporation of isoleucine in a protein: AUU, AUC, and AUA.

 (a) What sequences of bases in DNA give rise to these codons?
 (b) What are the corresponding anticodons in tRNA?

19.33. Which of the following structural features would contribute to water-solubility of a protein?

 (a) rich in glutamic acid
 (b) rich in valine
 (c) formation of bundles of helices
 (d) formation of a helix folded into a sphere
 (e) combination with glucose
 (f) combination with cholesterol (page 908)

19.34. Meat can be tenderized by soaking it overnight in a marinade of vinegar, spices, and sugar.

 (a) Why is the meat tenderized in this process?
 (b) Would a water solution of sucrose alone have the same tenderizing effect?

19.35. Match each of the following enzymes (or class of enzymes) with a reaction on the right that it might catalyze. (There may be more than one correct answer.)

(a) a phosphatase	(1) $Fe^{2+} \longrightarrow Fe^{3+}$
(b) a glycosidase	(2) cleavage of the *N*-terminal amino acid from a protein
(c) α-1,4-glucan 4-glucanohydrolase	(3) sucrose $\longrightarrow$ fructose + glucose
(d) an aminopeptidase	(4) glucose 6-phosphate $\longrightarrow$ glucose
(e) an oxidase	(5) cellulose $\longrightarrow$ glucose
(f) phosphohexose isomerase	(6) amylose $\longrightarrow$ glucose
	(7) glucose 6-phosphate $\longrightarrow$ fructose 6-phosphate

19.36. Which of the following compounds would be most likely to inhibit the enzymatic incorporation of nicotinic acid (3-pyridinecarboxylic acid) into NAD^+?

 (a) pyridine (b) 3-pyridinesulfonic acid (c) 3-methylpyridine
 (d) adenine (e) acetic acid (f) malonic acid

19.37. The tripeptide pyroglutamylhistidylprolinamide (page 870) is subjected to complete hydrolysis with HCl and H_2O. What is the structure of each product?

19.38. A peptide containing one equivalent each of tyr, ile, gly, arg, and cyS had no *C*-terminal amino acid and no *N*-terminal amino acid. Explain.

19.39. Suggest a reason for the differences in the isoelectric points of   (a) lysine (9.74) and histidine (7.59), and   (b) lysine and arginine (10.76).

19.40. An amino acid with an (R), or D, configuration at the α carbon cannot be enzymatically incorporated into proteins. Many D-amino acids are oxidized in a reaction catalyzed by **D-amino acid oxidase**.

$$\underset{R}{\overset{CO_2H}{H-C-NH_2}} + \tfrac{1}{2}O_2 \longrightarrow \underset{R}{\overset{CO_2H}{C=O}}$$

This enzyme-catalyzed reaction is very slow for L-amino acids, D-glutamic acid, D-lysine, and D-aspartic acid. The reaction is fast for D-proline, D-alanine, D-methionine, and D-tyrosine.

(a) Would you say that the differences in reaction rate are due to steric hindrance or to electronic effects?
(b) Would you predict that the binding site on the enzyme is polar or nonpolar?
(c) Which would be likely to undergo this enzymatic oxidation at a faster rate, D-cysteine or D-arginine?

19.41. In sickle-cell anemia, valine replaces a glutamic acid unit of normal hemoglobin. Glutamic acid has two codons (GAA and GAG) while valine has four (GUU, GUC, GUA, and GUG). Postulate the genetic change that accounts for sickle-cell anemia.

19.42. The complete hydrolysis of an acyclic nonapeptide yields a mixture of ala, asp, glu, gly, leu, lys, phe, tyr, and val. Terminal-residue analysis shows the *N*-terminal amino acid to be val and the *C*-terminal amino acid to be gly. Partial enzymatic hydrolysis with *chymotrypsin,* which catalyzes preferential cleavage at the carbonyl groups of phe or tyr, yields a pentapeptide and a tetrapeptide, among other products. The tetrapeptide is partially hydrolyzed to three dipeptides. One dipeptide contains ala and gly; the second contains asp and tyr; and the third contains asp and ala. Partial hydrolysis of the nonapeptide with *trypsin,* which catalyzes cleavage at the carbonyl group of lys, yields a pentapeptide and a tetrapeptide. The amino acid content of the tetrapeptide is glu, leu, lys, and val. What are the two possible structures of the nonapeptide? How could they be differentiated?

19.43. In mammals, arginine undergoes enzymatic hydrolysis to urea $(H_2N)_2C=O$, which is excreted in the urine, and a basic amino acid called *ornithine*. What is the structure of ornithine?

19.44. Treatment of ornithine (see Problem 19.43) with an alkaline solution of H_2N-CN yields arginine. Suggest a mechanism for this reaction. (*Hint:* Your first step should be to assign $\delta+$ and $\delta-$ charges to the atoms.)

19.45. *Glutathione* is a tripeptide found in most living cells. Partial hydrolysis yields cyS, glu, gly, glu–cyS, and cyS–gly.

(a) What is the amino acid sequence in glutathione?
(b) It has been discovered that glutamic acid forms a peptide link in glutathione with the *side chain* carboxyl group, rather than the carboxyl group adjacent to the amino group. What is the structural formula of glutathione?

CHAPTER 20

Lipids and Related Natural Products

A **lipid** is defined as a naturally occurring organic compound that is insoluble in water, but soluble in nonpolar organic solvents such as a hydrocarbon or diethyl ether. This definition sounds as if it might include many types of compounds, and indeed it does. The various classes of lipids are related to one another by this shared physical property; but their chemical, functional, and structural relationships, as well as their biological functions, are diverse. We will discuss here the classes usually thought of as lipids: fats and oils, terpenes, steroids, and a few other compounds of interest. (Line formulas are generally used for terpenes and steroids, as the following examples show. These formulas were introduced in Section 9.16.)

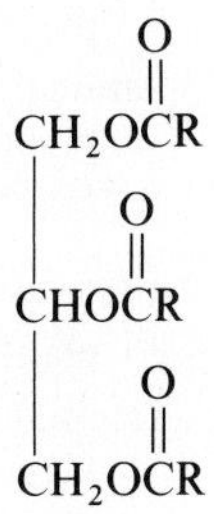

a fat, or triglyceride

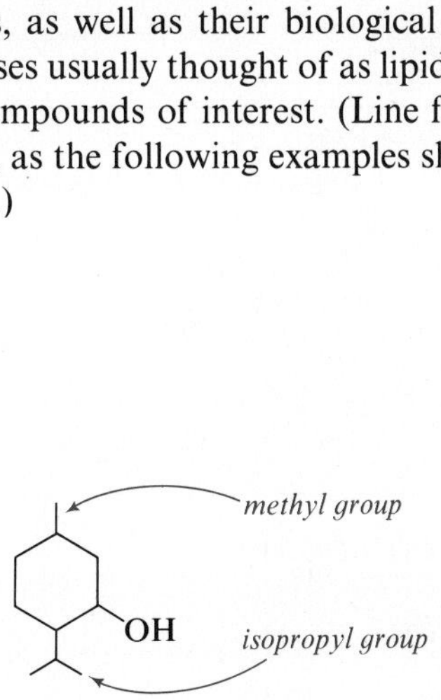

menthol

a terpene

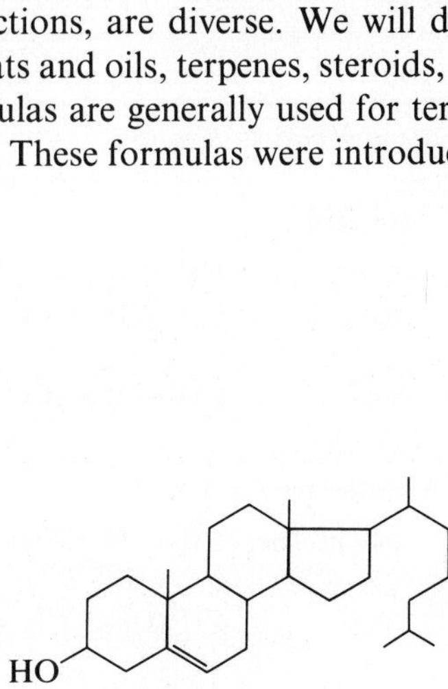

cholesterol

a steroid

SECTION 20.1.

Fats and Oils

Fats and oils are **triglycerides**, or **triacylglycerols**, both terms meaning "triesters of glycerol." The distinction between a fat and an oil is arbitrary: at room temperature a fat is solid and an oil is liquid. Most glycerides in animals are fats, while those in plants tend to be oils; hence the terms *animal fats* (bacon fat, beef fat) and *vegetable oils* (corn oil, safflower oil).

The carboxylic acid obtained from the hydrolysis of a fat or oil, called a **fatty acid**, generally has a long, unbranched hydrocarbon chain. Fats and oils are often named as derivatives of these fatty acids. For example, the tristearate of glycerol is named tristearin, and the tripalmitate of glycerol is named tripalmitin.

$$
\begin{array}{l}
CH_2O_2C(CH_2)_{16}CH_3 \\
| \\
CHO_2C(CH_2)_{16}CH_3 \ + 3\,H_2O \ \xrightarrow{\ H^+\ } \\
| \\
CH_2O_2C(CH_2)_{16}CH_3
\end{array}
\qquad
\begin{array}{l}
CH_2OH \\
| \\
CHOH \qquad + 3\,CH_3(CH_2)_{16}CO_2H \\
| \\
CH_2OH
\end{array}
$$

tristearin   glycerol   stearic acid
(glyceryl tristearate)   *a fatty acid*

a typical fat

Fatty acids can also be obtained from **waxes**, such as beeswax. In these cases, the fatty acid is esterified with a simple long-chain alcohol.

$C_{25}H_{51}CO_2C_{28}H_{57}$ $C_{27}H_{55}CO_2C_{32}H_{65}$ $C_{15}H_{31}CO_2C_{16}H_{33}$
in beeswax   *in carnauba wax*   cetyl palmitate
in spermaceti

Most naturally occurring fats and oils are *mixed* triglycerides—that is, the three fatty-acid portions of the glyceride are not the same. Table 20.1 lists some

TABLE 20.1. Selected fatty acids and their sources

Name of acid	Structure	Source
Saturated:		
butyric	$CH_3(CH_2)_2CO_2H$	milk fat
palmitic	$CH_3(CH_2)_{14}CO_2H$	animal and plant fat
stearic	$CH_3(CH_2)_{16}CO_2H$	animal and plant fat
Unsaturated:		
palmitoleic	$CH_3(CH_2)_5CH{=}CH(CH_2)_7CO_2H$	animal and plant fat
oleic	$CH_3(CH_2)_7CH{=}CH(CH_2)_7CO_2H$	animal and plant fat
linoleic	$CH_3(CH_2)_4CH{=}CHCH_2CH{=}CH(CH_2)_7CO_2H$	plant oils
linolenic	$CH_3CH_2CH{=}CHCH_2CH{=}CHCH_2CH{=}CH(CH_2)_7CO_2H$	linseed oil

representative fatty acids, and Table 20.2 shows the fatty-acid composition of some plant and animal triglycerides.

TABLE 20.2. Approximate fatty-acid composition of some common fats and oils

| | *Composition* $(\%)^a$ | | | |
Source	*palmitic*	*stearic*	*oleic*	*linoleic*
corn oil	10	5	45	38
soybean oil	10	—	25	55
lard	30	15	45	5
butter	25	10	35	—
human fat	25	8	46	10

^{a} Other fatty acids are also found in lesser amounts.

Almost all naturally occurring fatty acids have an *even* number of carbon atoms because they are biosynthesized from the two-carbon acetyl groups in acetylcoenzyme A.

The hydrocarbon chain in a fatty acid may be saturated or it may contain double bonds. The most widely distributed fatty acid in nature, oleic acid, contains one double bond. Fatty acids with more than one double bond are not uncommon, particularly in vegetable oils; these oils are the so-called *polyunsaturates*.

The configuration around any double bond in a naturally occurring fatty acid is *cis*, a configuration that results in the low melting points of oils. A saturated fatty acid forms zigzag chains that can fit compactly together, resulting in high van der Waals attractions; therefore, saturated fats are solids. If a few *cis* double bonds are present in the chains, the molecules cannot form neat, compact lattices, but tend to coil; polyunsaturated triglycerides tend to be oils. Figure 20.1 shows models of the two types of chains.

Triglycerides are one of the three principal foodstuffs, carbohydrates and proteins being the other two. As an energy source, triglycerides are the most efficient: they provide 9.5 kcal/gram, while the proteins provide 4.4 kcal/gram and the carbohydrates provide 4.2 kcal/gram.

In an organism, ingested fats are hydrolyzed into monoglycerides, diglycerides, fatty acids, and glycerol, all of which can be absorbed through the intestinal wall. The organism (1) uses these hydrolyzed or partially hydrolyzed fats as raw materials to synthesize its own fats; (2) converts the fatty acids to other

A saturated triglyceride:

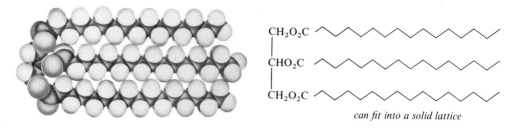

An unsaturated triglyceride:

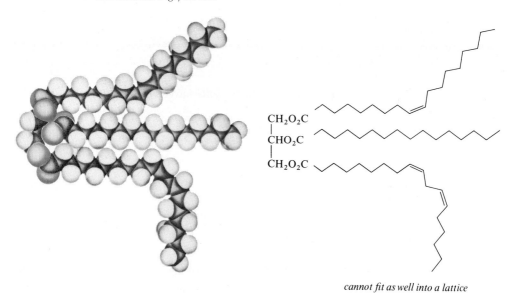

FIGURE 20.1. The shapes of saturated and unsaturated triglycerides. Adapted from William H. Brown and Judith A. McClarin, *Introduction to Organic and Biochemistry,* *3rd ed.* (Willard Grant Press, Boston, 1981).

compounds such as carbohydrates or cholesterol; or (3) converts the fatty acids to energy.

$$
\begin{array}{c}
CH_2O_2CR \\
| \\
CHO_2CR \\
| \\
CH_2O_2CR
\end{array}
\xrightarrow[\text{enzymes}]{H_2O}
\begin{array}{c}
CH_2O_2CR \\
| \\
CHO_2CR \\
| \\
CH_2OH
\end{array}
+
\begin{array}{c}
CH_2O_2CR \\
| \\
CHOH \\
| \\
CH_2OH
\end{array}
+
\begin{array}{c}
CH_2OH \\
| \\
CHOH \\
| \\
CH_2OH
\end{array}
+
RCO_2H
$$

　　　　　　　　　　a diglyceride　*a monoglyceride*　glycerol　*a fatty acid*

SECTION 20.2.

Soaps and Detergents

The word *saponify* means "make soap." Saponification of an ester with NaOH yields the sodium salt of a carboxylic acid (Section 13.5C). Saponification of a triglyceride yields a salt of a long-chain fatty acid, which is a **soap**. American pioneers used beef or pork fat and wood ashes (which contain alkaline salts, such as K_2CO_3) to make soap. (It was reported by Julius Caesar that Teutonic tribes of his era also made soap this way.)

Saponification:

$$
\begin{array}{c}
CH_2O_2C(CH_2)_{16}CH_3 \\
| \\
CHO_2C(CH_2)_{16}CH_3 \\
| \\
CH_2O_2C(CH_2)_{16}CH_3
\end{array}
+ 3\,NaOH
\xrightarrow{\text{heat}}
\begin{array}{c}
CH_2OH \\
| \\
CHOH \\
| \\
CH_2OH
\end{array}
+ 3\,CH_3(CH_2)_{16}CO_2^-\;Na^+
$$

　　　　tristearin　　　　　　　　　　　　　glycerol　　　　*a soap*

sodium stearate

A molecule of a soap contains a long hydrocarbon chain plus an ionic end. The hydrocarbon portion of the molecule is hydrophobic and soluble in nonpolar substances, while the ionic end is hydrophilic and water-soluble. Because of the hydrocarbon chain, a soap molecule as a whole is not truly soluble in water. However, soap is readily suspended in water because it forms **micelles**, clusters of hydrocarbon chains with their ionic ends facing the water (see Figure 20.2).

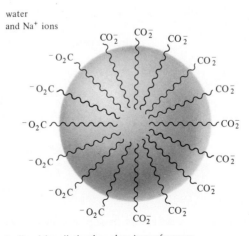

FIGURE 20.2. A micelle of the alkylcarboxylate ions of a soap.

The value of a soap is that it can emulsify oily dirt so that it can be rinsed away. This ability to act as an emulsifying agent arises from two properties of the soap. First, the hydrocarbon chain of a soap molecule dissolves in nonpolar substances, such as droplets of oil. Second, the anionic end of the soap molecule, which is attracted to water, is repelled by the anionic ends of soap molecules protruding from other drops of oil. Because of these repulsions between the soap–oil droplets, the oil cannot coalesce, but remains suspended.

in soapy water, oil droplets repel each other because of similar charges of soap's carboxylate groups

A disadvantage of soaps is that they form insoluble salts (bathtub ring) with Ca^{2+}, Mg^{2+}, and other ions found in hard water. ("Softening" water involves exchanging these ions for Na^+.)

$$2\ CH_3(CH_2)_{16}CO_2Na + Ca^{2+} \longrightarrow [CH_3(CH_2)_{16}CO_2]_2Ca + 2\ Na^+$$

sodium stearate $\qquad\qquad\qquad\qquad$ calcium stearate

insoluble

Most laundry products and many toilet "soaps" and shampoos are not soaps, but **detergents**. A detergent is a compound with a hydrophobic hydrocarbon end plus a sulfonate or sulfate ionic end. Because of this structure, a detergent has the same emulsifying properties as a soap. The advantage of a detergent is that most metal alkylsulfonates and sulfates are water-soluble; detergents do not precipitate with the metal ions found in hard water.

One of the first detergents in common use was a highly branched alkylbenzenesulfonate. The alkyl portion of this compound is synthesized by the polymerization of propylene and is attached to the benzene ring by a Friedel–Crafts alkylation reaction. Sulfonation, followed by treatment with base, yields the detergent.

a detergent

Although the microorganisms in septic tanks or sewage-treatment plants can break down continuous-chain alkyl groups into smaller organic molecules, they cannot degrade branched chains. The reason for this difference in biodegradability is that long-chain hydrocarbons are degraded two carbons at a time by way of a keto ester. Branching interferes with the formation of the ketone group, and thus blocks the entire sequence. (FAD, NAD$^+$, and HSCoA, shown in the following equation, are discussed in Sections 13.8 and 19.14B.)

$$RCH_2CH_2\overset{O}{\overset{\|}{C}}SCoA \xrightarrow[-2H]{FAD} RCH=CHCSCoA \xrightarrow{H_2O}$$

$$\underset{RCHCH_2CSCoA}{\overset{OH\ \ \ O}{\overset{|\ \ \ \ \|}{}}} \xrightarrow[-2H]{NAD^+} RCCH_2CSCoA \xrightarrow{HSCoA} RCSCoA + CH_3CSCoA$$

two carbons removed

To prevent the build-up of detergents in rivers and lakes, present-day detergents are designed with biodegradability in mind. One type of biodegradable detergent is an alkylbenzenesulfonate with a continuous-chain, rather than a branched-chain, alkyl group. Another type of biodegradable detergent is a continuous-chain alkylsulfate.

$$CH_3(CH_2)_{16}CH_2OSO_3^-\ Na^+$$

an alkylsulfate detergent

STUDY PROBLEM

20.1. Write equations for a reaction sequence that would convert triolein into a sodium alkylsulfate detergent.

SECTION 20.3.

Phospholipids

Phospholipids are lipids that contain phosphate ester groups. **Phosphoglycerides**, one type of phospholipid, are closely related to the fats and oils. These compounds usually contain fatty-acid esters at two positions of glycerol with a phosphate ester at the third position.

$$
\begin{array}{cc}
\underset{R'CO-C-H}{\overset{O\ \ \ CH_2OCR}{\overset{\|\ \ \ \ \ |}{}}} & \text{quaternary N} \\
\underset{CH_2OPOCH_2CH_2\overset{+}{N}(CH_3)_3}{\overset{O^-}{\overset{|}{}}} & \\
\overset{\|}{O} &
\end{array}
\qquad
\begin{array}{cc}
\underset{R'CO-C-H}{\overset{O\ \ \ CH_2OCR}{\overset{\|\ \ \ \ \ |}{}}} & \text{primary N} \\
\underset{CH_2OPOCH_2CH_2\overset{+}{N}H_3}{\overset{O^-}{\overset{|}{}}} & \\
\overset{\|}{O} &
\end{array}
$$

a lecithin,
or phosphatidylcholine   *a cephalin,*
or phosphatidylethanolamine

Lecithins and **cephalins** are two types of phosphoglyceride that are found principally in the brain, nerve cells, and liver of animals and are also found in egg yolks, wheat germ, yeast, soybeans, and other foods. These two types of compounds are similar to each other in structure. Lecithins are derivatives of choline chloride, $HOCH_2CH_2N(CH_3)_3{}^+$ Cl^-, which is involved in the transmission of nerve impulses. Cephalins are derivatives of ethanolamine, $HOCH_2CH_2NH_2$.

Other classes of phospholipids are represented by **plasmalogens**, which have vinyl ether groups instead of ester groups at carbon 1 of glycerol, and **sphingolipids**, of which sphingomyelin is an example. Sphingomyelin is a phosphate ester, not of glycerol, but of a long-chain allylic alcohol with an amide side chain.

$$\begin{array}{c} O \quad CH_2OCH{=}CHR \\ \parallel \; \mid \\ R'COCH \\ \mid \\ \quad\quad O^- \\ \quad\quad \mid \\ CH_2OPOCH_2CH_2\overset{+}{N}H_3 \\ \parallel \\ O \end{array}$$

a plasmalogen

$$\begin{array}{c} CH_3(CH_2)_{12}\diagdown \quad \diagup H \\ C \\ \parallel \\ C \\ H \diagup \quad \diagdown CHOH \\ \mid \\ \quad\quad O \\ \quad\quad \parallel \\ CHNHC(CH_2)_{22}CH_3 \\ \mid \\ \quad\quad O^- \\ \quad\quad \mid \\ CH_2OPOCH_2CH_2\overset{+}{N}(CH_3)_3 \\ \parallel \\ O \end{array}$$

sphingomyelin

a sphingolipid

STUDY PROBLEM

20.2. Hydrolysis of sphingomyelin yields phosphoric acid, choline, a 24-carbon fatty acid, and *sphingosine*. What is the structure of sphingosine?

Like soaps and detergents, phospholipids contain long-chain, hydrophobic hydrocarbon groups. The phosphate–amine end of a phospholipid is hydrophilic. Because of these structural features, the phospholipids are excellent emulsifying agents. In mayonnaise, it is the phosphoglycerides of the egg yolk that keep the vegetable oil emulsified in the vinegar.

Interest in phospholipids is high because of their occurrence in nerve cells and the brain. Some biological functions of these compounds are known, but there is still much to learn about their role in biological systems. It is known, for example, that phospholipids are important in the action of cell membranes. These membranes are formed from proteins associated with a bilayer, or double layer, of phosphoglyceride molecules with their hydrophobic ends pointing inward and their hydrophilic ends pointing outward. This bilayer helps form a barrier that allows selective passage of water, nutrients, hormones, and wastes in and out of the cell (see Figure 20.3).

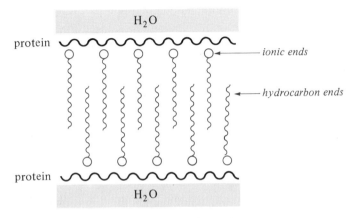

FIGURE 20.3. The bilayer of phospholipids in a cell membrane.

It is thought that the sphingolipids such as sphingomyelin contribute strength to the myelin (nerve cell) sheath by the intertwining of their hydrocarbon chains. Phospholipids are also thought to act as electrical insulation for the nerve cells. The myelin sheaths of people with multiple sclerosis (and some other diseases that affect this membrane) are deficient in the long hydrocarbon chains.

SECTION 20.4.

Prostaglandins

One of the more exciting areas of biochemical research today is that of the **prostaglandins**. These compounds were first discovered in semen, and it was recognized that they were synthesized in the prostate gland (hence the name). We now know that prostaglandins are found throughout the body and are also synthesized in the lungs, liver, uterus, and other organs and tissues.

Just how the prostaglandins act is still unknown. It is thought that they are moderators of hormone activity in the body, a theory that explains their far-reaching biological effects. For example, administration of remarkably small doses of some prostaglandins stimulates uterine contractions and can cause abortion. Imbalances in prostaglandins can lead to nausea, diarrhea, inflammation, pain, fever, menstrual disorders, asthma, ulcers, hypertension, drowsiness, or blood clots. Aspirin is known to interfere with the biosynthesis of prostaglandins. Presumably, the diminished rate of prostaglandin formation ultimately results in the reduction of a fever and in the relief of pain.

Although the structures of the prostaglandins are not particularly complex, their structures were not determined until 1962. The prostaglandins are 20-carbon carboxylic acids that contain cyclopentane rings. They are biosynthesized from the 20-carbon unsaturated fatty acids (one reason unsaturated fats are necessary in a good diet).

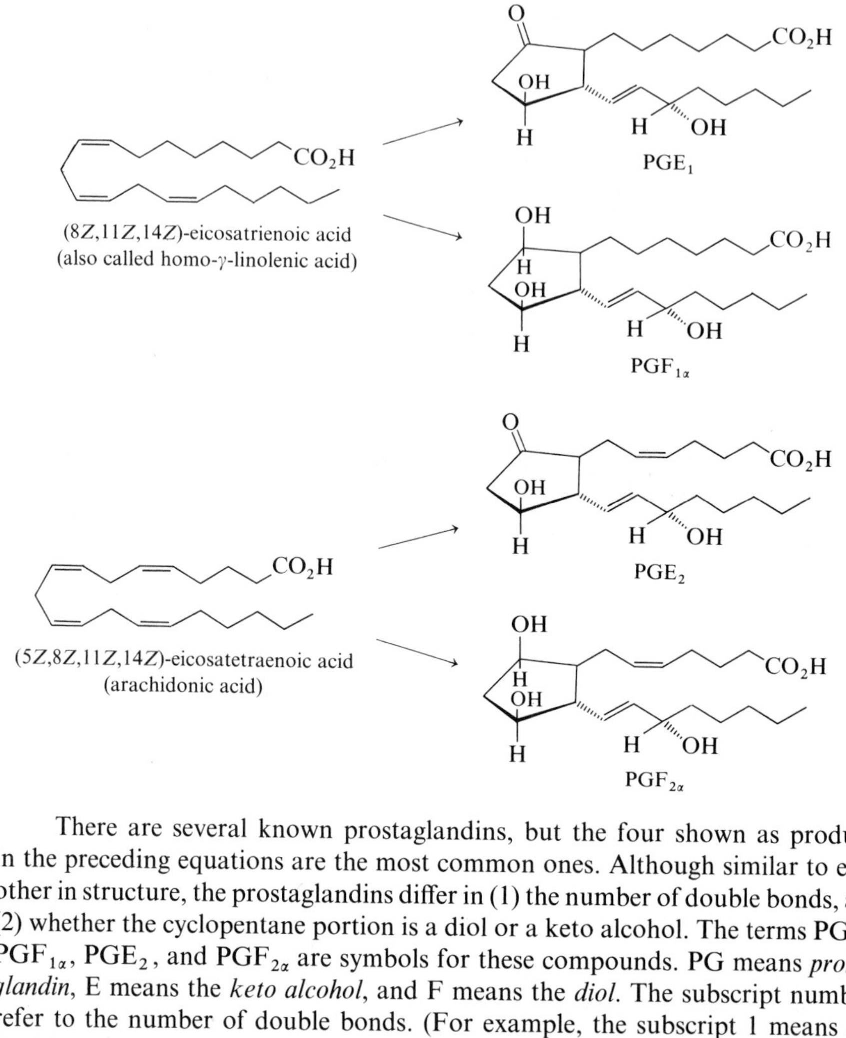

(8Z,11Z,14Z)-eicosatrienoic acid
(also called homo-γ-linolenic acid)

PGE$_1$

PGF$_{1\alpha}$

(5Z,8Z,11Z,14Z)-eicosatetraenoic acid
(arachidonic acid)

PGE$_2$

PGF$_{2\alpha}$

There are several known prostaglandins, but the four shown as products in the preceding equations are the most common ones. Although similar to each other in structure, the prostaglandins differ in (1) the number of double bonds, and (2) whether the cyclopentane portion is a diol or a keto alcohol. The terms PGE$_1$, PGF$_{1\alpha}$, PGE$_2$, and PGF$_{2\alpha}$ are symbols for these compounds. PG means *prostaglandin*, E means the *keto alcohol*, and F means the *diol*. The subscript numbers refer to the number of double bonds. (For example, the subscript 1 means *one* double bond.) The subscript α refers to the configuration of the OH at carbon 9 (*cis* to the carboxyl side chain).

The biosynthesis of the prostaglandins probably proceeds by a free-radical mechanism. Abstraction of a doubly allylic hydrogen, followed by an allylic rearrangement of one double bond, leads to attack by a hydroxyl radical at carbon 15 of the fatty acid.

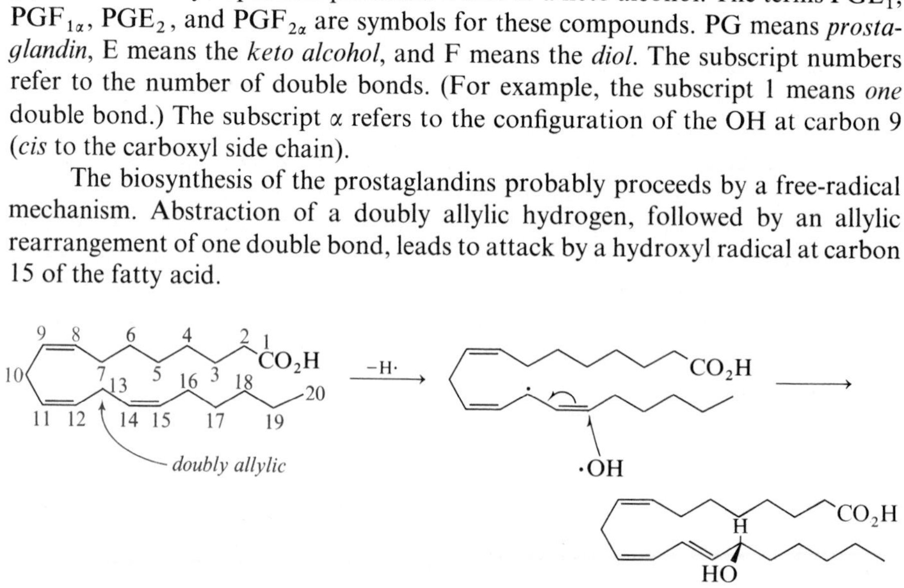

Oxygen adds to carbons 9 and 11, and the cyclopentane ring is formed. Oxidation of this peroxy intermediate leads to the keto alcohol, while reduction leads to the diol.

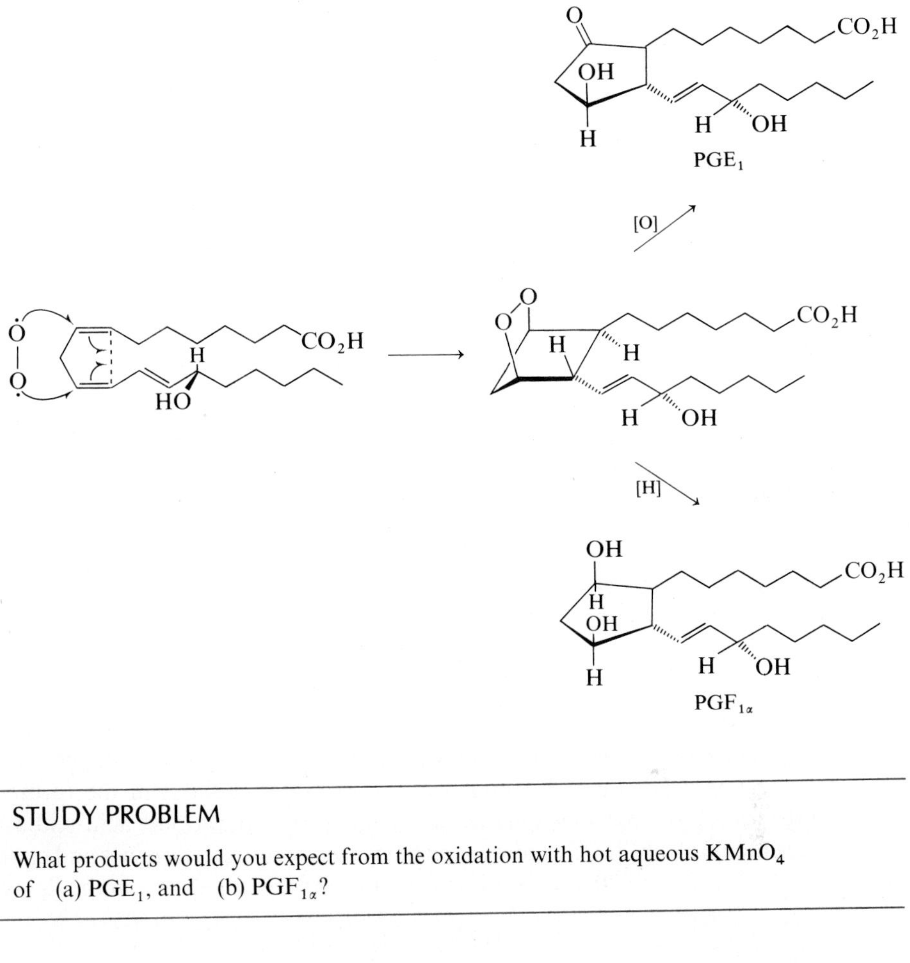

PGE₁

[O]

[H]

PGF₁ₐ

STUDY PROBLEM

20.3. What products would you expect from the oxidation with hot aqueous KMnO₄ of   (a) PGE₁, and   (b) PGF₁ₐ?

SECTION 20.5.

Terpenes

The odorous components of plants that can be separated from other plant materials by steam distillation are called **essential oils**. Many essential oils, such as those from flowers, are used in perfumes. Most essential oils are mixtures of **terpenes**, a class of natural products found in both plants and animals. The name comes from *turpentine*, which is rich in terpenes.

All terpenes appear to have been constructed by the *head-to-tail joining of isoprene skeletal units*. (The *head* is the end closer to the methyl branch.) Terpenes may contain two, three, or more isoprene units. Their molecules may be open-chain or cyclic. They may contain double bonds, hydroxyl groups, carbonyl groups, or other functional groups. A terpene-like structure that contains elements other than C and H is called a **terpenoid**.

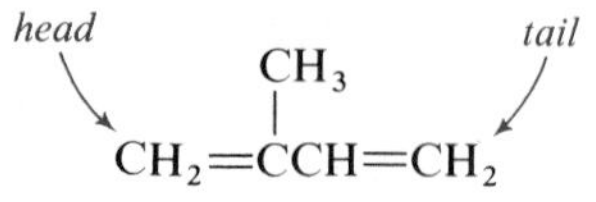

isoprene

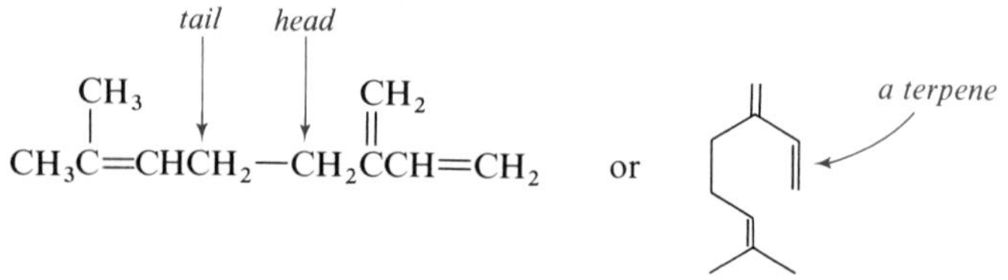

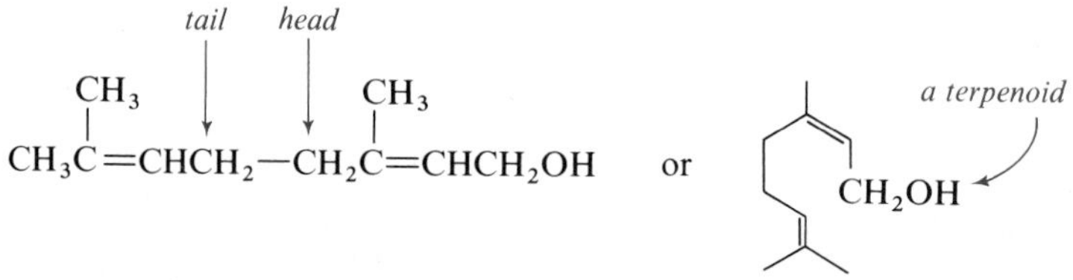

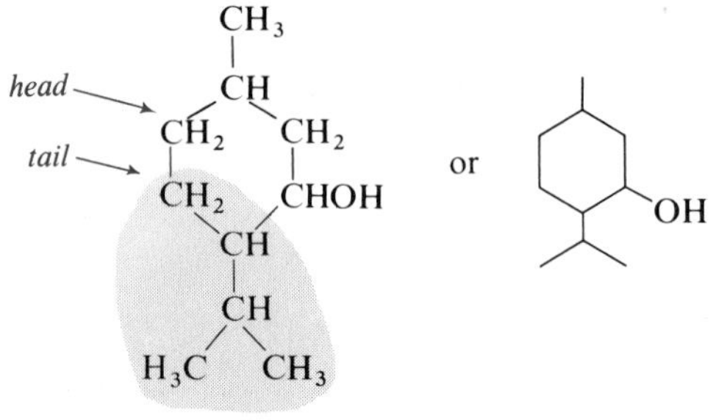

a terpene   or

a terpenoid   or

or

Although the idea is appealing, terpenes do not arise from polymerization of isoprene. The first step in terpene biosynthesis is an enzymatic ester condensation of the acetyl portions of acetylcoenzyme A. Intermediates in the formation of terpenes are the pyrophosphates (diphosphates) of mevalonic acid and a pair of isopentenyl alcohols. An abbreviated biosynthetic route to terpenes and cholesterol is found in Figure 20.4. (The phosphate groups are not shown in the figure.)

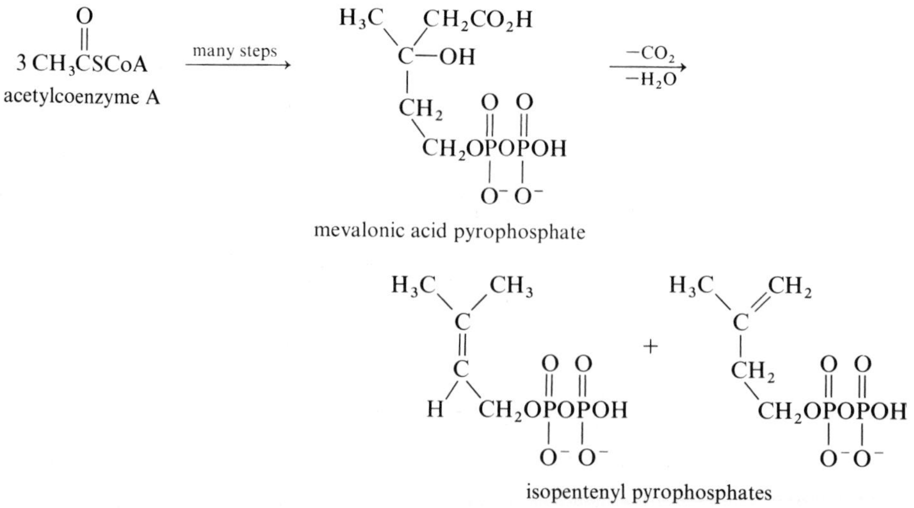

mevalonic acid pyrophosphate

isopentenyl pyrophosphates

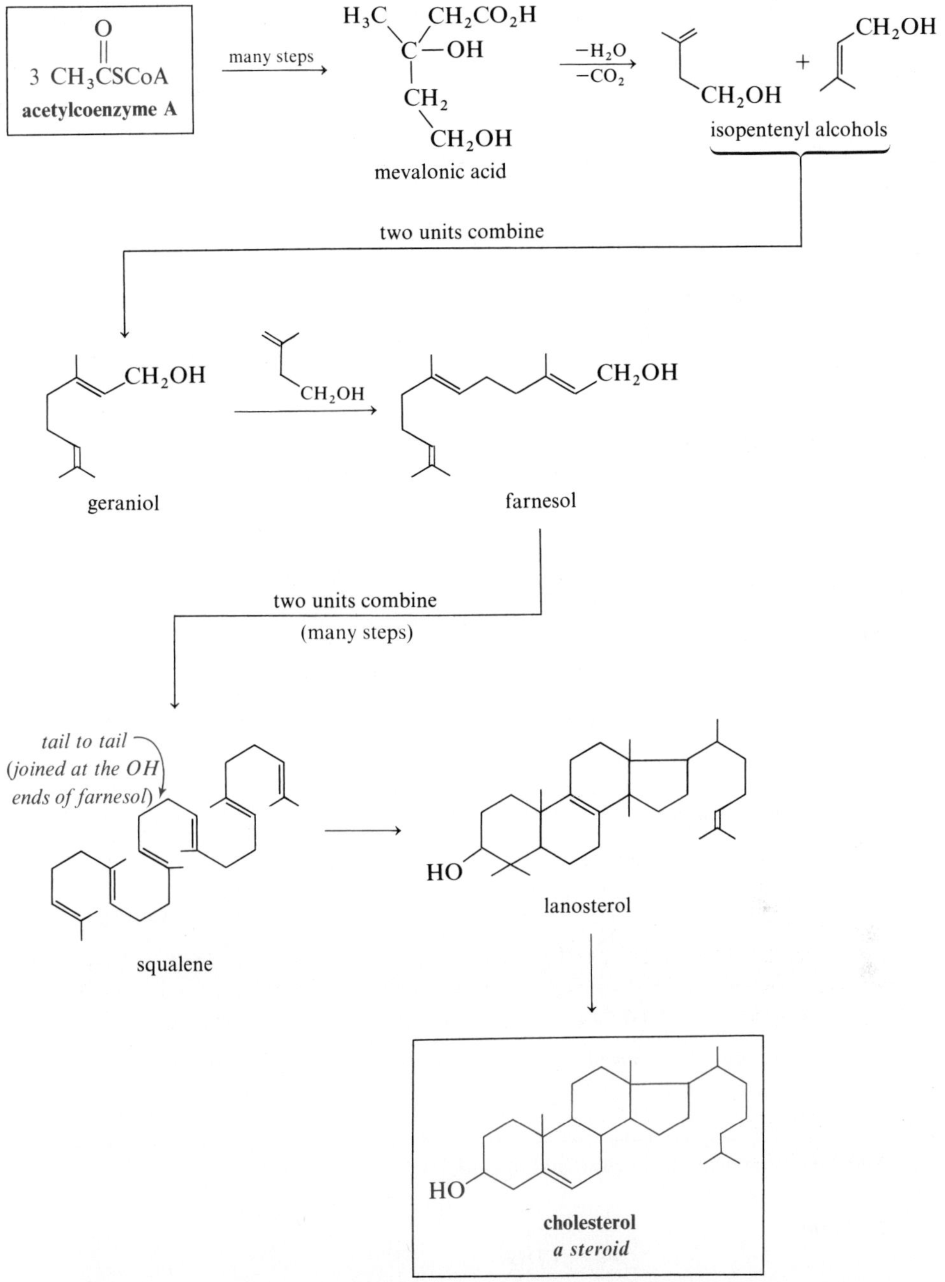

FIGURE 20.4. Abbreviated biosynthetic path from acetylcoenzyme A to cholesterol. (The conversion of squalene to lanosterol, including stereochemistry, is shown in Figure 20.5.)

Terpenes are categorized by the *number of pairs of isoprene units* they contain:

monoterpenes:	two isoprene units
sesquiterpenes:	three isoprene units
diterpenes:	four isoprene units
triterpenes:	six isoprene units
tetraterpenes:	eight isoprene units

A. Monoterpenes

Monoterpenes, with skeletons that contain only two isoprene units, are the simplest of the terpenes. Yet, even monoterpenes exhibit a variety of structures. (Although some terpenes and terpenoids, such as geraniol, are found in a variety of organisms, we have indicated sources especially rich in the following compounds.)

Acyclic monoterpenes:

geraniol

in roses

citral (geranial)

in lemongrass

Cyclic monoterpenes:

limonene

in citrus fruits

menthol

in mint

camphor

from camphor trees

α-pinene

in turpentine

SAMPLE PROBLEM

Show the isoprene units in citral and in camphor.

Solution:

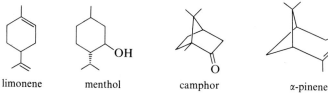

STUDY PROBLEM

20.4. Show the isoprene units in menthol and α-pinene.

Bridged ring systems, such as in camphor and α-pinene, are not uncommon in terpenes. Camphor is an example of a compound with a bicyclo[2.2.1]heptane skeleton. The prefix **bicyclo-** refers to a system in which two rings share two or more carbon atoms between them (that is, fused rings). The numbers [2.2.1] represent the number of carbon atoms in each "arm" of the ring system. (An arm constitutes the carbon atoms between bridgehead carbons.)

bicyclo[2.2.1]heptane
(norbornane)

Bridged systems are invaluable for mechanistic studies because the rings are frozen into one conformation. For example, the six-membered ring in camphor is frozen into a boat form, rather than the usual chair form.

Unless the ring size is quite large, bridged rings must have *cis* ring junctures. Overlap of orbitals to form a *trans* ring juncture would be sterically impossible.

Another restriction of bridged rings, called **Bredt's rule**, is that the bridgehead carbon cannot be doubly bonded. A model of a bridged system such as in camphor or pinene will demonstrate that the geometry is wrong for *p*-orbital overlap. Bredt's rule does not apply to fused rings that share only two carbons or to large rings that can accomodate a *trans* double bond.

stable   no double bond here

B.   Higher terpenes

We have shown several examples of monoterpenes, which are common in plants. Some higher terpenes of interest are *squalene* (found in yeast, wheat germ, and shark liver oil) and *lanosterol* (a component of lanolin, which is obtained from wool fat). Both these compounds are intermediates in the biosynthesis of steroids. The conversion of squalene to lanosterol is an important one that will be discussed shortly. (See Figure 20.4 or 20.5 for the structures of these two compounds.) Natural rubber is a polyterpene that we have already discussed (Section 9.17B).

Carrots contain an orange-colored tetraterpene called *carotene*. (If a person eats too many carrots, the deposition of carotene will color his skin orange. However, time is a cure for this condition.) Carotene can be cleaved enzymatically into two units of vitamin A. (The role of vitamin A in vision will be discussed in Chapter 21.)

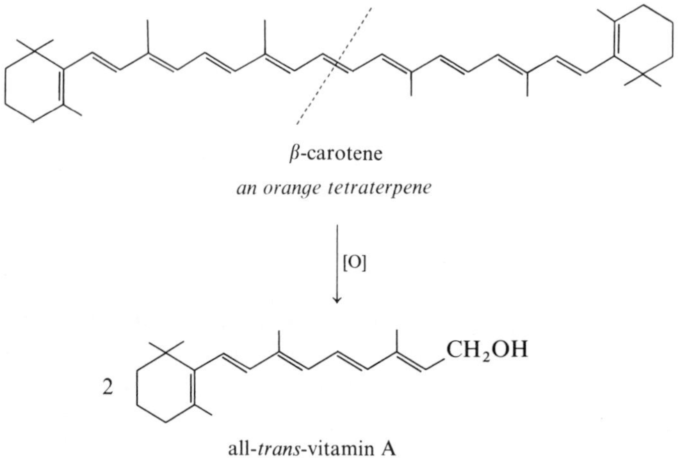

β-carotene

an orange tetraterpene

[O]

all-*trans*-vitamin A

C.   Reactions of terpenes

Terpenes undergo reactions typical of their functional groups. For example, addition to alkene double bonds can occur.

SAMPLE PROBLEM

Predict the product of addition of HCl to limonene. (Remember Markovnikov's rule.)

Solution:

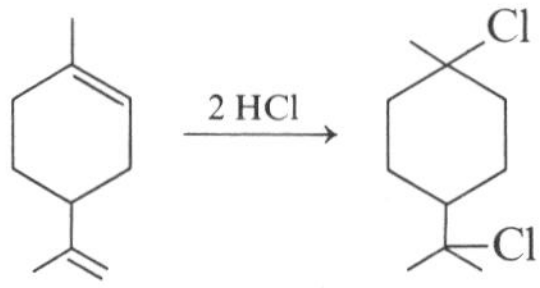

Terpene structures contain secondary and tertiary carbon atoms and often contain double bonds. Addition of H^+ to a double bond can lead to a carbocation, which then can undergo rearrangement to a more stable carbocation (Section 5.6G). Rearrangements of terpene skeletons are very common. For example, the addition of HCl to α-pinene does not yield a product with the α-pinene ring system intact. Instead, bornyl chloride, with a bornane ring system, is obtained.

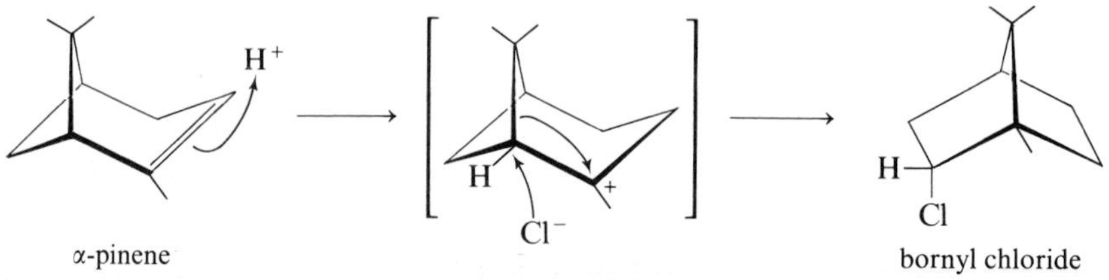

α-pinene   bornyl chloride

The sequence of reactions leading from squalene to lanosterol, which occurs in the biosynthesis of steroids, shows many of the features of terpene reactions. This sequence is shown in Figure 20.5. The first step is an enzymatic epoxide formation. Next, the epoxide oxygen is protonated and the epoxide ring is opened,

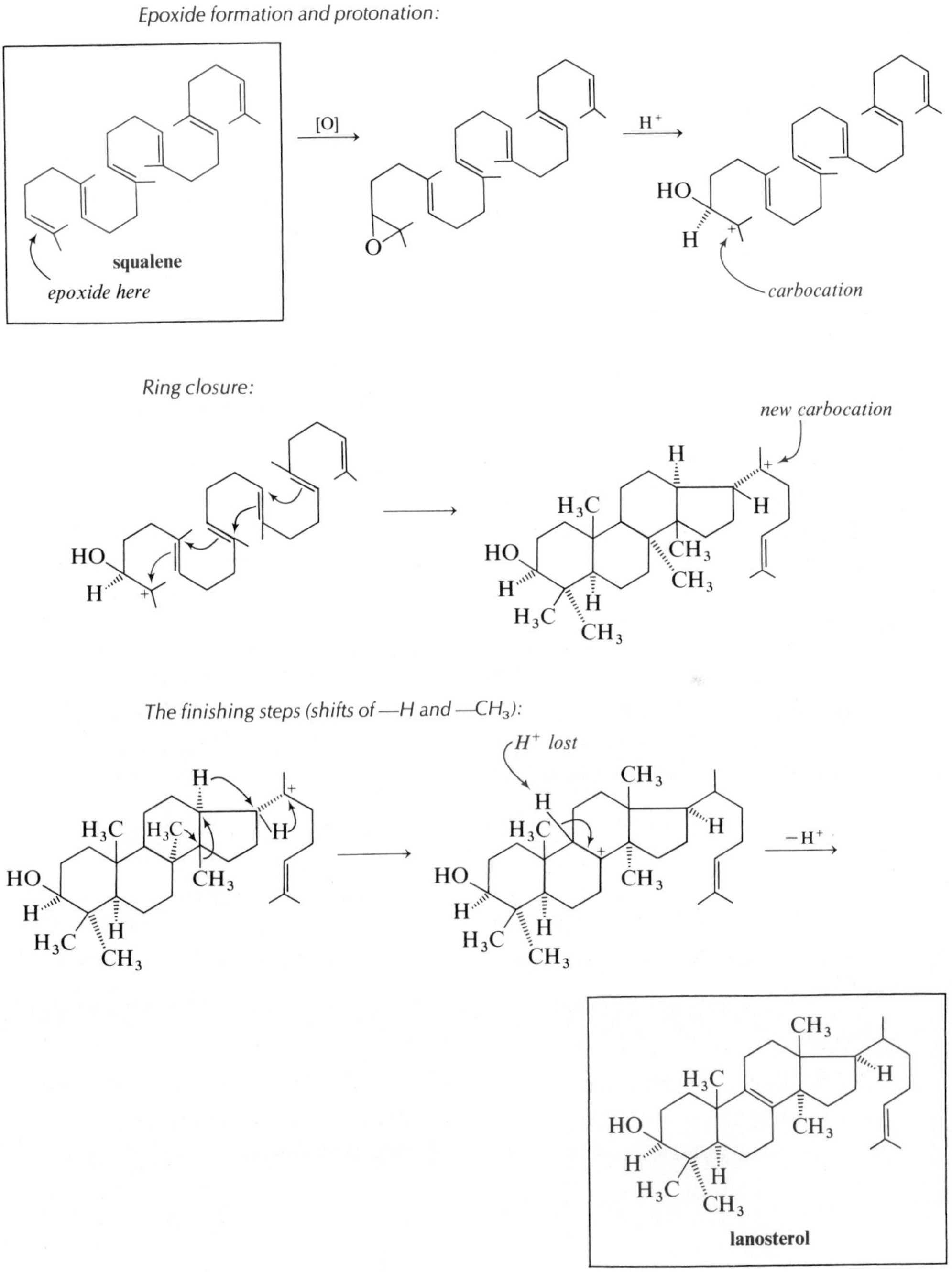

Epoxide formation and protonation:

Ring closure:

The finishing steps (shifts of —H and —CH₃):

FIGURE 20.5. The conversion of squalene to lanosterol.

leaving a carbocation. The formation of this carbocation sets in motion a concerted series of electron shifts that results in the ring closures of *four rings* and yields
the steroid ring system. The ring closures are followed by hydride and methyl
shifts to yield lanosterol. It is amazing that this whole sequence is catalyzed by
only one enzyme, squalene oxide cyclase!

SECTION 20.6.

Pheromones

Humans communicate by talking, using sign language, painting pictures, or
writing letters (or books). But insects and some animals communicate chemically.
A chemical secreted by one individual of a species that brings forth a response in
another individual of the same species is called a **pheromone** (from the Greek
phero, "carrier"). Insect pheromones called alarm pheromones signify danger.
Pheromones are also used to recruit others; for example, a pheromone secreted
by one bee helps alert other bees to the location of a food source. Insect sex
attractants are another class of pheromones.

Extremely small quantities of a pheromone can elicit the desired response.
A typical female insect may carry only 10^{-8} gram of sex attractant, yet that is
enough to attract over a billion males from miles away! A male gypsy moth can
"smell" a female at a distance of 7 miles.

Pheromone chemistry is an exciting and vigorous field of research that
includes natural-product isolation, structure identification, synthesis, and
biological studies. It is an area of study that also is of immediate practical value.
For example, sex-attractant pheromones have been used for insect control. In
some cases, male insects may be lured by a sex-attractant pheromone, trapped,
and then sterilized and released to mate unproductively with the females.

Many insect pheromones are not complex in structure. Geraniol and citral,
both terpenes, are recruiting pheromones for honeybees, while isoamyl acetate
(not a terpene) is a bee alarm pheromone. (Isoamyl acetate is also the principal
odorous component of banana oil.)

$$CH_3\overset{\displaystyle O}{\overset{\|}{C}}OCH_2CH_2CH(CH_3)_2$$

3-methylbutyl acetate
(isoamyl acetate)

geraniol   citral

The following compounds are sex attractants for a few different species of
insects:

$$cis\text{-}(CH_3)_2CH(CH_2)_4\overset{\displaystyle O}{\overset{\diagdown\diagup}{C}}HCH(CH_2)_9CH_3$$

gypsy moth

$$cis\text{-}CH_3(CH_2)_3CH=CH(CH_2)_6\overset{\displaystyle O}{\overset{\|}{O C}}CH_3$$

cabbage looper

boll weevil

$$cis\text{-}CH_3(CH_2)_{12}CH=CH(CH_2)_7CH_3$$

house fly

STUDY PROBLEM

20.5. The formula for the alarm pheromone for one species of ant is $C_7H_{14}O$. When treated with I_2 and NaOH, this pheromone yields iodoform and *n*-hexanoic acid. What is the structure of the pheromone?

SECTION 20.7.

Steroids

We have mentioned steroids several times. Now let us take a more detailed look at these compounds. **A steroid** is a compound that contains the following ring system. The four rings are designated *A*, *B*, *C*, and *D*. The carbons are numbered as shown, starting with ring *A*, progressing to ring *D*, then to the angular (bridgehead) methyl groups, and finally to a side chain if it is present.

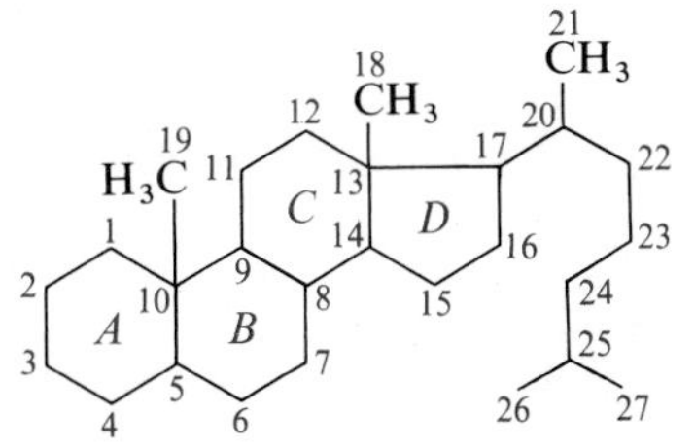

cholestane

Many steroids may be named as derivatives of this structure, which is called **cholestane**. (Steroid nuclei with different side chains also have names; however, we will not present them here.)

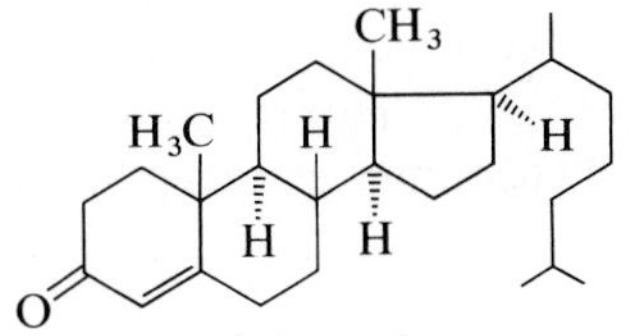

4-cholesten-3-one

a steroid with a cholestane skeleton

A. Conformation of steroids

Recall from Section 4.5B that the more stable isomer of a 1,2-dialkylcyclohexane is the one with both substituents *equatorial*. A *trans*-1,2-disubstituted cyclohexane can exist in a preferred conformation in which both substituents are equatorial, while the *cis*-isomer must have one axial substituent and one equatorial substituent. The *trans*-isomer is therefore the more stable one.

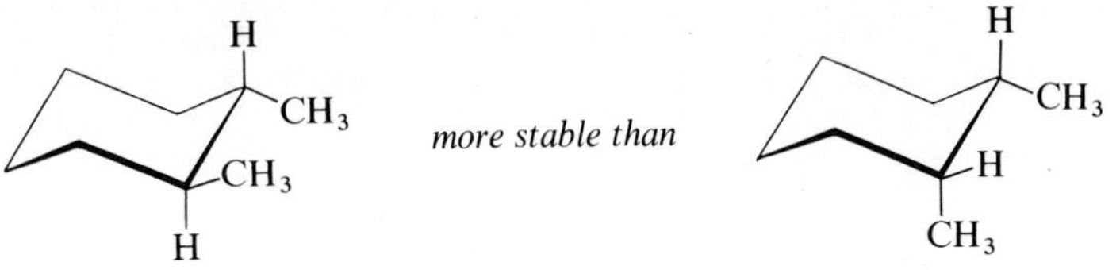

more stable than

trans (e,e)   *cis (a,e)*

One ring fused to another in the 1,2-positions can be *cis* or *trans*. The following structures represent *cis-* and *trans-*decalin.

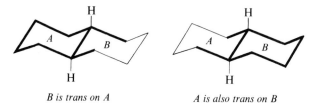

To verify that these two formulas represent *cis* and *trans* ring junctures, make the models. Alternatively, check the hydrogens on the ring-juncture carbons. They, too, must be *cis* at a *cis* ring juncture or *trans* at a *trans* ring juncture.

Note that when ring *B* is *trans* on ring *A*, ring *A* is also *trans* on ring *B*.

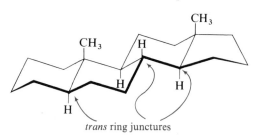

B is trans on A   *A is also trans on B*

As may be seen in the formulas for *trans-* and *cis-*decalin, the *trans* ring juncture is *e,e*, while the *cis* ring juncture is *a,e*. The *trans*-isomer is more stable than the *cis*-isomer by about 3 kcal/mole. The steroid nucleus contains three ring junctures (*A/B, B/C,* and *C/D*). In nature, these are usually the more stable *trans* ring junctures (but we will encounter an exception later in this chapter).

trans ring junctures

SAMPLE PROBLEM

Tell whether each of the following ring junctures is *cis, trans,* or neither:

(a)   (b)

(c)

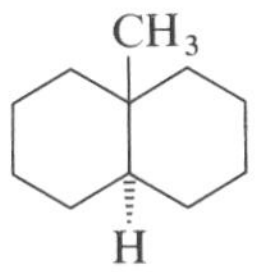

(d) 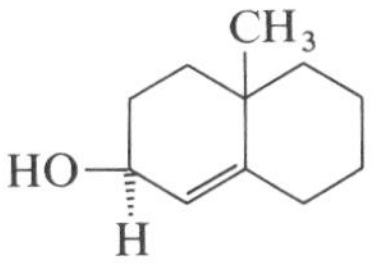   HO—

CH$_3$

CH$_3$

H

H

Solution: (a) *trans*; (b) neither (why not?); (c) *trans*; (d) neither.

Groups substituted on a steroid ring system may be *below* or *above* the plane of the ring as the ring system is generally drawn. A group that is below the plane (*trans* to the angular methyl groups) is called an α group, while one that is above the plane (*cis* to the angular methyl groups) is called a β group. (Do not confuse this use of α and β with their use in carbohydrate chemistry, where they refer to groups only at the anomeric carbon. In both classes of compounds, however, β denotes a group projected upward in the usual conformational formula, while α denotes a group projected downward.)

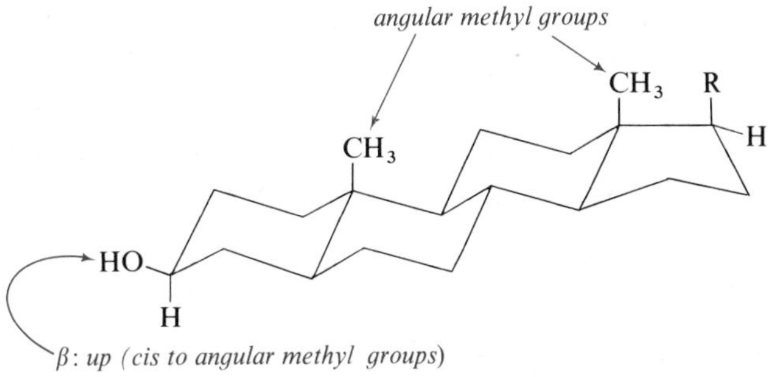

angular methyl groups

CH$_3$ R

CH$_3$

H

HO

H

β: up (cis to angular methyl groups)

The terms α and β may be used in the names of steroids to designate the stereo-chemistry of substituents. In the name, α or β immediately follows the position number for the substituent.

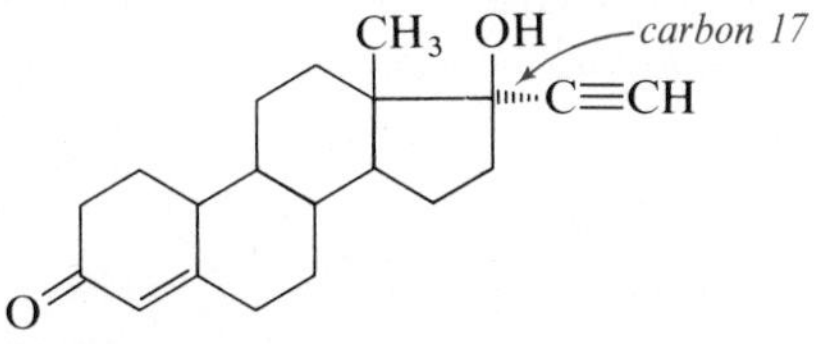

CH$_3$ OH   carbon 17

C≡CH

O

17α-ethynyl-17β-hydroxy-4-estren-3-one

an oral contraceptive, usually called norethinodrone

STUDY PROBLEMS

20.6. Draw conformational formulas for:

(a) cholestane   (b) 3α-cholestanol   (c) norethinodrone

20.7. Which is the more stable isomer, 3α-cholestanol or 3β-cholestanol? Why?

B.   Some important steroids

Cholesterol is the most widespread animal steroid and is found in almost all animal tissues. Human gallstones and egg yolks are especially rich sources of this compound. Cholesterol is a necessary intermediate in the biosynthesis of the steroid hormones; however, since it can be synthesized from acetylcoenzyme A, it is not a dietary necessity. High levels of blood cholesterol are associated with arteriosclerosis (hardening of the arteries), a condition in which cholesterol and other lipids coat the insides of the arteries. Whether or not the level of cholesterol in the blood can be controlled by diet is still a topic of controversy.

cholesterol
(5-cholesten-3β-ol)

A steroid related to cholesterol, *7-dehydrocholesterol*, which is found in the skin, is converted to *vitamin D* when irradiated with ultraviolet light. This reaction is discussed in Section 17.5.

Cortisone and **cortisol** (hydrocortisone) are two of 28 or more hormones secreted by the adrenal cortex. Both these steroids alter protein, carbohydrate, and lipid metabolism in ways not entirely understood. They are widely used to treat inflammation due to allergies or rheumatoid arthritis. Many related steroids with a carbonyl or hydroxyl group at carbon 11 have similar activity.

cortisone

cortisol

Sex hormones are produced primarily in the testes or the ovaries; their production is regulated by pituitary hormones. The sex hormones impart secondary sex characteristics and regulate the sexual and reproductive functions. Male hormones are collectively called **androgens**; female hormones, **estrogens**; and pregnancy hormones, **progestins**.

In pregnant females, the presence of progesterone suppresses ovulation and menstruation. Synthetic progestins, such as norethynodrel (Enovid), are used to suppress ovulation as a method of birth control.

Androgens:

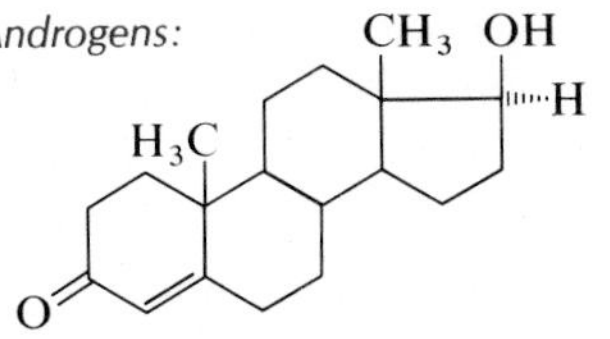

testosterone

the principal male hormone

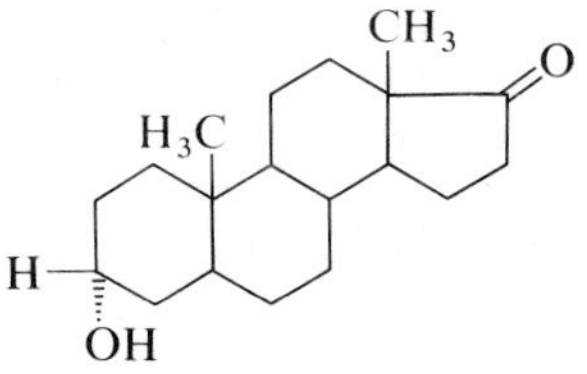

androsterone

a metabolized form of testosterone

Estrogens:

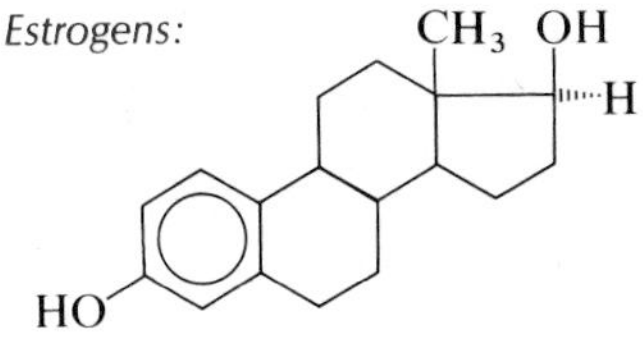

estradiol

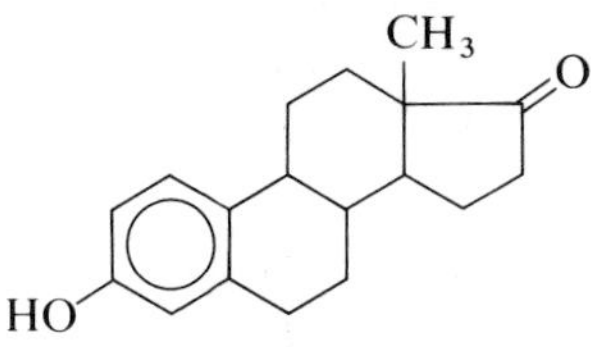

estrone

Progestins:

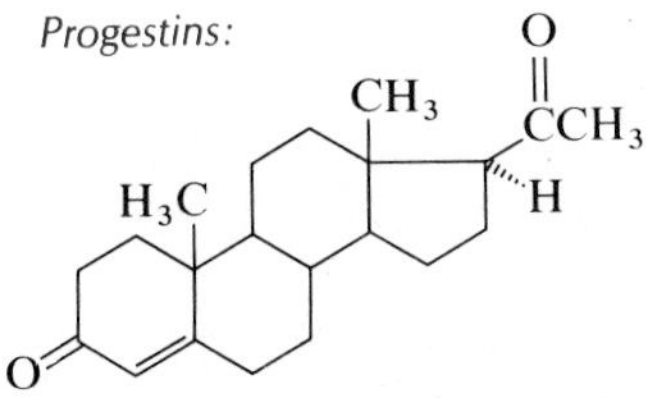

progesterone

suppresses ovulation

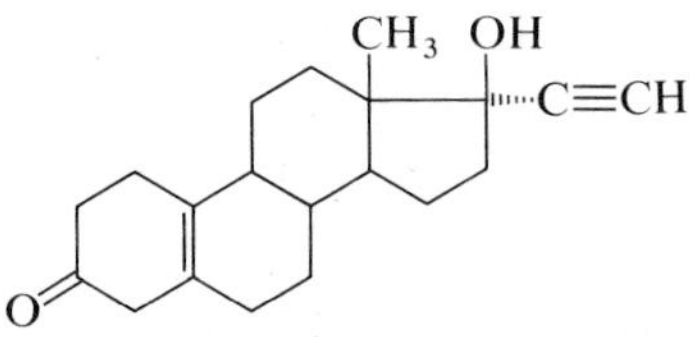

norethynodrel

a synthetic progestin

Bile acids are found in bile, which is produced in the liver and stored in the gall bladder. The structure of cholic acid, the most abundant bile acid, follows. Cholic acid, as well as other bile acids, has a *cis A/B* ring juncture instead of the usual *trans A/B* ring juncture.

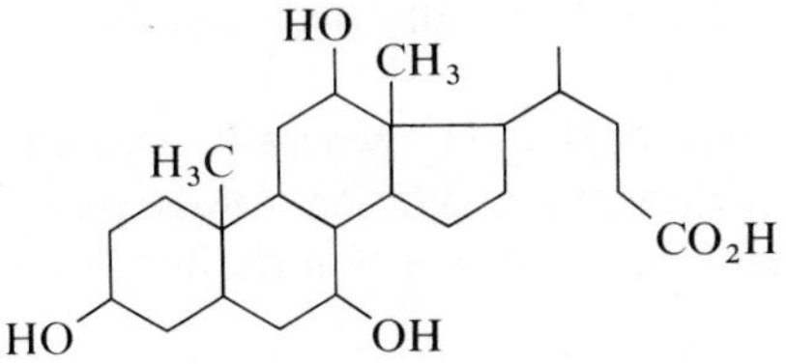

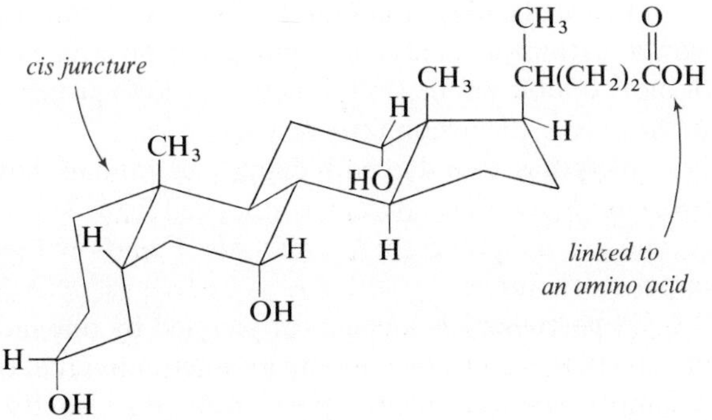

cholic acid

Bile acids are secreted into the intestines in combination with sodium salts of either glycine or taurine ($H_2NCH_2CH_2SO_3H$). The bile acid–amino acid link is an amide link between the carboxyl group of the bile acid and the amino group of the amino acid. In this combined form, the bile acid–amino acid acts to keep lipids emulsified in the intestines, thereby promoting their digestion.

$$RCO_2H + H_2NCH_2CO_2^- \ Na^+ \xrightarrow[-H_2O]{enzymes} \overset{\overset{\textstyle O}{\|}}{R}CNHCH_2CO_2^- \ Na^+$$

cholic acid   sodium salt of   *an emulsifying agent*
                     glycine

STUDY PROBLEM

20.8.   What structural features of cholic acid combined with glycine allow it to act as an emulsifying agent?

Summary

Fats and **oils** are triglycerides: triesters of glycerol and long-chain fatty acids. Generally, oils (liquid) contain more unsaturation than fats (solid). Fatty acids contain even numbers of carbon atoms and *cis* double bonds.

A **soap** is the alkali-metal salt of a fatty acid. A **detergent** is a salt of a sulfonate or sulfate that contains a long hydrocarbon chain. Naturally occurring **waxes** are esters of fatty acids and long-chain alcohols.

Phosphoglycerides, such as lecithins and cephalins, generally contain glycerol esterified with two fatty acids and a phosphatidylamine. Other phospholipids may differ in structure, but all contain long hydrocarbon chains plus a phosphate group and an amino group.

Prostaglandins, thought to be moderators of hormone activity, are 20-carbon carboxylic acids containing a cyclopentane ring plus hydroxyl groups, one or more double bonds, and sometimes a keto group. They are biosynthesized from unsaturated, 20-carbon fatty acids.

Terpenes, found in both plants and animals, have the skeletons of diisoprene, triisoprene, etc. They are not synthesized from isoprene, but from acetylcoenzyme A, and are the precursors of steroids. Functional groups may also be present in terpene structures.

A **pheromone** is a chemical secreted by one individual of a species (notably insects) that elicits a response in another individual of the same species. Generally of simple structure, pheromones are used to signify danger or food, or to act as sex attractants.

Cholesterol, cortisone, the sex hormones, and the bile acids are all **steroids**, compounds that contain the following ring system:

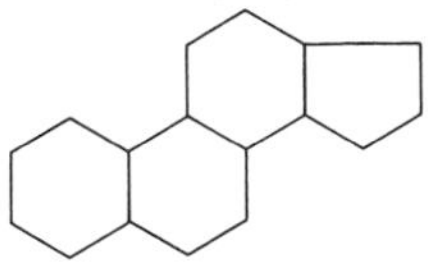

STUDY PROBLEMS

20.9. A mixed triglyceride contains two units of stearic acid and one unit of palmitoleic acid. What are the major organic products when this triglyceride is treated with:

(a) an excess of dilute aqueous NaOH and heat?
(b) H_2, copper chromite catalyst, heat, and pressure?
(c) bromine in CCl_4?

20.10. What are the major organic products of ozonolysis, followed by oxidative work-up, of each of the following fatty acids?　(a) palmitic acid;　(b) palmitoleic acid;　(c) linoleic acid; (d) linolenic acid.

20.11. How would you distinguish chemically between:　(a) tripalmitin and tripalmitolein; (b) beeswax and beef fat;　(c) beeswax and paraffin wax;　(d) linoleic acid and linseed oil; (e) sodium palmitate and sodium *p*-decylbenzenesulfonate;　(f) a vegetable oil and a motor oil?

20.12. Hydrolysis of *trimyristin*, a fat obtained from nutmeg, yields only one fatty acid, myristic acid. This same acid can be obtained from 1-bromododecane (the 12-carbon, continuous-chain alkyl bromide) and diethyl malonate in a malonic ester synthesis. What are the structures of myristic acid and trimyristin?

20.13. Which of the following compounds would show detergent activity?

(a) CH_3OSO_3Li

(b) $CH_3(CH_2)_4\overset{\displaystyle CH_3}{\overset{|}{CH}}(CH_2)_2OSO_3K$

(c) 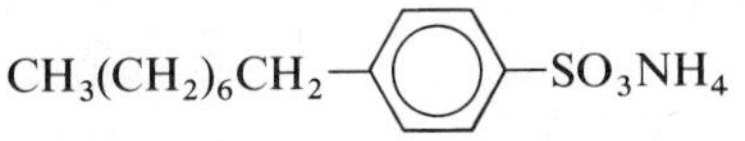 $CH_3(CH_2)_6CH_2$—⟨○⟩—SO_3NH_4

(d) $CH_3(CH_2)_{16}CH_2OH$

20.14. A fat of unknown structure is found to be optically active. Saponification, followed by acidification, yields two equivalents of palmitic acid and one equivalent of oleic acid. What is the structure of the fat?

20.15. Starting with tristearin as the only organic reagent, show by equations how you would prepare a wax.

20.16. Write the structures of the products of complete hydrolysis of each of the following phosphoglycerides with dilute aqueous NaOH:

$$CH_3(CH_2)_{16}CO_2CH_2$$

(a) $\quad CH_3(CH_2)_7CH{=}CH(CH_2)_7CO_2\overset{|}{C}H \quad O^-$

$$\underset{\underset{O}{\parallel}}{\overset{|}{C}}H_2OPOCH_2CH_2\overset{+}{N}(CH_3)_3 \quad Cl^-$$

$$CH_3(CH_2)_{16}CO_2CH_2$$

(b) $CH_3(CH_2)_7CH{=}CH(CH_2)_7CO_2\overset{|}{C}H \quad O^-$

$$\underset{\underset{O}{\parallel}}{\overset{|}{C}}H_2OPOCH_2CH_2\overset{+}{N}H_3 \quad Cl^-$$

20.17. *Cerebrosides* are glycosphingolipids that are found primarily in the sheaths of nerve cells. What fatty acid and sugar would be isolated from hydrolysis of the following cerebroside in acidic solution?

$$CH_3(CH_2)_{12}CH$$
$$\parallel$$
$$HC$$
$$|$$
$$CHOH \quad O$$
$$\qquad\qquad \parallel$$
$$HOCH_2 \qquad | \qquad$$
$$HO\!\!-\!\!O\ OCH_2CHNHC(CH_2)_{22}CH_3$$
$$OH$$
$$OH$$

20.18. Indicate the isoprene units and classify each structure as a monoterpene, sesquiterpene, or diterpene:

(a) $\qquad$ (b)

caryophyllene   in oil of balsam
(oil of cloves)

20.19. Write formulas for:

(a) bicyclo[3.2.0]heptane   (b) bicyclo[3.1.1]heptane   (c) bicyclo[3.2.1]octane

20.20. Name the following ring systems:

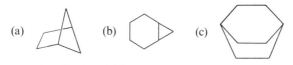

(a) $\qquad$ (b) $\qquad$ (c)

20.21. Predict the possible products of E2 elimination of the following alkyl chlorides:

(a) *cis*-1-chloro-2-methylcyclohexane   (b)

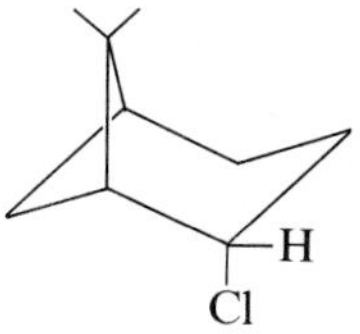

20.22. The structure of the aggregating pheromone of the bark beetle follows:

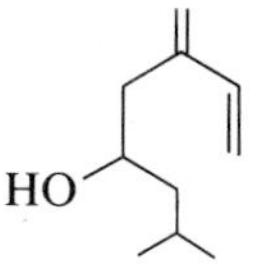

What are the organic products of its reaction with: (a) excess Br_2 in CCl_4; (b) excess gaseous HBr (no O_2 or peroxides);   (c) hot $KMnO_4$ solution?

20.23. Tell whether the substituents or the ring junctures in the following compounds are *cis* or *trans*:

(a) 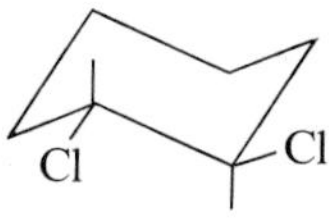   (b) 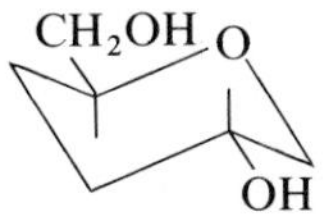   (c)

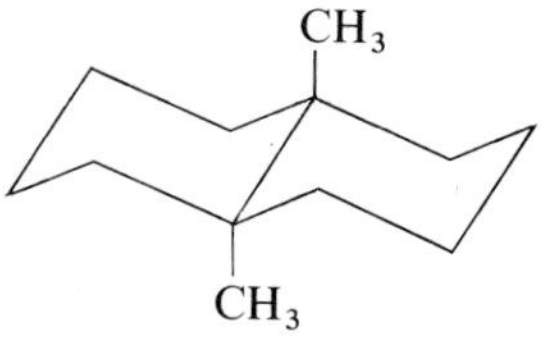

(d) 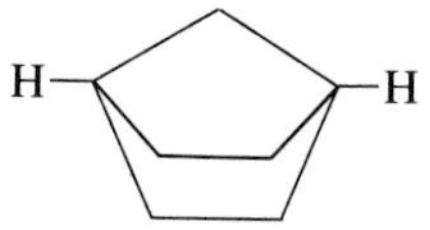   (e) 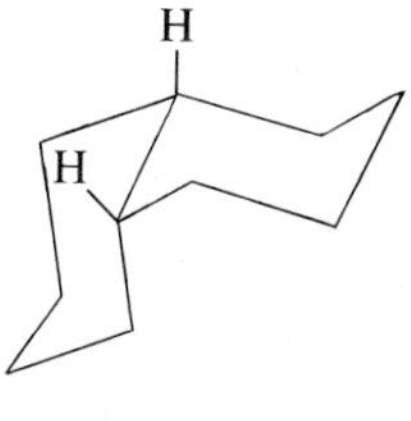   (f)

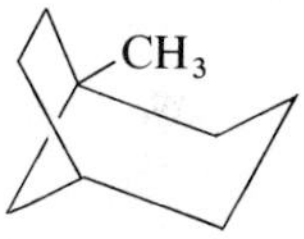

20.24. Which structure of each of the following pairs is more stable? Why?

(a)

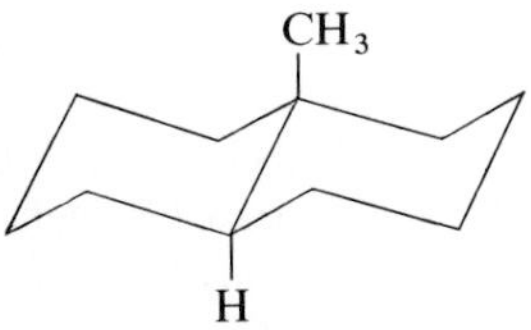

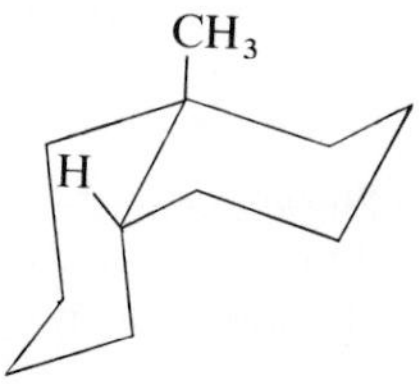

(1)   **(2)**

(b)

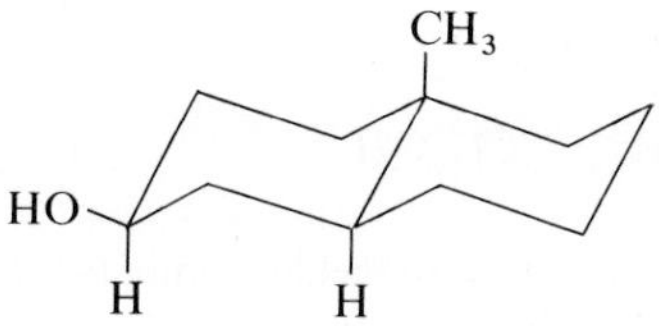

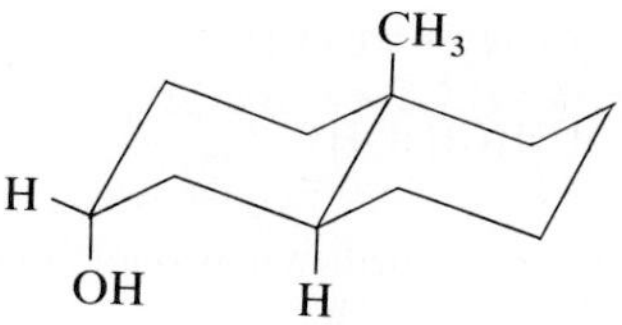

(1)   **(2)**

20.25. Draw the structure of each of the following steroids. (*Acetoxy-* means CH_3CO_2—.)

    (a)   5-cholesten-3α-ol
    (b)   1-cholesten-3-one
    (c)   4,6-cholestadien-3β-ol
    (d)   3β-acetoxy-6α,7β-dibromocholestane

20.26. How would you separate a mixture of estradiol and testosterone?

20.27. The structure of the bile acid *desoxycholic acid* is shown below. Redraw the structure showing the conformation of the ring system.

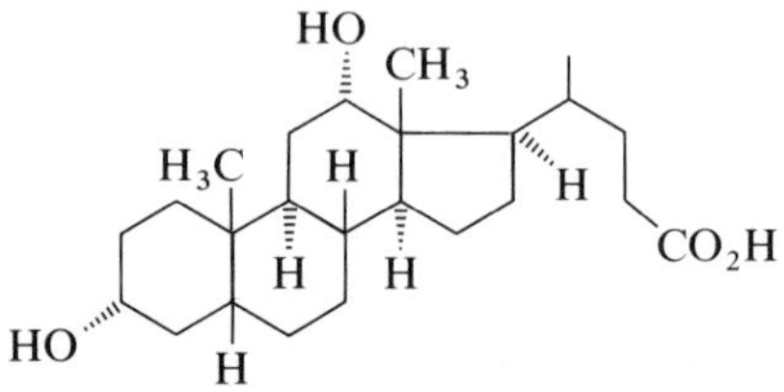

20.28. When cholic acid is treated with acetic anhydride, one monoacetate is formed preferentially. What is the structure of this monoacetate and why is it formed preferentially?

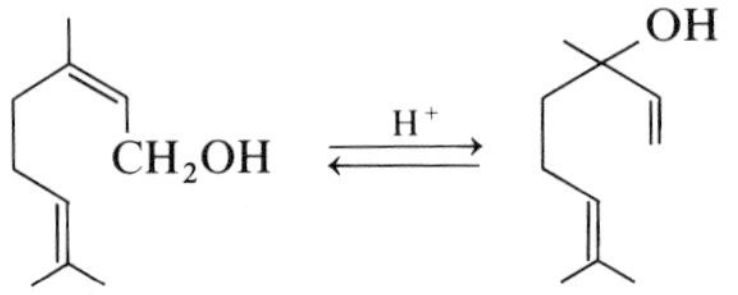

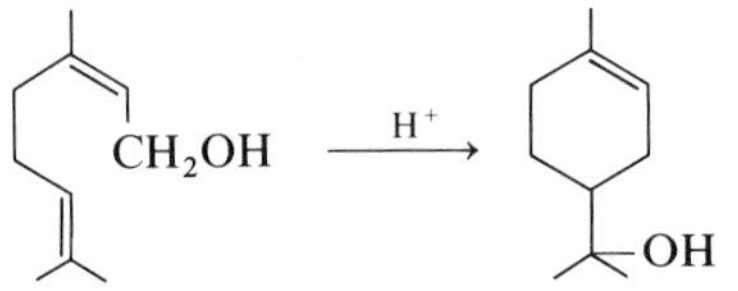

20.29. Suggest mechanisms for the following terpene rearrangements:

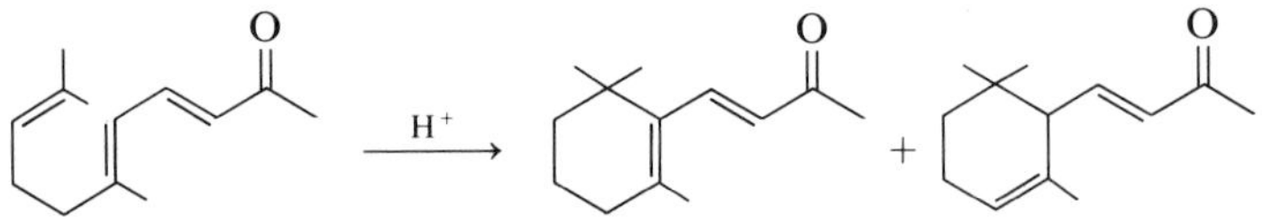

20.30. A biochemist prepared the following samples of mevalonic acid labeled with carbon-14. The three mevalonic acids were fed to a series of plants. Later, the isopentenyl alcohols from the plants were isolated. In which position would the carbon-14 be found in each of the product isopentenyl alcohols?

20.31. In a plant, two isopentenyl alcohol molecules combine to yield geraniol (see Figure 20.4). Identify the position of carbon-14 in geraniol arising from the labeled mevalonic acid in Problem 20.30(b).

20.32. *Citronellal* ($C_{10}H_{18}O$) is a terpenoid that undergoes reaction with Tollens reagent to yield citronellic acid ($C_{10}H_{18}O_2$). Chromate oxidation of citronellal yields acetone and $HO_2CCH_2CH(CH_3)CH_2CH_2CO_2H$. What is the structure of citronellal?

20.33. When cholesterol is treated with Br_2, a dibromide **I** is formed. When **I** is heated, another dibromide **II** is formed. **I** and **II** are isomers that differ only in the ring juncture between rings *A* and *B* (one dibromide has a *trans A/B* ring juncture; the other has a *cis A/B* ring juncture).

(a) Give the conformational structures of **I** and **II**.
(b) Show the mechanism for the formation of **I**.
(c) Explain why heating causes the ring juncture to change.

20.34. The synthesis of the acyclic monoterpene *alloocimene* was accomplished by the following route. What is the structure of alloocimene? (Show the structures of the intermediate products in your answer.)

$$\overset{\displaystyle O}{\underset{\displaystyle CH_3CCH_2CH_3}{\|}} \xrightarrow{NaC\equiv CH} \textbf{A} \xrightarrow{C_2H_5MgBr} \textbf{B}$$

$$\xrightarrow[\text{(2) } H_2O]{\text{(1) } (CH_3)_2CHCHO} \textbf{C} \xrightarrow[\text{poisoned Pd}]{H_2} \textbf{D} \xrightarrow[\text{heat}]{H^+} \text{alloocimene}$$

20.35. The terpenoid *pinol* ($C_{10}H_{16}O$) can be obtained by the treatment of the dibromide of α-terpineol with $NaOC_2H_5$. What is the structure of pinol? (*Hint*: Consider the stereochemistry of the intermediates.)

α-terpineol

20.36. Suggest a synthesis for α-terpineol from readily available, open-chain starting materials.

CHAPTER 21

Spectroscopy II:
Ultraviolet Spectra,
Color and Vision, Mass Spectra

In Chapter 8, we discussed the absorption of infrared and radiofrequency radiation by organic compounds and how the absorption can be used in structure identification. In this chapter, we will consider the absorption of **ultraviolet (uv) and visible light** by organic compounds. Ultraviolet and visible spectra are also used in structure determination. More important, the absorption of visible light results in vision; we will also discuss this topic, along with colors and dyes. Last, we will introduce **mass spectra**, which arise from fragmentation of molecules when they are bombarded with high-energy electrons.

SECTION 21.1.

Ultraviolet and Visible Spectra

The wavelengths of uv and visible light are substantially shorter than the wavelengths of infrared radiation (see Figure 8.2, page 313). The unit we will use to describe these wavelengths is the *nanometer* ($1 \text{ nm} = 10^{-7}$ cm). The visible spectrum spans from about 400 nm (violet) to 750 nm (red), while the ultraviolet spectrum ranges from 100 to 400 nm.

The quantity of energy absorbed by a compound is inversely proportional to the wavelength of the radiation:

$$\Delta E = h v = \frac{hc}{\lambda}$$

where ΔE = energy absorbed, in ergs
  h = Planck's constant, 6.6×10^{-27} erg-sec
  v = frequency, in Hz
  c = speed of light, 3×10^{10} cm/sec
  λ = wavelength, in cm

Infrared radiation is relatively low-energy radiation. Absorption of infrared radiation by a molecule leads to increased vibrations of covalent bonds. Molecular transitions from the ground state to an excited vibrational state require about 2–15 kcal/mole.

Both uv and visible radiation are of higher energy than infrared radiation. Absorption of ultraviolet or visible light results in **electronic transitions**, promotion of electrons from low-energy, ground-state orbitals to higher-energy excited-state orbitals. These transitions require about 40–300 kcal/mole. The energy absorbed is subsequently dissipated as heat, as light (see Section 21.7), or in chemical reactions (such as isomerization or free-radical reactions).

The wavelength of uv or visible light absorbed depends on the ease of electron promotion. Molecules that require *more energy* for electron-promotion absorb at *shorter wavelengths*. Molecules that require *less energy* absorb at *longer wavelengths*. Compounds that absorb light in the visible region (that is, colored compounds) have more-easily promoted electrons than compounds that absorb at shorter uv wavelengths.

absorption at 100 nm (uv) ⟶ 750 nm (visible)

increasing ease of electronic transition

STUDY PROBLEM

21.1. Which would have the more-easily promoted electrons, anthracene (colorless) or Tyrian purple?

anthracene   Tyrian purple

A uv or visible spectrophotometer has the same basic design as an infrared spectrophotometer (see Figure 8.5, page 316). Absorption of radiation by a sample is measured at various wavelengths and plotted by a recorder to give the spectrum. (Figure 21.1 shows a typical uv spectrum.)

Since energy absorption by a molecule is quantized, we might expect that the absorption for electronic transitions would be observed at discrete wavelengths as a spectrum of lines or sharp peaks. This is not the case. Rather, a uv or visible spectrum consists of broad absorption bands over a wide range of wavelengths. The reason for the broad absorption is that the energy levels of both the ground state and the excited state of a molecule are subdivided into *rotational and vibrational sublevels*. Electronic transitions may occur from any one of the sublevels of the ground state to any one of the sublevels of an excited state (Figure 21.2). Since these various transitions differ slightly in energy, their wavelengths of absorption also differ slightly and give rise to the broad band observed in the spectrum.

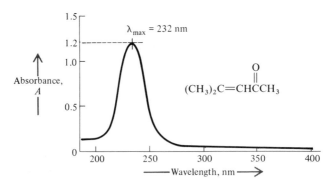

FIGURE 21.1. Ultraviolet spectrum of mesityl oxide, 9.2×10^{-5} M, 1.0-cm cell.

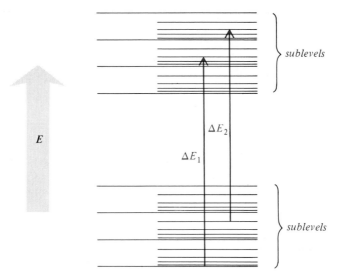

FIGURE 21.2. Schematic representation of electronic transitions from a low energy level to a high energy level.

SECTION 21.2.

Expressions Used in Ultraviolet Spectroscopy

Figure 21.1 shows the uv spectrum of a dilute solution of mesityl oxide (4-methyl-3-penten-2-one). The spectrum shows the scan from 200–400 nm. (Because absorption by atmospheric carbon dioxide becomes significant below 200 nm, the 100–200 nm region is usually not scanned.) The wavelength of absorption is usually reported as λ_{max}, the wavelength at the highest point of the curve. The λ_{max} for mesityl oxide is 232 nm.

The absorption of energy is recorded as **absorbance** (not transmittance as in infrared spectra). The absorbance at a particular wavelength is defined by the equation:

$$A = \log \frac{I_0}{I}$$

where A = absorbance
I_0 = intensity of the reference beam
I = intensity of the sample beam

The absorbance by a compound at a particular wavelength increases with an increasing number of molecules undergoing transition. Therefore the absorbance depends on the electronic structure of the compound and also upon the concentration of the sample and the length of the sample cell. For this reason, chemists report the energy absorption as **molar absorptivity ε** (sometimes called the *molar extinction coefficient*) rather than as the actual absorbance. Often, uv spectra are replotted to show ε or $\log \varepsilon$ instead of A as the ordinate. The $\log \varepsilon$ value is especially useful when values for ε are very large.

$$\varepsilon = \frac{A}{cl}$$

where ε = molar absorptivity
A = absorbance
c = concentration, in M
l = cell length, in cm

The molar absorptivity (usually reported at the λ_{max}) is a reproducible value that takes into account concentration and cell length. Although ε has the units M^{-1} cm^{-1}, it is usually shown as a unitless quantity. For mesityl oxide, the ε_{max} is $1.2 \div (9.2 \times 10^{-5} \times 1.0)$, or 13,000 (values taken from Figure 21.1).

SAMPLE PROBLEM

A flask of cyclohexane is known to be contaminated with benzene. At 260 nm, benzene has a molar absorptivity of 230, and cyclohexane has a molar absorptivity of zero. A uv spectrum of the contaminated cyclohexane (1.0-cm cell length) shows an absorbance of 0.030. What is the concentration of benzene?

Solution:
$$c = \frac{A}{\varepsilon l} = \frac{0.030}{230 \times 1.0} = 0.00013 M$$

SECTION 21.3.

Types of Electron Transitions

Let us consider the different types of electron transitions that give rise to ultra-violet or visible spectra. The ground state of an organic molecule contains valence electrons in three principal types of molecular orbitals: **sigma (σ) orbitals**; **pi (π) orbitals**; and **filled, but nonbonded, orbitals (n)**.

Both σ and π orbitals are formed from the overlap of two atomic or hybrid orbitals. Each of these molecular orbitals therefore has an antibonding σ^* or π^* orbital associated with it. An orbital containing n electrons does not have an antibonding orbital (because it was not formed from two orbitals). Electron transitions involve the promotion of an electron from one of the three ground states (σ, π, or n) to one of the two excited states (σ^* or π^*). There are six possible transitions; the four important transitions and their relative energies are shown in Figure 21.3.

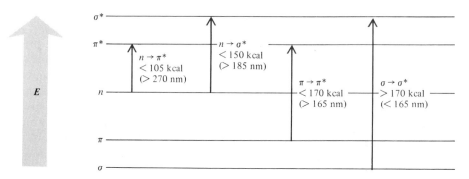

FIGURE 21.3. Energy requirements for important electronic transitions. (The corresponding wavelengths are in parentheses.)

The most useful region of the uv spectrum is at wavelengths longer than 200 nm. The following transitions give rise to absorption in the nonuseful 100–200 nm range: $\pi \rightarrow \pi^*$ for an isolated double bond and $\sigma \rightarrow \sigma^*$ for an ordinary carbon–carbon bond. The useful transitions (200–400 nm) are $\pi \rightarrow \pi^*$ for compounds with conjugated double bonds, and some $n \rightarrow \sigma^*$ and $n \rightarrow \pi^*$ transitions.

A.   Absorption by polyenes

Less energy is required to promote a π electron of 1,3-butadiene than is needed to promote a π electron of ethylene. The reason is that the energy difference between the HOMO (Highest Occupied Molecular Orbital) and the LUMO (Lowest Unoccupied Molecular Orbital) for conjugated double bonds is less than the energy difference for an isolated double bond. Resonance-stabilization of the excited state of a conjugated diene is one factor that decreases the energy of the excited state.

Because less energy is needed for a $\pi \rightarrow \pi^*$ transition of 1,3-butadiene, this diene absorbs uv radiation of longer wavelengths than does ethylene. As more conjugated double bonds are added to a molecule, the energy required to reach the first excited state decreases. Sufficient conjugation shifts the absorption to wavelengths that reach into the visible region of the spectrum; a compound with sufficient conjugation is colored. For example, lycopene, the compound responsible for the red color of tomatoes, has eleven conjugated double bonds. (Line formulas, such as the following one, are described in Section 9.16A.)

lycopene
$\lambda_{max} = 505$ nm

Table 21.1 lists the λ_{max} values for $\pi \rightarrow \pi^*$ transitions of a series of aldehydes with increasing conjugation. Inspection of the table reveals that the position of absorption is shifted to longer wavelengths as the extent of the conjugation increases. Generally, this increase is about 30 nm per conjugated double bond in a series of polyenes.

TABLE 21.1. Ultraviolet absorption for some unsaturated aldehydes

Structure	λ_{max}, nm
$CH_3CH=CHCHO$	217
$CH_3(CH=CH)_2CHO$	270
$CH_3(CH=CH)_3CHO$	312
$CH_3(CH=CH)_4CHO$	343
$CH_3(CH=CH)_5CHO$	370

STUDY PROBLEMS

21.2. List the following all-*trans*-polyenes in order of increasing λ_{max}:

(a) $CH_3(CH=CH)_{10}CH_3$ (b) $CH_3(CH=CH)_9CH_3$
(c) $CH_3(CH=CH)_8CH_3$

21.3. The λ_{max} for Compound (a) in Problem 21.2 is 476 nm. Predict the λ_{max} values for Compounds (b) and (c).

B.   Absorption by aromatic systems

Benzene and other aromatic compounds exhibit more-complex spectra than can be explained by simple $\pi \rightarrow \pi^*$ transitions. The complexity arises from the existence of *several* low-lying excited states. Benzene absorbs strongly at 184 nm ($\varepsilon = 47,000$) and at 202 nm ($\varepsilon = 7,000$) and has a series of absorption bands between 230–270 nm. A value of 260 nm is often reported as the λ_{max} for benzene because this is the position of strongest absorption above 200 nm. Solvents and substituents on the ring alter the uv spectra of benzene compounds.

The absorption of uv radiation by aromatic compounds composed of fused benzene rings is shifted to longer wavelengths as the number of rings is increased because of increasing conjugation and greater resonance-stabilization of the excited state.

benzene
$\lambda_{max} = 260$ nm

naphthalene
$\lambda_{max} = 280$ nm

phenanthrene
$\lambda_{max} = 350$ nm

anthracene
$\lambda_{max} = 375$ nm

naphthacene
$\lambda_{max} = 450$ nm
(yellow)

pentacene
$\lambda_{max} = 575$ nm
(blue)

coronene
$\lambda_{max} = 400$ nm
(yellow)

STUDY PROBLEM

21.4. Suggest a reason why coronene absorbs at a *shorter* wavelength than does naphthacene.

C. Absorption arising from transitions of *n* electrons

Compounds that contain nitrogen, oxygen, sulfur, phosphorus, or one of the halogens all have unshared *n* electrons. If the structure contains no π bonds, these *n* electrons can undergo only $n \rightarrow \sigma^*$ transitions. Because the *n* electrons are of higher energy than either the σ or π electrons, less energy is required to promote an *n* electron, and transitions occur at longer wavelengths than $\sigma \rightarrow \sigma^*$ transitions. Note that some of these values are within the usual uv spectral range of 200–400 nm (Table 21.2). The π^* orbital is of lower energy than the σ^* orbital; consequently, $n \rightarrow \pi^*$ transitions require less energy than $n \rightarrow \sigma^*$ transitions and often are in the range of a normal instrument scan.

The *n* electrons are in a different region of space from σ^* and π^* orbitals, and the probability of an *n* transition is low. Since molar absorptivity depends on the number of electrons undergoing transition, ε values for *n* transitions are low, in the 10–100 range (compared to about 10,000 for a $\pi \rightarrow \pi^*$ transition).

A compound such as acetone that contains both a π bond and *n* electrons exhibits both $\pi \rightarrow \pi^*$ and $n \rightarrow \pi^*$ transitions. Acetone shows absorption at 187 nm ($\pi \rightarrow \pi^*$) and 270 nm ($n \rightarrow \pi^*$).

TABLE 21.2. Ultraviolet absorption arising from $n \rightarrow \sigma^*$ transitions

Structure	λ_{max}, nm	ε
$CH_3\ddot{O}H$	177	200
$(CH_3)_3\ddot{N}$	199	3950
$CH_3\ddot{C}l:$	173	200
$CH_3CH_2CH_2\ddot{B}r:$	208	300
$CH_3\ddot{I}:$	259	400

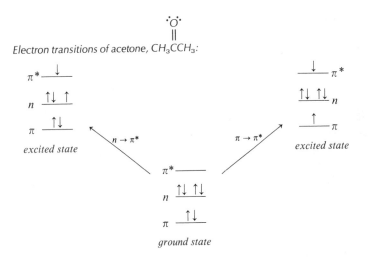

Electron transitions of acetone, CH_3CCH_3:

SECTION 21.4.

Color and Vision

Color has played a significant role in human society ever since people first learned to color clothes and other articles. Color is the result of a complex set of physiological and psychological responses to wavelengths of light of 400–750 nm striking the retina of the eye. If all wavelengths of visible light strike the retina, we perceive white; if none of them do, we perceive black or darkness. If a small range of wavelengths hits the eye, then we observe individual colors. Table 21.3 lists the wavelengths of the visible spectrum with their corresponding colors and complementary colors, which we will discuss shortly.

Our perception of color arises from a variety of physical processes. A few examples of how light of a particular wavelength may be directed to the eye follow: (1) The yellow-orange color of a sodium flame results from the **emission of light** with a wavelength of 589 nm; the emission is caused by excited electrons returning to lower-energy orbitals. (2) A prism causes a **diffraction of light** that varies with the wavelength; we observe the separated wavelengths as a rainbow pattern. (3) **Interference** results from light being reflected from two surfaces of a very thin film (e.g., soap bubbles or bird feathers). The light wave reflected from the farther surface is reflected out of phase with the reflection from the nearer surface, resulting

TABLE 21.3. Colors in the visible spectrum

Wavelength, nm	Color	Complementary (subtraction) color
400–424	violet	green-yellow
424–491	blue	yellow
491–570	green	red
570–585	yellow	blue
585–647	orange	green-blue
647–700	red	green

in wave interference and cancellation of some wavelengths; hence, we see color instead of white.

The fourth, and most common, process that leads to color is the **absorption of light of certain wavelengths** by a substance. Organic compounds with extensive conjugation absorb certain wavelengths of light because of $\pi \to \pi^*$ and $n \to \pi^*$ transitions. We do not observe the color absorbed, but we see its **complement**, which is reflected. A complementary color, sometimes called a **subtraction color**, is the result of subtraction of some of the visible wavelengths from the entire visual spectrum. For example, pentacene (page 922) absorbs at 575 nm, in the yellow portion of the visible spectrum. Thus, pentacene absorbs the yellow light (and, to a lesser extent, that of the surrounding wavelengths) and reflects the other wavelengths. Pentacene has a blue color, which is the complement of yellow.

Some compounds appear yellow even though their λ_{max} is in the ultraviolet range of the spectrum (for example, coronene, page 922). In such a case, the tail of the absorption band extends from the ultraviolet into the visible region and absorbs the violet to blue wavelengths. Figure 21.4 depicts the spectrum of such a compound.

A.   Mechanism of vision

The human eye is an amazingly intricate organ that converts photons of light into nerve pulses that travel to the brain and result in vision. The mechanism of the eye is remarkably sensitive. About one quantum of light energy is all that is necessary to trigger the mechanism resulting in a visual nerve pulse. We can detect

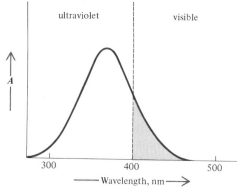

FIGURE 21.4. A compound with a λ_{max} in the uv region may also absorb light in the visible region.

as few as five quanta of light. (For comparison, a typical flashlight bulb radiates about 2×10^{18} quanta per second.)

The eye contains two types of photoreceptor—the **rods** and the **cones**. The cones contain pigments and are responsible for color vision and for vision in bright light. Animals that lack cones are color-blind. The rods are responsible for black and white photoreception and for vision in very dim light. While more is known about rods than cones, much is still to be learned—for example, how the nerve impulse is generated.

In the rods, light is detected by a reddish-purple pigment called **rhodopsin**, or **visual purple** ($\lambda_{max} = 498$ nm). Rhodopsin is formed from an aldehyde called *11-cis-retinal* and a protein called *opsin*. These two components of rhodopsin are bonded together by an imine link between the aldehyde group of 11-*cis*-retinal and an amino group in opsin. As is frequently the case with protein complexes, opsin has a shape that holds the 11-*cis*-retinal in a pocket. Compounds with other shapes do not fit into this pocket. In the combined form, the imine bond joining 11-*cis*-retinal and opsin is protected by the rest of the opsin molecule and is not readily hydrolyzed.

When a photon of light (*hv*) strikes rhodopsin, the rhodopsin undergoes a series of transformations to intermediates called **metarhodopsins**. In this process, the nerve impulse is generated, and the double bond at carbon 11 of the 11-*cis*-retinal is converted from *cis* to *trans*. All-*trans*-retinal does not fit in the pocket of opsin, and the imine bond becomes exposed when the isomerization occurs. The imine link (of metarhodopsin II) is thus cleaved, and the all-*trans*-retinal is released.

How does the isomerization of the *cis* double bond at carbon 11 occur? The process is not yet entirely understood. Possibly the absorption of light causes the promotion of one pi electron from the bonding orbital to an antibonding orbital. In this excited state, there is no overlap between the two *p* orbitals, the energy barrier for rotation is lower, and rotation can occur fairly easily. (However, newer evidence supports the theory that the isomerization step is a concerted reaction with no intermediate, initiated by protonation of the imine nitrogen.)

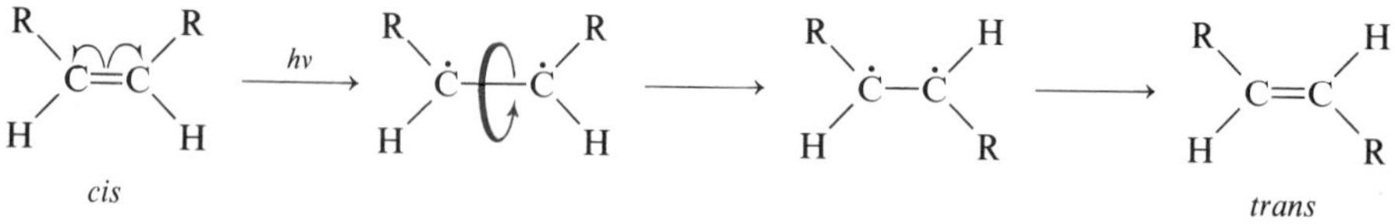

Rhodopsin must be regenerated for vision to be sustained. One way this occurs is by a *photo-induced reconversion* of metarhodopsin I back to rhodopsin. (This reaction is the reverse reaction of the first step in the flow equation shown on page 925.) After hydrolysis of the imine, all-*trans*-retinal also can be photo-isomerized to 11-*cis*-retinal, which can then reform rhodopsin.

In dim light, these reverse reactions are less likely to occur, and the all-*trans*-retinal undergoes a multistep *chemical conversion*. In this process, the all-*trans*-retinal undergoes enzymatic reduction to all-*trans*-vitamin A, which is transported to the liver where it is isomerized to 11-*cis*-vitamin A. The 11-cis-vitamin A then is transported back to the eye, where it is oxidized to 11-*cis*-retinal, which can again combine with opsin. (Figure 21.5 summarizes the chemistry of the visual cycle.)

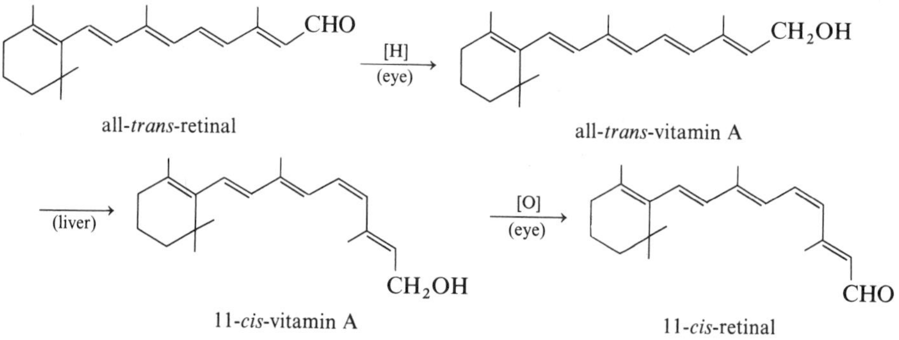

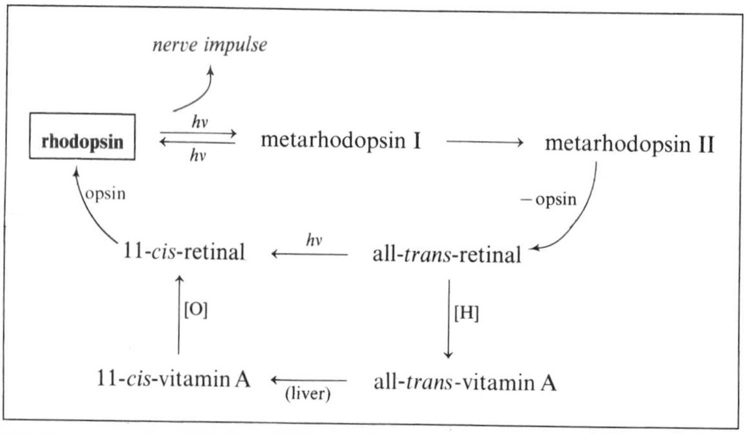

FIGURE 21.5. Summary of transformations in the visual cycle.

SECTION 21.5.

Colored Compounds and Dyes

Nature abounds with color. Some colors, such as those of hummingbird or peacock feathers, arise from light diffraction by the unique structure of the feathers. However, most of Nature's colors are due to the absorption of certain wavelengths of visible light by organic compounds.

Before the theories of electronic transition were developed, it was observed that some types of organic structures give rise to color, while others do not. The partial structures necessary for color (unsaturated groups that can undergo $\pi \to \pi^*$ and $n \to \pi^*$ transitions) were called **chromophores**, a term coined in 1876 (Greek *chroma*, "color," and *phoros*, "bearing").

Some chromophores:

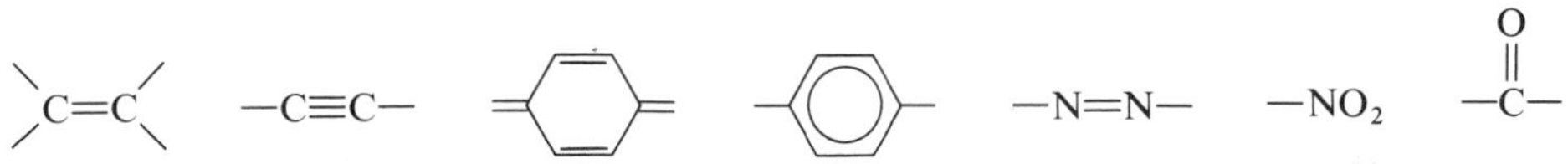

It was also observed that the presence of some other groups caused an intensification of color. These groups were called **auxochromes** (Greek *auxanein*, "to increase"). We now know that auxochromes are groups that cannot undergo $\pi \to \pi^*$ transitions, but can undergo transitions of n electrons.

Some auxochromes:

$$-OH \quad -OR \quad -NH_2 \quad -NHR \quad -NR_2 \quad -X$$

In the following discussion of natural colored compounds and dyes, note the presence of these chromophores and auxochromes.

A.   Some natural colored compounds

Naphthoquinones and **anthraquinones** are common natural coloring materials. *Juglone* is a naphthoquinone that is partly responsible for the coloring of walnut hulls. *Lawsone* is similar in structure to juglone; it is found in Indian henna, which is used as a red hair dye. A typical anthraquinone, carminic acid, is the principal red pigment of *cochineal*, a ground-up insect (*Coccus cacti L.*) that is used as a red dye in food and cosmetics. Alizarin (page 929) is another red dye of the anthraquinone class.

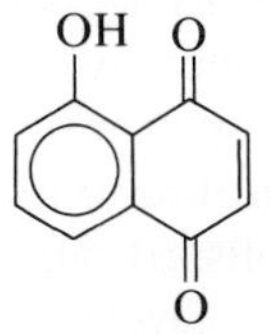

juglone 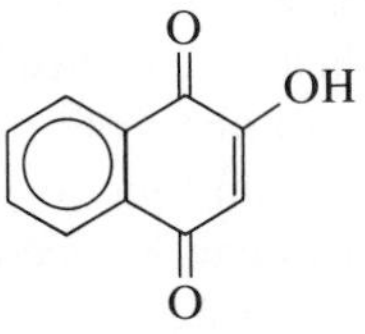   lawsone 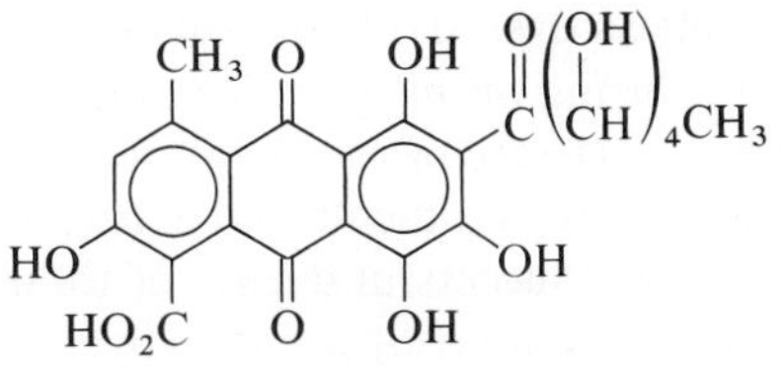   carminic acid

Most red and blue flowers owe their coloring to glucosides called **anthocyanins**. The nonsugar portion of the glucoside is called an *anthocyanidin* and is a type of *flavylium salt* (see below). The particular color imparted by an anthocyanin depends, in part, on the pH of the flower. The blue of cornflowers and the red of roses are due to the same anthocyanin, *cyanin*. In a red rose, cyanin is in its phenol form. In a blue cornflower, cyanin is in its anionic form, with a proton removed from one of the phenolic groups. In this respect, cyanin is similar to acid–base indicators, which will be discussed in Section 21.6.

cyanin

in red roses

cyanidin choride

a flavylium salt

The term flavylium salt comes from the name for *flavone*, which itself is a colorless compound. The addition of a 3-hydroxyl group leads to flavonol, which is yellow (Latin, *flavus*, "yellow").

flavone

colorless

flavonol

yellow

B.   Dyes

A **dye** is a colored organic compound that is used to impart color to an object or a fabric. The history of dyes goes back to prehistoric times. *Indigo* (page 929) is the oldest known dye; it was used by the ancient Egyptians to dye mummy cloths. *Tyrian purple* (page 917), which was obtained from *Murex snails* found near the city of Tyre, was used by the Romans to dye the togas of the emperors. *Alizarin* (page 929), also called *Turkey red*, was obtained from the roots of the madder plant and was used in the 1700's and 1800's to dye the red coats of British soldiers.

There are numerous colored organic compounds; however, only a few are suitable as dyes. To be useful as a dye, a compound must be *fast* (remain in the fabric during washing or cleaning). To be fast, a dye must, in one way or another, be bonded to the fabric. A fabric composed of fibers of polypropylene or a similar hydrocarbon is difficult to dye because it has no functional groups to attract dye molecules. Successful dyeing of these fabrics can be accomplished, however, by the incorporation of a metal–dye complex into the polymer. Dyeing of cotton (cellulose) is easier because hydrogen bonding between hydroxyl groups of the glucose units and groups of the dye molecule hold the dye to the cloth. Polypeptide fibers, such as wool or silk, are the easiest fabrics to dye because they contain numerous polar groups that can interact with dye molecules.

A **direct dye** is a dye that is applied directly to cloth from a hot aqueous solution. If the fabric to be dyed has polar groups, such as those found in polypeptide fibers, then the incorporation of a dye with either an amino group or a strongly acidic group will render the dye fast. *Martius yellow* is a typical direct dye. The acidic phenol group in Martius yellow undergoes reaction with basic side chains in wool or silk.

Martius yellow

A **vat dye** is a dye that is applied to fabric (in a vat) in a soluble form and then is allowed to undergo reaction to an insoluble form. The blue coats supplied by the French to the Americans during the American Revolution were dyed with *indigo*, a typical vat dye. Indigo was obtained by a fermentation of the woad plant (*Isatis tinctoria*) of Western Europe or of plants of the *Indigofera* species, found in tropical countries. Both types of plants contain the glucoside *indican*, which can be hydrolyzed to glucose and *indoxyl*, a colorless precursor of indigo. Fabrics were soaked in the fermentation mixture containing indoxyl, then were allowed to air dry. Air oxidation of indoxyl yields the blue, insoluble indigo. Indigo is deposited in the *cis* form, which undergoes spontaneous isomerization to the *trans* isomer.

indoxyl   *cis*-indigo   *trans*-indigo

A **mordant dye** is one that is rendered insoluble on a fabric by complexing or chelation with a metal ion, called a **mordant** (Latin *mordere*, "to bite"). The fabric is first treated with the metal salt (such as one of Al, Cu, Co, or Cr), then treated with a soluble form of the dye. The chelation reaction on the surface of the fabric results in a fast dye. One of the oldest mordant dyes is *alizarin*, which forms different colors depending upon the metal ion used. For example, alizarin gives a rose-red color with Al^{3+} and a blue color with Ba^{2+}.

alizarin–aluminum chelate

Azo dyes are the largest and most important class of dyestuffs, their numbers running into the thousands. In azo-dyeing, the fabric is first impregnated with an aromatic compound activated toward electrophilic substitution, then is treated with a diazonium salt to form the dye. This diazonium coupling reaction is discussed in Section 10.14.

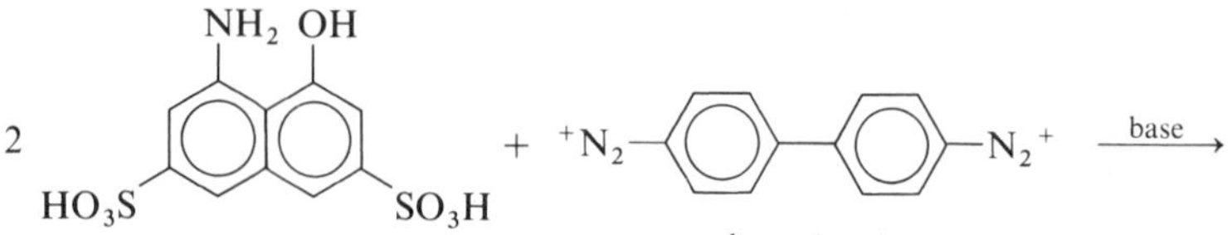

"H-acid"

*a key intermediate for
hundreds of azo dyes*

a diazonium ion

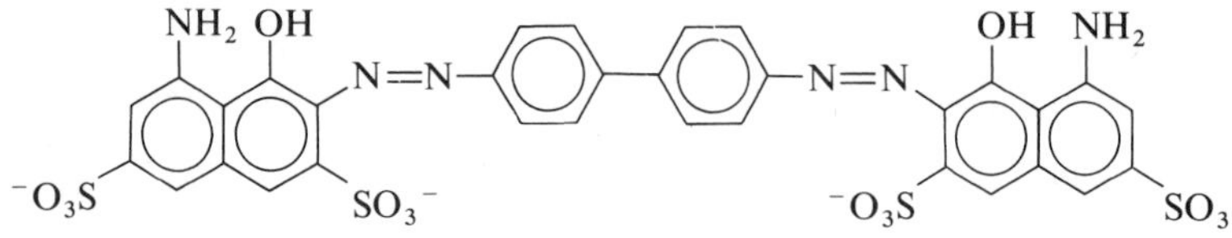

Direct Blue 2B

SECTION 21.6.

Acid–Base Indicators

An **acid–base indicator** is an organic compound that changes color with a change in pH. These compounds are most frequently encountered as titration endpoint indicators. Test papers, such as litmus paper, are impregnated with one or more of these substances.

Two typical indicators are *methyl orange* and *phenolphthalein*. Methyl orange is red in acidic solutions that have pH values less than 3.1. It is yellow in solutions with pH values greater than 4.4. Phenolphthalein, on the other hand, changes color on the alkaline side of the pH range. Up to pH 8.3, phenolphthalein is colorless. At pH 10, it is red. In strongly alkaline solutions, it again becomes colorless.

Indicators change color because the chromophoric system is changed by an acid–base reaction. In acidic solution, methyl orange exists as a resonance hybrid of a protonated azo structure; this resonance hybrid is red. The azo nitrogen is not strongly basic, and the protonated azo group loses the hydrogen ion at about pH 4.4. The loss of the proton changes the electronic structure of the compound, resulting in a change of color from red to yellow. Figure 21.6 shows spectra of methyl orange at two different pH values.

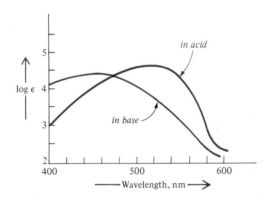

methyl orange
yellow in base

red in acid

FIGURE 21.6. Visible spectra of methyl orange in acidic and alkaline solutions (pH 1 and 13, respectively).

The commercial value of phenolphthalein is that it serves as the active ingredient in "candy" and "gum" laxatives. However, phenolphthalein is also one of the best-known titration indicators. In acidic solution, phenolphthalein exists as a colorless lactone. In the lactone, the center carbon is in the sp^3-hybrid state; consequently, the three benzene rings are isolated, not conjugated.

At pH values greater than 8.3 (alkaline solution), a phenolic hydrogen is removed from phenolphthalein, the lactone ring opens, and the center carbon becomes sp^2-hybridized. In this form, the benzene rings are in conjugation, and the extensive pi system gives rise to the red color that is observed in mildly alkaline solution. Figure 21.7 shows these reactions.

In strongly alkaline solution, the center carbon of phenolphthalein is hydroxylated and is converted to the sp^3 state. This reaction isolates the three pi systems again. At high pH values, phenolphthalein is colorless. Again, refer to Figure 21.7.

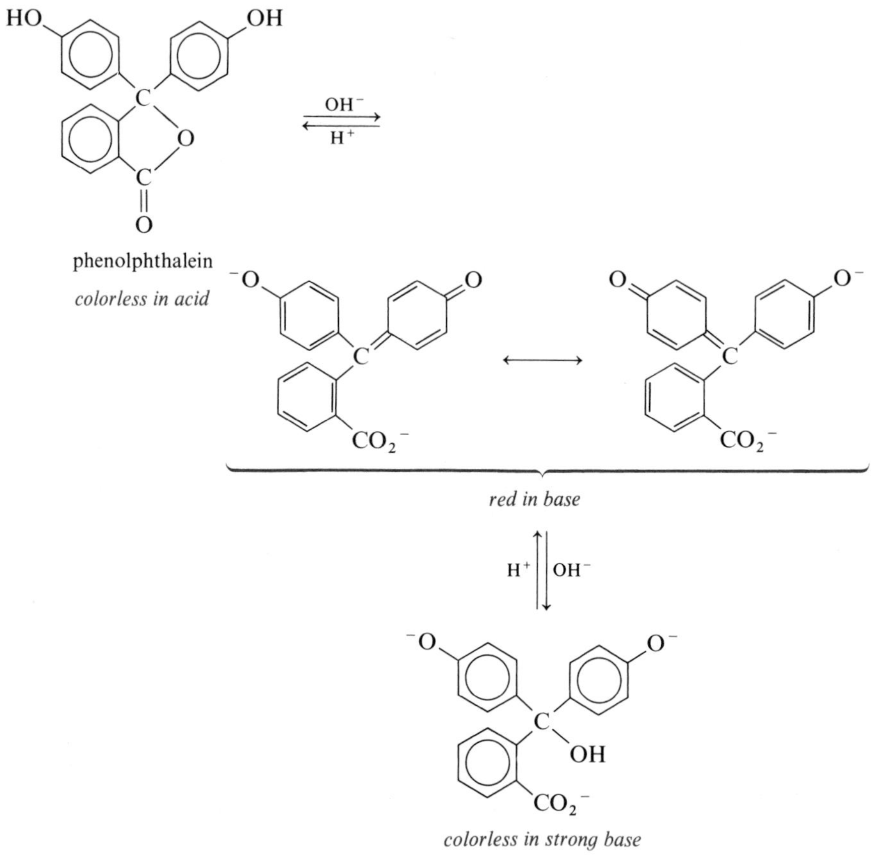

FIGURE 21.7. The acid–base reactions of phenolphthalein.

STUDY PROBLEM

21.5. One of the following indicators is blue-green at pH 7; the other is violet. Which is which? Explain your answer.

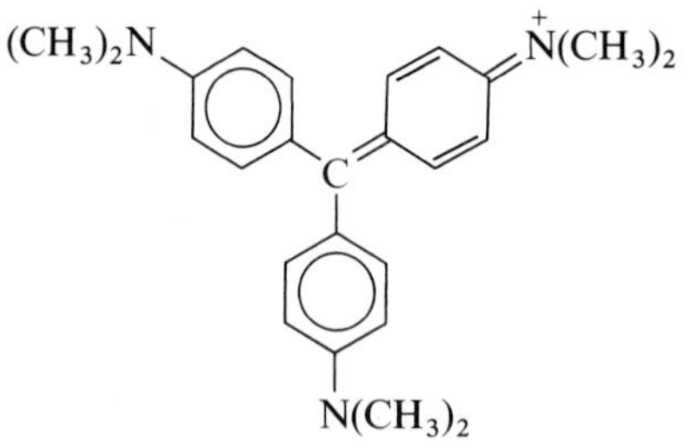

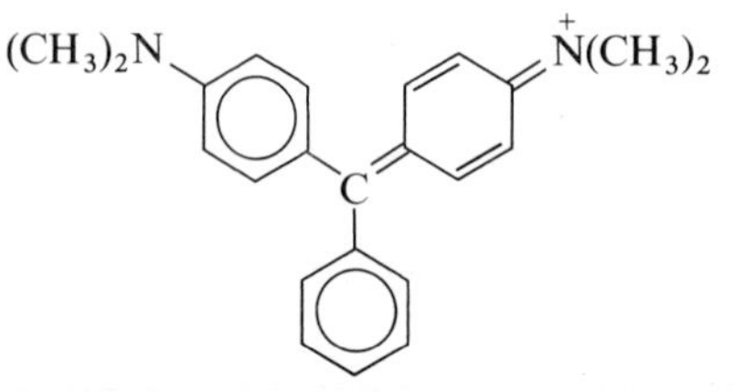

Fluorescence and Chemiluminescence

When a molecule undergoes absorption of ultraviolet or visible light, an electron is promoted from the ground state to an **excited singlet state**. (*Singlet* refers to a state in which electrons are *paired*; see the footnote on page 407.) Immediately after promotion (on the order of 10^{-11} sec), the electron drops to the lowest-energy excited singlet state. One of the ways in which an excited molecule with electrons in this lowest-energy singlet state can return to the ground state is to lose its energy as light. This process is also fast (10^{-7} sec). The energy lost by this emission of light is slightly *less* than the energy that was initially absorbed. (The difference results in increased molecular motion.) Consequently, the wavelength of light emitted is slightly longer than that which was absorbed.

A compound that absorbs light in the visible range appears colored. When the same compound emits light of a different wavelength, it appears two-colored, or **fluorescent**. An example of a fluorescent compound is **fluorescein**, which has been used as a marker for airplanes downed at sea. In aqueous solution and in the presence of light, fluorescein appears red with an intense yellow-green fluorescence.

fluorescein

Some fluorescent compounds, called **optical bleaches**, are used as fabric whiteners. These are colorless compounds that absorb ultraviolet light just out of the visible range, then emit blue-violet light at the edge of the visible spectrum. This blue-violet color masks yellowing of the fabric.

Two optical bleaches:

Blankophor R

Calcofluor SD

Chemiluminescence is a phenomenon in which (1) a chemical reaction generates products that contain excited molecules, and (2) the return of these excited products to the ground state is accompanied by the emission of light. A familiar example of chemiluminescence is the light of the firefly, caused by the enzymatic oxidation of *firefly luciferin.*

firefly luciferin

SECTION 21.8.

Mass Spectrometry

Most of the spectral techniques we have discussed arise from absorption of energy by molecules. **Mass spectrometry** is based on different principles. In a mass spectrometer, a sample in the gaseous state is bombarded with electrons of sufficient energy to exceed the first ionization potential of the compound. (The ionization potentials of most organic compounds are in the range of 185–300 kcal/mole.) Collision between an organic molecule and one of these high-energy electrons results in the loss of an electron from the molecule and the formation of an organic ion. The organic ions that result from this high-energy electron bombardment are unstable and fly apart into smaller fragments, both free radicals and other ions. In a typical mass spectrometer, the *positively charged fragments* are detected. The **mass spectrum** is a plot of **abundance** (the relative amounts of the different positively charged fragments) versus the **mass-to-charge ratio (*m/e*)** of the fragments. The ionic charge of most particles detected in a mass spectrometer is $+1$; the m/e value for such an ion is equal to its mass. Consequently, from a practical standpoint, the mass spectrum is a record of particle mass versus relative abundance of the particles.

How a molecule or ion breaks into fragments depends upon the carbon skeleton and functional groups present. Therefore, the structure and mass of the fragments give clues about the structure of the parent molecule. Also, it is frequently possible to determine the molecular weight of a compound from its mass spectrum.

Let us introduce mass spectrometry using methanol as an example. When methanol is bombarded with high-energy electrons, one of the valence electrons is lost. The result is an **ion radical**, a species with one unpaired electron and a charge of $+1$. An ion radical is symbolized by $^{+}\cdot$. The ion radical that results from abstraction of one electron of a molecule is called the **molecular ion** and is symbolized $M^{+}\cdot$. The mass of the molecular ion is the molecular weight of the compound. The molecular ion of methanol has a mass of 32 and a charge of $+1$. Its mass-to-charge ratio (m/e) is 32. (In the following example, the half-headed arrow signifies the loss of one electron from a methanol molecule.)

$$e^- + CH_3\overset{..}{\underset{..}{O}}H \xrightarrow{\ -2e^-\ } CH_3\overset{..}{O}H^+, \quad \text{usually written} \quad [CH_3OH]^{\ddagger}$$

the molecular ion of methanol, m/e = 32

A molecular ion can undergo fragmentation after it has been formed. In the case of $[CH_3OH]^+$, the ion radical can lose a hydrogen atom and become a cation: $[CH_2=OH]^+$. This fragment has a *m/e* of 31. In the mass spectrum of methanol (Figure 21.8), peaks for the particles with *m/e* values of 31 and 32 are evident. (The fragments that give rise to the other peaks in this mass spectrum will be discussed in Section 21.11.)

$$
\left[\begin{array}{c} H \\[2pt] H\!:\!\overset{\displaystyle H}{\underset{\displaystyle H}{C}}\!:\!\overset{..}{\underset{..}{O}}\!:\!H \end{array} \right]^{\ddagger}
\longrightarrow H\cdot +
\left[\begin{array}{c} H \\[2pt] \overset{\displaystyle H}{\underset{\displaystyle H}{C}}\!:\!:\!\overset{..}{\underset{..}{O}}\!:\!H \end{array} \right]^{+}
$$

a cation, m/e = 31
(not an ion radical because all electrons are paired)

As may be seen in Figure 21.8, a mass spectrum is presented as a bar graph. Each peak in the spectrum represents a fragment of the molecule. The fragments are scanned so that the peaks are arranged by increasing *m/e* from left to right in the spectrum. The intensities of the peaks are proportional to the relative abundance of the fragments, which in turn depends on their relative stabilities. By convention, the tallest peak in a spectrum, called the **base peak**, is given the intensity value of 100%; lesser peaks are reported as 20%, 30%, or whatever their value is relative to the base peak. The base peak sometimes arises from the molecular ion, but often arises from a smaller fragment.

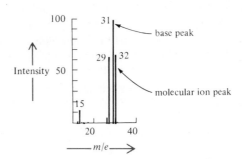

FIGURE 21.8. Mass spectrum of methanol, CH_3OH.

STUDY PROBLEMS

21.6. Write the formula for each product and tell whether it is a cation, an ion-radical, or a free radical:

(a) CH_4 minus one e^- (b) $[CH_4]^{\ddagger}$ minus H·
(c) $[CH_3CH_2]^+$ minus H·   (d) $[CH_3CH_2]^{\ddagger}$ minus H·

21.7. Write the formulas for the molecular ions of (a) CH_4 and (b) $CH_3CH_2CH_3$.

21.8. Give the m/e value for each of the following particles:

(a) $[CH_4]^{\ddagger}$ (b) $[(CH_3)_2CH]^+$ (c) $[O_2]^{\ddagger}$ (d) $[H_2O]^{\ddagger}$

SECTION 21.9.

The Mass Spectrometer

A diagram of a common type of mass spectrometer is shown in Figure 21.9. The sample is introduced, vaporized, and allowed to feed in a continuous stream to the *ionization chamber*. The ionization chamber (as well as the entire instrument) is kept under vacuum to minimize collisions and reactions between radicals, air molecules, and so forth. In this chamber, the sample passes through a stream of high-energy electrons, which causes the ionization of some of the sample molecules into their molecular ions.

After its formation, a molecular ion can undergo fragmentation and re-arrangement. These processes are extremely rapid (10^{-10}–10^{-6} sec). The longer-lived particles can be detected by the ion collector, but a shorter-lived particle may not have a sufficient lifetime to reach the ion collector. In some cases, the molecular ion is too short-lived to be detected, and only its fragmentation products exhibit peaks.

As the ion radicals and other particles are formed, they are fed past two electrodes, the *ion accelerator plates*, which accelerate the positively charged particles. (The neutral and negatively charged particles are not accelerated and are removed continuously by the vacuum pumps.) From the accelerator plates, the positively charged particles pass into the *analyzer tube*, where they are de-flected into a curved path by a magnetic field.

The radius of the curved path depends upon the particle's velocity, which in turn is dependent on the magnetic field strength, the accelerating voltage, and the m/e of the particle. At the same field strength and voltage, the particles of higher m/e have a path with a wider radius, while the lower-m/e particles have a path of smaller radius (see Figure 21.10). The continuous flow of positively charged particles through the analyzer tube therefore forms a pattern: higher-m/e particles

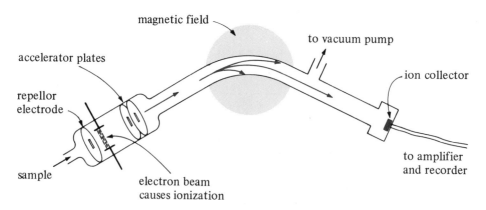

FIGURE 21.9. Diagram of a mass spectrometer.

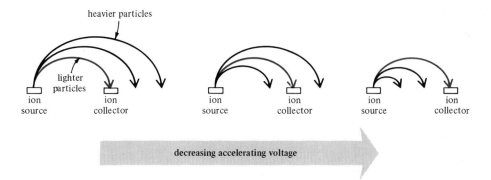

FIGURE 21.10. As the accelerating voltage is decreased, particles of successively higher
m/e hit the ion collector.

with a larger radius and lower-*m/e* particles with a smaller radius. If the accelerating voltage is slowly and continuously decreased, the velocities of all the particles decrease, and the radii of the paths of all the particles also decrease. By this technique, particles of successively higher *m/e* are allowed to strike the detector. Figure 21.10 shows the effects of decreasing the accelerating voltage on the paths of particles with three different *m/e* values. (The same effect can be obtained by increasing the magnetic field strength instead of decreasing the accelerating voltage.)

SECTION 21.10.

Isotopes in Mass Spectra

A mass spectrometer is so sensitive that particles differing by 1.0 mass unit give separate signals. The molecular weight of CH_3Br is 94.9 (15.0 atomic mass units for CH_3 and 79.9 for Br). However, the mass spectrum of this compound (Figure 21.11) does not show one molecular ion peak at *m/e* = 94.9; instead two peaks are observed, one at *m/e* = 94 and the other at *m/e* = 96. The reason for the two peaks is that naturally occurring Br exists in two isotopic forms, one with atomic mass 79

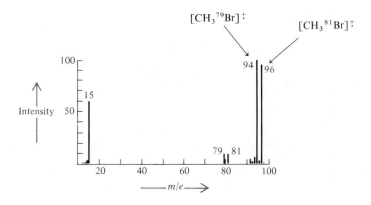

FIGURE 21.11. Mass spectrum of bromomethane, CH_3Br.

and the other with atomic mass 81. When we calculate the molecular weight of a bromine compound, we use a weighted average of these two isotopic masses (79.9). Because a mass spectrometer detects particles containing each of these isotopes as individual species, we cannot use average atomic masses when dealing with mass spectra, as we do when calculating the stoichiometry of a chemical reaction.

Table 21.4 lists the common natural isotopes and their relative abundance. Naturally occurring bromine exists as a 50.5%–49.5% mixture of bromine-79 and bromine-81, respectively. Particles of the same structure containing Br give a pair of peaks of approximately the same intensity that are 2.0 mass units apart. *The particle containing the lower-mass isotope is considered the molecular ion*; the peak for the other particle is called the **M + 2 peak** (molecular ion plus two mass units).

Naturally occurring chlorine is a mixture of 75.5% chlorine-35 and 24.5% chlorine-37. The particle containing chlorine-35 is considered the molecular ion, while the particle containing chlorine-37 gives rise to the M + 2 peak, which has an intensity approximately one-third that of the molecular ion peak.

Most elements common in organic compounds (except for Cl and Br) exist in nature as predominantly one isotope. For example, carbon is 98.89% carbon-12, and hydrogen is 99.985% hydrogen-1. For this reason, we generally assume for mass spectral purposes that all of C is carbon-12 and ignore the tiny proportion that is carbon-13. The presence of the isotopes of the common elements explains the multitude of small peaks surrounding a large peak in a mass spectrum (for example, the small peaks around the peaks at $m/e = 94$ and $m/e = 96$ in Figure 21.11).

STUDY PROBLEMS

21.9. Calculate the m/e value for the molecular ion of each of the following compounds: (a) ethane; (b) 1,2-dichloroethane; (c) ethanol; (d) *p*-bromophenol.

21.10. Figure 21.12 contains mass spectra of four compounds. Which of these compounds contains Br? Which contains Cl?

TABLE 21.4. Natural abundance of some isotopes

Isotope	*Abundance, %*	*Isotope*	*Abundance, %*	*Isotope*	*Abundance, %*
^1H	99.985	^2H	0.015		
^{12}C	98.89	^{13}C	1.11		
^{14}N	99.63	^{15}N	0.37		
^{16}O	99.76	^{17}O	0.04	^{18}O	0.20
^{32}S	95.0	^{33}S	0.76	^{34}S	4.2
^{19}F	100				
^{35}Cl	75.5			^{37}Cl	24.5
^{79}Br	50.5			^{81}Br	49.5
^{127}I	100				

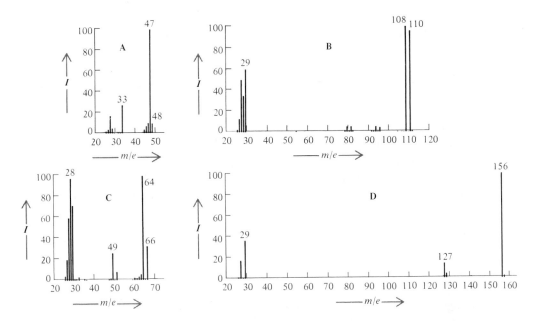

FIGURE 21.12. Mass spectra for Problem 21.10.

SECTION 21.11.

Ionization and Fragmentation in Mass Spectra

In the mass spectrometer, the first reaction of a molecule is the initial ionization—the abstraction of a single electron. The loss of an electron gives rise to the molecular ion. From the peak for this ion radical, which is usually the peak that is farthest to the right in the spectrum, the molecular weight of the compound may be determined. (Remember, it is the exact molecular weight for a molecule containing single isotopes and not an average molecular weight.)

The question arises, "Which type of electron is lost from the molecule?" This question cannot be answered with accuracy. It is believed that the electron in the highest-energy orbital (the "loosest" electron) is the first to be lost. If a molecule has n (unshared) electrons, one of these is lost. If there are no n electrons, then a pi electron is lost. If there are neither n electrons nor pi electrons, the molecular ion is formed by loss of a sigma electron.

$$CH_3\ddot{O}H \longrightarrow [CH_3OH]^{\ddagger} + n\ e^-$$

$$CH_3CH::CH_2 \longrightarrow [CH_3CH-CH_2]^{\ddagger} + \pi\ e^-$$

$$H:CH_3 \longrightarrow [CH_4]^{\ddagger} + \sigma\ e^-$$

After the initial ionization, the molecular ion undergoes fragmentation, a process in which free radicals or small neutral molecules are lost from the molecular ion. A molecular ion does not fragment in a random fashion, but tends to form the *most stable fragments possible*. In equations showing fragmentation, it is common practice not to show the free-radical fragments because they are not detected by the mass spectrometer.

Let us reconsider the mass spectrum of methanol in Figure 21.8. The spectrum consists of three principal peaks at $m/e = 29$, 31, and 32. The structures of the fragments may often be deduced from their masses. The M^+ peak (at 32) of methanol arises from loss of one electron. The peak at 31 must arise from loss of an H atom (which has a mass of 1.0). The peak at 29 must arise from an ion that has lost two more H atoms. What about the minor peak at 15? This peak arises from the loss of $\cdot OH$ from the molecular ion.

$$[CH_2{=}OH]^+ \xrightarrow{\ -H_2\ } [CH{\equiv}O]^+$$
$$m/e = 31 \qquad\qquad m/e = 29$$

$$CH_3OH \xrightarrow{\ -e^-\ } [CH_3OH]^{\ddagger}$$
$$m/e = 32$$

with $-H\cdot$ leading to the $m/e = 31$ fragment and $-\cdot OH$ leading to

$$[CH_3]^+$$
$$m/e = 15$$

Could other fragmentation patterns occur? For example, could the molecular ion lose H^+ to become $[CH_3O]^\cdot$? It possibly could, but we do not know, because only the positively charged particles are accelerated and detected.

A.   Effect of branching

Branching in a hydrocarbon chain leads to fragmentation primarily at the branch because secondary ion radicals and secondary carbocations are more stable than primary ion radicals and primary carbocations. Carbocation stability is a more important factor than free-radical stability. For example, a methylpropane molecular ion yields predominantly an isopropyl cation and a methyl radical (not the reverse).

$$\left[CH_3\text{—}CH\text{—}CH_3 \right]^{\ddagger} \longrightarrow CH_3CH^+ + \cdot CH_3 \quad \text{and very little} \quad CH_3CH + {}^+CH_3$$
(with CH_3 substituents)

STUDY PROBLEM

21.11. What would be the molecular ion and the principal positively charged fragments arising from ionization of: (a) 2-methylpentane;   (b) 2,2-dimethylpropane;   (c) 1-pentene?

B.   Effect of a heteroatom or carbonyl group

Consider another spectrum, that of *N*-ethylpropylamine (Figure 21.13). The molecular ion has $m/e = 87$. Fragmentation of this molecular ion takes place alpha to the nitrogen atom and yields fragments with $m/e = 58$ (loss of an ethyl

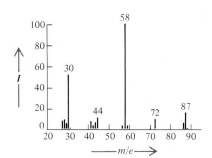

FIGURE 21.13. Mass spectrum of *N*-ethylpropylamine, $CH_3CH_2NHCH_2CH_2CH_3$

group) and 72 (loss of a methyl group). This type of fragmentation is called **α-fission** and is common in both amines and ethers.

fission here

$[CH_3 - CH_2NHCH_2 - CH_2CH_3]^+$

$m/e = 87$

$\xrightarrow{-\cdot C_2H_5}$ $[CH_3CH_2NH = CH_2]^+$

$m/e = 58$

$\xrightarrow{-\cdot CH_3}$ $[CH_2 = NHCH_2CH_2CH_3]^+$

$m/e = 72$

The reason for α-fission is that the cation formed in this reaction is resonance-stabilized:

$$[R - CH_2 - \ddot{N}HR]^+ \xrightarrow{-R\cdot} [CH_2 = \overset{+}{N}HR \longleftrightarrow \overset{+}{C}H_2 - \ddot{N}HR]$$

A similar fragmentation occurs at a bond adjacent to a carbonyl group (or α to the oxygen). Again, the resultant cation is resonance-stabilized.

$$\left[\begin{array}{c} \ddot{O}: \\ \| \\ RC - R \end{array}\right]^+ \xrightarrow{-R\cdot} [R\overset{+}{C} = \ddot{O}: \longleftrightarrow RC \equiv \overset{+}{O}:]$$

STUDY PROBLEM

21.12. The mass spectrum of an ether is shown in Figure 21.14. What is the structure of the ether?

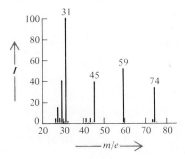

FIGURE 21.14. Mass spectrum for Problem 21.12.

C. Loss of a small molecule

Small stable molecules, such as H_2O, CO_2, CO, and C_2H_4, can be lost from a molecular ion. An alcohol, for example, readily loses H_2O and shows a peak at 18 mass units less than the peak of the molecular ion. This peak is referred to as the **M − 18 peak**. In many alcohols, elimination of H_2O is so facile that the molecular-ion peak is not even observed in the spectrum. The spectrum of 1-butanol (Figure 21.15) is a typical mass spectrum of an alcohol.

$$[CH_3CH_2CH_2CH_2OH]^{\ddagger} \xrightarrow{-H_2O} [CH_3CH_2CH=CH_2]^{\ddagger}$$

$$M^{\ddagger}, m/e = 74 \qquad\qquad M-18, m/e = 56$$

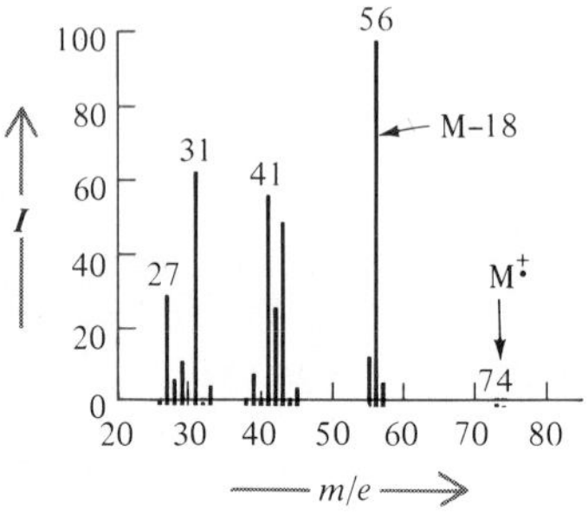

FIGURE 21.15. Mass spectrum of 1-butanol.

D. McLafferty rearrangement

When there is a hydrogen atom γ to a carbonyl group in the molecular ion, a **McLafferty rearrangement** may occur. In this rearrangement, an alkene is lost from the molecular ion.

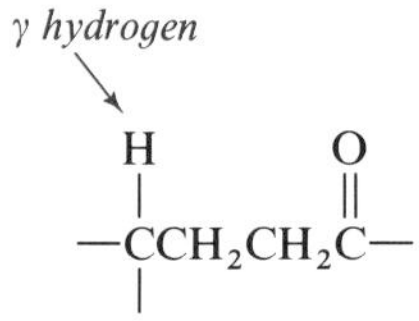

γ hydrogen

$$\underset{|}{\overset{\text{H}}{\underset{|}{-\text{C}}}}\text{CH}_2\text{CH}_2\overset{\overset{\text{O}}{\|}}{\text{C}}-$$

Rearrangement:

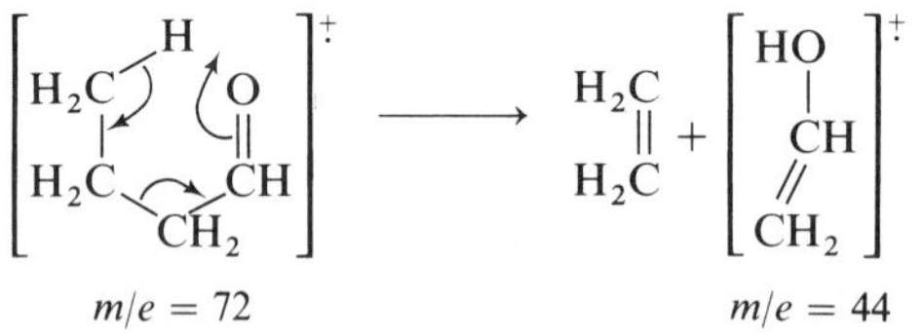

$$m/e = 72 \qquad\qquad m/e = 44$$

STUDY PROBLEM

21.13. Predict the m/e values for the products of the McLafferty rearrangement of the following compounds:

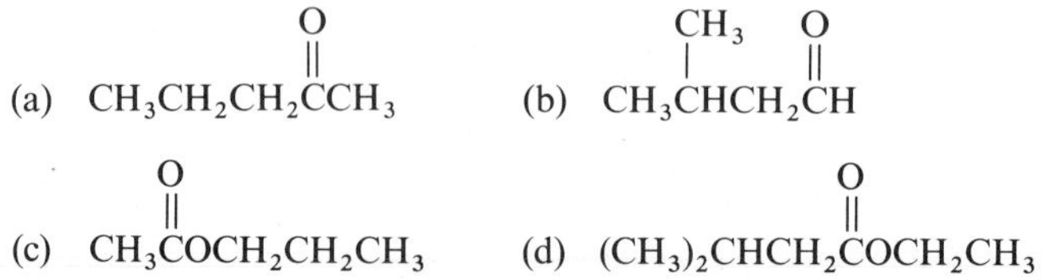

(a) $CH_3CH_2CH_2\overset{\overset{\text{O}}{\|}}{\text{C}}CH_3$
(b) $CH_3\overset{\overset{\text{CH}_3}{|}}{\text{CH}}CH_2\overset{\overset{\text{O}}{\|}}{\text{C}}H$

(c) $CH_3\overset{\overset{\text{O}}{\|}}{\text{C}}OCH_2CH_2CH_3$
(d) $(CH_3)_2CHCH_2\overset{\overset{\text{O}}{\|}}{\text{C}}OCH_2CH_3$

Summary

Absorption of ultraviolet (200–400 nm) or visible (400–750 nm) light results in **electronic transitions**, promotion of electrons from the ground-state orbitals to orbitals of higher energy. The wavelength λ of absorption is inversely proportional to the energy required. The uv or visible spectrum is a plot of **absorbance A** or **molar absorptivity** ε vs λ, where $\varepsilon = A/cl$. The position of maximum absorption is reported as λ_{max}.

The important electronic transitions are $\pi \to \pi^*$ for conjugated systems and $n \to \pi^*$. Increasing amounts of conjugation result in shifts of λ_{max} toward longer wavelengths. Compounds that absorb at wavelengths longer than 400 nm are colored; the apparent color is the **complementary color** of the wavelength absorbed.

Vision is made possible by the conversion of 11-*cis*-retinal in rhodopsin to all-*trans*-retinal, which in turn is reduced to all-*trans*-vitamin A. The regeneration of 11-*cis*-retinal is accomplished by enzymatic conversion of all-*trans*-vitamin A to 11-*cis*-vitamin A and then to 11-*cis*-retinal.

Dyes are colored compounds that adhere to fabric or other substance. An **acid–base indicator** is a compound that undergoes a color change in a reaction with acid or base. The color change arises from a change in the conjugated system and thus in the wavelength of absorption.

TABLE 21.5. Summary of some fragmentation patterns in mass spectra

Fragmentation	*Reaction type*
Alkanes:	
$[R_2CH\!\dotplus\!CH_3]^{\ddagger} \xrightarrow{\;-\cdot CH_3\;} R_2\overset{+}{C}H$	σ-bond fission to most stable carbocation
Amines and ethers:	
$[R\!\dotplus\!CH_2-NR'_2]^{\ddagger} \xrightarrow{\;-R\cdot\;} CH_2=\overset{+}{N}R'_2$	α-fission
$[R\!\dotplus\!CH_2-OR']^{\ddagger} \xrightarrow{\;-R\cdot\;} CH_2=\overset{+}{O}R'$	α-fission
Carbonyl compounds:	
$\left[\begin{smallmatrix} O \\ \| \\ RC\dotplus Y \end{smallmatrix}\right]^{\ddagger} \xrightarrow{\;-Y\cdot\;} R\overset{+}{C}=O$	α-fission
$\left[\begin{smallmatrix} H & & O \\ \| & & \| \\ R_2CCH_2\dotplus CH_2CY \end{smallmatrix}\right]^{\ddagger} \xrightarrow{\;-R_2C=CH_2\;} \left[\begin{smallmatrix} OH \\ \| \\ CH_2=CY \end{smallmatrix}\right]^{\ddagger}$ where Y = H, R', OH, OR', etc.	McLafferty rearrangement[a]
Alcohols:	
$\left[\begin{smallmatrix} OH \\ \| \\ R_2CHCR_2 \end{smallmatrix}\right]^{\ddagger} \xrightarrow{\;-H_2O\;} [R_2C=CR_2]^{\ddagger}$	loss of H_2O

[a] If a ketone or ester undergoes rearrangement and if Y contains a γ hydrogen, two types of rearrangements may be observed.

Fluorescence and **chemiluminescence** are the results of emission of light as excited molecules return to the ground state.

A **mass spectrum** is a graph of **abundance** versus **mass-to-charge ratio** (m/e) of positively charged particles that result from bombardment of a compound with high-energy electrons. Removal of one electron from a molecule of the compound results in the **molecular ion**. The molecular ion can lose atoms, ions, radicals, and small molecules to yield a variety of fragmentation products. Table 21.5 summarizes some of the fragmentation patterns.

STUDY PROBLEMS

21.14. A chemist prepared 1,3,5-hexatriene and 1,3,5,7-octatetraene, placed them in separate flasks, but neglected to label them. How could the two compounds be differentiated by uv spectroscopy?

21.15. Which of the following pairs of compounds could probably be distinguished from each other by their uv spectra? Explain.

(a) $CH_3CO_2CH_2CH_3$, $CH_3CH_2CO_2CH_3$

(b)

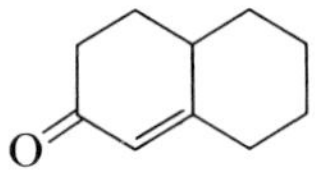

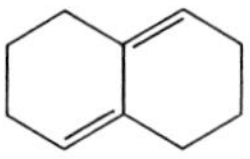

(c) $CH_3CH{=}CHCH_2\overset{\overset{\displaystyle O}{\|}}{C}CH_3$, $CH_3CH_2CH{=}CH\overset{\overset{\displaystyle O}{\|}}{C}CH_3$

(d)

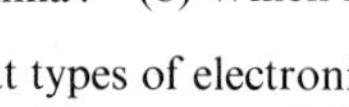

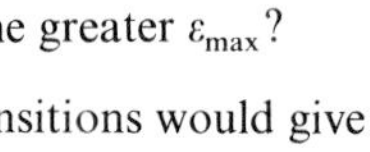

21.16. Calculate the molar absorptivities of the following compounds at the specified wavelengths:

(a) adenine ($9.54 \times 10^{-5} M$ solution, 1.0-cm cell), absorbance of 1.25 at 263 nm
(b) cyclohexanone ($0.038 M$ solution, 1.0-cm cell), absorbance of 0.75 at 288 nm

21.17. 3-Buten-2-one shows uv absorption maxima at 219 nm and 324 nm.   (a) Why are there two maxima?   (b) Which has the greater ε_{max}?

21.18. What types of electronic transitions would give rise to uv absorption in each of the following compounds?

(a) 2,4-octadiene (b) 2-cyclohexenone (c) aniline (d) formaldehyde

21.19. Rank the following compounds in the order of increasing λ_{max}:   (a) benzene; (b) biphenyl ($C_6H_5{-}C_6H_5$);   (c) styrene ($C_6H_5CH{=}CH_2$);   (d) stilbene ($C_6H_5CH{=}CHC_6H_5$).

21.20. A chemist has a liquid compound of unknown structure with the formula C_5H_8; the uv spectrum shows a λ_{max} at about 220 nm. What are the likely possible structures of this compound?

21.21. Suggest a reason why the λ_{max} for *trans*-stilbene ($C_6H_5CH{=}CHC_6H_5$) is 295 nm, while that for *cis*-stilbene is 280 nm.

21.22. Suggest a mechanism for the photochemical isomerization of *trans*-stilbene to *cis*-stilbene.

21.23. *Indophenol blue* is a vat dye that is oxidized to an insoluble blue dye after application to fabric. Give the structure of the oxidized form.

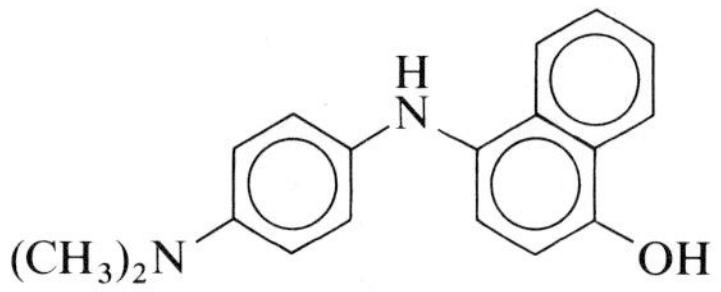

reduced form of Indophenol blue

21.24. As a solution of *Crystal violet* is acidified, it turns from violet to blue, then to green, and finally to yellow. Write equations that could explain these color changes.

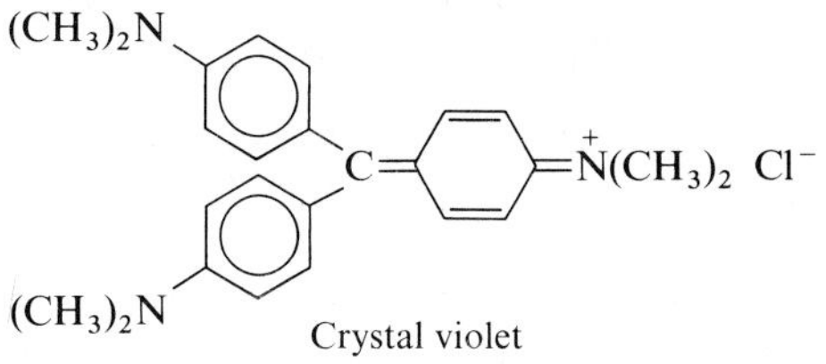

Crystal violet

21.25. *Phenol red* is an acid–base indicator that is yellow at pH 6 but red at pH 9. Draw the structures (and resonance structures) for the two forms of this compound.

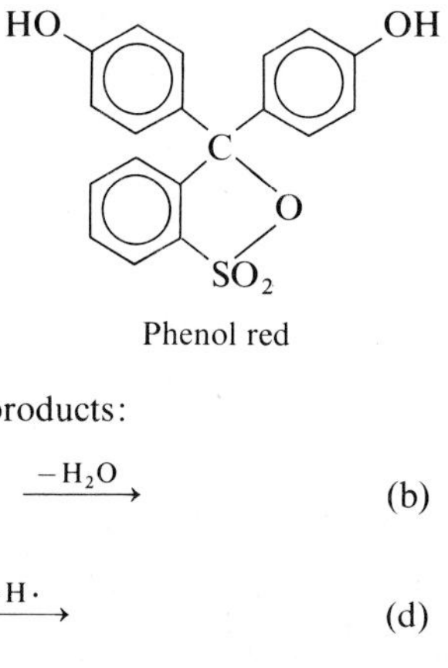

Phenol red

21.26. Give the structures of the products:

(a) $[CH_3CH_2CH_2OH]^{\ddagger} \xrightarrow{-H_2O}$

(b) 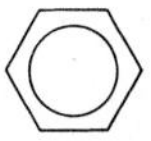 $\xrightarrow{-e^-}$

(c) 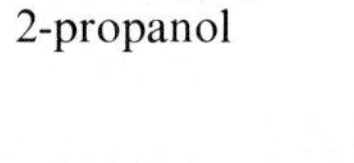 $\left[\langle \rangle -OH \right]^{\ddagger} \xrightarrow{-H\cdot}$

(d) $[(CH_3)_2CHCl]^{\ddagger} \xrightarrow{-Cl\cdot}$

21.27. For each of the following compounds, predict the structures and m/e values for the molecular ion and likely positively charged fragmentation products:

(a) ethyl isopropyl ether
(b) ethyl isobutyl ether
(c) 2-chloropropane
(d) 2,5-dimethylhexane
(e) 2-propanol
(f) 4-cyclopentylbutanal
      (McLafferty fragment only)

21.28. Suggest structures and fragmentation patterns that account for the following observed peaks in the mass spectra:

(a) *n*-butane, $m/e = 58, 57, 43, 29, 15$
(b) benzamide, $m/e = 121, 105, 77$
(c) 1-bromopropane, $m/e = 124, 122, 43, 29, 15$

21.29. What reactants would you need to prepare the yellow azo dye *Crysamine G*?

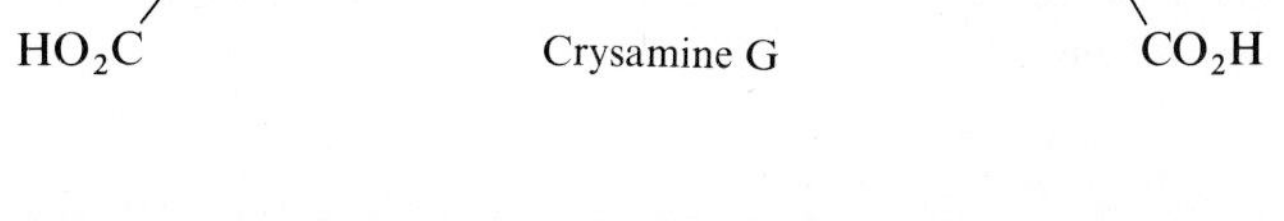

Crysamine G

21.30. In uv spectra, the presence of an additional double bond in conjugation adds about 30 nm to the λ_{max}. From the following observed values for λ_{max}, state how the *degree of substitution* on the sp^2 carbons of a polyene affects the position of the λ_{max}.

Structure	λ_{max}, nm
$CH_2{=}CHCH{=}CH_2$	217
$CH_3CH{=}CHCH{=}CH_2$	223
$CH_3CH{=}CHCH{=}CHCH_3$	227

$CH_2{=}CCH{=}CHCH_3$
    $\mid$
    CH_3
	227

21.31. From your answer in Problem 21.30, predict the λ_{max} values for the following polyenes:

(a)   (b)

21.32. Predict the λ_{max} values where indicated:

Structure	λ_{max}, nm	Structure	λ_{max}, nm
$CH_2{=}CHCCH_3$ (with O double bond)	219	$(CH_3)_2C{=}CHCCH_3$ (with O double bond)	(a)
$CH_3CH{=}CHCCH_3$ (with O double bond)	224	$(CH_3)_2C{=}CCCH_3$ (with O double bond, CH_3)	(b)
		$CH_2{=}CHCH{=}CHCCH_3$ (with O double bond)	(c)

21.33. A ketone ($C_6H_{12}O$) gives a positive iodoform test and shows principal peaks in the mass spectrum at the following m/e values: 100, 85, 57, and 43. Which of the following compounds are compatible with these data?

(a) $CH_3CH_2CCH_2CH_2CH_3$ (with O double bond)
(b) $CH_3CCH_2CH_2CH_2CH_3$ (with O double bond)

(c) $CH_3CCHCH_2CH_3$ (with O double bond, CH_3)
(d) $CH_3CCH_2CH(CH_3)_2$ (with O double bond)

21.34. A compound contains only C, H, and O. The infrared spectrum shows strong absorption at 1724 cm^{-1} (5.8 μm), 1388 cm^{-1} (7.2 μm), and 1231 cm^{-1} (8.1 μm) (plus other minor absorption). The nmr spectrum shows only one singlet at 2.1 ppm. The mass spectrum has principal peaks at 58 m/e and 43 m/e. What is the structure of the compound?

21.35. The infrared, nmr, and mass spectra for Compounds A through D are shown in Figures 21.16 through 21.19. From the spectra, deduce the structure of each compound.

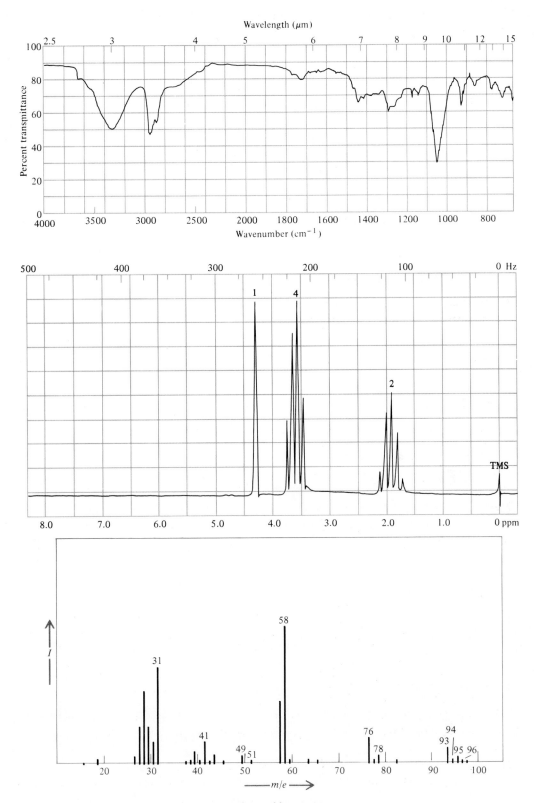

FIGURE 21.16. Spectra for Compound A, Problem 21.35.

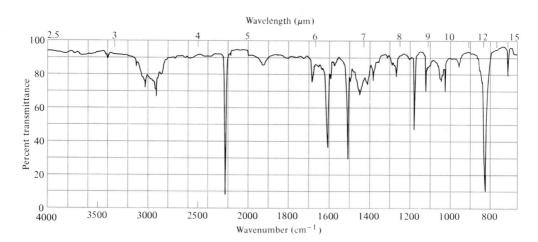

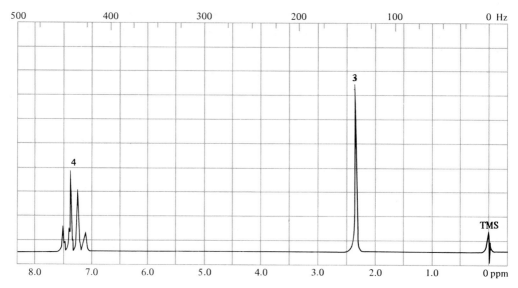

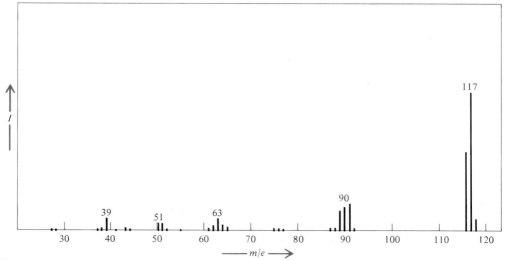

FIGURE 21.17. Spectra for Compound B, Problem 21.35.

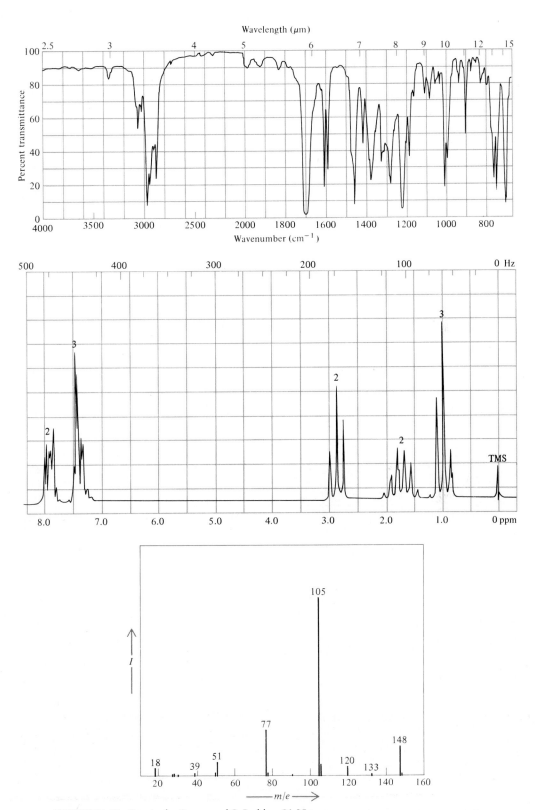

FIGURE 21.18. Spectra for Compound C, Problem 21.35.

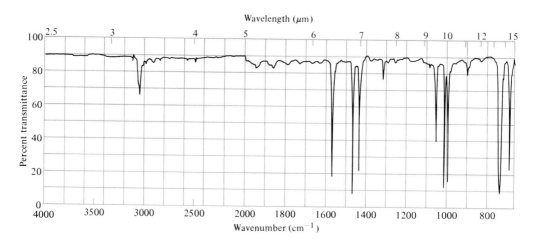

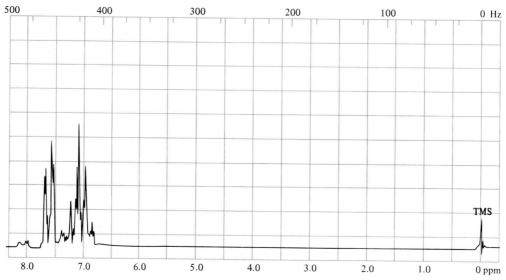

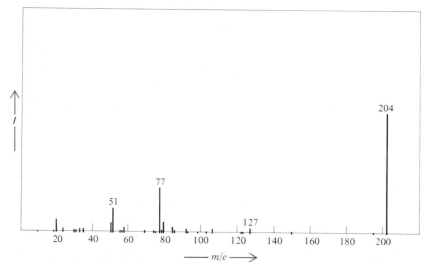

FIGURE 21.19. Spectra for Compound D, Problem 21.35.

APPENDIX

Nomenclature of Organic Compounds

A complete discussion of definitive rules of organic nomenclature would require more space than can be allotted in this text. We will survey some of the
more common nomenclature rules, both IUPAC and trivial. The following
references contain more detail:

IUPAC Nomenclature of Organic Chemistry, Sections A, B, & C, 2nd Ed.,
    Butterworths, London, 1971. Now available from Pergamon Press,
    Elmsford, New York. [Also may be found in *Pure Appl. Chem. II* (1–2)
    (1965).]
Chemical Abstracts Service, *Naming and Indexing of Chemical Substances for*
    Chemical Abstracts *during the Ninth Collective Period* (*1972–1976*)
    (*January–June, 1972*), American Chemical Society, Columbus, Ohio, 1973.
Chemical Abstracts Service, *Combined Introductions to the Indexes to Volume 66*
    (*January–June, 1967*), American Chemical Society, Columbus, Ohio, 1968.
A. M. Patterson, L. T. Capell, and D. F. Walker, *The Ring Index, 2nd Ed.,*
    American Chemical Society, Washington D.C., 1960; *Supplement I*
    (*1957–1959*), 1963; *Supplement II* (*1960–1961*), 1964; *Supplement III*
    (*1962–1963*), 1965.
R. S. Cahn, *J. Chem. Ed.,* **41**, 116 (1964).

Alkanes

The names for the first thirty continuous-chain alkanes are listed in Table A1.

Branched alkanes.   In naming an alkane with alkyl substituents, the longest
continuous chain is considered the parent. The parent is numbered from one end to

TABLE A1. Names of some continuous-chain alkanes

Molecular formula	Name	Molecular formula	Name
CH_4	methane	$C_{16}H_{34}$	hexadecane
C_2H_6	ethane	$C_{17}H_{36}$	heptadecane
C_3H_8	propane	$C_{18}H_{38}$	octadecane
C_4H_{10}	butane	$C_{19}H_{40}$	nonadecane
C_5H_{12}	pentane	$C_{20}H_{42}$	eicosane
C_6H_{14}	hexane	$C_{21}H_{44}$	heneicosane
C_7H_{16}	heptane	$C_{22}H_{46}$	docosane
C_8H_{18}	octane	$C_{23}H_{48}$	tricosane
C_9H_{20}	nonane	$C_{24}H_{50}$	tetracosane
$C_{10}H_{22}$	decane	$C_{25}H_{52}$	pentacosane
$C_{11}H_{24}$	undecane	$C_{26}H_{54}$	hexacosane
$C_{12}H_{26}$	dodecane	$C_{27}H_{56}$	heptacosane
$C_{13}H_{28}$	tridecane	$C_{28}H_{58}$	octacosane
$C_{14}H_{30}$	tetradecane	$C_{29}H_{60}$	nonacosane
$C_{15}H_{32}$	pentadecane	$C_{30}H_{62}$	triacontane

the other, the direction being chosen to give the lowest numbers to the substituents. The entire name of the structure is then composed of (1) the numbers of the positions of the substituents; (2) the names of the substituents; and (3) the name of the parent.

Alkyl substituents.   The names of the alkyl substituents (also called *branches*, or *radicals*) are derived from the names of their corresponding alkanes with the ending changed from *-ane* to *-yl*. For example, CH_3CH_2— is ethyl (from ethane). Multiple substituents are placed in alphabetical order, each preceded by its respective number and like substituents grouped together. Some common branched alkyl substituents have trivial names (see Table A2).

TABLE A2. Trivial names for some common alkyl groups

Structure	Name
$CH_3CH_2CH_2$—	*normal-*, or *n*-propyl[a]
$(CH_3)_2CH$—	isopropyl
$CH_3CH_2CH_2CH_2$—	*n*-butyl[a]
$(CH_3)_2CHCH_2$—	isobutyl
$CH_3CH_2CH(CH_3)$—	*secondary-*, or *sec*-butyl
$(CH_3)_3C$—	*tertiary-*, *tert-*, or *t*-butyl
$CH_3CH_2CH_2CH_2CH_2$—	*n*-pentyl (or *n*-amyl)[a]
$(CH_3)_2CHCH_2CH_2$—	isopentyl (or isoamyl)
$(CH_3)_3CCH_2$—	neopentyl

[a] The use of *n*- (to denote a continuous chain) is optional.

$$CH_2CH_3$$
$$|$$
$$CH_3CH_2CHCH_2CH_2CH_3$$
① ② ③ ④ ⑤ ⑥

3-ethylhexane

$$CH_3 \quad CH_2CH_3$$
$$| \qquad |$$
$$CH_3CHCH_2CHCH_2CHCH_2CH_3$$
① ② ③ ④ ⑤ ⑥ ⑦ ⑧
$$|$$
$$CH_3$$

4-ethyl-2,6-dimethyloctane

Alkenes and alkynes

Unbranched hydrocarbons having one double bond are named in the IUPAC system by replacing the ending *-ane* of the alkane name with *-ene*. If there are two or more double bonds, the ending is *-adiene*, *-atriene*, etc. The chain is numbered to give the lowest possible numbers to the double bonds. (The *lower* number of the two carbons joined by the double bond is used to give the position.)

① ② ③ ④ ⑤
$$CH_3CH=CHCH_2CH_3$$

2-pentene

① ② ③ ④ ⑤
$$CH_2=CHCH=CHCH_3$$

1,3-pentadiene

Unbranched hydrocarbons having one triple bond are named by replacing the ending *-ane* of the alkane name with *-yne*. If there are two or more triple bonds, the ending is *-adiyne*, *-atriyne*, etc. The chain is numbered to give the lowest possible numbers to the triple bonds. Again, the lower number is used to give the position. For example, $CH_3CH_2C{\equiv}CH$ is 1-butyne.

Unbranched hydrocarbons with both double and triple bonds are named by replacing the ending *-ane* of the name of the alkane with the ending *-enyne*. When necessary, position numbers are inserted. The double bond is assigned the lowest number; for example, $CH_2=CHC{\equiv}CCH_3$ is 1-penten-3-yne. Trivial names for some alkenes and alkynes are found in Table A3.

Branched alkenes and alkynes.   In the IUPAC name of a branched alkene or alkyne, the parent chain is the longest chain that contains the maximum number of double or triple bonds. (This may or may not be the longest continuous chain in the structure.)

TABLE A3. Trivial names for some alkenes and alkynes

Structure	Name
$CH_2=CH_2$	ethylene
$CH{\equiv}CH$	acetylene
$CH_2=C=CH_2$	allene
$CH_2=CHCH_3$	propylene
$CH_3C{\equiv}CH$	methylacetylene
$(CH_3)_2C=CH_2$	isobutylene
$CH_2=C(CH_3)CH=CH_2$	isoprene

3,4-dipropyl-1,3,5-heptatriene

Several unsaturated substituents have trivial names; some of these are listed in Table A4, along with their systematic names.

In complex compounds, the symbol Δ is sometimes used to denote double bonds.

is $\Delta^{4,6}$-cholestadiene

Geometric isomers. There are two methods for naming geometric isomers. One method employs the prefixes *cis-* (same side) or *trans-* (opposite sides) to designate the geometric isomers.

cis-2-butene   *trans*-2-butene

The other method employs (E) (groups with higher priority, according to the Cahn–Ingold–Prelog system, on opposite sides) or (Z) (groups with higher priority on the same side) for designation of the geometric isomers. The priority rules are found in Section 4.1B (page 115).

TABLE A4. IUPAC names for some unsaturated groups

Structure	Name
$CH_2=$	methylene
$CH_2=CH-$	ethenyl (vinyl)[a]
$CH_3CH=$	ethylidene
$CH\equiv C-$	ethynyl
$CH_2=CHCH_2-$	2-propenyl (allyl)[a]
$CH_3CH=CH-$	1-propenyl
$(CH_3)_2C=$	isopropylidene
	2-cyclopenten-1-yl

[a] Names in parentheses are trivial.

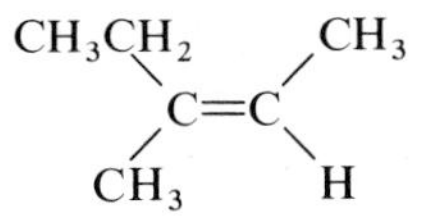

$$CH_3CH_2 \quad CH_3$$
$$C=C$$
$$CH_3 \quad H$$

$$CH_3 \quad CH_3$$
$$C=C$$
$$CH_3CH_2 \quad H$$

(Z)-3-methyl-2-pentene 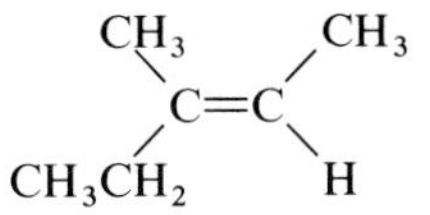   (E)-3-methyl-2-pentene

ethyl has priority over methyl; methyl has priority over H

Cyclic hydrocarbons

Cycloalkanes and cycloalkenes. The names of saturated monocyclic hydrocarbons are formed by attaching the prefix *cyclo-* to the name of the alkane with the same number of carbon atoms.

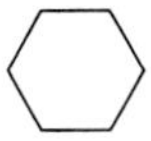

is cyclohexane

A ring is considered the parent unless there is a longer chain attached, in which case the longer chain is the parent.

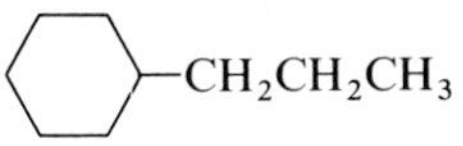

—$CH_2CH_2CH_3$ —$CH_2(CH_2)_5CH_3$

propylcyclohexane 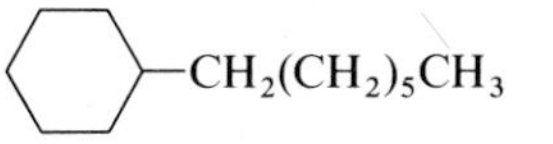   1-cyclohexylheptane

Alkyl substituents are named as prefixes and are given the lowest possible position numbers.

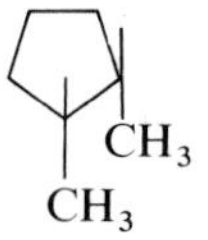

cis-1,2-dimethylcyclopentane

CH_3
CH_3

Unsaturated monocyclic hydrocarbons are named by changing the ending from *-ane* to *-ene* (*-adiene*, etc.). The ring is numbered to give the lowest numbers possible to the double bonds. With both alkyl substituents and double bonds, the ring is numbered so that the double bonds receive the lowest possible numbers.

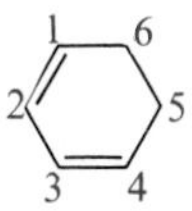

CH_3

1,3-cyclohexadiene 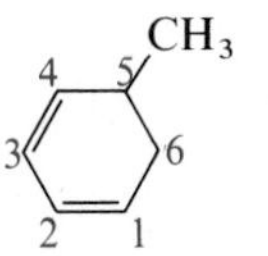   5-methyl-1,3-cyclohexadiene

Some common terpene ring structures have individual names.

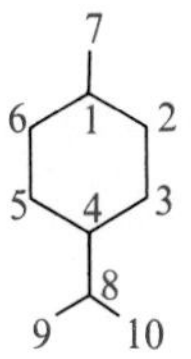

menthane 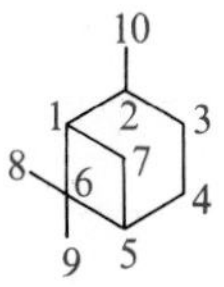   pinane 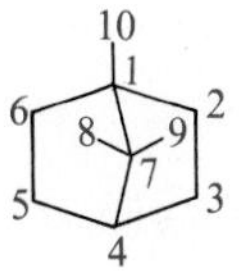   bornane 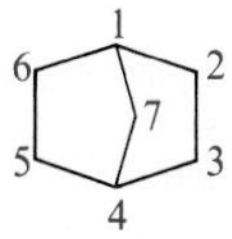   norbornane

Aromatic hydrocarbons.   Aromatic hydrocarbons are generally referred to by their trivial names (see Table A5). Systems composed of five or more linear fused benzene rings are named by a Greek-number prefix followed by *-cene*. (The prefix denotes the number of fused rings.)

pentacene

The aromatic system is considered the parent unless a longer chain is attached.

TABLE A5. Names for some arenes and aryl groups

Structure	Name	Structure	Name
Arenes:		**Aryl groups:**	
	benzene	C_6H_5—	phenyl-
	naphthalene	$C_6H_5CH_2$—	benzyl-
	anthracene	CH_3—⟨⟩—	p-tolyl-
	phenanthrene		1-naphthyl- (α-naphthyl-)
⟨⟩—CH_3	toluene	$C_6H_5CH{=}CHCH_2$—	cinnamyl-
⟨⟩—$CH(CH_3)_2$	cumene		
⟨⟩—$CH{=}CH_2$	styrene		
	o-xylene		
	mesitylene		

$$C_6H_5CH_2CH_2CH_3 \qquad C_6H_5CH_2(CH_2)_5CH_3$$

propylbenzene   1-phenylheptane

If there are only two substituents on a benzene ring, their positions may be designated either by prefix numbers or by *o-*, *m-*, or *p-* (*ortho*, *meta-*, or *para-*). If there are more than two substituents, numbers must be used.

o-dibromobenzene   *m*-dibromobenzene   *p*-dibromobenzene   1,2,3-triethylbenzene

The principal group, or a group that is part of the parent (for example, the CH_3 in toluene), is always considered to be attached to position 1 on the ring. The numbers of substituents are chosen to be as low as possible.

m-chlorophenol   *p*-nitrotoluene   3,5-dimethylstyrene

Heterocycles

Some common heterocycles are listed in Table A6. In monocyclic heterocycles with only one heteroatom, that atom is considered position 1. Other ring systems are numbered by convention (see Table A6).

To name monocyclic compounds with one or more heteroatoms, prefixes may be used: *oxa-* ($-O-$), *aza-* ($-NH-$), *thia-* ($-S-$). For unsaturated rings, the ring size is designated by suffixes: *-ole* means five, and *-ine* means six (used only with rings that contain nitrogen). *Example:* An *oxazole* is a five-membered ring containing O and N. In numbering, O has priority over N.

1,3-oxazole   1,3,5-triazine

Heteroatoms in chains

The *oxa, aza-* system may also be used for naming aliphatic compounds. This method is called *replacement nomenclature*. All the atoms in the chain are numbered in such a way that the heteroatoms receive the lowest possible numbers. The parent is considered to be the alkane that has the same number of carbon atoms as the total number of atoms in the continuous chain (heteroatoms, but not hydrogens, included).

$$\overset{1}{C}H_3\overset{2}{O}\overset{3}{C}H_2\overset{4}{C}H_2\overset{5}{O}\overset{6}{C}H_2\overset{7}{C}H_2\overset{8}{O}\overset{9}{C}H_2\overset{10}{C}H_3$$

2,5,8-trioxadecane

TABLE A6. Names of some common heterocycles

Structure	Name	Structure	Name
	furan		pyrimidine
	4*H*-pyran (γ-pyran)		morpholine
	pyrrole		thiophene
	pyrazole		indole
	imidazole		carbazole
	pyridine		purine
	piperidine		quinoline
	piperazine		isoquinoline

Putting names together

The prefixes. In the IUPAC system, alkyl and aryl substituents and many functional groups are named as prefixes on the parent (for example, iodomethane). Some common functional groups named as prefixes are listed in Table A7. Other prefix names are sometimes used for carbonyl groups, hydroxyl groups, etc. These are mentioned under their specific headings.

Like treatment of like things. Names should be as simple as possible and as consistent as possible. Two identical substituents should be treated alike, even

TABLE A7. Some common functional groups named as prefixes

Structure	Name	Structure	Name
—OR	alkoxy-[a]	—F	fluoro-
—NH$_2$	amino-	—H	hydro-[b]
—N=N—	azo-	—I	iodo-
—Br	bromo-	—NO$_2$	nitro-
—Cl	chloro-	—NO	nitroso-

[a] *methoxy-*, *ethoxy-*, etc., depending upon R.

[b] *Hydro-* is a prefix used to designate a hydrogenated derivative of an unsaturated parent. *Perhydro-* means completely hydrogenated.

4a,8a-dihydronaphthalene   perhydrophenanthrene

though a few other rules may be broken. Although C$_6$H$_5$CH$_3$ would be methylbenzene or toluene (rather than phenylmethane), (C$_6$H$_5$)$_2$CH$_2$ is called diphenylmethane.

Prefixes to designate like things. In simple compounds, the prefixes *di-*, *tri-*, *tetra-*, *penta-*, *hexa-*, etc. (not italicized) are used to indicate the number of times a substituent is found in the structure: e.g., dimethylamine for (CH$_3$)$_2$NH or dichloromethane for CH$_2$Cl$_2$.

In complex structures, the prefixes *bis-*, *tris-*, and *tetrakis-* (not italicized) are used: *bis-* means two of a kind; *tris-*, three of a kind; and *tetrakis-*, four of a kind.

[(CH$_3$)$_2$N]$_2$ is bis(dimethylamino)- and not di(dimethylamino)-

The prefix *bi-* is used for (1) "double" molecules, and (2) bridged hydrocarbons.

biphenyl   bicyclo[2.2.0]hexane

a double molecule   *a bridged hydrocarbon*

Order of prefixes. Prefixes are listed in *alphabetical order* (ethylmethyl-). In alphabetizing, a prefix denoting the number of times a substituent is found (di-, tri-, etc.) is disregarded. Ethyldimethyl- is the correct alphabetical order, even though *d* comes before *e*.

When to drop a vowel. In the case of *conjunctive names* (names that are formed by combining two names), vowels are not elided, but are maintained. For example, the *e* in indoleacetic acid is retained (not indolacetic acid).

If there are two successive suffixes, the vowel at the end of the first suffix is dropped (propenoic acid, not propeneoic acid), unless it is followed by a consonant (propanediol, not propandiol).

Nomenclature priority of functional groups. The various functional groups are ranked in priority as to which receives the suffix name and the lowest position number. A list of these priorities is given in Table A8. These are *not* the same priorities that are used for (*E*) and (*Z*) or (*R*) and (*S*).

$$\begin{matrix} & O \\ & \| \\ CH_3\overset{}{C}\overset{|}{C}HCH_3 \\ & | \\ & NH_2 \end{matrix}$$ is 3-amino-2-butanone (not 2-amino-3-butanone)

The principal functional group is indicated at the end of the name. A name with only one functional ending is preferred. *Example:* $HOCH_2CH_2CO_2H$ is 3-hydroxypropanoic acid. In names that must have two endings, the terminal ending refers to the principal functional group (see Table A8). *Example:* $CH_3CH{=}CHCO_2H$ is 2-butenoic acid.

TABLE A8. Nomenclature priority[a]

Structure	Name
$-N(CH_3)_3{}^+$ (as one example)	onium ion
$-CO_2H$	carboxylic acid
$-SO_3H$	sulfonic acid
$-COX$	acid halide
$-CONR_2$	amide
$-CN$	nitrile
$-CHO$	aldehyde
$-CO-$	ketone
ROH	alcohol
ArOH	phenol
$-SH$	thiol
$-NR_2$	amine
$-O-O-$	peroxide
$-MgX$ (as one example)	organometallic
$\underset{/}{\overset{\backslash}{C}}{=}\underset{\backslash}{\overset{/}{C}}$	alkene
$-C{\equiv}C-$	alkyne
R—, X—, etc.	other substituents

[a] Highest priority is at the top.

Numbering of the parent. The parent is numbered so that the principal function receives the lowest number (see preceding section). Greek letters are reserved for trivial names and may not correspond directly to the IUPAC numbers. *Alpha* (α, the first letter of the Greek alphabet) means *on the nearest carbon*, which is often number 2 in the systematic name.

$$\text{Br}$$
$$|$$
$$CH_3CHCO_2H$$

2-bromopropanoic acid (IUPAC)
α-bromopropionic acid (trivial)

Numbering of substituents and use of parentheses. If two numbering systems are required for complete identification of all atoms in the molecule, primes (') are often used for one of the systems to prevent confusion.

Alkyl substituents are numbered separately from the parent chain, beginning at the point of attachment. In these cases, a prime is not necessary provided that parentheses are used to enclose complex prefixes.

$$H_2N-\!\!\bigcirc\!\!-CH_2CH_2Br$$

p-(2-bromoethyl)aniline

$$CH_3CHClCH_2$$
$$|$$
$$CH_2\!=\!CHCH\!=\!CHC\!=\!CH_2$$

2-(2-chloropropyl)-1,3,5-hexatriene

Configuration around a chiral carbon. A chiral carbon atom with four different groups attached can have either an (R) or an (S) configuration. For purposes of designating the chiral carbon as either (R) or (S), the structure is placed so that the lowest-priority group is in the rear. (See Section 4.1B for determination of priority, which is *not the same* as nomenclature priority.) The direction from the highest-priority group to the second-highest-priority group is then determined. If the direction is *clockwise*, the chiral carbon is (R); if it is *counterclockwise*, then the carbon is (S). If the structure has only one chiral carbon, (R) or (S) is used as the first prefix in the name. If the molecule has more than one chiral carbon, the designation of each chiral carbon and its position number is enclosed in parentheses in the prefix—e.g., $(2R,3R)$-dibromopentane.

H in rear

(R)-2-butanol

(S)-2-butanol

Carboxylic acids

There are four principal types of names for carboxylic acids: (1) IUPAC; (2) trivial; (3) carboxylic acid; and (4) conjunctive.

IUPAC names. Except for acids of one to five carbons and some fatty acids, aliphatic monocarboxylic acids are named by the IUPAC system. The longest chain containing the —CO_2H group is chosen as the parent, and the chain is numbered starting with the carbon of the —CO_2H group as position 1. The name is taken from the name of the alkane with the same number of carbons, with the final -*e* replaced by -*oic acid*.

$$CH_3CH_2CH_2CH_2CH_2CO_2H$$

hexanoic acid

Substituents are designated by prefixes. A double bond is designated as a suffix preceding -oic acid.

$$CH_2{=}CHCH_2\overset{\overset{\displaystyle CH_3}{|}}{C}HCH_2CO_2H$$

3-methyl-5-hexenoic acid

TABLE A9. Trivial names of some monocarboxylic acids

Structure	Name of acid	Structure	Name of acid
Saturated chain:		**Other functionality[b]:**	
HCO_2H	formic	$CH_3COCH_2CO_2H$	acetoacetic
CH_3CO_2H	acetic	$CH_3CH(OH)CO_2H$	lactic
$CH_3CH_2CO_2H$	propionic	CH_3COCO_2H	pyruvic
$CH_3(CH_2)_2CO_2H$	butyric	$CH_3COCH_2CH_2CO_2H$	levulinic
$CH_3(CH_2)_3CO_2H$	valeric		
$CH_3(CH_2)_{10}CO_2H$	lauric	**On rings:**	
$CH_3(CH_2)_{12}CO_2H$	myristic	$C_6H_5CO_2H$	benzoic
$CH_3(CH_2)_{14}CO_2H$	palmitic		
$CH_3(CH_2)_{16}CO_2H$	stearic		salicylic
Unsaturated chain[a]:			
$CH_2{=}CHCO_2H$	acrylic		
$CH_2{=}C(CH_3)CO_2H$	methacrylic		2-naphthoic
trans-$CH_3CH{=}CHCO_2H$	crotonic		
			nicotinic

[a] For unsaturated fatty acids, see Table 20.1, page 888.
[b] For α-amino acids, see Table 19.1, page 852.

Trivial names.   Table A9 gives a list of commonly encountered trivial names for carboxylic acids.

Carboxylic acid names.   A carboxylic acid name is used when a $-CO_2H$ group is attached to a ring. The name is a combination of the name of the ring system with the suffix *-carboxylic acid*. The carboxyl group is considered attached to position 1 of the ring unless the ring system has its own unique numbering system. (The carbon of the $-CO_2H$ group is not numbered, as it is in an IUPAC name.)

cyclohexanecarboxylic acid   2-pyridinecarboxylic acid

Conjunctive names.   Conjunctive names are combinations of two names: in the following examples, the name of the ring plus the name of the acid.

cyclohexaneacetic acid   indole-2-acetic acid

Diacids and polycarboxylic acids.   Diacids may be named systematically as *-dioic acids*: for example, $HO_2CCH_2CO_2H$ is propanedioic acid. Trivial names are commonly used. Some of these are listed in Table A10.

Polycarboxylic acids of the aliphatic series containing more than two carboxyl groups are named by the carboxylic-acid nomenclature. The longest chain to which the greatest number of carboxyl groups are attached is chosen as the parent. If there is unsaturation, then the double or triple bonds are included in the chain if possible.

$$HO_2CCH_2CHCH_2CO_2H \quad \text{is 1,2,3-propanetricarboxylic acid}$$
$$| $$
$$CO_2H$$

Sulfonic acids.   Sulfonic acids are named by adding the ending *-sulfonic acid* to the name of the rest of the structure.

$$CH_3CH_2SO_3H \qquad\qquad C_6H_5SO_3H$$

ethanesulfonic acid   benzenesulfonic acid

Acid anhydrides

Acid anhydrides are named from the names of the component acid or acids with the word *acid* dropped and the word *anhydride* added. (Either IUPAC or trivial acid names may be used.)

TABLE A10. Trivial names of some diacids

Structure	*Name of acid*
Aliphatic:	
HO_2CCO_2H	oxalic
$HO_2CCH_2CO_2H$	malonic
$HO_2C(CH_2)_2CO_2H$	succinic
$HO_2C(CH_2)_3CO_2H$	glutaric
$HO_2C(CH_2)_4CO_2H$	adipic
$HO_2C(CH_2)_5CO_2H$	pimelic
$HO_2C(CH_2)_6CO_2H$	suberic
$HO_2C(CH_2)_7CO_2H$	azelaic
$HO_2C(CH_2)_8CO_2H$	sebacic
cis-$HO_2CCH{=}CHCO_2H$	maleic
trans-$HO_2CCH{=}CHCO_2H$	fumaric
$HO_2CCH(OH)CH(OH)CO_2H$	tartaric
Aromatic:	
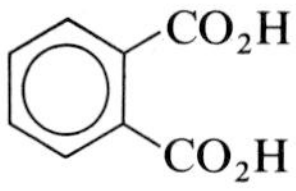	phthalic
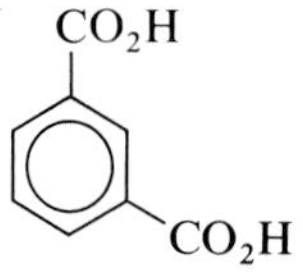	isophthalic
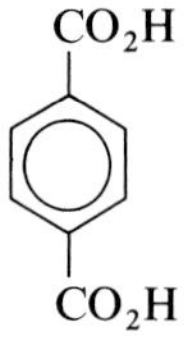	terephthalic

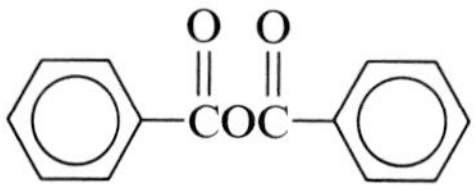

benzoic anhydride

$CH_3\overset{O}{\overset{\|}{C}}O\overset{O}{\overset{\|}{C}}CH_2CH_3$

acetic propionic anhydride

Cyclic anhydrides are named from the parent diacid.

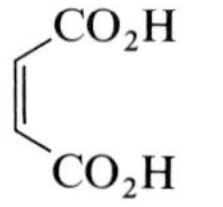

maleic acid

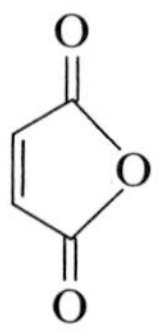

maleic anhydride

Acid halides

Acid halides are named by changing the ending of the carboxylic acid name from *-ic acid* to *-yl* plus the name of the halide. An ending of *-yl halide* on the name of a *diacid* implies that both carboxyl groups are acid-halide groups.

$$
\underset{\text{acetyl chloride}}{\overset{\displaystyle O \atop \displaystyle \|}{CH_3CCl}}
\qquad
\underset{\text{benzoyl bromide}}{\overset{\displaystyle O \atop \displaystyle \|}{C_6H_5CBr}}
\qquad
\underset{\text{succinyl chloride}}{\overset{\displaystyle O \qquad O \atop \displaystyle \| \qquad \|}{ClCCH_2CH_2CCl}}
$$

Alcohols

The names of alcohols may be (1) IUPAC; (2) trivial; or, occasionally, (3) conjunctive. *IUPAC names* are taken from the name of the alkane with the final *-e* changed to *-ol*. In the case of polyols, the prefix *di-*, *tri-*, etc. is placed just before *-ol*, with the position numbers placed at the start of the name, if possible.

$$CH_3CH_2CH_2CH_2CH_2OH$$

1-pentanol

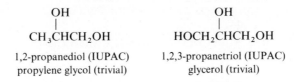

1,4-cyclohexanediol

In cases where confusion is possible, the number precedes *-ol*. *Example*: $CH_3CH{=}CHCH_2OH$ is 2-buten-1-ol.

If higher-priority functionality is present, or in complex molecules, the prefix *hydroxy-* should be used. *Example*: $CH_3CH(OH)CH_2CO_2H$ is 3-hydroxybutanoic acid.

Trivial names are generally composed of the name of the alkyl group plus the word *alcohol*.

$$(CH_3)_3COH \qquad\qquad C_6H_5CH_2OH$$

tert-butyl alcohol   benzyl alcohol

Conjunctive names are used principally with structures in which ring systems are attached to an aliphatic alcohol.

pyridine-3-methanol

Polyols. Structures with two OH groups on adjacent carbons (1,2-diols) are sometimes given trivial *glycol* names: the name of the *alkene* (not the alkane) from which the diol could be formed, plus the word *glycol*. *Glycerol* and *glycerin* are trivial names for 1,2,3-propanetriol.

$$
\underset{\begin{array}{c}\text{1,2-propanediol (IUPAC)}\\ \text{propylene glycol (trivial)}\end{array}}{\overset{\displaystyle OH \atop \displaystyle |}{CH_3CHCH_2OH}}
\qquad\qquad
\underset{\begin{array}{c}\text{1,2,3-propanetriol (IUPAC)}\\ \text{glycerol (trivial)}\end{array}}{\overset{\displaystyle OH \atop \displaystyle |}{HOCH_2CHCH_2OH}}
$$

Phenols

Phenols are those compounds in which an OH is attached directly to an arene ring. In these cases, phenol (or naphthol, etc.) is considered the parent.

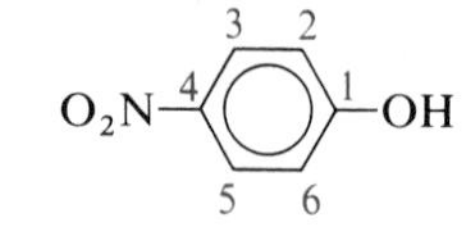

p-nitrophenol

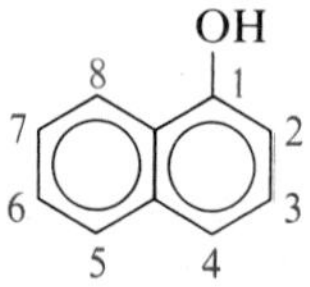

1-naphthol (IUPAC)
α-naphthol (trivial)

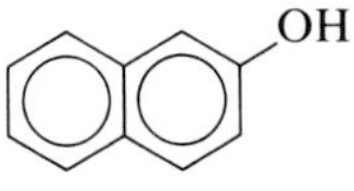

2-naphthol (IUPAC)
β-naphthol (trivial)

Many phenols and substituted phenols have trivial names.

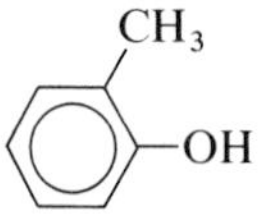

o-cresol 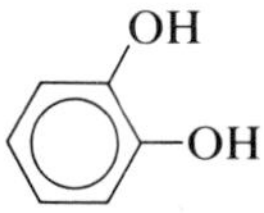   pyrocatechol 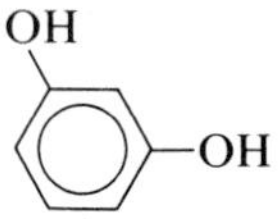   resorcinol 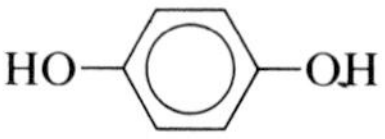   hydroquinone

Aldehydes

Aldehydes may be named by the IUPAC system or by trivial aldehyde names. In the IUPAC system, the *-oic acid* ending of the corresponding carboxylic acid is changed to *-al*.

$$CH_3CH_2CH_2CH_2CH_2CHO$$

hexanal

In trivial names, the *-ic* (or *-oic*) *acid* ending is changed to *-aldehyde*.

$$CH_3CHO \qquad C_6H_5CHO \qquad$$

acetaldehyde   benzaldehyde   cyclohexanecarboxaldehyde

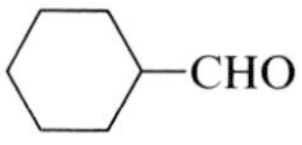

Some aldehydes have specific trivial names:

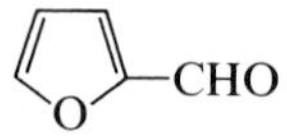

$$HOCH_2CH(OH)CHO$$

2-furaldehyde
or furfural

glyceraldehyde

Amides

In both the IUPAC and trivial systems, an amide is named by dropping the *-ic* (or *-oic*) *acid* ending of the corresponding acid name and adding *-amide*.

$$CH_3(CH_2)_4\overset{\overset{O}{\|}}{C}NH_2 \qquad CH_3\overset{\overset{O}{\|}}{C}NH_2$$

hexanamide (IUPAC)   acetamide (trivial)

Substituents on the amide nitrogen are named as prefixes preceded by *N-* or *N,N-*. $C_6H_5CONHCH_3$ is *N*-methylbenzamide, and $C_6H_5CON(CH_3)_2$ is *N,N*-dimethylbenzamide. *N*-Phenylamides have trivial names of *anilides*.

$$CH_3\overset{\overset{O}{\|}}{C}NHC_6H_5 \qquad C_6H_5\overset{\overset{O}{\|}}{C}NHC_6H_5$$

acetanilide   benzanilide

Sulfonamides are named by attaching the ending *-sulfonamide* to the name for the rest of the structure. *Example:* $C_6H_5SO_2NH_2$ is benzenesulfonamide.

$$H_2N-\overset{4}{\underset{5}{\overset{3}{\bigcirc}}^{2}_{6}}{}^{1}-SO_2NH_2$$

p-aminobenzenesulfonamide
(sulfanilamide)

Amines

Amines are named in two principal ways: with *-amine* as the ending and with *amino-* as a prefix. Unless there is a functional group of higher priority present, the *-amine* ending is used. Note that the largest group attached to the nitrogen is considered the parent or part of the parent amine.

Ending names:

$$(CH_3)_2NH \qquad CH_3NHCH_2CH_3$$

dimethylamine   *N*-methylethylamine

larger group

Prefix names:

$$\overset{NH_2}{\underset{CH_3\overset{|}{C}HCH_2OH}{}} \qquad \overset{NHCH_3}{\underset{CH_3\overset{|}{C}HCH_2OH}{}}$$

2-amino-1-propanol   2-(methylamino)-1-propanol

Polyamines may be named as *di-* or *triamines*, etc.

$$H_2NCH_2CH_2NH_2$$

1,2-ethanediamine

Some arylamines have their own names.

$$C_6H_5NH_2$$
aniline

$$CH_3-\!\!\left\langle\bigcirc\right\rangle\!\!-NH_2$$
p-toluidine

The prefix *aza-* is sometimes used to identify a nitrogen in a chain or ring; see page 957.

Amine salts. Amine salts are named as *ammonium salts* or (in simple cases) as amine hydrochlorides, etc.

$$(CH_3)_3NH^+ \ Cl^-$$
trimethylammonium chloride
(trimethylamine hydrochloride)

Cyclic salts often are named as *-inium salts*.

$$C_6H_5NH_3^+ \ Br^-$$
anilinium bromide
(aniline hydrobromide)

$$\left\langle\bigcirc\right\rangle\!\overset{+}{N}H \ ^-O_2CCH_3$$
pyridinium acetate

Esters and salts of carboxylic acids

Esters and salts of carboxylic acids are named as two words in both systematic and trivial names. The first word of the name is the name of the substituent on the oxygen. The second word of the name is derived from the name of the parent carboxylic acid with the ending changed from *-ic acid* to *-ate*.

$CH_3(CH_2)_4CO_2CH_3$	methyl hexanoate (IUPAC)
$CH_3CO_2CH_2CH_3$	ethyl acetate (trivial)
$CH_3CH_2CO_2Na$	sodium propanoate (or sodium propionate)
$CH_3-\!\!\left\langle\bigcirc\right\rangle\!\!-SO_3Na$	sodium *p*-toluenesulfonate (or sodium tosylate)

Ethers

Ethers are usually named by using the names of attached alkyl or aryl groups followed by the word *ether*. (These are trivial names.)

$$CH_3CH_2OCH_2CH_3$$
diethyl ether

$$CH_3O-\!\!\left\langle\bigcirc\right\rangle$$
cyclohexyl methyl ether

In more complex ethers, an *alkoxy-* prefix may be used. (This is the IUPAC preference.)

$$\overset{\displaystyle OCH_3}{\underset{\displaystyle \text{}}{\vert}}$$
$$CH_3CH_2CH_2CHCH_2CO_2H$$
3-methoxyhexanoic acid

Sometimes the prefix *oxa-* is used; see page 957.

Ketones

In the *systematic* names for ketones, the *-e* of the parent alkane name is dropped and *-one* is added. A prefix number is used if necessary.

$$\underset{\text{2-pentanone}}{CH_3\overset{\displaystyle O}{\overset{\|}{C}}CH_2CH_2CH_3} \qquad \underset{\text{3-penten-2-one}}{CH_3\overset{\displaystyle O}{\overset{\|}{C}}CH=CHCH_3} \qquad \underset{\text{2,4-pentanedione}}{CH_3\overset{\displaystyle O}{\overset{\|}{C}}CH_2\overset{\displaystyle O}{\overset{\|}{C}}CH_3}$$

In a complex structure, a ketone group may be named in the IUPAC system with the prefix *oxo-*. (The prefix *keto-* is also sometimes encountered.) Contrast the use of oxo- with that of oxa- (an ether).

$$CH_3\overset{\displaystyle O}{\overset{\|}{C}}CH_2CO_2H \qquad \text{is 3-oxobutanoic acid}$$

In *trivial* names, the alkyl groups attached to the C=O group are named, and the word *ketone* is added. Methyl ethyl ketone ($CH_3COCH_2CH_3$) is a common example. The $CH_3CO—$ group is sometimes called the *aceto-*, or *acetyl*, group, while the $C_6H_5CO—$ group is the *benzo-*, or *benzoyl*, group. Also encountered are *-phenone*, *-naphthone*, or *-acetone* endings, where one of these groups is an attachment to the ketone carbonyl group.

$$\underset{\text{acetoacetone}}{CH_3\overset{\displaystyle O}{\overset{\|}{C}}CH_2\overset{\displaystyle O}{\overset{\|}{C}}CH_3} \qquad \underset{\text{acetophenone}}{CH_3\overset{\displaystyle O}{\overset{\|}{C}}C_6H_5} \qquad \underset{\text{benzophenone}}{C_6H_5\overset{\displaystyle O}{\overset{\|}{C}}C_6H_5}$$

Some ketones also have specific trivial names: CH_3COCH_3 is called acetone.

Organometallics and metal alkoxides

Organometallic compounds, those with C bonded to a metal, are named by the name of the alkyl or aryl groups plus the name of the metal.

$$\underset{\text{triethylaluminum}}{(CH_3CH_2)_3Al} \qquad \underset{\text{phenylmagnesium bromide}}{C_6H_5MgBr}$$

Organosilicon or boron compounds are often named as derivatives of the metalloid hydrides. SiH_4 is silane; $(CH_3)_4Si$ is tetramethylsilane. BH_3 is borane; $(CH_3CH_2)_3B$ is triethylborane.

Sodium or potassium alkoxides are named as salts: the name of the cation plus the name of the alcohol with the *-anol* ending changed to *-oxide*. Salts of phenols are called *phenoxides*. (An alternate ending is *-olate*.)

$$\underset{\substack{\text{sodium ethoxide} \\ \text{(sodium ethanolate)}}}{CH_3CH_2ONa} \qquad \underset{\substack{\text{potassium phenoxide} \\ \text{(potassium phenolate)}}}{C_6H_5OK}$$

Glossary of some prefix symbols used in organic chemistry

(+)	dextrorotatory
(−)	levorotatory
(±)	racemic
α-	alpha: (1) on the adjacent carbon; (2) refers to configuration of carbon 1 in sugars; (3) refers to configuration of substituents on steroid ring systems
aldo-	aldehyde
allo-	closely related
andro-	relating to male
anhydro-	denoting abstraction of water
antho-	relating to flowers
anthra-	relating to coal or to anthracene
anti-	on opposite faces or sides
β-	beta: (1) opposite to that of α in configuration; (2) second carbon removed from a functional group or a heteroatom
bi-	twice or double
bisnor-	indicating removal of two carbons
chromo-	color or colored
cis-	on the same side of a double bond or ring
cyclo-	cyclic
Δ-	double bond
D-	on the right in the Fischer projection (see Section 18.3A)
d-	dextrorotatory
de-	removal of something, such as hydrogen (*dehydro-*) or oxygen (*deoxy-*)
dextro-	to the right, as in dextrorotatory
dl-	racemic
(*E*)-	on the opposite sides of a double bond
endo-	(1) on or in the ring and not on a side chain; (2) opposite the bridge side of a ring system:

	(3) attached as a bridge within a ring, as 1,4-*endo*-methyleneanthracene
epi-	(1) the 1,6-positions of naphthalene; (2) epimeric; (3) a bridge connection on a ring, as 9,10-epidioxyanthracene
erythro-	related in configuration to erythrose
exo-	(1) on a side chain attached to a ring; (2) on the bridge side of a ring system (see *endo-*)
gem-	attached to the same atom
hemi-	one half
hydro-	(1) denotes presence of H; (2) sometimes relating to water
hypo-	indicating a low, lower, or lowest state of oxidation
i-, iso-	(1) methyl branch at the end of the chain; (2) isomeric; (3) occasionally *i-* is used for *inactive*

L-	on the left in the Fischer projection (see Section 18.3A)
l-	levorotatory
leuco-	colorless or white
levo-	to the left, as in levorotatory
m-, *meta-*	(1) 1,3 on benzene; (2) closely related compound, as metaldehyde
meso-	(1) with a plane of symmetry and optically inactive; (2) middle position of certain cyclic organic compounds
n-	*normal*: continuous chain
neo-	one C connected to four other C's
nor-	(1) removal of one or more C's (with H's); (2) structure isomeric to that of root name, as norleucine.
o-, *ortho-*	1,2 on benzene
oligo-	few (units of)
p-, *para-*	(1) 1,4 on benzene; (2) polymeric, as paraformaldehyde
per-	saturated with, as in *perhydro-*, or *peroxy-*
peri-	(1) 1,8 on naphthalene; (2) fusion of ring to two or more adjoining rings, as perixanthenoxanthene.
pheno-	relating to phenyl or benzene
poly-	many (units of)
Ψ-, pseudo-	has a resemblance to
pyro-	indicating formation by heat
(*R*)-	clockwise configuration around a chiral carbon
(*R*)(*S*)-	racemic
s-	abbreviation for *secondary-* or *symmetrical-*
(*S*)-	counterclockwise configuration around a chiral carbon
seco-	denoting ring cleavage
sec-	abbreviation for *secondary-*
sym-	symmetrical
syn	on the same face or side
t-, *tert-*	abbreviation for *tertiary-*
threo-	related in configuration to threose
trans-	on opposite sides of a double bond or ring
uns-, *unsym-*	unsymmetrical
v-, *vic-*	vicinal: on adjacent C's
(*Z*)-	on the same side of a double bond

Answers to Problems

The answers to the problems within the textual material are given here. For the answers to the chapter-end problems, see the *Study Guide* that accompanies this text.

CHAPTER 1

1.1. (a) 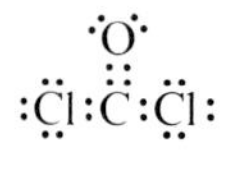 $:\ddot{\text{C}}\text{l}:\overset{\cdot\cdot}{\underset{\cdot\cdot}{\text{C}}}:\ddot{\text{C}}\text{l}:$ (b) $\overset{\ddot{\text{C}}\text{l}:}{\underset{\ddot{\text{C}}\text{l}:}{:}}\text{C}::\text{C}\overset{:\ddot{\text{C}}\text{l}:}{\underset{:\ddot{\text{C}}\text{l}:}{}}$ (c) $\overset{\text{H}}{\underset{\text{H}}{}}\text{C}::\text{C}\overset{:\ddot{\text{Br}}:}{\underset{\cdot\text{H}}{}}$ (d)  $\text{H}:\text{C}::\text{C}:\ddot{\text{F}}:$

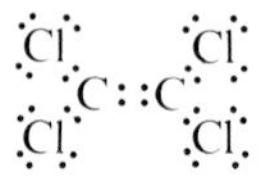

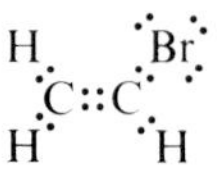

1.2. (a) neutral molecule, no formal charges
 (b) ion: $\overset{0}{\text{C}}\text{H}_3\overset{+1}{\text{N}}\text{H}_3$

1.3. (a) all atoms 0; (b) $\text{CH}_2\overset{+1}{=}\text{N}\overset{-1}{=}\text{N};$ (c) $\text{CH}_3\text{O}\overset{\overset{\displaystyle\text{O}^{-1}}{|}}{\underset{\underset{\displaystyle\text{O}^{-1}}{|}}{\text{S}}}\overset{+2}{}\text{OH},$ other atoms 0

1.4. (a) $(\text{CH}_3)_2\text{CHCH}_2\text{Cl};$ (b) $\text{CH}_3\text{CHCl}_2;$ (c) $\text{CH}_3(\text{CH}_2)_3\text{CHClCH}_2\text{Cl};$
 (d) $(\text{CH}_3)_2\text{C}=\text{C}(\text{CH}_3)_2;$ (e) $\text{CH}_2(\text{CN})_2$

1.5. (a)

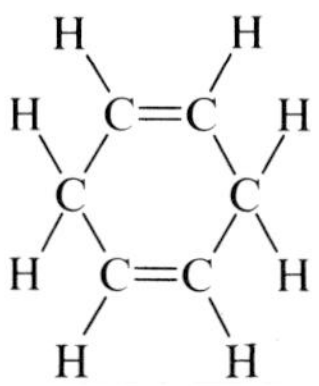

(b)

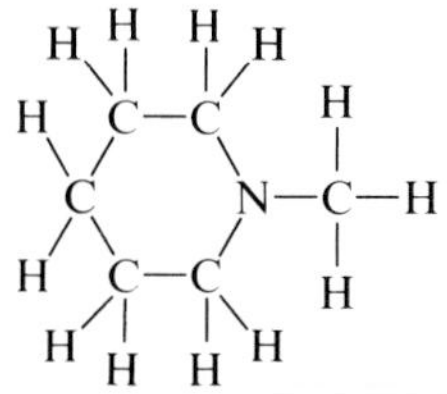

(c)

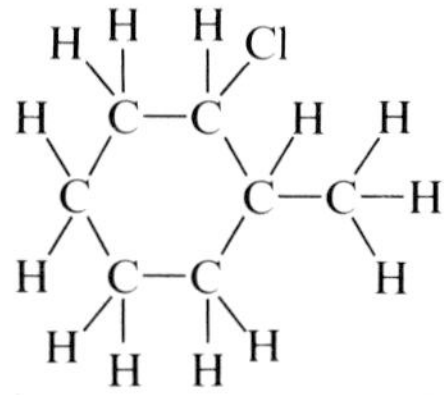

1.6. (a) [pentagon] (b) [fused bicyclic] (c) [5-membered ring with O] (d) [cyclobutane with two CH_3 groups on one carbon and a CH_3] (e) [cyclopentene]

1.7. (a) liberates more energy

1.8. (a) $\overset{\delta+}{C}H_3 - \overset{\delta-}{B}r;$ (b) $CH_3\overset{\overset{\displaystyle \overset{\delta-}{O}}{\|}}{C} - \overset{\delta-}{O} - H$
$\quad\quad\quad\quad\quad\quad\quad\quad\underset{\delta+}{}\quad\quad\underset{\delta+}{}$

1.9. (a) $CH_3OH;$ (b) $CH_3O-H;$ (c) CH_3Cl

1.10. (a) $CH_3\overset{\longrightarrow}{C\equiv N};$ (b) $(CH_3)_2\overset{\longrightarrow}{C=O};$ (c) none; (d) [cyclohexane ring]$\overset{\longrightarrow}{=O}$

1.11. (a) $CH_3CH_2CH_2\ddot{N}H_2 ---:NH_2CH_2CH_2CH_3$

(b) $CH_3\ddot{O}H---:\underset{\cdot\cdot}{O}CH_3,$ $CH_3\ddot{O}H---:\overset{\overset{\displaystyle H}{|}}{O}H_2,$ $CH_3\underset{\cdot\cdot}{\ddot{O}}:---H_2\ddot{O},$
$\quad\quad\quad\quad\quad\quad\quad\quad\quad\quad\quad\quad\quad\quad\quad\quad\quad\quad H_2\underset{\cdot\cdot}{O}:---H_2\underset{\cdot\cdot}{O}:$

(c) none (d) $(CH_3)_2\ddot{O}:---H_2\ddot{\underset{\cdot\cdot}{O}},$ $H_2\ddot{O}:---H_2\ddot{O}:$

1.12. (a) base; (b) acid; (c) base; (d) acid; (e) base

1.13. (a) $CH_3CH_2\overset{\overset{\displaystyle \cdot\ddot{O}\cdot}{\|}}{C}\ddot{O}H + H\ddot{O}H \rightleftharpoons CH_3CH_2\overset{\overset{\displaystyle \cdot\ddot{O}\cdot}{\|}}{C}\underset{\cdot\cdot}{\ddot{O}}:^- + H\overset{\overset{\displaystyle H}{|}}{\underset{+}{O}}H$

(b) [cyclohexyl]$-\overset{\overset{\displaystyle \cdot\ddot{O}\cdot}{\|}}{C}\ddot{O}H + {}^-:\ddot{O}H \longrightarrow$ [cyclohexyl]$-\overset{\overset{\displaystyle \cdot\ddot{O}\cdot}{\|}}{C}\underset{\cdot\cdot}{\ddot{O}}:^- + H\ddot{O}H$

(c) $(CH_3)_2\ddot{N}H + H\ddot{O}H \rightleftharpoons (CH_3)_2\overset{+}{\ddot{N}}H_2 + {}^-:\ddot{O}H$

(d) [piperidine ring]$\ddot{N}\cdot$ + $CH_3\overset{\overset{\displaystyle \cdot\ddot{O}\cdot}{\|}}{C}\ddot{O}H \rightleftharpoons$ [piperidine ring]$\overset{+}{N}\overset{H}{\diagup}$ + $CH_3\overset{\overset{\displaystyle \cdot\ddot{O}\cdot}{\|}}{C}\underset{\cdot\cdot}{\ddot{O}}:^-$
$\quad\quad\quad\quad\quad\underset{H}{}\quad\quad\quad\quad\quad\quad\quad\quad\quad\quad\quad\quad\quad\quad\quad\quad\underset{H}{}$

1.14. (a) < (b) < (c)

1.15. (c) < (b) < (a)

1.16. (a) $CH_3\overset{\overset{\displaystyle O}{\|}}{C}CH_3$, Lewis base; H^+, Lewis acid
(b) $(CH_3)_3C^+$, Lewis acid; Cl^-, Lewis base

(c) $CH_3\overset{\overset{\displaystyle O}{\|}}{C}OCH_3$, Lewis acid; $^-OCH_3$, Lewis base

CHAPTER 2

2.1. (a) 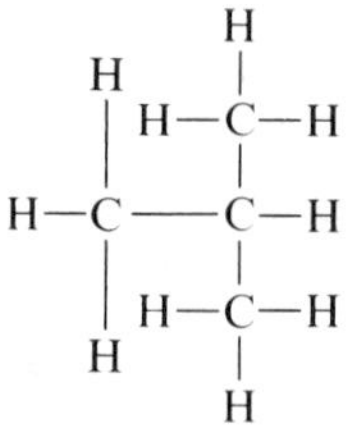 (b) 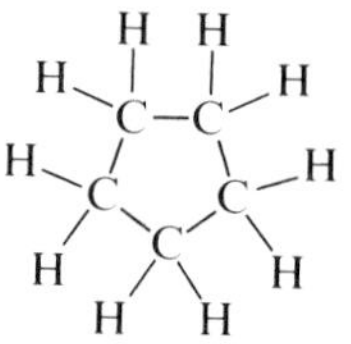

All C—C bonds are sp^3–sp^3 and all C—H bonds are sp^3–s.

2.2. (a) 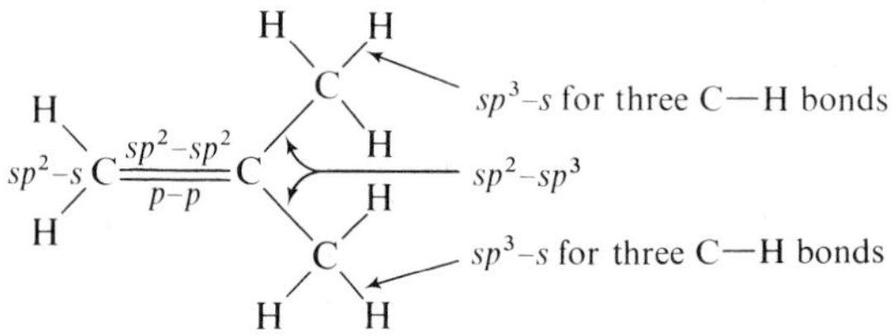
sp3–s for three C—H bonds
sp2–sp3
sp3–s for three C—H bonds
sp2–sp2
p–p
sp2–s

(b) 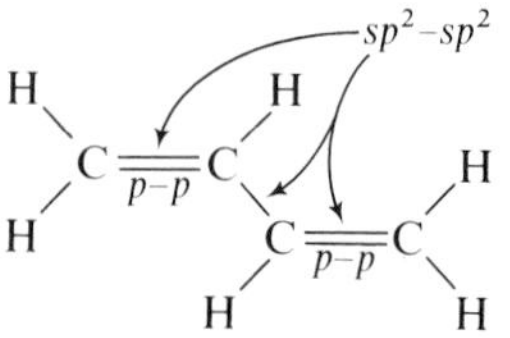
sp2–sp2
p–p
p–p

All C—H bonds are sp^2–s.

(c) 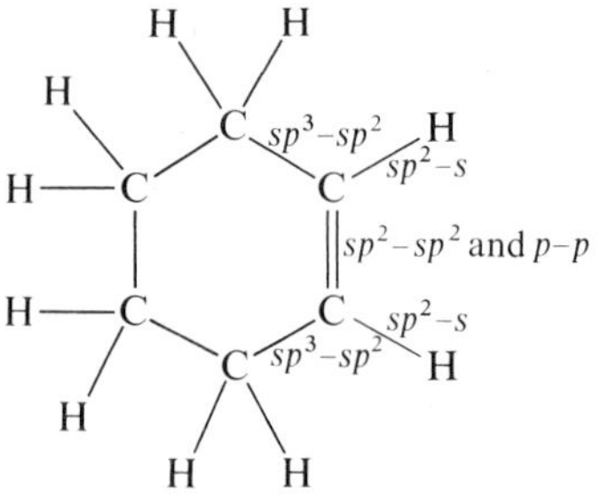
sp3–sp2
sp2–s
sp2–sp2 and p–p
sp2–s
sp3–sp2

All other C—C bonds are sp^3–sp^3 and all other C—H bonds are sp^3–s.

2.3. 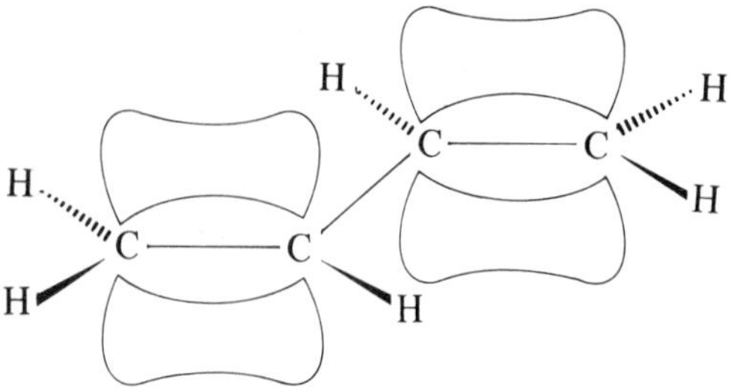

All C and H atoms are in one plane; the pi bonds are above and below this plane.

2.4. 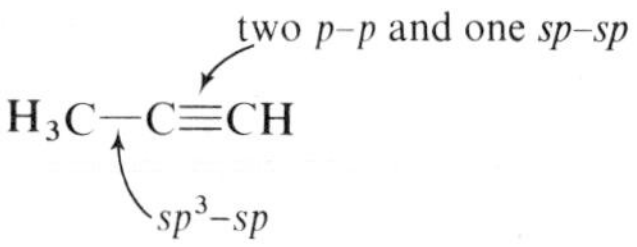
two p–p and one sp–sp
H3C—C≡CH
sp3–sp

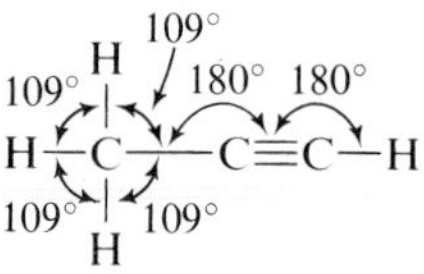
109°
109°
180°
180°
109°
109°

2.5.   (a) (2), (1);   (b) (2), (1);   (c) (1), (2), (3)

2.6.   (a) $H_2C\!=\!C\,HCH_2\,\overset{\overset{\textstyle O}{\|}}{C}H$;   (b) ◯―NH_2 ;   (c)

2.7. RCO_2H

2.8.   (a) $H\!-\!\overset{\overset{\textstyle H}{|}}{\underset{\underset{\textstyle H}{|}}{C}}\!-\!\overset{..}{N}\!-\!\overset{\overset{\textstyle H}{|}}{\underset{\underset{\textstyle H}{|}}{C}}\!-\!\overset{..}{N}\!-\!H$ All C—H and N—H bonds are sp^3–s,
  and all C—N bonds are sp^3–sp^3

$H\!-\!C\!-\!H$
$|$
H

(b) $sp^2\!-\!sp^2$ $sp^2\!-\!s$
and $p\!-\!p$ $\overset{..}{N}\!-\!H$ H $sp^3\!-\!sp$ All unmarked C—H and N—H bonds are sp^3–s.
$H\!-\!\overset{..}{N}\!-\!C\!-\!N\!-\!C\!-\!C\!\equiv\!N:$
$|$ $|$ $|$ $sp\!-\!sp$ and two $p\!-\!p$
H H H
$sp^3\!-\!sp^2$ $sp^3\!-\!sp^3$

(c)   All C—C and C—N bonds are sp^3–sp^3;
  all C—H and N—H bonds are sp^3–s

2.9.   The following structures are not the only possible answers:

(a) $CH_3CH_2CH_2\overset{\overset{\textstyle O}{\|}}{C}H$ (b) $CH_3\overset{\overset{\textstyle O}{\|}}{C}CH_2CH_3$ (c) $CH_3CH_2CH_2\overset{\overset{\textstyle O}{\|}}{C}OH$

(d) $CH_3\overset{\overset{\textstyle O}{\|}}{C}CH_2\overset{\overset{\textstyle O}{\|}}{C}OH$

2.10.   Five conjugated double bonds, no isolated double bonds.

2.11.   (a) $CH_3CH\!=\!CHCH\!=\!CHCH\!=\!CHCH_3$
  (b) $CH_2\!=\!CHCH_2CH\!=\!CHCH\!=\!CHCH_3$
  (c) $CH_2\!=\!CHCH_2CH\!=\!CHCH_2CH\!=\!CH_2$

2.12.   (a)   (b)

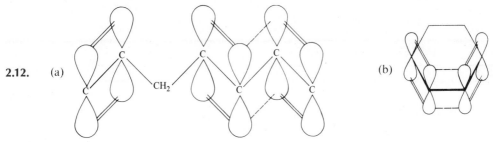

2.13. $CH_2{=}CH\overset{\overset{\displaystyle O}{\|}}{C}CH_3$

2.14. 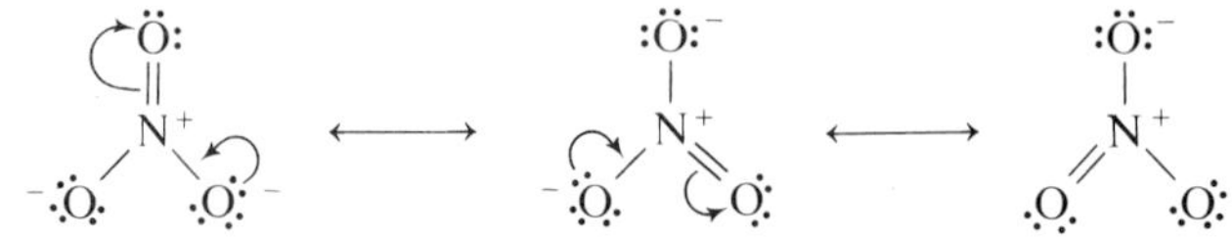

2.15. (a) The right-hand structure is the major contributor because each atom has an octet.
(b) The left-hand structure.

2.16. (a)  ; *equivalent*

(b) 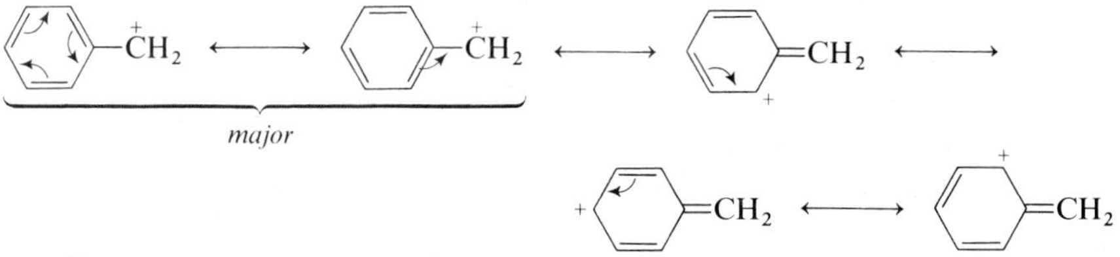

(c) 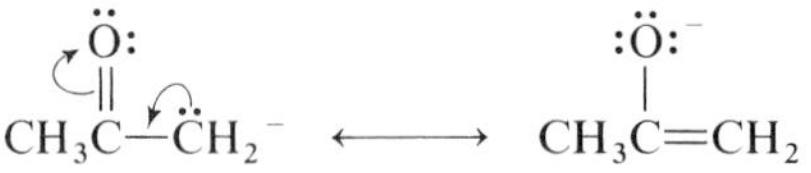
major

(d) 
major

(e) 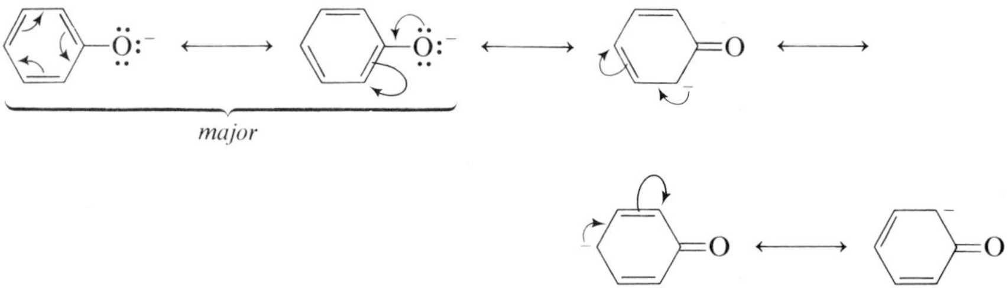

(f) 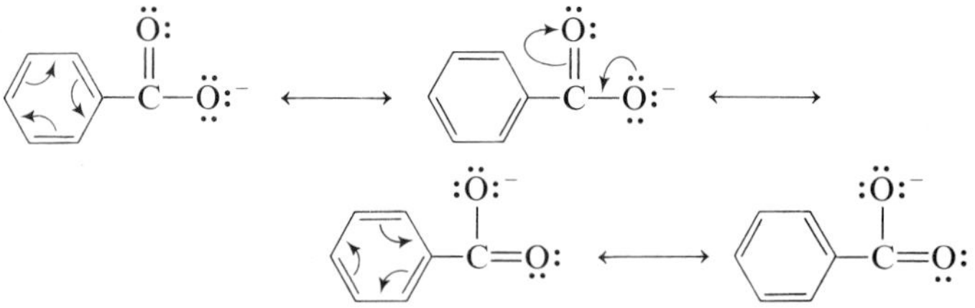
equivalent

In (f), note that the negative charge is not delocalized by the benzene ring.

CHAPTER 3

3.1. (a) $CH_3CH_2\overset{\overset{\displaystyle O}{\|}}{C}CH_2CH_3$ (b) 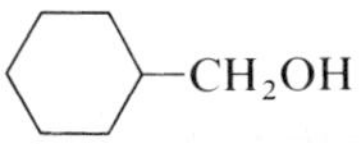 (c) $CH_3\overset{\overset{\displaystyle O}{\|}}{C}CH_3$

3.2. (a) $HOCH_2\overset{\overset{\displaystyle O}{\|}}{C}H$; (b) ⬡$-CH_2OH$; (c) $CH_3CH_2CH_2CH_2\overset{\overset{\displaystyle O}{\|}}{C}H$

3.3. (a) C_nH_{2n+2} is an open-chain alkane.
(b) C_nH_{2n} contains either one double bond or one ring.
(c) C_nH_{2n-2} contains either one triple bond, one double bond plus one ring, two double bonds, or two rings.

3.4. (a) $CH_3CH_2CH_2-$⬡; (b) $(CH_3)_2CHCH_2-$⬡;

(c) $CH_3CH_2CH_2\overset{\overset{\displaystyle C(CH_3)_3}{|}}{C}HCH_2CH_2CH_2CH_3$

3.5. (a) Cl_2CHCH_2Cl; (b) $Cl-$⬡$\overset{\overset{\displaystyle Cl}{|}}{}-NO_2$

3.6. (a) propene; (b) cyclohexene; (c) 1,3-cyclohexadiene; (d) 1-chloro-1-propene

3.7. (a) ⬠ (b) $CH\equiv CC\equiv CCH_3$ (c) ▱^{CH_3}

3.8. (a) 2-pentanol; 4-methylcyclohexylamine

(b) ^{CH_3}⬡$-OH$; $CH_3CH_2CH_2CH_2NHCH(CH_3)_2$

3.9. $CH_3CH_2CH_2CH_2\overset{\overset{\displaystyle O}{\|}}{C}H$; $CH_3\overset{\overset{\displaystyle O}{\|}}{C}CH_2CH_2CH_3$ or $CH_3CH_2\overset{\overset{\displaystyle O}{\|}}{C}CH_2CH_3$;

$\quad\quad$ pentanal $\quad\quad\quad\quad$ 2-pentanone $\quad\quad\quad\quad\quad$ 3-pentanone

$CH_3CH_2CH_2CH_2\overset{\overset{\displaystyle O}{\|}}{C}OH$
pentanoic acid

3.10. (a) $CH_3(CH_2)_4\overset{\overset{\displaystyle O}{\|}}{C}O-$⬡; (b) $CH_3CH_2CH_2\overset{\overset{\displaystyle O}{\|}}{C}OC(CH_3)_3$

3.11. (a) 6-nitro-3-hexanone; (b) 3-bromo-1-phenyl-1-propanol

---

CHAPTER 4

4.1. (a) (*E*)-2-chloro-1-fluoro-1-butene; (b) (*Z*)-2,3-dichloro-2-butene

4.2. (a) (*E*); (b) (*Z*); (c) (*E*); (d) (*Z*)

4.3.

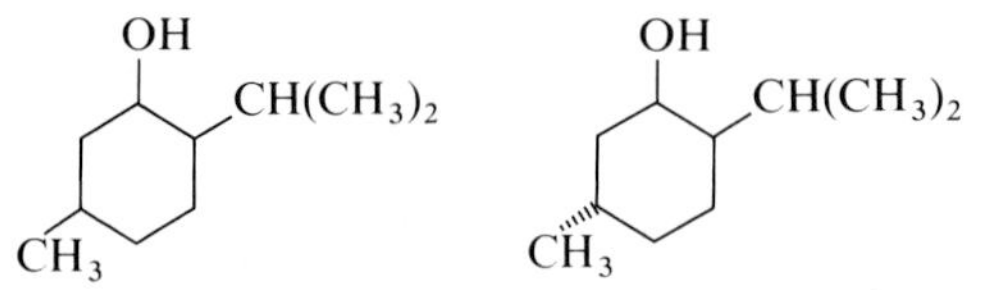

all *cis*

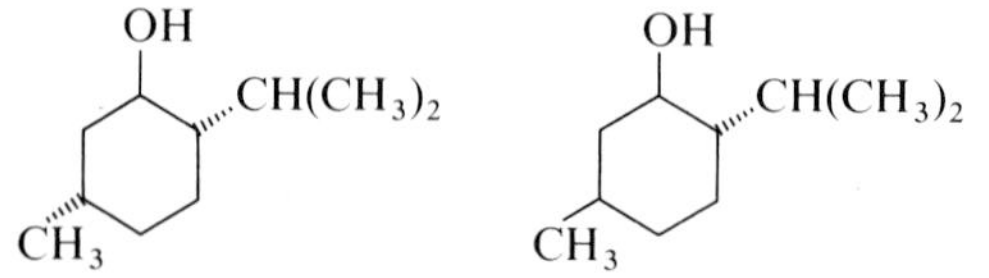

4.4. (a)

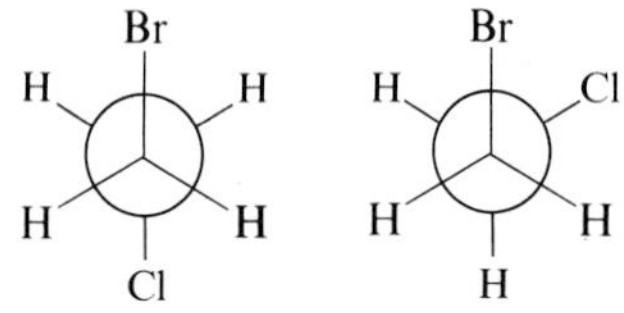

anti   *gauche*

(b)

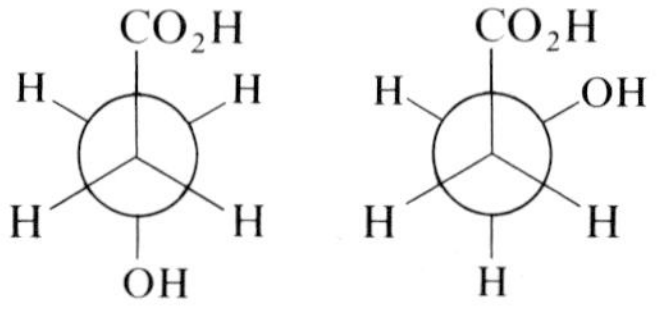

anti   *gauche*

4.5. (c) and (d)

4.6. (a) *trans e,e*; (b) *cis a,e*; (c) *trans e,e*; (d) no geometric isomers;
(e) *trans e,e*; (f) *trans a,e*

4.7. (a) none; (b) $CH_3CH_2\overset{*}{C}HBr$; (c) $C_6H_5CH_2\overset{*}{C}HC_6H_5$ (with CH_3 and F substituents respectively)

4.8. (a)

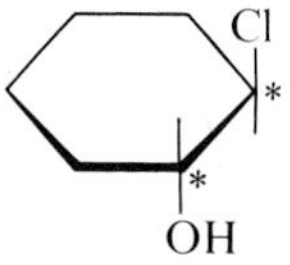

(b) none   (c)

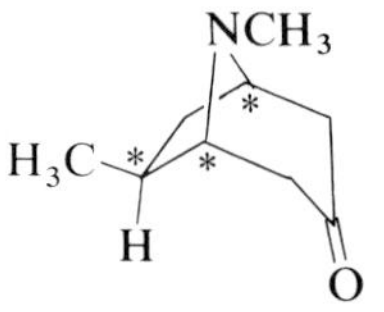

4.9. (a)

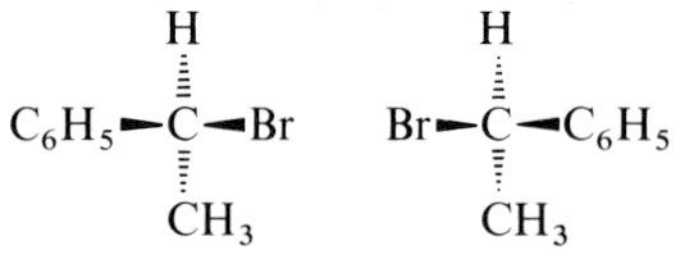

dimensional

(b)

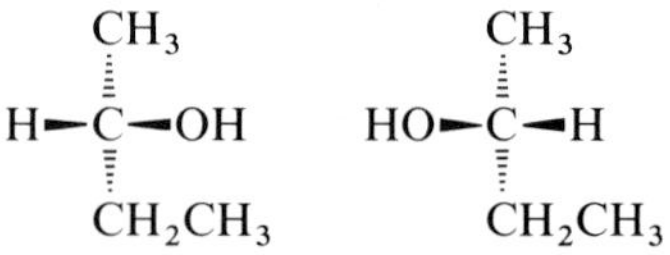

dimensional

4.10. (a)

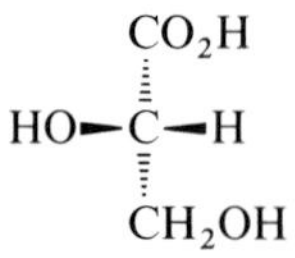

(b)

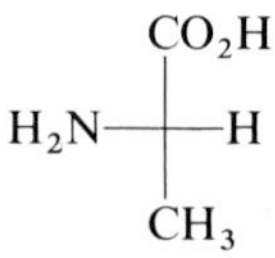

4.11. (a) (*R*); (b) (*S*); (c) (*R*); (d) (*S*)

4.12. (a)

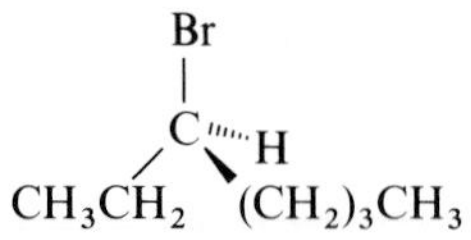

(b)

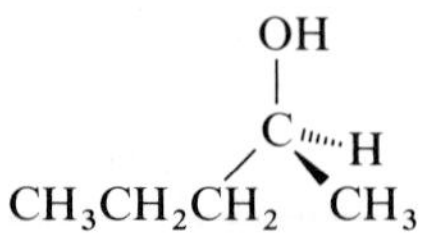

4.13. (a) four; (b) two; (c) eight

4.14.

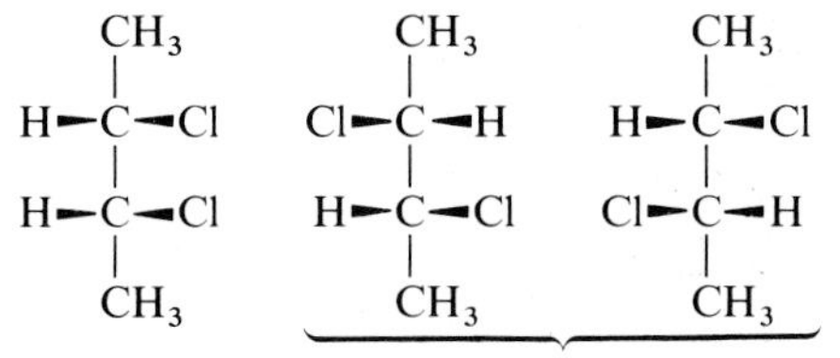

meso   enantiomers

4.15. (a) (*R*): H$_2$N $-$ | $-$ H   (*S*): H $-$ | $-$ NH$_2$

with CH$_3$ top and CH$_2$C$_6$H$_5$ bottom for both

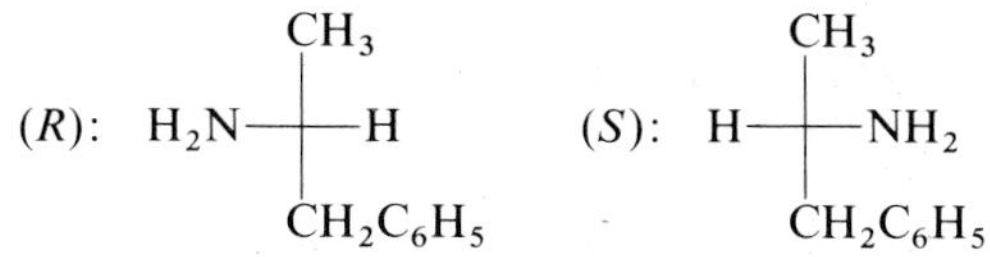

(b) CH$_3$CHCO$_2$H (OH on CH) + H$_2$N $-$ C $-$ H (CH$_3$ top, CH$_2$C$_6$H$_5$ bottom) $\longrightarrow$

(*R*) (*S*)   (*R*)

H $-$ C $-$ OH (CO$_2^-$ top, CH$_3$ bottom) H$_3$N$^+$ $-$ C $-$ H (CH$_3$ top, CH$_2$C$_6$H$_5$ bottom) + HO $-$ C $-$ H (CO$_2^-$ top, CH$_3$ bottom) H$_3$N$^+$ $-$ C $-$ H (CH$_3$ top, CH$_2$C$_6$H$_5$ bottom)

(*R*,*R*)   (*S*,*R*)

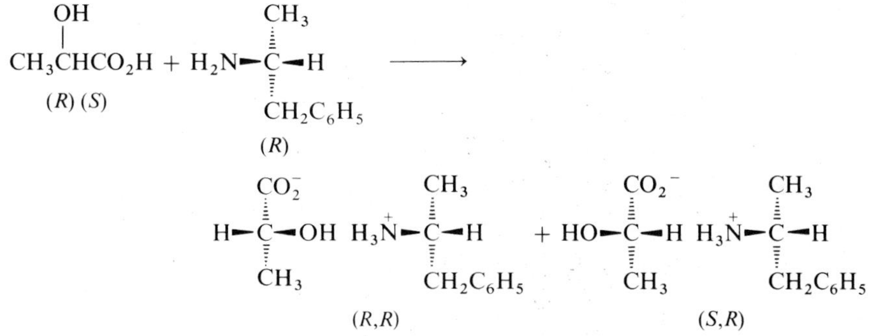

CHAPTER 5

5.1. (a) 2-iodo-2-methylpropane, *t*-butyl iodide
(b) 2-chloropropane, isopropyl chloride
(c) 1-iodo-2-methylpropane, isobutyl iodide

5.2. (a) Br$_2$CHCH$_2$CH$_2$CH$_3$ (b) CH$_3$CHClCH=CH$_2$ (c) FCH$_2$CH$_2$OH

5.3. (a) (CH$_3$CH$_2$)$_2$CHCl + CH$_3$O$^-$ $\longrightarrow$

(CH$_3$CH$_2$)$_2$CHOCH$_3$ + CH$_3$CH=CHCH$_2$CH$_3$

(b) ⬡$-$Br + CH$_3$O$^-$ $\longrightarrow$ ⬡$-$OCH$_3$ + ⬡(ene)

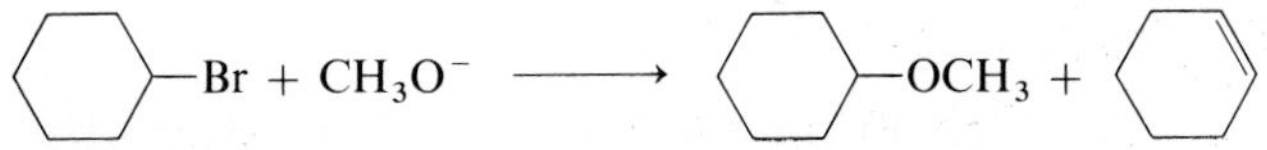

(c) ⬡$-$CH$_2$I + CH$_3$O$^-$ $\longrightarrow$ ⬡$-$CH$_2$OCH$_3$

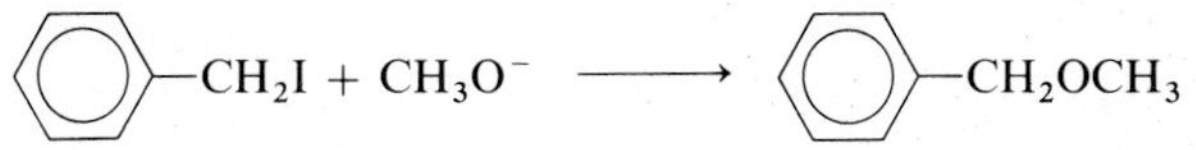

5.4. (a) CH$_3$CH$_2$CH$_2$O$^-$ + BrCH$_2$CH$_2$CH(CH$_3$)$_2$ ⟶

CH$_3$CH$_2$CH$_2$Br + $^-$OCH$_2$CH$_2$CH(CH$_3$)$_2$ ⟶ product

(b) ⬡$-$I + $^-$CN $\longrightarrow$ product

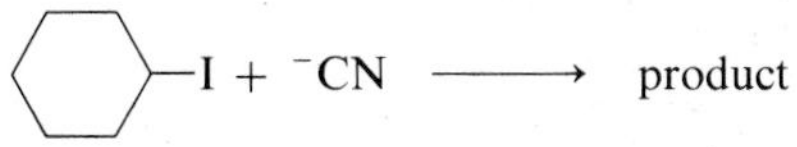

5.5. $NC^- +$ CH_3CH_2 $\overset{H}{\underset{CH_3}{\overset{(S)}{C}}}-Br$ $\longrightarrow$ $\overset{(R)}{NC-C}\overset{H}{\underset{CH_3}{}}CH_2CH_3$ $+ Br^-$

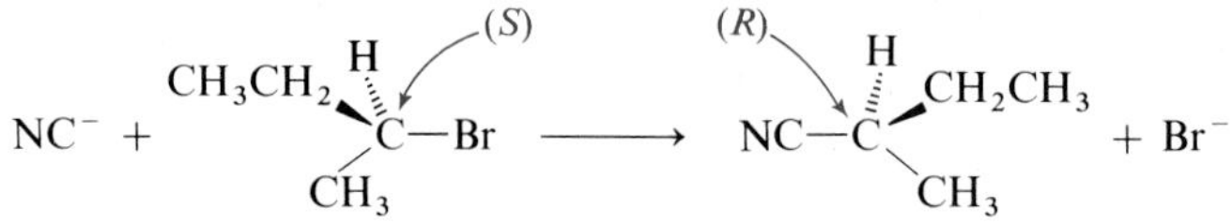

5.6.   (a) 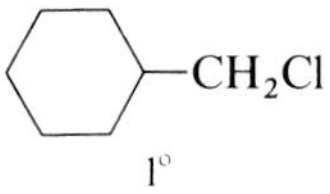 ⬡$-CH_2Cl$ (b) $(CH_3)_2CHCH_2CHClCH_3$ (c) 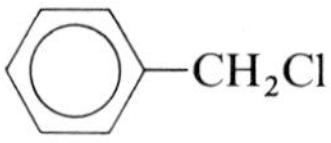 ⬡$-CH_2Cl$
$\qquad\quad$ $1°$ $\qquad\qquad\qquad\qquad\qquad$ *less hindered*

5.7.   (a) $(CH_3CH_2)_3COH + HI$ (b)  ⬡$\overset{CH_3}{\underset{OCH_3}{}}$ $+ HCl$

5.8.   (b), (a), (c)

5.9.   (a) no observed rearrangement products

$\qquad$ (b) $(CH_3)_2\overset{H}{\overset{\curvearrowright}{C}}CHCH_2CH_3$ $\longrightarrow$ $(CH_3)_2\overset{+}{C}CH_2CH_2CH_3$

5.10. $(CH_3)_2CHCHBrCH_2CH_3 + (CH_3)_2CBrCH_2CH_2CH_3$

5.11. (a) 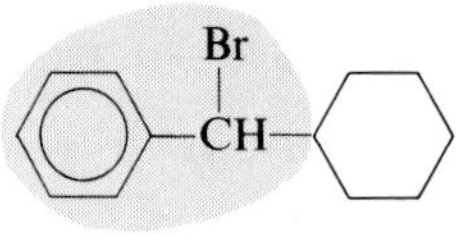 ⬡$-\overset{Br}{\underset{}{CH}}-$⬡   (b) 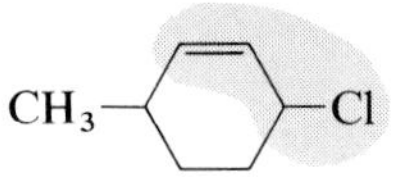 CH_3-⬡$-Cl$

5.12. (a) allylic;   (b) vinylic;   (c) aryl;   (d) benzylic

5.13. (a) 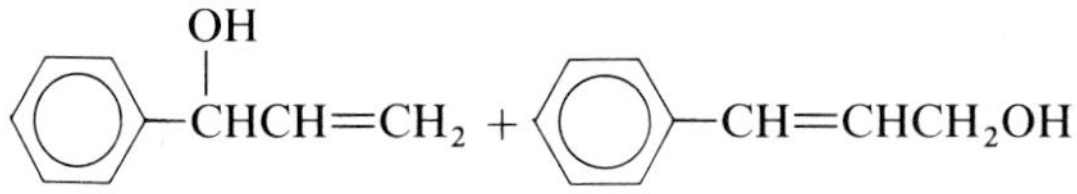 ⬡$-\overset{OH}{\underset{}{CH}}CH=CH_2 +$ ⬡$-CH=CHCH_2OH$

$\qquad$ (b) 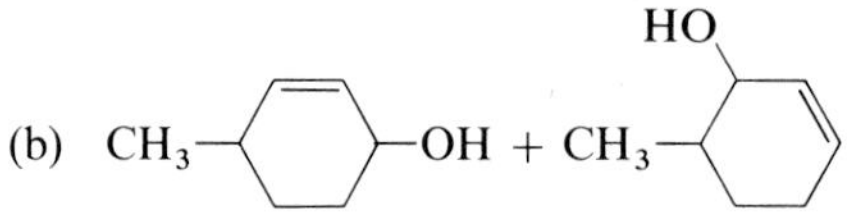 CH_3-⬡$-OH + CH_3-$⬡$\overset{HO}{}$

5.14. $(CH_3)_2\overset{Br}{\underset{}{C}}CH_2CH_3$ $\xrightarrow[-HBr]{CH_3CH_2OH}$

$\qquad\qquad\qquad$ $(CH_3)_2\overset{OCH_2CH_3}{\underset{}{C}}CH_2CH_3 + (CH_3)_2C=CHCH_3 + CH_2=\overset{CH_3}{\underset{}{C}}CH_2CH_3$

5.15. (a) $CH_2=CHCH_2CH_3 + CH_3CH=CHCH_3$
$\qquad\qquad\qquad\qquad\quad$ *cis* and *trans*

$\qquad$ (b) 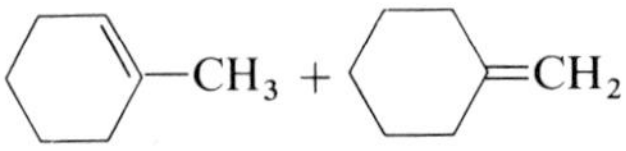 ⬡$-CH_3 +$ ⬡$=CH_2$

5.16. ⬡$-CH_3$

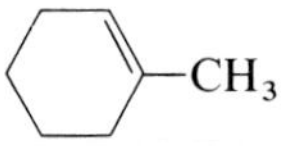

5.17.

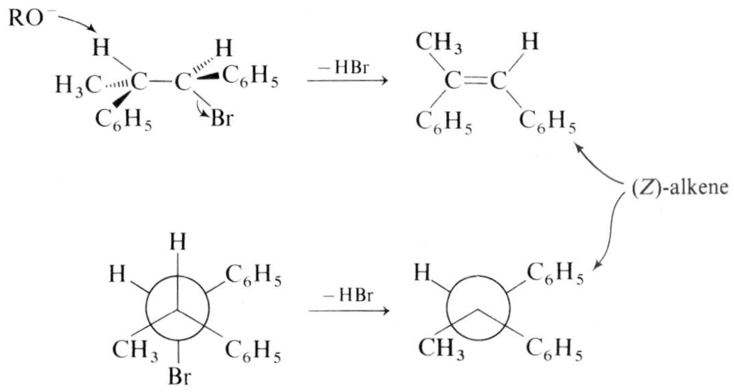

RO⁻

H₃C⋯C—C⟨C₆H₅ with H, H, C₆H₅, Br substituents —HBr→

CH₃, H / C₆H₅, C₆H₅ — (Z)-alkene

H / H, C₆H₅, CH₃, C₆H₅, Br —HBr→ H / H, C₆H₅, CH₃, C₆H₅

5.18.

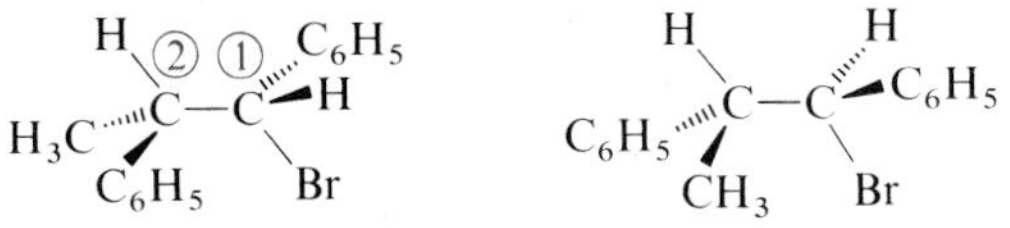

H₃C⋯C—C⟨C₆H₅ with H ② ①, H, C₆H₅, Br

(1R,2S)

C₆H₅⋯C—C⟨C₆H₅ with H, H, CH₃, Br

(1S,2R)

5.19. The enantiomers yield (Z)-1-bromo-1,2-diphenylethene (*trans*-phenyls); the *meso* form yields the (E)-isomer (*cis*-phenyls).

5.20.

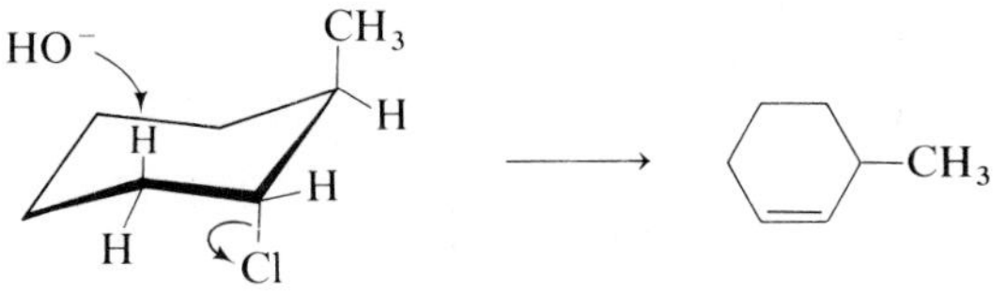

HO⁻, CH₃, H, H, H, H, Cl ⟶ ⬡—CH₃

Only one H can be abstracted. only E2 product

5.21. (a) $(CH_3)_2CHCHBrCH_2CH_3$ —HBr→

$(CH_3)_2C=CHCH_2CH_3$ + $(CH_3)_2CHCH=CHCH_3$

Saytseff *Hofmann*

(b) ⬡ with CH₃, Cl —HCl→ ⬡—CH₃ + ⬡=CH₂

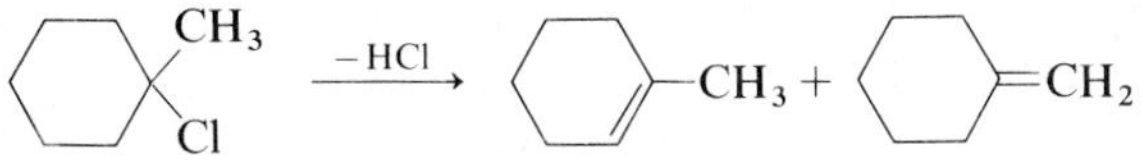

Saytseff *Hofmann*

5.22. (a) $CH_3CH_2CH_2CH=CH_2$; (b) *trans*-$CH_3CH=CHCH_3$

5.23. CH_3S^- is the better nucleophile because S is larger and more polarizable than O.

5.24. (a) ⬡—O⁻ + $CH_3CH_2CH_2I$ ⟶

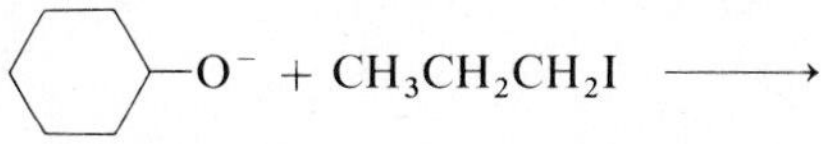

(b) $C_6H_5CH_2Br$ + CH_3S^- or $C_6H_5CH_2S^-$ + CH_3I ⟶

(c) CH_3CH_2Br + ⁻$OCH_2CH_2CH_3$ ⟶

(d) $C_6H_5CO_2^-$ + $CH_2=CHCH_2Br$ ⟶

(e) $C_6H_5O^-$ + CH_3CH_2Br ⟶

(f) $(CH_3)_2CHCH_2CHBrCH_3$ $\xrightarrow[CH_3OH]{^-OCH_3}$ or $(CH_3)_2CHCHBrCH_2CH_3$ $\xrightarrow[(CH_3)_3COH]{^-OC(CH_3)_3}$

In (c), the reaction shown would have a slightly faster rate than that of $CH_3CH_2O^-$ + $CH_3CH_2CH_2Br$.

CHAPTER 6

6.1. (a) $H:\overset{\cdot\cdot}{\underset{\cdot\cdot}{O}}\cdot$ (b) $H:\overset{H}{\underset{H}{C}}:\overset{\cdot\cdot}{\underset{\cdot\cdot}{O}}\cdot$ (c) $H:\overset{H}{\underset{H}{C}}:\overset{H}{\underset{H}{C}}\cdot$

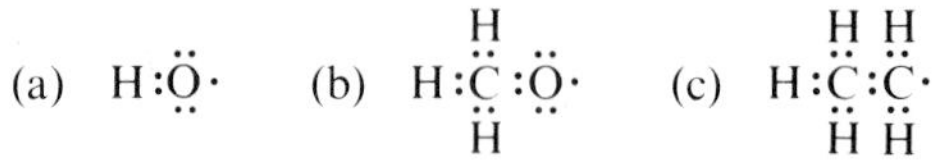

6.2. *initiation:* $Cl_2 \xrightarrow{h\nu} 2\,Cl\cdot$

propagation: ⬡ $+ Cl\cdot \longrightarrow$ ⬡· $+ HCl$

⬡· $+ Cl_2 \longrightarrow$ ⬡$-Cl + Cl\cdot$

termination: ⬡· $+ Cl\cdot \longrightarrow$ ⬡$-Cl$ or other combination of two radicals

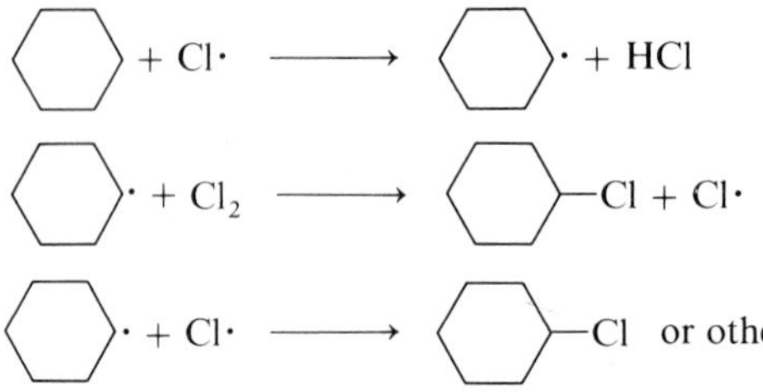

6.3. $CHCl_3 + Cl\cdot \longrightarrow \cdot CCl_3 + HCl$
$\cdot CCl_3 + Cl_2 \longrightarrow CCl_4 + Cl\cdot$

6.4. nine

6.5. not racemic, $CH_3\overset{\overset{\displaystyle Cl}{|}}{C}HCHCH_2Cl$

       $(R)(S)$ CH_3 (R)

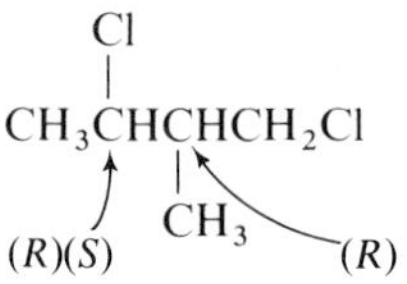

6.6. 9 to 1

6.7. (a)  $CH_2{=}CH{-}\dot{C}H_2 \longleftrightarrow \dot{C}H_2{-}CH{=}CH_2$

(b)

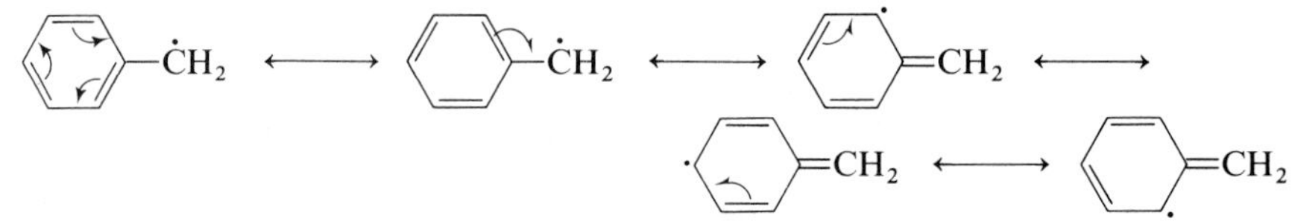

6.8. (b), (c), (a), (d)

6.9. (a) $C_6H_5CHBrCH_2CH_2CH_3$;   (b) 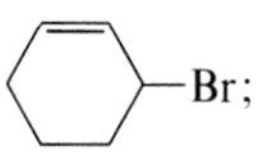 ⬡$-Br$;   (c) $C_6H_5CH{=}CHCH_2Br$

6.10.

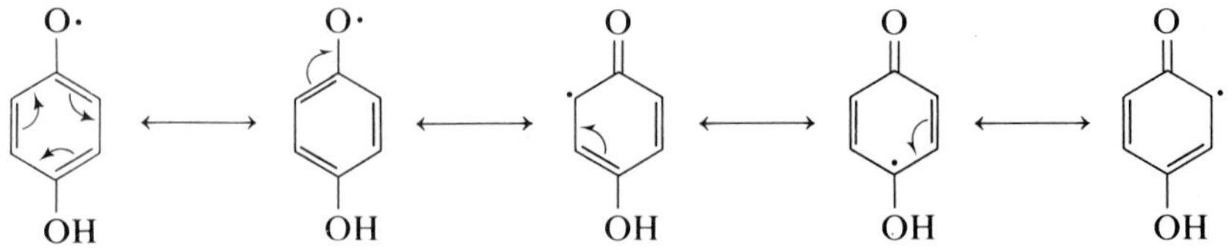

6.11. The reaction is faster because a 3° hydrogen is abstracted more readily than a 2° hydrogen.

$CH_3CH_2O\overset{2°}{C}H_2CH_3$ $(CH_3)_2CHO\overset{3°}{C}H(CH_3)_2$

6.12.

$$(CH_3)_2C\overset{C\equiv N}{\underset{}{|}}-N=N-\overset{C\equiv N}{\underset{}{|}}C(CH_3)_2 \xrightarrow{heat} 2(CH_3)_2\overset{C\equiv N}{\underset{}{|}}C\cdot \ + :N\equiv N:$$

The product radical is 3° and "allylic" and is thus relatively stable. The other product is N_2, which is easily lost from a molecule because of its stability.

6.13. $R\cdot \ +$ [naphthalene ring with $\overset{..}{N}HC_6H_5$ substituent] $\longrightarrow$ $RH \ +$ [naphthalene ring with $\overset{..}{N}C_6H_5$ substituent]

relatively stable free radical
(resonance-stabilized by both the
phenyl and naphthyl groups)

6.14.   (b), (c), (d)

6.15.   [cyclohexyl]—Br $\xrightarrow{Mg}$ [cyclohexyl]—MgBr $\xrightarrow[(2)\ H_2O,\ H^+]{(1)\ CH_3CHO}$ [cyclohexyl]—$\overset{OH}{\underset{}{|}}$CHCH$_3$

6.16.   (a)   [cyclohexyl]—Br $\xrightarrow[(2)\ CuI]{(1)\ Li}$ ([cyclohexyl])$_2$CuLi $\xrightarrow{CH_3CH_2CH_2Br}$ product

(b) $(CH_3)_3CBr \xrightarrow[(2)\ CuI]{(1)\ Li} [(CH_3)_3C]_2CuLi \xrightarrow{CH_3CH_2CH_2Br}$ product

6.17.   (a) and (d), which both contain acidic hydrogens that would destroy any Grignard reagent formed.

6.18.   [cyclohexyl]—Br $\xrightarrow{Mg}$ [cyclohexyl]—MgBr $\xrightarrow{D_2O}$ product

6.19.   (a) $(C_6H_5)_2CH_2 \xrightarrow{NBS} (C_6H_5)_2CHBr \xrightarrow[ether]{Mg}$

$(C_6H_5)_2CHMgBr \xrightarrow[(2)\ H_2O,\ H^+]{(1)\ CH_3CHO}$ product

(b) $C_6H_5CH_3 \xrightarrow{NBS} C_6H_5CH_2Br \xrightarrow[CH_3OH]{Na^+\ ^-OCH_3}$ product

(c)   Br—[benzene ring]—CH$_3$ $\xrightarrow[ether]{Mg}$ BrMg—[benzene ring]—CH$_3$ $\xrightarrow{D_2O}$

D—[benzene ring]—CH$_3$ $\xrightarrow{NBS}$ D—[benzene ring]—CH$_2$Br $\xrightarrow{KCN}$ D—[benzene ring]—CH$_2$CN

6.20.   (a) $(CH_3)_2CHCH_2Br \xrightarrow[ether]{Mg} (CH_3)_2CHCH_2MgBr \xrightarrow[(2)\ H_2O,\ H^+]{(1)\ CH_3\overset{O}{\overset{||}{C}}CH_2CH_3}$ product

(b)   [cyclohexyl]—Br $\xrightarrow[ether]{Mg}$ [cyclohexyl]—MgBr $\xrightarrow[(2)\ H_2O,\ H^+]{(1)\ HCHO}$ product

---

CHAPTER 7

7.1.   (a)   2,4-dimethyl-3-pentanol;   (b)   3-methyl-1,2-cyclohexanediol

7.2.   (a) $CH_3\overset{CH_3}{\underset{OH}{\overset{|}{\underset{|}{C}}H}C(CH_2CH_3)_2}$ (b) $HOCH_2\overset{CH_3}{\underset{CH_3}{\overset{|}{\underset{|}{C}}}CH_2\overset{OH}{\overset{|}{C}H}CH_2CH_3}$

7.3. (a) 3-cyclopenten-1-ol; (b) 4-hydroxy-1-cyclohexanone; (c) 2-bromo-1-ethanol

7.4. (a) $CH_3CH_2OC_6H_5$; (b) and (c) $CH_3\overset{\underset{\displaystyle |}{OH}}{C}HCH_2CH_3$; (d) $(CH_3)_2\overset{O}{\overbrace{C\!-\!CH_2}}$

7.5. (a) $C_6H_5Br \xrightarrow[\text{ether}]{Mg} C_6H_5MgBr \xrightarrow[\text{(2) } H_2O, H^+]{\text{(1) HCHO}} C_6H_5CH_2OH$

(b) $CH_3(CH_2)_3Cl \xrightarrow{OH^-} CH_3(CH_2)_3OH$

(c) $CH_3CH_2Br \xrightarrow[\text{ether}]{Mg} CH_3CH_2MgBr \xrightarrow[\text{(2) } H_2O, H^+]{\text{(1) } C_6H_5CHO} C_6H_5\overset{\underset{\displaystyle |}{OH}}{C}HCH_2CH_3$

7.6. (a) $CH_3CH{=}CHCH_3 \xrightarrow{H_2O, H^+}$ $CH_3\overset{\underset{\displaystyle |}{OH}}{C}HCH_2CH_3$

$CH_3\overset{\overset{\displaystyle O}{\|}}{C}CH_2CH_3 \xrightarrow[\text{(2) } H_2O, H^+]{\text{(1) NaBH}_4}$

(b) image with CH_3—cyclopentene—CH_3 $\xrightarrow{H_2O, H^+}$ and ketone $\xrightarrow[\text{(2) } H_2O, H^+]{\text{(1) NaBH}_4}$ giving cyclopentanol with OH, CH_3 groups

7.7. *trans*

7.8. (a) $HCl + ZnCl_2$, racemization; (b) PCl_3, inversion; (c) $SOCl_2 + R_3N$, inversion, or $SOCl_2$ + an ether, retention

7.9. *trans*, because it is more stable

7.10. (a) $(C_6H_5)_2CHCHO$; (b) $(C_6H_5)_2\overset{\overset{\displaystyle O}{\|}}{\underset{\underset{\displaystyle CH_3}{|}}{C}}CCH_3$

7.11. (a) and (b), $C_6H_5O^- + Na^+$

7.12. The nitro group helps share the negative charge in the anion.

7.13. (a) $CH_3CH_2CH_2\overset{\overset{\displaystyle O}{\|}}{C}OCH_3$; (b)

naphthalene with $CH_2O\overset{\overset{\displaystyle O}{\|}}{C}CH_3$ substituent

7.14. $H\overset{\cdot\cdot}{\underset{\cdot\cdot}{O}}NO + H{-}\overset{\cdot\cdot}{\underset{\cdot\cdot}{O}}NO \underset{\longleftarrow}{\overset{-NO_2^-}{\rightleftharpoons}} H_2\overset{+}{O}{-}NO \underset{\longleftarrow}{\overset{-H_2O}{\rightleftharpoons}} \overset{+}{N}O$

$CH_3CH_2\overset{\cdot\cdot}{\underset{\cdot\cdot}{O}}H + \overset{+}{N}O \longrightarrow CH_3CH_2\overset{+}{\underset{\underset{H}{|}}{\overset{\cdot\cdot}{O}}}NO \underset{\longleftarrow}{\overset{-H^+}{\rightleftharpoons}} CH_3CH_2\overset{\cdot\cdot}{\underset{\cdot\cdot}{O}}NO$

7.15. (a) $CH_3CH_2CH_2OH + TsCl \xrightarrow{-HCl} CH_3CH_2CH_2OTs$

     (b) $(R)\text{-}CH_3\overset{\overset{\displaystyle OH}{|}}{C}H(CH_2)_3CH_3 + TsCl \xrightarrow{-HCl} (R)\text{-}CH_3\overset{\overset{\displaystyle OTs}{|}}{C}H(CH_2)_3CH_3$

7.16. $CH_3\overset{\overset{\displaystyle \cdot\cdot\overset{\displaystyle O:}{\|}}{\|}}{\underset{\underset{\displaystyle :O:}{\|}}{S}}{-}\overset{\cdot\cdot}{\underset{\cdot\cdot}{O}}:^- \longleftrightarrow CH_3\overset{\overset{\displaystyle :\overset{\cdot\cdot}{O}:^-}{}}{\underset{\underset{\displaystyle :O:}{\|}}{S}}{=}\overset{\cdot\cdot}{O}: \longleftrightarrow CH_3\overset{\overset{\displaystyle \overset{\cdot\cdot}{O}:}{\|}}{\underset{\underset{\displaystyle :\overset{\cdot\cdot}{O}:^-}{|}}{S}}{=}\overset{\cdot\cdot}{O}:$

7.17. $(S)\text{-}CH_3\overset{\overset{\displaystyle OH}{|}}{C}H(CH_2)_5CH_3$

7.18. (b), (c), (a)

7.19. (b), (a), (c)

7.20. (a) cyclopentane ring $=O$; (b) $C_6H_5CO_2H$

7.21. $CH_3\overset{\overset{\displaystyle O}{\|}}{C}H + H\overset{\overset{\displaystyle O}{\|}}{C}CH_2OCH_3$

7.22.

$$\begin{array}{c}
\overset{\overset{\displaystyle O}{\|}}{C}H \\
| \\
H{-}C{-}OH \\
| \\
HO{-}C{-}H \\
| \\
H{-}C{-}OH \\
| \\
H{-}C{-}OH \\
| \\
CH_2OH
\end{array} \xrightarrow{HIO_4} 5\ H\overset{\overset{\displaystyle O}{\|}}{C}OH + H\overset{\overset{\displaystyle O}{\|}}{C}H$$

to HCHO

7.23. (a)

naphthalene with O^- substituent $+ CH_3I \longrightarrow$;   (b) $(CH_3)_2CHO^- + CH_3CH_2Br \longrightarrow$

7.24.

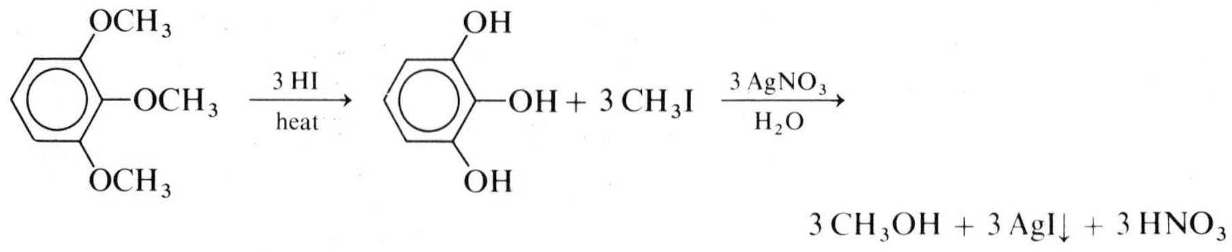

$$3\,CH_3OH + 3\,AgI\downarrow + 3\,HNO_3$$

7.25. (c)   **7.26.** (a)

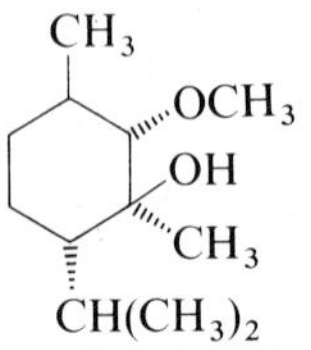

(b)

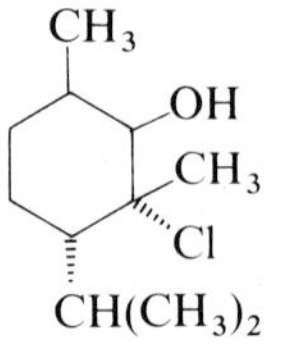

7.27. (a)   DMSO forms hydrogen bonds with H_2O: $(CH_3)_2S=\overset{..}{\underset{..}{O}}:---H_2O$

(b)   The anion is resonance-stabilized:

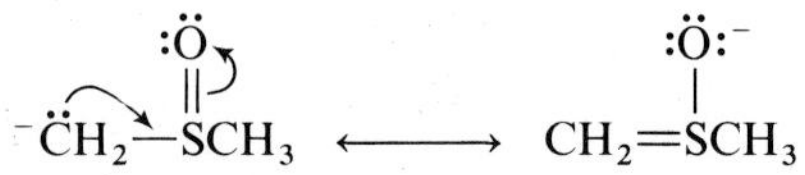

7.28. $CH_3CH_2OH \xrightarrow{H_2CrO_4} CH_3CO_2H \xrightarrow[heat]{CH_3CH_2OH,\ H^+} CH_3CO_2CH_2CH_3$

7.29.

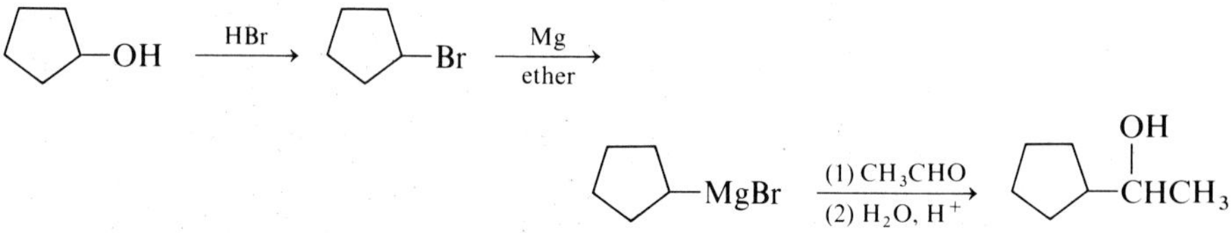

7.30. $(CH_3)_2CHCH_2OH \xrightarrow[\text{or } PCl_3]{SOCl_2} (CH_3)_2CHCH_2Cl \xrightarrow[ether]{Mg} (CH_3)_2CHCH_2MgCl$

CHAPTER 8

8.1. (a) $1600\ cm^{-1}$;   (b) $200\ nm$;   (c) $60,004\ Hz$

8.2. (a) $1667\ cm^{-1}$;   (b) $12.5\,\mu m$;   (c) $1.5\ \mu m$ **8.3.** I (d);   II (b)

8.4. $CH_3CH_2OCH_2CH_3$ **8.5.** I, ester;   II, ketone

8.6. (a)   two;   (b)   seven;   (c)   ten;   (d)   two   **8.7.** (a), (c), (d), (e)

8.8. (a)   one;   (b)   one   **8.9.** (a)   two;   (b)   three;   (c)   four

8.10.

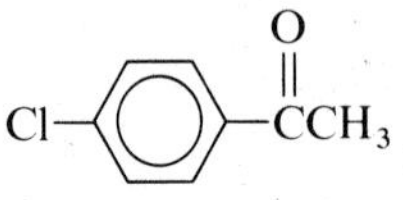

8.11.   Two Cl atoms have a greater electron-withdrawing power than one Cl atom.

8.12. $\underset{\textit{singlet, 7.15}}{\overset{\textit{triplet, 2.9}}{C_6H_5CH_2CH_2O_2CCH_3}}$

triplet, 2.9 ⟶ (arrow to CH₂)
triplet, 4.3 ⟶ (arrow to CH₂)
singlet, 7.15 ⟶ (arrow to C₆H₅)
singlet, 2.0 ⟶ (arrow to CCH₃)

8.13. I (e); II (d); III (a)   **8.14.** $BrCH_2CH_2CH_2\overset{\overset{\displaystyle O}{\|}}{C}OCH_2CH_3$

8.15. $(CH_3)_2CHOH$ **8.16.** $C_6H_5CH_2OH$ **8.17.** $CH_3CH_2CHClCO_2H$

CHAPTER 9

9.1. (a) $\underset{ClCH_2CH_2}{\overset{CH_3}{C}}=\underset{CH_3}{\overset{CH_2CH_3}{C}}$ (b) $\underset{H}{\overset{CH_2=CH}{C}}=\underset{H}{\overset{CH_3}{C}}$

(c) ⬡—C≡CH   (d) $C_6H_5C\equiv CC_6H_5$

9.2. (a) 2-methyl-1,4-pentadiene; (b) 2-propyn-1-ol

9.3. I contains C=C; II contains C≡C

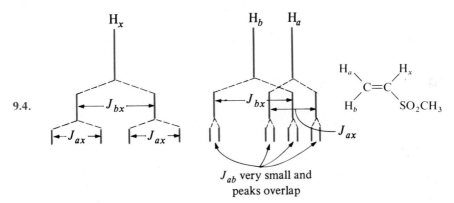

9.4.

J_{ab} very small and peaks overlap

9.5. In *p*-chlorostyrene, the chemical shifts are the same for the protons vicinal to the vinylic group and for the protons vicinal to the Cl. However, in styrene, the five aryl protons do not all have the same chemical shift.

9.6. (a) ⬡ ; (b) $CH_2=C(CH_3)_2$; (c) $CH_3CH=C(CH_3)_2$

9.7. (a) $CH_3C\equiv CH \xrightarrow{Na} CH_3C\equiv C^-\ Na^+ \xrightarrow{CH_3CH_2Br} CH_3CH_2C\equiv CCH_3$

(b) $CH_3C\equiv C^- \xrightarrow{C_6H_5CHBrCH_3} C_6H_5\overset{\overset{\displaystyle CH_3}{|}}{C}HC\equiv CCH_3$

9.8. $CH\equiv CH$ $\xrightarrow[-CH_3CH_3]{1\,CH_3CH_2MgBr}$ $CH\equiv CMgBr$ $\xrightarrow[(2)\,H_2O,\,H^+]{(1)\,CH_3CH_2\overset{\displaystyle O}{\overset{\displaystyle \|}{C}}H}$ $CH_3CH_2\overset{\displaystyle OH}{\underset{\displaystyle |}{C}}HC\equiv CH$

9.9. (a) $CH_3CH_2\overset{+}{C}HCH_3$ $\xrightarrow{Br^-}$ $CH_3CH_2\overset{\displaystyle Br}{\underset{\displaystyle |}{C}}HCH_3$

(c) $(CH_3)_2\overset{+}{C}CH_2CH_3$ $\xrightarrow{Br^-}$ $(CH_3)_2\overset{\displaystyle Br}{\underset{\displaystyle |}{C}}CH_2CH_3$

9.10. $CH_3CH\!=\!CH_2$ $\xrightarrow{H^+}$ $[CH_3\overset{+}{C}HCH_3]$ $\xrightarrow{CH_3CH_2OH}$

$CH_3CH_2\overset{+}{\underset{\displaystyle CH_3\overset{\displaystyle |}{C}HCH_3}{O}}H$ $\underset{-H^+}{\rightleftharpoons}$ $CH_3CH_2OCH(CH_3)_2$

9.11. (a) $CH_3CH_2CCl(CH_3)_2$; (b) $CH_3CH_2\overset{\displaystyle CH_3}{\underset{\displaystyle |}{C}}HCBr(CH_3)_2$

9.12. (a) $(CH_3)_2CHCH_2Br$; (b) $(CH_3)_2CBrCH_3$

9.13. $CH_3CH_2CH\!=\!CH_2$ $\xrightarrow{H^+}$ $[CH_3CH_2\overset{+}{C}HCH_3]$ $\xrightarrow{HSO_4^-}$ $CH_3CH_2\overset{\displaystyle OSO_3H}{\underset{\displaystyle |}{C}}HCH_3$

9.14. (a) $CH_3CH_2\overset{\displaystyle OH}{\underset{\displaystyle |}{C}}HCH_3$; (b) $(CH_3)_2\overset{\displaystyle OH}{\underset{\displaystyle |}{C}}CH(CH_3)_2$; (c) $(CH_3)_2\overset{\displaystyle OSO_3H}{\underset{\displaystyle |}{C}}CH_2CH_3$

9.15. (a) $(CH_3)_3CCH\!=\!CH_2$ $\xrightarrow[-\,^-O_2CCH_3]{Hg(O_2CCH_3)_2}$ $\left[(CH_3)_3\overset{\displaystyle \cdots}{C}H\underset{+}{-CH_2}\overset{HgO_2CCH_3}{}\right]$ $\xrightarrow[-H^+]{H_2O}$

$(CH_3)_3\overset{\displaystyle HgO_2CCH_3}{\underset{\displaystyle OH}{\underset{\displaystyle |}{C}H-\overset{|}{C}H_2}}$ $\xrightarrow{NaBH_4}$ $(CH_3)_3\overset{\displaystyle}{\underset{\displaystyle OH}{\underset{\displaystyle |}{C}CHCH_3}}$

(b) $(CH_3)_3CCH\!=\!CH_2 + H_2O$ $\xrightarrow{H^+}$ $(CH_3)_2\overset{\displaystyle OH}{\underset{\displaystyle |}{C}}CH(CH_3)_2$

9.16. $CH_2\!=\!\overset{\displaystyle CH_3}{\underset{\displaystyle |}{C}}CH_2CH_3$ $\xrightarrow[-\,^-O_2CCH_3]{Hg(O_2CCH_3)_2}$ $\overset{CH_3CO_2Hg}{\underset{+}{CH_2}}\!\!-\!\!\overset{CH_3}{\underset{}{C}}CH_2CH_3$ $\xrightarrow[-H^+]{CH_3OH}$

$\overset{CH_3CO_2Hg}{CH_2}\!\!-\!\!\overset{CH_3}{\underset{\displaystyle OCH_3}{C}}CH_2CH_3$ $\xrightarrow{NaBH_4}$ $CH_3\overset{CH_3}{\underset{\displaystyle OCH_3}{C}}CH_2CH_3$

9.17.

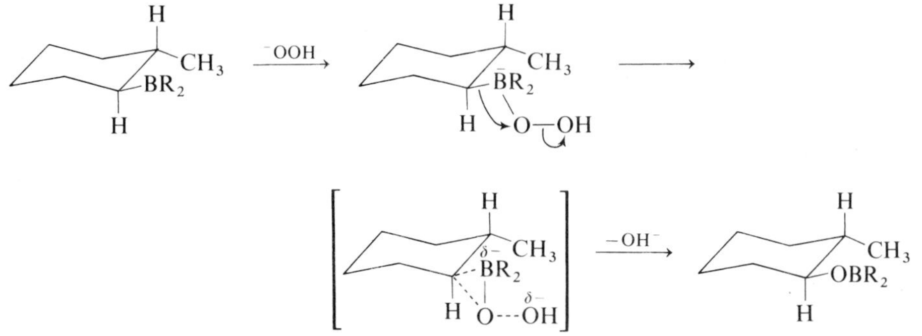

trans transition state
(cannot yield a *cis* alcohol)

9.18. (a)

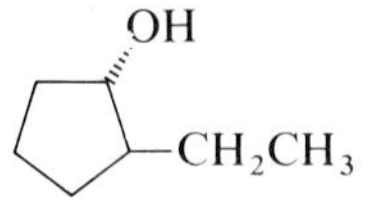

trans (and racemic)

(b)

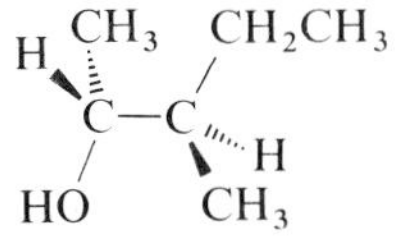

(2R,3R)-3-methyl-2-pentanol
and its (2S,3S) enantiomer

9.19. (a) $CH_3CHBrCHBrCH_3$; (b) $(CH_3)_2CBrCH_2Br$; (c) $(CH_3)_2CBrCHBrCH_3$

9.20. (a)

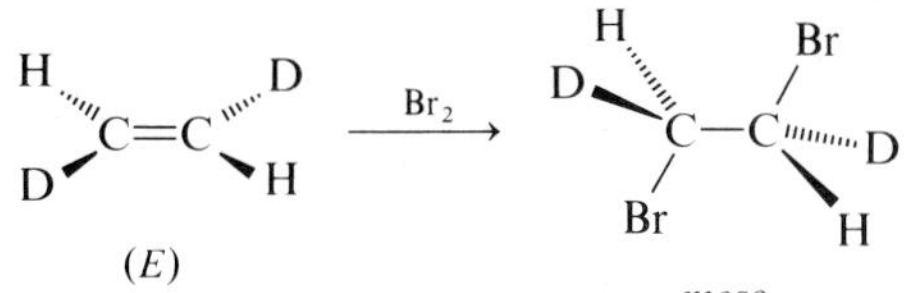

(E)   *meso*

(b)

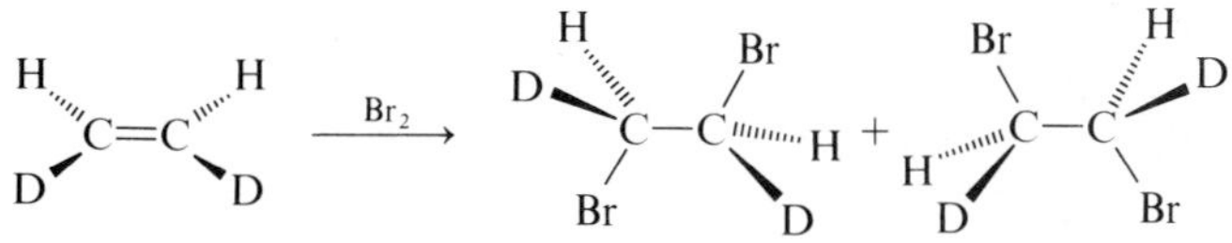

(Z)   (1S,2S)   (1R, 2R)

9.21. No, because the intermediate contains Br.

9.22.

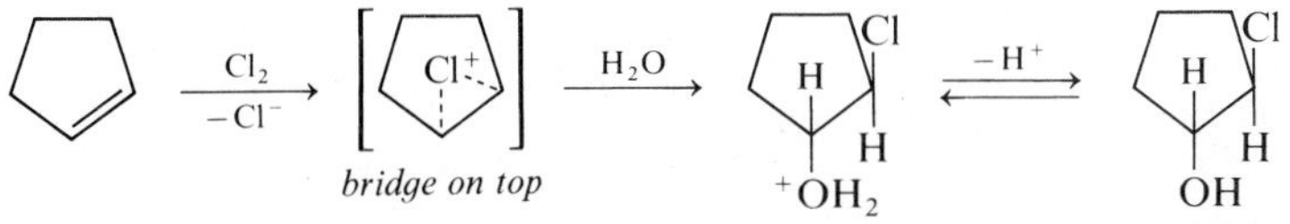

bridge on top

9.23. (a)

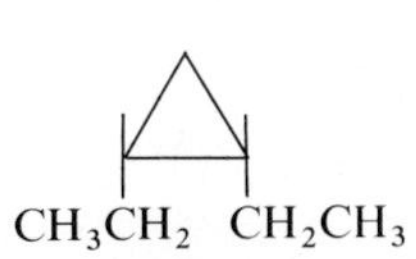

CH_3CH_2 CH_2CH_3

(b) and (d):

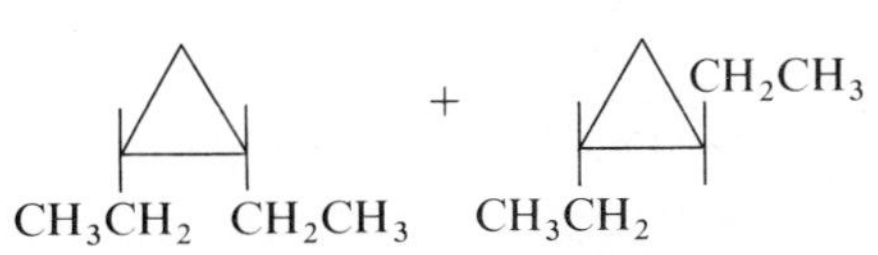

CH_3CH_2 CH_2CH_3 CH_3CH_2

(c)

CH_3CH_2

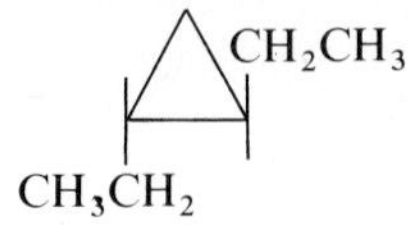

9.24. (a) 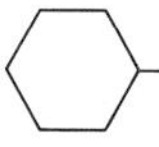 —CH=CH$_2$ + CHBr$_3$ + ⁻OR ⟶ product + ROH + Br⁻

(b) 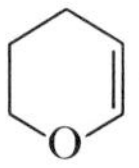 + CHCl$_3$ + ⁻OR ⟶ product + ROH + Cl⁻

9.25. (a) The difference in energy is greater between *cis-* and *trans-*(CH$_3$)$_3$CCH=CHCH$_2$CH$_3$ because of the greater steric hindrance in the *cis*-isomer. (b) The difference in energy is greater between *cis-* and *trans-*ClCH=CHCl because the *cis*-isomer contains greater dipole–dipole repulsions between the two chlorine atoms.

9.26. (a) CH$_3$CH—C(CH$_3$)$_2$ (b) 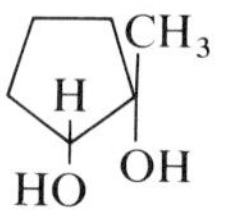 Both products are racemic.
$\quad\quad\quad\quad$ |$\quad$ |
$\quad\quad\quad\quad$ OH OH

9.27. (a) 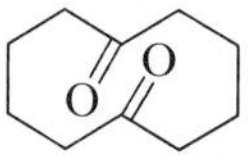 in either case

(b) 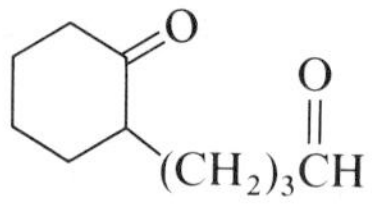 from reductive, 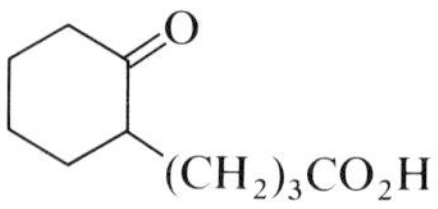 from oxidative

(c) OHC—CHO + OHC(CH$_2$)$_4$CHO from reductive
$\quad\quad\quad\quad\quad\quad\quad\quad\quad$ HO$_2$C—CO$_2$H + HO$_2$C(CH$_2$)$_4$CO$_2$H from oxidative

(d) OHC(CH$_2$)$_4$CHO from reductive, HO$_2$C(CH$_2$)$_4$CO$_2$H from oxidative

9.28. CH$_3$CHBrCHBrCH$_3$; CH$_3$CHClCHClCHClCHClCH$_3$

9.29. [CH$_3$CH$_2$ĊH—CH=CHCH$_3$ ⟷ CH$_3$CH$_2$CH=CHĊHCH$_3$] ⟶
$\quad\quad\quad\quad\quad\quad\quad\quad\quad$ CH$_3$CH$_2$CHICH=CHCH$_3$ + CH$_3$CH$_2$CH=CHCHICH$_3$

9.30. (a) 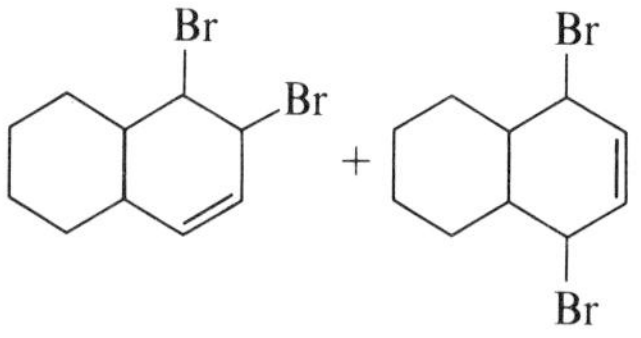 + (b) 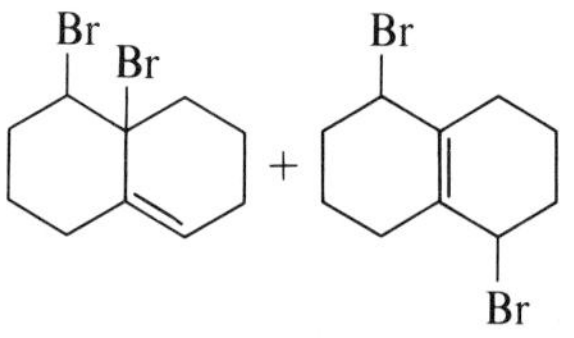 +

9.31. the *trans*-isomer, because it is more stable

9.32. **A**, BrCH$_2$CH=CHCH$_2$Br; **B**, BrCH$_2$CHBrCH=CH$_2$

9.33. (a) s-*trans*; (b) s-*cis*; (c) s-*trans*.   Only (a) is interconvertible.

9.34. (a) 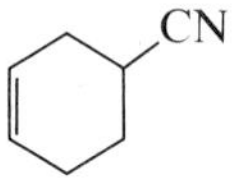 (b)

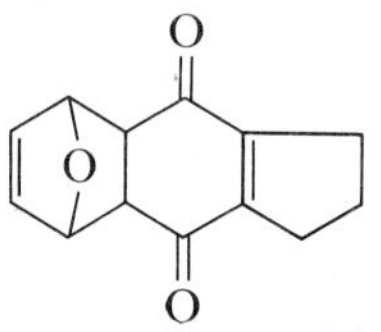

9.35. + $CH_3CH_2O_2CC\equiv CCO_2CH_2CH_3$ $\xrightarrow{\text{heat}}$

9.36.

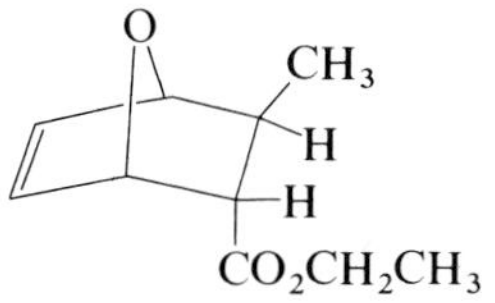

9.37. and

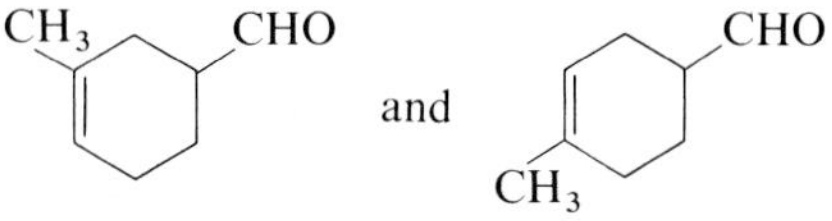

9.38. $\text{-}(CF_2CF_2)_x\text{-}$ **9.39.** $CH_2{=}CHCN$ **9.40.**

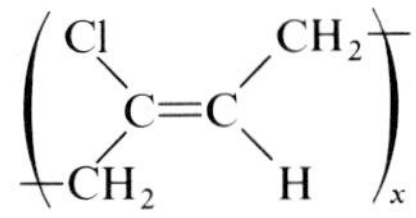

9.41. (a) $CH_3\overset{\overset{\displaystyle O}{\|}}{C}CH_2CH_3$ $\xrightarrow[\text{(2) } H_2O,\, H^+]{\text{(1) } (CH_3)_2CHMgBr}$ $CH_3\underset{\underset{\displaystyle CH(CH_3)_2}{|}}{\overset{\overset{\displaystyle OH}{|}}{C}}CH_2CH_3$ $\xrightarrow{H_2SO_4}$ product

(b) $CH_3CH_2CH{=}CH_2$ $\xrightarrow{Br_2}$ $CH_3CH_2CHBrCH_2Br$ $\xrightarrow{NaNH_2}$ product

(c) $\xrightarrow{\text{heat}}$ $\xrightarrow[\text{Pt}]{H_2}$ product

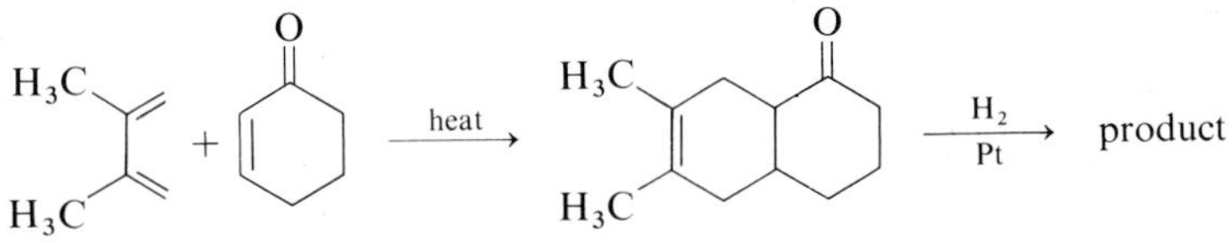

CHAPTER 10

10.1. (a)   (b)   (c) $CH_3CH_2\text{-}$$\text{-OH}$

(d) -CH_2CH_2OH (e) -CH_2Br

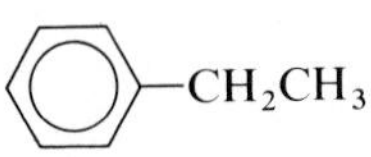

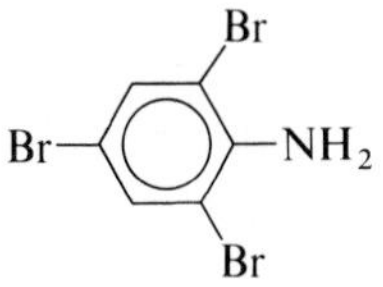

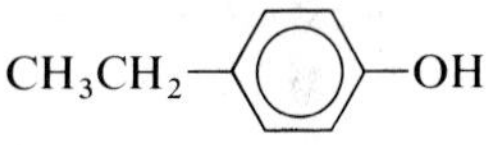

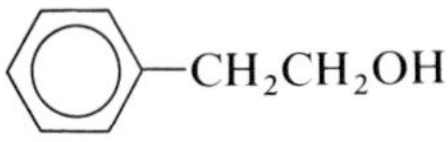

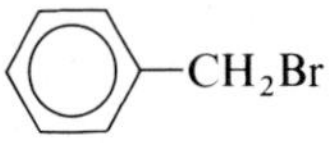

10.2. A, *p*-iodoanisole; B, *p*-chloroaniline

10.3. not aromatic   **10.4.** (a), (c)

10.5. (a) $C_6H_5\underset{\underset{\displaystyle CH_3}{|}}{C}HCH_2CH_3$;   (b) $C_6H_5C(CH_3)_3$;   (c) $C_6H_5\underset{\underset{\displaystyle CH_2CH_3}{|}}{C}(CH_3)_2$;
(d) $(C_6H_5)_2CH_2$

10.6. $C_6H_6 + R\overset{+}{C}{=}\ddot{O}\!:$ $\longrightarrow$ $\xrightarrow{-H^+}$

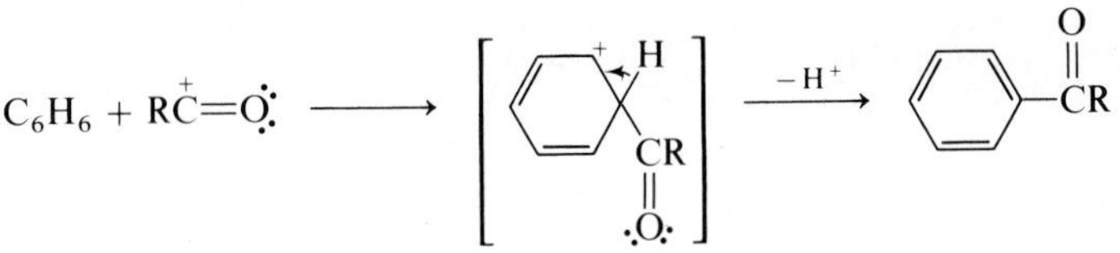

10.7. (a) $C_6H_5\overset{\overset{\displaystyle CH_3}{|}}{\underset{\underset{\displaystyle CH_3}{|}}{C}}CH_2CH_3$ (b) $C_6H_5CH_2\overset{\overset{\displaystyle CH_3}{|}}{C}HCH_2CH_3$

10.8. Two parts *o*-, two parts *m*-, and one part *p*-.

10.9. (a) —Cl + O$_2$N— —Cl   (b) —NO$_2$

(c) —OH + HO$_3$S— —OH

(d) —CH$_3$ + CH$_3$CH$_2$— —CH$_3$

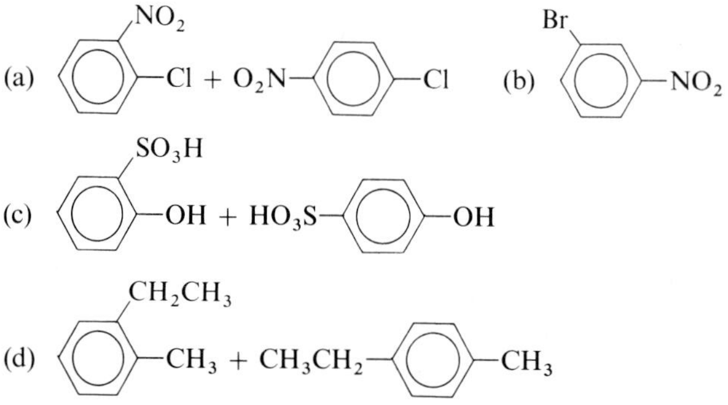

10.10.

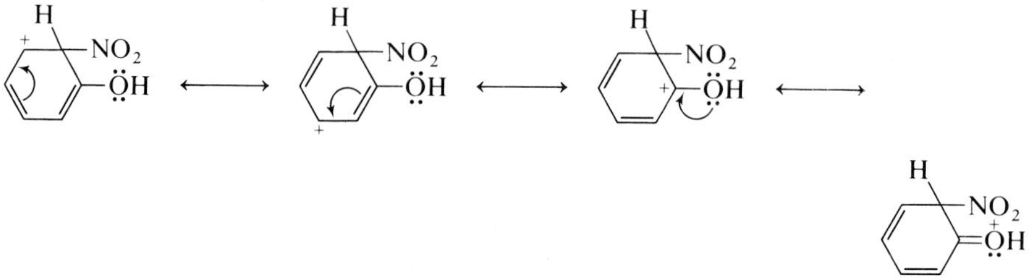

10.11. The unshared electrons of the nitrogen are delocalized by the carbonyl group and are less available for donation to the ring. Because the amide nitrogen is partially positive, it also exerts a greater electron-withdrawal than an amine nitrogen.

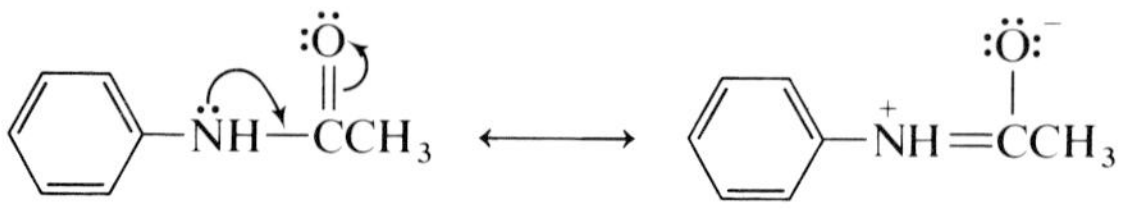

10.12. (a) $(CH_3)_2CH$— —Br + —Br

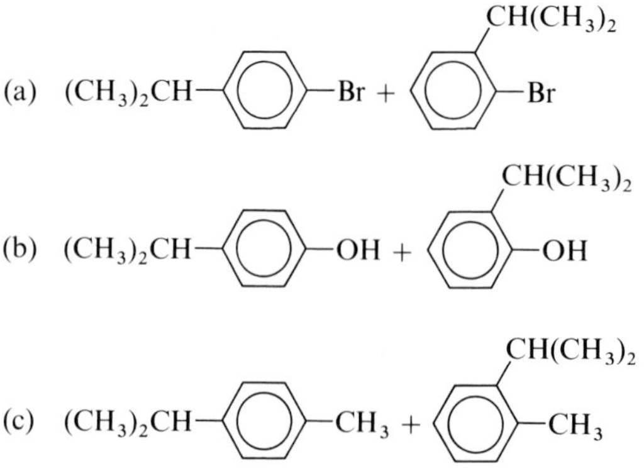

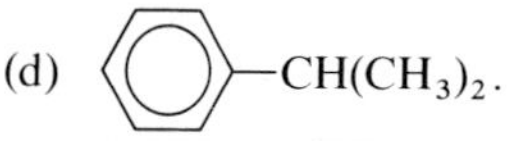

(b) $(CH_3)_2CH$— —OH + —OH

(c) $(CH_3)_2CH$— —CH$_3$ + —CH$_3$

(d) —CH(CH$_3$)$_2$.   Relative rates: (b) > (c) > (d) > (a).

10.13. (a) Cl—⬡—NH₂ + Cl-⬡(Cl)(Cl)—NH₂

(b) ⬡(CH₃)(NO₂)—OCH₃ + CH₃—⬡(NO₂)—OCH₃

(c) ⬡—C(=O)NH—⬡(NO₂) + ⬡—C(=O)NH—⬡—NO₂

because the right-hand ring is activated and the left-hand ring is deactivated.

10.14. ⬡–ĊH₂ ⟷ ⬡=CH₂ ⟷ ⁺⬡=CH₂ ⟷ ⬡=CH₂

10.15. ⬡(CO₂H)(CO₂H)

10.16. (a) naphthalene-OH $\xrightarrow{OH^-}$ naphthalene-O⁻ $\xrightarrow{C_6H_5CH_2Br}$ product

(b) naphthalene-OH + $C_6H_5\overset{O}{\overset{\|}{C}}O\overset{O}{\overset{\|}{C}}C_6H_5$ ⟶ product

10.17. (a) naphthalene-NH₂ $\xrightarrow[0°]{\substack{NaNO_2 \\ HCl}}$ naphthalene-N₂⁺ $\xrightarrow[KCN]{CuCN}$ product

(b) CH_3—⬡—NO_2 $\xrightarrow[(2)\ OH^-]{(1)\ Fe,\ HCl}$ CH_3—⬡—NH_2 $\xrightarrow[0°]{\substack{NaNO_2 \\ HCl}}$

CH_3—⬡—$N_2^+\ Cl^-$ $\xrightarrow{Cu_2Br_2}$ product

(c) C_6H_6 $\xrightarrow{\substack{HNO_3 \\ H_2SO_4}}$ $C_6H_5NO_2$ $\xrightarrow[(2)\ OH^-]{(1)\ Fe,\ HCl}$ $C_6H_5NH_2$ $\xrightarrow[0°]{\substack{NaNO_2 \\ HCl}}$

$C_6H_5N_2^+\ Cl^-$ $\xrightarrow[heat]{H_2O,\ H^+}$ C_6H_5OH $\xrightarrow[(2)\ CH_3I]{(1)\ OH^-}$ $C_6H_5OCH_3$ $\xrightarrow{C_6H_5N_2^+\ Cl^-}$ product

10.18.

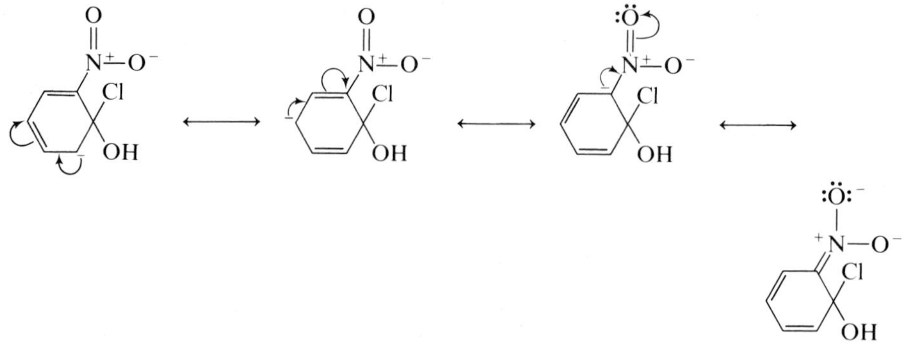

10.19. (a)

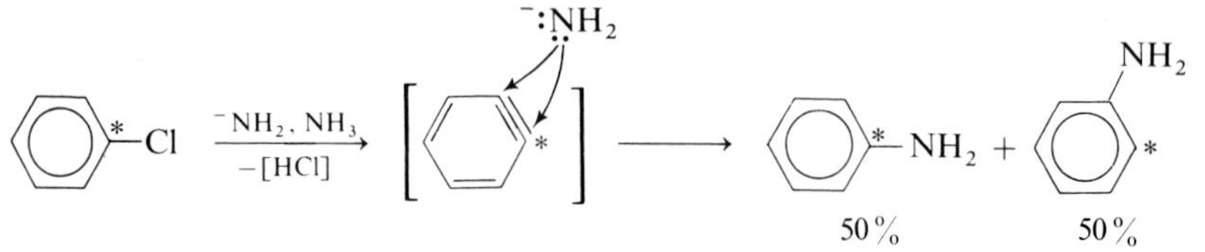

50%   50%

(b)

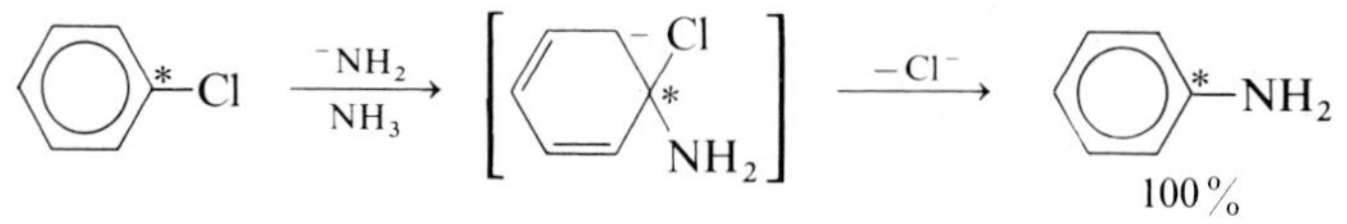

100%

10.20. (a) 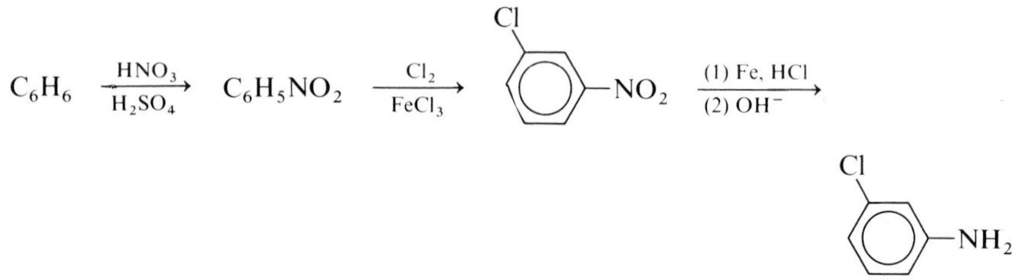 C_6H_6 $\xrightarrow[H_2SO_4]{HNO_3}$ $C_6H_5NO_2$ $\xrightarrow[FeCl_3]{Cl_2}$

$\xrightarrow[(2)\ OH^-]{(1)\ Fe,\ HCl}$

(b) $C_6H_6 + ClC(CH_2)_3CH_3$ $\xrightarrow{AlCl_3}$ $C_6H_5C(CH_2)_3CH_3$ $\xrightarrow[HCl]{Zn/Hg}$

$C_6H_5(CH_2)_4CH_3$

(c)   excess $C_6H_6 + ClCH_2CH_2Cl$ $\xrightarrow{AlCl_3}$ $C_6H_5CH_2CH_2C_6H_5$

(d)   excess $C_6H_6 + CHCl_3$ $\xrightarrow{AlCl_3}$ $(C_6H_5)_3CH$

CHAPTER 11

11.1. (c)   **11.2.** C_6H_5CHO

11.3. (a) cyclopentanone; (b) 2-pentanone; (c) pentanal;
(d) 2-chlorocyclopentanone.

11.4. (a) because of electron-withdrawal by the Cl atoms

11.5. (a)

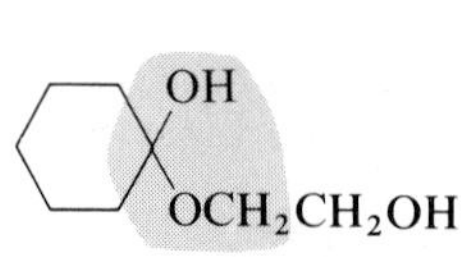

hemiketal

(b)

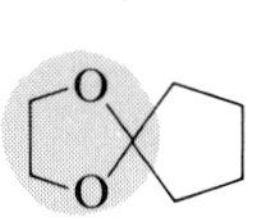

ketal

(c)

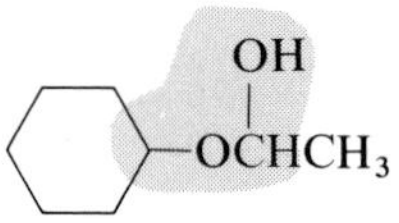

hemiacetal

11.6. (a) [cyclohexanone]$=O + HOCH_2CH_2OH$ (b) [cyclopentanone]$=O + HOCH_2CH_2OH$

(c) [cyclohexane ring]$-OH + \overset{\overset{\displaystyle O}{\|}}{H}CCH_3$

11.7. The hemiacetal carbocation is more stabilized because it is resonance-stabilized.

$$\underset{\overset{\displaystyle |}{\underset{\displaystyle :OR}{\overset{+}{C}}}}{\overset{+}{R}CH} \longleftrightarrow \underset{\overset{\|}{\underset{\displaystyle {}^+OR}{}}}{RCH}$$

11.8. (a) $H_3C-\overset{\overset{\displaystyle OH}{|}}{}\overset{\overset{\displaystyle O}{\|}}{}-CH_3 \rightleftharpoons H_3C-$[ring]$\overset{OOH}{}CH_3$

(b) $HOCH_2-\overset{\overset{\displaystyle OH}{|}}{}\overset{\overset{\displaystyle O}{\|}}{}-CH_2OH \rightleftharpoons$

[furanose ring] $HOCH_2-$ ... CH_2OH with OH, OH $+$ $HO-$[pyranose ring]... CH_2OH with OH, OH

11.9. $CH_3CH_2OH \xrightarrow[\text{pyridine}]{CrO_3} CH_3CHO \xrightarrow[CN^-]{HCN} CH_3\overset{\overset{\displaystyle OH}{|}}{C}HCN \xrightarrow{H_2O, H^+}$

$$CH_3\overset{\overset{\displaystyle OH}{|}}{C}HCO_2H$$

11.10. (a) $CH_3\overset{\overset{\displaystyle \ddot{O}:}{\|}}{C}H + \underset{\underset{\underset{\displaystyle Na^+}{}}{:\overset{..}{O}:}}{:SOH} \rightleftharpoons CH_3\overset{\overset{\displaystyle :\ddot{O}:^-}{|}}{C}H \underset{O=S=O}{\underset{\displaystyle :\underset{..}{O}-H}{}} \xrightleftharpoons{\text{proton transfer}} CH_3\overset{\overset{\displaystyle :\overset{..}{O}H}{|}}{C}H \underset{O=S=O}{\underset{\displaystyle :\underset{..}{O}:^- Na^+}{}}$

(b) (1) Extract the ether solution with an aqueous, alkaline solution to remove the carboxylic acid as RCO_2^- Na^+.

(2) Extract the ether solution with aqueous $NaHSO_3$ solution to remove the aldehyde as $R\overset{\overset{\displaystyle }{}}{C}HSO_3^-$ Na^+.
$\overset{\overset{\displaystyle |}{}}{OH}$

(3) Regenerate the acid by acidification of its solution. Regenerate the aldehyde by treatment of its solution with acid or base. The ketone is still in the ether solution.

11.11. An arylamine yields a product in which the double bond is in conjugation with the aromatic ring.

11.12. A pair of stereoisomers:

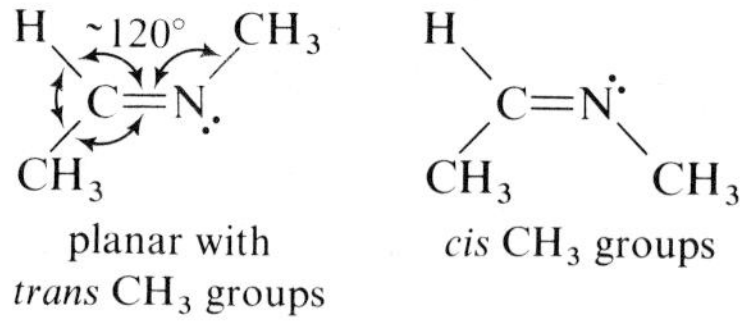

planar with
trans CH$_3$ groups   *cis* CH$_3$ groups

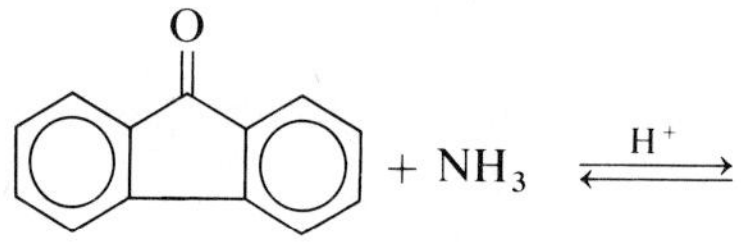

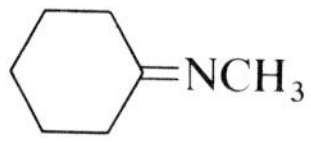

11.13. (a)

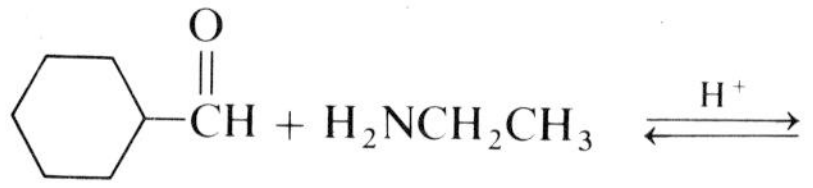

(b)

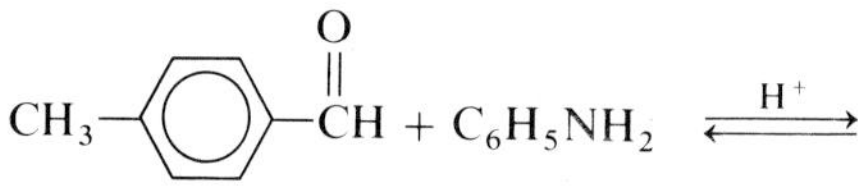

(c) $(CH_3)_2C=O + H_2N-$$-CH_3 \; \xrightarrow{H^+}$

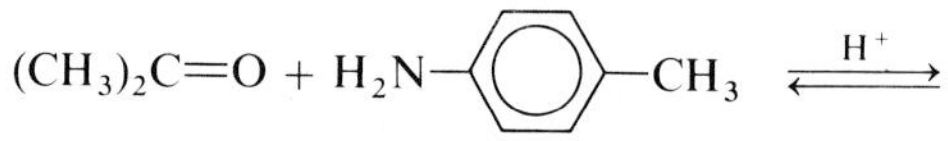

(d) $+ NH_3 \; \xrightarrow{H^+}$

11.14. (a) $=NCH_3$ (b) $-N(CH_3)_2$ (c)

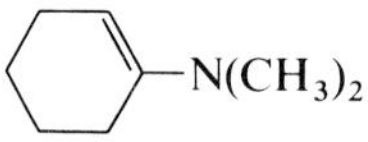

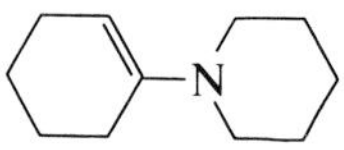

11.15.

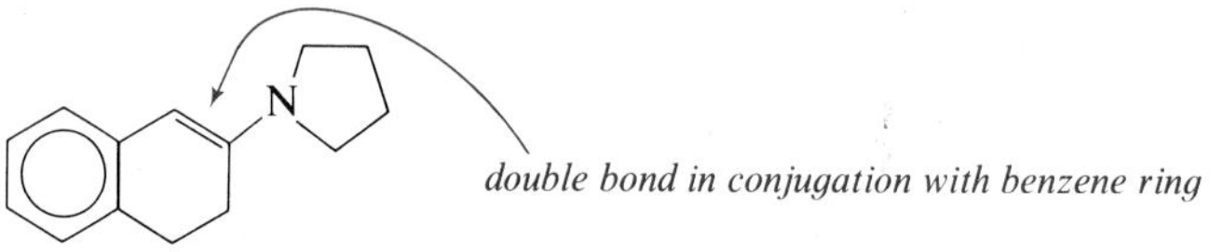

double bond in conjugation with benzene ring

11.16. (a)   (b)   (c)

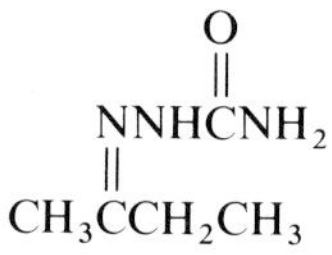

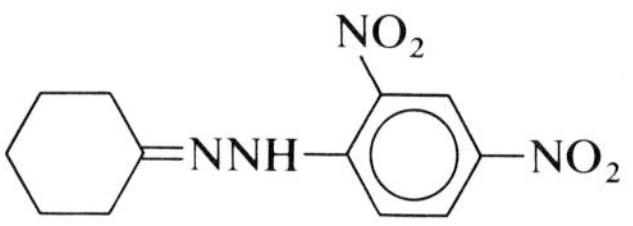

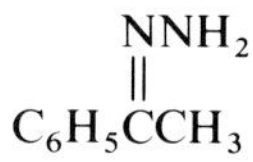

11.17. (a) $=O + (C_6H_5)_3P=CH_2 \longrightarrow$ (b) $-CH_3$

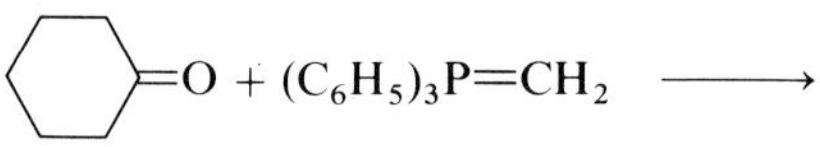

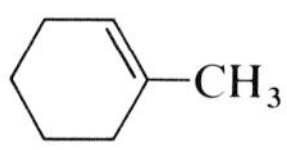

11.18. (a) $C_6H_5CHO, CH_3CH_2CH_2Br$ or $C_6H_5CH_2Br, CH_3CH_2CHO$

(b) $=O$, $Br-$

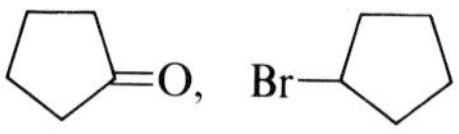

(c) , $BrCH_2CH_2CH_2CH_2Br$ or , $OHCCH_2CH_2CHO$

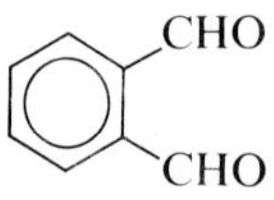

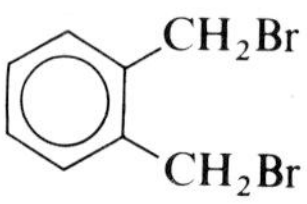

11.19. (a) C_6H_5MgBr; (b) $CH_3CH_2CH_2MgBr$

11.20. (a), (b), and (d) contain acidic protons.

11.21. $CH_3(CH_2)_4CHO$ $\xrightarrow[\text{(2) } H_2O,\ H^+]{\text{(1) } CH_3MgI}$

CH_3CHO $\xrightarrow[\text{(2) } H_2O,\ H^+]{\text{(1) } CH_3(CH_2)_3CH_2MgBr}$ $\longrightarrow$ $CH_3(CH_2)_4\overset{\overset{\displaystyle OH}{|}}{C}HCH_3$

$CH_3(CH_2)_4\overset{\overset{\displaystyle O}{\|}}{C}CH_3$ $\xrightarrow[\text{heat, pressure}]{H_2,\ Ni}$

11.22. (a) CH_3CH_2CHO $\xrightarrow[\text{(2) } H_2O,\ H^+]{\text{(1) } NaBH_4}$

(b) $\xrightarrow[\text{(2) } H_2O,\ H^+]{\text{(1) } NaBH_4}$ **(c)** $\xrightarrow[\text{(2) } H_2O,\ H^+]{\text{(1) } NaBH_4}$

11.23.

$$
\begin{array}{c}
CH_2OH \\
H\!-\!C\!-\!OH \\
HO\!-\!C\!-\!H \\
H\!-\!C\!-\!OH \\
H\!-\!C\!-\!OH \\
CH_2OH
\end{array}
$$

11.24. (a) $C_6H_5\overset{\overset{\displaystyle \ddot{O}:}{\|}}{C}-\overset{H}{C}H-\overset{\overset{\displaystyle :\ddot{O}}{\|}}{C}CH_3$ $\longleftrightarrow$ $C_6H_5\overset{\overset{\displaystyle \ddot{O}:}{|}}{C}-CH=\overset{\overset{\displaystyle :\ddot{O}:^-}{|}}{C}CH_3$ $\longleftrightarrow$

$C_6H_5\overset{\overset{\displaystyle :\ddot{O}:^-}{|}}{C}=CH-\overset{\overset{\displaystyle :\ddot{O}}{\|}}{C}CH_3$

(b) $CH_3\overset{\overset{\displaystyle \ddot{O}:}{\|}}{C}-CH-\overset{\overset{\displaystyle :\ddot{O}}{\|}}{C}CH_3$ $\longleftrightarrow$ $CH_3\overset{\overset{\displaystyle \ddot{O}:}{|}}{C}-CH=\overset{\overset{\displaystyle :\ddot{O}:^-}{|}}{C}CH_3$ $\longleftrightarrow$

$CH_3\overset{\overset{\displaystyle :\ddot{O}:^-}{|}}{C}=CH\overset{\overset{\displaystyle :\ddot{O}}{\|}}{C}CH_3$

11.25. (a) tautomers; **(b)** resonance structures; **(c)** tautomers;
(d) resonance structures

11.26. (a) $CH_3\overset{\overset{\displaystyle OH}{|}}{C}=CH\overset{\overset{\displaystyle O}{\|}}{C}H$ $\rightleftharpoons$ $CH_3\overset{\overset{\displaystyle O}{\|}}{C}CH_2\overset{\overset{\displaystyle O}{\|}}{C}H$ $\rightleftharpoons$ $CH_3\overset{\overset{\displaystyle O}{\|}}{C}CH=\overset{\overset{\displaystyle OH}{|}}{C}H$

(b) $CH_3\overset{\overset{\displaystyle OH}{|}}{\underset{\underset{\displaystyle CH_3}{|}}{C}}=C\overset{\overset{\displaystyle O}{\|}}{C}CH_3$ $\rightleftharpoons$ $CH_3\overset{\overset{\displaystyle O}{\|}}{C}\underset{\underset{\displaystyle CH_3}{|}}{C}H\overset{\overset{\displaystyle O}{\|}}{C}CH_3$ $\rightleftharpoons$ $CH_3\overset{\overset{\displaystyle O}{\|}}{C}\underset{\underset{\displaystyle CH_3}{|}}{C}=\overset{\overset{\displaystyle OH}{|}}{C}CH_3$

(c) $CH_3\overset{\overset{\displaystyle OH}{|}}{C}$ $\rightleftharpoons$ $CH_3\overset{\overset{\displaystyle O}{\|}}{C}$ $\rightleftharpoons$ $CH_3\overset{\overset{\displaystyle O}{\|}}{C}$

11.27.

$$\underset{\substack{|\\ \text{C}=\text{O}\\ |\\ \text{CH}_2\text{OH}}}{\text{CH}_2\text{OPO}_3\text{H}^-} \rightleftharpoons \left[\underset{\substack{|\\ \text{COH}\\ \|\\ \text{CHOH}}}{\text{CH}_2\text{OPO}_3\text{H}^-}\right] \rightleftharpoons \underset{\substack{|\\ \text{HCOH}\\ |\\ \text{CHO}}}{\text{CH}_2\text{OPO}_3\text{H}^-}$$

intermediate enediol

11.28. Ketone and OH$^-$, second-order kinetics

11.29. (1) $\underset{\substack{\|\\ \text{O}}}{\text{CH}_3\overset{\text{O}}{\text{C}}\text{CH}_2\text{Br}} + \text{OH}^- \underset{-\text{H}_2\text{O}}{\rightleftharpoons} \left[\text{CH}_3\overset{\text{O}}{\underset{\|}{\text{C}}}-\bar{\text{C}}\text{HBr} \longleftrightarrow \text{CH}_3\overset{\text{O}^-}{\underset{|}{\text{C}}}=\text{CHBr}\right]$

stabilized

(2) $\text{CH}_3\overset{\text{O}}{\underset{\|}{\text{C}}}\overset{\frown}{\text{C}}\text{HBr} + \text{Br}\overset{\frown}{-}\text{Br} \longrightarrow \text{CH}_3\overset{\text{O}}{\underset{\|}{\text{C}}}\text{CHBr}_2 + \text{Br}^-$

The anion in step (1) is stabilized by electron withdrawal by the electronegative Br atom.

11.30. (a) $\text{CH}_3\overset{\text{+OH}}{\underset{\|}{\text{C}}}\text{CH}_3$

(b) The rates would be the same because neither Br_2 nor I_2 is involved in the rate-determining step.

11.31. (a), (c)

11.32. (a) (b) $(\text{CH}_3)_2\text{CH}\overset{\text{O}}{\underset{\|}{\text{C}}}\text{CH}_3$ (c) $\text{CH}_3\overset{\text{O}}{\underset{\|}{\text{C}}}-\text{⟨⟩}-\overset{\text{O}}{\underset{\|}{\text{C}}}\text{CH}_3$

11.33. (a) (b) (c) (d)

11.34. (a) $\text{CH}_3\overset{\text{O}}{\underset{\|}{\text{C}}}\text{CH}_3 \xrightarrow[(2)\ \text{H}_2\text{O, H}^+]{(1)\ \text{C}_6\text{H}_5\text{MgBr}} \text{CH}_3\overset{\text{OH}}{\underset{\underset{\text{C}_6\text{H}_5}{|}}{\overset{|}{\text{C}}}}\text{CH}_3 \xrightarrow[\text{heat}]{\text{H}^+} \text{product}$

(b) $\xrightarrow{\text{HCN}}$ $\xrightarrow{\text{H}_2\text{O, H}^+}$ product

(c) $\text{CH}_3(\text{CH}_2)_4\text{CHO} \xrightarrow[(2)\ \text{H}_2\text{O, H}^+]{(1)\ \text{⟨⟩—MgBr}} \text{CH}_3(\text{CH}_2)_4\overset{\text{OH}}{\underset{|}{\text{CH}}}-\text{⟨⟩} \xrightarrow[\text{heat}]{\text{H}_2\text{CrO}_4}$

$\text{CH}_3(\text{CH}_2)_4\overset{\text{O}}{\underset{\|}{\text{C}}}-\text{⟨⟩} \xrightarrow{(\text{C}_6\text{H}_5)_3\text{P}=\text{CH}_2} \text{product}$

CHAPTER 12

12.1. (a) propenoic acid; (b) pentanedioic acid; (c) 2-bromopropanoic acid

12.2. (a) α-bromopropionic acid; (b) β-hydroxypropionic acid

12.3. (a)

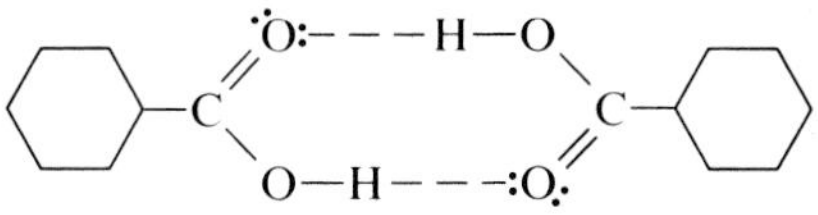

(b)

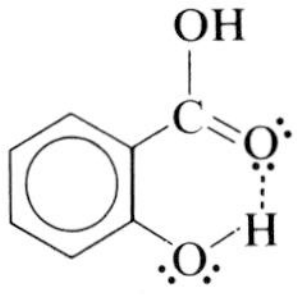

and 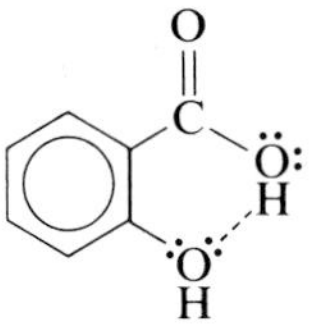 plus hydrogen bonds between molecules of *o*-hydroxybenzoic acid

(c) 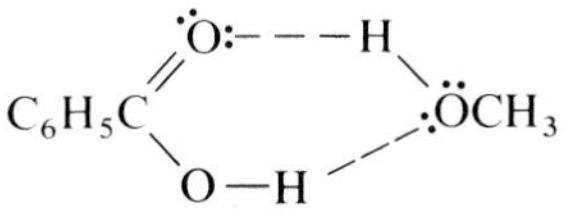 C_6H_5C plus hydrogen bonds within and between benzoic acid molecules and between methanol molecules.

12.4. (a) $HOCH_2CH_2Cl \xrightarrow{KCN} HOCH_2CH_2CN \xrightarrow{H_2O,\ H^+} HOCH_2CH_2CO_2H$

(b) $(CH_3)_3CCH_2Cl \xrightarrow[ether]{Mg} (CH_3)_3CCH_2MgCl \xrightarrow[(2)\ H_2O,\ H^+]{(1)\ CO_2}$

$(CH_3)_3CCH_2CO_2H$

(c)

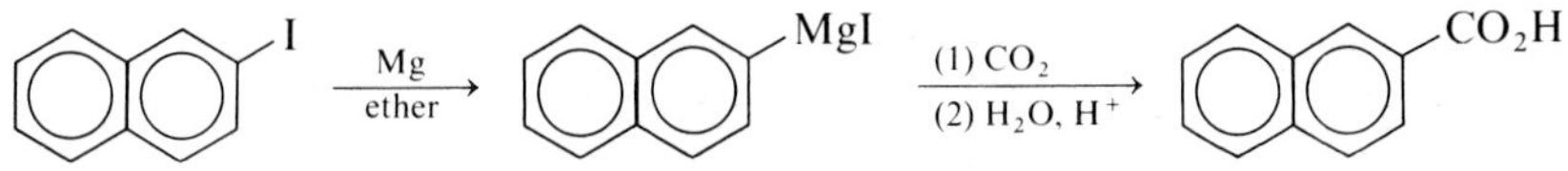

(d) $C_6H_5CH_2Br \xrightarrow{KCN} C_6H_5CH_2CN \xrightarrow{H_2O,\ H^+} C_6H_5CH_2CO_2H$

or $C_6H_5CH_2Br \xrightarrow[ether]{Mg} C_6H_5CH_2MgBr \xrightarrow[(2)\ H_2O,\ H^+]{(1)\ CO_2} C_6H_5CH_2CO_2H$

12.5. Dissolve the aldehyde in diethyl ether, extract the ether solution with dilute aqueous $NaHCO_3$ to remove the acid, and then evaporate the ether to recover the aldehyde.

12.6. Extract with aqueous $NaHCO_3$, which removes the acid. Extract with aqueous NaOH, which removes the naphthol. The octanol remains in the ether. The acid and phenol may be regenerated by acidification.

12.7. 76.5 **12.8.** (a) 59.0; (b) 64.0

12.9. (a) bromoacetic acid; (b) dibromoacetic acid; (c) 2-iodopropanoic acid

12.10. (a) weaker; (b) 3,5-dinitrobenzoic acid

12.11.

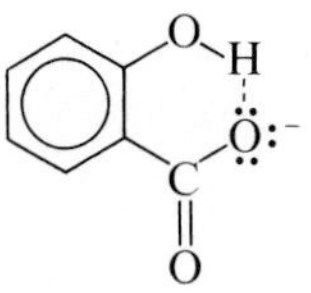

stabilized by hydrogen bonding 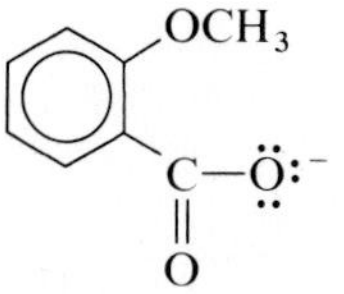   *not stabilized by hydrogen bonding*

12.12. (d), (c), (a), (b)

12.13.

$$CH_3\overset{O}{\overset{||}{C}}OH \;\underset{}{\overset{H^+}{\rightleftharpoons}}\; \left[CH_3\overset{{}^+OH}{\overset{||}{C}}OH \longleftrightarrow CH_3\overset{OH}{\overset{|}{C}}\overset{+}{O}H \right] \;\overset{CH_3{}^{18}OH}{\rightleftharpoons}\;$$

$$\left[\begin{matrix} OH \\ | \\ CH_3COH \\ | \\ CH_3{}^{18}\overset{+}{O}H \end{matrix} \right] \;\underset{then\,+H^+}{\overset{-H^+,}{\rightleftharpoons}}\; \left[\begin{matrix} OH \\ | \\ CH_3C\!-\!\overset{+}{\underset{..}{O}}H_2 \\ | \\ {}^{18}OCH_3 \end{matrix} \right] \;\overset{-H_2O}{\rightleftharpoons}\;$$

$$\left[\begin{matrix} :\overset{..}{O}H \\ | \\ CH_3C^+ \\ | \\ {}^{18}OCH_3 \end{matrix} \longleftrightarrow \begin{matrix} {}^+OH \\ || \\ CH_3C \\ | \\ {}^{18}OCH_3 \end{matrix} \right] \;\overset{-H^+}{\rightleftharpoons}\; CH_3\overset{O}{\overset{||}{C}}{}^{18}OCH_3$$

12.14. (a) CH_3—⬡—$CO_2CH(CH_3)_2$ (b) $CH_3CH_2O_2C$—⬡—$CO_2CH_2CH_3$

(c) $(R)\text{-}CH_3CO_2\underset{\underset{CH_3}{|}}{C}HCH_2CH_3$

12.15. [lactone structure] =O   **12.16.** (a) [cyclohexene]—CH_2OH (b) HO—[cyclohexane]—CH_2OH

cis and *trans*

12.17. [phthalic anhydride structure]   **12.18.** $\left[\begin{matrix} O & & O \\ || & & || \\ C(CH_2)_4C\!-\!O \end{matrix}\right]_x$

12.19. $HO\overset{O}{\overset{||}{C}}\underset{\underset{CH_2CH_3}{|}}{\underset{CH}{\curvearrowleft}}C\!=\!O \overset{-CO_2}{\longrightarrow} HO\overset{O}{\overset{||}{C}}{=}CHCH_2CH_3 \rightleftharpoons HO\overset{O}{\overset{||}{C}}CH_2CH_2CH_3$

12.20. (a) CH_3CO_2H (b) $CH_3\overset{O}{\overset{||}{C}}CH_2CH_3$ (c) $CH_3CH_2\overset{O}{\overset{||}{C}}CH_2CH_2CH_3$

(d) no reaction

12.21. $CH_3CH_2CH\!=\!CHCO_2^- \;{}^+MgBr + C_6H_6$

12.22. $CH_3CH_2CH_2CO_2H + CH_3OH \;\underset{heat}{\overset{H^+}{\searrow}}$

$CH_3CH_2CH_2CO_2CH_3$

$CH_3CH_2CH_2CO_2H \;\underset{(2)\ CH_3I}{\overset{(1)\ OH^-}{\nearrow}}$

12.23. (a) $CH_3CH_2CO_2H$ $\xrightarrow[\text{(2) H}_2\text{O}]{\text{(1) LiAlH}_4}$ $CH_3CH_2CH_2OH$ $\xrightarrow{\text{HBr}}$ $CH_3CH_2CH_2Br$ $\xrightarrow[\text{ether}]{\text{Mg}}$

$CH_3CH_2CH_2MgBr$ $\xrightarrow[\text{(2) H}_2\text{O, H}^+]{\text{(1) CO}_2}$ product

(b) $ClCH_2CH_2CO_2H$ $\xrightarrow[\text{(2) KCN}]{\text{(1) neutralize with OH}^-}$ $NCCH_2CH_2CO_2^-$ $\xrightarrow{\text{H}_2\text{O, H}^+}$ product

(c) C_6H_5⬡ $\xrightarrow[\text{(2) H}_2\text{O}_2\text{, H}^+]{\text{(1) O}_3}$ $C_6H_5CH{\Large\langle}^{CO_2H}_{CO_2H}$ $\xrightarrow[-\text{CO}_2]{\text{heat}}$ product

---

CHAPTER 13

13.1. Because of both resonance-stabilization and the inductive effect of the more electronegative oxygen, the carbonyl carbon is more positive than the vinyl carbon.

$$CH_3\overset{\overset{\displaystyle\ddot{O}:}{\|}}{C}Cl \quad\longleftrightarrow\quad CH_3\overset{\overset{\displaystyle:\ddot{O}:^-}{|}}{\underset{+}{C}}Cl$$

13.2. $CH_3\overset{\overset{\displaystyle O}{\|}}{O}CCH_2CH_2CN$

13.3. (a) $CH_3CH_2CO_2H + SOCl_2 \longrightarrow CH_3CH_2\overset{\overset{\displaystyle O}{\|}}{C}Cl + HCl + SO_2$

(b) $3\,HO_2C-CO_2H + 2\,PCl_3 \longrightarrow 3\,Cl\overset{\overset{\displaystyle O}{\|}}{C}-\overset{\overset{\displaystyle O}{\|}}{C}Cl + 2\,H_3PO_3$

(c) $3\,C_6H_5CH_2CO_2H + PCl_3 \longrightarrow 3\,C_6H_5CH_2\overset{\overset{\displaystyle O}{\|}}{C}Cl + H_3PO_3$

(*Note:* Either $SOCl_2$ or PCl_3 could be used in each case.)

13.4. $CH_3CH_2CH_2\overset{\overset{\displaystyle:\ddot{O}:}{\|}}{C}Cl + :\ddot{O}H^- \longrightarrow \left[CH_3CH_2CH_2\overset{\overset{\displaystyle:\ddot{O}:^-}{|}}{\underset{\underset{\displaystyle:OH}{|}}{C}}\ddot{C}l:\right] \xrightarrow{-Cl^-}$

$$CH_3CH_2CH_2\overset{\overset{\displaystyle:\ddot{O}}{\|}}{\underset{\underset{\displaystyle OH}{|}}{C}} \xrightarrow{OH^-} CH_3CH_2CH_2CO_2^- + H_2O$$

13.5. Either or both positions in the carboxyl group:

(a) ⬡$-\overset{\overset{\displaystyle O}{\|}}{C}-Cl + H_2{}^{18}O \longrightarrow$ ⬡$-\overset{\overset{\displaystyle O}{\|}}{C}-{}^{18}OH$

(b)

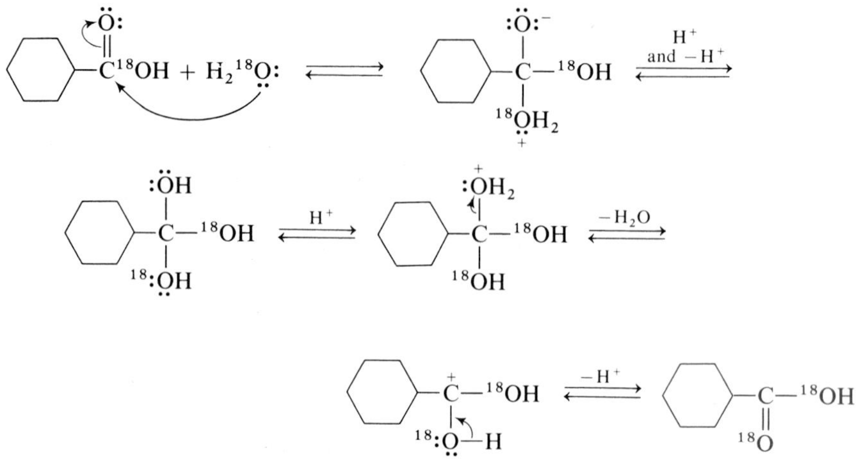

(c)

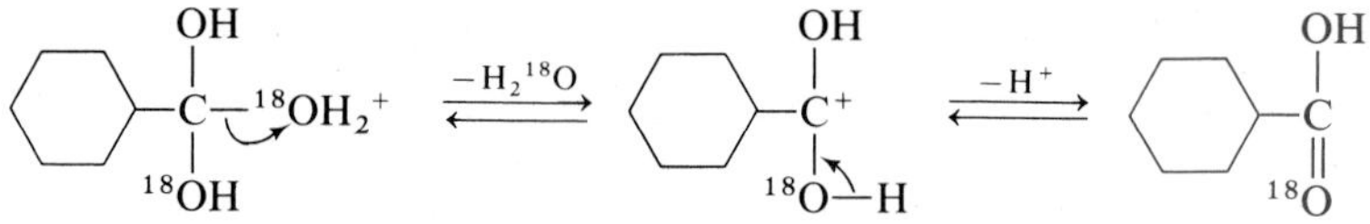

13.6.

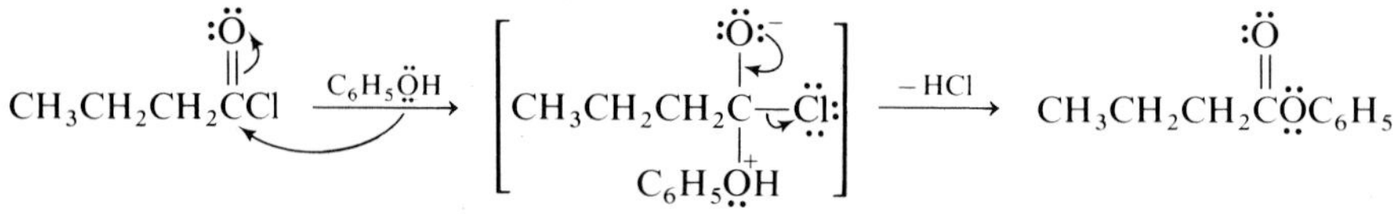

$$CH_3CH_2CH_2CCl \xrightarrow{C_6H_5\ddot{O}H} \left[CH_3CH_2CH_2C\!-\!\ddot{C}l: \right] \xrightarrow{-HCl} CH_3CH_2CH_2C\ddot{O}C_6H_5$$

13.7. $CH_3CO_2^-$ Na^+ + Cl^- + unreacted CH_3NH_2

13.8.

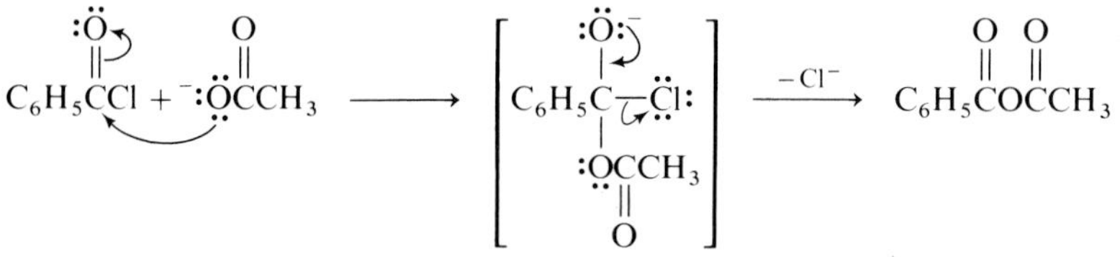

$$C_6H_5CCl + {}^-\!:\!\ddot{O}CCH_3 \longrightarrow \left[C_6H_5C\!-\!\ddot{C}l: \right] \xrightarrow{-Cl^-} C_6H_5COCCH_3$$

13.9. $C_6H_6 \xrightarrow[\text{AlCl}_3]{\overset{O}{\overset{\|}{ClCCH(CH_3)_2}}} C_6H_5CCH(CH_3)_2 \xrightarrow[\text{HCl}]{\text{Zn/Hg}} C_6H_5CH_2CH(CH_3)_2$

13.10. (a) $(CH_3)_2CHCl \xrightarrow[\text{(2) CuI}]{\text{(1) Li}} LiCu[CH(CH_3)_2]_2 \xrightarrow{\overset{O}{\overset{\|}{C_6H_5CCl}}} \text{product}$

$C_6H_5Br \xrightarrow[\text{(2) CdCl}_2]{\text{(1) Mg}} (C_6H_5)_2Cd \xrightarrow{\overset{O}{\overset{\|}{ClCCH(CH_3)_2}}} \text{product}$

(b) CH_3I $\xrightarrow[\text{(2) CuI}]{\text{(1) Li}}$ $LiCu(CH_3)_2$ $\longrightarrow$ product

CH_3I $\xrightarrow[\text{(2) CdCl}_2]{\text{(1) Mg}}$ $(CH_3)_2Cd$ $\longrightarrow$ product

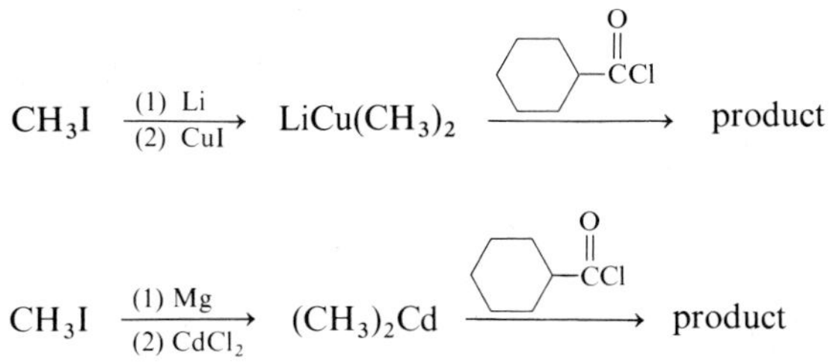

(*Note:* There are other possible combinations of reagents, but remember that secondary cadmium reagents are unstable.)

13.11. (a) $(CH_3)_2CClCO_2H$ (b) $C_6H_5CHBrCO_2H$

13.12.

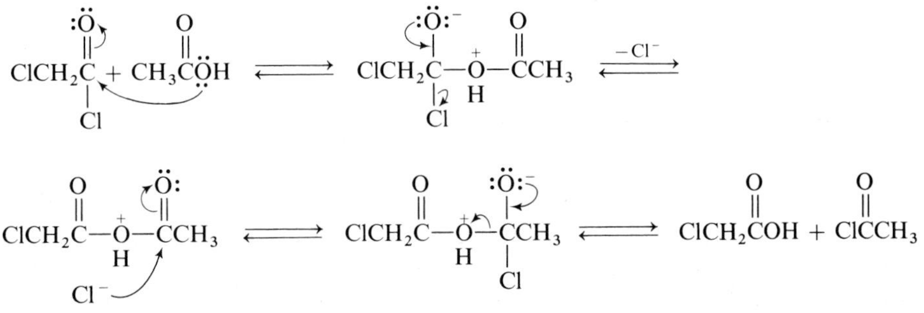

13.13. (a) pentanoic anhydride
(b) acetic benzoic anhydride or benzoic ethanoic anhydride

13.14.

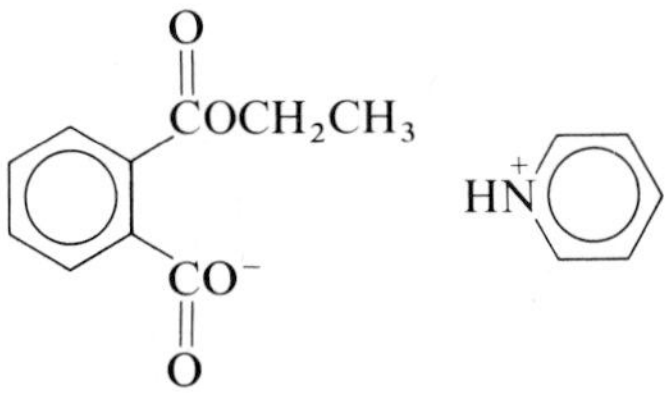

13.15.

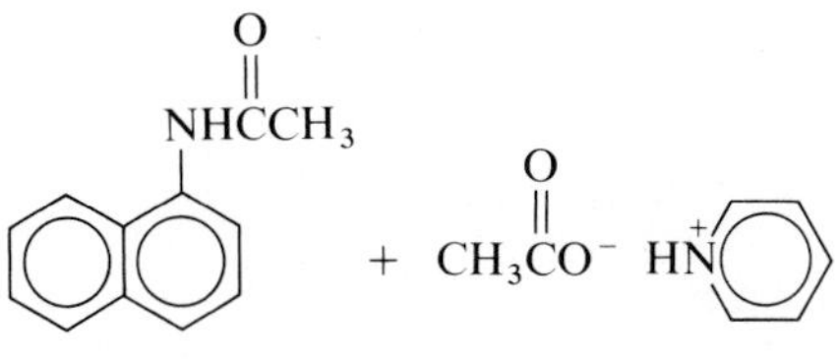

13.16.

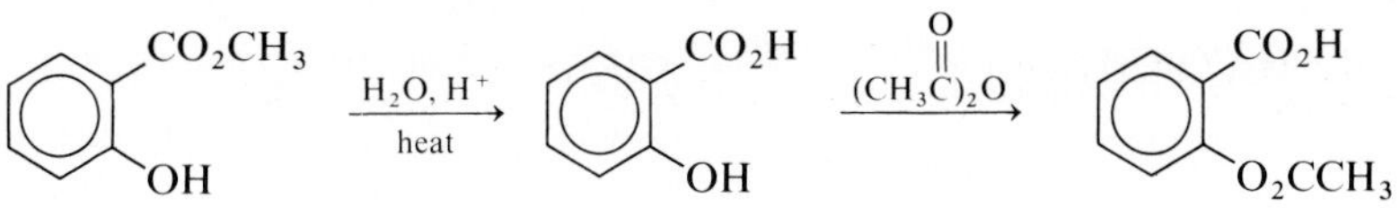

13.17. $CH_3CO_2H + H^{18}OCH_2CH_3$ **13.18.** $CH_3CO_2^- + H^{18}OCH_2CH_3$

13.19. (a) $C_6H_5CO_2CH_2CH_3 + OH^- \longrightarrow C_6H_5CO_2^- + HOCH_2CH_3$

(b) $=O + 2\,OH^- \longrightarrow HOCH_2CH_2OH + CO_3^{2-}$

(c) $+ 2\,OH^- \longrightarrow$ $+ H_2O$

13.20. (a) $\overset{\displaystyle :O:}{\overset{\displaystyle \|}{CH_3COCH_2CH_3}} \xrightleftharpoons{H^+} \left[\overset{\displaystyle \overset{+}{O}H}{\overset{\displaystyle \|}{CH_3COCH_2CH_3}} \longleftrightarrow \overset{\displaystyle :\ddot{O}H}{\underset{+}{CH_3COCH_2CH_3}} \right]$

$\xrightleftharpoons{CH_3OH} \left[\overset{\displaystyle :\ddot{O}H}{\underset{\displaystyle \overset{+}{CH_3\ddot{O}H}}{CH_3\overset{|}{\underset{|}{C}}CH_2CH_3}} \right] \xrightleftharpoons[\text{transfer}]{\text{proton}} \left[\overset{\displaystyle :\ddot{O}-H}{\underset{\displaystyle CH_3\ddot{O}\ \ H}{CH_3\overset{|}{\underset{|}{C}}-\overset{+}{\ddot{O}}CH_2CH_3}} \right]$

$\xrightleftharpoons{-H^+} \overset{\displaystyle O}{\overset{\displaystyle \|}{CH_3COCH_3}} + HOCH_2CH_3$

(b) $\overset{\displaystyle :\ddot{O}}{\overset{\displaystyle \|}{CH_3COCH_2CH_3}} \xrightleftharpoons{^-OCH_3} \left[\overset{\displaystyle :\ddot{O}:^-}{\underset{\displaystyle \overset{|}{O}CH_3}{CH_3\overset{|}{\underset{|}{C}}-\ddot{O}CH_2CH_3}} \right] \rightleftharpoons$

$\overset{\displaystyle \ddot{O}:}{\overset{\displaystyle \|}{CH_3COCH_3}} + {}^-:\ddot{O}CH_2CH_3$

13.21. $\overset{\displaystyle CH_3CHCO_2H}{\underset{\displaystyle OH}{|}} \xrightarrow{SOCl_2} \overset{\displaystyle O}{\overset{\displaystyle \|}{\underset{\displaystyle \underset{Cl}{|}}{CH_3CHCCl}}}$

13.22. $\overset{\displaystyle CH_2O_2C(CH_2)_{14}CH_3}{\underset{\displaystyle CH_2O_2C(CH_2)_{14}CH_3}{\overset{|}{\underset{|}{CHO_2C(CH_2)_{14}CH_3}}}} \xrightarrow[(2)\,H_2O]{(1)\,LiAlH_4} \overset{\displaystyle CH_2OH}{\underset{\displaystyle CH_2OH}{\overset{|}{\underset{|}{CHOH}}}} + 3\,HOCH_2(CH_2)_{14}CH_3$

13.23. (a) $CH_3CH_2CO_2CH_3 \xrightarrow[(2)\,H_2O,\,H^+]{(1)\,2\,CH_3MgI} \overset{\displaystyle OH}{\underset{}{CH_3CH_2\overset{|}{C}(CH_3)_2}}$

(b) $CH_3CH_2CO_2CH_3 \xrightarrow[(2)\,H_2O,\,H^+]{(1)\,2\,CH_3CH_2MgBr} \overset{\displaystyle OH}{\underset{}{CH_3CH_2\overset{|}{C}(CH_2CH_3)_2}}$

13.24. $\overset{\displaystyle O}{\overset{\displaystyle \|}{HCOCH_2CH_3}} \xrightarrow[(2)\,H_2O,\,H^+]{(1)\,2\,CH_3CH_2MgBr} HOCH(CH_2CH_3)_2$

13.25.

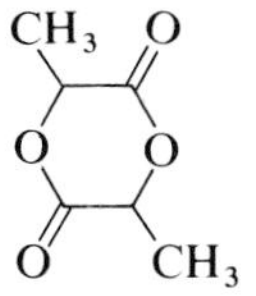

13.26. (a) $\xrightarrow{OH^-}$ x $H_2NCHCO_2^-$ (b) $\xrightarrow{H_2O,\ H^+}$ x $H_3\overset{+}{N}CHCO_2H$
$\qquad\qquad\qquad\qquad\quad$ | $\qquad\qquad\qquad\qquad\qquad\qquad$ |
$\qquad\qquad\qquad\qquad\quad$ CH$_3$ $\qquad\qquad\qquad\qquad\qquad\qquad$ CH$_3$

13.27. (a)

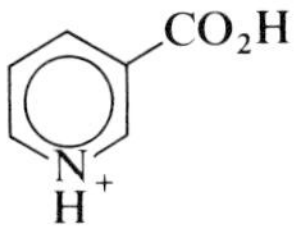

$+$ NH$_4^+$ (b) 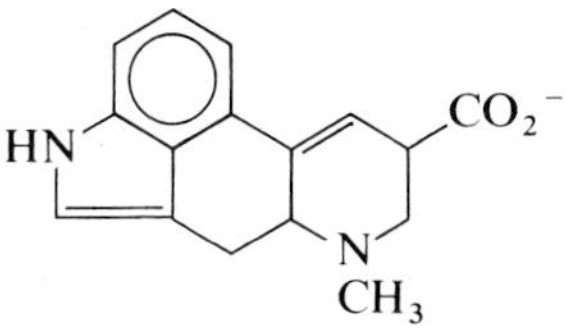 $+$ HN(CH$_2$CH$_3$)$_2$

(c) $\ ^-O_2C(CH_2)_3NHCH_3$

13.28. (a)

(b)

13.29.

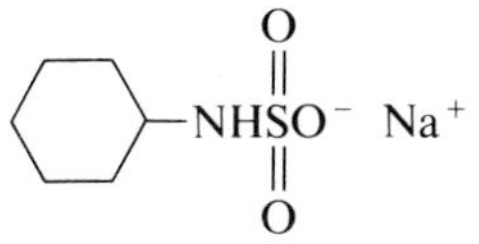

13.30. (a) CH$_3$CH$_2$CN; (b) CH$_3$CH$_2$CH$_2$CN

13.31. (a) $-$CO$_2$H $\xrightarrow{SOCl_2}$ $-\overset{\overset{O}{\|}}{C}Cl$ $\xrightarrow{2\ (CH_3)_2NH}$ product

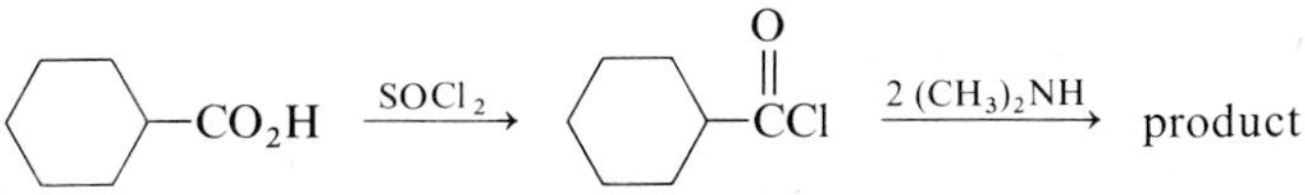

(b) (CH$_3$)$_2$CHCH$_2$CO$_2$CH$_3$ $\xrightarrow[\text{(2) } H_2O,\ H^+]{\text{(1) } 2\,CH_3CH_2CH_2MgBr}$ product

(c) CH$_3$CH$_2$CH$_2$OH $\xrightarrow{H_2CrO_4}$ CH$_3$CH$_2$CO$_2$H $\xrightarrow{SOCl_2}$ CH$_3$CH$_2\overset{\overset{O}{\|}}{C}Cl$

CH$_3$CH$_2$CH$_2$OH $\xrightarrow{HBr}$ CH$_3$CH$_2$CH$_2$Br $\xrightarrow{KCN}$ CH$_3$CH$_2$CH$_2$CN

$\xrightarrow[\text{(2) } H_2O]{\text{(1) } LiAlH_4}$ CH$_3$CH$_2$CH$_2$CH$_2$NH$_2$ $\xrightarrow{CH_3CH_2\overset{\overset{O}{\|}}{C}Cl}$ product

CHAPTER 14

14.1. (a)

14.2. (a) $CH_3CH_2CHO + CH_3CH_2O^- \rightleftharpoons CH_3\overset{-}{C}HCHO + CH_3CH_2OH$

(b) $(CH_3)_2CHCO_2CH_2CH_3 + CH_3CH_2O^- \rightleftharpoons$

$(CH_3)_2\overset{-}{C}CO_2CH_2CH_3 + CH_3CH_2OH$

(c) no reaction

(d) 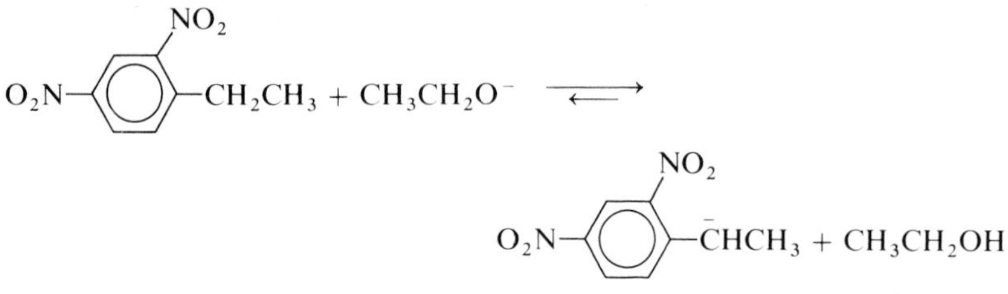 O_2N-⟨ring, NO$_2$⟩$-CH_2CH_3 + CH_3CH_2O^- \rightleftharpoons$

O_2N-⟨ring, NO$_2$⟩$-\overset{-}{C}HCH_3 + CH_3CH_2OH$

(e) $C_6H_5CH_2NO_2 + CH_3CH_2O^- \rightleftharpoons C_6H_5\overset{-}{C}HNO_2 + CH_3CH_2OH$

(f) $CH_2(CO_2H)_2 + 2\,CH_3CH_2O^- \rightleftharpoons CH_2(CO_2^-)_2 + 2\,CH_3CH_2OH$

14.3. Transesterification, as well as enolate formation:

$$CH_2(CO_2C_2H_5)_2 + 2\,CH_3O^- \rightleftharpoons CH_2(CO_2CH_3)_2 + 2\,C_2H_5O^-$$

14.4. (a) $CH_3CH_2CH_2CH(CO_2C_2H_5)_2$ (b) $CH_3CH_2\overset{\overset{\displaystyle CH_3}{\displaystyle |}}{C}(CO_2C_2H_5)_2$

14.5. (a) $CH_3CH_2CH_2CH(CO_2C_2H_5)_2 \xrightarrow[\text{heat}]{H_2O,\ OH^-}$

$CH_3CH_2CH_2CH(CO_2^-)_2 \xrightarrow{H^+} CH_3CH_2CH_2CH(CO_2H)_2$

(b) $(CH_3)_2\overset{*}{C}(CO_2C_2H_5)_2 \xrightarrow[\substack{\text{heat} \\ -CO_2}]{H_2O,\ H^+} (CH_3)_2CHCO_2H$

14.6.

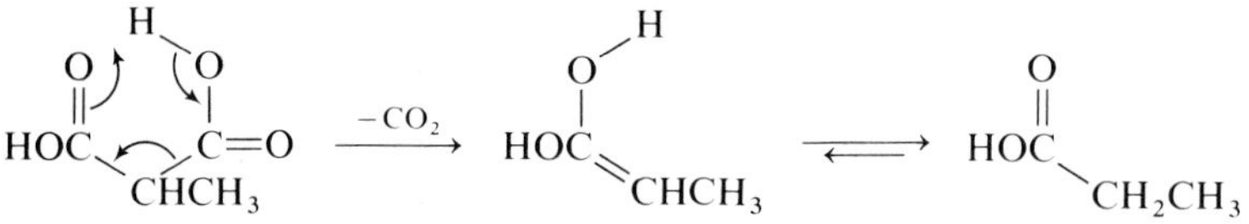

an enol

14.7. $CH_3\overset{\overset{\displaystyle O}{\displaystyle ||}}{C}\overset{-}{C}HCO_2C_2H_5\ Na^+;$ $CH_3\overset{\overset{\displaystyle O}{\displaystyle ||}}{C}\underset{\underset{\displaystyle CH_2CH_2-⟨ring⟩}{\displaystyle |}}{C}HCO_2C_2H_5$

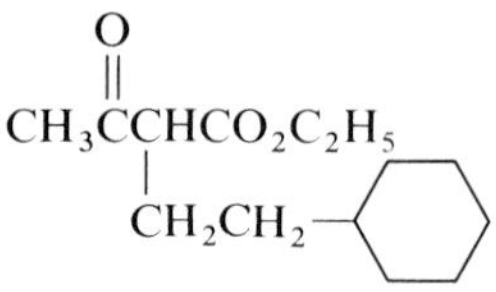

14.8. (a) $CH_3\overset{\overset{\displaystyle O}{\displaystyle ||}}{C}CH_2\overset{\overset{\displaystyle O}{\displaystyle ||}}{C}OC_2H_5 \xrightarrow[(2)\ CH_2=CHCH_2Br]{(1)\ NaOC_2H_5}$

(b) $CH_3\overset{O}{\overset{\|}{C}}CH_2\overset{O}{\overset{\|}{C}}OC_2H_5$ $\xrightarrow[\text{(2) } NaOC_2H_5,\text{ then } CH_3CH_2Br]{\text{(1) } NaOC_2H_5,\text{ then } CH_3CH_2Br}$ $CH_3\overset{O}{\overset{\|}{C}}\underset{\underset{CO_2C_2H_5}{|}}{C}(CH_2CH_3)_2$ $\xrightarrow[\text{heat}]{H_2O,\ H^+}$

14.9. (a) $CH_2(CO_2C_2H_5)_2$ $\xrightarrow[\text{(2) } CH_3(CH_2)_3Br]{\text{(1) } NaOC_2H_5}$ $CH_3(CH_2)_3CH(CO_2C_2H_5)_2$ $\xrightarrow[\text{(2) } H^+,\text{ cold}]{\text{(1) } OH^-,\ H_2O,\text{ heat}}$

(b) $CH_2(CO_2C_2H_5)_2$ $\xrightarrow[\text{(2) } NaOC_2H_5,\text{ then } CH_3I]{\text{(1) } NaOC_2H_5,\text{ then } CH_3I}$ $(CH_3)_2C(CO_2C_2H_5)_2$ $\xrightarrow[\text{heat}]{H_2O,\ H^+}$

(c) $CH_3\overset{O}{\overset{\|}{C}}CH_2CO_2C_2H_5$ $\xrightarrow[\text{(2) } CH_3CH_2CH_2Br]{\text{(1) } NaOC_2H_5}$ $CH_3\overset{O}{\overset{\|}{C}}\underset{\underset{CH_2CH_2CH_3}{|}}{C}HCO_2C_2H_5$ $\xrightarrow[\text{(2) } CH_3I]{\text{(1) } NaOC_2H_5}$

$CH_3\overset{O}{\overset{\|}{C}}-\overset{CH_3}{\underset{\underset{CH_2CH_2CH_3}{|}}{\overset{|}{C}}}CO_2C_2H_5$ $\xrightarrow[\text{heat}]{H_2O,\ H^+}$

(d) $CH_2(CO_2C_2H_5)_2$ $\xrightarrow[\text{(2) } BrCH_2CO_2C_2H_5]{\text{(1) } NaOC_2H_5}$

(e) $CH_2(CN)_2$ $\xrightarrow[\text{(2) } Br(CH_2)_4Br]{\text{(1) } NaOC_2H_5}$ $Br(CH_2)_4CH(CN)_2$ $\xrightarrow{NaOC_2H_5}$

$BrCH_2(CH_2)_3\overset{..}{C}(CN)_2$ $\longrightarrow$

(f) $\xrightarrow[\text{(2) } (CH_3)_2CHCH_2Br]{\text{(1) } NaOC_2H_5}$

14.10. $\left[\begin{array}{cc} CH_3\overset{O}{\overset{\|}{C}} & CH_3\overset{O}{\overset{\|}{C}} \\ (CH_3)_2\overset{|}{C}CH=\overset{+}{N} & (CH_3)_2\overset{|}{C}CH-\overset{..}{N}{}^+ \end{array}\right]$ $\xrightarrow{H_2O}$

$CH_3\overset{O}{\overset{\|}{C}}$
$(CH_3)_2\overset{|}{C}CH-\overset{..}{N}$
$\underset{+}{\overset{|}{O}H_2}$
$\underset{\xrightarrow[\text{transfer}]{\text{proton}}}{\rightleftharpoons}$
$CH_3\overset{O}{\overset{\|}{C}}$
$(CH_3)_2\overset{|}{C}CH-\overset{+}{N}H$
$\overset{|}{O}-H$
$\xrightarrow[\text{proton transfer}]{\text{elimination and}}$

$CH_3\overset{O}{\overset{\|}{C}}$
$(CH_3)_2\overset{|}{\underset{\underset{O}{\|}}{C}}CH + H_2\overset{+}{N}$

14.11. A cyclic amine has less steric hindrance around the nitrogen than an open-chain amine.

14.12.

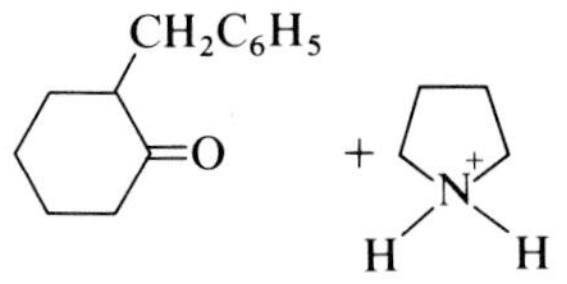

14.13. (a)

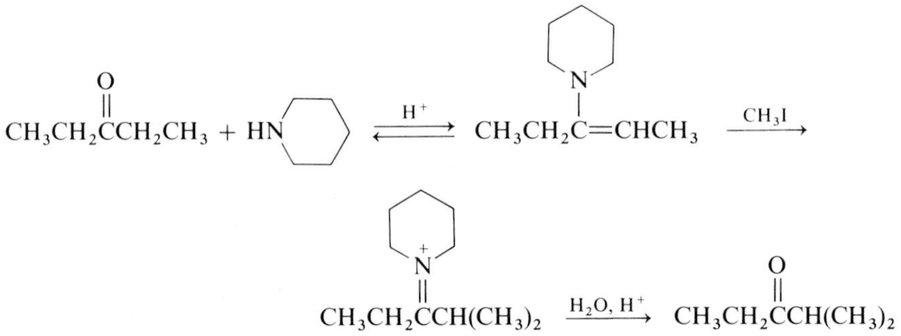

(b)

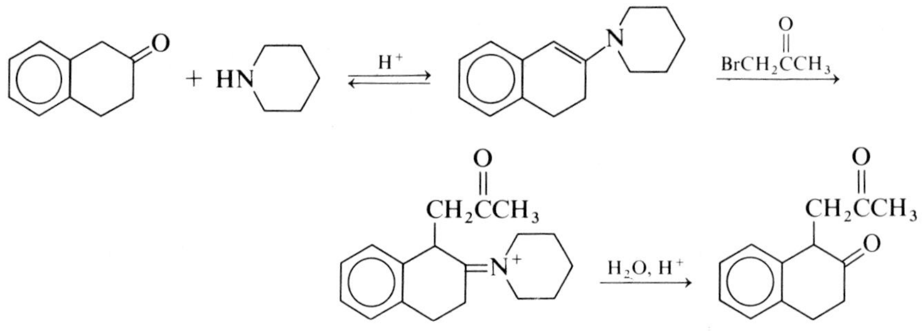

14.14. (c) and (e) because they contain α hydrogens

14.15. (a)

(b)

(c)

14.16. (a) $C_6H_5CH=CCHO$ with CH_2CH_3 substituent

(b)

14.17. (a) $C_6H_5CHO + CH_3CH_2CHO$ $\xrightarrow{OH^-}$

(b)

(c) —CHO + CH₃CHO $\xrightarrow{\text{OH}^-}$

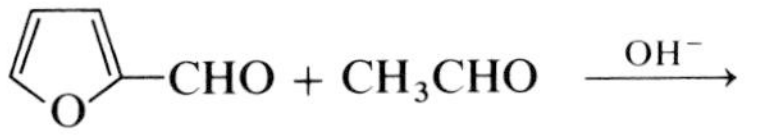

(d) OHC(CH₂)₅CHO $\xrightarrow{\text{OH}^-}$

14.18. (a) (CH₃CH₂)₂C=O + CH₂CN $\xrightarrow[\text{heat}]{\text{NH}_4\text{O}_2\text{CCH}_3 \atop \text{CH}_3\text{CO}_2\text{H}}$
    |
    CO₂C₂H₅

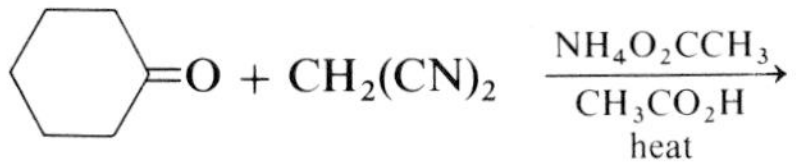

(b) =O + CH₂(CN)₂ $\xrightarrow[\text{heat}]{\text{NH}_4\text{O}_2\text{CCH}_3 \atop \text{CH}_3\text{CO}_2\text{H}}$

14.19. (a)
$$\underset{\underset{\text{C}_6\text{H}_5}{|}}{\text{C}_6\text{H}_5\text{CH}_2\overset{\overset{\text{O}}{||}}{\text{C}}\text{CHCO}_2\text{C}_2\text{H}_5}$$
(b) $\text{C}_6\text{H}_5\text{CH}_2\overset{\overset{\text{O}}{||}}{\text{C}}\text{CH}_2\text{C}_6\text{H}_5$

14.20. (a)

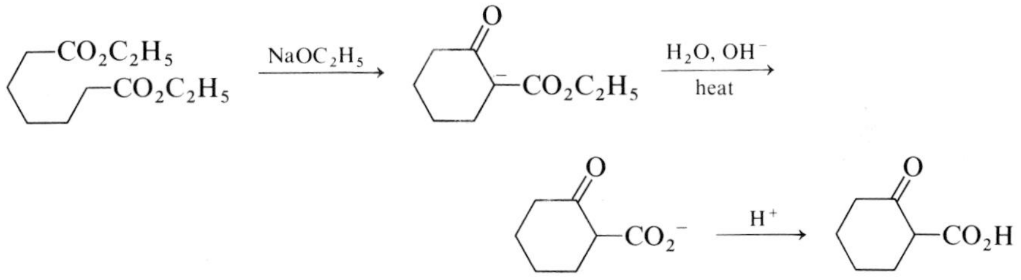

(b)

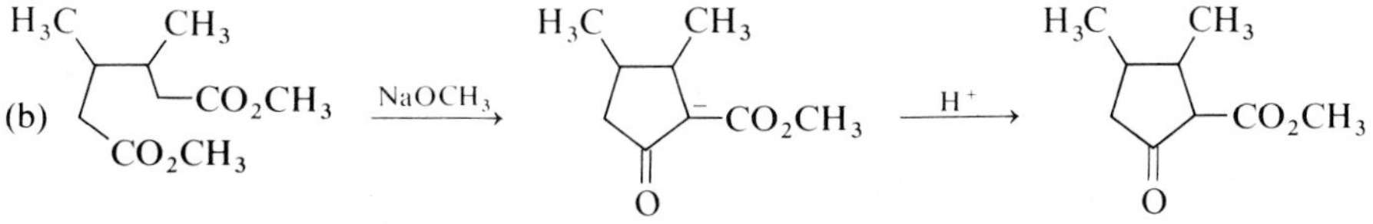

14.21. (a) $\text{C}_2\text{H}_5\text{O}_2\text{C}\overset{\overset{\text{O}}{||}}{\text{C}}\text{CH}_2\text{CO}_2\text{C}_2\text{H}_5$ (b)

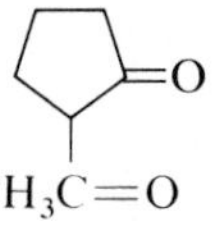

14.22. (a)

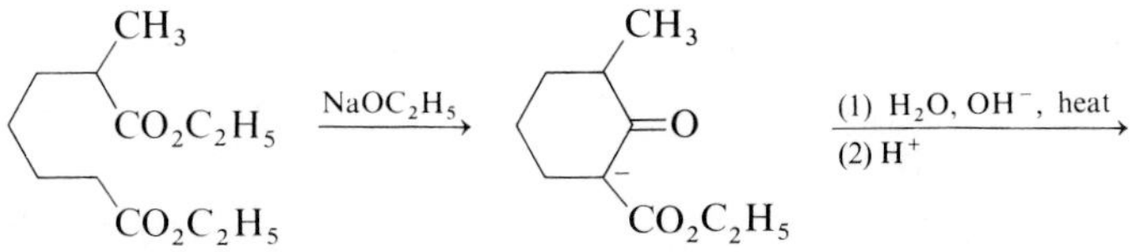

(b) C₆H₅CO₂C₂H₅ + CH₃CO₂C₂H₅ $\xrightarrow{\text{NaOC}_2\text{H}_5}$ $\text{C}_6\text{H}_5\overset{\overset{\text{O}}{||}}{\text{C}}\overset{-}{\text{C}}\text{HCO}_2\text{C}_2\text{H}_5$ $\xrightarrow[\text{(2) H}^+]{\text{(1) H}_2\text{O, OH}^-,\text{ heat}}$

(c) C₆H₅CO₂C₂H₅ + CH₂(CN)₂ $\xrightarrow[\text{(2) H}^+]{\text{(1) NaOC}_2\text{H}_5}$

(d) (CH₃)₃CCO₂C₂H₅ + CH₃$\overset{\overset{\text{O}}{||}}{\text{C}}$CH₃ $\xrightarrow[\text{(2) H}^+]{\text{(1) NaOC}_2\text{H}_5}$

14.23.

$$2 \overset{\overset{\text{O}}{\|}}{CH_3CSCoA} \xrightarrow{\hspace{1cm}} \overset{\overset{\text{O}}{\|}\hspace{0.5cm}\overset{\text{O}}{\|}}{CH_3CCH_2CSCoA} + HSCoA$$

↖ *acidic*

14.24.

14.25. (a) $CH_3CH(CO_2C_2H_5)_2 + CH_3CH_2CH{=}CHCO_2C_2H_5 \xrightarrow{NaOC_2H_5}$

$$\underset{\overset{\displaystyle |}{CH_3C(CO_2C_2H_5)_2}}{CH_3CH_2CHCH_2CO_2C_2H_5} \xrightarrow[\text{(2) H}^+]{\text{(1) H}_2O,\ OH^-,\ heat}$$

(b) $CH_2(CO_2C_2H_5)_2 + CH_3CH{=}CHCO_2C_2H_5 \xrightarrow[\text{(2) H}_2O,\ H^+]{\text{(1) NaOC}_2H_5}$

$$\underset{\overset{\displaystyle |}{CH(CO_2C_2H_5)_2}}{CH_3CHCH_2CO_2C_2H_5} \xrightarrow[\text{heat}]{H_2O,\ H^+}$$

(c)

$+\ CH_2(CO_2C_2H_5)_2 \xrightarrow[\text{(2) H}_2O,\ H^+]{\text{(1) NaOC}_2H_5}$ $\xrightarrow[\text{heat}]{H_2O,\ H^+}$

CHAPTER 15

15.1. (a) 1°; (b) 2°; (c) salt of 3°; (d) 1°; (e) quaternary; (f) 3°

15.2. (a) 1,4-butanediamine; (b) 1,6-hexanediamine;
(c) (2-methylpropyl)amine, or isobutylamine;
(d) *N,N*-dimethylethylamine; (e) *p*-chloro-*N*-methylaniline

15.3. (c) contains a chiral N, while (d) contains a chiral C. Each exists as a pair of enantiomers.

15.4. (a) $(CH_3)_2NH{-}{-}{-}:N(CH_3)_2$
 $\phantom{(CH_3)_2NH{-}{-}{-}:N}\overset{|}{H}$

(b) $(CH_3)_2NH{-}{-}{-}:N(CH_3)_2,\quad (CH_3)_2NH{-}{-}{-}:\overset{..}{O}H_2,\quad (CH_3)_2\overset{..}{N}{:}{-}{-}{-}H_2O,$
 $\phantom{(CH_3)_2NH{-}{-}{-}:N(CH_3)_2,\quad (CH_3)_2NH{-}{-}{-}:\overset{..}{O}H_2,\quad (CH_3)_2N}\overset{|}{H}$
 $\phantom{(CH_3)_2NH{-}{-}{-}:N(CH_3)_2,\quad (CH_3)_2NH{-}{-}{-}:\overset{..}{O}H_2,\quad (CH_3)}H_2\overset{..}{O}{:}{-}{-}{-}H_2O$

15.5.

15.6. (a) (1) $CH_3CH_2CH_2Br$, (2) H_2O, OH^-
(b) (1) $CH_2{=}CHCH_2Br$, (2) H_2O, OH^-
(c) (1) $C_6H_5CH_2Br$, (2) H_2O, OH^-

15.7.

$\xrightarrow{\text{BrCH(CO}_2\text{C}_2\text{H}_5)_2}$

$\xrightarrow{\text{NaOC}_2\text{H}_5}$

$\xrightarrow{\text{(CH}_3)_2\text{CHCH}_2\text{Br}}$

$\xrightarrow[\text{heat}]{\text{H}_2\text{O, H}^+}$ product

15.8. (a)

$\xrightarrow[\text{heat, pressure}]{\text{H}_2, \text{Ni, NH}_3}$

(b) $\text{BrCH}_2\text{CH}_2\text{CH}=\text{CH}_2 \xrightarrow{\text{KCN}} \text{NCCH}_2\text{CH}_2\text{CH}=\text{CH}_2 \xrightarrow[\text{(2) H}_2\text{O}]{\text{(1) LiAlH}_4}$ product

(c) $\text{C}_6\text{H}_5\text{CO}_2\text{H} \xrightarrow{\text{SOCl}_2} \text{C}_6\text{H}_5\overset{\overset{\displaystyle O}{\|}}{\text{C}}\text{Cl} \xrightarrow{2(\text{CH}_3)_2\text{NH}} \text{C}_6\text{H}_5\overset{\overset{\displaystyle O}{\|}}{\text{C}}\text{N(CH}_3)_2$

$\xrightarrow[\text{(2) H}_2\text{O}]{\text{(1) LiAlH}_4} \text{C}_6\text{H}_5\text{CH}_2\text{N(CH}_3)_2$

15.9. (a) $\text{CH}_3\text{CH}_2\overset{\overset{\displaystyle O}{\|}}{\text{C}}\text{CH}_3 \xrightarrow[\text{heat, pressure}]{\text{H}_2, \text{Ni, NH}_3} \text{CH}_3\text{CH}_2\overset{\overset{\displaystyle NH_2}{|}}{\text{C}}\text{HCH}_3 \longleftarrow$

$\xrightarrow[\text{heat}]{\text{H}_2\text{O, OH}^-}$

(b)

$+ \text{CH}_3\text{CH}_2\overset{\overset{\displaystyle Br}{|}}{\text{C}}\text{HCH}_3 \longrightarrow$

15.10. (a) $(R)\text{-C}_6\text{H}_5\text{CH}_2\overset{\overset{\displaystyle CH_3}{|}}{\text{C}}\text{HNH}_2$

(b) $\text{H}_2\text{NCH}_2\text{CH}_2\text{CH}_2\text{NH}_2$

15.11. (a) piperidine;   (b) piperazine

15.12. There is no conjugation between the nitrogen of benzylamine and the aromatic ring because the nitrogen is not attached to an sp^2 carbon, but to an sp^3 carbon.

15.13. *p*-Methylaniline is a stronger base than aniline because the methyl group is *electron-releasing*. *p*-Nitroaniline is a weaker base than aniline because the nitro group is *electron-withdrawing*.

15.14.

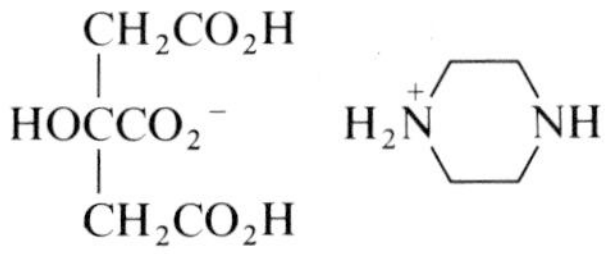

$$\underset{\overset{|}{CH_2CO_2H}}{\overset{CH_2CO_2H}{HOCCO_2^-}}$$

$$H_2\overset{+}{N}NH$$

15.15. $CH_3CH_2CH_2CH_2NH_2$

$$CH_3CH_2CH_2CH_2\overset{+}{-}N\equiv N$$

$\xrightarrow{-N_2} CH_3CH_2CH_2CH_2Cl$

$\xrightarrow{-N_2} CH_3CH_2CH_2CH_2\overset{+}{OH_2} \xrightarrow{-H^+} CH_3CH_2CH_2CH_2OH$

Cl⁻ or H_2O

$$CH_3CH_2CHCH_2\overset{+}{-}N\equiv N$$

$\xrightarrow{-N_2}$

$CH_3CH_2\overset{+}{C}HCH_3$

$\xrightarrow{-H^+} CH_3CH=CHCH_3 + CH_3CH_2CH=CH_2$

$\xrightarrow{Cl^-} CH_3CH_2CHClCH_3$

$\xrightarrow[-H^+]{H_2O} \underset{\overset{|}{OH}}{CH_3CH_2CHCH_3}$

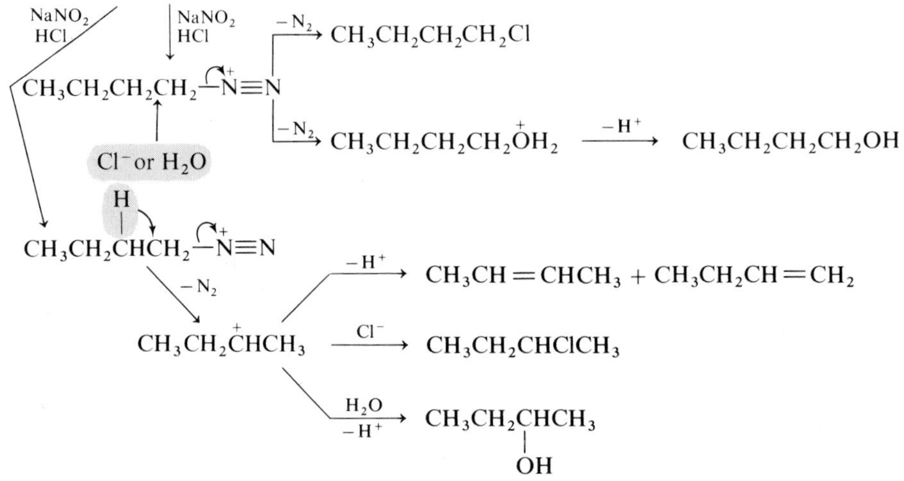

15.16. The benzenediazonium ion is resonance-stabilized.

15.17. (a) $CH_3CH_2CH_2CH_2N(CH_3)_2 + CH_2{=}CH_2 + H_2O$

(b) $C_6H_5CH{=}CH_2 + (CH_3)_2NCH_2CH_3 + H_2O$

15.18. (a) 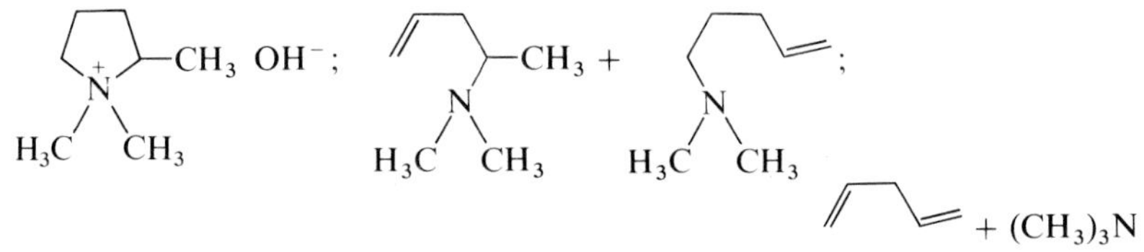 [structure] $-CH_3$ OH⁻; [structure] $-CH_3$ + [structure] ;

[structure] $+ (CH_3)_3N$

(b) 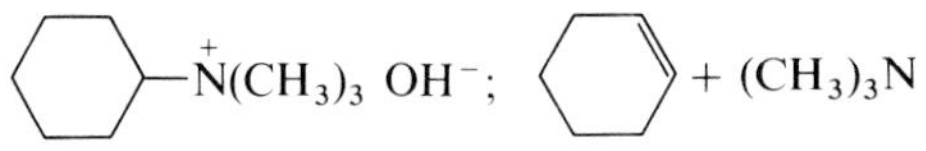 [structure] $-\overset{+}{N}(CH_3)_3$ OH⁻; [structure] $+ (CH_3)_3N$

15.19. [structure]

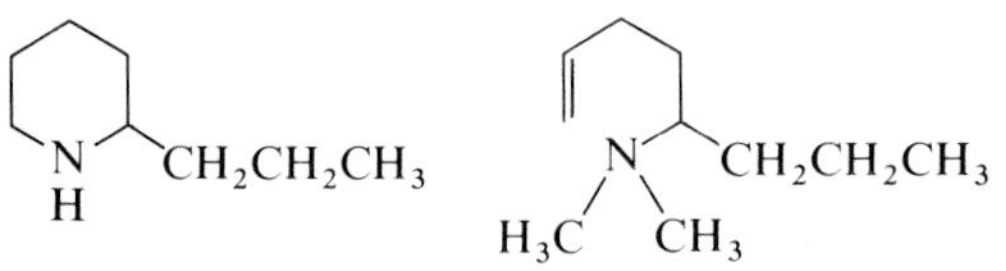

15.20. (a) 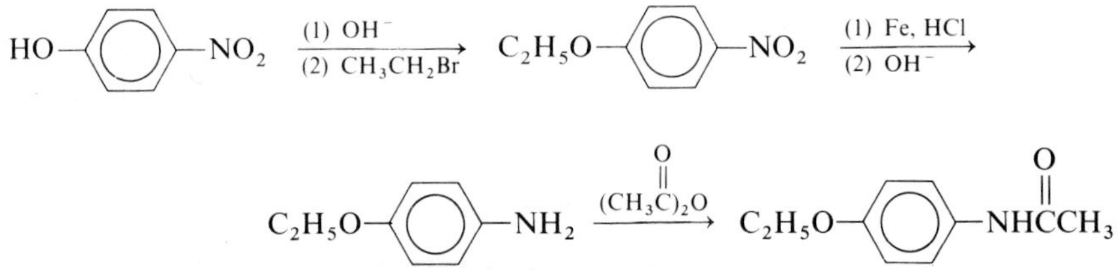 $HO-\!\!\bigcirc\!\!-NO_2 \xrightarrow[(2)\ CH_3CH_2Br]{(1)\ OH^-} C_2H_5O-\!\!\bigcirc\!\!-NO_2 \xrightarrow[(2)\ OH^-]{(1)\ Fe,\ HCl}$

$C_2H_5O-\!\!\bigcirc\!\!-NH_2 \xrightarrow{(CH_3\overset{\overset{O}{\|}}{C})_2O} C_2H_5O-\!\!\bigcirc\!\!-NH\overset{\overset{O}{\|}}{C}CH_3$

(b) 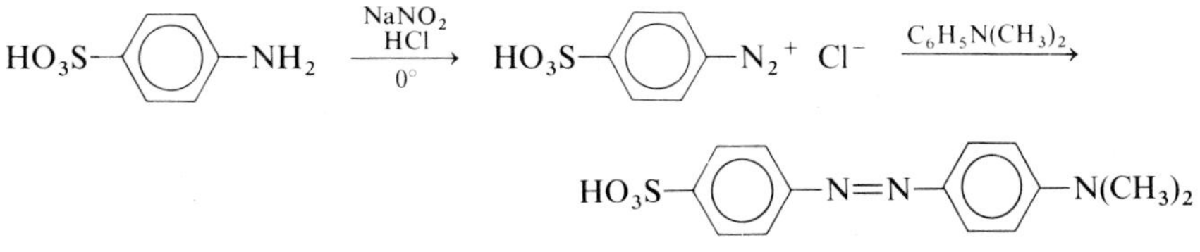 HO_3S—⟨benzene ring⟩—NH_2 $\xrightarrow[\substack{0°}]{\substack{NaNO_2 \\ HCl}}$ HO_3S—⟨benzene ring⟩—$N_2{}^+$ Cl^- $\xrightarrow{C_6H_5N(CH_3)_2}$

HO_3S—⟨benzene ring⟩—$N{=}N$—⟨benzene ring⟩—$N(CH_3)_2$

(c) ⟨cyclohexyl⟩—Br $\xrightarrow[\substack{(3)\ H_2O,\ H^+}]{\substack{(1)\ Mg,\ ether \\ (2)\ CO_2}}$ ⟨cyclohexyl⟩—CO_2H $\xrightarrow{SOCl_2}$ ⟨cyclohexyl⟩—$\overset{\overset{\displaystyle O}{\|}}{C}Cl$

⟨cyclohexanone⟩=O $\xrightarrow{\text{⟨piperidine, NH⟩}}$ ⟨enamine⟩—N⟨piperidine⟩ $\xrightarrow[\substack{(2)\ H_2O,\ H^+}]{\substack{(1)\ \text{⟨cyclohexyl⟩}-CCl}}$ product

---

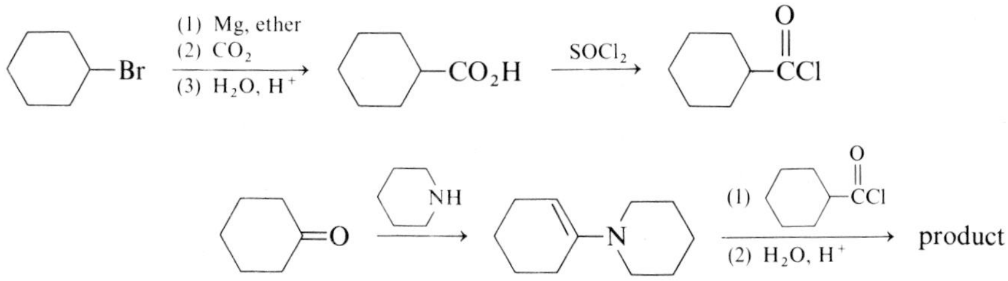

CHAPTER 16

16.1.

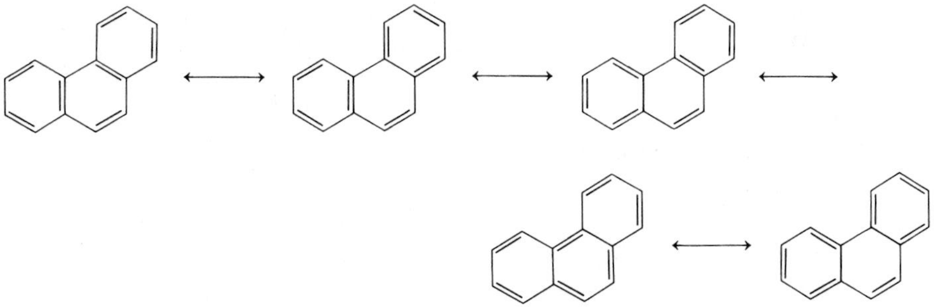

Four of the five resonance structures show a C9–C10 double bond.

16.2.   electrophilic attack

16.3.

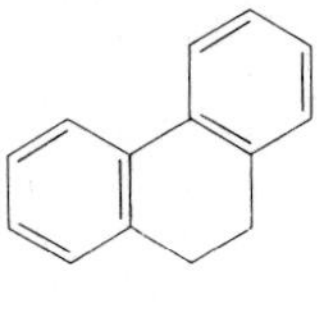

16.4.   (a)

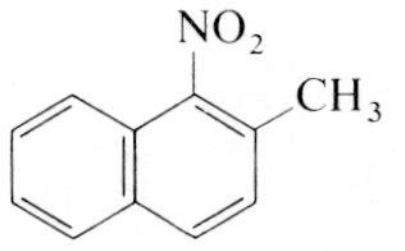

(b)

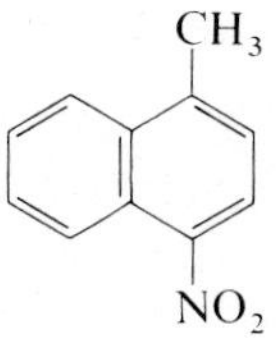

(c)

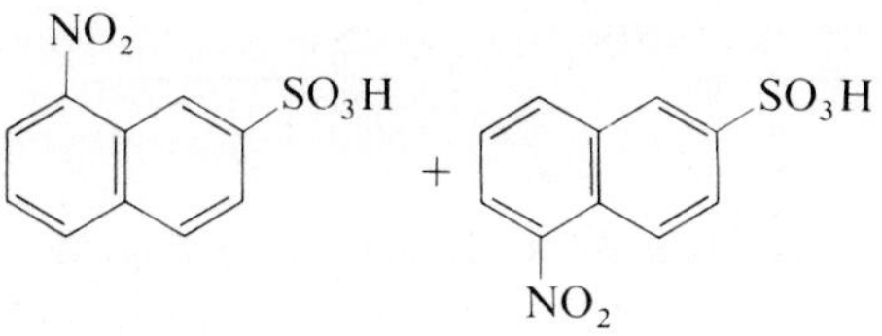

$+$

(d)

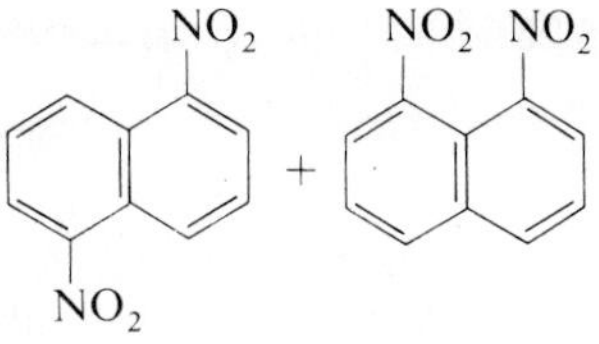

$+$

16.5. (a)

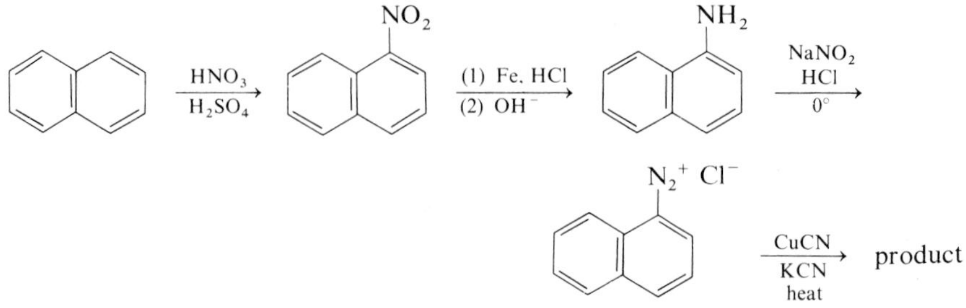

(b)

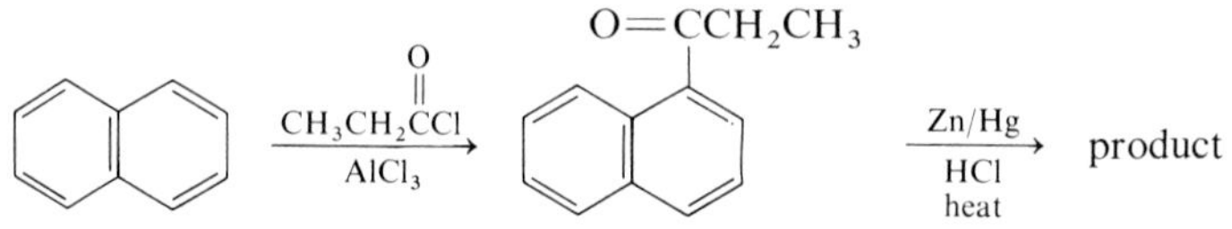

(c)

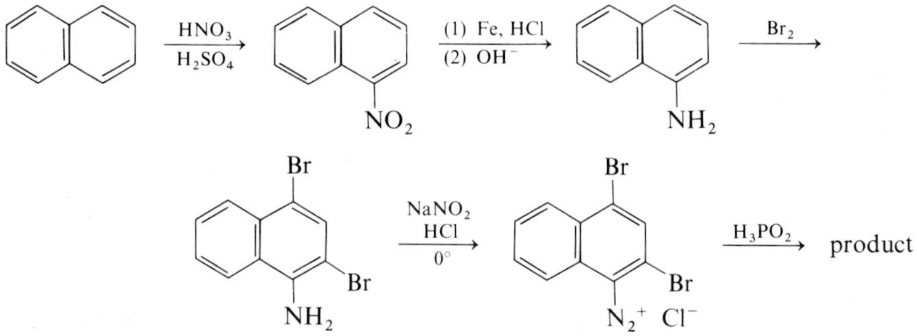

16.6.

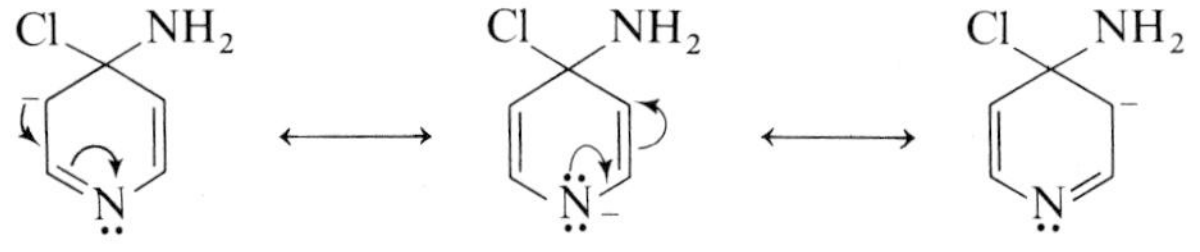

16.7.

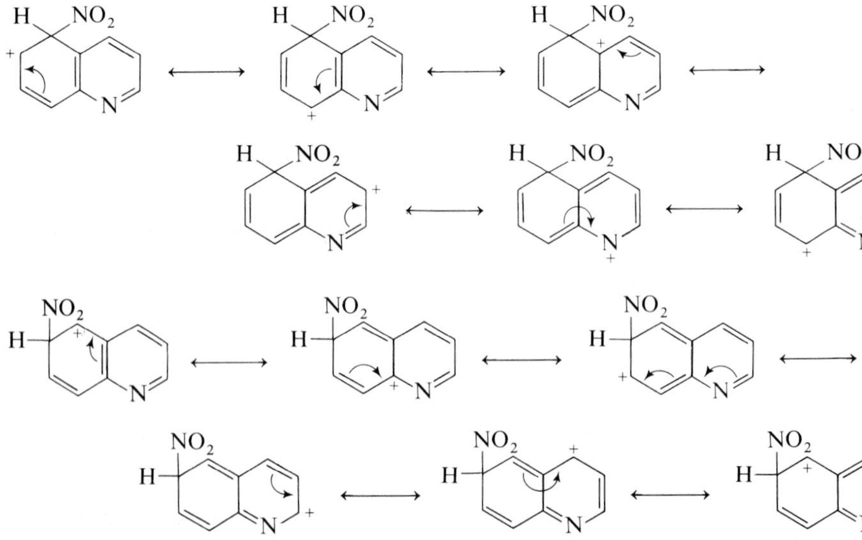

The 5-intermediate has three resonance structures with one aromatic ring, while the 6-intermediate has only two.

16.8. 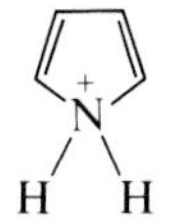   or sp^3 not aromatic

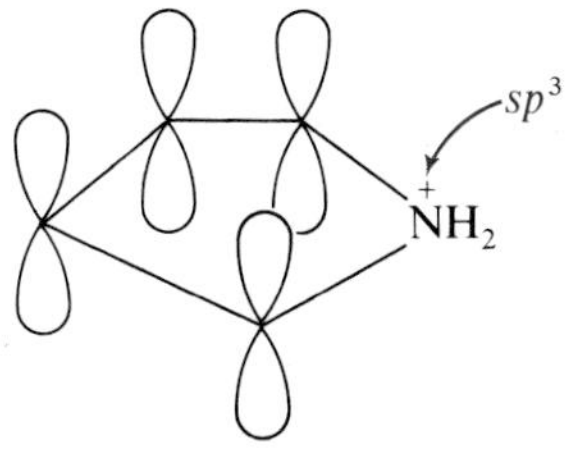

16.9. (a) 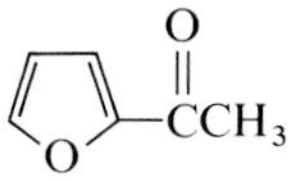 (b) 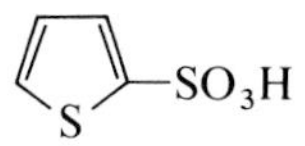 —SO_3H

(c) 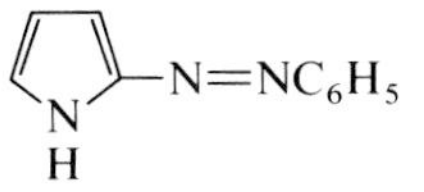 —$N{=}NC_6H_5$ (d) 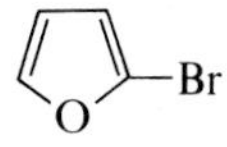 —Br

16.10.   + HO_2CCHCH_2OH

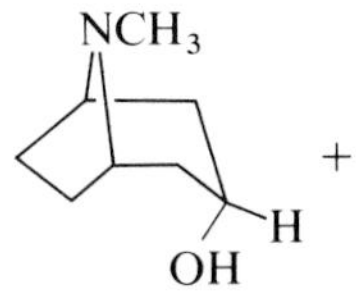

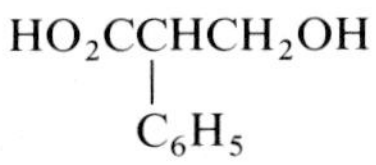

tropine   tropic acid

16.11. Compound C can be converted to Compound F, without the need for blocking the amino group, by the following scheme:

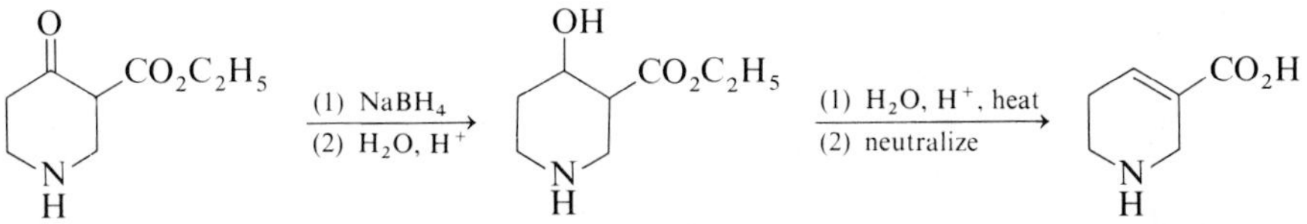

$\xrightarrow[\text{(2) H}_2\text{O, H}^+]{\text{(1) NaBH}_4}$

$\xrightarrow[\text{(2) neutralize}]{\text{(1) H}_2\text{O, H}^+\text{, heat}}$

16.12.

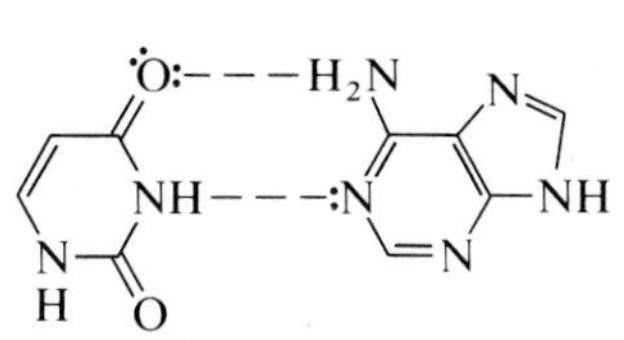

CHAPTER 17

17.1.

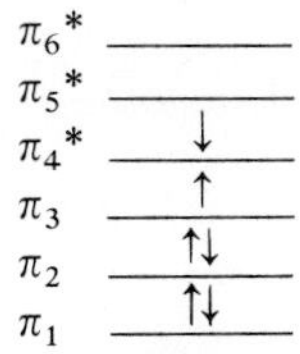

17.2. The double bonds in 1,4-pentadiene are not conjugated; thus, their π orbitals are independent of each other.

first C=C: π_2^* ⎯⎯⎯ π_1 ⎯↕⎯   second C=C: π_2^* ⎯⎯⎯ π_1 ⎯↕⎯

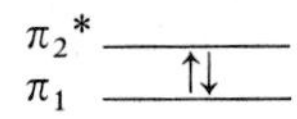

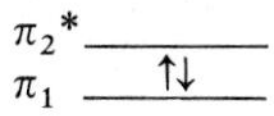

17.3.   [6 + 4]

17.4.   (a) $2\ C_6H_5CH{=}CHCO_2H$ $\xrightarrow{\ hv\ }$ (b) $\begin{array}{l} CH_2CH{=}CH_2 \\ | \\ CH_2CH{=}CH_2 \end{array}$ $\xrightarrow{\ hv\ }$

In (a), either *cis*- or *trans*-cinnamic acid (3-phenylpropenoic acid) could be used. Other stereoisomers (and structural isomers) of the product might be obtained as by-products.

17.5.   A photo-induced [4 + 2] cycloaddition cannot occur when *either* the dienophile or the diene is excited.

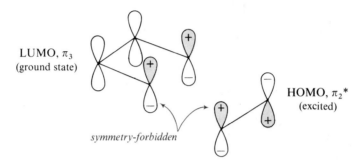

LUMO, π_3
(ground state)

HOMO, $\pi_2{}^*$
(excited)

symmetry-forbidden

17.6.   A [6 + 4] cycloaddition reaction is thermally induced because this reaction path is symmetry-allowed.

π_3, HOMO of
the triene

$\pi_3{}^*$, LUMO of
the diene

17.7.

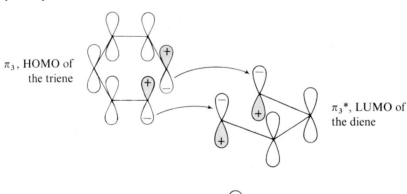

symmetry-forbidden

17.8.   (a)

CH$_3$

""CH$_3$

trans

(b)

CH$_2$CH$_3$

H

CH$_2$CH$_3$

H

(3*E*, 5*Z*, 7*Z*)

17.9.   (a) [3,3];   (b) [1,5]

17.10. (a) can proceed by a [1,5] sigmatropic shift, and (b) is a [3,3] sigmatropic shift. Both proceed readily.

17.11. Previtamin D contains a $(4n + 2)\, \pi$ system. The thermal reaction proceeds by disrotatory motion, and the photo-induced reaction proceeds by conrotatory motion.

thermal:

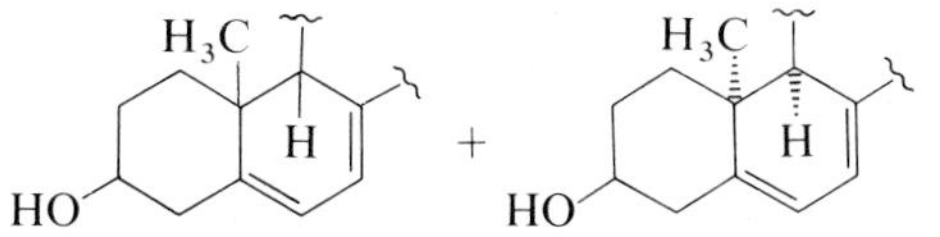

photochemical:

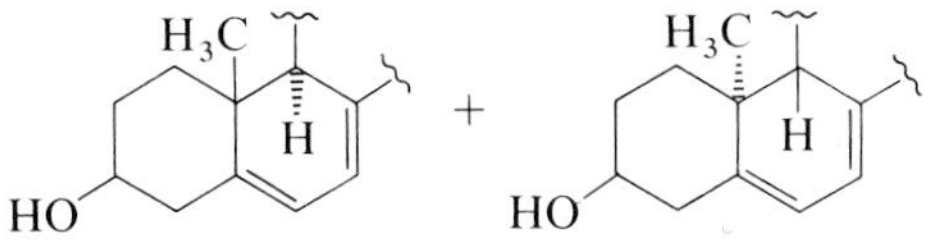

CHAPTER 18

18.1. (a) aldotetrose; (b) pentulose; (c) aldohexose

18.2. (a) (R) and D; (b) (S) and L; (c) $(3S,4R)$ and D

18.3. L-(+)- and *meso*-tartaric acid

18.4. (a) four; (b) four; (c) three

18.5. (a)

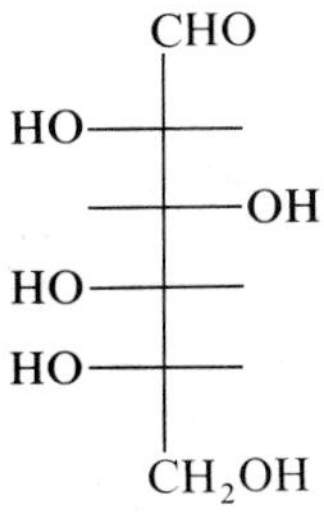

(b)

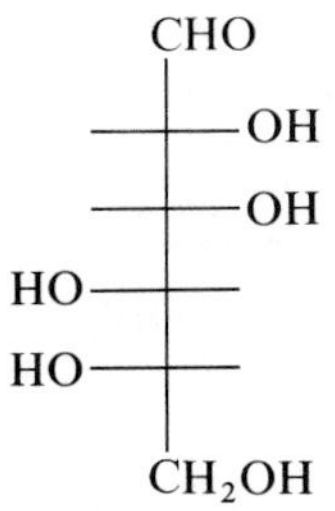

18.6. L-idose

18.7.

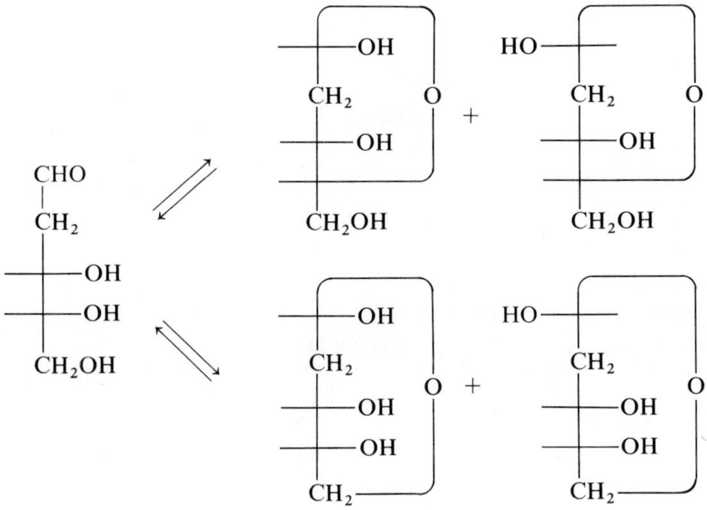

18.8. α: 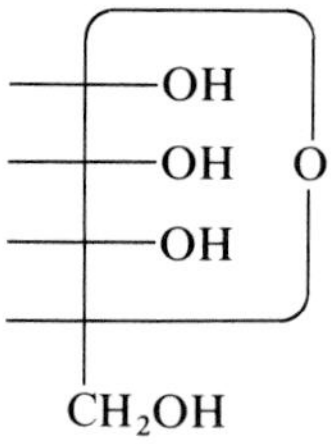 β:

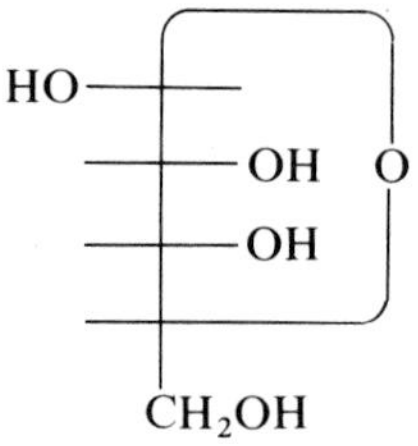

18.9.

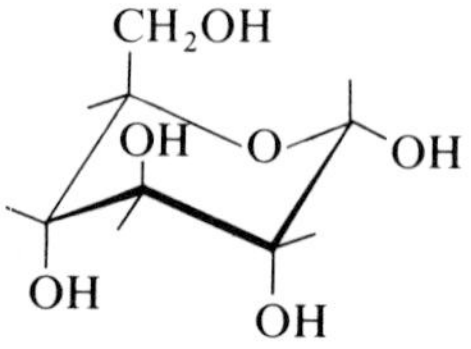

18.10. α:

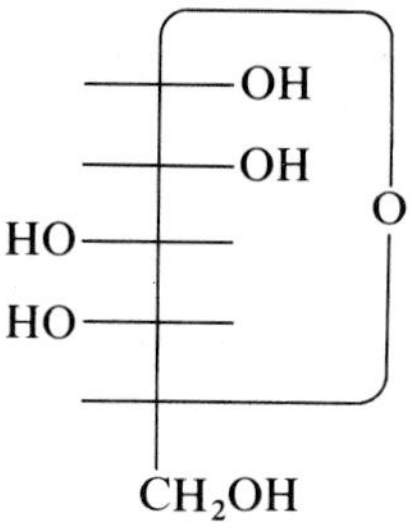

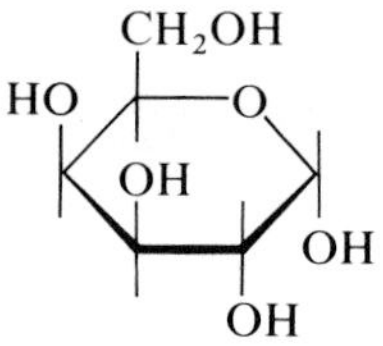

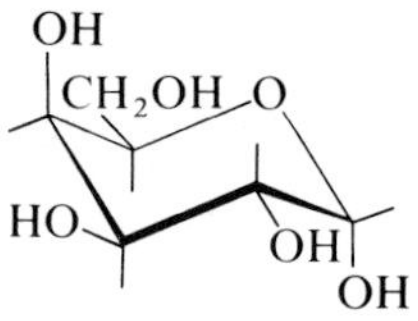

β:

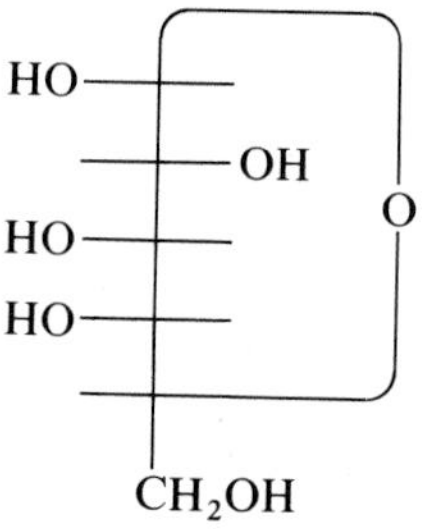

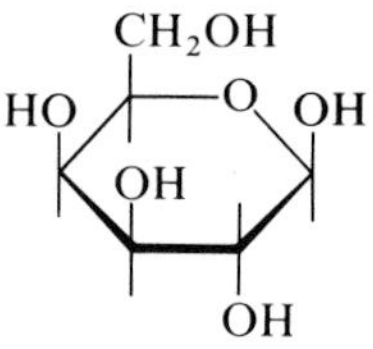

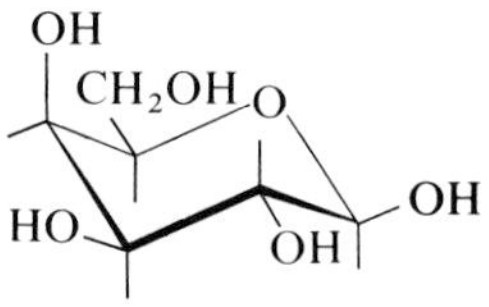

18.11. (a) 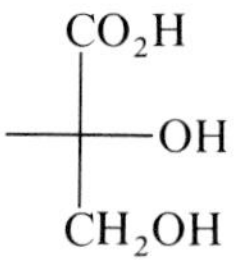 (b) 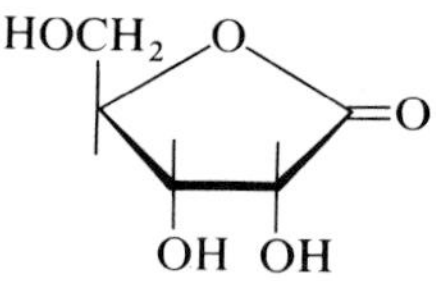 (c) none

(d)

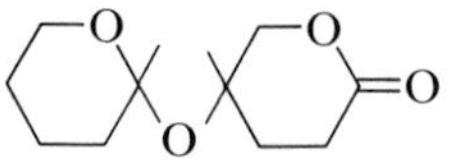

18.12. allose and galactose

18.13. (a) 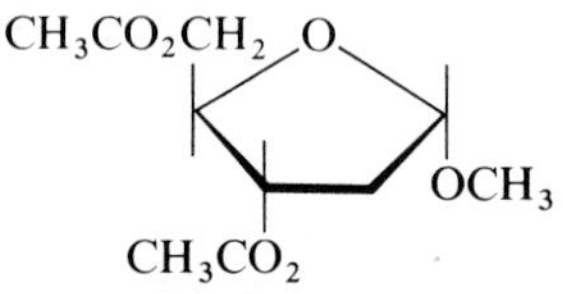 (b)

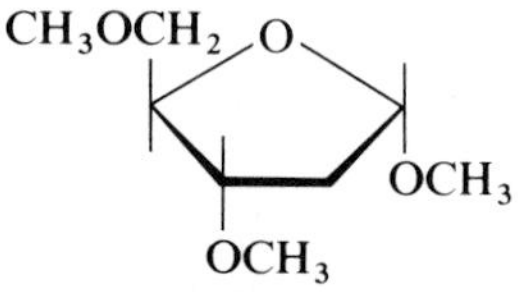

18.14.

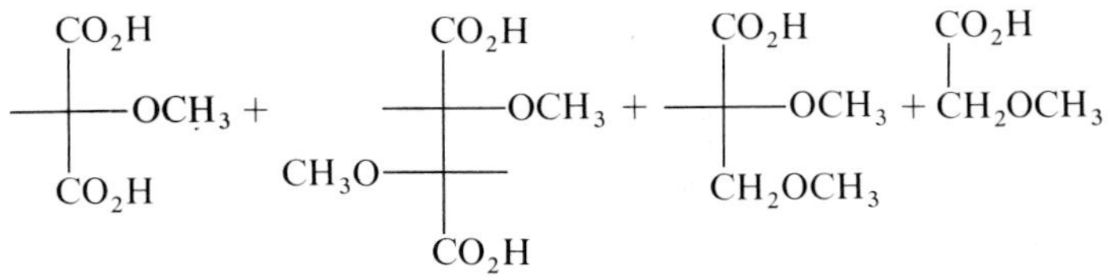

$$\begin{array}{c} CO_2H \\ | \\ -\!\!-\!\!-OCH_3 \\ | \\ CO_2H \end{array} + \begin{array}{c} CO_2H \\ | \\ CH_3O-\!\!-\!\!-OCH_3 \\ | \\ CO_2H \end{array} + \begin{array}{c} CO_2H \\ | \\ -\!\!-\!\!-OCH_3 \\ | \\ CH_2OCH_3 \end{array} + \begin{array}{c} CO_2H \\ | \\ CH_2OCH_3 \end{array}$$

18.15.

$$\begin{array}{c} CH_2OH \\ \cdots\cdots \\ CHOH \end{array} \longrightarrow \left\{ \begin{array}{c} HCHO \longleftarrow formaldehyde \\ + \\ CHO \\ \\ OHC \quad CHO \quad OCH_3 \end{array} \right.$$

$$\longrightarrow \left\{ \begin{array}{c} CH_2OH \\ OHC \quad CHO \quad OCH_3 \\ + \\ HCO_2H \longleftarrow formic\ acid \end{array} \right.$$

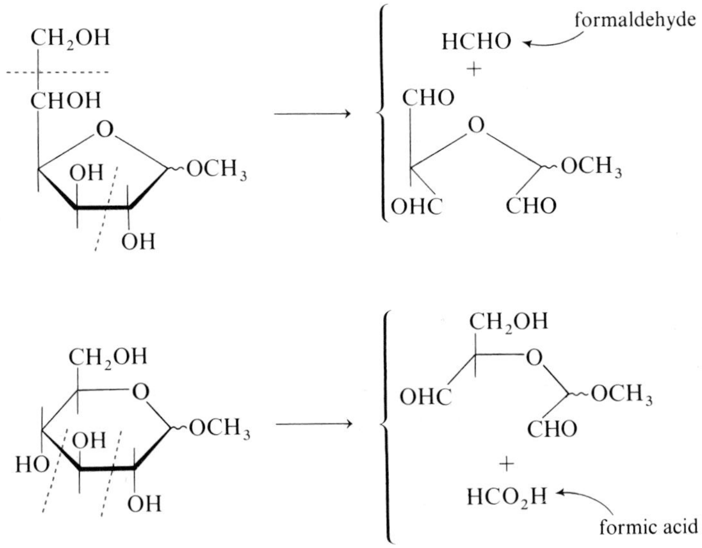

18.16. (a) 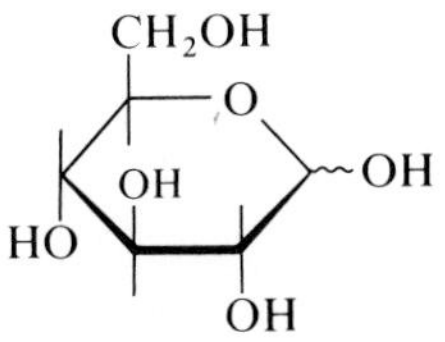 , (b)

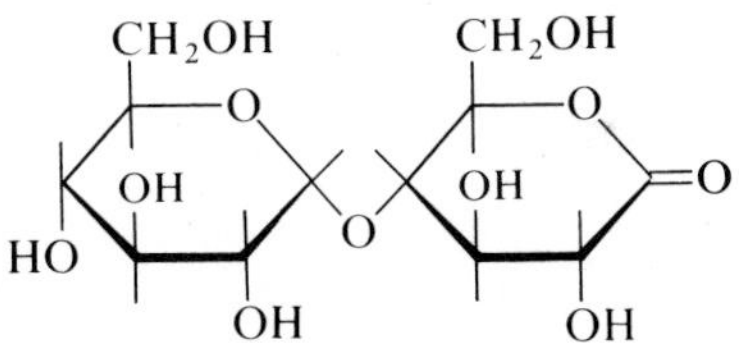

18.17. (a), (b), and (e):

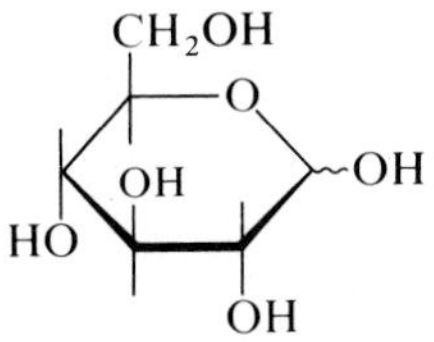

(c) 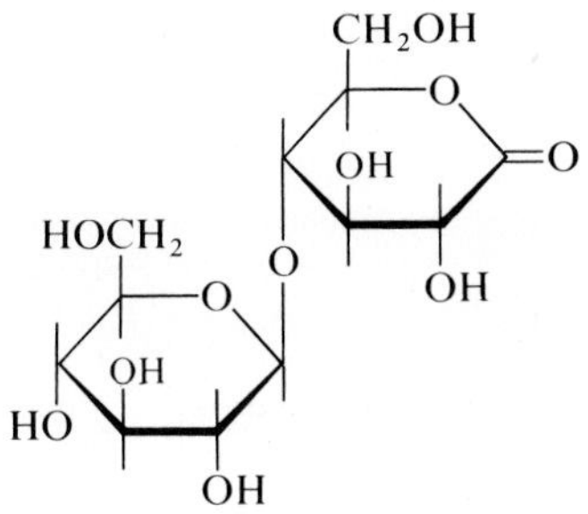 (d)

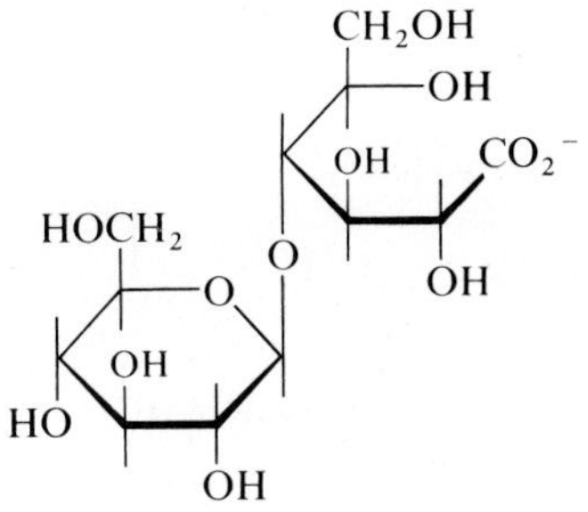

18.18. β-Cellobiose is more stable because each substituent on each ring is equatorial.

18.19. (a)

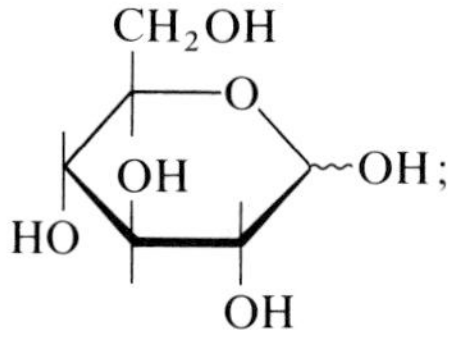

~OH; (b) no reaction; (c) no reaction;

(d)

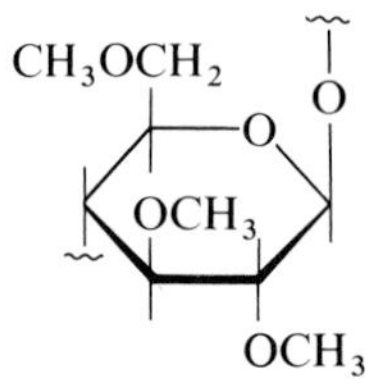

18.20. (a)

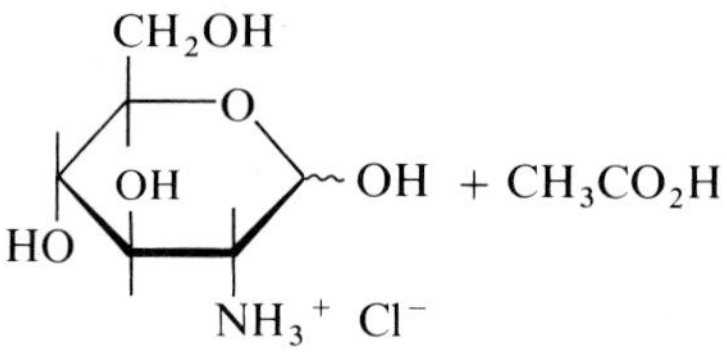

OH + CH$_3$CO$_2$H

(b)

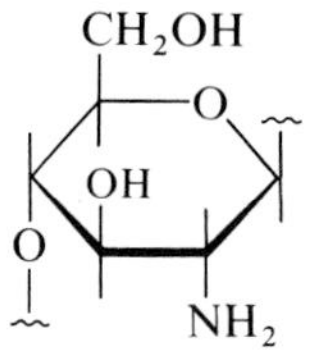

+ CH$_3$CO$_2^-$

CHAPTER 19

19.1. (a) acidic; (b) near-neutral; (c) near-neutral; (d) basic;
(e) near-neutral

19.2. HO$_2$CCH$_2$CH$_2$CHCO$_2^-$ Na$^+$
$\quad\quad\quad\quad\quad\quad\quad\quad$|
$\quad\quad\quad\quad\quad\quad\quad$NH$_2$

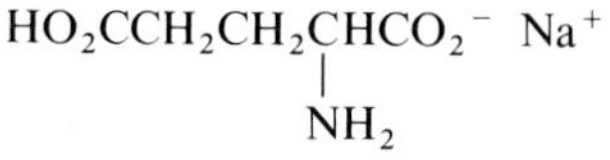

19.3. (a)

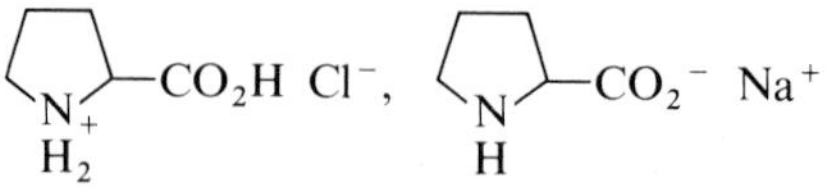

—CO$_2$H Cl$^-$,

—CO$_2^-$ Na$^+$

(b) HO—⟨○⟩—CH$_2$CHCO$_2$H Cl$^-$, $^-$O—⟨○⟩—CH$_2$CHCO$_2^-$ 2 Na$^+$
$\quad\quad\quad\quad\quad\quad\quad\quad\quad\quad$|$\quad\quad\quad\quad\quad\quad\quad\quad\quad\quad\quad\quad\quad\quad\quad$|
$\quad\quad\quad\quad\quad\quad\quad\quad\quad$NH$_3^+$$\quad\quad\quad\quad\quad\quad\quad\quad\quad\quad\quad\quad\quad\quad\quad$NH$_2$

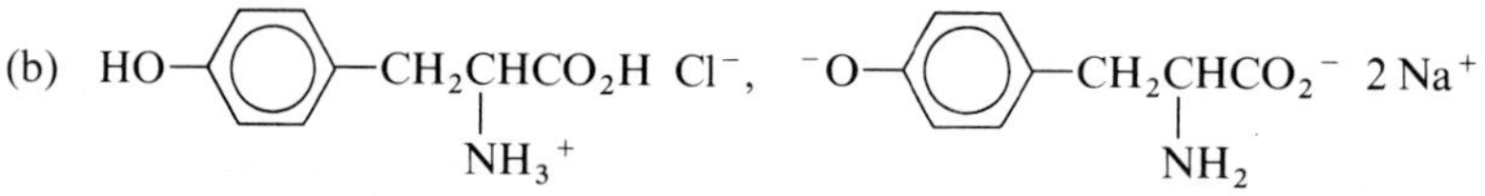

19.4. Like the nitrogen in pyrrole (Section 16.9), the nitrogen in tryptophan has no unshared bonding electrons; therefore, tryptophan is a neutral amino acid.

19.5.

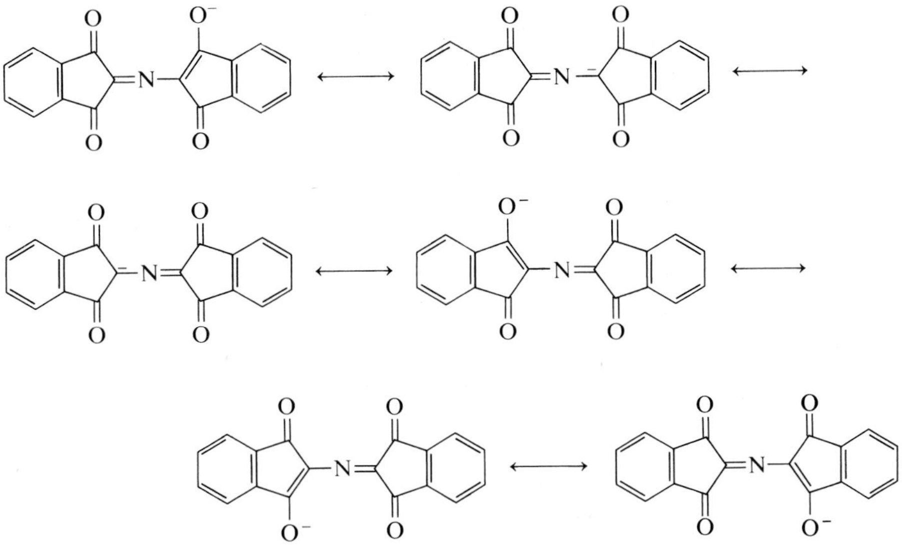

19.6. ala–gly–phe, ala–phe–gly, gly–phe–ala, gly–ala–phe, phe–ala–gly, phe–gly–ala

19.7.

$$C_6H_5CH_2O\overset{O}{\overset{||}{C}}-NHCH_2\overset{O}{\overset{||}{C}}-\underset{\underset{CH_2C_6H_5}{|}}{NHCHCO_2H} \xrightarrow{ClCO_2C_2H_5}$$

$$C_6H_5CH_2O\overset{O}{\overset{||}{C}}-NHCH_2\overset{O}{\overset{||}{C}}-\underset{\underset{CH_2C_6H_5}{|}}{NHCH}\overset{O}{\overset{||}{C}}OCO_2C_2H_5 \xrightarrow{CH_3CH(NH_2)CO_2H}$$

$$C_6H_5CH_2O\overset{O}{\overset{||}{C}}-NHCH_2\overset{O}{\overset{||}{C}}-\underset{\underset{CH_2C_6H_5}{|}}{NHCH}\overset{O}{\overset{||}{C}}-\underset{\overset{|}{CH_3}}{NHCH}CO_2H \xrightarrow{H_2,\ Pd} \text{gly–phe–ala}$$

19.8.

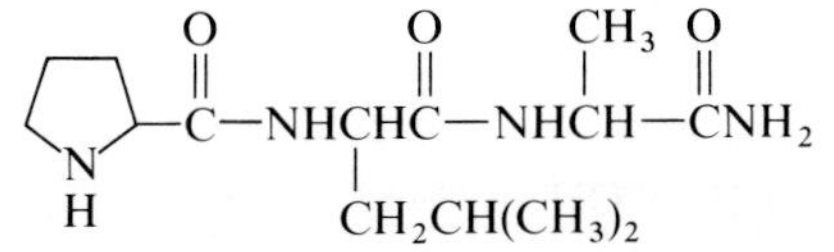

19.9. The enzyme catalyzes: (a) transfer of an acetyl group from one molecule to another; (b) conversion of phenylalanine to tyrosine by hydroxylation of the benzene ring; (c) dehydrogenation of an ester or a salt of pyruvic acid.

CHAPTER 20

20.1.

$$\begin{array}{l} CH_2O_2C(CH_2)_7CH{=}CH(CH_2)_7CH_3 \\ | \\ CHO_2C(CH_2)_7CH{=}CH(CH_2)_7CH_3 \\ | \\ CH_2O_2C(CH_2)_7CH{=}CH(CH_2)_7CH_3 \end{array} \xrightarrow[\text{heat, pressure}]{\text{excess } H_2\text{, catalyst}}$$

$$\begin{array}{l} CH_2OH \\ | \\ CHOH \\ | \\ CH_2OH \end{array} + 3\,CH_3(CH_2)_{16}CH_2OH \xrightarrow{H_2SO_4} CH_3(CH_2)_{16}CH_2OSO_3H$$

$$\xrightarrow{\text{NaOH}} CH_3(CH_2)_{16}CH_2OSO_3Na$$

20.2.

$$\textit{trans}\text{-}CH_3(CH_2)_{12}CH{=}CH\overset{\overset{OH}{|}}{C}H\underset{\underset{NH_2}{|}}{C}HCH_2OH$$

20.3. (a) and (b):

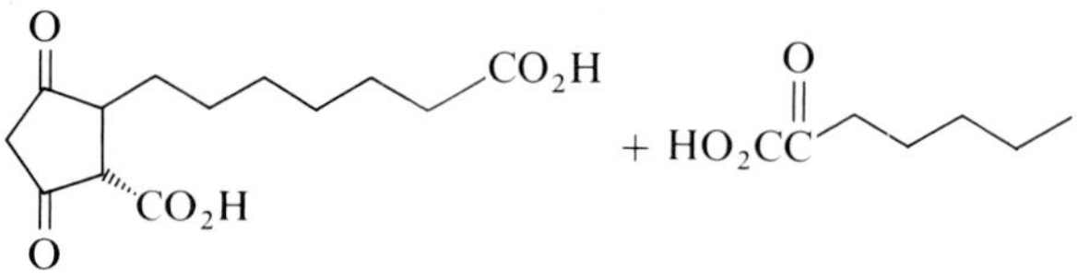

20.4.

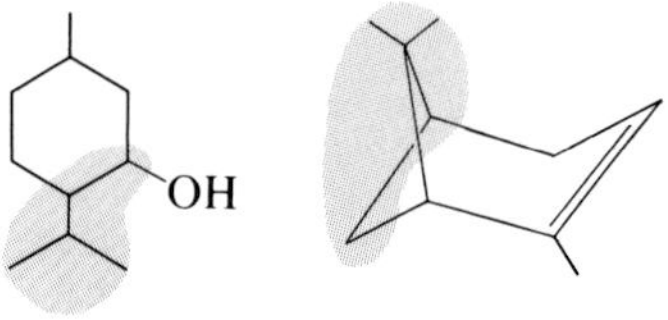

20.5. $CH_3(CH_2)_4\overset{\overset{O}{\|}}{C}CH_3$

20.6. (a)

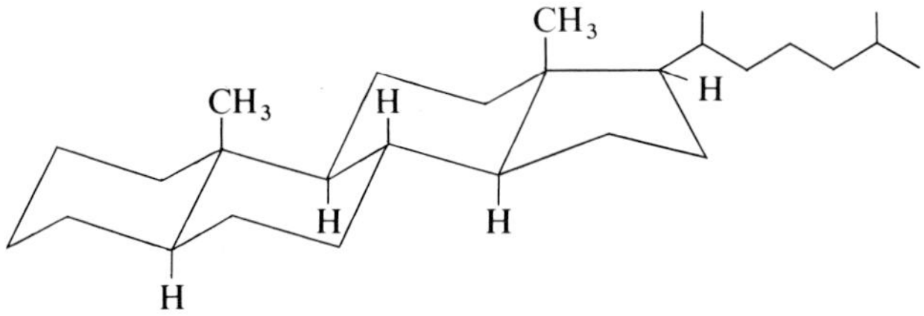

(b)

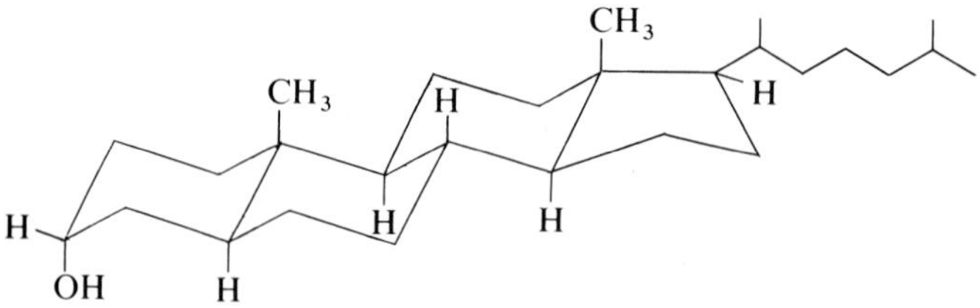

(c)

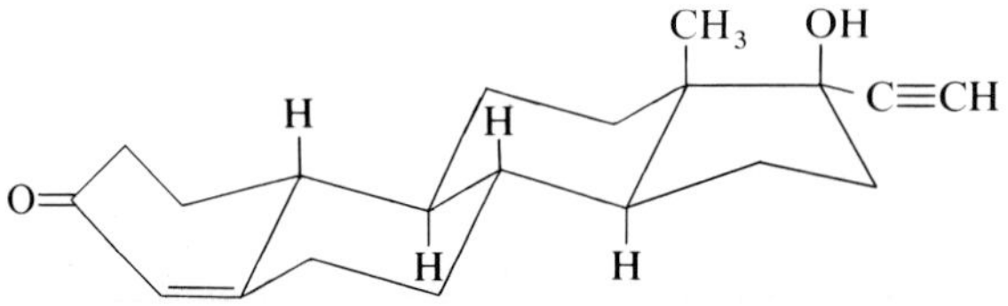

20.7. 3β-Cholestanol is more stable because its OH is equatorial.

20.8. The steroid ring system is a large, hydrophobic end, while the glycine portion contains the hydrophilic $-CO_2^-$ Na^+.

CHAPTER 21

21.1. Tyrian purple   **21.2.** (c), (b), (a)   **21.3.** (b) 446 nm; (c) 416 nm

21.4. The entire ring system of coronene has 24 pi electrons ($4n$, and not $4n + 2$). Therefore, not all the pi electrons are involved in the aromatic pi cloud. It is thought that only the peripheral carbons are part of the aromatic system.

21.5. A compound that appears violet absorbs at a shorter wavelength (about 570 nm) than one that appears blue-green (about 650 nm), and therefore is the compound with less delocalization. The structure in (b), with only two $-N(CH_3)_2$ groups, has less delocalization and is the violet-colored compound. The structure in (a), therefore, is the blue-green compound.

21.6. (a) $[CH_4]^{\ddagger}$, ion-radical; (b) $[CH_3]^+$, cation;
(c) $[CH_2CH_2]^{\ddagger}$, ion-radical; (d) $[CH_2CH_2]^+$, cation

21.7. (a) $[CH_4]^{\ddagger}$; (b) $[CH_3CH_2CH_3]^{\ddagger}$

21.8. (a) 16; (b) 43; (c) 32; (d) 18

21.9. (a) 30; (b) 98; (c) 46; (d) 172

21.10. B contains Br, and C contains Cl.

21.11. (a)

$$\left[\begin{array}{c} CH_3CHCH_2CH_2CH_3 \\ | \\ CH_3 \end{array}\right]^{\ddagger} \quad \begin{array}{c} \xrightarrow{-[CH_3]^{\cdot}} \left[\begin{array}{c} CHCH_2CH_2CH_3 \\ | \\ CH_3 \end{array}\right]^+ \\[2em] \xrightarrow{-[CH_3CH_2CH_2]^{\cdot}} \left[\begin{array}{c} CH_3CH \\ | \\ CH_3 \end{array}\right]^+ \end{array}$$

$[CH_3]^+$ and $[CH_3CH_2CH_2]^+$ might also be observed.

(b)

$$\left[\begin{array}{c} CH_3 \\ | \\ CH_3CCH_3 \\ | \\ CH_3 \end{array}\right]^{\ddagger} \xrightarrow{-[CH_3]^{\cdot}} [(CH_3)_3C]^+$$

(c) $[CH_2{=}CHCH_2CH_2CH_3]^{\ddagger} \xrightarrow{-[CH_3CH_2]^{\cdot}} [CH_2{=}CHCH_2]^+$

21.12. $CH_3OCH_2CH_2CH_3$ **21.13.** (a) 58; (b) 44; (c) 60; (d) 88 and 102

Index